15th International Conference on Industrial Manufacturing and Metallurgy

IOP Conference Series: Materials Science and Engineering
Volume 966

Online
18-19 June 2020

Editors:

Victor Goman
Maria Mironova

ISBN: 978-1-7138-4787-8
ISSN: 1757-8981

Printed from e-media with permission by:

Curran Associates, Inc.
57 Morehouse Lane
Red Hook, NY 12571

Some format issues inherent in the e-media version may also appear in this print version.

Institute of Physics
Dirac House, Temple Back
Bristol BS1 6BE UK

Phone: 44 1 17 929 7481
Fax: 44 1 17 920 0979

techtracking@iop.org

Additional copies of this publication are available from:

Curran Associates, Inc.
57 Morehouse Lane
Red Hook, NY 12571 USA
Phone: 845-758-0400
Fax: 845-758-2633
Email: curran@proceedings.com
Web: www.proceedings.com

TABLE OF CONTENTS

Preface

Commitees

Peer review declaration

The study of the relationship between stock value and financial performance of iron and steel enterprises using artificial intelligence techniques 1
V V Potanin, O Yu Sidorov

Comparative evaluation of effectiveness indexes of column packings in benzol absorbers........ 7
E V Nogovitsyna

Substantiation of parameters of underground geotechnology in the development of complex structural deposit of various grades of ores 13
A M Mazhitov, P V Volkov

Counterflow step-type bulk material cooler 19
A S Shishkin, S F Shishkin

Calculation of a closed grinding cycle in a jet mill with two classifiers........ 25
D S Barakovskikh, S F Shishkin

Influence of liquid nitriding on the properties of the electric steel 10 31
V A Korotkov, D V Isakov

Nickel sorption from solutions with high salt concentration 37
O Yu Makovskaya, O B Kolmachikhina, V G Lobanov, S E Polygalov

Technology for carrying out hydraulic fracturing using a new material 'nitinol' 43
A E Verisokin, A P Yanukyan, D Yu Serdyukov, L G Zhulina

Analysis of single- and double-threaded rolling for production of grinding balls of alloy steel grades 48
I Galimyanov

Study of possibility of cerium reduction from slags of CaO-SiO_2-Ce_2O_3–15%Al_2O_3–8%MgO system........ 54
A A Babenko, L A Smirnov, A G Upolovnikova, R R Shartdinov

Destruction electrospark deposits during Joule heating of base metal 60
G E Trekin, O I Shevchenko

Effect of basicity and chromium oxide on the viscosity of boron-containing slags 66
A A Babenko, R R Shartdinov, A G Upolovnikova, A N Smetannikov, L Yu Mikhailova

The use of additional filler wire to solve the tandem GMAW issues 71
I O Polevoy, A M Fiveyskii

Double-sided arc welding with multiple electrodes of vertical joints of steel tanks........ 77
O I Filippov, V A Erofeev, S I Poloskov, A Yu Melnikov

Application of surfacing using consumable electrode with an additional filler wire to ensure the required operational properties of pipeline valves 83
S S Poloskov, D S Buzorina, V A Erofeev

Influence of technological parameters on the vanadium valence form in slags 89
A V Klimov

Technology for fixing mine workings with friction anchors of the SZA type with a new type of anticorrosive coating 95
P V Volkov, R V Kulsaitov, A M Magitov

Kinetics of vanadium reduction from slags by solid carbon 100
A V Klimov

Modelling of the thermal state and the melting loss of a graphite electrode in the conditions of the evaporative cooling in the arc furnace 105
M V Bystrov, I M Yachikov, I V Portnova

Application of hydrohammers of heavy class in the development of marble deposits in mining operations 111
I A Pytalev, D V Domozhirov, N V Ugolnikov, A A Prochorov

The microstructure and hardness of casting a solid brake disc after late graphitizing modification 117
D A Boldyrev, R R Dema, O B Kalugina

Improving the production technology of drilling and blasting operations by blasting of high ledges 122
N V Ugolnikov, D V Domozhirov, N G Karaulov, A A Prochorov

Microstructure of impurity phase inclusions in gray cast iron and their composition 128
D A Boldyrev, R N Amirov, S P Nefed'ev

Influence of blast furnace slag type on physical and mechanical properties of Portland cement 133
F L Kapustin, T S Balueva

Prospects for the use of peat in metallurgy 139
T B Ykonovskaya, A I Zhigulskaya

Research on the influence of the strip tension on contact stress and cold rolling force with a strain-hardening 144
K Ilin, G Baranov

The effect of enrichment on quartz sand properties 150
L F Mufteeva, I A Pavlova, E P Farafontova

The use of technogenic raw materials to produce a high-alumina chamotte 155
A A Getman, I A Pavlova, E P Farafontova

Nanofine spatial dissipative structures with azimutal lattice curvation 159
V B Malkov, O V Chemezov, V G Pushin, A V Malkov, B V Shulgin, O V Malkov

Phenomenon of relaxation of inhomogeneous elastic rotational curvature of the lattice of nanofine spatial dissipative structures of hexagonal selenium 164
V B Malkov, O V Chemezov, V G Pushin, A V Malkov, B V Shulgin, O V Malkov

Thermal stability of hafnium bronze subjected to dynamic channel angular pressing 169
A Stolbovsky, V Popov, E Popova, R Falahutdinov, S Murzinova, E Shorokhov

The use of regression models to describe the temperature coefficient values of linear expansion and technological temperatures of $CaSO_4$-KPO_3-$Na_2B_4O_7$ low-melting glass system ... 174
A Stolbovsky

Physico-mechanical properties of the sinter of various chemical composition ... 179
S K Sibagatullin, A S Kharchenko, V I Sysoev, D N Gushin

The effect of wear-resistant coatings composition based on niobium nitride on the structural parameters, mechanical properties and efficiency of the cutting tool ... 185
V P Tabakov, L V Khudobin, A V Chikhranov, Y A Dolzhenko

Research features of opening of refractory arsenopyrite concentrates by using low-temperature roasting ... 189
N O Zvontsov, D A Rogozhnikov

Estimation of selective reduction of iron and phosphorus from manganese ores of different genesis ... 195
N Y Kosdauletov, V E Roshchin

Obtaining of niobium-containing ferroalloys from the Russian ore raw materials ... 201
V I Zhuchkov, O V Zayakin, L Yu Mikhailova

The problems of application of domestic ore raw materials in the production of ferroalloys ... 206
V I Zhuckov, L I Leontiev, O V Zayakin, V Ya Dashevsky

Scaling resistance of deposited metal by flux-cored wire 30N8Kh6M3STYu ... 211
E N Eremin, A S Losev, I A Ponomarev, S A Borodikhin, A E Matalasova

Method analysis of surveying of mineral storages in open-pit mining ... 217
E A Romanko, D V Domozhirov, O S Kolesatova, N V Litvinenko

Influence of technological depressurization on the operation of a cascade pneumatic separator ... 221
V B Ponomarev, I P Fesun

Simulation of patterns of crystallizing ingot macrostructure formation ... 227
O Yu Sidorov

Methodology of designing an air cascade separator ... 233
V B Ponomarev, A S Shishkin

Investigation of structure and properties of low alloy steel obtained by wire arc additive manufacturing under various fluxes ... 238
G E Trekin

Synthetic raw materials - possibility to increase refractory materials resistance ... 244
K G Zemlyanoi, A R Khafizova

A development of computer vision elements for processing the results of tensile tests of specimens and building diagrams of ultimate plasticity ... 248
M V Erpalov

The possibility evaluation of zinc sulphate forms extraction from sludge of PJSC EVRAZ NTMK with following metalloflux obtaining ... 254
O Yu Sheshukov, M A Mikheenkov, I V Nekrasov, D K Egiazaryan

Reducing the carbonitride formation risk with increasing titanium dioxide content in the blast furnaces charge of JSC EVRAZ NTMK ... 258
O Yu Sheshukov, M A Mikheenkov, I V Nekrasov, D K Egiazaryan

The effect of thermoplastic processing on the structure and mechanical properties of metastable austenitic steel 263
N N Ozerets

Effect of frictional treatment with a dense cubic boron nitride indenter on the micromechanical properties of the NiCrBSi–Cr_3C_2 coating 269
N N Soboleva, A V Makarov, E P Nikolaeva, P A Skorynina, I Yu Malygina

Destruction of cement stone modified with a plasticizer and a nanoscale aluminum oxide 275
E S Gerasimova, E D Trestsova

Utilization of final tailings concentration of titano-magnetite ores in Portland cement production 280
F L Kapustin, N N Bashkatov

Utilization of ferrochrome slag in concrete production 286
N N Bashkatov

Decoding variant for dilatometric measurements of metal samples heating and cooling 291
S B Mikhailov, V A Sharapova, N A Mikhailova, A E Karelova

The effect of elastic deformation on the minor hysteresis loops of low carbon steel 297
A N Mushnikov, E S Gorkunov, S M Zadvorkin, L S Goruleva, K D Kryucheva

Analysis of the influence of the parameters of the power supply network on the operating modes of the electric drives of the stands of the hot rolling mill 1700 of ArcelorMittal Temirtau JSC 302
I V Breido, V M Druzhinin

Optimization of the ion-plasma nitriding of structural steel 308
V R Baraz, E A Yurchenkov, E A Ishina

Technologies for remote lidar sensing of the surface to detect low concentrations of hydrocarbons in Siberia and the far north 313
S L Leshchenko

Iron and steel industry of Russia in the XXI century 319
S V Dokuchaev

Resource-saving potential of brick scrap from demolition of buildings 324
I K Domanskaya

Study of the main metallurgical characteristics of iron ore raw materials (sinter and pellets) 329
G Yu Vitkina, A N Dmitriev, R V Alektorov

Improvement in vanadium-containing titanomagnetite processing technology 335
A V Kushnarev, K V Mironov, S A Zagainov, A A Forshev

Plasma hardening of corrosion-resistant steel 341
M V Mironova

Graphite flake cast iron surface hardening 346
M V Mironova

Evolution of the structure of precipitation-hardening steel during heat treatment 351
E N Safonov

Research of features of structure formation in case of surface steel hardening 356
E N Safonov

Sulfur partition by process stages of metallurgical production of JSC EVRAZ NTMK 362
M V Savelyev, O Yu Sheshukov, A A Metelkin, O I Shevchenko, A S Tkachev, S V Shmakov

Calculation of sulfur removal in ladle furnace unit by means of ionic theory of slags 368
M V Savelyev, O Yu Sheshukov, A A Metelkin, O I Shevchenko, D K Egiazar'yan

Research into in-service deterioration of ball-rolling rolls 374
V Ju Rubcov, O I Shevchenko, A S Spirina, N A Pankova

Development of new roll pass designs for ball-rolling rolls with continuously variable parameters 380
V Ju Rubcov, O I Shevchenko

Ultrasound effect on electrospark cementation process 386
O I Shevchenko

Raising of energy efficiency of thermal performance of layered metallurgical units using layered type of natural gas combustion 392
V I Matyukhin, A V Matyukhina, V A Dudko

Use of temporary gas flow to improve the efficiency of metallurgical units operation 398
V I Matyukhin, A V Matyukhina, S Ya Zhuravlyov, N V Yamshanova

Development of radial-shear rolling mill special stands for continuous cast billets deformation 404
R V Iskhakov, Yu V Gamin, M V Kadach, A S Budnikov

Determination of cutting forces based on DMG MORI CTX300 ecoline CNC lathe drive power data 410
A P Starostin, D M Elkind

Development of the identifier system of indicators for improving the information potential of the holding 414
E V Dolzhenkova, L V Iurieva

Ejector jet mill with external pressure 420
D S Barakovskikh, S F Shishkin

Development of tool selection process for milling of pocket features 426
T A Gordeeva, S S Kugaevskii, A P Starostin

Development of super-professional competencies of the production personnel in the context of university supplementary education programs 432
Yu G Myslyakova, M V Kurashova

To the issue of transformation of 'old industrial area' geobrand through example of Sverdlovsk Region 437
Yu G Myslyakova, M V Kurashova, M M Shcherbinin

Role of management of industrial enterprises in formation of labor migration trend in region 443
M V Kurashova, Ye A Nadeina, Yu G Myslyakova

Difficulties of succession pipeline in large industrial enterprises of Nizhny Tagil in studies of tendencies of competitiveness of higher technical education 448
M M Shcherbinin, V V Potanin, S E Chetverikov

Development of algorithm for excavation control 454
O A Lukashuk, K Y Letnev, A P Komissarov

Informative modeling of subjective reality for intellectual anthropomorphic robots 458
A G Dolganov, K Y Letnev

Arranging mining machinery systems in a single classification by the types of raw peat being extracted 464
D R Iakupov, S L Ivanov, P V Ivanova, E K Permyakova

Reconditioning and hardening of mining equipment hydraulic cylinder working surfaces by using of double-electrode cladding 469
S I Zverev, A M Fiveyskii

Patterns of form error inheritance during machining 473
E V Pegashkina

The study of the influence of deviations of the arc energy parameters on the defects formation during automatic welding of pipelines 478
A E Filyakov, M A Sholokhov, S I Poloskov, A Yu Melnikov

Application of double electrode welding as the way to increase welding productivity of fillet welds of bridge structures 484
A A Mosin, M A Sholokhov, D S Buzorina

Evaluation of the resulting quality level index for interregional passenger automobile transport 490
V V Epifanov, M Y Obshivalkin, K A Generalova (Lukon'kina)

Organizational factors for manufacturing defects reduction in small batch production 496
I V Ershova, S G Baranchikova, E V Cherepanova, N N Chistyakov

Modern concept analysis of failure modes and consequences for quality assurance of products and processes 500
E V Kalinin, L L Kalinina, O S Norkina

Advanced standard operating procedures for survey of buildings and structures of metallurgical production 505
M S Kuznetsov, V G Dubinina, Yu V Patrushev

Instrument ofthe work breakdown accounting in the high-tech production 511
O O Podoliak, Yu N Markova, T A Mineeva, N A Kuznetcova

Comparative analysis of cutting fluids 517
I V Ershova, N E Kalinina, E V Cherepanova, D F Fayskhanov

Problem of model usage in mechanical engineering 522
V F Pegashkin, N U Ignatova

Marketing aspect in the 'Value Engineering' in the new products development of machine-building enterprises 527
A D Murukina, L M Tipner, M A Prilutskaya, N A Kalinina

Utilization aspects in a life cycle contract (the example of public transport in Russia) 533
N A Amosov, E Y Kuznetsova, M A Prilutskaya, G R Kukushkina

Upgrading of induction motors with energy efficiency enhancement at the electrical repair shops and sites of the running industrial companies 539
R G Mugalimov, A R Mugalimova, R A Zakirova

Processes in spark electric discharge in air, nitrogen and oxygen 545
I M Piskarev, N A Aristova

Determination of the reduction coefficients in a continuous finishing group of stands in a hot rolling mill using artificial neural network 550
O R Latypov, S I Platov, R R Dema, N V Urtsev

The research on the levelling process of thick slabs based on the electric motor parameters monitoring 554
S Yu Belyaev, A A Fedulov, A S Alexandrova, E Yu Raskatov, T V Meshchaninova

Multi-criteria model for the development of industrial logistics 560
A N Rakhmangulov, S N Kornilov, D V Aleksandrin, N O Shevkunov

Research and educational network of ground stations for receiving and processing information from educational satellites 566
A A Spiridonov, V A Saechnikov, D V Ushakov, A G Kezik, V E Cherny, A P Verstakovskaya, V E Evchik

Temperature calculation for linear induction motor in transport application with multiphysics approach 572
F N Sarapulov, V Goman, G E Trekin

Research in Autodesk Inventor weldment performance cultivator frame 578
V V Telegin, I V Telegin

Synchronous reluctance machines: performance evaluation with and without ferrite magnets 583
Mohamed N Ibrahim, Kotb B Tawfiq, E M Rashad, Peter Sergeant

Development of the temperature prediction algorithm for smart thermostat 589
S I Korolev, R M Muhutdinov, O Yu Sidorov, V Goman

Comparison of minimum energy consumption indicators for electric motors powered directly from the mains at part loads 594
V Goman

Sensitivity of robust systems controlling first order objects 600
M A Volkov, D V Isakov

Research of a bi-directional DC-DC converter integrated in electric car power installation 606
K H Gulyamov, B N Sharifov, A H Ghulomzoda, M Kh Safaraliev, R M Yunusov

Estimation of the operational characteristics and determination parameters of the vertical radial-axial hydraulic turbine type 400 612
A A Gulakhmadov, S Sh Qurbonalizoda, Z V Kobuliev, B N Ghoziev

Mathematical modelling of mutual electromagnetic influences of related power transmission lines in a transition process mode 618
S M Asanova, S M Suerkulov, M Kh Safaraliev, D U Asanova, R S Milutin

Method for designing drop-of-wire recognition systems on sections of undistorted two-wire power transmission lines 624
S M Asanova, J S Ahyoev, N Askarbek, S M Suerkulov, D U Asanova, M Kh Safaraliev

Automation of movement control for modern main tank T-90 630
M A Badrtdinov, D V Yudincev

Using of polymeric materials to defeat of lightly armored and easily vulnerable targets. Theory and experiments 635
E A Khmelnikov, K V Smagin, T E Zavodova

Developing manufacturing execution system with predictive analysis 641
V V Potekhin, AH Bahrami, B Katalinič

Epistemic responsibility of industrial enterprise engineer 647
N U Ignatova

The additive technology to obtain a three-dimensional model of the 81Cu-19Ni alloy 653
V R Baraz, S S Gerasimov, I A Gruzdeva, V P Shveikin, A V Chemesova, A O Ovchinnikova, I A Veretennikova

Tribological tests of the improved piston mechanism of the axial piston hydraulic machine 658
S Stazhkov, A Kuzmin, V Elchinskiy, N Yakovenko

Review of the features of augmented reality application in the training of operators and maintenance staff 663
S V Byvaltsev

Multi-criteria optimization of metal pressure processes using experiment planning methods 668
S V Byvaltsev

Increase in linear motion steadiness of tractor-trailers using stabilizing towing couplers 674
Y N Stroganov, E M Pampura

Study of the influence of the design of the turning tool on the removal of thermal energy from the cutting zone 680
Y L Liberman, S V Lukinskikh, A V Smirnov

Using of polymeric materials to defeat of lightly armored and easily vulnerable targets. Mathematical modeling 686
E A Khmelnikov, K V Smagin, T E Zavodova

Deep drawing failure modes 692
AV Nezamaev, V V Karzhavin

Foundry production digitalization 696
V M Gruzman

Interactional experience with the employer during a selection of educational content in a single-industry town 702
R A Karelova

Possibilities of an artificial neural network use to control oxygen consumption in a converter shop 706
R A Karelova

Possibilities of information technologies application to production problems solving by math modeling methods 711
V M Gruzman

Investigation of skew piercing and upsetting effect on wheel steel mechanical properties 715
A V Fomin, V P Romanenko, A N Nikulin, A A Sevastjyanov

Corporate culture as one of the key factors of effective industrial enterprise development 719
E A Putilova, A V Shutaleva

Automated tool support system of a large machine-building enterprise 724
A Bogoyavlensky, A Sirotyuk

Influence of thermal oxidation on the properties of titanium alloys 730
V I Golubev, V V Medison, D S Pastukhov, V F Pegashkin

The use of homomorphic image processing to analyze coke grading 736
V N Krouglov, A G Gruh, A L Gapak, I Khurelchuluun

Optimization of roll calibration for flange shape rolling. Groove space 742
D L Shvarts, A M Mikhaylenko, E I Ustinova

Application of laser interferometry to analyze cutting tool state during machining 748
I A Efimovich, I S Zolotukhin

Adapting standard maintenance approaches for mining excavators to actual operating condition 755
D A Shibanov, S L Ivanov, K A Safronchuk, V I Knyazkina

Optimization of alternate profile component part machining 761
L V Borshova, V F Pegashkin

Author Index

IOP Conf. Series: Materials Science and Engineering **966** (2020) 011001 doi:10.1088/1757-899X/966/1/011001

Preface

The 15th International Conference on Industrial Manufacturing and Metallurgy (ICIMM 2020)

The conference was organized at the Nizhny Tagil technological Institute (branch) of the Ural Federal University (Russia). The conference has been taking place since 1990 every 2 years.

The main aim of the conference was to improve the communication between scientists and industry representatives for the introduction of new ideas and technologies. In 2020 International scientific committee included scientists and practitioners from Italy, France, Germany, Belgium, Egypt, Kazakhstan, Russia. The main partner and co-organizer of the conference is EVRAZ-NTMK JSC – large metallurgical enterprise in Nizhny Tagil, which celebrated its 80th Anniversary in June 2020. The industrial partners of the conference are Danieli, Primetals, SMS Group, Horst Wiesinger Consulting, Paul Wurth, Mishima Kosan, Exolaunch GmbH, Iridium Ltd.

Nizhny Tagil Technological Institute (branch) of the Ural Federal University was founded in 1944. Over the years the Institute has trained more than 14500 engineers. Now NTI is a large multidisciplinary higher educational institution in Nizhny Tagil. The departments of higher professional education train metallurgists, mechanical engineers, economists, IT specialists, mechatronic, electric drives and industrial automation, maintenance of equipment, civil engineering, deference industry engineers not only for the enterprises of the city, but of the whole Ural region.

Unfortunately, in 2020, the world has found itself in the state of pandemia which is unprecedented. It has naturally had a dramatic impact on the science and education activities which have been switched to online mode completely. The conference was not an exclusion, and on the 18-th of June 2020 it was successfully conducted online. 25 selected reports were presented in real time via video communication service and all other reports were presented via virtual platform – Nizhny Tagil technological Institute (branch) of the Ural Federal University YouTube channel. Online presenters had 10 minutes interval for report and 5 minutes for Q&A session after report. Also, an opportunity for discussion using comments on the YouTube channel was realized by all attendees.

About 400 attendees from the following countries have participated in the conference: Russia, Belarus, Kazakhstan, Kirgizstan, Tadzhikistan, Mongolia, China, Belgium, Egypt, Germany, Austria, France, Italy. They represented more than 50 universities, research institutions and industrial enterprises.

On behalf of the Ural Federal University we would like to thank all authors, reviewers, technical staff, heads of sections by whose efforts the conference became valuable and meaningful.

Editors-in-chief

Victor Goman and Maria Mironova

IOP Conf. Series: Materials Science and Engineering **966** (2020) 011002 doi:10.1088/1757-899X/966/1/011002

International scientific committee

IOP Conf. Series: Materials Science and Engineering **966** (2020) 011002 doi:10.1088/1757-899X/966/1/011002

Local Organization Committee

Chair - Vladislav Potanin, director of the Nizhny Tagil Technological institute (NTI) of Ural Federal University

Co-Chair – Maria Mironova, PhD, Assoc. prof., vice-director NTI

Co-Chair – Victor Goman, PhD, Assoc. prof., director of the Research&Innovation Center, Chair of the Mechatronics and robotics program, IEEE member

Vladimir Pegashkin, Chair of the General Engineering department

Evgeniy Khmelnikov, Chair of the Special Engineering department

Evgeniy Safonov, Dr., Prof. of the General Engineering department, director of the post-graduate school NTI

Oleg Shevchenko Dr., Prof., chair of the Metallurgical Technologies department

Vyacheslav Gruzman, Dr., Prof. of the Metallurgical Technologies department

Oleg Sidorov, Dr., Prof., senior researcher, Research&Innovation Center

Grigory Trekin, Phd, Assoc. prof., researcher, Research&Innovation Center

Riya Karelova, PhD, Assoc. prof., Chair of the Information technology program

Oleg Cheremnikh, PhD, Assoc. prof of the General Engineering department

Kseniya Pykhteeva, PhD. Assoc. prof. of the Metallurgical Technologies department, NTI young scientist's coordinator, Research&Innovation Center

Conference Secretary - Tatyana Andreeva, M.Sc., Research&Innovation Center

IOP Conf. Series: Materials Science and Engineering **966** (2020) 011003 doi:10.1088/1757-899X/966/1/011003

Peer review declaration

All papers published in this volume of *IOP Conference Series: Materials Science and Engineering* have been peer reviewed through processes administered by the Editors. Reviews were conducted by expert referees to the professional and scientific standards expected of a proceedings journal published by IOP Publishing.

- **Type of peer review:** single-blind (82,1% of papers), double-blind (17,9% of papers)
- **Conference submission management system:** via e-mail and cloud services
- **Number of submissions received:** 259
- **Number of submissions sent for review:** 218
- **Number of submissions accepted:** 140
- **Acceptance Rate (Number of Submissions Accepted / Number of Submissions Received X 100):** 54%
- **Average number of reviews per paper:** 1.18
- **Total number of reviewers involved:** 24
- **Any additional info on review process:** normally single-blind review method was used. Double-blind review was applied on request of the first reviewer or editors-in-chief
- **Contact person for queries:** Victor Goman, v.v.goman@urfu.ru, +79126059692

IOP Conf. Series: Materials Science and Engineering **966** (2020) 012001 doi:10.1088/1757-899X/966/1/012001

The study of the relationship between stock value and financial performance of iron and steel enterprises using artificial intelligence techniques

V V Potanin[1, 2] and O Yu Sidorov[1]

[1] Ural Federal University named after the First President of Russia B N Yeltsin, Nizhniy Tagil Technological Institute, 59, Krasnogvardeyskaya str., Nizhniy Tagil, 622000, Russia

E-mail: [2] v.v.potanin@urfu.ru

Abstract. Study of the relationship between finance performance and market equity value of Russian public companies, such as PJSC Mechel, PJSC Novolipetsk Iron and Steel Works, PJSC Magnitogorsk Iron and Steel Works, PJSC Severstal using the artificial neural network (ANN) has been carried out. Data of annual financial statements, stock-exchange equity prices, USD/RUB, EUR/USD exchange rates, RF CB base rate were used as initial data. The number of instructional examples was equal to 15 and there were 16 input values included into each example. In the end, ANN calculated the ratio of the market value of one share to equity capital per share. ANN training was carried out using the backpropagation method and the conjugate gradient method. The average values of discrepancies between the actual value of the ratio of market value to equity capital and calculated value using ANN were 6.0, 2.5, 8.2 % for NLMK, Severstal and MMK, respectively. The use of ANN has shown the overestimated value of equity stock-exchange quotations of PJSC Mechel in relation to the market averages.

1. Introduction

The use of artificial neural networks (ANNs) is especially reasonable in case when the studied value depends on myriad of factors and this dependence is implicit and stochastic. This is often demonstrated when studying financial and economic phenomena.

The paper [1] considers the possibility of using ANN and semantic analysis to predict the corporate bankruptcy based on its annual financial statements.

The paper [2] introduces the possibility for prediction of financial time-series using combined ANN with layers of different types. The paper shows [3] that trade strategies based on the classification models of ANNs generate higher profit than buy-and-hold strategies. The paper [4] uses the recurrent neural network for prediction of stock international market indexes. A neural network model was used in papers [5–7] to predict stock value, stock market trends, and relationships with technical analysis using Japanese candlesticks.

The ability of ANN to predict daily quotations of NASDAQ stock exchange index for two types of input data was investigated in the study [8]: four previous days and nine previous days. The paper [9] analyses the neural network model which predicts the highest Sharpe ratio of investment portfolio.

The paper [10] presents the effects of four types of ANN for prediction of common-stock value. It has been noted that the convolutional ANN was able to predict the values of New York Stock Exchange Index based on training with the use of values of India Stock Exchange Index.

The ability of ANN to predict Gazprom stock prices (5-minute data) and Sberbank shares (daily data) was checked in the studies [11, 12]. The significant differences between predicted and actual values [11] were obtained in some cases for Gazprom shares. Acceptability of the obtained results was noted for Sberbank shares [12].

The paper [13] considers ANN for prediction of profitability of Finam Narodnyi hybrid unit investment trust. The possibility of correct prediction of the participation unit value trend and the impossibility of accurate prediction for the near-term tone of the participation unit value are noted.

The performed review shows the possibility and rationale of the studies regarding the use of neural network technologies to describe the financial and economic features of companies.

2.Problem statement and initial data

This paper aims at assessment of the market value of companies PJSC Mechel (Mechel) and JSC 'EVRAZ Nizhniy Tagil Iron and Steel Works' (NTMK) using ANN based on annual statement data as well as assessment of the market value of companies PJSC Novolipetsk Iron and Steel Works (NLKM), PJSC Magnitogorsk Iron and Steel Works (MMK), PJSC Severstal (Severstal).

Data of accounting statements (balance sheet, profit and loss statement), shown on the web site [14–18] for 5 years (2014–2018) was selected as initial data. The exception is Mechel, for which complete data is available only for 2015–2018.

In addition, equity quotations and their number of companies NLMK, MMK, Severstal [19], data on the external economic environment were used as input data: RF CB base rate, USD/RUB and EUR/USD exchange rates [20]. The latest data were selected with a forward shift of 3 months, i.e. the beginning of April in the year following the reporting one. This is due to the fact that reporting is prepared later than the end of the reporting period.

The initial data were specified: consolidated balance sheet items were divided into balance sheet total (BST); aggregates of profit and loss statement were divided into income (I); external economic data was corrected to the interval (0; 1).

The specified initial data (instructional example) for NLMK for 2018 is given in Table 1 as an example.

Table 1. Example of specified initial data (NLMK, according to the results of 2018).

No.	Input data for ANN	Value
1	Fixed assets/BST	0.0566
2	Current assets/BST	0.434
3	Capital and reserves/BST	0.581
4	Long-term liabilities/BST	0.0164
5	Short-term liabilities/BST	0.255
6	Initial cost/I	–0.671
7	Gross profit/I	0.0329
8	Sales profit/I	0.023
9	Profit before tax/I	0.0276
10	Net profit/I	0.0239
11	Profit before tax per share	0.02276
12	Equity capital per share (EC1)	0.05608
13	USD/RUB	0.654176
14	EUR/USD	0.11235
15	RF CB base rate	0.0775
16	Output target value for ANN: Market stock value/EC1	0.312

IOP Conf. Series: Materials Science and Engineering **966** (2020) 012001 doi:10.1088/1757-899X/966/1/012001

The number of instructional examples was equal to 15 and 16 input values (a set of training data contained 240 values) were included into each example. In the end, ANN calculated the ratio of the market value of one share to equity capital per share.

3. ANN description and its training

Artificial neural network (ANN), the structure of which is given in Figure 1, was used for simulation. Logistic function was used as compressive function [21]. The number of neurons in the hidden layer was selected to be equal to 15, and in the outer layer – 2 (one neuron was backup).

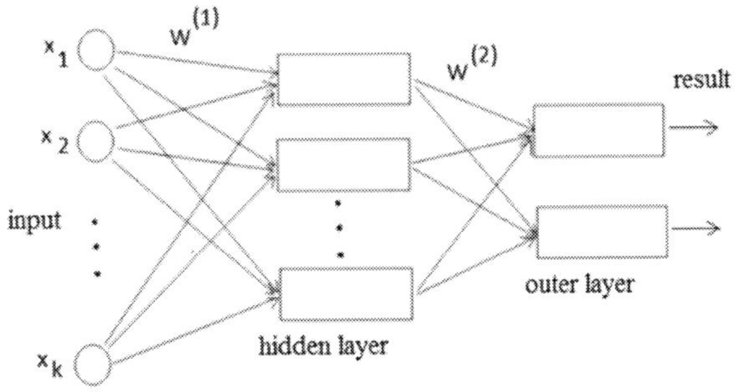

Figure 1. ANN structure.

For ANN training the backpropagation method was used (see, for example, [21]), within the framework of which functionality of type was minimized

$$E(w) = \frac{1}{2}\sum_{t=1}^{T}\sum_{l=1}^{n}(y_{lt} - d_{lt})^2 \rightarrow \min$$

where T – number of instructional examples, which the given result d_{lt} corresponds to (these are the values of the Market stock value/EC1 – see Table 1); n – number of neurons in the output layer ($n = 2$); y_{lt} – egress of neural network during the training.

Minimization parameters were weight matrices $w^{(1)}$ and $w^{(2)}$. The searching of elements of neural network weight matrices was carried out using the conjugate gradient method (see, for example, [22]). Within the framework of this approach it is necessary to determine the partial derivatives of the functionality $E(w)$ with respect to the elements of weight matrices $w_{ij}^{(1)}$ and $w_{ij}^{(2)}$. The following was obtained

$$\frac{\partial E}{\partial w_{jk}^{(2)}} = \sum_{t=1}^{T}\left(y_{kt}^{(2)} - d_{kt}\right)\cdot(1 - F(S_{kt}^{(2)}))\cdot F(S_{kt}^{(2)})\cdot y_{jt}^{(1)},\quad j = 1...n,\ k = 1...p \tag{1}$$

$$\frac{\partial E}{\partial w_{ij}^{(1)}} = \sum_{t=1}^{T}\sum_{l=1}^{p}\left(y_{lt}^{(2)} - d_{lt}\right)\cdot\left(1 - F\left(S_{lt}^{(2)}\right)\right)\cdot F\left(S_{lt}^{(2)}\right)\cdot\left(1 - F\left(S_{jt}^{(1)}\right)\right)\cdot F\left(S_{jt}^{(1)}\right)\cdot x_{it}\cdot w_{jl}^{(2)} \tag{2}$$

$i = 1...m,\ j = 1...n$

Here, the following designations were used

$$S_{jt}^{(1)} = \sum_{i=1}^{m} x_{it}\cdot w_{ij}^{(1)},\ j = 1...n;\quad y_{jt}^{(1)} = F(S_{jt}^{(1)}) = F\left(\sum_{i=1}^{m} x_{it}\cdot w_{ij}^{(1)}\right)$$

$$S_{lt}^{(2)} = \sum_{j=1}^{n} y_{jt}^{(1)}\cdot w_{jl}^{(2)},\ l = 1...p;\quad y_{lt}^{(2)} = F(S_{lt}^{(2)}) = F\left(\sum_{j=1}^{n} y_{jt}^{(1)}\cdot w_{jl}^{(2)}\right)$$

Formulas (1)–(2) determine the partial derivatives of the objective function $E(w)$ on the weight of neurons of the second (outer) and first (hidden) layers.

The following calculations were carried out within the framework of the conjugate gradient method [22]:

- set x_0 and calculate $p_0 = -g_0$ at the first step;
- calculate $p_k = -g_k + \beta_k \cdot p_{k-1}$, $\beta_k = \dfrac{\|g_k\|}{\|g_{k-1}\|}$, $x_k = x_{k-1} + \eta \cdot p_k$ at the step k.

Where g_k – gradient vector, the components of which are determined by formulae (1) and (2); x_k – vector, containing values of the elements of weight matrices $w_{ij}^{(1)}$ and $w_{ij}^{(2)}$; η - parameter.

The calculations are completed if vector components g_k become less than the specified accuracy of calculations. The result of using ANN is expressed with the following scheme

$$\begin{pmatrix} x_1 & \dots & x_m \end{pmatrix} \cdot w^{(1)} \rightarrow \begin{pmatrix} F\left(S_1^{(1)}\right) \\ \vdots \\ F\left(S_n^{(1)}\right) \end{pmatrix} = \begin{pmatrix} y_1^{(1)} \\ \vdots \\ y_n^{(1)} \end{pmatrix} \rightarrow \begin{pmatrix} y_1^{(1)} & \dots & y_n^{(1)} \end{pmatrix} \cdot w^{(2)} \rightarrow \begin{pmatrix} F\left(S_1^{(2)}\right) \\ \vdots \\ F\left(S_p^{(2)}\right) \end{pmatrix} = \begin{pmatrix} y_1^{(2)} \\ \vdots \\ y_p^{(2)} \end{pmatrix}$$

ANN training was carried out with an allowable error of 0.00001 and was equal to 204,744 iterations of the conjugate gradient method.

4. Results and discussion

Figure 2 shows the results of ANN training, at which designation 'Market value/Capital' characterizes the ratio of the market value (according to data of stock trading) to equity capital ('Capital and reserves' indicator according to data of balance sheet). The average values of discrepancies between the actual value of MV/EC and calculated value by means of ANN were 6.0, 2.5, 8.2 % for NLMK, Severstal and MMK, respectively.

The trained ANN will be used for assessment of market indicators of the companies in the same industry (ferrous metallurgy) – NTMK and Mechel. It should be noted that NTMK shares are not listed at stock exchanges; therefore, calculation of its market value can be carried out through the involvement of ANN trained according to the analogue companies. The accounting statement analysis [17] of Mechelcompany shows its spotty economic status, therefore, the assessment of its market value using the trained ANN brings us to the conclusion on conformity of Mechel equity stock-exchange quotations to the same values for NLMK, Severstal, MMK.

Calculation results shown in Figure 3 indicate that the MV/EC indicator for NTMK is in the range of 2.86–4.90. The calculated (ANN) and actual indicators of MV/EC for Mechel (Figure 3) are substantially different from each other, especially for 2017 and 2018. This brings us to the conclusion that the level of Mechel equity stock-exchange quotations in this period is overestimated in relation to the market averages.

Figure 4 shows the actual market values of NLMK, Severstal and MMK enterprises, as well as the market values of NTMK calculated using ANN. These data bring us to the conclusion that NTMK is at the level of leading iron and steel enterprises of the Russian Federation according to the economic indicators.

IOP Conf. Series: Materials Science and Engineering **966** (2020) 012001 doi:10.1088/1757-899X/966/1/012001

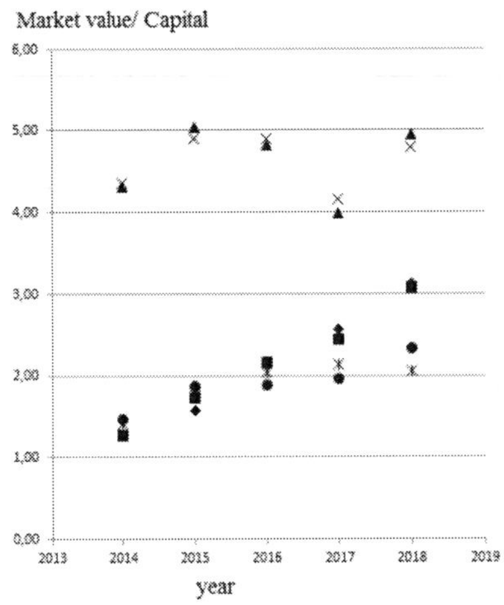

Figure 2. Results of ANN training: rhomb, square are actual values and results of ANN for NLMK; triangles, dashes are actual values and results of ANN for Severstal; asterisks, circles are actual values and results of ANN for MMK.

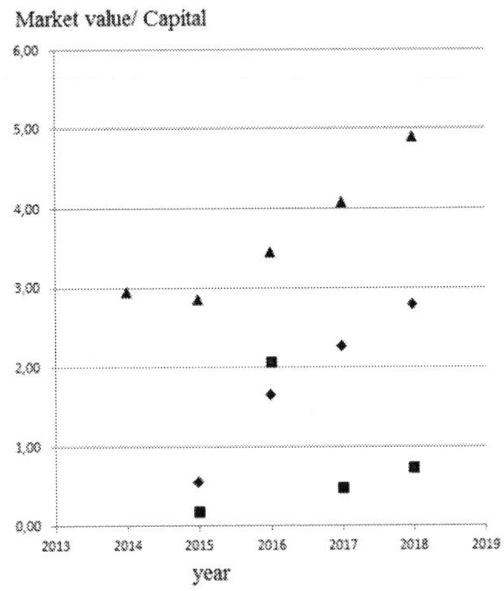

Figure 3. Actual values and calculation results. Triangles are calculation for NTMK using ANN; rhombs, squares are actual values and results of ANN for Mechel

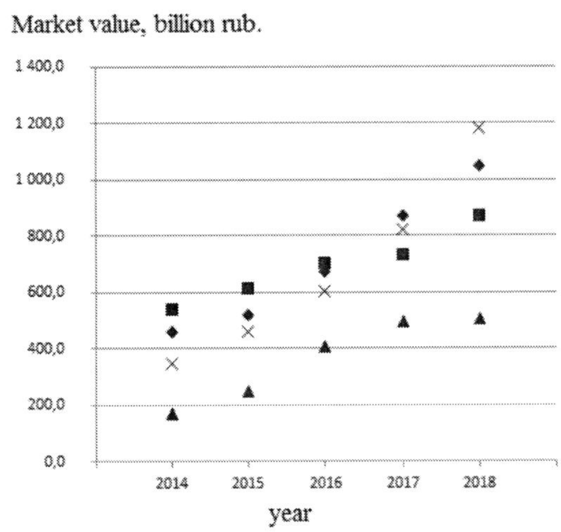

Figure 4. Market value change trends. Rhombs, squares, triangles are actual data for NLMK, Severstal, MMK, respectively; dashes are calculated values for NTMK, obtained using ANN.

5. Conclusions

1. The artificial neural network for assessment of the market value of iron and steel enterprises of the Russian Federation has been composed.

2. The use of artificial neural network brings us to the conclusion that the level of equity stock-exchange quotations of PJSC Mechel is overestimated in relation to the market averages within the period of 2017–2018.

3. The calculated market values of JSC 'EVRAZ Nizhniy Tagil Iron and Steel Works' indicate that the company is at the level of leading iron and steel enterprises of the Russian Federation according to the economic indicators.

The calculations were performed with the program in C#.

IOP Conf. Series: Materials Science and Engineering **966** (2020) 012001 doi:10.1088/1757-899X/966/1/012001

References

[1] Makeeva E Yu and Arshavsky I V 2014 Integration of Neural Networks and Semantic Interpretation for Bankruptcy Prediction *Journal of Corporate Finance Research* **4 (32)** pp 130–41
[2] Podvalnyy E S and Masloboyshchikov E V 2011 Features of Neural Network Prediction of Financial Time Series *Bulletin of Voronezh State Technical University* pp 25–9
[3] David E and Suraphan Th 2005 The use of data mining and neural networks for forecasting stock market returns *Expert Systems with Applications* **29** pp 927–40
[4] Hsieha T-J, Hsiaob H-F and Yeh W-Ch 2011 Forecasting stock markets using wavelet transforms and recurrent neural networks: An integrated system based on artificial bee colony algorithm *Applied Soft Computing* **11** pp 2510–25
[5] Pakdaman M, Hamidreza N, Homa T and Hashemi B 2010 Stock Market Value Prediction Using Neural Networks *International Conference on Computer Information Systems and Industrial Management Applications (CISIM)* pp 132–6
[6] Long J, Chen Zh, He W,Wu T and Ren J 2020 An integrated framework of deep learning and knowledge graph for prediction of stock price trend: An application in Chinese stock exchange market *Applied Soft Computing* **91** 106205
[7] Jasemi M, Kimiagari A M and Memariani A 2011 A modern neural network model to do stock market timing on the basis of the ancient investment technique of Japanese Candlestick *Expert Systems with Applications* **384** pp 3884–90
[8] Moghaddama A H, Moghaddamb M H and Esfandyaric M 2016 Stock market index prediction using artificial neural network *Journal of Economics, Finance and Administrative Science* **21** pp 89–93
[9] Vukovic D,Vyklyuk Y, Matsiuk N and Maiti M 2020 Neural network forecasting in prediction Sharpe ratio: Evidence from EU debt market *Physica A: Statistical Mechanics and its Applications* **542** 123331
[10] Hiransha M, Gopalakrishnan E A, Vijay Krishna Menona and Soman K P 2018 NSE Stock Market Prediction Using Deep-Learning Models *Procedia Computer Science* **132** pp 1351–62
[11] Lesik I A 2015 Forecasting Stock Price Growth Using Feedforward Neural Networks *Software & Systems* **2 (110)** pp 70–4
[12] Filin V V and Ledeneva T M 2012 The Neural Network Prediction of Share Price by Nonlinear Dynamic's Methods *Bulletin of Voronezh State Technical University* pp 57–60
[13] Butusov O B, Meshalkin V P, Nikiforova O P, Smoller A V and Nigmatullin R M 2011 Information System for Forecasting the Profitability of Unit Investment Trusts Using Backpropagation Neural Network *Journal of Applied Informatics* **1 (31)** pp 12–6
[14] https://www.audit-it.ru/buh_otchet/4823006703_pao-novolipetskiy-metallurgicheskiy-kombinat (accessed on 01.06.2020)
[15] https://www.audit-it.ru/buh_otchet/7414003633_pao-magnitogorskiy-metallurgicheskiy-kombinat (accessed on 01.06.2020)
[16] https://www.audit-it.ru/buh_otchet/3528000597_pao-severstal (accessed on 01.06.2020)
[17] https://www.audit-it.ru/buh_otchet/7703370008_pao-mechel (accessed on 01.06.2020)
[18] https://www.audit-it.ru/buh_otchet/6623000680_ao-evraz-nizhnetagilskiy-metallurgicheskiy-kombinat (accessed on 01.06.2020)
[19] https://ru.investing.com/equities/nlmk_rts (accessed on 01.06.2020)
[20] https://cbr.ru/ (accessed on 01.06.2020)
[21] Wasserman F 1992 *Neural Computing: Theory and Practice* (Moscow: Mir)
[22] Himmelblau D M 1972 *Applied Nonlinear Programming* (McGraw-Hill Book Company: The University of Texas)

IOP Conf. Series: Materials Science and Engineering **966** (2020) 012002 doi:10.1088/1757-899X/966/1/012002

Comparative evaluation of effectiveness indexes of column packings in benzol absorbers

E V Nogovitsyna

Ural Federal University named after the First President of Russia B N Yeltsin, Nizhniy Tagil Technological Institute, 59, Krasnogvardeyskaya str., Nizhniy Tagil, 622000, Russia

E-mail: helen_nog@mail.ru

Abstract. Coke-oven gas is produced during coking of charge coal in a coking chamber and it constitutes a mix of different substances. A composition of different components in the gas depends on a composition of the coal charge and conditions of the coking. At a recovery plant, aromatic hydrocarbons are retrieved from the coking gas by absorption oil in packed absorbers. Mass-transfer apparatus demand the equability of the phase distribution. Hydrodynamic conditions, appearing inside contacting apparatus, determinate the effectiveness of the mass-transfer, which means that unbiased evaluation of any given mass-transfer apparatus can be made while simultaneous considering its hydrodynamic and mass-transfer indicators. The different types of column packing have been investigated in this study. The calculations of benzol absorbers have been performed basing on the initial data using the standard methodologies for the recovery plant of JSC EVRAZ NTMK coke and by-product enterprise. For comparison of the results obtained, the complex indexes of effectiveness, that simultaneously take into account the mass-transfer rate, the device capacity, the features of packings and energy cost, have been used.

1. Introduction

The most part of by-product coke plants is characterized by excessive losses of raw benzol with a reverse coke gas, which explains the practical interest in studying the technology of raw benzol recovery from the coke gas and in studying technical means ensuring effectiveness of an absorption process. After the final gas coolers, the coke gas contains 32–35 g/m^3 of aromatic hydrocarbons. The basis of the latter is benzol itself and its homologues, that are toluene, xylenes and trimethylbenzenes [1].

The effectiveness of the aromatic hydrocarbons recovery from the coke gas depends on: compositions of the gas and absorption oil, temperature and pressure, hydrodynamic conditions of phase motion in the apparatus. The composition of the coke gas depends on the coal charge composition and conditions of coking. The composition of the absorption oil is controlled by the central laboratory. The absorbing capacity of oils decreases with increase of the temperature and growth of the molecular weight of an absorbent and its viscosity. Furthermore, the absorbing capacity of the oil significantly decreases in the case of water influx [2].

It is known that if the temperature reduces, the solubility of gases in the absorbent increases. At the temperature of 10°C, the viscosity of the oil increases sharply, and precipitation is also possible, which contaminates the packing and increases the hydraulic resistance of the absorber. For that reason, the process is conducted at the optimum temperature of the gas of 25°C. To avoid the condensation of water vapor from the coke gas and water influx of the wash oil, the temperature of the latter is kept by 2–3 degrees above the gas temperature [3].

The hydrodynamic conditions of the phase motion depend on selection of the contact devices in the absorber and they considerably determine a surface of inter phase contact, an operating mode of apparatus and its hydraulic resistance. Thus, the intensity of the mass-transfer at the time of absorption depends on selection of a packing. The data on influence of a packing surface on an extent of benzol hydrocarbons absorption is given in [4].

The upcoming reconstruction of the recovery plant of JSC EVRAZ NTMK coke and by-product enterprise, aimed at consolidation of two departments, makes this study relevant, since during the reconstruction it is possible to provide an option of substitution of an old packing in the benzol absorbers.

2. Main part

Due to the petrochemical industry development, a large number of new types of packings of regular and irregular type from various manufacturers have recently came to the market, including: EPC 'Ingehim'; 'Sulzer Chemtech'; RPC 'Kedr-89'. Detailed description of the packings, recommendations on their work conditions, calculation and empirical data on their operation are introduced in separate scientific articles [5–8] and monographs [9–11].

The following factors are essential in terms of the packing selection in the by-product coke industry: low hydraulic resistance, ability to operate steadily under strongly varying gas loads, ability to remove deposited sludge from the packing surface quickly and on the cheap, high corrosion resistance and mechanical strength [12].

In order to calculate the parameters of the absorbers, the following values [12], that are compliant with the data [1] and exceed the average performance indicators of recovery plant No. 3 in the second half of 2018, have been taken as the initial data (Table 1).

Table 1. Initial data for calculation.

Gas production rate, m^3/h	100.000
Content of BH in the gas at the inlet to absorber, kg/m^3	$32.5 \cdot 10^{-3}$
Content of BH in the gas at the outlet from absorber, kg/m^3	$3 \cdot 10^{-3}$
Content of BH in the washing oil at the inlet to absorber, %	0.25
Temperature inside the apparatus, °C	30
Gas pressure at the inlet to absorber, Pa	$0.12 \cdot 10^6$
Density of the coke gas, kg/m^3	0.453
Density of the wash oil, kg/m^3	1.010

Currently a Z-shaped packing and a wood grid packing are used in the absorbers at JSC EVRAZ NTMK coke and by-product plant. This study investigates more up-to-date types of regular and irregular packings offered by the manufacturers 'Sulzer Chemtech' [13], LLC EPC 'Ingehim' [14] and others considered in details in [9].

In the industry, irregular packings with a lower cost are still used to a much greater extent, especially at large-tonnage plants with large-diameter mass-transfer apparatus for realization of absorption-desorption processes. Irregular packing Ingehim-2000 and Leva have been chosen basing on the results of evaluation [9], where it is shown that they have the highest relative effectiveness in regard to the Raschig ring packing.

Packing Ingehim-2000 (Figure 1) is made of a sheet or strap of corrosion-resistant or carbon steel with the thickness from 0.3 to 2 mm. This packing is applicable to the polluted and corrosive environments. A surface of the sheet is exposed to special metalwork enhancing liquid spreading on the packing.

Packing Leva (Figure 2) is a metal saddle packing in the shape of semi-ring with rectangular perforation and petals bent inwards. This packing forms packed layers with a quite loose structure, due to which it has high capacity and low hydraulic resistance.

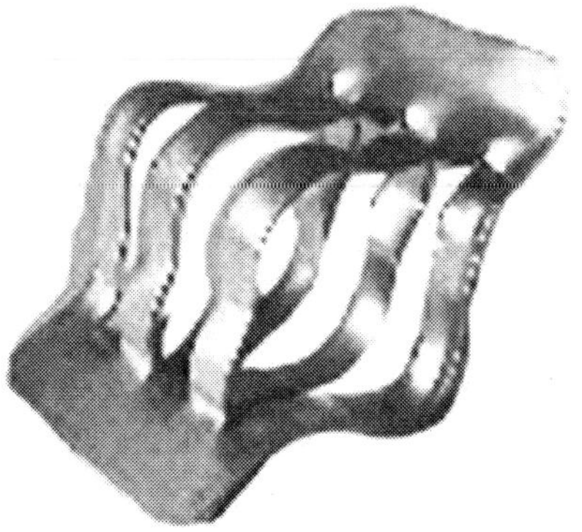

Figure 1. Element of packing Ingehim-2000.

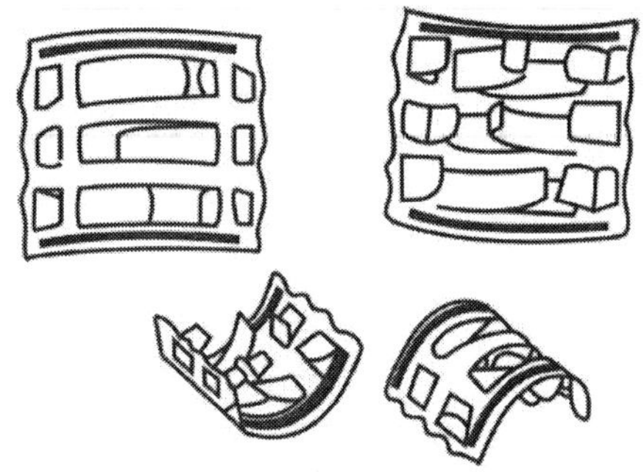

Figure 2. Elements of packing Leva.

Regular packings have a number of advantages compared to irregular ones: absence of liquid hold-up spots at points of contact between elements of an irregular packing, presence of theoretic possibility of 100% wetting of the packing surface with a liquid. Presence of regular-shaped channels contributes to lower hydraulic resistance per height unit of a packed layer and greater capacity. An enlarged surface of phase contact leads to an increase in the effectiveness [9].

Regular block packings have recently gained common use due to significant simplification of works on set up of the packings into the absorber. In terms of its properties, a block packing is as good as other types of regular packings [15]. The disadvantages of the regular block packings include sensitivity to equality of distribution of irrigation over the entire area of the packing and high cost.

Figure 3. Packing Mellapak.

Figure 4. Packing Vakupak.

Figure 5. Packing IRG.

Packing Mellapak produced by 'Sulzer Chemtech' (Figure 3) consists of inter-bedded layers of a perforated metal sheet of a light thickness (0.15–0.25 mm). Corrugation in adjoining layers of the packing are turned in opposite directions.

Indigenous packing Vakupak (a joint development project of VNIINEFTEMASH with 'Apparate- und Anlagenbau Germania' company [16,17]) (Figure 4) constitutes itself a package of vertical plates fabricated by a horizontal corrugation method. The corrugation has bent and downward directed protrusions with holes that provide an assigned distance between adjoining plates during production of a packing. The liquid flows down the corrugation in the form of a film, and the gas moves from the bottom upwards in gaps between corrugated sheets. The protrusions of the corrugation provide intensive internal mixing and breakdown of liquid and vapor flows [13].

Packing IRG (Ingechim Regular Gofr; EPC 'Ingehim') (Figure 5) constitutes itself a package of vertically installed metal corrugated plates. The corrugation of the plates is located at an angle to the horizon. The package is arranged in a way that a corrugation of adjoining plates is located crosswise.

The parameters of the packings necessary for calculation of the absorbers are presented in Table 2.

Table 2. Characteristics of packings.

Parameter	Regular packings			Irregular packings	
	Vakupak [17]	Mellapak [18]	IRG [9]	Ingehim-2000 [14]	Leva [9,19]
Specific surface area of a packing a, m^2/m^3	115.0	92.0	162.0	62.2	118.0
Void volume ε, m^3/m^3	0.980	0.954	0.980	0.967	0.980
Equivalent diameter of a packing d_e, m	0.035	0.042	0.024	0.056	0.033

The calculations have been performed according to the standard methodology suggested in [12]. The results show (Table 3) that the main regularities are fulfilled: the higher the mass-transfer coefficient value, the smaller the required mass-transfer surface. The total height of a packing is calculated taking into account data on the specific surface (Table 2).

Table 3. Parameters of absorbers with different packing types.

Parameter	Regular packings			Irregular packings	
	Vakupak	Mellapak	IRG	Ingehim-2000	Leva
Mass-transfer coefficient $K_s \cdot 10^{-4}$, kg/($m^2 \cdot s$)	0.79	0.35	0.54	1.14	0.56
Mass-transfer surface $F \cdot 10^5$, m^2	12.85	28.16	18.24	8.55	17.39
Diameter of an absorber d, m	3.2	3.8	3.6	3.2	4
Height of a packing H, m	129	99	107	143	117
Hydraulic resistance of an absorber ΔP, Pa	8,224	7,686	6,029	4,784	3,572
Number of absorbers n, pcs	4	3	3	4	4

Thus, the mass-transfer coefficient of an absorber with packing Ingehim-2000 has a maximum value, however, the total height of the packing turns out to be larger than in other cases due to the small specific surface. The optimum total height of the packing is ensured by the contact devices like Mellapak and IRG, which allows conducting retrieval of the aromatic hydrocarbons in three sequentially installed absorbers.

3. Results and discussion

To evaluate the operation effectiveness of different column packings of absorption apparatus, the authors [4, 20] suggest to use value I as a summarizing indicator of intensity. It comprises the capacity and intensity characteristics of the inter-phase exchange:

$$I = \frac{w}{h}$$

where w is the maximum allowed gas velocity in the cross-section of the apparatus, m/s; h is the height of mass transfer unit, m.

To compare the energy losses for gas passing through the absorber [4,20], it is allowed to use index $N/\Delta P$ (kPa^{-1}), where N means a number of mass transfer units.

Table 4. Indexes of operation effectiveness of column packings.

Parameter		Regular packings			Irregular packings	
		Vakupak	Mellapak	IRG	Ingehim-2000	Leva
Limiting gas velocity ω_{lmt}, m/s		8.4	5.7	7.1	8.1	5.3
Height of mass transfer unit h, m		2.08	2.29	2.34	2.22	2.91
Indexes of intensity	I	4.05	2.49	3.03	3.65	1.83
	$N/\Delta P$	0.43	0.46	0.58	0.73	0.98

When comparing different options for equipping the absorber, a technical solution capable of providing the maximum limiting velocity and $N/\Delta P$ at a minimum height of transfer unit is admitted optimal. Under otherwise equal conditions, an option with the greatest value I is selected [20].

With regard to the results given in Tables 3 and 4, based on the total of parameters, packing IRG can be recommended to use in the absorbers: the plant will be comprised of three absorbers, likewise Mellapak, but in contrast to it, packing IRG has higher indexes of intensity. Packing Leva has the highest values $N/\Delta P$, which corresponds to the minimum energy cost, but is inferior in all other indicators.

References

[1] *Coke Chemical Engineer Guide* 2009 ed Ye T Kovalev (Kharkov: Publishing House 'INZhEK') vol **3** p 432 [In Russian]

[2] Sabirova T M 2018 *Basic Principles of Chemical Products Recovery and Processing Technology: Study Guide* (Yekaterinburg: Publishing House of Ural University) [In Russian]

[3] Grebenyuk A F, Korobchanskiy V I, Vlasov G A and Kaufman S I 2002 *Recovery of Chemical Products of Coking* P2 (Donetsk: Eastern Publishing House) [In Russian]

[4] Brodskiy E V, Babitsin S M and Gudimenko S V 1986 Prospects for Intensification and Increasing the Efficiency of Benzol Scrubbers *Coke and Chemistry* **9** pp 32–6 [In Russian]

[5] Pushnov A S, Kagan A M, Berengarten M G, Ryabushenko A S and Stremyakov A V 2008 Regular packing for heat and mass exchange processes by the direct contact of phases *Engineering Mechanics* **1** pp 13–7 Available online: http://www.engineeringmechanics.cz/pdf/15_1_013.pdf (accessed on 1 June 2020)

[6] Dyakonov G S, Farakhov M I, Maryakhin N N, Yasaveev M Kh, Shigalov I M and Laptev A G 2000 Design of New Irregular Packing and its Hydrodynamic Research *Heat and Mass Transfer Processes and Devices of Chem. Tech. Interuniversity Subject Collection of Scientific Works* (Kazan: KSTU) pp 239–48 [In Russian]

[7] Farakhov T M, Basharov M M and Shigapov I M 2011 Hydraulic Properties of New Highly Irregular Heat and Mass Exchange Attachments *Oil and Gas Engineering* **2** pp 192–207 [In Russian]

[8] Leontiev V S 2012 Innovative Development of High-Intensity Mass Exchange Units to Upgrade Rectification Complexes *Oil and Gas Engineering* **1** pp 178–86 [In Russian]

[9] Kagan A M, Laptev A G, Pushnov A S and Farakhov M I 2013 *Contact Packings in Industrial Heat-and-Mass Transfer Apparatuses* ed A G Laptev (Kazan: Otechestvo) [In Russian]

[10] Reinhard Billet 1995 *Packed towers in processing and environmental technology* (Wiley-VCH)

[11] Sokol B A, Chernyshev A K, Baranov D A, Berengarten M G and Levin B V 2009 *Packings of Mass Transfer Columns* ed D A Baranov (Moscow: Infokhim) [In Russian]

IOP Conf. Series: Materials Science and Engineering **966** (2020) 012002 doi:10.1088/1757-899X/966/1/012002

[12] *Basic Processes and Apparatuses of Chemical Technology: Engineering Guide* 1983 ed Yu I Dytnerskiy (Moscow: Chemistry) [In Russian]

[13] *Information Bulletin of Sulzer Chemtech Regular Packing for Absorption and Rectification Processes.* Available online: https://www.sulzer.com/-/media/files/products/separation-technology/distillation-and-absorption/brochures/structured_packings.ashx?la=en (accessed on 1 June 2020)

[14] *Information Bulletin of LLC EPC 'Ingehim' Column Equipment.* Available online: https://ingehim.ru/files/kolonnoe-oborudovanie.pdf (accessed on 1 June 2020) [In Russian]

[15] Muzafarova A R and Yemelyanycheva Ye A 2016 Classification, Basic Requirements and Design Features of Modern Packings *Bulletin of Technical University* **2** pp 63–7 [In Russian]

[16] Lebedev Yu N, Chekmenyov V G and Vikhorev M B 2015 Import-Substituting Equipment for Oil and Gas Processing and Petrochemical Industries *Chem. Eng.* **1** pp 26–31 [In Russian]

[17] Skoblo A I, Molokanov Yu K, Vladimirov A I and Shchelkunov V A 2000 *Processes and Equipment of Oil and Gas Processing and Petrochemistry* (Moscow: Nedra-Business Center) [In Russian]

[18] Ratovskiy Yu Yu, Lebedev Yu N and Chekmenyov V G 2004 Vakupak and Kedr Packing for Vacuum Towers on AVT Units *Chemistry and Technology of Fuels and Oils* **1** pp 55–6 [In Russian]

[19] USSR Authorship Certificate No. 1375302 1987 *Packing for heat and mass exchange apparatuses* Available online: https://patents.su/4-1375302-nasadka-dlya-teplomassoobmennykh-apparatov.html (accessed on 1 June 2020) [In Russian]

[20] Umnov M A, Khomin V G, Brodskiy E V and Kolivashko A S 2003 Optimal Parameters of Absorption Devices *Coke and Chemistry* **5** pp 27–31

IOP Conf. Series: Materials Science and Engineering **966** (2020) 012003 doi:10.1088/1757-899X/966/1/012003

Substantiation of parameters of underground geotechnology in the development of complex structural deposit of various grades of ores

A M Mazhitov [1,2] and P V Volkov [1]

[1] Nosov Magnitogorsk State Technical University, 38, Lenin street, Magnitogorsk, 455000, Russia

E-mail: [2] artur.mazhitov@yandex.ru

Abstract. Analysis of mining production activities at the current stage of field development shows an acute deficit in mineral raw materials. At present, most enterprises have exhausted all their prospective reserves and inevitably return to the reserve. Such sites have as a rule a low content of the useful component, are composed of different-grade ores and are located in difficult mining and geological and technical conditions.Their development requires significant capital investment and accordingly this leads to a decrease in the profitability of production. In order to maintain production capacity and extend the life of the field, companies are forced to ensure their development, despite a decrease in overall profits.The search for possible technologies for mining these areas of deposits is an urgent and priority direction for many enterprises. So, in the presented report are the optimal variants of the development of deposits shown, it is represented by different-grade ore bodies, most of them are stacked off-balance ores. The presented solutions make it possible to increase the life of the enterprise by increasing the reserves of previously substandard ores.

1. Introduction

After working off the main reserves of copper pyrite deposits, as a rule, there are reserves dispersed throughout the mine field in the form of separate lenticular low-power ore bodies with complex morphology and low content of the useful component. The company is forced to work out such areas, both under the condition of compliance with the requirements for full recovery of reserves, and to maintain production capacity and extend the life of the field [1–5].

In modern mining, we have accumulated vast experience in working out low-power and dispersed ore bodies that represent the main reserves of the Deposit. The question of working off of newly discovered, left 'or later' complex ore bodies stays unaffected and highly relevant at the present time.

The solution of this question is made on the example of the development of the I-th Deposit of the Kamagansky field (Republic of Bashkortostan), represented by a group of lenticular ore bodies with a capacity from 1–1.5 m to 20 m and an angle of incidence from 0 to 30° [6,7]. In total, 15 ore bodies were identified, lying in tiers in the range of 47.4 to 726.5 m, the dimensions along the stretch from 50 to 320 m, in the cross – from 20 to 160 m. Ores (both balance and off-balance) are mostly thickly interspersed, in the form of debris, rarely there are separate intervals of solid ores with a content of copper from 0.68 to 2.9 % and zinc from 1.0 to 2.3 %. Ore bodies have interlayers of waste rocks.

The operational reserves of the Deposit (up to the border of the mining lease) are 703.45 thousand tons.

At present, the ores of the I-th Deposit are the necessary raw material base for replenishing the outgoing capacity of the Kamagansky field and maintaining the stable operation of the enterprise for

IOP Conf. Series: Materials Science and Engineering **966** (2020) 012003 doi:10.1088/1757-899X/966/1/012003

the coming years. Therefore, realization of research on the feasibility study of underground geotechnologies for the development of the I-th Deposit is a very relevant issue at this stage of the development of the enterprise.

Due to the scatter of ore bodies along the mining lease, the development of the Deposit requires significant capital investment and as a result, will lead to unprofitable operation of the mine. However, write-off of these reserves is not possible due to the relatively large volumes of reserves and the amount of metal. The conducted research has shown the need for joint development of the I-th Deposit and the main reserves of the field. At the same time, there is an increase in profit due to the prolongation of term of development of the field.

2. Research methods

The paper uses a comprehensive research method, that includes along with design solutions, analysis and generalization of the practice of developing dispersed long-lying ore bodies, analytical research, methods of mathematical statistics and technical and economic analysis, as well as economic and mathematical modeling.

3. The main part

An acute shortage of raw materials for production of mining operations forces enterprises to solve issues related to the development of areas of the field that were once left in the subsurface, due to the complexity of their occurrence, or low profitability of development. Currently, this issue is relevant and requires a comprehensive scientific and technical justification of options and parameters of geotechnical development of such sites.

So, because of depletion of operational reserves by the development of the Oktyabrsky field sites (ore bodies) scattered across the mine field are putting into commercial operation. As an example, we can give design solutions for the development of reserves in the Tashkulinsky area [8,9]. Technical and economic modeling of options for opening and working out this section predetermined the maximum use of existing mining and capital workings. At the same time, the option of geotechnology with the lowest possible time for commissioning of the site and the procedure for mining ore bodies was determined. First of all, ore bodies 31 and 34, located between the horizons of 220–300 m, are worked out. Ore bodies located above the horizon of 220m are worked out in a retreating order-28, 29, 33, 30 and 27.

Thus, the use of existing workings of the Oktyabrsky underground mine by opening up the reserves of the Tashkulinsky area allowed to reduce the volume of mining and capital works.

Design solutions for refining the ore bodies of the Kamagansky Deposit also prove the feasibility of putting previously unprofitable sections into operation [10].The design took into account the main requirement for the development of this type of field, which is a quick return on investment, and therefore a minimum time to achieve the design capacity and intensity of development of the field.

A striking example of commissioning an unprofitable section of the field is the design solution for underground geotechnology for the development of 'non-core' sections of the I-th Deposit of the Kamagansky field, represented by dispersed low-power ore bodies.

The Kamagansky field is a classic implementation of sequential combined geotechnology. All underground reserves are opened from the quarry by three tunnels located on different horizons. Reserves subject to underground development are located in the North-Western (section of ore bodies (O.B) 3 and 5) and North-Eastern (O.B 12) sides of the quarry. At present O.B. 12 has been fully worked out and O.B. 3 and 5 are being finalized.

After detailed operational exploration the contours of the O.B. 12 were clarified, whose callout was not developed by open pit mining in the southern part of the field.

The I-th Deposit under consideration is located a little to the West of O.B 12. A peculiarity of the site is the longline occurrence of numerous low-power ore bodies with the content of a useful component, that varies in a wide range. Development of the site is complicated by the large indentation of the array by tectonic disturbances and numerous cracks of the ordinary order, as well as the close location of the upper ore body to the surface (less than 50 m).Moreover, in the zone of influence of underground mining operations is an oil depot, which belongs to the second category of

IOP Conf. Series: Materials Science and Engineering **966** (2020) 012003 doi:10.1088/1757-899X/966/1/012003

protection.

Geomechanical calculations and mathematical modeling have shown that the predicted values of surface deformations do not exceed the permissible values for this category of protected objects, so there is no need to move oil base buildings beyond the displacement zone [11].

For the purpose of effective implementation of underground geotechnology was zoning of ore bodies of the I-th Deposit according to the elements of occurrence and the content of the useful component conducted and intervals of application of underground mining systems were established.

The presence of protected objects has predetermined the use of development systems with a full laying of the developed space or leaving the regular ore and the artificial pillars by the low power of the ore bodies.

The final choice of development systems was made by economic and mathematical modeling of options for each ore body, based on the criteria of maximum profit and minimum cost [12,13].

Based on the empirical dependencies on the comparative evaluation of mining systems given in V V Grigoriev's dissertation [14], relevant indicators were calculated for various variants of mining systems and a rational variant was selected for each ore body:

- camera development system with a laying (ore bodies with a capacity of more than 10–12 m at any angle of incidence);
- room-and-pillar development with a single-layer excavation and laying at the power of the ore body 2–6 m and 6–10 m (ore bodies and clippings with a capacity of 2 to 6 m with a drop angle of up to 30°);
- room-and-pillar development with leaving of the ore and the artificial support pillars (ore bodies with off-balance and substandard reserves);
- room-and-pillar development with delivery of ore by force of explosion (sections of ore bodies with an angle of incidence greater than 30–35°).

Variants of the quarrying schemes were subjected to technical and economic modeling:

- two existing tunnels of the horizon: 140 m and 90 m;
- tunnels of the horizon 140 m and elevator rising
- two vertical barrels from the surface;
- tunnels of the horizon: 140 m and crosscut ‘Novy Sibay–Kamagan’.

Calculations have shown that in order to reduce the capital costs of dispersed ore bodies of long-range occurrence, an opening scheme is necessary with maximum use of existing capital workings, as well as conditional division of the Deposit into sections and their phased commissioning with separate airing of each as mining operations develop. Thus, the I-th Deposit of the Kamagansk field should be divided into three independent sections by depth:

- reserves located in the floor of 140–50 m (from the surface to the bottom of the quarry);
- reserves-140–330 m (from the bottom of the quarry to the borders of the mountain branch);
- reserves located below the mountain branch.

The first complex (starting). As the main opening development is tunnel on the horizon 140 m used, it serves for fresh air supply, rock mass issuance, people's entrance and exit and delivery of materials. From the tunnel behind the zone of displacement on the lower section (140–330 m)for cleaning operations passes an inclined slope on horizon 50/140 m, that connect the operational horizons of the site. The Elevator rising horizon of 0/120 m serves for the release of polluted air and as a second mechanized exit. The method of airing is injection, using the existing fan installation at the portal of the horizon 140 m ad it, which includes a VZP-16 fan. Fresh air is fed through the tunnel to the inclined exit, washes the areas of treatment works and through the Elevator rising 0/120 m is issued to the surface.

Second complex. After commissioning of the first starting complex, it is planned to build an underground mine below the bottom of the quarry to the mountain branch (330 m).It is planned to use the stand bore of the ‘NovySibay’ section of the Sibay Deposit as an air supply. For this is the existent crosscut ‘NovySibay – Kamagan’ used. From the existent crosscut passes inclined crosscut 210/330 m to the site of the I-th deposit.

The sections of the treatment works are connected to each other by an inclined exit horizon 140/330 m. As an emergency exit, it is planned to pass the Elevator rising horizon 140/330 m. The

IOP Conf. Series: Materials Science and Engineering **966** (2020) 012003 doi:10.1088/1757-899X/966/1/012003

airing method is pressure-type. Fresh air is supplied through the stand bore of the 'NovySibay' section along an inclined crosscut at horizon of 330 m. Then it is fed to the operating areas of the ore bodies along an inclined entrance, washes the treatment works and is issued through the 140 m horizon adit in the quarry.

The third complex (final). The opening of ore bodies (No 26 and 27) located below the mining branch is carried out by deepening the inclined exit and Elevator rising. The opening of the I-th Deposit of the Kamagansky Deposit is shown in Figure1.

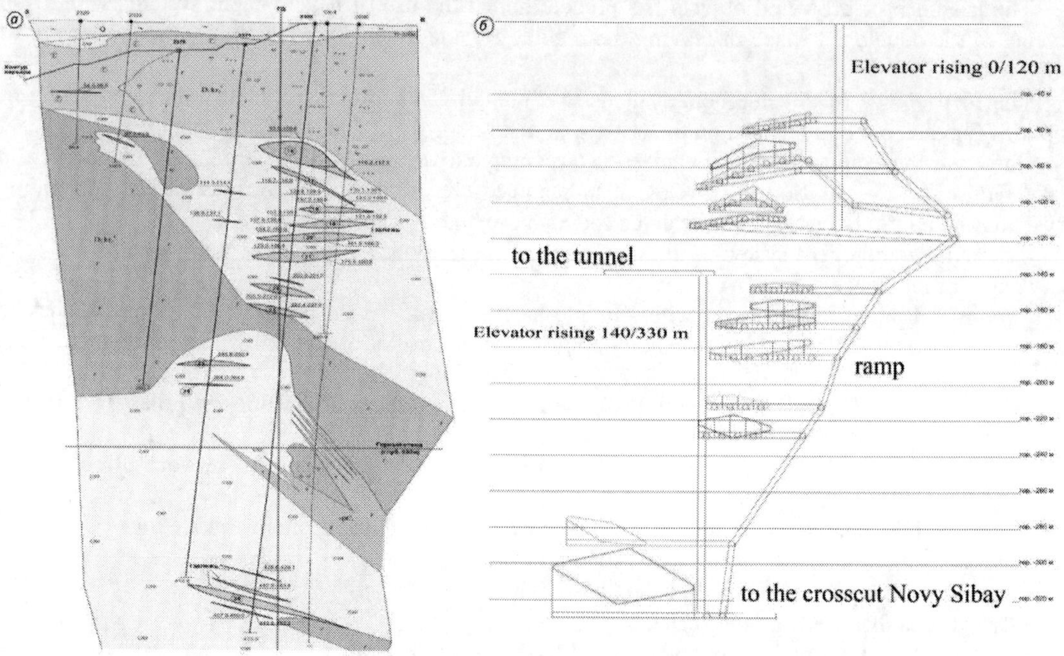

Figure 1. Geological section (a) and opening scheme of the I-th Deposit (b).

Due to the small capacity of the ore bodies was ore preparation from delivery drifts from an inclined slope accepted, that was passed in a supine side outside the zone of influence of the East Sibai tectonic fracture. An inclined slope is passed as close as possible to the dredging area in order to reduce the length of the field delivery drift [15]. Every ore body is a dredging site and the height of the floor (sub-floor) can vary depending on the morphology and conditions of occurrence, but not more than 40 m.

Preparatory work is to carry out on each excavation site mining complexes, which are necessary for production of ore reserves by the accepted development systems.

In General, the preparation of ore bodies of the Deposit is the same and consists in carrying out delivery and ventilation-laying drifts on the flanks of the ore body and brought together by drilling and delivery orts. The location of the ventilation and stowage drift is selected taking into account the difference in marks between the delivery drift. Orts are drawn along the axis of the worked out camera. Ore bodies with relatively small layers of empty rocks between them are worked out in one section. At the same time, several mining sites may be in development. The development of reserves in the lower-lying area can be started after the front of the treatment excavation in the upper section is shifted beyond the displacement zone, built at the appropriate angles, but not less than 20 m relative to the upper border of the ore body in the lower section.

The final detailed technical and economic calculation of the accepted variant of underground geotechnologies of dispersed ore bodies of long-range occurrence, which is the most optimal, showed the inefficiency of mining the reserves of the I-th ore Deposit of the Kamagansky field (at the dressing stage). In this regard, a variant of simultaneous mining of the I-th Deposit and the reserves of the lens out of O.B. 12 was analyzed. Technical and economic modeling indicators are shown in Figure 2 (profit indicators are taken as conventional units for the purpose of non-disclosure of trade secrets and are provided as a comparative analysis).

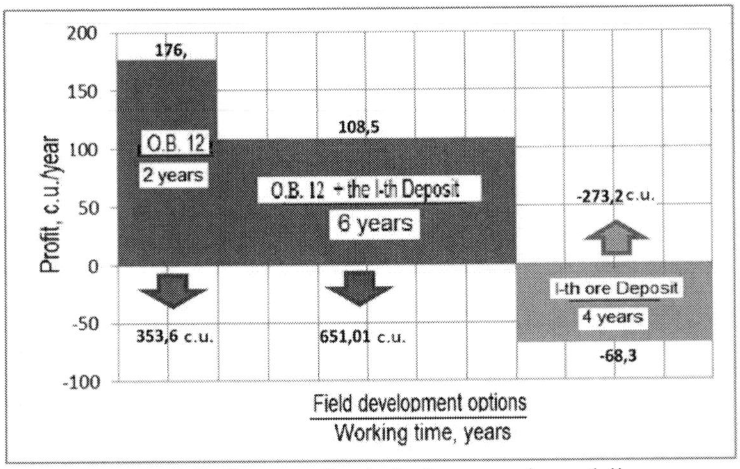

Figure 2. Indicators of technical–economic modeling.

Based on the technical and economic modeling the loss from the development of the I-th Deposit will be 273.2 conventional units. The feasibility of joint mining of the I-th Deposit and the undeveloped quarry of the southern clipping of O.B. 12 is clear. The annual profit will be 108.5 cu for a six-year period of development. Total profit is more than 651 conventional units. Individual development of the O.B. 12 increases the annual profit of the enterprise (up to 176.8 cu), but the term of development is reduced to 2 years.

Thus, the development of an unpromising site of low-power dispersed ore bodies is possible under the condition of maximum use of existing mining and capital workings. At the same time, it is necessary to divide the Deposit into operational sections with individual opening and airing method, and gradually introduce the sections of the Deposit into development. In case of tiered occurrence of dispersed ore bodies, in connection with the opening of the Deposit to the full depth, all ore bodies, both balance and off-balance, are subject to excavation. This solution allows to increase the volume of commercial ore of the site at a constant level of capital investment. It is possible to increase the profitability of the site by parallel treatment excavation of the main reserves of the field.

4. Conclusion

The paper contains a solution to an actual scientific and practical problem, which consists in the development and scientific justification of the parameters of the technology for working out a complex-structure Deposit, represented by a group of closely spaced ore bodies of long-range occurrence, which is important for the science and practice of the mining industry.

Main scientific results, practical conclusions and recommendations:

1. It is established that in the case of long-line occurrence of close ore bodies, the Deposit is worked out without combining them into a single ore zone, and it is advisable to work out all the ore bodies regardless of the quality of the ore. Opening of the Deposit is made taking into account the maximum use of existing mining and capital workings and preferred ore preparation.

2. By selecting of development systems is preliminarily zoning of the ore bodies by occurrence elements and content of a useful component produced. Zoning is performed to pre-select a particular version of the development system. The final choice of technology is made by economic and mathematical modeling of the mining of each ore body according to the indicators of maximum profit and minimum cost.

3. A rational order for the development of the mining front has been defined in order to minimize initial capital investment, which consists in dividing the Deposit vertically into sections with separate airing and putting them into operation in turn.

4. The technical and economic assessment showed that the development of deposits with a similar

structure and morphology is appropriate for maintaining the production capacity of the enterprise only in conjunction with the development of more promising areas of the field. Individual mining of the Deposit is preferable at higher prices for raw materials and the dollar exchange rate.

References

[1] Wang D S, Chang J P, Yin Z M and Lu Y G 2014 Deformation and failure characteristics of high and steep slope and the impact of underground mining *Transit Development in Rock Mechanics-Recognition, Thinking and Innovation rd: Proceedings of the 3 ISRM Young Scholars Symposium on Rock Mechanics. USA* pp 451–7

[2] Golik V I, Komashchenko V I and Kachurin N M 2016 On the problem of underground mining of ore deposits in the Central Federal district *Izvestiya Tula state University. Earth science.* iss **4** pp 127–39

[3] Jang H, Topal E and Kawamura Y 2015 Decision support system of unplanned dilution and ore-loss in underground stoping operations using a neuro-fuzzy system *Applied Soft Computing* vol **32** iss **C** pp 1–12

[4] Golik V I, Komashchenko V I, Morkun V and Gaponenko 2017 Improving the effectiveness of explosive breaking on the bade of new methods of borehole charges initiation in quarries *Metallurgical and Mining Industry* no **7** pp 383–7

[5] Wang G, Li R, Carranza E J M et al 2015 3D geological modeling for prediction of subsurface Mo targets in the Luanchuan district, China *Ore Geology Reviews* vol **71** pp 592–610

[6] *Technical and economic assessment of the development of the first Deposit of the Kamaganskoye field by underground method* 2015 (Magnitogorsk: LLC 'Uralgeoproekt')

[7] Tatarko N I 2012 *Operational calculation of reserves of copper-crushed ores of the first Deposit of the Kamaganskoye Deposit* (as of 01.01.2013) OTV.ISP (Ufa: LLC 'Bashkirgeologiya')

[8] Ilimbetov A F 2002 *Substantiation of a rational method for controlling rock pressure when working out dispersed ore bodies (on the example of the Oktyabrsky copper-coal Deposit)*: dis. ... cand. tech. science: 25.00.22 (Magnitogorsk)

[9] Rylnikova M V, Korneev S A, Mazhitov AM and Korneeva V S 2014 Selection of technological schemes for processing separate semi-underground ore bodies of the Oktyabrsky Deposit. *Surveyor's Bulletin* no **2 (100)** pp 15–9

[10] Rylnikova M V, Korneev S A, Mazhitov AM and Korneeva V S 2014 Substantiation of methods of development and development systems of low-power ore bodies of the Kamagansky copper-pyrite Deposit *Mining journal* no **5** pp 86–90

[11] Mazhitov A M 2016 Assessment of the impact of underground mining operations on the stress-strain state of the Kamagansky field near-barrier massif / *Actual problems of mining* no **1** pp 29–35

[12] Mossakovsky Ya V 2006 *Economics of mining industry: Textbook for universities* (Moscow: Moscow state mining University Press)

[13] Shestakov V A 2003 *Design of mining enterprises: Textbook for students.* 2nd ed. (Moscow: MGGU Publishing house)

[14] Grigoriev V V 2010 *Substantiation of systems for developing instrument reserves of copper-coal deposits in the design of combined geotechnologies*: dis. ... Cand. tech. science: 25.00.22, 25.00.21 / Grigoriev Vladimir Veniaminovich (Magnitogorsk)

[15] Mazhitov A M and Meshcheryakov E Yu 2013 Determination of parameters and indicators of an adaptive variant of the development system with an area-end release for conditions of shallow deposits mining / *Bulletin of the Magnitogorsk state technical University named after G I Nosov* no **2 (42)** pp 5–8

Counterflow step-type bulk material cooler

A S Shishkin[1, 2] and S F Shishkin[1]

[1] Department of equipment and automation of silicate productions of the Institute of new materials and technologies, Ural Federal University named after the First President of Russia B N Yeltsin, 19, Mira str., Yekaterinburg, 620000, Russia

E-mail: [2]a.s.shishkin@urfu.ru

Abstract. Counterflow cooling is extensively used in many industries to recover the heat from hot materials coming out of kiln. The method for calculating a counterflow step-type bulk material cooler is proposed. It is based on a matrix model of heat balance for individual cooling stages. The proposed equation system allows calculating the required number of steps and airing flow rate. The method was used to design a cyclone heat exchanger for an alumina calcination furnace and a counterflow louvered proppant cooler. It allows taking into account the degree of heat transfer incompleteness at individual stages and minimizing air flow for cooling. Examples of industrial introduction are given. The applicability of the method for counterflow heaters calculating is noted. On the basis of the offered method, step counterflow coolers with louvers were developed and introduced. Such devices and calculation methods can be used to heat the material before being fed into the furnace. The use of step-type counterflow heat exchanger makes it possible to utilize the heat of bulk material leaving the furnace and the heat of exhaust gases.

1. Introduction

In many industries, the task of cooling or heating bulk material is important. To obtain the maximum economic effect, it is advisable to use the heat of the exhaust gases and the thermal energy of the material leaving the furnace. Technically, this task is achieved through the installation of appropriate heat exchangers – heaters and coolers of bulk material. This approach allows saving fuel, increase productivity, i. e. creating efficient thermal installations.

A number of research papers have been published on energy analysis of the kiln heat exchangers in cement and alumina industry. Fans [1] has undertaken a study of clinker cooler efficiency. It is contented that the efficiency of the clinker heat exchanger depends substantially on the granulometry of the clinker. Rasul et al. [2] studied the thermal performance of a cement plant. They suggested using the exhaust gas leaving from kiln system for drying the raw material. Ziya et al. [3] have considered the operation of cement kiln system used in cement manufacturing. They proposed a mathematical model of new heat recovery exchanger to utilize the heat from the kiln system. Touil et al. [4] have applied the concept of exergy analysis to a clinker cooler of a cement production facility. Their research shows the importance of the inlet temperature ratio and the number of stages cross-current contacting. Taweel et al. [5] have reviewed an analysis of energy and exergy based on the clinker temperature profile. Their study allows predicting the temperatures of exhaust air and estimate of waste heat recovery from the clinker cooler system.

Another example is the cyclone calcination furnace for alumina production [6]. In this device, the heat of the exhaust gas is used to dry and heat the material entering the calciner. At the same time, air

IOP Conf. Series: Materials Science and Engineering **966** (2020) 012004 doi:10.1088/1757-899X/966/1/012004

is used to cool the outgoing alumina, which passes through a four-stage countercurrent cooler, and then enters the combustion. Thus, heat losses with flue gases and output material are minimized. In dry cement kilns, four-stage cyclone heat exchangers are often used to heat a raw meal.

An important problem is the cooling of bulk material at low temperatures, for example from 300–200 °C to 80–70 °C. Consider this problem. An effective way to solve it is to implement a countercurrent stepwise cooling scheme. The design scheme of the cooler is shown in figure 1.

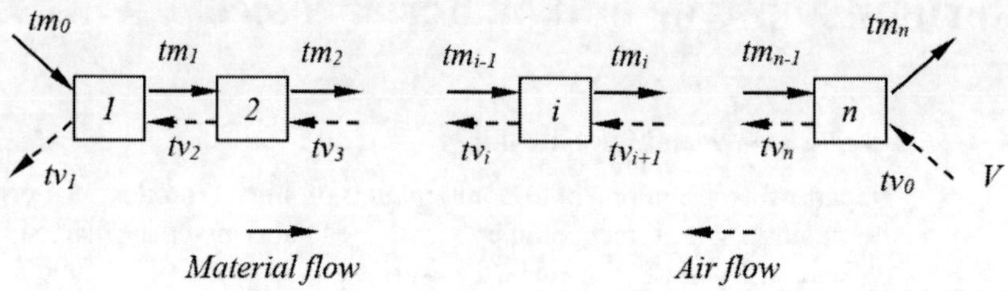

Figure 1. Design scheme of countercurrent cooler.

The following notation is used in figure 1: t_{mi}, t_{vi} – material and air temperatures at the exit of the *i-th* stage; $t_{mi\text{-}1}$, t_{vi+1} – the temperature of the material and air at the entrance to the *i-th* stage.

The material enters the first stage of the heat exchanger and moves in the direction of the last stage. Air enters the last stage of the cooler and moves counter currently in the direction of the first stage. Temperatures of the bulk material t_{m0} and air t_{v0} at the cooler inlet and also the flow rate of the bulk material G are known. It is necessary to calculate the air flow rate V, m^3/s, in order to provide a given final temperature t_{mk} of the bulk material at the outlet of the cooler.

2. The method of calculating the step-type counterflow heat exchanger

In the general case, specific heat of air c_v and bulk material c_m depend on temperature, therefore, the system of heat balance equations for n sections cooler will have next view:

$$\begin{cases} Gc_m(t_{m0})t_{m0} + Vc_v(t_{v2})t_{v2} = Gc_m(t_{m1})t_{m1} + Vc_v(t_{v1})t_{v1} + Q_1 \\ Gc_m(t_{m1})t_{m1} + Vc_v(t_{v3})t_{v3} = Gc_m(t_{m2})t_{m2} + Vc_v(t_{v2})t_{v2} + Q_2 \\ Gc_m(t_{m2})t_{m2} + Vc_v(t_{v4})t_{v4} = Gc_m(t_{m3})t_{m3} + Vc_v(t_{v3})t_{v3} + +Q_3 \\ \dots \\ Gc_m(t_{mi\text{-}1})t_{mi-1} + Vc_v(t_{vi+1})t_{vi+1} = Gc_m(t_{mi})t_{mi} + Vc_v(t_{vi})t_{vi} + Q_i \\ \dots \\ Gc_m(t_{mn\text{-}1})t_{mn-1} + Vc_v(t_{v0})t_{v0} = Gc_m(t_{mn})t_{mn} + Vc_v(t_{vn})t_{vn} + Q_n \end{cases} \quad (1)$$

The left side of each equation represents the heat input (enthalpy flux) to this section, the right side represents heat consumption, Q_i is the heat loss through the wall to the environment. For the cooler, at the first stage of calculations, it can be accepted that the heat capacity of the material and air is constant in a given temperature range, and heat losses can be neglected. The n equation system of heat balance for cooler sections will take next form:

$$\begin{cases} Gc_m t_{m0} + Vc_v t_{v2} = Gc_m t_{m1} + Vc_v t_{v1} \\ Gc_m t_{m1} + Vc_v t_{v3} = Gc_m t_{m2} + Vc_v t_{v2} \\ Gc_m t_{m2} + Vc_v t_{v4} = Gc_m t_{m3} + Vc_v t_{v3} \\ \cdot\ \cdot\ \cdot \\ Gc_m t_{mi-1} + Vc_v t_{vi+1} = Gc_m t_{mi} + Vc_v t_{vi} \\ \cdot\ \cdot\ \cdot \\ Gc_m t_{mn-1} + Vc_v t_{v0} = Gc_m t_{mn} + Vc_v t_{vn} \end{cases} \tag{2}$$

For fine bulk material, for example, alumina, the heat exchange process is completed quite quickly and it can be assumed that the temperatures of the material and air at the outlet of each stage are equal. For coarse bulk material, the heat exchange process doesn't have time to complete, and the following relation can be written for an arbitrary stage of the cooler:

$$t_{vi} = \eta t_{mi}, \tag{3}$$

where η is the coefficient of incompleteness.

The heat transfer incompleteness coefficient η for coarse bulk material is determined by calculation or experimentally for a specific device. Taking into account dependence (3), the equation system will take the form (4). This linear n equation system (4) contains n unknown material temperatures t_{m1}, t_{m2}, ... t_{mi}, ..., t_{mn}, and can be solved by any known method. Thus, the temperatures of the bulk material at the outlet of each stage will be found. Then, using relation (3), all air temperatures t_{v1}, t_{v2}, ... t_{vi}, ..., t_{vn} are found at the outlet of each stage.

$$\begin{cases} Gc_m t_{m0} + Vc_v \eta t_{m2} = Gc_m t_{m1} + Vc_v \eta t_{m1} \\ Gc_m t_{m1} + Vc_v \eta t_{m3} = Gc_m t_{m2} + Vc_v \eta t_{m2} \\ Gc_m t_{m2} + Vc_v \eta t_{m4} = Gc_m t_{m3} + Vc_v \eta t_{m3} \\ \cdot\ \cdot\ \cdot \\ Gc_m t_{mi-1} + Vc_v \eta t_{mi+1} = Gc_m t_{mi} + Vc_v \eta t_{mi} \\ \cdot\ \cdot\ \cdot \\ Gc_m t_{mn-1} + Vc_v \eta t_{m0} = Gc_m t_{mn} + Vc_v \eta t_{mn} \end{cases} \tag{4}$$

Let us introduce the following notation:

$$a = Gc_m;\ \mathrm{b} = Vc_v \eta;\ d = Vc_v$$

Then, the equation system (4) consisting of a different number of stages will take the corresponding form. In the case of one stage, the temperatures of material and air at the cooler outlet are found by next dependences:

$$t_{m1} = \frac{at_{m0} + dt_{v0}}{a+b} \quad \text{and} \quad t_{v1} = \eta t_{m1}$$

Two-stage cooler:

$$\left\{ \begin{bmatrix} -(a+b) & b \\ a & -(a+b) \end{bmatrix} \cdot \begin{bmatrix} t_{m1} \\ t_{m2} \end{bmatrix} = \begin{bmatrix} -at_{m0} \\ -dt_{m0} \end{bmatrix} \right.$$

Three-stage cooler:

$$\left\{\begin{bmatrix} -(a+b) & b & 0 \\ a & -(a+b) & b \\ 0 & a & -(a+b) \end{bmatrix} \cdot \begin{bmatrix} t_{\mathrm{m1}} \\ t_{\mathrm{m2}} \\ t_{\mathrm{m3}} \end{bmatrix} = \begin{bmatrix} -at_{\mathrm{m0}} \\ 0 \\ -dt_{\mathrm{v0}} \end{bmatrix}\right.$$

Further, it is not difficult to generalize to n steps. The temperatures of the material t_{m0} and the air t_{v0} at the inlet, as well as the matrix coefficients a, b, d are known. Having solved the system, the temperature distribution of the material at the cooler outlet of each stage can be found. The proposed model allows at the design phase to select the required number of a cooler stages and the required air flow to cool the material to the selected temperature.

3. Commercial tests

Based on this approach, several counter-flow bulk material coolers have been designed and are successfully operating for various tasks. As an example, a countercurrent cyclone alumina cooler on a cyclone calcination furnace [8] and a countercurrent louvered proppant cooler can be given. Consider these examples in more detail. Figure2shows a photograph that explains the principle of operation of a laboratory counterflow louver cooler. Material moves downward through inclined louvers, and air moves upward through the screens, cooling the material. Thus, a countercurrent three-stage cooling system is implemented. Figure 3 shows photograph of industrial counterflow louver proppant cooler operating at the factories of FORES LLC.

Determination of the required air flow for cooling and the number of cooler stages is performed using the proposed methodology. Figure 4 shows calculated final temperatures of the material and air at the outlet of the three-stage cooler, depending on the air flow. The temperature of material and air at the cooler inlet were set t_{m0} = 250 °C and t_{v0} = –10 °C, which corresponds to the data of industrial tests of a three-stage proppant cooler. The cooling capacity was G = 17 t/h. The temperature measurements for individual stages show that degree of incompleteness of the heat transfer process was η = 0.8. The material was supplied with a temperature of 240–270 °C and cooled to 77–90 °C. The air flow rate at the cooler inlet was V = 15000–15500 m^3/h. The air temperature at the cooler inlet was –10 °C. Calculated and experimental data are in satisfactory agreement with each other, as follows from the figure 4.

Figure 2. Counterflow coolers with louvers. The movement of the material in the laboratory model.

Figure 3. Industrial counterflow cooler with a capacity of 17 t/h.

Figure 5 shows the calculated graphs dependence of material and air temperature at the cooler outlet against the number of heat transfer stages. As follows from the graph, three stages are sufficient for cooling proppants under given conditions.

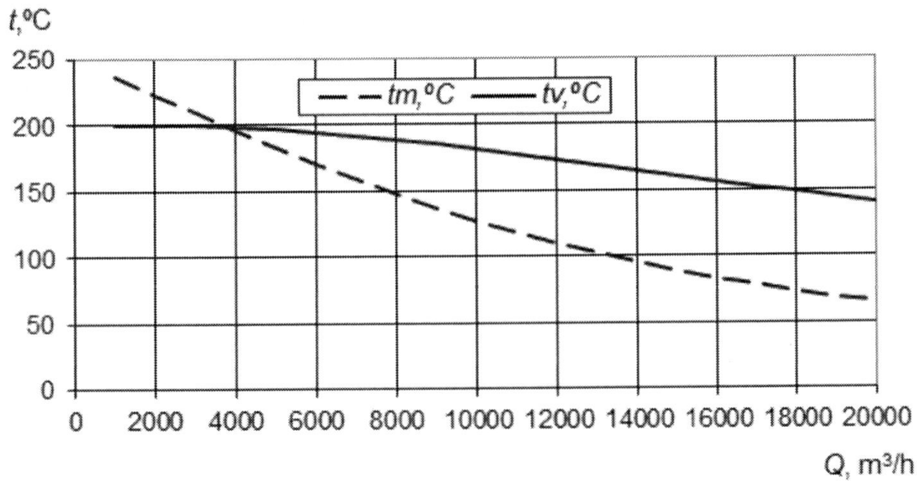

Figure 4. Air temperature at the cooler inlet t_{v0} = –10 °C, material t_{m0} = 250 °C, productivity G = 17000 kg/h, η = 0.8).

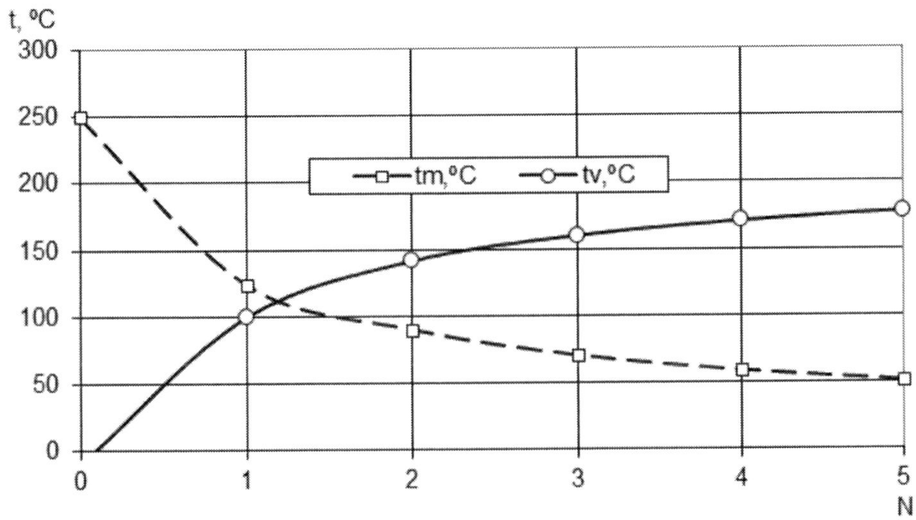

Figure 5. The dependence of material and air temperature at the cooler outlet against the number of heat transfer stages N. Air consumption V = 15000 m^3/h, G = 17 t/h, t_{m0} = 250 °C, t_{v0} = –10 °C, η = 0.8.

4. Conclusions

The proposed calculation method allows determining the required number of stages and air flow for cooling to the temperature setpoint. On the basis of the offered method, step counterflow coolers with louvers were developed and introduced. Such devices and calculation methods can be used to heat the material before being fed into the furnace. This will allow you to utilize the heat of the exhaust gases.

IOP Conf. Series: Materials Science and Engineering **966** (2020) 012004 doi:10.1088/1757-899X/966/1/012004

References

[1] Fans M 2017 Optimization of clinker cooler performance with understanding of the heat exchanger challenges *Cement International* **15 (2)** pp 56–61

[2] Rasul M, Widianto W and Mohanty B 2005 Assessment of the thermal performance and energy conservation opportunities of a cement industry in Indonesia *Appl. Therm. Eng.* **25** pp 50–65

[3] Ziya S, Zuhal O and Hikmet K 2010 Mathematical Modeling of Heat Recovery from Rotary Kiln *Appl. Therm. Eng.* **30** pp 817–25

[4] Touil D, Belabed H, Frances C and Belaadi S 2005 Heat exchange modeling of a grate clinker cooler and entropy production analysis *International Journal of Heat and Technology* **23 (1)** pp 61–8

[5] Taweel T J B, Sokolova E, Sergeev V, Solovev D Energy and Exergy 2018 Analysis of Clinker Cooler in the Cement Industry *IOP Conf. Series: Materials Science and Engineering* **463** 032101

[6] Shishkin S 2015 Cyclone calcination furnace *New refractories* **4** pp 12–5

[7] Shishkin S and Shishkin A 2012 *The solution of engineering problems in EXCEL* (Ekaterinburg: Ural Federal Universiry Publishing) p 365

[8] Shishkin A and Shishkin S 2016 *Examples of calculations of aerodynamic processes for processing bulk materials in Excel* (Ekaterinburg: Ural Federal Universiry Publishing) p 409

IOP Conf. Series: Materials Science and Engineering **966** (2020) 012005 doi:10.1088/1757-899X/966/1/012005

Calculation of a closed grinding cycle in a jet mill with two classifiers

D S Barakovskikh[1, 2] **and S F Shishkin**[1]

[1] Department of equipment and automation of silicate productions of the Institute of new materials and technologies, Ural Federal University named after the First President of Russia B N Yeltsin, 19, Mira str., Yekaterinburg, 620000, Russia

E-mail: [2] d.s.barakovskikh@urfu.ru

Abstract. The matrix calculation method is successfully used for calculating the closed cycle in a jet mill. In the traditional closed-cycle scheme of a jet mill, a dynamic centrifugal classifier is used for fine separation of the grinding product. The circulation load can be up to 5 or more. This involves using a high-performance classifier the cost of which will be a significant share of the total installation cost. In addition, the dynamic classifier rotor wear is proportional to the quantity of material, i.e. performance. To eliminate these shortcomings a closed loop scheme with two consecutive classifiers is proposed within the framework of the matrix model. Moreover, the first classifier must be of the pass-through type and the second is of the dynamic one. The use of this scheme in mica production allowed reducing the circulation load on the second centrifugal dynamic classifier for fine separation of the material by two times. Equations that allow determining the granulometric composition and mass yield of separation products, the multiplicity of circulation, the load on the mill and classifiers are obtained. The mathematical model proposed permits to calculate the optimal separation limits of each classifier.

1. Introduction

The matrix model based on material balance using narrow classes of fineness is considered in many investigations about grinding. In [1], the fracture matrix was used to predict the size of limestone sand particles after particles destruction under one-dimensional compression. In [2], matrices were used to calculate the sizes of hydrargillite particles during grinding in a bead mill. The matrix model was also used to predict serpentinite separation products in the air classification process [3].

This calculation method is successfully used in general and for calculating a closed grinding cycle in a jet mill in particular and allows calculating the granulometric composition of the finished product and the circulation load on the mill and classifier [4–6].

In accordance with the matrix model [7–8] granulometric compositions are interpreted using the concept of an arithmetic vector **r** whose components are partial remnants of narrow classes $\{r_1, r_2, r_3, ..., r_i, ..., r_n\}$. The grinding process in an open cycle is described by the matrix equation (1) which represents the material balance using narrow classes of fineness.

$$\begin{pmatrix} a_{11} & a_{12} & \cdots & a_{1n} \\ 0 & a_{22} & \cdots & a_{2n} \\ . & . & . & . \\ 0 & 0 & 0 & a_{nn} \end{pmatrix} \cdot \begin{pmatrix} r_{z1} \\ r_{z2} \\ \vdots \\ r_{zn} \end{pmatrix} = \begin{pmatrix} r_1 \\ r_2 \\ \vdots \\ r_n \end{pmatrix}, \quad (1)$$

IOP Conf. Series: Materials Science and Engineering **966** (2020) 012005 doi:10.1088/1757-899X/966/1/012005

where r_{zi}, r_i – the particles content of the i-th class of fineness in the initial and crushed products respectively; a_{ij} – the portion of the i-th class of fineness transferred to the smaller j-th class (i-row, j-column).

Formally, equation (1) is a transformation of the vector of the initial composition $\mathbf{r_z}$ into the vector of the finished product $\mathbf{r}$. Since a single narrow r_{zi} class can only pass to smaller classes as a result of grinding, the matrix **A** is triangular. Matrix **A** is determined experimentally. For example, for a jet mill the r_{zi}-th narrow class of fineness is crushed separately and the a_{ij} part that has passed to the smaller j-th class ($I <= j$) is determined.

When considering a closed grinding cycle, the material balance equation (1) becomes more complex since it is necessary to take into account the emerging circulation flows which depend on the selected scheme.

2. Calculation of a closed grinding circuit with two classifiers

The traditional closed-cycle grinding scheme which includes a jet mill and a classifier is widely used in practice [9–10]. With fine grinding the circulation load can be up to 5 or more. In this case, for example, if the initial feed capacity is 1 t/h the classifier must have a production capacity of 5 t/h. Since dynamic centrifugal classifiers are used for fine grinding it is obvious that the cost of a high-performance classifier will be a significant share of the total cost of a closed-loop installation. In addition, it is known that the dynamic classifier rotor wearout is proportional to the amount of material i.e. production capacity. A new closed-loop scheme with two consecutive classifiers is proposed to eliminate these shortcomings (Figure 1).

To build a balanced mathematical model of this scheme it is necessary to replace and calculate two classifiers with one combined cascade classifier (Figure 2).

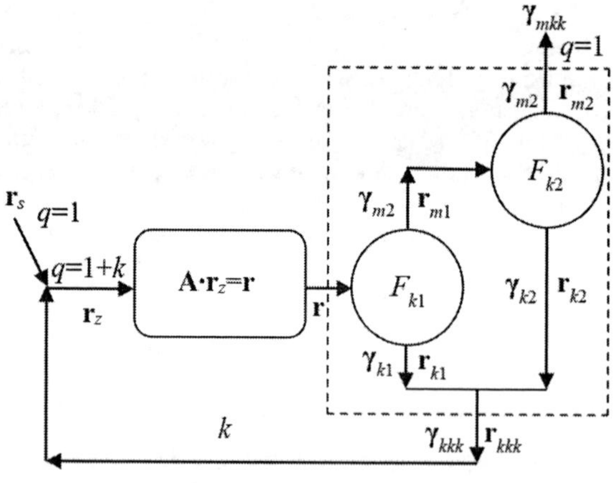

Figure 1. Calculation scheme of a closed-loop with two classifiers.

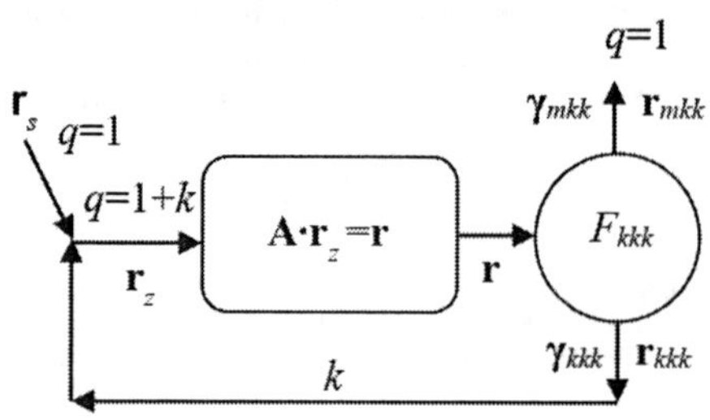

Figure 2. Combined cascade of two classifiers.

The following symbols are used in the figures:

$\mathbf{r}_s$, $\mathbf{r}_m$, $\mathbf{r}_k$ – vectors of granulometric composition of the initial material, fine and coarse separation products respectively;

$\mathbf{r}_z$, $\mathbf{r}$ – vectors of the granulometric composition of the material entering the mill input and classifier input respectively;

$\mathbf{r}_{m1}$, $\mathbf{r}_{m2}$, $\mathbf{r}_{k1}$, $\mathbf{r}_{k2}$, $\mathbf{r}_{mkk}$, $\mathbf{r}_{kkk}$ – vectors of granulometric composition of fine and coarse separation products for the first, second classifiers and the combined cascade respectively;

q – the mass flow of the material received at the input and output q=1;

k – the multiplicity of circulation;

A – destruction matrix for the jet mill;

F_{k1}, F_{k2}, F_{kkk} – a function of the degree of fractional extraction of narrow fineness classes into a coarse separation product for the first, second classifiers and the combined cascade respectively;
γ_{m1}, γ_{m2}, γ_{k1}, γ_{k2}, γ_{mkk}, γ_{kkk} – mass output of fine and coarse separation products for the first, second classifiers and the combined cascade respectively.

The theory of combined cascade classifiers is well developed [11–14]. It follows from this theory that several interconnected classifiers can be replaced by a single combined cascade classifier that has the appropriate degree of fractional extraction F_{kkk}. The function of the degree of fractional division into a coarse product for a combined cascade classifier that implements sequential resorting of a fine product will be described by the function:

$$F_{kkk} = 1 - F_{mkk} = 1 - (1 - F_{k1})(1 - F_{k2}) = F_{k1} - F_{k1} \cdot F_{k2} + F_{k2} \quad (2)$$

If the function of the degree of fractional extraction in a coarse separation product is approximated by Plitt dependence then F_{k1} and F_{k2} will have the form in equation (3):

$$F_{k1}(x_j) = \frac{(x_j/c_1)^{p_1}}{1 + (x_j/c_1)^{p_1}} \text{ и } F_{k2}(x_j) = \frac{(x_j/c_2)^{p_2}}{1 + (x_j/c_2)^{p_2}}, \quad (3)$$

where x_j is the average size of fineness class j; c_1, c_2 – is the size of the boundary grain (x_{50} separation boundary) for classifier 1 and 2; p_1, p_2 is the separation efficiency parameter for classifier 1 and 2.

The mass yield of coarse and fine products for the combined cascade is determined by the formulas:

$$\gamma_{kkk} = \sum_{j=1}^{n} r_j F_{kkk}(x_j) = \sum_{j=1}^{n} r_j \left(F_{k1}(x_j) - F_{k1}(x_j) \cdot F_{k2}(x_j) + F_{k2}(x_j) \right), \quad (4)$$

$$\gamma_{mkk} = \sum_{j=1}^{n} r_j \left[1 - F_{kkk}(x_j) \right] \quad (5)$$

In this case the multiplicity of circulation is found by a known dependence:

$$k = \frac{\gamma_{kkk}}{100 - \gamma_{kkk}} \quad (6)$$

The load on the mill and the first classifier by weight as a percentage of power is found by dependence:

$$q_n = 1 + k \quad (7)$$

The load on the second classifier by weight as a percentage of power will be lower and it found by dependence:

$$q_2 = q_n \frac{\gamma_{m1}}{100} \quad (8)$$

The granulometric composition of the finished product can be calculated using the formula:

$$r_{mkkj} = \frac{100 \cdot (1 - F_{kkk}(x_j)) \cdot r_j}{100 - \gamma_{kkk}} \quad (9)$$

The granulometric composition of the material (partial residues of class j size) entering the grinding process can be determined by the dependence:

$$r_{zj} = \frac{r_{sj} + r_{kkkj} k}{1 + k} \quad (10)$$

Expressing k in terms of γ_k it turns out:

$$r_{zj} = \frac{100-\gamma_{kkk}}{100} r_{sj} + \frac{\gamma_{kkk}}{100} r_{kkkj} \tag{11}$$

The granulometric composition of a coarse separation product is determined by the dependence:

$$r_{kkkj} = 100 F_{kkkj} \frac{r_j}{\gamma_{kkk}} \tag{12}$$

Given this dependences the granulometric composition of the material entering the grinding process can be determined:

$$r_{zj} = \frac{r_{sj}(100-\gamma_{kkk})}{100} + F_{kkkj} r_j \tag{13}$$

After conversion system (1) will look like:

$$\begin{cases} \sum_{j=1}^{n} c_1 F_{kkkj} r_j - \sum_{j=1}^{n} a_{1j} F_{kkkj} r_j + r_1 = b_1 \\ \sum_{j=1}^{n} c_2 F_{kkkj} r_j - \sum_{j=1}^{n} a_{2j} F_{kkkj} r_j + r_2 = b_2 \\ \quad . \; . \; . \qquad . \; . \; . \qquad . \; . \; . \\ \sum_{j=1}^{n} c_n F_{kkkj} r_j - \sum_{j=1}^{n} a_{nj} F_{kkkj} r_j + r_n = b_n \end{cases} \tag{14}$$

where the coefficients b_i and c_i are:

$$b_i = \sum_{j=1}^{n} a_{ij} r_{sj}; \; c_i = \frac{b_i}{100} \tag{15}$$

Thus, the equations obtained allow to calculate the granulometric compositions of products and the circulating load on the mill and both classifiers. This model was used in the design of a closed-loop mica jet grinding plant with two classifiers at JSC «Malyshevskoye rudoupravlenie».

3. Commercial tests

The scheme of the jet grinding plant for mica grinding is shown in figure 3. The unit capacity is 300 kg/h. The scheme uses two consecutive classifiers. The first classifier is a centrifugal static and through, the second one is a centrifugal dynamic with a rotating rotor.

The scheme includes: 1 – raw material hopper; 2 – screw feeder; 3 – receiving hopper; 4 – rotary valve for compressed air supply; 5 – receiver; 6 – pressure gauge; 7 – jet mill; 8 – supersonic nozzle; 9 – acceleration tube; 10 – grinding chamber; 11 – barrier; 12 – through classifier (D = 1000 mm); 13 – sleeve valve; 14 – centrifugal classifier of type 'SELECTOR-500/1500' (D = 500 mm, n = 960 rpm, N = 12 kW, variable-frequency drive); 15 – cyclone of type TSN-15 (D=800 mm); 16 – cyclone of type SDK TSN-33 (D = 1200 mm); 17 – rotary valve; 18 – diaphragm; 19 – bag filter (F_f= 120 m^2); 20 – fan of type VR-132-30-10,2 (version 5, n = 1650 rpm, N = 45 kW, variable-frequency drive).

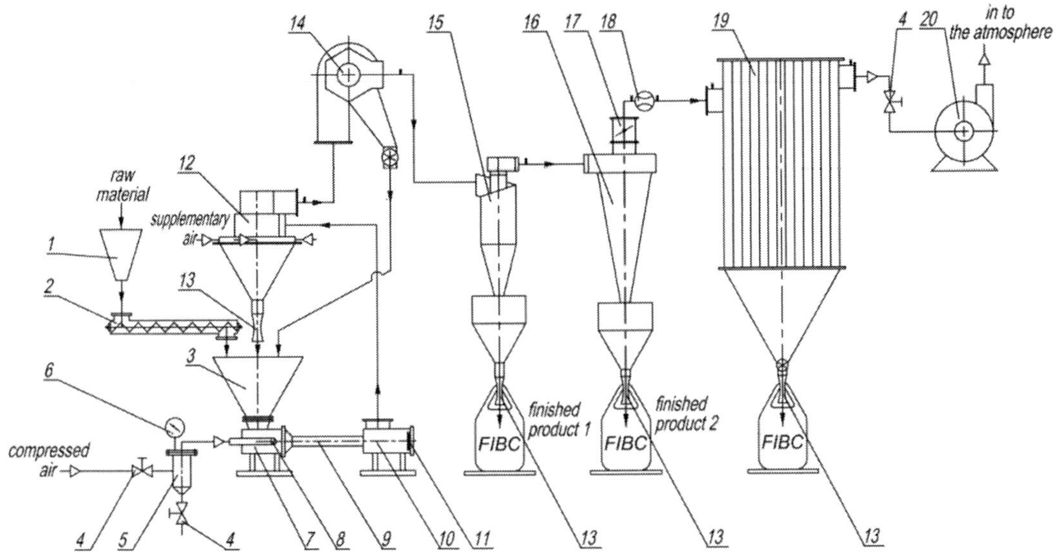

Figure 3. Technological scheme of closed-loop jet grinding with two classifiers for mica grinding at JSC 'Malyshevskoye rudoupravlenie'.

The granulometric compositions of the raw material and finished mica products are shown in table 1.

Table 1. Granulometric composition of products.

Sieve	Raw material			Cyclone 1 (TSN-15)			Cyclone 2 (SDK-TSN-33)			The combined product		
X (mkm)	*R* (%)	*D* (%)	*r* (%)	*Rm*1(%)	*Dm*1 (%)	*rm*1 (%)	*Rm*2(%)	*Dm*2 (%)	*rm*2 (%)	*Rm*3 (%)	*Dm*3 (%)	*rm*3 (%)
1000	1.4	98.6	1.4	0.0	100.0	0.0	0.0	100.0	0.0	0.0	100.0	0.0
630	7.6	92.4	6.2	0.0	100.0	0.0	0.0	100.0	0.0	0.0	100.0	0.0
400	19.3	80.7	11.6	0.2	99.8	0.2	0.0	100.0	0.0	0.2	99.8	0.2
315	26.9	73.1	7.6	0.2	99.8	0.0	0.0	100.0	0.0	0.2	99.8	0.0
200	47.4	52.6	20.5	0.8	99.2	0.6	1.4	98.6	1.4	0.9	99.1	0.7
160	57.4	42.6	10.0	1.6	98.4	0.8	2.2	97.8	0.8	1.7	98.3	0.8
100	73.7	26.3	16.3	6.5	93.5	4.9	2.9	97.1	0.6	6.1	93.9	4.4
63	83.5	16.5	9.8	14.0	86.0	7.5	4.3	95.7	1.4	13.0	87.1	6.8
40	93.8	6.2	10.2	35.1	64.9	21.1	10.2	89.8	5.9	32.4	67.6	19.5
0	100.0	0.0	6.2	100.0	0.0	64.9	100.0	0.0	89.8	100.0	0.0	67.6

From the presented data it is follow that using a closed circuit with two classifiers it is possible to obtain three products of different fineness: from a cyclone 15 of the TSN-15 type – a fraction of 0–63 microns (content of particles <63 microns is 86 %); from a cyclone 16 of the SDK-TSN-33 type – a fraction of 0–40 microns (content of particles <40 microns is 89.8 %); or a combined cyclone product. As a result of jet grinding the bulk mass of mica is reduced from 672.8 kg/m^3 to 377.0 kg/m^3 (product 1) and 303.6 kg/m^3 (product 2).

The multiplicity of circulation of the first classifier was $k_1 = 2.1$, the second one was $k_2 = 1.1$. As a result, the total productivity load on the first classifier and the mill was $q_1 = 4.2$, on the second classifier was $q_2 = 2.1$, thus it was possible to reduce the load on the dynamic classifier for fine separation of material by 2 times.

The separation efficiency of the first pass classifier according to the Eder-Mayer indicator ($E = 100 \cdot x_{75}/x_{25}$) was $E_{k1} = 12.69$ % at the separation boundary $c_1 = 34.5$ microns. The efficiency of the second dynamic centrifugal classifier – E_{k2} = 20.34 % at the separation boundary c_2 = 21.7 microns. The total efficiency of the combined cascade E_{kk} = 23.67 % at the separation boundary c_{kk} = 12.4 microns.

The relatively low separation efficiency is due to the fact that the mica particles have a flat shape. As a result, the mass force and the resistance force are proportional to the particles size that objectively does not allow qualitative size separation of mica by air classification. However, despite the limited separation efficiency the installation allows to adjust the fineness of the finished product by changing the x_{50} separation boundaries on the first and second classifiers.

4. Conclusions

A mathematical model of the grinding process in a jet mill with two classifiers is proposed. The obtained equations allow calculating the granulometric compositions of separation products, circulation multiplicity, load on the mill and classifiers. Using two classifiers in a closed circuit (the first static, the second dynamic) allows reducing the circulation load (capacity) on the centrifugal dynamic classifier by 2 times that is confirmed by the results of industrial tests. Reducing of the load on the dynamic classifier enables to reduce capital costs and reduce the cost of repairing or replacing the rotor.

References

[1] Tong C X, Zhang S and Sheng D 2018 A breakage matrix model for calcareous sands subjected to one-dimensional compression *4th Geo Shanghai Int. Conf. on Fundamentals of Soil Behaviours* (Singapore: Springer) pp 17–24

[2] Berthiaux H, Varinot C and Dodds J 1996 Approximate calculation of breakage parameters from batch grinding tests *Chem. Eng. Sci.* **51** pp 4509–16

[3] Ponomarev V, Dzuser V, Danilov V and Litvinov A 2010 Analytical forecasting of the process of serpentinite air classification for JSC 'Russian magnit' *Refractories and Technical Ceramics* **9** pp 12–5

[4] Shishkin S 2003 Calculation of a closed grinding cycle for a jet mill *Bulletin of UGTU-UPI. Construction and education* **6** pp 279–83

[5] Zhukov V, Otwinovski H, Smirnov S and Kaniowski P 2009 Computational and experimental studies of processes in fluidized bed jet mills *Bulletin of ISCTU* vol **4** (Ivanovo: Ivanovo State Chem.-Technol. University) pp 74–77

[6] Serebryanik I 2015 Modeling of jet grinding process taking into account the peculiarities of mica deformation mechanics *Fundamental and applied problems of engineering and technology* vol **6** (Oryol: Oryol State University) pp 39–44

[7] Gupta A and Yan D *Mathematical modelling of comminution processes. Mineral processing and operations* ed 2 (Amsterdam: Elsevier) chapter 11 pp 317–55

[8] Linch J 1981 *Crushing and grinding cycles* (Moscow: Nedra) p 343

[9] Shishkin S 1991 Calculation of the grinding process in a closed cycle *University News. Chem. and Chemical Technology* **24** pp 117–9.

[10] Zhukov V, AndreevA, Otwinovski H and Urbaniak D 2006 Mathematical model and method for calculating the dynamic classifier *University News. Chem. and Chemical Technology* vol **49 (5)** (Ivanovo: Ivanovo State Chem.-Technol. University) pp 99–102

[11] Chokhonelidze A, Braun A and Forgor L 2014 Development of a matrix model of a closed circuit of cement grinding *Naukovedeniye* **5 (24)** p 107

[12] Barskiy E 2010 Critical regimes of two-phase flows with a polydisperse solid phase *Fluid Mechanics and Its Applications* vol **93** (New York: Springer) p 348

[13] Mizonov V 1989 *Aerodynamic classification of powders* (Moscow: Chimiya) p 160

[14] Ponomarev V 2015 *Calculation and design of equipment for air separation of bulk materials* (Ekaterinburg: Ural Federal University Publishing) p 79

IOP Conf. Series: Materials Science and Engineering **966** (2020) 012006 doi:10.1088/1757-899X/966/1/012006

Influence of liquid nitriding on the properties of the electric steel 10

V A Korotkov[1, 2] and D V Isakov[1]

[1] Ural Federal University named after the First President of Russia B N Yeltsin, Nizhniy Tagil Technological Institute, 59, Krasnogvardeyskaya str., Nizhniy Tagil, 622000, Russia

E-mail: [2] vk@udgz.ru

Abstract. Liquid nitriding (carbonitriding) effect on hardness, wear resistance and magnetic properties of electric steel grade 10 is reviewed here, as well as on the electric contact resistance. It was found that hot-rolled steel 10 loses its hardness as a result of annealing. Surface carbonitriding of specimens generates an improved layer of higher hardness, high wear resistance and increased electrical resistance. The specimens carbonitrided have contact electric resistance higher than that of hot-rolled specimens after descaling. The magnetic permeability of 'grade 10' steel does not practically change after annealing and carbonitriding, while the magnetic permeability corresponds to the magnetic field strength in a very typical way with typical parameters for carbon steels. Traditional methods of processing electrical steels, carried out in order to achieve such a complex effect, combine heterogeneous processes, such as electroplating, vacuum, plasma, high - temperature annealing, coating-are very expensive and difficult to implement. Carbonitration is more accessible for practical use and has a significantly lower cost. The results obtained can be used in the design of magnetic circuits for various electric devices.

1. Introduction

Electric steels are used for the manufacture of magnetic conductors of electric machines and devices. A significant part of them is technically pure iron with a low hardness of ~HB130 (HV130). After some machining and peening, its grains get refined and some residual stresses are induced, and as a result, the magnetic permeability reduces and the coercive force increases. To improve (restore) electromagnetic properties of electric steels, annealing is practiced [1]. Another special thing about electric steel is dielectric coatings applied on the surface, in order to reduce Foucault currents and electromagnetic losses during operation [2, 3]. In addition, some parts made of electric steel need working surface friction wear protection e.g. electromagnetic clutches. For this purpose, the surface is coated with a precipitated layer of 3 to 5 microns of chemical nickel (in sodium thiophosphate baths) and then boron carbide powder is sprayed. A solid high wear resistant boride and nickel boride (~HV2015) layer is formed during follow-up annealing [4]. This hardening is definitely effective as the hardness is raised more than by 15 times (HV2015/HV130) but is achieved with a complex of heterogeneous processes (galvanic, plasma, vacuum), which makes it expensive and of little feasibility.

Liquid nitriding or carbonitriding can be considered a promising practice that improves the mechanical and electromagnetic properties of electric steel. It is carried out at ~570 °C, which is high enough to eliminate peening hardening, residual stresses and, as a result, improve the electromagnetic

IOP Conf. Series: Materials Science and Engineering **966** (2020) 012006 doi:10.1088/1757-899X/966/1/012006

properties. In addition, a hard coating is formed on the surface, which increases the wear resistance and increases the durability of parts [5].

The purpose of this research is influence of liquid nitriding (carbonitriding) on electromagnetic and hardness properties of no-alloyed electric steel grade 10.

2. Material, processing and preparation of specimens

Specimens for this research were cut mechanically from 5mm hot-rolled steel sheets. Chemical composition, weight %: C – 0.08, Si – 0.17, Mn – 0.17, Cr – 0.04, Ni – 0.02, Cu – 0.05, S – 0.006, P – 0.012. Mechanical properties per certificate: tensile strength is 405 MPa, yield strength is 290 MPa, percent elongation is 39.5 %. Electromagnetic and mechanical properties were determined using specimens 'as received' and after annealing at 800 °C and 950 °C. Annealing conditions and hardness before and after annealing is shown in Table 1. Then all the specimens were carbonitrided and reexamined. Carbonitriding (liquid nitriding) was performed in an industrial bath with the following sizes: d 450 mm, h 900 mm with molten potassium cyanide Ch (KNCO) at 575 °C.

Table 1. Annealing conditions for steel 10 specimens and their hardness.

Protective annealing environment	Annealing temperature cycle	Hardness, HRB TR-5006 device	
		Before annealing	After annealing
Vacuum: residual pressure (4–8)10×10^{-4} Pa	heating from 400°C to 800 °C for 1 hour soaking at 800 °C for 4 hours cooling in the furnace from 800 °C to 400 °C, then in the air for 4 hours	90	78
Argon purging inside the furnace, 20 l/min	heating from 400 °C to 950 °C for 2 hours soaking at 950 °C for 3 hours cooling in the furnace from 950 °C to 400 °C, then in the air for 15 hours		62

3. Mechanical tests

The specimens were measured for hardness by the HRB scale before and after annealing with TR-5006 device. It was found (Table 1) that the initial hardness HRB 90 decreased to HRB 78 during annealing at 800 °C, and even more after annealing at 950 °C i. e. to HRB 62. Specimen surface hardness was measured with PMT-3M device after short-term carbonitriding for 10 minutes at various loads on the indenter. The curves of Figure 1 show: the hardness of the steel carbonitrided 'as received' was HV102, with low indenter load (10 g), which is slightly higher than HV 77–89 obtained at loads = 20 g. This difference is explained by the fact that the carbonitration layer thickness is related to the indenter penetration depth [4, 5]. Therefore, the values measured represent an integral hardness of the carbonitrided layer and that of the base metal under it, which is not as solid.

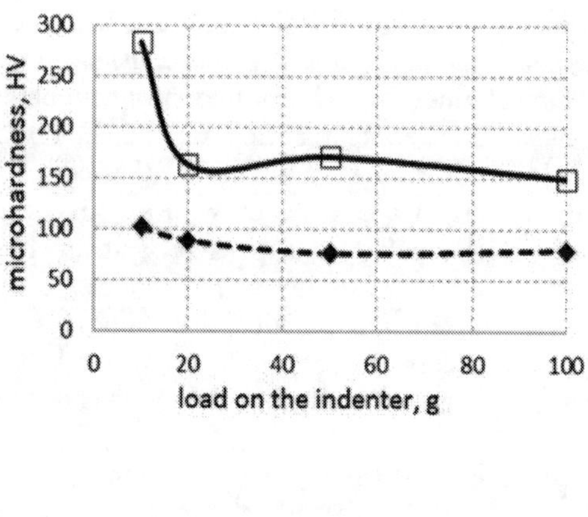

Figure 1. Microhardness of specimen surface
- -♦- - carbonitrided 'as received' (without annealing)
—□— carbonitriding after annealing at 800°C

It is obvious that the higher the indenter load is, the deeper it penetrates, and the greater is the effect of the solid carbonitrided layer on the resulting hardness. Hence, the hardness value at low indenter load is higher than at higher loads. Pre-annealing at 800 °C caused increased hardness of the surface carbonitrided (Figure 1). It is logical to assume that the diffusion processes were easier for them during carbonitriding, together with increased thickness of the hardened layer. This is also indicated by the hardness HV102 increased to HV283 at a low load of 10 g per indenter.

Figure 2. L-shaped hardness and magnetic permeability specimens.

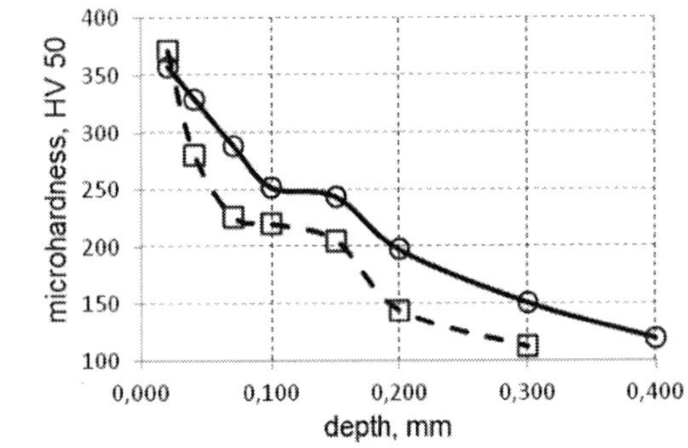

Figure 3. Hardness distribution from the surface after 3 hour carbonitriding
—○— as-received - -□- - annealing 950°C

The steel 10 was tested for magnetic permeability using a magnetic conductor made up of L-shaped specimens shown in figure 2. Some of the specimens were annealed (950 °C) before carbonitriding for 3 hours and some were carbonitrided 'as received'. Then some samples were cut off from the L-shaped specimen ends designed for this purpose and used to produce cross-section hardness test specimens. See Figure 3 for measurement results.

Figure 3 shows that the cross-section hardness near the surface is ~HV 360. This value is near the value obtained when measuring the surface hardness of a specimen carbonitrided after annealing (Figure 1). L-shaped specimens hardness goes down away from the surface, but it is slightly higher on a specimen without annealing than that of specimens annealed before carbonitriding. This is probably due to the higher hardness of the base metal whereon the thin carbonitrided layer is located (Table 1).

Since the late 90's of the last century, UZIT-3-type ultrasonic hardness measurement devices have become widespread in the industry. Their determine hardness by the resonant frequency of a magnetostrictive rod with a diamond pyramid indented in the measured surface. Besides, the indentation force is set low at the device, so the mark depth is ~12 microns [6]. This type of device has an advantage over stationary hardness meters that is it can be used for hardness measurement of parts in the production process directly covering up to 100 % of the product output.

In this regard, the hardness of steel 10 was measured after carbonitriding with the UZIT-3 device. It was found that the hardness of specimens annealed at 950 °C is HRC60 after 3 hours of carbonitriding, which is equal to HV~800 by Vickers, and HRC56 (HV~700) for specimens 'as received'. These values are higher than the hardness readings of the microhardomer shown in Figure 3, and therefore need to be explained.

The area available for measurement was maximized so as to refine the carbonitrided layer hardness values. For this purpose, the surface of the specimen was polished at an angle of ~8 degrees. As a result, the visible sector of the hardened layer on the cut (Figure 4) increased by ~8 degrees, and the indenter had no problem to reach it.

In addition, the indenter load was reduced by half (25 g), which minimized the probability for the indenter to penetrate through the thin solid carbonitride layer to the base metal, which is softer. A FUTURE TECH FM-300 (Japan) microhardness meter was used to make measurements under control of Thixomet software. The hardness distribution through the carbonitrided layer depth is shown in Figure 5. In addition, the specimen measurement values were multiplied by 0.13. According to the curve, the highest hardness HV_{25} 800 to 1000 occurs up to 15 microns deep and indicate the carbonitride layer thickness. The underlying diffusive nitrogen saturated zone of the metal 35 microns deep has its hardness stabilized at ~HV_{25} 530. Thus, it is established that the actual hardness of the thin solid carbonitride layer (15 μm) of on the surface of the steel 10 is HV_{25} 800 to 1000, and its lower values shown in Figure 3 are explained by the indenter penetration in the softer base metal below.

Figure 4. Carbonitrided layer on the metallographic specimen polished at 8° to the surface.

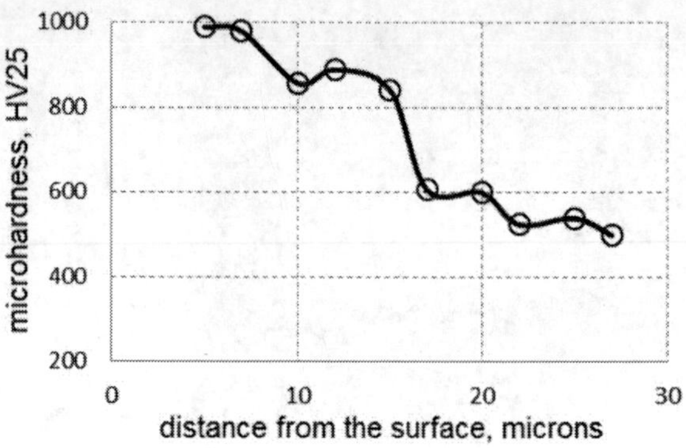

Figure 5. Hardness distributed through the depth of the specimen carbonitrided for 3 hours after annealing at 950 °C.

The carbonitriding effect on the steel 10 wear resistance can be described by the results obtained by dry disk-and-pad friction on the Mi-1M friction machine. It is established that the carbonitrided steel 10 disk with normalized steel 45 pad wears ~10 times lower than the normalized 45 steel disk in the fist stage, and ~20 times lower every 5 minutes in the 4th stage of the test.

4. Magnetic properties examination

The following specimens (Figure 1) were made of grade 10 steel in four different ways: 1) hot-rolled; 2) after annealing 950 °C (without carbonitriding); 3) after carbonitriding of hot-rolled specimens; 4) after annealing 950 °C and carbonitriding.

The magnetic steel properties before and after carbonitriding were examined using the ammeter and wattmeter method [8]. The magnetic conductor is made of two L-shaped elements that form a closed loop when matched. Two windings were placed on the magnetic conductor i.e. generator (primary) winding and measuring (secondary) winding. Sinusoidal 50 Hz voltage was applied to the generator winding, and magnetic field strength, magnetic induction and magnetic permeability was determined from the measurements of the primary winding current, secondary winding voltage and active power consumption. Current and voltage were metered with M320 and M890F multimeters; active power consumed was metered with a D539 wattmeter. The supply voltage was adjusted with a laboratory transformer (LATR).

The magnetic induction amplitude value was calculated by the equation from [9], assuming sinusoid nature of the magnetic flow:

$$B_m = \frac{\sqrt{2}E_2}{2\pi f S W_2}, \tag{1}$$

where E_2 – effective EMF value induced in the secondary winding, f – supply voltage frequency, S – magnetic conductor cross sectional area, W_2 – turn number of secondary (test) winding.

When real currents and voltages are replaced with equivalent sinusoidal currents and voltages (equivalence determined by equality of RMS values) and also with account of active magnetic conductor losses, and provided that winding and magnetic conductor parameters are reduced to the primary winding, finally, the magnetic force amplitude equation will take the following form:

$$H_m = \sqrt{2} \cdot \frac{W_1}{l_{cp}} \cdot \sqrt{I_1^2 - \left(\frac{P_{CT}}{E_2}\frac{W_2}{W_1}\right)^2} \qquad (2)$$

where I_1 is a current value in the primary (generator) winding, l_{cp} is an average magnetic conductor length, W_1 and W_2 are winding turn numbers, $P_{CT} = P_1 - I_1^2 R_1$ are magnetic conductor losses, P_1 is active power input, R_1 is active primary windings resistance.

Relative magnetic permeability μ is defined as a ratio of the magnetic induction amplitude to the magnetic force amplitude and magnetic constant: $\mu = B_m/(H_m \mu_0)$.

The magnetic characteristic determining method stated makes possible to adequately evaluate the heat treatment and carbonitriding effect on steel magnetic properties.

Electric properties were measured by ten points with primary winding currents ranging from 0.05 A to 0.3 A, which corresponds to the magnetic force from 250 A/m to 2000 A/m. The primary winding was heating up in the measurements and its active resistance varied from 127 Ohms to 140 Ohms.

The magnetic permeability dependences on the magnetic force turned out to be typical for carbon steels (figures 6, 7) [10, 11]. The representative peak of the curve $\mu(H)$ is observed in the magnetic force range of 300 to 400 A/m. The magnetic conductor gets saturated and the limit magnetic induction values are generated starting from 1000 A/m and higher.

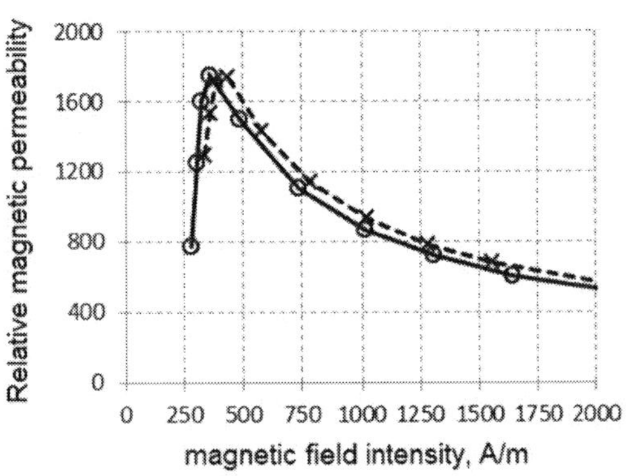

Figure 6. Relative magnetic permeability of specimens without carbonitriding
—o— without annealing (initial state)
---x--- annealing 950 °C.

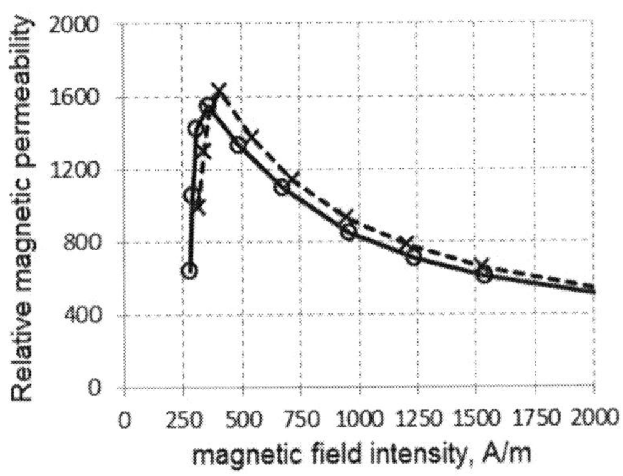

Figure 7. Relative magnetic permeability of specimens after carbonitriding
—o— carbonitriding without annealing
---x--- carbonitriding after annealing 950 °C.

The curves show that the magnetic properties of the steel are practically the same in its initial state, after annealed 950 °C and after carbonitriding; some minor differences are within the tolerance. It can be concluded that 4-hour annealing at 950 °C and carbonitriding at 575 °C have practically no effect on the magnetic properties of the hot-rolled steel 10.

5. Contact resistance examination

A specimen (L-shaped, Figure 1) for contact resistance measurement was placed between the same L-shaped specimens (covers) polished to the shining metal in a hot-rolled state. The measuring device electrodes were pressed against the covers with a force of 30 N. Thus, the measurement pattern used four contact (transient) resistances: 2 between the electrode and the cover and 2 between the cover and the specimen. SEW 6237 DLRO milliohm-meter was used for the measurements. Three series of

IOP Conf. Series: Materials Science and Engineering **966** (2020) 012006 doi:10.1088/1757-899X/966/1/012006

experiments were made with the following specimens: 1 is a hot-rolled specimen polished to the shining metal; 2 is a hot-rolled specimen with a surface scale layer; 3 is a carbonitrided specimen without surface polish.

The results achieved show the following: scales on the specimen surface make its contact resistance 59 times higher (1059/18), and a carbonitrided layer make it 3.6 times higher (65/18). The high hot-rolled specimen resistance is explained by a dense oxide film (scale) formed in the heating, combined with high rolling contact pressure. On the contrary, the carbonitrided layer, as it well known [5, 6], has a porous thin (5 microns) structure that with a dielectric property of a chemical carbonitride.

6. Conclusion

- The electrotechnic hot-rolled steel 10 relaxes its hardness from HRB 90 to HRB 62–78 as a result of annealing.
- The steel 10 carbonitrided in the hot-rolled state for 10 minutes generates a hardened layer on the surface with a hardness of 283 HV10. This steel annealed and carbonitrided for 3 hours generates a hardened layer harder than that of these steels in hot rolled condition: HRC 60 and HRC 56, accordingly (UZIT-3 device). The measurements with deeper indenter penetration into the surface (PMT-3M device), give the opposite result (Figure 3), as it is much more influenced by the base metal, which looses the hardness as a result of annealing.
- The solid carbonitrided surface brings a significant (10–20 times) wear reduction to the steel 10 under dry sliding friction on the normalized steel 45.
- Annealing 950 °C and carbonitriding at 575 °C has practically no effect on the hot-rolled steel 10 magnetic properties.
- The contact electrical resistance of carbonitrided specimens is 3.6 times higher compared to the resistance of hot-rolled specimens cleaned from scale. Besides, the porous structure of the carbonitrided surface is better for dielectric varnish coating that has to adhere to it.
- The results obtained can be used in the design of magnetic circuits for various electric devices.

References

[1] Guliayev A P, Guliayev A A 2011 *Metallurgy* (Moscow: Alians) p 644 [In Russian]

[2] Kopylov I P, Klokov B K and Morozkin V P edited Kopylova I P *Electric machine design* 2005 4th ed (Moscow: Higher school) p 767 [In Russian]

[3] Guseva S V, Lykasov A A and Tepliakov Yu N 2016 *Electric insulation manufacturing practice for silicon transformer steels* Patent RU 2600606

[4] Tarasov A N, Gorbachiov Yu M and Murashko V M 2002 *Method for manufacturing and heat treatment of parts made of soft magnetic steels for magnetic systems of electric low-thrust jet engines* Patent RU 2181777

[5] Prokoshkin D A 1984 *Chemical and thermal treatment of metals: Carbonitration* (Moscow: Metallurg, Machinebuilding) p 240 [In Russian]

[6] Korotkov V A 2015 Carbonitriding for steelmaking equipment wear protection *Metallurgist* no **12** pp 63–8

[7] *UZIT-3 ultrasonic hardness tester operating manual* 2000 RE 427113-002-20872624-00.

[8] Antonov V G, Petrov L M and Schelkin A P 1986 *Material magnetic parameter measurement tools* (Moscow: Energoatomizdat) p 216 [In Russian]

[9] Demirchian K S, Neiman L R, Korovkin N V and Chechurin V L 2003 *Theory of electrotechnics: In 3 volumes. College textbook vol 2* 4th ed (St. Peterburg: Piter) p 576

[10] Kazandzhal L B 2000 *Magnetic properties of electric steels and alloys* (Moscow: OOO 'Science and technology') p 224 [In Russian]

[11] Gerasimov V G 2007 *Reference book of electrotechnics: in 4 volumes. Vol. 1: General. Materials for electrotechnics* 9th ed (Moscow: Energoizdat) p 440 [In Russian]

IOP Conf. Series: Materials Science and Engineering **966** (2020) 012007 doi:10.1088/1757-899X/966/1/012007

Nickel sorption from solutions with high salt concentration

O Yu Makovskaya[1, 2], O B Kolmachikhina[1], V G Lobanov[1] and S E Polygalov[1]

[1] Ural Federal University named after the first President of Russia B N Yeltsin, 620002, 19, Mira str., Yekaterinburg, Russia

E-mail: [2] o.i.makovskaia@urfu.ru

Abstract. The object of the study is salt concentrates formed as a result of treatment of concentrated acidic and alkaline wastewater of galvanic processes, realized on Ural Optical & Mechanical Plant (Russia). Salt concentrates are to be disposed of as low-hazard substances, but periodically the presence of nickel ions in them was detected. Concentration of nickel ions reaches up to 0.3 g/L, the total salt content reaches 200 g/L. The possibility of sorption purification of salt concentrates from nickel on the aminocarboxylicionite Lewatit TP 207 was studied, and the capacitance characteristics of the ionite were determined. It is shown that a high salt background does not reduce the ion exchange capacity of ionite for nickel. The dynamic capacity of Lewatit TP 207 before the breakthrough was 0.155 g/g. Nickel desorption is carried out with 20% sulfuric acid. The maximum concentration of nickel in the eluates was 25.484 g/L. Such eluates can be returned to the nickel-plating bath. The technology of periodic ion-exchange purification of salt concentrates is proposed.

1. Introduction

Galvanic production is an integral part of instrument-making, mechanical and other large enterprises. Electroplating is used to protect metals from corrosion during operation and at the same time to give a good appearance to parts.

One of the urgent problems of galvanic production is generation of the large amount of wastewater. The enterprises construct local treatment facilities to neutralize these waters and reduce the consumption of fresh water. At the neutralization station, water comes from the parts washing baths and spent contaminated electrolytes. They contain heavy non-ferrous metal ions, acids, alkalis, various additives, as well as mineral oils and dyes. It is advisable to create at the water treatment sites such technological chain, in which the water circulation system tends to be closed, and forming waste can be returned to the process.

The existing techniques for wastewater treatment are based on methods provide metals to be converted into insoluble compounds (hydroxides). The resulting solids are disposed of. Non-ferrous metals contained in sediments are lost, and enterprises are forced to pay environmental fines for the dumping of toxic substances.

Typically, galvanic wastewater treatment technology consists of a combination of several methods. The most commonly reagent treatments [1], coagulation-flotation methods [2], electrochemical methods (electroflotation, electrodialysis, electrolysis, etc.) [3–5], membrane technologies (ultrafiltration, nanofiltration, reverse osmosis), sorption processes are used [6].

The aim of the present study is to investigate the sorption of nickel by ion exchange from concentrated salt concentrates, which are waste from the treatment facilities of the galvanic section of the Ural Optical & Mechanical Plant.

In 2014 the treatment facilities have been reconstructed and currently operating according to the following scheme. Galvanic wastewater is directed to the neutralization station via separate collectors. Plastic receiver tanks are installed outside the building. Wash water, concentrated alkaline and acidic effluents, chrome effluents, concentrated alkaline and acidic aluminum-containing effluents are taken in separate tanks, from where they are fed for further processing.

Wash waters are averaged, then mixed with chrome effluents and fed to an electrocoagulator. The preliminary adjustment of the pH value of HCl or NaOH to the value of hydrate formation of chromium was carried out. In the electrocoagulator, Cr^{6+} is reduced to Cr^{3+}, as well as sorption of heavy metals on the surface of electrolytically obtained precipitates is occurred.

From the electrocoagulator, an aqueous suspension of metal hydroxides is sent to a vertical parallel plate clarifier, from where the thickened suspension enters to the filter press, and the clarified effluents pass sequentially through grain-loaded filters, carbon filters and a coal particle recovery unit.

After pretreatment clarified wash water is fed to the unit of a two-stage reverse osmosis. In the process of reverse osmosis, the stream is divided into two parts: the filtrate (water) and the permeate (stream enriched with salts). Permeate is taken to the evaporation unit, and purified water is used for topping up the electroplating baths.

Concentrated galvanic effluents are fed to the reactor to adjust the pH level (pH=10) and then sent to a filter press. Concentrated aluminum-containing effluents are adjusted separately (pH=5) and fed to a vacuum filter.

The clarified effluents and permeates of the reverse osmosis unit are directed to the evaporation unit. The products of the evaporation unit are distilled water and liquid salt concentrate for disposal.

Salt concentrates contain up to 0.3 g/L of nickel. To purify them from nickel using sorption by ion exchangers is proposed. It is known that resins containing carboxyl groups are capable of selectively absorbing nickel from dilute solutions [7, 8]. In particular, Lewatit TP 207 is used to remove nickel [9, 10].

2. Methods and materials

Lewatit TP 207 (Lanxess, Germany) is a weakly acidic, macroporouscation exchange resin with the iminodiacetate functional groups used for the selective sorption of heavy metal cations from aqueous solutions, the total ion exchange capacity of 2.2 meq/mL, the particle size of 0.4–1.25 mm and the thermal stability of up to 80 °C. In the experiments, a solution containing 250 mg/L Ni was used. Density of solution is 1.2 g/mL, pH = 4.40. Salt background presented by Na_2SO_4, NaCl, $NaNO_3$, Na_3PO_4, Na_3BO_4, NH_4F.

2.1. Ionite pretreatment

Before to they use ionite was washed with NaOH and HCl (7.5 %) to remove organic and inorganic impurities and then washed several times with deionized water. The resin was finally converted to the H^+- form.

2.2. Batch experiments

To perform sorption under static conditions, 50 mL of the solution was brought into contact with 0.5 g of resin. The duration of contact of the resin with the solution was 24 hours. The filtrates were separated from the ion exchanger. The nickel concentration was determined by atomic absorption spectrometry.

The sorption capacity q_t [mg/g] of the appropriate ions was determined from the following formula:

$$q_t = (C_0 - C_t) \times \frac{V}{m},$$

where C_0 is the initial concentration of As (V) ions, [mg/L]; C_t is the concentration of As(V) ions at time t, [mg/L]; V is the volume of the solution, [L]; m is the mass of the ionite, [g].

2.3 Column experiments

In the column experiment, sorption was performed in a fixed-bed glass column with 1.0 cm internal diameter and 25 cm height, packed with 5 g of Lewatit TP 207 in H. The glass wool beads were added to improve the flow distribution.

The sorption duration was 78 hours; 2.59 L of a solution was passed. The filtrates were taken into 50 mL volumetric flasks.

2.4. Effect of $NaNO_3$ concentration on nickel sorption

The main component in salt concentrates is $NaNO_3$. The concentration of sodium nitrate is 194.78 g/L. A series of experiments was performed to study the effect of $NaNO_3$ content on nickel sorption. In 50 mL volumetric flasks, a $NiSO_4$ model solution with a nickel content of 250 mg/L and a $NaNO_3$ solution with a concentration of 0, 10, 20, 50, 100, 200 g/L were added. Contact duration 24 hours.

2.5. Desorption

Desorption was carried out with a 20% solution of sulfuric acid. The desorbent was supplied at the same rate as the solution for sorption.

3. Results and discussion

The results are shown in the table 1.

Table 1. Effect of $NaNO_3$ concentration on Lewatit TP207 capacity for nickel.

C($NaNO_3$), g/L	Mass of ionite, g	C_t(Ni), mg/L	q_t, mg/g
0	0.5	27	22.30
10	0.505	30.2	21.76
20	0.505	26.9	22.09
50	0.505	22	22.57
100	0.5	23.55	22.65
200	0.5	21.88	22.81

From the obtained results it follows that changes in the salt background does not significantly affect the capacity of the resin.

Experimental breakthrough curves for adsorption of Ni(II) on Lewatit TP207 is shown in figure 1.A breakthrough point was observed after passing 1.15 L of concentrate after 34 hours. After passing through 2.5 L of solution, the filtrate concentration was equal to the initial one.

The dynamic capacity of the ion exchanger before the breakthrough point was 0.155 g/g.

After desorption, the ion exchanger was washed with 125 mL of distilled water. The nickel concentration in the wash water is 34.6 mg/L. The concentration of nickel in the eluates was determined by AAS.

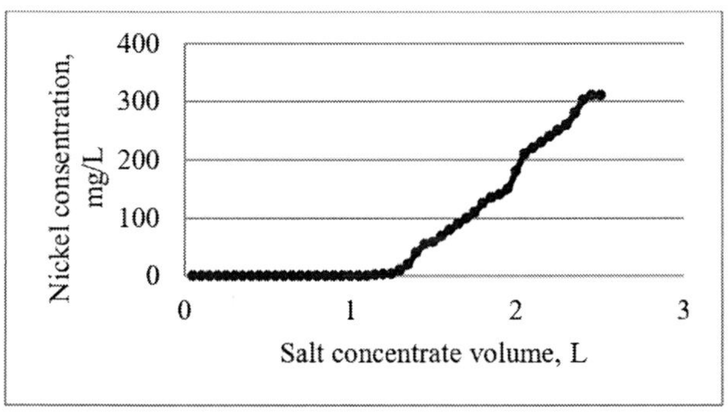

Figure 1. Sorption breakthrough curve.

The output desorption curve is shown in the figure 2.

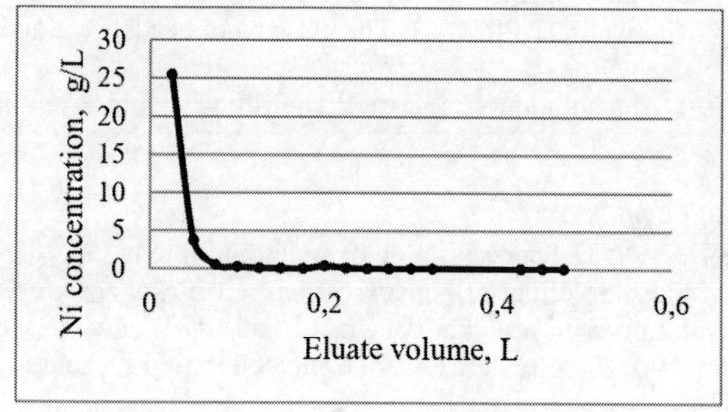

Figure 2. The output desorption curve.

In the richest eluate fraction the Ni concentration is 25.484 g/L. The concentration of Zn, Fe, Cu in several portions of eluates were analyzed by AAS. The results of the analysis are shown in table 2.

Table 2. The content of impurities in the eluate.

Eluate volume, L	Concentration, g/L		
	Zn	Cu	Fe
0.025	1.47	0.56	3.073
0.05	0.23	0.191	0.86
0.075	0.036	no data	0.63
0.1	0.005	no data	0.67
0.125	0.005	no data	0.83
0.15	0.036	no data	0.8
0.175	0.001	no data	0.77
0.2	0.001	0.002	no data

The resulting concentrations of impurities make it possible to return the eluates to the process chain, as an additive to the electrolyte in nickel baths, not exceeding the values established by the technological instruction.

Since salt concentrates have a temperature near 80 °C after the evaporation unit, the effect of temperature on nickel sorption has been studied.

There is practically no information in the literature concern the effect of temperature on the absorption of nickel by resins. There is evidence of the effect of temperature on nickel sorption by activated carbon, bentonite clays, minerals etc. [11, 12]. The obtained regularities cannot be fully applied to synthetic ion exchangers.

To study the sorption process at elevated temperature, the solution was passed through a thermostatic column.

Salt concentrate with pH=4.98; ρ=1.334 g/mL; C_{Ni}=189 mg/L, temperature 60 °C. Duration of sorption was 11 hours, 1.975 L of salt concentrate was passed through column. Solution rate is 0.18 L/hour.

Although the ion exchanger was loaded into the column in a fully swollen state, its volume decreased while passing the heated solution. Breakthrough of nickel ions was observed in the first portion of filtrate. Thus, it is not recommended to supply the salt concentrate directly to the sorption purification without preliminary cooling.

Currently, about 40 m^3 per year of salt concentrate is formed at the enterprise; the average nickel concentration is 0.310 g/L. It is supposed to treat the concentrate according to the scheme shown in figure 3.

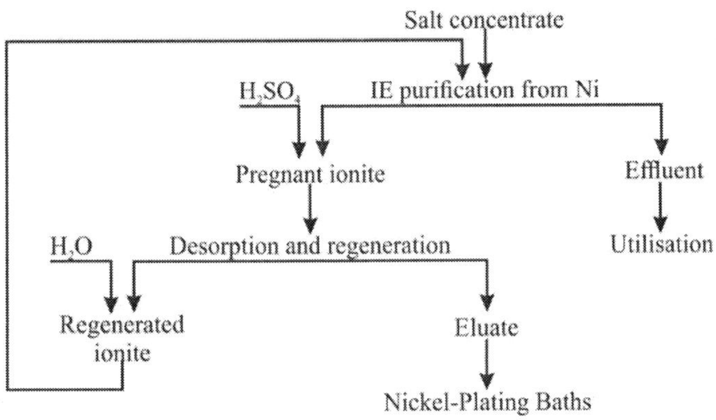

Figure 3. Flowsheet for the processing of salt concentrates.

Since the volume of generated effluents is small and require cooling, the best option would be to carry out sorption treatment periodically, as salt concentrates accumulate.

4. Conclusion

According to the results of the experiments, the working exchange capacity (before the breakthrough point) over nickel for the Lewatit TP 207 ion exchange resin was 0.155 g/g. Considering that about 40 m^3 of salt concentrate is formed per year, it is proposed to provide for a periodic operation mode of the installation. The degree of desorption, it is advisable to take 90 %. Periodically, complete desorption of the ion exchanger will be necessary.

References

[1] Arshad H, Iram J and Nadeem A Kh 2014 Characterization and treatment of electroplating industry wastewater using Fenton's reagent *Journal of Chemical and Pharmaceutical Research* **6(1)** pp 622–7

[2] Filatova E G, Soboleva A A, Dudarev V I and Pomazkina O I 2012 Electrocoagulation waste water treatment galvanic manufacture from nickel ions *Modern problems of science and education* **2**

[3] Benvenuti T, Rodrigues M A S, Moura Bernardes A and Zoppas Ferreira J 2014 Electrodialysis Treatment of Nickel Wastewater. *Electrodialysis and Water Reuse. Topics in Mining, Metallurgy and Materials Engineering* ed. Moura Bernardes (Berlin: Heidelberg)

[4] Li C L, Zhao H X, Tsuru T et al 1999 Recovery of spent electroless nickel plating bath by electrodialysis *J. Membr. Sci.* **157 (2)** pp 241–9

[5] Orhan G, Arslan C, Bombach H et al. 2002 Nickel recovery from the rinse waters of plating baths. *Hydrometal.* **65 (1)** pp 1–8

[6] Kharlamov D A et al. 2017 The post treatment of galvanic wastewaters from chromium (VI), copper (II) and nickel (II) ions using magnetic composite materials *Journal of Fundamental and Applied Sciences* **91S** pp 1811–19

[7] Bobkova L A, Kozik V V, Odnostorontseva T V et al 2012 Sorption selectivity of cobalt (II), nickel (II), and copper (II) ions by KB-2E macroreticular carboxylic cation exchanger from aqueous-salt solutions of alkali metals *Russ. J. Appl. Chem.* **85** p 1052

[8] Rudnicki P, Hubicki Z, Kołodyńska D. 2014 Evaluation of heavy metal ions removal from acidic waste water streams *Chemical Engineering Journal* **252** pp 362–73

[9] Halle O and Rossoni D 2005 A new (monodisperse) chelating resin designed for mining application: Lewatit MonoPlus TP207 *Proc. of Int. Nickel/Cobalt Conf. ALTA* (ALTA Metallurgical Services: Melbourne)
[10] Dizge N, Keskinlera B and Barlaset H 2009 Sorption of Ni (II) ions from aqueous solution by Lewatitcation-exchange resin *Journal of Hazardous Materials* **167** pp 915–26
[11] Alakhras F, Al-Abbad E, Alzamel N O, Abouzeid F M and Ouerfell N 2018 Contribution to Modelling the Effect of Temperature on Removal of Nickel Ions by Adsorption on Nano-Bentonite *Asian Journal of Chemistry* **30 (5)** pp 1147–56
[12] Scheckel K G and Sparks D L2001Temperature effects on nickel sorption kinetics at the mineral-water interface *Soil Sci. Soc. Am. J.* **65** pp 719–28

IOP Conf. Series: Materials Science and Engineering **966** (2020) 012008 doi:10.1088/1757-899X/966/1/012008

Technology for carrying out hydraulic fracturing using a new material 'nitinol'

A E Verisokin[1, 4], A P Yanukyan[2], D Yu Serdyukov[3] and L G Zhulina[1]

[1] Institute of Oil and Gas, NCFU, 1, Pushkin Str., Stavropol, 355009, Russia
[2] Tyumen industrial University, 49/3, Respubliki Str., Tyumen, 625000, Russia
[3] NGDU Rechitsaneft, RUP PO Belorusneft, 43, Lenin Str., Rechitsa, 247500, Belarus

E-mail: [4]verisokin.aleksandr@mail.ru

Abstract. The paper examines the causes of proppant backflow from a fracture to a well, and presents the main technologies for proppant control. The consequences of the removal of the fracture filler into the wellbore are determined. It has been noted that proppant backflow is most damaging to wells operated by electric centrifugal pumps. Existing methods of hydraulic fracturing are investigated, their disadvantages are indicated. A method has been developed for hydraulic fracturing with injection of compressed springs made of metal with the effect of 'memory' and restoring their shape under reservoir temperature stress. In the first stage, the remote section of the crack is filled with a fine fraction of ceramic proppant, and at the final stage of crack attachment, compressed springs made of nitinol material are fed. Using of material with a shape memory will allow the proppant packing to be compacted and the proppant to be 'locked' in the fracture.

1. Introduction

In the process of hydraulic fracturing (fracking) through the action of high pressure on the reservoir, a system of cracks is created into which granular material (proppant) is transported to fix the crack in the open state after relieving the overpressure [1]. Often there is a spontaneous proppant exit from the created hydraulic fracture beyond the reservoir (to water-saturated horizons) and hydraulic fracturing design optimization [2].

It is known that proppant backflow from a fracture into a well occurs after hydraulic fracturing during initial treatment and in some cases even after a well is fully developed [3]. One of the reasons for the removal of granular material from the fracture into the well is the destruction of the filler particles when pressed into the fracture, and the uneven distribution of particles in the fracture [4].

The filling of cracks with proppant is characterised by high permeability, high fluid velocities and the manifestation of inertial forces [5]. It is also noted that the wings of the crack work less efficiently, respectively, and the removal of proppant mainly comes from the central part of the crack. The backflow of proppant from the fracture to the well leads to a decrease in the conductive width of the fracture as a result of a decrease in the wedging effect, up to its complete collapse (Figure 1).

Due to the removal of the fracture filler, potentially high well flow rates are reduced, proppant plugs are formed on the bottom, which requires additional well flushing and the overhaul period of the electric centrifugal pump (ECP) is also reduced.

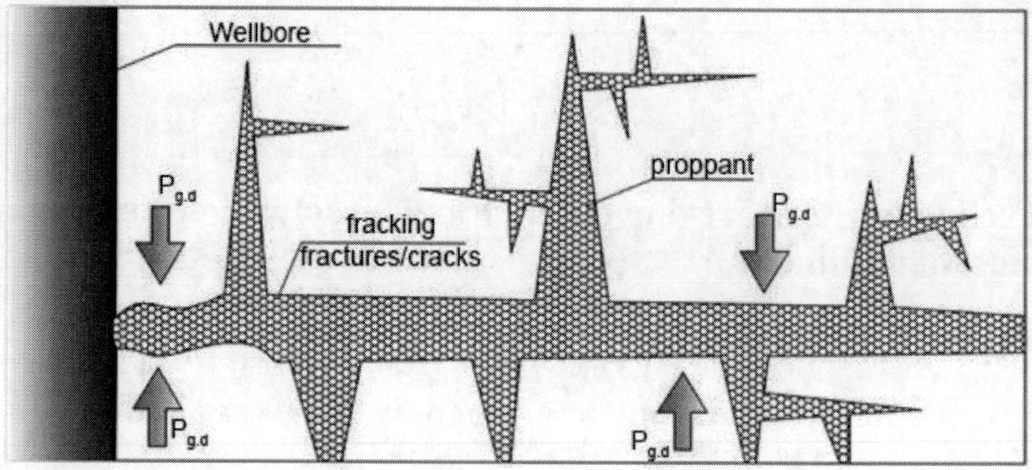

Figure 1. Reduction of the conductive width of the crack.

An analysis of the causes of failure of submersible pumps confirms that proppant back flow (PBF) is most harmful for wells operated by electric centrifugal pumps. ESPs with which a significant part of oil producing wells are equipped in our country are quickly clogged with proppant after hydraulic fracturing and fail ('ESP wedge' occurs) [6]. Because of this, the ESP, which is first installed in the well after hydraulic fracturing, is called a 'sacrificialpump' (Figure 2).

Figure 2. The negative effect of the back flow of the crack filler into the well on an ECP, clogged with proppant, during the working stage.

2. Materials and methods

To solve the problem, methods of hydraulic fracturing were investigated [7–8]. The analysis showed that the problem of fixing and holding the proppant in the fracture has not been solved at the present time. The considered methods have the following disadvantages:

- the use of a proppant, without taking into account the size of the fracture, makes it difficult to keep the proppant particles in the crack;
- there is no reliable filtration layer during the formation of the proppant crack frame;
- the occurrence of various fluids in the well, such as acids, gels and destructors leads to the destruction of the polymer and a decrease in the strength of the proppant itself, with its destruction and the formation of mobile abrasive particles in the stream;

- during transportation of the composite proppant along the axial channel of the drill pipe string, their adhesion inevitably occurs to the inner surface, with the formation of a proppant plug;
- the use of granular material of the same fractional composition to fill a hydraulic fracture along its entire length stimulates the process of particle migration into the well, since at the outlet into the wellbore, the flow rate of the formation fluid sharply and repeatedly increases.

In connection with the presence of an urgent problem, the technology of hydraulic fracturing with the injection of pressed springs from nitinol at the final stage of crack attachment is proposed. Nitinol is an alloy of titanium (45%) and nickel (55%) in the proportion of 45 to 55%, which has high corrosion and erosion resistance. It is unusual that this alloy has a memory effect. If the item is heated to red heat, then it will return this form. After cooling to room temperature, the part can be deformed, but when heated, it will return to its original shape. The response temperature range depends on the chemical composition of the alloy and can be changed depending on the required indicators. The shape of the element can be set in the laboratory using the recommended heat treatments. This behaviour is due to the fact that this material is an intermetallic compound and, when quenching, the mutual arrangement of atoms is ordered, which leads to shape memory. Super elasticity manifests itself during the transition upon heating from one structural internal state to another. Upon reaching the value of phase transformation, the alloy as a spring takes its original form.

3. Results and discussions

The proposed method of hydraulic fracturing includes running of the pipe string with the packer into the wellbore, the overlap of the annulus above the roof of the producing formation, the supply of hydraulic fracturing fluid along the pipe string with the creation of overpressure with a crack fixer, in the form of proppant particles of the calculated fractional composition, holding time. The dimensions of the granules used to fill remote zone of reservoir andnear wellbore zone of hydraulic fracture are determined in such a way as to prevent the migration of proppant particles by the flow of reservoir fluid. At the first stage, the remote section of the crack is filled with a fine fraction of ceramic proppant, and at the second stage, a large fraction of particles formed from a metal having a memory effect and changing their shape is supplied, with the formation of a filtration layer in the hydraulic fracture when exposed to formation temperature (figure 3).

To ensure the fixation of proppant particles in the near – wellbore zone, we determine its diametrical dimensions by the formula:

$$d_{\text{prop}} = \frac{V_{\text{CT}}}{V_{\text{f}}} \cdot d_{\text{p}},$$

where d_{prop} – is the diameter of the proppant particles at the outlet of the fracture into the wellbore, mm; V_{CT} – formation fluid flow rate at the outlet to the wellbore, m/s; V_{f} – the rate of filtration of the reservoir fluid at a remote section of the fracture, mm/s; d_{p} – the diameter of the proppant particles at a remote section of the fracture, mm.

When the proppant particles are diametrically sized to fill a crack in a remote area within d_{p} = 0.2–0.4 mm, the grain size in the near-well zone should be d_{prop} = 2–3 mm.

To ensure the supply of such proppant particles to the fracture, it is necessary to select the composition of the working fluid and the speed of transportation in the interval of the reservoir. Displacement fluid is pumped into the volume of the tubing string, ensuring that the suspension is completely displaced into the fracture. The formation of the filtration layer by springs from nitinol in the borehole zone of the hydraulic fracture is carried out in a chaotic arrangement.

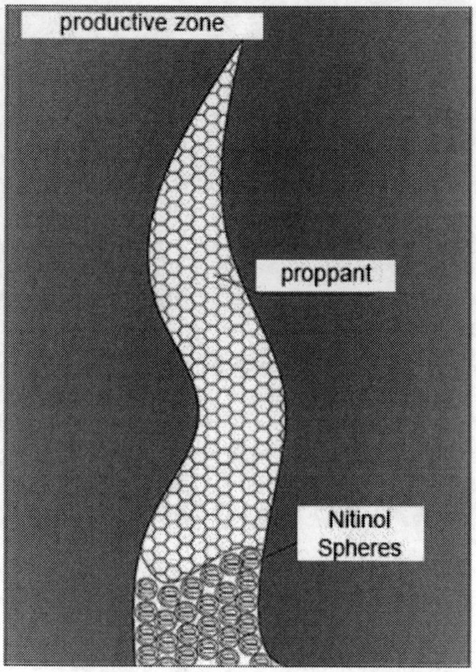

A – proppant granule from the material 'nitinol' in the hydraulic fracture in an unopened state

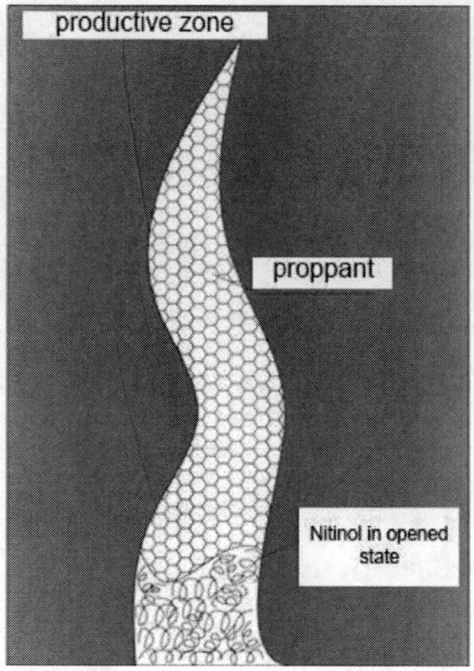

B – proppant from the material 'nitinol' in the hydraulic fracture in the open state after exposure to temperature

Figure 3. Using 'nitinol' to fix hydraulic fracture of the formation.

The conductivity of such a material can be tested at the developed stand [9]. 'Nitinol' has a better holding capacity compared to proppant or sand. Furthermore, the injection of material with shape memory at the final stage of the crack attachment will allow the proppant packing to be compacted and, in turn, to block the proppant in the crack (Figure 4).

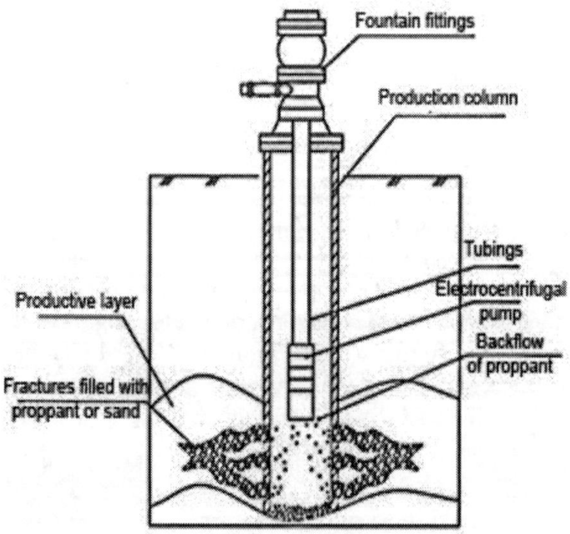

Proppant backflow using existing technologies

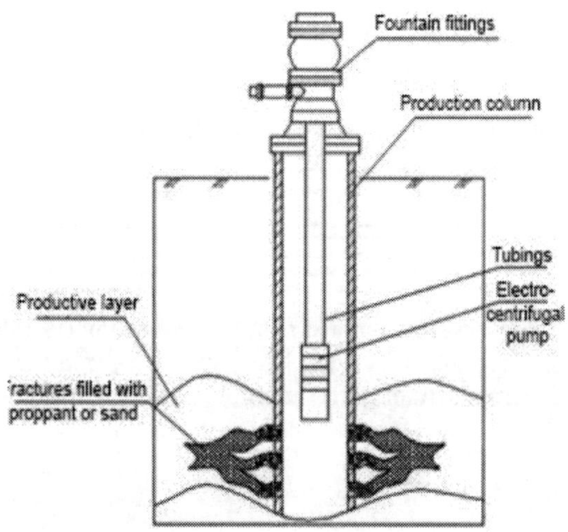

Lack of proppant back flow when using nitinol

Figure 4. Schematic diagram of the proposed crack attachment technology.

4. Conclusion

The proposed technology for strengthening cracks after hydraulic fracturing received a positive decision on the grant of a patent of the Russian Federation [10].

The application of this method in practice will obtain the following technical results:

- the delivery of proppant particles to the peripheral part of the fracture while maintaining their position in the process of reservoir fluid filtration during well development and hydrocarbon production;

- the formation of a highly permeable screen of the filter layer of the fracture in the near-wellbore zone in the presence of a large proppant fraction with optimal placement of hydraulic fracturing in the fracture;

- the possibility of forming a highly permeable screen in the borehole zone in the form of segments of helical coil springs made of metal having the 'memory' effect and restoring their shape due to the formation temperature.

References

[1] Verisokin A, Zinovieva L, Grab A and Serikov D 2018 The mechanism of deformation processes that occur during hydraulic fracturing in horizontal wells *Construction of oil and gas wells on land and at sea: scientific-technical* **10** pp 50–3

[2] Aslanyan I, Minakhmetova R, Trusov A, Lutfullin A, Khabibrahmanov A, Aflyatunov R and Khabipov R 2018 Determination of proppant removal zones by spectral noise metering *Oil industry* **5** pp 68–71

[3] Economides M and Nolte K 2000 *Reservoir stimulation 3rd ed.* John Wiley Sons (Schlumberger)

[4] Devyatov A 2010 Modeling hydraulic fractures using local grinding of a hydrodynamic grid in the case of a full-scale field development scenario *Oil field* **5** pp 32–6

[5] Gilmiev D 2013 Hydrodynamic model of fluid filtration in the reservoir in the presence of hydraulic fractures *Oil industry* **7** pp 108–10

[6] Keybal A and Keybal A 2009 Reasons for proppant backflow to the wellbore after hydraulic fracturing of the reservoir *Drilling and Oil* **11** pp 48–52

[7] Akimov O 2008 The potential of proppant consolidation technologies to increase the efficiency of hydraulic fracturing *Oil industry* **11** pp 31–3

[8] Demichev S and Otradnykh O 2008 Experimental studies on the fixation of proppant in hydraulic fractures *Drilling and Oil* **12** pp 19–21

[9] Verisokin A 2018 The proppant test procedure for hydraulic fracturing *Science and Technology in the gas industry* **2** pp 62–9

[10] Verisokin A and Serikov D 2019 Technology of hydraulic fracturing using titanium nickelide *Territory of oil and gas* **9** pp 20–4

IOP Conf. Series: Materials Science and Engineering **966** (2020) 012009 doi:10.1088/1757-899X/966/1/012009

Analysis of single- and double-threaded rolling for production of grinding balls of alloy steel grades

I Galimyanov

Engineering office JSC EVRAZ NTMK, 1, ul. Metallurgov, Nizhny Tagil, 622025, Russia

E-mail: Ilyas.Galimyanov@evraz.com

Abstract. JSC EVRAZ NTMK has mastered the technology of production of grinding balls with high hardness of surface and normalized hardness at the depth of 0.5 of a ball radius (4th hardness group) and normalized volume hardness (5th grade hardness) in accordance with Russian state standard GOST 7524–2015. Currently rolling of balls with the rated diameter of 100 mm is performed using double-threaded rolls. Single-threaded pass design is used for rolling of the 120-mm grinding balls. Double-threaded pass design increases a feed angle of rolls. The higher angle increases tangential stresses. 100-mm balls produced through the double-threaded pass design where the flanges excessively wear, show specific defects on a surface of the balls. This article describes the materials and results of the study of rolling of the grinding balls with the diameter of 100 mm and 120 mm produced via single- and double-threaded designs. Lower tangential loads applied to the balls during single-threaded rolling affects their macro structure, their hardening layer thickness.

1. Introduction

The main consumers of the grinding balls are domestic and foreign mining and processing combines (Ukraine, Kazakhstan, Poland, Mongolia, etc.). Many domestic and foreign consumers doing processing of hard ores of non-ferrous metals demand high strength, hardness and wear resistance of the grinding balls [1].

For this purpose, JSC EVRAZ NTMK has mastered the technology of production of grinding balls with high hardness of surface and normalized hardness at the depth of 0.5 of a ball radius (4th grade hardness) and normalized volume hardness (5th grade hardness) in accordance with [2].

For this purpose, a grinding ball plant has been commissioned. The plant includes: feedstock storage area; charging area; re-heating furnaces area; mill area; balls hardening area; tempering area (stress relieving); finished products shipment and storage area; change parts preparation area; water treatment and cooling system. A walking hearth re-heating furnace is divided along the length into continuous, heating and soaking zones. Cross ball rolling mill. Rolls with the gaged variable pitch screw thread are the working tool of SPSh 60 to 120.

The balls hardening area is intended for heat treatment of the balls with the rated diameter of 60, 70, 80, 90, 100, 110 and 120 mm, of hardness category 1–5 provided that tempering is done in a tempering furnace. An equipment set of the balls hardening section includes a hardening tank, a hardening drum, a driving station and an input chute.

In terms of structure, the hardening drum constitutes a central frame on which twenty-four through pipes are fixed along the circumference of the circle, with screw augers, having a screw line direction opposite to the drum rotation, installed inside (Figure 1). Tempering area (stress relieving). The tempering furnace is a conveyor furnace with circulatory atmosphere of direct convective heating.

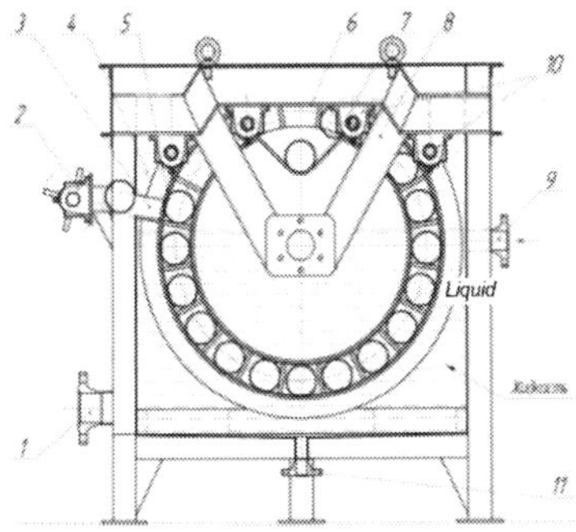

Figure 1. Structure main elements: 1 – fluid removal; 2 – dosing mechanism; 3 – input chute; 4 – container; 5 – external stationary part of drum; 6 – moving part of drum; 7 – internal moving part of drum; 8 – ball; 9 – fluid feed line; 10 – roll; 11 – fluid drainage.

2. Methods

Rolling of the balls from billets of different accuracy level, keeping or removing a web in a gorge requires rolls of different pass designs. A screw gage for the balls rolling consists of two sections that are the forming one and finishing one [2]. The forming section provides bite of a billet and its gradual reduction into a ball connected to the billet body via the web. The forming is performed by the flanges having a constantly increasing height, by which a diameter of the web is changed. A metal displaced from the webs into the ball provides radial deformation and increase of a ball diameter compared to a feedstock diameter. Therefore, the feedstock diameter is 1 to 2 mm smaller than that of a ball [4]. Correction of groove filling may be performed by adjusting the feed angle of the rolls a, which is normally 2 to 4° at the ball-rolling mills. A groove length (number of turns) is measured in degrees of of the flat groove screw line. If the groove total length is 900–1350°, the forming section takes 360-540°, and a sizing section, at which diameters of billets are aligned prior to forming and an elongated finishing section, takes up to 810°. When rolling the balls of bearings, only sizing of the ball is performed at the finishing section [5]. This section has a constant pitch and a threading profile corresponding to a profile of a ball being rolled, as well as a constant angle of threading rise equal to the feed angle of rolls a. One ball is rolled in one revolution of the rolls [6, 7].

When rolling the grinding balls, a roll is distinguished by the presence of the sizing section aligning the diameters of the billets prior to forming and the elongated finishing section (up to 810°). A small axial displacement of the rolls breaks off the web at the beginning of the finishing section, and then rotates the ball along the axis perpendicular to a rolling axis, the flanges cut off the webs and the vertex projections are rolled smooth [8, 11]. To increase the production output of the mill, double- and triple- threaded screw grooves are used along with the single- threaded ones, which allows getting 2–3 balls for one revolution of the rolls. The average production output of mills is 1–6 balls per second, up to 50–60 million of units/year (10–150 ths. tons/year) [9].

Forming of products during transverse rolling in the screw grooves is carried out by gradual reduction of a rotating feedstock. An allowable value of a single reduction for half a revolution of the feedstock is determined by conditions of its bite by the rolls [10]. Thus, to ensure the bite of the feedstock by the rolls, the reduction shall not exceed the allowable value.

As the operating experience shows, in addition to the single bite, loosening of the metal during the transverse rolling in the screw grooves is affected by: width of the groove flange, tension of the metal in a part of the feedstock being reduced, temperature of the rolling, rotation speed of the rolls, difficulty in an axial metal flow and the presence of an excess metal in the groove.

Currently rolling of the balls with the rated diameter of 100 mm is performed using the double-threaded rolls (Figure 2). Rolling of the grinding balls with the diameter of 120 mm is performed using the single-threaded design (Figure 3). The feed angle of the rolls is 4.5–5°, since it is equal to an average angle of inclination of the flanges, due to application of the double- threaded design in the process of rolling of the balls [11]. With the pitch of 227.387 (for the two- thread design) and the roll diameter of 700 mm, the theoretical angle is B = 5.7° (Table 1). When working with such angles, the tangential stresses increase due to an increase in the feed angle (Figure 4) [12].

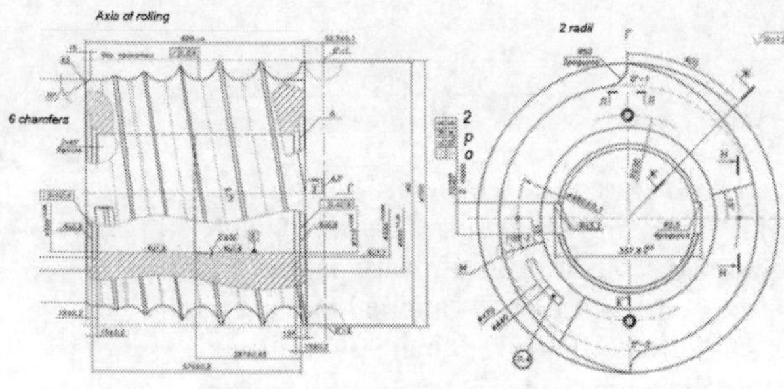

Figure 2. Roll with double- threaded design for rolling of the balls with the diameter of 100 mm.

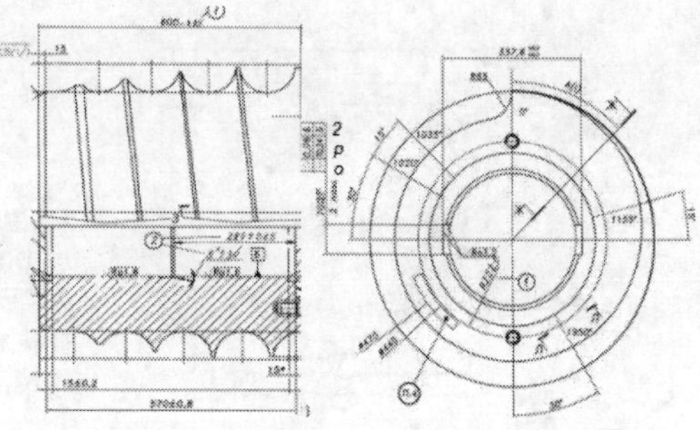

Figure 3. Roll with single- threaded design for rolling of the balls with the diameter of 120 mm.

Table 1. The estimated parameters.

RPD	Diameter of the cold ball (nom.), mm	Design distances from the roll axis to the rolling axis Dv (parallel rolls), mm Nominal value	MIN-output Support	Cut	Distance between the roll flanges in the gorge, mm	K, mm	Feed angle, ∝°, deg.
∅120 single-threaded	124.8	700	698.0	694.5	17.7	13 (10.3 %)	2.5–3.5
∅100 double-threaded	104.0	700	698.0	694.5	13.0	10 (9.5 %)	4–5

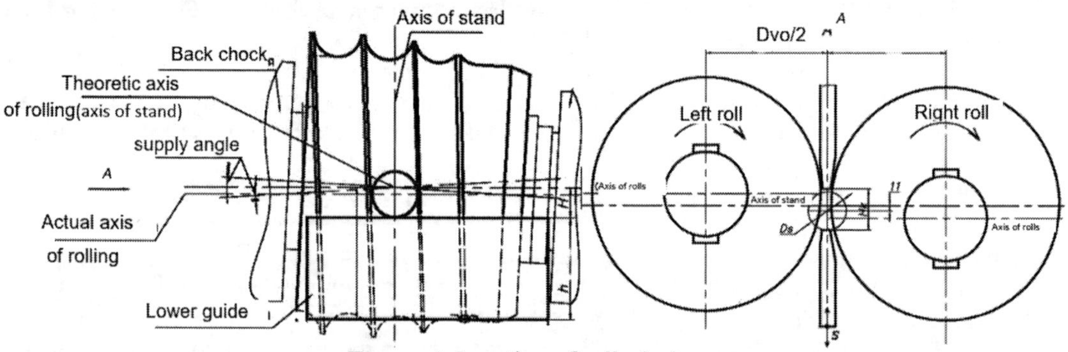

Figure 4. Location of rolls during setup.

When reprofiling, a ratio of the pitch to the diameter of the roll increases even more, which pass to the greatest tangential loads, which reduce the efficiency of useful forces during rolling. All this pass to an increase in loads necessary for the formation of the ball, and consequently to an increase in load on the rolling motor. In this regard, when rolling the ball with the diameter of 100 mm with a standard hourly production output, the loads on the rolling motor are at the average of 85–90 % of the nominal loads, and, for example, when rolling the ball with the diameter of 120 mm with an hourly production output, the loads on the rolling motor are 70–75 % from the nominal ones (Figure 5, 6). In addition, when the tangential loads are increased, distribution of a contact pressure is moved from the top of the flange to a side surface, which leads to intensive wear of the rolls.

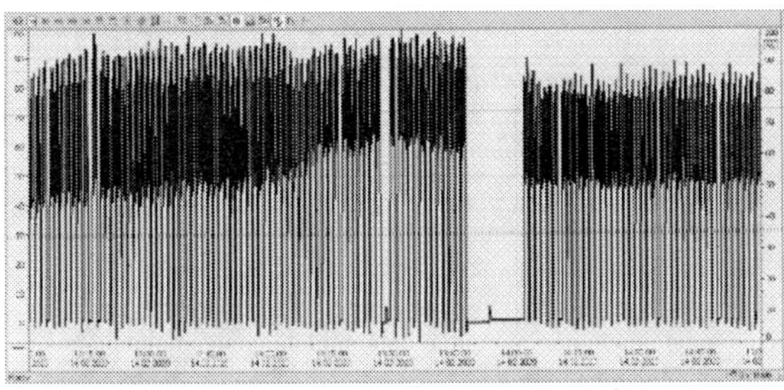

Figure 5. Rolling motor loads during rolling of 100 mm ball.

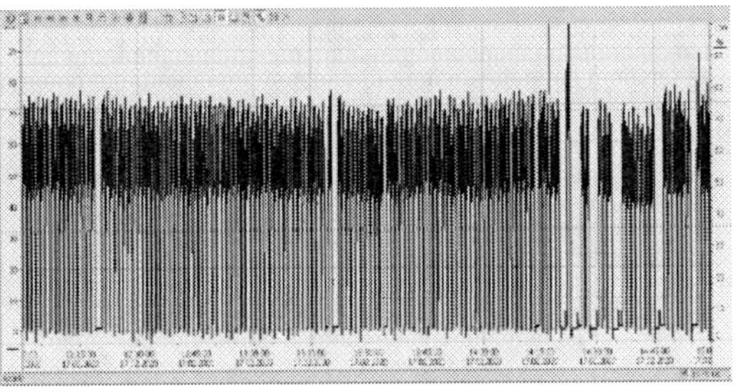

Figure 6. Rolling motor loads during rolling of 120 mm ball.

During the double-pass rolling of the grinding balls with the diameter of 100 mm, the large feed angle of the rolls causes increased wear of the flanges not only in the normal (perpendicularly to the roll axis – along the tops of the flanges), but also in the tangential (lengthwise the roll axis – along the side of the flanges) directions (Figure 7). This factor significantly reduces the consumption rate of the rolls.

Figure 7. The roll wear at the ball web break area.

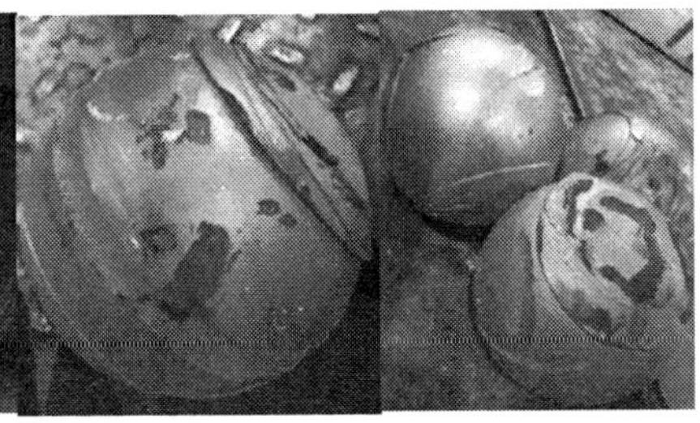

Figure 8. The grinding balls with the diameter of 100 mm affected by roll flange wear.

Production of the balls with the diameter of 100 mm on the rolls with the double- threaded design with excessive wear of the flanges causes unique defects on a surface of the balls, and dimensions of these defects may exceed maximum allowable deviations (Figure 8).

During the single- threaded rolling of the balls with the diameter of 120 mm, the smaller feed angle of the rolls causes decrease of wear of the flanges and the consumption rate of the rolls respectively.

Currently the possibility of production of the balls with the rated diameter of 100 mm on single-threaded rolls is considered, an increase of durability of the rolls is planned respectively.

3. Results

For this purpose, it is required to order development of design and manufacture a set of rollers and blocks. When comparing the methods of single- threaded and double- threaded rolling of the balls, the following factors are revealed:

– When using the single- threaded design, the feed angle will be reduced by 2 times, which will pass to a decrease in the tangential loads.

– During the single- threaded design of the forming section, the feedstock will be reduced in 3 full roll revolutions instead of 1.5 revolutions on a current design, which will pass to more gentle reduction regimes and allow reducing the load on the rolling motor and decreasing wear of the rolls.

– Decreasing the load on the rolling motor, as well as reducing an angle of spindles to 2.5 degrees, will reduce the total engine torque by 2 times, which will allow increasing a number of revolutions from 40 rpm currently used to 80 rpm and will keep the machine production output at the level of 80 balls per minute for the single- threaded design in general.

IOP Conf. Series: Materials Science and Engineering **966** (2020) 012009 doi:10.1088/1757-899X/966/1/012009

Reduction of the tangential stresses on the balls with the diameter of 100 mm during transfer to the single-threaded design influences on the macro structure, the value of hardening layer and hardness of balls (Figure 9). Reduction of the tangential stresses shall decrease a risk of crack of the balls [13].

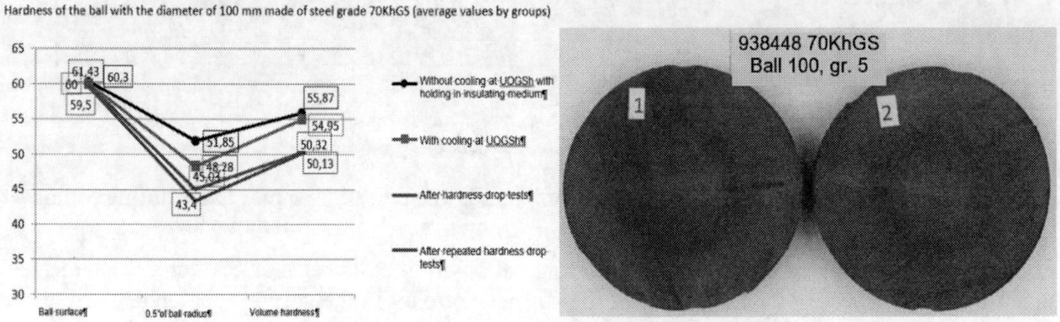

Figure 9. Hardness and the micro structure of the balls with the diameter of 100 mm made of steel grade 70KhGS.

Unacceptable defects and stress cracks were not detected in the course of control of the micro structure of the grinding balls with the diameter of 120 mm made of steel grade 70KhGS (Figure 10). The depth of the hardened layer was 18–31 mm.

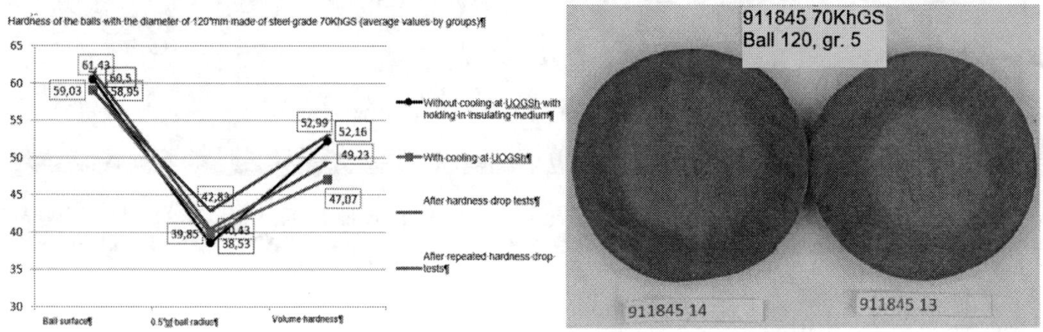

Figure 10. Hardness and the micro structure of the balls with the diameter of 120 mm made of steel grade 70KhGS.

According to the results of certification of the grinding balls, compliance with the technical requirements of the 5th hardness group according to [2] was confirmed.

Subject to agreement, pilot batches were sent to the consumer PJSC Polyus to test the use of the grinding balls during grinding of non-ferrous metal ores. The grinding balls were loaded into ball crushers in cooperation with specialists of JSC EVRAZ NTMK. According to the results of using the grinding balls of the 5th hardness group according to [2], the consumer received improved results on specific consumption rate of the balls and increased the scope of supply.

4. Conclusions

1. When rolling the ball with the diameter of 100 mm with the standard hourly production output (19.5 t/h), during the double- threaded rolling, the loads on the rolling motor are at the average of 85–90 % of the nominal loads, and, for example, when rolling the ball with the diameter of 120 mm with the hourly production output (22.1 t/h), during the single- threaded rolling, the loads on the rolling motor are 70-75% from the nominal ones.

2. Reduction of the tangential stresses on the balls with the diameter of 100 mm during transfer to the single- threaded design influences on the macro structure, the value of hardening layer and hardness of balls.

3. Decreasing the load on the rolling motor, as well as reducing an angle of spindles to 2.5 degrees, will reduce the total engine torque by 2 times, which will allow increasing the number of revolutions from 40 rpm currently used to 80 rpm and will keep the machine production output at the level of 80 balls per minute for the single- threaded design in general.

4. By means of innovative developments, the composition of advanced equipment allows achieving a stable technological process for producing the grinding balls that meet the technical requirements of the 5th hardness group according to [2].

5. JSC EVRAZ NTMK commissioned a unique grinding ball plant. The technology of industrial production of balls with the diameter from 60 mm to 120 mm with high hardness of surface and normalized hardness at the depth of 0.5 of a ball radius (4th hardness group) and normalized volume hardness (5th hardness group) as per [2], has been developed.

References

[1] Stalinskiy D V, Rudyuk A S and Solennyi V K 2017 Selection of Material and Technology for Heat Treatment of Grinding Balls Operating Mainly in Abrasion Wear Conditions *Steel* no **6** pp 64–9

[2] Russian state standard GOST 7524-2015 *Steel grinding balls for ball mills. Specifications.*

[3] Umanskiy A A, Golovatenko A V, Shchukin A G and Simachev A S 2019 Study of Influence of Parameters of Hardening Micro Structure of Grinding Balls, Produced by Cross-Rolling Method, on their Operational Characteristics *Production of Rolled Products* no **3** pp 34–9

[4] Chizh Ye V and Abenova M B 2016 Selection and Analysis of Technologies for Production of Grinding Bodies *Theory and Technology of Metallurgical Production* no **2** pp 26–33

[5] Vdovin K N, Foktistov N A, Abenov M B, Kulikov V D and Kondratyev I S 2015 Quality of Grinding Balls Made by Different Methods *Theory and Technology of Metallurgical Production* no **1 (16)** pp 78–81

[6] Rubtsov V Yu and Shevchenko O I 2018 Development of Production of Grinding Balls of the 5 th Hardness Group in Conditions of JSC EVRAZ NTMK *In Collected Book: Important Problems of Today's Science, Engineering and Education: Abstracts of 76th Int. Scientific and Technical Conference* (Magnitogorsk: Publishing House of Nosov Magnitogorsk State Technical University) vol **1** pp 117–8

[7] Rubtsov V Yu 2018 Operating Range of Ball Mill Parameters *Young Population and Science: Materials of Int. Scientific and Training Conference (May 25, 2018): in 2 volumes* (Nizhny Tagil: NTI (branch) UrFU) vol 1 pp 18–23

[8] Rubtsov V Yu 2017 Influence of Roll Body Dimensions of Ball Quality *Important Problems of Today's Science, Engineering and Education* vol **1** pp 90–3

[9] Rubtsov V Yu and Shevchenko O I 2017 The Application of the Variable Speed of the Rolls Rotation during the Transverse Rolling in order to Improve their Service Life *Ferrous Metallurgy: Bulletin of Institute 'Chermetinformatsiya'* no **11** (1415) pp 72–7

[10] Hu Z, Wang B and Zheng Z 2018 Research and industrialization of near-net rolling technology used in shaft parts *Frontiers of Mechanical Engineering* **13** pp 17–24

[11] Tomczak J, Pater Z and Bulzak T 2014 Designing of screw impressions in the helical rolling of balls *Archives of Civil and Mechanical Engineering* **14 (1)** pp 104–13

[12] Pater Z 2014 Analysis of the helical-wedge rolling process for producing a workholding bolt *Metalurgija* **53 (4)** pp 653–6

[13] Berti G A, Quagliato L and Monti M 2015 Set-up of radial–axial ring-rolling process: Process worksheet and ring geometry expansion prediction *Int. Journal of Mechanical Sciences* **99** pp 58–71

[14] Tomczak J, Pater Z and Bulzak T 2018 The flat wedge rolling mill for forming balls from heads of scrap railway rails *Arch. Metall. Mater.* **63** pp 5–12

IOP Conf. Series: Materials Science and Engineering **966** (2020) 012010 doi:10.1088/1757-899X/966/1/012010

Study of possibility of cerium reduction from slags of CaO-SiO_2-Ce_2O_3– 15%Al_2O_3– 8%MgO system

A A Babenko[1], L A Smirnov[1], A G Upolovnikova[1, 2] and R R Shartdinov[1]

[1] Institute of Metallurgy of the Ural Branch of the Russian Academy of Sciences, 101, Amundsen street, Yekaterinburg, 620016, Russia,

E-mail: [2] upol.ru@mail.ru

Abstract. By means of theoretical and experimental studies it is shown that it is possible to reduction cerium from the slags of the CaO–SiO_2–Ce_2O_3 system containing 15%Al_2O_3 and 8 % MgO, by aluminum dissolved in metal, at temperatures of 1550 and 1650°C. The results of mathematical modeling are presented graphically as 'composition – equilibrium content of cerium in metal' diagrams. It is found that depending on temperature of metal, basicity of slag and the content of cerium oxide, from 0.055 to 16 ppm cerium passes into metal containing 0.06 % of carbon, 0.25 % of silicon and 0.055 % of aluminium. Positive influence of the temperature factor, basicity of slags and the content of cerium oxide in a studied range of chemical composition on process of cerium reduction is explained from the perspective of phase structure of formed slags and thermodynamics of reactions of cerium reduction. The possibility of reduction of cerium from slags of CaO–SiO_2–Ce_2O_3 system, containing 15 % Al_2O_3 and 8 % MgO, has been experimentally confirmed. It has been shown that at basicity of 5 and 4 % of Ce_2O_3 in slag up to 16 ppm cerium passes into metal during 10 minutes exposure.

1. Introduction

Due to the increase in construction of long gas main pipelines in Russia the research and development of composition of high-strength pipe steels and innovative technological solutions for their production are becoming more and more important for the domestic metallurgy. The promising way of solving the problem is steel microalloying with boron [1–6]. Microalloying of low-carbon boron steel is used to increase the set of mechanical, technological and operational properties while maintaining economically alloyed chemical composition. However, the conditions under which pipelines are laid and operated are characterized, as a rule, by cold climate and high seismic activity, therefore along with high strength the pipe steels have high requirements to impact toughness at low temperatures and deformation ability. However, the ductility of low-carbon steels is related to the impact toughness, the increase in which is accompanied by a decrease in both relative elongation and contraction. One of the directions for solving the problem of maintaining of high strength and providing plasticity of low-carbon steel is organization of the process of modification of metal with rare-earth metals (REM). Positive influence of REM on ductility, impact toughness and resistance to cyclic cracking of pipe steel has been repeatedly confirmed by studies [7–11].

It is known that ferroalloys are used to modify steel with REM. It leads to an increase in the cost of steel. One of the directions of the problem solving can be REM reduction from oxide systems. It is known that introduction of Ce_2O_3 into the refining slag reduces the activity of harmful Al_2O_3 inclusions in the metal, melting temperature and slag viscosity, as well as increases the desulfurization

IOP Conf. Series: Materials Science and Engineering **966** (2020) 012010 doi:10.1088/1757-899X/966/1/012010

degree [12–18]. In addition, the possibility of reduction of cerium from the slags of the studied oxide system and its dissolution in steel in an amount up to 4ppm is noted, thus providing the effect of steel modification. The microstructure of the molten metal is better refined and consists of ferrite and a small amount of perlite [13–16].

Theoretical and experimental studies of the possibility of cerium reduction by aluminium dissolved in metal from the slags of the $CaO–SiO_2–Ce_2O_3$–15%Al_2O_3–8%MgO system at temperatures of 1550 and 1650 °C are presented.

2. Materials and methods

The thermodynamic modeling of cerium reduction from slags of the $CaO–SiO_2–Ce_2O_3$ system containing 15% Al_2O_3 and 8% MgO, by aluminum dissolved in metal at temperatures of 1550 and 1650°C was performed using HSC 8.03 Chemistry (Outokumpu) software package based on Gibbs energy minimization and variational principles of thermodynamics using simplex planning grid method [19–22]. The principle of the method of simplex planning grids is to build a planning matrix, where the basicity of slag and Ce_2O_3 content in slag are changed (Table 1) with the following construction, on the basis of thermodynamic calculations, of mathematical model describing dependence of equilibrium Ce_2O_3 content in metal on slag composition, presented as diagrams 'the chemical composition of slags of the $CaO–SiO_2–Ce_2O_3$ system containing 15% Al_2O_3 and 8% MgO–equilibrium content of cerium in metal' at temperatures of 1550°C and 1650°C (Figures 1 and 2). In diagrams blue lines indicate isolines of equilibrium content of cerium. Thin black lines indicate the basicity of slag (B=CaO/SiO_2), the figures indicate their values.

Table 1. Slag composition and the results of thermodynamic modeling of the equilibrium cerium content in metal.

№	Slag index	Slag composition, mas. %					[Ce], ppm		B=CaO/SiO_2
		CaO	SiO_2	Ce_2O_3	Al_2O_3	MgO	1550°C	1650°C	
1	Y_1	50.7	25.3	1	15	8	0.055	0.085	2
2	Y_2	63.3	12.7	1	15	8	1.89	2.68	5
3	Y_3	58.3	11.7	7	15	8	11.7	16.1	5
4	Y_4	46.7	23.3	7	15	8	0.42	0.64	2
5	Y_{13}	59.1	16.9	1	15	8	0.72	1.03	3.5
6	Y_{132}	56	16	5	15	8	3.47	4.94	3.5
7	Y_{22}	60	12	5	15	8	8.77	12.2	5
8	Y_{12}	54.9	21.1	1	15	8	0.23	0.34	2.6
9	Y_{121}	53.2	20.8	3	15	8	0.65	0.95	2.6
10	Y_{21}	61.6	12.4	3	15	8	5.43	7.61	5
11	Y_{131}	57.5	16.5	3	15	8	2.09	3.00	3.5
12	Y_{41}	48	24	5	15	8	0.29	0.44	2
13	Y_{31}	54.5	15.5	7	15	8	4.85	6.87	3.5
14	Y_{42}	49.4	24.6	3	15	8	0.17	0.26	2
15	Y_{32}	50.5	19.5	7	15	8	1.62	2.36	2.6
16	Y_{122}	51.9	20.1	5	15	8	1.13	1.66	2.6

Experimental studies on reduction of cerium from slag were carried out on the high-temperature unit, made on the basis of the Tamman resistance furnace. Slag from the 58%CaO–12%SiO_2–4%Ce_2O_3–15%Al_2O_3–8%MgO system was used as an experimental slag. Steel containing 0.06 % carbon, 0.25% silicon and 0.05% aluminum was used as metal. A 90 g sample of metal and 9 g slag (10 % of the metal weight) was loaded into a corundum crucible and installed in a furnace, heated and melted. The molten metal and slag were heated up to 1650°C and kept in the crucible for 10 min. Four parallel experiments were carried out with metal hot soak under the slag of the abovementioned composition in the crucible for 10 min.

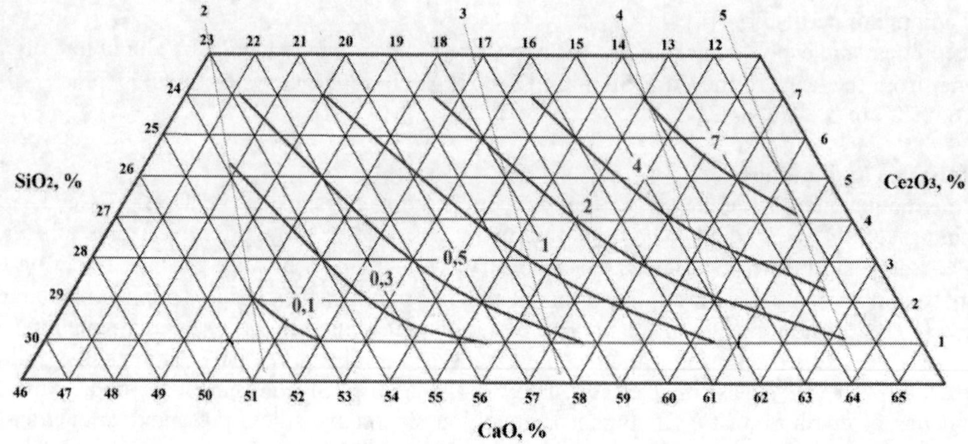

Figure 1. Diagram of the equilibrium content of cerium in metal hot-soaked under the slag of the $CaO–SiO_2–Ce_2O_3$ system containing 15% Al_2O_3 and 8% MgO at the temperature of 1550°C.

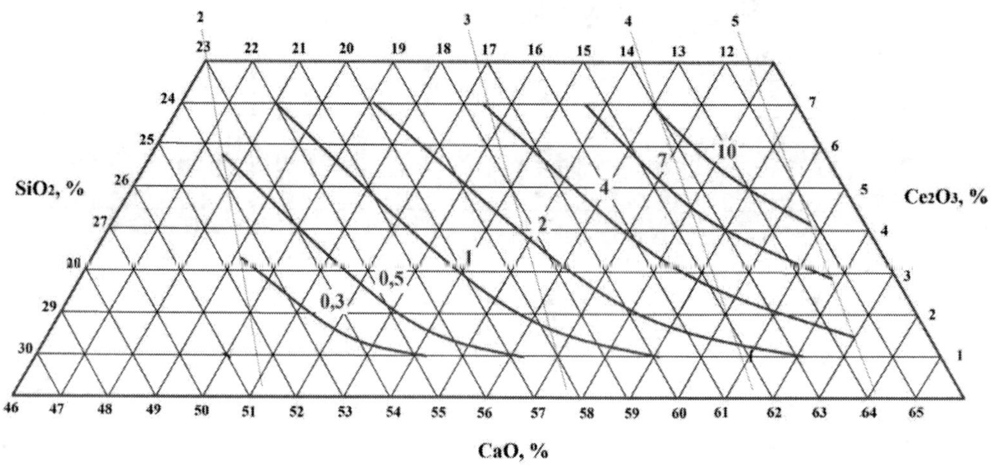

Figure 2. Diagram of the equilibrium content of cerium in metal hot-soaked under the slag of the $CaO–SiO_2–Ce_2O_3$ system containing 15% Al_2O_3 and 8% MgO at the temperature of 1650°C.

Results and discussion

Depending on temperature of metal, basicity of slag and the content of cerium oxide at the expense of the chemical reactions presented in Table 2, from 0.055 to 16 ppm cerium passes into metal. Thus hot soak of the metal containing 0.06 % C, 0.25 % Si and 0.055 % Al, under slag basicity of 2.0, containing 1.0 % cerium oxide, up to 0.055 ppm cerium passes into metal at temperature of 1550 °C. Increase of temperature of system to 1650°C is accompanied by insignificant increase in cerium concentration reaching no more than 0.085 ppm. At increase to 7.0 % of cerium oxide concentration in slag with basicity of 2.0 more essential increase of the content of cerium in metal is observed, reaching in a range of temperatures 1550–1650°C 0.4–0.6 ppm, accordingly.

The growth of basicity of slag favourably affects development of process of reduction of cerium. The increase in basicity of slag from 2 to 5 leads to the increase of the equilibrium content of cerium in the metal from 0.1 to 7 ppm at a temperature of 1550°C (Figure 1). Thus, with temperature increase to 1650°C the equilibrium content of cerium in metal increases and changes in a considered range of basicity of slag on the average from 0.3 up to 10 ppm (Figure 2).

Table 2. Gibbs energy change for chemical reactions of cerium reduction.

№	Chemical reactions	ΔG, kJ/mol	
		1550°C	1650°C
1	$2Ce_2O_3 + 6Al = CeAlO_3 + 2CeAl_2$	–173	–164
2	$CeAlO_3 + 2Al + 3CaO = Ca3Al2O6 + 2CeAl_2$	–83	–84
3	$CeAl_2 + 1.5SiO_2 = Al_2O_3 + Ce + 1.5Si$	–93	–91
4	$3CeO_2 + 4Al + 6CaO = 3Ce + 2Ca_3Al_2O_6$	–226	–233

According to the data given in the Table 2 aluminum dissolved in metal interacts with cerium oxides Ce_2O_3, CeO_2 and cerium aluminate $CeAl_2O_3$ by reactions 1, 2 and 4 with the formation of cerium aluminide $CeAl_2$, tri-calcium aluminate and metallic cerium. Cerium aluminide $CeAl_2$ interacting with silicon oxide SiO_2 by the reaction 3 provides the formation of aluminum oxide Al_2O_3 in the slag, cerium and silicon in the metal. However low concentrations of free cerium oxides Ce_2O_3, CeO_2, which do not exceed 0.009 and 0.005%, and free calcium oxide CaO, which do not exceed 2.5 %, in slags formed in the region of reduced to 2.0 basicity and containing no more than 1.0% of cerium oxide, do not provide proper development of the reactions given in Table 2 and, as a consequence, do not provide high concentrations of cerium in metal as well (Table 1). Observed insignificant increase in concentration of cerium in metal with increase in temperature of metal and concentration of free cerium oxides Ce_2O_3, CeO_2, reaching 0.05 and 0.007 %, apparently is caused by development of reactions 2 and 4. Increase in basicity of formed slags to 5.0 is accompanied by increase in concentration of free calcium oxide CaO to 24 % and, as consequence, increase in equilibrium concentration of cerium in metal (Table 1) at the expense of development of reactions 2 and 4.

Hot soak of metal containing 0.06% carbon, 0.25% silicon and 0.05% aluminum under slag containing 58% CaO, 12% SiO_2, 15% Al_2O_3, 8% MgO and 4% Ce_2O_3 is accompanied by reduction of cerium, the concentration of which after 10 minutes of hot soak approaches 16 ppm cerium. At the same time, experimental data practically confirm the results of thermodynamic modeling.

3. Conclusions

It is found that from 0.055 to 16 ppm cerium content passes in the steel, containing 0.06% carbon, 0.25 % silicon and 0.05% aluminum, depending on the metal temperature, basicity of slag and cerium oxide. Thermodynamic modeling together with the method of simplex planning grids allowed to obtain new data on the equilibrium content of cerium in a metal containing 0.06% C, 0.25% Si, 0.05 % Al, hot soaked under the slag of the $CaO–SiO_2–Ce_2O_3–15\%Al_2O_3–8\%MgO$ system in a wide range of chemical compositions at temperatures of 1550°C and 1650°C. The results of thermodynamic modeling have been experimentally confirmed.

Acknowledgments

The work was supported by RFBR grant № 19–08–00825.

References

[1] Ghali S N, El-Faramawy H S and Eissa M M 2012 Influence of Boron Additions on Mechanical Properties of Carbon Steel *Journal of Minerals and Materials Characterization and Engineering* **11** pp 995–9

[2] Murari F D, Vasconcelos da Costa e Silva A L and de Avillez R R 2015 Cold-rolled multiphase boron steels: microstructure and mechanical properties *Journal of Materials Research and Technology* **4 (2)** pp 191–6

[3] Łętkowska B, Dziurka R and Bała P 2015 The analysis of phase transformation of undercooled austenite and selected mechanical properties of low-alloy steel with boron addition *Archives*

of civil and mechanical engineering **15 pp** 308–16

[4] Lyakishev N P, Pliner Yu L and Lappo S I 1986 *Borsoderzhashchiestalii splavy [Boroncontaining steels and alloys]* (Moscow: Metallurgiia) [in Russian]

[5] Babenko A A, Smirnov L A, Upolovnikova A G and Sychev A V 2020 Theoretical Bases and Technology of Steel Exhaustive Metal Desulfurization and Direct Microalloying with Boron Beneath Basic Boron-Containing Slags *Metallurgist* **63** pp 1259–65

[6] Babenko A A, Zhuchkov V I, Smirnov L A, Upolovnikova A G, Selminskich N I and Sychev A V 2017 Formirovanie osnovnyh borsoderzhashchih shlakov – perspektivnoe napravlenie vnepechnoj desul'furacii i pryamogo mikrolegirovaniya nizkouglerodistoj stali borom [The formation of the basic boron-containing slags is the perspective direction of the ladle desulfurization and direct microalloying of the low carbon steel with the boron] *Ferrous metallurgy. Bulletin of Scientific, Technical and Economical Information* **9 (1413)** 50-4 [in Russian]

[7] Petryna D Yu, Kozak O L, Shulyar B R, Petryna Yu D and Hredil M I 2013 Influence of alloying by rare-earth metals on the mechanical properties of 17G1S pipe steel *Materials Science* **48 (5)** pp 575–81

[8] Makarchenko V D and Kindrachuk M V 2014 Vliyanie ceriya na mekhanicheskiei korrozionnye svojstva nizkolegirovannyh trubnyh stalej [The effect of cerium on the mechanical and corrosive properties of low-alloy pipe steels] *Kompressionnoe I energeticheskoe mashinostroenie* **3** 24-9 [in Russian]

[9] Torkamani H, Raygan Sh, Garcia-Mateo C, Rassizadehghani J, Palizdar Y and San-Martin D 2018 Evolution of pearlite microstructure in low-carbon cast microalloyed steel due to the addition of La and Ce *Metallurgical and materials transactions A* **49** pp 4495–508

[10] Smirnov L A, Rovnushkin V A, Oryshchenko A S, Kalinin G Yu and Milyuts V G 2016 Modification of Steel and Alloys with Rare-Earth Elements Part 1 *Metallurgist* **59** pp 1053–61

[11] Smirnov L A, Rovnushkin V A, Oryshchenko A S, Kalinin G Yu and Milyuts V G 2016 Modification of Steel and Alloys with Rare-Earth Elements. Part 2 *Metallurgist* **60** pp 38–46

[12] Xiaohong Y, Hu L, Guoguang C, Chengchuan W and Bin W 2011 Effect of refining slag containing Ce_2O_3 on steel cleanliness *Journal of rare earths* **29 (11)** pp 1079–83

[13] Long H 2011 Measurements of physical properties of Ce contained refining slags and their effects on steel cleanliness *Master thesis of University of Science and Technology Beijing* **32** p 169

[14] Guo M X and Suito H 1999 Effect of dissolved cerium on austenite grain growth in an Fe - 0.20 mass % C - 0.02 mas % P alloy *ISIJ International* **39 (11)** p 1169

[15] Wu C, Cheng G and Long H 2014 Effect of Ce_2O_3 and CaO/Al_2O_3 on the Phase, Melting Temperature and Viscosity of CaO-Al_2O_3-10 Mass % SiO_2 Based Slags *High Temp. Mater. Proc.* **33 (1)** pp 77–84

[16] Feifei H, Bo L, Da L, Ligang L, Ting D, Xuejun R, Qingxiang Y 2011 Effects of rare earth oxide on hardfacing metal microstructure of medium carbon steel and its refinement mechanism *Journal of rare earths* **29 (6)** pp 609–13

[17] Wang L J, Wang Q, Li J M and Chou K C 2016 Dissolution mechanism of Al_2O_3 in refining slags containing Ce_2O_3 *J. Min. Metall. Sect. B-Metall.* **52 (1)** pp 35–40

[18] Anacleto N M, Lee H-G, Hayes P C 1993 Sulphur Partition between CaO-SiO_2-Ce_2O_3 Slags and Carbon-saturated Iron *ISIJ International* **33 (5)** pp 549–55

[19] Kim V A, Nikolaj E I, Akberdin A A and Kulikov I S 1989 *Planirovanie eksperimenta pri issledovanii fiziko-himicheskih svojstv metallurgicheskih shlakov. Metodicheskoe posobie [The design of the experiment in the study of the physicochemical properties of metallurgical slag. Method book]* (Alma–Ata: Nauka) [in Russian]

[20] Kim V A, Akberdin A A, Kulikov I S et al. 1980 Ispol'zovanie metoda simpleksnyh reshetok dlya postroeniya diagram tipa sostav – vyazkost' [Using the simplex lattice method for

constructing composition-viscosity diagrams] *Izvestiya. Ferrous Metallurgy* **9** p 167 [in Russian]

[21] Babenko A A, ZHuchkov V I, Sychev A V and Upolovnikova A G 2016 Ispol'zovanie metoda simpleksnyh reshetok dlya postroeniya diagram sostav-vyazkost' shulakov sistemy $CaO-SiO_2-Al_2O_3-MgO-B_2O_3$ [Using of simplex lattices method for diagramming composition-viscosity of the slag system $CaO-SiO_2-Al_2O_3-MgO-B_2O_3$] *Butlerov Communications* **48 (11)** pp 40–4 [in Russian]

[22] Babenko A A, Zhuchkov V I, Leont'ev L I, Upolovnikova A G and Konysheva A A 2017 Ravnovesnoe raspredelenie bora mezhdu metallom sistemy Fe–C–Si–Al I borsoderzhashchim shlakom [Equilibrium distribution of boron between metal of Fe–C–Si–Al system and boron slag] *Izvestiya. Ferrous Metallurgy* **60(9)** pp 752-8 [in Russian]

IOP Conf. Series: Materials Science and Engineering **966** (2020) 012011 doi:10.1088/1757-899X/966/1/012011

Destruction electrospark deposits during Joule heating of base metal

G E Trekin[1, 2] and O I Shevchenko[1]

[1] Ural Federal University named after the First President of Russia B N Yeltsin, Nizhniy Tagil Technological Institute, 59, Krasnogvardeyskaya str., Nizhniy Tagil, 622000, Russia

E-mail: [2] trekin1963@yandex.ru

Abstract. Electrospark coating deposited on the surfaces of machine parts is a common surface hardening method. Hard materials are normally used for depositing to improve wear resistance or other properties. In this investigation electrospark deposition intended for use as micro-alloying of welds by deposition on to the surface of the base metal and in this case it remelts completely. Intense heating and destruction occurs before remelting, which changes the chemical composition of the melt. The aluminum and titanium coatings is shown to deteriorate at 500 to 650 °C generating a liquid phase due to eutectic available in both metals in this temperature range. Most of titanium and aluminum oxidized at that. It is found that a solid layer, appears after electrospark depositing using both coatings. For aluminum coating, after high-temperature heating, a thicker white layer appears with the same microhardness level as that of the base metal, and a thin layer of partial decarburizing. Fusion sectors are detected in the titanium coating that are harder than the base metal and have a dendritic structure, followed by a thin zone of complete decarburization and quite an extended zone of partial decarburization.

1. Introduction

Metal deposition on the surface during electric spark discharging (electrospark deposition – ESD) occurs as a result of the cathode sputtering effect. In a molt pool on the anode appears cavern due to displacement by the spark discharge pressure. This molten area also receives, most likely, liquid drop going off the cathode. When the frequency of spark discharges is high, the melting zones of each discharge overlap and quite a solid coating is generated with cracks, since the cooling rate is very high. A zone, similar to a welding heat affected zone (HAZ), develops in the volume under the deposited layer. Intensive evaporation of metal from the spark discharge zone results in splashing and loss of the electrode metal [1–3]. Hard wear-resistant alloys are the most used for electrospark deposition [4, 5]. Layers of other materials deposited are used as scale and corrosion-resistant etc. [6–8]. Deposited aluminum and titanium were used as a micro-alloying layer for subsequent remelting, since the layer is about tenths of a millimeter thick and has cracks. The aluminum and titanium microalloying effect is beneficent for welded metal properties [9–12], so in this paper we studied changes in the near-surface volume after electrospark deposition and high-temperature heating before welding melting.

2. Materials and study methods

The electrospark deposition were conducted in the air environment using an experimental setup, designed in the Department of Metallurgical technology of NTI (branch) of UrFU, in the parameters that make formation (Table 1).

Table 1. Electrospark deposition parameters.

Electrode metal	d_{el} (mm)	I_{av} (A)	U_{av} (V)	E_{1imp} (kJ)	v_{imp} (Hz)	ω_{el} (rpm)	V_{dep} (cm/s)
Al	40	18–22	75–80	4.85	300	50	6
Ti	20						

d_{el} – electrode diameter; I_{av} – average current; U_{av} – average voltage; E_{1imp} – pulse energy; v_{imp} – pulse rate; ω_{el} – electrode rotation rate; V_{dep} – electrode traveling speed relative to workpiece.

The base metal are plates of the steel 20 (Table 2) with the cross section 3×25 mm and 250 mm long. The ESD layer was put on all planes through the entire length of the specimen.

Table 2. The nominal chemical composition of Steel 20, mass % [13].

C	Mn	Si	P	S	Cr	Ni	Cu	As
			not above					
0.17–0.24	0.3–0.65	0.17–0.37	0.25	0.04	0.035	0.25	0.25	0.08

The metallographic study was performed with a Zeiss Observer D1m microscope running under Thixomet software. Etching was carried out with Nital (6 %) solution for 20 s at the room temperature.

Microhardness indentation were made using a Future Tech FM 300 testing machine by the Vickers pyramid for a load 50 g over 50 s, at right angle to the specimen thickness. The results were processed by means of Excel and MathCAD software.

The distribution of chemical elements through the modified layer depth was investigated with a Phenom PRO scanning electron microscope.

The temperature was registered with K–type thermocouple. The thermal heating cycle was registered by connecting the thermocouple to a Zet 410 amplifier and a Zet 210 analog-to-digital converter (ADC). The current and voltage were recorded using the same setup equipped with current transformers and signal rectification and smoothing circuits.

The specimen was heated under fluxes AN 348 A and FSA ChT A 650 20/80 (table 3, 4) using a transformer TSD-1000-3 in parameters that make possible melting of the flux and heating of the workpiece above the flux melting temperature. and then it was cooled in water (Figure 1).

3. Results and discussion

The microstructure of the surface after electrospark deposition is a white high hardness layer with a lot of straight cracks. The aluminum layer surface is uneven, the titanium layer is more even (Figure 2 a, 3 a, 4 a, b). The thickness of the layer applied is 40–80 μm for an aluminum electrode, and about 30 μm for a titanium electrode. Behind the layer applied there is a HAZ with a ferritic structure having the thickness equal for both electrodes of about 100 μm (Figure 2 a, 3 a).

It is found, that the deposited layer completely disappears after Joule heating and cooling, and is replaced with a white low hardness layer. This is a dense layer with relatively smooth borders and a few pores and inclusions, for aluminum coating under both fluxes. A thin partial decarburizing layer is observed at some distance from the white layer, looking like a thicker ferrite border around the quenched structure areas (Figure 2 b, c and 4 c). This is an uneven thickness layer with melting areas of dendrite structure, for titanium coating. Complete and partial decarburizing zones occur behind the melting sectors, of which on zone is narrow, about 200 μm, and the other is of about 1 mm deep (Figure 3 b,c, 4 d).

The aluminum distribution through the thickness of the modified layer shows that aluminum diffuses approximately 68 μm deep after short electrothermal exposure and its maximum concentration in the near-surface layer is 6–12 at. % (Figure 5 c). Titanium is distributed unevenly, in the melting area from 2 to 5 at. % with peaks up to 20 at. % (Figure 5 a, b), only peaks are also in other sections up to 20 at . %.

Table 3. The nominal chemical composition of AN-348-A flux, mass %.

Al_2O_3	MnO	CaO	MgO	CaF_2	Fe_2O_3	S	P
40–44	≤ 6	31–38	≤ 12	≤ 7	3–6	0.5–2.0	≤ 0.12

Table 4. The nominal chemical composition of FSA ChT A 650-20/80 flux, mass %.

The sum of oxides	Al_2O_3+CaO+ MgO	Al_2O_3 min	CaF_2
%	40	20	22

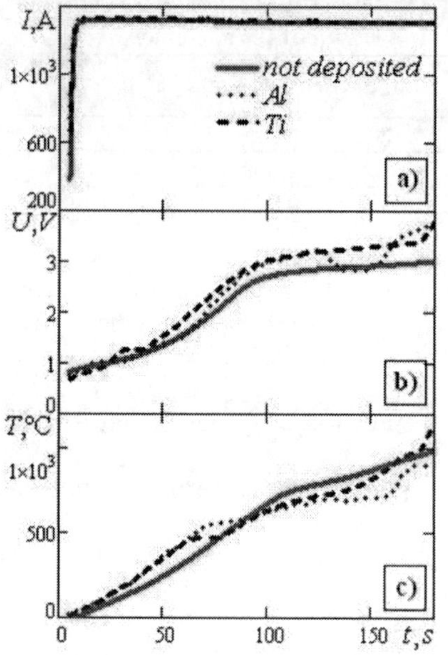

Figure 1. Changes in current (a), voltage (b) and temperature (c) during Joule heating under flux AN 348-A. Chemical element characters show the material of the electrode at ESD.

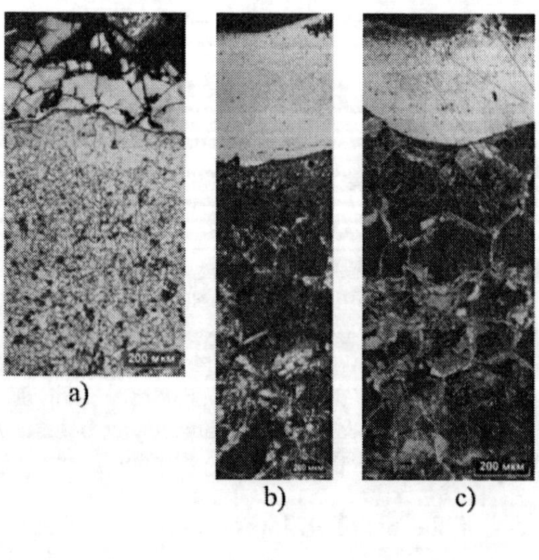

Figure 2. Steel 20 surface layer microstructure aluminum deposit, after: a) – ESD; b) – ESD and Joule heating under FSA ChT A 650 20/80 flux; c) – ESD and Joule heating under AN 348-A flux, µm.

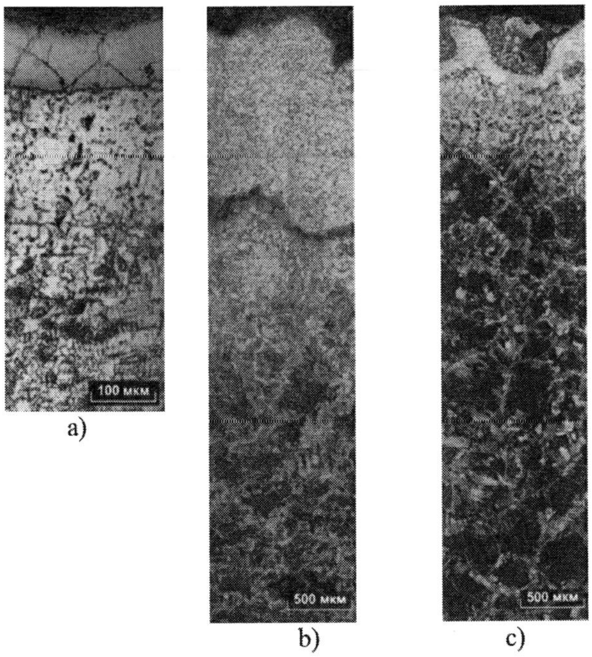

Figure 3. Steel 20 surface layer microstructure titanium deposit after: a) – ESD; b) – ESD and Joule heating under FSA ChT A 650 20/80 flux; c) – ESD and Joule heating under AN 348-A flux, мкм = μm.

Iron to aluminum and iron to titanium diagram review [14–16], shows that intermetallides are likely to occur in the electrospark deposited layer, as they are detected during deposition of these materials [17], which is also confirmed high coating hardness and cracks. Generation nitrides and oxides should also not be ruled out, since the coating was deposited in an air environment. Small volumes of relatively pure electrode material are very likely. For example, the titanium content in its peak value reaches 72 at. % (Figure 5 a). When heated, the base metal interacts in the manner of diffusion penetration: on the base metal part, for both electrodes applied, it is α a solid solution with γ a region above 900 °C, on the aluminum part, eutectic is at 652 °C, on the titanium part, eutectic also is at 590 °C. Therefore, heating of both electrospark layers has to generate a liquid phase, which is actually observed, and the aluminum coating becomes smooth, while the titanium generates local melting sectors with a dendritic structure. This is confirmed by Figure 2 c, where the heating curves change in the range 500–650 °C for samples with an electrospark layer.

With increasing temperature afterwards, for the aluminum layer, the interaction develops in the form of diffusion path, oxidation on the surface side, local melting and subsequent dissolution on the fusion line side, so an aluminum oxide film and a thin film of solid solution with about 5 at. % get in the weld. Temperature increased in the titanium coating most likely results in titanium binding of carbon, which is evidenced by the decarburized zones that occur after electro-thermal exposure. The average content of titanium is about 7 at. % in the thin layer. Assuming, that the thickness of the titanium containing layer is approximately the same before and after high-temperature heating, then about 10 % of titanium is transferred to the base metal i.e. most of the titanium is transferred into the melt as an oxide.

Since the main structural changes occur before the melting point of the fluxes, their influence is basically weak, so the results differ insufficiently.

4. Summary

Thus, when the electrospark deposited layers by aluminum and titanium electrodes are heated to high temperature, the surface layer melts followed by diffusion redistribution of elements. A white deposit with a thin partial decarburized layer is generated in place of the aluminum electrospark coating.

IOP Conf. Series: Materials Science and Engineering **966** (2020) 012011 doi:10.1088/1757-899X/966/1/012011

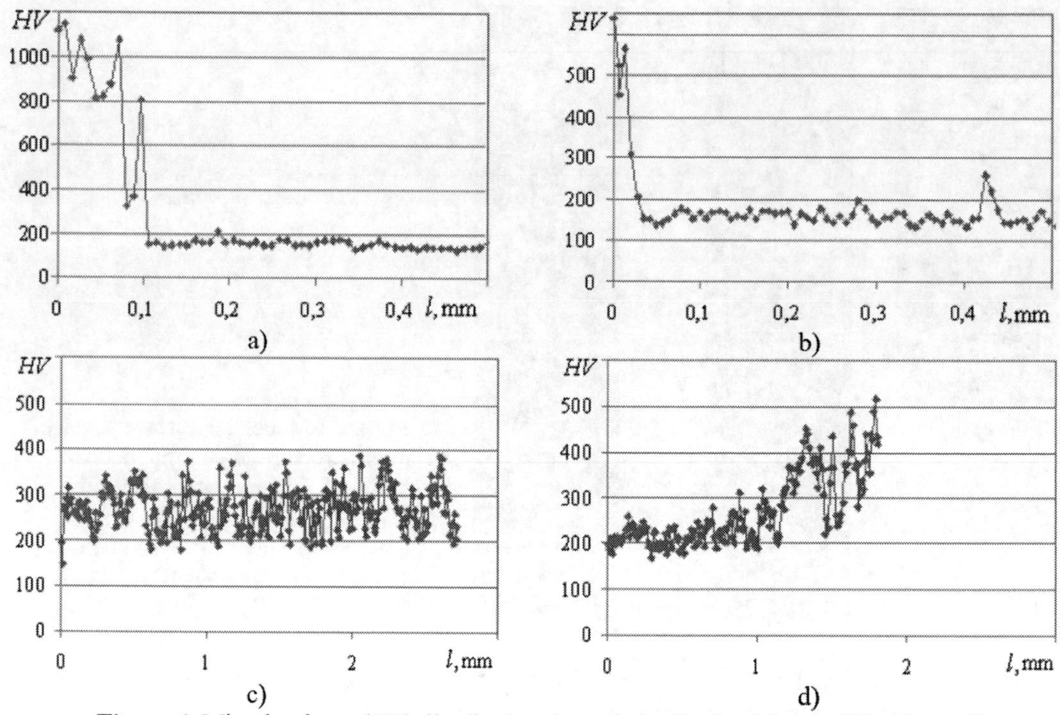

Figure 4. Microhardness (*HV*) distribution through the depth of the modified layer (*l*)
a, c – aluminum electrode; b, d – titanium electrode; a, b – after ESD;
c, d – after ESD and Joule heating under FSA ChT A 650 20/80 flux.

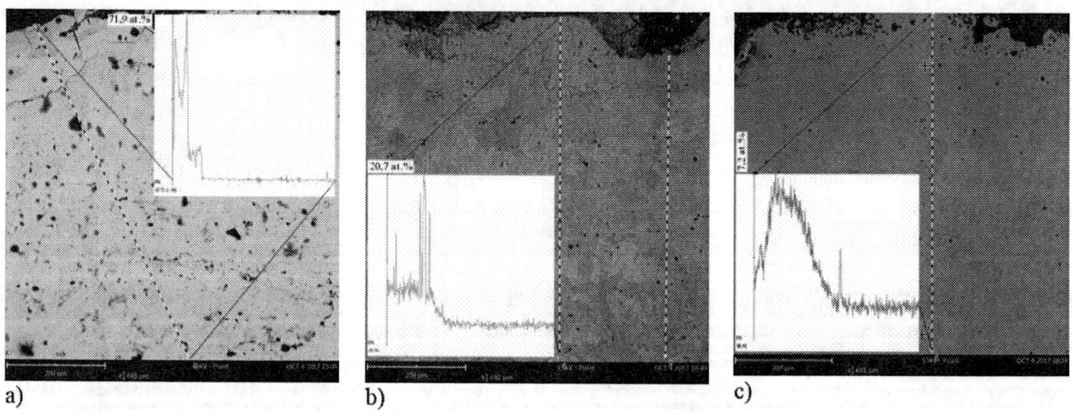

Figure 5. Titanium and aluminum distribution through the depth of the modified layer.
a, b – titanium electrode; c – aluminum electrode; a – after ESD; b, c – after ESD and Joule heating.

Local melting areas are generated in place of the titanium electrospark coating, followed by a thin complete decarburizing zone and about 1 mm of partial decarburizing. Most of the titanium and aluminum brought to the surface is oxidized and only max 10% remain in the electrospark modified layer after high-temperature heating.

References

[1] Ivanov V I, Burumkulov F Kh, Verkhoturov A D, Gordiyenko P S, Panin Ye S and Konevsov L A 2013 Formation of the surface layer on a low-carbon steel in electrospark treatment *Welding International* vol **27** issue **11** pp 903–6

[2] Sahin O and Ribalko A V 2011 Electrospark Deposition: Mass Transfer, Mass Transfer - Advanced Aspects Ed. Hironori Nakajima *InTech* Available online: http://www.intechopen.com/books/mass-transfer-advanced-aspects/electrospark-deposition-mass-transfer (accessed on 01.06.2020)

[3] Thamer A D, Hafiz M H and Mahdi B S 2012 Mechanism of Building-Up Deposited Layer during Electro-Spark Deposition. *Journal of Surface Engineered Materials and Advanced Technology* **2** pp 258–63

[4] Çakir A, Yilmaz M S, Ribalko A and Korkmaz K 2015 A Study on Modifcation of Micro-Alloy Steel Surfaces with Different Hard Materials Via Electro-Spark Deposition Method. *Proceedings of the 4th International Congress APMAS 2014, April 24-27, 2014, Fethiye, (Turkey) ACTA PHYSICA POLONICA A* vol **127** no **4** pp 1410–13

[5] Nikolenko S V, Verkhoturov A D, Syui N A and Kuz'michev E N 2016 Influence of electrospark discharge parameters on roughness and microabrasive wear of steel 45 surface after ESA by TiC-based electrodes. *Surface Engineering and Applied Electrochemistry* vol **52** issue **4** pp 342–9

[6] Jamnapara M I, Frangini St, Avtani D U and Nayak V S 2012 Microstructural studies of electrospark deposited aluminide coatings on 9Cr steels *Surface Engineering* vol **28** pp 700–4

[7] Jamnapara M I, Frangini St, Alphosa J, Chauhan N I and Mukherjee S 2015 Comparative analysis of insulating properties of plasma and thermally grown alumina films on electrospark aluminide coated 9Cr steels *Surface & Coatings Technology* **266** pp 146–50

[8] Kozyr A V, Konevtsov L A and Konovalov S V 2018 Research of heat resistance of the multilayer coating after electro-spark alloying of C45 steel Cr-Ni alloys *Letters on Materials* vol. **8** Issue **2** p 140-145.

[9] Bhadeshia H K and Svenson L E 1993 Modelling the evolution of microstructure in Steel weld metal *Mathematical Modelling of Weld Phenomena* eds H Cerjak, K E Easterling, Institute of materials (London) pp 109–82

[10] Trekin G E and Shevchenko O I 2015 Effect of aluminum electrospark sublayer and tempering temperature on the structure and qualities of low-carbon low-alloy welds *Bulletin of PNRPU Mechanical engineering, material science* vol **17** no **1** pp 112–8

[11] Wang J, Li G and Xiao A 2011 A Bainite-Ferrite Multi-Phase Steel Strengthened by Ti-Microalloying *Materials Transactions* vol **52** no **11** pp 2027–31

[12] Jorge J C F, Bott I S, Sauza L F G, Mendes M C, Araujo L S and Evans J M 2019 Mechanical and microstructural behavior of C-Mn steel weld deposits with varying titanium contents *J. Mater. Res. Techn.* **8 (5)** 4659–71

[13] Sorokin V G et al 2001 *Steels and alloys. Grade list: Ref.* ed. V G Sorokin and M A Gervasiev (Moscow: Intermet Engineering) p 608

[14] *Binary Alloy Phase Diagrams* 1986 ed by T B Massalski, J L Murray, L H Bennett and H Baker (ASM, Metals Park)

[15] Li X, Scherf A, Heilmaier M and Stein F 2016 The Al-Rich Part of the Fe-Al Phase Diagram *Journal of Phase Equilibria and Diffusion* vol **37** no **2** pp 162–73

[16] Okamoto H *Desk Handbook: Phase Diagrams for Binary Alloys* 2nd ed. ASM International® Materials Park, Ohio 44073-0002.

[17] Pyachin S A and Burkov A A 2015 Formation of intermetallic coatings by electrospark deposition of titanium and aluminum on a steel substrate *Surface Engineering and Applied Electrochemistry* **51:2** pp 118–24

IOP Conf. Series: Materials Science and Engineering **966** (2020) 012012 doi:10.1088/1757-899X/966/1/012012

Effect of basicity and chromium oxide on the viscosity of boron-containing slags

A A Babenko[1], R R Shartdinov[1, 2], A G Upolovnikova[1], A N Smetannikov[1] and L Yu Mikhailova[1]

[1] Institute of Metallurgy of the Ural Branch of the Russian Academy of Sciences, 101, Amundsen street, Yekaterinburg, 620016, Russia

E-mail: [2] rr.shartdinov@gmail.com

Abstract. The main diluent for chromium-containing slags is CaF_2. However, fluorspar disadvantages make it necessary to search for a replacement. As a substitute, boron oxide allows to improve slag physical properties and environmental situation. The paper presents the experimental study results of the chemical composition and temperature influence on viscosity of slags of $CaO–SiO_2–Cr_2O_3$ system containing 8% MgO, 3% Al_2O_3 and 6% B_2O_3. It was found that a slag free of chromium oxide with basicity of 1.0 has a sufficiently high fluidity of 0.2–0.6 Pa·s in a wide temperature range of 1200–1350°C, due to high concentration of low-melting phases, reaching 22%, and only 11 % high-melting. A slag with 18% Cr_2O_3 and the same basicity retains a low viscosity of 0.1–1.0 Pa·s but at higher temperature of 1450–1570 °C due to increase in high-melting compound content to 27%. The viscosity of a slag with basicity of 2.5 without chromium oxide is 0.07–1.0 Pa·s in a narrow temperature range of 1650–1700°C and 0.14–1.0 Pa·s for a 'shorter' slag with 18% Cr_2O_3 and same basicity in even narrower range of 1650–1670°C. These slags have the highest high-melting phase content of about 50%.

1. Introduction

Currently the main method for the production of low-carbon stainless steel in the world is argon oxygen decarburization (AOD), which consists of 2 periods: oxidation and reduction. The purpose of the reduction period of smelting at AOD units is both the reduction of chromium from slag to metal and desulfurization of steel, but due to the fact that during the oxidation period a significant part of chromium is oxidized and passes into slag in the form of Cr_2O_3, the slag viscosity increases significantly [1] and the refining ability decreases [2]. The authors [2–4] note that low viscosity slags are in a narrow range of compositions, therefore the possibility of reducing the viscosity of these slags with the addition of fluorspar was studied [5, 6]. However fluorspar has many disadvantages. Aggressive effect on the lining, inconstancy of influence on the properties of slag due to the formation of volatile fluorides, which are environmentally harmful [7–10]. Boron oxide may become an alternative. B_2O_3 has a low melting point and also forms low-melting eutectics with the main components of the slag, for example, $MgO·B_2O_3$ (melting point 988°C), $CaO·B_2O_3$ (melting point 1100 °C) and others. This oxide is actively used in blast furnace production, ladle metallurgy and in continuous casting of steel [11–16]. Nevertheless the number of studies studying the effect of B_2O_3 on the physicochemical properties of chromium-containing slag [17] is extremely limited.

Slags of this process are a complex multicomponent system, the physicochemical characteristics of which determine the course of smelting, separation of metal and slag, etc. [2,18], ultimately determining the quality of steel. Therefore, the aim of the work is to study the viscosity of slags of the CaO–SiO_2–Cr_2O_3–MgO–Al_2O_3–B_2O_3 system with a basicity of 1.0–2.5, containing 0–18% Cr_2O_3, 8% MgO, 3% Al_2O_3 and 6% B_2O_3 (in this expression and hereinafter indicated by mass %).

2. Materials and methods

Synthetic slags, the compositions of which are shown in Table 1, were smelted in a resistance furnace in molybdenum crucibles from reagent grade oxides, previously calcined for 2–3 hours at a temperature of 800°C (B_2O_3 at a temperature of 100°C). The viscosity of the slag test samples was measured on an electrovibration viscometer in molybdenum crucibles in a stream of purified argon. The temperature of the slag was measured using a tungsten-rhenium thermocouple.

Thermodynamic modeling of the phase composition of the prototypes was carried out using the HSC Chemistry 8.03 software package, which allows one to calculate the equilibrium compositions and quantities of the products formed according to the Gibbs energy minimization algorithm based on the concepts of metallurgical melts as ideal associated solutions [19].

The obtained temperature dependences of slag viscosity are presented graphically in Figure 1. Table 2 contains the results of the calculation of the phase composition.

Table 1. The chemical composition of the experimental slag.

No	Slag composition, %						
	CaO	SiO_2	Cr_2O_3	MgO	Al_2O_3	B_2O_3	$B = (CaO)/SiO_2$
1	41.50	41.50	0	8	3	6	1.0
2	59.29	23.71	0	8	3	6	2.5
3	46.43	18.57	18	8	3	6	2.5
4	32.50	32.50	18	8	3	6	1.0

3. Results and discussion

As it can be seen from Figure 1, most of the studied slags (No. 1, 2, 3) are the so-called 'short' slags with a pronounced inflection point. This is explained by the phase composition of the studied oxide system, namely, the content of a large number of such high-temperature compounds as $2CaO \cdot SiO_2$ and $CaO \cdot Cr_2O_3$ (Table 2).

The first slag (slag No. 1), not containing Cr_2O_3, having a basicity of 1.0 and containing 6.0 % B_2O_3, is characterized by a significant concentration of low-melting phases $CaO \cdot B_2O_3$, $2CaO \cdot B_2O_3$ and $CaO \cdot MgO \cdot 2SiO_2$ (up to 22% in total) with only 11% of high-melting phases $2CaO \cdot SiO_2$, CaO, MgO, and has a viscosity of 0.2–0.6 Pa·s in the range of 1200–1350°C. Moreover, the overwhelming number of phases $CaO \cdot SiO_2$, $CaO \cdot MgO \cdot SiO_2$, etc. accounts for the second interval (62%), which ensures a rapid speed of crystallization despite the low basicity.

An increase in the basicity of slag without Cr_2O_3 (slag No. 2), containing 6.0% boron oxide, to 2.5 is accompanied by a decrease to 5% in the fraction of low-melting phases ($2CaO \cdot B_2O_3$) and an increase in the fraction of medium- and high-melting compounds to 43 and 50 % respectively, which

Table 2. Slag phase composition.

Phases, %	$t_{melting}$, °C	Slags 1	2	3	4
Low-melting phases					
$CaO \cdot B_2O_3$	1130	4	0.1	0.4	4.5
$2CaO \cdot B_2O_3$	1280	9	5	9	8
$CaO \cdot MgO \cdot 2SiO_2$	1391	9	0.03	0.2	8
Medium-melting phases					
$2CaO \cdot MgO \cdot 2SiO_2$	1454	4	0.4	1	3
$3CaO \cdot B_2O_3$	1460	1	14	9	1
$3CaO \cdot 2SiO_2$	1460	13	8	7	6
$CaO \cdot MgO \cdot SiO_2$	1503	8	5	8	7
$CaO \cdot SiO_2$	1540	23	3	5	22
$MgO \cdot SiO_2$	1557	5	0.1	0.5	5
$3CaO \cdot MgO \cdot 2SiO_2$	1575	2	6	5	1
SiO_2	1710	6	0.03	0.1	5
High-melting phases					
$2CaO \cdot SiO_2$	2130	10	40	21	6
$CaO \cdot Cr_2O_3$	2170	0	0	20.5	7
$MgO \cdot Cr_2O_3$	2350	0	0	0.004	0.005
Cr_2O_3	2435	0	0	3	13
CaO	2570	0.2	6	2	0.2
MgO	2852	1	4	4	1

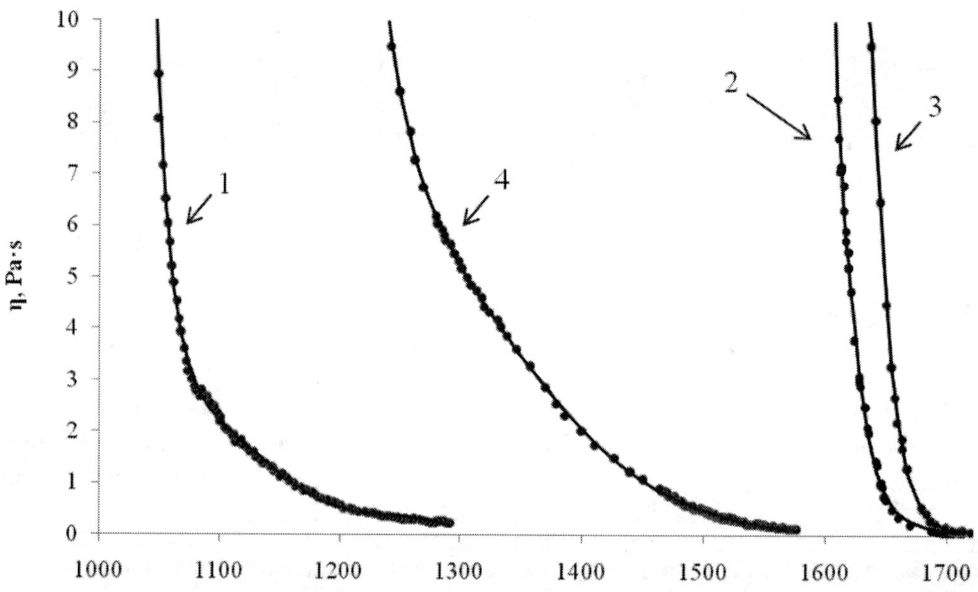

Figure 1. Temperature dependence of slag viscosity. 1–4 – slag numbers.

Causes a shorter crystallization at higher temperature (slag No. 2). Despite the increase in the fraction of the low-melting phase $2CaO \cdot B_2O_3$ to 9% in slag No. 3, which has the same basicity but contains 18% Cr_2O_3, an increase in the fraction of high-melting phases $2CaO \cdot SiO_2$, CaO, $CaO \cdot Cr_2O_3$, Cr_2O_3 and MgO to 53% provides it similar nature of the dependence of viscosity curve on temperature. In this case slag No. 2 and No. 3 have a viscosity of 0.07–1.0 Pa·s in the range of 1650–1700°C and up to 1.0 Pa·s at a temperature of 1670°C, respectively.

An increase in the basicity of slag without Cr_2O_3 (slag No. 2), containing 6.0% boron oxide, to 2.5 is accompanied by a decrease to 5% in the fraction of low-melting phases ($2CaO \cdot B_2O_3$) and an increase in the fraction of medium- and high-melting compounds to 43 and 50 % respectively, which causes a shorter crystallization at higher temperature (slag No. 2). Despite the increase in the fraction of the low-melting phase $2CaO \cdot B_2O_3$ to 9% in slag No. 3, which has the same basicity but contains 18 % Cr_2O_3, an increase in the fraction of high-melting phases $2CaO \cdot SiO_2$, CaO, $CaO \cdot Cr_2O_3$, Cr_2O_3 and MgO to 53% provides it similar nature of the dependence of viscosity curve on temperature. In this case slag No. 2 and No. 3 have a viscosity of 0.07–1.0 Pa·s in the range of 1650–1700°C and up to 1.0 Pa·s at a temperature of 1670°C, respectively.

Slag No. 4 is 'long' and doesn't have a pronounced inflection point due to low basicity and a large number of glass-forming oxides SiO_2, B_2O_3. The presence in the slag of the studied oxide system of 18 % chromium oxide, acting at a basicity of 1.0 or more as a network-former, increases the degree of slag polymerization [17]. With a slight decrease relative to slag No. 1, of the content of low-melting phases $CaO \cdot B_2O_3$, $2CaO \cdot B_2O_3$, $CaO \cdot MgO \cdot 2SiO_2$ up to 21%, the fraction of high-melting $2CaO \cdot SiO_2$, CaO, $CaO \cdot Cr_2O_3$, Cr_2O_3 and MgO increases up to 27% at the expense of compounds with an medium melting point. That leads to heterogenization of the oxide system and viscosity up to 0.1–1.0 Pa·s at increased temperature of 1450–1570°C.

4. Conclusions

As a result of a study of the physical properties of the slags of the studied oxide system it was found that boron oxide provides the formation of low-melting phases and a sufficiently high fluidity of the slags of the $CaO - SiO_2 - Al_2O_3 - MgO - B_2O_3$ system even at a high content of chromium oxide. The addition of 18% chromium oxide, acting at a basicity of 1.0 or more as a complexing agent, increases the degree of slag polymerization and the fraction of high-temperature phases in the slags of the studied oxide system, accompanied by their heterogenization and viscosity increase.

Acknowledgments

The work was carried out according to the state assignment for IMET UB RAS.

References

[1] *Atlas shlakov: sprav. izd. [Atlas of slag. Reference book]* 1985 (Moscow: Metallurgiia) [in Russian]

[2] Tokovoj O K 2015 *Argonokislorodnoe rafinirovanie nerzhaveyushchej stali: monografiya [Argon oxygen decarburization: monography]* (Chelyabinsk: Izdatel'skij centr YuUrGU) [in Russian]

[3] Wu T, Yuan F and Zhang Y 2018 Viscosity measurements of CaO-SiO_2-CrO Slag *ISIJ International* **58 (2)** 367-9

[4] Liu Zh, Dekkers R, Blanpain B and Guo M 2019 Experimental Study on the Viscosity of Stainless Steelmaking Slags *ISIJ International* **59 (2)** 404-11

[5] Magidson I A, Morozov A S, Sidorenko M F et al 1973 Vyazkost' hromistyh shlakov [The Viscosity of chromium-containing slags] *Izvestiya. Ferrous Metallurgy* **11** 61-4 [In Russian]

[6] Kalicka Z, Kawecka-Cebula E and Pytel K 2009 Application of the Iida Model for Estimation of Slag Viscosity for Al_2O_3-Cr_2O_3-CaO-CaF_2 Systems *Arch. Metall. Mater.* **54 (1)** 179-87

[7] Dyudkin D A and Kisilenko V V 2010 *Vnepechnaya metallurgiya stali [Ladle metallurgy of steel]* (Moscow: Teplotekhnik) [In Russian]

[8] Akberdin A A, Kulikov I S, Kim V A et al 1987 *Fizicheskie svojstva rasplavov sistemy CaO-Al_2O_3-SiO_2-MgO-CaF_2 [Physical properties of CaO-Al_2O_3-SiO_2-MgO-CaF_2 melt system]* (Moscow: Metallurgiia) [in Russian]

[9] Povolockij D Ya, Roshchin V E, Gribanov V P et al 1982 Vliyanie SiO_2 na letuchest' shlakov

sistemy Al_2O_3-Al_2O_3-CaF_2 [Effect of SiO_2 on the volatility of slags of the Al_2O_3-Al_2O_3-CaF_2 system] *Izvestiya. Ferrous Metallurgy* **8** 39–42 [In Russian]

[10] Zajcev A I, Mogutnov B M and Shahpazov E H 2008 *Fizicheskaya himiya metallurgicheskih shlakov [Physical chemistry of metallurgical slags]* (Moscow: Interkontakt Nauka) [In Russian]

[11] Wang H, Zhang T, Zhu H, Li G, Yan Y and Wang J 2011 Effect of B_2O_3 on Melting Temperature, Viscosity and Desulfurization Capacity of CaO-based Refining Flux. *ISIJ International* **51 (5)** 702–6

[12] Akberdin A A, Kireeva G M and Medvekovskaya I A 1986 Vliyanie B_2O_3 na vyazkost' shlakov sistemy CaO-Al_2O_3-SiO_2 [The effect of B_2O_3 on the slag viscosity of the CaO-Al_2O_3-SiO_2 system] *Izvestiya AN SSSR. Metally [Proceedings of the USSR Academy of Sciences. Metals]* **3** 55-6 [In Russian]

[13] Babenko A A, Istomin S A, Protopopov E V, Sychev A V and Ryabov V V 2014 Viscosity of CaO-SiO_2-Al_2O_3-MgO-B_2O_3 slags *Steel in Transl.* **44 (2)** 110–2

[14] Wang H, Li G, Li B, Zhang X and Yan Y 2010 Effect of B_2O_3 on Melting temperature of CaO-Based Ladle Refining Slag *ISIJ International* **17 (10)** 18-22

[15] Akberdin A A, Kireeva G M, Medvedovskaya et al 1986 *Atlas diagramm vyazkosti borsoderzhashchih shlakov [Atlas of viscosity diagrams of boron-containing slag]* (Dep. in VINITI) [In Russian]

[16] Wang W, Cai D and Zhang L 2018 A Review of Fluorine-free Mold Flux Development *ISIJ International* **58 (11)**1957–64

[17] Li Q, Yang Sh, Zhang Y, An Zh and Guo ZhCh 2017 Effects of MgO, Na_2O, and B_2O_3 on the viscosity and structure of Cr_2O_3-bearing CaO-SiO_2-Al_2O_3 slags *ISIJ International* **57 (4)** 689-96

[18] Povolockij D Ya and Gudim Yu A 1988 *Proizvodstvo nerzhaveyushchej stali [Stainless steel production]* (Chelyabinsk: Izd-voYuUrGU) [In Russian]

[19] Vatolin N A, Moiseev G K and Trusov B G 1994 *Termodinamicheskoe modelirovanie v vysokotemperaturnyh neorganicheskih sistemah [Thermodynamic modeling in high temperature inorganic systems]* (Moscow: Metallurgiia) [In Russian]

IOP Conf. Series: Materials Science and Engineering **966** (2020) 012013 doi:10.1088/1757-899X/966/1/012013

The use of additional filler wire to solve the tandem GMAW issues

I O Polevoy[1, 3] and A M Fiveyskii [1, 2]

[1] SHTORM Co. Ltd., Ekaterinburg, 620100, Russia

[2] Ural Federal University named after the First President of Russia B N Yeltsin, 19, Mira str., Yekaterinburg, 620000, Russia

E-mail: [3] ip@shtorm-its.ru

Abstract. A review of the studies showed that the existing technologies for tandem gas metal arc welding (GMAW) of butt welds have a number of problems that do not allow to increase productivity while maintaining low heat inputto base material and stability of the welding process. The possible solution of this problem by feeding of additional filler wire (AFW) in between the welding electrodes is researched in this study. Influence of addition of AFW into the arc and into the weld pool for single-arc GMAW on deposition rate and geometric properties of the welds are researched and the main problems of these methods are showed in the study. The effect of addition of AFW on the temperature of weld pool and heat input to base metal and solidification rate of the weld pool are showed in the study too. The main disadvantages of using pulsed welding modes for providing the stability of tandem welding process are described and the addition of AFW in between the welding electrodes for stabilization of trailing arc is justified. It is concluded that the further complex studies of effects of AFW on tandem GMAW are needed.

1. Introduction

Recently an increase in the productivity of welding processes only by increasing the specific heat input became non optimal, since this leads to deterioration in the weld metal formation and deterioration in the mechanical properties of the heat affected zone (HAZ). Moreover, the limitation of the specific heat input in favor of increase in the welding speed leads to reduction of the cross-sectional area of welds and, accordingly, to increase in the number of passes, lower productivity and an increased risk of defect formation [1].

To solve the problems encountered in various areas of welding production, multi-wire welding methods are widely used [2]. One of the most common methods of multi-wire welding is tandem gas metal arc welding (GMAW), which is actively used in the manufacture of ship panels, mining equipment and in bridge building and allows to achieve high process performance and reduce heat input to the base metal, on the one hand, due to the increased deposition rate, and on the other, due to an increase in welding speed, in some cases more than doubled.

A distinctive feature of tandem welding is the use of welding wires isolated from each other, which are fed through separate feeding mechanisms. The wires should be powered from separate welding sources [3], and during the welding process, control of the arcs can be carried out independently [3] or according to a certain control scheme. Due to this, it is possible to set the mode parameters for each wire separately, which allows to configure the welding arcs for various tasks. The most common

IOP Conf. Series: Materials Science and Engineering **966** (2020) 012013 doi:10.1088/1757-899X/966/1/012013

option is to use a main arc to ensure the required penetration depth and to set the geometric characteristics of the weld pool, and the trailing arc in order to ensure the maximum melting coefficient, while maintaining high quality of the weld surface, avoiding increased spatter [4], and also create the optimum temperature field in the weld pool, for improving the geometric characteristics and structure of the weld metal [5].

The main problems of this welding method are the insufficient productivity of the welding process compared to other multi-wire methods, the increased heat input to the base metal compared to single-arc welding, and the insufficient stability of the welding process. One of the solutions to these problems may be the use of an additional filler wire (AFW).

2. The use of AFW to increase productivity

In various studies, the use of AFW is often considered as a means of increasing the deposition rate, as well as influencing the geometric properties of welds. Typically, single-arc GMAW methods with AFW feed either into the arc [6] or into the weld pool are investigated.

2.1. The addition of AFW into the arc

The schematic of the method of welding with the addition of AFW into the arc is presented in Figure 1. According to the researches [7, 8], when feeding AFW into the arc, it becomes possible to increase the deposition rate by 100–140% compared to single-arc welding.

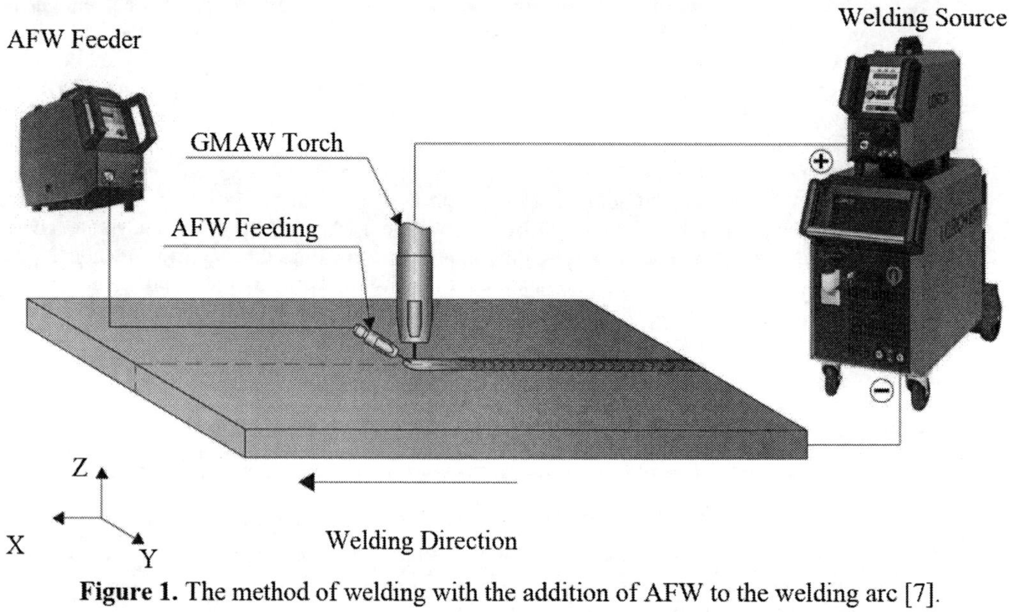

Figure 1. The method of welding with the addition of AFW to the welding arc [7].

The main problem of this welding method is low stability of the welding arc and increased spatter [7] with volumes of feed of the AFW from 60 to 80% of the supply volume of the welding wire. The solution to this problem is to further increase the relative feeding volume of the AFW. In this case, the arc completely passes to the AFW instead of the base metal due to the fact that the arc tends to flow along the path of the lowest resistance [7, 9]. A schematic of the arc transition to the surface of the AFW is presented in Figure 2.

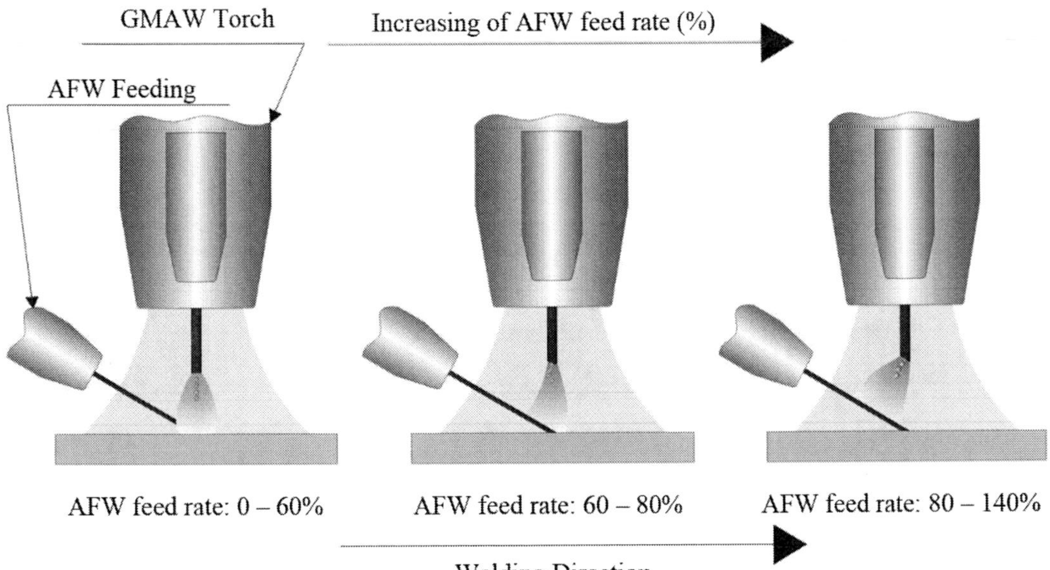

Figure 2. A schematic of the arc transition to the surface of the AFW with an increase in the volume of its supply.

However, with an increase in the AFW feed speed, the problem of reducing the penetration by half arises: the penetration values reduce from 2 mm to 1 mm at a current of 250 A, and from 3 mm to 1.5 mm at a current of 350 A, due to the deviation of the welding arc, and also the displacement of the cathode spot to the surface above the surface of the base metal, which reduces the pressure of the arc on the base metal. The percentage of inclusion of the base metal also decreases from 45% to 10% when welding at a current of 250 A and from 43% to 15% when welding at a current of 350 A [8]. Based on the obtained research results, we can conclude that this welding method is suitable for surfacing of large volumes of metal with small penetration.

2.2. The addition of AFW into the weld pool

Since the ability of the weld pool to melt the AFW is lower than that of the welding arc, the method in which the AFW is supplied into the weld pool is not characterized by high deposition rate and, accordingly, by different geometric characteristics of the weld pool [1]. So, the maximum achievable deposition rate is 50%. With a further increase in the volume of addition of AFW to the weld pool, periodic non-melting and jamming of the AFW in the crystallized weld pool can be observed [8].

The main advantage of this method in comparison with the method of feeding AFW to the welding arc is the absence of a significant reduction in the penetration depth [10].

The schematic of this welding procedure is presented in Figure 3.

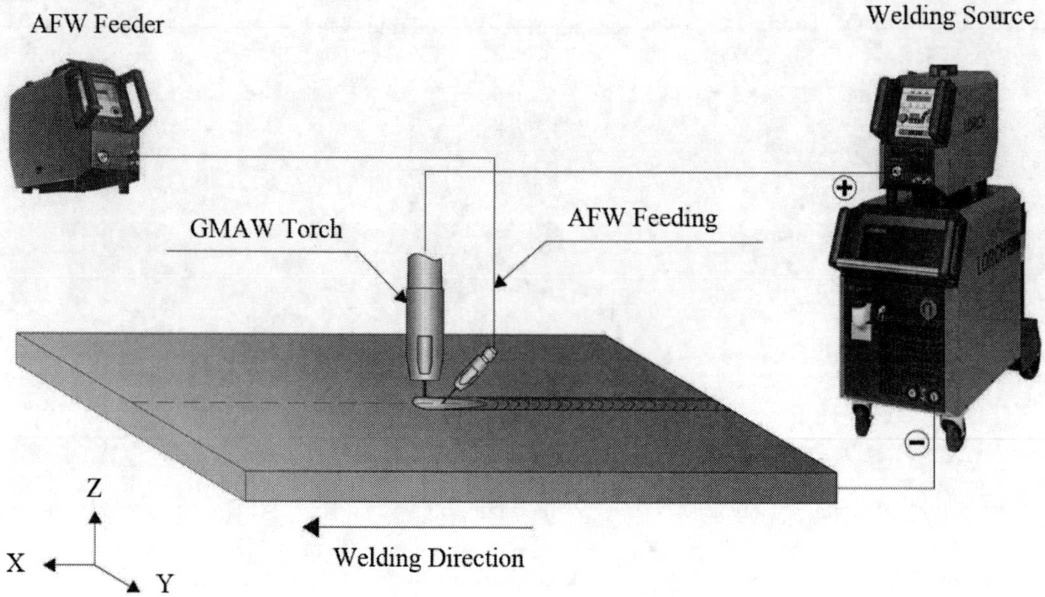

Figure 3. The schematic of the welding method with the addition of AFW into the weld pool.

3. The effect of AFW on heat input

The main reason for the decrease of welds quality during welding without AFW, is most commonly considered to be overheating of the metal of the weld pool and HAZ due to high heat input [1, 10–12].

According to the study [11], the addition of AFW into the arc leads to decrease of the weld pool temperature. The addition of AFW allowed reducing of the maximum temperature recorded by thermocouples from about 1550 °C to about 1400 °C when feeding the AFW to the welding wire in a volume ratio of 1 to 1.

A decrease in the temperature of the weld pool is also indicated by a decrease in its thermal influence on the base metal. As a result of the research [12], a decrease in the width of the coarsened grain zone in the HAZ to 50% was observed when butt welds were welded by GMAW with AFW as compared to the values obtained by GMAW without AFW [12]. The results of study [10] showed that addition of AFW with a volume of 50% of the volume of the welding wire into the weld pool decreases the angular distortion of the workpiece from 3.28° to 2.82°, which also indicates a decrease in heat input from the weld pool to the base metal.

Also, it is possible to influence the cooling rate of the weld pool depending on the place of the addition of AFW without changing the volume of added AFW. So, in the study [13], it is argued that the closer the AFW is added to the solidification zone, the smaller volume is required to increase the solidification rate. As the AFW approaches the welding arc, the melting rate of the AFW increases and the solidification rate decreases. As a result, due to a change of the AFW feed location, it is possible to control the cooling rate of the weld pool and thus influence the mechanical properties of the welding joint.

4. The main disadvantages of using pulsed welding modes

The stability of the tandem welding process should be determined by the ability of the welding arc not to be interrupted due to the magnetic influence of the arc nearby [14], not to violate gas shield quality due to periodic arc deviations [15], not to be oscillated between the filler and base metal. Low stability of the process leads to increased spatter, as well as instable weld pool formation [14, 16].

At present, the main research devoted to the leveling of the magnetic interaction of arcs are considered using pulsed welding modes as a solution to the problem. The most stable is the process in which pulses are provided synchronously and with a minimum phase difference. With an increase in the phase difference of the pulses, the number of interruptions of one of the arcs increases, which leads to an unstable seam width, increased spatter, and violation of gas shield quality [15, 17]. Studies in which more complex configurations of pulses are researched [18, 19] are contradictory and cannot be used unquestionably.

Pulse welding with synchronous pulses and with a small phase difference has a number of significant disadvantages and limitations. Thus, a stable welding process can be achieved only if the welding currents of both arcs are approximately equal, and if an increase in heat input into the base metal is allowed with a decrease in the penetration depth and an increase in the weld width [15].

5. The main conclusions and lines for further research

If it is necessary to ensure a large penetration depth with a high deposition rate and low heat input into the base metal, as well as to stabilize the tandem GMAW process, AFW feeding in between the welding wires can be used. With this welding method, a significant decrease in the penetration depth under the first arc should not be observed, as well as a significant increase in the deposition coefficient due to the influence of the heat of the second arc on the AFW. The arrangement of the wires and tandem GMAW process with AFW is shown in Figure 4.

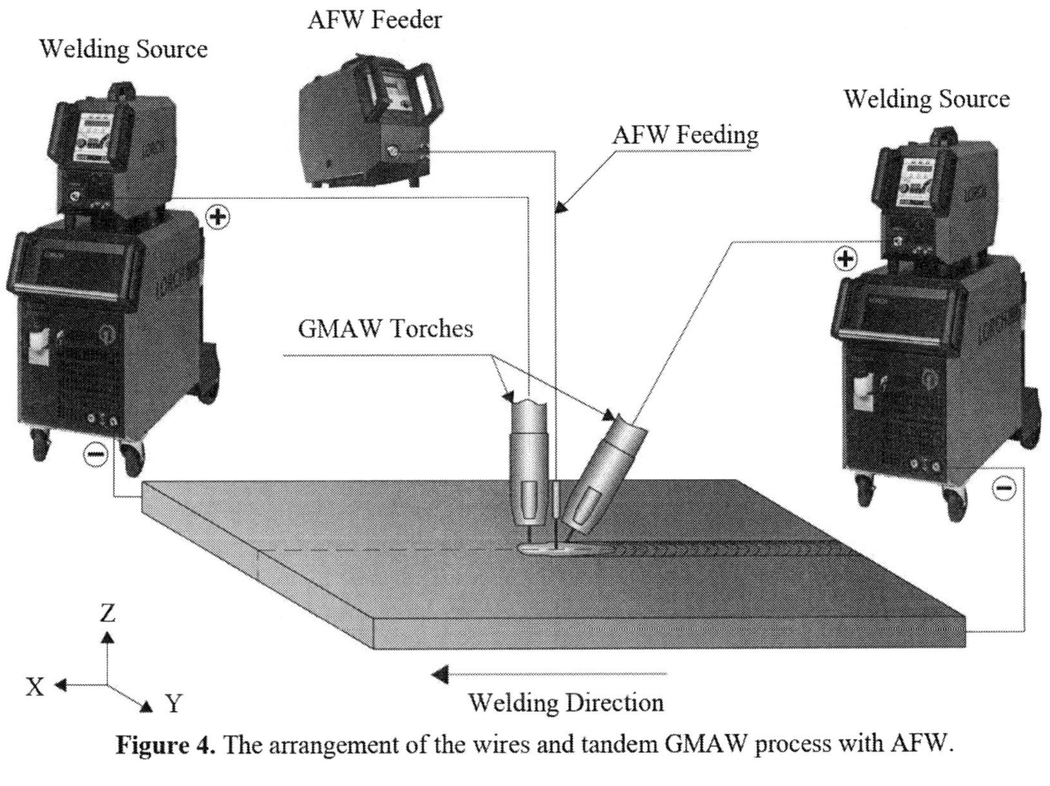

Figure 4. The arrangement of the wires and tandem GMAW process with AFW.

According to a few studies [20], an increase in the feed rate of the AFW to the weld pool, leads to stabilization of the cathode spot on the surface of the weld pool due to its constant cooling and liquid metal flow interruption at the location of the AFW feed. Thus, it becomes possible to stabilize the trailing arc and prevent its oscillations and interruption due to magnetic interactions, as well as to increase the deposition rate due to the heating of the AFW by the trailing arc. Existing studies of this phenomenon cannot provide a fuller picture, since they were carried out under conditions in which the AFW was added to the weld pool tail, at a great distance from both arcs, and studies of tandem welding with the addition of AFW in between welding wires in order to stabilize the welding process has not yet been carried out.

Thus, further studies should be aimed at a comprehensive research of the tandem GMAW process with the addition of the AFW between the welding wires in order to determine the applicability of this welding method in industrial production, as well as to possibly eliminate the use of pulsed modes and simplify the control of the welding process.

Acknowledgement
The assistance provided by the translator of the present article Ms. Natalia Karamysheva was greatly appreciated.

References

[1] Aleshin N P, Yakushin B F, Kobernik N V and Kilev V S 2018 *Svarochnoe Proizvodstvo* **10** pp 3–13
[2] Fiveyskiy A, Mosin A, Zverev S, Polevoy I 2019 *MATEC Web Conf.* **298** 00072
[3] Moinuddin S Q and Sharma A 2019 *Advances in Welding Technologies for Process Development* vol 1 ed J J Vora and V J Badheka (Boca Raton: CRC Press) pp 1–23
[4] Goecke S, Hedegård J, Lundin M and Kaufmann H 2001 *Svetsaren* **2-3** pp 24–8
[5] Hairulin T V and Stolbov V I 2012 *Fizika I Himiya Obrabotki Materialov* **4** pp 62–8
[6] Sholohov M A 2009 *Issledovanie i razrabotka tekhnologii mekhanizirovannoj svarki v zashchitnyh gazah stalej s ponizhennym soderzhaniem vrednyh primesej. Avtoref. dis. kand. tekhn. nauk* [in Russian] (Ekaterinburg: USTU-UPI named after the first President of Russia B N Yeltsin) pp 1–24
[7] Ribeiro R A, Dos Santos E B F, Assunção P D C, Braga E M and Gerlich A P 2019 *Weld. J.* **98** pp 135–49
[8] Ribeiro P P C, Dos Reis J R, Fonseca M G, Loayza C R L, Assunção P D C, Filho A A C and Braga E M 2017 *24th ABCM International Congress of Mechanical Engineering* (Curitiba: ABCM)
[9] Ribeiro R A, Assunção P D C, Dos Santos E B F, Filho A A C, Braga E M and Gerlich A P 2019 *Materials* **12 (3)** p 335
[10] Zhilin P L, Konischev B P and Lebedev S A 2014 *Trudi NGTU im R E Alekseeva* **5** pp 381–7
[11] Günter K, Bergmann J P, Zhang C, Rosenberger M and Notni G 2018 *Weld. J.* **97** pp 99–107
[12] Marques L F N, Santos E B F, Gerlich A P and Braga E M 2016 *Sci. Technol. Weld. Joi.* **22** pp 87–96
[13] Meshkov V V 2012 *Vestnik SGTU* **1 (63)** pp 48–53
[14] Ueyama T, Ohnawa T, Tanaka M and Nakata K 2007 *Sci. Technol. Weld. Joi.* **6** pp 523–9
[15] Chen D, Chen M and Wu C 2015 *J. Mater. Process. Technol.* **225** pp 45–55
[16] Dongsheng W, Xueming H, Xiaoli Y and Fang L 2017 *Int. J. Manuf. Technol.* **88** pp 219–27
[17] Schnick M , Wilhelm G, Lohse M, Füssel U and Murphy A B 2011 *J. Phys. D Appl. Phys.* **44** pp 185–205
[18] Yao P, Xue J, Zhou K, Wang X and Zhu Q 2016 *J. Mater. Process. Technol.* **229** pp 111–20
[19] Kaiyuan W, Peimin X, Zhao L, Min Z and Zhuoyomg L 2020 *J. Manuf. Process.* **49** pp 423–37
[20] Xiang T, Li H, Wei H L and Gao Y 2015 *Int. J. Adv. Manuf. Technol.* **83** pp 1583–93

IOP Conf. Series: Materials Science and Engineering **966** (2020) 012014 doi:10.1088/1757-899X/966/1/012014

Double-sided arc welding with multiple electrodes of vertical joints of steel tanks

O I Filippov[1, 5], V A Erofeev[2], S I Poloskov[3] and A Yu Melnikov[4]

[1] Transneftstroy Co. Ltd, Moscow, 127254, Russia
[2] Tula State University, 92, Lenina Avenue, Tula, Russia, 300600
[3] Bauman Moscow State Technical University, 5, ul. Baumanskaya 2-ya, Moscow, 105005, Russia
[4] SHTORM Co.Ltd, Ekaterinburg, 620100, Russia

E-mail: [5] foi_1@rambler.ru

Abstract. The increase of the welding productivity is still of immediate interest, which in turn requires the development and implementation of more high-performance welding methods that ensure a consistently high quality of welded joints. One of the ways to further increase the productivity of arc welding processes in the construction of tanks is the industrial use of automatic double-sided arc welding with multiple electrodes in separate weld pools.To establish the relationship between the parameters of the welding mode, energy characteristics, as well as the distance between the torches and their angle of inclination relative to the groove, a physical and mathematical model of the production of the welding joint when four arcs participate in pairs from both sides of the joint has been developed. Thus, computer modeling showed that during weld formation with the movement of the electrode down (vertical down), the possibility of supplying each pair of arcs with pulsed current should be taken into account. During groove filling with the movement of the electrodes up (vertical up), their lateral oscillations and delays at the edges should be carried out according to an aperiodic law in order to minimize the effects of 'magnetic blow'.

1. Introduction

Taking into account a number of advantages, including the simplicity of manufacturing, operation and maintenance, occupied space, vertical steel tanks for storing oil and oil products are one of the most popular and widespread varieties of tanks and vessels.Therefore, the Russian Federation has developed an extensive program for the construction of new tanks and maintaining high operational requirements for already constructed tanks. Unfortunately, tanks for receiving, storing, processing and dispensing of oil and oil products are dangerous industrial facilities, accidents at which can lead to serious material, environmental and social consequences. Therefore, for their construction, high requirements are imposed on all stages of their production, including the quality of welding works, since the analysis showed that the safe operation of the tanks is significantly affected by defects in welded joints [1].

The design of the tank includes a bottom, a shell and a fixed roof. Assembly and welding of metal structures of the bottom and shell of tanks is carried out on a previously prepared foundation slab and base. Metal structures can be made in two ways: by rolling or sheet assembly. It is also possible to produce them by a combined method. However, the accumulated experience of building tanks by the rolling method convincingly showed that such method is the most appropriate for the production of tanks with a shell thickness of not more than 8 mm; therefore, sheet-metal assembly and welding [2] became the main method for construction of large-capacity tanks. Currently, fusion arc welding is the most commonly used method for tanks assembling. Welding of inlets, ventilation pipes, manholes is performed by mechanized welding with a consumable electrode in active gases and gas mixtures. When performing the most time-consuming operations, which include production of a bottom corner weld connecting the bottom to the shell of the tank, welding of horizontal and vertical seams during sheet assembly, both mechanized and automatic welding are used [3, 4]. It is known [5, 6], that vertical seams are one of the weakest structural and technological units of tanks, since they account for about 15% of the total number of detected defects.

For now, the used technologies have reached their maximum performance. This occurs due to restrictions on the size of the weld pool and heat input into the welded joint. Productivity increase of welding works is still of immediate interest.

In our opinion, a promising way to further increase the productivity of arc welding during the construction of tanks is the industrial use of automatic two-arc double-sided welding in separate weld pools [7], since it is quite problematic to keep a weld pool of large volume and mass during production of vertical joints.Significant advantage of the proposed welding method is the fact that with the simultaneous exposure of a pair of arcs on both sides of the joint, additional phenomenon are achieved: during root welding, it is strengthened by an additional layer of electrode metal, and during filling of the grooves, thermal cycling of the weld and heat-affected zone (HAZ) is performed.

However, such complicated technological methods, as well as the spatial position of the weld pool, affect the distribution and amount of energy input, and, consequently, the quality of the joints [8]. Therefore, for the industrial use of the proposed technology of automatic double-sided arc welding with multiple electrodes of vertical joints, it is necessary to define scientifically-based requirements for technologies and equipment. In this case, it is necessary to establish the relationship of a large number of parameters: welding speeds and wire feed speed; energy characteristics of the process and heat input during welding; values of voltages and currents of arcs; parameters of current pulses and lateral oscillation of the welding torches; the distance between the torches and their angle of inclination relative to the groove.

2. Materials andMethods

To solve this class of problems, it is necessary to create a physical and mathematical model of the process, its computer implementation, and assessment of its adequacy. In the case of positive results, further research is possible to determine the range of acceptable values of welding parameters and their optimization, to study the sensitivity of the process to instability of parameters in the neighborliness of optimal values. The necessary source data for modeling the features of this type of multi-arc welding of vertical joints will be the thermophysical properties of steel and electrode wires, the thickness and geometric characteristics of the beveling, the location of the torches relative to the joint, the welding and feeding speeds of the electrodes, the diameters of the electrode wires.

Since in the proposed welding type four arcs participate in pairs from both sides of the joint, based on their interaction, the model will determine the equilibrium conditions of the weld pools in the modeling space limited by physical phenomena, Figure 1.

Based on preliminary work, it was determined that the formation of the root of the seam should be carried out with the synchronous movement of the torches down (vertical down), while filling the grooves with their movement up (vertical up). Therefore, the formation of the weld root was ensured by welding vertical down, with the location of each pair of arcs from different sides of the joint, Figure 1b. In this case, the second arc of each one-sided pair (Figure 1a) should be placed at a distance L_{ef}, to exclude the fusion of the weld pool. When forming a two-sided seam with synchronous exposure of arcs, electrode wires with a diameter of d_e (Figure 1b) are fed at a speed v_e on both sides of the X-groove (Figure 1c). Welding torches are placed at an angle β to the longitudinal section of the weld (Figure 1b), which ensures the highest electrodynamic pressure of the welding arcs and contributes to better retention of the melt of the weld pool. For root formation, it is advisable to use pulsed welding [9, 10], which initiates the fusion of the groove faces and retention of the weld pool in the groove, as well as minimizing the effect of magnetic blow on the wandering of arcs and spatter of electrode metal.When filling out a groove, to reduce the likelihood of a weld pool dripping, it is advisable to weld with lateral oscillation of the electrode (Figure 1d) and delays at the toes of the grooves, which reduce the length of the weld pool while keeping the power of the arc and reduce the likelihood of lack of penetration at the toes of the groove. Thus, the simultaneous activity of all four arcs from different sides of the joint provides a significant increase in the productivity of welding of vertical joints of steel tanks.

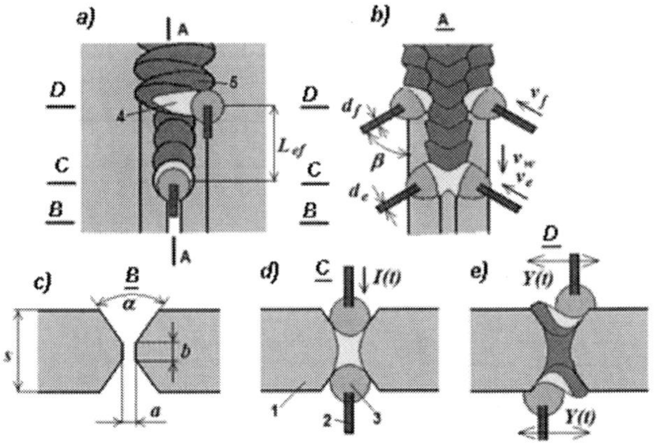

Figure 1. The schematic of the implementation of welding of vertical joints (a, b), cross section of the joint (c), and position of electrodes relative to the joint (g): 1 – parts of a groove weld; 2 – electrode wire; 3 – arc; 4 – weld pool; 5 – weld seam.

Considering these features of the multi-arc welding process, an area is selected for modeling physical phenomena, covering the edges to be welded and the weld pool formation zones. An important feature is the uncertainty in the structure of the modeling area, determined by the formation of the weld pool and seam. This causes the need to adjust the structure of an area depending on the results of the model equations. The morphology of the area is expressed within a Cartesian coordinate system of x, y, z-coordinates by the membership function of the points belonging to different zones, the properties and processes in each are different. Initially, the structure of the modeling area is determined by the geometry of the beveling (zone M_m). As the groove is filled with electrode metal, the molten zones of the weld pool M_L, solid-liquid metal M_s, and weld metal M_w arise. The listed zones are in contact with the gas environment G. The affiliation of a point to one of the zones is determined by the state of the metal. For example, points of a metal with a temperature below the solidus temperature belong to the M_m, zone, and above the liquidus temperature, they belong to the M_L zone. Internal interband surfaces are marked as intersections of the sets of points belonging to the corresponding zones. For example, the location of points $(x,y,z)_L$ of the surface of the weld pool in a metal is described by the intersection of the sets M_L and M_s : $(x,y,z)_L \in M_L \cap M_s$.

The location of the surface of the weld pool separating the metal M from the surrounding gas environment G is determined not by temperature, but by the pressure equilibrium on the surface $Z(x,y)$ of the liquid metal M_L, therefore it is additionally described by the function $Z(x,y) \in M_L \cap G$.

The sizes of the areas and the location of the interface between them is unknown, and they must be determined during the simulation, depending on the results obtained when solving a system of equations describing the physical phenomena that occur during welding. The need for continuous adjustment of the arrangement of the interface surfaces of the selected areas during the solution, as well as a change in the arc current and their location relative to the joint, makes it necessary to solve the nonstationary problem, which considers the change in the thermodynamic state of the metal under the thermal and power effects of the welding arc. This is more convenient to carry out in a fixed Cartesian coordinate system with a central plane of symmetry, in which the metal of the welded joint is stationary, and the arcs move with the welding speed v_w in the direction of the x coordinate. Accordingly, the coordinates of the centers of arcs are defined as:

$$x_{a1}=x_0+v_w t;\ x_{a1}=x_{al}-x_a;\ y_{a1}=0;\ y_{a2}=y(t) \tag{1}$$

IOP Conf. Series: Materials Science and Engineering **966** (2020) 012014 doi:10.1088/1757-899X/966/1/012014

where x_0 – arc initiation point; t – time since the arc initiation; $y(t)$ – function that describes the arc oscillations across the joint.

If double-sided symmetrical joint is welded with two pairs of arcs from different sides, the process is symmetrical. This allows us to limit ourselves to modeling processes on one side of the joint. At the same time, our three-dimensional analysis of heat transfer is based on the simultaneous use of two successive heat sources to account for the input heat from two separate welding arcs in the same plane.

The algorithm for virtual studying the technology of multi-arc welding of vertical joints involves establishing the spatial arrangement of the arcs in the groove, solving the heat equation and determining the location of the weld pools, solving the equilibrium equation of weld pools to achieve mass balance and spent electrode metal, determining the geometry of the welds and their quality indicators.

3. Results

In accordance with the proposed sequence of actions, we performed a computer simulation of the welding process of sheets of steel 09G2 (analogy 7Mn6 – DIN, Germany) with a thickness of 30 mm with two-sided symmetrical beveling. Shielding gas is a gas mixture. When conducting research we used electrode wire Sv-08G2S (analogy 13Mn6 – DIN, Germany) with a diameter of 1.2; 1.6 and 2.0 mm. Vertical seam, the root of the seam was welded with the movement of the torches down (vertical down), while filling the grooves with their movement up (vertical up). Stick out of 20 mm. Distance between arcs is of 30 mm.

Figure 2 shows the result of modeling the formation of the weld root during pulsed welding and electrode wire with a diameter of 1.2 mm.

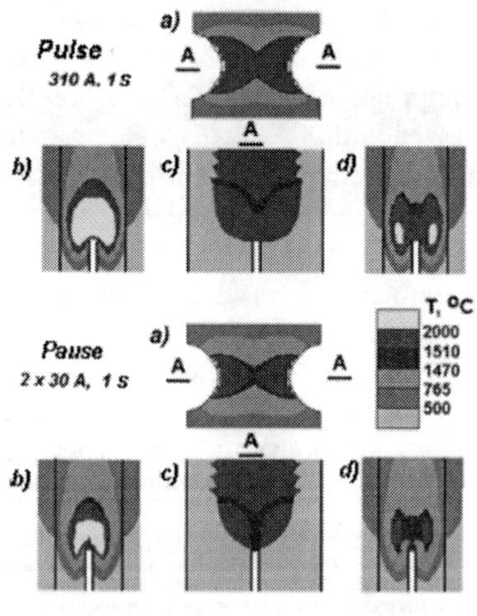

Figure 2. The result of modeling the temperature distribution at the end of the pulse and at the end of the pause during welding the root with two pulse arcs a) the temperature distribution in the weld cross section; b) on the surface of the seam; c) in the plane of symmetry; g) on the initial surface of the beveled edges.

During a root pass, it is important to ensure reliable penetration. It was found that with an enlarged root opening (up to 4 mm), more stable penetration is provided by a pulsed arc mode. An arc voltage of 30 V was chosen to obtain an arc of 3 mm in length. The accepted values are: 5.9 V – anode voltage, 6.7 – cathode voltage, 5.2 V / mm – potential gradient in the arc column. The pulse duration is 1 s, the pulse current is 310 A, the pause duration is 1 s, and the pause current is 30 A.

4. Discussion

It was found that by the end of the pulse, the mass of the weld pool increases to 12 g, and by the end of the pause it decreases almost by half – to 7 g. At the end of the pause, the weld pool partially crystallizes and lack of fusion is possible at the root of the seam (Figure 3d). But at a low welding speed, a current pulse provides a weld pool of sufficient length for complete penetration. In this case, the overlap in the center of the joint exceeds 60%. The pulsed mode increased the minimum penetration width of the root, Figure 3a, to 7 mm (versus 5 mm in continuous mode). However, the doubled arc current causes a strong heating of the electrode stick out, which forces us to recommend using a larger diameter wire (1.6 mm) during formation of root welds. In addition, with a pulsed arc, the surface of the seam has recesses, Figure 3c, which may impair the formation of subsequent layers when filling the grooves. In this case, the depth of the cavities on the seam surface is estimated to be about 1.6 mm. A significant drawback of welding with a pulsed arc with a consumable electrode is that when the arc current changes, the length of the arc gap also changes, which somewhat reduces the capabilities of pulsed technologies.

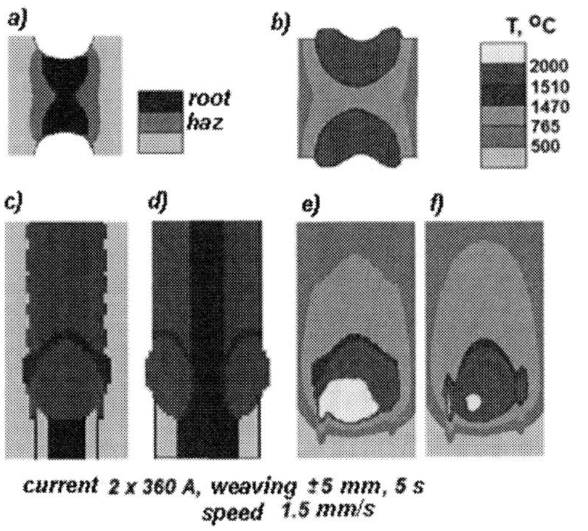

Figure 3. The simulation results of the welding of the filling passes with lateral oscillation of the electrode: a) the cross-section of the groove after the root pass; b) the maximum temperature distribution; c) temperature distribution on the weld surface; d) in the plane of symmetry; e), f) on the initial surface of the edges.

Since the root opening has a considerable width (14 mm), welding must be performed with lateral oscillations to fill it. The oscillation amplitude was set to ±3 mm, the period was 2–4 s, the delay at the edges was 0.2 s. At a speed of 1.5 mm/s, the oscillation step in the groove was 3 mm. The diameter of the electrode wire was increased to 2.0 mm, since the of the electrode stick out with a wire diameter of 1.6 mm was heated above 500 °C. It was determined that when using an electrode wire with a diameter of 2.0 mm, the temperature of stick outs does not exceed 320 °C, with a welding current of 360 A.

The simulation results of the filling passes are shown in Figure 3.

With virtual filling of the groove, an almost round weld pool with a diameter of about 16 mm is formed with the indicated welding parameters. Despite its significant weight (12 g), it is stable, since it is almost completely immersed in the groove and is held by the electrodynamic force of the arc, the value of which is estimated at 0.15 N. Nevertheless, the surface of the seam is formed with a slight convexification of about 2 mm, in the middle of which a hollow (meniscus) is located. This form of seam is allowed by regulatory documentation.

5. Conclusion

Thus, computer modeling of the features of double-sided arc welding with multiple electrodes showed that during weld formation with the movement of the electrode down (vertical down), the possibility of supplying each pair of arcs with pulsed current should be taken into account. Pulse technologies reduce the penetration sensitivity to the joint gap during the formation of the root welds. During groove filling with the movement of the electrodes up (vertical up), their lateral oscillations and delays at the edges should be carried out according to an aperiodic law in order to minimize the effects of 'magnetic blow'.

The presented results show that the proposed method for welding vertical joints has acceptable parameters, provides the required formation of the seam and can be used during tanks assembling. In addition, the use of the presented physical and mathematical model for virtual reproduction of the process of automatic double-sided welding of vertical joints will solve all main problems in the development of technologies and the necessary equipment for welding vertical joints of tanks with thick shells for storing large volume oil and oil products.

Acknowledgement
The assistance provided by the translator of the present article Ms. Natalia Karamysheva was greatly appreciated.

References

[1] Safina I, Kauzova P and Gushchin D 2016 Otsenka tekhnicheskogo sostoyaniya rezervuarov vertikalnykh stalnykh [In Russian] *Tech Nadzor* **3** pp 40–1
[2] Bezborodov Yu, Shram W and Kravtsova E 2015 *Rezervuary dlya priyema, khraneniyai otpuska nefteproduktov* [In Russian] (Krasnoyarsk) p 108
[3] Kah P and Martikainen J 2011 Current trends in welding processes and materials: Improve in effectiveness *Reviews on Advanced Materials Science* **30 2** pp 198–200
[4] Sakamoto R, Kobayashi K, Iijima T, Nishimura Y and Mizo Y 2010 Development of vertical submerged arc welding method for aboveground LNG storage tank construction *Ishikawajima-Harima Engineering Review* **43 2** pp 55–62
[5] Kunina P, Velichko E, Prikhodko M and Nizhnik A 2017 Razrabotka sovremennykh metodik diagnostiki svarnykh soyedineniy rezervuarov dlya khraneniya neftiinefteproduktov metodom ul'trazvukovogo kontrolya [In Russian] *Territoriya Neftegaz* **1** pp 20–4
[6] Goncharov N, Nesterov G and Yushin A 2019 Issledovaniye defektov svarnykh shvov rezervuarov vertikalnykh stalnykh [In Russian] *Bezopasnost' truda v promyshlennosti* **3** pp 69–74
[7] Filippov O 2020 Puti sovershenstvovaniya svarochnykh rabot pri stroitel'stve rezervuarov dlya khraneniya nefti i nefteproduktov [In Russian] *Blanking Productions in Mechanical Engineering* **17 7** (in the editorial portfolio)
[8] Cho D, Na S, Cho M and Lee J 2013 A study on V-groove GMAW for various welding positions *Journal of Materials Processing Technology* **213 9** pp 1640–52
[9] Scotti A, Larson U and Norrish J 1991 Bead instability of mechanized P-MIG [pulsed MIG] welding in vertical-up position *International Journal of the Joining of Materials* **3 1** pp 18–24
[10] Fiveyskii A, Buzorina D, Melnikov A and Filyakov A 2016 Sovershenstvovaniye vysokoproizvoditelnykh protsessov impulsnoydugovoy svarki v zashchitnykh gazakh [In Russian] *Svarkai Diagnostika* **6** pp 53–5

IOP Conf. Series: Materials Science and Engineering **966** (2020) 012015 doi:10.1088/1757-899X/966/1/012015

Application of surfacing using consumable electrode with an additional filler wire to ensure the required operational properties of pipeline valves

S S Poloskov[1], D S Buzorina[2, 3, 5] and V A Erofeev[4]

[1] Moscow Region State University, 24, ul. Veri Voloshinoy, Mytishchi, 141014, Moscow Region, Russia

[2] Ural Federal University named after the First President of Russia B N Yeltsin, 19, Mira str., Yekaterinburg, 620000, Russia

[3] SHTORM Co.Ltd, Ekaterinburg, 620100, Russia

[4] Tula State University, 92, Lenina Avenue, Tula, Russia, 300600

E-mail: [5] darja.buzorina@yandex.ru

Abstract. A promising way to ensure the required operational properties of pipeline valves with a significant increase in productivity and quality of surfacing during its manufacture is the introduction of anti-corrosion surfacing technologies using a consumable electrode with the supply of an additional filler wire. A solution to the optimization problem of the proposed surfacing technology is complicated by a large number of its parameters, therefor the task was solved in stages. At the first stage, the parameters of the arc interaction between the electrode and filler wire was determined, and at the second – the specific features of the formation of the surfacing bath with such interaction. A virtual study of the process of surfacing with Inconel 625 wires in argon on steel 09G2S was carried out. The influence of the thermal effect of the deposition mode parameters on the main metal is determined. It has been established that, due to the fact that the heat flux in the main metal is created mainly by the droplet stream of the deposited metal from the main arc at high currents, the diameter of the electrode wire guaranteeing its burning should be larger than the filler wire.

1. Introduction

Often, in the manufacture of parts and assemblies of pipeline valves, cheaper unalloyed steels are used, and the required anticorrosive properties of the sealing surfaces are provided by surfacing with alloys and high alloys. Therefore, surfacing is currently a priority in ensuring the required operational properties of sealing and contact surfaces of pipeline valves. Depending on the operating conditions of the pipeline valves, iron-based surfacing materials with the addition of chromium, nickel, cobalt, niobium, for example Inconel 625 are used.

Improvement of productivity and the quality of work performed can be provided by implemented automation of surfacing processes. Automatic surfacing with a non-consumable electrode and filler wire ensure process continuity with effective regulation of the substrate metal fractions in the deposited layer. However, it requires constant monitoring of the condition of the non-consumable electrode, since by wearing-out of its working surface, the melting ability of the arc decreases sharply. Conversely, with more efficient automatic surfacing with a consumable electrode, even with the addition of a filler wire, it is more difficult to achieve a minimum penetration of the base metal. Given the capabilities of automation, such promising technique could be a high-tech process of robotic surfacing with a consumable electrode with the supply of an additional filler wire to the front of the weld pool to shield the thermal effect of the arc [1, 2]. And with a science-based selection of materials for the electrode and filler wire and surfacing modes, it is possible to obtain the required chemical composition of the joints [3].

2. Materials andMethods

A solution to the optimization problem of the proposed surfacing technology is complicated by a large number of its parameters (speed of surfacing and wire feed, their diameters, voltage and arc current, amplitude, period and shape of vibrations, initial temperature of base material) and many quality indicators (dimensions and shape of the deposited layer, penetration depth and the thickness of the deposition, the chemical composition of the deposited metal, the properties of the deposited layer). Therefore, such a multi-level task should be solved in stages. At the first stage, the parameters of the arc interaction between the electrode and filler wire should be determined, and at the second – the specific features of the formation of the surfacing bath with such interaction.

At present, among modern methods of studying welding and surfacing processes that reduce the time and number of experiments, physical and mathematical modeling is distinguished. In this regard, we will first consider modeling the arc interaction during surfacing, and then we will perform modeling of the features of the formation of the molten pool taking into account the arc interaction of the electrode and filler wires.

First of all, we outline the modeling space for arc interaction during surfacing with a consumable electrode with an additional filler wire. A rational way of feeding the electrode wire (see Figure 1) is to feed it perpendicular to the plane of the base material, and the filler wire at an angle α so that its axis intersects the axis of the electrode at a given distance L_2 from the contact tip, and the second contact tip is located at a distance L_2 from the point of intersections. In this case, the first arc *a* burns between the electrode wire *1* and the base material *3*, and the second arc *b*– between the consumable electrode and the filler wire *2*. During modeling, the distance x_e from the point of intersection of the axes of the wires to the base material must be taken into account. Moreover, in our opinion, to exclude cases of welding ('freezing') of the non-melted filler wire in the solidified metal of the weld pool tail, it is advisable to feed the wire into the welding pool head.

In addition to the location of the arcs, the feed rates and the diameters of the electrode v_e, d_e and filler wires v_f, d_f must also be taken into account in the the modeling space. For the main arc (*a*), the open circuit voltage U_e, is set, and for the additional arc (*b*) – current I_f. It should be noted that the current I_e of the main arc is self-alignment due to the effect of self-regulation of its length l_e, in return, the voltage of the additional arc is determined by its length l_f.

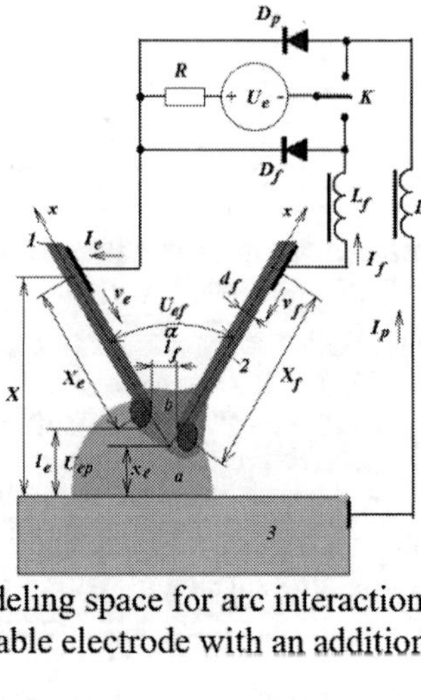

Figure 1. The modeling space for arc interaction during surfacing with a consumable electrode with an additional filler wire.

3. Results

During arc interaction, the control actions are the feed speeds of the electrode and filler wires, the open circuit voltage of the power source and the relative switch closure rate, which controls the voltage ratio of the main and additional arcs. These control actions determine the currents of the arcs, their lengths and the distribution of the process power. Additional parameters that determine the regulation results are the diameters of the wires, their stick outs, the angle between them.

Figure 2 shows the effect of the feed speed of an electrode wire with a diameter of 1.6 mm from Inconel 625 alloy on the current I_e of the main arc, the power P_k released in the cathode spot, the power P_e of the heat flow of the droplets of the electrode metal, the temperature T_e of the electrode stick out, and the length l_e of the main arc.

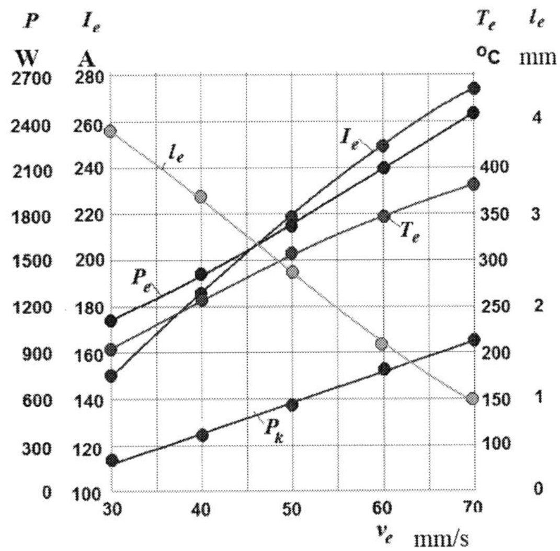

Figure 2. Dependence of the main arc parameters on the feed rate v_e of an electrode with a diameter of d_e = 1.6 mm from Inconel 625 alloy at a stick out of l_e = 10 mm, a supply voltage of U_e = 32 V, pulse ratio t_e/τ = 0.5, and an additional arc current of I_f = 108 A.

During virtual research, it was found that an increase in the feeding speed v_e of the electrode wire at a fixed current I_f of the additional arc, the current I_e, flowing through the electrode increases almost in proportion to the speed. The deviation from proportionality is explained by a noticeable increase in the temperature T_e of the electrode stick out. For the same reason, the power P_e, that is heating the electrode increases at a faster rate than the arc current I_e. The power P_k, released in the cathode spot on the surface of the base material is almost proportional to the feed rate, but much less than the thermal power P_e of droplet stream. The length l_e of the main arc decreases linearly as the electrode feed rate increases. At the increase in the feeding speed v_e of the electrode wire at a fixed current I_f of the additional arc, the current I_e, flowing through the electrode increases almost in proportion to the speed.

The parameters of the additional arc are also important. Figure 3 shows the effect of the feed speed v_f of a filler wire with a diameter of 1.2 mm from Inconel 625 alloy on the current I_f of the additional arc, the power P_k, released in the cathode spot, the power P_f of the heat flow of the droplets of the filler wire metal, the temperature T_a of the heated electrode stick out, and the length l_a of the main arc.

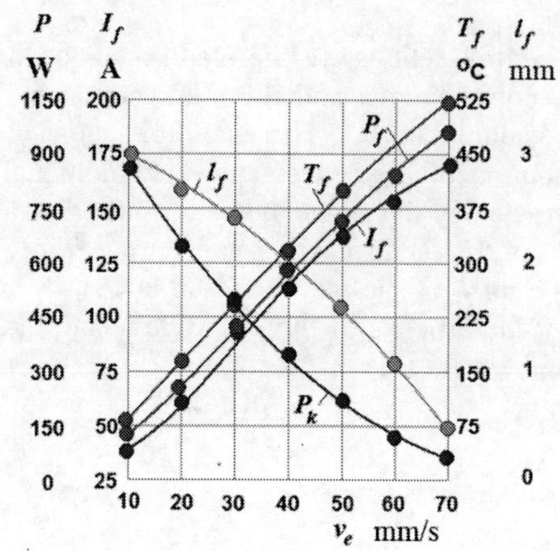

Figure 3. Dependence of the additional arc parameters on the feed rate v_f of a filler wire with a diameter of d_e = 1.2 mm from Inconel 625 alloy at a stick out of l_e = 10 mm, a supply voltage of U_e = 32 V, pulse ratio t_e/τ = 0.5, and a main arc current of I_e = 185 A.

4. Discussion

Heat flows of arcs and droplets of electrode and filler metal cause melting and formation of a molten pool with stirring. The pressure of the arc and the droplet stream extrudes the melt, forming a curved surface, the shape of which provides a balance between the electrodynamic pressure of the arc plasma, the pressure of the droplet stream, the gravitational and internal pressures in the melt and the capillary pressure determined by the curvature of the surface. Therefore, the physical phenomena during the formation of the molten pool is determined by the location of the active arc spot on the molten pool (Figure 1) and the droplet streams of the electrode and filler material, which changes in accordance with the given path of the torch [4]. It is the shape of this surface at the solidification boundary that forms the surface of the deposited layer.

Technological parameters of the process of formation of the molten pool are: grades of alloys of the base metal and deposited layer; diameter and feed rate of electrode and filler wires; parameters of power supplies of electric arcs; characteristics of lateral oscillation of the welding torch.

Therefore, the results of numerical modeling of the formation of the molten pool should be drawn with regard to the heat and mass transfer of the arcs and the geometric characteristics of the surfacing (Figure 4), including the minimum Y_0 and maximum Y_m width, thickness Z_H and area S_H of the cross section of the deposited layer, as well as depth Z_L and penetration S_L area of the base material.

According to [5], during surfacing with oscillations of the heating source, heat flows are unevenly distributed alongside the deposited layer and have minimum and maximum values. Given this circumstance, the geometrical characteristics should also include the dimensions of the Y_{850}, Z_{850}, S_{850} heat-affected zones, which for a steel base material are determined by the by the ultimate arrangement of isotherms of phase transformation temperatures (850 °C) and deep tempering (500 °C).

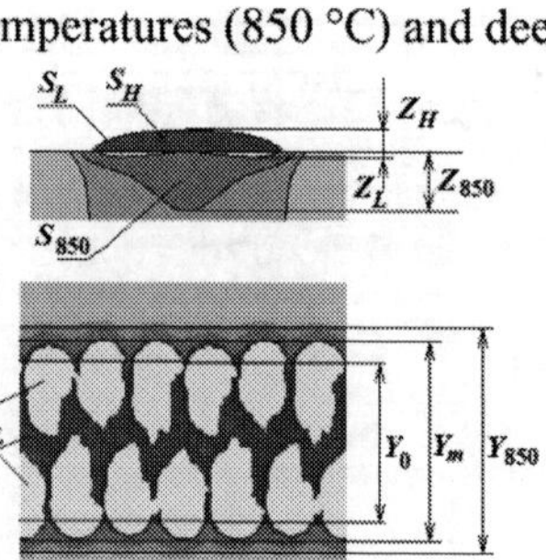

Figure 4. Geometrical characteristics of surfacing.

The presented procedures for the interaction of arcs and the values of the geometrical characteristics of surfacing allow sufficiently accurate determination of the structural transformations in the heat-affected zone and the deposited layer, its geometric shape and characteristic dimensions.

As a test example, we consider the process of surfacing with Inconel 625 wires in argon on steel 09G2S (analogy 13Mn6 – DIN, Germany) with athickness of 20 mm. The following parameters were set: diameter of the electrode wire 1.6 mm at a stick out of 10 mm and a feed speed of 40 mm/s, diameter of a filler wire 1.2 mm at a stick out of 20 mm and a feed speed of 44 mm/s. The open circuit voltage of the main arc power source is set to 15.5 V, the arc current heating the filler wire is 100 A. The deposition speed is 5 mm/s, the amplitude of lateral oscillations is 7.5 mm, the period is 1.5 s, the delay in the extreme positions is 0.15 s. The result of solving the equations of the model is shown in Figure 5.

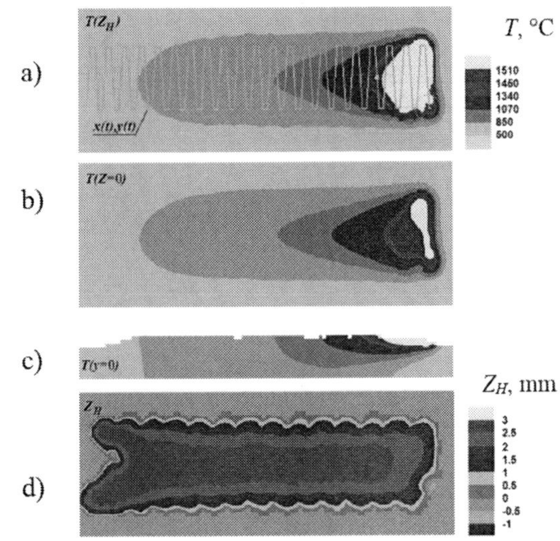

Figure 5. The result of solving the equations of the molten pool formation: a) temperature distribution $T(Z_H)$ and trajectory $x(t)$, $y(t)$ of the electrode center on the metal surface Z_H; b) on the surface of the steel base material; c) along the plane of symmetry in the direction of surfacing; d) topographic image of the surface of the deposited layer.

As a result of numerical modeling, the following steady-state surfacing process parameters are obtained:

- electrode current 240 A, current on the deposited surface 140 A;
- components of the heat flow into the metal: cathode spot power 646 W, droplets of electrode wire 946 W, filler wire 636 W;
- average values of the volume of liquid metal in the molten pool: steel 14.4 mm^3, Inconel 237.4 mm^3;
- maximum dimensions of the deposited layer: width 18.8 mm, height 1.9 mm, penetration depth of the steel base material 0.8 mm.

5. Conclusion

The obtained results show that the main heat source is the heat of droplets of the electrode and filler material, the heat flow of which (1582 W) is significantly greater than the heat release in the cathode spot (646 W). The distribution of this power over the metal surface is determined mainly by the amplitude of lateral oscillation of the electrode (±7.5 mm), which determines the width of the deposited layer (18 mm). With a small period of lateral oscillation (1.5 s) and a low deposition rate,

IOP Conf. Series: Materials Science and Engineering **966** (2020) 012015 doi:10.1088/1757-899X/966/1/012015

the width of the deposited layer varies slightly, since the molten pool covers almost the entire width of the deposited layer. The arc pressure on the molten pool is insignificant due to the small arc current burning between the electrode and the surface of the molten pool (140 A), which eliminates the formation of a crater in the melt, leading to a deeper penetration of the steel base material. The presented data indicate that it is possible to form a deposited layer with minimal penetration (not more than 0.8 mm) and only partial contact of the layer with the base material (37%). Most of the liquid Inconel (59%) is in contact with steel in a solid-liquid state, since the Inconel melting point (1340 °C) is lower than the solidus temperature of the steel (1450 °C). This provides a small iron content in the deposited layer, which is estimated by the ratio of the volumes of molten steel and the deposited Inconel value of less than 6%.

As part of the implementation of the Industry 4.0 concept, the creation of models and prototypes using the capabilities of wire arc additive manufacturing (WAAM – Wire Arc Additive Manufacturing) using wire for layer-by-layer deposition of material is given the most serious attention [6]. In view of this, the research results can be used in the additive hardfacing production using industrial robots.

Acknowledgement
The assistance provided by the translator of the present article Ms. Natalia Karamysheva was greatly appreciated.

References

[1] Zhang Y M, Jiang M and Lu W 2014 Double electrodes improve GMAW heat input control *Weld. J.* **83** 39–41

[2] Poloskov S S 2019 Problemy naplavki uplotnitel'nykh poverkhnostey truboprovodnoy armatury i puti ikh resheniya [In Russian] *Vestnik DGTU* **19** 349–56

[3] Schreiber F, Allebrodt B and Erpel T 2018 Hardfacing material solutions for high performance coatings in wear and corrosion applications *Advanced Technologies and Materials* **43** 21–5

[4] Poloskov S S, Erofeev V A and Sholohov M A 2019 Fiziko-matematicheskaya model' dugovogovzai modey s tviya pri naplavke plavyashchimsya elektrodom s dopolnitel'noy prisadochnoy provolokoy [In Russian] *Svarka i diagnostika* **6** 27–31

[5] Lebedev V A and Novikov S V 2016 Raschet parametrov avtomaticheskoy dugovoy naplavki tsilindricheskikh detaley s kolebatelnymi dvizheniyami svarochnogo instrumenta [In Russian] *Zagotovitelnye proizvodstva v mashinostroenii* **5** 6–12

[6] Ding D, Pan Z, Cuiuri D and Li H 2015 Wire-feed additive manufacturing of metal components: technologies, developments and future interests *The International Journal of Advanced Manufacturing Technology* **811** 465–81

Influence of technological parameters on the vanadium valence form in slags

A V Klimov

Institute of New Materials and Technologies, Ural Federal University named after the First President of Russia B N Yeltsin, 19, Mira str., Yekaterinburg, 620000, Russia

E-mail: a.v.klimov@urfu.ru

Abstract. The multivalence of vanadium in oxide melts is considered. It is shown that the vanadium oxidation state in the slags is determined by the slag and metal compositions as well as the oxidation potential of the gas phase. Using a polymer model, vanadium valence form fractions were calculated depending on technological parameters. Calculations show that in an air atmosphere, vanadium in the slag should be predominantly in the tetravalent and trivalent state, and in contact with the molten iron, mainly in the trivalent and partly divalent state. Upon contact with pig iron, VO oxide becomes the main form of vanadium in the slag. Slag basicity and temperature have a significant effect on the relationship of vanadium forms. The suggested design-basis using the polymer model can be useful in accomplishment of the thermodynamic calculations for the vanadium equilibrium distribution between metal and slag in the iron and steel production. It allows us to describe the dependence of the vanadium oxidation state on the free oxygen activity in the slag and the gas phase oxidation potential.

1. Introduction

Thermodynamic calculations of the vanadium distribution between the metal and slag are complicated by the vanadium multivalence, as it can be present in oxide melts in two-, three-, four- and pentavalent forms. In addition, the vanadium valency determines not only the equilibrium distribution, but also the kinetics of its transition between phases. Information on vanadium-containing oxide melts is mainly based on the results of bulk, petrographic and phase analysis of solidified slags. The slag composition effect on the form of vanadium existence is obvious.

Various experimental data are presented on the effect of the blast furnace and steelmaking slag composition and, in particular, their basicity ($B=\%CaO:\%SiO_2$) on the vanadium distribution coefficient between oxide and iron-based metal melts. For slags formed during the vanadium pig iron production [1], this coefficient decreases with an increase in the basicity of the oxide melt, while the opposite effect is observed for ferrous steel-making slags [2]. A weak dependence of the vanadium oxides activity on the basicity of slag is observed in the low-alloyed vanadium steels production in electric arc furnaces [3]. The relationship between the distribution coefficient of vanadium and the iron oxides content in the slag during the smelting of low alloy steel in industrial units is clearly manifested [4, 5]. The data presented indicate the trivalent vanadium state in the slag and the decisive role in its distribution of the reaction

$$\frac{3}{2}(FeO) + [V] = \frac{3}{2}[Fe] + \frac{1}{2}(V_2O_3) \tag{1}$$

According to [5, 6], the value of the vanadium trioxide activity coefficient $\gamma_{V_2O_3}$ in acid, neutral and basic steelmaking slags when normalized to pure solid vanadium trioxide is 2.6–2.8, which indicates the invariance of the valence state and a limited V_2O_3 solubility in the studied systems. However, the data represented by the authors for acid slags can also be linearized in the coordinates N_{VO}/a_V vs a_{FeO} which indicates the possibility of the divalent vanadium significant amount presence in acid slags.When normalizing to pure solid VO, its activity coefficient was 0.42 and 0.31 at 1873 and 1923 K, respectively.

The investigation of the vanadium valence state and the activity of its oxides in slag melts was carried out by the authors [7]. Two groups of experiments were performed in a furnace with a controlled atmosphere: with pure slags of the $CaO–SiO_2–Al_2O_3–V_2O_5$ system and these slags in contact with iron-carbon melts of various compositions. The gas phase consisted of a various $CO–CO_2$ mixtures, equivalent to the partial oxygen pressure from $1.5\cdot10^{-6}$ to $2\cdot10^{-10}$atm.The presence of two main vanadium valences was found: V^{5+} and V^{4+} at P_{O_2} more than 10^{-8}atm, V^{4+} and V^{3+} at P_{O_2} less than 10^{-8} atm. An increase in the CaO content promoted an increase in the share of vanadium with a higher valency. The results of the experiments could be affected by insufficient exposure time to achieve equilibrium and the oxidation of vanadium in the slag samples during sampling.

Experiments on the cathodic deposition of vanadium from oxide melts on solid electrodes showed [8] that, in the absence of a reducing agent and an air atmosphere, slag contain mainly V^{4+} form. In the atmosphere ($CO+N_2$) the V^{3+} form is stable. Pentavalent vanadium in the slag is unstable and goes into lower valence states. Upon contact with a carbon-saturated molten iron, vanadium is stable in the trivalent state. In a reducing atmosphere in the presence of solid vanadium, the main vanadium form in the slag is V^{2+}.

The presence of vanadium in the metal melt probably provides a disproportionation reaction at the phase boundary

$$2V^{3+} + V^0 = 3V^{2+} \tag{2}$$

An X-ray diffraction study of the oxide films formed during oxidation on the metal surface showed that at a vanadium concentration in the metal of less than 0.2% vanadium spinel of variable composition is formed. V_2O_3 is formed in the range from 0.2 to 0.3% V. At V > 0.3%, the product of vanadium oxidation, according to the authors of [9], should be V_2O_2 oxide. A slightly different concentration range V was found in [10]. According to these data, V_2O_2 appears at V > 0.9%. There is an opinion [11,12] that in the Fe–V–O system at low temperature only FeO–VO solutions of variable composition can be in equilibrium with the metal melt. A similar point of view is shared by the authors of [13, 14]. The given point of view is based on studies of the products of iron deoxidation by vanadium. However, according to the review [15], depending on the O/V ratio in the metal, the products of deoxidation are V_2O_5 (V < 0.1–0.2%), V_2O_3 (with V contents from 0.15–1.8 to 10%) and VO (from 0.3–4 to 10% V in the metal).

The presence in iron of such strong reducing agents as carbon and silicon leads to the reduction of transition metal ions in oxide melts, especially with a high SiO_2 content, to a divalent state [16]. This vanadium valence state is confirmed by laboratory data on the smelting of vanadium-containing cast iron. Figure 1 shows that the equilibrium between the metal and vanadium slag is well described by reaction

$$(VO) + [C] = [V] + \{CO\} \tag{3}$$

Thus, there is still no clear insight of how various technological parameters affect the form of vanadium in the slag, which makes it difficult to select the optimal process conditions by calculation.

2. Calculation design

The polymer model [17] makes it possible to estimate the effect of slag composition on the distribution of vanadium by valency. We assume that tetravalent vanadium is present in the melt in the form of the VO^{2+}cation, and pentavalent is in the form of the VO_4^{3-}anion [18].

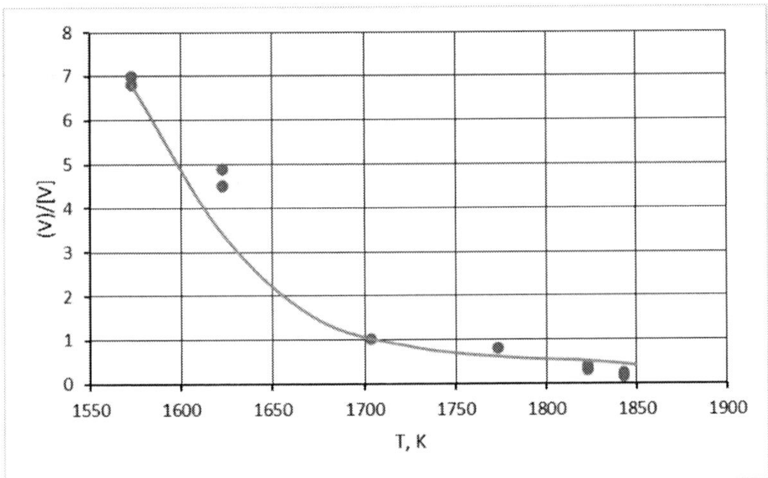

Figure 1. Vanadium distribution between slag and pig iron. Points are experimental values. Curve – calculation values for reaction (3).

Then the redistribution of vanadium valencies in oxide melts can be described by the equations:

$$VO_4^{3-} + O^- = VO^{2+} + 3O^{2-} + \tfrac{1}{2}\,O^o + \tfrac{1}{4}\,O_2 \qquad (4)$$

$$VO^{2+} + O^- = V^{3+} + O^{2-} + \tfrac{1}{2}\,O^o + \tfrac{1}{4}\,O_2 \qquad (5)$$

$$V^{3+} + O^- = V^{2+} + \tfrac{1}{2}\,O^o + \tfrac{1}{4}\,O_2 \qquad (6)$$

Here O^- is the non-bridging oxygen ion, O^{2-} is the free oxygen of the slag, O^o is the bridging oxygen atom. The suggested record takes account of the dependence of the vanadium valence forms on the partial pressure of oxygen and the oxide melt composition. The relationship between the various forms of oxygen in the slag is determined by the reaction

$$2O^- = O^o + O^{2-} \qquad (7)$$

The model makes it possible to determine the ionic fractions of oxygen various forms in slags of various compositions and use them in the calculation. From equations (4)–(6) it follows that

$$\frac{a_{VO^{2+}}}{a_{VO_4^{3-}}} = K_4 \frac{a_{O^-}}{a_{O} \cdot a_{O^0}^{1/2} \cdot P_{O_2}^{1/4}} \qquad (8)$$

$$\frac{a_{V^{3+}}}{a_{VO^{2+}}} = K_5 \frac{a_{O^-}}{a_{O^{2-}} \cdot a_{O^0}^{1/2} \cdot P_{O_2}^{1/4}} \qquad (9)$$

$$\frac{a_{V^{2+}}}{a_{V^{3+}}} = K_6 \frac{a_{O^-}}{a_{O^0}^{1/2} \cdot P_{O_2}^{1/4}} \qquad (10)$$

The equilibrium constants K_4–K_6 were found by combining the polymerization constant of reaction 7 and the equilibrium constants of reactions

$$V_2O_{5lq} = 2VO_{2lq} + \tfrac{1}{2}\,O_2 \qquad (11)$$

$$2VO_{2lq} = V_2O_{3lq} + \tfrac{1}{2}\,O_2 \qquad (12)$$

$$V_2O_{3lq} = 2VO_{lq} + \tfrac{1}{2}\,O_2 \qquad (13)$$

K_{11}–K_{13} values were calculated from reference data [19]. Due to the absence of reliable data on the vanadium activity coefficients in slags, we equate its activity to mole fractions. Let us assume by convention that in the vanadium subsystem

$$N_V^{5+} + N_V^{4+} + N_V^{3+} + N_V^{2+} = 1 \quad (14)$$

Solving equations (8)–(10) and (14) simultaneously, can be found the fractions of all vanadium valence forms depending on the slag composition and the gas phase oxidative potential.

3. Calculation results

Figure 2 shows the results of the calculations for slag of the CaO–SiO_2 system at 1600 °C. It was believed that the concentration of vanadium oxides in the slag is deficient to affect the distribution of oxygen in three forms.

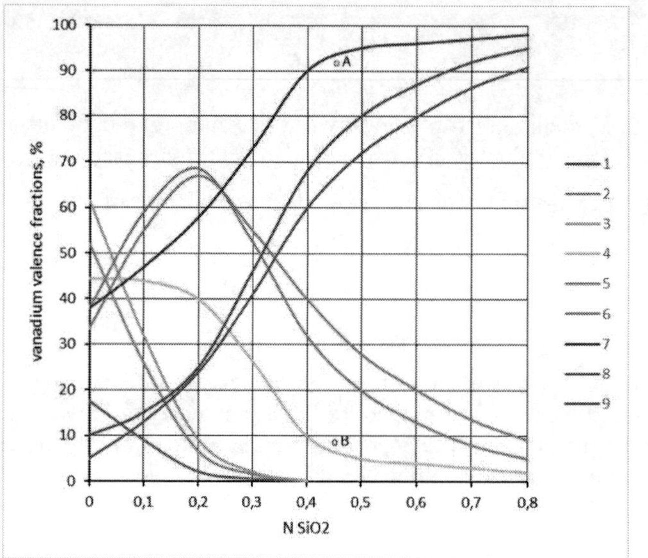

Figure 2. Dependence of the fractions of penta- (1–3), tetra- (4–6) and trivalent (7–9) vanadium on the SiO_2 content in the CaO – SiO_2 melt at a temperature of 1600 °C and an oxygen partial pressure of 0.001 (1,4,7), 0.21 (2,5,8) and 1.0 atm (3,6,9).

It can be seen that in silicate melts of the CaO-SiO_2 system, the main forms of vanadium are trivalent and tetravalent. Appreciable fractions of 5-valent vanadium are present with a molar fraction of SiO_2 of less than 0.3. The fraction of divalent vanadium even at P_{O_2} = 0.001 atm. varies from 0.066 to $7.75 \cdot 10^{-4}$ % and at P_{O_2} equal 1 atm. it is $3.20 \cdot 10^{-3}$–$6.49 \cdot 10^{-5}$ %. The fraction of tetravalent vanadium changes through a maximum at the level of 50–70 %, the position of which in the diagram depends on the partial oxygen pressure. In slags with a basicity of 0.93 or lower, the largest fraction is V^{3+}. The stable existence of trivalent and tetravalent vanadium at very close concentrations to the calculated values was confirmed experimentally for an argon atmosphere with an oxygen impurity of 0.001 atm. (points A and B).

To extend the proposed approach to other oxide systems, we evaluated the effect of the free oxygen concentration in the slag on the vanadium valence forms. This parameter can be obtained using the polymer model for most metallurgical systems for which polymerization constants are known. In addition, the authors of [20] suggested using the common logarithm of a ratio of the free oxygen ionic fraction to the polymerization constant as an indicator of the slag basicity. The authors have shown that this indicator plays the same role as pH in aqueous solutions. The dependence of vanadium fractions on the basicity index in the air atmosphere is shown in Figure 3.

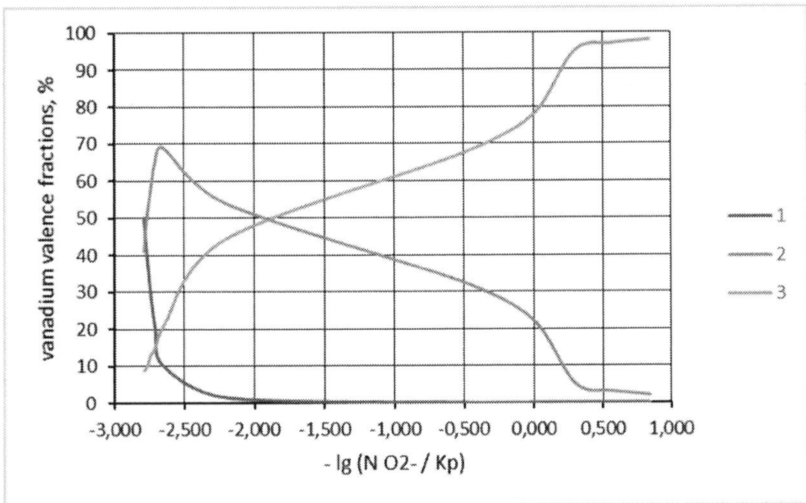

Figure 3. The dependence of vanadium valence fractions on the basicity index. Penta- (1), tetra- (2) and trivalent (3) vanadium.

Realization of similar calculations using metallic iron and FeO as reagents showed (Figure 4) that pentavalent vanadium practically disappears in the slag contacted with molten iron in the air atmosphere, and tetravalent fraction is greatly reduced. At the same time, a considerable amount of divalent vanadium appears, which increases with a decrease in basicity.

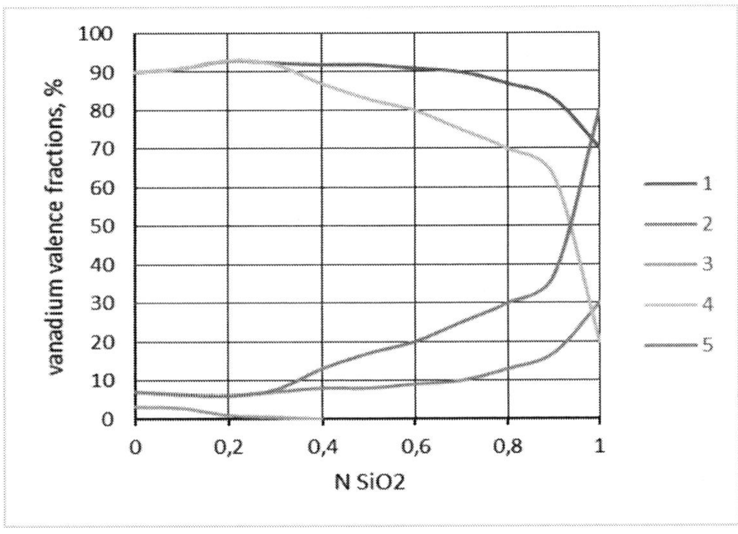

Figure 4. Vanadium valence forms in systemsFe–CaO–SiO_2 (1, 2, 3) and Fe–FeO–SiO_2 (3,4,5) at 1600 °C in air. 1,4 – V^{3+}; 2, 5 – V^{2+}; 3 – V^{4+}.

Upon contact with the molten pig iron, as was shown earlier (see Figure 1), the divalent form of vanadium, apparently, becomes predominant.

The results do not contradict the data on the oxidation state of vanadium obtained by X-ray spectroscopy and High-Temperature Mass Spectrometry [21–25].

IOP Conf. Series: Materials Science and Engineering **966** (2020) 012016 doi:10.1088/1757-899X/966/1/012016

4. Conclusions

Thus, the polymer model provides an effective estimate tool for the prevailing vanadium valence forms in slags depending on the process conditions. In particular, using the polymer model, the influence of the slag composition and the gas phase oxidation potential on the distribution of vanadium oxidation state is shown.

References

[1] Reznichenko V A and Shabalin L I 1986 *Titanomagnetites. Deposits, metallurgy, chemical technology* [in Russian] (Moscow: Nauka)
[2] Kramarov A D 1964 *Steel production in electric furnaces* [in Russian] (Moscow: Metallurgy)
[3] Rovnushkin V A, Toporishev G A, Rakovsky F S, Smirnov L A and Zhirnova L B 1979 *Proceedings of Higher Educational Institutions. Ferrous metallurgy* [in Russian] **10** **11**
[4] Smirnov L A, Deryabin Yu A, Filippenkov A A and Rakovsky F S 1985 *Production and use of vanadium slag* [in Russian] (Moscow: Metallurgy)
[5] Rovnushkin V A, Bokovikov B A, Bratchikov S G Arzamastsev E I, Amdur A M and Tyutyukov S A 1988 *Cox-free processing of titanomagnetite ores* [in Russian] (Moscow: Metallurgy)
[6] Zhirnova L B 1983 Interaction of vanadium slag with iron-based alloys, Ph.D dissertation [in Russian] Ural Polytechnic Institute, Yekaterinburg
[7] Rovnushkin V A, Nikiforov S V and Toporishchev G A 1989 *Metally* [in Russian] **5** **5**
[8] Klimov A V 1983 *Physicochemical Investigations of Metallurgical Processes* [in Russian] (Sverdlovsk: UPI) **11** 112
[9] Anderson J S and Khan A S 1970 *Less-Common Metals* **2** 209
[10] Casalot A 1972 *Materials Research Bulletin* **9** 903
[11] Mikhailov G G 1974 *Proceedings of Higher Educational Institutions. Ferrous metallurgy* [in Russian] **10** **5**
[12] Mikhailov G G and Lushnikova T V 1974 *Collection of scientific papers of the Chelyabinsk Polytechnic Institute* [in Russian] (Chelyabinsk: CHPI) **147** 10
[13] Karasev R A, Polyakov A Yu and Samarin A I 1952 *Izv. USSR Academy of Sciences* **6** 1313 [in Russian]
[14] Narita K and Kayama S 1971 *Physicochemical fundamentals of steel production* [in Russian] (Moscow: Nauka) 26
[15] Janke D and Fisher W A 1976 *Stahl und Eisen* **8** 398
[16] Charlu T V and Kleppa O J 1973 *High temperature science* **4** 260
[17] Novikov V K 1987 *J. Melts* **1** **6** 21
[18] Nagano M, Kato A, Motida I and Seyama T 1970 J. *Ceram. Soc. Japan* **904** 401
[19] Kubashevsky O and Olkok S B 1982 *Metallurgical thermochemistry* [in Russian] (Moscow: Metallurgy)
[20] Novikov V K and Nevidimov V N 2006 *Polymer nature of molten slag* [in Russian] (Ekaterinburg: UPI)
[21] Chaurand P, Rose J, Briois V, Salome M, Proux O and Nassif 2007 *J. Phys. Chem.* **111** 1501–10
[22] Song W, Li H, Zhang Y and Wang Y 2014 *Yejin Fenxi Metallurgical Analysis* **34** 27–31
[23] Zhdanov P A, Seregina I F, Osipov K B, Bolshov M A, Skryleva E A, Volkov A I and Seregin A N 2016 *Industrial laboratory. Diagnostics of materials* [in Russian] **82** 313-18
[24] Duchesnel M A, Nakano J, Hu Y, MacLennan A, Hughes R W, Bennett J and Nakano A 2016 *Proceedings of The 10th International Conference on Molten Slags, Fluxes and Salts (MOLTEN16)* (The Minerals, Metals & Materials Society) 1405–12
[25] Wang L, Teng L, Chou K-C and Seetharaman S 2013 *Metall. Mater. Trans.* **44** 948–53

IOP Conf. Series: Materials Science and Engineering **966** (2020) 012017 doi:10.1088/1757-899X/966/1/012017

Technology for fixing mine workings with friction anchors of the SZA type with a new type of anticorrosive coating

P V Volkov[1, 2], R V Kulsaitov[1] and A M Magitov[1]

[1] Department of Mineral Deposits, Nosov Magnitogorsk State Technical University, 39, Lenina av., Magnitogorsk, 455000, Russia

E-mail: [2]wolf1709@rambler.ru

Abstract. The issues of ensuring the stability of mine workings in weak rocks due to the use of combined roof support based on anchors with a new type of anti-corrosion coating on the example of mining enterprises of the Russian Federation are considered. As a result of the work, the quality indicators of man-made mine waters of mining enterprises of the Russian Federation were determined, the conditions of the mine atmosphere that contribute to the intensification of metal corrosion were identified, an experiment was conducted in the research laboratory to determine the corrosion resistance of self-fixing anchors, the most suitable anchor coatings for their use in aggressive mine environments were identified, The technology of fixing with anchors with a new type of anti-corrosion coating is proposed. Also shown are the conditions for the use of galvanized, paint and varnish coatings, as well as various protective materials, the scope of their application, the advantages, and disadvantages of these coatings in mine environments are indicated. The prospects for the use of new types of coating are shown and directions for further research in this direction are indicated.

1. Introduction

At present, in the underground mines of the Russian Federation, opening vertical as well as horizontal mining and rifling workings are carried out on unstable rocks and ores. Tunneling in such conditions is associated with the presence of metasomatic sites, ore masses with sulfur occurrences, and waters of high mineralization, which negatively affect the durability and reliability of metal structures and rapidly reduce their service life. In such areas, it is recommended to use anchor supports only with a corrosion-resistant coating.

2. Relevance, scientific significance of the issue with a brief overview of references

Self-fastening anchor support (SZA) is a widespread use when fixing mine workings in underground mines – this is a type of shaft support fixed in holes and boreholes due to frictional interaction. SZA is intended for fastening the roof and sides of underground mine workings of ore and non-metallic deposits, including those developed by the open-pit method, for fastening the sides and slopes of open pit mines. SZA is allowed to be used both independently and as part of a reinforced combined lining.

The use of combined fastening methods based on anchors in conjunction with shotcrete in rocks of low stability in domestic and foreign mines is becoming more and more popular.

In this regard, the results of studies conducted at the mining enterprises of Australia and Canada [7–9], which have experience in attaching workings using fiber-shotcrete, as well as studies [10] by scientists of the Russian Federation, are of great interest[3].

IOP Conf. Series: Materials Science and Engineering **966** (2020) 012017 doi:10.1088/1757-899X/966/1/012017

The main functions of the combined lining based on anchors and shotcrete are to form bonds between the lining and the rock, which creates a bearing rock beam, as well as holding pieces of rock between the anchors, while ensuring increased resistance of the massif to destruction and weakening from the effects of the mine atmosphere. The choice of parameters and construction of the lining is based on the analysis of several factors: cost, bearing capacity, operational readiness, expediency and laboriousness of the installation, load (static or dynamic), etc. [1–6].

Of the currently known achievements in the field of applying anti-corrosion coatings for objects operating in high humidity conditions, is the use of surface-active substances (surfactants), as well as inhibited coatings.

For anticorrosive protection of metal surfaces of objects operating in a humid environment, coatings with a gum composition, in particular, based on nairite NT, have spread.

The experience of the Ukrtsvetmetremont trust in protecting the metal and reinforced concrete structures with the gumming composition showed that their durability (under conditions of air humidity more than 75% and exposure to chlorine, sulfuric anhydride) increased by 3–5 times.

The experience of VO Soyuzkhimremoborudna on the protection against corrosion of metal structures using epoxy-coal compositions with high chemical resistance is interesting. The disadvantage of this method is that the start of operation of the structure can be made 7–10 days after coating.

Since it is necessary to ensure the preservation of the functional qualities of metal structures over the estimated life cycle, especially when used in a mine atmosphere, this is an urgent issue; during operation, the metal lining is in conditions most favorable for the occurrence and development of corrosion; annually up to 100 thousand tons of rolled metal is used for fastening mine workings; as a result of the influence of physical and chemical factors, about 10% of the metal is lost per year, as a result of which the metal lining loses its bearing capacity; self-fixing anchor support (SZA) has a core thickness of up to 3–4 mm, in connection with this, the issue of anticorrosion protection during their operation in mine conditions is especially acute [10–14].Therefore, the development of technology for fixing mine workings with friction anchors of the SZA type with a new type of anti-corrosion coating is an urgent task.

3. Problem statement

As mining operations develop, along with the deterioration of mining and geological conditions, we see the problem of stability of mining operations during the design life at minimal cost.

The experience gained during the use of self-fixing anchor supports in the mines of the Russian Federation indicates the need for protection the supports from the influence of aggressive environments of the mine atmosphere and the main factor – corrosive groundwater.

The purpose of this work is to find ways to corrosion protection of self-fixing anchor lining.

At a visual examination:

1) the location of the lining was analyzed relative to the factors of an aggressive environment - casing and washing of the lining structure with aggressive mine waters;

2) an overview of the design of anchors;

3) studied and analyzed design materials for the use of lining - passports for fixing mine workings with self-fixing anchor lining.

During visual examinations, the strength of the supporting structures and their water permeability are assessed by a thorough inspection with fixing in the log of observations, delaminations, cracks, caverns, traces of filtration and corrosion damage.

Measurement of metal corrosion is carried out by electrical methods, by weighing a corrosion film taken from a unit area of a metal, or by weighing a metal purified from corrosion. Also, to determine the corrosion of the metal supports in places, a device is used to measure the thickness of the metal, cleaned of the corrosion layer, which is produced using a dial indicator with an accuracy of 0.01 mm.

The main problems of using metal anchor support in mine conditions:

- 90–100% humidity;

IOP Conf. Series: Materials Science and Engineering **966** (2020) 012017 doi:10.1088/1757-899X/966/1/012017

- constant temperature 10–15°C;
- the presence in the soil and in mine water of lithoautotrophic– iron-oxidizing bacteria;
- wandering currents;
- low pH;
- mineralization of the liquid phase.

4. Theory

The hydrogeological conditions of the Ural deposits are characterized by the presence of groundwater of alluvial-deluvial deposits, weathering crust and rock zone. The water of the weathering crust and rock is hydraulically interconnected to form a single aquifer. The depth of active groundwater circulation is 80–150 m in tectonic zones, along contacts with dikes – up to 250 m. The water cut of the field is characterized by significant unevenness. Hydrogeological conditions for the development of deposits are classified as medium complexity. Pore-fractured (to depths of 6–70 m), fractured (to depths of 100–120 m), fractured-vein waters of tectonic fault zones (at depths below 120 m, up to 350–400 m) are developed in the deposits. The rate of filtration of fractured waters is 0.14–0.58 m/day.

According to the code of rules SP.28.13330.2012, insulation coatings for structures in soils must meet the requirements of Russian State Standard GOST 9.602 for medium-aggressive and highly aggressive environments, the following anti-corrosion coatings should be used:

• thermal diffusion zinc coatings with a thickness of 45–60 microns with overlapping with paint coatings of groups 2 and 3;

• hot zinc coatings, 60–100 microns thick with overlappings with paint and varnish coatings of groups 2 and 3;

• gas-thermal zinc or aluminum coatings, 120–160 microns thick, with overlappings by paint and varnish coatings of 2,3,4 groups;

• chemically resistant paint coatings of 2,3,4 groups;

• gas-thermal zinc coatings, 250–300 microns thick;

• insulation coatings in conjunction with electrical protection;

• electrochemical protection in liquid media and bottom soils.

In the research laboratory, the experiment was carried out in two stages:

Stage I included surface preparation before applying an anti-corrosion coating.

At the second stage of research, the corrosion resistance of self-fixing anchors, with different types of protection, when exposed to aggressive environments was determined.

5. Practical significance, offers and introduction results

Laboratory studies have shown that an innovative polymer coating applied using the new technology can provide reliable protection of the anchor fastener against corrosion for more than 10 years, which reduces the cost of maintaining the workings in a stable state.

Today, large-scale pilot-industrial tests of the first samples of polymer-coated anchor fasteners have been carried out at the enterprises of Apatit JSC, UGMK-Holding LLC, and PF MMC Norilsk Nickel, positive results have been obtained that allow us to carry out work on amendments to the regulatory – technical documentation.

6. Conclusions

As a result of the data obtained, it can be concluded that for acid mine waters, where the pH is below 4.5–5, a galvanized surface cannot be used, there is a strong etching of iron and the loss of metal mass during the month is 4.04%. This is because zinc in this case intensifies the process of iron etching due to the electrochemical action of the Zn-Fe galvanic pair.

Also, SNiP 2.03.11–85 stipulates that for medium and highly aggressive environments a galvanized coating.

Positive results were shown in coatings: epoxy and epoxypolyester. They withstand acidic basement and mine water, the loss of metal mass in two months of the experiment is 1.07%–1.37 % and 0.03 % for mine water when exposed to basement water, as well as coatings based on low-pressure polyethylene.

The cost of fixing 1 meter was made for the following types of supports: arch metal lining from the special profile of SVP; reinforced composite lining of reinforced concrete anchors and spray concrete reinforced with a metal mesh; reinforced combination support using self-fixing anchors with a new type of corrosion protection.

In the calculations, the use of the following equipment was taken into account: drilling holes – Boomer 282 drilling rig; UKK grid hinge and installation of SVP – PDM Toro – 301; spraying concrete – Aliva.

Calculations have shown that, compared with the technology used for fastening arch metal supports from the SVP special profile (cost 64.212 rubles/m^2) and reinforced combined support from reinforced concrete anchors and spray concrete reinforced with metal mesh (cost 36.630 rubles/m^2) reinforced combination lining with the use of self-anchoring anchors (cost 32622 rub/lm) will reduce the cost of fixing 1 lm output by 47% and 12%, respectively. The results obtained show the possibility of reducing the cost of securing workings in rocks below the average stability category. Reinforced combined lining with the use of self-fixing anchors is technologically and economically feasible in rocks of medium and lower stability category, and also allows you to effectively use this type of lining in order to reduce water manifestations in underground mine workings, as well as in case of hardening of unstable rocks.

References

[1] *Guidelines for the design of underground mine workings and the calculation of lining* 1983 (Moscow: Stroytekhizdat)

[2] Karetnikov V N, Kleimenov V B and Nordihin A G 1989 *Fastening capital and preparatory workings, reference* (Moscow: Nedra)

[3] Neugomonov S S, Volkov P V and Zhirnov A A 2018 Mounting Low Stability Rocks with Reinforced Combined Support on the Basis of Friction SZA Anchors *Gorny Zhurnal (Mining Journal)* **2** pp 31–4

[4] All-Union Building Standards 126-90 (Ministry of Transport of the USSR) Fastening workings with spray concrete and anchors in the construction of transport tunnels and subways. Standards for the design and production of work

[5] Masayev Yu A, Masayev V Yu and Filina L D 2015 Recent Developments in Area of Mounting and Enhancing Stability of Rock Outcrops in Mine Workings *Bulletin of KuzGTU* 1 41–44

[6] Zubkov A A, Zubkov A V, Kutlubayev I M and Latkin V V 2016 Enhancing Structure and Technology of Mounting Supports with Friction Fastening *Gorny Zhurnal (Mining Journal)* 5 pp 50–3

[7] Dyomin V F, Dyomina T V et al 2015 Evaluation of Stress-Deformed State of Rock Masses Around Mine Working Mounted with Anchor Support *Mining Information&Analytical Bulletin* **10** pp 70–6

[8] Zubkov A A, Latkin V V, Neugomonov S S and Volkov P V 2014 Advanced Methods of Mine Working Mounting at Underground Mines *Mining Information&Analytical Bulletin. Certain Articles (Special Issue)* **10** pp 106–17

[9] Kalmykov V N, Volkov P V and Latkin V V 2016 Analysis and Justification of Parameters of Resin-Grouted Anchor Support at Conduct of Pilot Testing under Conditions of Safyanovsky Underground Mine *Relevant Mining Issues* **2** pp 27–35

[10] TamilSelvi M, Thandavamoorthy T. S. 2015 Energy Absorption Characteristics of Steel, Polypropylene and Hybrid Fiber Reinforced Concrete Prisms *Int. J. of Earth Sci. and Eng.* **8** **2** 372-376

[11] Kalmykov V N, Meshcheryakov E Yu and Volkov P V 2011 Analysis and Justification of Parameters of Wellfield Module 'Cleaning Activities' at Development of Resources in

Border Areas of Quarries *Bulletin of Magnitogorsk Nosov State Technical University* **4** pp 5–8

[12] Kalmykov V N, Grigoryev V V, Volkov P V 2010 Searching for Options of Deposit Development Systems to Excavate Near Edge Zones at Combined Geotechnology *Bulletin of Magnitogorsk Nosov State Г И Technical University* **1** pp 17–22

[13] Yeremenko V A, Razumov Ye A, Zayatdinov D F 2012 Modern Technologies of Anchorage *Mining Information&Analytical Bulletin* **12** pp 38–45

[14] Kalmykov V N, Gibadullin Z P, Zubkov A A, Neugomonov S S, Volkov P V and Pushkarev Ye I 2013 Development of Technology of Mechanized Mounting of Mine Workings Using 'Liquid' Shotcrete at Underground Mines of JSC 'Uchalinskiy MPP' *Mining Information&Analytical Bulletin* **54** pp 64–70

[15] Lushnikov V N, Yeremenko V A et al 2014 Mounting Mine Workings Under Deformed and Bump Hazardous Rock Masses *Gorny Zhurnal (Mining Journal)* **4** pp 37–44

[16] Gibadullin Z R and Volkov P V 2009 Method of Evaluation of Options for Ore Transportation at Mining of Marginal Resources *Bulletin of Magnitogorsk Nosov State Technical University* **3** pp 11–3

[17] Nguyen Viet Dinh 2014 Defining Stability of Mine Workings with Combined Support on the Basis of Numerical Simulation *Mining Information&Analytical Bulletin* **1** pp 325–9

[18] Kalmykov V N, Volkov P V and Meshcheryakov E Yu 2009 Development of Integrated Processing Schemes of Intensive Deposit Development of Near Edge Quarry Areas *Combined Geotechnology: Integrated Development and Preservation of Subsoil Riches of Yekaterinburg, June 22–26, 2009. Materials of International Scientific and Technical Conference: Collection of Papers. 2009, Publishing House: MSTU* pp 31–3

[19] Kalmykov V N, Neugomonov S S and Volkov P V 2019 Features of Fastening Very Unstable Rocks with Combined Support on the Basis of SZA and Shotcrete. *IOP Conf. Series: Earth and Environmental Science* **272** 022108

[20] Kalmykov V N, Latkin V V, Zubkov A A, Neugomonov S S and Volkov P V 2015 Technological Peculiarities of Building Up Reinforced Combined Support at Underground Mines *Mining Information&Analytical Bulletin* 4 (Special Issue **15**) pp 63–9

[21] Kalmykov V N, Kotik M V and Volkov P V 2019 Evaluation of the Stressed State of the Rock Massif and the Renovated Shaft Support of the Skipova Shaft of Sibay Branch of Uchalinsky Ore Mining and Processing Enterprise *IOP Conf. Series: Earth and Environmental Science* **272** 032070

IOP Conf. Series: Materials Science and Engineering **966** (2020) 012018 doi:10.1088/1757-899X/966/1/012018

Kinetics of vanadium reduction from slags by solid carbon

A V Klimov

Institute of New Materials and Technologies, Ural Federal University named after the First President of Russia B N Yeltsin, 19, Mira str., Yekaterinburg, 620000, Russia

E-mail: a.v.klimov@urfu.ru

Abstract. The laboratory investigation results of the vanadium reduction are presented. Experiments were performed with $CaO–SiO_2–Al_2O_3–MgO–V_2O_3$ slags in an air atmosphere. The solid carbon was used as a reducing agent. The reduction kinetics was studied by sampling. It has been shown that reduction proceeds to VC carbide. The process rate is inversely proportional to the slag viscosity. The formation of a carbide film on the carbon surface leads to a sudden, by an order of magnitude, reduction in the rate. The process of vanadium reduction from oxide melts by solid carbon is electrochemical in nature and is limited by the stage of carbon anodic oxidation. The influence of slag viscosity on the process kinetics is explained by the fact that it is controlled by the growth rate and removal of gas bubbles. The appearance of the carbide film, apparently, transfers the reduction process into the intra diffusion mode, when the transfer of carbon and vanadium through the carbide film is limiting.

1. Introduction

To predict the progress of technological processes for the vanadium reduction it is necessary to have a physicochemical model of ones. From the physicochemical point of view, the question of the vanadium reduction kinetics from liquid slags by solid carbon-containing materials is practically studied insufficiently. So, the reduction of V_2O_5 and V_2O_3 from liquid slags with solid carbon was investigated by the gravimetric method [1, 2]. The experiments were carried out on slag (wt.%) 35CaO, 30SiO2, $10Al_2O_3$, 10MgO, $15CaF_2$. Within 15–20 minutes after the experiment started, the process slowed down significantly. A change in the V_2O_3 slag content affected the rate much more strongly than a change in the V_2O_5 content. The first order of reaction for V_2O_5 was found, which was confirmed in [3]. Although the authors themselves [1, 2] expressed the opinion that the reduction process is limited by the stage of carbon anodic oxidation, and therefore the concentration of vanadium is not included in the kinetic equation. Reduction of vanadium oxide by carbon in stainless steel slags with Fe and Cr oxides was investigated by the authors [4]. In general, the vanadium reduction process by solid carbon continues to remain insufficiently studied, in particular, the influence of the slag composition on the reduction rate has not been disclosed. Although an attempt was made to evaluate the effect of temperature and slag composition on the process rate [5]. Our investigations have revealed a number of evidences that make it possible to reason the mechanism of vanadium reduction from slag by solid carbon and to describe the rate of the process depending on the oxide melt viscosity.

2. Experimental procedure and materials

The initial compositions of the investigated slags are presented in Table 1. A certain amount of V_2O_3 (slags 1–4) or V_2O_5 (slags 5–8) was added to them. The experiments were carried out in graphite crucibles. A thoroughly mixed mixture of a given composition oxides a particle size of less than 0.1 mm was placed in the crucible. Part of the crucible was isolated from the slag by an inserted corundum tube, which ensured the constancy of the working surface during foaming of the slag. The furnace was heated at about 100° above the temperature set for the experiment. The temperature was controlled by the tungsten-rhenium thermocouple. The crucible with the mixture was heated in the upper part of the furnace and lowered into the working zone. In this case, it took two minutes to melt the slag and establish the required temperature. After that, sampling was started by freezing on a tungsten rod. Quenching of the sample was carried out under water.

Table 1. The initial compositions of the investigated slags (wt.%).

No.	CaO	SiO_2	Al_2O_3	MgO	t, ^0C
1	45	35	20	–	1550
2	50	30	20	–	1550
3	55	25	20	–	1550
4	35	45	20	–	1550
5	41	23	28	8	1500
6	44	24	24	8	1550
7	32.5	18	43.5	6	1600
8	43	–	51	6	1530

3. Results and discussion

Figure 1 shows that the V_2O_3 concentration changed linearly with time. In this case, two main parts are distinguished on the curve: in the initial period of the experiment, there is a strong decrease in the V_2O_3 slag concentration, and then the process sharply slows down so that a bend appears on the ($\%V_2O_3$) vs τ line. This type of dependence is possible if the reducing agent surface is blocked by the reaction products. Indeed, at the slag – crucible interface, a vanadium carbide film was formed, mainly consisting, as shown by X-ray diffraction studies, of VC carbide. As can be seen from the Table 2 the appearance of a carbide film significantly reduces the vanadium reduction rate. The presence of one linear part for slag 8 is explained by a low reduction rate. The amount of vanadium which was reduced during the experiment to carbide was insufficient to completely close the crucible reaction surface.

A comparison of the process rates in experiments 1–4 with ones in experiments 5–8 shows that they are practically independent of the added vanadium form and the initial vanadium concentration in the slag. Apparently, under conditions of experience at high temperatures, the transition of vanadium to the lower valence states was in time. Another reason for this fact may be that the limiting stage of the process is not related to the discharge of vanadium ions.

To identify the rate dependence on the slag composition, according to reference data [6, 7], the viscosity of the slag used in the experiments was estimated. The influence of V_2O_3 and V_2O_5 additives on slag viscosity was neglected. Figure 2 shows the dependence of the vanadium rate in both sections on viscosity in the coordinates υ_v, $1/\mu$. It turned out that the reduction rate in the first linear part is inversely proportional to viscosity, and in the second linear part after the bend, it is independent of viscosity. The absence of a connection between the reduction rate and the dynamic slag characteristics indicates that in the latter case the process is limited by reasons not related to slag, in particular, probably, with the diffusion of carbon and vanadium through carbide phase.

The dependence of the reduction rates on the slag viscosity in the initial sections is related to the anodic oxidation of carbon as a limiting stage. As was shown by us in [8], the rate of anodic oxidation of solid carbon in viscous slag can be determined by the rate of growth and removal of gas bubbles, which is inversely proportional to the melt viscosity.

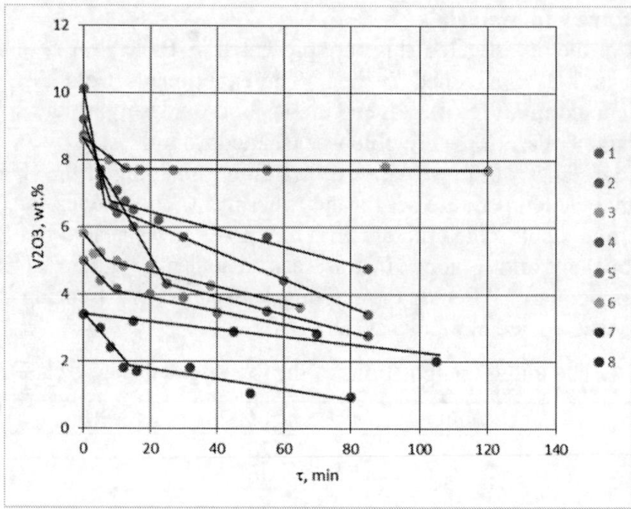

Figure 1. V_2O_3 concentration change over time during reduction by crucible carbon (line numbers correspond to the numbers of slags).

Table 2. Vanadium reduction rates, mol/(cm²·s).

No.	$\upsilon \cdot 10^8$ for the first part	$\upsilon \cdot 10^8$ for the second part
1	37.3	1.38
2	14.7	7.45
3	87.2	0
4	58.5	6.94
5	82.2	7.79
6	57.0	8.65
7	56.5	5.92
8	1.14	–

The presented results differ significantly from the case when vanadium is reduced simultaneously by an iron-carbon melt and crucible carbon. As shown in Figure 3, in this case, a line bend associated with carbide formation is not observed up to a deep (90–95%) reduction of vanadium. Concentration curves quietly reach equilibrium. These facts can be explained if we assume that the reduction of vanadium from slag by solid carbon is limited by the anode half reaction

$$C + O^{2-} - 2e = CO \quad (1)$$

In the metal melt presence in the system, this reaction goes at the metal-slag interface, or carbon oxidation goes at the metal-graphite interface, so blocking of the crucible surface by carbides does not affect the process rate.

The initial reduction rate of vanadium by a saturated iron-carbon melt in terms of a mole of vanadium is approximately twice as high as that of solid carbon.

When expressing the viscosity μ in Pa·s, the equation of the rate, expressed in mol/(cm²·s), has the form

$$v = \left(\frac{28{,}879}{\mu} - 27{,}572\right) 10^{-8} \quad (2)$$

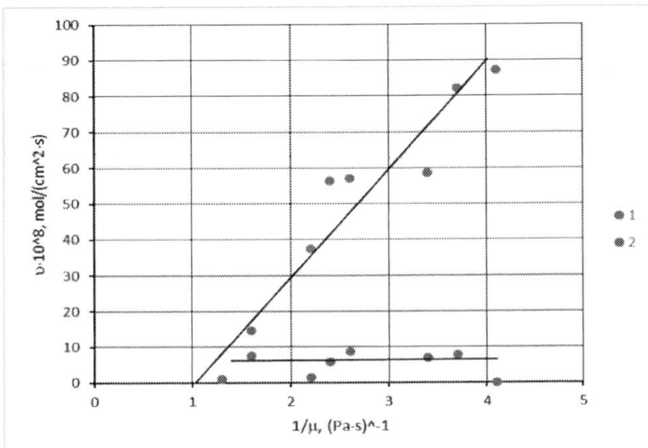

Figure 2. Dependence of the vanadium reduction rate on slag viscosity.

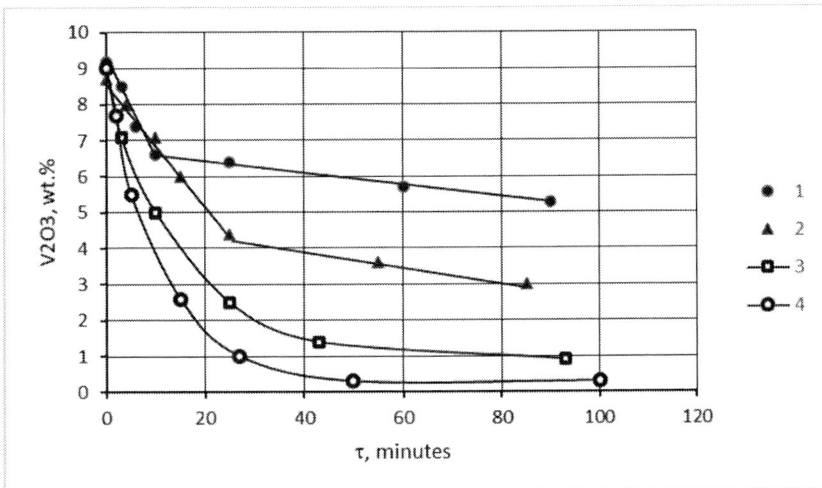

Figure 3. Change in V_2O_3 concentration with time in slags (wt.%) 35 CaO, 45 SiO_2, 20 Al_2O_3 (1,3) and 50 CaO, 30 SiO_2, 20 Al_2O_3 (2,4) with the V_2O_3 addition in graphite crucibles without (1, 2) and in the presence of metal (3, 4).

4. Conclusions

Thus, the presented results make it possible to discuss in more detail the mechanism of vanadium reduction from liquid slags by solid carbon, and also indicate the decisive role of slag viscosity in the process kinetics.The latter is explained by the fact that the process is controlled by the growth and removal rate of gas bubbles during anodic oxidation of carbon.

References

[1] Lyamkin S A 1973 *Kinetics of joint metals reduction from slag by solid carbon* Ph.D dissertation [in Russian] Ural Polytechnic Institute, Yekaterinburg.
[2] Boronenkov V N 1974 Kinetics of metal reduction from oxide melts, Ph.D dissertation [in Russian] Ural Polytechnic Institute, Yekaterinburg.
[3] Vostryakov A A, Lepinskykh B M and Neuymin A D 1971 *Transactions of IMET UFAN USSR* [in Russian] (Sverdlovsk: UFAN RAN) **25** p 134
[4] Mantha Divakar, Marten Görnerup and Ashok Kumar Lahir 2001 *Steel Research* **72** **2** pp 40–3

[5] Cai B, Hu B, Zhao H, Guo H and Liu J 2012 *CNKI Journal Hebei Metallurgy*
[6] Voskoboinikov V G, Dunaev N E et al. 1975 Pr*operties of liquid blast furnace slags: reference book* [in Russian] (Moscow: Metallurgy)
[7] Mazurin O V, Streltsina M V and Shvaiko-Shvaikovskaya T P 1977 *Properties of glasses and glass-forming melts: Reference book* [in Russian] (Leningrad: Nauka).
[8] Klimov A V 2011 *Transactions of XIII Russian Conference 'Structure and Properties of Metallic and Slag Melts. Vol. 3'* (Ekaterinburg: Ural Branch of the Russian Academy of Sciences) p 44

IOP Conf. Series: Materials Science and Engineering **966** (2020) 012019 doi:10.1088/1757-899X/966/1/012019

Modelling of the thermal state and the melting loss of a graphite electrode in the conditions of the evaporative cooling in the arc furnace

M V Bystrov[1], I M Yachikov[2] and I V Portnova[2, 3]

[1] Public Joint-Stock Company 'Machine Building Plant im. M I Kalinin', 18, Kosmonavtov avenue, Yekaterinburg, 620017, Russia
[2] Nosov Magnitogorsk State Technical University, 39, Lenina av., Magnitogorsk, 455000, Russia

E-mail: [3] otvetsek@magtu.ru

Abstract. The cost of the graphite electrodes amounts greater proportion in the final value of the finished product. This calls for finding opportunities to reduce the maintenance costs of AC and DC furnaces used in foundries. The aim of the study is the creation of a mathematical model and the analysis of the efficiency of the evaporative cooling method in reducing the amount of the graphite electrodes used in small capacity arc furnaces. The mathematical model and the computer program allow determining the thermal state of the electrode and the melting loss of the graphite, the model is developed for a primarily cylindrical electrode and takes into account: the current passing through an electrode, the time of the electrode being in the electrified furnace, and the parameters of the water evaporative cooling. The paper presents the received data about the melting loss of the graphite per ton of steel against the water consumption during the evaporative cooling. The graphite consumption in the arc furnaces of 10–12 tons' capacity is reduced by 40 %. The recommended water consumption for the evaporative cooling of electrodes for all types of foundry furnaces is 0.1–0.2 m^3/h.

1. Introduction

When melting metal in electric arc furnaces, some of the most expensive consumable materials are the graphite electrodes (GE). The expenditures on them amount up to 30 % of the total cost of the finished metal output, while their share in the prime cost of the foundry alloys is higher than the share of the consumed electric energy [1]. Besides, the price policy in the sphere of consumable materials and the finished products is under the influence of the world economic situation [2]. All this requires a search for the ways of reducing maintenance costs on graphite electrodes.

The main factors, determining the rate of consumption of graphite electrodes, are [3, 4]:
- the oxidation of the side and end surfaces (37 %);
- the thermal and mechanical fracture (35 %);
- the oxidation of the end parts (7 %);
- the field emission (10 %);
- the unsafe structures of the electrode clamps and nipples (3 %);
- the bending and impact loads (5 %);
- the graphite fracture caused by the liquid metal bath (3 %).

It is obvious, that one of the factors significantly influencing the amount of used graphite electrodes is their thermal and mechanical fracture and the subsequent oxidation of the side and end surfaces, determined by the electrode temperature and the availability of cooling.

There are various methods and ways of contributing to the prolongation of an electrode life: the strict limits in the use of the technological modes (working should be carried out with regard to the rating power at less current consumption, with an increased demand for making the charging basket and so on); putting a protective coating on the side surface of an electrode. There are several ways of reducing the destructive thermal impact on an electrode. Some of them are the use of a hollow graphite electrode (the inert gas supply through the axial passage results in the electrode cooling and additional arc stabilization), spraying of the coolant (water) under the electrode clamp onto the side surface of graphite electrodes [2, 5–7].

The drawbacks of the methods of putting protective coatings are: the high cost of the applied coating materials, the necessity of the additional technological operations connected with the coating process. Besides, the coating is not durable enough and requires new covering after every casting operation. The paper [8] demonstrates that the consumption of graphite electrodes mainly depends on the uneven thermal load during a casting operation. To reduce the temperature of the graphite electrodes, a forced cooling can be used, such as feeding an inert gas through the central axial passage of an electrode [2, 9]. It should be noted, that as the mathematical modelling points to the ineffective use of gas cooling through the axial passage, which allows removing only an insufficient amount of heat from the electrode, as the gas has a small heat capacity and quickly achieves the temperature comparable with the temperature of the electrode itself [9].

There is a well-known way of reducing the used amount of the electrodes by evaporative cooling. This method is considered to be the most promising and is used in powerful and UHP electric arc furnaces both in Russia and abroad [8–10]. Nowadays, it is mainly applied in large capacity steel making electric furnaces. For instance, in Russia the described system of the evaporative cooling is used in EAF-180, while in Byelorussia it is applied in EAFs-100 [10].

The system of evaporative cooling consists of a spray ring, fixed under the electrode clamp, to spray water onto its side surface (Figure 1). The water control takes place by means of a pressure regulator and valves. Water can be continuously sprayed up or down onto the outer periphery of a graphite electrode at 10–35° angle with regard to the level [11]. The system of electrode's water cooling can be fixed under the coupling band of the electrode clamp [10]. Our observations point out that the spray rings reduce the used amount of electrodes and contribute to the increase of the service life of the electrode clamp and the surrounding isolation [12].

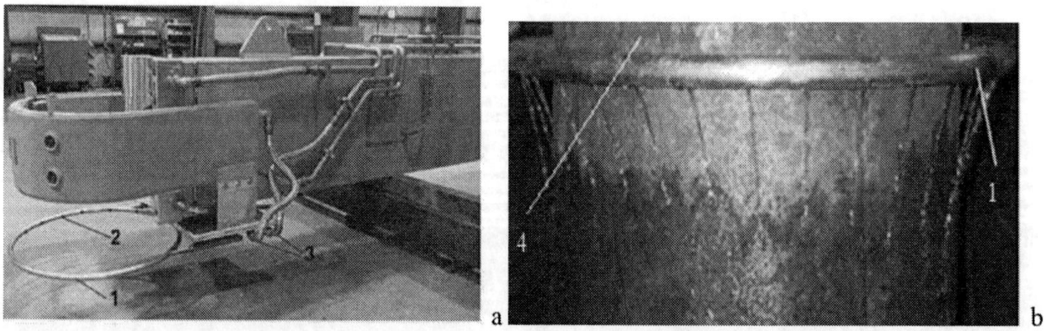

Figure 1. The design of the evaporative cooling system: a – a general view [15]; b – a spray ring around a graphite electrode; 1 – a ring around a graphite electrode; 2 – water spray nozzles for the water supply to an electrode; 3 – tubes for the supply and discharge of water used for the evaporative cooling; 4 – a graphite electrode.

The paper raises the issue of the effectiveness of the application of the forced evaporative cooling in small capacity electric arc furnaces. They are widely used in iron and steel founding and have an integrated work-cycle, including both the oxidation and the reduction periods. In doing so, there appear some problems connected with the choice of an efficient design, parameters of the evaporative cooling system and the valuation of its economic efficiency. In many ways, the solutions of these problems are determined by the knowledge of the parameters of cooling: the appropriate pressure and water consumption, the speed of water flow down the electrode, the time of switching the water supply on when the electrodes are still cold, etc.

The aim of the study is a mathematical model development aimed at evaluating the efficiency of the evaporative cooling in terms of reducing the amount of used graphite electrodes in small capacity electric arc furnaces (EAF).

2. The mathematical model of the thermal state and the melting loss of a graphite electrode

The mathematical model of the thermal state of an electrode with the diameter $D = 2R$ and the height H includes the heat equation in the cylindrical coordinate system (regarding the axial symmetry and auto thermal heat supply sources) [9]

$$\frac{\partial t}{\partial \tau} = \frac{1}{c \cdot \gamma}\left[\frac{\partial}{\partial r}\left(\lambda \frac{\partial t}{\partial r}\right) + \frac{\lambda}{r}\frac{\partial t}{\partial r} + \frac{\partial}{\partial z}\left(\lambda \frac{\partial t}{\partial z}\right)\right] + \frac{q_v}{c \cdot \gamma} \tag{1}$$

where $0 < r < R$, $0 < z < H$; c, γ is the thermal capacity and the density of the electrode material respectively; $q_v = 16 \cdot I_a^2 \rho / (\pi^2 \cdot D^4)$ is the volume power of auto thermal heat supply sources in a voltage carrying electrode; λ, ρ, c, γ are the heat conduction coefficient, the electric resistivity, the thermal capacity and the density of electrode material respectively; I_a is the arc current (the root-mean-square value for the alternative current).

The boundary problem involves a number of boundary conditions. The preliminary conditions are $t(r,z) = t_0$. Firstly, the evaluation of the heat flux from the arc at the electrode tip is carried out. Because the electric arc is more often at the end of the electrode tip, the heat flux q_a receives a higher value and is determined by the parabolic law, $q_a(r) = \eta_a \bar{q} + 3 \cdot (1-\eta_a)\bar{q}(r/R)^2$, where $\bar{q} = 4 \cdot I_a U_e / (\pi \cdot D^2)$ – is an average of the heat flux arriving at the electrode tip from the electric arc; U_e – is the near-electrode voltage drop; $\eta_a \approx 0.3$ is the fraction of the heat capacity of the arc, generated in the middle of the electrode tip.

We also set the third class boundary conditions of the heat exchange with the surrounding furnace atmosphere due to the convection q_{conv} and the heat exchange with the working wall surfaces of the furnace caused by the radiation q_{rad} (the water cooling is not applied) as well as the third class boundary conditions of the heat exchange with the cooling water, fed to the side surface of the electrode.

The heat-mass exchange in water moving along the side surface of an electrode is rather complicated, that is why a number of assumptions are taken into account:

- the gas is not saturated, being away from the liquid, at the same time there appears a substance flow, directed away from the surface where the evaporation takes place;
- the heat flux from the electrode surface and the surrounding atmosphere is directed to the liquid;
- the temperature of the water heat is not taken into account until it reaches the saturation point;
- all the heat absorbed by the liquid is spent on its evaporation and releases with the vapour (adiabatic process of evaporation);
- the water flows along the vapour film with the constant speed W_0;

The heat-balance equation is written regarding these assumptions,

$$\alpha_s(t_w - t_s) + q_{env} = \psi J \tag{2}$$

where q_{env} is the heat flux of radiation and convection on the surface of water by the furnace; ψ is the heat of water vaporization; J is the mass of the evaporating water per surface unit; α_s is the surface heat transfer coefficient; t_w is the water temperature; t_s is the temperature of the side surface of the electrode.

The depth of the water film δ changes along the electrode due to the evaporation,

$$\frac{d\delta(y)}{d\tau} = -\frac{J(y)}{\gamma_w} \tag{3}$$

where $J(y)$ is the steam- mass flow along the electrode; γ_w is the water density; τ is the time when the water is in the furnace; $y = H - z$.

To solve the differential equation (3), a boundary condition is added, $\delta|_{y=0} = \delta_0$, where $\delta_0 = Q_w / (\pi D \cdot W_0)$ is the depth of the water film in the area of its feeding with a volumetric water discharge Q_w. Knowing the thermal state of an electrode one can determine the linear velocity of the graphite melting loss,

$$\upsilon(T) = \frac{V(T)}{3600 \cdot \gamma_g}, \text{ m/c} \tag{4}$$

where $V(T)$ is the velocity of graphite melting loss due to the temperature, γ_g is the graphite density.

To calculate the thermal melting loss of the graphite electrode, the experimental dependence of the speed of the graphite melting loss (due to the oxidation and thermo-mechanical fracturing) on its temperature in the open air $V(T)$ is used [5, 13]. Regarding the empirical data, an approximation equation is received,

$$V(T) = \xi \cdot 10^{a \lg(T)^2 + b \lg(T) + c}, \text{ kg/(m}^2\text{ h)} \tag{5}$$

where $a = 4.9627; b = -23.836; c = 27.167$; T is the temperature, K, $\xi = 0.3 - 1$ is the parameter determining the resistance of the graphite electrode to the thermal loss.

The mass melting loss of graphite from the side surface of the electrode ($r = R$),

$$Q_b = \frac{dm}{d\tau} = \int_F V(T) dF = 2\pi R \int_0^H V(T) dz, \text{ kg/h.}$$

By analogy, the mass melting loss of graphite from the lower tip of an electrode ($z = 0$),

$$Q_t = \frac{dm}{d\tau} = \int_F V(T) dF = 2\pi \int_0^R V(T) r dr, \text{ kg/h.}$$

The total velocity of graphite melting loss Q_Σ and the mass loss Δm at a time τ_K from one graphite electrode, $Q_\Sigma = Q_b + Q_t$, $\Delta m = \int_0^{\tau_k} V_\Sigma d\tau$.

The energy emitted in the arc of the DC EAFs is determined as $P_\partial = I_\partial \cdot U_\partial$, while the energy in the three arcs of three phase AC EAFs is determined as $P_a = 3 \cdot I_a^* U_a^*$, where I_a = 16 kA, the ampere rating of an arc of the DC EAF; I_a^* is the active ampere rating of the arc of three phase AC EAF. An approximate ratio of currents passing through the graphite electrodes of tree phase AC EAF and DC EAF $I_a^* = 1.2 I_a / 3$. Is calculated regarding the same energy of the furnaces under consideration.

A computer program 'Modelling of the thermal state of an electrode of EAF' is developed to provide the calculation under the given mathematical model. It allows determining the thermal state and the graphite melting loss for the preliminary cylindrical electrode in dependence on the current passing through the electrode and the time of its being in the electrified furnace as well as the parameters of the water cooling.

3. The results of computer modelling

The foundry engineering uses arc furnaces with the capacity of 0.5–25 tons. To carry out the computer modelling, the performance specifications of EAF, used for shaped casting with the capacitance from 0.5 to 25 tons are taken from the work [14]. The following performance specifications of electric arc furnaces are used in the presented calculations: t_0 = 20 °C; the near-electrode voltage drop in the arc is 8 V; the emissivity factor of lining is 0.93; it working temperature is 1400 °C; the temperature of gases in the furnace is 1200 °C; the duration of heat is 1 hour. The thermophysical properties of a graphite electrode are: c = 2.1 kJ/(kg·K); γ = 1700 kg/m³; the emissivity factor of the graphite is 0.71; ρ = 5 microOhm·sm; λ = 120 W/(m·°C). The parameters of the water, cooling an electrode are: the preliminary temperature is 20 °C; ψ = 2.3 MJ/kg; the emissivity factor of the water is 0.5; γ_w = 1000 kg/m³; W_0 = 0.3 m/s; the evaporative cooling time-on after inserting e cold electrodes in a furnace is 60 seconds.

Figure 2 shows the received dependences of the discharge intensity of graphite per ton of steel on the consumption of water, fed for the evaporative cooling in EAFs with the capacitance from 0.5 to 25 tons. It is established that for almost all furnaces there is a significant decrease in the melting loss of graphite electrodes. It is clear that the less the furnace capacitance, the higher the melting loss of graphite electrodes. If the water consumption is over 0.2 m³/h the melting loss of electrodes decreases insufficiently.

Figure 3 shows the dependence of the graphite consumption on the furnace capacitance, provided the presence or absence of evaporative cooling with the water consumption of 0.1 m³/h. The decrease of graphite loss for the furnaces with the capacitance of 0.5–6 tons averages 28 %, for the furnaces with the capacitance of 10–12 tons it amounts to 40 %, for the furnaces of 25 tons it is 11.5 %. It is established that the water supply to the system of evaporative cooling is mostly efficient after current switching in the 3–5 minutes interval.

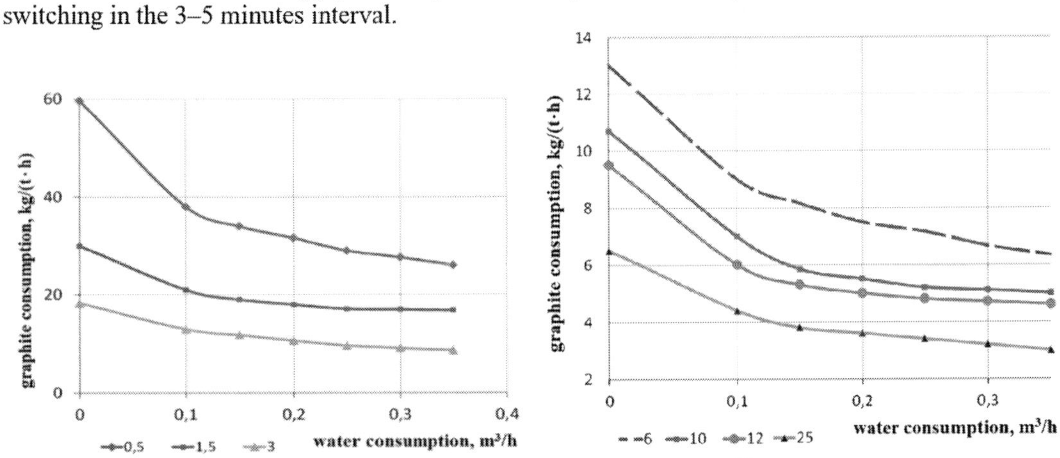

a) b)

Figure 2. The dependence of the graphite consumption per ton of steel on the consumption of water, fed for evaporative cooling
a – for EAFs with the capacitance from 0.5 to 3 tons; b – for EAFs with the capacitance from 6 to 12 tons.

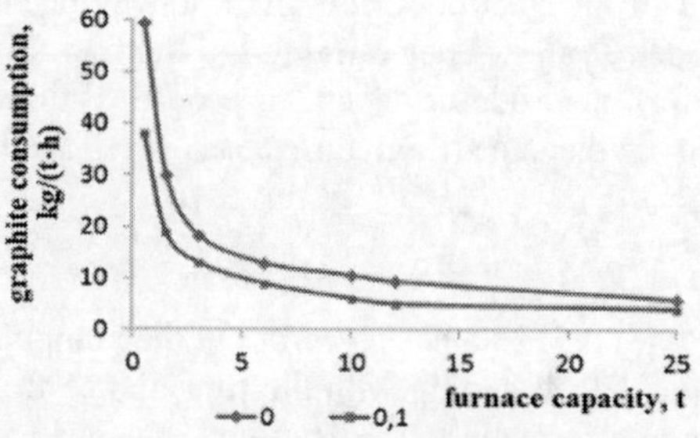

Figure 3. The dependence of the graphite consumption on the furnace capacitance: 1 – with the use of cooling, 2 – with the use of evaporative cooling at the water rate of 0.1 m^3/h.

4. Conclusion

1. A mathematical model of the thermal state and the melting loss of graphite electrodes has been created for the working at the nominal current of three phase AC EAFs and DC EAFs of small capacitance.

2. It is estimated that the use of the evaporative cooling in arc electric furnaces with the capacity of 0.5–25 tons decreases the temperature of the graphite electrode mostly in its upper area, which results in the less amount of the used graphite electrodes due to the less oxidation of their side surface and the thermo-mechanical fracture.

3. The use of the evaporative cooling of graphite electrodes can be recommended for reducing their consumption both in the existing and planned electric arc furnaces with the capacity of 3–12 tons. The recommended water consumption in cooling electrodes for all types of foundry furnaces should amount to 0.1–0.2 m^3/h.

References

[1] Grudnickij O M, Ishakov R A-R and Korobov V K 2011 *Ways to reduce the specific consumption of graphitized electrodes on electric steelmaking furnaces* (Lit'e i metallurgija) no 1 pp 100–1

[2] Yachikov I M, Portnova I V and Bystrov M V 2019 *Mathematical and software systems in the industrial and social fields* vol **7** no **1** pp 24–9

[3] Brusa E 2014 Thermomechanical coupling effect of graphite electrodes upon the electric arc furnace dynamics *Meccanica* vol **49** pp 2979–90

[4] Yachikov I M, Portnova I V and Bystrov M V 2019 *Materials Science Forum* vol **946** pp 444–9

[5] Schwabe W E 1971 *The Mechanics of Consumption of Graphite Electrodes in Electric Steel Furnaces* (Electric Furnace Proceedings) pp 140–5

[6] Dyskina B Sh and Kabanova T V 2014 *Advances in Chemistry and Chemical Technology* vol **10** pp 39–41

[7] Stopar K, Kovačič M, Kitak P and Pihler J 2017 *Materials and Manufacturing Processes* vol **32** pp 1189–200

[8] Yachikov I M, Portnova I V and Bystrov M V 2018 *Proceedings of the Fifteenth Congress of Steelmakers* (Yekaterinburg) pp 180–6

[9] Filippov V V, Matochkin V A, Tishchenko V A and Gulyaev M P 2002 *Casting and metallurgy* no **1** pp 80–3

[10] Jachikov I M, Portnova I V and Bystrov M V 2018 *Modern scientific achievements of metallurgical heat engineering and their implementation in industry, Sat. doc. II International scientific-practical. conf.* pp 203–8

[11] Nakamoto Y 1990 Patent 4941149 US.

[12] Chakrabarti A K 2006 *Steel Making* (PHI Learning Pvt. Ltd)

[13] Okorokov N V 1971 *Arc Furnaces* (Moscow)

[14] Bogdanov S P 2006 *Calculation of arc steel-smelting furnaces* (St. Petersburg)

Application of hydrohammers of heavy class in the development of marble deposits in mining operations

I A Pytalev[1], D V Domozhirov[1, 2], N V Ugolnikov[1] and A A Prochorov[1]

[1] FGBOU VO 'Magnitogorsk state technical University named G I Nosov', 38, Lenin street, Magnitogorsk, 455000, Russia.

E-mail: [2] dvd1975@mail.ru

Abstract. By development marble deposits in mining operations one of the main technological processes, that ensures minimal losses and dilution with maximum output of commercial products is the process of preparing rocks for excavation. Marble extracted for the production of microcalcite is prepared for excavation by drilling and blasting at most mining enterprises. This inevitably leads to overgrinding in the charging chamber zone and, as a result, to a decreasing in quality indicators (yellowness and whiteness), significant losses of minerals (0–20 mm fraction) up to 30% and a low percentage of high-grade crushed stone in the commodity balance. The paper considers a variant of improving the production technology on the example of the Eleninsky marble Deposit (Chelyabinsk region) by using of heavy-class hydrohammers, which allows to increase the efficiency and completeness of the development of balance reserves by reducing by 5–10% overgrinding and increasing output of a commercial product high grades (40–200 mm).

1. Introduction

Production of mineral fillers from marble on an industrial scale in Russia began relatively recently, at the beginning of the 21st century.

As of 2020, the Russian industry is provided with domestic fillers by only 60%. The deficit is filled with fillers from Turkish, Finnish and European manufacturers.

The main commodity product for the production of marble fillers is crushed stone of the following fractions:

- 1 fraction – 40–200 mm;
- 2 fraction – 20–40 mm;
- 3 fraction – 5–20 mm;
- secondary dropout (0–5 mm) – production of weighting material of drilling fluids.

Currently, marble is prepared for excavation by drilling and blasting, as a result the loss of minerals is increasing significantly and the percentage of output of high-grade crushed stone in the commodity balance is decreasing, which ensures the production of high-quality fillers [1–3].

A similar scheme of work is applied at the Eleninsky marble Deposit. The article presents a variant of changing the process of marble extraction technology, which allows reducing the loss of minerals and increasing the yield of high-grade commercial product at the Deposit under consideration.

Eleninsky marble Deposit is located in Kartalinsky district of the Chelyabinsk region, 3.5 km South-West of village Eleninka. Mineral fillers used in the production of high-quality plastics, rubber and technical products, paint products, modified dry building mixes, office paper, etc. are made from

Eleninsky marble (Figure 1). Eleninsky marble is characterized by high whiteness, low yellowness and very low Fe_2O_3 content.

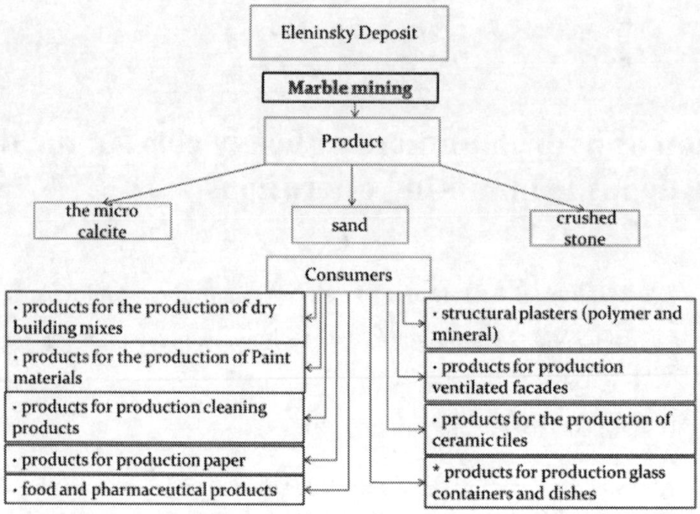

Figure 1. Products and main consumers of the Eleninsky marble Deposit.

2. Research methods

The paper uses a comprehensive research method that includes, along with technological solutions, analysis and generalization of the practice of developing marble deposits, analytical research, methods of technical and economic analysis, as well as economic and mathematical modeling.

3. The main part

As a result of the analysis of the fractional composition of the exploded mineral of the Eleninsky marble Deposit, which is received for further processing, it is established:

- mineral resources, including dilution, received on the DSU:
  - primary dropout (mud fractions), 0–20 mm – 18%
- mineral received for crushing (100%):
  - 40–200 mm – 52%
  - 20–40 mm – 28%
  - 5–20 mm – 6%
  - secondary dropout, 0–5 mm – 14%

It should be noted that in the composition of the mineral received for crushing, 28% is a finished product, 20% is a re-ground product, and only 52% is a raw material suitable for obtaining a liquid commodity product. In the process of crushing and grinding (20–200 mm), the finished product is obtained. The quarry uses drilling and blasting (DBW), which is a traditional method of preparing the rock mass for the extraction of solid minerals.

The assessment of options for preparing marble for excavation was carried out on the basis of economic and mathematical modeling of the use of DBW and mechanical destruction according to the criteria for obtaining the maximum profit P_i of the mining enterprise and achieving the minimum cost C_i of finished products [1, 2, 3, 4]:

$$P_i = Ц_i - \sum_{1}^{n} C_{g_i} \rightarrow \max , \qquad (1)$$

$$C_i = \frac{\sum C_{g_i}}{V_i} \rightarrow \min \qquad (2)$$

Tables 1 and 2 show the annual commodity balance and the cost of work on existing parameters in preparation for the extraction of mineral resources using DBW.

Table 1. Annual commodity balance in preparation for marble excavation with the help of DBW.

Appellation	%	Tons	Sales price (wholesale) RUB/*t* (without VAT)	Profit, RUB. (without VAT)
Annual productivity of the quarry for mineral resources, taking into account dilution	100	750000.00	–	–
Primary dropout (0–20 mm)	18	135000.00	–	–
		CSP		
Mineral resources for crushing	100	615000.00	–	–
40–200 mm	52	319800.00	400	127920000.00
20–40 mm	28	172200.00	160	27552000.00
5–20 mm	6	36900.00	160	5904000.00
Secondary dropout (0–5mm)	14	86100.00	100	8610000.00
TOTAL:		615000.00		169986000.00

Table 2. The cost of DBW on existing parameters.

Appellation	%	Tons	Cost of work, RUB/*t* (without VAT)	The cost of DBW per year, RUB (without VAT)
Volume of loosening of DBW	100	750000.00	27.5	20750000,00
Primary dropout(0-20 mm)	18	135000.00	–	–
The oversize (+ 200 mm)	10	75000.00	51	3825000.00
TOTAL:				24575000.00
Output of commercial products		615000.00	–	–
The cost of DBW in terms of one ton of commodity products			39.96	

In order to reduce losses during re-grinding of minerals [1, 2, 7], a variant with modified parameters of drilling and blasting operations (BWR) was considered and several experimental explosions were made. During the experiment, the specific consumption of explosives decreased (up to 10%) and the grid of wells expanded (up to 15%).

As a result, it was possible to reduce the loss of mineral resources by 8%, but at the same time the output of oversized material (+200 mm) was increased almost 5 times. Taking into account the cost of destruction of oversized by hydraulic hammer, the cost of drilling and blasting operations in terms of one ton of commercial products increased by 39.7% (table 3).

Table 3. Cost of drilling and blasting operations during experimental explosions.

Appellation	%	Tons	Cost of work, RUB/*t* (without VAT)	The cost of DBW per year, RUB (without VAT)
Volume of loosening of DBW	100	750000.00	27.5	20625000.00
Primary dropout (0–20 mm)	10	75000.00	–	–
The oversize (+200 mm)	49	367500.00	51	18742500.00
TOTAL:				39367500.00
Output of commercial products		675000.00		
The cost of DBW in terms of one ton of commodity products			58.32	

Thus, the use of this option with the changed parameters of the DBW only partially solves the problem. At the same time, the cost of BWR in terms of one ton of commercial products is increasing significantly.

As an alternative, we consider changing the process of preparing rocks for excavation at the Eleninsky marble Deposit [10–12], in order to reduce the loss of minerals and increase the yield of the commodity product without increasing the cost of finished products.

The annual capacity of the quarry for marble extraction is adopted in accordance with the technical project and is 280 thousand m^3 in a dense body.

The capacity of overburden rocks and useful thickness, the geological structure (Figure 2) and mining conditions of the Deposit determine the conduct of mining operations by the open method.

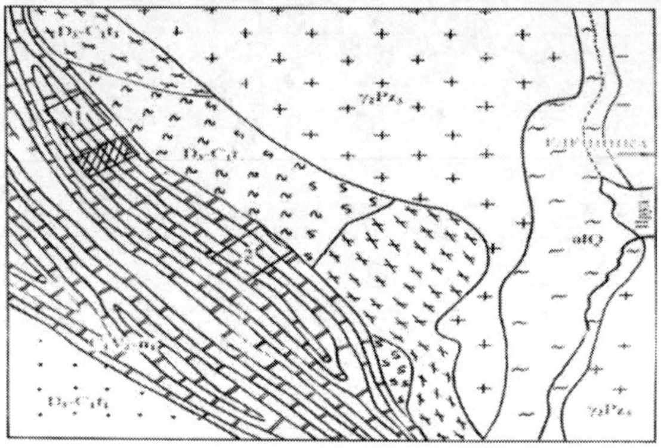

Figure 2. Geological map of the Eleninsky marble Deposit.

The option of non-explosive preparation is considered and the choice of the development system according to N. V. Melnikov is made. The site provides for the use of a transport development system that includes a four-process cyclical technology:

- mechanical loosening of rock mass (heavy hydraulic hammer);
- excavation and loading operations (excavator type mechanical shovel);
- transportation (motor transport);
- dumping and storage (external bulldozer).

On the basis of LLC 'UK TEK cars' data management preparation of rocks for excavation heavy hydraulic hammer on the limestone deposits in the quarry LLC 'Vostochnye Berniki' and graphite on LLC 'Taiginsky GOK' was performed techno-economic modelling (table 4 and 5) [6, 7, 12, 13], the results of which is the decision to carry out pilot testing hydraulic hammer for the preparation of the minerals on Eleninsky field of marbles (Figure 3).

Figure 3. Preparation of marble for excavation with the help of a heavy hydraulic hammer of the company 'Volvo' at the Eleninsky marble Deposit.

Table 4. Cost of work by preparing marble for excavation with a hydraulic hammer.

Appellation	%	Tons	Cost of work, RUB/t (without VAT)	Costs for hydraulic hammer in year, rbl. (without VAT)
Annual productivity of the quarry for mineral resources, taking into account dilution	100	750000.00	51	38250000.00
Primary dropout (0–20 mm)	5	37500.00		–
Output of commercial products	–	712500.00		–
The cost of mining operations in terms of one ton of commodity products			53.68	

Table 5. Loss of mineral resources and yield of commercial product in preparation for excavation with a hydraulic hammer.

Appellation	%	Tons	Profit, RUB (without VAT)
Annual productivity of the quarry for mineral resources, taking into account dilution	100	750000.00	–
Primary dropout (0–20 mm)	5	37500.00	–
		CSP	
Mineral resources for crushing	100	712500.00	
40–200 mm	70	498750.00	199500000.00
20–40 mm	18.5	131813.00	21090080.00
5–20 mm	7.5	53438.00	8550080.00
Secondary dropout (0–5 mm)	4	28500.00	2850000.00
TOTAL:		712500.00	231990160.00

Techno-economic modeling has shown (table 6) that the use of hydraulic hammer heavy class in the mining of marble leads to lower operating costs for the preparation of rocks for excavation by reducing losses, for the conditions on Eleninsky Deposit, at 97500 tons per year and increase the commercial yield of 16 %, including varieties.

Table 6. Technical and economic comparison of options for preparing a mineral for excavation.

Appellation	Hydrohammer		DBW		EFFECT	
	Tons	Sales price RUB/t without VAT	Tons	sales price RUB/t without VAT	Tons	Ruble per year
Mineral Resources For crushing	712500.00	–	615000.00	–	97500.00	–
40–200 mm	498750.00	199500000.00	319800.00	127920000.00	178950.00	71580000.00
20–40 mm	131813.00	21090080.00	172200.00	27552000.00	–40387.00	–6461920.00
5–20 mm	53438.00	8550080.00	36900.00	5904000.00	16538.00	2646080.00
Secondary dropout (0–5mm)	28500.00	2850000.00	86100.00	8610000.00	–57600.00	–5760000.00
TOTAL:	712500.00	231990160.00	615000.00	169986000.00	97500.00	62004160.00

4. Conclusion

The paper describes the solution of an actual scientific and practical problem, which consists in the development and scientific justification of the technology of mechanical preparation of rocks for excavation in the development of marble deposits using a heavy-class hydraulic hammer, which is important for the science and practice of the mining industry.

IOP Conf. Series: Materials Science and Engineering **966** (2020) 012020 doi:10.1088/1757-899X/966/1/012020

Main scientific results, practical conclusions and recommendations:

1. It is established, that by using the traditional method of preparing marble for excavation, namely DBW, leads to its over-grinding in the charging chamber zone and as a result to a decrease in quality indicators (yellowness and whiteness), significant losses of minerals (0–20 mm fraction) up to 30 % and a low percentage of high-grade crushed stone in the commodity balance.

2. Change of parameters of drilling and blasting operations by reducing specific consumption of explosives (10%) and expansion of the grid wells (up to 15 %) in terms of Eleninsky Deposit leads to a reduction of losses of minerals 8%, but increases the output of the oversized (+200 mm)in 5 times, and sledsvtie growth in the unit cost of marketable products on 39.7%.

3. Technical and economic modeling has shown that the use of a heavy class hydraulic hammer in the development of marble deposits reduces the operating costs of preparing marbles for excavation by reducing losses at the Eleninsky field by 97.500 tons per year and increasing the yield of commercial products by 16 %, including by grades.

References

[1] Safronov V P, Zaitsev Yu V and Safronov V V 2015 Efficiency of equipment usage for realization of explosion-free technologies of limestone and dolomite mining *News of TulGU. Technical Sciences* issue **7** pp 89–99

[2] Ligotsky D N 2015 Working out of contact zones by using hydrohammers to reduce the level of mineral losses *Mining information and analytical Bulletin* no **3** pp 7–13

[3] Domozhirov D V, Simakov D B, Nosov I I, Nosov V I and Golyak S A 2016 Increasing the competitiveness and safety of blasting operations by using emulsion EXPLOSIVES on the basis of mobile technological complexes *Mining information and analytical Bulletin (scientific and technical journal)* no **36** pp 35–42

[4] Mossakovsky Ya V 2006 *Economics of mining industry: Textbook for universities* (Moscow: Moscow state mining University Press) p 525

[5] Shestakov V A 2003 *Design of mining enterprises: Textbook for students. Higher educational.* 2nd ed (Moscow: MGGU Publishing house) p 800

[6] Domozhirov D V, Simakov D B and Zurkov I E 2004 Ways of improving the competitiveness of mining enterprises in modern economic conditions *Mining informational and analytical Bulletin* no **10** pp 82–5

[7] Mitusov A A, Reshetnikova O S, Mitusov P E and Lagunova Yu A 2016 Research of hydrohammer for rock crushing *Mining information and analytical Bulletin* no **11** pp 125–34

[8] Belin V A, Gorbonos M G, Mangush S K and Ekvist B V 2015 New technologies for conducting explosive works *GIAB. Separate issue 1. Proceedings of the international scientific Symposium 'miner's Week-2015'* pp 87–101

[9] Kutuzov B N 2014 Explosives reference: in 2 CH (Moscow: Mining) CH. P. Technique, technology and safety of explosive works p 304

[10] Pershin G D, Golyak S A, Domozhirov D V, Karaulov N G, Pshenichnaya E G and Ulyakov M S 2015 Modern methods of preparing granite blocks for excavation. Advances in modern natural science *Actual problems of modern science, technology and education* pp 655–8

[11] Kyzyrov K B, Mitusov A A and Reshetnikova O S 2018 Design researches of hydrohammer parameters for mining and construction industries *Mining information and analytical Bulletin* no **9** pp 220–6

[12] VOLVO GROUP MAGAZINE 2016 no **5** pp 52–5

[13] Federal norms and rules in the field of industrial safety 'Safety rules for mining operations and processing of solid minerals'. 2013 Rostechnadzor Order no 599 of December 11

IOP Conf. Series: Materials Science and Engineering **966** (2020) 012021 doi:10.1088/1757-899X/966/1/012021

The microstructure and hardness of casting a solid brake disc after late graphitizing modification

D A Boldyrev[1, 3], R R Dema[2] and O B Kalugina[2]

[1] Togliatti State University, 14, Belorusskaya Street, Togliatti City, 445020, Russia
[2] Nosov Magnitogorsk State Technical University, 38, Lenin Street, Magnitogorsk, 455000, Russia

E-mail: [3]denis.boldyrev@vaz.ru,

Abstract. The features of graphite formation in hypoeutectic and hypereutectic gray cast iron are considered. It is shown that this process in hypoeutectic gray cast irons is easier to control than in hypereutectic, because the formation of a predominantly rectilinear uniformly distributed graphite with a smaller size range and the presence of other distributions is achieved by an one-stage operation of ladle graphitization. The structure of graphite in hypereutectic gray cast irons is characterized by a large variety of its shapes, sizes and distributions, and it is also less technologically stable and controllable. Its characteristics can be approximated to the structure of hypoeutectic gray cast iron due to the obligatory second stage of graphitization. The microstructure and hardness of casting a solid brake disc after late graphitizing modification in a mold are investigated. The content of interdendritic distributions of graphite PGr8, PGr9 and short inclusions of graphite PGd45-90 (GOST 3443-87) was minimized due to the operation of the secondary (late) graphitizing modification. And Brinell hardness of the solid brake disc meets production requirements.

1. Introduction

The process of graphite formation and its features during crystallization of hypoeutectic gray cast irons (with a degree of saturation Sc <1) has been studied in sufficient detail [1–3]. This type of cast iron is quite susceptible to graphitizing modification. In such cast iron, a graphite structure can be formed at a relatively low consumption of graphitizing modifier (<0.3% for ladle modification using the 'under a single batch' method) with the following generally recognized as the most favorable characteristics [4, 5]: PGf1 – PGr1 – PGd (45–180) – PG (6–10) according to Russian state standard GOST 3443–87.

The process of graphite formation of hypoeutectic gray cast irons is quite easily controlled and an one-stage ladle graphitizing modification is quite sufficient to obtain the above structure. Figure 1 shows a typical graphite structure from a hypoeutectic gray cast iron.

During crystallization of hypereutectic gray cast irons (with a degree of saturation Sc > 1), the process of graphite formation is more complicated, since it is less controlled by graphitizing modification. The favorable graphite structure (during an one-stage graphitizing ladle modification) is not achieved and has the following characteristics: PGf1,2 – PGr1,3,5 – PGd (45–350) – PG (10–12) according to GOST 3443–87 [6–9]. That is, a swirling shape also appears along with a rectilinear lamellar structure. Both colony of lamellar graphite and its branch distribution are added to the uniform distribution, and the range of lengths of graphite inclusions increases in the direction of their

IOP Conf. Series: Materials Science and Engineering **966** (2020) 012021 doi:10.1088/1757-899X/966/1/012021

elongation. Below are the most characteristic microstructures of hypereutectic gray cast iron (Figure 2).

Figure 1. Typical structure in hypoeutectic gray iron castings (rectilinear lamellar structure).

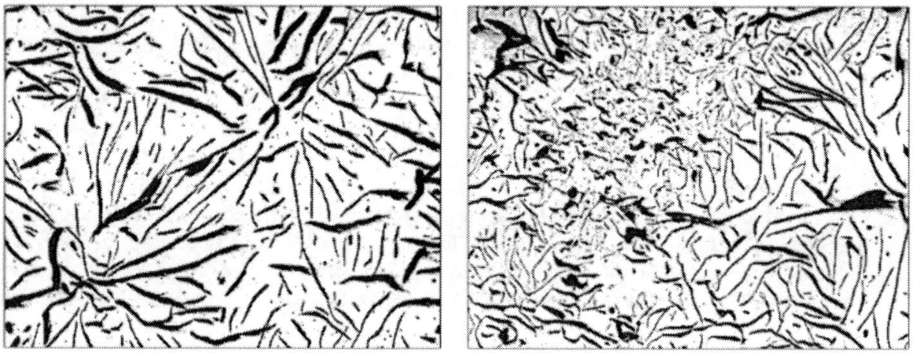

a b

Figure 2. Typical structures in hypereutectic gray iron castings: a – rectilinear; b – swirl.

This qualitative difference in the microstructure of graphite is explained by the crystallization features of hypoeutectic and hypereutectic gray cast irons. During crystallization of hypereutectic gray cast iron, the initiation of graphite formation occurs after and due to the introduction of a graphitizing modifier, which creates crystallization centers of graphite inclusions in quasi-eutectic zones locally saturated with silicon. A sufficient amount of a graphitizing modifier with a certain set of active elements (barium, calcium, strontium) completely removes the overcooling of the cast iron melt and allows forming a uniform distribution of PGr1 of rectilinear graphite inclusions PGf1 of an average length of 90–180 microns. A qualitative difference between the crystallization process of hypereutectic gray cast iron is the initial separation of primary graphite from the melt. The morphology of primary graphite is first and foremost formed by the introduced graphitizing modifier. Primary graphite after graphitization of cast iron in the ladle is a rectilinear plate PGf1 with uniform PGr1 and/or branch PGr5 distribution. However, the graphitizing effect of the initial modification to create a favorable graphite structure during eutectic transformation and lower temperatures, similar to the structure of hypoeutectic gray cast iron, is insufficient. As a result, colonies of PGr3 are formed in later crystallizing zones. Thus, the process of uniform structure formation of graphite after the eutectic transformation is violated, as evidenced by its crowding and thickening in the form of Pgr3 colonies, that is, it becomes uncontrollable. If the technology is not complied with regarding the consumption of the graphitizing modifier and/or the method of introducing it into the ladle, the described unfavorable

IOP Conf. Series: Materials Science and Engineering **966** (2020) 012021 doi:10.1088/1757-899X/966/1/012021

graphite structure is supplemented by interdendritic (non-directional point PGr8 and directed lamellar PGr9) distributions.

2. Methods and Materials

The object of the study was the casting of solid brake disc from gray cast iron (the composition is shown below). Castings of brake discs are obtained as follows. The melt of hypereutectic iron (C 3.92 %, Si 2.00%, Mn 0.72%, P 0.02%, S 0.08%, Cr 0.29%, Ni 0.09%, Cu 0.60%, Sn 0.01%) is obtained in an induction crucible furnace and poured into a holding induction crucible furnace (holding furnace). The Barinoc® 75 modifier with a fraction of 2–6 mm was used during graphitizing ladle modification in an amount of 0.15% (1.5 kg per 1 ton of molten iron in the ladle), the chill hard spots was 1.5–1.6 mm (chill test). The molten iron temperature in the ladle was 1420 °C, the ladle casting time lasted 8 minutes. Lump Fs75-3 of 70–110 g was used for secondary (late) graphitizing modification.

The hardness was measured using a stationary Brinell hardness tester HB-3000B. The microstructure of cast iron was studied using a Metam RV-21 metallographic microscope for visual observation of the microstructure of metals, alloys, and other opaque objects in reflected light. Magnification of the microscope from 50 to 1000.

3. Result and Discussion

The approximation of the graphite structure of hypereutectic gray cast iron to hypoeutectic in terms of achieving its uniform distribution of PGr1 can be realized in one of two ways:

1) carrying out late graphitizing modification to the stream during the pouring of the mold, which is universally recognized as the most effective way;

2) carrying out, along with the primary graphitizing ladle modification, late graphitizing modification with cast or pressed intra-shaped inserts placed in a special socket located in the smudge sump under the sprue, or with a briquette from pressed siftings of ferrosilicon in the sprue well.

However, for the implementation of the first way, a specialized metering device is necessary, which is a complex, high-tech equipment due to its automated operation and control. There is also a restriction on the type of automatic molding production line – it should be with step-by-step conveyor movement (the automatic supply of the modifier from the dispenser does not work satisfactorily while filling the casting molds with molten iron from the ladle on lines with continuous conveyor movement).

The second way is technologically simple and convenient. However, cast inserts for all their effectiveness are quite expensive (≈1 euro/pc.) due to their making in complex chill molds, and pressed ones are characterized by instability of dissolution. The most technologically simple and economical way to carry out the operation of late graphitizing modification is to put lumpy ferrosilicon (usually Fs75 grade – GOST 1415–93) into a casting mold with a mass of ~0.05% of its metal consumption (mass of all castings together with a gate system and a sprue). According to this technology, a batch of pilot castings of brake discs was made. Figure 3 shows a photograph of a solid brake disc, and table 1 shows the results of a study of the microstructure and Brinell hardness of the casting in zones No. 1–4.

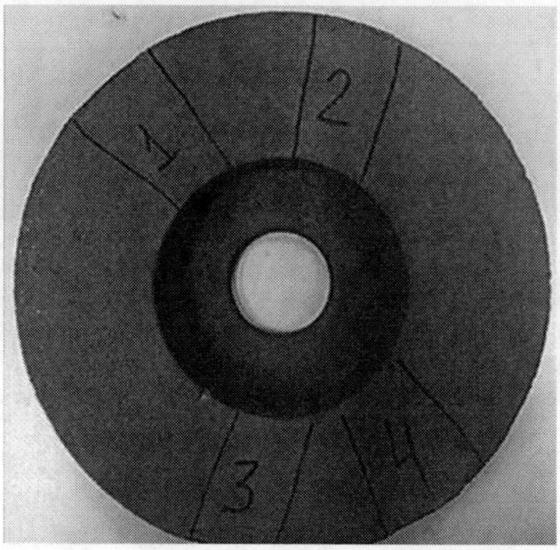

Figure 3. The solid brake disc casting with research zones No. 1–4.

Table 1. The microstructure and Brinell hardness of the solid brake disc casting.

No. zone	$HB_{5/750/10}$	Metal base,%			Graphite
		Perlite	Ferrite	Cementite	
1	187, 182, 179, 179, 185 Average 182	Basis	1	0	Non-massive zone PGr8 + PGr9 <5%, core PGd750. PGd45-90 <10% in the contact zone of the disc and brake pads and, partially, in the machining allowance zone, in the remaining zones PGd180–350
2	179, 179, 182, 179, 187 Average 181	Basis	1	0	Non-massive zone PGr8 + PGr9 <5%, core PGd750. PGd45-90 <10% in the contact zone of the disc and brake pads and, partially, in the machining allowance zone, in the remaining zones PGd180–350
3	179, 179, 177, 179, 187 Average 180	Basis	1	0	Non-massive zone PGr8 + PGr9 <5%, core PGd750. PGd45-90 <10% in the contact zone of the disc and brake pads and, partially, in the machining allowance zone, in the remaining zones PGd180–350
4	187, 187, 187, 180, 187 Average 186	Basis	1	0	Non-massive zone PGr8 + PGr9 <5%, core PGd750. PGd45-90 <10% in the contact zone of the disc and brake pads and, partially, in the machining allowance zone, in the remaining zones PGd180–350
Brake disc casting requirements					
	170–217	Basis	< 5	< 5	PGd180–350, Non-massive zone PGr8 + PGr9 <5%, core PGd759 or PGr7 <10%. Allowed in the contact zone of the disc and brake pads PGd180 <= 20 %, in other zones PGd180 <= 40%

Below are photos of microstructures obtained in zone No. 1 (Figure 4).

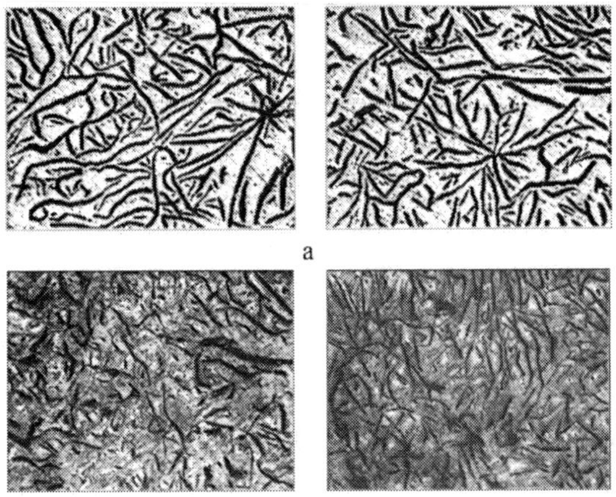

Figure 4. The microstructure of cast iron brake disc (enlarged ×100):
a – not etched; b – etched.

4. Conclusion

Thus, hypereutectic cast iron of the brake disc casting according to microstructure and Brinell hardness complies with the manufacturing requirements. The content of interdendritic distributions of graphite PGr8, PGr9 and short inclusions of graphite PGd45–90 was minimized due to the operation of the secondary (late) graphitizing modification in the microstructure of cast iron upon reaching the actual length of most graphite inclusions - no more than PGd350. The structure of hypereutectic gray cast iron with such characteristics provides the best combination of its antifriction and thermophysical properties.

Acknowledgments

This work is carried out within a framework of the government order (No. FZRU-2020-0011) of the Ministry of Science and Higher Education of the Russian Federation

References

[1] Vozdvizhenskij V I, Grachev V A and Spasskij V V 1984 *Foundry alloys and their melting technology in mechanical engineering* (Moscow: Mashinostroenie)
[2] Joseph R D 1996 *ASM Specialty Handbook: Cast Irons* (ASM International)
[3] Girshovich N G 1978 *Handbook of iron casting* (Moscow: Mashinostroenie)
[4] Truhov A P and Malyarov A I 2004 *Foundry alloys and smelting* (Moscow: Akademiya)
[5] Olawale J O, Ibitoye S A and Oluwasegun K M 2016 *IJSER* **7** p 397
[6] CHajkin V A, Boldyrev D A and CHajkina N V 2007 *Vestnik of NMSTU* **4** p 53
[7] Boldyrev D A, Shapovalov A N, Nefedev S P, Dema R R, Kononov V N, Kharchenko M V, Suvorova E V and Markova I Yu 2019 *J. Chem. Technol. Metall.* **54** p 348
[8] Boldyrev D A, Dema R R and Kalugina O B 2019 *Materials Today: Proc.* **11** p 510

IOP Conf. Series: Materials Science and Engineering **966** (2020) 012022 doi:10.1088/1757-899X/966/1/012022

Improving the production technology of drilling and blasting operations by blasting of high ledges

N V Ugolnikov[1, 2], D V Domozhirov[1], N G Karaulov[1] and A A Prochorov[1]

[1] FGBOU VO 'Magnitogorsk state technical University named G I Nosov', 38, Lenin street, Magnitogorsk, 455000, Russia

E-mail: [2] ugkit@mail.ru

Abstract. In modern economic conditions, the issue of improving the efficiency of mining enterprises is very relevant. One of the key processes isdrilling and blasting operations (DBO), which determine the efficiency of the entire complex of mining operations. One of the main methods of controlling the explosion energy is the diameter of the blast wells. According to the recommendations of prof. B N Kutuzov the diameter of blast wells is correlated with the fracturing and strength of the exploding rocks. With increasing blockage and rock strength to achieve the required quality of explosive crushing preference is given to drilling equipment with a small diameter of wells. By blasting of high ledges, especially large-block rocks with a strength coefficient of more than 15 on the M M Protodyakonov scale; the use of high-performance drilling equipment for drilling small-diameter blast wells is difficult. This is due to the fact that the calculated value of the resistance line along sole (RLAS) does not pass the safety condition for drilling the first row of wells. In this regard, the paper proposes and justifies the use of extensions of the lower parts of wells using mechanical expanders.

1. Introduction

In quarries of construction, materials for drilling explosive wells with a diameter of up to 200 mm in rocks of medium and strong strength are used machines with submerged pneumatic hammers. There are mainly machines of domestic and foreign production [1, 2]:

- URB-2A with a diameter of 120–190mm blast wells;
- Sandvik (D25KS, D245S) with a diameter of blast holes 127–203 mm;
- TAMROCK (PANTERA 800, PANTERA 900, PANTERA 1100, PANTERA 1500) with the diameter of blast wells 115–152 mm;
- Atlas Copco (ROC 460 PC, ROC F7, ROC F9, ROC L6, ROC L7) with blast hole diameters of 105–165 mm;
- FURUKAWA (HCR 1000-1200-1500, PCR200) with the diameter of blast wells 65–150 mm.

This drilling equipment is used in quarries with a ledge height of 10 meters or less. By increasing the height of the ledge to 12–15 m or more, drilling the first row of wells in large-block rocks of medium and above average strength with a small diameter does not provide a condition for safe drilling, since the calculated values of the resistance line along the sole do not pass the condition for safe drilling of the first row of wells.

In this situation are the following methods and technologies for drilling and blasting operations recommended [2–4]:

- use of larger diameter wells (if this type of drilling equipment is available);
- the use of boiler charges (difficulty in creating boiler cavities, uneven crushing) or mechanical extensions along the length of the well (increased consumption of drilling the first row of wells);

IOP Conf. Series: Materials Science and Engineering **966** (2020) 012022 doi:10.1088/1757-899X/966/1/012022

- use of inclined wells drilled parallel to the slope of the ledge (presence (absence) of this type of drilling equipment at the enterprise; difficulty in ensuring the safety of inclined wells with large volumes of drilling and blasting operations; difficulty of loading and installation of the explosive network, especially in the absence of mechanization; high wear of drilling equipment);
- increase of the charge energy of explosives (type of explosives, specific consumption of explosives);
- use of paired-converged wells (increased drilling expenditure of the first row of wells).

Currently a lot of mechanical expanders are designed for blast wells, which can be installed on existing drilling equipment as well as on auxiliary equipment and allow drilling small-diameter blast wells up to 250 mm or more.

The analysis of the considered methods of explosive crushing intensification in the conditions of Krutorozhinsky field of gabbro-diabase son PAO 'Orsk quarry management' has shown the applicability of mechanical expansion of explosive wells to ensure the safe drilling of the first row of wells [5, 6]. Therefore, the problem emerges of substantiating the parameters of expanded well cavities.

2. Methods of research

To determine the rational parameters of the location of blast wells, we suggest using the principle of self-similarity, based on taking into account the critical displacement rates of the array in the zone of action of adjacent borehole charges [5–11]. The proposed principle of self-similarity is well established in calculating the parameters of the location of borehole charges in underground conditions [7].

3. Main part

The study of rational parameters of drilling and blasting operations was carried out in conditions of Krutorozhinsky field of gabbro-diabases on PAO 'Orsk quarry management'. The parameters of the quarry and development system are shown in Table 1.

Table 1. Main parameters of the quarry and development system.

Parameters	Value
Design depth of the quarry, m	80–200
Length of the quarry on top, m	1870
Width of the quarry on top, m	800
Design height of the ledge, m	15
The angle of slope of the working ledge, the degree	75
The minimum width of a working platform, m	34–60

At the Krutorozhinsky quarry, the following types of explosives are used in waterlogged, partially waterlogged and dry wells (Table 2).

Table 2. Characteristics of the used explosives.

Characteristics for types explosives	Heat of explosion, kJ/kg	Volume of gaseous explosion products, l/kg	Bulk density, g/cm^3	Detonation speed, m/s
Grammonit 79/21	4291	895	0,85–0,90	3200–3600
Grammonit 30/70	3977	800	0,85–0,90	3800–4500
AS-25P	3200	910	1.00	4600–4800
Arsenit ГБ-1	4100–4200	820–850	0,75–0,85	4500–5100
Arsenit ГБ-2	4270–4750	850–860	0,75–0,85	4800–5500
Arsenit ГБ-4	4300–4500	790–820	0,75–0,80	5000–5200
Fortis	3260	968	1.00	5100
MS-U-1	5700–5800	710–895	0.65–0.75	6700

For these types of explosives were the rational values of the resistance line along the sole at the height of the ledge of 15 m and the diameters of wells of 150 and 190 mm for rocks of different strength and fracturing determined (Table 3).

Table 3. Values of the rational resistance line on the sole, m.

Diameter of well, mm	The fortress of rocks on the horizon							
	f=8	f=10		f=12		f=14		f=16
	Fracture category							
	II	III	IV	III	IV	III	IV	V
explosives – MS-U-1 (K_{VV}=0.83)								
150	6.1	4.8	4.1	4.9	4.3	4.8	4.1	3.5
190	6.7	5.4	4.5	5.5	4.7	5.3	4.6	3.9
explosives – Arsenit GB-2(K_{VV}=0.98)								
150	6.7	5.3	4.5	5.4	4.7	5.3	4.5	3.8
190	8.2	6.6	5.5	6.7	5.8	6.5	5.6	4.7
explosives – Arsenit GB-4 (K_{VV}=1)								
150	6.6	5.3	4.4	5.4	4.6	5.2	4.5	3.8
190	8.2	6.5	5.5	6.6	5.7	6.4	5.6	4.7
explosives – Grammonit 79/21 (K_{VV}=1)								
150	6.3	5.0	4.2	5.1	4.4	4.9	4.3	3.6
190	7.7	6.2	5.2	6.3	5.4	6.1	5.3	4.4
explosives – Arsenit GB-1 (K_{VV}=1.05)								
150	6.5	5.2	4.3	5.2	4.5	5.1	4.4	3.7
190	8.0	6.3	5.3	6.5	5.6	6.3	5.4	4.6
explosives – Grammonit 30/70 (K_{VV}=1.09)								
150	6.0	4.8	4.0	4.9	4.2	4.7	4.1	3.4
190	7.4	5.9	5.0	6.0	5.2	5.8	5.1	4.3
explosives – Fortis (K_{VV}=1.21)								
150	6.3	5.0	4.2	5.1	4.4	5.0	4.3	3.6
190	7.8	6.2	5.2	6.3	5.5	6.1	5.3	4.5
explosives – emulsion composition AS-25P (K_{VV}=1.25)								
150	7.0	5.6	4.7	5.7	4.9	5.5	4.8	4.0
190	8.6	6.9	5.8	7.0	6.1	6.8	5.9	5.0

The minimum allowable value of the resistance line along the sole under the conditions of safe placement of drilling equipment for drilling the first row of wells is 6.0 m.

According to calculations the mechanical expansion of blast wells should be used only in large-block rocks with a strength coefficient of more than 10 on the M M Protodiakonov scale.

To determine the rational parameters of mechanical expansions of blast wells, we propose the use of the principle of self-similarity [5–7, 12], which is based on the account of the critical displacement rates of the array in the zone of action of adjacent borehole charges [13, 14].

The voltages generated by the explosion of a borehole charge are determined by [5–7, 12]:

$$\sigma_{\text{сж(р)}} = \rho_0 C_p V_{\text{сж(р)}} \tag{1}$$

where $\sigma_{\text{сж(р)}}$ – acting compressive (tensile) stresses, Pa;

ρ_0 – specific mass of rock, kg /м^3;

C_p – speed of spread of a longitudinal elastic wave in mountain (sample), m/s;

$V_{\text{сж(р)}}$ – the displacement rate of the mountain under the action of compressive (tensile) voltages, m/s.

Hence the critical displacement rate of the mountain is defined as: [5–7]:

$$V_{\text{кр}}^{\text{сж(р)}} = \frac{\sigma_{\text{сж(р)}}}{\rho_0 C_p} K_{\text{д}} \tag{2}$$

where $K_{\text{д}}$ – coefficient of dynamism.

The mountain will be destroyed if the resulting array offset rates exceed or equal the critical values:

$$V_{\text{сж(р)}} \geq V_{\text{кр}}^{\text{сж(р)}} \tag{3}$$

where $V_{\text{сж(р)}}$ – the displacement rate of the mountain under the action of compressive (tensile) dynamic voltages, m/s;

$V_{кр}^{сж(р)}$ – minimum (critical) values of the displacement rate of the mountain, when destruction occurs, due to compressive (tensile) voltages, m/s.

The resulting explosion displacement rates of the array at this point are determined by [5–7]:

$$V_{сж(р)} = k_v \bar{r}^{-v} \tag{4}$$

where k_v – seismic proportionality coefficient depending on elastic parameters of the destroyed rocks [9, 10].

$$k_v = \sqrt[3]{\frac{C_p}{9\rho_0}\left(\frac{1+\mu}{1-\mu}\right)^2} = \sqrt[3]{\frac{C_p}{\rho_0}\left(1-\frac{4C_p^2}{3C_p^2}\right)^2} \tag{5}$$

where μ – coefficient Puasson's;

$v = 2{,}25$ – indicator of degree;

$\bar{r}$ – equivalent reduced distance, $\mathrm{m/kg}^{1/3}$.

Equivalent reduced distanceis determined by [6, 12]:

$$\bar{r} = \frac{r_n}{\sqrt[3]{Q_{eqv}}} \tag{6}$$

where r_n– distance from the test point to the center of gravity of the equivalent charge, m;

Q_{eqv} – the equivalent charge weight, kg.

According to rock properties of Krutorozhinsky field the critical displacement rates of the mountain (m\s) are shown in Figure 1, at which the destruction of the mountain occurs.

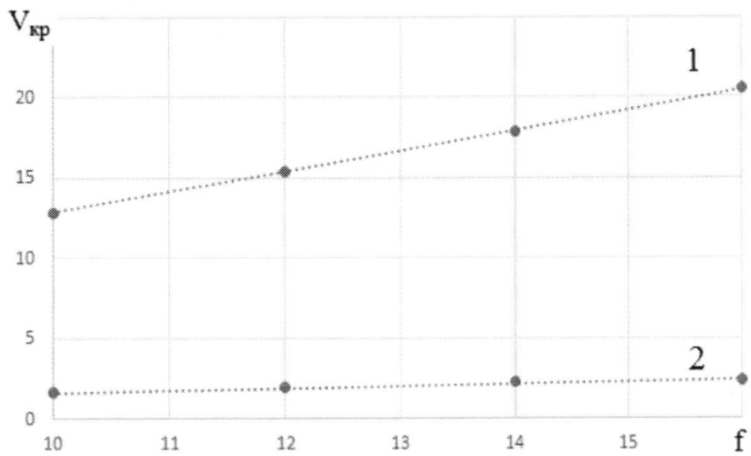

Figure 1. The value of the critical displacement rates of the array from the rock fortress:
1 –critical speed under the action of compressivevoltages;
2–critical speed under the action of tensile voltages.

Mechanical expanders with a diameter of 250 mm were used to calculate the parameters of borehole charge extensions.

Depending on the strength of the rocks and the corresponding critical displacement rate of the array, the optimal value of the reduced resistance line along the sole (m/kg^3) is determined Figure 2.

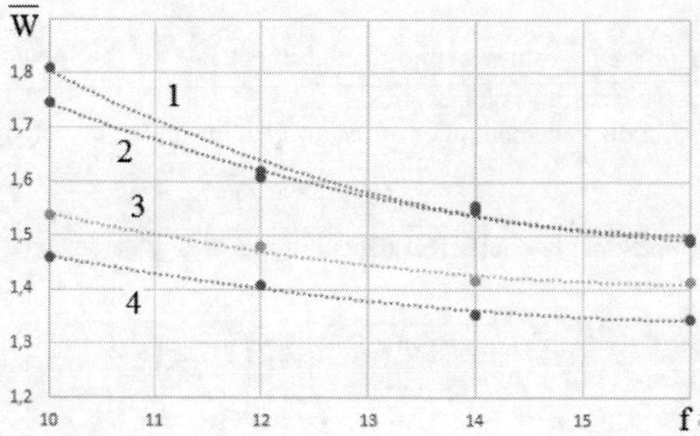

Figure 2. Dependence of the reduced RLAS from the coefficient of rock strength
1 – Grammonit 79/21; 2 –MS-U-1; 3 – Arsenit G-1; 4 – emulsion composition AS-25P.

The obtained dependencies are approximated by a polynomial equation of the form:

$$\overline{W} = \alpha f^2 - \beta f + \gamma \tag{7}$$

where $\overline{W}$ – rational reduced value of the resistance line along the sole, m/kg³;
f – the strength coefficient;
α, β, γ – empirical coefficients, the value of which is shown in Table 4.

Table 4. Value of empirical coefficients.

Type of explosives	α	β	γ
Grammonit 79/21	0.008	0.26	3.60
MS-U-1	0.005	0.17	2.96
Arsenit G-1	0.003	0.11	2.30
Emulsion composition AS-25P	0.002	0.09	2.08

The length of mechanical expansion of borehole charges (lp, m) is shown in Figure 3.

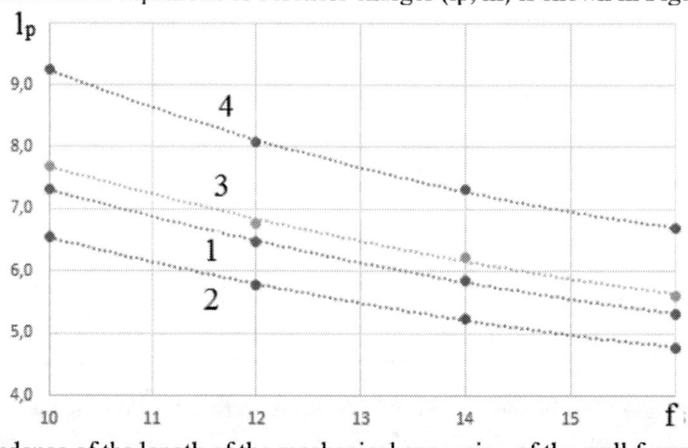

Figure 3. Dependence of the length of the mechanical expansion of the well from the rock strength
1 – Grammonit 79/21; 2 – MS-U-1; 3 – Arsenit G-1; 4 – emulsion composition AS-25P.

The paper defines the parameters of drilling and blasting operations using mechanical extensions of the lower parts of wells, with a ledge height of 15 m and a well diameter of 150 mm for large-block rocks of medium and high strength.

4. Conclusion

The application of a method for determining the parameters of mechanical extensions of borehole charges based on the principle of self-similarity, based on the comparison of critical displacement

IOP Conf. Series: Materials Science and Engineering **966** (2020) 012022 doi:10.1088/1757-899X/966/1/012022

rates of the array allows:

1.Increase the yield of blasted rock mass from one linear meter of wells, especially in medium and hard-to-explode rocks by 15–26%, by increasing the grid of wells, in comparison with inclined wells and wells with a diameter of 250 mm, both when exploding to a free surface, and when interacting with adjacent charges, exploding with deceleration;

2. Reduce the specific consumption of explosives by 8–23% in large-block hard rocks compared to wells with a diameter of 250 mm.

References

[1] Momeni A, Karakus M, Khanlari G R and Heidari M 2015 Effects of cyclic loading on the mechanical properties of a granite *International Journal of Rock Mechanics and Mining Sciences* vol **77** July pp 89–96

[2] Ugolnikov N V 2018 *Management of explosion energy* (Magnitogorsk: FGBOU VO 'MGTU named of G. I. Nosov')

[3] Belin V A, Gorbonos M G, Mangush S K and Ekvist B V 2015 New technologies for conducting explosive works *GIAB Separate issue 1. Proceedings of the international scientific Symposium 'Miner's Week-2015'* pp 87–101

[4] Gorodnichenko V I 2012 Expansion of wells for blasting of ore in underground mining operations *GIAB* no **7** pp 28–31

[5] Ugolnikov N V and Domozhirov D V 2019 Ensuring the safety of drilling and blasting operations by blasting paired-converged wells of high ledges in quarries *News of Tula state University. Earth science* no **3** pp 332–43

[6] Ugolnikov N V and Domozhirov D V 2019 Substantiation of rational parameters of a pair-of closely spaced wells in the quarries of non-metallic building materials *Combined geotechnologies: Transition to a new technological structure: Collection of articles on the results of the international conference* (Magnitogorsk: Magnitogorsk state technical University. G. I. Nosov University) pp 361–9

[7] Egemberdiev R I, Ugolnikov N V, Yusupov Kh A and Stolpovskikh I N 2019 Substantiation of parameters of extensions of borehole charges by breaking with fans *Mining information and analytical Bulletin* no **11** pp 48–58

[8] Pergament V, Melnikov Iv, Suraev V, Melnikov I, Vassiliev K, Kotik M and Shevtsov N 2014 Ensuring seismic safety of the explosive works and evaluation of the consequences related to technogenic and natural seismic events *Minnodeloi Geologia* (Bulgaria: 1000 Sofia) no **1–2** pp 57–63

[9] Pergament V Ch 1971 Accounting the spatial dispersion of the EXPLOSIVE charge in explosive problems *Engineering methods for controlling the action of an explosion. collection of proceedings* (Magnitogorsk: MGMI) issue **89** pp 3–14

[10] Pergament V Ch 1971 Critical speeds and parameters of drilling and blasting operations *Engineering methods for controlling the action of an explosion. collection of proceedings* (Magnitogorsk: MGMI) issue **89** pp 40–8

[11] Pergament V, Malarov I, Firstov P and Gitterman Y 1998 Experimental evaluation of near-source seismic effects of quarry blastes *XXVI General Assembly of the European Seismological Commission (ESC)* (TEL AVIV) p 29

[12] Belin V A, Kholodilov A N and Gospodarikov A P 2017 Methodical bases of forecasting of seismic action of mass explosions *Mining journal* no **2** pp 66–8

[13] Cunningham C V B 2011 *Control over Blasting Parameters and Its Effect on Quarry Productivity* (Rondebosch: AECI Explosives and Chemical Limited)

[14] Alenichev I A 2018 The reaction of the rock mass in the quarry space to dynamic effects in the production of blasting *Mining information and analytical bulletin* no **7** pp 189–95

Microstructure of impurity phase inclusions in gray cast iron and their composition

D A Boldyrev[1, 3], R N Amirov[2] and S P Nefed'ev[2]

[1] Togliatti State University, 14 Belorusskaya Street, Togliatti City, Samara Region, 445020, Russia
[2] Nosov Magnitogorsk State Technical University, 38, Lenin Street, Magnitogorsk, 455000, Russia

E-mail: [1]denis.boldyrev@vaz.ru

Abstract. It is quite difficult to control the content of carbide-forming elements in the production of casting 'pressure plate' of gray pearlitic iron grade SCh25 (GOST 1412–85). Therefore, it is necessary to conduct research and analysis of foreign inclusions, the basis of which are 'uncontrolled' carbide formers. The analysis of the content of 'uncontrolled' carbide-forming elements on the formation of zones with an interdendritic distribution of graphite types D, E (ASTM A247) castings of a pressure plate is carried out. The elemental composition of impurity inclusions is determined. It has been established that 'uncontrolled' carbide-forming elements are represented in cast iron in complex carbide-nitride and sulfide-phosphide phases. A conclusion is drawn on the effect of carbide formers on the formation of areas with an interdendritic distribution of graphite. All types of these phases are concentrated to a greater extent within the zones of interdendritic distributions of lamellar graphite D and E. This indirectly confirms the influence of 'uncontrolled' carbide-forming elements on their formation.

1. Introduction

The manufacturer is faced with problems of inconsistency of the structure and material properties of castings with the established regulatory requirements during the development of new foundry products. One of the reasons for such deviations is already observed at the beginning of the processor chain and is associated with the unsatisfactory quality of the charge materials used for melting [1–3]. This is manifested in the presence of 'uncontrolled' carbide-forming elements entering the molten iron from waste high-strength steels and noticeably worsening the morphology of the graphite phase in the manufacture of cast iron castings [4–8].

This problem manifested itself, in particular, when the 'Pressure plate' casting of pearlite gray cast iron grade SCh25 (Russian state standard GOST 1412–85 [9]) was put into production. It is relevant to conduct research and analysis of the presence, size, distribution and elemental composition of foreign inclusions (mainly uncontrolled carbide formers) to correct the 'infected' steel component of charge materials. Also relevant is the study of their influence on the formation of adverse interdendritic distributions of graphite.

2. Methods and Materials

Electron microscopic images and X-ray spectra were obtained using a Carl Zeiss Sigma scanning electron microscope (Germany) (resolution of the electron column at an optimal WD of 1.3 nm; range of movement: X – 125 mm; Y – 125 mm; Z – 50 mm; tilt 0 –90°; 360° rotation).The microscope is equipped with an EDAX analytical system (USA) with an Apollo detector and a Hikari back-scattered electron detector.

The object of the research was the casting 'Pressure plate' made of gray cast iron of pearlitic class grade SCh25.

Part of the electron microscopic photographs and the results of X-ray spectral microanalysis are shown in Figures 1–4. Figures 1, 2 show the sections of the middle part of the casting, Figures 3, 4 show the edge sections of the casting.

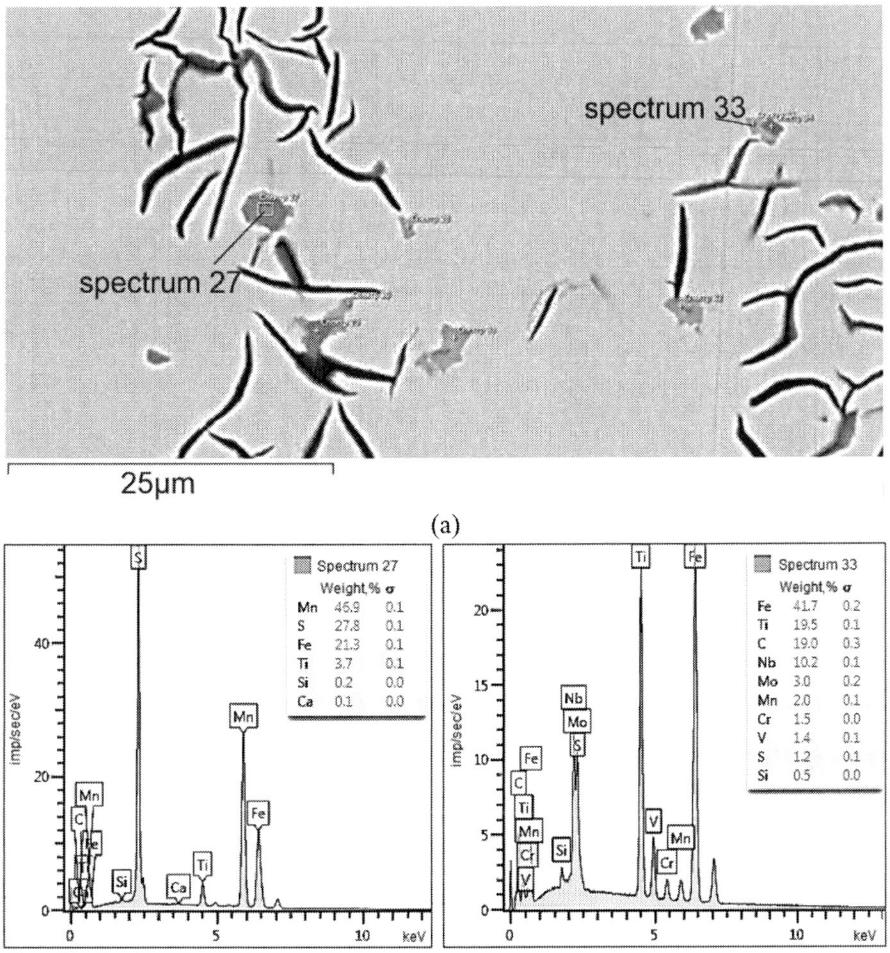

Figure 1. Electron microscopic image of the middle part of the casting(a) and X-ray spectra of sulfide (b) and carbide (c) inclusions (marked by the rectangle and the cross).

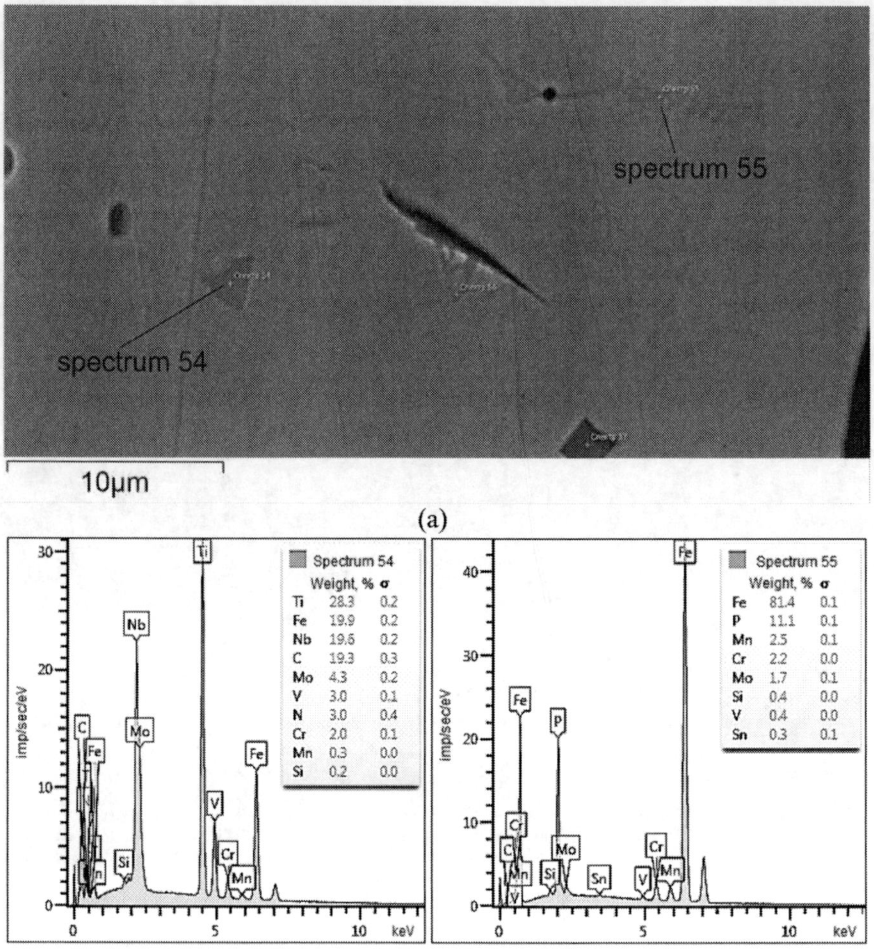

Figure 2. Electron microscopic image of the middle part of the casting (a) and X-ray spectra of carbonitride inclusion (b) and phosphide eutectic (c)(marked by crosses)

The degree of gray in these images characterizes the chemical composition of the test substance (the lighter - the greater the sequence number of the constituent elements according to the periodic table).The numerical results of the weight content of elements in the above X-ray spectral analysis are of a qualitative nature.

3. Result and Discussion

According to the results of electron microscopy analysis using x-ray spectral microanalysis, it was found that 'uncontrolled' carbide-forming elements such as zirconium, vanadium, titanium, tungsten, niobium and molybdenum coming from charge materials are represented in cast iron in complex carbide-nitride and sulfide-phosphide phases.

In complex phases of a compact form based on complex ferromanganese sulfides, such chemical elements as (together and separately) vanadium, titanium, molybdenum, niobium, tungsten and zirconium are concentrated. Separate phases of this type are sulfide-phosphide, because additionally contain phosphorus. In isolated cases, phosphide eutectic inclusions with dissolved uncontrolled carbide-forming impurity elements are found, morphologically different from the rest of the inclusions and presented as fragments of a broken network.

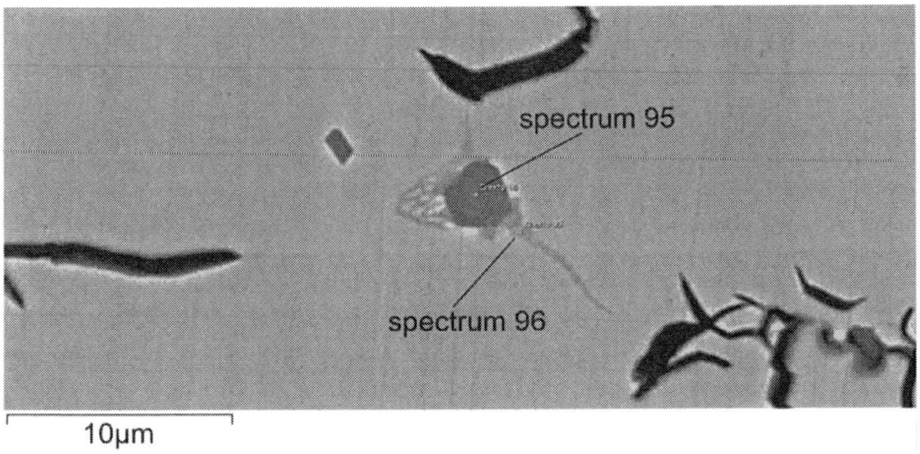

(a)

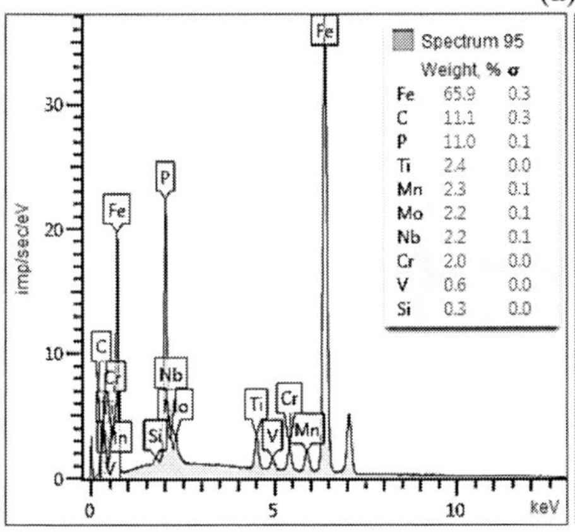

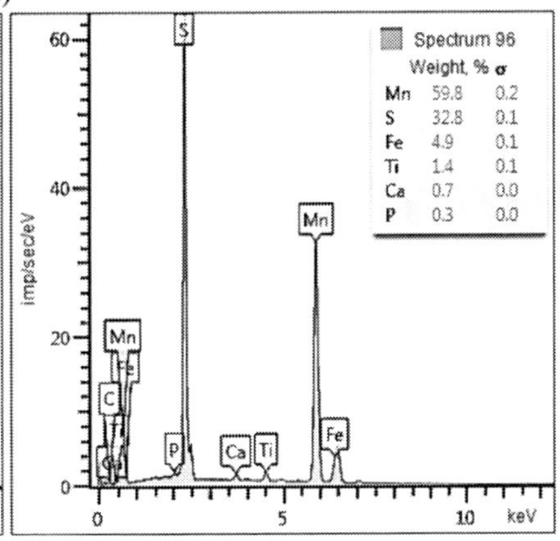

(b) (c)

Figure 3. Electron microscopic image of the edge of the casting (a)and X-ray spectra of the phosphide eutectic (b) and sulfide inclusion (c)(marked by crosses)

In complex carbide phases, mostly elongated and irregular shape, such chemical elements are concentrated in various combinations as vanadium, titanium, molybdenum, niobium, tungsten and zirconium. The individual phases of this type are carbonitrides, because additionally contain nitrogen.

In some cases, both types of these inclusions are combined into a conglomerate in the form of dark sulfide of a compact form with a light elongated carbide periphery.

4. Conclusion

All types of carbide-nitride and sulfide-phosphide phases are concentrated to a greater extent within the zones of interdendritic distributions of lamellar graphite PGr8 and PGr9 (GOST 3443–87 [10]) (D and E according to ASTM A247 [11]) both in the extreme parts of the sample and in its core. This indirectly confirms the influence on their formation of 'uncontrolled' carbide-forming elements that increase the tendency of molten iron to chill hard spots.

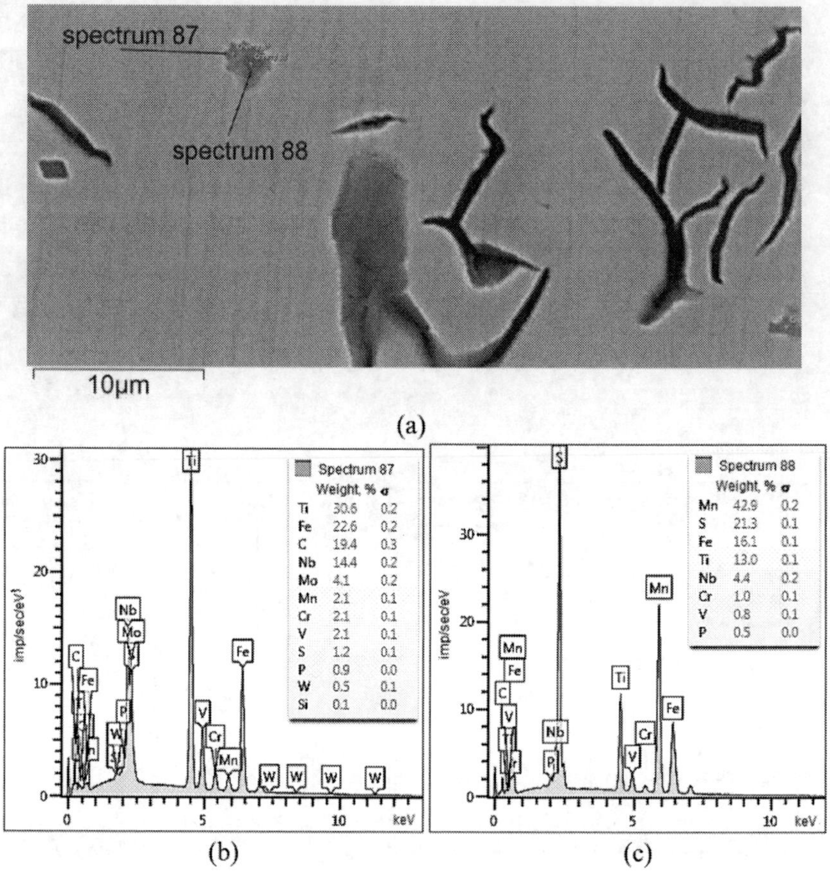

Figure 4. Electron microscopic image of the edge of the casting *a* and X-ray spectra of the carbide *b* and sulfide inclusions *c* (marked by crosses)

Acknowledgement
This work is carried out within a framework of the government order (No. FZRU-2020-0011) of the Ministry of Science and Higher Education of the Russian Federation

References

[1] Joseph R D 1996 *ASM Specialty Handbook: Cast Irons* (ASM International)
[2] Vozdvizhenskij V I, Grachev V A and Spasskij V V 1984 *Foundry alloys and their melting technology in mechanical engineering* (Moscow: Mashinostroenie)
[3] Truhov A P and Malyarov A I 2004 *Foundry alloys and smelting* (Moscow: Akademiya)
[4] Girshovich N G 1978 *Handbook of iron casting* (Moscow: Mashinostroenie)
[5] Olawale J O, Ibitoye S A and Oluwasegun K M 2016 *IJSER* **7** p 397
[6] CHajkin V A, Boldyrev D A and CHajkina N V 2007 *Vestnik of NMSTU* **4** p 53
[7] Boldyrev D A, Dema R R and Kalugina O B 2019 *Materials Today: Proc.* **11** p 510
[8] Boldyrev D A, Shapovalov A N, Nefedev S P, Dema R R, Kononov V N, Kharchenko M V, Suvorova E V and Markova I Yu 2019 *J. Chem. Technol. Metall.* **54** p 348
[9] GOST 1412-85 *Flake graphite iron for castings. Grades* (Moscow: IPK Izdatel'stvo standartov)
[10] GOST 3443-87 *Cast iron castings with graphite of different form. Methods of structure determination* (Moscow: IPK Izdatel'stvo standartov)
[11] ASTM A247 *Standard Test Method for Evaluating the Microstructure of Graphite in Iron Castings* (Committee A04 on Iron Castings)

IOP Conf. Series: Materials Science and Engineering **966** (2020) 012024 doi:10.1088/1757-899X/966/1/012024

Influence of blast furnace slag type on physical and mechanical properties of Portland cement

F L Kapustin[1] and T S Balueva[2]

[1] Department of materials sciences in construction of the Institute of new materials and technologies, Ural Federal University, 28 Mira Street, Yekaterinburg, 620002, Russia
[2] CJSC 'Nevyansky Tsementnik', bldg. 1, Lenina Street, Cementnyy, Neviansky District, Sverdlovskaya Region, 624173, Russia

E-mail: [1]f.l.kapustin@urfu.ru, [2]balueva@eurocem.ru

Abstract. Physical properties, grain, chemical and mineral compositions of blast-furnace slags selected from the dumps of JSC 'EVRAZ Nizhny Tagil metallurgical plant' and PJSC 'Nadezhdinski Metallurgical Plant' were studied. Also, the same parameters of granulated blast-furnace slag of JSC 'Mechel-Materialy' currently used by cement plants in the Ural region as a mineral additive during Portland cement grinding were investigated. By chemical composition all of them are acidic. In contrast to granulated slag which mainly contains the vitreous phase dump slags are well crystallized and mainly contain the minerals okermanite, mervinite and monticellite. Dump slags have a Student's t-criterion of more than 15.0, i.e. they can be used as a mineral additive during Portland cement grinding. The introduction of 20 % of the studied slags reduces the water demand of Portland cement of joint grinding and accelerates the setting of the cement paste, reduces the bending and compression strength compared to the additive-free cement. Slag-containing cements after 7 and 28 days of hardening have similar values of compressive strength: 27.2–29.2 and 41.1–43.0 MPa respectively, they are normally hardening and have a strength class of 42.5. Chemical, physical and mechanical properties of these cements meet the requirements of Russian Standard 31108–2016.

1. Introduction

At cement plants in Russia granulated blast furnace slags (GBFS) are mainly used as a mineral additive in grinding Portland cement. According to Russian Standard 31108–2016 [1] Portland cements of types CEM II/A and CEM II/B can contain mineral additive, i.e. GBFS in an amount of 6–20 % and 21–35 % respectively as the main component. Granulated blast furnace slags used in the cement industry must meet the requirements of Russian Standard 3476–74 in terms of their composition and properties [2]. The introduction of GBFS increases water demand, lengthens setting, reduces the heat of cement hydration and improves resistance of cement stone to aggressive environments [3–5].

Recently the GBFS volumes of usage by cement plants has decreased significantly due to not only reduction in blast-furnace production of cast iron but also an increase of CEM I Portland cement use in monolithic construction [6]. In addition, the slag cost including transportation expenses increased to 1,000 Russian rubles (RUR) per ton while the average price of Portland cement is 4,000 RUR per ton. It is also necessary to take into account that the wet granulation slag has a high humidity (up to 25 %) and its drying requires a large fuel consumption which increases the cost of production of slag cements [7]. Taking into account these factors cement plants in the Ural region are investigating the feasibility of cheaper and more accessible rocks or man-made waste as a mineral additive for grinding cement such as fly ash from Reftinskaya State District Power Plant, ash-and-slag mix from Verkhnetagilskaya

State District Power Plant, and dump blast furnace slags (DBFS) from nearby metallurgical plants [8–11].

The aim of this work is to study the influence of the type of blast furnace slag formed at Ural metallurgical plants on the grindability and physical and mechanical properties of Portland cement.

2. Raw materials and research methods

The following raw materials were used in the work:

- cement clinker produced by CJSC 'Nevyansky Tsementnik' (Nevyansk) having a saturation coefficient KN = 0.92, silicate module n = 1.9, alumina module p = 1.1, mineral composition, %: 61.5 C_3S; 13.1 C_2S; 6.6 C_3A; 15.7 C_4AF; 0.3 CaO_{free};
- gypsum-anhydrite stone of the Ergachinsky Deposit containing 36.5 % $CaSO_4{\cdot}2H_2O$ and 60.0 % $CaSO_4$, which meets the requirements of Russian Standard 4013–82 [12];
- GBFS of JSC 'Mechel-Materialy', Chelyabinsk (sample 1);
- DBGS of JSC 'EVRAZ Nizhny Tagil metallurgical plant', Nizhny Tagil (sample 2);
- DBGS of PJSC 'Nadezhdinski Metallurgical Plant', Serov (sample 3).

Grain composition, bulk and true density of blast furnace slags were determined using standard methods and the specific surface area of crushed slag was determined using the air permeability method on the PSH-11M device.

Slag and clinker grindability was studied in a laboratory drum mill with ball loading when the ratio of the material (of fraction less than 5 mm)/grinding body is equal to 1/9, measuring the period of grinding to a residue on sieve No. 008 equal 5.0 %. The chemical composition of slag samples was determined according to Russian Standard 5382–91 [13], their X-ray phase analysis was performed on a Miniflrex 600 diffractometer (Rigaku Technologies, Inc.) in the range of 2θ from 0 to 90° with automatic decoding using the ICDD PDF2 database (Japan).

The possibility of using of blast furnace slags as a mineral additive for Portland cement grinding was evaluated by their activity characterized by a statistical assessment of the significance of differences in compressive strength of samples with the addition of slag and quartz sand after steaming, i.e. by the Student's t-criterion, according to Russian Standard 25094–94 [14].

The hydraulic activity of slags with an MgO content of less than 10 % in accordance with Russian Standard 3476–74 was evaluated by the quality coefficient (K) calculated using the formula:

$$K = (CaO + Al_2O_3 + MgO)/(SiO_2 + TiO_2). \quad (1)$$

According to the values of the quality coefficient GBFS is classified to three grades: class I includes slags with a K of more than 1.65, class II – more than 1.45, and class III – at least 1.20. Also, the content of aluminum, magnesium, manganese and titanium oxides is limited in the blast furnace slags.

Grinding of Portland cement with 20 % slag and 6 % gypsum-anhydrite stone was carried out in a laboratory two-chamber ball mill. The grinding fineness, normal density and setting time of cement paste, specific surface area, bending and compression strength of cements with different types of slag were determined according to Russian Standard 30744–2001 [15] and the bending and compression strength of the cement mortar after steaming – according to Russian Standard 310.4–81 [16].

3. Results and discussion

3.1. Composition and properties of blast furnace slags

The results of determining the grain composition and density of samples of blast furnace slags are presented in table 1. It was found that the JSC 'Mechel-Materialy' GBFS contains mainly porous vitrified particles less than 5 mm in size (88.7 %). The slag sample of JSC 'EVRAZ Nizhny Tagil metallurgical plant' consists of up to 78.3 % broken stone with a 5–20 mm grains content and the sample of PJSC 'Nadezhdinski metallurgical plant' is represented by broken stone containing 92.0 % of particles of 10–40 mm in size. The slag of PJSC 'Nadezhdinski metallurgical plant' has the highest true density and the granulated slag of JSC 'Mechel-Materialy' has the lowest one.

Table 1. Grain composition and properties of blast furnace slags.

Sample number	Slag manufacturer	The contents of the grains, %, on the sieve, mm				Density, kg/m^3	
		20–40	10–20	5–10	<5	Bulk	True
1	JSC 'Mechel-Materialy'	–	8.5	2.8	88.7	1250	2860
2	JSC 'EVRAZ Nizhny Tagil metallurgical plant'	2.1	35.6	42.7	19.6	1410	2980
3	PJSC 'Nadezhdinski metallurgical plant'	60.3	31.7	7.2	0.8	1320	2880

According to the chemical composition the studied blast furnace slags differ mainly in the content of calcium, magnesium and titanium oxides (table 2) and are acidic (basicity modulus less than 1). GBFS (sample 1) and dump blast furnace slag (sample 3) have a quality coefficient of more than 1.45 and belong to grade 2 and the dump blast furnace slag (sample 2) does not meet the requirements of Russian Standard 3476–74 in the amount of TiO_2 (more than 4 %) (table 3). The activity assessment of blast furnace slags by the Student's t-criterion showed that GBFS (sample 1) has the highest value (51.0) and dump blast furnace slag (sample 2) has the lowest value (15.7). Dump blast furnace slag (sample 3) has a t-criterion equal to 18.6. Thus, the studied dump blast furnace slags have a t-criterion value of more than 15.0, i.e. they can be used as a mineral additive when Portland cement grinding.

Table 2. The chemical composition of blast furnace slags.

Sample number according to Table 1	Chemical composition (mass %)								
	SiO_2	Al_2O_3	Fe_2O_3	CaO	MgO	SO_3	MnO	TiO_2	Other
1	34.56	11.72	0.88	38.20	7.60	0.84	0.63	1.21	4.36
2	35.45	12.67	1.78	32.33	6.88	0.88	0.57	6.41	3.91
3	35.93	11.72	1.18	41.36	3.90	0.63	0.68	1.01	3.59

Table 3. Compliance of blast furnace slags properties with Russian Standard 3476 requirements.

Indicators	Requirements of the standard for 1, 2, 3 grades	Actual values for slag samples		
		1	2	3
Quality factor	≥ 1.65; 1.45; 1.20	1.61	1.24	1.54
Al_2O_3 quantity (mass %)	≥ 8; 7.5; –	11.72	12.67	11.72
MgO quantity (mass %)	≤ 15	7.60	6.88	3.90
TiO_2 quantity (mass %)	≤ 4	1.21	6.41	1.01
MnO quantity (mass %)	≤ 2; 3; 4	0.63	0.57	0.68

X-ray phase analysis of slag samples showed that GBFS (sample 1) is mainly represented by a glassy phase and a small content of the okermanite ($2CaO{\cdot}MgO{\cdot}2SiO_2$) and spinel ($MgO{\cdot}Al_2O_3$) minerals (Figure 1, a). In contrast to granulated slag, both samples of dump blast furnace slags are well crystallized and contain the following minerals: okermanite, mervinite $3CaO{\cdot}MgO{\cdot}2SiO_2$ and monticellite $CaO{\cdot}MgO{\cdot}SiO_2$ and the slag (sample 2) also contains perovskite and glass phase in a small amount (Figure 1, b, c). Thus, all three samples of blast furnace slag differ by the content of the vitreous phase and olivine group minerals which do not hydrate in the composition of Portland cement stone [17].

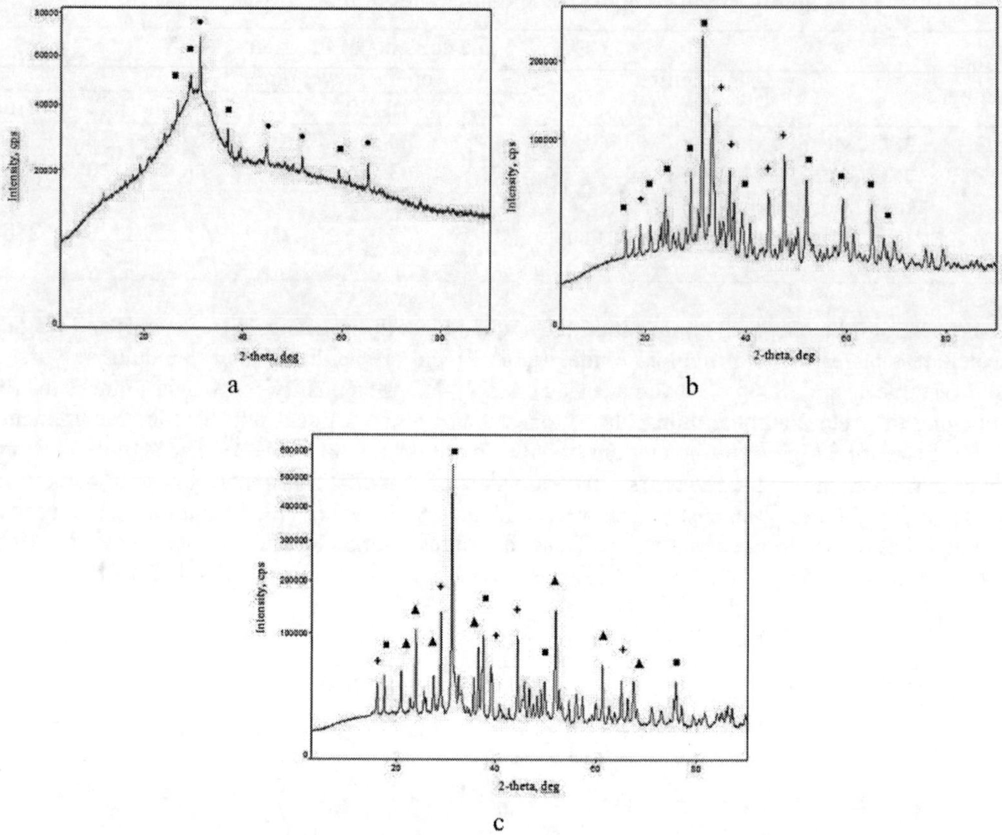

Figure 1. Diffraction pattern of samples of blast-furnace slag:
a – sample 1; b – sample 2; c – sample 3;
● – spinel; ■ – akermanite; + – merwinite; ▲– monticelli.

3.2. Influence of slag additives on the properties of Portland cement

It is known that the structure and mineral composition of metallurgical slags affects not only their activity but also their grindability [18]. The grindability of all three samples of the studied blast furnace slags as well as Portland cement clinker was determined. It was found that slags are crushed much worse than clinker with a grinding time of 80 minutes. The samples can be ranked according to their grinding time: the longest is 180 minutes for sample 3, then sample 2 and sample 1 (GBFS) with grinding times of 120 and 140 minutes respectively.

Four compositions of Portland cement were prepared by joint grinding to study possibility of using of slags as a mineral additives: the first – without the addition of slag, from the second to the fourth – with 20 % of slags. The results of determining their physical and mechanical properties are presented in Table 4. It is found that the additive-free cement and Portland cement with GBFS are ground 15 % faster than the compositions containing samples of dump blast furnace slags. All cements had the same grinding fineness and similar values of specific surface area.

The cements obtained have different water demand: composition with the sample 3 slag has the lowest normal density equal to 26.4 %; cement with sample 2 slag has the highest normal density equal to 28.2 % having an increased specific surface area unlike other Portland cements. Compared to additive-free Portland cement all three samples of blast furnace slags accelerates the setting of cement paste and most strongly – with the introduction of DBFS of sample 2: initial setting time has decreased 1.8 times, the final setting – by 55 %. DBFS of sample 3 has the minimal influence on the cement setting.

Table 4. Results of physical and mechanical tests of Portland cements.

Cement sample number	Slag sample number	Fineness (mass %)	Specific surface area (m^2/kg)	Normal consistency of cement paste (%)	Setting time (min)		The strength (MPa)			
							Bending, after (days)		Compression, after (days)	
					initial	final	7	28	7	28
1	–	2.5	324	27.4	215	310	7.9	8.5	38.6	47.3
2	1	2.6	315	27.0	155	260	6.5	7.9	29.2	43.0
3	2	2.5	337	28.2	120	200	6.6	8.3	27.2	41.1
4	3	2.7	327	26.4	190	315	6.1	7.3	28.1	42.1

It was found that the introduction of all three samples of blast furnace slag into Portland cement in an amount of 20 % reduces the bending and compression strength compared to additive-free cement: cement with DBFS (sample 2) has minimal decrease of bending strength by 28 days of water hardening (by 2.4 %), and cement with DBFS (sample 3) has the largest strength decrease (by 16.4 %). However, all cements with slags have similar values of compressive strength after 7 and 28 days of hardening, i.e. 27.2–29.2 and 41.1–43.0 MPa respectively. By physical and mechanical properties all Portland cements containing blast furnace slag meet the requirements of Russian Standard 31108–2016; they are normally hardening and have a strength class of 42.5. In addition, they also meet its requirements in terms of chemical parameters (mass loss during ignition, insoluble residue, and chloride ion content).

4. Conclusions

The composition and properties of dump blast furnace slags of JSC 'EVRAZ Nizhny Tagil metallurgical plant' and PJSC 'Nadezhdinski metallurgical plant' and granulated blast furnace slag of JSC 'Mechel-Materialy' used by cement plants in the Ural region as a mineral additive for grinding Portland cement were studied. By chemical composition all of them are acidic. Unlike granulated slag, the dump ones are well crystallized and mainly contain such minerals as akermanite, merwinite and monticellite. Dump slags have a Student's t-criterion of more than 15.0, i.e. can be used as a mineral additive when grinding Portland cement. Blast furnace slags are milled by 1.5–2.2 times longer than Portland cement clinker; the titanium slag is especially harder to mill. The introduction of 20 % of the studied slags reduces the water demand of Portland cement of joint grinding, accelerates the setting of the cement paste, reduces the bending and compression strength compared to the additive-free cement. Slag-containing cements have similar values of compressive strength after 7 and 28 days of hardening, i.e. 27.2–29.2 and 41.1–43.0 MPa respectively. They are normally hardening and have a strength class of 42.5 and meet the requirements of Russian Standard 31108–2016 [1] by chemical, physical and mechanical properties.

References

[1] Russian Standard 31108–2016: Cements for General construction Specifications (Moscow: Standardinform) [In Russian]
[2] Russian Standard 3476–74 Slags blast furnace and electric-phosphoric granulated for manufacturing of cement (Moscow: Standards publishing house) [In Russian]
[3] Breitenbücher R, Bäcker J, Kunz S, Ehrenberg A and Gerten C 2018 Optimizing the Acid Resistance of Concrete with Granulated Blast-Furnace Slag *International Conference on Concrete Repair, Rehabilitation and Retrofitting MATEC Web Conf.* **199** pp 1–7
[4] Suresh D and Nagaraju K 2015 Ground Granulated Blast Slag In Concrete – A Review IOSR *Journal of Mechanical and Civil Engineering* **12(4)** pp 76–82
[5] Dvořák K, Gazdič D, Hájková I and Magrla R Comparison of Selected Properties of Portland Cement Based Materials and Alkali Activated Materials Based on Granulated Blast Furnace Slag *Materials Science Forum* **865** pp 107–113

[6] Kiyutsina O 2018 Trends and prospects of development of monolithic construction in Russia *Cement and its Applications* **2** pp 28–29 [In Russian]

[7] Paschenko A, Serbin V and Starchevskaya E 1985 *Binding materials* (Kiev: Vishcha shkola) [In Russian]

[8] Kapustin F, Ufimtsev V, Vishnevskii A, Fomina I, Kapustin A and Zemlianoi K 2019 Use of fly ash from Reftinskaya DPS in the production of construction materials and products *Proc. Int. Conf. 'Fundamental research and applied development of processes of processing and utilization of technogenic formations'* (Ekaterinburg: UrO RAN) pp 50–55 [In Russian]

[9] Kapustin A and Semerikov I 2011 Composition and properties of composite cement with the addition of ash and slag mix of HPS *Bulletin of SUSU* **16** (Chelyabinsk: SUSU) pp 38–40 [In Russian]

[10] Kapustin F, Ragozin D, Kuznetsov A, Semerikov I and Kapustin A 2010 Influence of blast furnace slag on cement quality according to Russian Standard 31108–2003 *Bulletin of SUSU* **15** (Chelyabinsk: SUSU) pp 22–25 [In Russian]

[11] Trofimov B and Shuldyakov K 2015 Application of low-active granulated blast furnace slag *Architecture, urban planning and design* **6** pp 38–45

[12] Russian Standard 4013−82 Gypsum and gypsum-anhydrite stone for the production of binding materials. Technical conditions (Moscow: Standards publishing house) [In Russian]

[13] Russian Standard 5382−91 Cements and materials of cement production. Chemical analysis methods (Moscow: Standards publishing house) [In Russian]

[14] Russian Standard 25094−94 Active mineral additives for cements. Test method (Moscow: Standards publishing house) [In Russian]

[15] Russian Standard 30744−2001 Cements. Test methods using polyfractionated sand (Moscow: Standardinform) [In Russian]

[16] Russian Standard 310.4−81 Cements. Methods for determining the bending and compressive strength (Moscow: Standards publishing house) [In Russian]

[17] Evtushenko E, Starostina I and Kravtsov E 1999 Processes of crystallization and activity of the domain granulated slag *Modern problems of construction materials science. Materials of the fifth academic readings of the RAASN* (Voronezh: RAASN) pp 130–133 [In Russian]

[18] Piachev V, Shkolnik Ya and Burlakov V 1985 Grindability of granulated blast furnace slags *Cement* **8** pp 8–9 [In Russian]

Prospects for the use of peat in metallurgy

T B Ykonovskaya[1, 2] and A I Zhigulskaya[1]

[1] Tver State Technical University, 22, pr. Afanasiy Nikitin embankment, Tver, 170026, Russia

E-mail: [1]tby81@yandex.ru

Abstract. A large number of metallurgical industry enterprises in the regions of Russia mainly use imported and rather expensive coking coal or natural gas for steelmaking. However, almost all regions of Russia have large reserves of peat, which can replace expensive hydrocarbon raw materials in metallurgical processes. For this reason, the use of peat raw materials (briquette, piece) in metallurgy is of scientific and practical interest. The properties of peat briquettes and pellets in the process of pyrometallurgical processing are set by the methods and technological parameters of their preparation. The article presents the results of a study of the influence of the type and quality properties of peat on the technological process of recovering peat compositions prepared on the basis of iron ore from the Bakchinsky deposit. Peat of the Vasyugan peat massif was used for the production of peat melting compositions. Experimental data showed that, for peat compositions, low-ash peat on the watershed is suitable. The use of peat melting compositions allows us to solve the problem of the return of ore dust and substandard ore in the smelting processes and thereby develop a rational low-waste technology.

1. Introduction

Ensuring the growth of metallurgical production in Russia is associated with the problem of meeting the demand for coking coal. Despite the increased pace of coal mining in Russia, its cost is increasing annually. It should be noted that coal is an exhaustible natural resource, and therefore, the reserves of high-quality coking coal in Russia are depleted. Coking coal reserves are estimated at 42.3 billion tons, of which more than 70% are accounted for by the Kuznetsk coal basin, while the estimated resources are about 390 billion tons.

The coking coal market structure is an oligopoly with elements of a local monopoly in its individual segments and for certain brands of coal. In addition, there is a rather strong ‘coal lobby’ thanks to which Russian coal mining companies receive financial support from the state in the form of tax and tariff benefits, as well as direct subsidies and soft loans for development and production modernization. Specialists of the Mining Research Institute named after A A Skochinsky determined the dependence of the price of coal on the price of oil products [1, 2], however, despite the drop in exchange prices for oil observed in March-April 2020, the prices of coking coal are stable and show a slight increase. Although prices on the global coal market are falling amid the economic crisis.

Demand for coking coal in the future will continue and even grow, due to the increasing complexity of mining and geological conditions for their development and the quality of coal raw materials. The main value of coking coal in producing coke for metallurgical processes in the blast furnace production of steel and other metals. The annual demand of the largest metallurgical plants in Russia for coking coal is 20.8 million tons. However, the development of new deposits requires significant investment. In addition, in the raw material component of the cost of pig iron, 40–50% is

IOP Conf. Series: Materials Science and Engineering **966** (2020) 012025 doi:10.1088/1757-899X/966/1/012025

accounted for by coke. In this regard, the search for an alternative to coking coal raw materials for the production of coke, which will replace expensive coal and improve the quality of steel products, becomes a very urgent task. In this regard, peat has several advantages. With a high yield of chemical products of pyrolysis, a blast furnace process on peat can produce high-quality cast iron, such as, for example, coal, which is used for smelting high-quality steel.

The qualitative properties of coal for the purposes of metallurgical processes are characterized by the level of sintering and the presence of impurities of such substances as phosphorus and sulfur. Therefore, metallurgical enterprises have certain requirements for the quality of coal coke used, among which the most important is the ash indicator. As you know, the higher the ash content, the worse the quality of coal coke for the blast furnace process. In this regard, for metallurgical processes, the ash content of the charge should be no more than 10%, the yield of volatile substances up to 30%, and the sulfur content not more than 1%.Since natural fuel is used for the production of coke, peat can be used for these purposes. Unlike hard coal, peat has a low phosphorus and sulfur content, is a common and renewable resource, with up to 90% of peat reserves concentrated in the North-West Federal District and Western Siberia, in the metallurgical regions of Russia. Peat is an organic fuel and a valuable chemical raw material for a wide range of industrial uses. Many geologists [3–5] hypothesize that peat is the initial stage of coal formation. For different types of peat and the types of their formation, the ash content varies widely; for high peat, the ash content is from 0.5 to 5%, for transitional peat from 6 to 10%, and for low-lying peat more than 10%. Therefore, for the production of peat coke in terms of ash content, higher peat of a watershed type of formation is more suitable. And from a technical and economic point of view, in order to reduce transport costs, stocks of low-ash peat should be very close to metallurgical production (up to 100 km.). This issue is especially relevant for the Tomsk region in whose territory the Vasyugan-Tymsky complex of a wide range of natural resources of peat, iron ore, oil and gas are located. Peat coke is suitable when used as a thinning element of the mixture, as well as for the agglomeration of iron ores. Peat coke, lump peat and milled peat can serve as blast furnace fuel for foundry purposes. Due to its high reactivity, peat is a good reducing agent for iron. This corresponds to one of the main directions of progress in ferrous metallurgy – the intensification of the blast furnace process through the improvement of the aerodynamic properties of charge materials by creating a ‘charge’ based on a powder of finely ground concentrate – an iron ore enrichment product. In addition, quite a lot of substandard, ‘waste’ rock accumulated in the territory of iron ore mining enterprises, which is characterized by a low content of iron ore and forms artificial technogenic deposits. This ‘waste’ breed can also be used as raw material for the ‘charge’ and the extraction of steel [6–8].

Along with coke, peat semi-coke is successfully used in metallurgical processes, as well as for the cementation of carbon steels, the manufacture of sorbents in a glass charge. In Germany, peat coke and semi-coke were used for smelting ferroalloys, for heat treatment in mechanical engineering, for the production of refractory bricks, and for the production of activated carbon and carbon disulfide. However, it should be noted that in foreign countries the use of peat in metallurgy is not widespread. The scientific interest in the use of peat is also extremely weak.

2. Materials and methods

For several years, scientists from Tver State Technical University conducted research on peat and ore materials obtained back in 1980 from the Vasyugan peat deposit and the Bakchinsky iron ore deposit, located in the Tomsk Region and bordering each other. In a problematic laboratory, the influence of the initial moisture content of peat, the type of peat, the intensity of its processing, the molding pressure, the degree of grinding of ore and fluxes and their amount, the coking regime and other factors on the strength of peat briquettes and pellets was studied.

The technique for producing peat briquettes and pellets is based on the theory of structure-forming processes in peat colloidal systems according to the principles of physicochemical mechanics of dispersed materials. This theory was developed by the founding scientists of the Peat School of Tver State Technical University: Afanasyev A F, Gamayunov N I, Volarovich M P, Churaev N V,

Lykov A V Research on this issue is ongoing [8–10]. According to a number of its properties, peat is a model of capillary-porous bodies, and these properties have a significant impact on the formation of the strength structure of peat briquettes and pellets. Peat is a dispersed multiphase system. The presence in it of hydrophilic and hydrophobic organomineral colloids, solutions and dispersions of low and high molecular weight compounds leads to the formation of various solid phase associates. The higher the density and strength of peat briquettes and pellets, the better the process of iron recovery. Granules were made from milled peat that passed through the crusher and former. The granules, dried to 40% humidity, were loaded into a coking shaft. Peat coke, similar in technological characteristics to charcoal, but possessing the strength of coal coke, practically does not introduce undesirable impurities into cast iron. At the same time, from 6 tons of peat, 1 ton of coke can be obtained. The object of study of this work is peat material restored at a temperature of 500–1100°C. Peat obtained in 1980 from site No. 3 of the Vasyugan deposit (Figure 1) was used to recover iron from ore. The samples relate to high peat, accounting for 24% of the total reserves of plot No. 3, and prevailing low-lying peat in the plot (52% of the reserves). Peat data are considered as raw materials for the production of peat materials (TPM). TPM consists of 70% ore and 30% admixtures of peat of the high and low types. The ore is a loose mass of brown color, consisting of goethite and hydrogetite [11–14].

Table 1. Peat characteristics of plot № 3 of Vasyugan field.

Sample number	Peattype	Type of peat	Degree of decomposition, R %	Ash content, A^c %
1	Lowland	Sedge-sphagnum	35	5.5
2	Upstream	Fuskum	35	1.4

Ore chips were laid out on a peat mass in an even layer of 1–2 cm. The resulting peat mass was passed through the screw twice and formed from it cylindrical samples with a diameter of 34–35 mm by extrusion in a special matrix. TPM samples were dried in a laboratory room to a constant weight, periodically measuring the size and mass of the moldings to calculate the shrinkage coefficient. The obtained TPMs were tested for their metallization upon heating at a constant rate of 5 °C/min. Samples were placed in a 300 ml heat-resistant steel retort and heated in a TEP-1 crucible electric furnace to temperatures of 500–1000 °C with an interval of 100 °C. Combined gas and decomposition products of TPM were discharged through a fitting in the lid of the retort into condensation equipment. Upon reaching the final heating temperature, the retort was removed from the furnace and cooled to room temperature. Given the pyrophoricity of the reduction products upon cooling in the retort, an excess nitrogen pressure was maintained for half a day. To heat TPM to a temperature of 1100 °C, a high-temperature tube furnace was used. The composition of iron ore of total iron contained about 38%, silicon oxide – 24%, calcium oxide – 0.64%, aluminum oxide – 5%, phosphorus pentoxide – 1.17%.

3. Discussion

The resulting cast iron contained silicon – 1.1%, vanadium – 0.06%, phosphorus – 0.5%, manganese – 021 %. The slag contains: silicon oxide – 43.82%, calcium oxide – 30.36%, aluminum oxide – 15 %, iron oxide – 3%, phosphorus pentoxide – 0.1% magnesium oxide – 2%. The close contact of the reducing agent and the peat ores, as well as the high volatility of peat, favorably affect the recovery of iron ore. It is shown that the rate of reduction of magnetite and brown iron ore in the presence of peat is higher than with other solid reducing agents. This circumstance can significantly reduce the process time and lower its temperature. All this opens up favorable prospects for the use of peat materials by the method of a problematic laboratory for direct (non-explosive) iron production or other metallurgical processes where a highly reactive reducing agent is required.

IOP Conf. Series: Materials Science and Engineering **966** (2020) 012025 doi:10.1088/1757-899X/966/1/012025

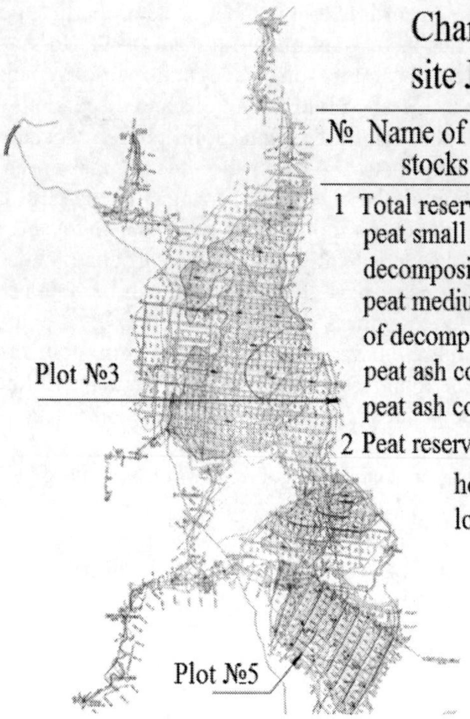

Characteristics of peat reserves in site №3 of the Vasyugan peat field

№	Name of peat stocks	Area, ha	Depth, m	Stocks, th.t
1	Total reserves	977	2	2036
	peat small degree of decomposition	329	0,80	201
	peat medium and hight degfee of decomposition	648	1,49	1835
	peat ash content up to 3%	648	1,49	1823
	peat ash content from 3% to 5%	68	1,2	12
2	Peat reserves by deposit type			
	horse	370	2,44	1144
	lowland	132	1,76	486

Figure 1. Map-scheme of the southern part of the Vasyugan peat massif (plots № 3 and № 5).

The complex chemical and metallurgical (blast furnace) process is based on the use of dry lump peat or 'mono-mixture' that has not undergone preliminary coking or thermal briquettes as blast furnace fuel. In this case, the blast furnace combines the process of smelting cast iron with the process of coking peat and, along with cast iron, produces high-calorie gas and other chemical products of peat pyrolysis: peat resin, gas gasoline.

A wide range of hydrocarbons and organic oxygen compounds can be obtained on the basis of carbon monoxide, which is the predominant component of blast furnace gas and water vapor, when these components are passed through the catalyst at a temperature of 220–280 °C. Gas can be used economically to generate electricity or for the synthesis of ammonia.

4. Conclusion

As a result, the degree of reduction of iron from TPM was 92%, the degree of metallization after firing was 94%. Experimental melting of porous iron gave a metal with a carbon content of 0.05%, sulfur and phosphorus less than 0.05%. In addition, more durable and dense TPMs are obtained when using peat. High thermal strength of briquettes and pellets from TPM is associated with the emergence of a metal frame in this case, with an increase in the degree of recovery, the thickness of the metal frame increases.

Peat for smelting cast iron can be either upland or lowland, provided that the ash content should not exceed 10% and the humidity not more than 30%. The resulting pressed peat piece from which coke is produced must have a mechanical strength of at least 2 MPa with a low content of fines and sulfur. As a disadvantage of peat coke in comparison with coal coke, there is not much mechanical strength. However, TPMs with various types of peat had good physical and mechanical characteristics both in cold and in heated states. Also, instead of iron ore, you can use its dust and get briquettes and pellets of TPM with the organization of milk waste technologies.

5. Results

In conclusion, we can say that the use of peat in metallurgy shows good prospects for this area. Shaping waste from metallurgical processes with peat makes it possible to obtain TPM briquettes and pellets with desired quality properties in ash, moisture, sulfur and phosphorus, and mechanical strength. The problem can arise only in the absence of a peat deposits base with specified geological characteristics of peat raw materials for the production of peat coke.

References

[1] Mikhailov A, Zhigulskaya A and Yakonovskaya T 2017 Strip mining of peat deposit mine / Planning and equipment selection: mpes 2017. *Proceeding of the 26th International Symposium* (Luleå, Sweden)

[2] Mikhailov A V, Zhigulskaya A I and Yakonovskaya T B 2017 Excavating and loading equipment for peat mining *IOP Conference Series: Earth and Environmental Science*

[3] Zhigulskaya A I et al 2016 Processing wood resources of peat deposits to produce environmentally friendly products *Actual problems of life safety and ecology: Sat. scientific tr II international scientific-practical conference with a scientific school for youth* (Tver: TSTU)

4] Zyuzin B F et al 2016 Complex use of peat and wood resources *Problems of the rational use of natural resources and sustainable development of Polesye: Sat. doc International scientific conference: in 2 vol*

[5] Yakonovskaya T B et al 2016 Rational use of peat resources in bioenergy of the Tver region. *Socio-economic and environmental problems of mining, construction and energy* ed. R.A. Kovaleva

[6] Gebler I V, Smolyaninov S I and Neznanov A G 1963 The influence of the degree of grinding of ore and fluxes on the quality of peat-based fuel-melting materials *Bulletin of the Tomsk Polytechnic University. Geo-Resource Engineering* vol **112**

[7] Smolyaninov S I and Denisov A M 1963 The effect of briquetting temperature and pressure on the mechanical properties of peat briquettes *Bulletin of the Tomsk Polytechnic University. Geo-Resource Engineering* vol **118**

[8] Smolyaninov S I 1974 Development of the fundamentals of the method of chemical and metallurgical use of peat *Bulletin of the Tomsk Polytechnic University. Geo-Resource Engineering* vol **118**

[9] Smolyaninov S I and Krinitsyn G G 1965 Fuel-melting materials from peat of low humidity *Bulletin of the Tomsk Polytechnic University. Geo-Resource Engineering* vol **126**

[10] Grevtsev N V, Aleksandrov B M and Plekhanov K A 2003 Prospects for the use of peat carbon-containing and ore-containing briquettes in metallurgical processes *Bulletin of the Ural State Mining and Geological Academy* vol **17**

[11] Patel N, Cameron I and Gordon Y 2016 Strategies for implementing direct reduction technologies in an integrated steel plant *Iron and steel technology* vol **13** no **4** pp 75–81

[12] Steinacker S and Antrekowitsch J. 2017 Hydro metallurgical treatment of a copper-iron-molybdenum alloy *World of metallurgy - erzmetall* vol **70** no **6** pp 319–26

[13] Ravisankar V, Venugopal R and Bhat H 2019 Investigation on beneficiation of goethite-rich iron ores using reduction roasting followed by magnetic separation *Mineral processing and extractive metallurgy: transactions of the institute of mining and metallurgy* vol **128** no **3**

[14] Huang Y, Zhang L, Wang M and Aindow M 2019 Fabrication of a multi-phase porous high-temperature Mo–Si–B alloy by in situ reaction synthesis *Powder metallurgy* vol **62** no **4**

IOP Conf. Series: Materials Science and Engineering **966** (2020) 012026 doi:10.1088/1757-899X/966/1/012026

Research on the influence of the strip tension on contact stress and cold rolling force with a strain-hardening

K Ilin[1, 2] and G Baranov[1]

[1] Ural Federal University named after the first President of Russia B N Yeltsin, 19, Mira St., Yekaterinburg, 620002, Russia

E-mail: [1]k.v.ilin@urfu.ru

Abstract. The influence of strip tension on contact stress and cold rolling force with strain-hardening has been investigated by using a numerical method. Dependences between roll pressure and strip tension for one rolling case have been studied. A comparison of the results provided by an introduced method to the already known methods allowed measurements of inaccuracy when the arc of contact has been replaced by chord and whenthe yield strength value for each step has been replaced by the average value of yield strength. Approaching to the real shape of the deformation zone and step-by-step calculations of strain-hardening increases magnitudes of average normal stress and rolling force. The highest difference between obtained results in comparison with the introduced method and already known methods reaches the largest value in the case with intensely hardening steels without strip tension. The increase of strip tension shows less difference between methods. The introduced method lets to calculate the rolling force for different metals depending on strip tension with more precision. Obtained results of the numerical analysis of the influence strip tension on the average normal stress and rolling force were summarized in the form of functions convenient for engineering calculations, with inaccuracy less than 5%.

1. Introduction

Strip tension one of the important factors defining energy intensity, rolling force and contact stresses during cold strip rolling. Contact stress during the rolling process depends on geometric parameters of deformation zone, mechanical characteristics of the material, interfacial friction, front and back tension [1]. Cold rolling is characterized by low interfacial friction and angle of bite less than 8 degrees. These facts are in consent with the application of strip theory for the metal stress analysis [2–10].

According to the method proposed by A Tselikov [1] for computing contacts stresses, the arc of contact replaced by chord and start strip tensions are taken into account via coefficients ξ_i:

$$\text{for backward slip zone: } p_A = \left(\frac{2\tau_S}{\delta}\right)\left[(\xi_0\delta - 1)\left(\frac{h_0}{h_x}\right)^{\delta} + 1\right] \tag{1}$$

$$\text{for forward slip zone: } p_B = \left(\frac{2\tau_S}{\delta}\right)\left[(\xi_1\delta + 1)\left(\frac{h_x}{h_1}\right)^{\delta} - 1\right] \tag{2}$$

where: $\tau_S = \frac{2\sigma_S}{\sqrt{3}}$ – pure shear stress; $\sigma_S = \frac{\sigma_{S0} + \sigma_{S1}}{2}$ – average yield stress; σ_{s0} and σ_{s1} – yield stress before and after the draft; $\delta = \mu \mathrm{Ctg}\left(\frac{\alpha}{2}\right)$; μ – friction coefficient; α – angle of bite; $\xi_0 = \frac{2\tau_S - \sigma_A}{2\tau_S}$; $\xi_1 = \frac{2\tau_S - \sigma_B}{2\tau_S}$; σ_A and σ_B – strip tension stress before and after deformation zone; h_0 – strip thickness before the draft respectively; h_1 – strip thickness after the draft; h_x – current value of strip thickness.

The average normal stress is defined by the equation:

$$p_{cp} = \frac{1}{l}\int_0^l p_x dx \tag{3}$$

where l – arc length.

To simplify calculations of average normal stress, the following equation [1] is used:

$$p_{cp} = \frac{1}{\Delta h}\left(\xi_0 2\tau_{S0}\frac{h_0}{\delta-2}\left[\left(\frac{h_0}{h_H}\right)^{\delta-2} - 1\right] + \xi_1 2\tau_{S1}\frac{h_1}{\delta+2}\left[\left(\frac{h_H}{h_1}\right)^{\delta+2} - 1\right]\right) \tag{4}$$

where Δh – metal draft; τ_{S0} and τ_{S1} – pure shear stress before and after the draft respectively; h_H – strip thickness at the neutral plane.

According to the method introduced by U Hessenberg and R Sims, average normal stress which includes the influence of strip tension is defined by the following equation:

$$p_{cp} = p'_{cp}\left(1 - 0,25\frac{\sigma_A - \sigma_B}{\tau_S}\right) \tag{5}$$

where p'_{cp} – average normal stress without strip tension.

A Korolev presented another simplified equation [8]:

$$p_{cp} = p'_{cp} - 0,5(\sigma_A + \sigma_B) \tag{6}$$

When average normal stress is defined by equations (1)–(3) and (5)–(6), and metal hardening has been taken into account via the average value of pure shear stress and the arc of contact replaced by chord, which may lead to imprecision [10].

As follows from equations (5)–(6), front and back strip tensions effect equally at rolling force. That inconsistent with already known experimental data, according to this data, back strip tension effects more at rolling force than front strip tension [1–2]. Already existing methods of assessment of strip tension influence usually connected with a particular case of the rolling process, which makes it difficult using them for engineering design [2, 11].

2. The new method for computing contact stress

The aim of this paper is the investigation of the influence of strip tension stress before and after deformation zone at contact stresses and rolling pressure during cold rolling metal strips with different parameters of strain-hardening. Determination of dependency of rolling pressure from strip tension, estimating imprecision from replacing arc of contact by chord and equivalent yield stress by average yield stress.

In this method arc of contact was replaced by a lot of numbers of small chords, while mechanic characteristics of metal in these spaces were set up as average value. Using equations (1) and (2), the recurrent relations between contact stresses at the beginning and the end of each space have been obtained:

for backward slip zone: $$p_{Ai} = \left(\frac{2\tau_{Si}}{\delta_i}\right)\left[(\xi_{Ai}\delta_i - 1)\left(\frac{h_i}{h_{i+1}}\right)^{\delta_i} + 1\right] \quad (7)$$

for forward slip zone: $$p_{Bi} = \left(\frac{2\tau_{Si}}{\delta_i}\right)\left[(\xi_{Bi}\delta_i + 1)\left(\frac{h_{i+1}}{h_i}\right)^{\delta_i} - 1\right] \quad (8)$$

where $\tau_{Si} = \frac{2\sigma_{Si}}{\sqrt{3}}$ – pure shear stress at i step; $\sigma_{Si} = \frac{\sigma_i + \sigma_{i+1}}{2}$ – average yield stress at i step; σ_i – yield stress at beginning of i step; $\xi_{Ai} = \frac{2\tau_{Si} \pm \sigma'_{Ai}}{2\tau_{Si}}$; $\sigma'_{Ai} = p_{Ai-1} - 2\tau_{Si}$; $\xi_{Bi} = \frac{2\tau_{Si} \pm \sigma'_{Bi}}{2\tau_{Si}}$; $\sigma'_{Bi} = p_{Bi-1} - 2\tau_{Si}$;

$\delta_i = \frac{\mu}{tg\varphi_i}$; $\varphi_i = \alpha - \frac{(1+2i')\alpha}{2\gamma}$ – the angle between the chord and rolling axis at i step; $i' = 0 \ldots i\text{-}1$; h_i – the height of the strip at beginning of i step; γ – number of chords;

Height of the strip has been defined by the equation:

$$h_i = h_1 + 2R\left(1 - \mathrm{Cos}\left(\frac{\alpha}{\gamma} i\right)\right)$$

where R – radius of roll; i change from γ to 0 for equation (7) and from 0 to γ for equation (8). The strain-hardening determination has been defined by the equation:

$$\sigma_{Si} = \sigma_{SO}\left(1 + m\varepsilon^n\right)$$

where m and n – coefficients of hardening curve; $\varepsilon = \ln\left(\frac{h_0}{h_i}\right)$ – deformation ratio.

According to the aforementioned method using equations (7)–(8), contact stress and average normal stress for different values of front and back strip tension for steel 65G and steel 10 were found, and the obtained results were compared to the results provided via equations (1)–(4).

Plots of contact stress for rolling with parameters: $R = 275$ mm; $\mu = 0.12$; $\lambda = 1.5$; $h_1 = 3$ mm are presented in figures 1–2, where plots 1 and 2 have been calculated by equations (1)–(2) and (7)–(8) respectively, without strip tension, and plots 3 and 4 have been calculated by equations (1)–(2) and (7)–(8) respectively, with strip tension $q_0 = q_1 = 0{,}5$; where: $q_0 = 0{,}5\left(\frac{\sigma_A}{\tau_{S0}}\right)$; $q_1 = 0{,}5\left(\frac{\sigma_B}{\tau_{S1}}\right)$.

The relationship between the average normal stress and strip tension has been measured by the coefficient $\upsilon = \frac{p'_{cp}}{p_{cp}}$. The value of p'_{cp} was defined by equation (3) using equations (7)–(8).

Plots presenting the dependence of ν from q_0 and q_1 for steel 65G for the aforementioned rolling parameters and different strip tensions are shown in figures 3–5. Plot 1 has been defined by equations (7)–(8), plot 2 by (1)–(2), and plot 3 by equation (4).

The same computing was made for steel 10. As the result, it has been established that the magnitude of ν coefficient for steel 65G without strip tension calculated by equations (7)–(8) was higher for about 19% than by equations (1)–(2) and 34% higher than coefficient magnitude defined by equation (4). For steel 10 difference between methods amounted to 10% and 24% respectively. Higher strip tension shows less difference between methods.

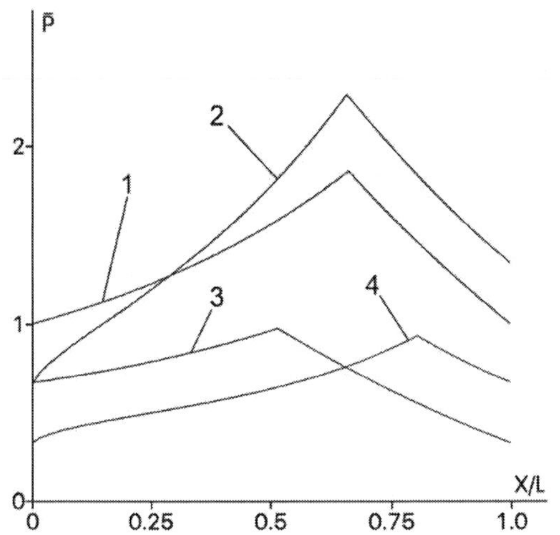

Figure 1. Plots of contact stress for steel 10 with $m = 1.86$ and $n = 0.64$.

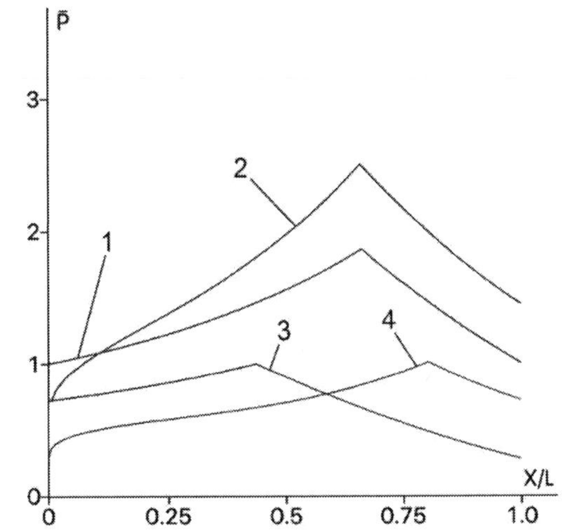

Figure 2. Plots of contact stress for steel 65G with $m = 2.2$ and $n = 0.35$.

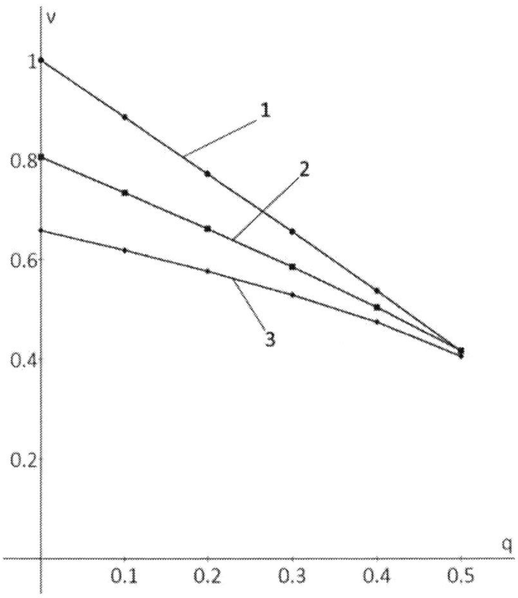

Figure 3. Plots of dependence ν from q_0 and q_1 for steel 65G.

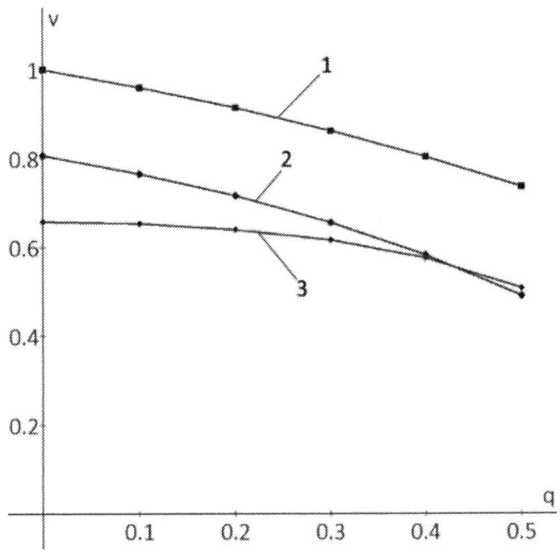

Figure 4. Plots of dependence ν from q with front tension of strip for steel 65G.

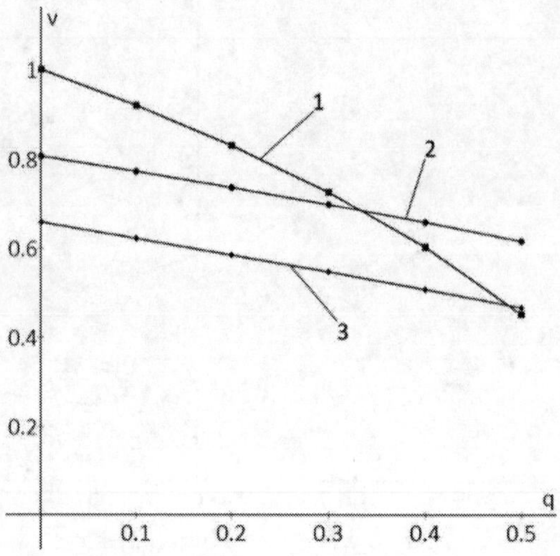

Figure 5. Plots of dependence ν from *q* with back tension of strip for steel 65G.

Analysis of influence rolling process parameters and exponent of hardening curve *n* on coefficient ν, made with using introduced model showed that λ in the range from 1.1 till 2.5 and δ in the range from 1.6 till 6 in the case when only front strip tension exists, inconsiderable effects on ν and dependence ν from $q_1$and may be approximated with inaccuracy less than 5% via equation $\nu = \nu_1 = 1 - 0{,}47 q_1^{1{,}15} n^{-0{,}23}$. In the case when only back strip tension exists influence *n* on ν may be neglected approximating results via equation $\nu = \nu_0 = 1 - q_0^{1{,}2} \delta^{0{,}07}$. In the case when front and back strip tension exists coefficient ν, with inaccuracy mentioned above, approximating via equation

$$\nu = \nu_0 - (1 - \nu_1)(1 - 1{,}5 q_0)$$

Obtained theoretical results well within already exist experimental data [1–2, 12] and provide considering the influence of strip tension with more precision.

3. Conclusions

The investigation of the influence of front and back strip tension at contact stresses and rolling forces during cold rolling of metal has been performed, plots of dependence rolling force from strip tension were made, and the obtained results were compared with results provided by already known methods. Approaching to the real shape of the deformation zone and step-by-step calculations of strain-hardening increases magnitudes of average normal stress and rolling force. The highest difference between obtained results in comparison with the introduced method and already known methods amounts to 34% in the case with intensely hardening steels without strip tension. The increase of strip tension shows less difference between methods. Obtained results of the numerical analysis of the influence strip tension on the average normal stress and rolling force were summarized in the form of functions convenient for engineering calculations.

References

[1] Tselikov A I, Nikitin G S and Rokotian S E 1980 *Theory of longitudinal rolling* (Moscow: Metallurgiya) p 320

[2] Vasilev Ya D 2012 Theoretical study of influence tension on energy efficiency of cold strip rolling process *Izv. Vuzov. Chernaya metalluriya* vol **6** pp 3–5

[3] Durnev V D and Nastich V P 2000 Influence of strip tension on the stress distribution in the deformation source in cold rolling *Russian Metallurgy (Metally)* vol **4** pp 55–59

[4] Inatovich U V and Gareeva L V 2018 Influence of strip tension on drafting schedule during cold

strip rolling *Proizvodstvo prokata* vol **11** pp 3–6
[5] Tselikov A I, Tomlenow A D and Zuzin V I 1982 *Theory of rolling: handbook* (Moscow: Metallurgiya) p 335
[6] Nikitin G S 2009 *Continuous Rolling Theory: schoolbook* (Moscow: Bauman Moscow State Technical University) p 399
[7] Treryakov A V 1966 *Theory, calculation and research of cold rolling mills* (Moscow: Metallurgiya) p 250
[8] Korolev A A 1969 *Design and calculation of machines and mechanisms of rolling mills* (Moscow: Metallurgiya) p 462
[9] Baranov G L 2015 Improved Calculation of the Contact Stress in Strip Rolling *Steel in translation* vol **45** issue **6** pp 447–52
[10] Baranov G L 2018 Influence of Strain Hardening on the Force in Cold Strip Rolling *Steel in translation* vol **48** issue **11** pp 739–44
[11] El-Bitar T A 1993 Computer program for the calculation of roll force and torque with strip tension in cold rolling *I&SM* vol **20 (5)** pp 87–96
[12] Chernov P P, Muhin U A and Bahaev K V 2005 Influence of interstand tension mode on specific energy consumption during cold rolling *Trudy VI Kongressa prokatchikov* (Lipetsk) pp 186–90

IOP Conf. Series: Materials Science and Engineering **966** (2020) 012027 doi:10.1088/1757-899X/966/1/012027

The effect of enrichment on quartz sand properties

L F Mufteeva[1], I A Pavlova[1, 2] and E P Farafontova[1]

[1] Institute of new materials and technologies, Ural Federal University named after the First President of Russia B N Yeltsin, 19, Mira str., Yekaterinburg, 620002, Russia

E-mail: [2] i.a.pavlova@urfu.ru

Abstract. The technological properties of quartz sand (without and after enrichment) by methods of emission spectral analysis, petrography, and X-ray phase analysis are investigated. The studied sand is a waste after extraction of titanium-containing components from raw materials. It is shown that the enrichment of quartz sand expands the field of its application. It has been established that quartz sand can be used as a filler of heavy, light, fine-grained, cellular and silicate concrete, mortar, and the preparation of dry construction mixtures. Also, sand can be claimed as a raw material component to produce ceramic tiles, porcelain stoneware, glazed ceramic products, wall ceramics, as well as proppants. Hydraulic enrichment leads to a decrease in the content of clay particles in the sand, which allows it to be used also for the construction of bases and coatings of roads and airfields. The integrated use of raw materials and waste allows solving the problem of creating non-waste and environmentally friendly technologies, which ensures the saving of natural raw materials, and on the other hand, allows it to be disposed of, improving the environmental situation.

1. Introduction

The quality of quartz sand used in various fields of industry is regulated by notations in accordance with the technical specifications. These requirements and quartz sand properties determine the technology and the necessity of enrichment. The final cost of the sand is determined by the method and degree of enrichment. The utilization of high-quality sands is limited by their high cost.

The most common impurities in the sands are associated with the presence of heavy and/or magnetic minerals (ilmenite, rutile, magnetite, hematite, goethite, etc.); weathered or layered minerals that absorbed iron (feldspar, mica); surface films covering quartz grains; clay; inclusions in the structure of quartz, etc. The enrichment of sand from clay impurities can be carried out by the dry method (air classification) or by the wet method (the hydraulic one). Dry enrichment methods, as a rule, are inferior in quality to sand enriched by the hydraulic method, especially on sands of complex mineral composition and sands with a high clay content.

Fine quartz sands are used, as a rule, in concrete production as a filler, in mortar production, in the production of ceramic materials as a non-shrinkage component, in the production of molding sands. The aim of this project is to determine the application of quartz sand in its natural form and to consider the effect of enrichment on the properties and applications.

Given the current state of consumption of natural raw materials, there is no doubt that the utilization and involvement of by-products in production is of value [1–3]. There are clay deposits in Tyumen region (Russia). Combined with extraction of titanium-containing components, the quartz waste comes into being [4]. It is important to find applications for this waste. It can be used in the construction industry [5–6]. The aim of this project is to study the possibility of using quartz waste of raw materials to produce dry building mixtures, proppants, cement, silicate brick, glass and glassware, fine ceramics and as molding sand [7–13]. Quartz sands are used to obtain cristobalite as a composite filler, coating material, surface finishing medium and structural ceramics [14].

2. Experimental procedure

The chemical composition of quartz sand was determined by emission spectral analysis with inductively coupled plasma using Optima 4300 DV optical emission spectrometer (Perkin Elmer, USA). The grain composition, modulus of sand fineness and real density using the pycnometric method were determined in accordance with [15]. Petrographic studies of refractory products were carried out in reflected light by optical microscope Polam R-311[15, 16].

3. Results and discussion

Unenriched sand visually contains large flinders more than 5 mm in size. When sifted, these pieces are easily destroyed, and the entire sample of sand passes through a No. 5 sieve. Therefore, the unenriched sand does not contain particles larger than 5 mm. The clay content in pieces of unenriched sand is 5% (for fine and very fine sands it should not exceed 1%, whereas in the enriched sand there are no such lumps. The clay-aggregated sand sample sare failed upon enrichment in water.

Table 1. Grain composition of quartz sand.

Quartz sand type	Size modulus	Fraction content (wt.%, mm)							
		2.5	1.25	0.63	0.315	0.16	0.1	0.004	0
unenriched	0.44	3.50	2.00	1.50	1.20	12.10	67.47	11.97	0.26
enriched	0.11	–	0.03	0.11	0.22	10.44	75.5	13.63	0.07

The full residue on sieve No. 063 for unenriched sand is 7.0% as well as for enriched one is 0.14 %, which corresponds to a group of very fine sands. The decrease in the sand particles content of more than 0.63 mm in the enriched sample is associated with the destruction of clay-aggregated sand particles.

Particle content less than 0.16 mm for unenriched sand is 79.7% and for enriched one is 89.13%, therefore, sands under discussion are very fine sands and can be used as fillers. The class of studied sands corresponds to II according to [17].

The chemical composition of the sand is given in Table 2. After enrichment, the content of Al_2O_3, Fe_2O_3, TiO_2 in sand decreases. It is associated with the removal of the clay part. The silica content in this case, as one would expect, increases.

Table 2. Chemical composition of quartz sand.

Quartz sand type	Weight content (%)									
	Al_2O_3	CaO	Cr_2O_3	Fe_2O_3	K_2O	MgO	Mn_2O_3	Na_2O	SiO_2	TiO_2
unenriched	4.274	0.239	0.014	1.179	2.075	0.107	0.0437	0.679	88.181	1.122
enriched	3.876	0.209	0.015	0.813	2.061	0.070	0.037	0.675	89.741	0.829

The content of dusty and clay particles, determined by the method of elutriation, in non-enriched sand is 5.6%, in enriched one is 0.46%, which corresponds to the requirements of [17] (up to 10%).

By the presence of organic impurities, unenriched and enriched sands are suitable for use in concrete and mortars, since the liquid above the unenriched and enriched sand samples is slightly colored. The optical density of the liquid over the unenriched sand is 0.044 and over the enriched one is 0.040 in comparison with a reference solution of 0.6–0.68.

As a result of petrographic studies of enriched sand, it was found that the studied samples are represented by alluvial sands and sandstones with a SiO_2 content of 95–98%, from milky white to

IOP Conf. Series: Materials Science and Engineering **966** (2020) 012027 doi:10.1088/1757-899X/966/1/012027

beige and red (sandstone) in color with admixtures of muscovite and biotite. Mica (muscovite and biotite) are concentrated mainly in fine sand fractions of 0.16–0.315 (3–6 and 1–3 wt. %, respectively) and 0.315–0.63 (3–5 and 2–4 wt. %, respectively). Here, white and transparent quartz is concentrated in well-rounded grains with a high degree of crystallization. The 0.63–1.25 mm fraction is represented by a mixture of white milk and partially transparent, well crystallized quartz and sandstone aggregates from red to dark gray, consisting of cemented small (0.05–0.1 mm) rounded grains forms. Cement is weak, non-hydrated, aggregates are not durable. Sandstone content is 10–13 wt. % There are individual particles of biotite fractions of less than 0.315 mm. A fraction of more than 1.25–2.5 mm consists entirely of fragile agglomerates of sandstone.

According to the X-ray diffraction data, unenriched quartz sand contains up to 5% montmorilonite, ~ 77% quartz, ~ 18% feldspar and muscovite in total. The mineral composition of enriched quartz sand is as follows: quartz is about 81%, as well as feldspar and muscovite in the amount of up to 19%.

The real density of the unreached sand sample was determined by the pycnometric method (Table 3). The voidness of enriched sand is increased, and the bulk density is decreased due to leaching offine clay fractions filling the voids between large sand particles. An increase in the true density of sand also indicates the leaching of less dense clay particles.

Table 3. Physical properties of quartz sand.

Quartz sand type	Real density ($kg \cdot m^{-3}$)	Bulk density($kg \cdot m^{-3}$)	Voidness (%)	Absolute humidity (%)
unenriched	2646.6	1338.90	49.4	0.5
enriched	2723.0	1275.18	53.3	0.1

The reactivity of unenriched and enriched sand is 37.66 and 37.24 $mmol \cdot L^{-1}$, respectively. It is in accordance with [17] (not more than 50 $mmol \cdot L^{-1}$). Consequently, the sands contain a permissible amount of amorphous varieties of silicon dioxide, soluble in alkalis (chalcedon, opal, flint, etc.) and sand can be used as a filler for concrete, as it is resistant to the chemical effects of alkali cement.

The content in unenriched and enriched sand of sulfide sulfur is 0.0043 and 0.0065%, respectively. The content of sulfate sulfur is 0.015% and 0.01%. According to [17], sulfur, sulfides, except pyrite (marcasite, pyrrhotite, etc.) and sulfates (gypsum, anhydrite, etc.) in terms of SO_3 should be not more than 1.0%, pyrite in terms of SO_3 – not more than 4% by weight). Therefore, both samples of sand are suitable for use as a filler for concrete and mortars.

For the suitability of quartz sand in road construction, the content of clay particles was determined by the swelling method according to [17]. An increase in the material volume in the sand was not observed. Therefore, the sands under study can be recommended for road construction.

To determine the suitability of sand as proppants, [18, 19] were used. According to the chemical composition (Table 2) in terms of MgO and Al_2O_3 content, these sands do not meet the requirements for either magnesia-quartz or aluminosilicate proppants. The minimum size of proppants according to [18, 19] corresponds to a size of 0.212 mm. The studied sands in the bulk consist of particles smaller than 0.16 mm (80–90%). Therefore, the sands under investigation cannot be used as proppants due to the discrepancy in grain and chemical compositions. But the studied sands can be recommended as silica-containing raw materials to produce aluminosilicate or magnesian proppants if necessary.

To determine the suitability of unenriched sand as molding one according to [20] the content of silicon dioxide should be at least 90.0%.With careful enrichment of the sand it is possible to bring this indicator to 90%,in the elutriated sand sample the content of silicon dioxide is 89.741%. The content of harmful impurities Fe_2O_3 decreases in the enriched sand sample from 1.179% to 0.813%. According to [20], the Fe_2O_3 content should be no more than 1%. Therefore, enriched sand according to this indicator meets the requirements. The content of alkaline and alkaline earth ($Na_2O + K_2O + + MgO + CaO$) according to [20] should be no more than 2%. In enriched sand it is 3.015%, in unenriched one is 3.1%. Therefore, the studied sands cannot be used as molding sand even after enrichment due to the high content of alkaline and alkaline earth oxides. Alkaline oxides are found in the sand in the form of feldspar and muscovite. Other enrichment methods are needed to remove such minerals.

To utilize studied sands in Portland cement production as a silicate additive, it does not require enrichment, since the clay component contained in unenriched sand will partially play the role of an alumina additive.

The use of quartz sand in the production of ceramic tile is possible due to the content in the sand of up to 19% of feldspars with muscovite. It is known that the fluxes are introduced into the body composition, which are used as feldspars and hydromica (in our case, muscovite). The sand enrichment in this case reduces the content of iron oxide, which allows the use of sand under investigation in the porcelain production.

4. Conclusion

The unenriched quartz sand belongs to the group of very fine sands (class II) with a particle size modulus of 0.44. The particle content of less than 0.16 mm is 79.7%, and particles of more than 0.063 mm are 7.03%. The sand does not contain large inclusions of more than 5 mm; it belongs to sands with a clay content of 5% in lumps and 5.6% of dust and clay particles. Sand is characterized by a low content of organic impurities, a bulk density of 1338.9 $kg \cdot m^{-3}$, a real density of 2646.6 $kg \cdot m^{-3}$ and a voidness of 49.4%. It has a low reactivity of 37.24 $mmol \cdot L^{-1}$. The content of sulfur compounds is negligible.

Enriched quartz sand belongs to the group of very fine sands (class II) with a particle size modulus of 0.11, a particle content of less than 0.16 mm is 89.2%, and particles of more than 0.063 mm are 0.14%. Sand does not contain large inclusions more than 5 mm. There is no clay in the lumps in the sand, and dusty and clay particles are practically absent (0.46%). Sand is characterized by a low content of organic impurities, a bulk density of 1275.2 $kg \cdot m^{-3}$, a real density of 2723 $kg \cdot m^{-3}$ and a voidness of 53.3%. It has a low reactivity of 37.66 $mmol \cdot L^{-1}$.

Thus, the studied sands can be recommended as a filler for heavy, light, fine-grained, cellular and silicate concrete, mortar, and preparation of dry mixes. Sand after enrichment can also be recommended for the construction of bases and coatings for roads and airfields.

References

[1] Ramdani S, Guettala A, Benmalek M L and Aguiar J B 2019 *J. Build. Eng.* **21** pp 302–11
[2] Kazanskaya L F and Smirnova O 2018 *Int. J. Civil Eng. Tech.* **9** pp 3006–12
[3] Okereafor U, Makhatha M, Mekuto L and Mavumengwana V 2020 *Minerals* **10** p 448
[4] Magenheimer S and Mazzullo J 1987 *J. Sedim. Petrol.* **57** pp 479–87
[5] Kapustin F L and Shul'meister T M 2017 *Glass Ceram.* **73** pp 3–4
[6] Pavlova I A, Kashcheev I D, Zemlyanoi K G and Shul'meister T M 2016 *Glass Ceram.* **72** pp 341–4
[7] Soliman N A 2017 *Construction and Building Materials* **145** pp 243–52
[8] Pavlova I A 2017 *Glass Ceram.* **74** pp 104–6
[9] Pavlova I A, Kutasheva S S and Farafontova E P 2019 *IOP Conf. Series Mat. Science Eng.* **687** 022022
[10] Pavlova I A and Sapozhnikova M A 2020 *Solid State Phenomena* **299** pp 3–7
[11] Arica N E 2000 *American Ceramic Society Bulletin* **79 (8)** pp 89–91
[12] Radin M S R M, Al-Gheethi A A S, Mohd S L A, Siti A B, Sabariah M and Amir H M 2019 *Pertanika J. Sci. & Technol.* **27** pp 1–9
[13] Cole W F 1991 *Materials Science Forum* **34–36** pp 1019–21
[14] Nurbaiti U and Pratapa S 2018 *Journal of Physics Conference Series* **983 (1)** 012014
[15] Russian state standard GOST 8735-88 *Sand for construction work. Testing methods* [in Russian]
[16] Russian state standard GOST 32723-2014 *Automobile roads of general use. Natural and crushed sand. Determination of mineralogical and petrographic composition* [in Russian]

[17] Russian state standard GOST 8736-2014 *Sand for construction works. Specifications* [in Russian]
[18] Russian state standard GOST R 54571-2011 *Magnesial-quarts proppants. Specifications* [in Russian]
[19] Russian state standard GOST R 51761-2013 *Aluminosilicate proppants. Specifications* [in Russian]
[20] Russian state standard GOST 2138-91 *Moulding sands. General specifications* [in Russian]

The use of technogenic raw materials to produce a high-alumina chamotte

A A Getman[1], I A Pavlova[1] and E P Farafontova[1, 2]

[1] Institute of new materials and technologies, Ural Federal University named after the First President of Russia B N Yeltsin, 19, Mira str., Yekaterinburg, 620002, Russia

E-mail: [2] ep.farafontova@urfu.ru

Abstract. When fused corundum is crushed, a finely dispersed powder is formed with an Al_2O_3 content of 93–95%, in the form of substandard material. It is advantageous to utilize this powder to obtain high-alumina chamotte with Al_2O_3 content of more than 62%. High-alumina aggregate (chamotte) was obtained by semi-dry technology with intermediate briquette molding from a mixture of corundum dispersed powder and enriched kaolin. Based on a final Al_2O_3 content of 65 wt. % the batch composition of the mass to produce a high alumina aggregate (chamotte) was calculated. From a mixture of dispersed corundum powder and kaolin moistened with a 5–7% technical lignosulfonate solution with a density of 1050 $kg \cdot m^{-3}$, a briquette was formed at a specific pressing pressure of 15–20 MPa. The briquette was fired at a temperature of 1700°C. The fired briquette had a strength of 164 to 193 MPa. The water absorption of briquettes was 11–12%. The phase composition of the briquette is represented mainly by corundum and mullite. The enriched kaolin did not show sintering effect on fine corundum.

1. Introduction

To produce mullite refractories with an Al_2O_3 content of 62–72%, the minerals of the sillimanite group are used: kyanite (distin), andalusite, sillimanite, etc [1–3]. These minerals have the same composition corresponding to the formula $Al_2O_3 \cdot SiO_2$ (62.9% Al_2O_3 and 37.1% SiO_2). After firing at appropriate temperatures, they all go over to mullite by the reaction: $3Al_2O_3 \cdot SiO_2 \rightarrow 3Al_2O_3 \cdot 2\ SiO_2 + SiO_2$. The theoretical yield of mullite is 86%, the remaining quantity of 14% is silicic acid, which passes into cristobalite during firing. Minerals of the sillimanite group are mainly found in quartz rocks, in which other impurities (muscovite and sericite) containing alkali are simultaneously present. Therefore, to produce high-alumina refractories, it is necessary to enrich it preliminary. The Al_2O_3 content in the enriched concentrate should be about 60%. The total fluxes (CaO, MgO, Na_2O, K_2O, Fe_2O_3) should not exceed 3%, which is very difficult to achieve. This is the main reason for the weak development of the production of high-alumina refractories from them, despite the significant reserves of this raw material. Deposits of kyanite that do not require enrichment are rare.

Another type of raw material to produce high alumina refractories is alumina hydrates: hydrargillite, boehmite, diasporas. These minerals are less likely than diasporas in the rock bauxite. Bauxites are sedimentary rocks, in addition to alumina hydrates, contain iron oxides (up to 20%). To produce high-alumina chamotte, low-iron bauxites (up to 5% Fe_2O_3) are suitable, the deposits of which are very rare. Due to the high content of iron oxides and other impurities, bauxite must also be pre-enriched.

The most common raw materials to produce mullite refractories are mainly technical alumina, pure refractory clays, kaolins and quartz. Mullite refractories are obtained in two stages: first, the synthesis of mullite, and then products from it. Technical alumina is produced in large quantities mainly for the production of aluminum by calcining artificial alumina hydrates. The content of Al_2O_3 in it reaches 99–99.8%. The mineral composition is represented by γ-alumina, which is represented by porous spherulites with a diameter of 20–70 μm, which makes it difficult to sinter and obtain dense products from it. Therefore, industrial alumina is usually ground to particles less than 3 microns in size. In addition, technogenic raw materials are used, for example high-alumina coal ash [4–5].

From technical alumina, electrocorundum is produced by melting at 2000–2400°C with a content of 97–99% Al_2O_3. Corundum is used in the production of corundum and mullite-corundum refractories. When fused corundum is crushed, a finely dispersed powder is formed with an Al_2O_3 content of 93–95%, in the form of substandard material. The low Al_2O_3 content is explained by the fact that the least stable phase in fused corundum is the glass phase, which is formed during the melting of alumina due to the content of impurities. This low-strength glass phase in comparison with crystals of corundum is transformed into a dust-like state during crushing first. Thus, it is of interest to utilize the resulting fine powder. Due to the low Al_2O_3 content, finely dispersed powder can be claimed in the production of high-alumina refractories with a lower Al_2O_3 content. It is advisable to utilize this powder, for example, to obtain high-alumina chamotte with an Al_2O_3 content of more than 62%. It is proposed to use kaolin of wet enrichment after filter pressing as a binding component. It is proposed to mix the obtained waste materials with a moisture content of 20–25% with finely dispersed corundum powder, form the briquette and, after drying, fire it.

2. Experimental procedure

Phase composition was determined by an X-ray phase method in a Miniflex 600 (Cu Kα-radiation, λ = 1.541862 Å, recording interval 3–90°, scanning step 0.02 deg), Rigaku, Carl Zeiss, Japan, with MiniFlex Guidance and PDXL Basic data treatment package program control and data collection. Identification of diffraction maxima was carried out using a JSPDS data bank. The chemical composition of quartz sand was determined by emission spectral analysis with inductively coupled plasma in an Optima 4300 DV(Perkin Elmer, USA) optical emission spectrometer.

The water absorption, open and total porosity and apparent density were determined according to [6]. The grain size composition of finely milled powders was determined by sedimentation analysis in a SLAD 2201 (Shimadzu Corp.) particle laser diffraction analyzer according to [7]. Sieve analysis was carried out in accordance with [8].

3. Results and discussion

To produce a high-alumina aggregate (chamotte) using briquette technology, plastic molding was used with kaolin from the Kyshtymskoye deposit. The chemical composition of kaolin and substandard fine powder of corundum are given in Table 1.

Table 1. The chemical composition of materials.

Material	SiO_2	Al_2O_3	CaO	Fe_2O_3	MgO	R_2O	TiO_2	LOI
Kaolin (%)	46.0-49.6	36.1-38.5	0.1-0.6	0.5-1.0	0.2-0.8	0.4-0.6	0.8	12.2-13.8
Fine substandard corundum powder (%)	0.05-0.3	93.0-95.0	0.2-0.24	0.13-0.5	0.44-0.62	0.3-0.55	2.2-5.0	–

The dispersed composition of kaolin is shown in Figure 1. The content of particles less than 2 microns is 8.15%. The maximum particle size is 112 microns. The main fractions are distributed uniformly and are in range of 60 to 2 microns. Particles larger than 60 microns are about 6%.

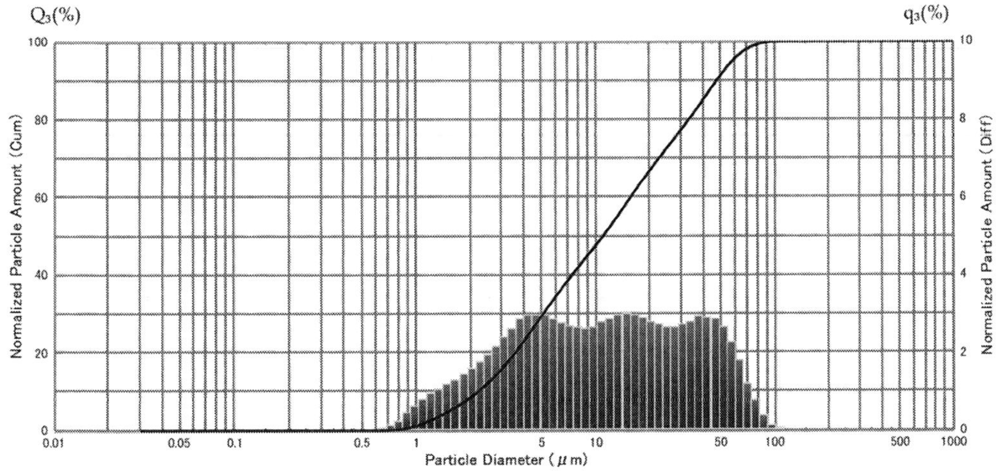

Figure 1. The dispersed composition of kaolin.

The grain composition of finely dispersed corundum is presented in Table 2. The main dusty fractions are particles smaller than 0.16 mm and more than 90%.

Table 2. The distribution of fine fractions in corundum.

Particle size (mm)	> 0.5	0.5–0.16	0.16–0.063	< 0.63
Fraction content (%)	3.07	4.52	51.45	40.96

According to the X-ray diffraction results, distribution of crystalline phases of corundum dust is represented by the following minerals (%): corundum89.5, dialuminumtitaniumoxide1.2, rutile 0.16, α-SiO_2 8.7.

To produce a high-alumina aggregate, the composition of the mixture with an Al_2O_3 content of 65 wt. % was calculated. To increase the density of the briquette and reduce the water absorption of the aggregate, the technology of pressing the briquette through 'false grain' was used [9]. Corundum dust was mixed with kaolin in a plastic way, dried at 105°C. The obtained material was crushed before passing through a mesh with a mesh size of 3 mm.

The strength of the briquette is affected by mass composition, moisture and pressing pressure. Samples with a relative humidity of 5, 7, 9% were formed from the obtained powder at various pressing pressures of 5, 10, 15, 20, 25 MPa. For wetting, the technical lignosulfonate solution with a density of 1.05 g·cm^{-3} was used. The samples were dried, and the compressive strength was determined. The results are presented in Figure 2. The briquettes molded at a pressing pressure of 5 MPa and at a moisture content of 5–9% do not have enough strength (4 MPa) [10]. When firing in a furnace, such briquette will crumble, that results in a decrease in the yield of the finished product. When the pressing pressure is increased to 10 MPa, the samples molded at a moisture content of 9 % are characterized by enough strength. A further increase in the pressing pressure to 15 MPa results in a strong briquette molded at a moisture content of 7–9%. With increasing pressing pressure above 20 MPa, the briquette molded at a relative humidity of 5–9% is also characterized by ample strength.

Chamotte firing is carried out in rotary or shaft furnaces. To reduce the dust and prevent the destruction of the granules during firing of the briquette, its transport strength should be at least 4 MPa [10]. The samples were fired at a temperature of 1700°C. The characteristics of the obtained chamotte are given in Table 3.

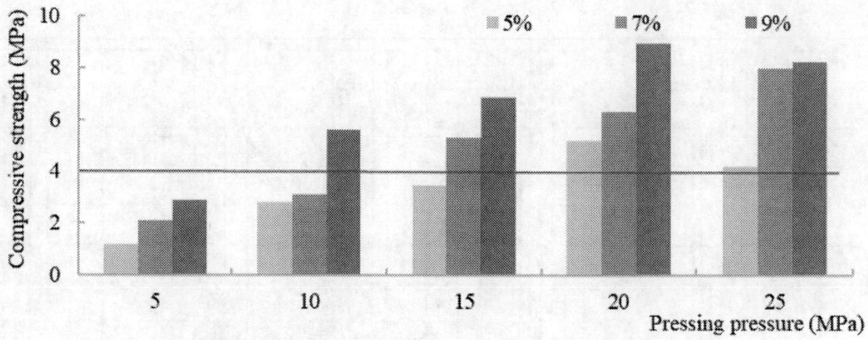

Figure 2. Dependence of compressive strength of unfired briquette on specific pressing pressure and relative humidity.

Table3. Post-firing properties of the samples.

Total shrinkage (%)	Water absorption (%)	Open porosity (%)	Apparent density ($kg{\cdot}m^{-3}$)	Compressive strength (MPa)
4–6	11–12	27–28	2310–2330	16.4–19.3

The distribution of crystalline phases of chamotte is as follows (%): corundum 35–38, mullite 46–50, cristobalite 6–7, quartz 4–6, β-Al_2TiO_5~4.

According to the results of this project at this stage, positive results were not reached on such properties as water absorption and density. The synthesis of mullite did not pass in full.

4. Conclusion

At this stage of project, high-quality high-alumina chamotte with an Al_2O_3 content of 65 wt.% with water absorption of less than 2% was not obtained. The enriched kaolin used with the contents of SiO_2 and Al_2O_3, respectively, 46–50% and 36–39% (the rest of the impurity) did not show sintering effect on finely dispersed corundum. To obtain chamotte with water absorption of less than 2%, it is necessary to turn 6–7% of cristobalite into mullite. In addition, it is necessary to introduce sintering components into the mixture and use the intergrinding of clay or kaolin with finely dispersed corundum powder.

References

[1] Degtyareva É V, Kabakova I I, Kotel'nikov G R et al 1976 *Refractories* **17** pp 235–40
[2] Kainarskii I S 1978 *Refractories* **19** pp 503–8
[3] Karklit A K and Katorgin G M 1995 *Refractories* **36** pp 122–6
[4] Zong Y., Li F, Chen W et al 2019 *SN Appl. Sci.* **1** 783 doi 10.1007/s42452-019-0814-8
[5] Gao Y, Liang K, Gou Y et al 2020 *Reviews in Chemical Eng.* doi 10.1515/revce-2019-0032
[6] Russian state standard GOST ISO 5017-2014 *Dense shaped refractory products. Determination of bulk density, apparent porosity and true porosity* [in Russian]
[7] Russian state standard GOST 8.777-2011 *State system for ensuring the uniformity of measurements. Disperse composition of aerosols and suspensions. Determination of particle size by laser diffraction* [in Russian]
[8] Russian state standard GOST 27707–2007 *Unshaped refractories. Methods for determination of grain composition* [in Russian]
[9] Kashcheev I D and Pavlova I A 2006 *Glass and Ceramics* **63** pp 86–8
[10] KashcheevI D, GlyzinaA É, Finkel'shtein A B et al 2019 *Refractories and Industrial Ceramics* **60** pp 362–4

IOP Conf. Series: Materials Science and Engineering **966** (2020) 012029 doi:10.1088/1757-899X/966/1/012029

Nanofine spatial dissipative structures with azimutal lattice curvation

V B Malkov[1], O V Chemezov[2, 5], V G Pushin[3], A V Malkov[4], B V Shulgin[2] and O V Malkov[4]

[1] Institute of High-Temperature Electrochemistry, Ural Branch of the Russian Academy of Sciences, 20, Academicheckaya str., Yekaterinburg, 620990, Russia
[2] Ural Federal University named after the First President of Russia B N Yeltsin, 19, Mira str., Yekaterinburg, 620000, Russia
[3] Institute of Metal Physics named after M N Mikheeva Ural Branch of the Russian Academy of Sciences, 18, Sophyi Kovalevskoy str., Yekaterinburg, 620108, Russia
[4] CJSC SPC 'Rosna', Yekaterinburg, Russia

E-mail: [5]o.v.chemezov@urfu.ru

Abstract. The nanofine spatial dissipative structures (NSDS) were characterized by transmission electron microscopy and electron diffraction. The NSDS obtained by thermogradiently processing an amorphous selenium film by unilateral heating of its lower surface at $T = 413$ K preliminarily. The results indicate that the NSDS of hexagonal selenium obtained in an amorphous film possess an azimuthal curvature of the lattice and a nonlinear fan-shaped system of flexural contours on their electron microscopic image. The lattice of the above NSDS undergoes elastic – plastic rotational curvature around three mutually perpendicular directions. The lattice rotation angles of hexagonal selenium NSDS reach: around [001], – 25°, around the direction perpendicular [001] and lying in the plane of the amorphous film – 32°, around the direction perpendicular to the first two ones and not lying in the plane of the amorphous film - azimuthal curvature of the lattice, – 35°. Thus, as a result of electron-microscopic and microdiffraction studies, it was found that NSDS of hexagonal selenium with azimuthal curvature of the lattice, causing the curvature of its habitus, are in a nonequilibrium state.

1. Introduction

It is known that the formation of dissipative structures can be realized by an explosive crystallization of amorphous substances [1]. Another way of forming such structures exists – with rotational curvature of the lattice by thermogradient processing of an amorphous film by unilateral heating of its lower surface [2, 3]. The rate of formation of the NSDS during the explosive crystallization creates difficulties in obtaining dissipative structures of a given size. Taking into account the above, the formation of NSDS of hexagonal selenium with rotational curvature of the lattice and given dimensions was carried out by the method of thermogradient processing of an amorphous film [2, 3].

2. Experimental

The study of NSDS of hexagonal selenium was carried out by electron microscopy and electron diffraction [5–11]. NSDS of hexagonal selenium with azimuthal curvature of the lattice are formed in an amorphous Se film during thermogradient processing of by heating its lower surface at $T = 413$ K

(Figure 1 a). A characteristic detail of the electron microscopic images of NSDS is the system of bending contours [4, 9–12]. The appearance of the system of bending contours is caused either by their bending as a whole or by the rotational curvature of their lattice [2, 3, 10–12]. Systems of nonlinear fan-shaped extinction contours are present on electron microscopic images of NPDS of hexagonal selenium with a curved habit (Figure 1 a).

Quenching of NSDS of hexagonal selenium by cooling in an air was performed after completion of the thermogradient processing of the amorphous Se film. In NSDS of hexagonal selenium formed in the amorphous film during quenching, the lattice structural defects are fixed. Lattice structural defects are atoms and vacancies displaced from their equilibrium positions. The concentration of the nonequilibrium structural defects, corresponding to the temperature of one-sided heating of the lower surface of the amorphous Se film, remains in NSDS formed in the amorphous film with a sharp decrease in temperature [2, 3, 13–15]. Quenching of the nonequilibrium structural defects allows the structure of NSDS to be 'frozen' and exist without exchange of both energy and matter with the environment. The NSDS formed in the amorphous Se film possess all the signs of nanofine crystals after quenching.

3. Results and Discussion

In microdiffraction studies [5–8] of NSDS of hexagonal selenium with curved habitus microelectron diffraction patterns were obtained from the central parts of NSDS (Figure 1 b) and from their peripheral parts: from the 'right' part (Figure 1 c) and from the 'left' part (Figure 1 d), respectively. Table 1 show the results of the calculation of microelectron diffraction patterns of the nanofine PDS of hexagonal selenium, respectively: from the central part (Figure 1 b), from the 'left' part (Figure 1 c) and from the 'right' part (Figure 1 c). Reflections in the microelectron diffraction pattern from the central part of NSDS of hexagonal selenium with azimuthal curvature of the lattice (Figure 1 b) indicate that prismatic planes of the first kind are in the reflecting position in this part of the NSDS. The results of the calculation of microelectron diffraction patterns of the peripheral parts of NSDS (Figure 1 c and Figure 1 d) allow determining their position in the reciprocal lattice by NSDS of hexagonal selenium with azimuthal curvature of the lattice. The position of microelectron diffraction patterns from the 'left' and 'right' parts of NSDS of hexagonal selenium in its reciprocal lattice is characterized by mirror symmetry.

Comparative analysis of microelectron diffraction patterns (Figure 1 b, 1 c and 1 d); and calculations performed using standard crystallographic formulas allow to establish that the lattice of the studied NSDS of hexagonal selenium with the curved habitus undergoes elastic - plastic rotational curvature around three mutually perpendicular directions – around [001], which coincides with the smallest size of NSDS; around the direction perpendicular to [001] and lying in the plane of the amorphous film and the azimuthal curvature around the direction perpendicular to the first two ones (Figure 1 a).

The elastic – plastic nature of the rotational curvature of the lattice of NSDS with azimuthal curvature of the lattice formed in the amorphous Se film is indicated, on the one hand, by the peculiar properties of their bending extinction contours. Indeed, the bending extinction contours of NSDS of hexagonal selenium (Figure 1 a), in their peripheral part, are discontinuous. On the other hand, high values of the angles of rotational curvature of the lattice of NSDS around three mutually perpendicular directions exclude the purely elastic nature of the rotational curvature of their lattice.

Thus, as a result of electron-microscopic and microdiffraction studies, it was found that NSDS of hexagonal selenium with azimuthal curvature of the lattice, causing the curvature of its habitus, are in a nonequilibrium state. The lattice of NSDS experiences elastic – plastic rotational curvature around three mutually perpendicular directions. The lattice rotation angles of these NSDS of hexagonal selenium reach accordingly: around [001], 250, around the direction perpendicular [001] and lying in the plane of the amorphous film – 320, around the direction perpendicular to the first two – azimuthal curvature of the lattice – 350.

IOP Conf. Series: Materials Science and Engineering **966** (2020) 012029 doi:10.1088/1757-899X/966/1/012029

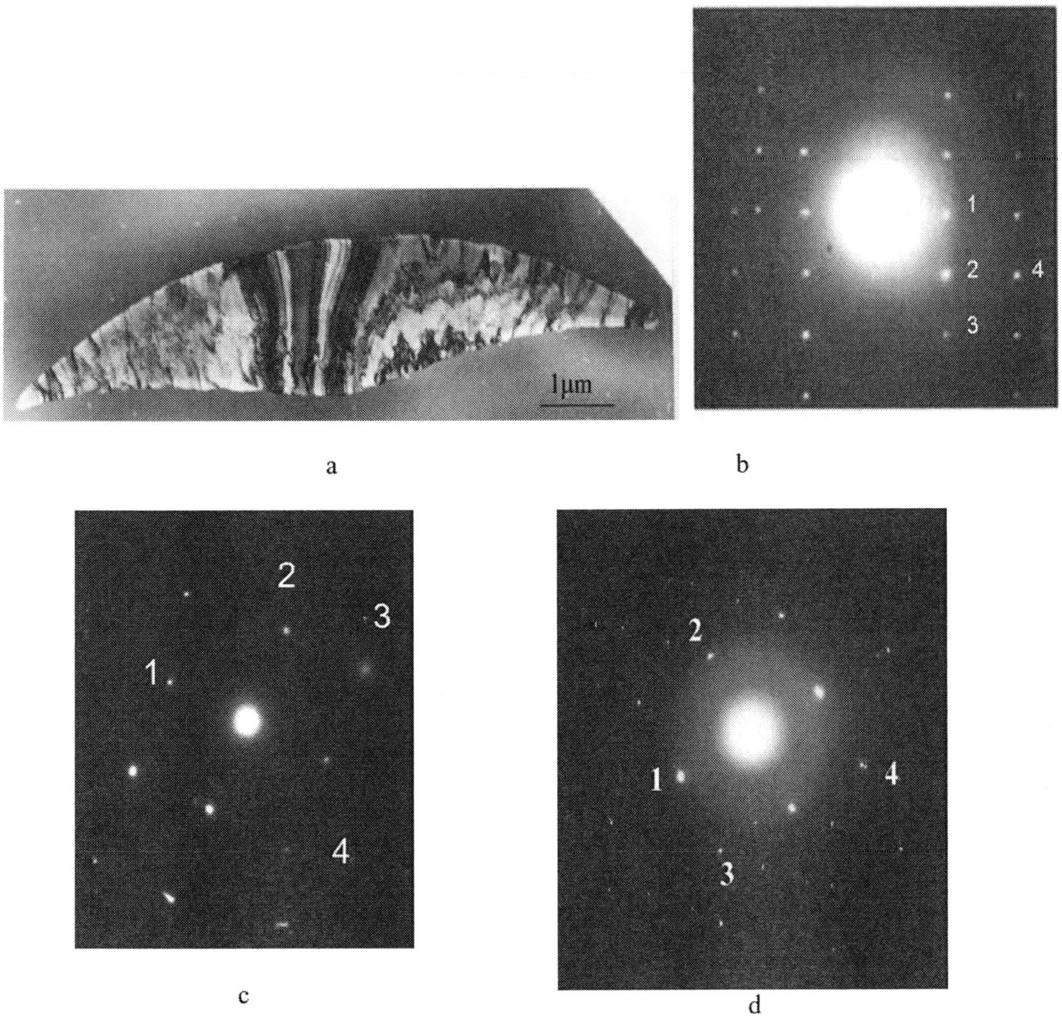

Figure 1. Electron-microscopic image of NSDS of hexagonal selenium with azimuthal curvature of the lattice (Figure 1 a). JEM-200CX JEOL Ltd, Japan. Microelectron diffraction pattern from the central part of NSDS of hexagonal selenium with azimuthal curvature of the lattice (Figure 1 b); prismatic planes of the first kind are in a reflective position. Microelectron diffraction patterns from the 'left' (Figure 1 c) and 'right' (Figure 1 d) of the peripheral parts of NSDS.

The rotation of the reciprocal lattice by NSDS of hexagonal selenium with azimuthal curvature of the lattice is the result of the rotation of selenium macromolecules – structural units that form a lattice NSDS. Accordingly cooperative movements of the structural units forming the lattice NSDS with azimuthal curvature of lattice are cooperative rotations of the selenium macromolecules around three mutually perpendicular directions

Table 1. Interplanar distances and Miller indices for a microelectron diffraction pattern of NSDS of hexagonal selenium with azimuthal curvature of the lattice: B) from the central part; C) from the 'left' part; D) from the 'right' part, respectively.

Part NSDS	Reflex numbers	Interplanar distances Experiment (Å)	Interplanar distances Theory (Å)	Miller indices
B	1	3.810	3.800	$10\bar{1}0$
	2	2.977	2.975	$10\bar{1}1$
	3	2.093	2.060	$10\bar{1}2$
	4	1.739	1.755	$20\bar{2}1$
C	1	2.977	2.975	$\bar{1}01\bar{1}$
	2	2.976	2.975	$01\bar{1}\bar{1}$
	3	2.186	2.167	$11\bar{2}0$
	4	2.046	2.060	$1\bar{1}02$
D	1	2.977	2.975	$\bar{1}011$
	2	2.977	2.975	$\bar{1}10\bar{1}$
	3	2.046	2.060	$0\bar{1}12$
	4	2.180	2.167	$2\bar{1}\bar{1}0$

4. Conclusion

Thus, the formation of NSDS of hexagonal selenium with azimuthal curvature of the lattice occurs in an amorphous film during its thermogradient treatment – one-sided heating of the lower surface of the amorphous film with subsequent quenching by cooling in air and is the result of all the necessary and sufficient conditions for the formation of spatial dissipative structures: a) NSDS with azimuthal curvature of the lattice are formed in a thermodynamically open system; b) this system is removed from an equilibrium; c) this system is described by nonlinear equations; c) the formation of NSDS of hexagonal selenium with azimuthal curvature of the lattice is the result of the cooperative movement of the structural units of their lattice – the cooperative rotation of selenium macromolecules around three mutually perpendicular directions [2, 3, 17–20].

Reference

[1] Shklovsky V A and Kuzmenko V M 1989 Explosive crystallization of amorphous substances *Advances in Physical Sciences* **157 2** pp 311–38 [In Russian]

[2] Malkov V B, Nikolaenko I V, Shveikin G P, Malkov A V, Pushin V G, Shulgin B V, Malkov O V and Plaksin S V 2017 Patent RU 2637396 [In Russian]

[3] Malkov V B, Nikolaenko I V, Shveikin G P, Pushin V G, Malkov A V, Malkov O V and Shulgin B V 2018 Formation of dissipative structures in an amorphous film *Reports of the Academy of Sciences* **478 5** pp 543–5

[4] Kveglis L I and Kashkin V B 2015 *Dissipative structures in thin nanocrystalline films* (Siberian Federal University: Prospekt) [In Russian]

[5] Hirsch P B, Howie R B and Nicholson D W 1968 *Electron microscopy of thin crystals* (London: Butterworths) pp 418–20

[6] Tomas G and Goringe M J 1979 *Transmission electron microscopy of materials* (New York-Chichester-Brisane-Toronto) pp 198–203

[7] Utevsky L M 1973 *Diffraction electron microscopy in metal science* (Moscow: Metallurgy) [In Russian]
[8] Andrews K V, Dyson D J and Keown S R 1968 *Interpretation of Electron Diffraction Patterns* (London) pp 39–56
[9] Malkov V B, Nikolaenko I V, Shveikin G P, Malkov A V, Pushin V G, Malkov O V and Shulgin B V 2014 Patent RU 2534719 [In Russian]
[10] Bolotov I E and Kolosov V Yu 1980 Bending of thin-film crystals of selenium, detected by extinction contours *Izv. USSR Academy of Sciences Ser. Physical* **44** **6** pp 1194–97 [In Russian]
[11] Bolotov I E and Kolosov V Yu 1982 Investigation of Crystals Based on Bend-Contour Arrangement 1 Relationship between Bend-Contour Arrangement and Bend Geometry *Phys. Stat. Sol.* (a) **69** **1** pp 85–96
[12] Kolosov V Yu 1982 *Abstract of dissertation* Electron-microscopic study of defects in crystals growing in amorphous films (Sverdlovsk: IFM) [In Russian]
[13] Damascus A and Deans J *Point defects in metals* ed. Lyubova B Ya (Moscow: Mir) 1966 [In Russian]
[14] Malkov V B, Nikolaenko I V, Shveikin G P, Malkov A V, Pushin V G, Shulgin B V, Malkov O V and Plaksin C V 2019 Patent RU 2687876 [In Russian]
[15] Bolotov I E and Prilepo V L 1983 On the vacancy mechanism of the occurrence of compressive stresses during crystallization of amorphous films *Phys. Stat. Sol. ABD* **1** pp k67- k70
[16] Prigogine I and Condepudi D 2002 *Modern thermodynamics. From heat engines to dissipative structures* (Moscow: Mir) [In Russian]
[17] Askhabov A M 1993 *Crystallogenesis and evolution of the crystal-medium system* ed. Yushkin N P (Saint Petersburg: Nauka) [In Russian]
[18] Nikolis G and Prigozhin I 1979 *Self-organization in nonequilibrium systems* ed. Chizmadzheva Yu A (Moscow: Mir) [In Russian]
[19] Mandelkern L 1966 *Crystallization of polymers* (Leningrad: Chemistry) [In Russian]
[20] Wunderlich B 1979 1984 *Physics of macromolecules* (Moscow: Mir) T 2, 3 [In Russian]

Phenomenon of relaxation of inhomogeneous elastic rotational curvature of the lattice of nanofine spatial dissipative structures of hexagonal selenium

V B Malkov[1], O V Chemezov[2, 5], V G Pushin[3], A V Malkov[4], B V Shulgin[2] and O V Malkov[4]
[1] Institute of High-Temperature Electrochemistry, Ural Branch of the Russian Academy of Sciences, 20, Academicheckaya str., Yekaterinburg, 620990, Russia
[2] Ural Federal University named after the First President of Russia B N Yeltsin, 19, Mira str., Yekaterinburg, 620000, Russia
[3] Institute of Metal Physics named after M N Mikheeva Ural Branch of the Russian Academy of Sciences, 18, Sophyi Kovalevskoy str., Yekaterinburg, 620108, Russia
[4] CJSC SPC 'Rosna', Yekaterinburg, Russia

E-mail: [5] o.v.chemezov@urfu.ru

Abstract. Nanofine rhomboid spatial dissipative structures (NRSDS) of hexagonal selenium were found and studied by means of transmission electron microscopy in amorphous Se films coated with nanofine layers of amorphous carbon, at a temperature of their thermo gradient treatment of 423 K. As a result, pictures were received of fan-shaped linear bending extinction contours on the electron-microscopic images of NRSDS. It was shown that in the above NRSDS take place continuous linear increase in the bend radius of the lattice along [001], due to continuous linear relaxation of the inhomogeneous elastic torsion of the lattice around [001]. The continuous linear relaxation of the elastic rotational curvature of the lattice around [001] in NRSDS of hexagonal selenium with inhomogeneous elastic rotational curvature of the lattice covers the entire nanofine rhomboid spatial dissipative structures, and not its part, as is the case with the formation of interblock boundaries in the nanofine rhomboid spatial dissipative structures.

1. Introduction

It is known that the formation of interblock boundaries take place with varying module of the misorientation vector ϖ in NRSDS of hexagonal selenium – nanofine crystals after the stage of quenching of NRSDS's [1–4]. This phenomenon is the result of relaxation of an inhomogeneous elastic rotational curvature of the lattice around [001] in one of NRSDS block [5–9]. At the same time, it is known that the formation of interblock boundaries in NRSDS, along which not only the modulus but also the sign of the misorientation vector changes, is a result of relaxation of the inhomogeneous elastic rotational curvature of the lattice around [001] in neighboring blocks of NRSDS [8, 9]. The driving force of relaxation processes in NRSDS's of hexagonal selenium, leading to the formation of interblock boundaries with a crystal – geometric parameters varying along the boundaries, is the tendency to decrease the energy of the inhomogeneous elastic rotational curvature of their lattice around [001] [5–7]. The relaxation process of the inhomogeneous elastic rotational curvature of the lattice of NRSDS around [001], which determine the formation of the interblock torsion boundaries, are realized only in the part of NRSDS's of hexagonal selenium [5–9].

2. Results and discussion

The method of transmission electron microscopy and electron diffraction method [10–13] used to study the amorphous films of selenium, after thermo gradient processing by unilateral heating of their lower surface at a temperature of 423 K. NRSDS of hexagonal selenium were discovered and investigated. Pictures of fan-shaped linear bending extinction contours are present on the electron microscopic images of the detected NRSDS. Let us consider the features of the mutual arrangement of bending extinction contours (Figure 1 a). It is noteworthy that the 'fan' of bending contours, the base of which is located in one of the blunt corners of the rhombus, is formed by rectilinear bending contours.

The bending contours deviate from the direction of the lowest formation speed towards the direction of the highest formation speed of NRSDS occur when moving take place along the direction of the lowest formation rate of NRSDS, which coincides with the short diagonal of a rhombus (Figure 1 a).

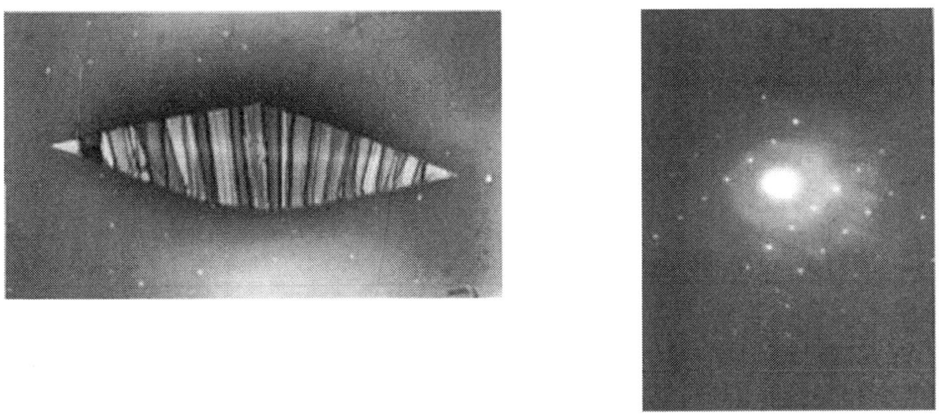

a b

Figure 1. a) NRSDS of hexagonal selenium with a linear fan-shaped system of bending contours; b) The microelectron diffraction pattern from the central part of NRSDS of hexagonal selenium. In a reflective position are prismatic planes of the first kind.

Bending contours deviating from the position parallel to [001] retain the shape of straight lines (Figure 1 a). The distance between the bending extinction contours, which differ only in the sign of the indices $hkl, \bar{h}\bar{k}\bar{l}$ increase in the direction of the opening of the 'fan' of the contours. The distance between the centers of pairs of bending contours, the appearance of which on the electron microscopic images of NRSDS due to reflection from pairs of planes with different indices $h_1k_1l_1, \bar{h}_1\bar{k}_1\bar{l}_1; h_2k_2l_2, \bar{h}_2\bar{k}_2\bar{l}_2$, increase in the same direction. The increase in the distances between bending contours that differ only in sign, and between the centers of pairs of bending contours with different indices, is also observed along the direction of the highest rate of formation of NRSDS. The similar change in the intercontour distances was previously observed for NRSDS of hexagonal selenium with inhomogeneous elastic rotational curvature of the lattice around [001] [14], which indicates the heterogeneity of the elastic rotational curvature of lattice in the studied NRSDS.

Electron diffraction patterns were obtained from the central part of the NRSDS of hexagonal selenium (Figure 1 b) in electron diffraction studies. The ones coincide with electron diffraction patterns obtained from the central part of NRSDS of hexagonal selenium with systems of bending contours parallel to the short diagonal of the rhombus and inhomogeneous elastic curvature of the lattice around [001] [14]. The conclusion follows from here that the initial state for the formation of

NRSDS with a linear fan-shaped system of bending extinction contours on electron microscopic images is the state of NRSDS of hexagonal selenium with inhomogeneous elastic rotational curvature of the lattice around [001] [14].

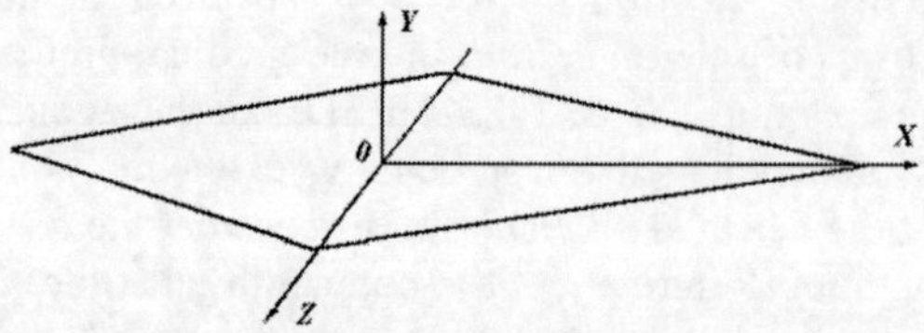

Figure 2. The coordinate system associated with NRSDS of hexagonal selenium.

It has been established that the lattice of NRSDS of hexagonal selenium (Figure 1 a) experiences an inhomogeneous elastic rotational curvature of the lattice around two mutually perpendicular directions – around [001], which coincides with the OZ axis (Figure 2) and around the direction of OX (Figure 2), which coincides with the long diagonal of the rhombus [15–18]. Calculations performed using standard crystallographic formulas show that the rotation of the lattice in the reference plane (001) reaches 18°, and the deviation of [001] from a position parallel to the plane of the film reaches 22°.The features of the patterns of bending extinction contours on the electron microscopic images of NRSDS indicates that the rotational curvature of the lattice of NRSDS with a linear fan-shaped system of bending contours has continuous progressive nature. For a more complete study of the organization of the lattice of NRSDS (Figure 1 a), we used the formulas of the method of bending extinction contours [19, 20].

As it was established above, for NRSDS of hexagonal selenium with a linear fan-shaped system of bending contours in the electron microscope image, the increase in the inter – contour distances in the direction of the opening of the 'fan' of contours is observed both between pairs of contours that differ only in signs $hkl; \overline{hkl}$ and between the centers of pairs of contours with different indices $h_1k_1l_1, \overline{h_1k_1l_1}; h_2k_2l_2, \overline{h_2k_2l_2}$. Let us consider how the lattice bending parameters of the above mentioned NRSDS are changed to realize the revealed change in the intercontour distances. For this purpose, we use the formulas for the distance between contours obtained in [19, 20]. The formula for the distance between the pair of contours that differ only in signs $hkl; \overline{hkl}$:

$$N^{(\beta)} = \frac{2r\cdot\sin\varphi_{UVW}}{1+\sin^2\beta\cdot ctg^2\gamma} \qquad (1)$$

The formula for the distance between the centers of pairs of contours, the appearance of which on the electron microscopic image of the crystal is due to reflections from planes with different indices $h_1k_1l_1, \overline{h_1k_1l_1}; h_2k_2l_2, \overline{h_2k_2l_2}$:

$$M^{(\beta)} = r\cdot\sqrt{1-\frac{1}{sin^2\beta\cdot ctg^2\gamma}} \qquad (2)$$

Since, under the conditions of our experiment, the quantities $sin\beta, ctg\gamma, sin\phi_{UVW}$ are constant, so far as formulas (1) and (2) can be transformed as follows:

$$N^{(\beta)} = r\cdot c_1'; M^{(\beta)} = r\cdot c_2'$$

Having marked, $\frac{1}{c_1'} = c_1, \frac{1}{c_2'} = c_2$, we will receive:

$$r = c_1 \cdot N^{(\beta)}; r = c_2 \cdot M^{(\beta)} \tag{3}$$

From formula (3), it is seen that the dependence between the bending radius of the crystal lattice r and distances, between contours that differ only in sign $hkl; \overline{hkl}$ and between the centers of the contour pairs, the appearance of which on the electron microscopic image of the crystal is due to reflections from planes with different indices $h_1k_1l_1, \overline{h_1k_1l_1}; h_2k_2l_2, \overline{h_2k_2l_2}$ is linear. It is clear that this nature of the change in the intercontour distances is due to an increase in the bend radius of the lattice of NRSDS in the direction of the opening of the 'fan' of the contours. Given the straightforwardness and continuity of bending contours present on the electron microscopic images of NRSDS and formulas (3) [17], it is clear that the increase in the radius of the rotational curvature of the lattice will occur according to a linear law. The calculations performed using the formulas of the bending contour method [19, 20] and formula (3) confirm the above conclusion.

3. Conclusion

Thus, in the process of studying NRSDS of hexagonal selenium the features of their lattice organization were revealed. The bending radius of the lattice of these NRSDS increases in direction of contour 'fan' opening according to the linear law. The initial state for the formation of NRSDS with a linear fan-shaped system of bending contours is the state of the NRSDS of hexagonal selenium with inhomogeneous elastic rotational curvature of the lattice around [001]. The increase in the bending radius of the lattice of NRSDS in the direction of contours those are similar to an 'open fan' take place according to the linear law. This phenomenon can be explained as the result of continuous linear relaxation of the elastic rotational curvature of the lattice around [001] in the NRSDS of hexagonal selenium with inhomogeneous elastic rotational curvature of the lattice. In this case, the continuous linear relaxation of the elastic rotational curvature of the lattice around [001] in NRSDS of hexagonal selenium with inhomogeneous elastic rotational curvature of the lattice covers the entire NRSDS's, and not its part, as is the case with the formation of interblock boundaries in NRSDS's.

Reference

[1] Prigogine I and Condepudi D 2002 *Modern thermodynamics From heat engines to dissipative structures* (Moscow: Mir) [In Russian]

[2] Nikolis G and Prigozhin I 1979 *Self-organization in nonequilibrium systems* ed Chizmadzheva Yu A (Moscow: Mir) [In Russian]

[3] Malkov V B, Nikolaenko I V, Shveikin G P, Malkov A V, Pushin V G, Shulgin B V, Malkov O V and Plaksin S V 2017 Patent RU 2637396 [In Russian]

[4] Malkov V B, Nikolaenko I V, Shveikin G P, Pushin V G, Malkov A V, Malkov O V and Shulgin B V 2018 Formation of dissipative structures in an amorphous film *Reports of the Academy of Sciences* **478 5** pp 543–5

[5] Bolotov I E, Kolosov V Yu and Malkov V B 1984 Application of Bend-Contour Method the Investigation of Block Formation in Thin-Film Crystals Budapest *Electron Microscopy* **2** pp 42–5

[6] Bolotov I E, Kolosov V Yu and Malkov V B 1986 Electron Microscope Investigation of Crystals Based on Bend-Contour Arrangement III Formation of Subgrain Boundaries in Dislocation-Free Crystals of selenium *Phys. Stat. Sol. (a)* **95** pp 377–83

[7] Bolotov I E, Kolosov V Yu and Malkov V B 1986 A new type of interblock boundaries in thin-film selenium crystals *Crystallography* **31 1** pp 204–6 [In Russian]

[8] Malkov V B, Malkov A V, Pushin V G, Streckalovskiy V N and Malkov O V 2012 Patent RU 2534719 [In Russian]

[9] Malkov V B, Nikolaenko I V, Shveikin G P, Pushin V G, Malkov A V, Malkov O V and Shulgin B V 2018 The mechanism of the formation of interblock boundaries, the effect of changing the sign of the vector of orientation along dangling interblock boundaries in

IOP Conf. Series: Materials Science and Engineering **966** (2020) 012030 doi:10.1088/1757-899X/966/1/012030

nanofine spatial dissipative structures *Reports of the Academy of Sciences* **481** **1** pp 30–2

[10] Hirsch P B, Howie A, Nicholson R B and Pashley D W 1965 *Electron Microscopy of Thin Crystals* (London: Butterworths) pp 418–20

[11] Tomas G and Goringe M J 1979 *Transmission electron microscopy of materials* (New York-Chichester-Brisbane-Toronto) pp 198–203

[12] Utevsky L M 1973 *Diffraction electron microscopy in metal science* (Moscow: Metallurgy) [In Russian]

[13] Andrews K V, Dyson D J and Keown S R 1968 *Interpretation of Electron Diffraction Patterns* (London) pp 39–56

[14] Malkov V B, Malkov A V, Malkov O V, Pushin V G, Shulgin B V and Agalakov S P 2008 The phenomenon of elastic rotational curvature of the lattice of nanofine crystals of hexagonal selenium around [001] in the framework of the asymmetric theory of elasticity *Collection of reports Kharkov Nanotechnological Assembly* **2** (Kharkov Ukraine) pp18–23

[15] Malkov V B, Nikolaenko I V, Shveikin G P, Malkov A V, Pushin V G, Malkov O V and Shulgin B V 2014 Patent RU 2534719 [In Russian]

[16] Malkov V B, Nikolaenko I V, Shveikin G P, Malkov A V, Pushin V G, Shulgin B V and Malkov O V 2018 The lattice symmetry of nanofine crystals isomorphism of the forward and backward spaces and the Curie symmetry principle *Reports of the Academy of Sciences* **479** **1** pp 18–20

[17] Malkov V B, Nikolaenko I V, Shveikin G P, Malkov A V, Pushin V G, Malkov O V and Shulgin B V 2015 Patent RU 2570106 [In Russian]

[18] Malkov V B, Nikolaenko I V, Shveikin G P, Malkov A V, Pushin V G, Shulgin B V, Malkov O V and Plaksin C V 2019 Patent RU 2687876

[19] Bolotov I E and Kolosov V Yu 1980 Bending of thin-film crystals of selenium detected by extinction contours *Izv. USSR Academy of Sciences ser. physical* **44** **6** pp1194–97 [In Russian]

[20] Bolotov I E and Kolosov V Yu 1982 Investigation of Crystals Based on Bend-Contour Arrangement 1 Relationship between Bend-Contour Arrangement and Bend Geometry *Phys. Stat. Sol. (a)* **69** **1** pp 85–96

IOP Conf. Series: Materials Science and Engineering **966** (2020) 012031 doi:10.1088/1757-899X/966/1/012031

Thermal stability of hafnium bronze subjected to dynamic channel angular pressing

A Stolbovsky[1, 2, 4], V Popov[2], E Popova[2], R Falahutdinov[2], S Murzinova[2] and E Shorokhov[3]

[1] Ural Federal University named after the first President of Russia B N Yeltsin, 19, Mira str., Yekaterinburg, Russia
[2] M N Miheev Institute of Metal Physics, UB RAS, 18, S. Kovalevskaya str., Yekaterinburg, 620108, Russia
[3] Russian Federal Nuclear Center – VNIITF, Snezhinsk, Russia

E-mail: [4] stolbovsky@imp.uran.ru

Abstract. Thermal stability of the structure of Cu-0.8%Hf alloy subjected to severe plastic deformation by 2 passes of dynamic channel angular pressing (DCAP) has been investigated. It is demonstrated that after such deformation the microstructure of the hafnium bronze is non-uniform, where areas of cell structure are neighboring with thin-lamellar structure of twinning type. Such structure indicates the action of two deformation mechanisms, dislocation sliding and twinning. It has been revealed that doping with hafnium has considerably affected susceptibility of copper to relaxation processes under the DCAP and following annealing. It has been found that the microstructure formed under the deformation can be retained at the annealing up to 500°C without noticeable changes, and the microhardness changes only slightly in this temperature range. At the same time, at heating up to 600°C and higher the relaxation processes are developed in the structure which is accompanied with a drastic drop of microhardness.

1. Introduction

In the last decades materials with ultrafine-grained (UFG) structure attract great attention of the researchers due to their unique physical and mechanical properties [1, 2]. Application of severe plastic deformation (SPD) enables obtaining bulk UFG materials demonstrating unusual mechanical behavior and a complex of unique properties which are often unachievable for coarse-grained materials.

A number of various techniques of SPD have been developed by now [3]. One of the most widely used of them is the so-called equal-channel angular pressing (ECAP) [4]. At the same time, one of the new promising methods of SPD is the dynamic channel angular pressing (DCAP), which has the scheme of deformation similar to that of the ECAP, but its main difference is the high deformation rate (10^4 – 10^6 s^{-1}). The DCAP was worked out at the Russian Federal Nuclear Center [5]. In contrast to ECAP, for which expensive pressing equipment is required, in the DCAP the energy of powder gasses is used instead of the press, which enables to subject massive materials to SPD with high speed and high productivity. Compared to the ECAP, in case of the DCAP not only the rate of plastic deformation is increased, but also the shock-wave impact is added, which increases the SPD influence on the structure and properties of materials. Thus, application of the DCAP makes it possible to change the properties of metals and alloys substantially by the formation such structures in them

IOP Conf. Series: Materials Science and Engineering **966** (2020) 012031 doi:10.1088/1757-899X/966/1/012031

which demonstrate the combination of high strength and plasticity, as was shown in the studies of evolution of the microstructure in aluminum, copper, nickel and bronze under the DCAP [6–8]. Besides, it was found that the UFG structure can be obtained by the DCAP by the lesser numbers of passes compared to the ECAP [7, 9].

However, when various techniques of SPD are applied, relaxation processes are developed under deformation along with the microstructure refinement, which causes an attainment of the so-called saturation state, when the strain increasing does not result in further refinement and strengthening, especially in pure metals [10-12]. One of the ways to overcome this limitation is doping [8, 13, 14]. That is why it is of great interest to study the behavior of doped metals under the SPD by particularly DCAP and following heating, especially if their behavior in pure state has already been studied [6, 15]. Thus, the main goal of the present study was to investigate evolution of microstructure of hafnium bronze obtained by the DCAP under the deformation and following annealing.

2. Materials and methods

Specimens for investigation were made of an alloy of commercially pure copper (99.9) with 0.8 wt. % of hafnium. After casting the specimens were forged at 600°C and annealed for their homogenization at 950°C for 1 h in the vacuum with following quenching in water. This treatment resulted in the polycrystalline microstructure with an average crystallite size of about 50 μm. The cylindrical specimens with the diameter of 10 mm and the length of 40 mm were fabricated from the as-obtained billets and subjected to DCAP at the rate of 10^4–10^5 s^{-1}. The initial speed of specimens was about 250 m/s, and the powder gases pressure was about 2 GPa. The pressing was carried out in a matrix consisting of two channels a the diameter of 10 mm intersecting under an angle of 90°, with an external rounding radius of 5 mm, which equals to a half of channels diameter, and an internal radius 1-2 mm. The DCAP was carried out in one and two passes, and after the first pass a specimen was turned by 180°. To study the thermal stability of the structure obtained the as-deformed specimens were annealed in the temperature range of 100-600°C in a high-vacuum furnace Varian for 1 h. The microstructure was studied in transmission electron microscope Tecnai G2 30 Twin. Microhardness was measured by a special unit adjusted to optical microscope Neophot-21 and calculated by the method described in detail in [14].

3. Results and discussion

As shown by metallographic studies (Figure 1), the structure of specimens after DCAP in transverse sections is quite uniform excluding narrow areas in upper (near internal radius of matrix) and lower (near external radius) parts, which is typical of specimens subjected to the dynamic channel angular pressing [6, 8, 13, 15]. As in [6, 13], the structure in the metallographic images is mainly stringy, with curved bands resulting from the high-speed deformation by DCAP.

The more detailed information on the structure evolution in hafnium bronze and mechanisms of deformation by DCAP was obtained by transmission electron microscopy. Figure 2 demonstrates electron micrographs and diffraction patterns of the structure after two passes of DCAP and annealing at 400°C.

According to TEM studies, the structure after 2 passes of DCAP is non-uniform, and it is seen that there are two deformation mechanisms, dislocation sliding and twinning. These mechanisms result if the formation of two types of microstructure, namely, areas of cell structure interchanging with thin-lamellar structure of twinning type. The sizes of some cells are up to 1 μm.

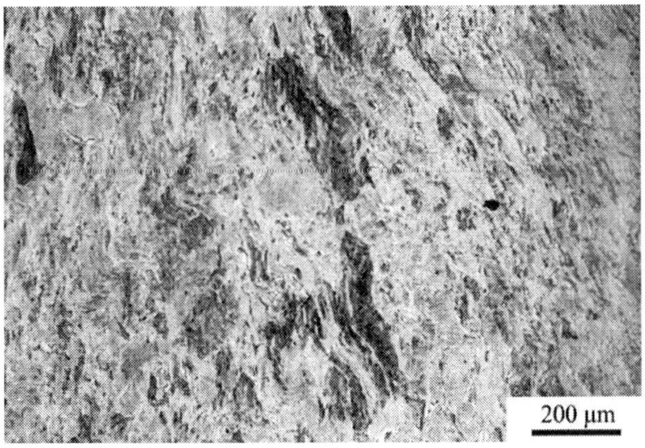

Figure 1. Structure of Cu-0.8Hf alloy after DCAP by 2 passes in transverse section

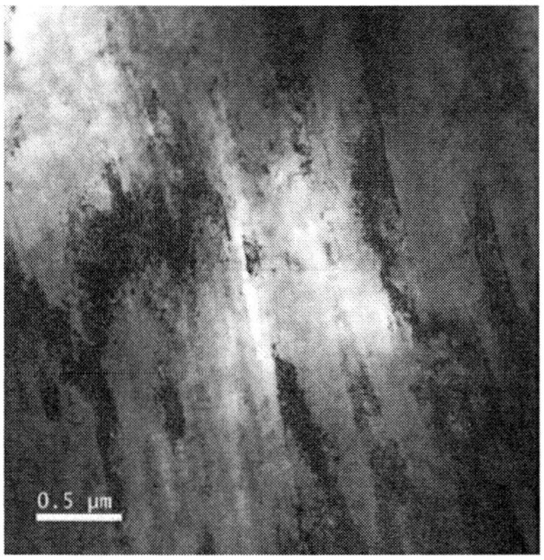

(a)

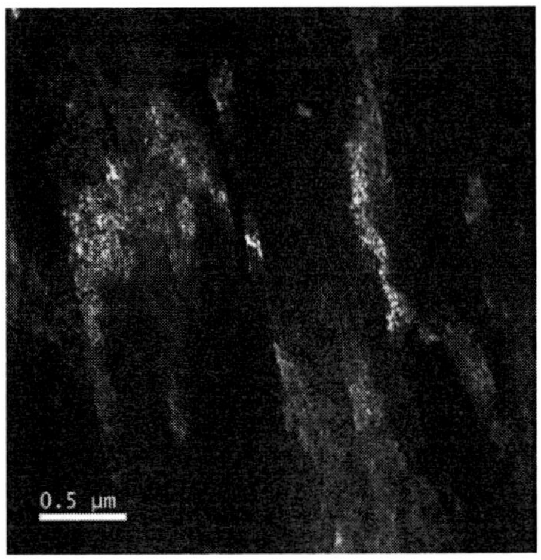

(b)

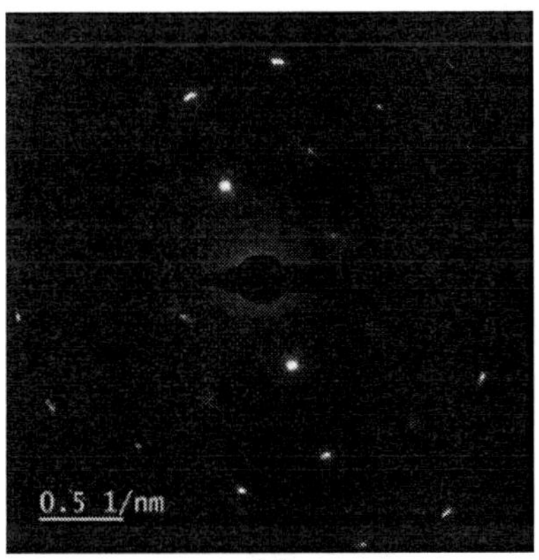

(c)

Figure 2. Microstructure of Cu-0.8Hf alloy after DCAP by 2 passes and annealing at 400°C for 1 h: (a) – bright-field image; (b) – dark-field image in $(111)_{Cu}$ reflection; (c) – electron diffraction pattern

Under the low-temperature annealing of the hafnium bronze the structure formed by DCAP does not undergo noticeable changes. Thus, after the annealing at 400°C the above-mentioned two types of structures – cell structure and thin-lamellar of twinning type – are present (Figure 2). The size of some cells is about 1 μm, and they have wide dislocation boundaries. Considerable defectiveness of cells observed directly after the deformation is retained after the annealing at this temperature. The areas with twins are also retained, and micro-twins belonging to one and the same twinning system form pili-ups, and it is confirmed by the selected area diffraction patterns (Figure 2c).

It is of great importance to emphasize that the structure observed directly after the deformation by DCAP is retained without noticeable changes at annealing in the temperature range up to 400°C and even up to 500°C, thus, in the hafnium bronze under consideration the structure is more stable than, for example, in tin bronze [13]. The enhanced thermal stability is obviously due to precipitation of ultra-dispersed particles of Cu_5Hf compound [14]. However, in the electron micrographs and electron-diffraction patterns of the bronze under consideration these intermetallic precipitates are not observed, which is maybe because of their high dispersity and location in highly-defective crystallite and cell boundaries. Nevertheless, the enhanced thermal stability of the structure is confirmed by microstructure measurements. Thus, in the quenched non-deformed state the microhardness of the alloy is about 700 ± 20 MPa. After the DCAP by 2 passes it increased up to 1800-2170 MPa dependently on the point of measurement, and these high values are retained at annealing up to 500°C. However, at annealing to 600°C the microhardness dramatically decreases and falls into the range of 1000-1200 MPa, which indicates the development of relaxation processes at this annealing temperature.

4. Summary

According to the data obtained in this study, after the deformation by 2 passes of DCAP the structure of hafnium bronze is non-uniform, with two types of microstructure – cell structure and lamellar structure of twinning type. Such structure indicates the action of two deformation mechanisms, dislocation sliding and twinning.

It has been established, that doping of copper by hafnium considerably affects the inclination of the structure formed under the DCAP to the development of relaxation processes. It is demonstrated that the microstructure formed under the deformation is retained at annealing in the temperature range up to 500°C without noticeable changes. The microhardness changes only slightly under the annealing up to 500°C compared to its values directly after the deformation and falls in the range of 1800-2170 MPa dependently on the area of the measurements. Under the annealing at 600°C the microhardness decreases considerably to the range of 1000-1200 MPa, which indicates the development of relaxation processes.

Acknowledgments

The work has been done under the State Assignment of FASO of Russia (topic "Function", No. AAAA-A19-119012990095-0) with partial support of the Program of Fundamental Research of Ural Branch of RAS (Project No. 18–10–2–37).

References

[1] Gleiter H 2000 *Acta Materialia* **48** 1-29.
[2] Valiev R, Zhilyaev A and Langdon T 2014 *Bulk Nanostructured Materials: Fundamentals and Applications* (Hoboken, New Jersey, USA: TMS, Wiley)
[3] Valiev R, Estrin Y, Horita Z, Langdon T, Zehetbauer M and Zhu Y 2006 *The Journal of The Minerals, Metals & Materials Society* **58 (4)** 33–9.
[4] Segal V 1999 *Materials Science and Engineering: A* **271** 322–33.
[5] Shorokhov E V, Zhgilev I N and Valiev R Z 2006 Russian Federation Patent No. 2283717
[6] Khomskaya I, Zel'dovich V, Frolova N, Kheifets A, Shorokhov E, and Zhgilev I 2008 *The Physics of Metals and Metallography* **105** 586–93.

[7] Popov V, Popova E, Kuznetsov D, Stolbovsky A, Shorohov E, Reglitz G, Divinski S and Wilde G 2014 *Defect Diffusion Forum* **354** 109–19.
[8] Zel'dovich V, Frolova N, Khomskaya I, Kheifets A, Dobatkin S, Shorokhov E and Nasonov P 2016 *The Physics of Metals and Metallography* **117** 74–82.
[9] Popov V, Popova E, Kuznetsov D, Stolbovsky A, Shorohov E, Nasonov P, Gaan K, Reglitz G, Divinski S and Wilde G 2013 *Materials Science and Engineering: A* **585** 281–91.
[10] Pippan R, Scheriau S, Taylor A, Hafok M, Hohenwarter A and Bachmaier A 2010, *Annual Review of Materials Research* **40** 319–43.
[11] Popova E, Popov V, Romanov E and Pilyugin V 2007 *The Physics of Metals and Metallography* **103** 407–13.
[12] Popov V, Stolbovsky A, Popova E and Pilyugin V 2010 *Defect Diffusion Forum* **297–301** 1312–21.
[13] Popov V, Stolbovsky A, Popova E, Falahutdinov R and Shorohov E 2017 *The Physics of Metals and Metallography* **118** 864–71.
[14] Shangina D, Bochvar N and Dobatkin S 2016 *Inorganic Materials: Applied Research* **7** 465–70.
[15] Khomshaya I, Zel'dovich V, Kheifets A, Shorokhov E, Frolova N, Nasonov P, Ushakov A and Zhgilev I 2011 *The Physics of Metals and Metallography* **111** 612–22.

IOP Conf. Series: Materials Science and Engineering **966** (2020) 012032 doi:10.1088/1757-899X/966/1/012032

The use of regression models to describe the temperature coefficient values of linear expansion and technological temperatures of $CaSO_4$-KPO_3-$Na_2B_4O_7$ low-melting glass system

A Stolbovsky[1, 2, 3]

[1] Ural Federal University named after the first President of Russia B.N. Yeltsin, 19, Mira str., Yekaterinburg, Russia
[2] M.N. Miheev Institute of Metal Physics, UB RAS, 18, S. Kovalevskoy str., Yekaterinburg, 620108, Russian

E-mail: [3] stolbovsky@imp.uran.ru

Abstract. Low-melting noncrystallizing glasses are used as solders for connections of various metallic and nonmetallic parts. However, theoretical data on glass formation and physical parameters of low-melting glasses are still limited. The use of modelling allows to simplify the selection of a glass system as a solder by the mathematical description of the dependencies of both that the temperature coefficient values of linear expansion and other specific temperatures, which are necessary for a practical application. Therefore, the fourth-order regression polynomial models are offered for describing dependences of the temperature coefficient values of linear expansion, the softening temperature and the glass transition temperature from the component concentrations of the $CaSO_4$-KPO_3-$Na_2B_4O_7$ low-melting glass system in the glass formation region. The model involver both pair and ternary interactions of the glass system's components. The analysis of the obtained results showed that the concentration of the CaSO4 component has the highest effect on the temperature coefficient values of linear expansion.

1. Introduction

At present, the application of a leak-free and reliable bonding of parts, which are made of various kind of materials with each other is required in modern instrument engineering. Low-melting noncrystallizing glasses are used as solders for connections of various metallic and nonmetallic parts [1]. The lower softening temperature of such solders allows to perform a junction at quite low temperatures in vacuum that will prevent the oxidation and deformation at soldering [1-2]. In addition, the creation of strong junctions between parts requires from solders that the temperature coefficient values of linear expansion (TCLE) must be comparable, the complete spreading temperature during heat treatment, as a rule, must not exceed above 700 °C and the softening temperature must be below 450–600°C as well [2].

Considerable experimental material concerning low-melting glasses in numerous glass-forming systems has already obtained [2-6]. The data about the properties such as TCLE, the softening temperature (T_s) and the glass transition temperature (T_g) for various glass systems is not generalized.

That leads to difficulties in choosing low-melting glasses for practical application in industry as solder.

At the same time, theoretical data on glass formation and physical parameters of low-melting glasses are still limited [3,7]. Therefore, the availability of data on TCLE and the technological temperatures for various compositions would allow to make an accurate selection of the composition with the most preferred properties for a specific application.

As represented in the work [8], the use of modelling allows to simplify the selection of a glass system as a solder by the mathematical description of the dependencies of both TCLE and other specific temperatures, which are necessary for a practical application.

In the previous study [9], the possibility of using the ternary $CaSO_4$-KPO_3-$Na_2B_4O_7$ low-melting glass system as a glass-solder material was shown. The aim of the present paper is to study the possibilities of using a regression model for predicting TCLE and the technological temperatures' values depending on the glass composition.

2. Materials and methods

The results of the study [9] devoted to the investigation of the $CaSO_4$-KPO_3-$Na_2B_4O_7$ system have been taken. According to the regression analysis recommendations [10-12], polynomial regression was chosen as a model for fitting the experimental data. The chi-squared (χ^2) test as the quality factor to estimate the fitting accuracy was chosen because χ^2 is a more sensitive characteristic in comparison with the coefficient of determination R^2 [12].

The general view of the N-order polynomial with pair and ternary interactions, which was used for fitting experimental data is represented by the following expression:

$$\begin{aligned}\alpha = C_0 + C_x x + C_y y + C_z z + C_{xy} xy + C_{xz} xz + C_{yz} yz + C_{xyz} xyz + \\ + C_{x2} x^2 + C_{y2} y^2 + C_{z2} z^2 + C_{xy2} x^2 y^2 + C_{xz2} x^2 z^2 + C_{yz2} y^2 z^2 + C_{xyz2} x^2 y^2 z^2 + \\ \dots \\ + C_{xN} x^N + C_{yN} y^N + C_{zN} z^N + C_{xyN} x^N y^N + C_{xzN} x^N z^N + C_{yzN} y^N z^N + C_{xyzN} x^N y^N z^N\end{aligned}, \quad (1)$$

where α is TCLE value ($\alpha \cdot 10^{-7}$ K^{-1}), various indexes – the calculated model coefficients, x is the concentration in mole fractions of $CaSO_4$, y – KPO_3 and z – $Na_2B_4O_7$, accordingly.

The same expression (1) was used for fitting T_s (°C) and T_g (°C) dependences as well.

The calculation was carried out in two stages. The first stage was the calculation of the full model with expression (1). At the second stage, coefficients, which were a few orders of magnitude less than 1 and other model coefficients as well, were excluded from the model to simplify one. Then the simplified model was recalculated to estimate accuracy. If it was necessary the process of simplifying was repeated.

3. Results and discussion

The regression analysis of experimental data allows to make the TCLE and technological temperatures' values interpolation between the experimental points and estimate the relationships between a glass composition and these dependent variables. Such approach will allow to consider the surface features in multidimensional space without deep theoretical studies.

As shown by the regression analysis of the experimental dataset, the TCLE, T_s and T_g dependences on components composition of the glass system are well described by means of the regression model in the form of a fourth-order polynomial in contrast to the second-order one for the $MgSO_4$-KPO_3-$Na_2B_4O_7$ system [8]. The results of approximation for the TCLE, T_s and T_g as calculated coefficients of the model are shown in Table 1. The numerical values are represented in exponential form. The sign in front of the model's coefficient indicates a direction of an influence.

It should be noted that suggested regression models do not have a physical basis. That is statistical description only. However, such statistical representation allows to estimate an influence of

IOP Conf. Series: Materials Science and Engineering **966** (2020) 012032 doi:10.1088/1757-899X/966/1/012032

components on the resulting value. That can be important for understanding when knowledge of physical foundations does not exist, or it is significant limited.

Moreover, the use of the models for practical application should be restricted in glass formation region [9]. It is important since the represented models does not take into account the limitation of the composition. In order to avoid an incorrect estimation of values of the TCLE and the technological temperatures, should use the following set of expressions for the components of the glass system:

$$\begin{cases} x \in (0.0;0.3) \\ y \in (0.0;0.8) \\ z \in (0.0;0.9) \\ x+y+z=1 \end{cases}, \quad (2)$$

where x is the concentration in mole fractions of $CaSO_4$, y – KPO_3 and z – $Na_2B_4O_7$, accordingly.

Table 1. Results of approximation.

Model's coefficients	$\alpha \times 10^{-7}$, K^{-1}	T_s, °C	T_g, °C
C_0	-	3.39E+02	-
C_x	-	-	3.81E+02
C_z	-	-	2.45E+02
C_{xy}	-	6.04E+02	7.62E+02
C_{yz}	3.76E+02	7.99E+01	4.99E+02
C_{xyz}	2.05E+03	-	-
C_{z2}	2.92E+02	1.31E+01	6.74E+01
C_{xy2}	-	1.65E+01	9.14E+02
C_{xz2}	2.13E+03	-	-
C_{y3}	2.09E+02	5.12E+01	2.27E+02
C_{xy3}	5.83E+03	-	-
C_{yz3}	-	-	7.77E+02
C_{xyz3}	1.03E+00	1.01E+00	-
C_{x4}	9.59E+03	-	-
C_{y4}	-	-	2.22E+02

To visually estimate α, T_s and T_g, colored ternary diagrams of these properties can be used. Figure 1 is represented the calculated values of the TCLE, T_s and T_g for the glass formation region of the $CaSO_4$-KPO_3-$Na_2B_4O_7$ system, which are indicated by colored areas on the ternary diagrams. It can be seen than the TCLE is variated in very wide range of $140\text{-}270 \cdot 10^{-7}$ K^{-1}. The values of the glass transition temperature are in the range of 270-390°C and the values of the softening temperature in the range from 350 to 490°C. These are lower values of T_s and T_g, which are usually required for glass solders [2].

In comparison with the $MgSO_4$-KPO_3-$Na_2B_4O_7$ system [8], the replacement of the $MgSO_4$ component expanded the range of the TCLE values. At the same time, the second-order regression polynomial model, which was used for the $MgSO_4$-KPO_3-$Na_2B_4O_7$ system, did not allow to describe the dependencies adequately. That can be caused by a different effect of Mg^{2+} and Ca^{2+} cations on the glass structure.

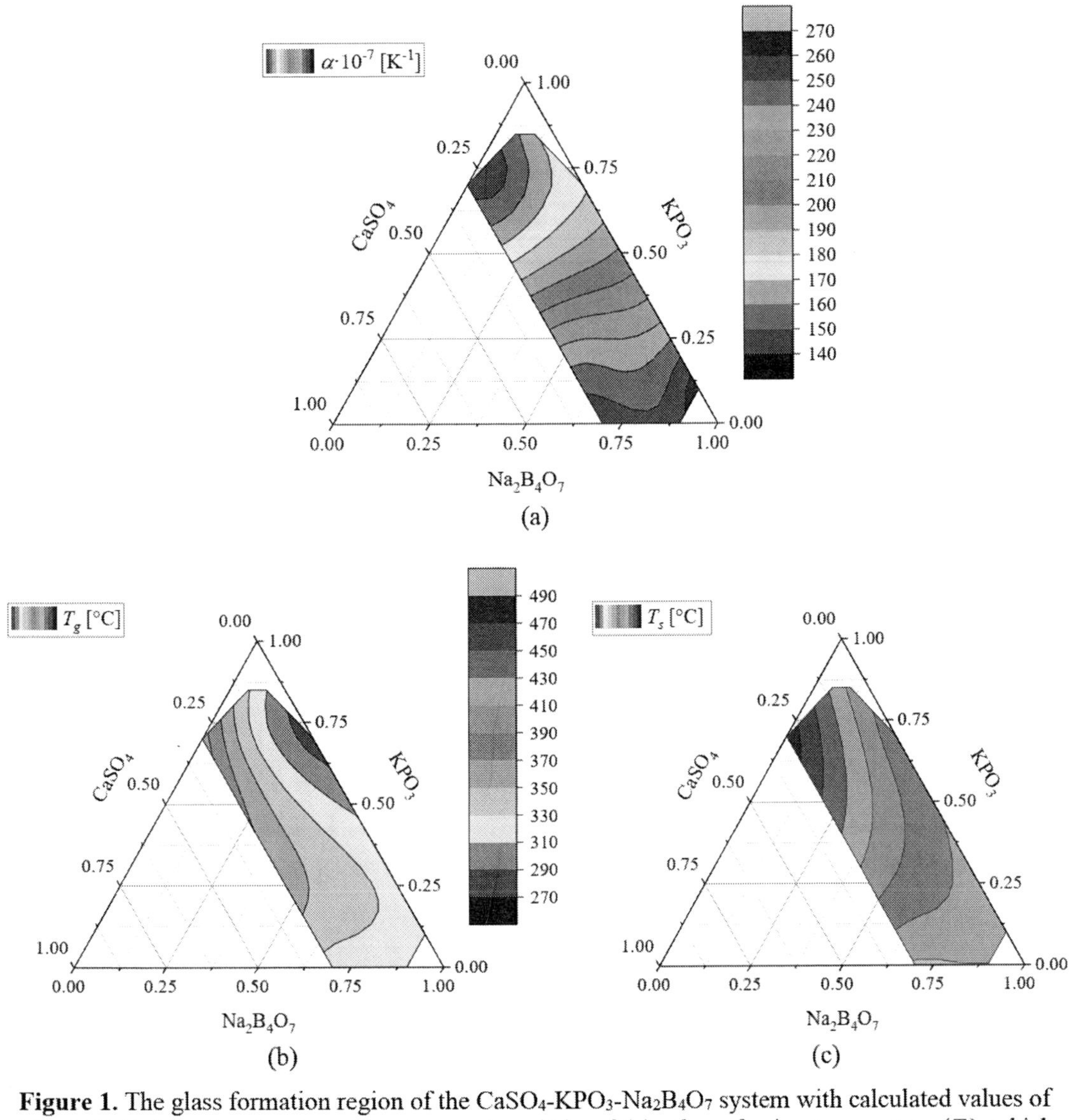

Figure 1. The glass formation region of the $CaSO_4$-KPO_3-$Na_2B_4O_7$ system with calculated values of (a) – TCLE; (b) – the glass transition temperature (T_g) and (c) - the softening temperature (T_s), which are represented by colored areas

Besides, it can be seen (Table 1) that, in the TCLE model, among the coefficients at the fourth-order magnitude of components, the coefficient corresponding to the $CaSO_4$ component is presented only. Moreover, this coefficient is the highest that may also indicate differences in the influence of Mg^{2+} and Ca^{2+} cations on the glass structure additionally.

4. Summary

It was established that the second-order regression polynomial model, which was used for the MgSO4-KPO3-Na2B4O7 system, did not allow adequately to describe the dependencies of TCLE and the technological temperatures' values of the $CaSO_4$-KPO_3-$Na_2B_4O_7$ system. The fourth-order regression polynomial model is offered for describing dependences of the TCLE, the softening temperature and the glass transition temperature from the concentration in mole fractions of $CaSO_4$, KPO_3 and $Na_2B_4O_7$ components of the glass system. The proposed model takes into account pair and ternary interactions of the glass system's components.

As shown by the analysis of the obtained results, in comparison with the $MgSO_4$-KPO_3-$Na_2B_4O_7$ system, the replacement of the $MgSO_4$ component with $CaSO_4$ allowed to expand the range of the TCLE values. In addition, it was found, the $CaSO_4$ component has the highest effect on the TCLE that may points out at different influence of Mg^{2+} and Ca^{2+} cations on the glass structure.

Acknowledgments

The research was carried out within the state assignment of Minobrnauki of Russia (theme "Function" No. AAAA-A19-119012990095-0).

References

[1] Bobkova N 2009 *Glass and Ceramics* **66** 206–209.
[2] Bobkova N and Trusova E 2012 *Glass and Ceramics* **68** 349-352.
[3] Mamoshin V 1996 *Glass and Ceramics* **53** 166-168.
[4] Vyatchina V, Perelyaeva L, Zuev M and Baklanova I 2009 *Glass Physics and Chemistry* **35** 580-585.
[5] Feng H, Zi-jun H, Jun-lin X, Shu-xia M and Ming-fang J 2016 *Journal of Central South University* **23** 1541-1547.
[6] Geodakyan D, Petrosyan B, Stepanyan S and Geodakyan K 2009 *Glass and Ceramics* **66** 381-384.
[7] Oganesyan M, Oganesyan R and Knyazyan N 2010 *Theoretical Foundations of Chemical Engineering* **44** 500-502.
[8] Stolbovsky A, Farafontova E and Vyatchina V 2019 *Materials Science Forum* **946** 331–335.
[9] Stolbovsky A, Murzinova S and Vyatchina V 2020 *Materials Science Forum* **989** 265-269.
[10] Draper N and Smith H 1998 *Applied Regression Analysis, 3rd Ed.* (New York: John Wiley & Sons, Inc.).
[11] Pardoe I 2013 *Applied Regression Modeling, 2nd Ed.* (New York: John Wiley & Sons, Inc.).
[12] Freund R, Wilson W and Sa Ping 2006 *Regression Analysis: Statistical Modeling of a Response Variable, 2nd Ed.* (Elsevier Inc.).

Physico-mechanical properties of the sinter of various chemical composition

S K Sibagatullin[1], A S Kharchenko[1], V I Sysoev[1, 3] and D N Gushin[2]

[1] Nosov Magnitogorsk State Technical University, 38, Lenin avenue, Magnitogorsk, Chelyabinsk Region, 455000, Russian
[2] PJSC 'MMK', Magnitogorsk, Chelyabinsk Region, Russia

E-mail: [3] viktor.sysoev.86@yandex.ru

Abstract. The paper includes data on the physico-mechanical properties of sinter of various origins including density (bulk, average, and true), porosity (open and total), critical angle of repose, internal and external angles of repose measured on dropping the material down onto a horizontal surface from a steel sheet inclined by 40°, 60°, and 80° to the horizon,angle of external friction on the steel surface, the strength properties of the sinter in a cold state and after reduction at a temperature of 500 °C.The investigated sinter had the following characteristics: an average density of 3.20–3.30 g/cm^3; a true density of 4.43–4.57 g/cm^3; an external angle of repose of 20.5–32.3 deg; a bulk porosity of 0.46–0.50 m^3/m^3; a degree of reduction by hydrogen at 500 °C of 3.3–4.8%. It demonstrated the following strength characteristics in the cold state: IR_{+5} = 92.0%, $D_{0.5\text{-}5}$ = 5.5%, $A_{-0.5}$ = 2.5%. The strength characteristics after low-temperature reduction in hydrogen atmosphere were as follows: IR_{+5} = 29.80 %; $D_{0.5\text{-}5}$ = 57.6 %; $A_{-0.5}$ = 12.6 %.

1. Introduction

Physico-mechanical and physico-chemical properties of iron ore raw materials loaded into a blast furnace determine the technical and economic indices of its operation [1–7]. The particle-size distribution of agglomerate, the reposeangles of the loaded materials, as well as their bulk density, affect the distribution of components of the blast furnace charge along the radius and around the circumference of the furnace top [8–12]. Therefore, reliable information about these indices makes it possible to effectively regulate the loading of the blast furnace [13–19]. The porosity of materials is of high importance for the course of heat and mass transfer processes, including the reduction of iron and other elements from oxides [20–22]. Information on the porosity of sinter helps to analyzethe movement of gases through the interparticle voids [23]. The strength characteristics of agglomerate, theshare of fine fractions of less than 5 mm or 5–10 mm in it, affect the gas permeability of the stock column [24,25]; therefore, a number of recommendations and criteria have been proposed for the current operating conditions of blast furnaces [26]. Loading of the agglomerate having reduced physical and mechanical characteristics, a high content of the fine fraction, intothe blast furnace operating with an upper limiting zone by gas dynamics, increases the mutual resistance of the charge and gas flows. This results in the suspension of burden and uneven run of the furnace, which will significantly reduce its productivity and raise the specific consumption of coke [27–29].

An analysis of the physico-mechanical characteristics of sinter makes it possible to develop a rational operating mode for a blast furnace. Using of such a mode provides the efficient application of agglomerated iron ore with various physico-mechanical properties and increases the technical and economic indices of smelting.

2. Test material

For laboratory study, the samples of iron ore sinter having various chemical composition were taken. The chemical composition of the investigated samples of sinter are presented in Table 1.

Table 1. Chemical composition of the investigated samples of sinter.

No. of sample	Element/oxide content, wt %												Bas.[a]
	Fe	FeO	S	SiO_2	CaO	MgO	Al_2O_3	TiO_2	P	Zn	Cr	MnO	
1	55.7	10.4	0.061	5.83	11.24	1.95	1.69	0.22	0.026	0.029	0.021	0.25	1.93
2	55.5	10.1	0.071	6.28	11.32	1.83	1.66	0.19	0.026	0.039	0.018	0.24	1.80
3	55.2	9.1	0.070	6.37	11.24	1.87	1.84	0.23	0.024	0.038	0.015	0.24	1.77
4	58.2	11.1	0.039	5.31	9.17	1.55	1.53	0.25	0.020	0.014	0.009	0.17	1.73

[a] Basicity.

As can be seen from Table 1, the sinter sample No. 4 has an increased total iron content (58.2 % versus 55.2–55.7% for the samples No. 1, 2, and 3), which is due to the more advanced production technology that allows the use of up to 90% of the fine concentrates having an increased iron content in the sinter charge. The sinter sample No. 4 is also characterized by a lower CaO/SiO_2 ratio (1.73 versus 1.77–1.93 for the samples No. 1–3) and an increased content of iron oxide FeO (11.1% versus 9.1–10.4 %f or the samples No. 1–3).

3. Results

3.1. Grain-size composition

The particle-size distribution was determined by sieve analysis [30]. The results of sieve analysis are given in Table 2.

According to the table, the sinter sample No. 4 is characterized by the highest value of the equivalent surface diameter of particles. It is 27.0 mm versus 10.1–12.1 mm for the samples No. 1–3. The weighted-mean diameter of particles of the sinter sample No. 4 is also significantly higher than the weighted-mean diameter of particles of the other sinter samples and amounts to 34.7 mm compared to 17.9–23.9 mm for the samples No. 1–3. The sinter sample No. 4 is also characterized by the higher uniformity of particle sizes, which amounts to 0.78 against 0.44–0.62 for the samples No. 1–3.

3.2. Density and porosity

The determination of bulk, average and true density of the sinter was conducted for the sinter samples No. 1–4 [31].

The connected porosity of the sinter was determined by its saturation with water under vacuum [32]. The total porosity was calculated by determining the average and true density [31].

The bulk porosity of the sinter was calculated by bulk (ρ_{blk}) and average (ρ_{avg}) densities:

$$\varepsilon = 1 - \rho_{blk}/\rho_{avr} \quad (1)$$

The results of determining the density and porosity of the sinter are given in Table 3. According to the Table 3, the sinter sample No. 4 has an average density of 3.20 g/cm^3, which is less than the average density of the samples No. 1, 2 and 3, the value of which is at the level of 3.27–3.30 g/cm^3. Moreover, the true density of the sinter sample No. 4 is higher than that of the samples No. 1, 2 and 3 by 2.9 % on average.

The increased uniformity of particle sizes by 33% (Table 2) of the sinter sample No. 4, compared with the samples No. 1, 2, and 3, provided the sinter sample No. 4 with the highest bulk porosity of 0.50. The content of the fraction of 0–5 mm in the samples No. 1, 2, and 3 in the range of 5.4–9.9% with a uniformity of particle sizes in the range of 0.44–0.62 provided bulk porosity of 0.46–0.47, which is less than that ofthe sinter sample No. 4.

IOP Conf. Series: Materials Science and Engineering **966** (2020) 012033 doi:10.1088/1757-899X/966/1/012033

Table 2. Granulometric composition of the sinter from the samples No. 1–4.

Index name	Fraction content, wt %			
No. of sample	1	2	3	4
Size range, mm:				
80–100	–	–	5.82	–
60–80	4.63	–	1.35	8.16
40–60	11.05	6.62	7.26	19.76
25–40	19.62	11.24	14.26	44.23
10–25	43.29	50.39	38.30	26.53
5–10	13.85	26.41	23.12	0.53
0–5	7.56	5.35	9.89	0.8
Total	100	100	100	100
Size range (fine fractions), mm:				
7–10	43.13	56.90	45.97	47.76
5–7	21.56	26.26	24.08	12.42
3–5	16.71	8.36	17.51	9.55
1–3	7.55	2.54	7.09	9.55
0.5–1.0	2.26	1.34	1.40	5.06
0–0.5	8.79	4.60	3.94	15.66
Total	100	100	100	100
Equivalent surface diameter of particles, mm	12.14	11.09	10.11	26.90
Weighted-mean diameter of particles, mm	23.94	17.89	23.13	34.68
Homogeneity of the particle diameters	0.51	0.62	0.44	0.78

Table 3. Density and porosity of the sinter samples No. 1–4.

Index name	Index value			
No. of sample	1	2	3	4
Density:				
bulk, t/m^3	1.75	1.75	1.75	1.6
average, g/cm^3	3.30	3.30	3.27	3.20
true, g/cm^3	4.46	4.44	4.43	4.57
Porosity, %				
unconnected	26.0	25.6	26.2	29.9
connected	14.2	11.8	10.7	9.2
Bulk porosity, m^3/m^3	0.47	0.47	0.46	0.50

3.3. Angles of repose and external friction

The angle of repose of bulk materials can be measured by a variety of methods the most widely used of which are: the tilting box method, the fixed funnel method, the revolving cylinder/drum method, the hollow cylinder method, and the tilting cylinder method. The difference between the angles of repose measured by various methods for the components of the blast furnace burden usually can be ignored. For the purposes of this study, the angle of repose was determined by the hollow cylinder method.

The outer and inner angles of repose of the sinter were determined by dropping it down from a height of 0.5 meters after moving along the surface of the steel sheet inclined at angles of 40°, 60°, and 80°to the horizon and having a length of 1 m. After the formation of an asymmetric pile both its angles to the horizon were determined characterizing external (the angle forming along the direction of movement of particles) and internal (the opposite angle) angles of repose correspondingly.

By raising the horizontally installed surface of the steel sheet with the sinter layer placed on it, the angle of external friction of the sinter on the steel surface was determined as the maximum angle at which the stationary state of the material is maintained.

The results of determining the angle of repose, the external and internal angles of repose in operating conditions and the angle of external friction on the steel surface are given in Table 4.

Table 4. Angles of repose and external friction for the sinter samples No. 1–4.

Index name	Index value			
No. of sample	1	2	3	4
Natural angle of repose, deg	32.3	34.3	33.0	33.6
Internal angle of repose (deg) determined by dropping the sinter down onto horizontal surface from the steel surface inclined to the horizon at an angle of, deg:				
40	41.9	38.3	37.6	27.6
60	39.2	35.2	34.6	22.5
80	39.7	42.6	36.8	25.2
Internal angle of repose (deg) determined by dropping the sinter down onto horizontal surface from the steel surface inclined to the horizon at an angle of, deg:				
40	25.8	20.5	33.5	21.0
60	24.6	27.6	25.3	17.5
80	32.3	34.9	28.3	21.8
External friction angle on the steel surface, deg	23	27	22	24

The strength of a suitable sinter expressed as resistance to impact loads and abrasion was determined by testing a sample weighing 200 g in a steel pipe with an internal diameter of 78.5 mm and a length of 1 m, rotating at a speed of 27 rpm, for 8 minutes. After testing the sample was sieved into three fractions, +5 mm, 0.5–5 mm, and –0.5 mm.The resistance of a suitable sinter to impact (IR_{+5}) was determined by the percentage of the fraction +5 mm, the destruction of a suitable sinter ($D_{0.5\text{-}5}$) – by the percentage of a fraction of 0.5–5 mm, abrasion of a suitable sinter ($A_{-0.5}$) –by the percentage of the fraction –0.5 mm [31].

The determination of the 'hot' strength of sinter, or the strength after reduction, was conducted as follows. The samples of sinter were crushed and sieved into a fraction of 10–15 mm and placed into an electric tube furnace. Reduction of the sample was carried out in the furnace by isothermal holding at a temperature of 500 °C in a stream of pure hydrogen (99.997%) for 1 hour. After cooling the sample in an inert medium (99.5% nitrogen) a drum strength test was conducted similar to the determination of the 'cold' strength of sinter.

The results of determining the 'cold' and 'hot' strength of the sinter are given in Table 5. The degree of reduction of the sinter tested for evaluation of the 'hot' strength is shown in Table 6.

Comparison of the indices of resistance to impact loads for a 'cold' suitable sinter from different plants shows that the sinter samples No. 1 (IR_{+5} = 92.93%) and No. 4 (IR_{+5} = 92.0%) have the highest values of impact resistance. The sinters No. 2 (IR_{+5} = 88.56%) and No. 3 (IR_{+5} = 88.38%) are less resistant to impact. The samples No. 3 ($D_{0.5\text{-}5}$ = 8.59%) and No. 2 ($D_{0.5\text{-}5}$ = 8.46%) are characterized by the greatest destruction in the cold state; the sinters No. 4 ($D_{0.5\text{-}5}$ = 5.50%) and No. 1 ($D_{0.5\text{-}5}$ = 5.05%) are less susceptible to destruction. The order of changing of abrasion for the investigated samples of sinters depending on the sample number is similar to the order of changing of destruction for these sinters. The sinter samples No. 3 ($A_{-0.5}$ = 3.03%) and No. 2 ($A_{-0.5}$ = 2.99%) are characterized by the greatest abrasion in the cold state while the sinters No. 4 ($A_{-0.5}$ = 2.50%) and No. 1 ($A_{-0.5}$ = 2.02%) demonstrate a much lesser degree of abrasion.

After reduction in hydrogen atmosphere, the strength characteristics of all the kinds of sinters studied in this work decrease to a different degree. Strength characteristics of the sinter sample No. 4were especially strongly reduced. The resistance of a suitable sinter to impact loads measured by the content of the +5 mm fraction decreased from 92.00% to 29.80%, i.e., about 3.1 times. The destruction of a suitable sinter measured by the content of fraction 0.5–5 mm increased from 5.50% to 57.58%, that is, about 10.5 times. The abrasion of a suitable sinter increased from 2.50% to 12.63%, that is, about 5.1 times. The relative drop in the corresponding strength characteristics of the other samples of sinter was significantly lower. Its values were in the following ranges: for impact resistance – from 1.4 to 1.8; for destruction – from 3.4 to 7.9; for abrasion – from 2.0 to 3.2. Data on the degree of reduction of the studied sinter samples, which are given in Table 6, indicate that the main reason for the sharp decrease in the strength characteristics of the sinter sample No. 4 after low-temperature reduction in hydrogen atmosphere is the increased reducibility of the material.

Table 5. 'Cold' and 'hot' strength of the sinter samples No. 1–4.

No. of sample	Strength indices, %					
	Resistance of a suitable sinter to impact (+5 mm)		Destruction of a suitable sinter (0.5–5 mm)		Abrasion of a suitable sinter (0–0.5 mm)	
	'Cold' strength	'Hot' strength	'Cold' strength	'Hot' strength	'Cold' strength	'Hot' strength
1	92.93	53.50	5.05	40.00	2.02	6.50
2	88.56	64.82	8.46	29.15	2.99	6.03
3	88.38	49.75	8.59	42.21	3.03	8.04
4	92.00	29.80	5.50	57.58	2.50	12.63

Table 6. Degree of reduction of the sinter tested for evaluation of the 'hot' strength and its FeO content.

No. of sample	Degree of reduction, %	FeO content, %
1	3.3	16.34
2	3.3	16.64
3	3.3	15.92
4	4.8	20.51

4. Conclusions

The differences in the physico-mechanical properties of the sinter sample No. 4, compared with the sinter samples No. 1–3, are as follows:

- its average density of 3.20 g/cm^{3}is higher than the average densities of the sinter samples No. 1, 2, and 3, the values of which are at the level of 3.27–3.30 g/cm^3;
- the true density is increased by 2.9%;
- the internal angle of repose of 34.6% and external angle of repose of 28.4% are lower than the ones for the other sinter samples;
- the highest bulk porosity –for the samples No. 1, 2, 3, and 4, its values are 0.47, 0.47, 0.46, and 0.50 m^3/m^3, respectively;
- the reducibility in hydrogen atmosphere at a temperature of 500 °C is increased by 1.5% (abs.) in comparison with the sinter samples No. 1, 2, and 3;
- high strength characteristics in the cold state, IR_{+5} = 92.00%; $D_{0.5\text{-}5}$ = 5.50%; $A_{-0.5}$ = 2.50%, and relatively low strength characteristics after low-temperature reduction in hydrogen atmosphere: IR_{+5} = 29.80%; $D_{0.5\text{-}5}$ = 57.58%; $A_{-0.5}$ = 12.63%, due to the high reducibility of the material.

References

[1] Dmitriev A N, Shumakov N S, Leontyev L I and Onorin O P 2005 *Fundamentals of Theory and Technology of Blast Furnace Smelting* (Ekaterinburg: Ural Branch of Russian Academy of Sciences, Institute of Metallurgy)

[2] Sibagatullin S K and Kharchenko A S 2012 *The Quality of the Charge Materials of Blast Furnace Smelting, Including Titanomagnetite and Siderite* (Magnitogorsk: NMSTU Publishing House)

[3] Babich A, Senk D and Gudenau H W 2009 *Ironmaking and Steelmaking* **36 (3)** pp 222–29

[4] Watakabe S, Takeda K, Nishimura H, Goto S, Nishimura N, Uchida T and Kiguchi M 2006 *ISIJ International* **3** pp 513–22

[5] Mousa E, Senk D and Babich A 2010 *Steel Research International* **81 (9)** pp 706–15

[6] Huatao Z and Minghua D 2012 *ISIJ International* **52 (12)** pp 2177–85

[7] Kashihara Y, Iwai Y, Sato T, Ishiwata N and Sato M 2015 *ISIJ International* **55** (6) pp 1237–44

[8] Sibagatullin S K, Kharchenko A S, Devyatchenko L D and Steblyanko V L 2017 *Journal of Chemical Technology and Metallurgy* **52 (4)** pp 694–701
[9] Kharchenko A S 2018 *Bulletin of Nosov Magnitogorsk State Technical University* **3** pp 46–56
[10] Zhao Y, Capo J C and McKnight S J 2011 *Iron & Steel Technology* **1** pp 52–61
[11] Jimenez J, Mochon J and S de Ayala J 2004 *ISIJ International* **44 (3)** pp 518–26
[12] Krivenko S V, Tarasov V P 2012 *Metallurgical Processes and Equipment* **30 (4)** pp 4–10
[13] Sibagatullin S K, Kharchenko A S, Savchenko G Yu and Beginyuk V A 2018 *CIS Iron and Steel Review* **16** pp 11–14
[14] Tovarovskii I G 2014 *Metallurgist* **58 (7–8)** pp 648–54
[15] Sibagatullin S K, Kharchenko A S 2018 *Steel in translation* **48 (10)** pp 624–30
[16] Sibagatullin S K, Harchenko A S and Logachev G N 2016 *The International Journal of Advanced Manufacturing Technology* **86 (1–4)** pp 531–37
[17] Sibagatullin S K, Harchenko A S, Teplykh E O, Stepanov E N, Mezin D A and Fetisov V B 2012 *Coke and Chemistry* **55 (2)** pp 62–5
[18] Nakano K, Sunahara K and Inada T 2010 *ISIJ International* **50 (7)** pp 994–9
[19] Teng Zh J, Cheng Sh S, Du P Yu and Guo X B 2013 *International Journal of Minerals, Metallurgy, and Materials* **20 (7)** pp 620–6
[20] Roshchin V E and Roshchin A V 2019 *CIS Iron and Steel Review* **17** pp 14–24
[21] Tovarovskii I G and Merkulov A E 2016 *Metallurgist* **60 (5–6)** pp 589–93
[22] Tomash A A *et al.*1999 *Reporter of the Priazovskyi State Technical University. Section: Technical Science* **8** pp 9–14
[23] Tarasov V P and Krivenko S V 201 *Steel in Translation* **44 (5)** pp 359–62
[24] Sibagatullin S K, Kharchenko A S, Chernov V P and Beginyuk V A 2017 *Chernye Metally* **8** pp 27–33
[25] Filatov S V, Titov V N, Kurunov I F, Lozovich A V and Zagainov S A 2018 *Metallurgist* **61 (9–10)** pp 844–48
[26] Tovarovskii I G 2014 *Steel in Translation* **44 (5)** pp 350–8
[27] Dmitriev A N, Zolotykh M O, Vitkina G Y and Gordon Y M 2016 *Proceedings of the Iron and Steel Technology Conference AISTech* **61 (9–10)** pp 3141–6
[28] Chukin M V, Sibagatullin S K, Kharchenko A S, Chernov V P and Logachev G N 2016 *CIS Iron and Steel Review* **12** pp 9–13
[29] Sibagatullin S K, Kharchenko A S, Savinov A S, Selivanov V N, Guschin D N and Mazur I P 2018 *Journal of Chemical Technology and Metallurgy* **53 (5)** pp 990–4
[30] GOST 27562-87 (Russian National Standard)
[31] GOST 25732-88 (Russian National Standard)
[32] GOST 2409-95 (Russian National Standard)

IOP Conf. Series: Materials Science and Engineering **966** (2020) 012034 doi:10.1088/1757-899X/966/1/012034

The effect of wear-resistant coatings composition based on niobium nitride on the structural parameters, mechanical properties and efficiency of the cutting tool

V P Tabakov[1], L V Khudobin[1], A V Chikhranov[1, 2, 3] and Y A Dolzhenko[2]

[1] Federal State Budgetary Educational Institution of Higher Education Ulyanovsk State Technical University, Ulyanovsk, 432027, Russia

[2] Federal State Budgetary Educational Institution of Higher Education Ulyanovsk Institute of Civil Aviation named after Chief Marshal of Aviation B P Bugaev, Ulyanovsk, 432071, Russia

E-mail: [3] chihranov@mail.ru

Abstract. Single-layer wear-resistant coatings based on niobium nitride deposited on a carbide cutting tool are considered in this article. The influence of one and two alloying elements on the change of the coating structure, mechanical properties and efficiency of the cutting tool was studied. It was revealed that in all cases, alloying with titanium, zirconium, aluminum and chromium leads to a change in the microstructure and an increase in the mechanical properties of the coatings and the inclusion of two elements into the composition of coatings increases mechanical properties to a greater extent. Research has shown that cutting tools with coatings based on niobium nitride are characterized by a decrease in the stress state of the tool base and an increase in the total compressive stresses in the coating. Compared to titanium nitride, the niobium nitride coating reduces the temperature on the front surface of the cutting tool, but for complex-alloy coatings, there is a slight increase in contact temperatures. Changing the heat-stressed state and increasing the mechanical properties leads to an increase in the crack resistance of such coatings. It is shown that the use of developed compositions of multi-element coatings can improve the efficiency of the cutting tool.

1. Introduction

Great importance is given today to improving the efficiency of the cutting tool. Therefore, most high-speed and carbide cutting tools are applied with wear-resistant coatings to improve performance [1–5]. This cutting tool has higher mechanical properties, crack resistance and, therefore, efficiency [6–10]. Coatings based on simple and complex nitrides of refractory elements, primarily titanium, zirconium, chromium, molybdenum, and aluminum nitrides, are most widely used among wear-resistant coatings. Coatings based on niobium nitride are particularly interesting among the coatings which have low thermal conductivity, which allows to reduce the heat flux into the cutting wedge of the tool [10, 11].

2. Methodology of Experiment

Coatings based on niobium nitride NbN, NbTiN, NbZrN, NbTiZrN, NbTiAlN and NbTiCrN with a thickness of 6 mkm were studied. These coatings were applied to MK8 carbide plates using the apparatus «Bulat-6». The chemical composition of coatings was determined using the method of quantitative x-ray diffraction analysis at the MAP-4 installation, taking into account ZAF corrections.

Lattice constants a and c and the residual compressive stresses $\sigma_0$were investigated by means of an x-ray diffractometer 'DRON-3M' [12]. Micro-hardness H_μ, elasticity modulus E, yield stress σ_y and fracture toughness K_{IC} of coatings were determined as in [10]. The calculation of the crack resistance of coatings was carried out according to the method of work [13]. The efficiency of carbide tool was determined during longitudinal turning of workpieces of structural steel 30KHGSA (alloying systems: 0.28–0.34% C, 0.8–1.1% Cr, 0.8–1.1% Mn, 0.9–1.2% Si).

3. Research results and discussion

The chemical composition of the coatings, the crystal lattice parameters a and c, the Burgers vector b and the length of the germ microcrack l_0 are presented in Table 1.

It was found that coatings based on niobium nitride are single-phase and have a tetragonal crystal lattice. The presence of titanium in composition of coatings leads to an increase in the values of Burgers vector b and the length of the germ microcrack l_0.

Values of the mechanical properties of the coatings are presented in table 2. Coatings of complex composition are characterized by greater microhardness and high values of elastic modulus and yield strength. Moreover, three-element coatings are characterized by a significant increase in fracture toughness: by 21–30% as compared to the coating of niobium nitride.

Table 1. Chemical composition and structural parameters of coatings.

Coating	Content of elements (% at)					a (nm)	c (nm)	b (nm)	l_0 (nm)
	Ti	Nb	Zr	Al	Cr				
TiN	100	–	–	–	–	0.4241	–	0.2999	0.5998
NbN	–	100	–	–	–	0.3015	0.5715	0.3015	0.6030
NbTiN	42.6	57.4	–	–	–	0.4399	0.8684	0.4399	0.8798
NbZrN	41.8	–	58.2	–	–	0.3007	0.5715	0.3007	0.6014
NbTiZrN	46.5	41.9	11.6	–	–	0.4372	0.8791	0.4372	0.8744
NbTiAlN	49.2	40.0	–	10.8	–	0.4449	0.8704	0.4372	0.8898
NbTiCrN	43.8	48.1	–	–	8.1	0.4458	0.8404	0.4458	0.8916

Table 2. Mechanical properties of coatings.

Coating	H_μ (GPa)	E (GPa)	K_{IC} (MPa·$m^{½}$)	σ_y (MPa)
TiN	26.2	327	8.76	9.53
NbN	29.8	467	9.78	9.67
NbTiN	31.3	422	10.16	10.7
NbZrN	32.1	428	9.97	10.61
NbTiZrN	32.9	496	11.84	11.12
NbTiAlN	32.1	455	12.76	10.91
NbTiCrN	33.4	448	12.69	11.42

Determination of the contact characteristics of the cutting process, temperatures and stresses acting in the cutting wedge of the tool was carried out by analytical calculation of the current unit loads and temperatures acting on the front and back surfaces of the cutting wedge of the tool according to the methods of [14, 15], followed by numerical calculation by the finite element method in a package ANSYS programs.

Calculations show a change in the thermal and stress state of cutting tools with coatings based on niobium nitride (table 3).

As it was shown in [10, 16], niobium nitride coatings have low thermal conductivity. This makes it possible to slightly reduce the temperature on the front surface of the cutting tool under the coating in the carbide base both by changing the contact characteristics of the cutting and by reducing the thermal conductivity of the coatings. NbN coating has the lowest maximum contact temperature T_{max}, while composite coatings have a temperature slightly higher (by 1.2–2.8 %) than titanium nitride.

Table 3. Heat-stressed condition and characteristics of crack resistance of coatings and tool base.

Coating	Tool base		Coating						
	T_{max} (°C)	σ_x (MPa)	σ_0 (MPa)	σ_1 (MPa)	σ_{therm} (MPa)	σ_Σ (MPa)	t_1 (min)	t_2 (min)	t_Σ (min)
TiN	1106	831	–775	513	–1910	–2172	1.63	7.28	8.91
NbN	1039	815	–2280	718	–3075	–3075	10.74	8.11	18.85
NbTiN	1120	779	–1497	620	–2709	–2709	6.55	12.16	18.71
NbZrN	1130	788	–2661	636	–1934	–1934	5.04	12.27	17.31
NbTiZrN	1132	736	–1704	689	–3021	–2729	11.36	15.76	27.12
NbTiAlN	1126	747	–1923	641	–2729	–3021	12.80	16.07	28.87
NbTiCrN	1137	751	–1941	534	–2819	–2819	9.83	16.58	26.41

At the same time, coatings with complex composition reduce the stress value in the tool base σ_x by 4–9 % as compared to NbN coating and by 5–12% as compared to TiN coating.

Taking into account the thermophysical properties of the coatings, the temperature and stress state of the cutting wedge of the tool were calculated the stresses arising in the coatings due to the difference with the tool base of their elastic modulus σ_1 and the thermal expansion coefficients σ_{therm} were calculated. The total stresses in coatings, taking into account the residual stresses σ_0 arising during deposition, allow us to state a more favorable picture for NbN coatings and three-element coatings of complex composition.

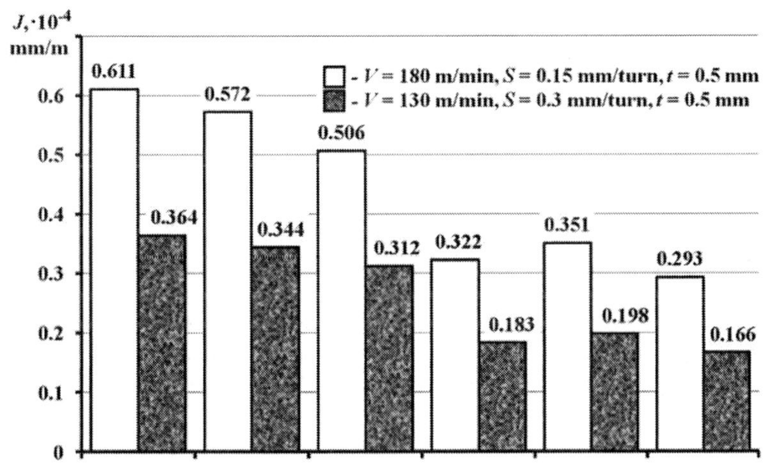

Figure 1. Influence of the coating composition on the wear intensity of the cutting tool with the wear-resistant coating.

Calculation of fatigue crack growth time by the method of [12] showed that niobium nitride-based coatings more effectively resist crack formation processes. High values of compressive stresses combined with a more favorable combination of thermal and stressed states provide the cutting tool with niobium nitride-based coatings an increase in the time t_1 before the crack development starts in the coating by 3.1–7.9 times. Different coatings resist differently to the development of a crack at the stage of its growth. In this case, coatings of a complex composition having high mechanical properties are characterized by longer time t_2 before a through crack is formed in the coating.

As a result, the total time before the occurrence of cracks starts in the coating (time of fatigue crack growth) t_Σ has the highest values for three-element coatings NbTiCrN, NbTiZrN and NbTiAlN. In comparison with NbN coating, the time value of t_Σ increases approximately by about 1.5 times for these coatings, and in comparison with TiN coating by 3–3.2 times.

The study of cutting tool efficiency with coatings showed a correlation between a decrease in the intensity of wear of the cutting tool and an increase in the crack resistance of coatings (Figure 1).Use of coatings of complex compositions allows to reduce wear intensity of cutting tool by 1.8–2.1 times compared to coatings of TiN and NbN and by 1.6–1.9 times compared to coating of NbTiN.

4. Conclusions

Conducted research has shown high efficiency of using cutting tools with developed coatings based on niobium nitride.

Acknowledgment

This paper is drawn up with the aid of the grant from the Russian foundation for basic research (RFBR) – the project № 18-48-730011.

References

[1] Mokritskiy B Ya 2010 *STIN* **11** pp 11–5
[2] Tabakov V P and Chikhranov A V 2009 *Russian engineering research* **10** pp 1047–53
[3] Tabakov V P and Chikhranov A V 2018 *Russian engineering research* **38** pp 105–9
[4] Volosova M A and Gurin V D 2013 *Journal of Friction and Wear* **3** pp 183–9
[5] Santecchia E, Hamouda A, Musharavati F, Zalnezhad E, Cabibbo M and Spigarelli S 2015 *Ceramics International* **41** pp 10349–79
[6] Vereschaka A A and Grigoriev S N 2017 *Wear* **378–379** pp 43–57
[7] Gadalov V N, Salnikov V G, Romanenko D N and Shirin I V 2011 *Mechanical Engineering Technology* pp 726–8
[8] Vaulina O Y, Ovechkin B B, Papchenko A V and Shvagrukova E V 2016 *IOP Conference Series: Materials Science and Engineering* **116 (1)** 012033
[9] Blinkov I V, Volkhonskii O A, Anikin V N et al 2012 *Prot. Met. Phys. Chem. Surf.* **48** pp 649–55
[10] Zhang S 2020 *Nanostructured Thin Films and Coatings: Mechanical Properties* (CRC PressINC) p 550
[11] Tabakov V P, Chikhranov A V and Dolzhenko Y A 2020 *IOP Conf. Series: Materials Science and Engineering* **709** 033096
[12] Gorelik S S, Rastorguyev L N and Skakov Yu A 1970 *X-ray Structure and Electron-Optical Analysis of Metals* (Moscow: Metallurgizdat) p 366 [In Russian]
[13] Chikhranov A V and Tabakov V P 2018 *Vestnik of P. A. Solovyov Rybinsk State Aviation Technical University* **2** pp 223–9
[14] Vasin S A,Vereschaka A S, Kushner V S 2001 *Thermomechanical approach to the system of relationships in cutting* (Moscow: BMSTU) p 488 [In Russian]
[15] Reznikov A N 1981 *Thermophysics of processes of mechanical processing of metals* (Moscow: Mashinostroenie) p 279 [In Russian]
[16] Saidakhmedov R H 2005 *Plasma coatings based on transition metal nitrides and carbides with adjustable stoichiometry* (Tashkent: Fan) p 226 [In Russian]

IOP Conf. Series: Materials Science and Engineering **966** (2020) 012035 doi:10.1088/1757-899X/966/1/012035

Research features of opening of refractory arsenopyrite concentrates by using low-temperature roasting

N O Zvontsov[1, 2] and D A Rogozhnikov[1]

[1]Department of Metallurgy of non-ferrous metals, Ural Federal University named after the First President of Russia B N Yeltsin, 19, Mira str., Yekaterinburg, 620000, Russia

E-mail: [2]zvontsovno@mail.ru

Abstract. The article describes an alternative pre-treatment method of refractory arsenopyrite concentrates using low temperature roasting. The process consists of carrying out oxidative roasting in the temperature range of 200–400 °C – relatively low for traditional process. The calculation results of the equilibrium composition of roasting products are presented on the example of a flotation gold-bearing concentrate of the Uderey's deposit in the HSC Chemistry 6 program. The main concentrate minerals containing gold arepyrite (FeS_2) and arsenopyrite (FeAsS). In addition, the concentrate of the Uderey's deposit is characterized by the presence of antimony in the form of Sb_2S_3. During the calculations of the equilibrium composition, the optimal conditions for low-temperature roasting were selected. It was found that under these conditions, the dissociation of higher sulfides occurs, arsenic goes into the form of trisulfide (As_2S_3) and some of the iron goes into the form of sulfate ($FeSO_4$). All this suggests that the process of low-temperature roasting allows to open the refractory concentrate and get a cinder, the further processing of which will not cause special problems: the non-volatile form of arsenic trisulfide will not cause complications when cleaning exhaust gases, and iron sulfate will simplify further leaching processes.

1. Introduction

The main reason for the refractoriness of arsenopyrite concentrates is the thin dissemination of gold in the main host minerals – arsenopyrite (FeAsS) and pyrite (FeS_2). In such minerals the gold is contained in a submicroscopic form, which does not allow extracting it without opening before. It is proved that, the refractoriness of gold associated with sulfides is the presence of not only native gold nanoparticles, but also solid solution, colloidal particles, and surface gold. The presence of preg-robbing carbon in the composition of ores and concentrates also leads to the manifestation of refractory gold [1–3].

The main objective of the pretreatment in such refractory materials is the oxidation of gold-containing minerals in order to destroy their crystal lattice and release gold particles. The most common methods are autoclave and bacterial oxidation, operations of thermal decomposition, hydrometallurgical opening with preliminary fine grinding [4–6]. However, these methods entail high costs for consumables, such as bacterial strains, leaching reagents, and others. Another quite effective way is smelting [7]. Its advantage is that, when the particles of the concentrate are melted, the release of gold particles occurs by itself and there is no need for pre-treatment. But this method also has its drawbacks: a significant proportion of arsenic and sulfur may be contained in the exhaust gases, which can also be expensive to clean.

Oxidative roasting is one of the most economical methods for pretreating refractory gold-bearing concentrates. However, to achieve the desired recovery, roasting is often operated in two or even three stages. Since single-stage roasting provides significant savings in both capital investment and operating costs, most other studies, by scientists through over the world, focus on optimizing conditions for single-stage roasting, which can provide gold recovery comparable to two-stage roasting processes [8–9].

In turn, the low-temperature roasting method make it possible for us not only to convert arsenic to the trisulfide form, but also to convert some of the iron to the sulfate form, which will significantly reduce operating cost when cleaning exhaust gases and at further stages of leaching of the calcined product.

2. Low temperature roasting

Roasting is called low temperature, since it's carried out at low temperatures relative to traditional oxidative roasting: in the range of 200–400 °C. At these temperatures, the reactions do not proceed as intensively as during traditional roasting (500–700 °C) [10], however, this avoids the melting of concentrate particles, a typical problem of roasting processes.

The main goal of roasting is the dissociation of higher concentrate sulfides (FeAsS and FeS_2) and the conversion of arsenic to non-volatile As_2S_3. Since arsenic in this form does not get into the exhaust gases and will not cause additional problems of cleaning them. The same can be said about further processing of the cinder - arsenic trisulfide can be easily eliminated by leaching with sulfuric or nitric acid [11].

The problem for such roasting is the search for optimal conditions to achieve the desired result. The main parameters affecting the efficiency of the process: temperature and the amount of oxygen. Even though temperature range is from 200 to 400 °C, the reactions proceed not so intensively, their speed can be compensated by the amount of air blast supplied. However, there is a risk of dramatic increase of the roasting temperature due to exothermic sulfur oxidation reactions. Therefore, when conducting low temperature roasting happened, it is necessary to constantly monitor the temperature and volume of oxygen supplied [12].

Below are the main reactions that occur during the low-temperature roasting, also the values of the Gibbs energy and the logarithms of the equilibrium constants calculated for the temperature range 200–450 °C (tables 1, 2) [13].

$$2FeS_2 + 1.5O_2 = Fe_2O_3 + 2S_2\uparrow \quad (1)$$

$$4FeAsS \xrightarrow{t,\ C} 4FeS + As_4\uparrow \quad (2)$$

$$2S_2 + As_4 = As_4S_4\uparrow \quad (3)$$

$$As_4S_4 + 7O_2 = 2As_2O_3\uparrow + 4SO_2\uparrow \quad (4)$$

$$2As_2O_3 + 4.5S_2 = 2As_2S_3\downarrow + 3SO_2\uparrow \quad (5)$$

$$2FeS_2 + 5O_2 = 2FeO + 4SO_2\uparrow \quad (6)$$

$$2SO_2 + O_2 = 2SO_3\uparrow \quad (7)$$

$$FeO + SO_3 = FeSO_4 \quad (8)$$

Table 1. Calculations of changes in Gibbs energy, kJ/mol, for reactions 1–8.

T,°C	Number of reaction, ΔG, kJ/mol							
	1	2	3	4	5	6	7	8
200	-284.546	-127.284	-258.215	-1755.855	-680.344	-1367.174	-108.743	-174.482
250	-291.197	-133.336	-241.730	-1747.596	-646.841	-1365.746	-99.278	-164.970
300	-297.898	-139.543	-225.364	-1739.152	-613.655	-1364.294	-89.824	-155.520
350	-304.653	-145.903	-209.105	-1730.553	-584.665	-1362.833	-80.388	-146.132
400	-311.465	-152.243	-192.945	-1721.820	-557.587	-1361.375	-70.973	-136.807
450	-318.337	-158.438	-176.877	-1712.970	-531.220	-1359.923	-61.582	-127.541

Table 2. Calculations of the logarithms of equilibrium constants, for reactions 1–8.

T,°C	Number of reactions, lnK_p							
	1	2	3	4	5	6	7	8
200	31.416	14.053	28.509	193.859	75.115	150.946	12.006	19.264
250	29.077	13.314	24.138	174.506	64.590	136.376	9.913	16.473
300	27.152	12.718	20.540	158.513	55.931	124.347	8.187	14.175
350	25.539	12.231	17.529	145.073	49.013	114.247	6.739	12.250
400	24.171	11.815	14.973	133.620	43.271	105.648	5.508	10.617
450	22.996	11.445	12.777	123.742	38.374	98.238	4.449	9.213

According to the results of calculating the values of the Gibbs energy change and the logarithms of the reaction equilibrium constants from tables 1 and 2, we can conclude that the thermodynamic probability of reactions 1–8 under the considered conditions of low-temperature firing is high.

To determine the desired parameters of the low-temperature roasting process, the equilibrium composition of the process products was calculated under certain conditions of temperature and blast volume. The calculations were made of using the gold-bearing concentrate of the Uderey's deposit as an example in the HSC Chemistry 6 program.

3. The calculation of the equilibrium composition

To determine the optimal roasting parameters, the 'Equilibrium compositions' module of the HSC Chemistry 6 application software package was used. The module is designed to calculate the equilibrium compositions of the contacting phases involved in technological processes. In our case, only two phases are considered: solid and gaseous. The solid phase consists of the initial concentrate and cinder formed during the process. The gaseous phase is represented by air blast and the resulting furnace gases, consisting mainly of sulfur dioxide (SO_2) [14].

Concentrate of the Uderey's deposit was taken as a starting material for calculations. Gold-bearing ores of this deposit, in addition to the main components in the form of pyrite and arsenopyrite, are characterized by the presence of antimony in the ore. The rational composition per 100 kg of this concentrate is shown in table 3.

Table 3. The rational composition of the concentrate of the Uderey's deposit.

	Fe	Cu	S	Sb	As	Si	Al	O	Other	Total
FeS_2	9.3	–	10.7	–	–	–	–	–	–	20.0
FeAsS	12.4	–	7.1	–	16.6	–	–	–	–	36.0
Sb_2S_3	–	–	1.7	4.3	–	–	–	–	–	6.0
SiO_2	–	–	–	–	–	7.5	–	8.5	–	16.0
Al_2O_3	–	–	–	–	–	–	10.6	9.4	–	20.0
Other	–	–	–	–	–	–	–	–	2.0	2.0
Total	21.7	0.0	19.4	4.3	16.6	7.5	10.6	17.9	2.0	100.0

Calculations of the equilibrium composition were carrying out at various temperatures and the amount of oxygen per 100 kg of concentrate. The calculation results at various temperatures are shown in the form of diagrams in Figures 1–3.

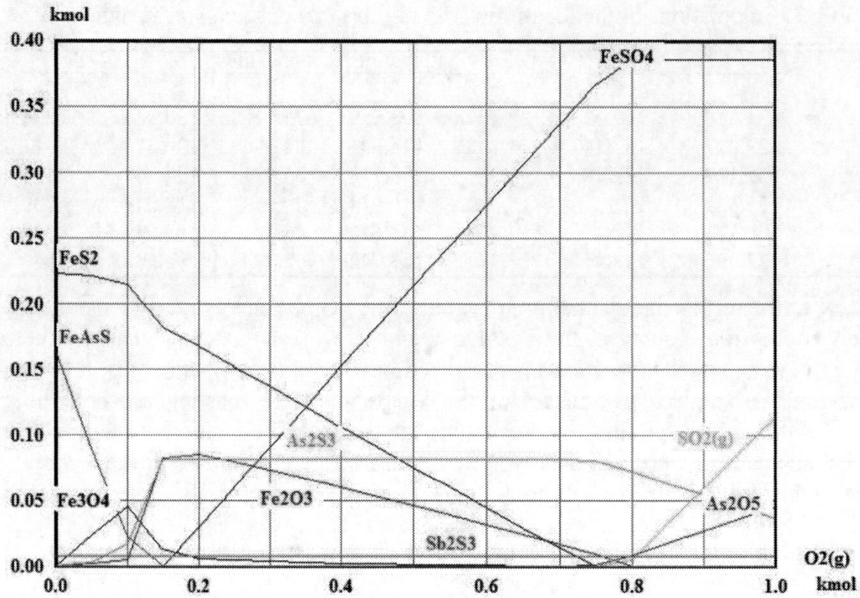

Figure 1. The equilibrium composition of the products of low-temperature roasting at 200 °C.

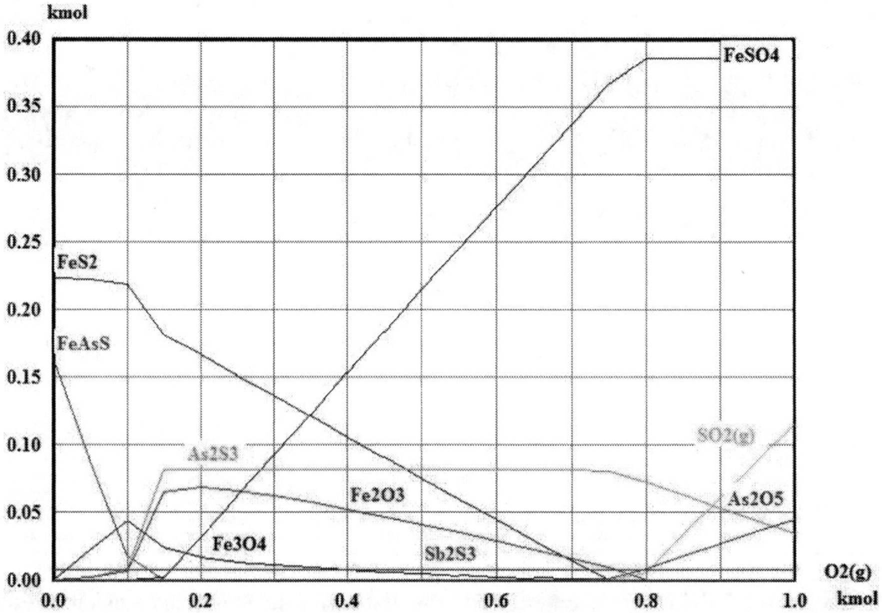

Figure 2. The equilibrium composition of the products of low-temperature roasting at 350 °C.

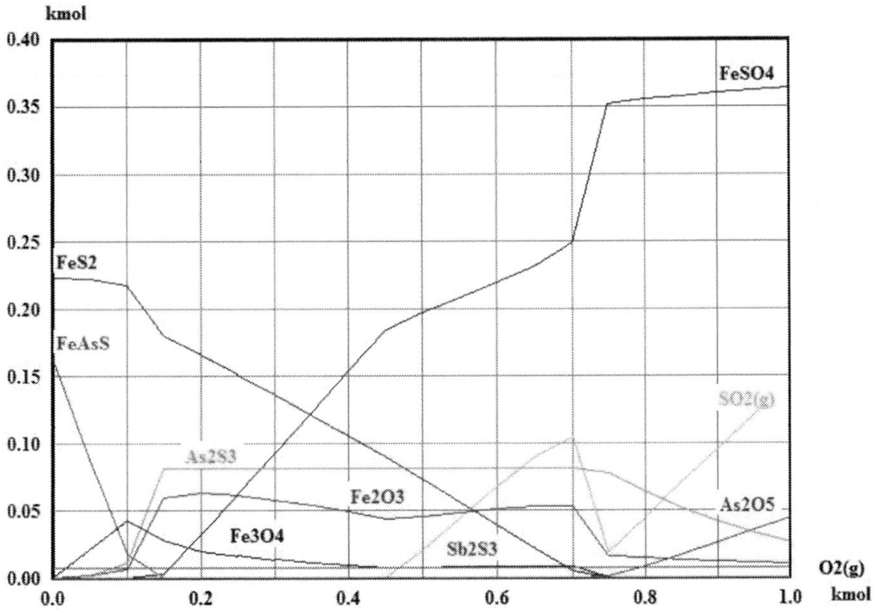

Figure 3. The equilibrium composition of the products of low-temperature roasting at 350 °C.

As shown in Figures 1–3, the most optimal flow rate is 0.6–0.7 kmol of oxygen per 100 kg of concentrate, since the maximum amount of arsenic in the form of trisulfide is maintain in this range. Also, with such an amount of oxygen, the maximum possible amount of iron sulfate is formed. With a further increase in such amount of oxygen, the formation of volatile arsenic pentoxide (As_2O_5) occurs, which contradicts the purpose of this roasting – to convert all arsenic to non-volatile form and leave it in the cinder. The calculation results at temperatures of 250 and 300 °C are comparable with the results of calculations at temperatures of 200 and 350 °C. The best result can be achieved at high temperatures of this roasting (Figure 2 and 3), since at temperatures of 350–400 °C oxidation reactions will proceed faster than at 200–300 °C, which will reduce operating cost.

As shown in Figure 3 at a temperature of 400 °C, the SO_2 curve begins to change and the sulfur content in the exhaust gases increases markedly with increasing oxygen, which indicates an increase in the intensity of exothermic sulfur oxidation reactions. In this case, there is a risk of allowing the melting of the cinder particles and thereby only complicate the further extraction of gold. So we can conclude that at temperatures more than 400 °C roasting is not advisable.

All figures show that the amount of Sb_2S_3 is presented in a straight line and does not change during the reaction. This means that antimony, known for its volatility and toxicity, will not affect the results of the roasting process in these conditions.

4. Conclusions

Calculations of the equilibrium composition were carried out in order to find the suitable conditions for carrying out roasting. Roasting under such conditions is intended to produce a cinder containing arsenic in the form of trisulfide, which is confirmed in Figures 1–3. Based on the calculation results, it can be concluded that the most optimal conditions for low-temperature roasting are temperatures in the range 350–400 °C and approximately 0.6–0.7 kmol of oxygen per 100 kg of concentrate.

In addition, Figures 1–3 show that even at 0.7 kmol O_2, complete dissociation of pyrite does not occur. This suggests that the opening did not occur completely and some of the gold will remain in a closed state and will be difficult to extract. However, the degree of opening of pyrite at a level of 90 % indicates a high degree of efficiency of low-temperature roasting as a pre-treatment of the concentrate.

It is also worth noting that the degree of opening of arsenopyrite is about 100%. Since the main goal of low-temperature roasting was to convert all arsenic to trisulfide, the goal can be achieved.

At this stage of analytical research, is a positive calculation result and the optimal parameters for roasting are selected, which is the basis for continuing research in this direction. In the future, it is planned to carry out calculations with the raw materials of other gold deposits, as well as conduct experiments to confirm analytical calculations.

References

[1] Zakharov B A 2013 *Gold: refractory ores: scientific publication* (Moscow: Ore and Metals) p 452

[2] Gulyashinov P A, Paleev P L and Gulyashinov A N 2016 Investigation of the roasting process of gold-bearing scorodite ore *Bulletin of Irkutsk State Technical University* vol **20** no **10** p 154–62 DOI: 10.21285 / 1814-3520-2016-10-154-162

[3] Bazhko V and Yahorava V 2017 Evaluation of ozonation technology for gold recovery and cyanide management during processing of a double refractory gold ore *The Journal of the Southern African institute of Mining and Metallurgy* vol **117** pp 749–56

[4] Litvinova N M 2008 *Improvement of technological methods for grinding refractory gold-bearing ores* (Russia, Khabarovsk: abstract of thesis ... PhD) p 21

[5] Polezhaev S. Yu 2015 *Increasing gold recovery in the technology of autoclave oxidation of concentrates of 'double tenacity' by preliminary heat treatment* (Saint Petersburg: Nat min. cheese. univ. 'Mountain', dis. ... PhD) p 132

[6] Leng F et al 2016 *Arsenic bioleaching in medical realgar ore and arsenic-bearing refractory gold ore by combination of Acidithiobacillusferrooxidans and Acidithiobacillusthiooxidans* (China, Benin: Tropical journal of pharmaceutical research) pp 1031–38

[7] Seitkan A and Redfern S A T 2016 *Processing double refractory gold-arsenic-bearing concentrates by direct reductive melting* (Minerals engineering, Elsevier) pp 286–302

[8] Zhang D et al 2016 Acid leaching decarbonization and following pressure oxidation of carbonic refractory gold ore *J. Cent. South Univ.* pp 1584–90 DOI: 10.1007/s11771-016-3212-z

[9] Lodeishchikov V V 1999 *Technology for the extraction of gold and silver from refractory ores: in 2 volumes* (Irkutsk: OJSC Irgiredmet) p 775

[10] De Michelis I et al 2013 Roasting and chlorine leaching of gold-bearing refractory concentrate: experimental and process analysis *International journal of mining science and technology* (Elsevier) pp 709–15

[11] Lidin R A, Molochkova V A and Andreeva L L 2000 *Chemical properties of inorganic substances: textbook. manual for universities. 3rd ed.* (Russia, Moscow: Chemistry) p 480

[12] Liu X et al 2019 Simultaneous removal of S and As from a refractory gold ore in a single stage O_2-enriched roasting process *The Minerals, Metals & Materials Society and ASM International* 2019 vol **50** B pp 1588–96

[13] Naboychenko S S, Ageev N G and Zhukov V P et al 2015 *Processes and apparatuses of non-ferrous metallurgy* (Yekaterinburg: GOU VPO USTU -UPI) p 700

[14] Ageev N G and Naboychenko S S *2016 Metallurgical calculations using the application package HSC Chemistry* (Yekaterinburg: Publishing House of the Ural University) p 124

IOP Conf. Series: Materials Science and Engineering **966** (2020) 012036 doi:10.1088/1757-899X/966/1/012036

Estimation of selective reduction of iron and phosphorus from manganese ores of different genesis

N Y Kosdauletov[1, 2] and V E Roshchin[1]

[1] South Ural State University (National Research University), 76, Lenina av., Chelyabinsk, 454080, Russia

E-mail: [2]kosdauletovn@susu.ru

Abstract. The possibility of selective reduction of iron from complex oxides by carbon monoxide gas is investigated. Brazilian ore rich in manganese oxide, ferromanganese ore from the Selezenskoye Deposit (Russia) and concentrate from the Zhayrem GOK (Kazakhstan) were used as samples. Experiments have shown that it is possible to selectively reduce iron from complex manganese ores to a metallic state not only with solid carbon, but also with carbon monoxide. Experimental results of solid-phase reduction showed that at a temperature of 900 °C. With an exposure time of 1.5 hours, the release of metallic iron is detected. At the same time, in the atmosphere with the metal phase, there is no presence of other elements other than iron in the metal. As the temperature rises to 1000 °C, the amount of metallic iron, as well as manganese and phosphorus, increases. At a temperature of 1000 °C, phosphorus turns into metal, despite the type of reducing agent. In all experiments, oxides of manganese, silicon and calcium remained in the oxide part.

1. Introduction

The ferrous metallurgy is the major consumer of manganese products in the world. Approximately 95 % of all the manganese ore is consumed during production of steel, mostly in the form of ferromanganese or silicomanganese, as well as in the form of other small-scale production associated with ferroalloys [1–2]. In addition to the metallurgical sphere, manganese is also used in other industries [3–4].

The confirmed supply of manganese ores in the world is 3.5 billion tons and the total amount is estimated to be 8.9 billion tons [5]. The results of various studies of preliminary reduction of metals from iron [6–7], chromium [8–11], and also manganese ores [12–14] have been recently published. Currently, there are large reserves of low-quality ferromanganese ores in the CIS countries, which after processing can be considered as an alternative raw material base for manganese production. In Kazakhstan at the Western Kamys deposit the reserves of oxidized manganese ores are estimated to be 700–800 thousand tons [15], and most of this raw material base is represented by the ferromanganese ores. In addition to the ores of the Western Kamys deposit, iron-manganese ores are also found in other deposits of central Kazakhstan. This refers to the deposits, e. g. Zhomart, Kamys, and Eastern Karazhal. These deposits have large amount of manganese ores with a high iron content [16]. Based on the literature data [17], it can be concluded that Kazakhstan has huge reserves of manganese ores; however, these ores are not always suitable for production of standard grade manganese alloys and only about 70 % of them are ferromanganese grades of ore. The iron manganese ores of the Middle Obuzha are known in Ukraine [18]. On the territory of Russia [19] in the Kemerovo region there are

IOP Conf. Series: Materials Science and Engineering **966** (2020) 012036 doi:10.1088/1757-899X/966/1/012036

iron-manganese ores of the Durnovsky deposit containing about 300 thousand tons, as well as the Selezensky deposit, represented as rich boulder ores of manganese and fragments of iron-manganese ores; the estimated reserves are 5 million tons. Currently, there are no efficient processing methods for such ores. A distinctive feature of the ferromanganese ores is their high content of both manganese and iron-containing minerals. The technological schemes which are developed for the mechanical enrichment of this type of ore do not allow sufficient separation of minerals of iron and manganese. In work [15] the author's study made it possible to establish an approximate temperature for firing iron-manganese ores in the presence of carbon, which was estimated to be 800–900 K. In works [20–21] the authors studied the features of the hydrometallurgical processing of iron-manganese ores using advanced ferrous-sulfate technology. All of these methods for removal of phosphorus and iron from composition of the ferromanganese ores have not yet been implemented in practice. Therefore, the development of theory and technology of processing is required for processing of ferromanganese ores.

2. The aim of this work

Was a comparative assessment of the possibility of selective reduction of metals from manganese ores using solid carbon and gaseous carbon monoxide as reducing agents.

3. Materials and methods

Three samples were used as the studied material: Brazilian ore rich in manganese oxide, ferromanganese ore of the Selezensky deposit (Russia) and concentrate of the Zhairemsky GOK (Kazakhstan). The carbothermic reduction was carried out in a sealed resistance furnace (Tamman furnace) with a graphite heater. Ground graphite electrodes were used as a reducing agent. Six corundum crucibles with samples were simultaneously placed in the working space of the furnace. In the first three crucibles, the concentrates of different manganese ores with fraction size range 0.1–0.4 mm and reducing agent in the form of powder were placed. The other three crucibles were filled with the same concentrates of different manganese ores, but without graphite. The furnace was closed with a lid in order to create a reducing atmosphere, heated to temperatures of 900 and 1000 °C and held for 1.5 or 3 hours. The temperature was controlled by a tungsten – rhenium thermocouple W-Re 5/20. The crucibles were cooled down to room temperature with furnace. The resulting reduction products were mechanically separated from the unreacted reducing agent and put into an epoxy resin and polished to make samples with cross sections for analysis. The chemical composition of the phases was determined using an Oxford INCA X – max 80 energy dispersive spectrometer mounted on a JEOL JSM – 7001F scanning electron microscope.

4. Results

The structure of ores after reduction at temperature of 900 °C and 1.5 hours reduction time in CO atmosphere is shown in Figure 1, and the chemical composition of the phases at the points of analysis in Table 1. After reduction the metal phase was detected on the surface and inside the ore grains (Figure 1 a, point 3, Figure 1 b, point 3, 4, Table 1). After initial stage of reduction the metal particles contained significant amount of iron. The adjacent oxide phase contained manganese oxides (Figure 1 a, point 1, 2, Figure 1 b, point 1, 2, Figure 1 c, point 1, 2, 3, Table 1), whereas phosphorus remained in the oxide phase during reduction in the atmosphere of CO gas. It should be noted that under these conditions of reduction the metal phases were observed only in the samples of two different types of ore (a and b); in the ore (c), such a change after reduction in the atmosphere of CO was not observed, since this ore contains higher amount of manganese.

During interaction of ore with solid carbon iron was reduced to a metallic state (Figure 2 a, point 2, 3, Figure 2 b, point 2, Figure 2 c, point 2, Table 2). After reduction the oxide phase of ores mostly contained oxides of manganese, silicon and calcium (Figure 2 a, point 1, Figure 2 b, point 1, 3, Figure 2 c, point 1, 3, Table 2).

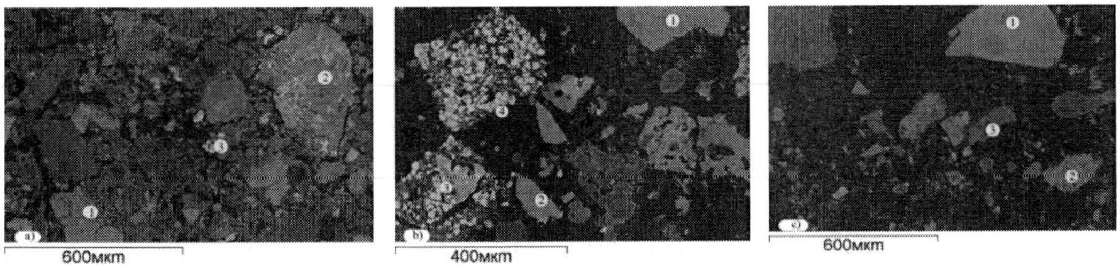

Figure 1. Reduction by CO at temperature of 900 °C and 1.5 hours reduction time: a – manganese concentrate Zhairemsky GOK, b – ferromanganese ore from Selezen deposit and c – Brazilian manganese ore.

Table 1. The content of elements after recovery, at. %.

Point of analysis	O	Al	Si	P	Mn	Fe	Cu	Ba
Spectrum 1a	26	0.0	1.8	0.0	72.6	0.0	0.0	0.0
Spectrum 2a	11	0.0	0.0	0.0	84.9	0.0	0.0	3.9
Spectrum 3a	0.0	0.0	0.0	0.0	0.0	99.6	0.4	0.0
Spectrum 1b	24	0.0	0.0	0.2	76.3	0.0	0.0	0.0
Spectrum 2b	25	0.0	0.0	0.1	74.5	0.0	0.0	0.0
Spectrum 3b	0.0	0.0	0.0	0.0	0.0	99	1	0.0
Spectrum 4b	0.0	0.0	0.0	0.0	0.0	98.8	0.6	0.0
Spectrum 1c	25	1.3	0.0	0.0	74	0.0	0.0	0.0
Spectrum 2c	30	2.6	0.0	0.0	65.5	2.2	0.0	0.0
Spectrum 3c	29	16.7	1.8	0.0	47.6	4.9	0.0	0.0

Figure 2. Reduction by solid carbon at temperature of 900 °C and 1.5 hours reduction time: a – manganese concentrate Zhairemsky GOK, b – ferromanganese ore place of birth Selezen and c – Brazilian manganese ore.

Table 2. The content of elements after reduction, at.%.

Point of analysis	O	Al	Mn	Fe	Ba	Cu
Spectrum 1a	25	0.0	74.9	0.0	0.0	0.0
Spectrum 2a	0.0	0.0	0.0	100	0.0	0.0
Spectrum 3a	0.0	0.0	3.2	96.8	0.0	0.0
Spectrum 4a	29	0.0	64.6	6.9	0.0	0.0
Spectrum 1b	28	2.3	65.6	0.0	4.1	0.0
Spectrum 2b	0.0	0.0	0.0	99	0.0	1
Spectrum 3b	28	0.0	52.7	17.8	1.5	0.0
Spectrum 1c	28	4.4	63.2	4.5	0.0	0.0
Spectrum 2c	0.0	0.0	0.0	98.5	0.0	1.5
Spectrum 3c	27	3.3	65.9	3.4	0.0	0.0

The experimental results after reduction at temperature of 1000 °C and holding time for 3 hours indicated that iron and phosphorus were reduced in the samples (Figure 3 a, point 1, Figure 3 b, point 1, 2, Figure 3 c, point 3, Table 3). However, under these conditions manganese was also present in small amounts together with iron in the metal phase. In the oxide phase after reduction by CO, individual phases of manganese monoxide and manganese were found in the silicate phase (Figure 3 a, point 2, 3, Figure 3 b, point 3, Figure 3 c, point 1, 2, Table 3). From the data in Table 3 it follows that in the atmosphere of CO gas in the metal phase the phosphorus content was 0.1–0.4 at. %, that contradicts to the results from Table 1, (Figure 1 a, point 3, Figure 1 b, point 3, 4).

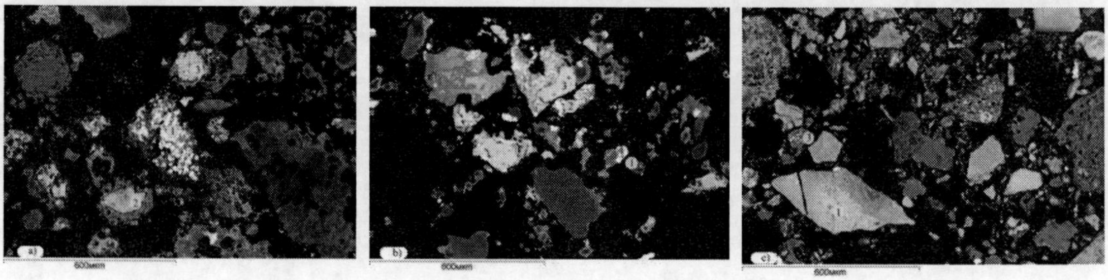

Figure 3. Reduction by CO at temperature of 1000 °C and 3 hours reduction time: a – manganese concentrate Zhairemsky GOK, b – ferromanganese ore from Selezen and c – Brazilian manganese ore.

Table 3. The content of elements after reduction, at. %.

Point of analysis	O	Al	Si	P	Mn	Fe	Cu	Ba
Spectrum 1a	0.0	0.0	0.0	0.1	0.0	99.9	0.0	0.0
Spectrum 2a	32	0.0	0.0	0.1	61.7	6	0.0	0.0
Spectrum 3a	33	0.0	4.9	0.0	45.4	16.4	0.0	0.0
Spectrum 1b	0.0	0.0	0.0	0.4	0.0	99	0.6	0.0
Spectrum 2b	0.0	0.0	0.0	0.4	8.9	90.5	0.2	0.0
Spectrum 3b	30	1.2	0.0	0.0	64.2	2.1	0.0	3
Spectrum 1c	31	3.6	0.4	0.5	59.2	4.9	0.0	0.0
Spectrum 2c	28	3.9	2.7	0.0	45.4	20	0.0	0.0
Spectrum 3c	0.0	0.0	0.0	0.4	3	95.2	1.4	0.0

At temperature of 1000 °C, the reduction of iron by solid carbon and in the atmosphere of CO occured in all samples. Metallic iron was formed after reduction, which also contained solution of manganese and phosphorus. It can be seen in (Figure 4 a, point 1, Table 4) that iron was almost completely reduced in contact with solid carbon, and in the samples (b and c) iron was present under these conditions in the oxide phase (Figure 4 b, point 2, 3, Figure 4 c, point 2, 3, Table 4).

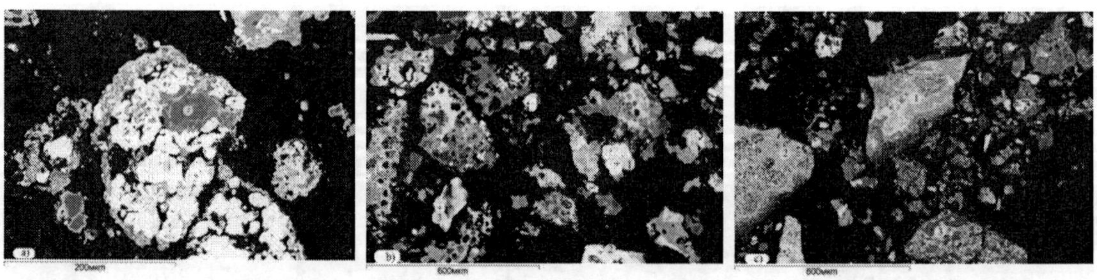

Figure 4. Reduction by solid carbon at temperature of 1000 °C and 3 hours reduction time: a – manganese concentrate Zhairemsky GOK, b – ferromanganese ore from Selezen and c – Brazilian manganese ore.

Table 4. The content of elements after reduction, at.%.

Point of analysis	O	Na	Al	Si	P	K	Mn	Fe	Ba	Cu
Spectrum 1a	0.0	0.0	0.0	0.0	0.1	0.0	0.0	99.9	0.0	0.0
Spectrum 2a	53	2	8.6	27.9	0.0	8.8	0.0	0.0	0.0	0.0
Spectrum 3a	52	1.8	3	29.2	0.0	3.2	6.6	0.0	4.5	0.0
Spectrum 1b	0.0	0.0	0.0	0.0	0.0	0.0	8.9	90.3	0.0	0.8
Spectrum 2b	29	0.0	0.0	0.0	0.0	0.0	62.5	2	6.2	0.0
Spectrum 3b	30	0.0	0.0	0.0	0.1	0.0	63	4	3.1	0.0
Spectrum 1c	31	0.0	0.0	0.0	0.3	0.0	68.5	0.0	0.0	0.0
Spectrum 2c	30	0.0	0.0	0.0	0.0	0.0	67.9	2.1	0.0	0.0
Spectrum 3c	32	0.0	2.9	0.0	0.0	0.0	32.8	32.2	0.0	0.0

5. Discussion

The experimental results of solid-phase reduction showed that solid-phase reduction of iron occurred at relatively low temperature (below the melting point of iron and oxide phase). After reduction at temperature of 900 °C and 1.5 hours holding time metallic iron was observed. As the metal phase was reduced in the atmosphere of CO, there were no other elements in the metal. When the temperature was enhanced to 1000 °C, the amount of metallic iron increased, as well as the amount of manganese and phosphorus. With increase in temperature and time of reduction, phosphorus transferred into metal phase together with iron. Thereby, phosphorus was evenly distributed between metal and oxide phases. At temperature of 1000 °C phosphorus transferred into metal despite the type of reducing agent used. In all the experiments, manganese remained in the oxide phase.

6. Conclusion

In laboratory conditions, the process of solid-phase selective reduction of iron from various manganese ores was tested in order to determine the possibility of separation of iron and manganese. The experiments confirmed the possibility of iron reduction from complex manganese ores to a metallic state not only by solid carbon, but also CO gas. According to the results of experiments at temperature of 900 °C in the atmosphere of CO gas it is possible to reduce iron selectively from ores (a and b). In the ore (c), such a change was not observed in the atmosphere of CO gas, since this ore contains higher amount of manganese. At temperature of 1000 °C phosphorus transfers into metal despite the type of a reducing agent used. In all the experiments the oxide part mostly contained oxides of manganese, silicon and calcium.

References

[1] Olsen S E, Tangstad M and Lindstad T 2007 *Production of Manganese Ferroalloys* 1st ed. (SINTEF and Tapir Academic Press: Trondheim, Norway) pp 11–69

[2] Sancho J P, Verdeja L F and Ballester A 2008 *Metalurgia Extractiva: Procesos de Obtención* 1st ed (Madrid: Editorial Síntesis) vol **2** pp 15–161

[3] O'Hara K D 2014 *Cave Art and Climate Change* 1st ed (Bloomington: Archway Publishing) p 64

[4] Freestone I C and Stapleton C P 2015 *Composition, technology and production of coloured glasses from Roman mosaic vessels. In Glass of the Roman World,* 1st ed. Bayley J, Freestone I, Jackson C, Eds (Oxford: Oxbow Books) pp 61–66

[5] Tolymbekova L B 2010 *Development of technology for smelting ferrosilicomanganese from pelletized high-silicon manganese raw materials (in Kazahstan)* pp 8–9

[6] Li K Q, Ni W, Zhu M, Zheng M J and Li Y 2011 Iron extraction from oolitic iron ore by a deep reduction process *Journal of iron and steel research international* vol **18** no **8** pp 9–13

[7] Kapelyushin Y, Xing X, Zhang J, Jeong S, Sasaki Y and Ostrovski O 2015 Effect of Alumina on the Gaseous Reduction of Magnetite in CO/CO 2 Gas Mixtures *Metallurgical and Materials Transactions B* vol **46** no **3** pp 1175–85

IOP Conf. Series: Materials Science and Engineering **966** (2020) 012036 doi:10.1088/1757-899X/966/1/012036

[8] Leikola M, Taskinen P and Eric R H 2018 Reduction of Kemi chromite with methane *Proceeding of 15th INFACON* (Cape Town, SAR)
[9] Anacleto N M, Solheim I, Sorensen B, Ringdalen E and Ostrovski O 2018 Reduction of chromium oxide and ore bymethane-containing gas mixtures *Proceeding of 15th INFACON,* (Cape Town, SAR)
[10] Zayakin O V and Zhuchkov V I 2018 Chromium recovery in an ore-smelting electric furnace *Proceeding of 15th INFACON* (Cape Town, SAR)
[11] Kucukkaragov C S and Eric R H 2018 Dissolution and reduction of chromite ore in ferrochromium slags *Proceeding of 15th INFACON* (Cape Town, SAR)
[12] Kumar M, Ranganathan S and Sinha S N 2007 Kinetics of different manganese ores *Infacon XI, Eleventh International Ferroalloys Congress* (New Delhi, India) pp 241–6
[13] Kalenga M K and Pan X 2018 Pre-reduction of a South African manganese ore: more insight on the formation of phases ed R T Jones, P den Hoed, & M W Erwee *Southern African Institute of Mining and Metallurgy* (Cape Town)
[14] Ostrovski O, Olsen S E, Tangstad M and Yastreboff M 2002 Kinetic Modelling Of MnO Reduction From Manganese Ore *Canadian Institute of Mining, Metallurgy and Petroleum* vol **41 (3)** pp 309–18
[15] Sirgetaev G E 2008 *Development and research of technologies for firing magnetic extraction of ferromanganese ore Western Kamys (Kazakhstan)* pp 12–20
[16] Miroshnichenko L A, Tilepov Z T, Pavlova Z N and Zhukov A V 1999 *Deposits of Manganese Kazakhstan* (Almaty) p 31
[17] Zhunusov A K and Tolymbekova L B 2016 *Metallurgical processing of manganese ores of the Tur and Western Kamys deposits* Fun of. Kazakhstan (Povlodar) p 3
[18] Yarosh T P, Kushniruk N V and Kozachenko L V 2011 Hydrometallurgical processing of ferromanganese ores *Zbagachennya of cinnamon copalins: Science-tech. Zb.* vol **45 (86)** pp 145–52
[19] http://geomar.ru/articles/mineral-resources/407-increase-reserves-manganese-ore-in-deposit-durnovskoye.html (accessed on 01.06.2020)
[20] Dzyuba O I and Yarosh T P 2007 Studying the thermodynamic features of sulfatizing roasting of ferromanganese ores *News of the Krivorizhsky Technological University* no **16** pp 84–7
[21] Yarosh T P and Zhugan E A 2008 Thermodynamic prerequisites for the chemical beneficiation of ferromanganese ores *News of the Krivorizhsky University of Technology* no **21** pp 86–9

IOP Conf. Series: Materials Science and Engineering **966** (2020) 012037 doi:10.1088/1757-899X/966/1/012037

Obtaining of niobium-containing ferroalloys from the Russian ore raw materials

V I Zhuchkov[1], O V Zayakin[1, 2] and L Yu Mikhailova[1]

[1] Institute of Metallurgy of Ural Division of Russian Academy of Science, 101, Amundsena str., Yekaterinburg, 620016, Russia

E-mail: [2] zferro@mail.ru

Abstract. The possibility of processing niobium-containing pyrochlore-apatite concentrates obtained from ores of Beloziminskoye deposit has been studied. A two-stage scheme for selective reduction of phosphorus into a metal phase to obtain low-phosphorus niobium-containing slag was suggested. The reduction of phosphorus and iron was carried out by a carbothermic method using coke and pig iron carbon as the reducing agent. Pig iron chip and magnetite product (80 % Fe_2O_3) were used as the precipitating agent. In laboratory and semi-industrial conditions the basic possibility of selective separation and sedimentation of phosphorus (to 91 %) and iron from a niobic concentrate without considerable reduction of niobium (up to 1–2 %), according to the two-stage scheme with minimum possible temperature of process during the first period (~ 1260 °C) and hot soak at the increased temperature (~ 1450 °C) during the second period were shown. Obtained low-phosphorus niobium-containing slag after crushing can be used for ferroniobium production according to existing technological schemes.

1. Introduction

Among the main problems of Russian ferroalloy production, the availability of ore raw materials is becoming increasingly important. This is due to deterioration of quality and decreasing of ore raw materials reserves with increase of demand for ferroalloys of high standards. Only a few types of alloys are produced at domestic plants from their own raw materials, and the major part of them is either imported from abroad or melted from imported raw materials [1–3].

One of the most strategically important elements of metallurgical production is niobium [4–6]. Consumption of ferroniobium by Russian steelmakers increased from 200–300 tons/year in the late 1990s to 4–6.5 thousand tons/year at present. However, its own production in the Russian Federation is only ~ 200 tons/year due to the absence of modern technological solutions for processing raw materials from available domestic niobium deposits [7,8].

In the Russian Federation a large area of niobium deposits (Beloziminsky, Veletagninsky, Shtikhinsky, etc.) has been explored, in the Chitinsky region – Katuginsky, in the Krasnoyarsk region – Tatar, Chuktugon, etc. All deposits differ from each other in their physical composition, which includes pyrochlore, apatite, monacite, vermiculite and other minerals [8].

According to [8], promising industrial sources of this raw material can be divided into three groups:

1. The Objects of the accelerated development. These include the Tatar deposit (Krasnoyarsk Krai) represented by industrial minerals: pyrochlor, apatite and vermiculite; Sallanlatva (Murmansk Region), minerals: lueshite, barite.

2. Objects of complex industrial development with associated pyrochlore. This is Katugin deposit (Chitinsky Region), minerals: gagarinite, pyrochlore, colombian, zircon, cryolyte.

3. The super-large and large objects with leading pyrochlore specialization and naturally alloyed Fe-Nb ores. These include deposits: Tomtorskoe (Saha-Yakutia), pyrochlore, monacite, crandallite, etc.; Veletagninsky (Irkutsk Region), pyrochlore, apatite; Beloziminsky (Irkutsk Region), pyrochlore, apatite, Colombian, Monacite.

The content of Nb_2O_3 in the ores varies from 0.1 to 1%, and there are also different amounts of phosphorus, thorium and other elements.

When enriching ores of these deposits the concentrates of different composition are obtained, Nb_2O_3 content in which varies from 15–25 to 40–50%, phosphorus up to 10–15% [9].

In this regard there are difficult tasks to develop for each type of Russian ore raw material effective processes for obtaining niobium ferroalloys acceptable for steelmakers.

In the work [10,11] the technology of smelting ferroniobium and niobium-containing alloys from concentrates of ore of Veletagninskoye deposit is considered. For processing pure niobium pentoxide the out-of-furnace aluminothermal melting is proposed. One-stage electro-furnace aluminothermal smelting is proposed for processing of pre-dephosphorized concentrates. Two-stage scheme is proposed for processing concentrates with high content of phosphorus. At the first stage dephosphoration is carried out by conversion of phosphorus into associated metal and production of low-phosphorus niobium-containing slag. In the second step a niobium-containing alloy is produced from the low-phosphorus niobium-containing slag.

2. Experimental part

IMET UB RAS studied the possibility of processing pyrochlore-apatite concentrates obtained from the ores of the Beloziminsky deposit. The scheme of enrichment of these ores provides for production of rough and conditioned concentrates.

The task of the research was to develop physical and chemical bases and technology of the process of obtaining niobium ferroalloys of rational composition with deep dephosphoration.

On the basis of the studies carried out it was concluded that pyroselection using three main reduction methods (carbo-, alumino- and silicothermy) can be used to obtain complex niobium-containing alloys with deep Dephosphoration.

Dephosphoration of ore concentrates of Beloziminskoye deposit was carried out by pyroselection method.

The process was studied with a conditioned Belozimine concentrate (Table 1) [12].

Table 1. Composition of niobium concentrate, %.

Material	Nb_2O_5	P_2O_5	SiO_2	Al_2O_3	CaO	MgO	Fe_2O_3	TiO_2
Conditioned concentrate	47.1	1.2	8.8	1.5	15.8	0.2	1.2	5.1

Composition of charge materials is given in Table 2.

Charge material size: conditioned concentrate – 0.3–1.2 mm, Coke – up to 3 mm, Ferrosilicon, silicoaluminium and aluminium– 0.5–2 mm, Pig iron chip – 0.2–0.4 mm, Lime, quartzite and hydrofluoric spar – 2–3 mm.

The difficulty of pyroselection is due to the fact that as a result of melting, it is necessary to leave niobium almost completely in the slag during deep reduction and transition of phosphorus in metal. At the same time selective reduction is a complex physical-chemical complex of processes of reducing various oxides, interaction of reduced metal globules with each other and with slag, coagulation of metal droplets, their deposition, etc.

Under laboratory conditions the physical and chemical characteristics of concentrates and the influence of various factors (amount and type of reducing agent, flux and precipitating agent, melt viscosity, process temperature, its duration) on the selective reduction of iron and phosphorus, ensuring the production of slag with Nb/P > 20, were studied.

The viscosity of the concentrate decreased sharply at 1350 °C, reaching < 0.1 N·s/m^2, the addition of fluorspar and alumina (the latter to a lesser extent) reduced the viscosity, the melting range decreased by 40–50°C.

Table 2. Chemical composition of charge materials, %.

Material name	C	A^c	V^r	W^a	Si	Al	Fe	SiO_2	CaO	CaF_2	Al_2O_3
Coke	84.6	7.6	2.7	5.1	–	–	–	–	–	–	–
Semi-coke	53.0	27.0	7.0	8.6	–	–	–	–	–	–	–
Aluminium	–	–	–	–	0.3	99.6	0.2	–	–	–	–
Pig iron chip	4.4	–	–	–	–	–	ost.	–	–	–	–
Lime	–	–	–	–	–	–	–	–	91.1	–	–
Quartzite	–	–	–	–	–	–	–	97.5	–	–	–
Fluorspar	–	–	–	–	–	–	–	–	–	92.0	–
Alumina	–	–	–	–	–	–	–	–	–	–	98.7

The feasibility of two-stage smelting was determined. Iron and phosphorus were reduced at 1260–1450 °C at the 1st stage. At the 2nd stage after completion of reduction reaction at the temperature of 1410–1450°C the melt viscosity was reduced and metal globules were precipitated.

As a precipitator, pig iron chip and magnetite product (80% Fe_2O_3) were tested in an amount of 2 to 15 and 3.6 to 26.8 g per 100 g concentrate, respectively (equal amount of Fe), and the amount of coke carbon and pig iron was taken based on the reduction of P and Fe (Figure 1 and 2).

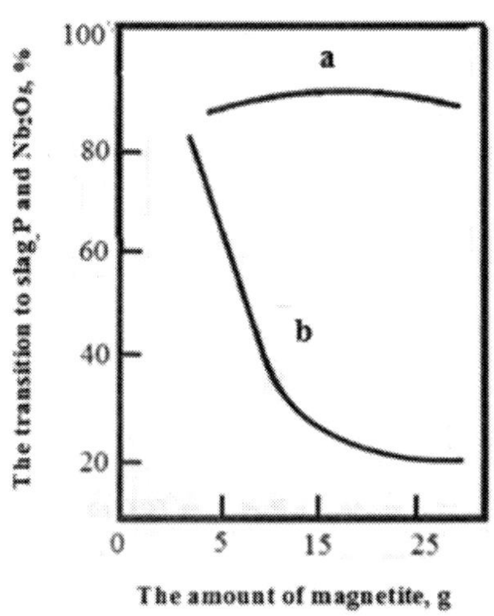

Figure 1. Transition of niobium (a) and phosphorus (b) to slag, depending on the amount of magnetite product.

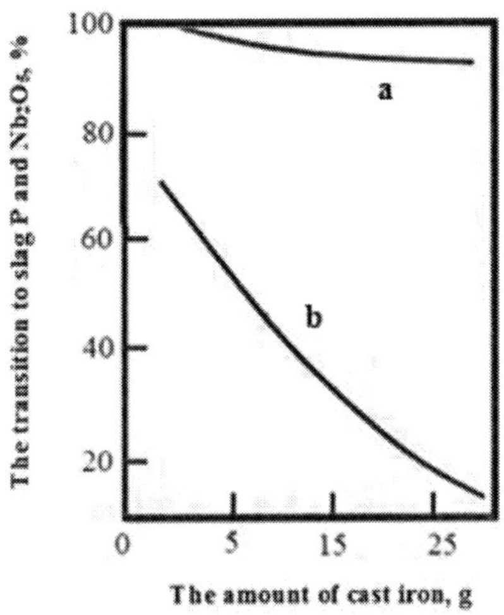

Figure 2. Transition of niobium (a) and phosphorus (b) to slag, depending on the amount cast iron in the charge.

In melts with magnetite product the dephosphoration of concentrate was quite complete, but niobium losses were also significant due to the increased amount of coke in the charge (on reduction of magnetite iron), reacting with oxides in the upper layers at the melt-coke boundary. In pig iron chip the phosphorus content decreased with increasing pig iron without significant transition of niobium to metal. The optimal amount of charge materials was determined experimentally: coke not less than stoichiometrically required for reduction of iron and phosphorus, pig iron chip ~ 15% of the concentrate weight.

It has been shown that it is possible by carbothermic reduction to selectively separate and precipitate phosphorus and iron from the conditioned niobium concentrate to the required limits with a slight reduction of niobium (up to 1–2%) according to a 2-step scheme: the mixture of concentrate,

IOP Conf. Series: Materials Science and Engineering **966** (2020) 012037 doi:10.1088/1757-899X/966/1/012037

coke and pig iron chip was being melted in an electric furnace at 1260–1300 °C for about 40 minutes, then the temperature was raised to 1410–1450 °C and was being held for 20–25 minutes. This scheme was recommended for further semi-industrial tests. The resulting dephosphorized slag after crushing can be used to melt [13, 14] ferroniobium or other niobium ferroalloys according to known technologies.

Semi-industrial experiments were carried out in a double-electrode electric furnace with a transformer capacity of 50 kVA lined with magnesite brick. 12 heats were performed.

Smelting showed that selective reduction of P and Fe (up to 91% P is removed into the metal) can be carried out in the electric arc furnace without significant reduction of niobium. Optimal for pyroselection was a step process with a minimum temperature in the first period (~ 1260 °C) and an increased (~ 1450 °C) in the second. Phosphorus removal was carried out to low values (~ 0.15% P in slag). An important condition for successful pyroselection in the electric arc furnace was the selection of electric parameters of smelting, as it depended on the degree of local overheats, the catch of elements, carbon participation in reduction. The best pyroselection results are obtained with an excess of carbon in the charge of 30% (from stoichiometrically required for reduction of P and Fe).

Industrial smelting on dephosphorization by pyroselection of conditioned concentrates was carried out in a furnace with a transformer capacity of 1100 kVA.

The raw material was a conditioned niobium concentrate (Nb_2O_5 = 38.0%; P_2O_5=4.5–6.4 %; Fe_2O_3 = 11.4–14.6%), pig iron chip and coke (68–73% C).

To achieve in slag a low phosphorus content (< 0.5%) upon condition of insignificant reduction of niobium it was necessary to have excess of reducer of 20–40% in furnace charge.

3. Conclusions

Satisfactory results were obtained. The electricity consumption per 1 ton of slag ranged from 1600–1900 to 970–1060 kWh. The content of phosphorus in the slag was quite low: 0.1–0.3% at Nb_2O_5 = 34–39%, about 99% of phosphorus passed into the metal and into the gas phase, reduction and transition of niobium to metal was less than 1%. A total of 3 tonnes of slag was produced.

Industrial technology showed the fundamental possibility of dephosphorisation by pyroselection in an electric furnace with satisfactory characteristics.

Obtained low-phosphorus niobium-containing slag can be used for ferroniobium production after crushing, according to existing technological schemes.

Acknowledges

Yu B Maltsev took part in the work.

The work was carried out according to the state assignment for IMET UB RAS.

References

[1] Zhuchkov V I, Leontiev L I and Dashevskiy V Y 2019 Situation and development of ferroalloy metallurgy in Russia *Transactions conf. Prospects for the development of metallurgy and mechanical engineering: FERROALLOYS* (Russia, Ekaterinburg: Institute of Metallurgy of Ural Division of Russian Academy of Science) pp 1–14

[2] Zhuchkov V I, Zayakin O V, Leont'ev L I et al 2008 Main Trends in the Processing of Poor Chrome Ore Raw Materials *Russ. Metall. (Metally)* **8** pp 709–12

[3] Gasik M I, Gladkikh V A, Zhdanov A V et al 2009 Calculation of the Value of Manganese Ore Raw Materials *Russ. Metall. (Metally)* **8** pp 756–58

[4] Nishimura I, Matsumara S *1995 Proc. Inter. Symp. Tantalum and Niobium* (Goslar. Germany) pp 355–69

[5] Tu Jui-Fan, Yang Kuo-Cheng, Chiang Lung-Jen et al 2017 The effects niobium and molybdenum Co-additions on bending property of hot stamping steels *SQJ* **46 (1)** pp 55–9

[6] Mayorov V G, Nikolaev A I 2002 Tantalum (V) and niobium (V) extraction by octanol *Hydrometallurgy* **66** pp 77–83

IOP Conf. Series: Materials Science and Engineering **966** (2020) 012037 doi:10.1088/1757-899X/966/1/012037

[7] Melent'yev G 2018 Prospects for providing our own rare-metal raw materials and development offerroniobium production in Russia *Transactions of Conf. Prospects for the development of metallurgy and mechanical engineering: FERROALLOYS* (Russia, Ekaterinburg: Institute of Metallurgy of Ural Division of Russian Academy of Science) pp 36–45 [In Russian]

[8] Mashkovtsev G A and Bakanova T V 2019 On Mineral Resources for Ferroalloy Production *Theoretical and practical conference with international participation and School for young scientists «FERROALLOYS: Development prospects of metallurgy and machine building based on completed Research and Development». Knowledge E Materials Science* pp 29–45

[9] Zhuchkov V, Vatolin N and Moiseyev G 1978 Study of the physicochemical characteristics of the processes of pyrometallurgical processing of phosphorous niobium-containing raw materials. Physical and chemical properties of metallurgical melts *Transactions of Institute of Metallurgy of the Ufa Scientific Center of the Academy of Sciences of the USSR* **31** pp 70–84 [In Russian]

[10] Stulov P Ye, Seregin A N and Pikalova V S 2012 Development of technology for smelting ferroniobium and niobium-containing alloys from ore concentrates of the Bolshetagninskoye deposit *Problems of Ferrous Metallurgy and Materials Science* **4** pp 5–11 [In Russian]

[11] Temnov A V and Pikalova V S Scenarios for the implementation of the mineral resource potential of complex rare-metal deposits of the Ziminsky ore district *Exploration and mineral protection* **7** pp 54–60 [In Russian]

[12] Lyakishev N P, Tulin N A and Pliner YU L 1981 *Doped Alloys and Steels with Niobium* (Moscow: Metallurgy) [In Russian]

[13] Lyakishev N P, Pliner YU L, Ignatenko G F and LappoS I 1978 *Alyuminotermiya* (Moscow: Metallurgy) [In Russian]

[14] Reznichenko V A 1967 *Metallurgy of tungsten, molybdenum and niobium* (Moscow: Nauka) [In Russian]

IOP Conf. Series: Materials Science and Engineering **966** (2020) 012038 doi:10.1088/1757-899X/966/1/012038

The problems of application of domestic ore raw materials in the production of ferroalloys

V I Zhuckov[1], L I Leontiev[2], O V Zayakin[1, 4] and V Ya Dashevsky[3]

[1] Institute of Metallurgy of Ural Branch of Russian Academy Sciences, Yekaterinburg, 101, str. Amundsena, 620016, Russia

[2] Presidium of the Russian Academy Sciences, 32a, Leninsky prospect, 119991, Moscow, Russia

[3] Institute of Metallurgy and Materials Science A A Baykova of Russian Academy Sciences, Moscow, 49, Leninsky Prospect, 119334, Russia

E-mail: [4]zferro@mail.ru

Abstract. There is a difficult situation with ferroalloy production in Russia – production has a raw material import dependence in the presence of our own large, although not quite high-quality, in accordance with world standards, mineral resource base. As rule, domestic ferroalloy raw materials are of low quality: a low content of leading elements (manganese, chromium ores), a high content of phosphorus (manganese, niobium ores), sulfur (manganese ores). The paper gives examples of a complex of physicochemical studies and the development of a number of alternative technologies for the smelting of new ferroalloy compositions with Mn, Cr, Nb, W. The scientific studies together with the development of the technology have been carried out to study the effect of ferroalloys composition on consumer properties. Because of the fact that the chemical composition of the ores of most new deposits cannot ensure the production of standard ferroalloys. By help deep physicochemical and technological studies it is possible to create new processes and combinations of different types of ferroalloys from non-traditional domestic mineral raw materials. The alloys are not inferior in terms of technical and economic indicators of products obtained from rich imported ore materials.

1. Introduction

Currently, the importance of ferroalloy production is constantly growing due to increasing demands on the quality of metal and the creation of new products types with unique properties. The development of the industry is prevented by the widening gap between the need for quantity and improvement of the quality of ore raw materials and also declining reserves of high-quality ores. At present, about 40 million tons of various ferroalloys are produced in the world, in Russia ~ 2 million tons [1].

The structure of ferroalloy production in the world and in Russia is different (figure). The world structure of ferroalloy production over the past decades has not undergone significant changes. It is consistent with their global consumption. The production of manganese alloys ranks first in the world. In individual countries the consumption of ferroalloys is balanced by their exports and imports, and production depends mainly on the available mineral resource base [1–4].

The structure of ferroalloy production in the Russian Federation predetermined by the reserves of domestic raw materials. Silicon alloys are produced mainly in Russia due to the lack of restrictions in the raw material base (49.4%), then manganese (25.8%) and chromium (16.4%) ferroalloys. The remaining ferroalloys account for ~ 9% of production [5].

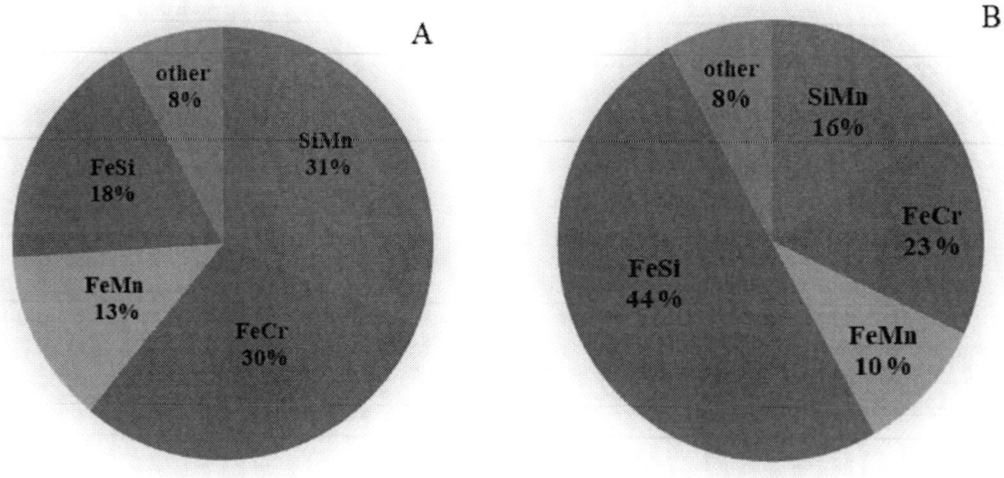

Figure. The structure of ferroalloy production in the world (A) and in Russia (B).

In addition to ferroalloys of the 'large' group (manganese, chromium, silicon), there are ferroalloys of the 'small' group, which include vanadium, niobium, molybdenum, titanium and others.

2. Results and Discussion

The main problem of ferroalloy production both in the world and in Russia is the availability of ore raw materials. A few types of alloys are produced at domestic plants from our own raw materials (ferrosilicon, vanadium alloys). The majority of ferroalloys are either imported from abroad or smelted from imported raw materials [6, 7].

The demand for manganese ferroalloys from the Russian steelmakers is ~ 650 thousand tons, which is met by imported ferro and silicomanganese. The rest is smelted from foreign raw materials (South Africa, Kazakhstan, Gabon). The volume of the consumption increased from 516 thousand tons in 2002 to 1276 thousand tons in 2018. Manganese ore mining was stopped in 2014. Now all supplies of manganese raw materials are carried out by import. Previously manganese ores deliveries were from Ukraine, but now it is ceased. Imports from Kazakhstan increased to 670 thousand tons/year, but then decreased to 128 thousand tons in 2018. New import flows of manganese concentrate from the Republic of South Africa (up to 843 thousand tons/year) and Gabon (up to 338 thousand tons/year) [8].

At the same time, we have about 290 million tons in categories A + B + C1 + C2 of manganese ores reserves. The estimated resources are more than 1 billion tons.

The largest manganese ore deposit in Russia is Usinskoye (Kemerovo Region), Porozhinskoye (Krasnoyarsk Territory), Parnokskoye (Komi Republic), Severouralskoye (Sverdlovsk Region). In the Republic of Sakha, rich (40–50% Mn) sulfide manganese ores have been discovered. The domestic manganese ores are mainly poor and high phosphorous. However, the numerous laboratory and industrial studies indicate the possibility of their wide use for obtaining different types of ferroalloys [7].

Russia fully provides itself with chromium ferroalloys in contrast with manganese alloys. The chromium ferroalloys are obtained mainly from foreign raw materials. Domestic ores are used in limited quantities (about 35%). The extraction of chromite ore in the Russian Federation is at the level of 400 thousand tons/year with the content of Cr_2O_3 in concentrate 37–39%. Balance reserves (A + B + C1 + C2) amount to 51.2 million tons, forecast resources – more than 540 million tons [9]. Ore mining is currently carried out at the deposits of Saranovskoye (Ural, Perm region) and Central (Polar Ural, chromite-bearing array Rai-Iz of the Yamalo-Nenets Autonomous Okrug).

IOP Conf. Series: Materials Science and Engineering **966** (2020) 012038 doi:10.1088/1757-899X/966/1/012038

Russian imports of chromium concentrates in different years range from 640 to 1110 thousand tons/year. The main part (94%) is supplies from Kazakhstan.

Domestic chromium ores are poorer than imported ores (38–39% Cr_2O_3 versus > 45% Cr_2O_3). Though the minerals are cheaper and can be used in larger amounts. The ores from different deposits has differ in chemical, phase, and particle size distribution. They require an individual technological approach when using them [6].

The production of ferroalloys of the 'small' group is also characterized by the presence of its own ore raw materials. The production volumes of ferroalloys (with Nb, Mo, W, etc.) do not satisfy own needs [9, 10].

According to G A Mashkovtseva [9], a large of niobium deposits explored In the Irkutsk Region (Beloziminskoe, Bolshetagninskoye, Zashikhinskoye, etc.), in the Chita region–Katuginskoye, in the Krasnoyarsk Territory – Tatarskoye, Chuktugonskoye and etc. All deposits differ from each other in material composition, which includes pyrochlore, apatite, monazite, vermiculite and other minerals.

A number of molybdenum deposits have been explored. In the Republic of Buryatia, this is Orekitkan and Malo-Oinogorsk, in the Chita region is Bugdain, and in the Kurgan region is Koklanovskoe and others.

Almost all of these deposits are not being developed. In respect to ferroalloy production, Russia has 'a paradoxical situation: the raw material dependence of production on imports in the presence of its own large, although not quite high quality, in accordance with world standards, mineral and raw material base' [10].

In addition to organizational and financial, one of the main reasons for this situation is the well-established conservative approach to smelting ferroalloys with using the same type of ore raw materials and obtaining standard guest products.

As a rule, domestic ferroalloy raw materials are of low quality. It has a low content of leading elements (manganese, chromium ores), a high content of phosphorus (manganese, niobium ores), sulfur (chromium, manganese ores). This requires a complex of physicochemical studies and the creation of a number of new alternative technologies.

The scientific studies together with the development of the technology have been carried out to study the effect of ferroalloys composition on consumer properties. Because of the fact that the chemical composition of the ores of most new deposits cannot ensure the production of standard ferroalloys.

It is necessary to conduct the scientific research together with the development of the technology to study the effect of ferroalloys composition on consumer properties. Because of the fact that the chemical composition of the ores of most new deposits cannot ensure the production of standard ferroalloys.

Based on the studies of the metallurgical characteristics of poor chromium ore materials, the physicochemical properties of the alloys obtained from them, and industrial scientific research [11–13], four main directions are formulated:

1. Selective reduction of ore components with separation into a metal intermediate with a low chromium content (~ 20%) and an oxide product with a high Cr_2O_3 content. The metal intermediate is suitable to the smelting of corrosion-resistant steel grades. The oxide is suitable for the smelting of high-grade grades of ferrochrome.
2. The rational additional charging of poor domestic chromium ores to rich imported ones to production of standard grades of ferrochrome.
3. Obtaining new alloys including commercial high-carbon ferrochrome with a low content of chromium, carbon and with a high content of silicon and also complex ferroalloys.
4. Obtaining foundry ferro- and silicochrome and their use in the smelting of refinery ferrochrome.

The directions allow to put into operation a number of known, but not used (or poorly used) previously deposits of poor chromium ores.

All presented directions have been investigated and, to varying degrees, brought to industrial testing and implementation.

IOP Conf. Series: Materials Science and Engineering **966** (2020) 012038 doi:10.1088/1757-899X/966/1/012038

It is necessary to know their service characteristics to obtain and use new ferroalloys. Changing and improving the characteristics of the alloys should be based on the study of their properties that affect the degree of assimilation of the leading components of the alloys.

Studies of the physicochemical characteristics of chromium ferroalloys and the development of rational compositions of their compositions were carried out.

The data obtained showed that ferrochrome with a high content of silicon and a low chromium content, as well as complex alloys with manganese can be obtained from poor domestic raw materials. They are characterized by more favorable physicochemical properties compared to traditional high-percent (~ 65% Cr) ferrochrome grade FeCr850 from the point of view of their application for steel processing. Laboratory experiments have shown that an increase in the silicon content in high-carbon ferrochrome to 10% leads to an increase in the degree of absorption of chromium in steel by 11%. In addition, standard ferrochrome contains less than 1% silicon and is used only for alloying steel with chromium. The proposed alloys can contain up to 10% Si and are used not only for alloying, but also for partial deoxidation of steel.

The prospects for using poor Russian manganese ore raw materials are largely associated with the study of the possibility of successful use of ferro- and silicomanganese with a low manganese content (40–55%) and the development of methods to reducing phosphorus in manganese ferroalloys. Mixing of poor manganese ore to rich ore also requires a founded scientific approach to choosing the optimal ratio of these two components of the raw material, taking into account the composition of the formed melt (basicity, viscosity, distribution coefficient of manganese), phosphorus content in the alloy, etc.

Significant economic interest is the increase in output in Russia for domestic use and for the export of tungsten products. In one form or another, tungsten is widely used in metallurgy and machine building, the electric, oil and chemical industries in pure form, alloy steel, hard alloys, etc. Tungsten is a material that is profitable for sale, the price of ferro-tungsten (65–75% W) is 15.000–35.000 USD/t [14]. Increasing the output of tungsten-containing products in the presence of significant reserves of raw materials (3rd place in the world) is an important state task [2, 8, 9].

Currently according to [9], the State Balance of Russia takes into account 52 primary and 40 placer accumulation deposits of tungsten. Inferred resources (WO_3) are estimated at 2.234 thousand tons, production in 2017 amounted to 3.500 tons. 929 tons of WO_3 were obtained from industrial waste. In 2017, the export of tungsten concentrates amounted to ~ 2000 tons, the import is 1470 tons.

Development of the Lermontovskoye (Kabardino-Balkaria), Holtoson (Buryatia), Bon-Gorkhonskoye (Transbaikalia) deposits is continue. Explored new deposits Agylkinskoye and Mikhailovskoye, in Yakutia, as well as a number of reserves in the Western Baikal region, Eastern Sayan and in the north of the Irkutsk region. It should be noted that in many domestic ores there is a high tungsten content (above 1% WO3). The enrichment of tungsten-containing ores is carried out mainly by gravity methods. This does not provide a stable quality of concentrates. The problem of cleaning concentrates from non-ferrous metal impurities such as copper, arsenic, lead, tin, bismuth, antimony, which are not removed by gravity enrichment, is extremely acute [15].

The concentrates obtained as a result of enrichment have significant fluctuations in the tungsten content and impurities due to the different composition of ores of different deposits and the methods of their enrichment. This creates difficulties in the development of technology for their processing into commercial products (metal tungsten, ferro-tungsten, etc.) Nevertheless, there is large arsenal of melting techniques. This including furnace and ladle melting; carbo-, silico- and aluminothermy reduction processes, pyroselecting and refining methods. This allows to developing and producing tungsten products of the required composition using rational technology.

3. Conclusion

Thus, we and other researchers have shown that it is realistically possible to create new processes and combinations of different types of ferroalloys from non-traditional domestic ore raw materials based on deep physicochemical and technological studies that are not inferior in terms of their technical and economic indicators of products obtained from imported materials. To successfully solve the problem of providing the ferroalloy industry with domestic ore raw materials, it is necessary to combine the research of geologists, enrichment and metallurgists, creating integrated teams and government projects under the leadership of the Russian Academy of Sciences and the Ministry of Natural Resources of the Russian Federation.

Acknowledgments
The work was carried out according to the government contract of IMET UB RAS and according to the RFBR project No. 19-03-00451.

Reference

[1] Boyarko G Yu and Khatkov V U 2018 Commodity flows of ferroalloys in Russia *Ferrous metals* **3** pp 60–3
[2] Mukhatdinov N Kh, Brodov A A and Kosyrev K A 2014 Strategy for the development of ferrous metallurgy in Russia for the period 2014–2020 and for the long term up to 2030 *Proc. of XIII Int. steelmakers congress* (Moscow: OJSC 'Seversky Pipe Plant') pp 18–22
[3] Zhuchkov V I, Zayakin O V, Leont'ev L I etc 2008 Main Trends in the Processing of Poor Chrome Ore Raw Materials *Russian Metallurgy (Metally)* **8** pp 709–12
[4] Hodina M A 2017 Russian market of manganese products *Prospect and protection of mineral resources* **2** pp 42–8
[5] *The market of ferroalloys in 2017–2018: production in Russia* 2018 Available online: www.metalresearch.ru/ferroalloys_market.html (accessed on 01.06.2020)
[6] Zayakin O V and Zhuchkov V I 2019 Promising Directions of the Application for Poor Raw Materials of the Ferroalloy Production *Proc. of Theoretical and practical conference with international participation and School for young scientists 'FERROALLOYS: Development prospects of metallurgy and machine building based on completed Research and Development'* (Knowledge E Materials Science) pp 70–6
[7] Zhuchkov V I, Smirnov L A, Zaiko V P and Voronov Yu I 2007 *Technology of manganese ferroalloys Part 1 High-carbon ferromanganese* (Yekaterinburg: Ural Branch of RAS) p 412
[8] Boyarko G Yu and Matveev B N 2020 The material and raw materials base of ferroalloys and the features of the development of their production in Russia. Part 1. Main ferroalloys *Ferrous metals* **1** pp 17–22
[9] Mashkovtsev G A and Bakanova T V 2019 On Mineral Resources for Ferroalloy Production 2019 *Proc. of Theoretical and practical conference with international participation and School for young scientists 'FERROALLOYS: Development prospects of metallurgy and machine building based on completed Research and Development'* (Knowledge E Materials Science) pp 29–45
[10] Melentyev G B 2018 Prospects for providing own rare-metal raw materials and development of ferroniobium production in Russia 2018 *Proc. of conf. 'Prospects for the development of metallurgy and mechanical engineering':* FERROSLAVY (Ekaterinburg: LLC AlfaPrint Publishing House) pp 36–45
[11] Gal'perin L L, Zayakin O V, Ostrovskii Ya I et al 2003 Production of high-carbon ferrochromium from various types of chromium ore *Steel in Translation* **33 (11)** pp 38–41
[12] Esenzhulov A B, Ostrovskii Ya I, Afanas'ev V I et al 2008 Russian Chromium Ore in Smelting High-Carbon Ferrochrome at OAO SZF *Steel in Translation* **38 (4)** pp 315–7
[13] Gasik M I, Gladkikh V A, Zhdanov A V et al Calculation of the Value of Manganese Ore Raw Materials *Russian Metallurgy (Metally)* **8** pp 756–8
[14] Boyarko G Yu and Matveev B N 2020 The material and raw materials base of ferroalloys and the features of the development of their production in Russia Part 2. Microalloying ferroalloys *Ferrous metals* **2** pp 15–20
[15] Pashkeev I U, Pashkeev K U and Mikhailov G G 2018 *Aluminothermy of ferro-tungsten* (Chelyabinsk: Publishing Center of SUSU) p 142

Scaling resistance of deposited metal by flux-cored wire 30N8Kh6M3STYu

E N Eremin[1, 2], **A S Losev**[1], **I A Ponomarev**[1], **S A Borodikhin**[1] **and A E Matalasova**[1]

[1] Department of Mechanical Engineering and Materials Science, Omsk State Technical University, 11, Mira av., Omsk, 644050, Russia

E-mail: [2] weld_techn@mail.ru

Abstract. The scaling resistance of a coating deposited by a flux-cored wire 30N8X6M3STYu subjected to surface oxidation at a temperature of 900 °C was studied. It was established that the main increase in the mass of scale for a given metal occurs in the first hours, and later this dependence is almost straightforward. The average weight gain of the metal scale of such a coating at 900 °C is 0.002 kg/(m^2·h), which is 16.5 times less than that of 30Kh2V8F steel. An X-ray analysis of a metal of the composition 30N8X6M3STYu showed that the scale consists not only of the oxide phase, but also of the remnants of non-oxidized grains of solid solution, nitrides and carbides. It was found that the basis of scale, along with hematite Fe_2O_3, magnetite Fe_3O_4, are various compounds of the main alloying elements such as Cr_2FeO_4, Fe_2NiO_4 and Fe_3N, which have good protective properties. The results show that the coating of steel 30N8Kh6M3STYu can be used for applying on the surface of parts operating at high temperatures.

1. Introduction

The efficiency of metallurgical production is largely determined by the operability of the hot metal deformation tool. In its turn, the operational reliability of the tool is directly dependent not only on its wear resistance, but also on the intensity of the scale formation process [1]. In addition, when heated, the metal surface interacts with oxygen, which leads to the formation of scale [2]. As a result, the chemical composition of the surface layer of the metal changes and its operational properties decrease [3]. Scale has abrasive properties due to its significant hardness, which is maintained at high temperatures of hot processing. Therefore, the presence of plentiful scale in the contact zone enhances tool wear. Therefore, the problem of increasing the operability of a tool that works not only in conditions of abrasive wear, but also at high temperatures is extremely important.

It is possible to increase the service life of the working surfaces of the tool by surfacing with new types of wear-resistant steels [4–6]. From this point of view, the processes of applying wear-resistant coatings on economically alloyed martensitic steels deposited by flux-cored wire to the working surfaces are widely used [6, 7]. In this regard, precipitation hardening nickel-chrome steels alloyed by molybdenum, silicon, titanium, and aluminum are promising [8, 9].

Previous studies of such steel 30N8Kh6M3STYu (alloying systems 0.03%C–8.0%Ni–6.0%Cr–3.0%Mo–1.0%Si–0.55%Ti–0.30%Al) found that it has a significant hardening effect as a result of aging and can be used for wear-resistant surfacing [10]. However, the features of the operation of such steel at high temperatures have not been studied.

In this regard, the features of scale formation and changes in the structure and phase composition of 30N8Kh6M3STYu steel as a result of exposure to high temperatures in the air were studied.

2. Objects and methods of research

The object of research was cast steel coating deposited by flux-cored wire based on steel 30N8Kh6M3STYu. To ensure the necessary carbon content in steel, high-carbon FeCr was introduced into the composition of the flux-cored wire. The calculated composition of the experimental flux-cored wire is shown in table 1.

Table 1. The calculated composition of the experimental flux-cored wire, %.

Ni	FeCr	Mo	FeSi	FeTi	FeAl	Na_2SiF_6	Fe	Fe-type
8.5	8	3.5	2.6	5.2	1.8	0.3	16.5	rest

For comparison, the metal obtained by surfacing with chromium-tungsten steel 30Kh2V8F, widely used for the manufacture of a hot metal deformation technological tool, was investigated.

Tests of the experimental composition of the coating metal for scale resistance (heat resistance) were carried out in a calm air atmosphere according to Russian state standard GOST 6130–71 [11]. An increase in the mass of samples at a temperature of 900 °C during a test period of 25 hours was chosen as a characteristic of scale resistance.

Metallographic studies of oxidized samples were performed using a JEOL JCM-5700 scanning electron microscope with a JED-2300 energy dispersive spectrometer.

X-ray phase analysis of oxidation products was carried out on a Shimadzu XRD-7000 multifunctional X-ray diffractometer. To process and analyze diffraction spectra, the Match! Software package version 3.8.1.151 was used.

3. Results of the experiments and discussion

The change in the mass gain of a metal deposited by a flux-cored wire of composition 30N8Kh6M3STYu and metal deposited by a wire 30Kh2V8F depending on the exposure time in an oxidizing atmosphere is shown in Figure 1.

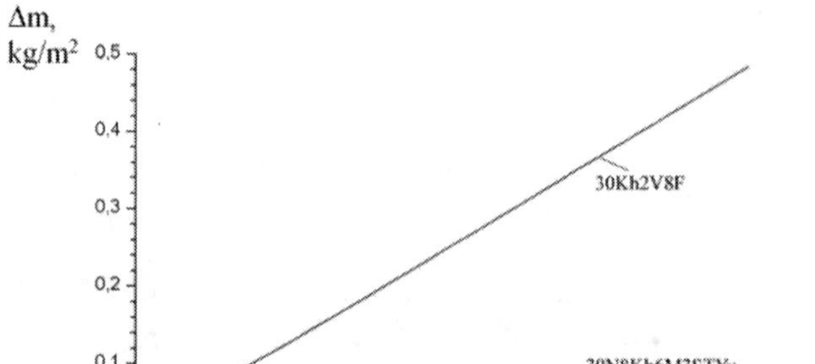

Figure 1. Change in the increase in mass Δm of the metal 30N8Kh6M3STYu and 30Kh2V8F from the exposure time at 900 °C.

The graph shows the jump in the increase in the mass of scale on the surface of the metal 30N8Kh6M3STYu occurs during the first hour. Over the next four hours, no increase in the mass of scale was observed. After five hours of testing, the growth rate of scale increases, but to a much lesser extent than in the first hour of testing, and in the future this dependence is almost straightforward. So, if in 1 hour of testing the increase in the mass of scale was about 0.0295 kg/m^2, then in 5 hours it was only 0.0216 kg/m^2, and after 25 hours only 0.0506 kg/m^2. Moreover, the thickness of the scale after 1 hour of testing at 900 °C is only 9.37–18.25 microns, and after 5 hours – 27.45–40.07 microns. On the surface of a 30N8Kh6M3STYu metal sample that passed 25 hours of testing, a uniform thin layer of light gray dense scale formed, the thickness of which remained virtually unchanged and amounted to 35.74–55.28 microns.

The scale formation indices of steel 30Kh2V8F are significantly worse than metal 30N8Kh6M3STYu. So for 1 hour of testing, the increase in the mass of scale was about 0.0397 kg/m^2, for 5 hours it was already 0.0986 kg/m^2, and after 25 hours 0.4858 kg/m^2. The thickness of the scale after 1 hour of testing at 900 °C is 44.17–52.29 microns, and after 5 hours – 70.07–81.17 microns. On the surface of a sample of 30Kh2V8F steel of the past 25 hours, the thickness of the oxide layer increased significantly and amounted to 491.67–519.37 microns.

To identify the causes of increased scale resistance of the metal deposited by flux-cored wire 30N8Kh6M3STYu, the structure and phase composition of the deposited metal and scale were studied.

The microstructure of the thin section of the lateral surface of the metal 30N8Kh6M3STYu with a scale obtained after holding for 25 hours with the location of the scan points is shown in Figure 2. The chemical composition of the scanned areas is presented in Table 2.

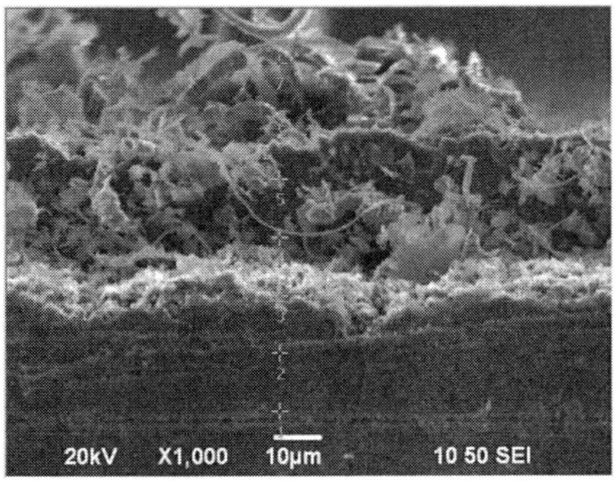

Figure 2. The lateral surface of a thin section of a metal of composition 30N8Kh6M3STYu with an oxide layer after a scale test at 900 °C – 25 hours with the location of the scan areas at points.

Table 2. The chemical composition of the scanned regions of the metal composition 30N8Kh6M3STYu after test for scale resistance at 900 °C – 25 hours.

Point No	N	O	Al	Si	Ti	Cr	Fe	Ni	Mo
1	0	0	0.04	1.52	0.21	6.12	83.89	8.25	3.84
2	0	0	0.65	2.69	0.98	5.33	76.28	8.72	6.2
3	1.53	4.07	2.12	4.9	3.4	4.34	62.63	7.2	13.83
4	5.52	9.8	0.24	8.1	3.89	12.48	63.67	0.61	5.7
5	0.41	7.52	0.13	0.33	0.71	3.11	89.93	0.85	0
6	0.63	0.52	0	0.33	0	10.56	95.57	0.14	0.26
7	2.05	1.26	0.12	0.49	0.16	7.8	92.4	0.76	1.95

Figure 2 shows that the scale is layered, with many outgoing cavities of various shapes and sizes. In the outer layer of scale, the amount and dispersion of residues of non-oxidized metal is much less than in the layer adjacent to the metal.

The results of chemical analysis show that the concentration in the metal of the main alloying elements Ni, Mo, Cr, Si, Ti, as they approach the transition layer (points 1, 2), remain at a high level. Oxygen and nitrogen appear in the transition layer (point 3), and the concentrations of all alloying elements reach very high values. As approach the surface of the scale (points 4, 5, 6), the concentrations of Ti, Al, Si, Ni, Mo decrease. Directly at the surface (point 7), a high concentration of Cr and Mo is observed.

IOP Conf. Series: Materials Science and Engineering **966** (2020) 012039 doi:10.1088/1757-899X/966/1/012039

The results obtained indicate the formation of a significant chemical microinhomogeneity in the deposited coating during oxidation. As approach the scale, metal depletion of the main alloying elements is observed. A rapid decrease in the concentration of the main alloying elements is associated with their active participation in diffusion processes due to high temperatures [12].

The characteristic structure of the surface part of the scale with the location of the scanning points is shown in Figure 3. The chemical composition of scanned objects is given in Table 3.

Figure 3. The surface of the oxide layer of the metal with the composition 30N8Kh6M3STYu after the scale resistance test at 900 °C – 25 hours with the location of the scan areas at points.

Table 3. The chemical composition of the scanned regions of the surface of the oxide layer of the metal with the composition 30N8Kh6M3STYu after a scale test at 900 °C – 25 hours.

Point No	N	O	Al	Si	Ti	Cr	Fe	Ni	Mo
1	5.34	19.13	0.51	0.1	0.39	12.21	48.45	12.92	0.95
2	0	19.79	1.76	1.11	1.16	8.78	55.77	11.63	0
3	6.76	12.59	0.73	0.27	0.28	11.18	55.61	12.58	0
4	4.4	15.04	1.45	0.58	1.37	14.55	50	11.99	0.62
5	8.16	13.79	1.14	0.29	0.34	14.33	52.14	9.81	0
6	5.16	14.93	0.75	0.37	0.57	12.78	54.13	11.11	0.2
7	5.14	13.14	1.11	1.75	0.11	11.14	55.92	9.74	1.95
8	9.74	18.44	1.13	0.18	0.26	10.2	51.68	8.37	0
9	5.7	14.45	0.61	0.27	0.28	5.93	64.09	8.67	0
10	5.24	21	1.67	0	0	8.77	51.67	10.42	1.23
11	9.72	14.83	0.61	0.09	0.16	6.78	63.87	2.98	0.96
12	5.64	15.88	0.99	0.3	0.46	8.8	57.67	10.26	0
13	5.64	17.15	1.46	2.2	0.33	5.32	60.85	7.05	0
14	2.86	15.09	0.66	0.59	0.24	6.49	71.31	2.76	0
15	13.04	17.21	1.27	0.04	0.29	10.13	49.02	9	0
16	8.32	16.76	0.86	0.51	0.49	8.76	55.75	8.55	0
17	8.88	20.7	1.21	0	0.56	10.03	49.04	9.34	0.24
18	0.84	10.1	0.44	0	0.95	8.93	62.47	14.55	1.72
19	9.3	19.14	1.37	0.23	0.53	12.21	46.05	11.17	0
20	6.86	12.04	0.66	0.23	1.54	10.72	59.08	7.98	0.89
21	7.2	18	0.63	0	0.38	8.41	63.01	1.22	1.15
22	10.26	21.92	1.3	0.83	0.45	11.62	43.86	9.49	0.27
23	0	21.79	1.29	0.51	0.15	9.75	55.25	10.19	1.07

IOP Conf. Series: Materials Science and Engineering **966** (2020) 012039 doi:10.1088/1757-899X/966/1/012039

It can be seen that the surface layer of the scale is a randomly entangled mixture of conglomerates of irregularly shaped particles with sizes up to 5 μm. All particles based on Fe, Cr, and Ni contain oxygen, and most of them also contain nitrogen. A number of particles contain all the main alloying elements (points 1, 4, 6, 7, 11, 20, 22, 23). Another series of particles contains mostly Fe, Cr, Ni, Ti, Al (points 2, 4, 5, 6, 13, 15, 17, 19, 20, 22). In some objects, a high concentration of Mo is observed (points 1, 4, 7, 10, 11, 18, 20, 21, 23).

The results of x-ray diffraction analysis of the metal scale composition 30N8Kh6M3STYu after exposure for 25 hours at a temperature of 900 °C are shown in table 4.

Table 4. The phase composition of the metal scale composition 30N8Kh6M3STYu after exposure at a temperature of 900 °C.

No	Phase designation and card number	Int.	Type of lattice	Lattice parameters
1	Fe_2O_3 (96-154-6384)	920	trigonal (hexagonal axis)	a = 5.0300 Å c = 13.7500 Å
2	Fe_3O_4 (96-900-2320)	820	cubic	a = 8.3517 Å
3	Cr_2FeO_4 (96-900-7326)	590	cubic	a = 8.3765 Å
4	Fe (96-900-6602)	420	cubic	a = 3.1590 Å
5	Fe_2NiO_4 (96-230-0296)	110	cubic	a = 8.35966 Å
6	Fe_3N 96-231-0870	105	trigonal (hexagonal axis)	a = 4.71600 Å c = 4.39400 Å
7	$Ni_{1,875}O_2$ (96-152-8875)	85	monoclinic	a = 5.1150 Å b = 5.9580 Å c = 5.9540 Å a = 125.190 Å
8	TiO_2 (96-900-8217)	44	tetragonal	a = 3.8040 Å c = 9.6140 Å
9	Ni_2SiO_4 (96-152-9934)	43	cubic	a = 7.9840 Å
10	$MoN_{0,506}$ (96-152-8389)	42	cubic	a = 4.16158 Å
11	$FeCO_3$ (96-901-5261)	40	trigonal (hexagonal axis))	a = 4.5270 Å c = 13.5700 Å
12	$A1_2O_3$ (96-152-8428)	40	hexagonal	a = 3.1120 Å c = 4.9880 Å
13	Mo_2C (96-153-9796)	40	orthorhombic	a = 4.7350 Å b = 6.0250 Å c = 5.2100 Å
14	Al_2FeO_4 (96-900-5171)	36	cubic	a = 8.1522 Å
15	$A1_2CO$ (96-153-7348)	34	hexagonal	a = 3.2000 Å c = 5.1000 Å
16	TiN (96-110-1082)	30	cubic	a = 4.2410 Å
17	Mo_3O (96-152-8029)	28	cubic	a = 5.5490 Å

The obtained results show that the basis of the metal scale of composition 30N8Kh6M3STYu is hematite Fe_2O_3, magnetite Fe_3O_4 and Cr_2FeO_4, as well as Fe_2NiO_4 and Fe_3N, and in small quantities $Ni_{1.875}O_2$, TiO_2, Ni_2SiO_4, $A1_2O_3$, $MoN_{0.506}$.

Thus, the presence of compounds with high protective properties in the scale composition determines the increased scale resistance of the deposited metal. The increase in the mass of scale is 16.5 times less, and its thickness is 11 times less than that of the metal deposited by wire 30Kh2V8F. The results obtained show that the metal deposited by flux-cored wire PP30N8Kh6M3STYu can be used for coating a tool operating at high temperatures.

4. Conclusion

1. The investigated coating of steel 30N8Kh6M3STYu is subject to slight surface oxidation. The average weight gain of the metal scale of such a coating at 900 °C is 0.002 kg/($m^2 \cdot h$).

2. The coating made of steel 30N8Kh6M3STYu can be used for applying to the surface of parts operating at high temperatures up to 900 °C.

Acknowledgments
This research was made possible through the Russian Science Foundation grant (project №17-19-01224).

References

[1] Tolstov I A, Pryakhin A V and Nikolaev A M 1990 *Improving the performance of the hot deformation tool* (Moscow: Metallurgy) p 143
[2] Teplyakov Y N 2017 *High temperature metal oxidation* (Chelyabinsk: Ed. Center of SUSU) p 215
[3] Oryshchenko A S 2011 *Heat resistant, heat resistant alloys* (St. Petersburg: Science) p 191
[4] Shekhter S Ya and Shvartser A Ya 1981 *Surfacing of parts of metallurgical equipment. directory* **2** pp 66–8
[5] Ryabtsev I A 2004 *Surfacing of machine parts and mechanisms* (Kiev: Ecotechnology) p 159
[6] Pokhodnya I K, Shlepakov V N, Maksimov S Yu and Ryabtsev I A 2010 *Automatic welding* **12** pp 34–42
[7] Malinov L S, Malinov V L 2007 *Economically alloyed martensitic alloys and hardening technologies* (Khar'kov: NSC KIPT) p 346
[8] Birman S R 1974 *Economically alloyed maraging steel* (Moscow: Metallurgy) p 208
[9] Sha W and Guo Z 2009 *Maraging steels: modelling of microstructure, properties and applications* (London: Elsevier Ltd) p 203
[10] Eremin E N, Losev A S, Ponomarev I A, Borodikhin S A and Matalasova A E 2019 *Journal of Physics: Conf. Series: Mechanical Science and Technology Update* **1260** pp 1–8
[11] Russian state standard GOST 6130-71 1990 *Metals. Methods of determining heat resistance* (Moscow: Standards publishing) p 53
[12] Anastasiadi G P 1991 *The formation of chemical microinhomogeneity in cast alloys* (St. Petersburg: Polytechnic) p 148

IOP Conf. Series: Materials Science and Engineering **966** (2020) 012040 doi:10.1088/1757-899X/966/1/012040

Method analysis of surveying of mineral storages in open-pit mining

E A Romanko[1, 3], D V Domozhirov[1], O S Kolesatova[2] and N V Litvinenko[1]

[1] Nosov Magnitogorsk State Technical University, Magnitogorsk, 38, Lenina av., Russia

[2] Technical University UGMK, Verkhnyaya Pyshma, 3, Uspensky av., 624091, Russia

E-mail: [3] Romanko_H@mail.ru

Abstract. The final products of mining enterprises are stored in open or closed areas-mineral storages. The duties of the surveyor service include monitoring of the volume of mineral, which is placed in storage. Requirements for the method of shooting and the accuracy of the received results are regulated by normative documents. Taking into account the introduction into production of modern geodetic technology survey of storage can be realized both methods of classical geodesy (the method of perpendiculars, total station survey) and with the use of GPS receivers, scanning laser systems, unmanned aerial vehicles or drones. In the article is given brief characteristic of specified survey methods and possible scope of their use. Based on the characteristic features of the storage: its form and type and taking into account possible method of measurement production (automated, non-automated) was developed method of selecting of survey. So, on the example of the open transshipment ore storage taking into account its form – medium and type – open and predominant use of automated methods of measurement production are survey methods recommended. Based on the results of the calculation of the storage volume was made the conclusion about compliance of received results with requirement of normative documents.

1. Introduction

Currently, the mining industry is experiencing an increase in the productivity of enterprises, this is due to the involvement of poorer ore reserves in the development. This leads to an increase in the intensity of mining operations, the need to place more of the final product of the enterprise and minerals in storages. One of the tasks of the survey service at the enterprise is the production of measurements of the volume of extracted ore mass, including in ore storages.

2. Material and methods

Measurements are made by the company's survey service in order to control the correctness and completeness of the Deposit development and execution of the mining plan. Periodicity of the surveying is determined by the amount of production and overburden volumes. At large and medium-sized mining enterprises the surveying is performed monthly, by small volumes-1 time per quarter. Independently of the used method of surveying, location of clear outlines, received by surveying points, bases of photographing or stereo pairs, must not exceed 1mm in plan, and for fuzzy outline – 1.5 mm. The height difference of pickets is allowed no more than 0.4 m [1–3].

IOP Conf. Series: Materials Science and Engineering 966 (2020) 012040 doi:10.1088/1757-899X/966/1/012040

The technology of surveying measurements consists in comparing of the volumes of residues of minerals in storage at start and at the end of the reporting period. Calculations of the volume of residues of minerals in storage are based on survey data [4–5].

Depending on the shape and complexity of the object, the presence of possible obstacles in the form of overpasses, galleries, communications, various technologies are used for the production of surveys – traditional and modern surveying technologies. Traditional methods include perpendicular and total station methods, while modern methods include satellite technologies, laser scanning, and photogrammetry (ground and air). These types of surveys are made using laser rangefinders, optical and electronic total stations, GPS receivers, laser scanners, unmanned aerial vehicles or drones [6–19].

The analysis of various technologies of surveying production is carried out, on these grounds is the systematization of technical means performed (Table 1) [8, 10–11].

Table 1. Surveying methods.

Way of surveying	Applied equipment	Advantages	Disadvantages
Classical geodesy: perpendicular, total station	Laser range finder, optical and electronic total station	good knowledge, low cost, reliability.	high labor intensity of surveying, not always ensuring the safety of surveying, low information content, depending on weather conditions.
Technologies global navigation satellite system (GNSS)	GPS receiver	measurement independence, efficiency, automation, high measurement accuracy	unstable satellite signal during surveying, it is not always possible to ensure the safety of surveying
Ground-based laser scanning	Scanning system	high measurement accuracy; safety of the performer by surveying; high surveying performance	extremely large excess volume of received information, high cost of work, dependence on weather conditions
Aerial survey	Unmanned aerial vehicle, drone	effective in hard-to-reach, unsafe and moving objects, large areas, fast-moving processes, reliability and sufficient accuracy	dependence on meteorological conditions, complex structure of organizational work, the need for large computing power of personal computers

Analyzing the data shown in the table, it should be noted that surveys using electronic total stations and geodetic GPS/GLONASS receivers are the most common for the production of surveys of various mining facilities. This is due to high performance and relatively simple processing of the received information.

The use of laser scanning and aerial survey systems using manned aircraft requires a large amount of material, organizational and time costs for the production of surveying. Therefore, their use is not profitable for small mining enterprises. At the same time are extensively unmanned aerial vehicles (UAVs) with a relatively low cost and an ability of operational deployment of equipment complex implementing. This will significantly reduce the labor intensity and cost of performing work on territories with an area of up to 1–2 km^2 [9, 15–19].

3. Results and discussions

The purpose of this work is consideration of the possibilities of modern geodetic equipment in the production of surveys of various mining facilities.

The object of research is an open transshipment storage of apatite ore.

To select the method of production of surveys on the object, their analysis was performed according to the following indicators: the shape and type of the object, the method of measurement. The results are shown in table 2.

Based on the performed analysis is it possible to draw the following conclusions:

1. By surveying on simple objects of closed and open type is it better to use the 1 and 2 methods of surveying.

2. Surveying of medium-sized objects is carried out using traditional and modern methods of surveying.

It is better to survey complex open-type objects using 3, 4, and 5 methods, by surveying of the closed-type storages, it is not possible to use GNSS technologies.

According to the proposed indicators object of the work has medium shape and belongs to the open type. In accordance with table 2 the following methods can be used for surveying: total station, GNSS, laser scanning and an aerial survey.

Table 2. Selecting of surveying method.

Way of surveying	Object form			Measurement method		Object type	
	simple	medium	difficult	non-automated	automated	opened	closed
1. Perpendiculars	+	–	–	+	–	+	+
2. Total station	+	+	–	+	+	+	+
3 GNSS	–	+	+	–	+	+	–
4. Laser scanning	–	+	+	–	+	+	+
5. Aerial survey	–	+	+	–	+	+	+

The surveying was performed using the following equipment:

- total station – electronic total station Leica TS06 plus Arctic 5";
- GNSS-two-frequency GNSS receivers operating in real time (Real Time Kinematic):mobile receiver (Leica GS08) from the base receiver (Leica GR10).The volume of mineral resources for these two methods of survey is determined by the method of vertical sections in the GIS GEOMIX software package.
- laser scanning with the Leica HDS4400 laser system with volume calculation in the I-Site Studio software;
- aerial survey - DJI PHANTOM 4 Pro with data processing in the Agisoft Metashape software package and volume calculation in the Credo master Plan.

The results of determining the volume of the mineral storage are shown in table 3.

Table 3. Results of volume determination.

Way of surveying	Volume of storage, m^3	Relative error in determining of volume, %
Total station	82087	2.6
GNSS	82190	2.5
Lasers canning	84718	1.0
Aerial survey	84845	0.9

To evaluate the accuracy of the definition of the storage volume a received value was compared with the data of operational accounting, which amounted to 86500 m^3.

4. Conclusion

Analyzing the results of calculating the volume of the mineral storage, it can be concluded that all methods, for the accuracy of determining the volume, meet the requirements of the Instructions [1]. By the final choosing of surveying method it is necessary to take into account the time spent on the production of works, the difference in the amount of capital and operating costs for each type of used equipment.

References

[1] *Protection of mineral resources and geological survey control. Instructions for the production of surveying works* (RD 07-603-03) 2003 ser 07 issue 15 Call. auth. M. State unitary enterprise 'Scientific and technical center for safety and industry of Gosgortehnadzor Russia' p 120

[2] *Regulations about geological and survey support of industrial safety and protection of mineral resources* (RD 07-408-01), 2001 approved by the resolution of Gosgortehnadzor of Russia

IOP Conf. Series: Materials Science and Engineering **966** (2020) 012040 doi:10.1088/1757-899X/966/1/012040

from 22.05.01 no 18

[3] *Protection of mineral resources and geological survey control: normative provision of objects protection from the harmful influence of mining operations and their conduct in dangerous zones: collection of documents* 2008 Feder. service for ecology, technologist. and an atom. supervision; A I Subbotin and others (Moscow: Industrial safety) p 212

[4] Ushakov I N 1989 *Mine surveying* (Moscow: Nedra)

[5] Ogloblin D N 1981 *Surveying* (Moscow: Nedra)

[6] Gusev V N, Naumenko A I, Volohov E M and Golovanov V A 2007 *Fundamentals of ground-based laser scanning* (Saint Petersburg) p 86

[7] Ignatov Yi M and Cigankov S A 2010 Method for building a digital mountain-geometric model of the structure of a mountain range for analyzing its structure using GIS technologies *Mining information and analytical Bulletin (scientific and technical journal)* **4** pp 91–6

[8] Woodget A S, Carbonneau P E, Visser F and Maddock I P 2014 Quantifying submerged fluvial topography using hyper spatial resolution UAS imagery and structure from motion photogrammetry *Earth Surface Processes and Landforms* vol **40 (1)** pp 47–64

[9] Liba N, Jarve I and Rand M 2013 Quality of orthophoto mosaics made with different methods *Zemes Ukio Mokslai* vol **20** no **2** pp 90–9

[10] Pravdina E and Lepikhina O 2017 Laser scanner data capture time management *ARPN Journal of Engineering and Applied Sciences* vol **12** no **5** pp 1649–61

[11] Rysbekov K, Huayang D, Kalybekov T, Sandybekov M, Idrissov K, Zhakypbek Y and Bakhmagambetova G 2019 Application features of the surface laser scanning technology when solving the main tasks of surveying support for reclamation *Mining of Mineral Deposits* **13 (3)** pp 40–8

[12] Gorova A, Pavlychenko A, Borysovs'ka O, and Krups'ka L 2013 The development of methodology for assessment of environmental risk degree in mining regions *Annual Scientific-Technical Collection – Mining of Mineral Deposit* pp 207–9

[13] Zarovnyaev B N, Shubin G V, Vasiliev I V and Varlamova L D 2016 Pitwall monitoring in deep surface mines using surface laser scanning *Gornyi Zhurnal* **9** pp 37–40 https://doi.org/10.17580/gzh.2016.09.07

[14] Kolesatova O S, Romanko E A, Litvinenko N V and Mavrin Y D 2019 Analysis of applied methods for monitoring deformations of quarry sides *Current problems of mining* **1** pp 3–7

[15] Siqueira H L, Junio J M, Matsubara E T, Eltner A, Colares R A and Santos F M 2019 The Impact of Ground Control Point Quantity on Area and Volume Measurements with UAV SFM Photogrammetry Applied in Open Pit Mines *International Geoscience and Remote Sensing Symposium (IGARSS)* 8897829 pp 9093–6

[16] Molnar A and Domozi Z 2017 Tracking Production Volumes of Open-Pit Mines with Photogrammetry Proceedings UKSim-AMSS *11th European Modelling Symposium on Computer Modelling and Simulation, EMS* pp 100–6 doi: 10.1109/EMS.2017.28

[17] Molnar A and Domozi Z 2018 Some Practical Problems of Photogrammetry Based Volume Control of Open-pit Mines *IEEE 16th World Symp. on Appl. Machine Intellig. and Informatics (SAMI)* pp 217–24

[18] Suh J and Choi Y 2017 Mapping hazardous mining-induced sinkhole subsidence using unmanned aerial vehicle (drone) photogrammetry *Environmental Earth Sciences* **76 (4)** doi: 10.1007/s12665-017-6458-3

[19] Lima R P, Coelho C, Vinueza G, Grassi J T and Castiglione L H G 2018 The use of RPAS – Remotely piloted aircraft systems in the topographic mapping for mining *REM Int. Eng. J.* vol **71** issue **2** pp 281–7

IOP Conf. Series: Materials Science and Engineering **966** (2020) 012041 doi:10.1088/1757-899X/966/1/012041

Influence of technological depressurization on the operation of a cascade pneumatic separator

V B Ponomarev[1,2] and I P Fesun[1]

[1]Ural Federal University named after the First President of Russia B N Yeltsin, 19, Mira str., Yekaterinburg, 620000, Russia

E-mail: [2]v.b.ponomarev@urfu.ru

Abstract. The article is devoted to assessing the impact of the technological depressurization of the powdered material dispenser on the work of the cascading pneumatic separator. The results of laboratory tests are presented. The article states that the magnitude of effectiveness and the separation boundary of powdered materials are reduced when working with a depressurized dispenser. The mathematical model of the cascading separator and the results of the computer experiment are presented. The article presents the dependence of the quantity of the product and the content of fine particle classes in the finished material on the amount of relative air flow through the powdered dispenser of the initial product. The change in the granulometric composition of the enriched quartz sand received is shown, depending on the amount of relative air consumption through the powdered dispenser at a fixed air velocity through the upper part of the separator. It was found that the maximum relative speed of air flow through the powder dispenser should be no more than 10–15 %.

1. Introduction

The factional processes of bulk materials in pneumatic separators are widely used in mineral enrichment [1–7]. Cascade classifiers are the most effective [7–8] because they are made of identical successive elements in the form of different sloping shelves, cones and impellers. Re-cleaning of powdered products consistently occurs in each of these cascades.

Each powdery materials separation system consists of the powdered dispenser, cyclones and a bunker with coarse material. The main problems of tightness arise in the powdered dispensers of the source material.

Unauthorized movement of air flow through the powder dispenser disrupts the operation of the separator. The flow rate of gas decreases in cascades below the powdered dispenser.

Determining the effect of the technological depressurization of the powdered dispenser on the magnitude of effectiveness of a cascading pneumatic powder separator is an urgent task.

2. Experiment

A laboratory experiment was conducted to assess the impact of technological depressurization in the powdered dispenser on the efficiency of the separator. During the experiment, a cascading separator was studied with the delivery of the initial product to the central section. The separator consisted of three cascading sections above and three cascading sections under the powdered dispenser. Each cascade consisted of an impeller and a perforated cone (Figure 1).

The experiment explored the possibility of obtaining enriched sand 0–5 mm with the minimization of dust fraction minus 0.16 mm in a cascading separator. The fractionation process was carried out at three different air flow velocities through an airtight installation and a depressurization powdered dispenser. In each series of experiments, the velocity flow was measured in the lower section of the device and in the powdered dispenser using a thermoanemometer. Total air flow was determined by a normal diaphragm located after cyclone. The relative air flow through the powdered dispenser in the experiments ranged from 6 to 23 %, while the rate of air velocity in the lower sections of the classifier ranged from 3 to 25 %.

The experiment showed that when the powdered dispenser is depressurized, the separation boundary and the overall effectiveness of fractionation of the bulk material is reduced (Figure 2, Table 1).

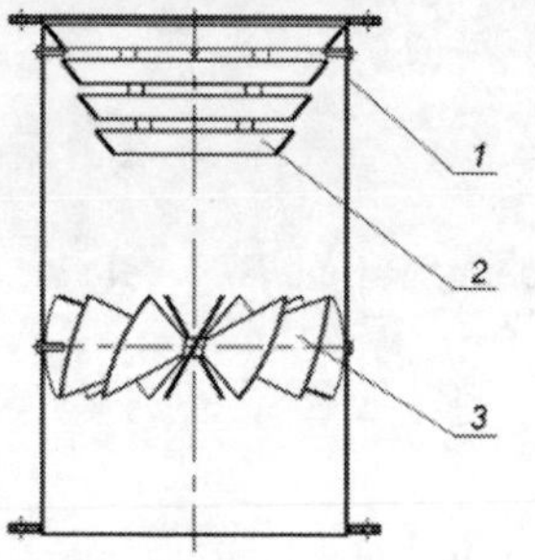

Figure 1. Single stage separator: 1–Housing, 2–multi Cone, 3–Impeller.

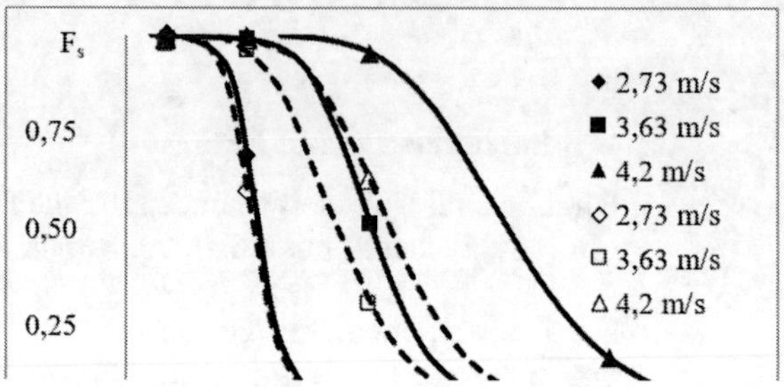

Figure 2. Functions of fractional separation at different air velocity: solid lines-separator without depressurization; dotted lines – with depressurization.

If the granulometric compositions of fine $r_S(x_i)$, coarse $r_L(x_i)$ and $r_B(x_i)$ material is known, the functions of fractional extraction for different w modes can be calculated [9–11]. If they are approximated to the Plitt function [12], the dependence of the airflow velocity through the classifier on the main characteristics of the fractionation process can be detected – the separation boundary x_{50w} and the sharpness p_w of separation. The fractional composition of the separation products is represented in Table 1.

Table 1. Fractional composition.

Sieve, mm	Source Material	A coarse product with an airtight feeder			A coarse product with a non-sealed feeder		
		2.73 m/s	3.63 m/s	4.2 m/s	2.73 m/s	3.63 m/s	4.2 m/s
5.00	0.07	0.07	0.09	0.10	0.07	0.08	0.09
2.50	23.11	26.62	31.12	37.12	26.34	29.49	32.16
1.25	16.84	19.40	22.67	27.00	19.19	21.49	23.43
0.63	25.66	29.56	34.32	34.42	29.25	32.47	34.93
0.32	17.97	20.61	11.72	1.35	20.37	15.90	9.28
0.16	10.36	3.73	0.08	0.01	4.78	0.56	0.11
–	6.00	0.00	0.00	0.00	0.00	0.00	0.00

Formula for the relative velocity of air flow through a depressurized dispenser:

$$\bar{V} = \frac{V_d}{V}, \qquad (1)$$

the relative separation boundary of the division will be defined as:

$$\bar{x}_{50} = \frac{x_{50d}}{x_{50}} \qquad (2)$$

where is V_d – the air flow through powdered dispenser, m³/h; V – the total separator air flow, m³/h; x_{50d} – the separation boundary with airtight supply, mm; x_{50} – the separation boundary with unsealed powdered dispenser, mm.

The formula of the relative separation boundary from the relative air flow rate for this experiment will be:

$$\bar{x}_{50} = 1 - 9{,}61\bar{V}^2 + 0{,}09\bar{V} . \tag{3}$$

3. Mathematical model

To determine the impact of technological depressurization on the efficiency of separation, consider the mathematical model of the regular cascade separator [13]. At the same time, the cascading classifier will be presented in the form of two separate devices and a power section.

The source material is fed into the power section, where it is perfectly mixed. The material is evenly distributed between the upper and lower devices. We will decompose the continuous process of fractionation of bulk material in a cascading separator. Let's imagine that at any given time there is a discrete number of parallel simple processes. Consider two main re-cleanings: in the first stage in the upper apparatus there is a separation from a fine product of random coarse particles, in the lower apparatus there is a de-dust of coarse material. In the second stage in the lower devices, a coarse product of the first stage of the upper devices is separated from fine particles. In the upper devices there is a clean-up of the fine product of the first stage of the lower devices. In order to implement the material balance and simplify the calculations, we believe that the unaccounted products of the separation of subsequent stages in each section fall into the relevant products of the separation process without additional re-cleaning.

Each separation cascade has its own function of factional separation. For a fine product, it can be approximated by the Plitt function:

$$\mathrm{F}_{\mathrm{S}j}(x_i) = \frac{1}{1+\left(\dfrac{x_i}{x_{50j}}\right)^{p_j}} . \tag{4}$$

The output of fine $\gamma_{\mathrm{S}j}$ and coarse $\gamma_{\mathrm{L}j}$ products from the cascade can be calculated as:

$$\gamma_{\mathrm{S}j} = \frac{\sum_i r_{\mathrm{B}j}(x_i)\mathrm{F}_{\mathrm{S}j}(x_i)}{100}; \quad \gamma_{\mathrm{L}j} = 1 - \gamma_{\mathrm{S}j} , \tag{5}$$

where $r_{\mathrm{B}j}(x_i)$ – is the content of i factions in a product that comes to be re-cleaned in the j-cascade.

The content of the i faction in the fine and coarse product of this cascade is calculated by the following formulas:

$$r_{\mathrm{S}j}(x_i) = \frac{r_{\mathrm{B}j}(x_i)\mathrm{F}_{\mathrm{S}j}(x_i)}{\gamma_{\mathrm{S}j}}; \quad r_{\mathrm{L}j}(x_i) = \frac{r_{\mathrm{B}j}(x_i)\left[1-\mathrm{F}_{\mathrm{S}j}(x_i)\right]}{\gamma_{\mathrm{L}j}} . \tag{6}$$

The scheme of material flows of the two stage recleaning of the shared material is presented on Figure 3. Here i^* is the power section; $\gamma_{\mathrm{S}j1}$, $\gamma_{\mathrm{S}j2}$ – amount of fine product j section of the first and second stages of cleaning; $\gamma_{\mathrm{L}j1}$, $\gamma_{\mathrm{L}j2}$ – number of coarse product j section of the first and second stages of re-cleaning. If the initial factional composition, the separation boundary and the separation efficiency of each cascade are known, the total fractional compositions of coarse materials and fine materials can be determined:

$$r_S(x_i) = \frac{r_{\mathrm{S}11}(x_i)\gamma_{\mathrm{S}11} + r_{\mathrm{S}12}(x_i)\gamma_{\mathrm{S}12} + \sum_{j=i^*+1}^{z}\left(r_{\mathrm{S}j1}(x_i)\gamma_{\mathrm{S}j1} + r_{\mathrm{S}j2}(x_i)\gamma_{\mathrm{S}j2}\right)}{\gamma_{\mathrm{S}11} + \gamma_{\mathrm{S}12} + \sum_{j=i^*+1}^{z}\left(\gamma_{\mathrm{S}j1} + \gamma_{\mathrm{S}j2}\right)}; \tag{7}$$

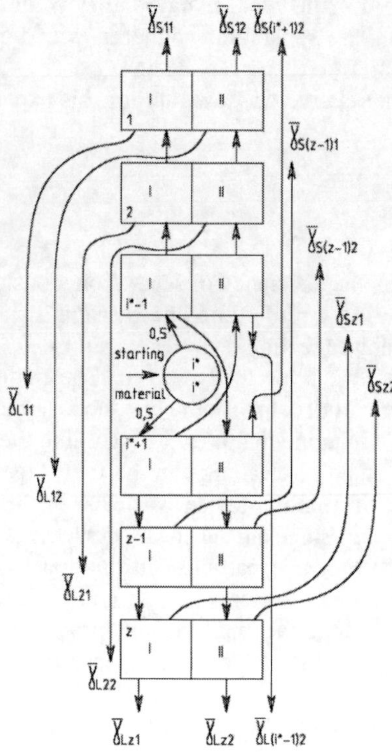

Figure 3. The scheme of material flows of the two stage recleaning of the bulk material.

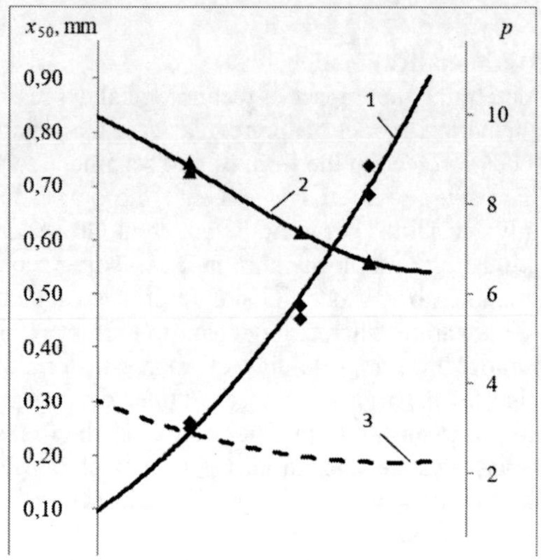

Figure 4. The dependence of the basic process parameters for the fractionation of air velocity: 1 – border; 2; 3 is the sharpness of separation of separator and a separate cascade. Solid lines – approximation of experimental data; dotted line-calculation.

$$r_{\mathrm{L}}(x_i)=\frac{r_{\mathrm{B}}(x_i)-r_{\mathrm{S}}(x_i)\left[\gamma_{\mathrm{S11}}+\gamma_{\mathrm{S12}}+\sum_{j=i^*+1}^{z}\left(\gamma_{\mathrm{S}j1}+\gamma_{\mathrm{S}j2}\right)\right]}{\gamma_{\mathrm{L}z1}+\gamma_{\mathrm{L}z2}+\sum_{j=1}^{j=i^*-1}\left(\gamma_{\mathrm{L}j1}+\gamma_{\mathrm{L}j2}\right)}; \quad (8)$$

Total output of fine material according to Figure 3:

$$\bar{\gamma}_{\mathrm{S}}=\bar{\gamma}_{\mathrm{S11}}+\bar{\gamma}_{\mathrm{S12}}+\sum_{j=i^*+2}^{z}\bar{\gamma}_{\mathrm{S}j1}+\sum_{j=i^*+2}^{z}\bar{\gamma}_{\mathrm{S}j2}+\bar{\gamma}_{\mathrm{S}(i^*+1)2}, \quad (9)$$

where is

$$\bar{\gamma}_{\mathrm{S11}}=\frac{1}{2}\prod_{j=1}^{i^*-1}\gamma_{\mathrm{S}j1}; \quad (10)$$

$$\bar{\gamma}_{\mathrm{S}j1}=\frac{1}{2}\gamma_{\mathrm{S}j1}\prod_{k=i^*+1}^{j}\gamma_{\mathrm{L}k1}; \quad (11)$$

$$\bar{\gamma}_{\mathrm{S12}}=\frac{1}{2}\gamma_{\mathrm{S}(i^*+1)1}\prod_{k=1}^{i^*-1}\gamma_{\mathrm{S}k2}; \quad (12)$$

$$\overline{\gamma}_{Sj2} = \frac{1}{2}\gamma_{Sj2}\gamma_{L(i^*-1)1}\prod_{k=i^*+1}^{j-1}\gamma_{Lk2}\,; \tag{13}$$

$$\overline{\gamma}_{S(i^*+1)2} = \frac{1}{2}\gamma_{L(i^*-1)1}\gamma_{S(i^*+1)2}\,. \tag{14}$$

Total output of coarse material:

$$\overline{\gamma}_L = 1-\gamma_S\,. \tag{15}$$

Given that the separation boundary is directly dependent on the flow velocity, it will be different in the proposed model for each of the two classifiers. The sharpness of the separation is determined by the design of the device and is the same for each cascade. If experimental data is inserted into the presented model, it will be possible to determine the dependence of the main technological parameters of the fractionation process on the velocity of airflow (Figure 4). If the air velocity varies from 2 to 4.5 m/s, the separation boundary for this classifier is described by the expression:

$$x_{50} = 0{,}053w^2 - 0{,}057w\,; \tag{16}$$

total indicator of the sharpness of the separation:

$$p = 0{,}18w^3 - 1{,}6w^2 + 3w + 8{,}8\,; \tag{17}$$

indicator of the sharpness of the separation of the individual cascade:

$$p = 0{,}23w^2 - 2{,}1w + 6{,}8\,. \tag{18}$$

4. Applying a mathematical model

Consider the application of the proposed method in order to calculate the granulometric composition of the enriched quartz sand fraction (0.7–5 mm). The source material is sand 0–5 mm. We will predict the operation of the air separator at varying amounts of the relative air flow through the depressurized feeder. The limit on class content minus 0.315 mm in enriched sand is no more than 1.5 %. In the mathematical model (Figure 3), the rate of separation of each section in the upper and lower devices will be the same and will be determined by equation (18). The separation boundary for each cascade in the top classifier will be determined by the expression (16). In the lower devices, the separation boundary will depend on the amount of relative depressurization (3) and, taking into account (16), will be equal to

$$x_{50d} = \left(0{,}053w^2 - 0{,}057w\right)\left(1 - 9{,}61\overline{V}^2 + 0{,}09\overline{V}\right). \tag{19}$$

Relative air flow rate through the powdered dispenser affects the velocity of air flow in the upper separator and the output of the finished product. Figure 5 shows this effect when performing a class content limit of minus 0.315 mm in enriched sand no more than 1.5 %. At a constant velocity of airflow through the upper classifier with an increase in the value of relative depressurization increases the content of classes minus 0.315 mm in enriched sand (Figure 6).

5. Conclusions

Laboratory tests and analysis of the mathematical model have shown that unauthorized air flow through the powdered dispenser reduces the efficiency of the powder fractionation process. With the limits on the content of fine fractions in the finished material to solve the problem, you need to increase the total air flow rate through the classifier, which leads to a decrease in the output of the product. With a fixed amount of air flow and depressurization, there are no established restrictions on the fractional composition of the finished material. As shown in Figure 5, the maximum relative air flow rate through the powdered dispenser should be no more than 10–15 %. At the same time, the total air flow rate increases by 3–6 %, and the reduction in the mass of output product is projected by no more than 2 %. The proposed mathematical model allows to carry out any technological calculations of cascading separators.

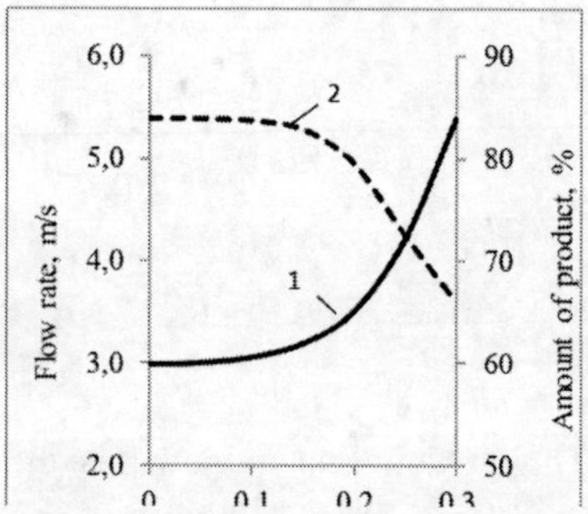

Figure 5. Dependence of relative air flow through the powdered dispenser on the velocity of airflow and output of the finished product: 1 – Air flow velocity, 2 – Output of the finished product.

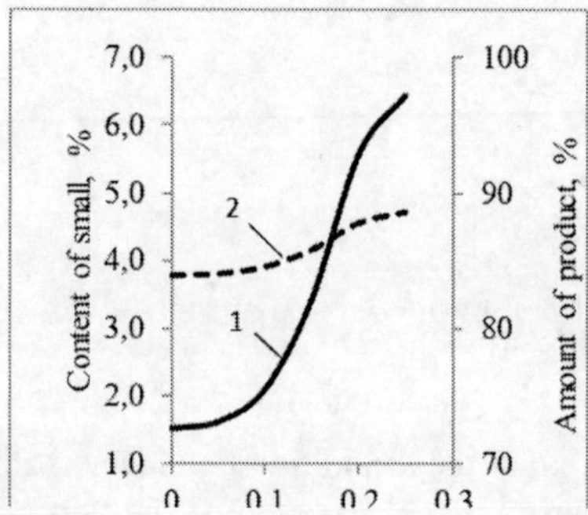

Figure 6. Effect of relative air flow through the powdered dispenser on class content minus 0.315 mm in enriched sand: 1 – Content minus 0.315 mm, 2 – Output of the finished product.

References

[1] Korsak L L and Lygach V N 2007 Gravity-centrifuge technology of concentrating minerals Russian *Journal of Non-Ferrous Metals* **48 (2)** pp 85–7

[2] Edilbayev A I 2011 Development of technologies for enrichment of high-clay ore and ore sands *Gornaya promyshlennost* **5 (99)** pp 76–9

[3] Galperin V I 2006 Aerial classification of bulk materials (about one little-known area of technology) part 3. Ways and devices. Gravitational air classifiers *Khimicheskaya promyshlennost segodnya* **3** pp 33–44

[4] Shapiro M and Galperin V 2005 Air Classification of Solid Particles: A Review *Chemical Engineering and Processing* **44** pp 279–285

[5] Ponomarev V B 2015 Dry processing of rock breaking waste *Gornyi Zhurnal* **12** pp 50–2

[6] Kapustin F L and Ponomarev V B 2016 Beneficiated sand production from rock material crushing screenings by means of air classifier *Obogashchenie Rud* (*Mineral Processing*) **4** pp 56–60

[7] Ponomarev V B and Dzyuzer V Ya 2010 The release of foreign impurities from silicon amorphous dioxide *Ogneupory i tekhnicheskaya keramika* **11–12** pp 44–8

[8] Kirsanov V A 2003 Distribution of air flow velocity in cascading air classifiers *Izvestiya vysshikh uchebnykh zavedeniy* **3** pp 65–70

[9] Tkachuk D M and Izmalkov V 2002 Equations of volume and fractional composition of the sands of the anti-enrichment cascade *Gornyy informatsionno-analiticheskiy byulleten* **7** pp 48–51

[10] Kuskov V B 2002 *Theory of separation processes* (Sankt-Peterburgskiy gornyy institut. SPb) p 55

[11] Shishkin A S 2009 Approximation of granulometric composition by finite distributions *Vestnik BGTU im V G Shukhova* **4** pp 78–81

[12] Ji Li, Chen Jiang, Kuang Shibo, Qi Zheng, Chu Kaiwei and Yu Aibing 2019 Prediction of separation performance of hydrocyclones by a PC-based model *Separation and Purification Technology* **211** pp 141–150

[13] Barsky E 2010 Critical regimes of two-phase flows with a polydisperse solid phase *Fluid Mechanics and Its Applications* **93** (New York: Springer) p 348

Simulation of patterns of crystallizing ingot macrostructure formation

O Yu Sidorov

Ural Federal University named after the First President of Russia B N Yeltsin, Nizhniy Tagil Technological Institute, 59, Krasnogvardeyskaya str, Nizhniy Tagil, 622000, Russia

E-mail: sidorov-ou-62@yandex.ru

Abstract. The electromagnetic, hydrodynamic, and temperature fields of a crystallizing cylindrical ingot with a diameter of 163 mm have been calculated using the finite-difference method. Crystallization has been simulated in an electromagnetic crystallizer. The crystallization rate has been calculated, and then the ingot areas have been divided into small crystals, dendrites and globular ones. Quantitatively, the relationship between the temperature field gradient during crystallization and the ingot structure can be expressed by the following rules: 1) elongated grains of about 10 mm – cooling rate of about $1.8 \cdot 10^{-4}$ kg/s; 2) elongated grains of about 20–30 mm – cooling rate of about $0.7 \cdot 10^{-4}$–$1.8 \cdot 10^{-4}$ kg/s; 3) globular grains with a size of 10–20 mm – cooling rate of less than $0.7 \cdot 10^{-4}$ kg/s. The possibility of simulation of the ingot structure during its crystallization based on the relationship between the crystallization rate and the grain size has been shown. The influence of the inductor current frequency on the structure of a crystallizing ingot has been simulated. It has been shown that an increase in the current frequency from 5 to 500 Hz leads to the expansion of the area with small crystals.

1. Introduction

A large number of papers have been devoted to experimental and theoretical studies of the electromagnetic field impact on metal systems. The electromagnetic field impact is expressed in grain refinement, leveling of the chemical composition with respect to the ingot volume [1]. In this case, the grain size can be related to the local cooling rate (see, for example, [2]).

It is noted in [3] that the electromagnetic field can change the temperature gradient, improve the coating microstructure, reduce cracks, and improve the coating microhardness. In the study [4], an assumption is made that growing dendrites are destroyed under the influence of a melt flow (under the influence of an electromagnetic field), and then some of the dendrite fragments are partially remelted. The possibility of concentration changes in the alloy composition under the influence of an electromagnetic field is shown in the papers [5, 6]. A change in the melt crystallization front under the influence of an electromagnetic field is shown in the paper [7], and the direction of growth of columnar crystals almost becomes parallel. The study [8, 10] provides data on the acceleration of purification of molten Al, Si from boron and phosphorus using a rotating electromagnetic field. It is noted in the paper [9] that the microstructure of Al-Zn-Mg-Cu alloy slabs was significantly improved by using an alternating magnetic field with industrial current frequency and frequency of 386 A and 50 Hz, respectively. The studies [11] for the aluminum alloy with silicon show the feasibility of electromagnetic impact during the assimilation of silicon in liquid aluminum. The feasibility of electromagnetic stirring during continuous casting of large-size slabs is shown in the paper [12]. The study [13] considers the concentration non-uniformity of the composition in relation to electromagnetic stirring. For a more effective impact of the electromagnetic field on a liquid or crystallizing molten metal, various schemes of inductors and their power modes have been proposed (see, for example, [13–18]).

2. Problem statement

This study aims at the simulation of an ingot structure (grain size) during its crystallization in a traveling electromagnetic field for various inductor current frequencies. To solve this problem, it is necessary to use a comprehensive approach that takes into account processes of various physical and chemical nature and various time constants. In this case, electromagnetic processes can be considered as quasi-steady. The hydrodynamic characteristics (speed and pressure) shall be recalculated at each time interval. Thermal processes are considered in dynamics.

An electromagnetic crystallizer for casting a brass cylindrical billet with a diameter of about 160 mm was selected as the study object [19]. The inductor winding coils are made of copper ribbon; the number of turns in the coil winding is 55, the winding arrangement is AAZZBBXXCCYY, and current density is 3 A/mm^2. A liquid alloy (brass) crystallizes in a water-cooled sleeve that separates the melt from the inductor.

3. Mathematical model

To solve field tasks, a cylindrical coordinate system was used, and the presence of cylindrical symmetry was assumed. In this case, the system of the electromagnetic field equations can be reduced to a single equation [20]

$$-\frac{\partial}{\partial r}\left(\frac{A_\phi}{\mu r}+\frac{1}{\mu}\frac{\partial A_\phi}{\partial r}\right)-\frac{\partial}{\partial z}\left(\frac{1}{\mu}\frac{\partial A_\phi}{\partial z}\right)+\beta\frac{\partial A_\phi}{\partial z}+\alpha\cdot\left(\frac{A_\phi}{r}+\frac{\partial A_\phi}{\partial r}\right)+j\cdot\omega\cdot\gamma\cdot A_\phi=J_\phi \tag{1}$$

Where $\beta = -\gamma \cdot w$and $\alpha = -\gamma \cdot v$; γ – medium specific electrical conductivity; v, w – components of medium motion speed along r and z axes, respectively.

Melt motion in a crystallizing ingot with cylindrical symmetry can be written as follows [20]

$$v = v^0 + \Delta t\left\{-v\frac{\partial v}{\partial r} - w\frac{\partial v}{\partial z} + \frac{1}{\rho}F_r - \frac{1}{\rho}\frac{\partial P}{\partial r} + v_t\left(\frac{\partial^2 v}{\partial r^2} + \frac{\partial^2 v}{\partial z^2} + \frac{1}{r}\frac{\partial v}{\partial r} - \frac{v}{r^2}\right)\right\}; \tag{2}$$

$$w = w^0 + \Delta t\left\{-v\frac{\partial w}{\partial r} - w\frac{\partial w}{\partial z} + \frac{1}{\rho}F_z - \frac{1}{\rho}\frac{\partial P}{\partial z} + v_t\left(\frac{\partial^2 w}{\partial r^2} + \frac{\partial^2 w}{\partial z^2} + \frac{1}{r}\frac{\partial w}{\partial r}\right)\right\} \tag{3}$$

In equations (2) and (3), F_r, F_z – components of the density of electromagnetic forces obtained in the calculation of the electromagnetic field according to equation (1); v_t – eddy kinematic viscosity.
To solve equations (2), (3), the finite-difference method with a staggered grid was used (see, for example, [21]).

The following equation was used for determination of temperatures in a moving medium [20]

$$\frac{\partial T}{\partial t} = -v\frac{\partial T}{\partial r} - w\frac{\partial T}{\partial z} + a_t\left(\frac{\partial^2 T}{\partial r^2} + \frac{1}{r}\frac{\partial T}{\partial r} + \frac{\partial^2 T}{\partial z^2}\right) + \frac{q_V}{\rho C_p} + \frac{L}{C_p}\frac{\partial \psi}{\partial t} \tag{4}$$

Where q_V – rate of internal heat sources; a_t – eddy diffusivity; L, ψ – crystallization heat and hard phase proportion, respectively; C_p, ρ – heat capacity and density, respectively.

4. Procedure for joint solving of field tasks

The specifics of simultaneous solving the above-mentioned tasks are their various time constants. During simulation, the following provisions have been adopted:

1) a change in the temperature and hydrodynamic fields is considered in dynamics;

2) the electromagnetic field is steady-state;

3) the hydrodynamic field is recalculated at each time interval in order to take into account the occurring solid phase in the melt;

To perform calculations, the finite-difference method was used, and an author's software package which allows for obtaining the necessary results with acceptable accuracy was created.

5. Simulation results

A fragment of the calculation area is shown in Figure 1. The liquid melt area was divided into 10 layers along the r axis (horizontal axis) and 56 layers along the z axis (vertical axis).

The melt motion speed practically does not affect the energy characteristics of the electromagnetic field, which follows from the calculations performed in the course of the study. Therefore, the electromagnetic field is calculated once.

The flow structure has a dual-vortex nature (Figure 2) with maximum speeds on the axis of symmetry ($r = 0$), which are about 0.74 m/s in the middle of the height. In the melt layer as close to the inductor as possible, the highest speed is about 0.4 m/s at the supply current frequency of 500 Hz.

```
666666666664444444444
666666666664444444444
666666666654444444444
111111111154444444444
111111111154444444444
111111111154222244444
111111111154222244444
111111111154AA2244444
111111111154AA2244444
111111111154222244444
111111111154222244444
111111111154AA2244444
111111111154AA2244444
111111111154222244444
111111111154222244444
111111111154ZZ2244444
111111111154ZZ2244444
111111111154222244444
111111111154222244444
111111111154ZZ2244444
111111111154ZZ2244444
111111111154222244444
111111111154222244444
111111111154BB2244444
111111111154BB2244444
```

Figure 1. Fragment of the calculation area:
1 – melt area; 2 – iron circuit;
A, Z, B – phase windings; 4 – air;
5 – crystallizer; 6 – heat insulation.

Figure 2. Fragment of the molten metal motion at the supply current frequency of 500 Hz. The field direction is from top to bottom.

The crystallization temperature range was 1,083–1,070 K. It was assumed that the initial melt temperature was 1,090 K. The melt crystallization begins about 100 s after the start of calculations. The first crystals appear in the metal layer closest to the inductor (since there is a water-cooled insert). In this layer, there is the largest temperature gradient from all areas with liquid metal.

The crystallization front moves from the metal layers closest to the inductor towards the ingot axis. It was found that the crystallization rate was in the range of 3–5 mm/min in the first 5–25 minutes after the start of simulation.

As it is known (see, for example, [1,2]), crystal growth occurs in the direction opposite to the direction of the heat flux, which in turn is determined by the temperature field gradient

$$\vec{q} = -\lambda \cdot \left(\frac{\partial T}{\partial r}\vec{e}_r + 0\vec{e}_\phi + \frac{\partial T}{\partial z}\vec{e}_z\right)$$

Let us try to establish the relationship between the temperature field gradient during crystallization and the solid metal alloy structure.

For the number of calculation steps of 540 (25 minutes), the fraction of the solid phase in each cell of the calculation area exceeds 0.75 Let us compare the experimentally observed structure of the metal ingot and the components of the temperature gradient $\frac{\partial T}{\partial r}$. In the outer layer with a thickness of 1/10 of the ingot radius, an equiaxial fine-grain structure is observed. In the inner layer (from the center to the periphery) with a thickness of 5/10 of the ingot radius, equiaxial large crystals are observed. Between these layers (4/10 of the ingot radius), columnar crystals directed along the radius are observed. We believe that the ingot grain size is related to the crystallization rate [2]. It has been found that quantitatively, the relationship between the temperature field gradient during crystallization and the ingot structure can be expressed in the following rules:

- elongated grains of about 10 mm – cooling rate of about $1.8 \cdot 10^{-4}$ kg/s;
- elongated grains of about 20-30 mm – cooling rate of about $0.7 \cdot 10^{-4}$–$1.8 \cdot 10^{-4}$ kg/s;
- globular grains with a size of 10-20 mm – cooling rate of less than $0.7 \cdot 10^{-4}$ kg/s.

Let us apply the obtained rules for simulation of a crystallizing ingot structure.

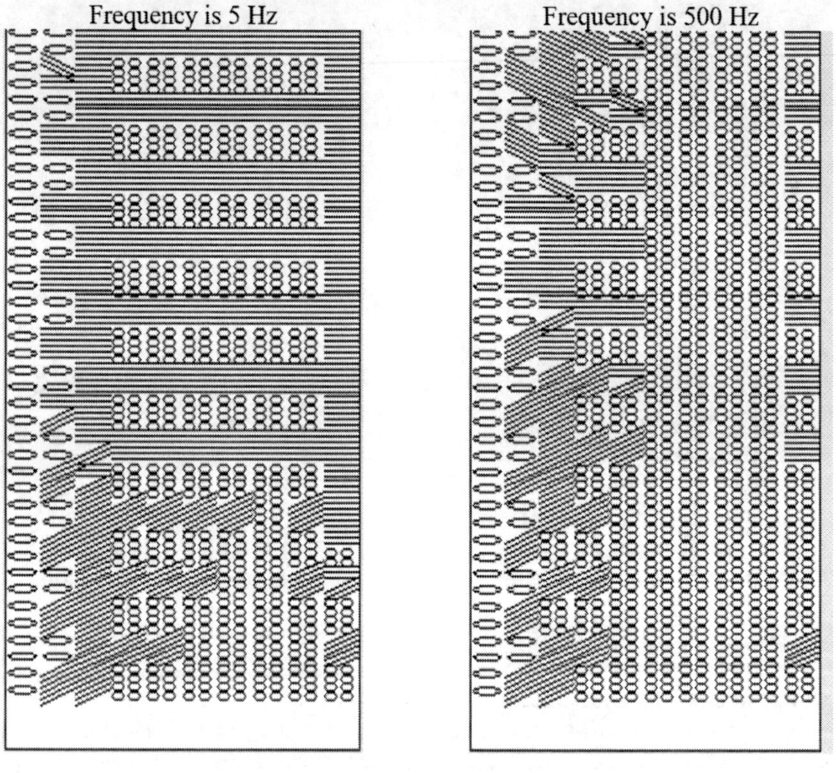

Figure 3. Result of simulation of the ingot axial section structure (a part of the ingot is shown).

– small crystals; – dendritic structure;
– large globular crystals.

Figure 3 shows the calculated structure of the crystallized ingot for two current frequencies in the inductor (*r=0* axis on the left; inductor on the right). From these results it follows that in the outer layer of the ingot, smaller elongated crystals (continuous small ellipses) form. Next to them and further to the center of the ingot, crystals in the form of dendrites (horizontal and inclined lines showing the direction of their growth) form. Then the formation of larger globular grains (large ellipses) located near the ingot axis begins.

The impact of the current frequency in the inductor is observed (Figure 3). At the frequency of 5 Hz, dendritic structures penetrate the ingot throughout the cross section. At the current with frequency of 500 Hz, a wide zone of small crystals appears and is about half the cross section. Further to the center of the ingot, there form crystals in the form of dendrites (horizontal and inclined lines). Then the formation of larger globular grains (large ellipses) located near the ingot axis begins.

6. Conclusion

- A quantitative relationship between the temperature field gradient, crystallization rate and grain size has been established.
- Molten metal crystallization in an electromagnetic field has been simulated.
- It has been shown that an increase in the current frequency from 5 to 500 Hz leads to the expansion of the area with small crystals.

References

[1] Samoilovich Yu A 1983 *System Analysis of an Ingot Crystallization* (Kiev: Naukova Dumka) [In Russian]

[2] Winegard W 1967 *An Introduction to the Solidification of Metals* (Moscow: Mir) [In Russian]

[3] Zhai L L, Ban C Y and Zhang J W 2019 Investigation on laser cladding Ni-base coating assisted by electromagnetic field *Optics & Laser Technology* **114** pp 81–8

[4] Xu Y, Wang T, Wang F and Wang E 2017 Influence of lower frequency electromagnetic field on dendritic crystal growth in special alloys *Journal of Crystal Growth* **468** pp 506–9

[5] Zhai L, Ban C, Zhang J and Yao X 2019 Characteristics of dilution and microstructure in laser cladding Ni-Cr-B-Si coating assisted by electromagnetic compound field *Materials Letters* **243** pp 195–8

[6] Chen X, Jia Y, Liao Q, Jia W, Le Q, Ning S and Yu F 2019 The simultaneous application of variable frequency ultrasonic and low frequency electromagnetic fields in semi continuous casting of AZ80 magnesium alloy *Journal of Alloys and Compounds* **774** pp 710–20

[7] Yang Y, Chen R, Guo J, Ding H and Su Y 2018 Numerical analysis for electromagnetic field Influence on heat transfer behaviors in cold crucible used for directional solidification *International Journal of Heat and Mass Transfer* **122** pp 1128–37

[8] Khatsayuk M Yu, Pervukhin M V and Golovenko Ye A 2011 MHD-Processes in the Liquid Phase of an Ingot Formed in a Magnetic Field *News of Higher Educational Institutions. Electromechanics* **1** pp 26–30

[9] Bana B, Zhang T, Lia J, Baia X, Pana X, Chena J and Tabaian S H 2018 Solidification refining of MG-Si by Al-Si alloy under rotating electromagnetic field with varying frequencies *Separation and Purification Technology* **202** pp 266–74

[10] Zoua Q, Hanb N, Shenb Z, Jiea J and Li T 2018 Effects of AlB2/AlP phase and electromagnetic stirring on impurity B/P removal in the solidification process of Al-30Si alloy *Separation and Purification Technology* **207** pp 151–7

[11] Nikitin K V, Amosov E A, Nikitin V I, Glushenkov V A and Chernikov D G 2015 Theoretical and Experimental Substantiation of Treatment of Aluminum-Based Melts by Pulsed Magnetic Fields *News of Higher Educational Institutions. Non-Ferrous Metallurgy* **5** pp 11–9

[12] Goman V, Sokolov I and Fedoreev S 2019 Modeling electromagnetic stirring processes during continuous casting of large-format slabs. *International Conference on Industrial Engineering, Applications and Manufacturing, ICIEAM 2019* (Institute of Electrical and Electronics Engineers Inc.) https://doi.org/10.1109/ICIEAM.2019.8743005

[13] Tsaplin A I 1986 Dynamics of the Circulation of a Crystallizing Continuous Ingot Liquid Core in the Traveling Field of an Inductor *Magnetic Hydrodynamics* **1** pp 127–31 [In Russian]

[14] Sarapulov F N, Sidorov O Yu and Timofeev V N 2001 Induction MHD-devices and mathematical simulation *Fifth international Conf. on Unconventional electromechanical and*

electrical systems (Miedzyzdroje, Poland, September 05-08) **1** p 165

[15] Sarapulov F N, Sarapulov S F, Sidorov O Yu and Sokunov B A 2006 Formation of Electromagnetic Impact on the Liquid Phase of a Crystallizing Ingot *Collection of Research Papers of the 12th International Ples Conference on Magnetic Fluids* (August-September, 2006. Ples, Russia. State Educational Institution of Higher Professional Education Ivanovo State Power University named after V I Lenin) p 407

[16] Gruzdeva I A, Sulitsin A V, Mysik R K and Sokunov B A 2006 Influence of Electromagnetic Stirring on the Quality of Tin Bronze Ingots *Proceedings of the International Research-to-Practice Conference. Features of Treatment and Use of Products Made of Heavy Non-Ferrous Metals* (Yekaterinburg: Ural Branch of the Russian Academy of Sciences) p 147

[17] Begalov V, Bychkov A, Sarapulov F and Sokunov B 2004 Poliphase schemes of inductors for electromagnetic treatment of smelts *Proceedings of HES-04* (Padua, Italy) p 79

[18] Erkenov N Kh 2002 Study of the Influence of Power Supply Systems on the Parameters of an Electromagnetic Crystallizer *News of Higher Educational Institutions. Electromechanics* **2** pp 28–31

[19] Sarapulov S F, Sokunov B A and Frizen V E 2003 Induction Agitator for Aluminum Alloys Stirring during Crystallization *Issue 38. Bulletin of USTU. Electromechanical and Electromagnetic Energy Converters and Controlled Electromechanical Systems: Collection of Articles. Part 2: Special Electrical Machines and Electromagnetic Devices. Energy Saving Issues. Educational Projects.* (Yekaterinburg: USTU) no 5 (25) p 171 [In Russian]

[20] Sidorov O Yu, Sarapulov F N and Sarapulov S F 2010 *Finite-Element and Finite-Difference Methods in Electromechanics and Electrotechnics* (Moscow: Energoatomizdat) [In Russian]

[21] Patankar S 1984 *Numerical Heat Transfer and Fluid Flow* (Moscow: Mir) [In Russian]

Methodology of designing an air cascade separator

V B Ponomarev[1, 2] and A S Shishkin[1]

[1]Ural Federal University named after the First President of Russia B N Yeltsin, 19, Mira str., Yekaterinburg, 620000, Russia

E-mail: [1]v.b.ponomarev@urfu.ru

Abstract. The calculation technique of industrial cascade separators based on experimental and analytical research process methods for air fractionation of bulk materials is considered. The main technological factors and process defining parameters used in calculation of separator design are presented. It is shown that for the design of an industrial separator, it is necessary to find a relationship between the initial granulometric distribution and the fractional requirements for the finished product from the concentration of the material flow and the air flow velocity to the complete cross-section of the separator. The experiment results of material flow concentration influence on division efficiency for various cascade separator constructions are given. Mathematical relationship of the air flow rate through the apparatus on the separation boundary and bulk concentration is proposed. The correlation of these parameters with the constructive sizes of the separator is shown. The initial particle size distribution of bulk, the set restrictions for fractional characteristics of the received powder and the approximating equations of fractional extraction function of fineness narrow classes in a large or small product based on experimental data are used for calculation of the equipment.

1. Introduction

Cascade air separators are widely used for solving various production tasks [1–7] related to processing and enrichment of bulk materials. This is because by their high efficiency of division [8] and ability to get fractions powdery materials in wide borders with low power expenses. To a large extent, their implementation at enterprises is difficult because of the low popularity of design techniques for such devices, the complexity of the transition from the mathematical model of the process to the overall dimensions and technological characteristics of the equipment. This problem is resolves in this paper.

2. Design technique

In our opinion, the most common at present are three types of cascade pneumatic separators: single-column separators with various bulk elements; multi–row separators combined over a common perforated grate and separators with an inclined grate.

In all cases, the structural analysis comes down to determining the cross-sectional area of separator or lattice. It was shown in [9] that separator efficiency increase practically stops if the number of cascades exceeds 6–7 units. For one-column separator this value is 6–7 pour out elements, for a combined one – 6 or 7 columns with the same number of overturning shelves. Knowledge of the area of one cell cascade allows us to determine the overall separator dimensions. The dimensions of inclined separator can be easily defined with knowledge of the grille cross-sectional area [9].

For given initial supply capacity Q and known flow concentration μ, it is possible to calculate the separator air flow V:

IOP Conf. Series: Materials Science and Engineering **966** (2020) 012043 doi:10.1088/1757-899X/966/1/012043

$$V = \frac{Q}{\mu} \tag{1}$$

In turn, the cross–sectional area S of the separator (grille) is defined as:

$$S = \frac{V}{w} \tag{2}$$

where w is velocity of the air flow to the total cross section of the separator or grille.

The flow concentration determines the ratio of the bulk material amount passing through the installation and the air flow, which in turn sets the dimensions of separator and associated equipment, and also the energy consumption per ton of the product obtained.

The optimum concentration value is the maximum value at which the requirements for the product are met, while observing the limitations on the "loss" of the valuable fraction (component) with the waste.

Therefore, to design an industrial separator, it is necessary to find the relationship between initial particle-size distribution and fractional requirements for the finished product from flow concentration of material and speed of the air flow to the total cross section of the separator.

Universal characteristic of the bulk material separation process is the function $F_{sm}(x_i)$ for fractionally extracting narrow classes xi into a fine product or the backward Tromp curve [10].
Knowing the function of fractional extraction, it is possible to determine the content of any fraction in fine or coarse product [2]:

$$r_{smi} = \frac{F_{sm}(x_i) r_{ini}}{\gamma_{sm}}; \tag{3}$$

$$\gamma_{sm} = \sum_i F_{sm}(x_i) r_{ini}; \tag{4}$$

$$r_{ci} = \frac{r_{ini} - r_{smi}\gamma_{sm}}{1-\gamma_{sm}}, \tag{5}$$

where r_{ini}, r_{smi}, r_{ci} – the content of the i-th fraction in the initial, fine or coarse product;
γ_{sm} – is the output of fine material.

The function of fractional extraction is characterized by two main indicators – the x_{50} separation boundary and the magnitude of effectiveness. To calculate the separation efficiency, for example, the Eder-Mayer point test $k_{75/25}$ [8, 10] is widely used. The function of fractional extraction can be approximated, for example, by the two-parameter Rosina-Ramler function [11]:

$$F_{sm}(x_i) = \exp\left(-\ln 2\left(\frac{x_i}{x_{50}}\right)^p\right) \tag{6}$$

or Plitte function [11]:

$$F_{sm}(x_i) = \frac{1}{1+\left(\frac{x_i}{x_{50}}\right)^p}, \tag{7}$$

and the exponent p is expressed in terms of $k_{75/25}$.

For the Plitte function [8]:

$$k_{75/25} = \left(\frac{1}{9}\right)^{\frac{1}{p}}. \tag{8}$$

For the Rosina – Ramler function:

$$k_{75/25} = \left(\frac{\ln 0,75}{\ln 0,25} \right)^{\frac{1}{p}}. \tag{9}$$

It is possible to model the fractionation process and calculate the parameters p and x_{50} by the selection method [11], using the formulas (3–5) and one of the approximations (6 or 7).
In [8, 9], the effect of the flow concentration on the separation boundary and efficiency by the Eder-Mayer criterion was investigated. It is noted that the efficiency curve has a certain zone of constancy and then a monotonic decrease with concentration increasing. Figure 1 shows a graph of the parameter p dependence on the flow concentration μ by the combined multi-row separator when fractionating the technical salt. The inverse relationship will have the form Figure 2.

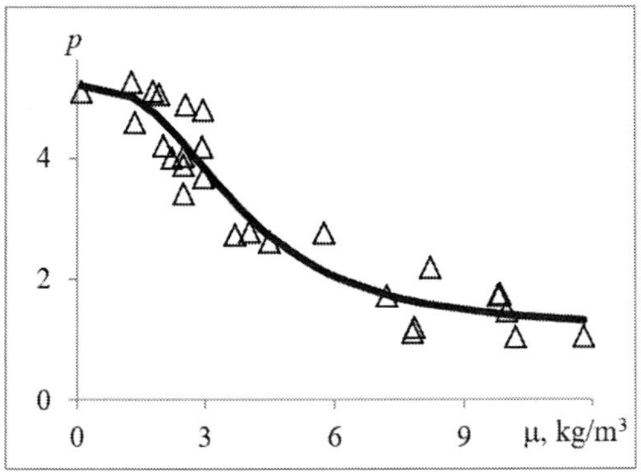

Figure 1. Dependence of the approximation parameter p of the Plate on the flow concentration.

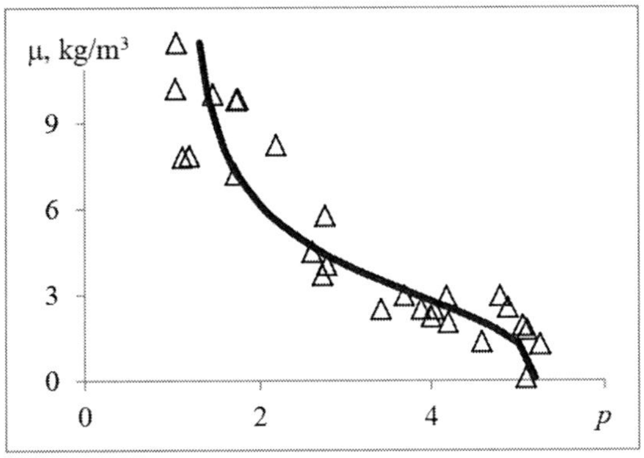

Figure 2. Dependence of the flow concentration on the parameter p of the Plitt approximation.

A mathematical approximation of this relationship looks like:

$$\mu = 3,82 \left(\frac{4,07}{p - 1,14} \right)^{\frac{1}{2,75}}. \tag{10}$$

Similar dependencies can be obtained for any other material on the laboratory model of the projected cascade separator. The separation boundary in the bulk material fractionation depends not only on the flow concentration, but also on the gas flow rate in the square

$$x_{50} = f\left(\mu, w^2\right). \tag{11}$$

Taking into account the analysis of dimensions, it is possible to propose an investigation of the following dependence

$$\frac{x_{50} g}{w^2} = f(\mu). \tag{12}$$

The Figure 3 shows the dependence of the parameter $\frac{x_{50} g}{w^2}$ on the flow concentration, which can be approximated by a parabola

$$\frac{x_{50} g}{w^2} = 4 \cdot 10^{-5} \mu^2 + 8 \cdot 10^{-4} \mu. \tag{13}$$

Separating the flow velocity from [equation (14)], we obtain

IOP Conf. Series: Materials Science and Engineering **966** (2020) 012043 doi:10.1088/1757-899X/966/1/012043

$$w=\sqrt{\frac{x_{50}g}{4\cdot10^{-5}\mu^2+8\cdot10^{-4}\mu}}=P(x_{50},\mu). \tag{14}$$

If you calculate the parameter for experimental values of the flow rate, you can draw a graph (Figure 4).

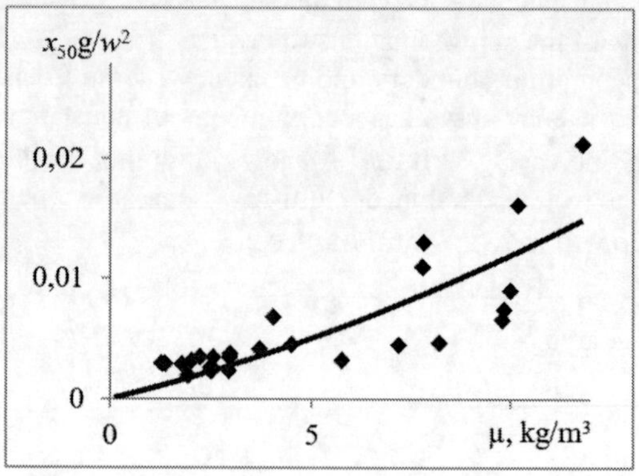

Figure 3. Dependence of the parameter $\frac{x_{50}g}{w^2}$ on the flow concentration.

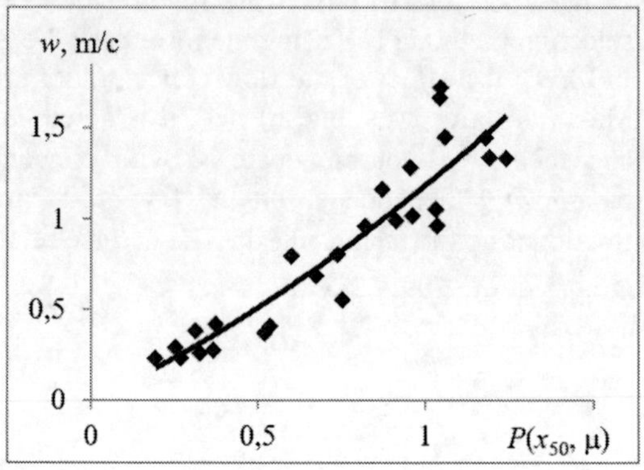

Figure 4. The dependence of the flow velocity on the parameter $P(x_{50},\mu)$.

Finally, the equation can be written as

$$w=\frac{0{,}33x_{50}g}{4\cdot10^{-5}\mu^2+8\cdot10^{-4}\mu}+0{,}86\sqrt{\frac{x_{50}g}{4\cdot10^{-5}\mu^2+8\cdot10^{-4}\mu}}. \tag{15}$$

3. Conclusion

Thus, it is possible to list the main stages of calculating the cascade pneumatic separator:

1. The mathematical model of the process of fractionation of bulk materials should be developed taking into account the requirements for the finished product. To calculate the fractional extraction function, we use formulas (6) or (7). Calculations are more conveniently carried out in the form of table 1

Table 1. Powder fractionation process model.

Fraction size x_i	r_{ini}	$F_{sm}(x_i)$	$F_{sm}(x_i)r_{ini}$	r_{smi}	r_{ci}
x_1	r_{in1}	$F_{sm}(x_1)$	$F_{sm}(x_1)r_{in1}$	r_{sm1}	r_{c1}
…	…	…	…	…	…
x_n	r_{inn}	$F_{sm}(x_n)$	$F_{sm}(x_n)r_{inn}$	r_{smn}	r_{cn}
		Output γ_{sm}	$\Sigma F_{sm}(x_i)\, r_{ini}$		

Determine the x_{50} separation boundary and the parameter p ensuring the production of the desired product, using the computer selection method.

2. Calculate the dependence of separation efficiency with the Eder-Mayer criterion on the flow concentration, according to laboratory experiment data.
3. Find the required value of material flow concentration, at which the p parameter value turns out to be no less than calculated by the mathematical model (Section 1).
4. Using laboratory experiment data and calculated flow concentration, we find the velocity of the air flow to the total cross section of the separator (grille).
5. Calculate the flow rate of air flow through the separator, the overall dimensions of the installation. Make selection of additional equipment and air blower.

IOP Conf. Series: Materials Science and Engineering **966** (2020) 012043 doi:10.1088/1757-899X/966/1/012043

If it is necessary to improve the efficiency of the separation, it is possible to compose a technological scheme with several clearings.

The results of this methodology have been successfully tested in the calculation and implementation of cascade separators at various industrial enterprises [2–4, 7, 12, 13].

References

[1] Avdeev N E 2005 *Prospective types of centrifugal and gravitational separators: theory and analysis of structures* (Voronezh State University, Voronezh) p 125

[2] Ponomarev V B 2015 Dry processing of rock breaking waste *Gornyi Zhurnal* **12** pp 50–2

[3] Kapustin F L and Ponomarev V B 2016 Beneficiated sand production from rock material crushing screenings by means of air classifier *Obogashchenie Rud* (*Mineral Processing*) **4** pp 56–60

[4] Shapiro M and Galperin V 2005 Air Classification of Solid Particles: A Review *Chemical Engineering and Processing* **44** pp 279–85

[5] Kleiv R A 2012 Value enhancement of the olivine process, dust through air classification *International Journal of Minerals, Metallurgy and Materials* **19** pp 185–91

[6] Felk A 2014 Fine-Milling and Air Classification of Ceramic Materials by the Dry Method *Glass and Ceramics* **71** pp 92–5

[7] Shishkin S F, Dzyuzer V Ya and Shishkin A S 2001 Air Classification of Sands for the Glass Industry *Glass and Ceramics* **58** pp 370–3

[8] Kirsanov V A, Avdeeva A A and Avdeev M N 2006 Analysis of basic characteristics of cascade pneumatic classifiers *Chemical and Petroleum Engineering* **42** pp 545–7

[9] Ponomarev V B 2017 *Calculation and design of equipment for air separation of bulk materials*, (Ekaterinburg) p 79

[10] Kirsanov V A, Kirsanov M V and Berdnik V M 2017 Analysis of criteria for assessing the efficiency of the process of separation of bulk materials *Actual problems of the humanities and natural sciences* pp 17–21

[11] Shishkin A S and Shishkin S F 2015 *Examples of calculations of aerodynamic processes of bulk materials processing in Excel* (available online on http://hdl.handle.net/10995/36127)

[12] Kapustin F L, Perepelitsyn V A, Ponomarev V B and Loshkarev A B 2017 Increase in the efficiency of the use of cuttings for crushing rock rocks *Physico-technical problems of developing minerals* **3** pp 103–8

[13] Barsky E 2009 Efficacy of separation of a pourable material *High Temperature* **47(6)** pp 822–8

Investigation of structure and properties of low alloy steel obtained by wire arc additive manufacturing under various fluxes

G E Trekin

Ural Federal University named after the First President of Russia B N Yeltsin, Nizhniy Tagil Technological Institute, 59, Krasnogvardeyskaya str, Nizhniy Tagil, 622000, Russia

E-mail: trekin1963@yandex.ru

Abstract. Wire arc additive manufacturing (WAAM) is vigorously explicating direction of mechanical engineering. When using submerged arc welding (SAW) it is possible to apply low-alloy steels with high efficiency and properties, that will allow to use it for manufacture of details and instrument. Microstructure, hardness, distribution of nonmetallic and chemical composition were investigated for SAW build up walls, in as-deposited condition, by low-alloy wire under fluxes with a different basicity. The characteristics of microstructure fixed by an optical microscopic metallography remain practically constant for all deposited weld metals. It is ferrite – martensite (bainite) structure, which has a high dispersibility. The exploration of hardness has shown formation of the partially quenched top with increased hardness and previous thermo- cyclic treated passes with lower hardness. When depositing SAW insignificant saturation of metal with silicium and manganese and a carbon drop occurs. The distribution of nonmetallics is most favorable for additive manufactured material under neutral and basic fluxes. As a result, a comparative research of fluxes was conducted, allowing to make a choice for additive manufacturing of low-alloy work pieces.

1. Introduction

The additive technologies allow to manufacture work pieces with the improved properties and ensure their economic expediency when the shape is irregular [1]. Most of the reviews do not give attention to low-alloy steels and submerged arc welding (SAW) [2, 3], because there is no its robotized equipment. The works using WAAM for workspaces from low-carbon and low-alloy steels, apply gas-metal arc welding [4, 5, 6]. The structure forming of deposited metal at arc additive manufacturing occurs at a crystallization and consequent technologically inevitable (automatic) thermo- cyclic treatment (ATCT) from the last passes to the previous ones [7, 8]. Thus ferrite – martensite (bainite) structure of different ratio of ferrite with a different morphology (acicular, polygonal granular etc.) is formed, and the martensite -austenitic constituents also detected [4]. The geometrical sizes of a building joint are determined by a diameter of a wire, welding parameters and surface-tension of a melt [9]. The chemical composition also depends on welding mode and saturation\expulsion of alloy elements in the process of chemical interaction between depositing metal and protective medium [10]. The set of metal properties obtained by arc additive manufacture is attractive enough for its use in industry. The metal has high level of residence to brittle fracture [11].

2. Materials and methods

The wall-shaped work pieces were built-up by means of SAW equipment ADC -1250 SF with a wire 30KhaGSA (Table 5) at diameter of 3 mm and standard mode (Table 1) under fluxes AN - 348A, FSA ChTA 650 -20/80 and UF-01 (Table 2–4), number of passes 51, as a base metal a low carbon steel plate with the thickness of 20 mm was used. The exploration of chemical composition was executed on Q2 - ION device.

Table 1. Submerget arc welding parameters.

Current (A)	Voltage (V)	Welding speed (m/hr)	Wire feed speed (m/min)	Electrode stick out (mm)	Inter-pass time (min)
540	44	40	0.9	30	50–60

Table 2. The nominal chemical composition (mass. %) and basicity index (BI_f) ofAN-348A flux.

SiO_2	MnO	MgO	CaF_2	CaO	Al_2O_3	Fe_2O_3	S	P	BI_f
41–44	34–38	5–7.5	4–5.5	<6.5	<4.5	<2	<0.15	<0.12	0,75

Table 3. The nominal chemical composition (mass. %) and basicity index (BI_f) of FSA ChTA 650 -20/80 flux.

Al_2O_3+ CaO+ MgO	Al_2O_3	CaF_2	BI_f
>40	>20	>22	1.2

Table 4. The nominal chemical composition (mass. %) and basicity index (BI_f) of UF -01 flux.

CaO+ MgO+ CaF_2+MnO	SiO_2	CaF_2	BI_f
>50	>20	>15	3.2

The metallographic study was performed with a Zeiss Observer D1m microscope running under Thixomet software. Etching was carried out with Nital (6 %) solution for 20 s at the room temperature. Hardness indentation were made using a DURA Jet testing machine by Rockwell C-method. The results were processed by means of Excel and MathCAD software.

3. Results and discussion

The shape of a single weld pass is crescent with a superficial weld penetration cog in the center. After a consequent deposit welding, pass by pass, a vertical wall of height 60–65 mm and width of 14–18 mm is formed without cracks. Because of low toughness of molten metal under flux, there is an increment of a thin layer with a thickness about 1 mm. The previous pass is remelted by 74 % and 1 mm thick strip remain here. The strip completely gets into a heat-affected zone and exposes to short thermal effect. After a deposit welding of consequent passes, there is triple remelting and auto thermo-cyclic treatment (ATCT) by short thermal heats with decreasing of maximal temperature from near melting point to a low tempering (Figure 1).

After SAW under fluxes AN-348A and FSA ChTA 650 -20/80 there remains about 40 % of carbon initially contained in a wire and about 60 % under a flux UF-01. The SAW building-up under all fluxes results in insignificant saturation of deposited metal with silicium. The significant saturation with manganese occurs only under flux AN-348A and FSA ChTA 650 -20/80 by 1.5 % and 0.79 %, for a flux UF-01 it is accordingly insignificant and makes about 0.2 %. The expulsion of a chromium is observed only for a flux AN-348A (table 5). Under fluxes AN-348A and FSA ChTA 650 -20/80 chemical composition with the least carbon content and greatest silicium and manganese content is formed. Under flux UF-01 – with the greatest carbon content and with the least silicium and manganese content is formed, with hardly varying of the chromium content.

The analysis of hardness distribution along the height of deposited metal showed, that the top passes have 16–32 % higher hardness in comparison with the bottom located passes, which passed auto thermo- cyclic treatment (figure 2, table 5).

IOP Conf. Series: Materials Science and Engineering **966** (2020) 012044 doi:10.1088/1757-899X/966/1/012044

The analysis of structure revealed a ferrite – martensite (bainite) structure in the top passes. It has a small volumetric fraction of ferrite for all investigated samples of a deposited metal, that stipulates higher hardness of the top pass. In bottom areas of a weld metal the fraction of ferrite was increased and dispersed perlite was most likely formed after high-temperature thermo – cyclic treatment affect.

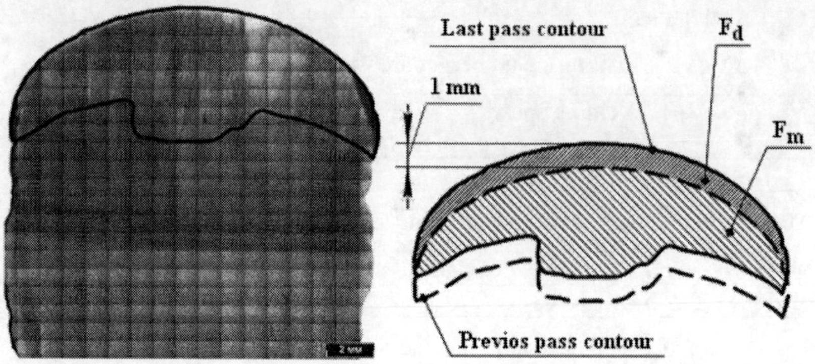

a) b)

Figure 1. Wire arc additive manufactured wall formation (flux AN-348A).
a – transversal section, b – passes contours reconstruction;
F_d – deposited metal area, F_m – melting area; ММ = mm

Table 5. The chemical composition and hardness of deposited metal (DM) and wire.

Flux	contents of elements (mass. %)				Hardness (HRC) after:	
	C	Si	Mn	Cr	deposition	deposition and ATCT
Wire 30KhaGSA	0.34	0.97	0.81	0.75	–	–
AN-348A	0.13	1.15	2.31	0.56	31	26
FSA ChTA 650 -20/80	0.14	1.20	1.59	0.72	31	21
UF-01	0.21	1.04	1.02	0.79	26	19

The structure of metal deposited under the low basic flux has the greatest dispersibility (Figure 4).

The estimation of nonmetallics was fabricated at small (×50) and large (×1000) magnification. First of them characterize particulate contamination with rather big inclusions, which are formed mainly at a stage of interaction of hydro dynamical currents of metal and molten flux in a welding bosh. Second of them originate in a liquid and solid condition, they grow or can be dissolved mainly in a solid condition.

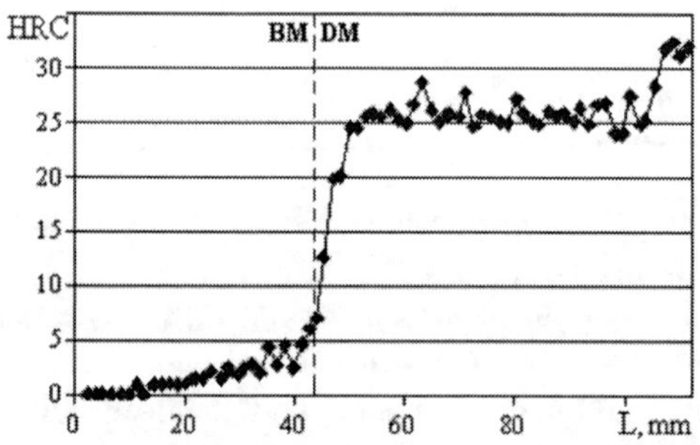

Figure 2. Hardness distribution (HRC) along wall height (L) (flux AN-348A): BM – basemetal (initial plate); DM – deposited weld metal.

The greatest inclusion volume fractioning during metal building up, deposited under the flux AN - 348A, is connected with its high saturation with oxygen. For flux FSA ChTA 650 -20/80 and UF-01 these magnitudes are less (Figure 5). The high fraction of big inclusions which got to polished section with the size from 30 up to 910 μm² is more probably of more hampered conditions floating of nonmetallics at the used conditions for fluxes AN - 348A and FSA ChTA 650 -20/80 if compare with flux UF-01 (Figure 4 b, d, f). At the large increase, the greatest fraction of conclusions up to 1 μm^2 for fluxes FSA ChTA 650 -20/80 and UF-01 and the significant fraction of conclusions with the sizes from 1,5 up to 18 μm^2 (Figure 4, d, f) is connected to distinctions in a melting temperature of fluxes. The interval of melting for welding fluxes strongly influences on an amount and aspect micro – flux contaminations, remaining in a weld metal. The components of a welding flux having a solidification point higher, than the weld metal are present in liquid metal of welding pool in the form of the smallest spherical particles and have time to be floated up from the pool before its crystallization. The acid and neutral fluxes have a melting temperature about 1350–1450 °C, therefore the amount of micro – slag particles in metal of a weld is more than at the use of basic fluxes.

Formation of the melting contour shape is not absolutely favorable under all considered fluxes, as the central melting cog is formed, that can increase probability of forming interlayers cracks. Therefore it is necessary to achieve a melting contour approaching to a direct line without the central local melting. It can be achieved diffusing electromagnetic effects application, welding with the modulated current, other technological effects and changing SAW modes.

The obtained structure is capable to show high properties as the metal was remelted three times and because of a thin remaining strip exposes practically homogeneous consequent pulse thermal effect, that in combination with temperature monitoring between passes leads to a high homogeneity of structure, that is confirmed by a small scatter of hardness and chemical composition. The areas of initial changes of chemical composition and final partial ATCT are approximately identical and consist of 10–12 passes.

The fraction of rather large inclusions with its area of more than 300 μm^2, if accepting particles as spherical, has the size of more than 19.5 μm, makes 0.079 for AN - 348A; 0.095 for FSA ChTA 650 -20/80 and 0.074 for UF-01. It shows that the content level of relatively large nonmetallics a little bit higher for FSA ChTA 650 -20/80, but in general is of the same order, therefore it is difficult to identify the best flux from this parameter. The analysis of the particle sizes revealed at high magnification shows, that their sizes are less than the dangerous size of 10 – 20 μm (38.5 – 78.5 μm^2), defined in sources [12, 13], for all used fluxes, but their density in metal, deposited under the flux AN - 348A is many times more than when we use two other fluxes. The most favorable is considered to be a dispersed structure with an acceptable amount of non-metallic inclusions. Flux FSA ChTA 650 -20/80 is the most suitable for this criterion.

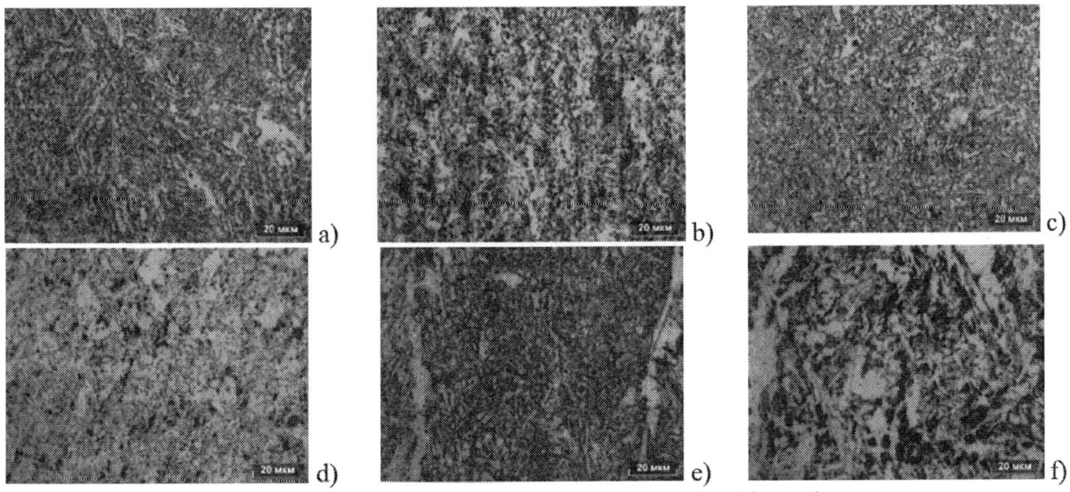

Figure 3. Optical micrographs of deposited weld metal.
flux: a, b- AN-348A; c, d - FSA ChTA 650 -20/80; e, f - UF -01
a, c, e – after deposition; b, d, f - after deposition and ATCT; scale in μm

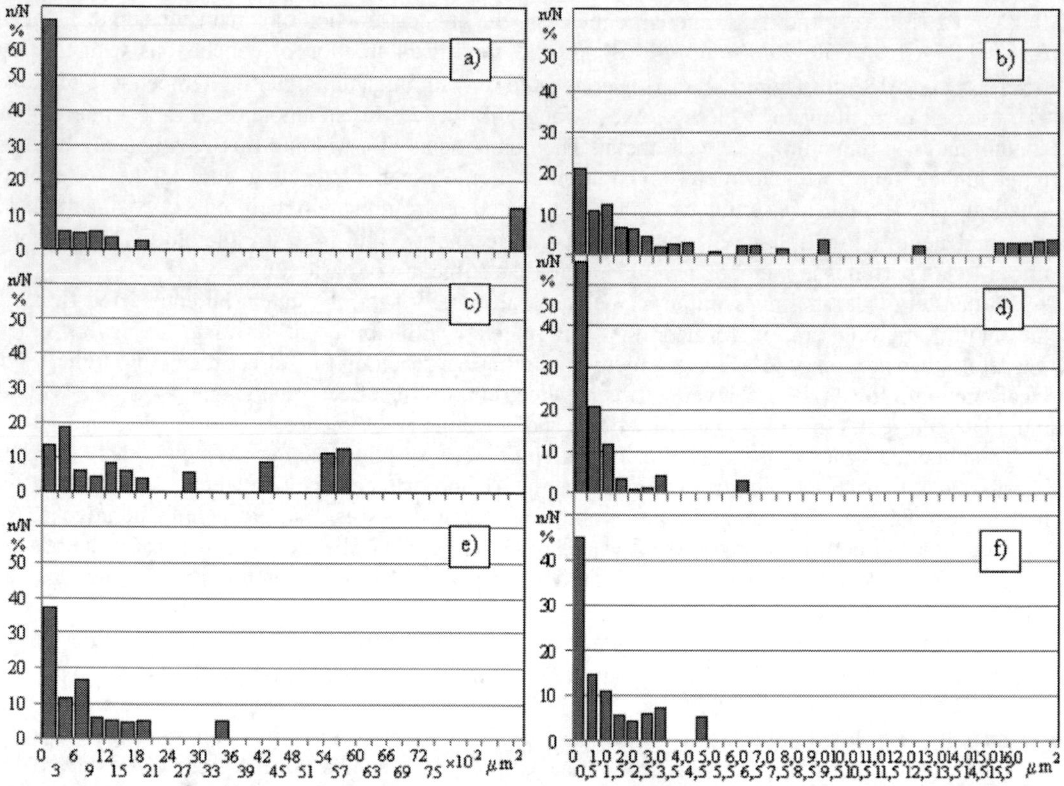

Figure 4. Histograms of the distributions non-metallic inclusions by area in the deposited weld metal. flux: a, b - AN-348A; c,d - FSA ChTA 650 -20/80; e,f - UF -01 measured, using magnification: a, c, e - ×50; b, d, f – ×1000

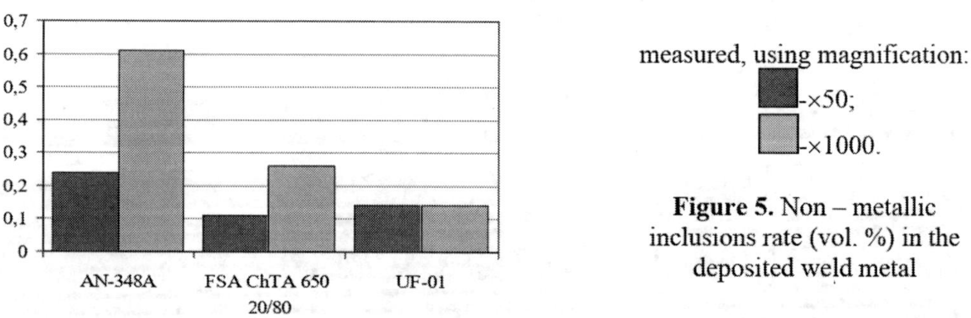

Figure 5. Non – metallic inclusions rate (vol. %) in the deposited weld metal

4. Summary

Thus, in the additive manufacturing of SAW under a flux by a wire with a diameter of 3 mm, using standard modes, a wall with a thickness of 14–18 mm forms. It has a high homogeneity and dispersibility of structure, except of the initial and final sections with the width of approximately the same as the thickness of the wall. When using 30KhaGSA wire under a layer of fluxes AN - 348A and FSA ChTA 650 -20/80 chemical composition with the least carbon content and greatest silicium and manganese content is formed. And under the flux UF-01 – with the greatest carbon content and with the least silicium and manganese content is formed, with hardly varying of the chromium content. The deposited metal has martensite – ferrite structure after a weld deposition, which stipulates higher hardness of the top passes. In bottom passes, hardness of 19 to 26 HRC is formed, which is stipulated with the growth of ferrite fraction, formation of a dispersible pearlite and tempering of a martensite because of auto thermo – cyclic treatment. The greatest contamination with non-metallic inclusions has metal, deposited under the flux AN - 348A and the least – under UF-01. Based on the results of the research flux FSA ChTA 650 -20/80 is recommended for the use. It is also necessary to improve the weld deposition technology, to remove a weld penetration cog in the center of the pass.

References

[1] Cunningham C R, Wikshaland S, Xu F, Kemakolam N, Shokrani E, Dhokia V and Newman S T 2017 Cost modelling and sensitivity analysis of wire and arc additive manufacturing *Procedia Manufacturing* vol **11** pp 650–7

[2] Williams S W, Martina F, Addison A C,.Ding J, Pardal G and Colegrove P 2016 Wire + Arc Additive Manufacturing *Materials Science and Technology* vol **32** no **7** pp 641–7

[3] Wu B, Pan Z, Ding D, Cuiuri D, Li H, Xu J and Norrish J 2018 A review of the wire arc additive manufacturing of metals: properties, defects and quality improvement *Journal of Manufacturing Processes* vol **35** pp 127–39

[4] Rodriges T A, Duarte V, Avila J A, Santos T G, Miranda R M and Oliveira J P 2019 Wire and arc additive manufacturing of HSLA steel: Effect of thermal cycles on microstructure and mechanical properties *Additive Manufacturing* vol **27** pp 440–50

[5] Nemani A V, Ghaffari M and Nasiri A 2020 Comparison of microstructural characteristics and mechanical properties of shipbuilding steel plates fabricated by conventional rolling versus wire arc additive manufacturing *Additive Manufacturing* vol **32** 101086

[6] Rafieazad M, Ghaffari M, Nemani A V et al 2019 Microstructural evolution and mechanical properties of a low-carbon low-alloy steel produced by wire arc additive manufacturing. *Int. J. Adv. Manuf. Technol* **105** pp 2121–34

[7] Xiong J, Li R, Lei Y and Chen H 2018 Heat propagation of circular thin-walled parts fabricated in additive manufacturing using gas metal arc welding *J. of Materials Processing Tech.* **251** pp 12–9

[8] Ding D, Pan Z, Cuiuri D, Li H, Duin S and Larkin N 2016 Bead modelling and implementation of adaptive MAT path in wire and arc additive manufacturing *Robotics and Computer-Integrated Manufacturing* **39** pp 32–42

[9] Rios S, Colegerove P A, Martina F and Williams S W 2018 Analytical process model for wire + arc additive manufacturing *Additive Manufacturing* **21** pp 651–7

[10] Kanjilal P, Pal T K and Majumdar S K 2006 Combined effect of flux and welding parameters on chemical composition and mechanical properties of submerged arc weld metal *Journal of Materials Processing Technology* **171** pp 223–31

[11] Shassere B, Nycz A, Noakes M V, Masuo C and Sridharan N 2019 Correlation of Microstructure and Mechanical Properties of Metal Big Area Additive Manufacturing. *Appl. Sci* **9 787**

[12] Krivonosova E A and Belinin D S 2017 Transformation non-metallic inclugions in steels after plasma heating. Bulletin of PNRPU *Mechanical engeneering, material science* vol **19** no **1** pp 58–77

[13] Krivonosova E A 2016 The features of affecting non-metallic phases on deformation behavior weld deposited layers. *Bulletin of PNRPU. Mechanical engeneering, material science* vol **18** no **1** pp 189–24

IOP Conf. Series: Materials Science and Engineering **966** (2020) 012045 doi:10.1088/1757-899X/966/1/012045

Synthetic raw materials - possibility to increase refractory materials resistance

K G Zemlyanoi[1, 2] and A R Khafizova[1]

[1]Chemical technology of ceramic and refractory department, Ural Federal University named after the First President of Russia B N Yeltsin, 19, Mira str., Yekaterinburg, 620000, Russia

E-mail: [2]kir77766617@yandex.ru

Abstract. The analysis of refractory materials and factors groups destruction types causing these destructions during operation of refractory materials is presented. One of the reasons preventing further increase of refractory resistance is shown to be natural barrier of impurities presence in natural raw materials for refractory production. Disclosed is a method for natural and/or man-made raw materials deeper enrichment by hydrochemical processing methods. This work is devoted to the chemical synthesis of raw materials enrichment for the production of refractory materials with increased resistance. The studies were carried out with the aim of developing a technology for producing high-purity aluminum powders with high physical and mechanical properties using raw materials from the Ural field and metakaolinite to produce high-purity alumina from aluminum sulfate. Thus, the authors tested the technology of complex processing of industrial and/or natural aluminosilicate materials waste using an extractable leaching solution of sulfuric acid - an ammonium hydrosulfate solution.

1. Introduction

The refractory materials production technology always develops together with the technologies' development of metallurgical processes. The refractories and lining quality is determined, and in turn determines the performance and efficiency of the thermal metallurgical units. Increasing requirements for metallurgical processes, both in terms of production intensity and metal quality, leads to increasing requirements for refractory materials and linings resistance in general [1–3], which are estimated by reducing the refractory per unit specific consumption of produced product.

Depending on the phase-structural conversion nature occurring during refractory operation in different environment and conditions, eight types of wear and tear are established:

1) high-temperature chemical corrosion;
2) thermal chipping;
3) structural strength loss (regeneration or 'aging' of the refractory);
4) melting;
5) pyroplastic deformation;
6) molten erosion;
7) gas phase sublimation and erosion;
8) mechanical destruction.

The first two types account for 80–90 % of refractory linings failure cases. In reality, there is usually simultaneous exposure to several types of wear and tear. The above phase-structural transformations type and intensity of the refractory during thermal aggregate linings operation are influenced by a large number of factors:

1) chemical and phase compliance of refractory is against an aggressive factor (melt, gas fluid composition, dust);

2) compliance of refractory structure with operating conditions (by density and strength, by gas permeability and melt permeability, heat conductivity and heat resistance);

3) compliance with the thermal unit standard operating mode;

4) refractory material thermal energy density (characterizes energy density in one cm^3 of crystalline phase taking into account its melting temperature, $(kJ\times deg)/cm^3$);

5) impurities composition and quantity in the refractory forming with aggressive chemical factor low-melting compounds and melts with low viscosity at the thermal unit working temperatures.

Compliance of the refractory material/product with some of these factors is established at the lining design and refractory type and appearance selection stages (1 and 2 factors); Impact of operating conditions (3 factors) depends only on the technological discipline of consumer. Thermal energy density is a complex indicator depending on the selected refractory phase type (such as chemical bonding, the crystal lattice parameters), the material and product obtaining method. The impurities composition and amount in the refractory depends on the raw material used (natural or synthetic) and producing method of the material and product (introduced in receiving process of the functional additives and the lighter/more fusible components removing possibility in the refractory production process).

If the first 4 factors are quite easily regulated and/or set during technological processes preparation and implementation in metallurgy and refractory products production and, theoretically, can be indefinitely improved as science develops, and the empirical observations mass increases, the 5th factor has a natural and quite obvious limit of development, to which practically all producers of refractory products have now approached. This is very well visible on the nomenclature of the main periclase and periclasocarbon refractory for black metallurgy: GOST 28874–2004 (reprinting GOST 28874–90) "Refractory. Classification" refers to periclase all refractory materials containing 85 % by weight MgO inclusive. According to GOST 4689–94 "Periclase refractory. Specification", content MgO of the refractory starts with 89 wt. %. As early as 5 years ago periclase and/or periclase carbon products with a content of 95 wt % MgO were considered qualitative. Now we use brand 97 periclase (97.0–97.5 wt. % MgO), also we speak about brand 98 desirability, and 99 [4–7] is better. But there is no natural raw material of this quality, given the basic genesis type of the magnesia raw material in nature (calcium carbonates into magnesium carbonates recrystallization in seawater) any magnesia raw material will necessarily contain some calcium ions in the solid solution in the magnesia lattice. And it is impossible to get rid of them by any mechanical and/or pyrotechnic enrichment methods. The same applies to iron (II) oxide forming solid solutions with magnesium oxide and, more broadly, to any natural raw material.

Thus, the chemical enrichment of natural and/or chemical synthesis of artificial raw materials for the production of refractory materials with increased resistance is now becoming increasingly apparent and relevant. And such examples already exist, namely, continuing a magnesian raw materials subject: a number of producers receives natural brines processing (sea water, natural bischofites) magnesium oxide with purity to 99.99 wt. % MgO (magnesia burned) for the chemical, electrometallurgical, electronic, ceramic industries, medicine. The main magnesium oxide producers from seawater are the USA, Netherlands, Japan, Mexico, Israel, Ireland, Jordan [8, 9]. And, although at present this technology cost is quite high, the hydrochemical enrichment or this technology (synthetic) production direction of basic refractory oxides (MgO, Al_2O_3, ZrO_2, Cr_2O_3) as well as carbon is seen as the only refractory materials increasing the chemical purity prospect. In addition, the hydrochemical method of enriching/producing pure refractory materials enables:

- still at the production stage cheaply and controllably alloy their structures, obtaining the required physical and chemical properties;
- involve in the clean materials production not only natural quality raw materials, but also technogenic wastes, simultaneously solving tasks of resource saving and anthropogenic impact reduction on the environment.

2. Materials and method

The authors have tested the technology of complex processing of industrial and/or natural aluminosilicate materials production wastes with the use of recoverable leaching sulphuric acid solution - ammonium hydrosulfate solution. The technology includes:

1) Sulphuric acid leaching of raw materials with ammonium hydrosulphate solution with extraction of iron III and aluminum into the solution by the following reactions:

$$2AlOOH + 6NH_4HSO_4 = Al_2(SO_4)_3 + 3(NH_4)2SO_4 + 4H_2O \; Fe_2O_3 + 6NH_4HSO_4 =$$

$$= Fe_2(SO_4)_3 + 3(NH_4)2SO_4 + 3H_2O;$$

2) Filtration of acid-resistant residue containing RO_2, R_2O_5 oxides for subsequent use as raw materials for the production of building materials, refractories, ceramics, abrasives, etc;

3) Iron III deposition from ammonia solution and separation of iron hydroxide III for further use in the production of pigments;

4) Aluminium precipitation from ammonia solution and separation of aluminium hydroxide for further use as a raw material;

5) Obtaining a pure solution of ammonium sulfate from which crystalline ammonium sulfate is extracted; according to the technology developed and tested at the pilot plant, it is thermally decomposed into ammonium hydrosulfate and ammonia by reaction:

$$(NH_4)2SO_4 = NH_4HSO_4 + NH_3 \uparrow.$$

6) Return of the obtained ammonium sulfate and ammonia into production.

The raw materials for the study were selected as kaolin waste from the Zhuravlinny Log field (Plast, Chelyabinsk Region) and Poletaevskoye field (Chelyabinsk) as raw kaolin. Phase composition of initial materials includes, wt. %: kaolinite $Al_4[Si_4O_{10}](OH)_8$ – 95.0–97.0; quartz SiO_2 – 1.0–2.0 and feldspars (microcline K, $Na(AlSi_3O_8)$) – 1.0–3.0.

The offered technology of hydrochemical processing, allowing to improve opening of initial raw materials and to increase the target product yield, consists of:

- Preliminary grinding of raw materials up to the full passage through the sieve 0.08 mm;
- Preparation of H_2SO_4 technical solution with content: 10, 15 и 20 %;
- Pelletizing of ground raw materials with addition of H_2SO_4 solution;
- Classification of the obtained granules with the separation of the fraction of 3–7 mm and the return of the remaining fractions to the stage of granulation;
- Heat treatment of the obtained granules at a temperature of 350–600 °C.

Phase composition of granulated products after heat treatment at 350 (and 600) °C is presented, wt. %: metakaolinite $Al_2O_3 \cdot 2SiO_2$ – 85.0–94.0 (14.0-66.0); quartz SiO_2 – 1.0–2.0 (13.0–33.0); hydroslides $Na,K,Al_2[AlSi_3O_{10}](OH)_2$ – 3.0–14.0 (43.0–58.0) and aluminum sulfate 0.0–2.0 (15.0–23.0). It has been established that in the process of heat treatment the materials granulated with 20 % sulphuric acid solution and heat-treated at the temperature of not less than 500 °C are subject to the greatest change; at the same time the products have a latent-crystalline, almost amorphous structure of metakaolinite $Al_2O_3 \cdot 2SiO_2$. The leaching process of the target products is based on the reaction:

$$Al_2O_3\,2SiO_2 + 3H_2SO_4 = Al_2(SO_4)_3 + 2SiO_2 + H_2O \quad (1)$$

in slow boiling mode (5–7 min) and natural cooling of the solution (50–60 min) with further decantation and washing of the sludge. Wash water is collected and directed to obtaining a leaching solution. The results of the leaching process are presented in Table 1.

Table 1. Leaching process results.

	Point No.		1	2	3	4	5	6
	Process				Leaching			
Reagents, g	Metakaolinite	10 % H_2SO_4	3	–	3		–	
		15 % H_2SO_4	–	3	–	300	450	500
	25 % H_2SO_4 solution			17.5		1748	1966	2185
Conditions	Boiling point, °C			–			90	
	Boiling point, min		50	90		50	60	90
	Excess solution H_2SO_4, %				10 %			
	Δ Drying Point, °C		207	212	258	212	306	297
	Drying Point, min		40	71	110	71	104	120
	Filter method				Decantation			
Mass of settled waters (master batch), g			375.6	308,9	301	1137	1949	3818
Mass of undissolved sediment, g		Practical	1.63	1.73	1.77	481	335	303
		Theoretical	1.62	1.62	1.62	270	270	270
Product yield (sludge), %			109	116	110	178	124	112
Additionally		pH initial/final	0.02/ 0.51	0.25/ 0.63	/0.75	/1.06	/0.94	/1.39

IOP Conf. Series: Materials Science and Engineering **966** (2020) 012045 doi:10.1088/1757-899X/966/1/012045

Based on the results of leaching, it can be concluded that the best way to achieve leaching is to use 15 % H_2SO_4 as the leaching agent.

The process of deposition of aluminum sulphate solution purified from iron oxides was performed at the temperature of solution 20–60 °C by step-by-step introduction of solution NH_4OH in the ratio 1 : 2 with pH control of the solution every 5–10 minutes to the final pH of the solution 4.0 – Table 2. After sedimentation the sludge settling, decantation and washing is performed. Drying and thermal treatment of hydroxide was carried out in laboratory furnaces at the temperature of 400–550 °C until the mass loss was stopped. The obtained product contains, wt. %: Al_2O_3 – 99.2; R_2O – 0.14; RO – 0.01; Fe_2O_3 – 0.05; SiO_2 – 0.08. The specific surface of the material was 5680 m^2/g, the average particle size was 2.7 μm.

Table 2. Precipitation results.

Point No.		1	2	3	4	5	6	7	8
Reagents, g	$Al_2(SO4)_3$ solution	588	250	500	500			100	
	NH_4OH solution	–	146	268	278	431	548	915	670
	10 % solution of NH_4NaSO_3	520	520	1040	1210	2420	2420	1815	1815
Excess 10% solution of NH_4NaSO_3			1.5			2		1.5	
Boiling point, °C					82			74	
Boiling point, min					120				
Δ Drying Point, °C		127–278	162–347	166–296	204–343	194–289	193–320	201–315	199–321
Drying Point, min		124	235	189	180	199	201	200	198
Filter method					Decantation				
Product					$Al(OH)_3$				
Undissolved sediment mass (master batch), g		–	680	–	–	–	–	–	–
Sludge mass actual, g		14	13	33	41	38	140		
Sludge mass theoretical, g		13	13	26	26	26	52		
Product yield, %		108	100	127	158	146	269		
pH after adding technical ammonia		–	4.15	4	–	–	–	–	3.7
pH master batch		–	5.66	–	–	–	–	4.01	–

3. Conclusion

The possibility of developing a technology based on the raw materials of the Ural field using metakaolinite to produce pure aluminium oxide and other products from aluminium sulphate, which will make it possible to create an energy- and resource-efficient technology to produce high-purity Al_2O_3 powders with high level of physical and mechanical properties, was demonstrated.

References

[1] Baask A, Dyubers D et al 2011 *Ferrous metals* vol **9** pp 32–39
[2] Smirnov A N 2014 Main trends in the development of the market of refractory materials [an electronic resource] access mode: http://steellab.com.ua/news/2014/01/01.php
[3] Axelrod L M 2017 *New refractory* vol **11** pp 3–13
[4] Visloguzova É A, Kashcheev I D et al 2013 *Refractories and Industrial Ceramics* vol **2** pp 83-87
[5] Goto K, Hanagari S et al 2013 Progress and Perspective of Refractory Technology *Nippon steel technical report* vol **104** pp 21–25
[6] Poirier J 2015 *Metall. Res. Technol* vol **4** p 20
[7] Smirnov A N 2014 Main trends in the development of the market of refractory materials access mode: http://steellab.com.ua/news/2014/01/01.php
[8] Shand M A 2006 *The chemistry and technology of magnesia* (John Wiley&Sons Inc) p 263
[9] Kramer D A *Minerals Year Book 2002 Magnesium Compounds*. (USGS, Reston. Virginia)
[10] Kashcheev I D, Zemlyanoi K G et al 2014 *New refractory* vol **4** pp 6–13
[11] Kashcheev I D and Zemlyanoi K G et al 2010 *New refractory* vol **10** pp 28–35
[12] Zemlyanoi K G and Stepanova K O 2017 *New refractory* vol **10** pp 31–37

IOP Conf. Series: Materials Science and Engineering **966** (2020) 012046 doi:10.1088/1757-899X/966/1/012046

A development of computer vision elements for processing the results of tensile tests of specimens and building diagrams of ultimate plasticity

M V Erpalov

Ural Federal University named after the first President of Russia B N Yeltsin, 19 Mira Street, Ekaterinburg, 620002, Russia

E-mail: m.v.erpalov@urfu.ru

Abstract. The use of specimens with a round notch allows expanding the application of tensile tests in order to study the ultimate plasticity of materials. However the main difficulty in processing the results of experiments is the need to measure the radius of curvature of the neck and the diameter in the minimum cross-section of the specimen. These dimensions determine a stress triaxiality value of the material in the center of the neck according to the Bridgman model or the other ones. This article is devoted to the development of elements of a computer vision system designed to measure the neck, which is formed both on cylindrical specimens and on specimens with a notch. To extract the contour of the specimen and to measure the forming neck, the video recording of the test process on a digital single-lens reflex (DSLR) camera is used. The paper proposes a method for recording the test process using the background, which allows to increase the accuracy of determining the contour of the specimen during subsequent image processing. The Sobel filter is used to extract the contour of the specimen, and a special neck profile equation allows to determine the neck dimensions.

1. Introduction

The ability to predict the moment of ductile fracture of materials is very important for solving practical problems of metal forming, as well as for creating adequate computer models of technological processes and products – digital twins. The value of the material ultimate plasticity is quantitatively characterized by the strain value accumulated up to the time of fracture. The plasticity of the material depends on temperature, strain rate, external hydrostatic pressure, Lode angle, and also the metal stresses value in the deformation zone. There are many criteria in literature that can be used to predict the ductile fracture moment [1–5]. Johnson-Cook and Cockcroft-Latham are the most popular of them. Despite the influence of many factors, the ultimate plasticity of materials is usually estimated in simple tests, which provide a uniform distribution of stresses and strains throughout the sample and a constant value of a stress state index. Such tests in particular include the extension of cylindrical specimens. However, after localization of deformation in the neck, the stress state of the metal becomes triaxial and must be taken into account to increase the accuracy of fracture models identification.

Another limitation of tensile tests of cylindrical specimens is the inability to change the stress triaxiality over a wide range of values, while it differs significantly for most technological processes of metal forming. For example, when pressing various profiles, compressive stresses prevail, and when drawing wires, tensile stresses prevail. The stress triaxiality values are defined as following:

$$\eta = \frac{\sigma}{\sigma_{eq}}, \tag{1}$$

where σ is the mean normal stress and σ_{eq} the equivalent von Mises stress. The use of specimens with different notches while studying the materials plasticity allows you to extend the range of stress triaxiality values [5–6]. However, the change in this parameter during tensile tests must also be taken into account to increase the accuracy of fracture models identification.

In [7], Bridgman showed a relationship between the stress triaxiality value in the minimum cross-section of specimen and the geometry of the neck:

$$\eta = \frac{1}{3} + \ln\left(1 + \frac{d}{4R}\right), \tag{2}$$

where d is the minimum cross-sectional diameter of the neck and R the radius of its curvature. In addition to the Bridgman model, other models of stress distribution in the neck are presented in the literature [8–9]. However, they also derive that the stress triaxiality is determined by the geometry of the neck. Thus, the need to take into account the stress triaxiality value as well as its change while identifying materials fracture models poses the problem of measuring the neck profile of cylindrical and notched specimens during the test process.

In works [10–11], La Rosa and Mirone have proposed a material-independent necking model without the need of experimental profile curvature measurement. However, this model does not take into account the influence of the hardening curve of a wide range of materials on the change in the neck. Therefore, the use of this model must be preceded by numerous experimental confirmations.

Focusing on the methods that use the experimental measurement of neck profile, several approaches have been proposed in [6, 12–13]. In [12], G'Shell et al. applied a tensile testing system with the computer-aided video extensometer and a true strain-rate controller. To analyze the neck profile, this system used frames of 512×512 pixels. In [6], Sancho et al. have used a single DSLR camera and an automated image processing algorithm for neck profile extraction. In order to optimize the sharpness of images, authors applied an additional light focused on specimens painted in white. In [13], they proposed an improved test procedure that used only one backlight placed behind the specimen.

In this paper, we propose our own approach to computer-aided measurement of the neck profile while testing cylindrical and notched specimens in order to determine the radius of neck curvature and the minimum diameter of the specimen in the neck.

2. The analytical description of the neck profile while testing notched specimens

The choice of an analytical expression for describing the functional dependence of the radial coordinate of points on specimen surface ρ on the corresponding axial coordinate z is an important step in the development of a computer-aided system aimed for constructing ultimate plasticity diagrams. Fitted equation of the neck profile allows you to calculate the radius of the neck curvature in the minimum cross-section with the use of the first curvature of a space curve at a point:

$$R = \frac{\left[1 + \left(\rho'(z_0)^2\right)\right]^{\frac{3}{2}}}{\left|\rho''(z_0)\right|}, \tag{3}$$

where $\rho'(z_0)$ and $\rho''(z_0)$ are the first and second derivatives of the neck profile equation, $z_0 = 0$ is the axial coordinate of minimum cross-section.

In work [10], La Rosa and Mirone have proposed fitting techniques that use the approximation of the neck profile by way of second or third order polynomials. Nevertheless, in this paper, it is proposed to use a special type of equation, which, as shown in [14], describes the profile of necked cylindrical specimen along the entire gauge length rather well:

$$\rho = \frac{d_1}{2} - \frac{d_1 - d}{2} \cdot \left(1 + \frac{z^2}{c}\right)^{-1}, \tag{4}$$

where ρ and z are radial and axial coordinates of points on surface of specimen respectively, c is the parameter sensitive to material properties, d – the minimum neck diameter and d_1 – the diameter of the specimen at the beginning of necking.

It should be noted that equation (4) does not allow approximating the entire surface of the notched specimen, in the smallest section of which the neck is formed. However, the graph of equation (4) has exactly two inflection points between which the neck is approximated by the proposed equation with high accuracy.

In this work, to estimate the accuracy of the neck profile approximation, the computer simulation has been performed in the Deform-2D software. We studied the shape change of the neck on standard five-fold specimens (according to the Russian state standard GOST no. 1497), as well as specimens having a radial notch according to figure 1. The dimensions of the notched specimens are listed in table 1.

Table 1. Characteristics of the analysed specimens.

Specimen number	Specimen characteristics	Notch acuity, d_0/R_0	Curvature, R_0 (mm)	Diameter, d_0 (mm)	Theoretical triaxiality, η_0
1	Smooth specimen	0	∞		0.333
2		0.5	12		0.451
3	Notched specimen	1	6	6	0.556
4		2	3		0.739

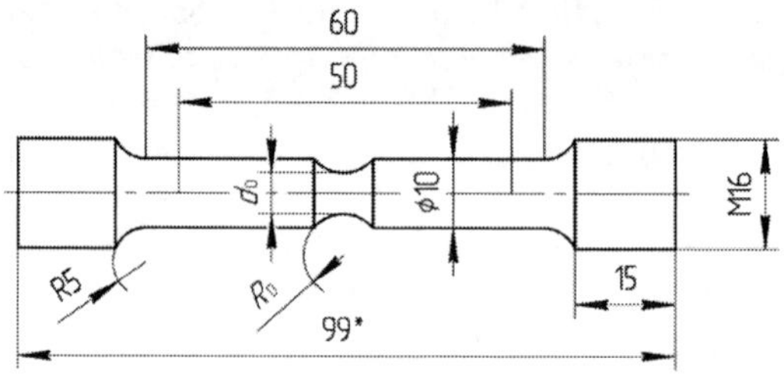

Figure 1. The geometry of specimens having the round notch.

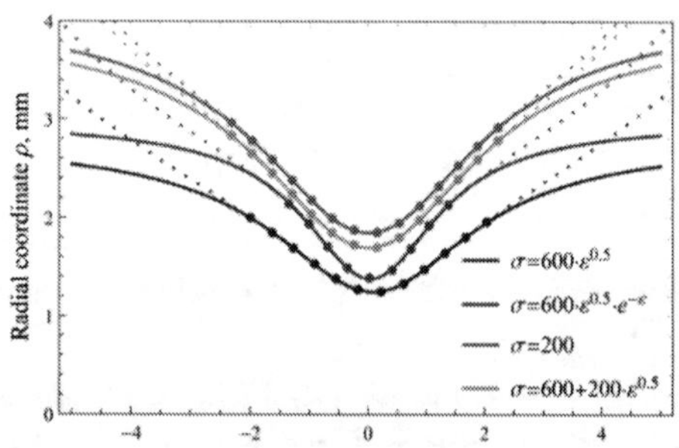

Figure 2. The comparison of fitted curves with FEM results for specimen with notch acuity 1, made from materials with different stress-strain curve.

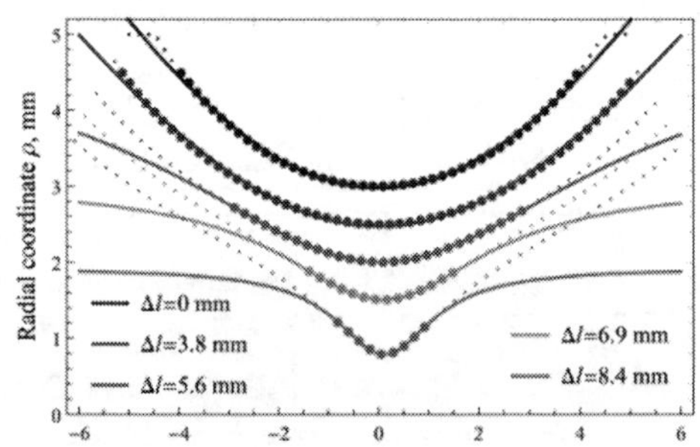

Figure 3. The comparison of fitted curves with FEM results for specimen with notch acuity 1, made from material with stress-strain curve $\sigma = 600 \cdot \varepsilon^{0.5}$ for different instants of test.

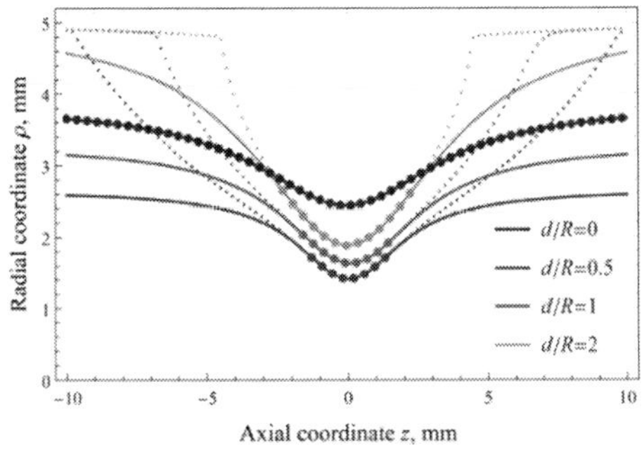

Figure 4. The comparison of fitted curves with FEM results for specimens with different notch acuity, made from material with stress-strain curve $\sigma = 600 \cdot \varepsilon^{0.5}$.

Figures 2–4 represent graphs of approximating curves according to equation (4). These curves are shown by continuous lines in comparison with the actual neck profiles that are shown by points and found during computer simulation. Large dots on these figures correspond to the points on the specimen surface that were only used to approximate the neck profile.

The results of computer simulation show that a high accuracy of the neck profile approximation is achieved when you use the only points on the specimen surface lied between the inflections. This is true in the case of using materials with different hardening curves (figure 2), for any instant of the test (figure 3), as well as when using specimens with different notch acuity (figure 4). Hardening curves of test materials are shown in figure 2.

The approximation accuracy was estimated as the maximum difference between the actual radial coordinate of any point on the specimen surface and the corresponding calculated value in accordance with the fitted equation (4). According to figure 2, the maximum difference was 0.018 mm; according to figure 3–0.11 mm; according to figure 4–0.05 mm. The maximum error occurs at the beginning of the test of the notch specimens, when the neck is still not evident. However, the error value is rather small.

3. The computer vision system for measuring the neck profile

The computer vision systems are being used today in a wide variety of real-world applications [15], which include, for example, optical character recognition, machine inspection, automotive safety, fingerprint recognition, face detection, etc. An automated edge detection is one of them which allows performing quantitative image analysis.

3.1. Neck profile extraction

The use of optical measurement systems is a good solution for obtaining neck size data during the test for a number of reasons. Firstly, you get the results of a neck measurement without the need to remove a specimen from the grips of a test setup. Secondly, the number and accuracy of measurements are determined by the algorithms for obtaining and processing images that are embedded in the computer vision system. At the same time, the influence of the human factor on the accuracy of the data obtained is reduced. Finally, the speed of processing the test results increases significantly.

In this work, we have developed a system for optical measurement of the neck profile, which is also formed on specimens with a notch. To obtain the coordinates of the points corresponding to the specimen surface, we use a single DSLR camera Canon 700D. In the case of testing specimens with a notch, when the neck size changes quickly, the test process is better to record in a video file. Extracting individual frames from a video file is carried out using the FFmpeg software. For automated processing of a large number of individual frames, a specialized program is written in the Wolfram Language.

Extracting the contour of the neck consists of several stages. At the first stage, on each frame, an area is defined on which a specimen is necked. At the second stage, the specimen axis is searched and the image is rotated so that the axis found coincides with the vertical. At the third stage, a filter is applied to separate the specimen from the background. Finally, the neck contour is extracted and the received data is processed.

One of the most difficult for the software implementation is the filtering algorithm that allows you to separate the specimen from the background. To increase the contrast of the image and the accuracy of the edge detection, a colored backdrop can be used. The program provides various background options – the commonly used green backdrop, the blue, red, the black or the white ones. In the last two cases, the recording of the test process can be carried out similarly to [6] and [13], when the specimen is painted or backlighted, respectively. Figure 5 represents an example of the own chroma key filter tuning. The filter is tuned by several manipulators that set the illumination levels of the background or specimen for each color channel separately or together (grayscale mode) depending on the user's choice.

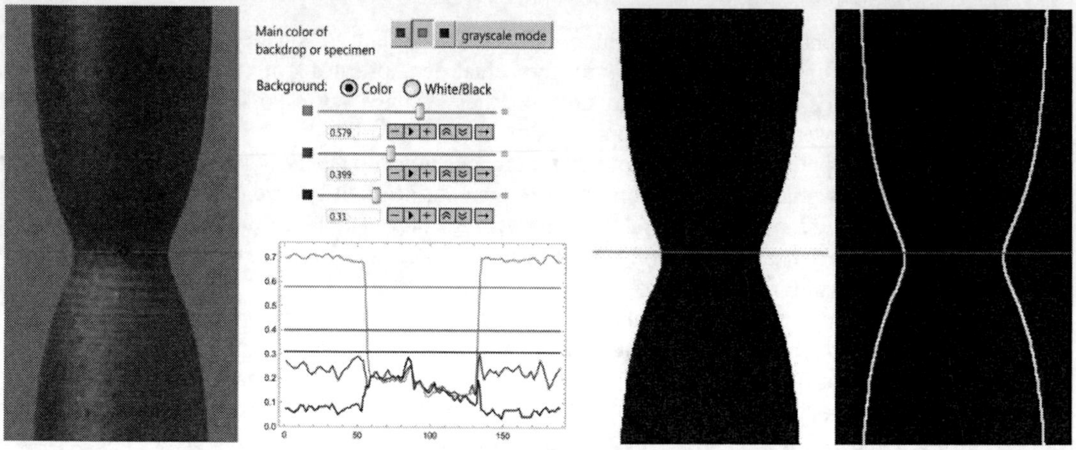

Figure 5. The tuning of the chroma key filter.

To extract neck profile, the image with a previously removed background is subjected to binarization, and then to processing by the gradient filter. We have considered the possibility of applying different gradient filters such as Prewitt, Sobel and Scharr, which differ in the coefficients of the convolution mask. It is established that the Sobel filter provides the most accurate result. In addition, a convolution mask with a size of 3×3 pixels provides white coloring of two pixels located around the actual specimen surface from different sides. One pixel is located on the side of the specimen, and the other one on the side of the background. Further, the approximation of white pixels using the proposed equation (4) and the least-squares technique is carried out in such a way that the fitted curve is located between the indicated white pixels, i.e. closest to the actual specimen surface. This approach is an alternative to the sub-pixel accuracy algorithm [13].

3.2. Refining the boundaries of the neck used for approximation

The results of computer simulation of tensile testing of the specimens with a round notch have shown that only those points on the specimen surface that are located between the nearest inflections should be used in order to achieve the most accurate neck profile approximation. The presence of only two inflection points in the graph of equation (4) makes it possible to carry out an automatic iterative search of the "neck boundaries" which is a part of the specimen surface only used for approximation. To do this, it is necessary to minimize the difference between the axial coordinates of the neck boundaries z_b and the axial coordinates of the inflection points:

$$z_b - \left(\frac{c}{3}\right)^{1/2} \to \min. \qquad (5)$$

In this work, the entire surface of the specimen part indicated on the frames previously as the region of interest is used for approximation in the first iteration. The found value of the parameter c, which is included in equation (4), allows to calculate the axial coordinates of the inflection points and compare them with the boundaries of the neck. For subsequent iterations of approximation, only those surface points are used that are between the calculated inflection points. The approximation procedure is repeated until the neck boundaries in the notch become inflection points of equation (4) with a given accuracy. Figure 6 presents an example of such an iteration.

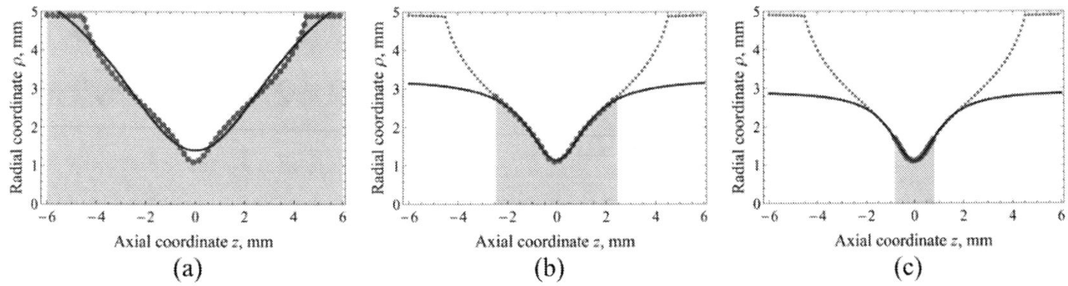

Figure 6. Automatic search for "neck boundaries" based on the coordinates of the inflection points: (a) the first iteration, (b) the second iteration, (c) the third iteration.

4. Conclusions

The measurement of the radius of curvature and the diameter of specimen in the minimum cross section of the neck is based on the approximation of experimental data using the proposed equation. The study has shown that the proposed equation is effective. To automate the processing of video records of the test process and the measuring of the neck, the program is developed in the Wolfram Language. The program for processing the test results includes algorithms widely used in computer vision systems. This program is able to determine the profile of the neck, which is formed both on cylindrical specimens and on specimens with a radial notch at any instant of the test. The found dimensions of the neck are necessary for building diagrams of the materials ultimate plasticity.

Acknowledgments

This work was supported by the grant MK–1878.2020.8 from the President of the Russian Federation for young scientists – candidates of sciences.

References

[1] Cockcroft M G and Latham D J 1968 *J. Inst. Metals* **96** 33–9
[2] Gurson A L 1977 *J. Eng. Mater. Technol.* **99** 2–15
[3] Mecking H and Kocks U 1981 *Acta Metall.* **29** 1865–75
[4] Johnson G R and Cook W H 1985 *Eng. Fract. Mech.* **21** 31–48
[5] Bai Y and Wierzbicki T 2008 *Int. J. Plasticity* **24** 1071–96
[6] Sancho A, Cox M J, Cartwright T, Davies C M, Hooper P A and Dear J P 2019 *Int. J. Solids Struct.* **176–177** 191–206
[7] Bridgman P W 1952 *Studies in Large Plastic Flow and Fracture, with Special Emphasis on the Effect of Hydrostatic Pressure* (New York: McGraw-Hill) p 362
[8] Davidenkov N N and Spiridonova N I 1946 *Proc. ASTM* **46** 1147–59
[9] Ostsemin A A 2009 *Strength of Materials* **41** 356–62
[10] La Rosa G Mirone G and Risitano A 2003 *Metall. Mater. Trans.* A **34A** 615–24
[11] Mirone G 2004 *Int. J. Solids Struct.* **41** 3545–64
[12] G'Sell C, Hiver J M, Dahoun A and Souahi A 1992 *J. Mater. Sci.* **27** 5031–9
[13] Sancho A, Cox M J, Aldrich-Smith G, Cartwright T, Davies C M, Hooper P A and Dear J P 2018 *EPJ Web Conf.* **183** 02063
[14] Erpalov M and Nukhov D 2020 *Mater. Sci. Forum* **989** 372–7
[15] Szeliski R 2010 *Computer Vision: Algorithms and Applications* (London: Springer) p 812

IOP Conf. Series: Materials Science and Engineering **966** (2020) 012047 doi:10.1088/1757-899X/966/1/012047

The possibility evaluation of zinc sulphate forms extraction from sludge of PJSC EVRAZ NTMK with following metalloflux obtaining

O Yu Sheshukov[1,2], M A Mikheenkov[1], I V Nekrasov[1,2] and D K Egiazaryan[1,2]

[1] IMET UB RAS, 101, Amundsena str, Ekaterinburg, 620016, Russia
[2] Ural Federal University named after the First President of Russia B N Yeltsin, 19, Mira str., Yekaterinburg, 620000, Russia

E-mail: o.j.sheshukov@urfu.ru

Abstract. EVRAZ NTMK JSC has significant amount of blast furnace sludges, that accumulated in dumps. Chemical and phase analysis of that sludges was carried out. It has shown significant amount of zinc sulphate, which prevents such sludge utilization in a blast furnace. Moreover, zinc sulphate requires a high temperature unit for recycling by common schemes. The possibility of zinc sulphate conversation into oxidized zinc with simultaneous metallized residue obtaining under the common for Waelz process temperatures was confirmed in laboratory. The obtained during laboratory experiments metallized residue is suitable for using in blast furnace as charge addition. The results of a thermodynamic analysis of the zinc sulphate conversion to oxide and the physicochemical basis for the extraction of such zinc forms are presented. Theoretical calculations and laboratory experiments confirmed that sludges with zinc sulphate could be suitable materials for blast furnace. But, sludges of different processes allow to obtain fired briquettes with different physical properties.

1. Introduction

A significant amount of sludge from the blast furnace shop has been accumulated on the EVRAZ NTMK JSC. Due to the impossibility of processing the open-hearth shop aspiration dust, it was also stored at the plant, despite the fact that the open-hearth production at the plant is closed. The accumulated reserves amount to several million tons. The easiest way to process these sludges is to mix them to raw materials of blast furnace. However, this processing method is not possible, due to zinc containing in these sludges, which leads to the formation of zinc crust in the upper part of the blast furnace. Zinc crust complicate the flow of the blast furnace process and their removal is possible only with overhaul of the blast furnace.

If zinc is removed from the sludge, then it is possible to use it in a blast furnace charge. It is desirable to increase the sludge metallization simultaneously with the zinc removal. However, the zinc removal by traditional methods [1, 2, 3] is difficult, due to zinc in the sludge is in sulfate form.

2. Materials and experimental method

To assess the possibility of removing zinc from the sludge of EVRAZ NTMK JSC, the company provided two samples of sludge: from the blast furnace shop and from the open-hearth furnace dust aspiration system. The chemical analysis is presented in Table 1. Figures 1 and 2 show the phase analysis of that sludges. Qualitative x-ray phase analysis was carried out on an XRD-7000 diffractometer (Shimadzu).

Table 1. The chemical composition of the provided sludges.

Sludge	Content, mass. %									
	CaO	SiO_2	Al_2O_3	MgO	MnO	V_2O_5	Fe_{all}	Zn	S	C
Blast furnace shop	6.9	6.2	0.3	2.2	0.37	0.7	55.2	5.1	1.1	9.8
Open-hearth furnace dust aspiration system	0.9	0.7	0.45	1.1	1.4	0.1	73.9	5.5	0.4	0

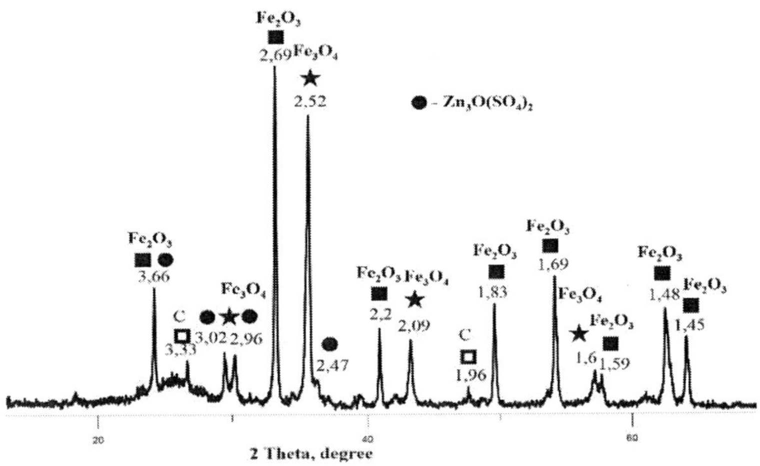

Figure 1. Phase composition of blast furnace shop sludge.

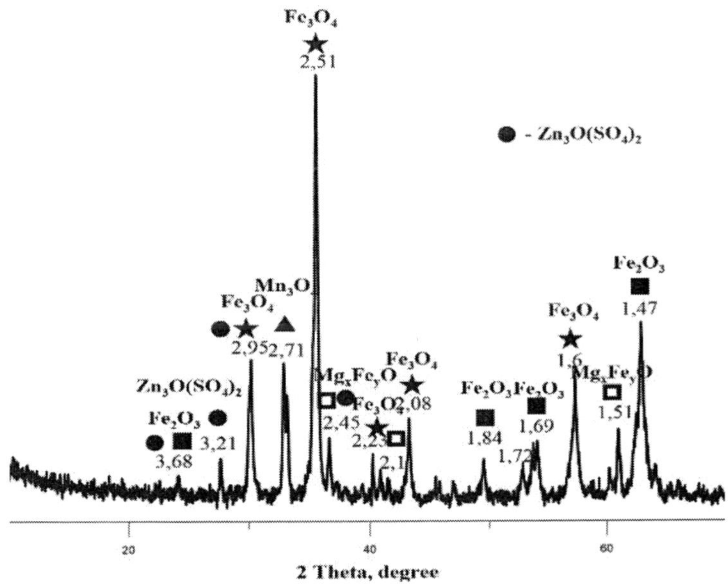

Figure 2. Phase composition of open-hearth furnace dust aspiration system sludge.

IOP Conf. Series: Materials Science and Engineering **966** (2020) 012047 doi:10.1088/1757-899X/966/1/012047

To determine the metallization possibility of the sludge and the removal of zinc according to the developed technology, limestone and coke were mixed with sludges. The raw material mixture was homogenized by co-grinding and pressed. Briquettes were fired according to the proved regime.

3. Thermodynamic calculations and experimental results

According to phase analysis, the sludge is mainly composed of iron oxides. Blast furnace sludge contains carbon in the coke form. Zinc in the sludge is present in complex sulphate form $ZnO(SO_4)_2$. The extraction of such zinc from sludge is difficult, due to under reducing conditions above 100 °C, sulphate zinc is converted to sulphide zinc by the reaction 1. Reduction of zinc sulphide by carbon is possible by the reaction 2.

$$ZnO(SO_4)_2 + 9C = ZnS + 9CO\uparrow, \quad (1)$$

$$2ZnS + C = 2Zn\uparrow + CS_2. \quad (2)$$

The results of thermodynamic calculations indicate that reaction 3 is possible only above 1900 °C. To realize such reaction in practice, the use of special high-temperature units will be required, which is not economically feasible.

To convert zinc from the sulphide form to the oxide form, we proposed adding limestone, magnesite, or dolomite to the raw mix before firing [4, 5]. When using limestone to decompose zinc sulphide and its sublimation, reaction 3 is proceeds.

$$2ZnS + 2CaCO_3 + C = 2Zn\uparrow + 2CaS + 3CO_2\uparrow. \quad (3)$$

The results of thermodynamic calculations indicate that reaction 3 is possible at above 1100 °C.

The composition of the raw mix was adjusted in accordance with the proved methodology and fired. The firing products chemical analysis are shown in table 2.

Table 2. The chemical composition of fired briquettes.

Sludge	Content, mass. %									
	CaO	SiO_2	Al_2O_3	MgO	MnO	V_2O_5	Fe_{all}	Zn	S	C
Blast furnace shop	19.9	20.7	4.8	5.9	0.18	0.7	50.5	0.14	2.7	0
Open-hearth furnace dust aspiration system	19.3	20.3	5.4	4.2	2.6	0.3	50.9	0.78	1.1	0

4. Results discussion

According to chemical analysis, the content of silicate materials increased in firing products, while total iron decreased. The zinc content decreased from 5 % to a tenth of a percent. After firing, blast furnace sludge briquettes dispersed, and open-hearth sludge briquettes retain their shape and are well metallized. The phase analysis for the open-hearth sludge briquettes after firing is shown in figure 3.

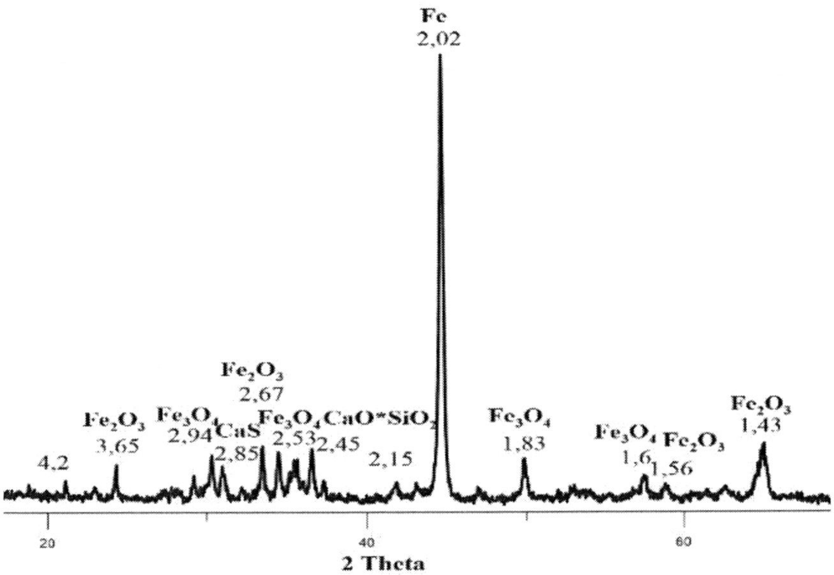

Figure 3. Phase composition of open-hearth furnace sludge briquettes after firing.

According to the Figure 3, it can be seen that the firing products contain a significant amount of metallic iron. The degree of metallization was 72 %. Zinc in the firing products is absent.

5. Conclusion

The test results indicate that the zinc extraction from the JSC EVRAZ NTMK sludge with subsequent metallization of remains is possible.

Acknowledgments

The reported study was funded by RFBR according to the research project № 18-29-24064\18.

References

[1] Global steel dust. Waelz Kiln technology: [electronic source]. United States US Global Steel Dust Inc. 109 N. Post Oak Lane #415 Houston, TX 77024: http://www.globalsteeldust.com/waelz_kiln_technology (accessed on 01.06.2020)

[2] Zinc Ox Korean Recycling Plant: [electronic source]. Crown House, High Street Hartley Wintney Hampshire RG27 8NW United Kingdom: https://zincox.com/korean-recycling-plant/ (accessed on 01.06.2020)

[3] Gregurek D, Peng Z and Wenzl C 2015 *The J. of the Minerals, Metals & Materials Society* vol **67** no **9** pp 1986–7

[4] Mikheenkov M A, Sheshukov O Yu, Nekrasov I V and Egiazaryan D K 2019 *Abstr. of the XXI Mendeleev Cong. on General and Applied Chemistry* (St. Petersburg) vol **3** p **80**

[5] Mikheenkov M A, Sheshukov O Yu, Nekrasov I V and Egiazaryan D K 2019 *Proc. of the Cong. with int. participation and the conf. of young scientists "Fundamental research and applied development of processes for processing and utilization of technogenic formations": "TECHNOGEN-2019"* – (Ekaterinburg: Ural Branch of the Russian Academy of Sciences) pp 368–72

IOP Conf. Series: Materials Science and Engineering **966** (2020) 012048 doi:10.1088/1757-899X/966/1/012048

Reducing the carbonitride formation risk with increasing titanium dioxide content in the blast furnaces charge of JSC EVRAZ NTMK

O Yu Sheshukov[1,2], M A Mikheenkov[1], I V Nekrasov[1,2] and D K Egiazaryan[1,2]

[1] IMET UB RAS, 101, Amundsena str, Ekaterinburg, 620016, Russia
[2] Ural Federal University named after the First President of Russia B N Yeltsin, 19, Mira str., Yekaterinburg, 620000, Russia

E-mail: o.j.sheshukov@urfu.ru

Abstract. Analyse of the blast furnace charge chemical content changing with the JSC EVRAZ NTMK raw materials base alteration was carried out. That expected change would result in TiO_2 content increasing in the blast furnace slags. That increasing would lead to carbonitride formation in the upper levels of the blast furnace, which reduce technical parameters of the blast furnace operation by reducing effective volume of the furnace and derange regular charge moving. Possible ways of that problem solving was shown. It was proposed the complex technology, that affect 3 technological stages: agglomeration, blast furnace process and out-of-furnace treatment of steel. Converting of TiO_2 into $CaO{\cdot}TiO_2$ allows to delay carbonitride formation due to increasing required temperature for that process. In that case, carbonitride formation would move to the tuyere zone of blast furnace, where it increase lining durability. The proposed technology was industrially tested, and the main idea was proved. Few problems were found during industrial tests, but theoretical solving of them was proposed.

1. Introduction

At present, JSC EVRAZ NTMK processes titanomagnetites of the Gusevogorskoe field. In the near future it is planned to develop the actual Kachkanarskoe field (AKF) with processing of its ores [1].

In table 1, according to [1], a comparative chemical composition of the initial ore and concentrates of vanadium-containing deposits of the Urals is given.

Table 1. The material chemical composition of some Ural vanadium-containing deposits.

Field	Content, mass. %					
	Ore			Concentrate		
	Fe	V_2O_5	TiO_2	Fe	V_2O_5	TiO_2
Gusevogorskoe	16.6	0.13	1.23	61.5	0.59	2.50
Kachkanarskoe	16.6	0.14	1.24	63.0	0.60	3.60
Medvedevskoe	24.0	0.25	7.00	60.0	0.7–0.8	10.15
Kopanskoe	36.7	0.45	9.90	54.60	0.7–0.9	8.13

In accordance with the table, after starting developing of the Kachkanarskoe field, the TiO_2 content in concentrate will increase from 2.5 to 3.6 mass. %. According to the same source, the IMET UB RAS carried out of blast furnace smelting calculations with using of Gusevogorskoe and Kachkanarskoe fields agglomerated concentrates of by the technology of JSC EVRAZ NTMK. The Results are summarized in figure 1 and table 2. According to the presented data, the TiO_2 content in the slag will increase from 10.53 % to 14.92 %.

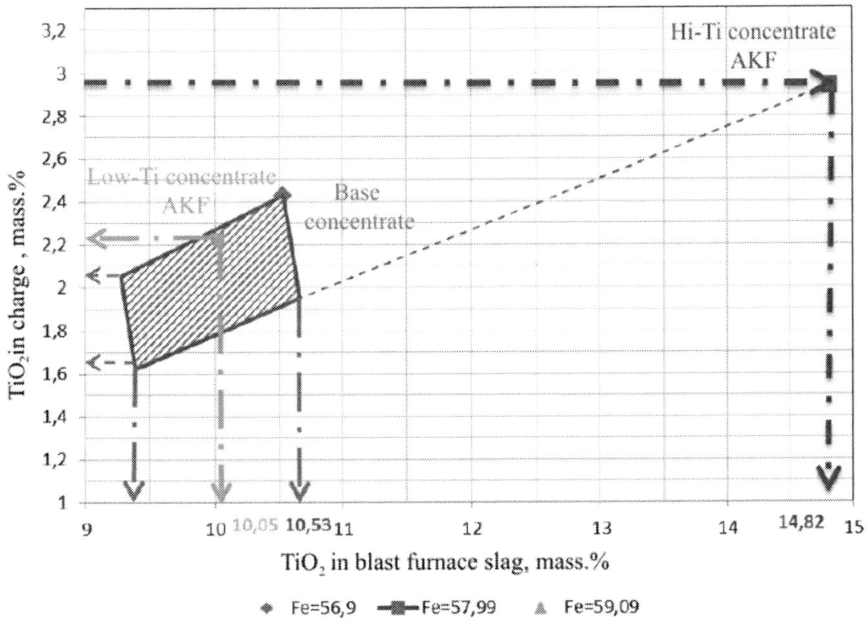

Figure 1. Effect of TiO_2 content in a blast furnace charge on TiO_2 content in slag.

Table 2. Effect of TiO2 content in a blast furnace charge on TiO2 content in slag.

Parameter Blast furnace charge	Base concentrate (Gusevogor skoe field)	Hi-Ti concentrate (AKF, main pit)	Low-Ti concentrate (AKF, west pit)
Agglomerate, %	38.3	38.3	38.3
Pellets, %	52.3	52.3	52.3
Iron-flux, %	9.4	9.4	9.4
Fe_{all}, %	56.9	57.99	59.09
TiO_2, %	2.43	2.94	2.25
Blast furnace slag			
TiO_2, %	10.53	14.92	10.05

Table 3 shows the comparative chemical composition of titanic blast furnace slag of JSC EVRAZ NTMK according to 2016 data and the forecast for the transition to high-titanium raw materials.

Due to the increase in the titanium dioxide content in the slag, the titanium carbonitride formation risk in a blast furnace increases [2]. The technology of titanomagnetite processing with vanadium extraction operating at EVRAZ-NTMK JSC is unique. It is implemented only in this enterprise in Russia. At present, the problem of preventing carbonitride formation is solved by introducing a significant amount of fluorite into the blast furnace slag and monitoring the blast furnace lining thickness [3, 4]. Such a path requires constant monitoring of the lining condition.

IOP Conf. Series: Materials Science and Engineering **966** (2020) 012048 doi:10.1088/1757-899X/966/1/012048

Table 3. The chemical composition of blast furnace slag.

Technology variant	Content, mass. %								Basicity*
	CaO	SiO_2	Al_2O_3	FeO	TiO_2	MgO	MnO	V_2O_5	
Regular technology	32.75	27.06	15.12	0.63	10.37	13.72	0.6	0.24	1.21
Forecasting technology	31.60	26.70	14.70	0.60	14.40	11.40	0.40	0.16	1.18

* – CaO/SiO_2

2. Phase content and thermodynamic calculations

We propose a technical solution to prevent carbonitride formation by adjusting the slag phase composition. In the CaO-SiO_2-TiO_2 diagram, shown in Figure 2, with Al_2O_3 content of about 20 %, the JSC EVRAZ-NTMK blast furnace slags are currently located in region A. With the titanium dioxide increasing, they will shift to region B.

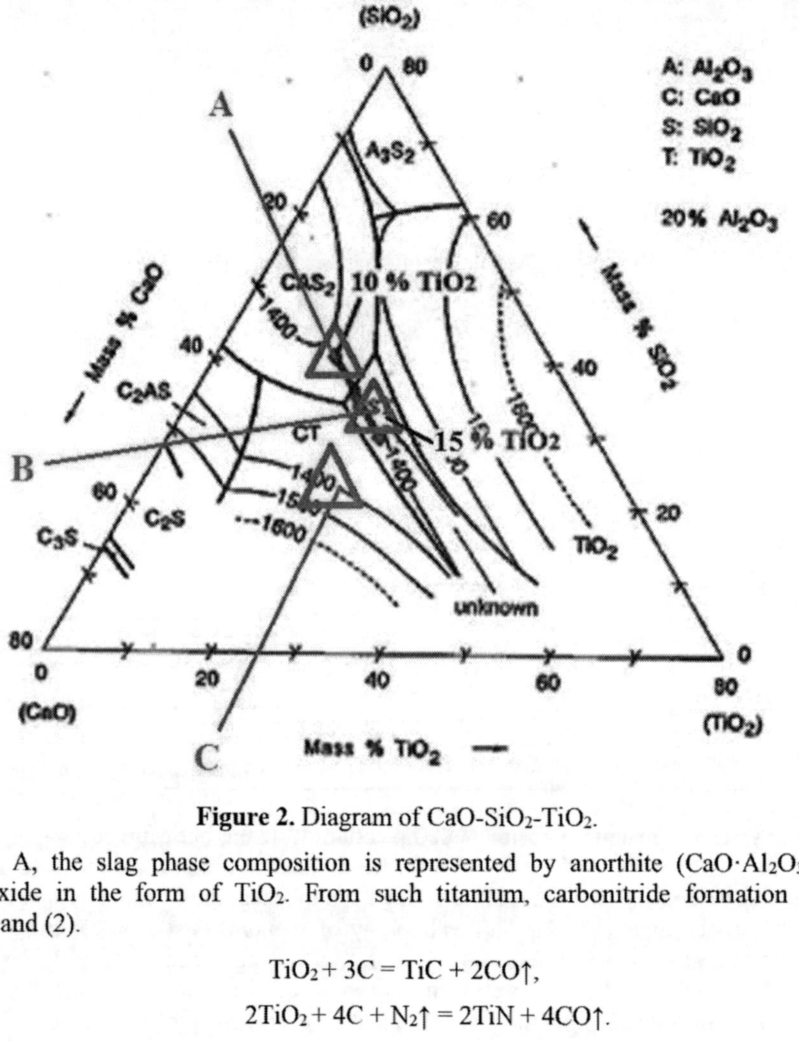

Figure 2. Diagram of CaO-SiO_2-TiO_2.

In region A, the slag phase composition is represented by anorthite ($CaO\cdot Al_2O_3\cdot 2SiO_2$) and titanium dioxide in the form of TiO_2. From such titanium, carbonitride formation proceeds by reactions (1) and (2).

$$TiO_2 + 3C = TiC + 2CO\uparrow, \quad (1)$$

$$2TiO_2 + 4C + N_2\uparrow = 2TiN + 4CO\uparrow. \quad (2)$$

Thermodynamic calculations indicate that reaction (1) is possible from 1300 °C, and reaction (2) – from 1200 °C. When slag compositions are moved to region B, titanium dioxide is converted to sphene ($CaO{\cdot}TiO_2{\cdot}SiO_2$). The titanium carbonitride formation becomes possible by reactions (3) and (4).

$$CaO{\cdot}TiO_2{\cdot}SiO_2 + 3C = TiC + 2CO\uparrow + CaO{\cdot}SiO_2, \quad (3)$$

$$2(CaO{\cdot}TiO_2{\cdot}SiO_2) + 4C + N_2\uparrow = 2TiN + 4CO\uparrow + 2(CaO{\cdot}SiO_2). \quad (4)$$

In this case, the risk of carbonitride formation remains, since according to thermodynamic calculations, the start the titanium carbo-nitrides formation by reactions (3) and (4) is shifted to a higher temperature region by only 100 °C (reaction (3) – 1400 °C, and reaction (4) – 1300 °C).

3. Proposed technology

To reduce that risk, it is proposed to convert titanium dioxide to perovskite ($CaO{\cdot}TiO_2$) at the agglomeration stage. In that case, carbonitride formation will proceed according to the reactions (5) and (6)

$$CaO{\cdot}TiO_2 + 3C = CaO + TiC + 2CO\uparrow, \quad (5)$$

$$2CaO{\cdot}TiO_2 + 4C + N2\uparrow = 2TiN + 2CaO + 4CO\uparrow. \quad (6)$$

Reactions (5) and (6) are possible only above 1600 °C, i.e. carbonitride formation will shift to the tuyere zone of blast furnace. Here it could provide a positive effect by formation of a refractory slag coating (skull).

At the same time, the basicity of blast furnace slags will increase, and they will shift to the region of perovskite existence (region C). An increase in the slag basicity will lead to an increase in their melting point and heterogeneity, which can have a negative effect on the domain process.

According to [5], blast furnace slags should contain manganese oxides, which destroy titanium carbides by reaction (7).

$$2MnO + TiC = 2Mn + TiO + CO\uparrow. \quad (7)$$

In the same work, it is noted that the MnO effectiveness in the titanium carbide formation prevention can be improved by reducing the slag viscosity. That could be achieved by addition of liquifying additives into the mixture. For example, it could be a boron-containing materials.

4. Technology verification

Our group in cooperation with EVRAZ-NTMK JSC conducted tests to replace limestone with slags of ladle-furnace steel processing (LF-slags) during agglomeration [6]. This replacement provided:

- increase the specific productivity of agglomeration machines by 4.3 %;
- increase the strength index $B_{+5\,mm}$ – by 0.3 %;
- reduce the consumption of coke breeze by 10 %.

However, these tests revealed the disadvantages of this method. LF-slags contain a significant amount of metal shots and large metal scraps. That hampers infusion of LF-slags into the agglomerate.

Our group have developed methods for the LF-slags complex stabilization [7]. In the case of using boron additives, that methods can not only stabilize the slag, but also reduce its viscosity. That will result in reducing of metal shots in LF slags. It is proposed to use processed slag during agglomeration. Along with the titanium dioxide conversion to perovskite, that approach will reduce the slag viscosity and heterogeneity.

5. Conclusion

The proposed technology affects three processes of JSC EVRAZ-NTMK: ladle-furnace steel treatment, agglomeration and blast furnace process; and allows to reduce the risk of carbonitride formation during the plant raw material base changing to high-titanium raw materials.

IOP Conf. Series: Materials Science and Engineering **966** (2020) 012048 doi:10.1088/1757-899X/966/1/012048

References

[1] Petukhov R V, Dmitriev A N et al 2014 Metallurgical evaluation of ores and concentrates of the Kachkanarskoe field. *Conf. proc. The creative heritage of Grum-Grzhimailo* (Ekaterinburg) pp 90–95.

[2] Smirnov L A, Deryabin Yu A and Shavrin S V 1990 *Metallurgical processing of vanadium-containing titanomagnetites* (Metallurgy) p 256 [In Russian]

[3] Dmitriev A 2019 *Mathematical Modeling of the Blast Furnace Process* (Cambridge Scholars Publishing) p 209

[4] Nisioka K, Ujisava Y and Takatani K 2018 Development of Mathematical Models for Blast Furnaces *Nippon Steel & Sumitomo Metal Technical Report* No **120** pp 69–75

[5] Tleugabulov B S 2009 Improving the slag regime of blast furnace smelting through the use of modifying additives *Abstract of the PhD dissertation*

[6] Zazhigaev P A, Sheshukov O Yu, Mikheenkov M A, Metelkin A A, Mironov K V, Saveliev M V, Forshev A A, Mamonov A L and Temnikov V V 2019 The use of out-of-furnace steel slag in agglomeration *Steel* no **12** pp 78–79

[7] Sheshukov O Yu, Mikheenkov M A, Nekrasov I V, Egiazaryan D K, Metelkin A A and Shevchenko O I 2015 *Issues of utilization of refining slags of steelmaking production* (Nizhny Tagil: NTI (branch) UrFU) p 153

IOP Conf. Series: Materials Science and Engineering **966** (2020) 012049 doi:10.1088/1757-899X/966/1/012049

The effect of thermoplastic processing on the structure and mechanical properties of metastable austenitic steel

N N Ozerets

Ural Federal University named after the First President of Russia B N Yeltsin, 19, Mira str., Yekaterinburg, 620000, Russia

E-mail: ozerets@mail.ru

Abstract. The patterns of structural transformations and the change in the mechanical properties of a metastable austenitic steel subjected to cold plastic deformation by drawing with high total degrees of compression and subsequent aging were studied in this work. The studies showed that after quenching, the use of intense plastic deformation by drawing revealed the high technological capabilities of the studied austenitic steel. Austenite in the steel is deformation-unstable and turns into deformation martensite during cold plastic deformation by drawing. Aging of deformed steel causes an additional increase in mechanical properties, due to the separation of the intermetallic phase NiAl from the bcc solid solution (strain martensite). It was revealed that almost carbon-free corrosion-resistant austenitic steel, as a result of correctly selected alloying, combines the advantages of three steels: metastable austenitic steels, Transformation-Induced Plasticity (TRIP)-steels and martensitic aging steels. As a result of using all possible hardening mechanisms, a high-strength state was achieved.

1. Introduction

The use of high-strength, corrosion-resistant steels makes it possible to obtain products with high service properties, to reduce metal loss, and to reduce the metal consumption of products. Martensitic, austenitic and austenitic-ferritic grades are used for the production of corrosion-resistant wire, which is employed in instrumentation, mechanical engineering, and the manufacture of medical instruments. The specific working conditions of most elastic components, as well as the core medical instrument, require the use of steels and alloys with a high level of strength, considerable elasticity, sufficient ductility and increased corrosion resistance [1].

The formation of a high-strength state in steels is achieved by choosing the appropriate alloying principles and obtaining the desired structural class of the material, as well as by combining various hardening mechanisms: solid-solution hardening, strain hardening in matrix phases without phase transitions, strain hardening due to $\gamma \rightarrow \alpha$ transformation and dispersion hardening with the release of intermetallic phases. In addition, the creation of a nanocrystalline state in the structure of the material leads to the formation of a fundamentally new complex of high physical and mechanical properties. It has been established [2–4] that nanostructured materials and alloys (i.e., ultrafine-grained materials with grain sizes $\leq$ 100 nm) exhibit very high hardness, strength, impact strength and wear resistance.

In the framework of this study, a new high-strength, corrosion-resistant steel of the austenitic class was developed. This is a steel in which all possible hardening mechanisms can be realized, and which possesses the nanocrystalline structure necessary for providing high mechanical properties on a wire designed for the manufacture of elastic elements [5].

The aim of this work is to study the structure and mechanical properties of austenitic, corrosion-resistant steel at each technological stage of obtaining high-strength wire: depending on the temperature of heating for quenching, after cold plastic deformation by drawing, and also as a result of post-deformation aging.

2. Research methodology

Corrosion-resistant steel of the austenitic class was selected as the material for study: 03Kh13N10K5M2YuT steel alloying systems 0.03 % C – 13 % Cr – 10 % Ni – 5 % Co – 2 % Mo – 0.8 % Al – 0.4 % Ti. The influence of quenching the heating temperature was studied at a temperature range of 800–1300 °C. Cold plastic deformation was carried out by drawing blanks from a diameter of 14.3 to 2.77 mm at room temperature. Samples were taken along the drawing route to study structural changes during deformations with compression ratios of 30 % (e = 0.39), 41 % (e = 0.52), 69 % (e = 1.15), 80 % (e = 1.60), 88 % (e = 2.17), 94 % (e = 2.32) and 96 % (e = 3.27). After deformation, aging was carried out at a temperature range of 300–700 °C for 1 h.

Mechanical tests were carried out on wire samples; the tensile strength (σ_s), yield strength ($\sigma_{0.2}$), relative narrowing (ψ) and elongation (δ) were determined. X-ray phase analysis was carried out on a DRON-2 diffractometer. The amount of the magnetic phase was determined on a Faraday scale. Hardness was measured using a Vickers instrument. Microhardness was measured on a Leco PC series automatic hardness tester with a programmable pitch and a load of 0.001 kg. Metallographic studies were performed on a Neophot optical microscope. Electron microscopic studies included a study of the fine structure of the steel and the determination of phase composition using a JAM 200-CX microscope.

3. Research results and discussion

In the framework of this work, a new high-strength, corrosion-resistant, austenitic carbon-free steel based on a Fe-Cr-Ni base, additionally alloyed with Co, Mo, Ti and Al, designed for the manufacture of high-strength wire of thin and thin sections, was developed. An optimal mode for producing high-strength wire from this new corrosion-resistant steel has been developed [6]. The optimal technological processes are:

1) obtaining a degree of supersaturation of the solid solution on large cross-section blanks by austenitization, followed by quenching in water. After hardening, an austenitic state with a deformation-unstable structure should be formed, which is the basis for obtaining high plasticity during drawing;
2) the subsequent deformation by drawing with large total degrees of compression, in combination with deformation transformation, should provide maximum fragmentation (defectiveness) of the structure, as well as a nanocrystalline structure, as the basis of a high-strength state;
3) further post-deformation aging should provide the necessary form and shape of reinforcing dispersed particles without a significant change in the phase composition of the matrix.

3.1. Quenching temperature selection

Microstructural studies of the steel after quenching from temperatures of 800–1300 °C showed that the steel structure consists of 100 % austenite and contains polyhedral grains with a large number of annealing twins (Figure 1). The higher the temperature, the larger the austenite grain is. Phase X-ray diffraction analysis confirms that after quenching from all the temperatures, the main phase is austenite; however, with an increase in the heating temperature to 1200 °C and higher, a bcc phase (δ-ferrite) appears.

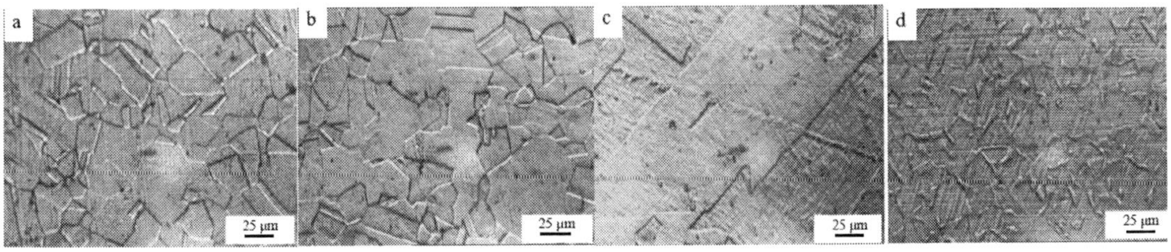

Figure 1. Microstructure of austenitic 03Kh13N10K5M2YuT steel after quenching from various temperatures: (a) 800 °C; (b) 1000 °C; (c) 1200 °C; (d) 1300 °C.

Increased strength after low-temperature quenching is associated with the presence in the γ-solid solution of a high-temperature intermetallic Ni_3Al type phase (Figure 2). The dissolution of this phase with an increase in the heating temperature for quenching to 1000–1100 °C leads to the formation of a homogeneous γ-solid solution, while strength decreases and ductility increases. A further increase in temperature to 1300 °C leads to an increase in hardness due to the appearance of a certain amount of δ-ferrite in the steel structure (see Figure 1, d) as a result of the $\gamma \rightarrow \delta$ transformation.

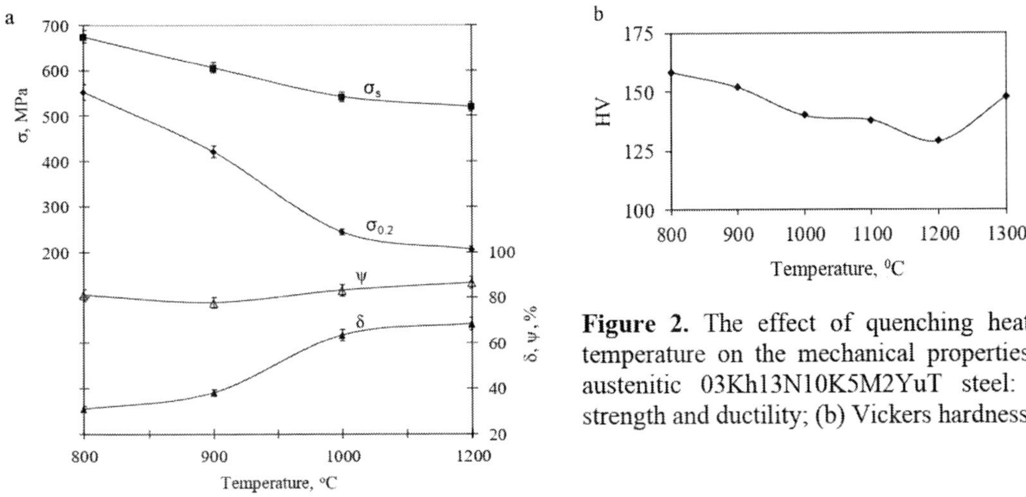

Figure 2. The effect of quenching heating temperature on the mechanical properties of austenitic 03Kh13N10K5M2YuT steel: (a) strength and ductility; (b) Vickers hardness.

Thus, from the point of view of choosing a temperature for quenching for further plastic deformation, the best is quenching from a temperature range of 1000–1100 °C in water. This is due, on the one hand, to the dissolution of excess intermetallic phases that reduce the aging effect and obtain a uniform austenitic structure and, on the other hand, to the absence of intensive grain growth and the formation of δ-ferrite. The mechanical properties of the studied hardened steel after quenching from 1000 °C are: σ_s = 540 MPa, $\sigma_{0.2}$ = 245 MPa, δ = 63 %, ψ = 83 %, and hardness HV = 143.

3.2. Cold plastic deformation by drawing

Cold plastic deformation by drawing is the second stage of the developed regime for producing high-strength wire from new corrosion-resistant steels. Drawing deformation with large total compression ratios, in combination with deformation transformation, should provide maximum fragmentation (imperfection), as well as a nanocrystalline structure, as the basis for a high-strength state.

An increase in the degree of cold plastic deformation to e = 2.32 made it possible to obtain high strength values (σ_s = 1500 MPa): the increase in strength was 960 MPa (Figure 3, a, b). High hardness values were also obtained (425 HV): the increase in hardness was 280 HV. As shown by the data from phase X-ray diffraction analysis (Figure 3, c), the austenite undergoes a martensitic transformation during cold plastic deformation: with a strain of $e = 2.32$, the amount of martensite is about 90 %.

The intensity of martensite formation was studied both by X-ray diffraction and magnetometric methods. The appearance of a martensite strain was observed at a sufficiently large strain ≈ 41%. With further drawing, the amount of martensite steadily increases.

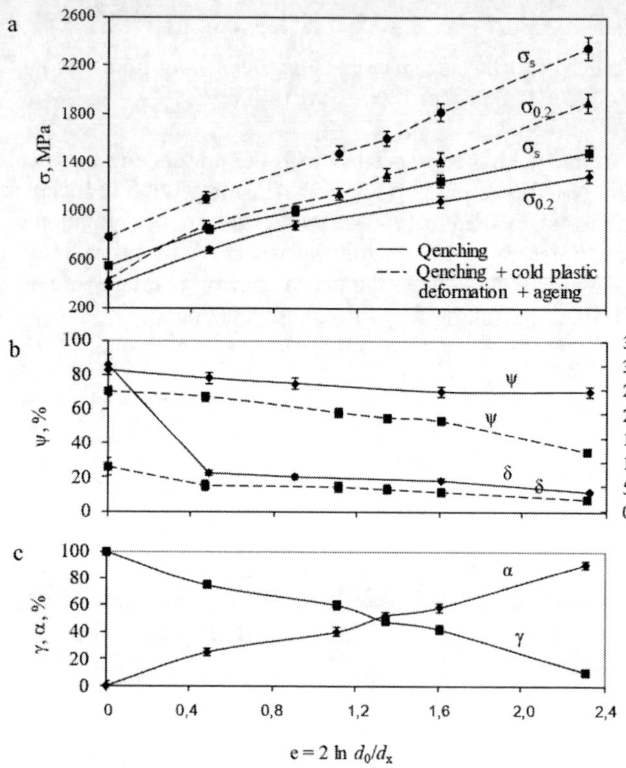

Figure 3. The effect of cold plastic deformation on the properties of hardened austenitic 03Kh13N10K5M2YuT steel: (a) strength properties; (b) plastic properties; (c) phase composition.

To determine the effect of the strain on the steel's structure, microstructural studies of the deformed wire samples were carried out. With moderate compressions (15–30 %), slip bands appear in individual grains. With greater deformation, the shape of the grains' changes: from equiaxed they become more fibrous and elongated along the drawing axis (Figure 4).

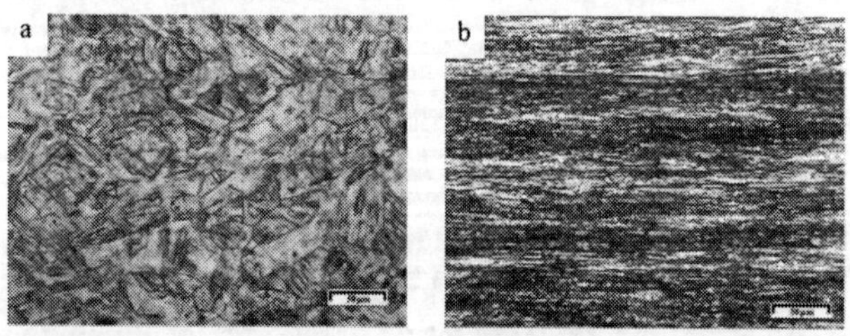

Figure 4. The microstructure of 03Kh13N10K5M2YuT steel after quenching and deformation at e = 0.52 (a) and e = 2.32 (b).

As a result of electron microstructural studies, it was found that, with a total strain of 88 % ($e = 2.17$), ring diffraction patterns appear (Figure 5, b) due to the misorientation of crystals at an angle greater than 5°. This is due to the fine-grained structure with reflections like α and γ phases. With a strain of ≈ 94 % (e = 2.32), martensite crystals with a submicroscopic size ≈100 nm and less occur (Figure 5, e-d).

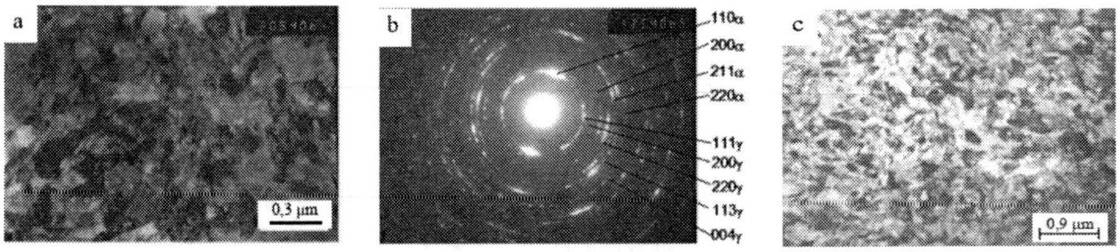

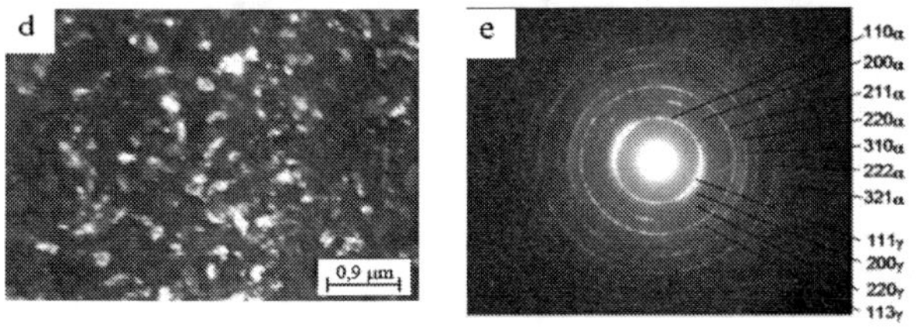

Figure 5. Microstructure of 03Kh13N10K5M2YuT steel after deformation e = 2.17 (ε = 88 %) (a-b) and deformations e = 2.32 (94 %) (c-e): (a), (c) bright field image; (b), (d) microdiffraction; (e) dark-field image in the reflection of the α-phase (110).

Thus, the extremely high ductility found in the steel is due to a martensitic transformation with optimal intensity, which ensures the formation of martensite submicrocrystals. The mechanical properties obtained by stretching the wire after cold plastic deformation $e = 2.17$ are: $\sigma_s = 1480$ MPa, $\sigma_{0.2} = 1200$ MPa, $\delta = 4$ %, $\psi = 70$ % and martensite crystal size 100–200 nm. After cold plastic deformation, the mechanical properties are: $e = 2.32$: $\sigma_s = 1500$ MPa, $\sigma_{0.2} = 1300$ MPa, $\delta = 3$ %, $\psi = 70$ % and martensite crystals 20–100 nm in size. The extremely high manufacturability of this steel should be noted, as it was able to stretch with very high total deformations without intermediate softening annealing.

3.3. The influence of post-deformation aging

We studied the possibility of further increasing the strength properties, and, in particular, the feasibility of using subsequent aging after quenching and cold plastic deformation. It was revealed that post-deformation aging of the 03Kh13N10K5M2YuT steel at 500 °C for 1 h leads to a significant increase in strength properties (Figure 3, dashed curves). This is a result of aging the bcc phase (strain martensite) with the release of a dispersed ordered intermetallic NiAl phase. The greater the amount of martensite formed in the process of deformation (drawing), the higher the increase in strength properties (σ_s, $\sigma_{0.2}$). Sufficiently high plasticity characteristics ($\delta = 3$ % and $\psi = 45$ %) are also maintained.

An electron microstructural study of pre-deformed samples of the 03Kh13N10K5M2YuT steel after aging at 500 °C for 1 h showed that the structural components of the deformed and aged steel are the bcc and fcc phases. The main mechanism of aging is the heterogeneous precipitation of the intermetallic phase from the bcc solid solution. In the structure of deformed and aged metastable austenitic steel (Figure 6), a ripple-type contrast is observed in electronic photographs due to the appearance of fine particles. Superstructural reflections belonging to particles of the intermetallic NiAl phase appear on electron diffraction patterns. An increase in temperature and aging time (650 °C, 2 h) leads to an increase in the precipitated particles of the NiAl phase (Figure 6, c).

Thus, the available data allow us to conclude that the presence of martensite is effective for achieving a high-strength state in unstable austenitic steel not only due to hardening from the use of plastic deformation, but also due to subsequent post-deformation aging. At the same time, when analyzing the causes of thermomechanical hardening, one cannot ignore the undoubted influence of the structural state of another solid solution, austenite. Deformation hardening of the latter can also contribute to a general change in the properties of a deformed metal during aging.

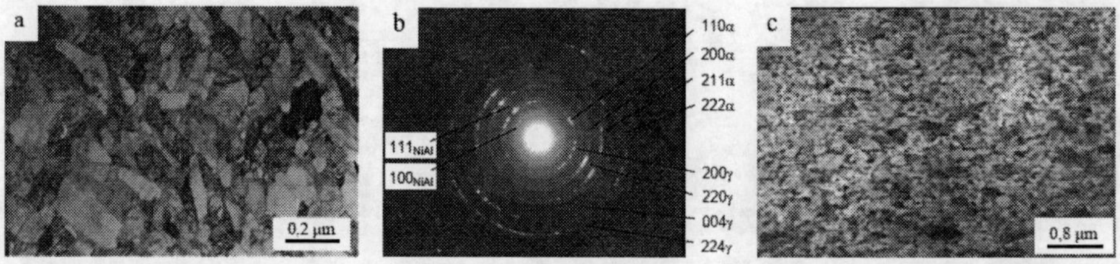

Figure 6. The structure of 03Kh13N10K5M2YuT steel after quenching from 1000 °C, deformation e = 2.32 and aging: at 500 °C, 1 h (a–b), at 650 °C, 2 h (c).

4. Conclusions

Cold plastic deformation is a reliable and easy way to achieve a high level of strength. The austenite of the steel under study is deformation-unstable and is almost completely transformed into deformation martensite during cold plastic deformation. The increase in strength reaches 900–1100 MPa (depending on the final diameter) with fairly good ductility. It is shown that the extremely high ductility found in this austenitic steel is due to the occurrence of a martensitic transformation with optimal intensity, which provides for the formation of 20–100 nm martensite submicrocrystals (nanostructures). The aging of deformed steel causes an additional increase in mechanical properties, which is associated with the decomposition processes of a supersaturated bcc solid solution (deformation martensite).

It was revealed that the phase responsible for the hardening of the deformed test steel during aging is an intermetallic NiAl phase. Thus, the available data allow us to conclude that almost carbon-free, corrosion-resistant austenitic steel, as a result of properly selected alloying, combines the advantages of three steels: metastable austenitic steels, TRIP-steels and martensitic aging steels. As a result of using all possible hardening mechanisms, a high-strength state was achieved.

References

[1] Järvenpää A, Jaskari M, Kisko A and Karjalainen P 2020 Processing and properties of reversion treated austenitic stainless steels *Metals* vol **10** Issue **2** pp 281 DOI: 10.3390/met10020281

[2] Abramova M M, Enikeev N A, Kim J G, Valiev R Z, Karavaeva M V and Kim H S 2016 Structural and phase transformation in a TWIP steel subjected to high pressure torsion *Materials Letters* vol **166** pp 321-324 DOI: 10.1016/j.matlet.2015.12.095

[3] Karavaeva M V, Abramova M M, Enikeev N A, Raab G I and Valiev R Z 2017 Effect of combined loading on the microstructure and microhardness of austenitic steel *Letters on Materials* vol **7** Issue **1** pp 29-33 DOI: 10.22226/2410-3535-2017-1-29-33

[4] Kim J G, Enikeev N A, Seol J B, Abramova M M, Karavaeva M V, Valiev R Z, Park C G and Kim H S 2018 Superior Strength and Multiple Strengthening Mechanisms in Nanocrystalline TWIP Steel *Scientific Reports* vol **8** Issue **1** pp 11200 DOI: 10.1038/s41598-018-29632-y

[5] Mal'tseva L A, Sharapova V A, Mal'tseva T V, Ozerets N N, Levina A V and Tsaplina Ye M 2011 Korrozionno-stoykaya austenitnaya stal' Patent 2430187 RU

[6] Maltseva L A, Sharapova V A, Maltseva T V, Levina A V, Ozerets N N and Zadvorkin S M 2016 High Strength Corrosion-Resistant Steels with Metastable Austenite for Elastic Elements and Springs for Demanding Applications *AIP Conference Proceedings* vol **1785** 040036 DOI: 10.1063/1.4967093

IOP Conf. Series: Materials Science and Engineering **966** (2020) 012050 doi:10.1088/1757-899X/966/1/012050

Effect of frictional treatment with a dense cubic boron nitride indenter on the micromechanical properties of the NiCrBSi–Cr_3C_2 coating

N N Soboleva[1,2,4], A V Makarov[3], E P Nikolaeva[3], P A Skorynina[1] and I Yu Malygina[1]

[1] Institute of Engineering Science, Ural Branch of the Russian Academy of Sciences, 34 Komsomolskaya St., Ekaterinburg, 620049, Russia
[2] Ural Federal University named after the First President of Russia B N Yeltsin, 19, Mira str., Yekaterinburg, 620000, Russia
[3] M N Miheev Institute of Metal Physics, Ural Branch of the Russian Academy of Sciences, 18 S Kovalevskoy St., Ekaterinburg, 620108, Russia

[4]E-mail: natashasoboleva@list.ru

Abstract. Coatings based on NiCrBSi alloys with Cr_3C_2 addition can effectively resist wear, corrosion, and oxidation at high temperatures. The use of frictional treatment by sliding indenters as finishing is a modern way to create high surface quality and to enhance the strength and wear resistance of the surface of parts. The article studies the characteristics determined by instrumented microindentation of the surface of a NiCrBSi–Cr_3C_2 laser clad coating subjected to frictional treatments with a sliding indenter made of dense cubic boron nitride DBN in the air at loads on the indenter of 350, 500 and 700 N, and after grinding. Frictional treatment in the entire considered range of loads contributes to an increase in strength performance, as well as parameters indicating an increased ability of the coating surface to resist elastoplastic deformation. The highest growth of the parameters is observed after frictional treatment at a load of 700 N.

1. Introduction

The formation of protective coatings on the surface of parts is an effective method to increase their durability and reliability. Coatings based on NiCrBSi alloys have good performance under conditions of wear, corrosion, elevated temperatures and are widely used worldwide at present [1, 2].

Gas powder laser cladding is a perspective method for obtaining coatings. The laser beam melts the filler material (powder) and simultaneously the thin surface layer of the base (substrate) during gas powder laser cladding [3]. Compared with other methods of applying surface layers, laser cladding forms coatings with increased hardness and uniformity, as well as with excellent metallurgical adhesion to the substrate [4, 5].

The creation of metal matrix composite coatings is a modern way to improve the performance of coatings. They consist of a rather soft matrix as well as hard reinforcing particles. NiCrBSi powders can be used as a metal matrix material for wear-resistant composite coatings owing to their excellent wear and corrosion resistance. In these alloys, an extremely thermal-resistant (up to 1000 °C) 'frame-like' structure of large chromium borides and carbides can be formed by high-temperature annealing [6–8]. This significantly expands the possibilities of using nickel-chromium laser clad coatings for high-temperature applications, in particular, in metallurgy [9]. In addition, the considered alloys have a rather low melting point and can serve as a plastic bond for reinforcing phases. Additives of carbides, borides, oxides, and other compounds are used as strengthening phases in NiCrBSi coatings. Coatings based on NiCrBSi with Cr_3C_2 addition can effectively resist wear, corrosion, and oxidation at high temperatures [10].

Laser clad coatings have significant roughness and undulation [11]. In most cases, these disadvantages are eliminated by grinding with abrasive wheels. However, grinding can provide 'burns' and microcracks on the surface, as well as critical tensile stresses, which can lead to a decrease in the service life of parts. The use of surface deformation by sliding indenters as finishing is a modern way to create high surface quality, favorable compressive stresses in the surface layer, and to enhance the strength and wear resistance of the surface of parts [12]. An effective method for obtaining a hardened NiCrBSi coating on metal parts has been proposed. It includes gas powder laser cladding and frictional treatment with a hemispherical indenter made of dense cubic boron nitride (DBN) [13]. The authors also showed that frictional treatments with a DBN indenter at loads of 350–700 N increase the microhardness of the NiCrBSi–Cr_3C_2 laser clad coating [14].

Instrumented microindentation is one of the methods for assessing the mechanical properties of nickel-based coatings [15–17]. It makes it possible to register diagrams during loading and unloading of the indenter. The method allows evaluating the mechanical properties of materials for which there is no opportunity to conduct standard tests for tensile, compression, and bending [18].

However, the instrumented microindentation method has not been previously used to evaluate the micromechanical characteristics of the NiCrBSi–Cr_3C_2 coating formed by laser cladding and subjected to frictional treatment. Therefore, the purpose of this work was to research the influence of frictional treatments with a DBN indenter at loads of 350–700 N on the micromechanical properties of the NiCrBSi–Cr_3C_2 laser clad coating.

2. Material and Experimental Procedure

The cladding material was two powders: 85 wt. % of the NiCrBSi with a dispersion of 40–160 μm and 15 wt. % of Cr_3C_2 with a dispersion of 50–150 μm. Chemical composition of NiCrBSi powder, wt. %: 14.8 Cr; 2.1 B; 2.9 Si; 0.48 C; 2.6 Fe; the rest is Ni. Two-layer cladding was carried out on plates made of low-carbon steel (0.2 % C) using a CO_2 laser in the continuous mode. Cladding was implemented in the following modes: radiation power 1.4–1.6 kW; the size of the laser spot on the surface 2–6 mm; powder consumption 2.9–3.8 g/min; speed 160–200 mm/min; shift 4.0–4.5 mm. An inert gas (argon) was used as a transport gas for the powder mixture. To exclude the undulation, grinding of the surface with a deposited coating was executed on a grinding machine where intensive cooling was provided. After grinding, the thickness of the NiCrBSi–Cr_3C_2 coatings was 0.7–1.1 mm.

Frictional treatment was performed in a laboratory setup on flat samples with the coatings after grinding their surface. An indenter made of dense cubic boron nitride DBN with hemispherical shape was reciprocating along the sample surface. The load P on the indenter was 350, 500, and 700 N. Frictional treatment was performed with a five-time scanning of the surface with the indenter in air at an speed of 0.013 m/s, a length of stroke of 17 mm, and displacement of an indenter was 0.1 mm per double stroke.

Instrumented microindentation with recording the loading and unloading diagrams was conducted using a measuring system Fischerscope HM2000 XYm. The indenter was a Vickers indenter, a maximum load on the indenter was 0.245 N. Micromechanical parameters were calculated according to ISO 14577 [19]. Loading/unloading time – 5 s. Holding time at maximum load – 20 s. The measurement error of microindentation characteristics from 10 measurements was determined by the standard deviation with a confidence probability of 0.95.

3. Experimental Results and Discussion

Figure 1 presents the diagrams of loading and unloading of the indenter during microindentation of the surfaces of samples with NiCrBSi–Cr_3C_2 coating after grinding, as well as after frictional treatment with the DBN indenter at loads of 350, 500, and 700 N.

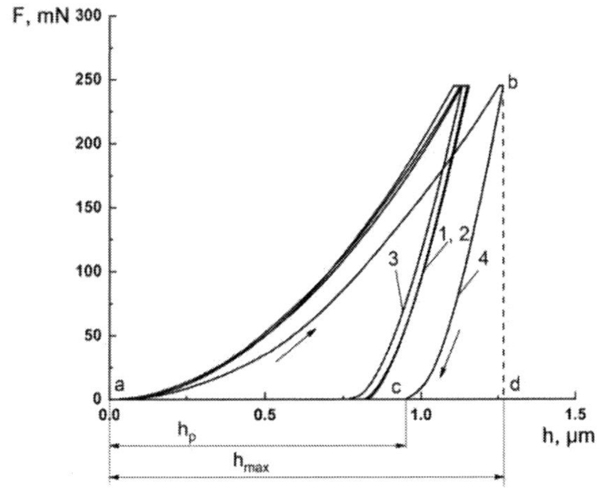

Figure 1. Experimental curves 'load F – displacement of the indenter h' during microindentation the surfaces of samples with NiCrBSi–Cr_3C_2 coating after frictional treatment with the DBN indenter in air at loads of 350, 500 N and 700 N (curves 1, 2 and 3, respectively) and after grinding (curve 4).

Table 1. The microindentation results of the samples with NiCrBSi–Cr_3C_2 laser clad coating after different treatments.

Surface treatment	h_{max} (μm)	h_p (μm)	HM (GPa)	H_{IT} (GPa)	E^* (GPa)	W_t (nJ)	W_e (nJ)
1 Frictional treatment, P=350 N	1.17±0.03	0.83±0.03	7.1±0.3	10.2±0.6	228±6	94.0±2.7	31.8±0.5
2 Frictional treatment, P=500 N	1.18±0.04	0.83±0.04	7.0±0.4	10.1±0.7	219±9	95.1±3.5	32.6±0.9
3 Frictional treatment, P=700 N	1.14±0.03	0.78±0.04	7.6±0.4	11.1±0.7	224±4	91.2±1.6	33.6±0.8
4 Grinding	1.26±0.05	0.93±0.06	6.1±0.5	8.4±0.8	229±10	103.5±6.3	28.9±0.9

Figure 1 shows that the frictional treatment of the coating has a significant effect on the type and position of the lines, shifting them towards lower values of the indenter displacements h during indentation in comparison with the measure of h for the surface after grinding.

The characteristics displayed in table 1 were determined using the curves of loading (a → b in figure 1) and unloading (b → c in figure 1), according to the method of Oliver and Farr [20]. They reflect the features of the mechanical behaviour of the considered coatings under elastoplastic deformation. The values in table 1 show that the frictional treatment of the NiCrBSi–Cr_3C_2 coating surface reduces the maximum h_{max} (to 1.14–1.18 μm) and the residual h_p (to 0.78–0.83 μm) indenter indentation depths compared to the values of the polished surface (h_{max} = 1.26 and h_p = 0.93 μm). The lowest values of the maximum and residual indentation depths during indentation are characteristic of the surface of the coating subjected to frictional treatment at a load of 700 N.

Frictional treatment with a DBN indenter at all loads used also increases the values of the hardness on the Martens scale HM, which takes into account both plastic and elastic deformation, compared to with the values after grinding. Furthermore, treatment at loads of 350–500 N increases the HM values by 15 %, and at loads of 700 N – by 25 % (see table 1). After frictional treatment, the indentation hardness values at the maximum load H_{IT} increase even more significantly – by 20–21 % after treatment at loads of 350–500 N and by 32 % after treatment at a load of 700 N (see table 1). An increase in H_{IT} indicates an increase in the resistance to constant deformation [19].

The contact elastic modulus when indenting the surface of the coating E^* after frictional treatment at various loads little (within the error) varies from the characteristics of the polished surface. It is known that the elastic modulus of metal materials can change during deformation [21–23]. However, in this case, the surface after grinding can also be considered as deformed. It is also important to note that the composite NiCrBSi–Cr_3C_2 coating after frictional treatment in air with a DBN indenter at a load of 350 N has a 13 % higher level of the contact elastic modulus ($E^* = 228 \pm 6$ GPa) (see table 1) than the NiCrBSi coating of the same composition, but without Cr_3C_2 additives ($E^* = 202 \pm 5$ GPa) [24]. In addition, the contact elastic modulus of NiCrBSi coatings of the same composition not subjected to frictional treatment was 195 ± 5 GPa [25]. This can be explained by the presence in the composite coating of 15 % Cr_3C_2 carbides with a significantly increased elastic modulus (E =370 GPa) [26]. Similarly, in [27], an increase in the elastic modulus of a composite material was observed with an increase in the amount of phase with a large elastic modulus.

Table 1 also shows that the values of the total mechanical work of indentation W_t, which is determined by the field under the load curve (triangle abd in figure 1) and consists of plastic deformation work and elastic reserve deformation work, inversely depend on the type of treatment compared to the values of hardness HM and H_{IT}, since the more hardened the material, the less it deforms under the indenter and, accordingly, the less work is spent on such deformation. The elastic reserve deformation work W_e, which is determined by the field under the unloading curve (triangle cbd in figure 1) and released when the applied load is removed, grows with increasing load on the indenter during frictional treatment.

The determinate values of the work (the total mechanical work of indentation W_t and the elastic reserve deformation work W_e) were used in the calculation according to the formula $(1-(W_e/W_t))100$ % of the plastic component of the work when indenting the surface of the coating (table 2). The largest (among the four considered coating states) plastic component of the indentation work (72 %) has a coating after grinding. Growth in the load on the indenter during frictional treatment successively reduces the plastic component from 66 % at a load of 350 N to 63 % at a load of 700 N.

Table 2 presents several calculated parameters determined by microindentation data depending on the type of surface treatment of the coating. The parameters are used to evaluate the resistance of material's surface to elastoplastic deformation under mechanical contact loading. The value of elastic recovery $R = ((h_{max} - h_p)/h_{max})\cdot 100$ % [28–30] and the ratio H_{IT}/E^* (specific contact hardness) [31] change in the same way during frictional treatment compared to the state after grinding. The values increase during frictional treatment at loads of 350–500 N and even more significantly at a load of 700 N. It is generally accepted [30, 32] that the elastic recovery of R and the ratio H_{IT}/E^* characterize the proportion of elastic deformation in the total deformation during indentation.

Table 2. The plastic component of the indentation work $(1-(W_e/W_t))\cdot 100$, elastic recovery R, the ratios H_{IT}/E^* и H_{IT}^3/E^{*2}, calculated from the microindentation results of the samples with NiCrBSi–Cr_3C_2 laser clad coating after different treatments.

Surface treatment	$(1-(W_e/W_t))\cdot 100$ (%)	R (%)	H_{IT}/E^*	H_{IT}^3/E^{*2} (GPa)
1 Frictional treatment, $P = 350$ N	66	29.1	0.045	0.020
2 Frictional treatment, $P = 500$ N	65	29.3	0.046	0.021
3 Frictional treatment, $P = 700$ N	63	31.6	0.050	0.027
4 Grinding	72	25.9	0.036	0.011

The values of the H_{IT}^3/E^{*2} ratio presented in table 2 are regarded to be a parameter of the material's resistance to plastic deformation after the beginning of the flow since the ratio is proportional to the flow stress P_y of the material [33]. The ratio H_{IT}^3/E^{*2} has a minimum value for the coating after grinding. Frictional treatment even at a minimum load of 350 N contributes to the growth of this indicator by 82 %. Treatment at a load of 700 N increases this indicator by 145 %.

It was previously shown [34] that there are coarse (50–100 μm) original chromium carbides Cr_3C_2 in the structure of the NiCrBSi–Cr_3C_2 laser clad coating. Frictional treatment with a DBN indenter at a load of 350 N smoothes the NiCrBSi matrix, and protruding initial chromium carbides Cr_3C_2 are remarked on the treated surface. The protruding wear-resistant framework on the wear surface made of the reinforcing phases plays a predominant role (compared with the metal matrix role) in the abrasion resistance of NiCrBSi coatings formed by laser cladding [35]. Therefore, it can be expected that the coating after frictional treatment at a load of 350 N will be characterized by high resistance to abrasive wear [14].

The large initial chromium carbides are partially removed from the surface of the NiCrBSi– Cr_3C_2 coating during frictional treatment at a load of 700 N [14]. The micromechanical properties determined by instrumented microindentation show that increasing the load during frictional treatment to 700 N (compared to treatment at 350–500 N) increases the hardness of the coating and its ability to resist plastic deformation. This indicates that the level of hardening of the metal matrix, in this case, makes a greater contribution to the overall strength properties of the coating measured during microindentation than their possible reduction due to some absence of coarse chromium carbides after frictional treatment at a load of 700 N.

Thus, in comparison with the coating after grinding, the coating after frictional treatments is characterized by both maximum deformation in the elastic region (i. e., by a delayed transition to the plastic stage of deformation) and an increased ability to resist to contact loads after the beginning of the plastic flow. To a greater extent, these features are manifested for the coating after frictional treatment at the highest load used. In [36], the results of microindentation for steel surfaces after frictional treatment correlate with wear mechanisms during sliding friction. The increased ability to resist elastoplastic deformation may indicate better wear resistance of the coating after frictional treatment at a load of 700 N under sliding friction conditions.

4. Conclusion

According to the data obtained with instrumented microindentation, frictional treatment of the surface of the composite NiCrBSi–Cr_3C_2 coating with an indenter made of dense cubic boron nitride in air at loads of 350–700 N causes a significant increase in strength characteristics (Martens hardness *HM* and indentation hardness at maximum load H_{IT}) and does not affect the level of the contact elastic modulus E^* compared to the parameters after grinding. Additive 15 wt. % of Cr_3C_2 carbide with a high modulus of elasticity leads to a 13 % increase in the value of the contact modulus of elasticity of the coating subjected to frictional treatment.

In consequence of frictional treatment, the plastic component of the indentation work decreases in all the studied modes, and the values of elastic recovery *R* as well as the H_{IT}/E^* and H_{IT}^3/E^{*2} ratios (compared to the corresponding parameters after grinding) increase. This indicates a slow transition to plastic deformation during contact loading of the coating, followed by a more significant resistance to the development of plastic flow.

Frictional treatment at the maximum load on the indenter (700 N), which results in some removal of large initial chromium carbides from the surface, but increases the hardness of the metal matrix, provides a maximum increase in the average values of strength characteristics measured during microindentation, as well as parameters indicating an increased ability of the coating surface after the considered treatment to resist to elastic and plastic deformation.

Acknowledgments

The work was performed with the financial support of the Grant from the President of the Russian Federation for young scientists MK-391.2019.8 and at the expense of funds making up the income from the trust management of the target capital for the development of UrFU, formed with the participation of UMMC-Holding Corp, as well as within the state order for IES UB RAS (AAAA-A18-118020790147-4) and IMP UB RAS (AAAA-A18-118020190116-6 and AAAA-A19-119070490049-8). The experimental research was performed on the equipment of the Plastometriya Collective Use Center of IES UB RAS.

References

[1] Günen A and Çürük A 2020 *JOM* **72**

[2] Garrido M A, Rico A, Gómez M T, Cadenas M, Fernández-Rico J E and Rodríguez J 2017 *J. Therm. Spray Technol* **26**
[3] Gao W, Chang C, Li G, Xue Y, Wang J, Zhang Z and Lin X 2019 *Optik* **178**
[4] Tamanna N, Crouch R and Naher S 2019 *Opt. Lasers Eng* **122**
[5] Li C, Zhang Q, Wang F, Deng P, Lu Q, Zhang Y, Li S, Ma P, Li W and Wang Y 2019 *Appl. Surf. Sci* **485**
[6] Makarov A V, Soboleva N N, Malygina I Yu and Osintseva A L 2013 Patent RU 2492980
[7] Makarov A V, Soboleva N N, Malygina I Yu and Osintseva A L 2015 *Met. Sci. Heat Treat* **57**
[8] Makarov A V, Soboleva N N, Malygina I Yu and Kharanzhevskiy E 2019 *J. Cryst. Growth* **525** 125200
[9] Makarov A V, Korobov Yu S, Soboleva N N, Khudorozhkova Yu V, Vopneruk A A, Balu P, Kotte L, Malygina I Yu, Burov S V and Stepchenkov A K 2019 *Lett. Mater* **4**
[10] Rojacz H, Zikin A, Mozelt C, Winkelmann H and Badisch E 2013 *Surf. Coat. Technol* **222**
[11] Singh R, Kumar D, Mishra S K and Tiwari S K 2014 *Surf. Coat. Technol* **351**
[12] Makarov A V and Korshunov L G 2019 *Phys. Met. Metallogr* **120**
[13] Soboleva N N, Makarov A V and Malygina I Yu 2019 Patent RU 2709550
[14] Soboleva N N, Makarov A V, Skorynina P A, Nikolaeva E P and Malygina I Yu 2019 *AIP Conf. Proc* **2167** 020345
[15] Gómez del Río T, Garrido M A, Fernández J E, Cadenas M and Rodríguez J 2008 *J. Mater. Process. Technol* **204**
[16] Serres N, Portha N and Machi F 2011 *Surf. Coat. Technol* **205**
[17] Xuan H F, Wang Q Y, Bai S L, Liu Z D, Sun H G and Yan P Ch 2014 *Surf. Coat. Technol* **244**
[18] Smirnov S V, Veretennikova I A, Smirnova E O and Pestov A V 2017 *Diagnostics, Resource and Mechanics of Materials and Structures* **6**
[19] Standard ISO **14577–1** 2002
[20] Oliver W C and Pharr J M 1992 *J. Mater. Res* **7**
[21] Benito J A, Jorba J, Manero J M and Roca A 2005 *Metall. Mater. Trans. A* **36**
[22] Makarov A V, Savrai R A, Schastlivtsev V M, Tabatchikova T I, Yakovleva I L and Egorova L Yu 2011 *Phys. Met. Metallogr* **111**
[23] Savrai R A, Makarov A V, Malygina I Yu, Rogovaya S A and Osintseva A L 2017 *Diagnostics, Resource and Mechanics of Materials and Structures* **5**
[24] Soboleva N N, Makarov A V and Malygina I Yu 2017 *Vektor nauki Tolyattinskogo gosudarstvennogo Universiteta* **4**
[25] Savrai R A, Makarov A V, Soboleva N N, Malygina I Yu and Osintseva A L 2016 *J. Mater. Eng. Perform* **25**
[26] Alekseev A G, Bovkun G A, Bolgar A S and Borisova A L, ed. 1986 *Properties, Preparation and Application of Refractory Compounds* (Moscow: Metallurgy Publ) p 928 [In Russian]
[27] Pugacheva N B, Michurov N S, Senaeva E I and Bykova T M 2016 *Phys. Met. Metallogr* **117**
[28] Page T F and Hainsworth S V 1993 *Surf. Coat. Technol* **61**
[29] Petrzhik M I and Levashov E A 2007 *Crystallogr. Rep* **52**
[30] Petrzhik M I, Filonov M R, Pecherkin K A, Levashov E A, Olesova V N and Pozdeev A I 2005 *Russ. J. Non-Ferr. Met* **6**
[31] Gorban V F and Pechkovskii É P 2010 *Powder Metall. Met. Ceram* **49**
[32] Cheng Y T and Cheng C M 1998 *Appl. Phys. Lett***73**
[33] Mayrhofer P H, Mitterer C and Musil *Surf. Coat. Technol* **174–175**
[34] Savrai R A, Makarov A V, Gorkunov E S, Soboleva N N, Kogan L Kh, Malygina I Yu, Osintseva A L and Davydova N A 2017 *AIP Conf. Proc* **1915** 040049
[35] Makarov A V, Soboleva N N and Malygina I Yu 2017 *J. Frict. Wear* **38**
[36] Makarov A V, Skorynina P A, Osintseva A L, Yurovskikh A S and Savrai R A 2015 *Obrab. Met (Met. Work. Mat. Sci)* **4**

Destruction of cement stone modified with a plasticizer and a nanoscale aluminum oxide

E S Gerasimova[1, 2] and E D Trestsova[1] (student)

[1] Department of materials sciences in construction of the Institute of new materials and technologies, Ural Federal University, 28, Mira str., Ekaterinburg, 620002, Russia

E-mail: [2]e.s.gerasimova@urfu.ru

Abstract. It is well known that hydration and hardening of Portland cement are associated with complicated physical and chemical processes, in particular the interaction of clinker minerals with water to form new hydrate phases of different composition. Sometimes the presence of various substances can affect these processes leading to self-destruction of the hydrated cement stone. The morphology and composition of damaged cement stone modified with an additive and with a nanoscale aluminum oxide were studied. It is shown that in the presence of the 'Linamix PC' plasticizer and a nanoscale aluminum oxide, well-crystallized $Ca(OH)_2$ plates with an average thickness of 1.5 microns and a length and width equal to 300–400 microns are formed. Also, a network of thread-like ettringite with a diameter of less than 1 micron on the surface of microcracks of cement stone is formed. Most likely the cement stone sample could be damaged by the forces of crystallization pressure during the growth of both $Ca(OH)_2$ and ettringite crystals.

1. Introduction

Hydration and hardening of Portland cement is a complex of physical and chemical processes that is accompanied by the formation of hydrate phases when clinker minerals interact with water. Crystals of calcium hydrosilicates, hydroaluminates and hydroalumoferrites, their solid solutions and complex compounds which differ by composition and structure are formed during hydration of the clinker minerals. A large variety of crystallohydrates in cement stone greatly complicates its study [1]. Also, the hydration and hardening processes are strongly affected by the presence of various plasticizers and other substances (nanoscale additives) [2–3]. Ettringite is the first stable hydration product that is formed by the reaction of C_3A, calcium sulfate (gypsum) and water [1, 4]. It is reported in [2, 4–7] that plasticizer molecules are usually adsorbed primarily on the surface of both C_3A and new formed ettringite. This leads to a change in its morphology. It was found in [8] that the composition and type of chemical functional additives affect the shape and size of ettringite. It is also known that as a result of the ettringite crystals growth under an increased content of gypsum destructive processes can occur in the cement stone leading to a decrease in strength [9].

The aim is to study the morphology and composition of damaged cement stone modified with an additive based on polyoxyethylene derivatives of unsaturated carboxylic acids and with nanoscale aluminum oxide.

2. Raw materials and mix preparation

The following materials were used in the work:

- Portland cement of CEM I 42.5 B type produced by JSC 'Sukholozhskcement' (Sverdlovsk region);

- aluminum nanooxide 'TG' which is obtained by thermal hydrolysis method developed at the Institute of chemistry of solid state of the Ural branch of the Russian Academy of Sciences (Ekaterinburg) [10];
- plasticizer 'Linamix PC' based on polyoxyethylene derivatives of unsaturated carboxylic acids produced by JSC 'Polyplast-URALSIB' (Pervouralsk).

The basis of the method for preparing the cement mix was taken from [11]. 0.3 % of Portland cement was introduced into 30 ml of water with addition of 0.5 % of nanooxide. The suspension obtained was treated with ultrasound for 30 minutes at a temperature of 30 °C in an 'Elmsonic S30N' ultrasonic bath. The suspension treated was then mixed with 1.0 % plasticizer and then injected into the remaining part of cement (99.7 %). Nanooxide and plasticizer were introduced in excess of 100 % of the cement mass. After combining all components, the mix was thoroughly mixed in a 'SL-1500' high-speed mixer for two minutes.

Small samples–cubes (2×2×2 cm) were formed from the mix obtained, at the age of 24 hours they were removed from the forms and then stored in high humidity conditions.

3. Results and discussion

After 28 days of storing the destruction and formation of crystals were observed in cement samples of the composition mentioned above. Figure 1 shows an image of one of the samples that were damaged. It can be seen that one side of the sample was exfoliated, macro- and microcracks appeared and small crystals were found on the surface of the rift and the sample itself. Some of the crystals were visible even at low magnification (figure 1, b).

a b

Figure 1. Sample of cement stone with destruction signs: a – general view of the sample; b – new formations on the sample surface at 25X magnification.

For a more detailed study micrographs of cement stone were made with the determination of the chemical composition at some points. The morphology of the samples and their chemical composition were studied using a JEOL JSM-6390LA scanning electron microscope equipped with an 'EDS Inca Enegy 250' X-ray spectrometer (figure 2–3).

Figure 2 shows well-crystallized plates arranged in order and parallel to each other. The thickness of formed plates on a cement stone sample with signs of destruction from 0.6 to 2 microns, on average – 1.5 microns. The length and width of the plates are approximately 300–400 microns.

The forming spherical particle that can be observed in figure 3 has a diameter of 30 microns. Smaller spherical particles attached to a large sphere have an average diameter of 2–5 microns. Thin fibers with a diameter of less than 1 micron were also detected; they are uniformly distributed on the cement stone sample surface and also in micro cracks (figure 3).

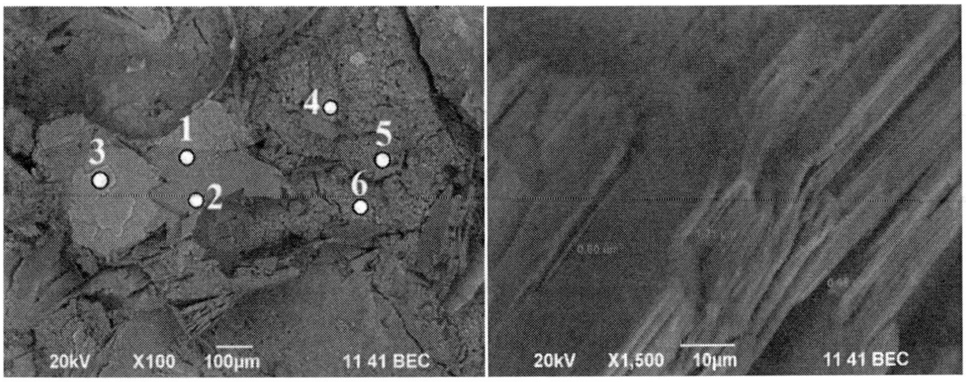

Figure 2. Structure of cement stone at different magnification.

The quantity of basic oxides obtained from microanalysis of points shown in the figures 2–3 is presented in table 1. It can be assumed that the distinct plates at the image (figure 2, points 1, 2 and 3) are $Ca(OH)_2$ which formed as a result of the hydration reaction of alite and belite [1]. The sphere (figure 3) can be formed by the main products of alite (C_3S) hydration which are calcium hydrosilicates (the ratio of CaO/SiO_2 is equal 3/1 according to table 1, points 13 and 14). The fibres on image (figure 3) are most likely ettringite.

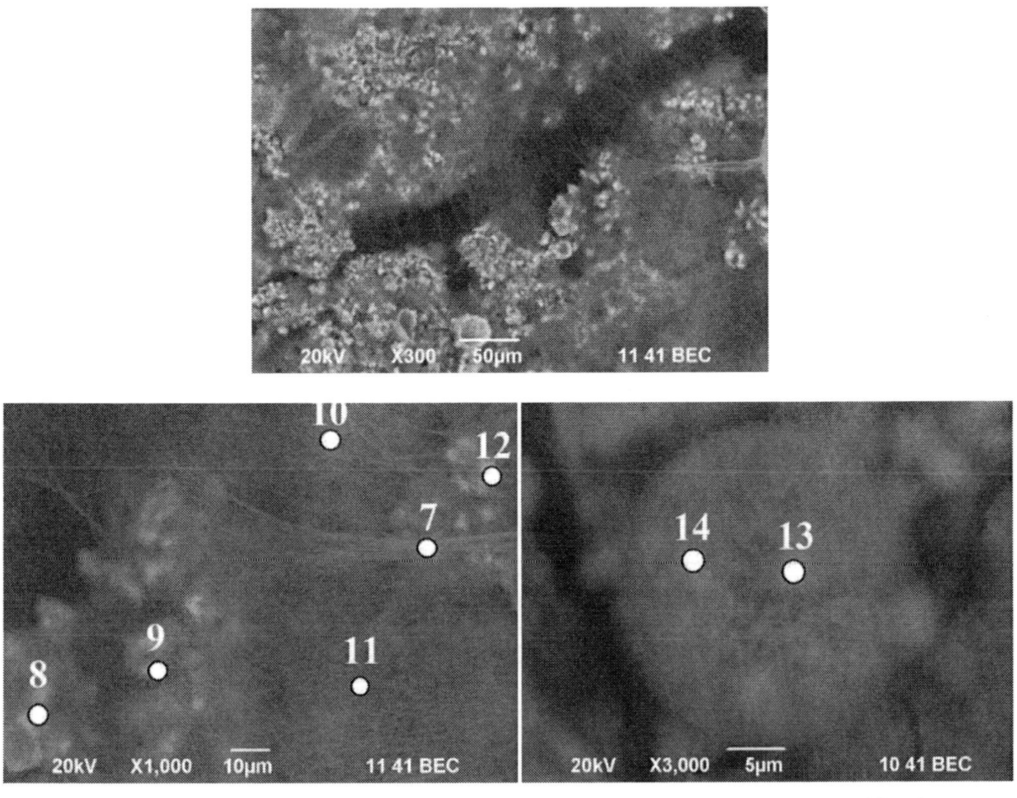

Figure 3. Fibres and sphere in the structure of cement stone at different magnification.

Chernyshov and Korotkikh reported in [12] about the effect of nano-self-reinforcement of the cement stone structure in which needle-fiber reinforcing elements (filamentous new formations of ettringite) are formed directly during the physical and chemical processes of hydration and hardening of the binder. The sizes of hydrate crystals formed in cement stone do not exceed several tens, less often hundreds of nanometers across [13]. However the situation is not so simple since the crystallization of long hydrates is accompanied by an increase in the crystallization pressure as a result of which the positive effects of nano-reinforcement of cement stone are 'superimposed' by the negative effects of its possible self-destruction from the action of such pressure [14]. Most likely the signs of such processes were encountered and recorded during the hardening of the cement stone containing the plasticizer 'Linamix PC' and aluminum nanoxide.

Table 1. Microanalysis results of cement stone at individual points.

Point number (at figures 2 and 3)	Quantity of basic oxides (%)						
	$Ca(OH)_2$	CaO	SiO_2	Al_2O_3	SO_3	FeO	MgO
1	100.00	–	–	–	–	–	–
2	97.10	–	2.90	–	–	–	–
3	100.00	–	–	–	–	–	–
4	–	85.07	6.85	1.44	1.14	5.50	–
5	–	69.32	23.27	2.30	–	5.11	–
6	–	76.21	3.51	6.02	3.16	11.10	–
7	–	61.66	13.86	8.68	14.73	2.88	–
8	–	59.51	27.36	2.54	1.42	3.61	–
9	–	60.06	14.61	8.50	14.30	2.53	–
10	–	60.91	22.81	6.00	7.13	3.14	–
11	–	59.19	25.49	5.08	5.47	3.92	0.85
12	–	59.68	19.93	6.84	8.69	3.41	1.44
13	–	60.19	23.19	5.32	5.95	4.37	0.98
14	–	56.35	27.11	6.46	6.69	3.39	–

4. Conclusions

The morphology and composition of damaged cement stone modified with an additive and with a nanoscale aluminum oxide were studied. It is shown that in the presence of 'Linamix PC' plasticizer and a nanoscale aluminum oxide, well-crystallized $Ca(OH)_2$ plates with an average thickness of 1.5 microns and a length and width equal to 300–400 microns are formed. Also, on the surface of microcracks of cement stone a network of thread-like ettringite is formed with a diameter of less than 1 micron. Most likely the cement stone sample could be damaged by the forces of crystallization pressure during the growth of both $Ca(OH)_2$ and ettringite crystals.

References

[1] Kuznetsova T V, Kudriashov I V and Timashev V V 1989 *Physical Chemistry of Binders* (Moscow: Vysshaia shkola) p 383

[2] Qianping R, Liu J, Yang Y, Shu X, Zhang J and Mao Y 2015 Effect of Molecular Weight of Polycarboxylate Superplasticizer on Its Dispersion, Adsorption, and Hydration of a Cementitious System *Journal of Materials in Civil Engineering* **28** 04015184 doi 10.1061/(ASCE)MT.1943-5533.0001460

[3] Li Z, Wang H, He S, Lu Y and Wang M 2006 Investigations on the Preparation and Mechanical Properties of the Nano-Alumina Reinforced Cement Composite *Materials Letters* **60(3)** 356–359 doi 10.1016/j.matlet.2005.08.061

[4] Prince W, Espagne M and Aïtcin P-C 2003 Ettringite Formation: A Crucial Step in Cement Superplasticizer Compatibility *Cement and Concrete Research* 635–641

[5] Kim B, Jiang S, Jolicoeur C and Aïtcin P-C 2000 The Adsorption of PNS Superplasticizer and Its Relation to Fluidity of Cement Paste *Cement and Concrete Research* **30** 887–893 doi 10.1016/S0008-8846(00)00256-8

[6] Arel H and Aydin E 2017 Effects of Ca-, Mg-, K-, and Na-lignosulfonates on the Behavior of Fresh Concrete *Construction and Building Materials* **157** 1084–1091 doi 10.1016/j.conbuildmat.2017.09.190
[7] Dalas F, Pourchet S, Rinaldi D, Nonat A, Sabio S and Mosquet M 2015 Modification of the Rate of Formation and Surface Area of Ettringite by Polycarboxylate Ether Superplasticizers during Early C_3A-$CaSO_4$ Hydration *Cement and Concrete Research* **69** 105–113
[8] Matiukhina O N and Evdokimova I V 2012 Ettringite Crystallization in the presence of Functional Additives for Dry Building Mixes *Dry Building Mixes* **6** 18–21
[9] Babkov V V, Sakhibgareev R R, Chuikin A E, Terekhov I G and Kabanets V V 2005 Hardening and destruction of cement stone under long-term conditions *Bulletin of BSTU* vol **9** (Belgorod: BSTU) pp 14–21
[10] Method for Producing Metal Oxide Nanoparticles Patent RU 2384522
[11] El-Gamal S.M.A, Abo-El-Enein S A, El-Hosiny F I et al 2018 Thermal Resistance, Microstructure and Mechanical Properties of type I Portland Cement Pastes containing Low-Cost Nanoparticles *Journal Therm. An. Calorim.* **131** 949–968 doi 10.1007/s10973-017-6629-1
[12] Chernyshov E M and Korotkikh D N 2009 Nanoreinforcement of Cement Stone Structures with the help of crystals of Ettringite for Concrete Fracture Strength increasing *Scientific Herald of the Voronezh State University of Architecture and Civil Engineering. Construction and Architecture* **2** 21–29 192 [In Russian]
[13] Sheykin A E 1974 *Structure, Durability and Cracking Resistance of Cement Stone* (Moscow: Stroyizdat) p 192 [In Russian]
[14] Beylin M I 1984 *Research and application of straining concrete and reinforced concrete constructions: collection of scientific labours of NIIZHB* (Moscow: Stroyizdat) p 15–23 192 [In Russian]

Utilization of final tailings concentration of titano-magnetite ores in Portland cement production

F L Kapustin[1,2] and N N Bashkatov[1]

[1]Institute of new materials and technologies, Ural Federal University named after the first President of Russia B N Yeltsin, 19, Mira str., Yekaterinburg, 620002, Russia

E-mail: [2]f.l.kapustin@urfu.ru

Abstract. When producing iron ore concentrate at Kachkanarsk integrated mining works 2 types of wastes are formed, namely: final tailings of dry magnetic separation (DMS) and final tailings of wet magnetic separation (WMS). Final tailings of wet magnetic separation (WMS) are pulled off or dumped into disposal area up to 45mln tons annually causing great damages to the environment. These final tailings are finely dispersed materials, their chemical composition is oxides of silicon, aluminium, calcium and iron in general. All this allows to use the wastes under consideration to produce Portland cement clinker replacing clayey and iron components in the raw material mixture completely and limestone partially. The replacement of clay with final tailings of ore enrichment are suitable to manufacture clinker raw material mixture allowing to use two-component raw mix instead of three-component and four-component one thus simplifying the technology of cement production and improving the ecological situation in Sverdlovsk region. The utilization of final tailings for concentration of titano-magnetite ores decreases fuel consumption when burning clinker compared to the traditional compositions. Experimental cement meets all the requirements of the Russian Standard 10178 as to their chemical and physic-mechanical properties and have grade 500.

1. Introduction

Mining industry all over the world is developing approximately 1.5 times quicker than other branches of industry using mineral raw materials. In this case the Russian Federation is not an exception which results in an invariable increase of both mining volumes and raw material enrichment.Therefore, the quantity of technological wastes is increasing, including slimes, final tailings and slags which are to be stored in disposal area.The availability of such dumps including proper activity of integrated mining works and mining benefication works (IMW) causes definite damage to environment, namely: air pollution, ground waters and land tenure. There are different suppositions as to utilization of integrated mining works wastes beginning from building materials, and articles manufactured from secondary raw materials for metallurgical industry [1–11].

As the final tailings of ore concentration are fine dense materials firstly, they were supposed to be used as fine aggregates in structural concrete. This was a very successful solution because the concretes produced have higher flowability and place mentability compared to concrete mixtures produced by using quartz sands. Besides the chemical reaction takes place on the border 'cement-aggregate grain' to form calcium hydrosilicates with low basility whichincreasestheconcretestrength [12]. However, dueto this the average strength of concrete density increases in the range of 60 to 250 kg/m^3, porosity decreases from 1.87 to 0.83 %, and water adsorbtion of concrete from 0.78 to 0.19 %.

According to the researches carried out in Ukraine the final tailings from IMW of Krivorozhsky basin containing up to 60–80% of quartz can be used to produce geopolymer binding materials without clinker. The cements in question feature higher durability in sulphatemedia, they can be effectively used to produce concretes with up to 50–75MPa [13]. In Novocherkassk polytechnic institute white cement was made delete it on the basis final tailings from Tyrkyauzsky IMW which were used both as an aluminosilicate component in the raw materials clinker mixture and as a mineral addition when cement was milled [14].

Kachkanarsky IMW concentration poor titanium-magnetite ores containing up to 16% of iron by using multistepped recycling which results in the formation of large quantity crushed and ground material wastes of two types namely: final tailings of dried magnetic separation (DMS) and final tailings ofwet magnetic separation (WMS) [15]. DMS final tailings are the result of crushed rock screening and their grain size is from 10 to 40mm. The particles more than 10mm in size are subjected to wet grinding and further to the first stage of magnetic separation. Then the material passes several stages of grinding and separation. After that WMS final tailings with up to 85% particles from 0.1–1.6mm in size are poured into settling tank. About 45mln tons of similar wastes in the form of slime are produced at IMW annually.

2. Raw materials and methods of investigation

Final tailings of WMS and DMS were used in this research at Kachkanarsky IMW. The mineral composition of DMS final tailings consists of up to 95%pyroxenites, up to 10% plagioclases and not more than 5% gabbro. WMS final tailings consists of up to 80–90% pyroxenes, from 3 to 20%olivines and up to 20% amphiboles. Besides plagioclases, titanomagnetites and ilmenites are also available in their composition as impurities. The chemical composition of final tailings used in experimental research are given in table1.

Table 1. Chemical composition of final tailings from Kachkanarsk IMW.

Waste type	Oxide content in mass%						
	LOI	SiO_2	Al_2O_3	Fe_2O_3	CaO	MgO	others
DMS	0.70	47.34	12.22	11.74	18.43	6.22	3.35
WMS	0.68	47.51	8.15	10.30	19.99	14.06	1.47

To prepare two-component mixtures limestone from Neviansky and Pashinsky deposit were used (table 2). Four-component raw meal from ZAO Neviansky cementnik and DMS final tailings from OAO Gornozavodsk Cement were used as control contents. To regulate cement setting gypsum anhydride stone from Ergachynsky deposit was used.

Table 2. Chemical composition of raw materials.

Materials	Oxide content in mass%						Modula		
	LOI	SiO_2	Al_2O_3	Fe_2O_3	CaO	MgO	*SC*	*n*	*p*
Limestone of Neviansky deposit	41.59	2.79	1.98	0.89	52.02	0.41			
Limestone of Pashinsky deposit	39.35	6.22	1.99	1.01	50.11	0.81			
Raw meal	32.86	14.18	3.84	3.25	43.73	1.21	0.91	2.00	1.18
Raw slime	34.87	13.72	3.56	2.79	42.79	1.62	0.94	2.16	1.28

Chemical composition of raw materials and mixes were determined by Russian State Standard 5282 'Cements and materials of cement production. Methods of chemical analyses'. Calculation of the raw materials content was done according to generally accepted methods for the saturation coefficient SC=0.94, the raw mix was milled to specific gravity of 300m^2/kgto produce clinker from laboratory and plant mixes briquettes were prepared which were burnt in the muffle kiln at 1450°C and is other mic availability for 30 minutes. The physic-mechanical propertiesof the experimental and control

IOP Conf. Series: Materials Science and Engineering **966** (2020) 012052 doi:10.1088/1757-899X/966/1/012052

cements were determined in accordance with Russian Standard 310 'Cement. Methods of testing' and the results obtained were compared with Russian Standard 10178 'Portland cement and Slag Portland cement. Technical conditions'.

3. Results of experiments and discussion

The chemical composition of WMS final tailings from Kachkanarsk IMW is characterized by stability of the same order kind as foe the raw materials used in cement industry. However, their chemical and mineral compositions vary considerably from the IMW final tailings of the south-western part of Russia and Kazakhstan, tested as raw materials to produce cement clinker [14, 16]. Kachkanar final tailings are characterized by high content of CaO, MgO, TiO_2 and SiO_2 and differ from the closest to their content final ore tailings of Sokolovo-Sarbysky deposit.It is found that the raw mixtures produced by using final tailings of Kachkanarsk IMW and limestones from Neviansky deposit are characterized by higher silicate modulus, but Pashinsky raw mixtures are characterized by higher silicate modulus (n) too and lower value of Al_2O_3 modulus (p) (table 3). Besides the latter contain the higher quantity of MgOwhich requires additional tests of cements as to the uniformity of volume changes when MgO expands.

Table 3. Composition and characteristics of tested mixtures containing WMS final tailings.

The origin of limestone	Components content, mass%		Modula			MgO quantity in clinker, mass%
	limestone	final tailings	*SC*	*n*	*p*	
Neviansky	75.05	24.95	0.94	2.15	1.19	5.62
Pashinsky	80.03	19.97	0.94	2.38	1.13	5.04

Taking into account that the raw meal used at ZAO Neviansky cementnik contains 79.36% of limestone, but the raw slime from OAO Gornozavodsk Cement consists of 85.57% of carbonate content, the use of the final tailings from Kachkanar IMW allows not only to exclude clayey and iron components, but also to decrease limestone consumption in the two-component mixtures. And availability of CaO in the final tailings bound in silicates will contribute to fuel consumption reduction which is used for the clinker formation processes.

The research of the raw mixtures sintering showed that CaO_f assimilation in the experimental mix using Neviansky limestone is taking place approximately in the same way as when the burning of the plant free lime (CaO_f) occurs in spite of high silicate modulus and raised quantity of MgO(table 4). The considerable content of CaO_f in the mix under study after burning at 950°C explains the acceleration of the limestone decarbonization when final tailings concentration is available. The tested raw mix using the limestone from Pashynsky deposit has shown a smaller quantity of CaO_f during burning at all temperatures compared to the raw slimes from OAO Gornozavodsk Cement that acknowledges the improvement of clinker burnability when WMS final tailings are introduced.

The research has shown that Portland cement produced by grinding of experimental clinker with a 5 % addition of gypsum-anhydride stone has physic-mechanical properties close to cements made of plant mixtures (table 5). However, they have a shorter period of setting and higher compressive strength. Experimental cements meet the Russian Standard 10178 requirements, as their properties and for ZAO Neviansky cementnik plant Portland cement has 500 grade as to strength, but for OAO Gornozavodsk Cement Portland cement in question has 400 grade.

Table 4. The influence of WMS final tailings on the burnability of raw mixtures.

Cement plant	Mixture type	CaO_f content, mass % after burning at temperatures°C					
		950	1150	1250	1350	1400	1450
Neviansky cementnik	plant	26.0	24.0	11.0	3.7	2.8	0.9
	experimental	32.0	21.0	7.0	3.0	2.5	0.8
Gornozavodsk Cement	plant	28.0	26.0	18.0	2.6	2.1	0.6
	experimental	28.0	17.0	13.0	2.0	1.6	0.5

Table 5. Physic-mechanical properties of Portland cement.

Cement plant	Charge type	Fineness of grinding, %	Normal thickness of cement paste, mass%	Setting times, hours-minutes		Compression strength after 28 days, MPa	
				beginning	ending	bending strength	compressive strength
Neviansky cementnik	plant	4.8	31.7	2-10	3-15	6.9	47.9
	experimental	9.7	28.0	1-55	2-50	6.8	49.6
Gornozavodsk Cement	plant	5.5	34.6	1-29	5-47	6.6	43.5
	experimental	8.4	29.3	1-08	4-30	6.4	44.1

With the aim to confirm the lab research results at Nizhnetagilsky cement plant proof-production tests were carried out to obtain Portland cementby burning twocomponent mixture consisting of limestone and DMS final tailings the chemical composition of this two-component mixture is given in table 1. The five-component raw mixture containing 78.9% in mass of limestone, 18.3% of clay, 1.6 % of sand, 0.7% of blast furnace granulated slag and 0.5% of pyrite cinder was used to burn clinker at the plant in the traditional manner. The experimental slime consisted of 74.0% of limestone and 24.5% of final tailings.

The preparation of the raw material slime was done according to the technology accepted at the plant with chemical composition correction in vertical basins and storage of finished slimes in the horizontal basin 4270m^3 of experimental slime have been prepared with characteristics shown in table 6. The recording of the rotary kilns work to burn plant slime were fixed within 24 hours and the experimental slime recording were fixed for 34 hours. The productivity of the rotary kilns for clinker burning was determined by means of slime consumption calculation taking into account its moisture and mass losses when calcining takes place.

Table 6. Composition and properties of raw material slims.

Raw material slims	Modula			Sieve residue No02, mass%	Humidity, mass%	LOI, mass%
	SC	*n*	*p*			
Plant	0.91	1.81	1.00	1.4	36.0	34.54
Experimental	0.91	1.99	1.04	1.3	30.7	30.89

When carrying out the experiments it was established that the productivity of the raw mills grinding either slime does not differ. Feeding the kilns with experimental slime was supported on the same level and the level was not changed when the common mixture was burnt, but at the same time the productivity concerning clinker burring increased due to less moisture and mass losses on calcining. So, the kiln productivity with ∅3.6 by 127m in size has increased by 30% and the kiln with ∅3.6 by 150m in size increased by 11%and the specific fuel consumption decreased by 21.4 % for the former and by 12.5% for the latter one. The difference in the results of the kiln work while burning clinker is apparently takes place due to carry-over of dust. It is quite possible that while introducing the experimental slime into the short kiln the ablation of material increases because of poor work of the chain curtain. The increase of speed and temperature of the kiln gases caused by gaseous fuel consumption growth when burning the mixture and also vacuum or rarefaction in dust chamber could contribute to this too. As to the long kiln these indicators decreased when two-component slime was used. The working parameters of the kiln with ∅3.6 by 150m are given in table 7.

Table7. The indicators of the kiln with ∅3.6 by 150 m dimentions to burn clinker.

Indicators	Raw material slims	
	plant	experimental
t° of off-gases	247	269
Vacuum in dust chamber, Pa	820	795
Fuel consumption, m^3/h	5465	5428
Productivity as to clinker	26.8	30.4
Specific fuel consumption, m^3/tclinker	203.7	178.3

Six samples of the experimental and plant clinkers have been chosen during proof-production tests. The samples of the experimental clinker possessed fairly close modula characteristics but differed from the plant clinker with higher quantity of MgO (table 8). Besides in 2 samples of the experimental clinker with *SC*=0.93 the CaO_f content was 1.46 and 2.39%, but in the rest contents the CaO_f content did not exceed 1%. Insufficient degree of burning of some experimental clinkers can be caused by high kiln productivity when the raw slime was burnt which consisted of final tailings and shortage of fuel necessary to burn these experimental clinkers. When *SC* of the experimental clinker is on 0.91, level the better burnability of the clinker is provided and accordingly the place. The plant and experimental clinkers with rationing CaO_f content have an activity about 50.0 MPa, but the raised quantity of $CaO_f$the activity decreases down to 44.1MPa.

Table 8. Modula characteristics and activity of clinkers.

Clinker	Modula			MgO quantity in clinker, mass%	Compression strength after 28 days, MPa
	SC	*n*	*p*		
Plant	0.91	1.81	0.95	1.64	50.6
Experimental	0.91–0.93	1.99–2.08	0.98–1.12	3.82–4.32	48.8–49.0

4. Conclusion

The final tailings of titanium-magnetite ores concentrationform in considerable quantity as aluminosilicate component of Portland cement raw mixture. But iron component is to be excludedfrom it. By using the final tailings in the composition of the two-component mixture allows to simplify the technology of the raw slime preparation to reduce its moisture and fuel consumption when burning clinker and also to raise the productivity of the rotary kiln which is confirmed with the proof-production tests.

References

[1] Makarov D V, Melkonyan R Y, Suvorova O V and Kumarova V A 2016 Perspectives of the use of the industrial wastes to obtain ceramic binding materials *Mining info-analytical bulletin* **5** pp 254–81

[2] Gafarova N E 2016 Application of shell rock into monolithic building *International journal of applied fundamental research* **6** pp 630–2

[3] Sverguzova S V, Starostina I V, Fomina E V, Porozhnuk L A, Denisova L V and Shykhiev I Y 2016 Obtaining of decorative plaster mixes by using final tailing to enrich ferric quartzites *Herald of Technological University* **1923** pp 144–8

[4] Melkonyan R G 2016 Ecological problems of the use of technogenic raw materials in the production of silicate materials *Mining info-analytical bulletin* **S1** pp 499–510

[5] Krasheninnikov O N, Belogurova T P, Latshuk V V and Puk A A 2007 Overburden rocks of Colsky peninsula deposit and getting broken stone on its base *Ecology of industrial production* **1** pp 64–73

[6] Belogurova T P and Krasheninnikov O N 2004 Utilization of overburden rocks of Hybinsky apatite-nepheline deposits in building *Building materials* **7** pp 32–5
[7] Absatarov S H, Loktionov S V and Fedorovsky U A 2007 Manufacture of crushed stone from overburden rocks at Lebedinsky IMW *Mining journal* **7** pp 56–58
[8] Chen Jia, Chen Tiejun and Zhang Yimin 2012 Vanadium tailings for high performance ceramsite synthesis *Metal Mine. In Chinese* **1427** pp 161–5
[9] Zengxiang Lu and Meifeng Cai 2012 Disposal methods in solid wastes from mines in transition from open-pit to underground mining *Procedia Environmental Sciences* **16** pp 715–21
[10] Zhou Lianbi 2007 Investigation and practice on mining land rehabilitation and ecological reconstruction in China *Nonferrous Metals* **259** pp 90–4
[11] Vyborov S G and Silin A A 2012 Perspectives of dump rocks as aluminium raw materials *Ukraine coal* **5** pp 31–7
[12] Chuprunenko E V and Larionova Z M 1976 Investigation of aggregates interaction with cement stone from Dnepropetrovsk IMW *Scientific works of Belgorod Technological Institute of binding materials* **21** pp 24–30
[13] Shevchenko O, Chehov V and Mysin V 1973 Slags lime clinkerless cement *Binding materials and constructions* **5** pp 20–1
[14] Florova R I 1975 Utilization of wastes from integrated mining works to produce cements *Scientific works of Novocherkask polytecnik Institute* **312** pp 87–8
[15] Prokofieva V V, Bozhenov P I, Syhachev A N and Eremmin H Ya 1986 Utilization of products to enrich iron in the building on the North (Leningrad: Stroyizdat) [In Russian]
[16] Sylumanov A T 1986 Binding materials made of industrial by-products (Moscow: Stroyizdat) [In Russian]

Utilization of ferrochrome slag in concrete production

N N Bashkatov

Institute of new materials and technologies, Ural Federal University, 19, Mira str., Yekaterinburg 620002, Russia

E-mail: n.n.bashkatov@urfu.ru

Abstract. Ferrochrome slag is waste resulted from the manufacture process of ligature melting. It is a porous material of dark grey-green colour with 2530 kg/m^3 density and the open pores size from 1mm to 5 mm. After crushing crushed slag possesses the following characteristics or properties, namely: the least grain coarseness is 20 mm, the largest one is 120mm, the needle and plane grain content is 25.5%, so it refers to ordinary group of crushed stone, its crushability grade is M800, attrition grade AI and frostproofness is F25. Screenings or riddlings obtained by crushing and fractioning of slag possess gradation factor 2.9 and it refers it coarse sand group of class II. Concrete mixes have been investigated by using granite aggregates and ferrochrome ones. It is determined that mixes containing ferrochrome aggregate are stiffer than the traditional ones. In this case strength characteristics of the experimental concretes are 1.7 times higher than those of the traditional concretes. When evaluating approximate cement consumption for concrete made of ferrochrome slag aggregates it was established that such concrete possesses strength of B30 class, but by using granite aggregate in the concrete mix result in strength of B20 class.

1. Introduction

Metallurgy and associated with it mining complex takes one of the leading places both in the mineral raw materials mined and the recycled one per ton of the finished products among other branches of industry. Slags, their chemico-mineralogical composition and physico-mechanical properties allow to consider them as a valuable raw material to produce building materials. They are formed as by-products at the main technological process stages of ferrous and non-ferrous metals production. Since the geological conditions of ferrous metals deposits worked out become more complicated from year to year and their regular content in mineral raw material decreases, the increase of mining volume and ores recycling require more and more substantial expenses. On the one hand the economy or saving every ton the mineral raw material is a worthwhile matter, on the other hand the damage is becoming increasing perceived because the complex utilization of the mineral raw material is not fully used. Therefore, more wide and complex utilization of the mineral raw material is one of the key problems of technical progress in non-ferrous metallurgy, its solution practically involves the increased level of metals production. All these problems result in razing waste volume both in metallurgical and mining complexes and consequently the pollution of environment. The issue of waste recycling of both complexes is of vital importance for all highly developed industrialized countries [1–5]. In Russia they are generally utilized as aggregates or fillers [6–9].

To transport waste (slags) into disposal area is not profitable, because presents great material expenses as to their storage. Metallurgical plants, as a rule, prefer to crush such slags and suggest to use them as coarse aggregate, but fine fractions resulting from crushing are removed and wasted. Cast crushed slag as to its physico-mechanical properties is as good as natural dense crushed stones for effective types of heavy concrete aggregates [10].

Ferro-chrome slag from OOO Kluchevskaya concentrating or dressing mill (KCM) is one of the waste of non-ferrous metallurgy resulted from ligature. There exist technologies of returning a number of wastes such as ferrous slags, melts and non-standard metals into production [11]. However, this fails to solve the recycling problem completely though the waste quantities are decreased. In this country there are 6 specialized mills to produce ferroalloy products, four of them are located on the territory of Sverdlovsk region [12]. This fact counts in favour of utilization such wastes urgently and without delay.

2. Raw materials and methodology

Ferrochrome slag from KCM is the material of dark grey-green colour, 2530kg/m^3 density, hardness 7–8 on Moss scale, water adsorbtion 2.63% and the developed surface on which there are open pores from 1 to 5mm in size which are distinctly seen. 3 laboratory slag samples have been investigated. These samples differ in chemical composition, but possessquite enough close values of basility M_b which is equal to 1.41 on the average (table 1). Because M_b is larger than 1, ferrochrome slag is referred to this group.

Table 1. Chemical composition of slag.

Sample No	Oxide content in mass%							M_b
	SiO_2	Al_2O_3	FeO	CaO	MgO	Cr_2O_3	others	
1	27.5	7.0	0.2	33.0	16.1	14.3	1.9	1.42
2	30.8	7.0	10.7	34.0	17.0	0.2	0.3	1.35
3	26.7	7.5	14.9	31.0	18.7	0.2	1.0	1.45

Portland cement CEM II/A-S manufactured by OAO Sycholozhcement, granite broken stone of 10-20 mm fraction and sand from granite crushing riddling of Shartash deposit have been used when heavy concrete content was selected (table 2). To obtain concretes ferrochrome slag was crushed up to the same fractions.

Table 2. Characteristics of aggregate.

Material	The largest grain size, ∅mm	The fineness modulus of sand	Bulk density, kg/m^3	Grain density, kg/m^3
Slag crushed stone	120	–	1260	2530
Slag sand	–	3.0	1640	2530
Crushed granite stone	20	–	1320	2600
Sand from granite crushed riddlings	–	2.1	1520	2620

The grain density of materials differs, but not so much. It is explained by the more developed surface of ferrochrome slag the open pores 1–5 mm in size are distinctly seen on it. The volume of open pores of granite crushed stone is 0.41%, but the volume of the crushed slag ones is 1.82%. Physico-mechanical testings of ferrochrome slag were carried out according to Russian Standard 8269.0 and 8735 to conform the requirements to Russian Standard 5578 'Crushed stone and sand for slag of ferrous and nonferrous metallurgy for concretes. Technical conditions' and Russian Standard 3344 'Slag crushed stone and sand for road building. Technical conditions'.

Calculation of concretes content was done according to the method of absolute volumes. Concrete mixes and concretes proper were examined in accordance with Russian Standard 1273.0 'Concretes. General requirements to the methods to determine density, moisture, water adsorbtion, porosity and water permeability', Russian Standard 10180 'Concretes. The methods of strength determination by the control samples', to conform the requirements to Russian Standard 26633 'Heavy and fine-grained concretes. Technical conditions'.

3. Results of experiments and discussion

As a coarse aggregate ferrochrome slag possesses the following characteristics including: needle and grain forms content is 25.5% and due to this content, it refers to the ordinary crushed stone group, the content of dust particles is 0.09 and the crushability grade is M800, the attrition grade is AI and the frost resistance is F25 (table 3). Thus, ferrochrome slag meets the requirements standard 3344 'Slag crushed stone and sand for road building. Technical conditions' and it can be recommended for utilization as coarse aggregate in road building.

Table 3. Physic-mechanical properties of crushed ferrochrome slag.

Quality indices	Value	Conclusion Russian Standard 3344–83
d the largest	120	meets
d the least	20	meets
Group of grains as to their from	ordinary	meets
Content of dust particles, mass %	0.09	meets
Water adsorption, %	1.44	is not normalized
Crushability, mass %	18.96	markM800
Attrition, mass %	22.13	markAI
Frost resistance, mass %	9.96	mark F25

Being the coarse aggregate for concrete ferrochrome slag waste as to its content of plane and needle form refers to the third group having the crushability mark M600, the other characteristics conform to table 3, speaking about its capability or practicability to be used as broken slag aggregate when producing concrete mix. Riddlings obtained as a result of slag crushing refers to the sand group of class II, it possesses the gradation factor 2.9 and meet Russian Standard 5578 requirements accordingly and it can be considered as slag sand.

Four concrete mix proportions or concrete mixes have been calculated in accordance with mark M350 which was predetermined or prescribed, and the cone sedimentation mobility is from 2 to 4 cm (table 4). The concrete mixes were distinguished by aggregates proper. Four aggregates were tested, namely:

– The first aggregate was the control one, it consisted of the granite crushed stone and its riddling after crushing.

– The second aggregate was made of the granite crushed stone and riddling from ferrochrome crushing.

–The third aggregate was made of ferrochrome slag and riddling from granite crushing.

– The fourth aggregate was made of ferrochrome slag and its riddling after crushing.

According to the calculation trial batches or trial mixes and their correction have been designed (table 5).

Placeability of concrete mix was assessed or evaluated by standard cone sedimentation. Due to this index all concrete mixes have placeability mark P1, but experimental concrete mixes are stiffer because their standard cone sedimentation is from 2.0 cm to 2.5 cm against 3.5 cm of the control concrete mix. This is explained by the fact that ferrochrome slag has more developed surface in comparison with the granite one, therefore, its water consumption is higher.

Comparing concrete mix No 1 with No 3 and No 4 ones it is seen and understandable that concrete production consumes more coarse aggregate and less fine one during the last years. This is explained by the fact that there exists the possibility to economize natural raw material when producing concrete. Besides it should be noted that the experimental mixes require less cement consumption which is the most expensive concrete mix component.

Table 4. Calculated contents of concrete mixes at W/C=0.51.

Content No	Material consumption per 1 m^3 of concrete mix, kg						Calculated density of concrete mix, kg/m^3
			granite		ferrochrome slag		
	water, l	cement	crushed stone	sand	crushed stone	sand	
1	170	335	1123	726	–	–	2354
2	170	335	1123	–	–	752	2380
3	170	335	–	776	1068	–	2349
4	170	335	–	–	1068	750	2323

Table 5. Factual contents of concrete mixes when W/C=0.51.

Content No	Material consumption per 1 m^3 of concrete mix, kg						Density of concrete mix, kg/m^3
			granite		ferrochrome slag		
	water, l	cement	crushed stone	sand	crushed stone	sand	
1	200	395	1132	640	–	–	2367
2	214	421	1123	–	–	752	2510
3	192	380	–	645	1165	–	2382
4	200	389	–	–	1165	623	2377

During the experiment is was established that concrete made of ferrochrome slagin the form of coarse and fine aggregates possesses, higher strength characteristics (table6). Since as it was indicated above slag has more developed surface and due to this cohesion of aggregate and binder or binding material improves. Due to water suction by porous aggregate cement slurry penetrates into open pores, as a result as if growing of cement stone and aggregates takes place. Water adsorbtion contributes to elimination of water films formation danger on the surface of aggregates which hinder cohesion.

Table 6. Physic-mechanical properties of concrete.

Content No	Density, kg/m^3	Compressive strength, MPa in a day			Class
		3	7	28	
1	2358	14.8	25.2	28.3	B20
2	2500	17.2	23.6	30.4	B20
3	2378	12.6	20.9	34.3	B25
4	2372	13.5	23.9	43.0	B30

4. Conclusion

Ferrochrome slag due to its physic-mechanical properties meets Russian Standard 3344–83 and 5578 requirements. It can be utilized both crushed stone in road building and coarse and fine aggregate in concrete production. Concretes made of ferrochrome slag conform Russian Standard 26633 and have higher density and strength than concretes made of traditional materials, they refer to B25-B30 class. The replacement of granite crushed stone by ferrochrome one results in Portland cement consumption reduction decreasing or reducing concrete cost price.

References

[1] Chen Jia, Chen Tiejun and Zhang Yimin 2012 Vanadium tailings for high performance ceramsite synthesis *Metal Mine. In Chinese* **1427** pp 161–5
[2] Zengxiang Lu and Meifeng Cai 2012 Disposal methods in solid wastes from mines in transition from open-pit to underground mining *Procedia Environmental Sciences* **16** pp 715–21
[3] Zhou Lianbi 2007 Investigation and practice on mining land rehabilitation and ecological reconstruction in China *Nonferrous Metals* **259** pp 90–4

[4] Sylumanov A T 1986 Binding materials made of industrial by-products (Moscow: Stroyizdat) [In Russian]
[5] Batalin B S and Kuryakova H B 2004 Sulphate activation of spontaneous disintegration ferro-vanadium slag *Izvestiya of higher schools* **9 (549)** pp 43–9
[6] Zhdanov A V, Zhuchkov V I, Dashevsky V Ya and Leontiev L I 2014 Utilization of wastes of ferroalloy production *Stal* **3** pp 92–8
[7] Galeev R R 2011 Utilization of metallurgical industry wastes as aggregates in the products made of nonplasticized PVC *Socially-economic and technical systems: research, design, optimization* **59 2** pp 18–24
[8] Ovchinnikov V P, Aksenova H A, Rozhkova O V and Fedorovskaya V A 2014 Utilization of blast-furnace slags to increase cohesion quality of fastening of high temperature wells *Theoretical and applied aspect of modern science* **3-1** pp 127–32
[9] Rytwin V M, Perepelytsa V A, Abyzov VA and Khvatov A V 2013 Practice of recycling and utilization of ferroalloy aluminothermal slags *Refractories and engineering ceramics* **10** pp 38–43
[10] Itskovich S M, Chymakov L D and Bazhenov U M 1991 Technology of concrete aggregates (Moscow: Higher school) [In Russian]
[11] Pozdnyakova E A 2011 Policy of energy saving in the frame of technology modernization of ferroalloy produduction *Economy of region* **2** pp 163–8
[12] Romanova O A and Pozdnyakova E A 2013 Development of ferroalloy production raw material: new tendency and economic possibilities *Economy of region* **1** pp 167–77

IOP Conf. Series: Materials Science and Engineering **966** (2020) 012054 doi:10.1088/1757-899X/966/1/012054

Decoding variant for dilatometric measurements of metal samples heating and cooling

S B Mikhailov[1, 3], V A Sharapova[1], N A Mikhailova[2] and A E Karelova[1]

[1] Ural Federal University named after the First President of Russia B N Yeltsin, 19, Mira str., Yekaterinburg, 620000, Russia
[2] Ural State University of Railway Transport (USURT), 66, Kolmogorova str., Ekaterinburg, 620034, Russia

E-mail: [3] s.b.mikhailov@urfu.ru

Abstract. Dilatometric measurements in the form of sample sizes registration with temperature changes are widely used in the practice of physical materials science. In recent decades, the hardware component of these measurements has substantially changed with a practically unchanged version of their use as applied to the characteristics of various phase transformations. The basic point, in this case, is the identification of the critical points temperature for the 'start/finish' transformations based on the analysis of dilatograms deviations from a tangents carried out in the temperature intervals near the break observed on the obtained experimental curves. However, in most cases, this information has a framework character due to the presence of deficiencies in the critical points temperatures determining. In some cases, to compensate for these shortcomings, results of the changes analysis in the first derivative of dilatometric curves are additionally used. A series of works carried out by the authors is studying the features at changing the temperature coefficient of linear expansion true values. An algorithm for obtaining this spectrum is prescribed.

1. Introduction

Dilatometric measurements are used to search for temperature ranges of structural and phase transformations in materials [1, 2]. For this, a technique is used to register the beginning and end of transformations. The determination of temperature points is performed by crossing the tangents drawn in the area of dilatometric curves inflections. The curves are obtained during the experiment by recording the thermal expansion/contraction of samples linear dimensions [3].

The technique is simple and clear. However, the technique neglects some experimental curves features. For example, the nonlinearity of dilatograms at inflection points, as well as the change in their slope within and outside the temperature ranges of the analyzed structural changes. In this case, part of the information about transformations features has been lost.

An additional method for decoding dilatograms is the analysis of the temperature coefficient of linear expansion of samples (LTEC) curves, obtained by differentiating the experimental curve. The most widely used is the calculation of the LTEC values relative to room temperature (20 °C) [4]. It is integral information on the variation of the thermal expansion coefficient within the temperature ranges from 20 °C to the upper temperatures. Unfortunately, these calculations used in the design calculations for building and engineering parts are accompanied by significant distortions of the 'true' values of thermal expansion coefficient [4]. It cannot unambiguously reflect the occurrence of structure transformations in materials upon thermal exposure to them.

The practice of using the 'true' values of the thermal expansion coefficient (true) [3, 4], obtained by numerical differentiation based on two adjacent points of experimental dilatograms, showed a higher information content in the unambiguous reflection of the desired structural and phase transformations occurring in the material of the samples subjected to heat. This attracted attention to the determination of the true values of LTEC (true) [6–11].

However, there is a need for a background 'zero' line, against which dilatometric effects can be distinguished. Similar lines are widely used in chromatography when working with curves that have various kinds of peaks that reflect the course of various processes. In the case of dilatometry, a variant of the background 'zero' line of LTEC heating ('DL-BACKGROUND') was proposed in a previous work [7]. These values are obtained by numerically differentiating the analyzed experimental dilatograms.

Namely, the partial contribution of only the thermal vibrations of atoms is used. This is a monotonically changing process of the same sign within the temperature range of exposure to the sample. In this case, the effects of structural and phase transformations accompanying the true thermal change in the size of the sample are discarded.

In turn, it is taken into account that the monotonicity of the LTEC curve is preserved only in the temperature limits of the existence of one type crystal lattice matrix structure in the investigated material.

The monotonic component of the true LTEC curves was allowed to be interpreted by individual logarithmic curves. It is this dependence, based on purely external signs, that most successfully reflects the change in the course of the temperature curves of the true values of the thermal expansion coefficient for pure substances of the periodic system. This fact is formulated on the basis of the experimental data contained in the work of Novikova [4]. Moreover, the concept of vacancy and quantum-mechanical corrections taking place above 500 °C and below room temperature, respectively, was introduced [5].

2. Research Methodology

The methodology allows one to interpret the monotonicity of the LTEC temperature variation in graphical form by a line constructed on one experimental point. And in the digital version, the technique allows you to display the equations of the logarithmic dependence with the pre-logarithmic factor (Aln). The value of the factor (Aln) is functionally related to the melting temperature of the analyte [4, Figure 2a]. This suggests the existence of a functional relationship between the energy of interaction of atoms in the structure of the analyte with the value of the selected number (Aln).

The presence of polymorphism suggests the use of two similar curves as applied to modifications of crystal lattices. Moreover, the available experimental data suggest that for the iron-carbon system the ratio of the coefficients (Aln) for the ferritic and austenitic components is in the ratio of about two to three, respectively.

The combination of two monotonous phase segments with an inserted closing straight line allows us to obtain a continuous 'total' broken curve of the true values of LTEC – 'LTEC' – BACKGROUND (total)'. Using the numerical integration of the obtained total curve 'LTEC – BACKGROUND (total)' allows us to obtain a curve with the dimension of the experimental dilatogram reflecting the 'expected' thermal change in sample size in its pure form. It was decided to use this curve directly as the background 'zero' line of the experimental dilatogram – 'DL-BACKGROUND'. Accordingly, any deviations from this line can be considered as dilatometric effects [12].

3. Results and discussion

The above analysis made it possible to distinguish three groups of dilatometric effects associated with thermal expansion as applied to heating a sample (in the annealed state) of the 15Kh13N2MFB alloy

IOP Conf. Series: Materials Science and Engineering **966** (2020) 012054 doi:10.1088/1757-899X/966/1/012054

[7]. The dilatometric effect of ferromagnetism in the form of an increase in the size of the sample takes place at temperatures below the Curie point (780 °C). Deviations from the background curve reflect the magnetostrictive effect. The existence of two mutually exclusive processes of changing the sample size in the temperature range of the $\alpha \rightarrow \gamma$ transformation takes place at higher temperatures.

Firstly, this is an intensive reduction in the size of the sample in accordance with generally accepted ideas. Secondly, a small effect of an increase in size was revealed, which experimentally does not contradict the process of transition from the "ferritic" level of LTEC to the "austenitic" one. Such a bifurcation is significant and obviously requires close attention.

The initial data were dilatograms of cooling from austenitization temperature 1050 ± 10 °C, (15 min) at a speed of 50 degrees per minute. The same heating and subsequent cooling curves were averaged before further processing. Two initial conditions were used before heating: after preliminary quenching at 1050 °C, 0.5 hours, cooling in air and subsequent tempering at 650 °C, 2.0 hours/air.

The initial averaged dilatograms of direct measurements are presented in Figure 1a, 1b during heating (1 heating) and subsequent cooling (1 cooling). A "countdown" on the temperature scale was used for cooling.

Differentiation by two adjacent experimental points was carried out in order to obtain the true values of the TEC [4]. The result is presented in the form of temperature curves of changes in the true values of the thermal expansion coefficient of heating (2heating) and cooling (2heating), respectively (Figure 1a, 1b). The values of the thermal expansion coefficient of cooling were calculated so as to obtain the opposite sign of the change in the thermal expansion coefficient of heating.

The logarithmic lines 3α (Figure 1c) and 3α (Figure 1d) are drawn with reference to the ferritic state of the sample material 'ferrite level of LTEC'. Similarly, the logarithmic lines 5γ (Figure 1c) and 5γ (Figure 1d) were drawn, but with respect to the austenitic state of the structure 'austenitic level of LTEC'.

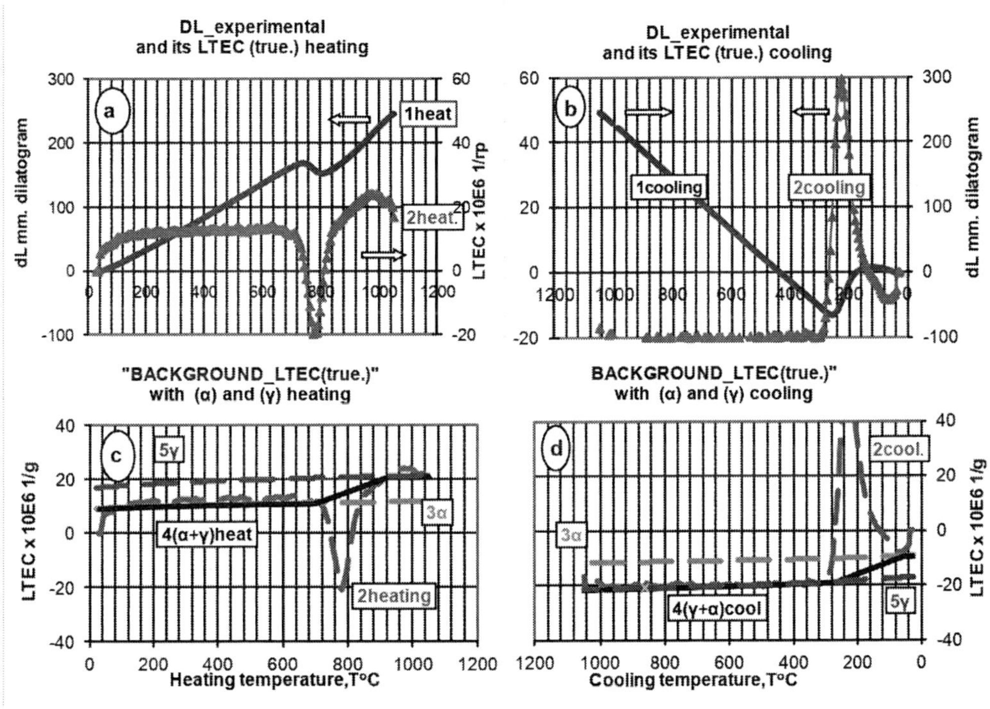

Figure 1. An example of processing averaged dilatograms of heating (a, c) and cooling

IOP Conf. Series: Materials Science and Engineering **966** (2020) 012054 doi:10.1088/1757-899X/966/1/012054

(b, d, with a countdown of temperature) of 15Kh13N2MFB alloy samples after quenching and subsequent tempering/.

For the ferritic state, this is the temperature range limited by the Curie point (750...780 °C). During cooling, this is the portion of the dilatogram from the minimum temperature of the manifestation of the vacancy correction zone (about 500 °C) up to the temperature of the onset of martensitic transformation (about 300 °C).

In this case, the values of the prelogarithmic factor (Aln) were for the ferritic state (α-Aln) = 1.65 (heating) and – 1.65 (cooling). For the austenitic state (γ-Aln) = 3.0 (heating) and – 3.0 (cooling). For almost pure iron, the absolute values of such a factor were (α-Aln) = 1.97 and (γ-Aln) = 3.27 [4, (Figure 2b)]. The introduction of 13 % chromium into the chemical composition of the alloy affected the decrease in the multiplier (Aln).

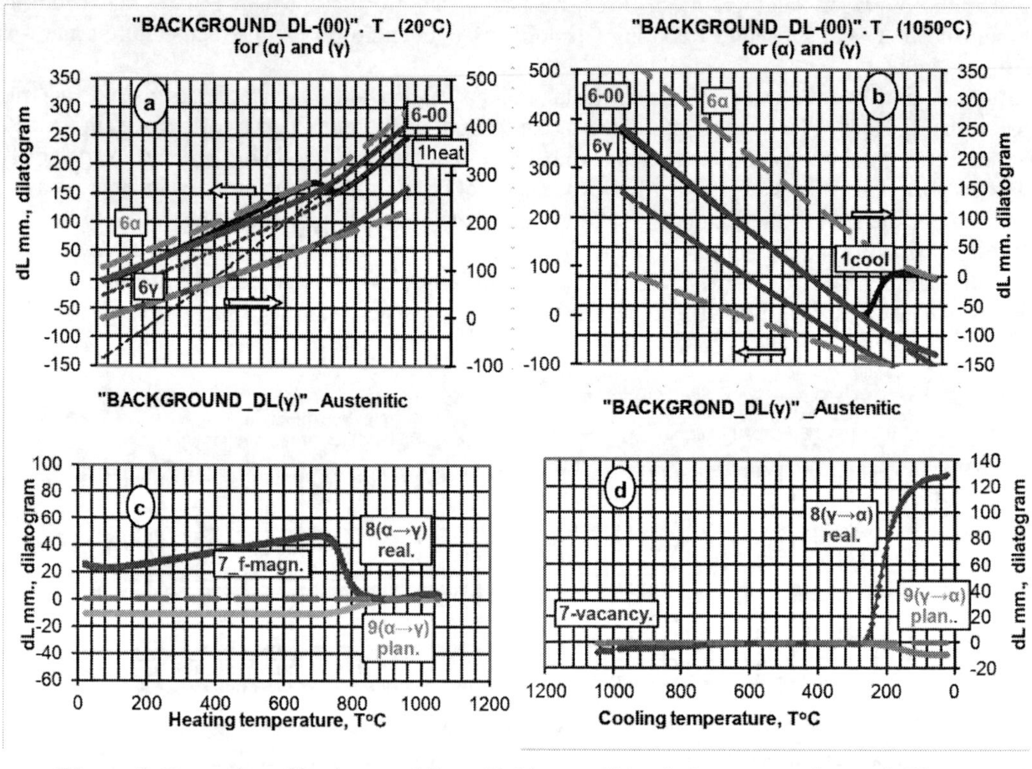

Figure 2. Examples of background lines (6-00, a and b) relative to experimental dilatograms (1 heating and 1 cooling). Type of dilatometric effects, reflecting the occurrence of thermal expansion processes (c and d) for the 'austenitic' line during heating (c) and cooling (d) of the 15Kh13N2MFB alloy.

It is assumed that broken lines (4 (α + γ) cooling and 4 (γ + α) heating) in terms of LTEC should reflect the process of monotonic thermal change in sample size in its pure form. Moreover, the sections of the experimental LTEC curves (curves with index 2) that do not coincide with the course of the constructed 'total' background curves (with index 4) are a result of thermal resizing in the material (Figure 1). It is these areas that are ignored in the process of forming the 'total' background lines of the LTEC (with an index of 4 in Figure 2).

The next step is the process of numerical integration of the obtained 'BACKGROUND-LTEC (total) curves' in order to obtain 'DL-BACKGROUND' curves with the dimension of experimental dilatograms. The result of this operation is shown in Figure 2 curves 6-00.

It should be noted that the result of numerical integration of the 4 ($\alpha + \gamma$) cooling and 4 ($\gamma + \alpha$) heating lines is a series of parallel curves of type 6-00, the location of which is regulated by the value of the free term of integration. In curves 6-00 in Figure 2, their location corresponds to the situation when the 'DL-BACKGROUND' line is drawn for the variant of coincidence of the values of the free integration terms when restoring the experimental dilatogram. In this case, the sought-for difference between the experimental dilatogram and the 'DL-BACKGROUND' line has a complex form, making it difficult to clearly identify the distinguished dilatometric effects of structural and phase transformations.

To accurately identify these effects, it is proposed, by selecting the value of the free term of numerical integration, to combine the 'DL-BACKGROUND' line with the characteristic sections of the experimental dilatogram. For Figure 2 a is the beginning of heating (line 6-α 'ferritic variant'), and for Figure 2 b – the beginning of cooling (line 6-γ 'austenitic variant'). These situations correspond to the single-phase state of the sample structure outside the temperature ranges of structural and phase transformations.

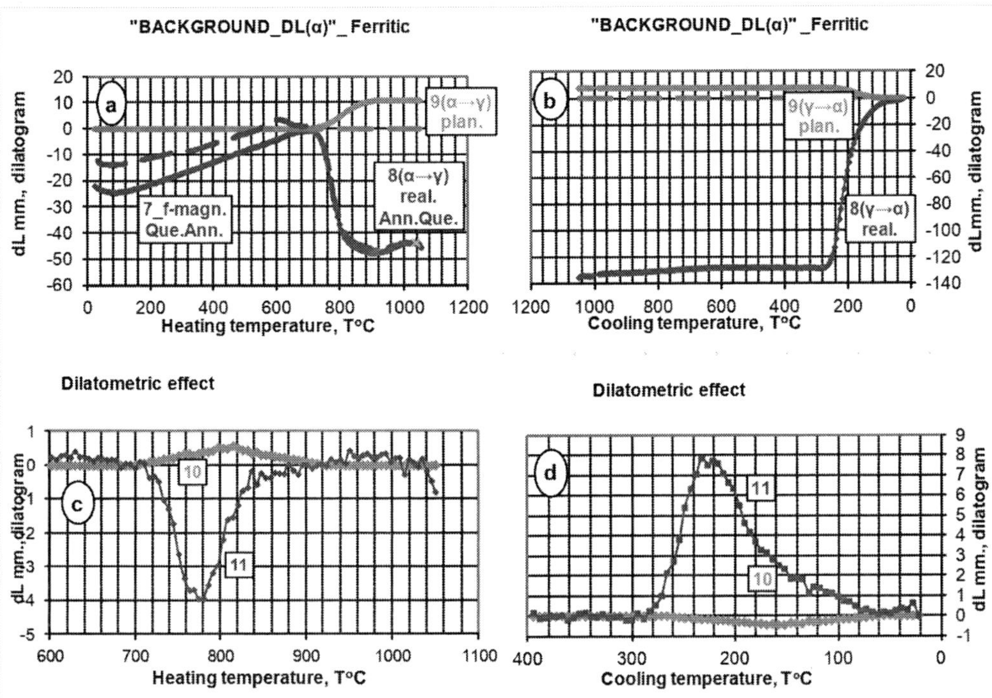

Figure 3. Dilatometric effects of thermal expansion (a and b) for the 'ferrite version' of the DL-FON line arrangement during heating (a) and cooling (b) of the 15Kh13N2MFB alloy. The result of preliminary differentiation of the pair effect sections during heating (c) and cooling (d) curves.

So, for Figure 2 a, 2 b, the 6-α zero line is drawn according to the 'ferritic version', through the Curie point upon heating and at the time of the 100 % ferrite phase upon cooling at the end point of the martensitic transformation. Accordingly, the 6-γ zero line is drawn according to the 'austenitic version'. We combine the line with the regions of the purely austenitic state of the sample structure, above the temperature range of the conversion of ferrite to austenite during heating and in the absence of vacancy formation from about 500 °C to the onset of martensitic transformation upon cooling. In these cases, the difference between the DL-FON zero line and the experimental dilatogram reflects the structural and phase transformations represented by curves 7, 8, in Figure 2 and 3.

IOP Conf. Series: Materials Science and Engineering **966** (2020) 012054 doi:10.1088/1757-899X/966/1/012054

In addition to curves 7 and 8, Figure 2 and 3 show a curve with index 9 (α-γ) expected. The formation of these curves is illustrated by the lower groups of lines in Figure 2 a and 2 b with dashed sections of curves. It is this difference that should reflect the 'expected' effect from the thermal change in the sample size at the recrystallization. At the same time, a large dilatometric effect 8 (α-γ) appears additionally to this effect as an additional result of the phase transformation in its temperature range.

4. Conclusion

The zero line as the base line separates the 'expected' effect of purely thermal expansion of the sample material from an additional set of dilatometric effects. Effects are represented by mirrored paired sections of 8-expect curves and 9th observation (Figure 2 and 3). This feature allows us to consider that this effect is not random.

A free integration term can establish various representations of these effects. In particular, for the 'ferritic' and 'austenitic' options. Moreover, when heating, it is preferable to use the 'ferritic' version (Figure 3 a), and when cooling, it is 'austenitic' (Figure 2 d). For both options, in the case of analysis by differentiation of the segments reflecting this pairedness, identical curves are obtained in the form of maxima presented in Figure 3 c and 3 d by curves with indices 10 and 11, respectively.

The dilatometer, in this case, plays the role of a DTA device, which has its own sample as a temperature sensor, detecting local changes in the temperature of the sample under the influence of absorbed or released heat of the structural or phase transformation itself against the background of general thermal effects by changing its size. In this case, the recorded maximum of the true LTEC curve should mark the temperature corresponding to the maximum rate of heat generation/absorption in a situation of an equal number of components involved in the transformation. The position and dimensions of the base of the maximum LTEC should mark the beginning and end of the temperature range of the investigated transformation.

References

[1] Ivasishin O M, Cherepin V T, Kolesnik V N and Gumenyak N M 2010 *Instrum. and Exp. Tech.* **53** **3** pp 457–60

[2] Cherepin V T 1986 *Experimental technique in physical metallography* (Kiev: Technique) p 280 [In Russian]

[3] Romanov P V, Radchenko V P 1960 *Austenite transformation during continuous cooling of steel: Atlas of thermokinetic diagrams. Part 1* (Novosibirsk: Publishing House Sib Sep USSR Academy of Sciences) p 51 [In Russian]

[4] Neymark B E 1967 P*hysical properties of steels and alloys used in power engineering: a reference book* (Moscow: Energy) p 238 [In Russian]

[5] Novikova S I 1974 *Thermal expansion of solids* (Moscow: Nauka) p 291 [In Russian]

[6] Mikhailov S B, Tsvetkova K Yu and Mikhailova N A 2013 *Actual problems of strength: Sat. abstracts 54 Int. conf.* (Yekaterinburg, FSBI IFM UB RAS) p 287

[7] Mikhaylov S B, Anashina A A and Mikhaylova N A 2018 *4th Int. Conf. on Ind. Engng.* pp 15–18

[8] Novikova S I 1982 *Meas. Tech.* pp 430–1 [In Russian]

[9] Basargin O V, Zakharov A I and Trush K V 1979 *Meas. Tech.* pp 1477–79 [In Russian]

[10] Sokolovskaya Y A and Berezovskaya V V 2015 *Russ. Metall* pp 211–215

[11] Mirzaeva D A, Sozykina S A, Makovetskii A N, Krasnotalova A O and Yusupova L I 2019 *Physics of Metals and Metallography* pp 770–74

[12] Teplukhina I, Golod V and Tsvetkov A 2018 *Letters on Materials* pp 37–41

IOP Conf. Series: Materials Science and Engineering **966** (2020) 012055 doi:10.1088/1757-899X/966/1/012055

The effect of elastic deformation on the minor hysteresis loops of low carbon steel

A N Mushnikov[1, 2], E S Gorkunov[1], S M Zadvorkin[1], L S Goruleva[1] and K D Kryucheva[1]

[1] Institute of Engineering Science, Ural Branch of the Russian Academy of Sciences, 34, Komsomolskaya St., Ekaterinburg, 620049, Russia

E-mail: [2] mushnikov@imach.uran.ru

Abstract. The effect of elastic tensile deformations on the magnetic permeability of a structural low-carbon steel 20K (0.23 % C, 0.20 % Si, 0.46 % Mn) at various magnetic states was experimentally studied. To obtain these parameters, the measurements of minor hysteresis loops, which begin in the demagnetized state and in the remanent magnetization state, were carried out. The maximum tensile stress equal to 250 MPa that corresponds to 90 % of the conditional yield strength of studied steel. Stress dependences of the initial and reversible magnetic permeabilities as well as differential magnetic permeability along the descending branch of the minor hysteresis loop, which starts in the remanent magnetization state, are non-monotonic. They have extremums at σ = 50–70 MPa. However, stress dependences of the sums of those parameters are monotonic. The relative changes reach 25 % at tensile stresses of 250 MPa which make the promise of using the sums of magnetic permeabilities as parameters for assessing the magnitude of tensile stresses.

1. Introduction

Studies of the dependences of the magnetic properties of ferromagnetic materials on the parameters of their stress state make it possible to create methods for assessing the stress-strain state of constructions made from such materials. Approaches associated with magnetization along return curves are common in magnetic non-destructive testing of ferromagnetic materials [1]. A number of papers demonstrate the possibility of using such parameter as a reversible magnetic permeability in the remanent magnetization state μ_{rev}. The permeability μ_{rev} depends not only on the mobility of the domain boundaries but also on the magnetic state of the material, which is determined by the concentration of various magnetic phases in the measured direction, the number and volume of domain boundaries. Dependences of μ_{rev} make it possible to calculate the magnitude of the induced magnetic anisotropy and residual stresses [2, 3]. Reversible permeability could be used to evaluate the remanent life of advanced ferritic steel [4]. However, the dependences μ_{rev} of low-carbon steels on elastic deformations can be non-monotonic [5].

To obtain a unique dependence, the ratio of magnetic permeabilities obtained in different magnetic states can be used. With increase of the tempering temperature of some steels the values of μ_{rev} and initial permeability μ_{init} vary ambiguously [6], and their changes qualitatively coincide, but the ratio μ_{rev} / μ_{init} changes monotonically with increasing tempering temperature. This is due to the fact that the concentration of magnetic phases at the demagnetized (field $H = 0$, induction $B = 0$) and remanent magnetization ($H = 0$, $B = B_r$) states depends on the tempering temperature differently. There is a high correlation of the ratio μ_{rev} / μ_{init} with the value of residual stresses in ferromagnetic materials. Also, the ratio of μ_{rev} / μ_{init} can be used to estimate the number of loading cycles of metal constructions operating under cyclic loads [7]. It is due to the significant residual compressive stresses appearing in a large number of grains along the tension axis under stress relief after cycling testing [8–11]. The effect of elastic stresses on the structural state has its own specifics and is of interest to study.

The aim of this work is to study the effect of elastic deformations on minor hysteresis loops, including those that begin in the remanent magnetization state, to identify parameters that can be used to develop methods for non-destructive evaluation of the stress-strain state of metal constructions.

2. Materials and experimental procedure

The studies were carried out on flat samples of structural low-carbon steel 20K (0.23 % C, 0.20 % Si, 0.46 % Mn). Magnetic hysteresis loops were measured directly during deformation on a bench consisting of a Tinius Olsen SL-60 hydraulic testing machine and a magneto-measuring complex Remagraph C-500.

Tensile stresses were limited by 250 MPa, which corresponds to 90 % of the conditional yield strength of steel 20 K. To calculate the errors in measured parameters, 10 measurements were performed at each loading point.

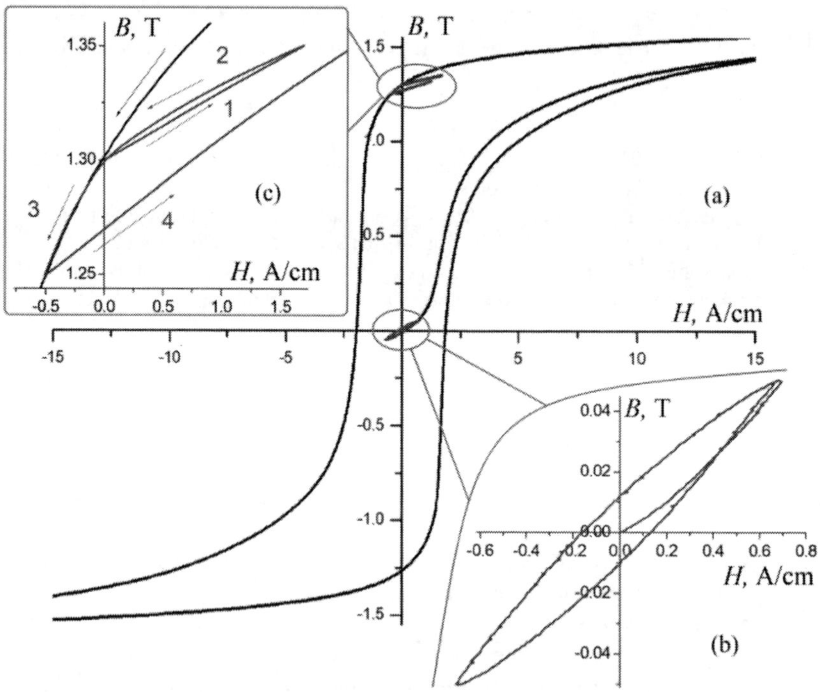

Figure 1. Hysteresis loops: (a) – major, (b) – minor in the demagnetized state, (c) – minor in the remanent magnetization state.

Minor hysteresis loops were measured in the induction range of ± 0.05 T from the starting point of measurement. Figure 1 shows the measured minor hysteresis loops and their location relative to the major hysteresis loop (Figure 1a) for steel 20K in the unloaded state. Parameters μ_{init} and μ_{rev} were calculated from the initial section of magnetization, where $B(H)$ is a linear dependence. The size of this section varies for different stress states. The differential magnetic permeability $\mu_{dif} = \mu_0 \times dB / dH$ was calculated on the descending branch of the hysteresis (μ_0 is a permeability of free space).

3. Results and discussion

The minor loop in the demagnetized state (Figure 1b) is a symmetric loop in weak fields. It can be described by the Rayleigh law, according to which irreversible magnetization processes are proportional to H^2. The minor loop in the remanent magnetization state is not symmetrical, as seen in Figure 1c. The following 4 steps can be noted (indicated by numbers in Figure 1c):

At step 1 the dependence of the magnetization on the field in weak fields is close to linear, and this region of 'linear dependence' is larger than the same area on the magnetization curve from the demagnetized state.

At step 2 the magnetization reversal curve does not coincide with the magnetization curve, but the residual induction returns to its original value B_r. This can be seen in more detail in Figure 2 – the changes of residual induction are close to zero and are independent of tensile stresses. The scatter of values is due to the measurement error and does not exceed 1.5 % of the magnetization range.

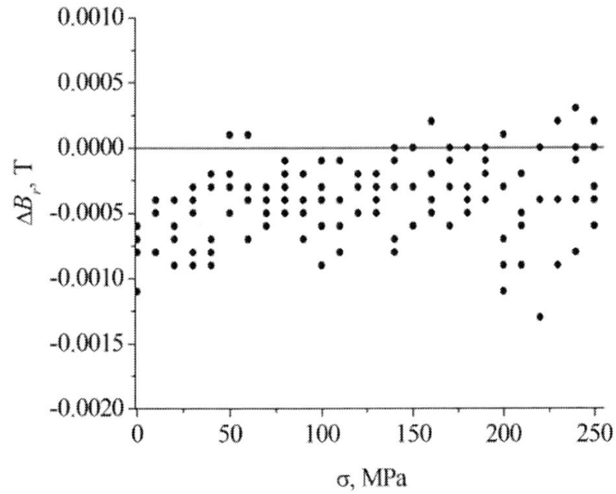

Figure 2. Changes in residual induction after a cycle 1–2 at various tensile stresses.

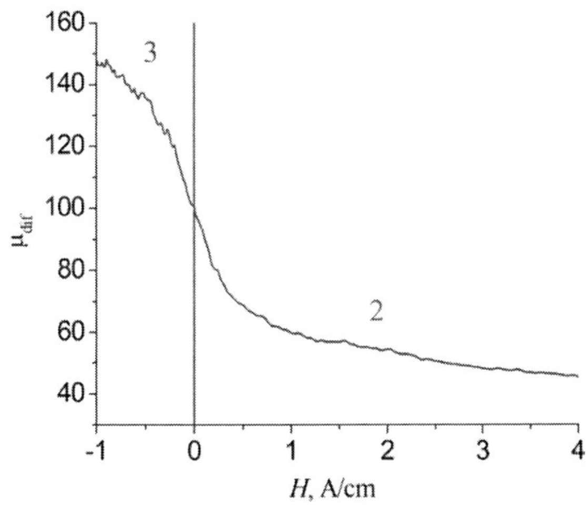

Figure 3. Field dependence of the differential magnetic permeability at steps 2 and 3.

At step 3 the magnetization reversal curve in negative fields coincides with a part of the descending branch of the major hysteresis loop, which is a consequence of the closure of cycle 1–2. The domain structure at the beginning of step 3 corresponds to the structure in the usual remanent magnetization state: negative fields that have not been previously passed are critical for the motion of domain walls through defects and their overcoming leads to irreversible changes in the magnetization. Therefore, an abrupt increase in the differential magnetic permeability occurs at the boundary between step 2 and step 3 (Figure 3).

At step 4 the return curve that begins with induction ($B_r - 0.05$ T) has a slightly higher permeability compared to μ_{rev}, but the cycle is not closed at the induction ($B_r + 0.05$ T).

Due to the fact that the range of the linear changes $B(H)$ in the remanent magnetization state is greater than in the demagnetized state, the obtained values of μ_{rev} have better repeatability compared to μ_{init}. For steel 20 K both of dependences $\mu_{rev}(\sigma)$ and $\mu_{init}(\sigma)$ are non-monotonic. They have extremums at $\sigma = 50–70$ MPa (Figure 4a). The effect of stress on permeability is different, which may be due to changes in the sign of magnetostriction at large values of induction, which is typical mainly for low alloy steels [12–15]. So, the ratio μ_{rev} / μ_{init} has a relatively high sensitivity to tensile stresses (Figure 5a). However, this ratio also non-monotonic varies from the value of tensile stresses, and therefore cannot be recommended as a parameter for evaluating the stresses for steel 20 K.

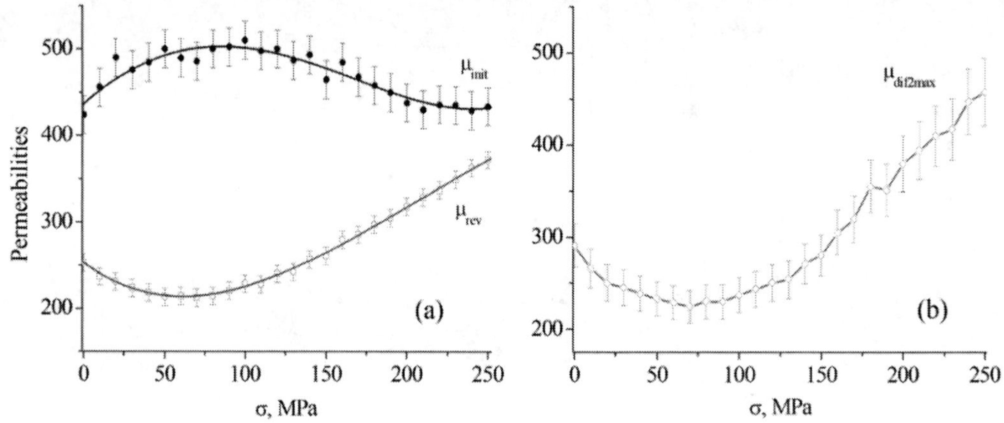

Figure 4. The influence of tensile stresses on the initial and reversible magnetic permeabilities (a) and the value of differential magnetic permeability at zero field (b).

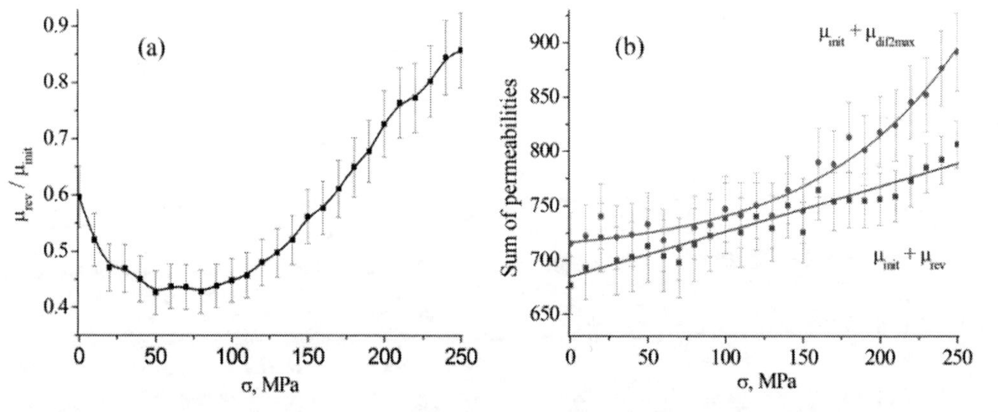

Figure 5. The influence of tensile stresses on the ratio (a) and the sum (b) of the permeabilities.

Figure 3 shows that at step 2, with field decreasing, the differential magnetic permeability monotonically increases, reaching a maximum value of $\mu_{dif2max}$ at $H = 0$. Figure 4b shows that the parameter $\mu_{dif2max}$ varies significantly from the action of large tensile stresses. The form of the dependence $\mu_{dif2max}(\sigma)$ is qualitatively similar to $\mu_{rev}(\sigma)$, and the largest differences are observed at $\sigma > 150$ MPa. These differences characterize the difference in the mobility of domain walls in the remanent state before and after exposure to a weak magnetic field.

Due to the fact that the extremum is the maximum in the dependence $\mu_{init}(\sigma)$, and it is the minimum in the dependences $\mu_{rev}(\sigma)$ and $\mu_{dif2max}(\sigma)$, the permeability sums can be used as parameters for assessing tensile stresses (Figure 5b). The sum of $\mu_{init} + \mu_{rev}$ linearly (correlation coefficient is 0.95) increases with increasing tensile stresses.

The dependence of the sum $\mu_{init} + \mu_{dif2max}$ can be divided into two parts. At the initial stage, tension has little effect on the amount of permeability and the changes fit into the measurement error. At the stress value of more than 150 MPa, a monotonic increase in the sum $\mu_{init} + \mu_{dif2max}$ occurs. The relative changes reach 25 % at tensile stresses of 250 MPa. Thus, the growth of $\mu_{init} + \mu_{dif2max}$ can be used as an indicator of a significant magnitude of tensile stresses (for steel 20 K, $\sigma = 150$ MPa corresponds to 0.5 yield strength).

However, when using such parameters, it must be taken into account that the relative error of the sum of permeabilities is the sum of the relative errors of the individual components, and the relative error of measurement of $\mu_{dif2max}$ is larger than for μ_{rev}.

4. Conclusion

The results of studies of the effect of tensile stresses on the characteristics of minor hysteresis loops of the steel 20 K obtained in a statically demagnetized and remanent magnetization states have shown the promise of using the sums of magnetic permeabilities as parameters for assessing the magnitude of tensile stresses.

Acknowledgments

The research was funded by RFBR, project number 20-48-660035. The experiments were carried out with the use of the equipment of the Plastometriya collective use center affiliated to IES UB RAS.

References

[1] Gorkunov E S 2015 *Diagnostics, Resource and Mechanics of materials and structures* **1** pp 6–40
[2] Stashkov A N, Kuleev V G, Shchapova E A and Nichipuruk A P 2018 *Russ. J. Nondestr. Test* **54** pp 855–60
[3] Kuleev V G, Stashkov A N and Nichipuruk A P 2019 *Phys. Metals Metallogr* **120** pp 632–8
[4] Ryu K S, Kim C S, Baek U B and Lee J S 2013 *J. Magn. Magn. Mat* **326** pp 257–60
[5] Makar J M and Tanner B K 1998 *J. Magn. Magn. Mat* **187** pp 353–65
[6] Gorkunov E S, Somova V M and Buldakova N B 1988 *Sov. J. Nondestr. Test.* **24** pp 42–8
[7] Gorkunov E S, Povolotskaya A M, Zadvorkin S M, Putilova E A, Mushnikov A N, Bazulin E G and Vopilkin A Kh 2019 *Russ. J. Nondestr. Test* **55** pp 827–36
[8] Abukus S 1977 *Jpn. J. Appl. Phys* **16** pp 1161–70
[9] Bulte D P and Langman R A 2002 *J. Magn. Magn. Mater* **251** pp 229–43
[10] Kuleev V G, Tsar'kova T P, Nichipuruk A P, Voronin V I and Berger I F 2007 *Phys. Metals Metallogr* **103** pp 131–41
[11] Gorkunov E S, Povolotskaya A M, Zadvorkin S M and Putilova E A 2017 *Russ. J. Nondestr. Test* **53** pp 636–43
[12] Mushnikov A N and Mitropolskaya S Yu 2016 *Diagnostics, Resource and Mechanics of materials and structures* **4** 57–70
[13] Jus A, Nowak P and Ginko O 2017 *Acta Phys. Pol. A* **131** pp 1084–6
[14] Lahyaoui O, Lanfranchi V, Buiron N, Martin F, Aydin U and Belahcen A 2019 *Int. J. Appl. Electromagn. Mech* **60** pp 299–312
[15] Gorkunov E S, Zadvorkin S M, Mushnikov A N, Smirnov S V and Yakushenko E I 2014 *J. Appl. Mech. Tech. Phys* **55** pp 530–8

Analysis of the influence of the parameters of the power supply network on the operating modes of the electric drives of the stands of the hot rolling mill 1700 of ArcelorMittal Temirtau JSC

I V Breido[1, 3] and V M Druzhinin[2]

[1] Doctor of Technical Sciences of Kazakhstan, Professor, Head of the department of automation of manufacturing processes, Karaganda State Technical University, 56, Nursultan Nazarbayev Avenue, Karaganda city, 100027, The Republic of Kazakhstan
[2] Doctoral student, Master of technical Sciences in 'Electric Power', Karaganda State Technical University, 56, Nursultan Nazarbayev Avenue, Karaganda city, 100027, The Republic of Kazakhstan

E-mail: [3] jbreido@mail.ru

Abstract. The finishing group of stands of the 1700 SRS-1 continuous mill of ArcelorMittal Temirtau JSC is characterized by high rolling speeds and high productivity. The use of DC motors as power sources for the main drives of the stands of high-power thyristor converter units significantly affects the nature of the operating modes of the power supply system and the quality of electricity on the buses of the SRS-1 substation. Experiments were performed to study the electricity grid supply substation in which the basic indexes of quality of electric energy and the built waveform voltage SAG during rolling. Voltage dips are observed throughout the rolling process and reach 10% of the rated network voltage. Due to the complexity of removing individual parameters of electric drives and conducting experiments on a real mill, an adequate model of the electric drive of the stand of the finishing group of the 1700 SRS-1 mill was developed and built.

1. Introduction

A modern metallurgical rolling mill is a complex set of various mechanisms that are different in purpose and operating modes [1].

The continuous broadband hot rolling mill of 1700 sheet rolling shop No. 1 (SRS-1) of Arcelor Mittal Temirtau JSC is characterized by high rolling speeds and productivity. At a high speed of rolling, the metal temperature does not have time to significantly decrease, as a result of which rolling is carried out at a high average temperature, which leads to the required quality of rolled products and an increase in the service life of the work rolls of the stands [2].

A feature of rolling in the finishing group of the mill stands is that it is carried out simultaneously in all stands of the group and the metal passes in one direction while it is crimped in each working stand [3].

The rolling process at the 1700 mill is accompanied by a significant level of electrical energy consumption and accounts for 90% of the total power consumption of the 110/10 kV two-transformer substation MSDS 1A with installed power transformers 63MVA [4].

Thyristor converters that feed the main drives of the mill stands significantly affect the operating modes of the power supply system and the quality of electricity on the busbars of the substation SRS 1.

The operating modes of the thyristor converters of the main drives of the 1700 mill are characterized by sharply variable consumption of electric energy, generation of higher harmonic oscillations into the electric network. This causes a discrepancy, voltage fluctuation of the network, distortion of the shape of the voltage curve, which is confirmed by the results of the experiments.

The rolling of energy-intensive profiles causes voltage drops at substation No. 6 with a voltage of 10 kV, which exceed the maximum permissible values of State Standard 32144–2013 [5]. As a result of tripping of protection against voltage surges and thyristor converters Simoreg from Siemens, an emergency shutdown of the main and auxiliary drives of the mill occurs.

2. Methods

In the course of experiments to study the power supply network of substation No. 6, from which the main drives of the mill 1700-SRS-1 are supplied with the help of the 'Resurs-UF' measuring device, the main indicators of the quality of electric energy are determined and voltage dips during rolling are constructed. The analysis of experiments showed that the average voltage drop exceeds the maximum permissible values of State Standard 32144-2013 [5] (over 10%). One of the registers is shown in Figure 1.

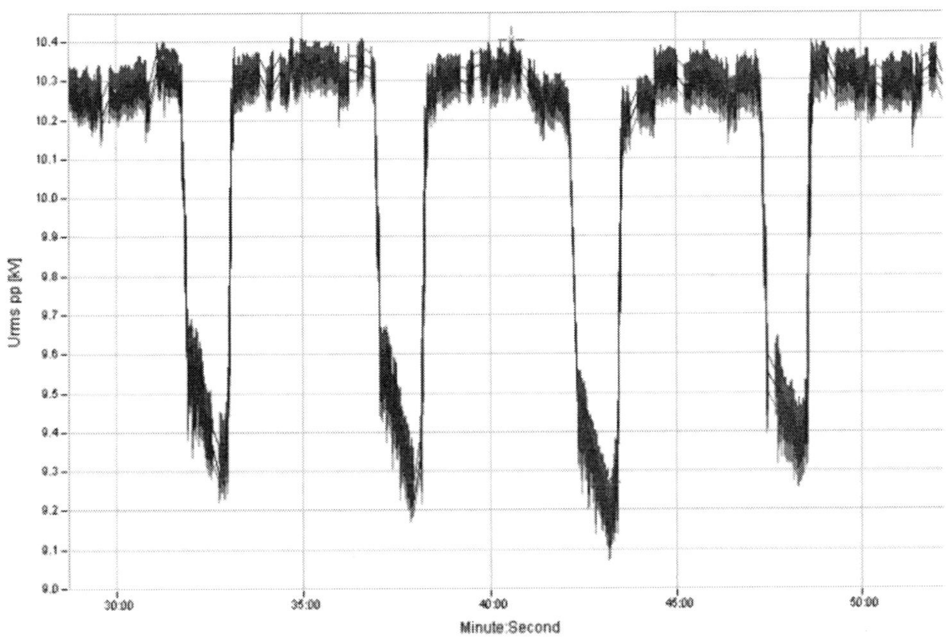

Figure 1. The voltage register on the busbars of the substation No. 6.

As can be seen from the register, voltage dips are observed during the entire rolling process. When the metal enters the rolls, a shock application of the load occurs, which causes a voltage drop, while the shock surge and load shedding are clearly traced on the register.

3. Results and discussion

Due to the difficulty of taking individual parameters of electric drives and conducting experiments on a real mill, a mathematical simulation model was developed on the example of an electric drive of stand No. 6 of the finishing group of the hot rolling mill 1700 SRS-1 of ArcelorMittal Temirtau JSC.

To build the model, a scheme for replacing a DC machine is made under generally accepted assumptions (the armature circuit and eddy currents are not taken into account). The excitation circuits are also not taken into account, since a DC machine with independent excitation is used and there is no control over the excitation circuit.

An equivalent substitution scheme of the electric machine armature circuit is shown in Figure 2 [6].

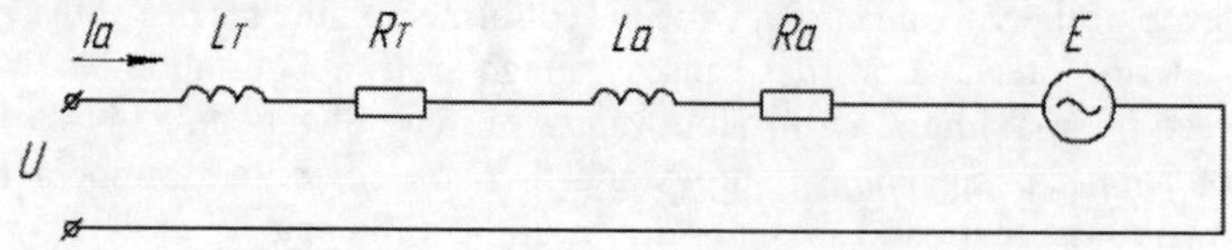

Figure 2. The equivalent substitution scheme of the armature circuit of a DC machine.

In Figure 2: U – the voltage of the supply network; R_T – the active resistance of the power transformer; R_a – active resistance of the electric machine armature circuit; L_T – inductive resistance of the power transformer; L_a – inductive resistance of the electric machine armature circuit; E – electromotive force (EMF) electric machine I_a – current of the armature circuit

The system of equations corresponding to the substitution scheme has the form:

$$\begin{cases} U = I_a R_a + L_a \dfrac{dI_a}{dt} + c\Phi\omega + I_a R_T + L_T \dfrac{dI_a}{dt} \\ U - I_a R_T - L_T \dfrac{dI_a}{dt} = I_a R_a + L_a \dfrac{dI_a}{dt} + c\Phi\omega \\ U - I_a(R_T + L_T p) = I_a(R_a + L_a p) + c\Phi\omega \end{cases} \quad (1)$$

where ω – the angular velocity of the electric machine; c – EMF coefficient at Φ = const; $c\Phi\omega$ – electric machine EMF; i_a – the current of the armature circuit; p – Laplace operator.

Based on the above system of equations (1), a simulation model of the electric drive of the stand of the hot rolling mill has been developed. The simulation model is a dual-circuit automatic control system, which is built on the principle of subordinate regulation.

The internal (subordinate) circuit is the current circuit, the external circuit is the speed circuit [7, 8]. The following notations are used on the model: '*Rw*1'is the proportional part of the speed controller; '*Rw*' is the integral part of the speed controller; '*Ri*1' is the proportional part of the current regulator; '*Ri*' is the integral part of the current regulator; 'Engine' is the electric part of the electric machine; 'Engine 1'is the mechanical part of the electric machine; Kw is the feedback coefficient of the speed; Ki is the feedback coefficient of the current; 'R_T', 'L_T' is the blocks that take into account the parameters of the supply network, namely the active and inductive resistance of the power transformer.

The structure and parameters of the model correspond to the electric drive and parameters of stand No. 6 of the finishing group of mill 1700 SRS-1 of ArcelorMittal Temirtau JSC. This model is implemented using a software package for solving MATLAB technical computing problems in the Simulink software simulation complex [9, 10].

The developed block diagram and model, taking into account the joint operation of the stand drive and power transformer, are presented in Figure 3.

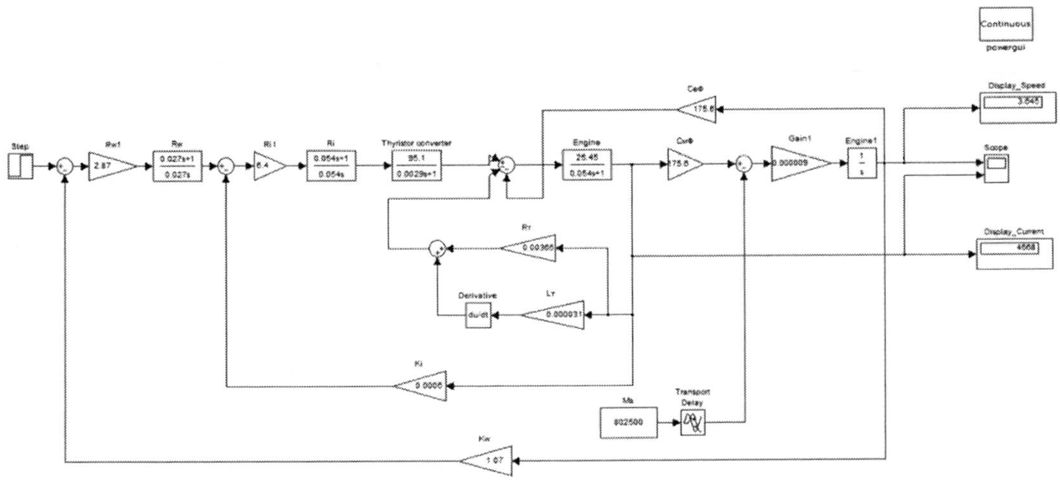

Figure 3. Structural diagram of the model taking into account the supply network.

Based on the simulation results, transient graphs were obtained, presented in Figure 4, and Table 1 shows the comparative data of the simulation results and the electric machine rated parameters.

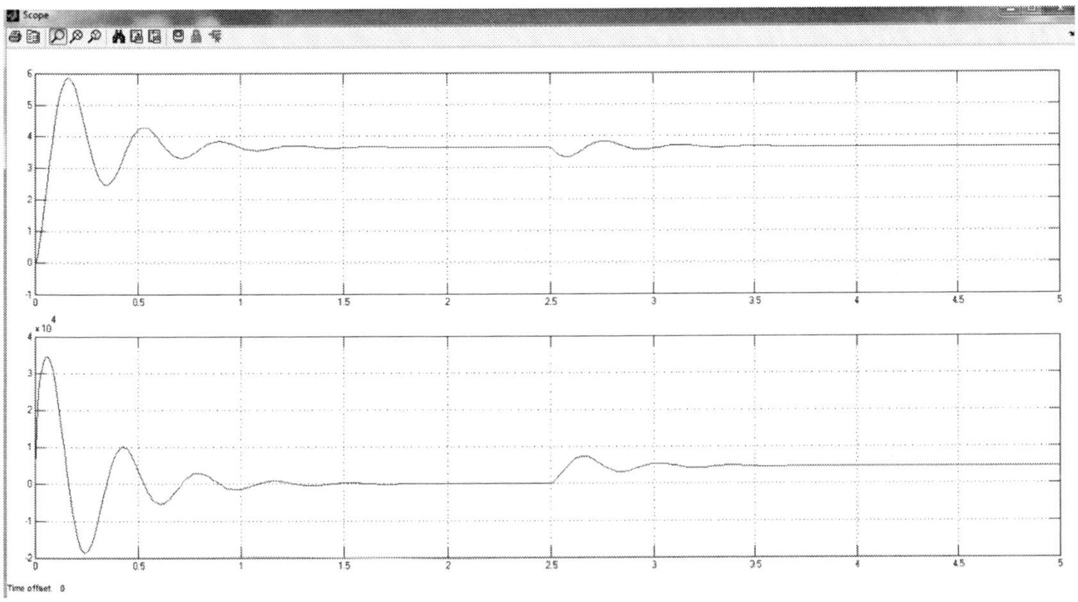

Figure 4. Transient graphs of the model taking into account the supply network.

Figure 4 above shows a graph of the transient in the velocity loop $w = f(t)$, rad/s; below is a graph of the transient in the current loop $I_a = f(t)$, A.

According to the results of processing data from real and simulation experiments, the discrepancy between the values of the speed of the electric machine and the current in the armature circuit obtained in a real network and in a simulation model is not more than 10%. This confirms the adequacy of the presented simulation model.

Table 1. The results of simulation and nominal data of the electric machine 2P25/105-3.15.

Parameter Name	Rated current of the electric machine armature circuit, I_a, A	Rated electric machine speed, w, rad/s
Rated parameters of the electric machine	4580	3.92
Simulation Results	4568	3.645
Discrepancy, Δ, %	0.262	7.02

Thus, transients (Figure 4) with a sufficient degree of accuracy reflect the real processes occurring in electric drives of stands of hot rolling mill 1700 of ArcelorMittal Temirtau JSC under shock application of load. The nature of these processes is important for continuous rolling, especially when rolling at high speeds and at small distances between stands.

4. Conclusions

The nature of the transient processes, with a surge in power as a result of the application of shock load, confirms a decrease in the quality of transient processes. In this case, the influence of the power transformer as an element of the supply network is obvious. According to the graphs presented in Figure 4, it can be seen that when the load occurs when the metal enters the crate, taking into account the influence of the supply network, the time of dynamic subsidence increases, while the nature of the transition process is delayed, which leads to a deterioration in the quality of rolled products.

Significant in time changing in the rolling speed leads to non-compliance with the temperature regime, the tension of the metal in the inter-stand spaces, which leads to the release of low-quality products [11].

The negative influence of the power supply network on the operation of power receivers of the hot rolling mill 1700 requires research and development of methods for controlling the electric drives of the stands of the finishing group in order to solve the problems of their electromagnetic compatibility with the power supply network.

References

[1] Hemakumar B, Aminur A M R B and Prasad M P R 2018 Performance Improvement in Hot Strip Rolling Mills by Improving Looper System. *2nd International Conference on Trends in Electronics and Informatics (ICOEI)* (Tirunelveli, India) IEEE INSPEC Accession Number: 18290923 pp 827–32 DOI: 10.1109/ICOEI.2018.8553709

[2] Radionov A A, Karandaev A S and Khramshin V R 2014 Speed and Load Modes of Rolling Hollow Billet at the Wide-Strip Rolling Mill *Proceedings of 2014 International Conference on Mechanical Engineering, Automation and Control Systems (MEACS)* p 5 DOI: 10.1109/MEACS.2014.6986841

[3] Radionov A A, Karandaev A S, Evdokimov A S, Andryushin I Yu, Gostev A N and Shubin A G 2015 Mathematical Modelling of the Interrelated Electric and Mechanical Systems of Continuous Sub-Group of the Rolling Mill Stands. Development of the Mathematical Model. *Bulletin of the South Ural State University. Ser. Power Engineering* part **1** vol **15** no **1** pp 59–73 DOI: 10.14529/power150108

[4] Druzhinin V M 2006 Analiz vliyanija parametrov seti elektrosnabzhenija naparametry j elektroprivodov stana 1700 LPC-1 AO 'Mittal Stil' *Trudy 5-oj mezhdunarodnoj nauchno – tehnicheskoj konferencii 'Energetika, telekommunikacii i vysshee obrazovanie v sovremennyh uslovijah'* (Almaty) pp 238–42 [In Russian]

[5] State standard (GOST) 32144-2013 Elektricheskaja energija. Sovmestimost' tehnicheskih sredstv elektromagnitnaja. Normy kachestva elektricheskoj energii v sistemah elektrosnabzhenija obshhego naznachenija. Vvedeno 2015–06–01 (Moskva: FGUP 'STANDARTINFORM') p 20

[6] Druzhinin V M 2007 Razrabotka imitacionnoj modeli elektroprivodov chistovoj gruppy stana 1700 LPC-1 AO 'Mittal Stil Temirtau' with uchetom parametrov seti elektrosnabzhenija *Trudy universiteta* no **4** (Karaganda: Izdatelstvo KarGTU) pp 70–3 [In Russian]

[7] Shokhin V V, Andreev A I and Morkovina P S 2017 Simulation of a Rolling Mill Using Physical and Mathematical Models. *Elektrotekhnicheskie sistemyi kompleksy* [*Electrotechnical Systems and Complexes*] no **2 (35)** pp 13–8 DOI 10.18503/2311-8318-2017-2(35)-13-18

[8] Chapman S J 2001 *MATLAB Programming for Engineers* (2nd Edition). (United States: Brooks/Cole Publishing Co. Div. of International Thomson Publ. Co) p 464

[9] Chernykh I V 2014 *Modelirovanie elektrotekhnicheskikh ustroistv v MATLAB, Sim Power System and Simulink* (Moscow: DMK Press) p 288 [In Russian]

[10] Terehin V B and Demen'tev Ju N 2015 *Komp'juternoe modelirovanie postojannogo and peremennogo toka v Simulink* (Tomsk: Izdatelstvo Tomskogo politehnicheskogo universiteta) p 307 [In Russian]

[11] Atrjaskin V F, Burlakov S A, Kulikov V I, Kolynjuk E P, Lozovoj V N and Sejsimbinov T S 1994 Sposob upravlenija gorjachej prokatkoj metallicheskih polos v chistovoj gruppe kletej nepreryvnogo stana. Preliminary Patent of the Republic of Kazakhstan no 446 15.03.1994

IOP Conf. Series: Materials Science and Engineering **966** (2020) 012057 doi:10.1088/1757-899X/966/1/012057

Optimization of the ion-plasma nitriding of structural steel

V R Baraz[1, 4], E A Yurchenkov[2] and E A Ishina[3]

[1, 3] Ural Federal University named after the First President of Russia B N Yeltsin, 19, Mira str., Yekaterinburg, 620000, Russia
[2] LLC 'Velesstroy', 114, Baku Commissars str., Yekaterinburg, 620135, Russia

E-mail: [4]vrbaraz@mail.ru

Abstract. Mathematical planning was used in order to determine the optimal mode for ion nitriding the structural steel 0.38C–2Cr–2Ni–Mo. This was justified by the requirement to preserve strictly limited functional parameters obtained as a result of diffusion saturation: hardness in the range of 450–650 HV in a layer 0.15–0.40 mm deep. According to of X-ray diffraction analysis in addition to reflections from the matrix phase (alloyed ferrite α-Fe), the reflexes of nitride compounds are recorded, namely, the γ'-phase with a FCC lattice (Fe_4N) and the ε-phase with a HCP lattice (Fe_3N) with a high nitrogen content. An analytical expression in the form of a linear function for the studied optimization parameters (microhardness and depth of nitration) was obtained by implementing factor planning and regression processing of the obtained data. This made it possible to assess the degree and direction of influence of the investigated factors on the optimization parameters under study.

1. Introduction

As is well known, ion-plasma nitriding is an effective form of diffusive nitrogen saturation for the surface layers of processed products [1–3]. It is carried out in a glow discharge excited on a cathode's surface in a nitrogen-containing gaseous medium (ammonia, nitrogen, etc.) under conditions of deep rarefaction. The saturation process is conducted in a steel container, the anode. Active gas ionization occurs due to the high voltage generated. Nitrogen ions enter the high-tension zone, accelerate to high speeds, and, colliding with the cathode, are embedded in its surface. The high kinetic energy possessed by the nitrogen ions is converted into thermal energy. As a result, the cathode is heated to a temperature of 450–600 °C in a short period of time (approximately 15–30 minutes), at which point diffusion occurs.

Unlike traditional gas nitriding in furnaces, this chemical-thermal treatment reduces the process' overall duration (by 1.5–2 times), makes it possible to regulate the process in order to obtain a nitrided layer with the desired properties, and is more environmentally friendly [4–5].

In particular, this technology is used for nitriding the structural steel 0.38C–2Cr–2Ni–Mo. Formally, this steel is not a classic example of an alloy steel intended for nitriding because it does not contain important elements like aluminum. At the same time, the processing modes currently used are unable to provide the desired combination of functional parameters. The diffusion layer has a hardness value higher than the regulated one and is characterized by insufficient depth. Therefore, in this investigation, it was necessary to find a mode of chemical-heat treatment that provides from this structural steel a hardness in the regulated range of 450–650 HV. It also needs to maintain a layer depth of 0.15–0.40 mm.

Thus, the main purpose of this research is to study the peculiarities of ion-plasma nitriding of 0.38C–2Cr–2Ni–Mo steel and determine the optimal regime for ion nitriding while taking into account whether the specified properties are obtained.

2. Experiment

The object of research is a standard structural steel 0.38C–2Cr–2Ni–Mo, the chemical composition of which is shown in Table 1.

Table 1. Chemical composition of 0.38C-2Cr-2Ni-Mo steel (mass. %).

C	Si	Mn	Ni	Cr	Mo	S	P	Cu
0.36	0.28	0.38	1.46	1.49	0.22	0.008	0.012	0.17

The research samples for ion-plasma nitriding were rectangular parallelepipeds with sides of 10 mm×10 mm×50 mm. In order to obtain a stable microstructure, the samples were annealed at a temperature of 840 °C for 4 hours. The standard heat treatment before nitriding was quenching and high-temperature tempering. Specifically for the studied steel, the quenching temperature was 880 °C with cooling in oil; the tempering temperature was in the range of 560–600 °C.

Chemical heat treatment was performed on an ion-plasma nitriding unit with a YON-50 glow discharge. The hardness of the nitrided surface layer was measured using a DuraScan automatic universal hardness tester that operates in micro-and macro-ranges and has a load of 100 MPa. The nitration's depth was determined on a transverse metallographic sample after etching with a 10 % solution of nitric acid using a metallographic complex, including a GX51 optical microscope and a software unit. A 'D2 PHASER' diffractometer in cobalt K_{α}-radiation was used for X-ray diffraction studies of phase composition. Phase analysis was carried out using the international diffraction data file ICDD PDF-4 (the International Centre for Diffraction Data) [6].

3. Results and discussion

Initially, the effect of the tempering temperature on the hardness and depth of the diffusion layer after ion-plasma nitriding was determined. Standard heat treatment was used for this steel: after quenching from 880 °C, tempering was carried out for two hours at temperatures of 560 and 600 °C. The preliminary ion-plasma nitriding of the hardened and tempered steel was carried out in a regulated regime: at 525 °C for 90 minutes.

The data acquired on the hardness and depth of the nitration are shown in Table 2.

Table 2. The hardness and depth of nitration after heat treatment in a regulated regime.

№	Quenching temperature, °C	Tempering temperature, °C	Hardness after tempering, HV_{10}	Hardness after nitriding, HV_{10}	Depth of nitration, mm
1	880	600	395	750	0.172
2	880	560	425	775	0.171

The metallographic (see box) and diffraction patterns of the nitrided layer are presented in Figure 1.

The microstructure is a characteristic hetero phase image with the presence of nitride particles, which are mainly spherical. X-ray phase analysis allowed us to identify the phase composition of the steel after ion-plasma nitriding. In addition to reflections from the matrix phase (alloyed ferrite α-Fe), the investigated nitrided layer γ'-phase with a FCC lattice (Fe_4N) and the more nitrogen-rich ε-phase with a HCP lattice (Fe_3N) were fixed.

The results of preliminary experiments have shown that it is possible to effectively influence hardness after nitriding by changing the tempering temperature after quenching. For this reason, it is necessary to conduct a more detailed investigation in which the influence of the factors can be analyzed over a fairly wide range. It seemed appropriate to conduct experiments using mathematical planning [7, 8].

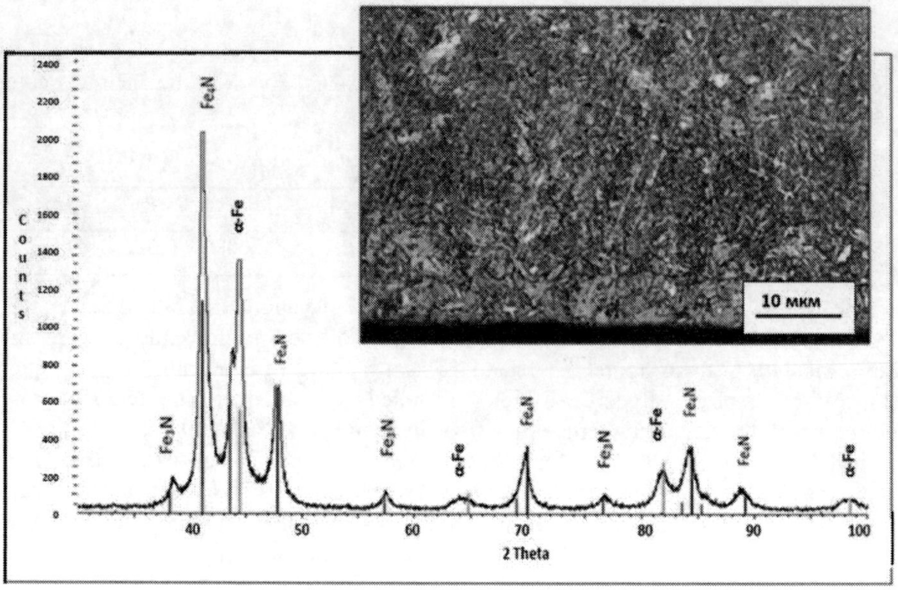

Figure 1. X-ray diffraction and metallographic patterns of 0.38C–2Cr–2Ni–Mo steel after ion-plasma nitriding.

The mathematical planning method is considered effective for solving problems when it is necessary to search for the optimal properties of the research object. The idea behind the method is to conduct a minimum number of experiments while simultaneously varying the values of independent variable factors according to specially formulated rules, to find the optimal region, and to obtain its mathematical expression.

In this article, the following variables were used as the studied factors:

a) the tempering temperature after quenching, °C;

b) the temperature of ion-plasma nitriding, °C;

c) the time for nitriding, min.

Parameters like hardness HV_{10} (y_1) and the depth of the nitration h (y_2) were used as optimization parameters. A complete three-factor type 2^3 experiment was carried out.

For the center of the plan (zero level), the standard parameters of 38Cr2Ni2Mo steel were used:

a) the tempering temperature after quenching, equal to 580 °C (x_1);

b) the nitriding temperature, equal to 525 °C (x_2);

c) the time for nitriding, equal to 60 min (x_3).

Thus, 9 experiments with the experiment in the center of the plan were conducted.

Table 3 shows data on the values of the discussed factors at zero level, the intervals of their variation, and the indicators at the upper and lower levels.

Table 4 provides the planning matrix based on the above factors and variation intervals, as well as the obtained values of the hardness HV_{10} and the depth of the layer h.

Standard statistical processing of the results was performed. After verification of the equation itself and its coefficients for adequacy, a regression dependence was obtained in the form of two linear functions. The final forms of the equations are presented, taking into account the established significance of the regression coefficients. Namely, this is the absence of the effects of paired and triple interaction, as well as the obvious insignificance of the effect of the tempering temperature on nitration depth.

Table 3. Factors and variation intervals.

	Tempering temperature, °C	Factors	Time for nitriding, min
Code	x_1	x_2	x_3
Zero level (0)	580	525	60
Intervals of variation (J)	20	25	30
Upperlevel (+1)	600	550	90
Lower level(–1)	560	500	60

Table. 4. The planning matrix.

Experiment number	x_1	x_2	x_3	Hardness, HV_{10}	Depth of nitration h, mm
1	560	500	30	590	0.072
2	600	500	30	525	0.079
3	560	550	30	640	0.115
4	600	550	30	575	0.139
5	560	500	90	670	0.144
6	600	500	90	605	0.139
7	560	550	90	700	0.191
8	600	550	90	645	0.165
9	580	525	60	650	0.123

Hardness, HV_{10}: $y_1=618-31x_1+21x_2+35x_3$; Depth of nitration, h, mm: $y_2=0.13+89x_2+118x_3$

Analysis of the regression equations allows us to evaluate the degree and direction of the influence of the investigated factors on the optimization parameters. The tempering temperature after quenching, as expected, has an effect on the hardness with the opposite sign. However, in general, the impact of all factors is approximately at the same level. The nitriding time has a slightly greater effect on nitration depth. The tempering temperature has no statistical effect, which is reflected in the absence of the x_1 indicator in the equation.

Since the experiment requires a certain combination of nitriding conditions and the required technological parameters (hardness in the range of 450–650 NV and nitration depth in the range of 0.15–0.40 mm), the most acceptable treatment mode is experiment 8 (Table 4).This regime involves tempering at 600 °C (for 2 hours) after quenching, followed by ion-plasma nitriding at 550 °C for 90 minutes. As a result of this treatment, the hardness reaches 645 HV_{10} and the depth of the nitration h-0.165 mm, which satisfies the required performance indicators.

A physical indicator like the diffusion coefficient (D) acquires an important functional value when performing diffusion saturation operations. Therefore, an empirical calculation of this characteristic was performed.

The diffusion coefficient D was calculated using formula [9,10]:

$$D = \frac{x^2}{2\tau}$$

where D – diffusion coefficient; x – average depth of the nitration, cm; τ –the time of diffusion, sec.

Computation was performed for the following experimental conditions:

1. For the center of the plan: the temperature is 525 °C, the saturation time is 60 min, and the depth of the nitration is 0.13 mm. The calculated value of the diffusion coefficient is D=2.2·10^{-8} cm²/ sec.

2. For experiment 8: the temperature is 550 °C, the saturation time is 90 min, and the depth of the nitration is 0.165 mm. D=2.5·10^{-8} cm²/ sec.

IOP Conf. Series: Materials Science and Engineering **966** (2020) 012057 doi:10.1088/1757-899X/966/1/012057

The obtained values of the diffusion coefficient are consistent with the available data from the literature describing the coefficient values for iron saturation at different temperatures and phase states [11].

4. Conclusion

The purpose of this work was to study changes in the hardness and depth of the nitration during nitrogen saturation of 0.38C–2Cr–2Ni–Mo steel in an ion-plasma nitriding unit, as well as the selection of optimal nitriding parameters according to specified criteria, namely certain ranges of the diffusion layer's hardness and depth.

To find the optimal regime for nitriding this steel in a glow discharge, experimental research based on mathematical planning was carried out. The hardness values HV_{10} and the depth of the nitration h were used as the studied optimization parameters. At the same time, it was necessary to choose the optimal saturation regime, which provides a regulated range of these characteristics within the limits of HV_{10} 450–650 and h 0.15–0.40 mm. As a result, the required regime of chemical heat treatment was determined. It includes quenching, tempering at 600 °C (for 2 hours), and subsequent ion-plasma nitriding at 550 °C for 90 minutes. The required results were attained: a hardness of 645 HV_{10} and a nitration depth of h 0.165 mm.

Qualitative phase analysis of the samples was carried out via X-ray diffraction. The reflexes observed after ion-plasma nitriding can be interpreted as being obtained by reflection from the matrix phase of α-Fe type FeCr (chromium-doped ferrite). In addition, reflections from the phases are recorded that correspond to nitrate compounds, namely the γ′-phase (Fe_4N) and the more nitrogen-rich ε- phase (Fe_3N).

References

[1] Filippov M, Kositsyna I and Gervasiev M 2009 *Surface treatment and coatings in mechanical engineering* [In Russian] (Yekaterinburg: UGTU-UPI Press) p 213
[2] Goncharenko I, Zolotukhin V and Gvozdev A 2006 *Fundamentals of steel heat treatment technology* [In Russian] (Tula: Grif and K) p 326
[3] Ramazanov K, Agzamov R and Budilov V 2016 *Ionic nitriding of steels in a low-pressure glow discharge* [In Russian] (Moscow: Innovative engineering) p 335
[4] Lakhtin Yu, Kogan Ya and Shpis G 1991 *Theory and technology of nitriding* [In Russian] (Moscow: Metallurgy) p 320
[5] Petrova L 2001 Internal nitriding of high temperature steels and alloys [In Russian] *Metal Science and Heat Treatment* vol **43** no **12** pp 11–7
[6] *ICDD. International Centre for Diffraction Data – PDF4+* 2016 (USA: Newtown Square PA)
[7] Babin A and Rakipov D 2014 *Organization and planning of the experiment* [In Russian] (Ekaterinburg: UrFU Press) p 112
[8] Adler Yu, Markova E and Granovsky Yu 1976 *Planning an experiment in the search for optimal conditions* [In Russian] (Moscow: Nauka) p 280
[9] Lakhtin Yu and Arzamasov B 1985 *Chemical heat treatment of metals* [In Russian] (Moscow: Metallurgy) p 255
[10] Lobanov L and Zorina M 2017 *Methods for the determination of diffusion coefficients* [In Russian] (Ekaterinburg: UrFU Press) p 100
[11] Merer H 2011 *Diffusion in solids* [In Russian] (Moscow: Intellect) p 536

IOP Conf. Series: Materials Science and Engineering **966** (2020) 012058 doi:10.1088/1757-899X/966/1/012058

Technologies for remote lidar sensing of the surface to detect low concentrations of hydrocarbons in Siberia and the far north

S L Leshchenko

Siberian Federal University, School of space and information technology
26, Akademika Kirenskogo Street, Krasnoyarsk, 660074, Russia

E-mail: lsl24@mail.ru

Abstract. The results of research in the field of remote sensing using the technologies of radar and laser detection of hydrocarbons raw materials aimed at the search, localization and topographic location of discovered oil and gas field, the determination of leaks and atmospheric pollution by hydrocarbon gases (e.g. methane) are presented. The main problems with sounding are revealed. To determine the extremely low concentrations of methane in the atmosphere, the main constraining factors were determined and the need to suppress side interference arising on the probing path of the laser beam was shown. To formulate requirements for the sensitivity of equipment and the accuracy of software calculations, the choice of a mathematical model of the path of the laser beam is justified. A modification of the known model is proposed to represent the lidar path based on the differential absorption method taking into account correction coefficients of molecular and aerosol absorption and scattering. Thus, the presented model is a set of methods for providing more accurate and localized data for the implementation of a hardware-software complex.

1. Introduction

A significant problem in the exploration of oil fields in the region of Siberia and the far north is the inaccessibility and extremely low temperatures, which reduces the rate of discovery of new deposits. A number of approaches are known for solving the indicated problem [1–3].

One of the most promising methods is the exploration method based on the use of remote lidar sensing technologies for the surface layer from an aircraft, which makes it possible to detect extremely low concentrations of hydrocarbons [4]. By detecting stable areas with a high content of suspended solids in the atmosphere, fixing local natural gas outlets, observing the spatial dynamics of surface leaks in real time, it becomes possible to topologically link the proposed deposits to the terrain.

Existing laser-location methods [5,6] are used everywhere for shooting and detecting gas pipeline leaks and solving topographic problems, which allows obtaining results in a wide range of subject areas. However, the known methods and hardware are designed to determine large concentrations of hydrocarbons (up to 1 ppm) [5]. However, for the localization of prospective deposits, it is necessary to determine the extremely low gas content in the surface space.

A solution to this problem can be obtained using a lidar based on the differential absorption method with high selectivity, which is provided by a set of technical solutions [7]. First of all, the functional composition of the lidar hardware should allow generating and efficiently receiving laser radiation at appropriate wavelengths that are close in their values to absorbed octanes and exclude the possibility of spurious solar illumination. Secondly, at the hardware and software level, it is necessary to solve filtering, noise reduction, and isolating the received useful signal with high selectivity. This requires the development of an adequate mathematical representation of the propagation path of the laser beam. Such a model will allow you to take into account the main factors affecting the accuracy of the processing of the useful signal, namely: type of landscape, signal dispersion, limitation of signal reception distance.

In order to solve the problems of signal scattering and develop a list of requirements for the developed complex, a mathematical apparatus is proposed that describes the path of the light beam.

2. Mathematical model

According to the authors of [4–6,8], the following representation of the lidar equation (1) is the most accurate for this case:

$$P_\lambda(R) = P_0\eta(\frac{c\tau}{2})A(R)\frac{\beta_\lambda(R)}{R^2}\exp[-2\int_0^R \alpha_\lambda(x)dx], \quad (1)$$

R is the distance to the scattering volume; c is the speed of light; τ is the laser pulse duration; P_λ is the power of the scattered radiation obtained during the time $t = \frac{2R}{c}$; P_0 is the laser power; η is the receiver efficiency (calibration constant); λ is the wavelength of the laser radiation; β_λ is the volume coefficient of backscattering of the atmosphere; A is the effective area of the receiver; α_λ – volume coefficient of attenuation (extinction) of the atmosphere.

The coefficients of molecular scattering (β_λ^{mol}) and attenuation (α_λ^{mol}) can be calculated with good accuracy based on the Rayleigh scattering theory [9], or using a specific model of the atmosphere [8]. For a correct estimation of the transmission of laser radiation by the atmosphere, information is needed on the physical and optical models of the atmosphere. One of the most complete for solving the tasks of remote sensing seems to be a physical model of the atmosphere developed at the V E Zueva SB RAS [4]. In addition to data on temperature, pressure, and concentration of H_2O and O_3 for various climatic zones, it contains information on the altitude distribution of CO_2, CO, CH_4, N_2O, NO, NO_2, as well as information on their standard deviations for various heights.

The theory of molecular scattering of Rayleigh light gives the following expression (2) for the volume coefficient of backward molecular scattering (in cm-1 * sr-1) in gases:

$$\beta_\lambda^{mol} = \frac{\pi^2(n^2-1)^2}{N^2\lambda^4}\frac{6+3\Delta}{6-7\Delta 0}, \quad (2)$$

N is the parameter of concentration of molecules; n is the refractive index; Δ is the degree of depolarization scattered radiation.

For mixture of atmospheric gases at altitudes up to 100 km, expression (2) describes the volume coefficient of backscattering, and will have form:

$$\beta_\lambda^{mol} = 5.45\cdot 10^{-28} N(\frac{550}{\lambda})^4,$$

where λ is taken in nanometers. Considering a single molecular scattering it has parameter Δ, which is associated with the polarization anisotropy of the molecules. For example, such as monatomic gases of the argon type, it is equal to zero (for air, Δ = 0.035, and for nitrogen Δ = 0.036) in the case of isotropic scattering centers.

The coefficients of aerosol scattering (β_λ^{aer}) and attenuation (α_λ^{aer}) can be determined either theoretically. It is based on the theory of aerosol scattering (Mie theory) [9], or experimentally (from lidar signals). It is based on algorithms of solving the lidar equation, which are detailed considered in a number of papers [10].

According to the theory of scattering of electromagnetic waves by aerosol particles in the approximation of dielectric spheres (Mie theory), one can obtain expressions for the volume scattering coefficients (3) and attenuation (4) at a given wavelength:

$$a^{aer}(n,\lambda)=\int_0^\infty \pi r^2 Q_{ext}(p,n,\lambda)f(r)\mathrm{d}r, \qquad (3)$$

$$\beta^{aer}(n,\lambda)=\int_0^\infty \pi r^2 Q_{scat}(p,n,\lambda)f(r)\mathrm{d}r, \qquad (4)$$

r is the radius; $f(r)$ are size distribution functions of aerosol particles; n' is the complex refractive index of the dielectric sphere; Q_{ext} and Q_{scat} are the scattering and attenuation efficiency coefficients; $p=\frac{2\pi r}{\lambda}$ is the relative particle size.

In [6], aerosol attenuation is described by the empirical expression (5):

$$\alpha^{aer}=\frac{3.912}{R_m}\left(\frac{\lambda}{0.55}\right)^q, \qquad (5)$$

where R_m is the meteorological range of visibility at $\lambda = 0.55$ µm.

The values of R_m for various visibility conditions are known [9], and the coefficient q is given by the following expression (6):

$$q=\begin{cases}0.585(R_m)^{1/3}\ R_m \le 6\ km\\ 1.36\ km \le R_m \le 5\ km\\ 1.6 R_m \ge 50\ km\end{cases}, \qquad (6)$$

When developing a hardware-software complex of a lidar for an aviation platform (Figure 2), it should be taken into account that the parameter $A(R)$ in equation (1) can be determined only by the characteristics of the receiver and transmitter. It depends on the spatial distribution of power in the laser beam, as well as the degree of overlap of the laser beam and the field of view of the telescope. This factor takes into account the influence of the shadow of the secondary mirror of the telescope, the aberration of the optical system, the inhomogeneity of the surface of the detector and the effective area of the telescope. When determining the size of the telescope mirror, it should be taken into account that the effective receiver area $A(R)$ is given by the following expression (7):

$$A(R)=\frac{A_0}{\pi W^2(R)}\times\int_{r=0}^{r_{max}}\int_{\psi=0}^{2\pi}\xi(R,r,\psi)F(R,r,\psi)r\mathrm{d}r\mathrm{d}\psi=A_0\xi(R), \qquad (7)$$

$F(R,r,\psi)$ is the spatial distribution function of the laser radiation intensity; $\xi(R,r,\psi)$ is the geometric probability coefficient; ψ is the azimuthal angle; $W(R)$ is the size of the laser spot at a distance R; A_0 is the area of the entrance aperture of the telescope; $\xi(R)$ is the function of the geometric factor (FGF) of the lidar, which takes into account the degree of interception of the laser beam reflected from the target.

The developed hardware-software complex for controlling lidar uses the principle of differential absorption (DP), based on the phenomenon of resonant absorption of laser radiation inside the absorption line of the studied gas. The gas concentration is calculated using beam signals at two close frequencies, one of which is inside the absorption line, and the other outside it.

Using lidar equation (1) for elastic scattering at wavelengths λ_{on} and λ_{off}, from the power ratios of the reflected signals P_{on} and P_{off} at two wavelengths, we obtain the following expression (8) for the concentration of the molecules under study.

$$N(R)=\frac{-1}{2\Delta\sigma}\frac{\partial}{\partial R}\left[\ln\frac{P_{on}(R)}{P_{off}(R)}\right]+\frac{1}{\Delta\sigma}\frac{\partial}{\partial R}\left[\ln\frac{\beta_{on}(R)}{\beta_{off}(R)}\right]-\frac{\alpha_{on}(R)-\alpha_{off}(R)}{\Delta\sigma}, \quad (8)$$

β_{on} and β_{off} are the coefficients of backscattering of the atmosphere at wavelengths λ_{on} and λ_{off}; α_{on} and α_{off} are the attenuation coefficients of the atmosphere for the waves λ_{on} and λ_{off},

$$\Delta\sigma=\sigma(\lambda_{on})-\sigma(\lambda_{off})=\sigma_{on}-\sigma_{off}, \quad (9)$$

$\Delta\sigma$ is the cross section of the DP (differential absorption); $\sigma(\lambda_{on})$ and $\sigma(\lambda_{off})$ are the effective absorption cross sections at two wavelengths. In the case of a surface sensing path, $\sigma(v)$ for a line with a central frequency v_0 is well described by the Lorentz contour [4–5] shown in expression (10):

$$\sigma_L(v)=\frac{S}{\pi}\frac{\gamma_L}{\gamma_L^2+(v-v_0)^2}, \quad (10)$$

S, γ_L are the intensity (12) and half-width (11) of the absorption line, respectively, which for vibrational-rotational lines at pressure p and temperature T have the following form:

$$\gamma_L(T,p)=\gamma_L^0\left(\frac{T_0}{T}\right)^l\frac{p}{p_0}, \quad (11)$$

$$S(T)=S_0\left(\frac{T_0}{T}\right)^m\exp\left[-E''\frac{hc}{kT}\left(\frac{1}{T}-\frac{1}{T_0}\right)\right], \quad (12)$$

γ_L^0 and S_0 are the Lorentz an half-widths and intensities at temperature T_0 and pressure p_0; E'' is the energy of the lower vibrational-rotational state; h is Planck's constant; k is the Boltzmann constant; m and l are the parameters of the temperature dependence of the half-width and intensity, respectively. The parameter l for various gases can vary from 0.5 to 1, the parameter $m = 1$ for linear molecules and 1.5 for nonlinear ones.

Expression (8) shows the possibility of restoring the continuous profile $N(R)$ during continuous recording of lidar signals. However, in practice, lidar signals will be recorded by the receiving equipment in the form of a discrete digital data array obtained at equal time intervals Δt, which corresponds to the spatial resolution $\Delta R=\frac{c\Delta t}{2}$. Then, from (8) for signals received from atmospheric objects remote along the sensing path at a distance of R and $R+\Delta R$, the expression for the average concentration (13–15) of the studied gas in the indicated range of distances follows:

$$N'=\frac{-1}{2\Delta\sigma\Delta R}\ln\left[\frac{P_{on}(R+\Delta R)\,P_{off}(R)}{P_{off}(R+\Delta R)\,P_{on}(R)}\right]-\frac{\alpha_{on}^{mol}-\alpha_{off}^{mol}}{\Delta\sigma}+B_S+E_a, \quad (13)$$

$$B_S=\frac{-1}{2\Delta\sigma\Delta R}\ln\left[\frac{P_{on}(R+\Delta R)\,P_{off}(R)}{P_{off}(R+\Delta R)\,P_{on}(R)}\right]-\frac{\alpha_{on}^{mol}-\alpha_{off}^{mol}}{\Delta\sigma}+B_S+E_a, \quad (14)$$

$$E_a=\frac{-\alpha_{on}^{aer}-\alpha_{off}^{aer}}{\Delta\sigma}, \quad (15)$$

When calibrating the transmitting and receiving equipment of the complex, the wavelengths should be chosen so that the coefficients of backscattering and attenuation of the atmosphere are independent of the wavelength. Therefore, the spectral interval between two wavelengths should be so small that the terms B_s and E_a can be neglected. Then the expression for concentration is simplified:

$$N' = \frac{-1}{2\Delta\sigma\Delta R} \ln\left[\frac{P_{on}(R+\Delta R)}{P_{off}(R+\Delta R)} \frac{P_{off}(R)}{P_{on}(R)}\right], \tag{16}$$

Based on the foregoing and the results of preliminary experiments [11–13] in the problem area, it was noted that the obtained equation (16) adequately describes the depend of the concentration of desired gas on the reflected radiation power at two wavelengths λ_{on} and λ_{off}. In this case, λ_{on} is inside the absorption line of the desired gas, and λ_{off} is outside it (Figure 1).

To assess the sensitivity of the developed equipment – the maximum allowable values of atmospheric gas concentrations N_{min}, which can be detected with a minimum measurement error of optical signals, it is convenient to use the formula proposed in [6]:

$$N_{\min}(R) = \frac{1}{\Delta\sigma\Delta R} \frac{2}{SNR(R)}, \tag{17}$$

SNR is the signal-to-noise ratio at a distance *R* from the lidar.

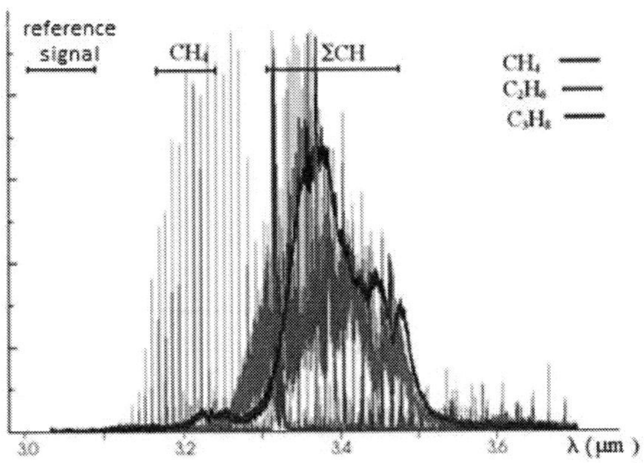

Figure 1. The spectra of the determined hydrocarbons and the spectral ranges selected for measuring their concentrations [14].

Thus, to increase the sensitivity of the device at a given spatial resolution ΔR, one should choose the most intense absorption lines of the gas under investigation with a maximum DP cross section $\Delta\sigma$.

3. Conclusions

The initial assumption that it is possible in principle to create a lidar to determine exactly the maximum concentrations of hydrocarbons in the surface layer is confirmed by simulation results. It was determined that the proposed architecture, the main parameters of the equipment and the differential absorption method are effective for remote sensing on paths not exceeding 200 m.

When setting the lidar to determine the octane group gases by the differential absorption method, wavelengths should be used for which the ratio $\frac{\Delta\sigma}{\Delta\lambda}$ is maximum, where $\Delta\lambda = \lambda_{on} - \lambda_{off}\,\Delta$. It can be seen from expression (17) that, at significant signal-to-noise ratios, the concentration of N_{min} is limited by the systematic error of the lidar, which can be from 10 to 20 %.

Currently, the mathematical apparatus proposed for calculating the concentration of saturated hydrocarbons along the lidar route requires adequate algorithmic representation and modeling, software development for a microprocessor-based lidar control system, and a series of field experiments.

IOP Conf. Series: Materials Science and Engineering **966** (2020) 012058 doi:10.1088/1757-899X/966/1/012058

References

[1] Bartholomew J, Lyman Ph, Weimer C and Ruppert L 2017 Airborne Active Sensing for Pipeline Leak Survey *AIAA SciTech Forum.*

[2] Wojcik M, Crowther B, Lemon R and other 2015 Development of Differential Absorption Lidar (DIAL) for Detection of CO2, CH4 and PM *Alberta Chemical and Biologial Sensing Technologies XII Proc. of SPIE* vol **9486** 94860K @ 2015 SPIE.

[3] Davenport V B and Ruth V L 1960 *Introduction to the theory of random signals and noise: a tutorial* (Moscow: publishing house of foreign literature) p 467.

[4] Zuev V E and Zuev V V 1992 *Remote optical sensing of the atmosphere* (St. Petersburg: Gidrometeoizdat) p 231 [In Russian].

[5] Mezheris R 1987 *Laser remote sensing* (Moscow: Mir) p 550 [In Russian].

[6] Collis R T, Hinckley E D, Inawa X et al 1979 *Laser control of the atmosphere* (Moscow: Mir) p 416.

[7] Nepomnyashchiy O V, Ten S F and Khabarov V A 2011 Mathematical and hardware support of a complex of geophysical studies for remote, aircraft sounding of the earth's surface *Aerospace Instrumentation* no **10** (Moscow: Nauchtekhizdat) p 38–43 [In Russian].

[8] Zuev V E and Krekov G M 1986 *Optical models of the atmosphere* (Leningrad: Gidrometeoizdat) p 256 [In Russian].

[9] Van de Hulst H C 1981 *Light Scattering by Small Particles* (New York: JWS Inc) p 470

[10] Kovalev V A and Eichinger W E 2004 *Elastic Lidar: Theory, Practice and Analysis Methods* (New York: JWS Inc) p 615 [In Russian].

[11] Nepomnyashchiy O V, Ten S F and Khabarov V A 2011 The method of dispersionless absorption measurement of methane concentration based on the hardware technique of a meter with an open optical channel *Information-measuring and control systems* no **2** vol **9** (Moscow: ZAO Publishing House 'Radio Engineering') p 3–7 [In Russian].

[12] Nepomnyashchiy O, Sirotinina N, Popov D and Leshenko S 2018 Lidar beam path model for measuring extremely low concentrations of hydrocarbons in the surface layer. *Anthology of scientific research papers 'Space Engineering, Technologies & Exploration'.* ECM Space Technologies GmbH, Berlin, Germany pp 186–190 [In Russian].

[13] Leshchenko S L, Popov D V and Nepomnyashchy D O 2018 The mathematical model of the lidar route *Mathematical methods in technology and technologies - MMTT.* Saratov State Technical University named after Gagarin Yu.A. Saratov pp 21–24 [In Russian].

[14] Roberto N 2005 Empirical relationships between extinction coefficient and visibility in fog *Appl. Opt.* vol **44**.

IOP Conf. Series: Materials Science and Engineering **966** (2020) 012059 doi:10.1088/1757-899X/966/1/012059

Iron and steel industry of Russia in the XXI century

S V Dokuchaev

Ural Federal University named after the First President of Russia B N Yeltsin, Nizhniy Tagil Technological Institute, 59, Krasnogvardeyskaya str., Nizhniy Tagil, 622000, Russia

E-mail: Dokuchaev-S-V@yandex.ru

Abstract. The author states that over the past twenty years, the iron and steel industry of Russia has held its position as one of the key industries of the Russian economy. Technological upgrading of production facilities has minimized the backwardness, however the rapid steel production growth in Asia has resulted in loosing business in the global steel market, while getting more dependent on the market situation. This article deals with the main trends, problems and collisions of the steel industry development in the first two decades of this century. The structure of Russian steel export is reviewed. This work is based on statistical and comparative analysis, ratings and forecasts, as well as theoretical postulates of Douglas North about the close relationship of market economy processes with political and social institutions of the country. The purpose of the article is to show that the industry is facing impending stagnation under the economic sanctions, export dependency of the Russian steel industry and worldwide steel overproduction. The following may be taken as the scientific results of the study: systematization, generalization and author's assessment of processes in the Russian steel industry and its development prospects, against the background of global trends in the industry.

Introduction

The steel industry is a key industry for the country. The high level of its export orientation makes it one of the basic sectors of the real economy and an important source of foreign currency in the Russian Federation. The goal of this research is to identify the main trends and collisions in the development of the Russian steel industry, that have occurred over the past two decades. To achieve this goal, the industry development was assessed basing on the analysis of the main steel industry performance indicators, both in Russia and worldwide.

By the beginning of the 21st century, the industry had finally adapted to market conditions. This process was quite painful and took the last decade of the twentieth century. In 2001, Russia was the 4th largest steel producer by the total output. The results of this period were:

A) Privatization and final ownership redistribution both in the industry and in the national economy; decentralization in the industry and getting of full independence in business activities by companies;

B) Corporatization of companies of the industry, establishing of several vertically integrated structures, which unite producers of feedstock and final products, and producing 80 % of the total steel product output [1];

C) Considerable output drop, both due to reduced sales market after the USSR collapse and due to the general output drop for all industries of Russia;

D) improvement (recovery) of financial indicators of companies of the industry after the default in August of 1998.

The steel industry of Russia had the following stable advantages: many local material suppliers and cheap labor.

At the same time there were serious problems in the industry:

1) Inefficient structure of most companies established in the course of privatization. All assets that a company had on its balance sheet were automatically included in its equity capital. As a result, the industry had a large share of non-core assets: there were many social facilities among them. More than 70 % of large companies in the industry were city-forming enterprises [2]. The high social burden of the industry made its products less competitive in the world market.

2) Serious dependency on the external market. Backward export structure: most products were semi-finished i.e. with low added value.

3) Necessity for the introduction of advanced technologies. Only Russian and Ukrainian companies continued to use open hearth furnace. Completion of the transition of the industry to the technological level of industrialized countries [3].

Main trends and collisions in the development of Russian steel industry in the first twenty years of the 21st century

Through 2001 to 2007, companies of the industry were increasing their production output both through technological and arrangement activities. The high monopolization level of the Russian steel industry plus availability of resources invited key market players to acquire foreign assets. Such acquisitions were made in: Ukraine (in particular, Evraz acquired 5 companies in Ukraine in 2007), Czech Republic, Italy, and USA. Not all acquisitions turned out to be smart and many of them were made just before the economic crisis in 2008, when assets were at the peak of their price. Some of the assets were depreciated and some had to be sold cheap. After 2014, the life of Russian companies in Ukraine is complicated, which also motivated them to reduce their assets in that land.

The crisis led to the output drop through 2008 to 2009, and then the slow growth began. The industry minimized its losses due to expanded supply to EU countries, but the level of earlier days was reached not in a day. The output performance of 2007 was beat only in 2018 for cast steel and in 2014 for rolled steel products [4]. The pipe demand was fluctuating and depending on progress of big transport projects in the oil and gas industry. Basically, that situation in the industry could be explained by low domestic demand and dependency on the external market, which had already become unsafe [5]. In fact, the crisis was replaced by stagnation, which was only growing after the economic sanctions against Russia were imposed in 2014.

During the period under review the positions of Russian steel industry was weakened in the global market since the growth of steel production in Asian countries was outstripping: over a period of a little more than a quarter of a century the steel production output was increased by almost 11.5 times in China, by 6 times in India and by more than 2.5 times in South Korea (as shown in Table 1).

Table 1. Steel production output trends (ths. tons) [6, 7].

Countries	1992	2000	2009	2018	Output change through 1992–2018
China	80935	127236	577070	928264	1146.93 %
India	18117	26294	63527	109272	603.15 %
USA	84322	101803	59384	86607	102.71 %
Japan	98132	106444	87534	104319	106.3 %
Russia	67029	59136	60011	72042	107.48 %
South Korea	28055	43107	48572	72464	258.29 %
Germany	39711	43376	32670	42435	106.86 %

Eventually, within the key steel producer group, Russia moved from the 4th place in 1992 to the 6th place in 2018 by absolute output, and its share in the global steel output decreased from 9.3 % to about 4 % as shown in Table 2.

Table 2. Russia's share change in percentage of global steel production output [6, 7].

1992	2000	2009	2018
9.3	6.98	4.85	3.97

Percentage change trends per various technologies of steel production worldwide through 2000-2018 are shown in table 3.

Table 3. Percentage per technologies of steel production in the period 2000/2018 [6, 7].

Country	Oxygen-blown converter	Electric furnace	Open hearth furnace	Other processes
China	62.6/88.4	15.9/11.6	1.4/–	19.8/–
India	49.8/45.2	40.9/54.8	9.3/–	–/–
USA	53/32	47/68	–/–	–/–
Japan	71.2/75	28.8/25	–/–	–/–
Russia	58.1/66.9	14.6/30.8	27.4/2.8	–/–
South Korea	57.2/66.6	42.8/33.4	–/–	–/–
EU	60.2/58.5	39.8/41.5	–/–	–/–

The industry sustained critical technological changes in the period under review. Buying and using advanced equipment mostly made by European manufacturers became a philosophy [8]. The Russian steel industry became less backward relating to the steel industry leaders with reference to the manner of production compared by its technological form. The progress was slowed down by the economic crisis of 2008: plans were frustrated for the development of thin casting systems and construction of dozens of mini-mills, and just about ten production facility enterprises were actually launched. The share of open-hearth production in the total volume of Russian steel output was minimized by an order and its final closure is planned to be carried out in 2020. Among the main steel producers only Russia and Ukraine remained.

Adding to inefficient structures of companies, the technological lag gave labor productivity indicators lower than in the PRS. In 2010, 1 employee in Russia produced only 11.8 % of the steel output performance in PRC i. e. it was 8.5 times lower [9].

The Russian export structure was slowly changing. It was determined mainly by demand in foreign markets, and to a lesser extent by our technological capabilities. The bulk of exported steel was in the form of feedstock (iron ore, pellets and scrap) and low-availability products (pig iron, iron and non-alloy steel semi-products), which together made up about 80 % of the total metal export in physical terms. The share of their products (hardware, pipes, and products of reprocessing) was not more than 20 % of the export at the end of the first decade of the 21st century [10, 11].

The industry has preserved its existing structure with prevailing large integrated plants having high capacity and high output of products with low added value, that was a consequence of development focused mainly on the world market. This trend of development met the state line of industrial policy i. e. keeping people in one-company towns employed to avoid social problems in such towns. In return, owners of vertical groups of companies could expect for priority access to the governmental financial support. The support included an extended list of imported production equipment not subject to VAT and preferential loans for big investment projects.

Lack of steady growth in domestic steel consumption resulted in greater export orientation of the industry. Export was a priority for the industry as only export could support mass steel production through all this time.

However, export growth was critically limited in the second decade of the 21st century. The following problems can be highlighted:

a) Protectionism in PRC, restricting their markets for foreign exporters; for Russia it means that the industry will have to focus on cast steel and rolled products with low added value, despite the stepwise upgrade of production technologies;

b) Sanctions of the industrial countries against Russian economy after 2014 that slowed down its development, reduced the domestic demand for steel and put additional obstacles in the way of Russian steel producers in the foreign markets;

c) Both rapid domestic production growth and steel overproduction in Asia, like worldwide. Steel product prices drop.

d) Environmental problems: Europe's transition to stricter environmental standards for the steel industry will also affect the Russian steel industry. Today, the European Union (with its New Green Deal) is considering a carbon tax on imports (border adjustment carbon tax) to support its environment-friendly production. That assumes that a special tax will be charged for steel imported to the European Union (produced using the conventional carbon-intensive technology). For European steel producers, it is about 25 Euro per one ton of carbon dioxide, which encourages them to switch to clean technologies. Russian enterprises will have to update production according to EU standards (including EU equipment to be bought) or incur additional expenses for that tax [12].

Automotive sheet and pipes of various diameters added to the product range are among positive development aspects of the industry in the second decade of the 21st century. Russia was a number 2 exporter of pipes and pipe production capacity increased and exceeded the domestic demand by much. However, the output growth due to these positions was short-lived. The automotive industry was hit hard by the ruble devaluation in 2014 and polymer pipes became a real alternative in the housing and public utility sector, so Gazprom had to abandon a number of its projects. The domestic demand could not consume the entire industry's output, although it showed some growth in 2017–2018. Competition also has increased in the world market of steel. The results of 9 months of 2019 show that exports has decreased by 11 % for flat products and by 15.6 % for long products compared to the same period of the previous year, and the structure of Russian steel export remained unchanged over the past years and had a high share of semi-products [13].

Apart from external factors that had a negative effect on the industry development, it is necessary to point out internal reasons that are not less, but even more important. This is the absence of institutional changes and related economic reforms in this country, and 'persistently ineffective solutions' chosen by the establishment [14].

Summary

The Russia's economic growth in the first decade of the 21st century was stimulated by high energy prices. However, this growth was not associated with institutional changes and was not accompanied by economic and political reforms. As a result, the steel industry growth as well as the entire Russian economy was short-lived. After the global economic crisis in 2008, many European countries and the United States introduced protective actions against steel producers from other countries. Steel production in Asia grew at a faster pace. The steel demand in the world during 2014–2018 was unstable: growing was followed by falling. With recessed Chinese economy growth, all these factors contributed to the steel overproduction growth in the world and led to a drop in the prices, which made life of Russian steel producers even more complicated. Steady performance of the industry may be associated with an increase of in metal consumption in the domestic market, but under the existing economic sanctions and the absence of institutional changes, this seems unlikely.

References

[1] Lugacheva L I and Musatova M M 2012 Consolidation and vertical integration of iron and steel companies *ECO* **8** pp 97–113

[2] Analytical report '*Problems and prospects of development of domestic ferrous metallurgy*' 2010 Interdepartmental analytical center pp 3–8 Available online: http://www.iacenter.ru/publication-files/130/110.pdf (accessed on 01.06.2020)

[3] Zapariy V V 2011 Ural metallurgy in the context of market reforms (end of XX–XXI century) *Historical and economic research* **2** pp 133–63

IOP Conf. Series: Materials Science and Engineering **966** (2020) 012059 doi:10.1088/1757-899X/966/1/012059

[4] RIA Rating Russia today 2019 Analytical bulletin *Metallurgy: trends and forecasts Results of 2018* Moscow **33** pp 4–6

[5] Shaybakova L F and Novosyolov S V 2017 Development of Ferrous Metallurgy in Russia: Trends, Special Features and Problems *Upravlenets* **5 (69)** pp 40–9

[6] World Steel Association 2019 *Steel Statistical Yearbook Concise version.* (Brussels) pp 3–4

[7] International Iron and Steel Institute/ Committee On Economic Studies December 2002 *Steel Statistical Yearbook* (Brussels) pp 10–2

[8] Tatarkin A I, Romanova O A, Korovin G B and Chenchevich S G 2015 Ural Orientation of Russian Metallurgy Innovative Development *ECO* **3 (489)** pp 79–97

[9] Ministry of Industry and Trade of Russia order dated 05.05.2014 On approval of the Strategy for the development of ferrous metallurgy in Russia for 2014 – 2020 and for the long term until 2030 and Strategies for the development of non-ferrous metallurgy in Russia for 2014-2020 and for the long term until 2030 **839** 7 Available online: http://www.consultant.ru/document/cons_doc_LAW_165502/ (accessed on 01.06.2020)

[10] Zinoveva N G 2018 Tendencies of export and import of steel industry products of Russia in 2017 – 2018 *Ferrous metallurgy Bulletion of scientific, technical and economic information* (Moscow) **12 (1428)** pp 5–15

[11] Zinoveva N G 2019 Tendencies of export and import of steel industry products of Russia in 2017 – 2018 *Ferrous metallurgy Bulletin of scientific, technical and economic information* (Moscow) **8** pp 903–12

[12] Shevelev L N and Brodov A A Energy saving, increasing energy efficiency and reducing greenhouse gas emissions in the iron and steel industry of Russia *Ferrous metallurgy Bulletion of scientific, technical and economic information* (Moscow) **2 (1418)** pp 3–7

[13] Zinoveva N G 2020 Tendencies of export and import of steel industry products of Russia within 9 months of 2019 *Ferrous metallurgy Bulletion of scientific, technical and economic information* (Moscow) **2** pp 107–13

[14] North D 1997 *Institutions, Institutional Change and Economic Performance* (Moscow: Foundation of economic thought 'Nachala') p 24

IOP Conf. Series: Materials Science and Engineering **966** (2020) 012060 doi:10.1088/1757-899X/966/1/012060

Resource-saving potential of brick scrap from demolition of buildings

I K Domanskaya[1, 2, 3]

[1]Department of Building Materials, Ural Federal University named after the First President of Russia B N Yeltsin, 19, Mira str., Yekaterinburg, 620000, Russia
[2]Department of Commodity Science and Expertise, Ural State University of Economics, Street 8 March/NarodnayaVolya, 62/45, Yekaterinburg, 620144, Yekaterinburg, Russia

Email: [3]doman-1@yandex.ru

Abstract. The composition and properties of crushing products of brick scrap of different sizes were studied. It is shown that a large (5–40 mm) and small (0.16–5.0 mm) fraction of brick scrap crushing products may be of interest for replacing natural aggregates, while a pulverized (0–0.16 mm) fraction has pozzolanic activity. The presence of cement masonry mortar in the composition of products for crushing scrap brick walls of buildings is the main obstacle to their use as a large aggregate for light concrete: the average density of crushed brick grains in this case was 890 kg/m^3, while for crushed cement grains – at least 1350 kg/m^3. With an increase in the degree of crushing of these wastes, their material composition and physical properties are averaged and therefore more suitable for reuse in construction, including as a fine aggregate for mortars.The chemical and mineral composition of brick scrap, studied on the example of a dusty fraction, allows us to predict its good prospects for replacing natural mineral raw materials in the production of firing materials, for example, Portland cement.

1. Introduction

The need for economical and careful attitude to natural resources is becoming more and more obvious. A particular concern of modern society is the depletion of non-renewable mineral resources, which are the raw material base of the most important sectors of material production, such as metallurgy, chemical industry of inorganic materials, construction and others. The importance of these problems, their specificity and possible solutions are reflected in UN documents as part of the Sustainable Development Agenda for the period until 2030 [1].

To ensure sustainable development, both social and technological innovations are developed and proposed, and the assessment of the sustainability of any material object is based on an assessment of its life cycle [2–7]. In a simplified form, the life cycle of an object consists of the following stages: design, manufacture, operation, and disposal (burial).On each of them, in one way or another, the object interacts with the environment, 'consumes and/or allocates' material and energy resources. It is obvious that from the point of view of preserving non-renewable natural raw materials, it is necessary to ensure the closed, 'cyclical' nature of the existence of products, regardless of their purpose.

Among the most material-intensive industries, metallurgy seems to have achieved the greatest success in product recycling [8,9]. The situation is much worse with the recycling of waste from the construction industry, which includes the production of building materials. According to various sources, about 20 million tons of construction and demolition waste is generated annually in the Russian Federation, and their utilization rate does not exceed 10–20 %. Publications by Yu M Bazhenov, M O Korovkin, V S Lesovik, S-A Yu Murtazayev and other authors indicate that there is a certain domestic experience in the processing and utilization of concrete scrap, while the use of construction scrap from the demolition of brick buildings has been studied worse.

The purpose of this work is to study the composition and properties of the products of crushing brick scrap of different sizes and determine the possible directions of their return to the construction industry to replace natural mineral raw materials.

2. Initial materials and research methods characteristic

For research, crushed scrap of wall structures from the demolition of a building built in the 60s of the 20th century (Yekaterinburg, Sverdlovsk region) was used. The maximum size of fragments is up to 150 mm (Figure 1).

Figure 1. Appearance of crushed scrap of wall structures.

The content of cement mortar in this sample was quite high – at least 30 % of the total mass of the brick scrap (usually, according to literature data, 10–15 %). So, 3 samples (types) of material was subjected to additional crushing: brick scrap in the natural state (BS), pre-selected fragments of ceramic brick (CB) and fragments of cement mortar (CM). After crushing, the resulting products were sieved on sieves and 3 fractions were isolated: particles with a grain size from 5 to 40 mm (large aggregate), from 0.16 to 5 mm (fine aggregate) and less than 0.16 mm (pulverized, or dust, fraction).

The study of the properties of granular materials was carried out in accordance with the requirements of regulatory documents: GOST 8267–93 'Crushed stone and gravel of solid rocks for construction works. Specifications'; GOST 32496–2013 'Fillers porous for light concrete. Specifications'; GOST 32021–2012 'Hard rock aggregates and fillers for production dry building mixes. Specifications'; GOST 8736–2014 'Sand for construction works. Specifications'.

Physical and mechanical parameters of the BS dust fraction were determined in accordance with GOST 30744–2001, a chemical composition – according to GOST 5382–91 'Cements and materials for cement production'. Chemical analysis methods, and its mineral composition was studied by x-ray-graphic method on a MiniFlex 600 diffractometer (Cuka-radiation, λ=1.541862 Å, survey interval – 3.00–90.00) 'Rigaku-Carl Zeiss' (Japan).

The pozzolan activity of the BS dust fraction was determined by the method [10].

3. Experiment results and discussion

The average grain density of crushed material is an important characteristic, since it determines whether it can be used as a filler in heavy or light concretes. According to GOST 8267–93, the grain density of aggregates intended for heavy concrete must be between 2000 and 3000 kg/m^3. In this case, the CM grains correspond to the required density, but the density of CB fragments, and therefore of all crushed BS, is less, therefore they are not suitable for heavy concrete and road construction (Table 1).

Table 1. Comparative characteristics of the components of crushed brick scrap.

Property	Brick scrap	Ceramic brick	Cement mortar	GOST 32496–2013 requirements
Bulk density, kg/m^3	940	890	1350	100–1200
Average grain density, kg/m^3	–	1800	2000	not rated
Fraction (grain size), mm	10–20	10–20	10–20	5–10 10–20 20–40
Mass loss during calcination, %	5.10	0,70	11.72	3

Analysis of the BS bulk density and its comparison with the requirements of GOST 32496–2013 shows that crushed brick scrap is suitable as a large aggregate for the production of light concrete (Figure 2). Its bulk density is comparable to such a porous artificial aggregate as aggloporite crushed stone (400–900 kg/m^3).

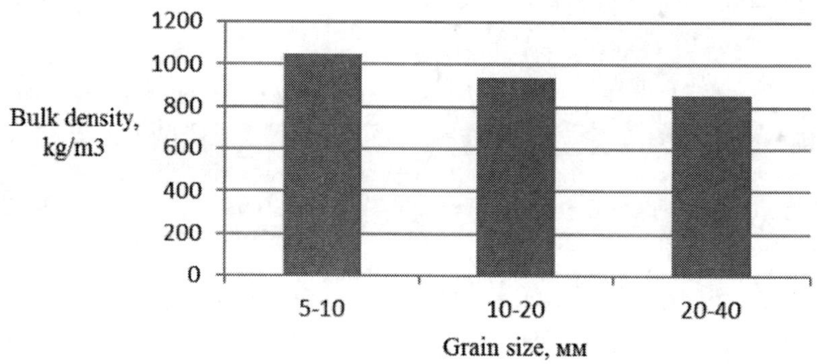

Figure 2. Bulk density of crushed brick scrap of different fractions

At the same time, the investigated batch of BS, due to the significant content of CM, does not meet the regulatory requirements for mass loss during calcination (Table 1). The compression strength in the BS cylinder (0.10–0.25 MPa) and its frost resistance (mass loss after 15 cycles of freezing and thawing was 12%, and no more than 8% is allowed) also do not meet the requirements of GOST 32496–2013. Therefore, the possibility of using crushed brick scrap as a large aggregate in light concrete can not be excluded, but in each case, additional research will be required.

Tests of the fine-grained BS fraction (0.16–5.00 mm) have shown that it meets the regulatory requirements (Table 2).

As for the powdery fraction of crushed BS, its output depends on the conditions and the number of crushing cycles of concrete scrap. According to the results of our research, it is from 3 to 20 %. The specific surface area of the dust was 300 m^2/kg. The chemical composition of the dust fraction, %: 53.68 SiO2; 21.72 CaO; 10.88 Al2O3; 7.17 F2O3; 6.28 loss of heat; 0,27 etc. X-ray-graphic analysis showed that the dust contains crystalline phases: the highest intensity among them is characterized by quartz reflections (4.25; 3.35; 2.45; 2.29). The presence of calcite cannot be excluded (3.85; 3.03; 2.45), tobermorite (3.05; 2.78; 1.82), anorthite, and other phases.

Table2. Fine aggregate characteristics.

Property	BS	GOST 32021–2012 requirements
Bulk density, kg/m^3	1400	1300–1600
Real density, kg/m^3	2510	2 000–28000
The fineness modulus	2.5	1.5–2.5
Grain content of lamellar (bream) and needle-shaped grains, %	19	15–35

The chemical and mineral composition of the dust obtained as a result of BS crushing seems to correspond to the average composition of BS. This suggests that such a material has prospects for use as a raw material component in the production of firing construction composites and Portland cement. There is a well-known experience of producing clinker and cement using crushed concrete waste [11].

It was found that the test sample of the dusty fraction has pozzolanic activity: the dough made from lime (80 %) and the test powder (20 %) when stored in air-wet conditions solidified after 72 hours (allowed for later than 7 days) [10]. Therefore, such an additive can be used as part of construction composites for partial replacement of Portland cement.

4. Conclusion

When choosing the optimal ways to dispose of brick scrap, you should take into account the heterogeneity of its composition due to the presence of cement mortar, the amount of which can be from 10 to 30 %. For this reason, the use of large fraction crushing products (particle size greater than 5 mm) in the form of crushed stone in the composition of concretes is unlikely. At the same time, the fine fraction (0.16–5.00 mm), due to its higher uniformity, can be successfully used to replace natural sand in construction composites after checking its compliance with the requirements of regulatory documents.

The dust formed when crushing brick scrap has pozzolanic activity, therefore, it can be used as part of construction composites for partial replacement of Portland cement.

The chemical and mineral composition of brick scrap (the total content of silicon and aluminum oxides is not less than 60%), studied on the example of a dusty fraction, allows us to predict its good prospects for replacing natural mineral raw materials in the production of firing materials, for example, Portland cement.

Thus, several resource-saving areas for using brick scrap from building demolition can be distinguished: for replacing crushed rocks, which act as aggregates in concrete and mortars, instead of siliceous rocks such as diatomite as a pozzolanic additive, as well as for partial replacement of clays and marls in high temperature synthesis technologies, for example, in Portland cement technology.

References

[1] Sustainable Development Goals Fund 2016. Goal 12: Responsible consumption and production. *United Nations Development Programme.* Available online: https://www.un.org/sustainabledevelopment/ru/sustainable-consumption-production/ (accessed on 01.06.2020)

[2] Leismann K, Schmitt M, Rohn H and Baedeker C 2013 Collaborative consumption: Towards a resource-saving consumption culture. *Resources* **2 (3)** pp 184–203

[3] Cook M 2004 *Understanding the potential opportunities provided by service-orientated concepts to improve resource productivity.* Tracy Bhamra; Bernard Hon (eds.). Design and Manufacture for Sustainable Development (John Wiley and Sons) p 125

[4] Vezzoli C, Ceschin F, Osanjo L, M'Rithaa M K M, Moalosi R, Nakazibwe V and Diehl J C 2018 Sustainable Product-Service System (S.PSS). *Designing Sustainable Energy for All. Green Energy and Technology.* (Springer, Cham)

IOP Conf. Series: Materials Science and Engineering **966** (2020) 012060 doi:10.1088/1757-899X/966/1/012060

[5] Robèrt K H, Schmidt-Bleek B, Aloisi De Larderel J, Basile G, Jansen J L, Kuehr R and Wackernagel M 2002 Strategic sustainable development - Selection, design and synergies of applied tools *J. of Cleaner Production* **10** (3) pp 197–214 https://doi.org/10.1016/S0959-6526(01)00061-010

[6] Frischknecht R, Jungbluth N, Althaus H J, Doka G, Dones R, Heck T and Spielmann M 2005 The ecoinvent database: Overview and methodological framework. *International Journal of Life Cycle Assessment.* https://doi.org/10.1065/lca2004.10.181.1

[7] Ortiz O, Castells F and Sonnemann G 2009 Sustainability in the construction industry: A review of recent developments based on LCA. *Construction and Building Materials* **23** pp 28–39 https://doi.org/10.1016/j.conbuildmat.2007.11.012

[8] Smirnov V V 2018 Recycling is a strategic direction for improving the efficiency of metallurgy in the Russian Federation *Accounting. Analysis. Audit* **5 (4)** pp 30–9 Available online: https://cyberleninka.ru/article/n/retsikling-strategicheskoe-napravlenie-povysheniya-effektivnosti-metallurgii-v-rossiyskoy-federatsii (accessed on 01.06.2020)

[9] Matinde E, Simate G S and Ndlovu S 2018 Mining and metallurgical wastes: a review of recycling and re-use practices. *J. S. Afr. Inst. Min. Metall.* vol. **118** no. **8**

[10] Butt Y M and Timashev V V 1973 *Practice on chemical technology of binding materials* [In Russian] (Moscow: Higher school Press) p 504

[11] Isore L 2018 Production of clinker and cement using crushed concrete waste *J. Cement and its application* **2** pp 50–3

IOP Conf. Series: Materials Science and Engineering **966** (2020) 012061 doi:10.1088/1757-899X/966/1/012061

Study of the main metallurgical characteristics of iron ore raw materials (sinter and pellets)

G Yu Vitkina[1, 2], A N Dmitriev[1] and R V Alektorov[1]

[1]Institute of metallurgy of the Ural branch of the Russian academy of sciences, 101, Amundsen street, Yekaterinburg, Russia

E-mail: [2] 20procents@mail.ru

Abstract. The paper presents the results of studies of the chemical, phase composition and metallurgical characteristics of titanomagnetite sinter. The iron ore used in blast furnaces of JSC 'EVRAZ NTMK' is titanomagnetite sinter obtained from ores of the Gusevogorsky deposit. Samples of sinter with different basicities as well as with addition of binding polymers in the amount of 300 and 500 g per ton of sinter were investigated. The results of industrial tests of the production and blast furnace smelting of sinter with different basicities and additives of binding polymers are presented on the example of the operation of blast furnaces no. 5 and 6 of JSC 'EVRAZ NTMK'. It was shown that an increase in the basicity of the sinter from 2.1 to 2.4 and the introduction of a polymer binder (in the amount of 500 g per ton of sinter) positively affect the complex of sinter metallurgical characteristics – durability after reduction, reducibility, softening and melting temperatures, and also decrease the coke rate in blast furnace smelting by 1.0–1.2%.

1. Introduction

The consumption of stocks of traditional magnetite raw materials forces us to reconsider our attitude to metal production methods and complex ore processing schemes, taking into account market demands and the use of regional resources. Ores with a low titanium content (Gusevogorsky deposit) are processed according to the metallurgical scheme, including the smelting of pig iron in blast furnaces of JSC 'EVRAZ NTMK'.

The use of Ural titanomagnetites consists in the efficient provision of high-quality raw materials for blast furnace production of JSC 'EVRAZ NTMK'. Raw materials processing technologies continue to develop continuously [1], therefore, those measures that are aimed at producing sinter with improved metallurgical characteristics are of particular interest [2–4].

2. Chemical and phase compositions of titanomagnetite sinter

Samples of titanomagnetite sinter with a basicity of 2.1 were selected for the study. The chemical composition of the studied samples of sinter is given in Table 1. The designation of the samples is as follows: 1 – sinter of current production (sinter machine no. 1); 2 – sinter (lower part of the pallet, sinter machine no. 1); 3 – sinter (middle part of the pallet, sinter machine no. 1); 4 – sinter (upper part of the pallet, sinter machine no. 1).

Figure 1 shows a typical diffraction pattern of samples of sinter. The main phase in sinter no. 1–4 is Fe_3O_4 (magnetite), Fe_2O_3 (hematite) is contained in a smaller amount, and Ca-containing silicate of a complex composition – $Ca_{2,3}Mg_{0,8}Al_{1,5}Fe_{8,3}Si_{1,1}O_{20}$ – is also well manifested (named SFCA in

IOP Conf. Series: Materials Science and Engineering **966** (2020) 012061 doi:10.1088/1757-899X/966/1/012061

literature [5–7]). The sinter contains γ-Ca_2SiO_4 (dicalcium silicate). In samples of sinter with a basicity of 2.1, the increase in the unstabilized γ-Ca_2SiO_4 phase goes from the top of the layer to the bottom. Since the presence of this phase leads to disintegration of the initial sinter, a similar pattern is expected during further experiments on reduction and durability after reduction.

Table 1. The chemical composition of the samples of sinter.

Probe	Fe_{total} (%)	FeO (%)	Basicity
1	54.18	10.44	
2	54.66	8.13	2.1
3	54.97	8.80	
4	55.87	10.5	
5	54.20	8.95	
6	54.31	8.30	
7	53.82	10.27	2.4
8	53.08	8.38	
9	52.60	8.22	
10	52.70	7.24	
11	53.00	8.42	2.4
12	52.20	10.50	
13	52.60	12.50	
14	52.30	9.67	

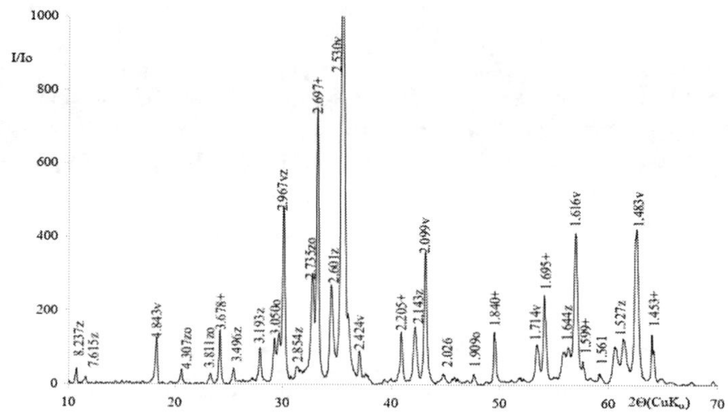

Figure 1. Diffractogramm of sinter of current production:
v – Fe_3O_4; + – Fe_2O_3; z – $Ca_{2,3}Mg_{0,8}Al_{1,5}Fe_{8,3}Si_{1,1}O_{20}$; o – γ-$Ca_2SiO_4$.

2.1 Sinter with basicity 2.4

Samples of titanomagnetite sinter with increased basicity of 2.4 were selected for the study. The chemical composition of the sinter is shown in Table 1. The designation of the samples is as follows: 5 – sinter (sinter machine no. 2); 6 – sinter (lower part of the pallet); 7 – sinter (middle part of the pallet); 8 – sinter (upper part of the pallet).

Figure 2 shows the characteristic diffractogramm of samples of the sinter. The main phase in sinter no. 5–8 is Fe_3O_4, Fe_2O_3 is contained in a smaller amount, and Ca-containing silicate of a complex composition – $Ca_{2,3}Mg_{0,8}Al_{1,5}Fe_{8,3}Si_{1,1}O_{20}$– is also well manifested. There are small amounts contain γ-Ca_2SiO_4 (dicalcium silicate). In sinter samples with a basicity of 2.4, the increase in the unstabilized γ-Ca_2SiO_4 phase is opposite from the bottom of the layer to the top.

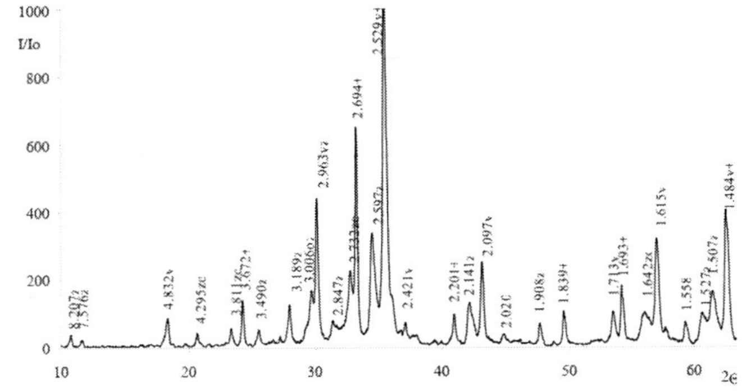

Figure 2. Diffractogramm of sinter with basicity 2.4:
v – Fe_3O_4; + – Fe_2O_3;z – $Ca_{2,3}Mg_{0,8}Al_{1,5}Fe_{8,3}Si_{1,1}O_{20}$; o – γ-$Ca_2SiO_4$.

2.2 Sinter with additives of binding polymers 300 and 500 g/ton of sinter

Samples of titanomagnetite sinter with increased basicity of 2.4 and additives of binding polymers in the amount of 300 and 500 g per ton of sinter were selected for the study.

The chemical composition of the sinter is shown in Table 1. The designation of the samples is as follows: 9 – sinter with the addition of 300 g/ton of sinter (upper part of the pallet); 10 – sinter with the addition of 300 g/ton of sinter (middle part of the pallet); 11 – sinter with the addition of 300 g/ton of sinter (lower part of the pallet); 12 – sinter with the addition of 500 g/ton of sinter (upper part of the pallet); 13 – sinter with the addition of 500 g/ton of sinter (middle part of the pallet); 14 – sinter with the addition of 500 g/ton of sinter (lower part of the pallet).

As a polymer additive, a line of additives of the 'Thermoplast SV' series of the company 'Polyplast – UralSib' specialized for mining and processing plants was used [8].

Figure 3 shows the diffractogramms of the studied samples of the sinter.

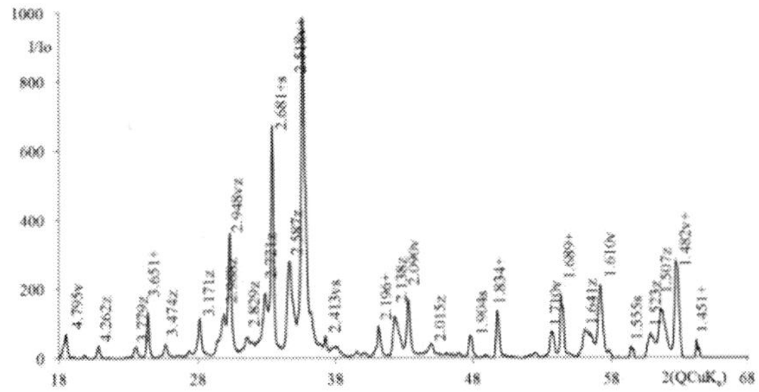

a

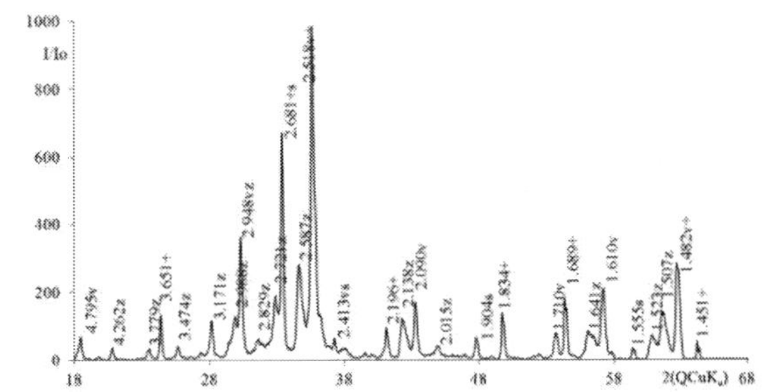

b

Figure 3. Diffractogramms of sinter with an increased basicity of 2.4 and additives of binding polymers (a – no. 10, b – no. 13): v – Fe_3O_4; + – Fe_2O_3; z – Ca-Mg-Al-Fe-Si-O; s – $CaO{\cdot}TiO_2$.

3. Metallurgical characteristics of sinter

The results of studies of Low Temperature Index ($LTD_{+6.3}$), also called Durability after reduction, in accordance with ISO 13930 [9], the temperature range of sinter samples in accordance with State Standard of Russian Federation no. 26517–85 [10], and sinter reducibility in accordance with State Standard of Russian Federation no. 17212–84 [11] are shown in Table 2.

On the whole, according to the samples taken from the middle and lower layers, the durability of the sinter (with a polymer additive content of 500 g per ton of sinter) significantly improved.

Table 2. Results of investigation of metallurgical characteristics of sinter probes.

Probe	$LTD_{+6.3}$ (%)	Reducibility (%)	Temperature of beginning of softening (°C)	Temperature of ending of softening (°C)	Temperature range (°C)
1	6.13	74.70	1180	1300	120
2	8.38	65.80	1200	1300	100
3	8.65	83.20	1220	1290	70
4	1.99	80.70	1240	1360	120
5	13.68	64.70	1140	1280	140
6	10.40	61.80	1200	1320	120
7	14.74	61.70	1260	1340	80
8	7.370	70.20	1210	1330	120
9	22.39	61.70	1200	1320	120
10	22.82	74.03	1160	1210	50
11	22.39	55.86	1120	1170	50
12	22.82	61.87	1260	1340	80
13	40.74	69.63	1170	1250	80
14	39.05	69.59	1130	1190	60

The reduction of the sinter with a polymer additive leads to the formation of metallic iron, wustite (FeO), partially magnetite (Fe_3O_4). Silicate phases are formed against the background of iron-forming phases: β-$2CaO \cdot SiO_2$ and $2CaO \cdot Al_2O_3 \cdot SiO_2$. Also, $3CaO \cdot Al_2O_3$ is weakly manifested in the samples.

4. Results of industrial tests of production and blast furnace smelting of sinter with different basicity and additives of binding polymers [12]

Tables 3–4 show some indicators of the quality of sinter and blast furnace (BF) smelting (BF no. 5, 6 of JSC 'EVRAZ NTMK') for the period April – October of 2015. This period can be conditionally divided into stages: April-May – stage 1 or basic (basicity of the sinter 2.1); June-July – stage 2 (basicity of the sinter 2.4); August-September – stage 3 (basicity of sinter 2.4 with a polymer additive); October – stage 4 (basicity of the sinter 2.4).

Coke rate at BF no. 5 (kg/t of pig iron) decreased 335.3 in April to 331.3, i.e. by 4.0 kg/t of pig iron (1.2 %), total fuel rate (kg/t of pig iron) decreased 483.2 in April to 478.5, i.e. 4.7 kg/t of pig iron (1.0 %). Coke rate at BF no. 6 decreased 390.5 in April to 386.7, i.e. 3.8 kg/t of pig iron (1.0 %), total fuel rate (kg/t of pig iron) fell 508.8 in April to 503.3, i.e. 5.5 kg/t of pig iron (1.1 %).

The metallurgical characteristics of the sinter changed in stages as follows (Table 5). Attention should be paid to the uneven distribution of the quality indicator along the height of the layer; the sinter from the upper layer has the least value. An increase in the basicity of the sinter also led to an increase in 'hot' durability, which is confirmed by published data. In particular, in paper [13–15], an increase in the hot durability of the sinter is explained by an increase in the amount of the SFCA phase with an increase in basicity.

The results show a decrease in the position of the cohesion zone down in height and a narrowing of the temperature range, which should generally have a positive effect on the operation of the blast furnace.

Table 3. Indicators of BF no. 5 and quality of sinter in 2015.

	April	May	June	July	August	September	October
Coke rate (kg/ton of pig iron)	335.3	335.7	331.3	332.9	343.8	331.3	345.7
Natural gas rate (m^3/ton of pig iron)	106.0	105.8	104.0	104.5	107.2	104.9	114.6
Pulverized coal fuel rate (kg/ton of pig iron)	73.6	69.4	77.8	78.9	68.4	76.5	56.5
Total fuel rate (kg/ton of pig iron)	483.2	479.2	480.5	483.6	486.0	478.5	482.9
Degree of CO (%)	51.8	51.4	51.9	51	50.3	50.3	50.2
Burden rate (%): sinter	38.5	38.7	35.6	39.7	39.4	38.3	39.4
pellets	53.4	51.2	51.4	52.6	54.1	54.7	55.8
iron flux	8.1	10.0	12.9	7.7	6.5	7.1	4.8
Sifting (%)	16.2	16.3	15.4	15.6	12.8	13.4	13.9
Durability +5 mm (%)	74.6	73.8	74.3	74.9	75.8	75.9	76.0

Table 4. Indicators of BF no. 6.

	April	May	June	July	August	September	October
Coke rate (kg/ton of pig iron)	390.5	384.9	376.3	387.7	383.5	386.7	406.3
Natural gas rate (m^3/ton of pig iron)	118.4	113.4	102.4	110.4	107.2	118.5	119.5
Pulverized coal fuel rate (kg/ton of pig iron)	29.3	30.3	53.3	38.2	40.4	28.2	6.6
Total fuel rate (kg/ton of pig iron)	508.8	500.3	503.6	507.8	502.5	503.3	505.4
Degree of CO (%)	50.1	51.1	50.0	50.7	50.6	50.8	50.8
Burden composition (%): sinter	38.1	38.6	37.5	38.8	38.1	38.1	40.7
pellets	52.1	50.7	52.5	51.7	52.3	51.5	53.9
iron flux	9.9	10.6	10.1	9.5	9.6	10.3	5.4

Table 5. Metallurgical characteristics of the sinter by stages (averaged values according to the measurement results).

	Stage 1	Stage 2	Stage 3	Stage 4
LTD+6.3 (%)	11.01	13.68	12.57	39.9
Reducibility (%)	74.75	64.74	64.90	69.61
Temperature of beginning of softening (°C)	1060	1140	1140	1150
Temperature of ending of softening (°C)	1200	1280	1190	1220
Temperature range (°C)	140	140	50	70

Conclusion

It was found that the reducibility of the sinter with a basicity of 2.4 compared to 2.1 is 10 % lower, which does not contradict the LTD data: the lower the reducibility, the higher the hot durability. According to LTD, sinter with a basicity of 2.4 shows higher values in all three quality indicators: +6.3, –3.15, –0.5. The temperature of the beginning and ending of softening of the sinter with a basicity of 2.4 compared with 2.1 increased by 80 °C. The softening interval has not changed.

The addition of a polymer binder in the amount of 500 g/ton to the sinter burden has a more significant effect on the quality of the sinter than the addition of 300 g/ton. The durability of the sinter during reduction significantly improved in all three indicators. So, $LTD_{+6.3}$ amounted to 39.9 % against 11.01 % in the base period. Attention should be paid to the uneven distribution of quality indicators along the height of the layer; the sinter from the upper layer has the least value. The reducibility of the sinter with the addition of a polymer binder in an amount of 300 g/ton, and especially in an amount of 500 g/ton, increased to 64.9–69.61 %. This indicator is close to recommended for the blast furnace process. The results of determining the softening and melting temperatures for sinter with polymer additives show a decrease in the cohesion zone down in height and a narrowing of the temperature range, which will generally have a positive effect on the operation of the blast furnace.

Acknowledgment
The work was carried out as part of the implementation of the State Task IMET UB RAS.

References
[1] Dmitriev A, Petukhov R, Vitkina G and Vyaznikova E 2019 *Defect and Diffusion Forum* **391** pp 215–20
[2] Vitkina G, Dmitriev A and Alektorov P 2019 Investigation of the metallurgical characteristics of titanium-containing mineral raw materials with different contents of titanium dioxide *Modern trends in the theory and practice of the extraction and processing of mineral and industrial raw materials* (Ekaterinburg: JSC Uralmekhanobr) pp 401–4
[3] Dmitriev A, Vitkina G, Petukhov R, Petrova S and Chesnokov Yu 2019 *Bulletin of scientific, technical and economic information. Ferrous metallurgy* **75 2** pp 154–66
[4] Dmitriev A, Petukhov R, Vitkina G and Vyaznikova E 2019 *Defect and Diffusion Forum* **391** 221-5
[5] Honeyands T, Manuel J, Matthews L, O'Dea D, Pinson D, Leedham J, Zhang G, Li H, Monaghan B, Liu X et al 2019 *Minerals* **9 6** pp 333–49
[6] Webster N A S, Pownceby M I, Madsen I C, Studer A J, Manuel J R and Kimpton J A 2014 *Metallurgical and Materials Transactions B: Process Metallurgy and Materials Processing Science* **456** pp 2097–105
[7] Nicol S, Chen J, Pownceby M I and Webster N A S 2018 *ISIJ International* **58 12** pp 2157–72
[8] Volkov D, Mamonov A, Shatokhin D, Parfenov A and Kormina I 2018 Using a polymer binder in the sintering *Science – Education – Production: Experience and development prospects (Mining and metallurgical production. Engineering and metal working Vol. 1*) (Nizhny Tagil: NTI UrFU) pp 37–48
[9] ISO 13930:2015 *Iron ores for blast furnace feedstocks – Determination of low-temperature reduction-disintegration indices by dynamic method* p 9
[10] Russian State Standard 26517-85 *Iron ores, agglomerates and pellets. Method for determining the temperature of the onset of softening and the temperature range of softening* (Moscow: Standartinform Publ.) p 8
[11] Russian State Standard 17212-84 *Iron ores, agglomerates and pellets. Method for determining of reducibility* (Moscow: Standartinform Publ.) p 9
[12] Dmitriev A, Vitkina G and Petukhov R 2017 *Pure and Applied Chemistry* **89 10** pp 1543–51
[13] Zhengwei Y, Guanghui L, Tao J, Yuanbo Z, Feng Z and Zhiwei P 2015 *ISIJ International* 55 **4** pp 907–9
[14] Li J, Zhang Z, Zhang M, Guo M and Wang X 2011 *Steel Research International* **82 6** pp 607–14
[15] Chen H, Chen G, Wu Y, Peng J, Srinivasakannan C and Jin C 2017 *JOM* **69 12** pp 2660–5

Improvement in vanadium-containing titanomagnetite processing technology

A V Kushnarev[1], K V Mironov[1], S A Zagainov[2, 3] and A A Forshev[1]

[1] EVRAZ NTMK, ul. Metallurgov, 1, Nizhnij Tagil, Sverdlovskaya oblast, 622025, Russia

[2] Ural Federal University named after the First President of Russia B N Yeltsin, 19, Mira str., Yekaterinburg, 620000, Russia

E-mail: [3] s.a.zagaynov@urfu.ru

Abstract. The article describes the modern technology of iron smelting from vanadium-containing titanomagnetite. The complex of research presented by the NTMK team and the leading research teams of the Ural Institute of metals, the metallurgical Institute of the Ural branch of the Russian Academy of Sciences, and the Ural Federal University is the basis for achieving high productivity of the blast furnaces EVRAZ NTMK. The presented research and development works are the basis for the technology of metallurgical processing of vanadium-containing titanomagnetite. The developed technology makes it possible to achieve high rates of melting intensity and total consumption of carbon contained in the fuel. Such indicators are achieved by reconstruction, namely by changing the line of the working space of the blast furnace. Modern loading devices were used, energy-efficient technology of coal-dust fuel intake was introduced, and the oxygen content in the air was increased. Considerable attention is paid to the analysis of the domain process and problems of domain melting of titanomagnetite. Particular attention is paid to the implementation of measures to reduce the content of titanium in cast iron.

1. Introduction

A package of research and development works performed by the NTMK team and by leading research teams of OJSC 'Ural Institute of Metals', Metallurgical Institute of Ural Branch of the Russian Academy of Sciences, and Ural Federal University is the basis for achieving high performance of blast furnaces at JSC EVRAZ NTMK. The development work was devoted to the metallurgical processing of vanadium-containing titanomagnetite [1, 2, 4]. Significant attention is devoted to the analysis of the blast-furnace process and problems of blast-furnace smelting of titanomagnetite [3, 6]. Materials provided by different information sources concerning the subject [16, 17] were reviewed in the study.

Special attention was given to the implementation of measures to reduce the occurrence of titanium in cast iron [5, 7, 11]. Specialists focus on the impact of base-to-acid ratio on crystallization properties of titanium-containing blast furnace slag [14–16].

2. Material and Research Methods

Thermodynamic analysis [5–7] showed that the main reaction of titanium oxide during reduction and transformation into cast iron is the reaction: $TiO + C = TiC + CO$. The reaction occurs with an increase in the volume of gas. Therefore, increasing pressure prevents the formation of titanium carbide. The equilibrium concentration of TiC in cast iron at a pressure of 0.40–0.45 is 0.7–0.5 %. The actual concentration of titanium in cast iron is 0.20 %. This is relevant in connection with the protection of furnaces from exposure to materials containing titanium [28].

In the course of the study, the melt movement conditions were analyzed during the melting of titanomagnetite [8–10] and recommendations were developed for stabilizing the profile of melting products during the melting of vanadium cast iron [12]. For blast furnaces, a specific capacity of more than 3.0 t / m^3 was achieved, which is the best result in the Russian Federation.

3. Results and Discussion

The raw material base of the complex is the Kachkanarskoye field. The specificity of vanadium smelting is related to the presence of titanium in the charge, which is partially dissolved in cast iron. This leads to the formation of refractory compounds. These compounds (nitrides and carbonitrides) under normal conditions lead to an increase in vanadium losses and the production of metal with slag, furnace failures, and complications when pouring melting products.

The equation for calculating the content of titanium in cast iron Δ[Ti] relative to the base value [Ti] is obtained from the analysis of physical and chemical processes that determine the extraction of titanium [7–15], taking into account the parameters that determine the change in the content of titanium in cast iron.

$$\frac{\Delta[TiO]}{[TiO]}=\frac{\Delta TiO_2^{mix}}{TiO_2^{mix}}+\frac{\Delta K_r}{K_r}-\frac{\Delta P}{P}-k_1\frac{\Delta(MnO)}{(MnO)}-k_2\frac{\Delta P_{day}}{P_{day}}-k_3\frac{\Delta CSR}{CSR}-k_4\frac{\Delta\eta}{\eta} \quad (1)$$

Where TiO_2^{mix} – content of TiO_2 in batch mixture, %; K_r – reaction equilibrium constant of titanium carbide formation, which is calculated with the use of mid-range temperature of melted slag in the lower section of a furnace; P – blast pressure, kPa; (MnO) – content of MnO in slag, %; CSR – coke hot strength indicator, %; η – slag viscosity, Pa-s; k_1-k_4 – factors, which were found with the use of mathematical model approach. A comparison of titanium content in cast iron calculated with the use of equation (1) with actual content is given in Figure 1.

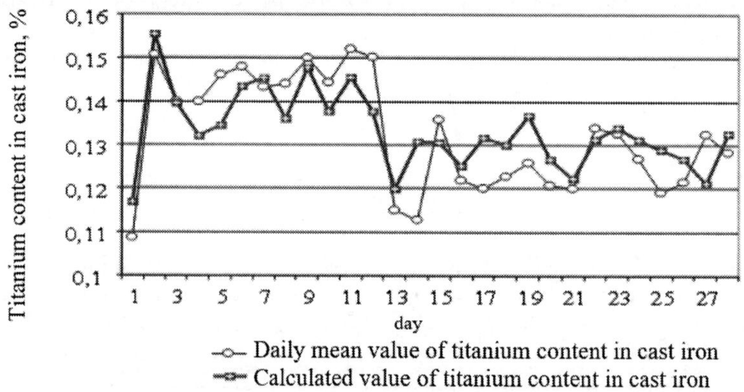

Figure 1. A comparison of titanium content in cast iron calculated with the use of equation (2) with actual content.

One of the major activities ensuring favorable conditions for the production of cast iron with low titanium content is high smelting rate. Analysis of the blast furnace work has shown that increasing of the furnace production rate is accompanied by an increase in the ratio of (TiO_2) in slag to titanium content in cast iron [Ti] (Figure 2).

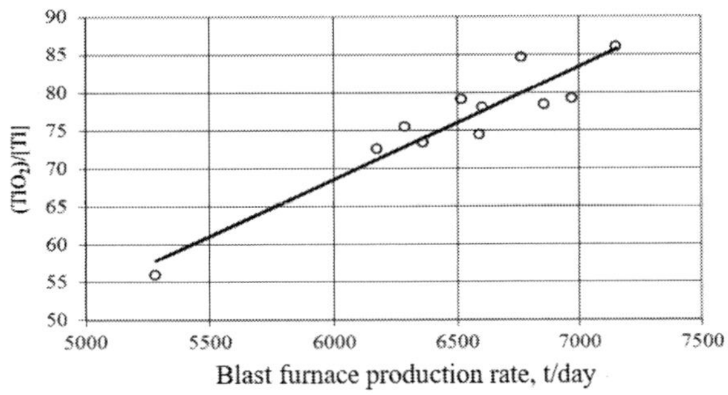

Figure 2. Influence of blast furnace production rate on titanium recovery when silicon content in cast iron is less than 0.09 %.

Work of blast furnace with high pressure under the blast-furnace top is provided by the use of bell-less top charger. Effectiveness of blast-furnace work with increased pressure under the blast-furnace top is given in Figure 3.

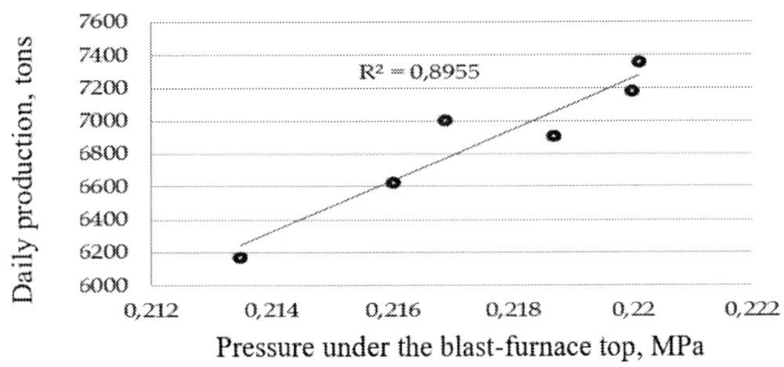

Figure 3. Daily production as a function of pressure under the blast-furnace top.

The increase of solubility of titanium in cast iron impedes the process of titanium carbides formation. The increase of solubility is achieved by the reduction of silicon content in cast iron. Manganese oxides also contribute to the destruction of titanium carbides. Currently, blast furnaces produce cast iron with silicon content ranging between 0.05–0.07 % and with the maximum allowable (according to requirements for ferrovanadium production) content of manganese in cast iron.

An opportunity to improve blast furnace production rate arose after the furnace line had been changed. On the basis of an analysis of data on the work of blast furnaces of an older design, it was believed that the increase of oxygen content in the air blast should not be accompanied by the increase of tuyere gas temperature [16].

Differential equation describing batch mixture temperature variation (t_{mix}) depending on the height of the furnace is as follows:

$$\frac{dt}{dh} = \frac{\alpha}{w_{mix} c_{app}} (t_g - t_{mix}), \tag{2}$$

Where α – heat transfer factor, W/m³ K; w_{mix} – speed of batch mixture movement, m/s; c_{app} – apparent heat capacity of a batch mixture, J/m³K; t_g – temperature of gas entering the bed.

Conditions, under which the theoretical flame temperature increases while the batch mixture temperature upon exiting the bed is maintained, are provided by the constancy of (2). Apparent heat capacity of a batch mixture depends on the amount of recovery additives, a specific yield of slag, and specific coke consumption. Calculations have shown that the dominant factors in heat transfer analysis are speed of batch mixture movement and theoretical tuyere gas temperature. In this case, the maintenance of (2) at the same level is achieved if the following conditions apply:

$$\frac{\Delta t_g}{t_g - t_{mix}} = \frac{\Delta w_{mix}}{w_{mix}}, \cdot \qquad (3)$$

As can be seen, the ratio (3) is equal to blast furnace production rate variation. Calculations have shown that for every 1 percent of the increase in the production rate the acceptable increase of theoretical tuyere gas temperature is 6.0–7.5 °C.

Taking account of the fact that an increase of oxygen content in the air blast of 1 % results in increase of production rate of 3.5–4.0 % it has been calculated that for each percent of the increase of oxygen content in the air blast the increase of theoretical flame temperature for 15–20 °C is acceptable.

During the analysis of the effectiveness of blast furnace work with the use of oxygen, the working period of the blast furnace no. 5 was divided on 3 periods (table 1). Annual averages were used. During the first period, the furnace had been working with the oxygen content of less than 26 % (the years 2006–2008). The second period is characterized by an increase of oxygen content in the air blast with a simultaneous increase in the natural gas consumption and with almost constant theoretical temperature. The second period is characterized by the implementation of blast-furnace smelting technology with the use of pulverized coal fuel.

Blast furnace work indicators with varying degrees of oxygen content in the air blast.

The analysis results have shown that it is possible, useful and effective to have an increased oxygen content in the air blast when smelting cast iron out of vanadium-containing titanomagnetite.

Forced work of blast furnaces is ensured by the range of measures aimed at increasing the hot strength of coke. The task has been solved through the choice of a batch mixture for carbonization and through the use of a petroleum additive.

Implementation of cast-iron smelting technology with the use of pulverized coal fuel was a significant factor ensuring the reduction of specific coke consumption [24, 25].

Injection of pulverized coal fuel into the blast furnace leads to changes in the work of the furnace and creates certain problems and hazards during smelting.

Theoretical analysis of hazards has shown that the main factor determining the effectiveness of blast-furnace work with the use of pulverized coal fuel is variations in gas tightness of a bed in the area of the viscoplastic state. Taking into account that pulverized coal fuel is a hydrocarbon additive its impact on specific consumption of coke and pressure differential in the lower part of the bin has been analyzed. In addition, it was assumed that 0.40–0.45 m^3 of oxygen per kilo of pulverized coal fuel, and 3.75 m^3 of natural gas per kilo of carbon were injected into the furnace.

The calculations have shown that there are two working modes for the blast-furnace with the use of pulverized coal fuel. In the first mode can be achieved increased performance of the furnace, and in the second mode, an increase of pulverized coal fuel consumption is achieved through the maximum possible reduction of specific consumption of coke while reducing smelting intensity.

Daily production as a function pulverized coal fuel consumption is given in Figure 4.

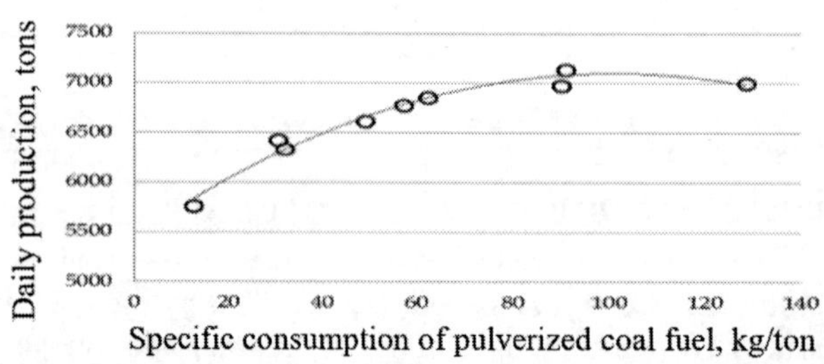

Figure 4. Daily production as a function of pulverized coal fuel consumption.

Obtained functions have shown that when using pulverized coal fuel, it is possible to choose such pulverized coal fuel and natural gas consumption rates at which either minimum coke consumption or maximum performance of the furnace can be achieved.

Nowadays the blast furnaces work in high-performance mode with optimum relationship between main fuel types (see Table 2).

Table 1. Main indicators of the blast furnaces work.

Indicator	BF no. 5	BF no. 7
Specific production rate, t/m^3 day	3.11	3.27
Specific natural gas consumption rate, м^3/ton of cast iron	115.1	114.0
Specific pulverized coal fuel consumption rate kg/ton of cast iron	46.8	66.3
Specific coke consumption rate kg/ton of cast iron	367.7	338.2

It should be noted that titanium containment in cast iron did not increase with the use of pulverized coal fuel (Figure 5).

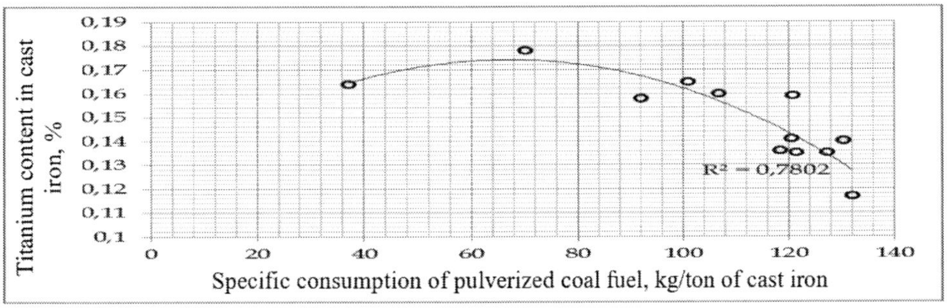

Figure 5. The titanium content in cast iron as a function of pulverized coal fuel consumption.

4. Conclusions

Modern technology of cast iron smelting out of vanadium-containing titanomagnetite has been implemented at JSC EVRAZ NTMK. The technology has made it possible to achieve one of the best indicators of smelting intensity and total fuel carbon consumption rate in the Russian Federation.

Such indicators were achieved through renovation – the line of blast furnace working space had been changed, modern charging devices have been used, energy-efficient technology of pulverized coal fuel injection has been implemented, and oxygen content in the air has been increased.

References

[1] Smirnov L A, Deryabin Yu A and Shavrin S V 1990 *Metallurgical Processing of Vanadium-containing Titanomagnetite* (Chelyabinsk: Metallurgy) p 256

[2] Volkov V V, Filippov V V and Gavrilyuk G G 2000 Experience of Titanomagnetite Processing in Blast Furnaces *Steel* no **11** pp 24–8

[3] Zakharov A I, Rudin V S, Filippov V V and Shavrin S V 2004 Analysis of the Blast-furnace Process in Conditions of Smelting of Low-siliceous Vanadium Cast Iron *Steel* no **4** pp 14–6

[4] Nosov S K, Filippov V V and Shavrin S V 2003 Problems of Blast-furnace Smelting of Titanomagnetite and Ways of Addressing the Problems *Steel* no **6** p 6–9

[5] Kushnarev A V, Shavrin S V, Zagainov S A and Sobianina O N *et al* 2007 Implementation of Measures to Reduce the Occurrence of Titanium in Cast Iron *Collected works of the International conference dedicated to quality improvement of metal used in the transport industry* (Nizhny Tagil: Publishing House of JSC NTMK) p 121–23

[6] Zagainov S A, Sobianina O N and Filippov V V *et al* 2008 Analysis of Blast-furnace Smelting in Production of Vanadium Cast Iron *Collected works of the International research-to-practice*

IOP Conf. Series: Materials Science and Engineering **966** (2020) 012062 doi:10.1088/1757-899X/966/1/012062

conference 'Improving the Quality of Education and Scientific Research' within the framework of the VII Satbayev Readings (Ekibastuz: Publishing House of Ekibastuz Institution of Engineering and Technology named after Academician K I Satbayev) pp 370–71

[7] Kushnarev AV, Shavrin S V, Zagainov S A and Sobianina O N *et al* 2008 *Collected works of the 3rd International conference 'Transmet-2007'* Implementation of Measures to Reduce the Occurrence of Titanium in Cast Iron (Yekaterinburg: Publishing House of USTU) pp 169–71

[8] Sobianina O N, Filatov S V and Zagainov S A 2008 Analysis of Conditions of Melt Movement in the Blast-furnace Well During Smelting of Titanomagnetite *News of Higher Educational Institutions. Ferrous Metallurgy* no **4** pp 69–70

[9] Sobianina O N, Filatov S V and Zagainov S A 2012 Analysis of Specificities of Titanium Recovery in Blast Furnace *Steel* no **3** pp 9–11

[10] Sobianina O N and Zagainov S A 2007 Specificities of Melt Movement in the Blast-furnace Well During Smelting of Titanomagnetite Ingoing Materials *Collected works of the XIII Reporting Scientific Conference of Young Scientists of USTU* (Yekaterinburg: Publishing House of USTU) pp 230–2

[11] Kushnarev A V, Shavrin S V, Zagainov S A and Sobianina O N *et al* 2007 Implementation of Measures to Reduce the Occurrence of Titanium in Cast Iron *Collected works of the International conference dedicated to quality improvement of metal used in the transport industry* (Nizhny Tagil: Publishing House of JSC NTMK) pp 121–3

[12] Zagainov S A, Onorin O P and Sobianina O N *et al* 2008 Development and Issuance of Recommendations on Stabilization of Smelting Products Profile During Vanadium Cast Iron Smelting *Bulletin 'Ferrous Metallurgy', JSC Chermetinformatsiya* no **7** p 70

[13] Jianlong Wu, Hui Chen, Jian Sun, Zhixing Zhao, Wei Wang and Yan Zhang 2019 Metallurgical properties and furnace protection practice of different titanium-bearing burdens, *Ironmaking & Steelmaking*

[14] Zhao W, Chu M and Liu Z *et al* Aug 2019 High-Temperature Interactions Between Vanadium-Titanium Magnetite Carbon Composite Hot Briquettes and Pellets Under Simulated Blast Furnace Conditions *Metallurgical and materials transactions b-process metallurgy and materials processing science* vol **50** (Edition 4) pp 1878–95

[15] Zhao W, Chu M and Wang H *et al* Jan 15 2019 Reduction behavior of vanadium-titanium magnetite carbon composite hot briquette in blast furnace process Powder technology vol **342** pp 214–23

[16] Hu M, Wei R and Gao L *et al* Mar 2018 Effect of the Basicity on the Crystallization Behavior of Titanium Bearing Blast Furnace Slag High temperature materials and processes vol **37** (Edition 3) pp 193–200

[17] Li W, Wang N and Fu G *et al* Feb 15 2018 Influence of TiO2 addition on the oxidation induration and reduction behavior of Hongge vanadium titanomagnetite pellets with simulated shaft furnace gases *Powder technology* vol **326** pp 137–45

IOP Conf. Series: Materials Science and Engineering **966** (2020) 012063 doi:10.1088/1757-899X/966/1/012063

Plasma hardening of corrosion-resistant steel

M V Mironova

Ural Federal University named after the First President of Russia B N Yeltsin, Nizhniy Tagil Technological Institute, 59, Krasnogvardeyskaya str., Nizhniy Tagil, 622000, Russia

E-mail: maria.mironova@urfu.ru

Abstract. The structure, phase composition and hardness of a 20X13 steel coating treated with two consecutive passes of a plasma arc are studied. It is shown that plasma treatment makes it possible to obtain martensite with a significant (30–50 %) share of permanent austenite in the structure of the local hardening zone. In this case, the hardness of the heat-strengthened blanket is 650–700 HB. The total depth of the local hardening zone with the modified structure reaches one millimeter. The study shows that a homogeneous structure with periodically changing phase composition and hardness develops in the surface layer of 20X13 steel, heat-treated plasma arc. in this case, the plasma arc must pass sequentially with a certain degree of overlap of local quenching zones. It is obvious that by adjusting the parameters of the processing conditions (the current of the plasma arc, the speed of its movement, the degree of overlap of local quenching zones), it is possible to purposefully form a certain structural-phase state of the surface layer with properties that best meet the operating conditions.

1. Introduction

Steel 20X13 is designed for manufacture of high-ductility critical parts, subjected to dynamic loads. Along with increased requirements for corrosion- and heat resistance, such parts, as a rule, shall ensure high wear resistance of the working surface in friction conditions. For the formation of the given set of properties after bulk heat treatment, subsequent local thermal strengthening of the working surface, e.g., high-frequency currents (HFC) hardening is usually provided. Given the technical and economic aspects of the rational use of high-frequency technology, surface strengthening by plasma hardening was analyzed.

The process of surface plasma tempering provides hardened cases with a depth of 0.5–2.5 mm with a fine-grained martensitic-austenitic structure, high microhardness, wear and crack resistance. Such properties are achieved by dispersion and increasing the local inhomogeneity of the structure of massive and plate martensite with different carbon content, increasing the density of dislocations, and preserving particles of carbide and retained austenite undissolved during heating in the surface layer, which helps to reduce crumbling and cracking at the breaking-in stage. The wear resistance of surfaces heat-treated with a plasma arc increases by 2–8 times; multiple layer-by-layer structural modification of the treated areas may occur; the hardening process is non-waste, environmentally friendly and can easily be integrated into flow charts for manufacturing and repairing machine parts as a finishing operation [1–6].

The use of plasma hardening for surface strengthening of parts made of steel 20X13 revealed a set of features due to a significant difference in its thermal and physical characteristics from the corresponding indices for ordinary carbon ones, and required the study of the structure, phase composition, and hardness of the heat affected zone.

2. Material and Research Methods

A 50×30×500 mm sample made of steel 20X13, containing 0.19 % carbon and 12.23 % chromium with the concentration of the remaining components according to Russian State Standard GOST 5632–72 was used for experiments. Initial heat treatment of the sample: oil quenching from 1050 °C, 770 °C tempering with air cooling. Areas of surface hardening were formed at the edges of the sample using transferred arc plasma gun [2] by two successive passes with overlapping local hardened regions to ~30 % of their width (Figure 1). Processing conditions: plasma arc current – 190–220 A, voltage – 48–50 V, plasma gun traveling speed – 2; 2.5 and 3 cm/s. The structure and hardness of the plasma hardened region was studied on sections using Olympus and Neophot-2 microscopes at magnifications of 50–1000^x. The hardness was measured with Duramin-2 instrument under a load of 4.9 N.

3. Results and Discussion

Figure 1 shows the cross section of a heat-strengthened surface layer formed by two successive passes of a plasma arc with a heat input of 0.4 MJ/m. At the same time, no surface washing is observed and the modification of the initial structure (secondary sorbite) occurs in solid.

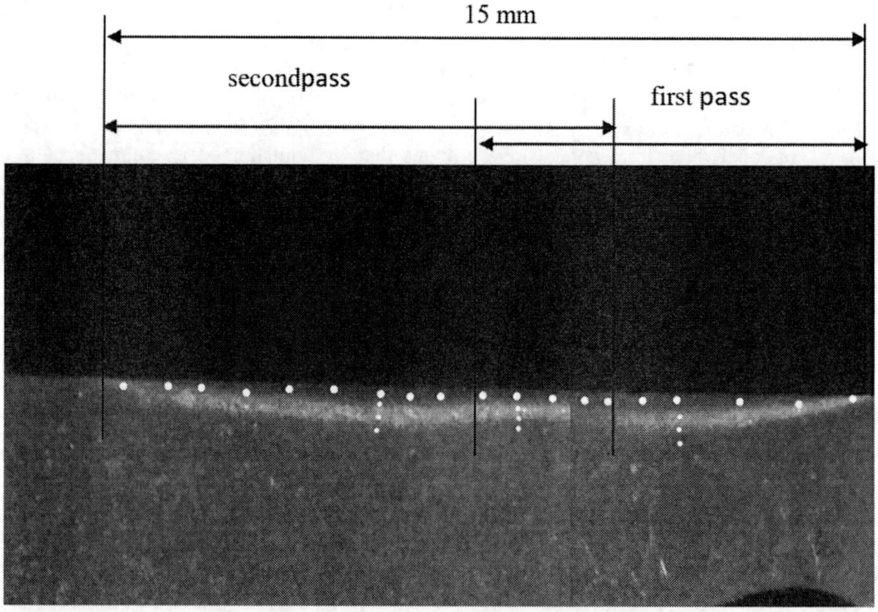

Figure 1. Cross section of a heat-strengthened surface layer formed by two successive passes of the plasma arc (dots indicate the hardness measurement diagram). ×10.

The process of plasma heat treatment is characterized by the supply of thermal energy from the surface and its distribution into the inner layers of the metal due to thermal conductivity. The microvolumes located at different distances from the surface are heated to different temperatures, and the heating and cooling rates of such microvolumes are also different. There is also no holding at a constant temperature, characteristic of volume heat treatment; phase transformations occur under conditions of flash heating and cooling and differ in various degrees of completeness; therefore, a wide range of structural conditions is observed in the heat-affected area (Figure 2).

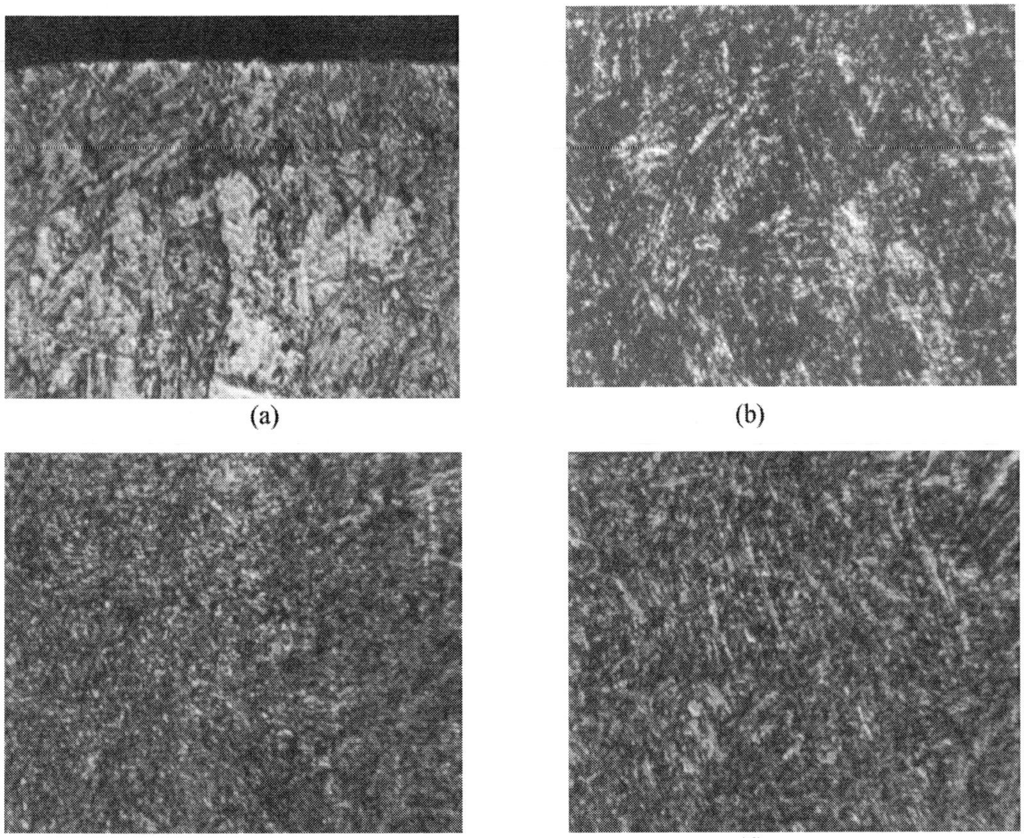

(a) (b) (c) (d)

Figure 2. Structure of the areas of the local zone of plasma hardening.×1000.

The temperature-time conditions for the existence of austenite under the conditions of the ongoing thermal cycle do not ensure its carbon homogenization and, upon rapid cooling, the stability of austenite in microvolumes with different carbon contents also turns out to be different. In areas with its high concentration, where the M_n point decreases, retained austenite is recorded. In areas with a low carbon content, where the M_n point is higher, martensite is formed. Since the martensite transformation progresses in fine-grain austenite, a structureless or typical massive martensite is formed, between the rods of which there are areas of retained austenite (Figure 2, a), the fraction of which near the surface can exceed 50 %. It can be noted that the nature of distribution of retained austenite in the martensitic matrix changes qualitatively in comparison with bulk hardening – the austenitic phase is evenly distributed in the form of thin interlayers between martensite crystals, which increases the plasticity of the integrated structure.

There is no doubt that the presence of metastable retained austenite in steels is desirable, since it allows the energy-absorbing process of transformation $\gamma_{ret} \rightarrow \alpha_m$ to be realized during operational loading. The martensite transformation occurring in the process of contact-impact or abrasive interaction of the working surface with a wear medium provides not only hardening, but also partial dissipation of the fracture energy in combination with stress relaxation during the formation of deformation martensite crystals. The wear resistance of the surface layer that underwent such a transformation increases many times. A similar effect also develops in the surface layer hardened by plasma quenching [3].

At a depth of more than 0.3 mm, the hardness decreases due to the formation of coarse-needled martensite with a decrease in the fraction of retained austenite (Figure 2, b); even lower, at a depth of more than 0.6 mm, an incompletely quenched area with the structure of finely divided troostite is observed (Figure 2, c), which smoothly transforms into the principal metal secondary sorbite structure (Figure 2, d).

Thus, according to experimental data, processing by a plasma arc with a heat input of 0.4 MJ/m allows to obtain martensite in the surface layer of steel 20X13 with a significant (30–50 %) fraction of retained austenite. The hardness of the surface layer in this case is 650–700 HV, the depth of the hardened area is 0.8 mm. Conditions with an increased value of the plasma arc current and a low velocity of its movement are characterized by increased energy input and, when processing hypoeutectoid carbon steels, they allow reaching maximum values of the depth and hardness of the hardened zone. In plasma hardening of the steel under consideration, the heat input of 0.4 MJ/m is apparently close to the limit one, since its increase to 0.45 MJ/m, even if providing a slight increase in the depth and hardness of the hardened zone, is accompanied by surface glazing, which is not always acceptable.

The change in hardness in the cross section of a heat-strengthened layer at a depth of ~ 0.3 mm (see Figure 1) is presented in Figure 3. The extreme points correspond to the hardness of steel not subjected to plasma treatment.

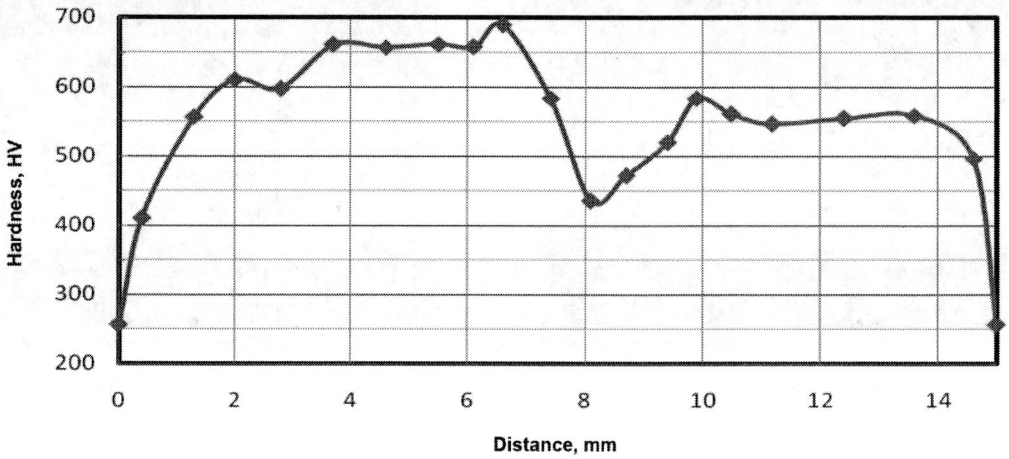

Figure 3. Change in hardness in the cross section of heat-strengthened layer at a depth of ~ 0.3 mm (see Figure 1).

The decrease in hardness in the area of overlapping of the local zones of plasma exposure is caused by tempering of steel hardened during the first pass, which occurs under conditions of reheating. When each subsequent pass forms new local quenching zone, in the area of preformed zone where the heating temperature exceeds Ac_3, rehardening occurs, as a result of which these parts do not differ in structural composition, however, according to [5], the dislocation density increases, as well as the degree of dispersion of the structure and hardness. In that part of preformed strengthening zone, where the heating temperature does not exceed Ac_1, high-speed tempering of the previously hardened structure takes place. Although the hardness in the tempering zone is lower than in the hardened zones, it significantly exceeds the level characteristic of the original structure (Figure 4).

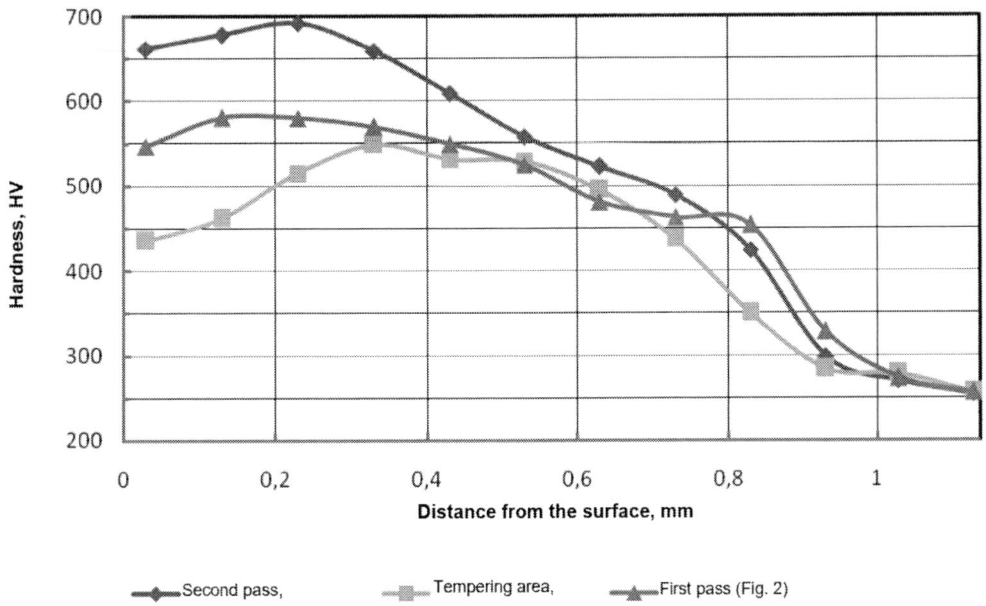

Figure 4. The change in hardness along the depth of the heat-strengthened layer.

Tempering zones with dispersed structure and reduced hardness serve as barriers to crack propagation, causing them to temporarily stop and branch out due to the higher fracture toughness of the structural components. The periodic alternation of hard and plastic layers in the structure of the working surface can effectively increase the operational life of parts operating under dynamic and thermomechanical loads [4].

Thus, a regular structure with periodically changing phase composition and hardness develops in a surface layer of steel 20X13 heat-treated with plasma arc, formed by successive passes of the plasma arc with a certain degree of overlap. Obviously, by adjusting the parameters of the processing conditions (plasma arc current, speed of its movement, degree of overlap of the local quenching zones), it is possible to purposefully form a certain structural-phase state of the surface layer with properties that most fully meet the operating conditions.

4. Conclusions

Processing by a plasma arc with a heat input of 0.4 MJ/m allows to obtain martensite in the surface layer of steel 20X13 with a significant (30–50 %) fraction of retained austenite. The hardness of the surface layer in this case is 650–700 HV, the depth of the hardened area is 0.8 mm.

A regular structure with periodically changing phase composition and hardness develops in a surface layer of steel 20X13 heat-treated with plasma arc, formed by successive passes of the plasma arc with a certain degree of overlap of local quenching zones.

References

[1] Korotkov V A 2019 *Russian Eng. Res.* **39 (3)** 234–6
[2] Safonov E N and Mironova M V 2018 *IOP Conf. Series: Materials Sci. and Eng.* **411 (1)**, 012069
[3] Korotkov V A 2015 *J. of Frict. and Wear* **36 (2)** 149–52
[4] Korotkov V A and Zlokazov M V 2014 *J. of Frict. and Wear* **35 (2)** 133–6
[5] Balanovsky A E 2015 *Strengt. Techn. and Coat.* **12** 18–30
[6] Korotkov V A, Ananyev S A and Shekurov A V 2014 *Welding Int.* **(28) 2** 140–2

IOP Conf. Series: Materials Science and Engineering **966** (2020) 012064 doi:10.1088/1757-899X/966/1/012064

Graphite flake cast iron surface hardening

M V Mironova

Ural Federal University named after the First President of Russia B N Yeltsin, Nizhniy Tagil Technological Institute, 59, Krasnogvardeyskaya str., Nizhniy Tagil, 622000, Russia

E-mail: maria.mironova@urfu.ru

Abstract. The influence of the operating parameters of electric arc heat treatment of the surface layer of graphite flake cast iron type СП-61 on the structure, microhardness of structural components and the size of the hardening zone is studied. It is shown that controlling the arc power, its speed and the degree of overlap of local hardening zones allows purposefully forming modified structural States with properties most suitable for specific operating conditions in the surface working layer of cast iron. Overlapping of local hardening zones allows purposefully forming modified structures with suitable properties for specific operating conditions in the surface working layer of cast iron rolls. According to the research results, regularities of formation of the structure and phase composition and properties of the surface layer of cast iron type СП-61 under the influence of electric arc heat treatment were obtained. Control of parameters of the processing mode (power and speed of the arc, the degree of overlap of local hardening zones) allows you to purposefully form modified structures with suitable properties for specific operating conditions in the surface working layer of cast iron.

1. Introduction

Hardening of the working surface of cast iron rolls with high and industrial frequency currents in order to form a bleached layer with increased wear resistance is used only in their primary production. When repairing the worn surface of the rolls, the hardness of the cooler layer formed during manufacturing is reduced. In this case, the output of the rolling mill is also reduced by reducing the inter-repair cycle and the diameter of the rolls.

To increase the service life of cast-iron rolls, it is promising to use heat treatment for their working surfaces with a concentrated energy flow. Hardening-type structures formed during high-speed heating and cooling have high hardness, wear resistance, and fracture resistance. Moreover, it is advisable to perform treatment after each repair machining using low-cost and high-efficiency methods with low labor intensity that are suitable for use in the conditions of metallurgical production. A method of surface heat treatment with argon shielded plasma (constricted) or conventional electric arc [1–9] is rational in terms of universality, availability, environmental friendliness and economic efficiency for extending the service life of cast rolls. Without changing the parameters of surface roughness, such treatment is easily integrated into the process of preparation and repair of rolls, being a finishing operation, low-cost, quite efficient and makes it possible to extend their service life.

The purpose of the work is to identify the basic laws of structure formation, as well as quantitative ratios of the phase composition, microhardness of structural components, the size of the heat-affected zone formed in roll-foundry iron, with parameters of the mode of surface heat treatment using direct electric arc. It is also necessary to evaluate the range and efficiency of controlling the characteristics of

IOP Conf. Series: Materials Science and Engineering **966** (2020) 012064 doi:10.1088/1757-899X/966/1/012064

the heat-treated working layer of the rolls by changing the parameters of the mode and determine the area of rational treatment modes.

2. Material and research methods

The study was carried out on samples cut from hot rolling rolls (СП-61 type graphite flake hypoeutectic mottled cast iron containing 3 % of carbon, 0.7 % of silicon and 0.5 % of manganese), the surface of which was heat-treated by electric arc in a range of mode parameters specified in Table 1.

Table 1. Parameters of treatment modes and characteristics of hardened layer.

Mode number	1	2	3
Arc current strength, A	250	250	350
Arc speed, m/h	200	200	300
Overlapping of local hardening zones during layer formation, % of the zone width	60	30	30
Maximum depth of the hardened layer, mm	0.60	0.45	0.54
Average depth of the hardened layer, mm	0.53	0.35	0.47
Peak-to-peak amplitude of the hardened layer depth, mm	0.12	0.17	0.22
Average depth of the ledeburitic interlayer, mm	0.10	0.04	0.06

The features of structure formation in the heat-affected zone were revealed using optical metallography, electron microscopy, durometry, and X-ray diffraction analysis. Microhardness of the structural components was determined using a Neophot-2 instrument under load of 0.78 N. X-ray diffraction patterns were recorded using a DRON-3.0 diffractometer in cobalt Kα-radiation. In order to obtain a continuous layer on the treated surface of a structure modified as a result of thermal action with increased microhardness of the components, the treatment was performed with sequential overlapping of the local heat-affected zones by 30–60 % of their width.

3. Results and discussion

The initial structure of cast iron is presented in Figure 1, c. Microhardness of cementite is 12,700 MPa, and perlite is 3,200 MPa.

It was found that heat treatment with a current of 250 A at arc speed of 200 m/h with the sequential formation of local hardening zones with 30 % overlapping ensures the maximum layer depth of 0.45 mm and the minimum of 0.28 mm (the average value of the modified layer depth is 0,35 mm). In this case, the ledeburitic interlayer on the surface has a thickness of ~0.04 mm. The option of local zones overlapping by 60 % of their width (mode 1) makes it possible to increase the maximum depth of the hardened layer to 0.60 mm, and the average value is 0.53 mm. The depth of the ledeburitic interlayer varies between 0.096–0.120 mm (the average value is 0.100 mm). It should be noted that this treatment option provides a hardened layer that is more uniform in depth – the difference in depth of different areas of the layer is 0.12 mm, while after treatment in mode 2 (30 % overlapping) the difference is 0.17 mm at a lower average depth. Treatment by current of 350 A at the speed of 300 m/h increases the depth of the layer compared to the second mode by almost 1.5 times (up to 0.47 mm), however, the peak-to-peak amplitude of the hardened layer depth in this case is 0.22 mm.

Figure 1 shows microphotographs of the structure of various areas of the heat-strengthened layer obtained by treatment in the first mode. The heat-affected zone manifests itself by the appearance of martensitic areas (Figure 1, d). In the transition zone, these areas border on areas in which perlite cementite experienced only coagulation. When moving towards the surface, the length of the martensite needles increases while at the same time their density decreases; the microhardness of martensite is 7,500 MPa. In the surface layer there are areas of ledeburite with a high degree of dispersion of the components. The areas within which the density of martensite needles is very low

IOP Conf. Series: Materials Science and Engineering **966** (2020) 012064 doi:10.1088/1757-899X/966/1/012064

(1–3 per austenite grain, Figure 1, a) are observed. The main part of the surface layer is represented by a ledeburite eutectic, the components of which are perpendicular to the surface (Figure 1, a, b).

X-ray diffraction analysis of the surface of heat-treated cast iron samples showed that cementite is textured in the studied treatment options. This is evident in the deviation of the intensity of the interference lines of this phase in comparison with its normal (perfectly crystalline) state. Figure 2, a shows that the interference lines (120), (130) and (140), in the formation of which the basal plane and planes close to it in orientation take part, have abnormally high intensity values. The latter fact indicates a directional arrangement of cementite crystals relative to the surface of the sample. Austenite and martensite do not have such a pronounced texture. The presence of cementite texture indicates the oriented growth of its particles, which can be implemented by melting the surface of the samples and subsequent directional crystallization.

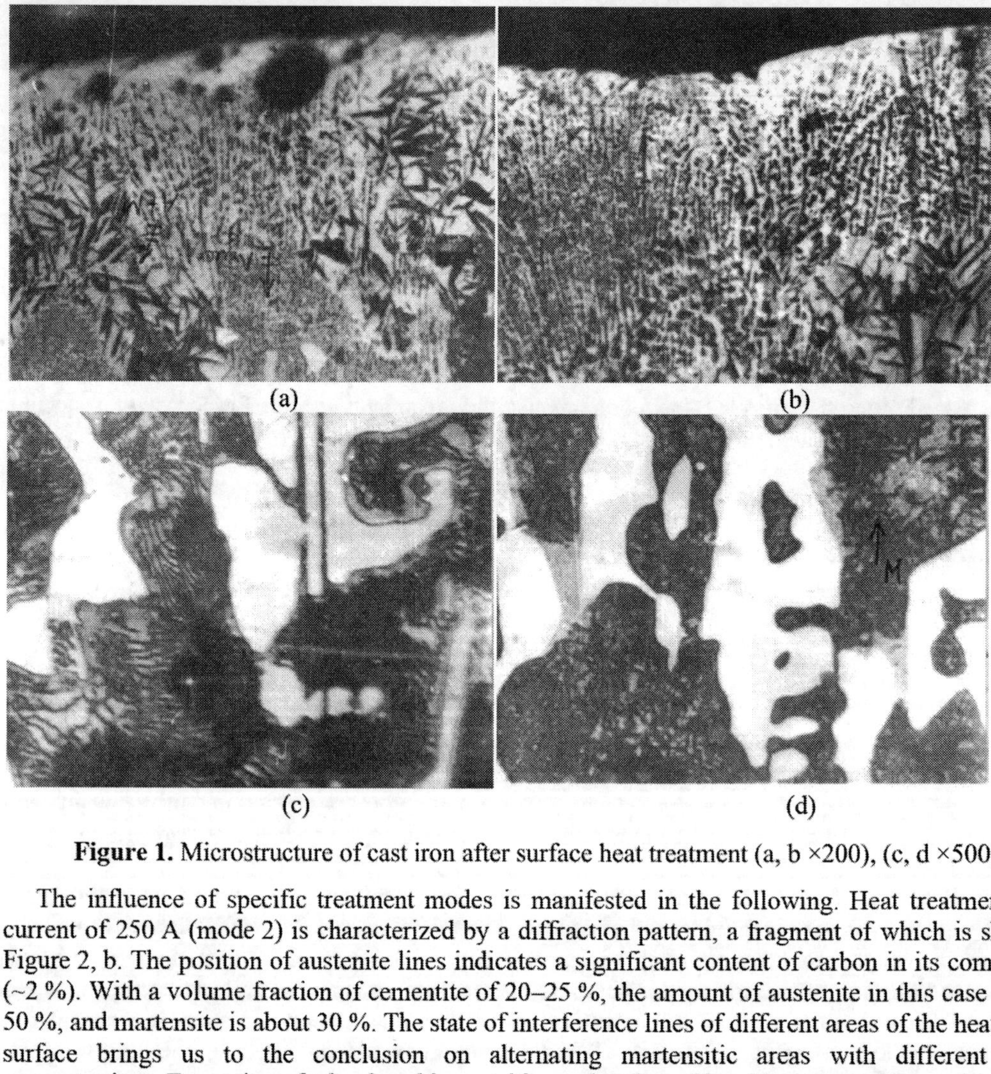

(a) (b) (c) (d)

Figure 1. Microstructure of cast iron after surface heat treatment (a, b ×200), (c, d ×500).

The influence of specific treatment modes is manifested in the following. Heat treatment using current of 250 A (mode 2) is characterized by a diffraction pattern, a fragment of which is shown in Figure 2, b. The position of austenite lines indicates a significant content of carbon in its composition (~2 %). With a volume fraction of cementite of 20–25 %, the amount of austenite in this case is about 50 %, and martensite is about 30 %. The state of interference lines of different areas of the heat-treated surface brings us to the conclusion on alternating martensitic areas with different carbon concentrations. Formation of a hardened layer with overlapping of local heat-treated zones by 60 % of their width is accompanied by a decrease in the degree of texturization of cementite and an increase in the volume fraction of austenite to 60–65 % due to a decrease in the amount of martensite. The state of martensite in this case is more stable than after treatment with less overlapping, (Figure 2, c).

Treatment using higher values of arc current and speed leads to an increase in the intensity of interference lines of cementite, the content of martensite also increases (up to 40 %), and the fraction of residual austenite decreases accordingly (Figure 2, d).

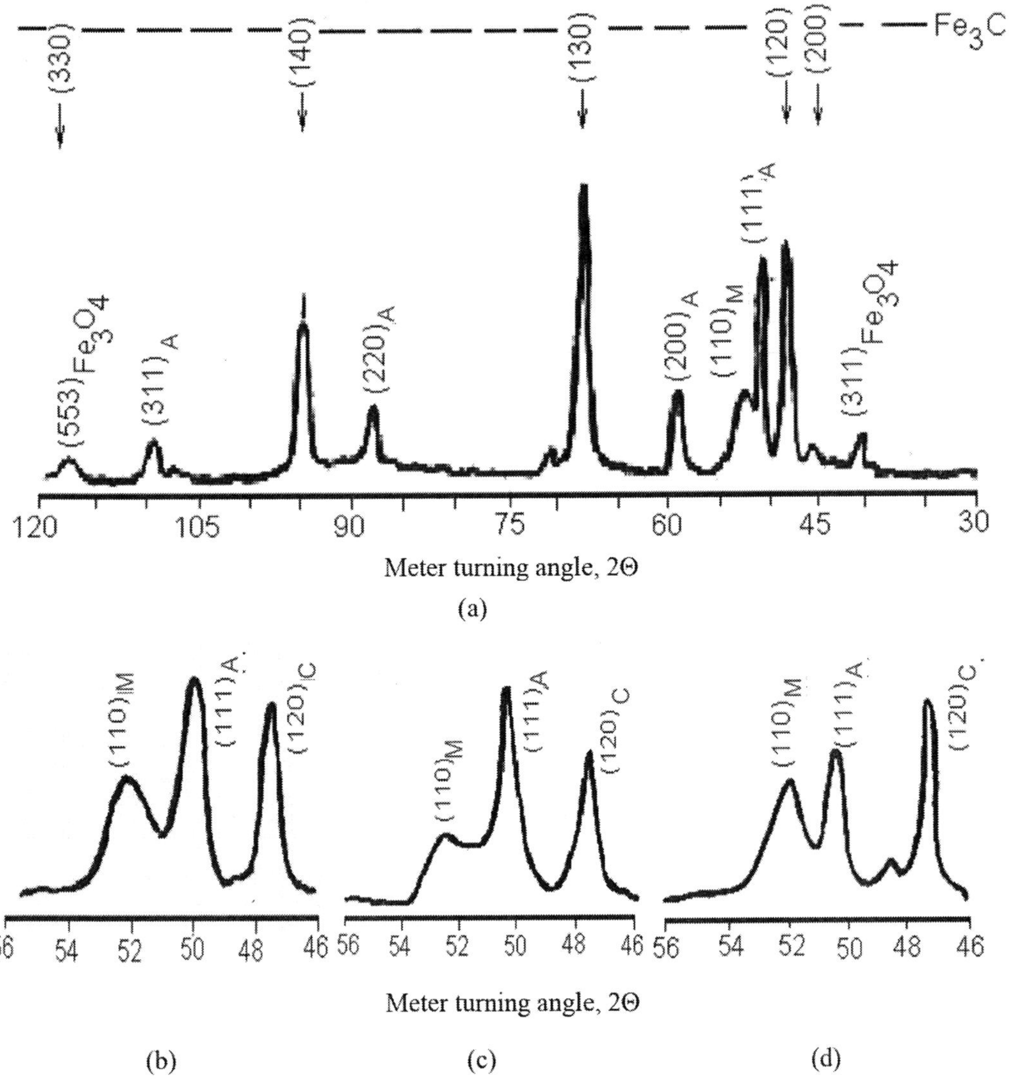

Figure 2. Areas of X-ray diffraction patterns from the surface of cast iron after surface heat treatment.

The main part of the surface layer is almost completely austenitic and ledeburitic; martensite does not reach the surface. Microhardness of ledeburitic areas is slightly inferior to microhardness of cementite and is commensurate with microhardness of martensitic areas. Massive martensite in the treatment option under consideration is found directly near the surface austenitic and ledeburitic layer; its depth can reach 0.2 mm.

The rest of the heat-affected zone is, in addition to cementite inclusions, a complex structural combination of austenite and martensite, the crystallites of which have a high degree of dispersion.

In general, various treatment options with a certain combination of mode parameters (arc heat input, its speed and degree of successive overlapping of local heat-affected zones) allows for obtaining various combinations of structural components with the corresponding variation of properties.

Thus, a relatively mild mode (arc current is 250 A, treatment speed is 200 m/h) with successive overlapping of hardening zones by 30 % of their width allows for obtaining a ledeburitic and austenitic structure with a relatively low (15–20 %) martensite content in the surface layer with a thickness of 10–30 μm. An increase in the degree of zones overlapping leads to the formation of a uniform ledeburitic layer with a thickness of up to 60 μm with a martensite content increasing to ~30 % on the surface. A further increase in the arc current and speed is accompanied by the formation of an uneven martensitic and ledeburitic layer on the surface.

With an increase in the arc current and the degree of overlapping of local heat-treated zones, the heat input to the treated material increases, while the depth of the heat-affected zone with the modified structure and the hardness of its surface also increase in the studied range of process mode parameters. This is associated with an increase in the duration of stay of near-surface volumes in the temperature range exceeding the transition points, with a corresponding increase in the stability of austenite upon subsequent cooling. In this case, structurally this is manifested by a decrease in the martensite fraction. An increase in the arc speed or a decrease in the degree of zones overlapping at a constant current value leads to a decrease in the specific heat flux through the treated surface and is accompanied by a decrease in the considered indicators.

Thus, changing the treatment mode parameters, primarily the arc power and speed, as well as degree of overlapping of local hardening zones makes it possible to purposively form modified structural states with properties that are the most suitable for particular operating conditions in the surface working layer of cast-iron rolls.

For example, the formation of a structure with a certain amount of residual austenite provides an opportunity for further strain hardening of the working layer under conditions of dynamic frictional contact during operational loading.

4. Conclusions

The analysis of the results of studies revealed the patterns of formation of the structure and phase composition and properties of the СП-61 type cast iron surface layer when exposed to electric-arc heat treatment.

Control of the treatment mode parameters, primarily the arc power and speed, as well as degree of overlapping of local hardening zones makes it possible to purposively form modified structural states with properties that are the most suitable for particular operating conditions in the cast iron surface working layer.

References

[1] Korotkov V A 2019 *Russian Eng. Res.* **39 (3)** 234–6
[2] Safonov E N and Mironova M V 2018 *IOP Conf. Series: Materials Sci. and Eng.* **411 (1)**, 012069
[3] Korotkov V A 2015 *J. of Frict. and Wear* **36 (2)** 149–52
[4] Korotkov V A and Zlokazov M V 2014 *J. of Frict. and Wear* **35 (2)** 133–6
[5] Korotkov V A, Ananyev S A and Shekurov A V 2014 *Welding Int.* **28 (2)** 140–2
[6] Safonov E N and Zhuravlev V I 1998 *Welding Int.* **12 (4)** 326–8
[7] Balanovsky A E 2015 *Strengthening Techn. and Coatings* **12** 18–30
[8] Korotkov V A 2016 *Metal Science and Heat Treatment* **58 (7-8)** 449–54
[9] Safonov E N 2005 *Metal Science and Heat Treatment* **47 (9–10)** 434–9

IOP Conf. Series: Materials Science and Engineering **966** (2020) 012065 doi:10.1088/1757-899X/966/1/012065

Evolution of the structure of precipitation-hardening steel during heat treatment

E N Safonov

Ural Federal University named after the First President of Russia B N Yeltsin, Nizhniy Tagil Technological Institute, 59, Krasnogvardeyskaya str., Nizhniy Tagil, 622000, Russia

E-mail: e.n.safonov@urfu.ru

Abstract. In order to identify the mechanism of the processes responsible for the formation of the properties of deposited steels of type 15N4M2D2G2AYuF and 20N4M2D2G2YuF, we studied the patterns of formation of their structure and phase composition as a result of heat treatment. The analysis of the results made it possible to recommend tempering at the temperature of 550 °C for two hours as the main mode of heat treatment of these steels. The development of the properties and structure of the deposited steel of the presented type is the result of multi-step interaction of several hardening mechanisms. The main role in the formation of the structure and properties is played by martensitic transformations, as well as solid-solution hardening by interstitial and substitutive atoms. At the first stage, the hardening is caused by the deposition of dispersed intermetallic compounds and carbides. In addition, dispersed particles of intermetallic phases are released at the early stages of aging and slow down the annihilation of dislocations in the matrix. With longer aging, carbides deposited at fixed dislocations contribute to hardening.

1. Introduction

The broad range of conflicting requirements is imposed on deposited steels designed for operation under cyclic thermomechanical effect conditions (high wear-resisting properties, cracking resistance, red hardness and heat resistance in combination with low hardness after deposition, good weldability, minimum content of expensive and scarce alloying elements). To one degree or another, these requirements can be satisfied only by combining different strengthening mechanisms that form the structure and properties of steel.

In most steels used for deposition of hot rolls, carbides are the main strengthening phase. Their quantity, composition and distribution, as well as the availability and velocity of carbide transformations under operating conditions, play a defining role in the process of high temperature wear.

Due to the features of crystallization of the welding pool and the effect of the thermal cycle on the underlying layers, significant structural and chemical heterogeneity of the deposited metal is observed, which determines the anisotropy of its properties, including uneven wear. Such heterogeneity in the deposited metal with carbide strengthening can be largely eliminated by tempering at 500–550 °C, as evidenced by an increase in wear resistance and a decrease in the banding of worn rolls [1, 2].

A structure with higher stability and resistance to high temperature wear can be formed in precipitation-hardening steel. In case of tempering in such steel, a fairly uniform precipitation of the disperse strengthening phase occurs, resulting in increased wear resistance at high temperatures. Precipitation-hardening steels have high heat resistance in the strengthened condition and fairly good machinability after deposition [3].

Production experience demonstrates that the unsatisfactory resistance of the hot working tool is associated with insufficient wear resistance and thermal fatigue resistance (heat resistance) of the used steels. One way to increase heat resistance is to reduce the carbon concentration, which is accompanied by an undesirable decrease in hardness and wear resistance. It is possible to compensate for this decrease by additional alloying with elements that cause precipitation strengthening due to the formation of intermetallic phases, or (and) through precipitation of the strengthening phases (nitrides, borides, oxides) directly from the crystallizing melt or austenite during cooling of the deposited metal.

The set of the properties of sparingly alloyed precipitation-hardening steels is formed by three main strengthening mechanisms – martensite transformation, solid-solution strengthening by interstitial and substitutional elements, and also strengthening by dispersed precipitates during tempering (ageing) of martensite.

In order to identify the mechanism of the processes responsible for the formation of the properties of the deposited metal of type 15N4M2D2G2AYuF and 20N4M2D2G2YuF as a result of heat treatment, we studied the patterns of formation of its structure and phase composition.

2. Procedure

The metallurgical survey was performed using Neophot-2 optical microscope. The X-ray diffraction method was used to determine the phase composition of the deposited metal, crystal lattice parameter, block size (coherent-scattering regions), and microdistortions ($\Delta a/a$). The composition of the phases was specified during the X-ray diffraction analysis of the extracted precipitation. Electron microscopic investigation of the deposited metal was performed with EMV-100L device using the method of thin foils in transmitted light at acceleration voltage of 100 kV.

3. Results and discussion

It was established that steel of type 15..20N4M2D2G2(A)YuF has a predominantly martensite structure after deposition and contains oxide (Al_2O_3), nitride (AlN) and carbide (MeC and Me_2C) phases (Table 1). In the main component of the structure, dislocation massive martensite, it was possible to state a high density of dislocations.

The martensite crystals have a fragmentary block structure. Retained austenite (up to 5 %) is located in martensitic packets at the boundaries between individual crystals in the form of thin layers, which helps to preserve the structural integrity upon conjugation of the martensite crystals of different orientations inside the packets (Figure 1, a). Therefore, the properties of metal of the studied composition in the state after deposition are formed as a result of martensite transformation, as well as solid-solution strengthening by interstitial and substitutional atoms.

Table 1. Mass fraction (%) and interstitial phase composition in the deposited steel.

Phase	15N4M2D2G2AYuF	20N4M2D2G2YuF
Al_2O_3	0.0566	0.0597
AlN	0.0880	0.0082
$(V_{0.8}Mo_{0.2})C$	0.051	0.10
Mo_2C	0.20	0.23

Figure 1, b shows the structure of the deposited metal of type 20N4M2D2G2YuF after two hours of ageing at 550 °C. It can be noted that there are no significant changes in the morphology of martensite compared to the initial structure. A high density of dislocations is maintained, against the background of which dispersed precipitates inside the crystals of the martensitic matrix are visible. Reflections associated with the presence of intermetallic (NiAl) and carbide (MeC, Me_2C) phases are observed in the microdiffraction pattern. According to the data of chemical and X-ray diffraction

IOP Conf. Series: Materials Science and Engineering **966** (2020) 012065 doi:10.1088/1757-899X/966/1/012065

analysis, the composition of the carbide phase is practically unchanged compared to the initial state, but its amount increases, Figure 2.

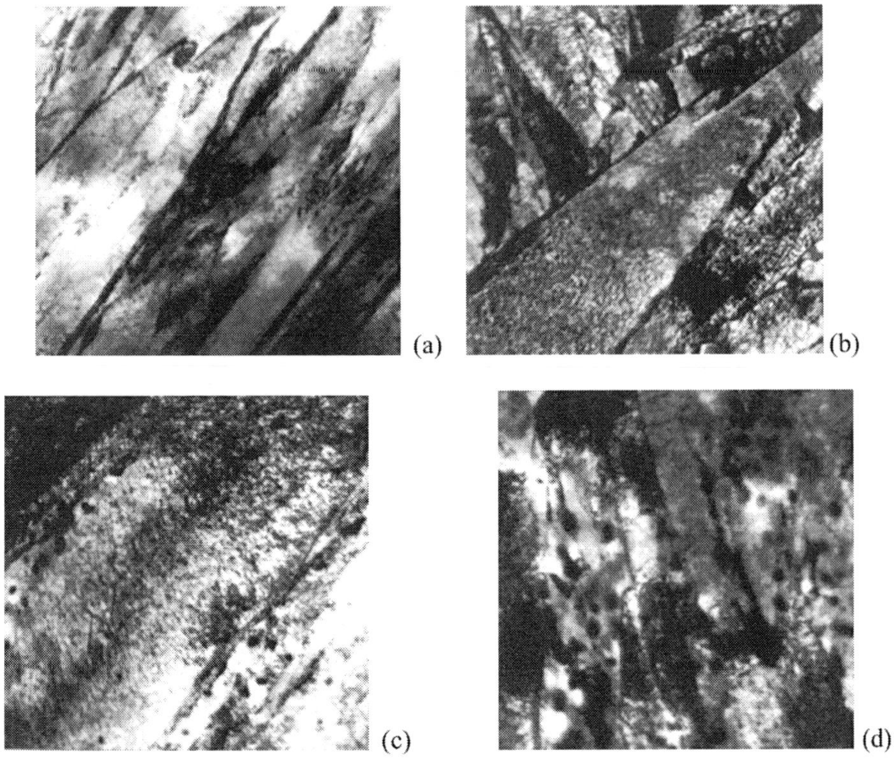

Figure 1. Structure of 20N4M2D2G2YuF steel:
(a) – after deposition, ×26000; (b) – after ageing at 550 °C, 2 hours, ×26000;
(c) – after ageing at 600°C, 1 hour, ×25000; (d) – after holding at 700°C, 1 hour, ×28000.

As it is known, the precipitation of particles of strengthening phases in the process of maraging leads to the generation of new defects in the crystalline structure (misfit dislocations) and the formation of a fine structure of alloys. The number of secondary dislocations is determined by the degree of misfit of matrix lattices and precipitates, the size and volume fraction of particles. The degree of non-conformance to the martensitic matrix increases in the row of carbides Fe_3C, Me_3C, Me_7C_3, $Me_{23}C_6$, Mo_2C, VC, NbC, and the ageing effect noted above and caused by the precipitation of strengthening phases increases in the same sequence. Radiographically, it manifests itself in broadening of diffraction maxima associated with the reduction of blocks (coherent-scattering regions) and increase in the level of microdistortions. In the deposited metal of type 15..20N4M2D2G2(A)YuF, the extreme values of microdistortions coincide with the peak of precipitation hardening (Figure 2), overageing leads to the coagulation processes. In particular, during the tempering of martensite, coagulation, in addition to coarsening, is structurally manifested in the shift of precipitation sites from dislocations to the boundaries of subgrains and grains, providing better conditions for particle growth.

Indeed, analysis of the structure shown in Figure 1, c makes it possible to note that after ageing of the deposited metal at 600 °C, precipitates of the strengthening phases coarsen, and they are located mainly near the boundaries of the martensite crystals. At this stage of ageing, the microdistortions in the martensite lattice are reduced to some extent, but sufficiently high dislocation density is maintained. Apparently, the stability of the martensite dislocation structure is the consequence of pinning of dislocations by dispersed precipitates.

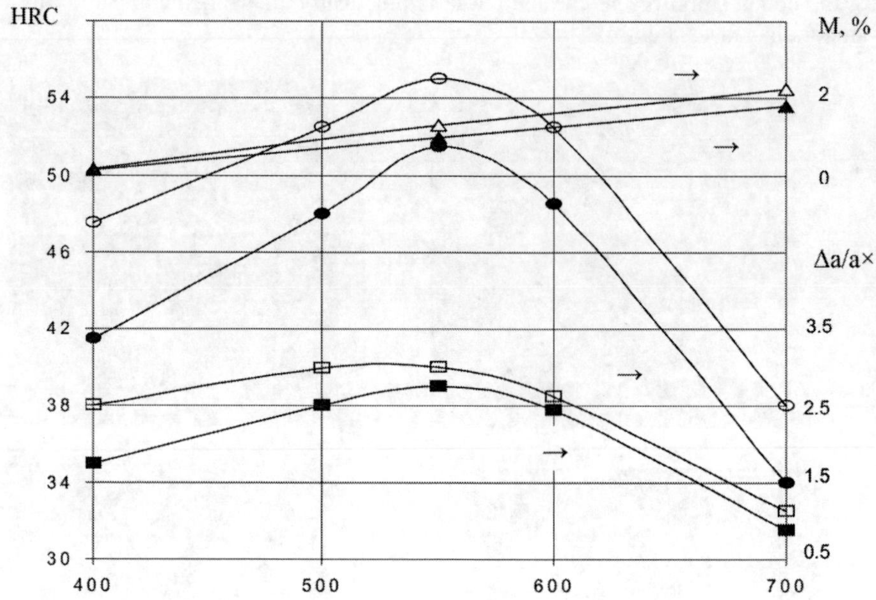

Figure 2. Change in hardness (○, ●), level of microdistortions (Δa/a) in the crystal lattice of the deposited metal (□, ■) and mass fraction of the carbide phases (M, %) – Δ, ▲ depending on the temperature of one-hour ageing: Δ, ○, □ – 20N4M2D2G2YuF; ▲, ●, ■ – 15N4M2D2G2AYuF.

The most significant changes in the morphology of the fine structure of martensite and precipitates are observed after holding of the deposited metal at 700 °C (Figure 1, d). The growth of particles of strengthening phases near the boundaries of the martensite crystals is accompanied by the significant decrease in the dislocation density, level of microdistortions and hardness of the deposited metal. This indicates the development of the stage of coagulation of precipitates, although the formation of the carbide phase is still in progress (Figure 2), and Me_6C carbide appears in its composition. Based on the comparison of the relative intensity of the interference lines and determination of the parameter of lattice of the specified carbide (a = 1.099 nm), it was concluded that its composition can be represented by formula Fe_3Mo_3C, with the metallic constituent presumably containing a certain quantity of vanadium (a_{table} = 1.104 nm).

4. Conclusions

The development of the properties and structure of the deposited steel of type 15..20N4M2D2G2(A)YuF during the heat treatment is the result of the multi-stage interaction of several strengthening mechanisms. Directly after deposition, the martensite transformation, as well as solid-solution strengthening by interstitial and substitutional atoms play the main role in the formation of structure and properties. At the first stage of holding at the temperature of 500–550 °C, strengthening is caused by the precipitation of dispersed intermetallic compounds of type Ni_3Al, NiAl and carbides of type $(V_{0.8}Mo_{0.2})C$, Mo_2C at dislocations during the solid solution decomposition. Moreover, dispersed particles of intermetallic phases are released at the early stages of ageing and slow down the annihilation of dislocations in the matrix. In case of longer ageing, carbides precipitated at pinned dislocations contribute to the strengthening. The further increase in tempering temperature to 600–700 °C leads to the predominant growth of particles of the strengthening phases near the boundaries of the martensite crystals, which is accompanied by the significant decrease in the

IOP Conf. Series: Materials Science and Engineering **966** (2020) 012065 doi:10.1088/1757-899X/966/1/012065

dislocation density, level of microdistortions and hardness of the deposited metal. This indicates the development of the stage of coagulation of precipitates and the associated softening of the deposited metal, regardless of increase in the mass fraction of the carbide phase in its structure.

The analysis of the results made it possible to recommend tempering at the temperature of 550 °C for two hours as the main mode of heat treatment of deposited steels of type 15..20N4M2D2G2(A)YuF.

References

[1] Rastegaev I A et al 2018 *J. of Friction and Wear* **39(5)** 422–6
[2] Korotkov V A 2015 *J. of Friction and Wear* **36(2)** 149–52
[3] Korotkov V A et al 2007 *Welding Int.* **21(7)** 540–2
[4] Feng C and Tahir I K 2008 *J. Mater. Sci.* **43(2)** 788–92
[5] Calik A 2009 *Int. J. Phys. Sci.* **4(9)** 514–8
[6] Daramola O O, Adewuyi B O and Oladele I O (2010) *J. Miner. Mater. Charact. Eng.* **9(08)** 693–708
[7] Kazakov Yu V et al 1991 *Welding Int.* **5(3)** 202–5

Research of features of structure formation in case of surface steel hardening

E N Safonov

Ural Federal University named after the First President of Russia B N Yeltsin, Nizhniy Tagil Technological Institute, 59, Krasnogvardeyskaya str., Nizhniy Tagil, 622000, Russia

E-mail: e.n.safonov@urfu.ru

Abstract. The geometric characteristics, microhardness and features of structure formation were studied in the heat-affected zone during the surface hardening of 09Г2, 20ФЛ, 45ХНМ, 65Г steels with direct current electric arc in argon. It is shown that the depth of zone with thermally modified structure is 0.7–1.2 mm with a tendency to decrease at maximum values of arc speed. Reheating of previously formed local hardening zone is accompanied by certain decrease in surface hardness, however, by changing the overlapping, the hardness of treated surface can be optimized with achieving of formation of regular structure. A regular gradient structure with periodically changing phase composition and microhardness is formed in the surface layer of steel that has been thermally treated with an electric arc. By changing the parameters of the processing conditions (electric arc current, speed of its movement, overlap of local hardening zones), it is possible to purposefully form a certain structural-phase state of the steel surface with properties that meet the operating conditions.

1. Introduction

The significant part of machine parts fails as a result of destruction of the most loaded surface layer during the wear under conditions of frictional contact. To increase the hardness, wear resistance and other operational properties of such parts, it is effective to use surface heat strengthening with concentrated energy sources [1–17]. In this case, physical property of metal strengthening is associated with the change in structural and phase state during the local heating of surface layers and their subsequent rapid cooling.

It should be noted that the main regularities of change in the structure and properties of iron-carbon alloys in the heat-affected zone after laser treatment are studied in most detail [1–15]. The use of electric or plasma arc as energy carrier makes it possible to increase performance of the process due to sufficiently high thermal power and heat flux density (104–106 W/cm^2) on the surface of hot spot with larger (10^{-4} against 10^{-8} cm^2) area [16, 17].

The essence of surface heat strengthening of iron-carbon alloys consists in heating of local area of part surface above the critical temperatures of phase transitions (Ac_1, Ac_3, A_{cm}) and subsequent cooling with high rate which ensures formation of hardening structures. As with usual heat treatment, the features of structural state obtained as a result of surface hardening are determined with degree of solution heat treatment of austenite during the heating, its duration and also initial alloy composition and structure. The final structural state and properties formed in the heat-affected zone after surface heating depend on the cooling rate within the temperature range of the least stability of austenite, composition and size of its grain, number of other factors determined by the parameters of thermal cycle. In this case, the effects of structure modification occurring under conditions of high-rate heating and cooling affect relatively large volumes of metal – usually the local plasma hardening zone is represented in cross section with characteristic contour of segment (Figure 1, a), 5–15 mm wide with maximum depth up to 2.5 mm which significantly exceeds the size of radiation zones.

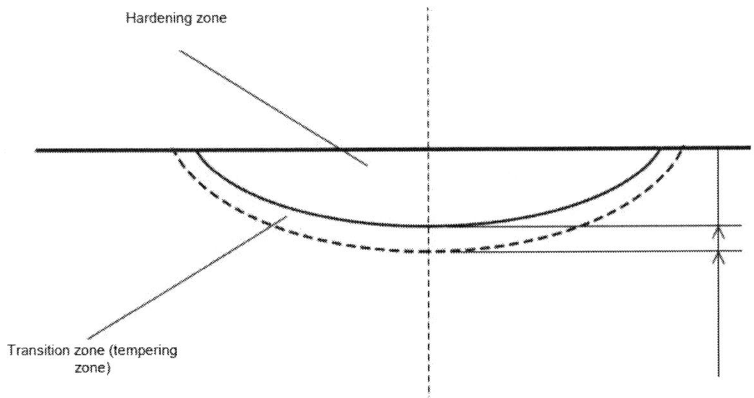

(a)

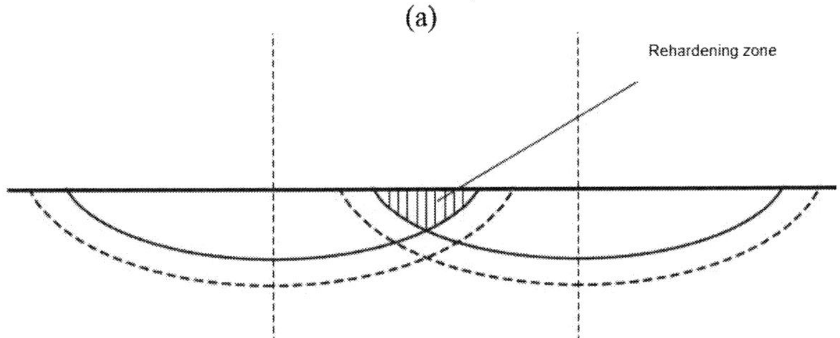

(b)

Figure 1. Cross section scheme of local zone of surface hardening (a) and heat-treated layer obtained by sequential formation of local zones.

Depending on the chemical composition and previous heat treatment of the part, under the influence of plasma heating, gradient structure is formed on its surface in the strengthening zone which composition and properties regularly vary in depth. In general, the following characteristic layers can be distinguished in the structure of this zone [1, B]:

– melting zone formed as a result of high-rate crystallization of molten metal of the surface layer. It is usually represented by dispersed dendrites elongated perpendicular to the surface near which they degenerate into cellular structure as during the micromelting. The phase composition for steel is usually a mixture of martensite and retained austenite;

– zone of hardening from solid state with a fine-grained martensitic-austenitic structure, high microhardness (8.000–10.000 MPa), wear and crack resistance. Such properties are achieved by dispersion and increasing the local inhomogeneity of the structure of massive and plate martensite with different carbon content, increasing the density of dislocations, and preserving particles of carbide and retained austenite undissolved during heating, which helps to reduce crumbling and cracking at the breaking-in stage;

– localized hardening zone transitional to the initial structure of base metal of the part. It contains a mixture of martensite, decomposition products of austenite and martensite, elements of the initial structure of base metal not subjected to phase transitions.

In case of surface hardening of parts subjected to volume heat treatment as well as when local zones of plasma hardening are overlapped in the areas heated below temperature Ac_1, tempering zone is formed (Figure 1, a).

The overall depth, features of composition and structure of strengthening zone with gradient structure are determined by a number of factors, which accounting completeness and degree affect the results of strengthening but the specified regularities are generally preserved.

The strengthening of surface of given area is achieved by sequential formation of local hardening zones in the form of extended strips (with gap, butt or overlapping between them) at each arc pass (Figure 1, b).

In this case, the number of passes shall be minimum essential and sufficient for strengthening of given surface of the product. The quality of strengthened layer is determined by center-to-center distance or overlapping of adjacent hardening zones. When each subsequent pass forms new local zone with overlapping, in the area of preformed zone where the heating temperature exceeds Ac_3, complete phase recrystallization and rehardening take place. When center-to-center distance is greater than the half-width of local hardening zone, there is no overlapping within two adjacent hardening zones and high-speed tempering takes place in that part of preformed zone where the reheating temperature does not exceed Ac_1.

2. Methods of research

The geometric characteristics, microhardness and features of structure formation were studied in the heat-affected zone during the surface hardening of 09Г2, 20ФЛ, 45ХНМ, 65Г steels with direct current electric arc in argon. Since strengthening of surface of given area was performed during the sequential formation of local hardening zones, the overlapping of these zones was 30 or 50 % of their width.

Arc current varied within 180–330 A, the rate of its movement was 80–180 m/h. Researches of phase composition were performed using optical metallography (Neophot-2), durometry (Leizt), X-ray diffraction analysis (DRON-3.0) in cobalt K_α-radiation. The profile and width of interference lines (111) of austenite and (110) α-phase were controlled, volume fraction of phases was evaluated with respect to integrated intensity of lines. Microhardness of the structural components was determined under load of 0.78 N.

3. Results and discussion of the results

It is found that the depth of zone with thermally modified structure within the studied range of parameters of the mode is 0.7–1.2 mm with a tendency to decrease at maximum values of arc speed. The initial structure of studied steels is ferritic-pearlitic, small amount of bainitic component is found in 09Г2 steel.

On the surface of local zone of heat treatment of 09Г2, 20ФЛ steels, formation of thin (~50 μm) low-etching layer with high (50–55 %) austenite content is observed that is unexpected for low-carbon steels. If pearlite colony, when heated at high rate, completely passes to the austenitic state, the carbon content in austenite shall be at the level of 0.8 % and the total amount of fixed carbon will be (0.5×0.8 % C) ~0.4 % that significantly exceeds its content in steel.

Apparently, increase in carbon concentration is consequence of microchemical inhomogeneity arising under conditions of flash heating of surface with electric arc. In the presence of structurally free ferrite, homogenization of austenite is delayed by increasing the carbon diffusion paths through ferritic grain. Therefore, even in the thin surface layer heated almost to the melting point, processes of carbon diffusion and leveling of austenite composition do not complete and zones of concentration inhomogeneity with increased carbon content remain in place of former perlite.

The Figure 2, d shows the change in the amount of austenite within strengthening zone successively formed by four passes on the surface of 20ФЛ steel. The austenite content periodically changes, in this case minimum values correspond to the boundary regions between zones and maximum concentration reaches 50–52 %.

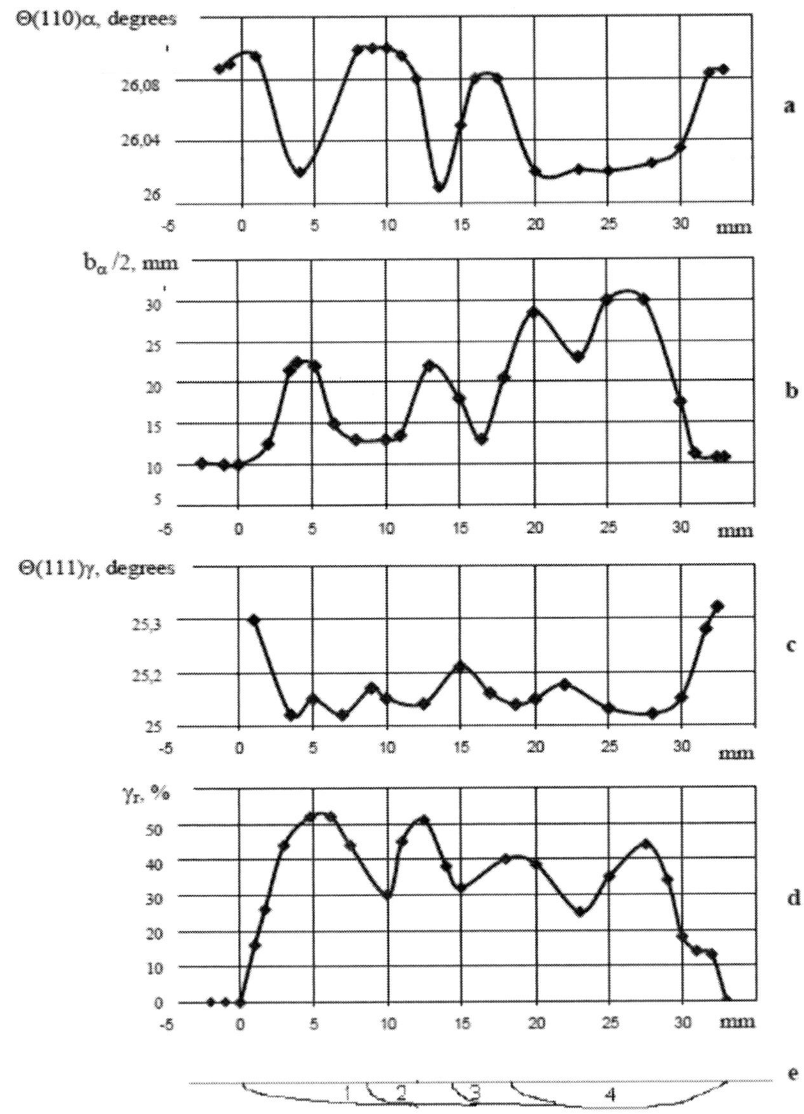

Figure 2. Change of position of interference lines of α – (a) and γ – (c) phases, half-width of line of α-phase (b), volume fraction of retained austenite γ_r (d) on the surface of 20ФЛ steel sample after arc heat treatment with sequential formation (1–4) of local hardening zones (e). Arc current is 210 A, rate of its movement is 72 m/h, overlapping of local hardening zones is 30 %.

The minimum values of reflection angle correspond to increased amounts of austenite (Figure 2, c), that is associated with the largest amount of carbon in composition of this phase. There is frequency in the change of half-width of interference lines of α-phase (Figure 2, b). The level of broadening observed in this case is consequence of not so much phase hardening as degree of tetragonal distortions of crystal lattice of this phase due to the significant amount of carbon in its composition, i. e. formation of carbon martensite. The additional fact confirming this conclusion is their displacement to the area of smaller reflection angles simultaneously with broadening of the lines associated with increase in the volume of elementary lattice (Figure 2, a).

After plasma treatment structure represented by martensite, retained austenite and carbides is formed on the surface of medium-carbon steels. The fraction of retained austenite reaches 40 % in 45XHM steel, 50 % in 65G steel (arc current is 300 A, treatment speed is 150 m/h).

When each subsequent pass forms new local zone, in the area of preformed zone where the heating temperature exceeds Ac_3, complete phase recrystallization and rehardening take place. As a result, these parts do not differ in structural composition, however, degree of dispersion of structure and depth of hardened zone increase. Microhardness is somewhat made even over the depth of zone, Figure 3, that can be explained by increase in degree of solution heat treatment of austenite over carbon as a result of increased time of retention of metal in the temperature range of existence of austenite.

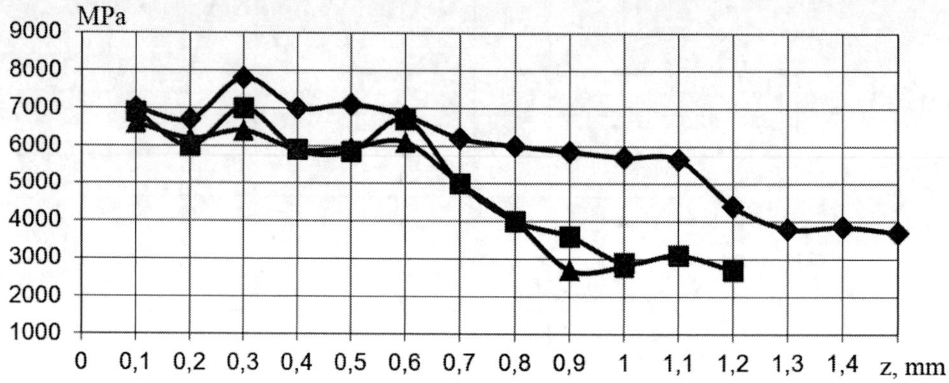

Figure 3. Change of microhardness over the depth of hardening zone (z) as a result of threefold electric-arc heat treatment of 65Г steel. Arc current strength is 300 A, arc speed is 60 m/h. The first pass is ▲, the second one is ■, the third pass is ♦.

High-speed tempering takes place in that part of preformed strengthening zone where the reheating temperature does not exceed Ac_1. Martensitic-austenitic structure of previously formed hardening zone in the areas of reheating is subject to decomposition with formation of bainitic and sorbitic structures. In place of carbon martensite particles of globular cementite are precipitated. Retained austenite of the surface layer is also partially decomposed with formation of sorbitic structure.

Since increase in treatment speed is accompanied by decrease in the specific heat flux through the treated surface, degree of solution heat treatment of austenite over carbon decreases during the heating. The martensite formed upon subsequent cooling, inheriting the microchemical inhomogeneity of austenite, has a lower carbon concentration, tetragonality and microhardness. Depth of strengthening zone also decreases. Treatment with increased heat input, when significant fraction of retained austenite is formed in the surface layer, makes it possible to obtain carbon martensite with microhardness of 6.100–8.300 MPa in the main part of strengthening zone. In this case the depth of strengthened zone reaches one millimeter.

It is known that periodic alternation of hard and soft layers can effectively increase the operational life of parts operating under dynamic and thermomechanical loads. In addition, presence of retained austenite on the surface treated with electric arc increases the incubation time of initiation and propagation of cracks, reduces the running-in time in friction pairs and allows for implementation of energy-absorbing process of its transformation to deformation martensite with partial dissipation of the fracture energy and simultaneous strain strengthening under conditions of dynamic impact of wear medium.

4. Conclusions

In a surface layer of steel heat-treated with electric arc which is formed by sequential formation of local hardening zones with certain overlapping, regular gradient structure is formed with periodically changing phase composition and microhardness.

IOP Conf. Series: Materials Science and Engineering **966** (2020) 012066 doi:10.1088/1757-899X/966/1/012066

By changing the parameters of the processing conditions (electric arc current, rate of its movement, overlapping of local hardening zones), it is possible to purposefully form a certain structural and phase state of steel surface with properties that meet the operating conditions.

References

[1] Anusha E, Kumar A and Shariff S 2019 *Optik* **163** 357
[2] Moradi M et al 2019 *Optik* **188**, 277–86
[3] Maharjan N et al 2019 *Surf. Coat. Technol.* **366** 311–20
[4] Oh S and Ki H 2017 *Appl. Therm. Eng.* **121** 951–62
[5] Liverani E et al 2017 *J. Manuf. Process.* **26** 262–8
[6] Guarino S, Barletta M and Afilal A 2017 *J. Manuf. Process.* **28** 266–71
[7] Khorram A et al 2019 *Opt. Laser Technol.* **119**, 105617
[8] Filep A et al 2014 *Adv. Mater. Res.* **996**, 538–43
[9] Jiang J, Xue L and Wang S 2011 *Surf. Coat. Technol.* **205** 5156–64
[10] Nath A K, Gupta A and Benny F 2012 *Surf. Coat. Technol* **206** 2602–15
[11] Liu A and Previtali B 2010 *Phys. Procedia***5** 439–48
[12] Kennedy E, Byrne G and Collins D N 2004 *J. Mater. Process. Technol.* **155–156** 1855–60
[13] Pashby I.R., Barnes S. and Bryden B.G. 2003 *J. Mater. Process. Technol.* **139** 585–8
[14] Skvarenina S and Shin Y C 2006 *Surf. Coat. Technol.* **201** 2256–69
[15] Ion J C 2002 *Surf. Eng.* **18** 14-31
[16] Safonov E N and Zhuravlev V I 1998 *Welding Int.* **12 (4)** 326–8
[17] Korotkov V A 2015 *J. of Frict. and Wear* **36 (2)** 149–52

Sulfur partition by process stages of metallurgical production of JSC EVRAZ NTMK

M V Savelyev[1, 5], O Yu Sheshukov[2, 3], A A Metelkin[4], O I Shevchenko[4], A S Tkachev[1] and S V Shmakov[1, 4]

[1] EVRAZ NTMK, ul. Metallurgov, 1, Nizhnij Tagil, Sverdlovskaya oblast, 622025, Russia
[2] Ural Federal University named after the First President of Russia B N Yeltsin, 19, Mira str., Yekaterinburg, 620000, Russia
[3] Metallurgical Institute of Ural Branch of the Russian Academy of Sciences, 101, Amundsena str., Yekaterinburg, 620016, Russia
[4] Ural Federal University named after the First President of Russia B N Yeltsin, Nizhniy Tagil Technological Institute, 59, Krasnogvardeyskaya str., Nizhniy Tagil, 622000, Russia

E-mail: [5] maxim.savelev@evraz.com

Abstract. The current level of development of the industry requires the manufacture of steel with high purity in relation to detrimental impurities. Sulfur is one of such impurities that significantly reduces the service properties, the content of which is strictly regulated in the finished product. Deep desulfurization may be achieved through development of cross-cutting technology of metal production as a single process including agglomeration and blast-furnace process and steelmaking. Sulfur partition by metallurgical process stages via conversion from raw materials of the blast-furnace process to steel continuously cast blank is studied in this work. It is shown that removal of sulfur from the half-finished product at the desulfurization plant will not provide the required sulfur content (less than 0.005 %) in the steel ingot. It is determined that the activity plan comprising treatment of the half-finished product at the desulfurization plant, smelting in the converter, treatment of steel at the ladle furnace unit and then in the vacuum vessel is required to obtain the final content of sulfur of less than 0.005%.

1. Introduction

Modern consumer of metallurgical products imposes increasingly demanding requirements to the produced steel quality. Sulfur is subjected to the strictest limitations on content in metal [1]. Significant number of theoretical and practical works is dedicated to deep desulfurization of metal in order to obtain low sulfur content in the finished products. This direction has special actuality for production of high-strength steels (pressure vessels, oil and gas pipes of large diameter [2–4]), sheet steels for deep drawing [5, 6], reinforcement steels [7, 8], as well as mass-produced steels [9, 10]. Steels with the regulated low sulfur content are required for aviation, power and transport engineering, construction and other industrial fields [11–13].

Normally increase of requirements to steel properties outstrips development of processing methods aimed at improvement of metal purity. As a result, it is necessary to search for combinations of efficient methods of refining for all types of process stages of metallurgical production. Currently, not a single type of metallurgical raw materials is used without preliminary preparation, which allows to carry out measures for sulfur removal at the stage of preparing the charge materials for blast-furnace smelting [14, 15]. Behavior of sulfur during agglomeration of iron-ore materials is well-studied, therefore, in case of correct process organization, a high sulfur removal rate will be achieved ensuring structural and other characteristics of agglomerated materials. Major part of work for removal of sulfur from metal is carried out in the blast-furnace process and during cast iron treatment afterwards. Continuous mode of desulfurization implemented in the blast furnace is the most efficient among all periodic processes offered alternatively. Cast iron quality becomes particularly important for the stage of converter operation where the possibilities of sulfur removal are limited [16–19].

Steel treatment at ladle furnace unit is an integral part of production of high-quality metal [16]; the technology provides induction of new high-basic free-running slag. Final concentration of sulfur in steel reduces together with increase of slag layer and its main basicity. Duration of argon blowing of the melt, increase of blowing rate and steel temperature result in reduction of sulfur concentration. Desulfurization conditions will deteriorate in case of increase of initial sulfur concentration and slag oxidation [17].

Special place of sulfur among other impurities included in steels is related to its impact on mechanical properties. Increase of sulfur content induces red shortness, decreases impact hardness, corrosion resistance, electric properties, and deterioration of plastic properties results in mass faulty production during stamping of deep-drawing hollow parts of sheet steel [1, 11, 18].

2. Theoretical data

Sulfur has unlimited solubility in the liquid iron (up to 38 % if its content in FeS is 36.5 %) and limited solubility in the solid iron: in γ-Fe at 1365 °C ~0.055%, and in α-Fe at 900 °C ≤0.015% and at room temperature ~0.0001 % [1]. At transition of metal from liquid state to solid state, solubility of sulfur in it sharply decreases and continues to decrease at cooling of solid metal from crystallization temperature to room temperature. That is, hardening and cooling of metal with high sulfur content will lead to eutectic and release of FeS iron sulfide from the solid solution.

Fe-FeS eutectic has a low melting point (985 °C) compared to the temperature of the solidus of steel. In the presence of oxygen due to the formation of oxysulphides, eutectic crystallization occurs at even lower temperature. Presence of liquid state along grain boundaries in combination with contraction stresses of ingot and deformations of forging, rolling, etc. will lead to formation of cracks. Such phenomenon has been termed the ‘red shortness’. Sulfur may also have a negative influence on impact hardness of steel at low temperatures (less than –30 °C), embrittling metal. High sulfur content leads to sharp anisotropy of properties in the finished rolled products (Figure 1) [1]. Anisotropy of mechanical characteristics which depend on the mutual orientation of sulphide inclusions (orientation of sample relating to direction of rolling) is mostly demonstrated at plastic properties of metal, for example, relative elongation, relative reduction and especially impact hardness.

However, the Figure 1 shows that if sulfur content is less than 0.005 %, impact hardness of steel in transverse samples is significantly lower than in transverse ones, i.e. metal according to its properties becomes homogeneous at full volume.

Consequently, sulfur content of the metal, which does not influence the homogeneity of its properties is taken as not more than 0.005 %. However, in order to guarantee the absence of influence, its content must be even less and must not exceed 0.0001 % – the limit of dissolution of this element in iron at standard conditions [1].

Finished steel in some quantity contains all impurities that are part not only of metallic melting stock, but also of the slag and gas phases and lining, so the sulfur content in the final product depends on both the technology of production and the raw materials used. Conversion in metallurgy usually means the processing of material in which the chemical or phase composition of the product significantly changes its physical and mechanical properties. In order to achieve the required results in removing sulfur from the metal, it is necessary to analyze the main sources of sulfur input to the melt, as well as technological factors affecting the desulfurization process for conversion in each process stage.

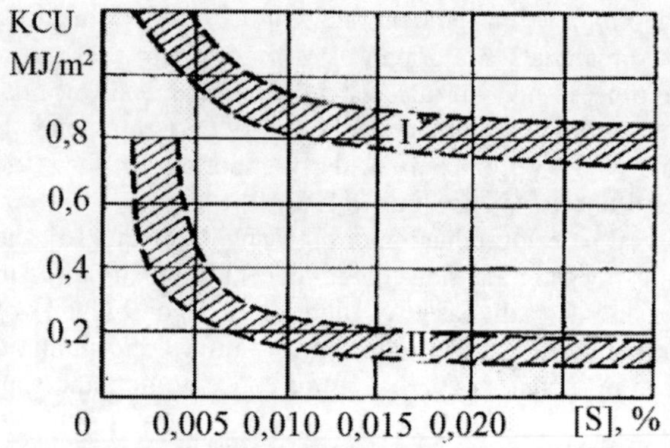

Figure 1. Dependence of impact hardness of aluminum deoxidized steel on content of sulfur determined for longitudinal (I) and transverse (II) samples of rolled products at room temperature.

3. Research results

The classical steel production scheme includes the following stages:

1. Obtaining of ore concentrate (ore production, sinter and pellets production.
2. Blast-furnace process (cast iron production).
3. External cast iron desulfurization.
4. Basic oxygen furnace process, steel production.
5. Steel processing in ladle furnace unit (LFU).
6. Metal processing at decreased pressure units.
7. Steel teeming in continuous-casting machines (CCMs).

At each stage, the sulfur content in metal will change, depending on the physical and chemical processes in the melting facility.

Depending on the chemical composition of cast iron, several special technological options are used to convert it into steel. JSC EVRAZ NTMK implemented vanadium conversion – processing of vanadium cast iron into steel and vanadium slag. The classic scheme is distinguished by the presence of a duplex process – a two-stage conversion. The first stage consists in extracting vanadium from cast iron and converting it as an oxide phase into slag. The result of this stage is the production of half-finished product (de-vanadium cast iron) and vanadium slag. After vanadium extraction, the molten metal with carbon content of 2.6–3.5 % goes to the converter for processing to obtain steel with a given chemical composition – this is the second stage. Between two stages of conversion, the half-finished product can be delivered to the desulfurization plant. Change of sulfur content in metal in stages of metallurgical conversion is given in Table 1.

Graphically the change of sulfur content and the degree of its change for conversion of production are presented in Figures 2 and 3.

The data presented in the figures show that the sulfur content after treatment at the desulfurization plant is 0.0068%, which is not sufficient to eliminate the effect of detrimental impurities on the service properties of the products. The minimum sulfur content in metal reaches the value of 0.0039% only with the complete metal treatment complex in the extra-furnace steel processing units – the ladle-furnace unit and the circulation vacuum vessel.

In order to study the sulfur partition (LS) between metal and slag in the LFU, let us compose a system of equations (1)–(3).

$$M \times [S]_0 + m \times (S)_0 = M \times [S]_f + m \times (S)_f \quad (1)$$

Where $(S)_0$ – initial sulfur content in slag (which is taken equal to zero); $[S]_0$ – initial sulfur content in metal; $[S]_f$ – final sulfur content in metal; $(S)_f$ – final sulfur content in slag; M – average heat weight (152.0 tons); m – mass of slag generated in the steel-pouring ladle of the LFU (1.98 tons).

IOP Conf. Series: Materials Science and Engineering **966** (2020) 012067 doi:10.1088/1757-899X/966/1/012067

Table 1. Sulfur partition in stages of manufacture of steel in case of different technology options.

Stages of manufacture of steel	Sulfur content in metal or raw materials, % wt.			
	Without treatment at the desulfurization plant		With treatment at the desulfurization plant	
	–	Treatment at the circulation degassing plant	–	Treatment at the circulation degassing plant
Obtaining of ore concentrate and fuel for blast furnace	Ore……0.006 Sinter and pellets (IORM)……0.009 Coke……0.59 Pulverized coal fuel……0.34			
Blast-furnace process	0.0249			
Converter process 1.	0.0244			
Desulfurization of half-finished product	–	–	0.0068	0.0068
Converter process 2.	0.0196	0.0206	0.0083	0.0085
LFU (beginning of treatment)	0.0192	0.0196	0.0084	0.0086
LFU (end of treatment)	0.012	0.0126	0.0046	0.0057
Circulation vacuum vessel	–	0.008	0.0039	0.0057
CCM	0.012	0.008	0.0039	0.0057

$$L_S = \frac{(S)_f}{[S]_f} \quad (2)$$

$$\lambda = \frac{m}{M} \quad (3)$$

whereλ – mass ratio of slag and metal.

Using equations 1–3, let us define the sulfur partition coefficient between metal and slag L_S:

$$L_S = \frac{\left(\frac{[S]_0}{[S]_f} - 1\right)}{\lambda} \quad (4)$$

Let us apply the formula 4 and the data of Table 1 for calculation of average values of the sulfur partition coefficient between metal and slag. Let us compare the calculated value L_S=47.0 with the data of sources [1] (L_S = 50–100) and [20] (L_S ≈50–250). Despite the 'satisfactory' final sulfur content in metal (less than 0.005%), the steel manufacture technology including such stages as the treatment of the half-finished product at the desulphurization plant and the treatment of steel in the LFU and vacuum vessel, has reserves for the removal of sulfur. By adjusting the composition of the LFU slag, it is possible to increase their desulphurization capacity.

4. Conclusion

Thus, the current level of development of the industry requires the manufacture of steel with high purity in relation to detrimental impurities. Sulfur is one of the impurities that significantly reduces the service properties and the content of which in the finished product is strictly regulated. Deep desulfurization may be achieved through development of cross-cutting technology of metal production as a single process including agglomeration and blast-furnace process and steelmaking. Sulfur partition by metallurgical process stages via conversion from raw materials of the blast-furnace process to steel continuously cast blank is studied in this work. It is shown that removal of sulfur from the half-finished product at the desulfurization plant will not provide the required sulfur content (less than 0.005 %) in the steel ingot. It was determined that the activity plan comprising treatment of the half-finished product at the desulfurization plant, smelting in the converter, treatment of steel at the ladle furnace unit and then in the vacuum vessel is required to obtain the final content of sulfur of less than 0.005%. According to the study of sulfur partition coefficient between metal and slug, despite the low final sulfur content in metal, this technology has reserves related to desulfurization achieved by correction of LFU.

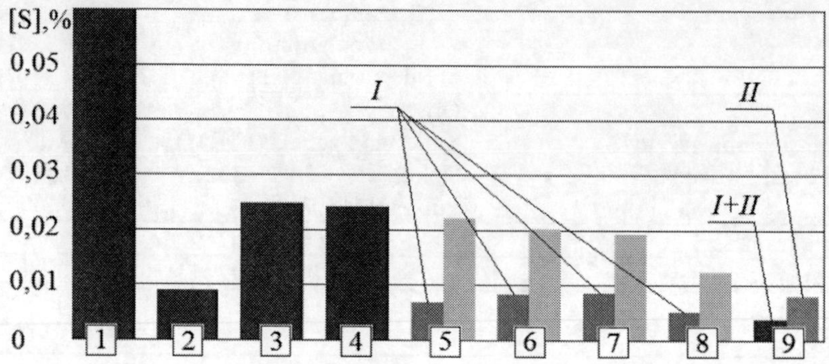

Figure 2. Diagram of sulfur partition in stages of manufacture of JSC EVRAZ NTMK, *I*– melts treated at the desulfurization plant, *II* – melts treated at the circulation vacuum vessel.Sulfur content in stages of manufacture: 1–in ore; 2 – in sinter; 3 – in cast iron; 4 – in half-finished product; 5 – desulfurization plant; 6 – at point of discharge from converter; 7 –deoxidation-alloying; 8 – treatment in LFU; 9 – treatment at RH.

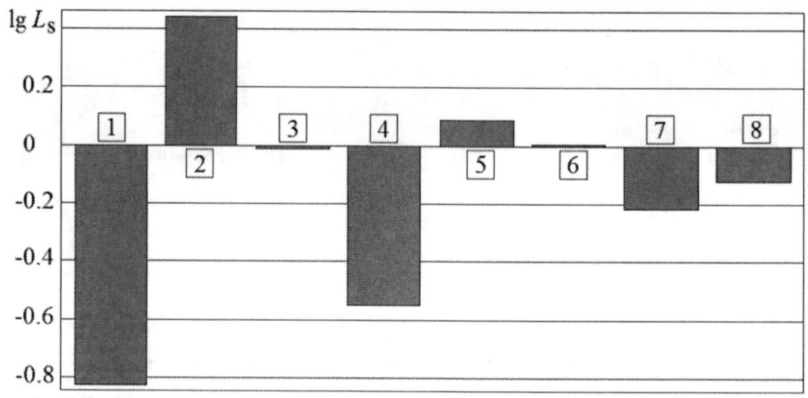

Figure 3. Changes in the sulfur content partition coefficient in stages of manufacture of JSC EVRAZ NTMK: 1 –Sintering production; 2 –Cast iron production; 3 –Vanadic conversion; 4 –Desulfurization of half-finished product; 5 –Steel conversion; 6 –Deoxidizing agents; 7 –Treatment at LFU; 8 –Treatment at RH.

References

[1] Bigeev A M and Bigeev V A 2000 *Metallurgiya stali. Teoriya i tekhnologiya plavki stali* (Magnitogorsk : MGTU) p 544 (in Russian)

[2] Ushakov A N, Bigeev V A, Stolyarov A M and Potapova M V 2019 Technology for production of pipeline ultra low sulfur steel *Chernye Metally* **12** pp 26–31

[3] Ushakov A N, Bigeev VA, Stolyarov A M and Potapova M V 2018 Ladle Desulfurization of Converter Low-Sulfur Pipe Steel *Metallurgist* **62 (7-8)** pp 667–73

[4] Ushakov A N, Avramenko V A, Bigeev V A, Stolyarov A M and Potapova M V 2018 Manufacture of Low-Sulfur Pipe Steel with Ladle Desulfurization of Cast Iron *Metallurgist* **61 (11-12)** pp 967–70

[5] Gorkusha D V, Grigorovich K V, Karasev A V and Komolova O A 2019 Content modification

of different types of nonmetallic inclusions during low-carbon if steel ladle treatment *Izvestiya Ferrous Metallurgy* **62 (5)** pp 345–52

[6] Ghosh P, Ghosh C and Ray R K 2009 Precipitation in interstitial free high strength steels *ISIJ International* **49 (7)** pp 1080–6

[7] Sheremet V A, Kekukh A V, Razdobreev V G, Kuvaev V N and Ivanov D A 2004 Influence of conditions of thermomechanical treatment on structures, mechanical and service properties of thermally strengthened reinforcing bars of A1000 class *Izvestiya Ferrous Metallurgy* **11** pp 40–3

[8] Sheremet V A, Kekukh A V, Razdobreev V G, Kuvaev V N and Ivanov D A 2004 Influence of the thermomechanical-treatment conditions on the structure and properties of thermohardened class-A1000 reinforcement *Steel in Translation* **34 (11)** pp 59–62

[9] Siyasiya C W and Stumpf W E 2008 The effects of hot rolling process and the nitrogen and sulfur content on the microstructural development of aluminium killed hot rolled low carbon strip steel *Journal of the South. African Inst. of Mining and Metallurgy* **108 (8)** pp 481–9

[10] Li Y, Zhu F, Cui F and Fang K, 2007 Analysis of forming mechanism of lamination defect of steel plate *Dongbei Daxue Xuebao/Journal of Northeastern Univ.* **28 (7)** pp 1002–5

[11] Yugov P I and Baeva L A 2004 Technological principles of the production of clean cold-resistant steels from pig iron with low contents of manganese and silicon *Metallurgist* **48 (7-8)** pp 307–10

[12] Pike T J 2000 Production and metallurgy perspective of modern constructional steels *Proceedings of Annual Techn. Session, Structural Stability Research Council* 2000 pp 14–30

[13] Tursunov N K, Semin A E and Sanokulov E A 2017 Study of dephosphoration and desulphurization processes in the smelting of 20GL steel in the induction crucible furnace with consequent ladle treatment using rare earth metals *Chernye Metally* **1** pp 33–40

[14] Inaba S and Kimura Y 2004 Behavior of sulfur in the carbon-bearing iron oxide pellet during heating *ISIJ International* **44 (12)** pp 2112–4

[15] Park J and Jung S 2016 Effects of various slag systems on metal/slag separation of CCA and slag composition on desulfurization and dephosphorization of iron nugget *Advances in Molten Slags, Fluxes, and Salts: Proceedings of the 10th International Conference on Molten Slags, Fluxes and Salts* pp 1025–30

[16] Serov Y V and Mikhalevich A G 1982 Effectiveness of conversion pig improvement to meet steelmaking requirements *Metallurgist* **26 (2)** pp 37–43

[17] Schrama F, Van den Berg B and Van Hattum G 2015 A comparison of the leading hot metal desulphurization methods *Proceedings of the 6th International Congress on the Science and Technology of Steelmaking, ICS 2015* pp 61–6

[18] Schrama F, Van den Berg B and Van Hattum G 2015 Electric steelmaking: Comparing hot metal desulphurisation methods *Steel Times International* **39 (4)** pp 42–9

[19] Bol'shakov V I, Shevchenko A F, Ye L D, Aleksandrov V A, Sen S T, Bashmakov A M and Trotsenko E A 2009 Rational ladle treatment for desulfurization of hot metal *Steel in Translation* **39 (4)** pp 326–33

[20] Umezawa K and Kajioka H 1977 Slag metal reactions and refining abilities of the ladle furnace *Tetsu-To-Hagane/Journal of the Iron and Steel Institute of Japan* **63 (13)** pp 2034–42

IOP Conf. Series: Materials Science and Engineering **966** (2020) 012068 doi:10.1088/1757-899X/966/1/012068

Calculation of sulfur removal in ladle furnace unit by means of ionic theory of slags

M V Savelyev[1, 5], O Yu Sheshukov[2,3], A A Metelkin[4], O I Shevchenko[4], and D K Egiazar'yan[2,3]

[1]EVRAZ NTMK, ul. Metallurgov, 1, Nizhnij Tagil, Sverdlovskaya oblast, 622025, Russia
[2]Ural Federal University named after the First President of Russia B N Yeltsin, 19, Mira str., Yekaterinburg, 620000, Russia
[3]Metallurgical Institute of Ural Branch of the Russian Academy of Sciences, 101, Amundsena str., Yekaterinburg, 620016, Russia
[4]Ural Federal University named after the First President of Russia B N Yeltsin, Nizhniy Tagil Technological Institute, 59, Krasnogvardeyskaya str., Nizhniy Tagil, 622000, Russia

E-mail: [5] maxim.savelev@evraz.com

Abstract. This article describes issues of sulfur removal in ladle furnace unit. First of all, metal desulfurization in ladle steel treatment units is achieved due to transition of sulfur to the slag phase. Coefficient of partition of sulfur between metal and slag is affected by sulfide capacity of slag, coefficient of activity of sulfur in the metal, oxidation potential of medium and reaction equilibrium constant of partition of sulfur between metal and slag. Temperature of liquid state in the ladle has significant impact on sulfide capacity of slag. Proposed calculation procedure based on provisions of ionic structure of slags allows evaluating a concentration of sulfur in steel on the basis of determination of coefficient of its partition between metal and slag. Optical basicity is suggested as criterion of refining property of slag, it has been shown that the amphoteric oxide Al_2O_3 is essential to calculation of this indicator. Its impact on sulfide capacity of slag is detected.

1. Introduction

After tapping of metal from the steelmaking vessel all further actions aimed at finishing of steel in terms ofchemical composition and temperature are performed in the ladle. Ladle metallurgy is intended to solve three main tasks: refining of steel on nonmetallic inclusions and detrimental impurities; correction of chemical composition by means of addition of corresponding master alloys; homogenization of metal by temperature and chemical composition [1, 2]. The current level of the industry development requires the manufacture of steel with high purity in relation to nonmetallic inclusions and detrimental impurities [3–6]. Sulfur is one of the impurities that significantly reduces the service properties of steel and the content of which is strictly regulated in the finished product. A great number of theoretical and practical researches have been devoted to processes of desulfurization, but this topic is still relevant at present.

It is known that metal desulfurization can be achieved due to transition of sulfur to the slag [1]. This process can be implemented using ladle furnace unit at the metal surface of which fresh slag is

IOP Conf. Series: Materials Science and Engineering **966** (2020) 012068 doi:10.1088/1757-899X/966/1/012068

built up. Slag is a multi-component oxide melt that interacts with metal and performs important technological functions [1, 11, 12]. At present, it is considered that slag has an ionic structure and consists of positively charged cations and negatively charged anions. Evidence of the ionic structure of slags are numerous X-ray diffraction studies of solidified slags, electrical conductivity of molten slags, presence of electrical charges in the boundary layers of metal and slag, high values of surface tension of slags, etc.

2. Theoretical data

In molten metallurgical slags, there are ions of the following groups [1, 13–15].

1. Cations Ca^{2+}, Mg^{2+}, Mn^{2+}, Fe^{2+}.
2. Anions O^{2-}, S^{2-}, SiO_4^{4-}, PO_4^{3-}, AlO_3^{3-}, FeO_2^{1-}.
3. More complex silica acid anions $(SiO_3^{2-})_n$,$Si_3O_9^{6-}$, $Si_4O_{12}^{8-}$, $Si_6O_{18}^{12-}$ etc. can form in acid slags [16–19].

Taking into account the presented data it is important to study metal desulfurization from the position of ionic theory of slag structure.

It is known that the sulfur partition coefficient between metal and slag is determined by the following formula [20]:

$$L_S = \frac{(S)}{[S]} = C_S \times \gamma_{[S]} \times \frac{1}{p_{\{O_2\}}^{0.5}} \times \frac{1}{K_{[S]}} \tag{1}$$

where C_S – sulfide capacity of slag;$\gamma_{[S]}$ – coefficient of activity of sulfur in the metal; $p_{\{O_2\}}^{0.5}$ – oxidation potential of medium; $K_{[S]}$ – reaction equilibrium constant of partition of sulfur between metal and slag.

Let us review each multiplier in the right part of the equation 1.

Reaction equilibrium constant of partition of sulfur between metal and slag according to data [1] is determined by the formula $\lg K_S = -\frac{6500}{T} + 2.625$ where T– temperature, K.

In order to calculate values of coefficient of activity of sulfur dissolved in liquid steel, according to data, it is allowed to use values of interaction parameters:

$$\lg \gamma_{[S]} = \sum_{i=1}^{n} e_S^i \times [X_i] = e_S^1 \times [X_1] + e_S^2 \times [X_2] + \cdots + e_S^i \times [X_i] \tag{2}$$

where e^i_S – parameter of interaction of corresponding metal element with sulfur; X – content of element in liquid iron, %.

According to data of [1, 21, 22] the content of FeO in metal and slag is one of the main factors determining the value of L_S.

$$L_S \approx \frac{1}{p_{\{O_2\}}^{0.5}}$$

The higher oxidation potential of medium, the smaller L_S, but this potential corresponds to the content of oxygen in the metal [O], i.e. L_S depends on oxidation of the metal. Thus, oxygen in form of oxide (FeO) in slag, and oxygen in metal [O] significantly reduce degree of desulfurization.

Sulfide capacity of slag C_S is one of the most important characteristics of refining property of the slags applied during extra-furnace steel processing. This value is determined as function of slag temperature and composition, i.e. this value is experimentally determined and thermodynamically evaluated.

One of the factors influencing on sulfide capacity is temperature of the process. Influence of the temperature on dependence of the sulfide capacity on the optical basicity was covered in the paper [21]. The data are provided in Figure 1.

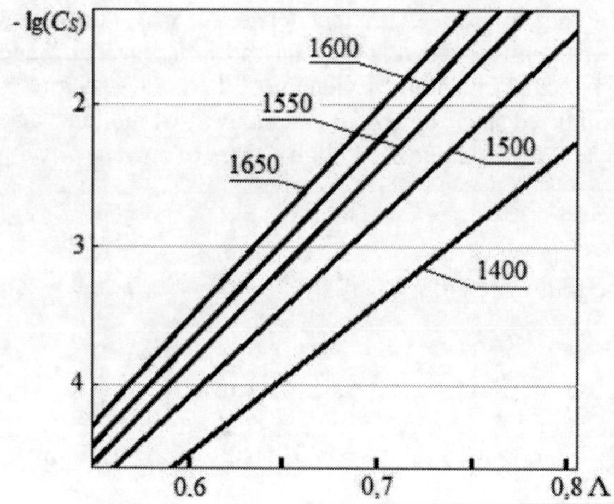

Figure 1. Dependence of the sulfide capacity lg(C_S) on the optical basicity (Λ)at different temperatures (°C).

Mathematical charts may be described using equation (3), with temperature intervals 1400–1650 °C and at optical basicity Λ not more than 0.75, error of provided formula is not more than 6 %.

$$\lg C_S = 14.3 \times \Lambda - 7.01 - \frac{9908.1}{T} \quad (3)$$

Optical basicity, which is characterized as the ability of oxygen anions which present in slag to give their electrons to ion acceptors (ions collected by probes [16–19]), is most often used as a criterion of basicity of oxide melts. For pure oxides, optical basicity (Λ) is connected with electronegativity of elements according to Poling (X_i) by the following ratio:

$$\Lambda_i = \frac{1}{1.36 \times (X_i - 0.26)} \quad (4)$$

Earlier papers identified that optical basicity of oxide Λ is connected with electronegativity of cation according to Poling (X_i) by the following expression:

$$\Lambda_i = \frac{0.75}{(X_i - 0.26)} \quad (5)$$

Use of equations 4 and 5 allows obtaining optical basicity for any multi-component systems consisting of nontransition (nonamphoteric) metals according to the following ratio:

$$\Lambda = \sum_{i=1}^{n}(X_i \times \Lambda_i) = X_{MO_1} \times \Lambda_1 + X_{MO_2} \times \Lambda_2 + \cdots + X_{MO_n} \times \Lambda_n \quad (6)$$

where X_i – equivalent fraction of anions introduced by this component; Λ_i – optical basicity of system component.

Optical basicity of main, acid and amphoteric oxides is determined. However, we found that Al_2O_3 may show both main and acid properties, i.e. optical basicity of this oxide will change depending on the slag composition, respectively.

The metallurgical slags built up in the steel-pouring ladle of the ladle-furnace unit contain oxides that absorb 'free' oxygen anions, such as SiO_2, to form a complex anion SiO_4^{4-}, and also contain amphoteric oxides, which can act both as donors and absorbers of free oxygen 'anions'. These parameters in classical formula 5 are not taken into account.

To determine optical basicity on the basis of the calculation formula (7) proposed by us, it is necessary to add together main oxides (m.o.) and subtract acid oxides (a.o.), influence of amphoteric oxide Al_2O_3 is taken into account additionally.

$$\Lambda = \left(\sum_{i=1}^{n}(X_i \times \Lambda_i)\right)_{\text{m.o.}} - \left(\sum_{i=1}^{n}(X_i \times \Lambda_i)\right)_{\text{a.o.}} + X_{Al_2O_3} \times \Lambda_{Al_2O_3} \quad (7)$$

3. Research results

In the steel-pouring ladle of the ladle–furnace unit (LFU) free-running highly basic slags with the following chemical composition are built up (Table 1):

Table 1. Composition of liquid state of slag of the ladle-furnace unit.

Values	Slag composition, % wt.					
	CaO	SiO_2	Al_2O_3	MgO	FeO	MnO
Range	45.0–61.9	10.0–30.2	1.8–29.6	2.1–9.8	< 1.0	< 1.0
Medium	54.0	22.1	13.5	7.3	–	–

To determine the influence of amphoteric oxide Al_2O_3on the final value of optical basicity (Λ), calculated by the formula (7), the melting parameters fixed on the ladle furnace unit were analyzed. The array of data was calculated for each melting: sulfur partition coefficient between metal and slag ($L_{S\ PRACT.}$); coefficient of activity of sulfur in the metal ($\gamma_{[S]}$); oxidation potential of medium $p_{\{O_2\}}^{0.5}$; reaction equilibrium constant of partition of sulfur between metal and slag ($K_{[S]}$); sulfide capacity of slag (C_S). Moreover, calculation of sulfide capacity was performed in compliance with the conditions of $L_{S\ PRACT.} = L_{S\ THEORET.}$, through variation of the optical basicity of oxide Al_2O_3 ($\Lambda_{Al_2O_3}$). Array of input data and results of calculation are provided in Table 2, influence of content of Al_2O_3 in the homogeneous constituent of slag on the optical basicity is shown in Figure 2.

Table 2. Indicators of analyzed processes of melting performed by the ladle-furnace unit and results of the calculation.

T,°C	L_S		C_s $\times 10^3$	$\gamma_{[S]}$	K_s	$P_{O_2} \times 10^{-8}$	$\Lambda_{Al_2O_3}$	Composition, % wt.			
	actual	calcul-ated						CaO	Al_2O_3	SiO_2	MgO
1557	11.11	11.88	0.10	1.05	390.73	2.27	0.50	50.34	14.16	22.44	11.2
1553	16.33	18.80	0.13	1.06	394.80	1.80	0.47	51.45	12.74	21.63	8.75
1589	28.04	28.72	0.37	1.06	360.20	3.75	0.44	53.95	15.98	19.80	9.25
1574	26.32	34.89	0.36	1.06	374.07	2.89	0.38	55.15	17.91	19.05	6.84
1576	27.53	32.82	0.32	1.02	372.17	2.66	0.34	54.04	16.15	19.06	8.58
1557	25.23	29.49	0.21	1.06	390.73	1.96	0.34	52.63	17.62	19.11	8.90
1583	40.39	40.52	0.43	1.03	365.65	3.01	0.33	55.68	19.38	18.02	6.38
1591	12.00	13.15	0.28	1.01	358.40	5.91	0.33	53.65	17.22	19.20	8.03
1557	23.31	27.34	0.23	1.05	390.73	2.27	0.31	52.52	18.77	18.35	8.86
1582	30.78	31.65	0.52	1.17	366.58	5.28	0.20	54.59	24.00	14.66	5.77
1558	20.05	20.18	0.16	1.06	389.72	2.11	0.20	51.91	19.04	17.98	8.93
1581	19.07	20.20	0.44	1.17	367.50	6.94	0.16	54.09	24.61	14.28	6.04

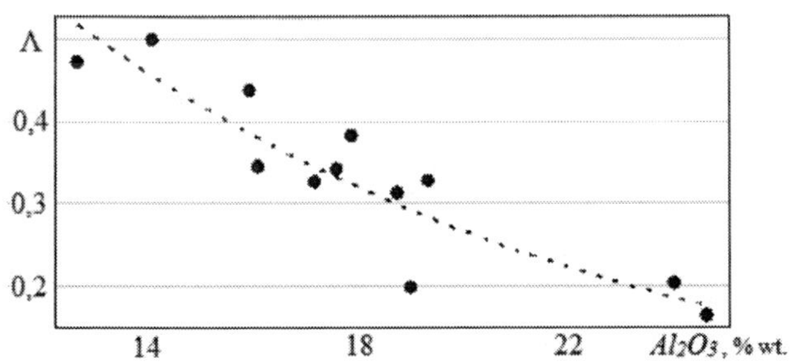

Figure 2. Dependence of optical basicity on the content of Al_2O_3 in the homogeneous constituent of slag.

Presented data show that with an increase in the content of Al_2O_3 in slag, the optical basicity of this oxide decreases. It can be determined by the following expression:

IOP Conf. Series: Materials Science and Engineering **966** (2020) 012068 doi:10.1088/1757-899X/966/1/012068

$$\Lambda_{Al_2O_3} = 1.65 \times e^{-0.0908 \times X} \quad (8)$$

where $\Lambda_{Al_2O_3}$ – optical basicity of oxide Al_2O_3; X – content of oxide in homogeneous constituent of metallurgical slag; e – base of the natural logarithm.

Therefore, optical basicity for heterogeneous slags built up using steel-pouring ladle of the ladle–furnace unit may be determined by the following expression:

$$\Lambda = (\textstyle\sum_{i=1}^{n}(X_i \times \Lambda_i))_{\text{m.o.}} - (\sum_{i=1}^{n}(X_i \times \Lambda_i))_{\text{a.o.}} + 1.65 \times e^{-0.0908 \times X} \quad (9)$$

where X_i – equivalent fraction of anions introduced by this component; Λ_i – optical basicity of system component; X – content of oxide in homogeneous constituent of metallurgical slag.

The optical basicity level determines sulfide capacity of slag, sulfur partition coefficient between metal and slag and, as a result, content of sulfur in metal.

4. Conclusion

Thus, metal desulfurization in ladle steel treatment units is achieved, first of all, due to the transition of sulfur to the slag phase. Coefficient of partition of sulfur between metal and slag is affected by sulfide capacity of slag, coefficient of activity of sulfur in the metal, oxidation potential of medium and reaction equilibrium constant of partition of sulfur between metal and slag. Temperature of liquid state in the ladle has significant impact on sulfide capacity of slag. Proposed calculation procedure based on provisions of ionic structure of slag allows to evaluate concentration of sulfur in steel on the basis of determination of coefficient of its partition between metal and slag. Optical basicity is suggested as criterion of refining property of the slag, it has been shown that the amphoteric oxide Al_2O_3 is essential to calculation of this indicator.

References

[1] Bigeev A M, Bigeew V A 2000 *Metallurgiya stali. Teoriya i tekhnologiya plavki stali* (Magnitogorsk: MGTU) p 544 [in Russian].

[2] Fandrich R, Lüngen H and Wuppermann C 2008 Actual review on secondary metallurgys *Revue de Metallurgie.Cahiers D'Informations Techniques* **105 (7-8)** pp 364–74

[3] Ushakov A N, Bigeev V A, Stolyarov A N and Potapova M V 2019 Technology for production of pipeline ultra low sulfur steel *Chernye Metally* **12** pp 26–31

[4] Fandrich R, Lungen H B and Wuppermann C 2008 Secondary metallurgy - State of the art and research trends in Germany *Stahl und Eisen* **128 (2)** pp 45–53

[5] Ushakov A N, Bigeev V A, Stolyarov A N and Potapova M V 2018 Ladle Desulfurization of Converter Low-Sulfur Pipe Steel *Metallurgist* **62 (7-8)** pp 667–73

[6] Ushakov A N, Avramenko V A, Bigeev V A, Stolyarov A N and Potapova M V 2018 Manufacture of Low-Sulfur Pipe Steel with Ladle Desulfurization of Cast Iron *Metallurgist* **61 (11-12)** pp 967–70

[7] Turkdogan E T 1983 Ladle deoxidation, urisation and inclusions in steel - 1. Fundamentals *Archiv fur das Eisenhuttenwesen* **54 (1)** pp 1–10

[8] Pluschkell W 1990 Metallurgical reaction techniques for adjusting very low contents of C, P, S and N in steel *Stahl und Eisen* **110 (5)** pp 61–70

[9] Jonsson L, Sichen D and Jönsson P 1998 A New Approach to Model Sulphur Refining in a Gas-stirred Ladle - A Coupled CFD and Thermodynamic Model *ISIJ International* **38 (3)** pp 260–7

[10] Cao Q, Pitts A and Nastac L 2018 Numerical modelling of fluid flow and desulphurisation kinetics in an argon-stirred ladle furnace *Ironmaking and Steelmaking* **45 (3)** pp 280–7

[11] Tursunov N K, Semin A E and Sanokulov E A 2016 Study of desulfurization process of structural steel using solid slag mixtures and rare earth metals *Chernye Metally* **4** pp 32–7

[12] Tursunov N K, Semin A E and Kotelnikov G I 2017 Kinetic features of desulphurization process during steel melting in induction crucible furnace *Chernye Metally* **5** pp 23–9

[13] Kapustin E A and Kharlashin P S 2001 Theoretical principles of metallurgical technology *Steelin Translation* **31 (12)** pp 15–8
[14] Konoplya V G and Kharlashin P S 2002 Thermodynamic and kinetic analysis of mass transfer of sulfur in slag-metal melts during application of low-sulfur cast iron *Izvestiya Ferrous Metallurgy* **4** pp 56–9
[15] Kharlashin P S, Kolomiytseva Y S, Grigoryeva M A and Baklanskiy V M 2010 Kinetics of desulfurization and resulfurization when low-sulfur steel making at the stage of oxidizing refining *Metallurgical and Mining Industry* **2 (4)** pp 267–70
[16] Novikov V K 1989 Development of a polymeric model of molten silicates *Melts Moscow* **1 (6)** pp 501–12
[17] Novikov V K and Maifat M V 1989 Application of polymeric model to the calculation of surface tension in multicomponent silicate melts *Melts Moscow* **2 (3)** pp 218–21
[18] Novikov V K, Spiridonov M A and Zinov'eva IS 1998 Thermochemical basicity indexes of oxides *Russian Journal of Physical Chemistry A* **72 (2)** pp 173–6
[19] Novikov V K, Spiridonov M A and Sangalova I S 2008 Chemical thermodynamics and elements of structure in oxide melts *Journal of Physics: Conference Series* **98 (1)** 012020
[20] Burmasov S P, Gudov A G, Yaroshenko Y G, Meling V V and Dresvyankina L E 2015 Mass transfer in the ladle refining of steel with gas mixing *Steel in Translation* **45 (9)** pp 635–9
[21] Sommerville I D 1986 Measurement, prediction and use of capacities of metallurgical slags. *Scaninject IV: 4th Int. Conference on Injection Metallurgy* (Lulea, Sweden) pp 8.1-8.21
[22] Sommerville I D and Masson C R 1992 Group optical basicities of polymerized anions in slags *Metallurgical Transactions B* **23 (2)** pp 227–9

IOP Conf. Series: Materials Science and Engineering **966** (2020) 012069 doi:10.1088/1757-899X/966/1/012069

Research into in-service deterioration of ball-rolling rolls

V Ju Rubcov[1, 2, 3], O I Shevchenko[1], A S Spirina[2] and N A Pankova[2]

[1] Ural Federal University named after the First President of Russia B N Yeltsin, Nizhniy Tagil Technological Institute, 59, Krasnogvardeyskaya str., Nizhniy Tagil, 622000, Russia

[2] EVRAZ NTMK, ul. Metallurgov, 1, Nizhnij Tagil, Sverdlovskaya oblast, 622025, Russia

E-mail: [3] uriylot@mail.ru

Abstract. This work features research findings on causes of deterioration occurring in ball-rolling rolls. The causes of such deterioration may originate not only in the excessive loads application and the use of bottleneck operating parameters but also in the specific process being implemented and the equipment used to heat-treat the rolling rolls. The deterioration occurring in roll flanges having a relatively small thickness was used to demonstrate that, to prevent the deterioration of ball-rolling roll flanges, the mill design must accommodate the minimum allowable flange widths obtained as a function of maximum allowable loads and the roll pass calculations. The next example of roll deterioration involved the use of a gas furnace to implement the heating stage of the quench hardening process for rolls made of 35KhGSA steel, which, as a result, developed roll deterioration. The use of rolls of the same steel grade, which were heat-treated in an arc furnace prior to being quenched, has been shown to deliver long-lasting positive results evidenced by many years of their operation. Rolls made of 5KhNM steel have also been shown to respond well to the heat treatment preceding the quenching process, but this time, the treatment was implemented in a gas furnace.

1. Introduction

The one component of a ball-rolling mill which is most susceptible to deterioration is the forming tool comprising: rolls and guides. Currently, the operating life cycle prescribed for a set of rolls before they require remachining (calculated in terms of cumulative output) is set at 600–2000 t depending on the section design being obtained and the roll pass design being used. The roll operates via flange penetration into the material, so it is the flange surface that is most susceptible to wear. However, roll failures may arise not only due to reaching of critical wear [1], but also in the process of deterioration of roll elements. The causes of such deterioration may originate in the roll material and roll pass design being used as well as in the heat treatment and operation conditions [2]. Deterioration-induced roll failures as well as early wear and tear thereof are among the factors resulting in unscheduled downtimes and forced roll changes [3], which is what makes research into these criteria a crucial task. This article discusses the causes of deteriorations processes that occur in ball-rolling rolls and actions to be taken to prevent them.

The rarest factor which, however, can still be encountered among the causes of deterioration of ball-rolling rolls is the use of bottleneck roll pass design configurations. The main challenge facing the designers of a ball-rolling mill is to ensure that their design meets the volume-per-time constancy principle, i. e. that, as the metal travels between the rolls, the volume constrained by the roll flanges (the neck portions included) is maintained constant at all times. According to the theory of A I Tselikov [4], rolls may be safely designed with a roll pass overfill allowance not exceeding 4 %. If the roll passes are underfilled, the metal will not come in contact with the flange, which can result in the rolled product passing over onto the bar; therefore, most roll pass configurations are initially designed with an insignificant overfill allowance. Exceeding of acceptable overfill leads to the increased roll wear, as well as to obtaining substandard balls up to occurrence of Mannesmann effect.

Another important factor to consider where deterioration is possible is the minimum thickness of the flange that projects outwardly from the roll surface and thus takes maximum loads. Scientists form Lublin University suggested several roll models where the flange thickness can increase, decrease or remain almost constant as the metal travels through the pass [6]. We also suggested our own roll design concepts featuring constant as well as variable flange width solutions equally compliant with the volume constancy requirement [7]. The most effective separation of the neck occurs when the flange width at the crossing points is minimal. However, it leads to the intensive deterioration within this zone and to further formation of strings in the rolling process. Summarizing the results obtained in research works [3, 6, 8–11], it can be confirmed that roll pass designs must be configured in such a way that the flange thickness would increase in particularly intensive zones, such as the neck separation zone. The differences between roll diameters and the effects of alternating forces that need to emerge to implement the ball twisting and neck separation process are also among the causes of the deterioration occurring in this zone. In addition, in the crossing zone, flange penetration into the metal is achieved by means of the cutoff roll only.

2. Research results

With a streamlined roll pass design solution, the overall wear and tear affecting the flanges will be characterized by the abrasion wear occurring along the working faces of the flanges. When, in particular, the choice of roll material was made in favor of 5KhNM steel and the hardness of the quench-hardened layer was relatively low (38–42 HRC), flange deterioration reached an extent where relatively large portions of it would break off.

Figure 1. Roll flange spalling occurring in the neck cutoff zone.

To determine the cause of such deterioration, a 30 × 250 mm sample of the deteriorated flange material that had broken off (Figure 1) from the roll (Figure 2) was taken to be put through a metallographic examination. The hardness of the sample after measurement was 31.5–38.5 HRC which is below the predefined one. The evaluation of microstructure was performed by means of Leica MEF4A optical microscope on 2 sections before and after etching process in 3% alcohol solution of hydrogen nitrate with x50–500 magnification.

The examination results showed that the base metal used in the roll flange had uniform microstructure and could be identified as tempered martensite (Figure 2, b).

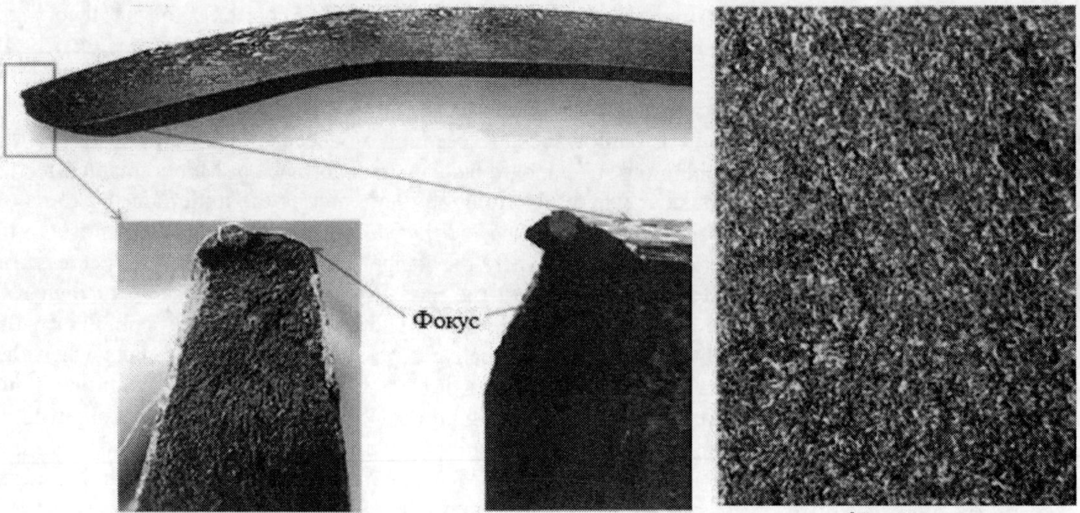

a) ×50 b) ×500

Figure 2. Microstructure of the base metal of the flange.

The metal of the rolling work material was welded onto the side surface of the flange and also on the top of it (Figure 3, a.) followed by the separation of a portion of the roll surface, a change in the roll metal structure and the emergence of microcracks (Figure 3, b). White layer of the martensite with the depth up to 0.35 mm was found by the microchemical analysis at the site of focus (Figure 3, c). Nonmetallics, liquation of sulfur and phosphorus at the site of deterioration are absent. In view of the foregoing, it was the adhesion of the hot metal being rolled that caused the deterioration of the flange, as evidenced by the metallic pickup on the side surface and the top of the flange.

In this case insignificant load increase led to the flange overheating because of its low thickness (9.5 mm) and to further deterioration. As a result, the flange thickness is important to the design of ball-rolling rolls.

Having excluded any causes attributable to roll pass design and mill setup configurations, other possible causes of critical deteriorations may be the heat treatment conditions used for the steel grades in question. Application of 35KhGSA steel grade in the conditions of JSC EVRAZ NTMK can be a remarkable example. Bodies of ball-rolling rolls made of the mentioned steel have been regularly used during several decades under conditions of the ball section of a Large-struction Rolling Shop where heating up for heat treatment has been carried out using the arc furnace. On a trial basis, the above-mentioned grade of steel was used at the site of ball section of Rail and Beam Shop instead of the 5KhNM steel used there. Heat treatment conditions were identical to the ones for a Large-struction Rolling Shop, but due to unit feasibility the heating was carried out in the gas furnace. The complete deterioration of the roll occurred in the process of the first charging of the rolls and trial rolling. Moreover, the cracks which were the cause of the deterioration, occurred not only along the stress concentrators (keyslots, etc.) but also in parallel to the axis along the whole perimeter (Figure 4). Fractures formed on both sides of punctured part of the roll are fragile, with focuses extending from the surface of the roll-pass. Cross cracks were found on the surface of the flange roll-pass.

Table 1. Chemical composition of the roll sample.

Name	Element content in %							
	C	Mn	Si	Cr	Ni	Mo	S	P
Roller sample	0.347	0.93	1.25	1.33	0.042	< 0.005	0.0045	0.011
GOST 4543-2016	0.32–0.39	0.8–1.1	1.1–1.4	1.1–1.4	< 0.3	< 0.3	< 0.025	< 0.025

IOP Conf. Series: Materials Science and Engineering **966** (2020) 012069 doi:10.1088/1757-899X/966/1/012069

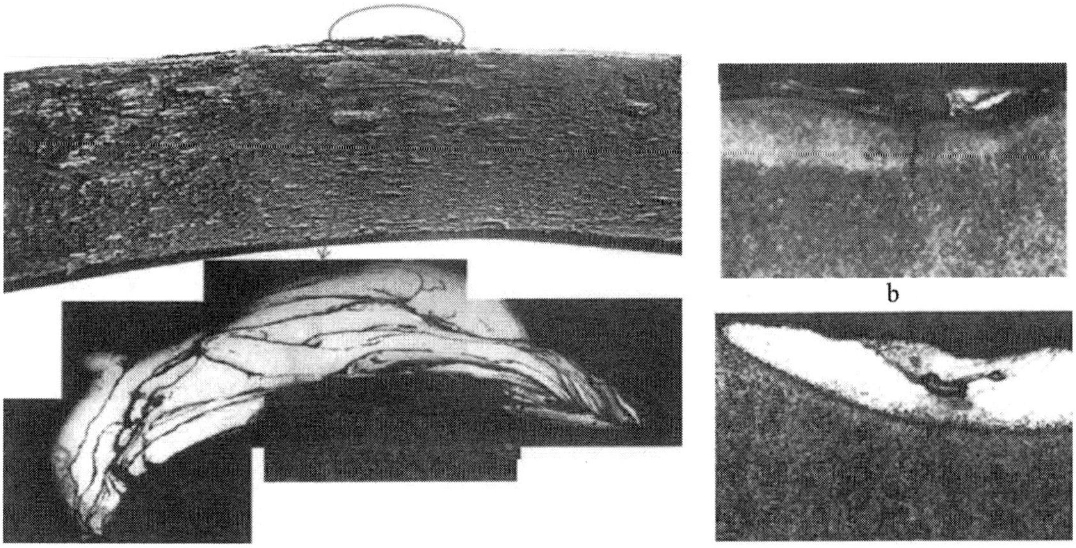

a c

Figure 3. Microstructure of the flange metal at the side of focus of deterioration:
a – welded metal of the rolling work material on the top of the flange (x50 magnification);
b – microcracks and changing of the structure (x100 magnification);
c – white layer of the martensite (x50 magnification).

Figure 4. Rolling roll fracture mode.

Chemical composition of the roll metal according to the chemical composition analysis meets requirements prescribed for 35KhGSA steel according to GOST 4543–2016 [12] (Table 1).

Macroanalysis of the cross section (Figure 5, a) performed by the deep etching method in hot 50% solution of hydrochloric acid showed that the following was found in the deterioration area:

–hardened case extended along the depression of roll-pass to the depth of 10 mm,

–through hardening of the flanges of the roll-pass,

–freckle-type segregation, 3 points according to GOST 10243–75 [13].

IOP Conf. Series: Materials Science and Engineering **966** (2020) 012069 doi:10.1088/1757-899X/966/1/012069

Hardness distribution over the roll section is uneven (Figure 5, b) and is from 48 HRC with gradual decreasing to 45 HRC on the surface. Hardness of the base metal measured by Brinell scale was 255 HB.

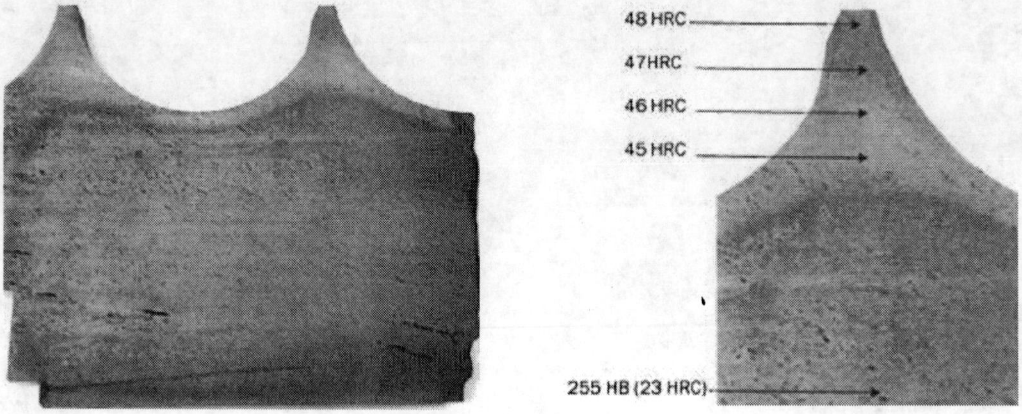

a b

Figure 5. Results of sample macroanalysis.

Microanalysis of the roll identified the presence of inadmissible decarburized layer with the depth up to 0.3 mm on the surface of the roll-pass. 15 mm deep thermal cracks and metal chipping along them were detected on one of the lateral sides (Figure 6, a).

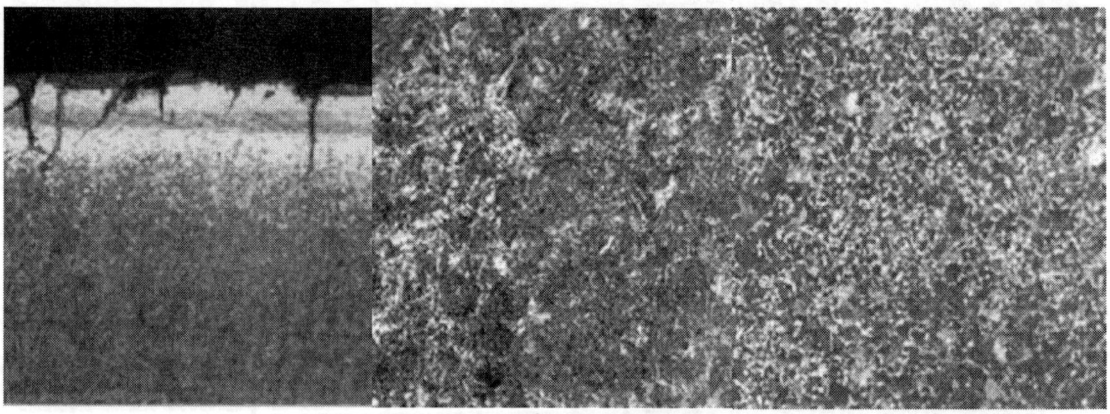

a b c

Figure 6. Sample microanalysis results:
a – hardened case x100, b – hardened case x500, c – base metal x100.

Microstructure of the hardened case extending through the roll-passes represents tempered martensite with sites of retained austenite (Figure 6, b). Structure of the base metal is hypopearlitic uneven-grained with No. 6–8 grain size according to GOST 5639–82 [14] (Figure 6, c).

As a result of above mentioned, it is possible to suggest that observation of time-temperature modes during heat treatment of ball-rolling rolls is not the only factor that has impact on their further structure including operability, but it is also equipment where heat treatment is performed. Under condition that 5KhNM steel used for ball-rolling rolls of Rail and Beam Shop is not subject to decarburization in the gas furnace, for 35KhGSA steel it was the key parameter that led to the rolls deterioration.

3. Conclusion

1. To avoid deterioration of ball-rolling rolls and their elements while designing apart from the type of the applied roll-pass it is necessary to count on minimum permissible flange width depending on the maximum allowable loads taken.

2. Rollers roll-pass must be calculated according to the condition of minimum possible overfill which will guarantee absence of circumferential loads if the metal is between neighboring roll-passes in the rolling process, and it will increase the rolls service life.

3. Except the structure of the ball-rolling rolls, their manufacturing technique is also important, as well as observation of heat treatment conditions and selection of equipment where it will be manufactured. For example, for a roll made of 5KhNM steel material, usage of gas furnace for heating for hardening is permissible but is impermissible for a roll made of 35KhGSA steel, as it was shown in this article.

References

[1] Chyla P, Pater Z, Tomczak P 2016 Numerical analysis of a rolling process for producing steel balls using helical rolls *Arch. Metall. Mater.* **61 2** pp 485–92

[2] Lam M M, Serov A I, Ternavskiy A A, Bazarova G S and Smyrnov Y N 2015 Heat treatment of rollers in the ShPS 30-80 ball-rolling mill at JSC DMPZ *Steel in Translation* **45 10** pp 784–8

[3] Rubtsov V Yu and Shevchenko O I 2019 *Harakter intensivnosti iznosa sharoprokatnyh valkov i kriterii otbrakovki* (Cause of wear rate of ball-rolling rolls and reject criteria) *Proceedings of the XII Congress of rollinger* **1** pp 187–91 [In Russian]

[4] Tselikov A I and Barbarich M V 1971 *Spetsial'nye prokatnye stany (Special Rolling Mills)* (Moscow: Metallurgiya) [In Russian]

[5] Kozhevnikova G V 2009 Uslovija ustojchivogo protekanija poperechnoj i poperechno-klinovoj prokatok (Conditions of stable flow of cross rolling and cross-wedge rolling) *Vestnik Belorussko-Rossijskogo universiteta - Bulletin of the Belarusian-Russian University* **1 (22)** pp 44–53 [In Russian]

[6] Pater Zb, Tomczak J, Bartnicki J and Bulzak T 2018 Thermomechanical Analysis of a Helical-Wedge Rolling Process for Producing Balls *Metals* **8 862** pp 1–14

[7] Rubtsov V and Shevchenko O 2018 Calibration of ball-rolling rollers with continuously variable pitch *Ferrous Metallurgy. Bulletin of Scientific, Technical and Economic Information* **8** pp 58–63

[8] Protasyev V B 2020 Design of Rolls For Hot Cross-helical Rolling of Billets Without Defects in The Axial Zone *Chernye Metally* **3** pp 42–7

[9] Yang S C and Chen C K 2001 The surface geometry of rollers with skew rolling of steel balls *Proc. Inst. Mech. Eng. C J. Mech. Eng. Sci.* **215** pp 523–32

[10] Shi X, Wang B and Li Z 2010 Optimization of pass parameters on two-grooves skew rolling for Al ball *Key Eng. Mater.* **419** pp 313–6

[11] Cao Q, Hua L and Qian D S 2015 Numerical simulation on hot helical rolling forming process of blank for bearing steel balls *Bearing* **1** pp 16–21

[12] GOST 4543-2016 *Metalloprodukcija iz konstrukcionnoj legirovannoj stali [Structural alloy steel products. Specifications]* (Moscow: Standartinform Publ.) p 53 [In Russian]

[13] GOST 10243-75 1985 *Stal'. Metody ispytanij i ocenki makrostruktury [Steel. Methods of test and estimation of macrostructure]* (Moscow: Izdatel'stvo standartov) p 37 [In Russian]

[14] GOST 5639-82 2003 *Stali i splavy. Metody vyjavlenija i opredelenija velichiny zerna. [Steels and alloys. Methods for detection and determination of grain size]* (Moscow: Izdatel'stvo standartov) p 38 [In Russian]

IOP Conf. Series: Materials Science and Engineering **966** (2020) 012070 doi:10.1088/1757-899X/966/1/012070

Development of new roll pass designs for ball-rolling rolls with continuously variable parameters

V Ju Rubcov[1, 2, 3] and O I Shevchenko[1]

[1] Ural Federal University named after the First President of Russia B N Yeltsin, Nizhniy Tagil Technological Institute, 59, Krasnogvardeyskaya str., Nizhniy Tagil, 622000, Russia
[2] EVRAZ NTMK, ul. Metallurgov, 1, Nizhnij Tagil, Sverdlovskaya oblast, 622025, Russia

E-mail: [3] uriylot@mail.ru

Abstract. This article outlines modern approaches to the creation of roll passes for helical rolling mills for the production of grinding balls. The key feature is the roll design with continually varying parameters of its elements. Such elements include: flange width varying along the length of the roll; flange height, with a nonlinear function of flange rising; pitch of the helical groove which maintains constancy of the volume of the metal being deformed. Particular attention is paid to a new conceptual pass design for ball-rolling rolls with a variable groove depth, the prospects of which are connected with uniformity of reduction and the production of balls not only with the correct geometry, but also with a more homogeneous structure. The article shows functions of a continuously varying pitch of a helical groove of rolls with a constant and variable flange width, as well as with a variable depth of the groove. A practical test was conducted on ball-rolling rolls for production of balls with a nominal diameter 120 mm with continuously varying pitch. Experience has shown positive results both in reducing loads and in increasing the dimensional accuracy.

1. Introduction

One of the well-known and productive methods for the production of grinding balls is helical rolling [1–2]. The deformation of the metal is carried out on a ball rolling mill in rolls with screw passes.

The workpiece is fed between rolls that rotate in the same direction and are located at certain angles relative to each other. After passing through the roll passes, the workpiece is gradually crimped and takes the form of balls, which are separated from each other by cutting the bridge with the roll flange [1] at the last moment. Pass design of ball-rolling rolls, including the design of profile elements, should allow implementation of an efficient technology and, provide the possibility of manufacturing rolling tools on existing metal-cutting equipment.

Initially, from the 50s of the 20th century, cutting screw passes in rolls of ball-rolling mills caused certain difficulties since they were produced on screw-cutting lathes.

At the tie, manufacturing of roll passes required adjusting of the calculated screw pitch according to the capabilities of the machines available. Usually, the screw parameters were provided by selection of pairs of interchangeable gears and changes could only be discrete. That is, the finished roll had a discretely varying pitch of the helical pass [1–4]. The advantage of such passes is the relative simplicity and accessibility of metal-cutting equipment for the production of rolls. The disadvantages are the uneven reduction during the transition from one step to another, causing peaks in the load on the rolling drive, as well as long manufacturing time of rolls of this type due to the constant set-up of the machine feed change gear train and the need for its precise tuning to the required step.

For the first time, a continuously changing nonlinear parameter in the form of a step was used in practice and described in [5–6]. This cutting method was provided through the use of semi-automatic copy machines. At present, the production of ball-rolling rolls is carried out at 4 or 5 axis machining centers with independent drives, which allow to obtain an open profile of any complexity formed by curves of the 1st and 2nd orders [7] using the 3-D model.

The combination of rational pass design and dynamic rolling mode improvement, as was shown by us in [8], unambiguously lead to an increase in the service life of the rolling tool.

In connection with the above, development of roll design with continuously changing parameters of the profile elements is an urgent task, as this will provide less wear on the working surfaces, reduce dynamic load peaks, uniform reduction will lead to a more homogeneous grain structure and a uniform temperature distribution on the surface of balls subjected to further heat treatment.

2. Variable parameters used in roll passes of ball-rolling rolls

Among all the variable parameters of profile elements of ball-rolling rolls, the main ones can be distinguished: the pitch of the helical line T_i, the height of the flanges H_i, the width of the flanges B_i (Figure 1).

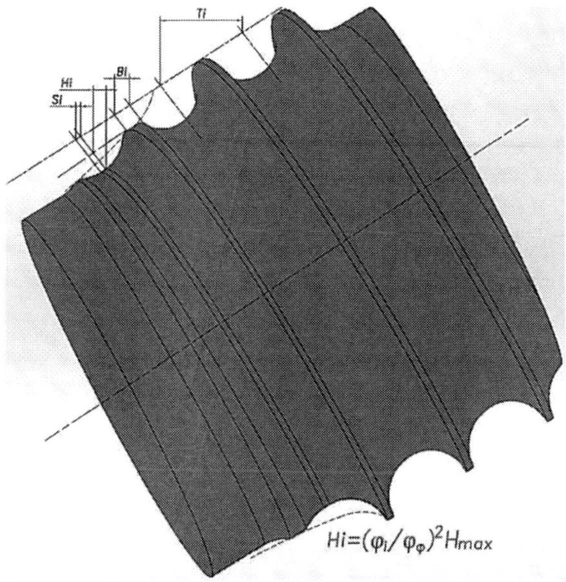

Figure 1. Ball rolling model with parabolic flange rising rule.
T_i – the pitch of the helix;
B_i – the flange width;
H_i – the flange height;
$H_{\max}$ – the maximum flange height (at the end of the forming section);
S_i – the magnitude of pass gauge expansion;
φ_i – the position angle along the helical line from the start of the flange;
φ_Φ – the angle at which the forming section ends (at which the flange has a maximum height $H_{\max}$).

The influence of the pitch varying on the rolling process and the quality of the products were analyzed in the following works [9–14]. In their works, authors of Lublin University of Technology [15] present the results of experimental studies using rolls with different flange widths. Three possible structural cases were studied where the flange width changes in the direction of the workpiece movement along the rolls: 1 – decreases, 2 – increases 3 –remain approximately constant. In the first case, wear of rolls in the workpiece entry area decreases, in the second case wear decreases at the crossing point of the two rolls, however, in both cases there is a significant difference in reductions at the surface of the ball, which leads to higher wear intensity of the rolls. The third case with approximately the same flange width along the entire length of the roll is recognized as the most acceptable. Judging by the data of [1, 4, 6], the width of the flange depends on the helical groove pitch T_i and the pass gauge spacing S_i (Figure 1). As a result, a design can be considered rational where the flange first slightly thins out, and then increases as the workpiece moves along the rolls, creating favorable conditions for separating the tail from the finished ball.

The rise of flanges is characterized by an increase in their height in the initial, forming, section of the roll. This is necessary to ensure gradual cutting into the workpiece. The function that determines the increase in height - the so-called rule of flange rising the can be linear, parabolic or hyperbolic [7]. At the moment, the linear rule has received the widest application. If a nonlinear rule is applied, the flange width will 'float' unevenly (Figure 1), which will lead to uneven metal adherence to the roll and instability of the rolling process.

3. New pass designs with continuously varying parameter

As noted above, the pitch of the helix must continuously vary to maintain a constant volume of metal between the rolls, which is limited by adjacent flanges, including the volume of metal in the tail. In our previously published works [7, 16], the pitch variance function Δt was derived for rolls with a continuously varying pitch and a constant flange width.

$$\Delta t = T_{mn} - \int_{R}^{0} \frac{\pi b(R^2 - x^2) + x^2 (R - \frac{x}{3})}{\pi R^2} dx \qquad (1)$$

where b_φ (b) is the flange width (tail length); R is the radius of the sphere of the conditional pass (the depth of the roll groove); T_{mn} is the main pass pitch (set during design taking into account the flange width).

For ball-rolling rolls with a linearly increasing flange width, the pitch variance function of a screw pass has the form equation (2).

$$\Delta t = T_{mn} - \int_{R}^{0} \frac{\pi b_\varphi (1 - x) + x^2 (R - \frac{x}{3})}{\pi R^2} dx \qquad (2)$$

Application of passes with a continuously varying pitch, calculated according to equations (1 and 2), eliminates jumps in loads during the rolling process.

A roll pass design with a variable depth of the groove has been developed to improve reduction uniformity and temperature distribution, as well as the formation of a more uniform structure of the balls.

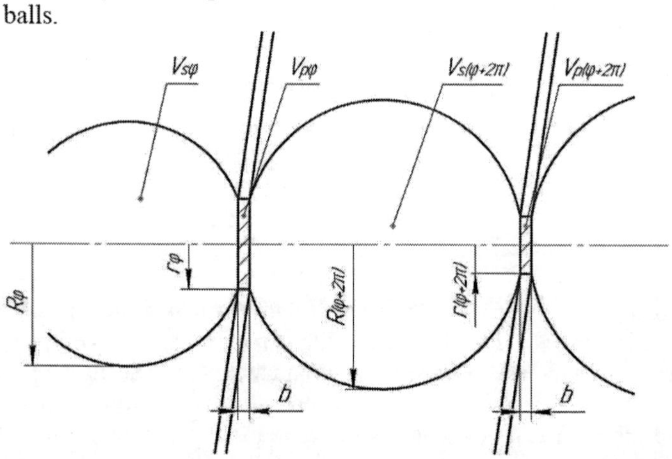

Figure 2. Section of a workpiece during rolling with a variable groove depth.

Given the strength conditions, the width of the flange was set, which is assumed to be constant. According to Figure 2, applying the partition into the basic geometric figures, we find the change in the radius ΔR which forms the ball.

From the conditions of constancy of metal volume passing through a screw pass, we compose equation (3)

$$V_{p\varphi} + V_{s\varphi} - V_{p(\varphi+2\pi)} - V_{s(\varphi+2\pi)} = 0 \tag{3}$$

where $V_{p\varphi}$ is the volume of the truncated tail of the conditional pass, $V_{s\varphi}$ is the volume of the truncated ball of the conditional pass, $V_{p(\varphi+2\pi)}$ is the volume of the truncated tail in the next turn of the conditional pass, $V_{s(\varphi+2\pi)}$ is the volume of the truncated ball in the next turn of the conditional pass.

When substituting the values of volumes of simple shapes that make up this system, it turns out:

$$\pi r_\varphi^2 b + [\frac{3}{4}\pi R_\varphi^3 - r_\varphi^2 (R_\varphi - \frac{r_\varphi}{3})] - \pi r_{(\varphi+2\pi)}^2 b - [\frac{3}{4}\pi R_{(\varphi+2\pi)}^3 - r_{(\varphi+2\pi)}^2 (R_{(\varphi+2\pi)} - \frac{r_{(\varphi+2\pi)}}{3}) - r_\varphi^2 (R_\varphi - \frac{r_\varphi}{3})] = 0 \tag{4}$$

where r_φ – is the tail radius of the conditional pass.

To derive the function of groove depth variance, we take $r_{(\varphi+2\pi)} = 0$, because conditionally, the flange is reduced to contact. By substituting this value in equation (2), we get:

$$\pi r_\varphi^2 b + \frac{3}{4}\pi R_\varphi^3 - \frac{3}{4}\pi R_{(\varphi+2\pi)}^3 = 0 \tag{5}$$

The change in groove depth along the helical line will be determined by the difference in the grooves of adjacent turns: $\Delta R = R_{(\varphi+2\pi)} - R_\varphi$, substituting this value in equation 5 and transforming, we obtain:

$$\Delta R = \sqrt[3]{\frac{4}{3} r_\varphi^2 b + R_\varphi^3} - R_\varphi \tag{6}$$

Because as the groove depth increases, the radius of the pass grows in all directions, the change in the radius of the pass will be equal to the change in the pitch of the helical line: $\Delta R = \Delta T$.

Given a variable value: $0 < r_\varphi < R$, where, r_φ is the radius of the previous position, and $R = R_\varphi$ is the radius of the workpiece, the pitch variance function will have the form:

$$\Delta t = \int_0^R (\sqrt[3]{\frac{4}{3} x^2 b + R^3} - R) dx \tag{7}$$

Thus, expressions (1), (2), (7) can be used to create models of rolls necessary for their manufacture.

4. Industrial trials of new pass designs

As noted above, the pitch as a production experiment, ball-rolling rolls were calibrated with a continuously varying pitch to obtain balls with a diameter of 120 mm.

The width of the flanges along the entire length of the roll is assumed constant and, based on the strength conditions, was 11 mm, in this regard, the calculation was carried out according to equation (1).

By integrating expression (1), 5 points are defined for tracing a curve to form the helical surface of the roll. The helical line of the roll with a continuously varying pitch is a kind of a second-order spatial curve, therefore, it is determined by its five points, and no four of them should lie on one straight line [17]. Based on these 5 points, a curve is built that is a helix with a continuously varying pitch; on the basis of this curve, a 3-D roll model is built. A set of rolls was made using this model on a 5-axis machining center (Figure 3). Results of test rolling showed that the load when using rolls with a continuously varying pitch during rolling decreased by 25–35%, in contrast to rolls with a discretely varying pitch.

Figure 3. Roll with constant flange width and continuously variable pitch.

Figure 4. Ball produced during the trial rolling.

The dimensions of the ball resulting from new rolls meets the requirements of Russian State Standard GOST 7524–2015 [18], and a significantly lower error is ensured. According to [18], the dimensions should be in the range: 125±5 mm. Actual results were 125±0.4 mm. An insignificant girdle up to 4 mm wide (Figure 4) is not a defect, but is required to prevent overflow of the pass, in the case of rolling a workpiece with a diameter at the upper tolerance limit. The produced batch of balls had no surface imperfections which could be rated as defective according to the standard.

References

[1] Tselikov AI and Barbarich MV 1971 *Spetsial'nye prokatnye stany (Special Rolling Mills)*(Moscow: Metallurgiya) [In Russian]

[2] Smirnov VS, Anisiforov VP, Vasil'chikov MV et al 1957 *Poperechnaya prokatka v mashinostroenii (Transverse Rolling in Manufacturing)* (Moscow: Mashgiz) [In Russian]

[3] Shor ER 1960 *Novye protsessy prokatki (New rolling processes)*(Moscow: Metallurgizdat) [In Russian]

[4] Peretyat'ko VN, Klimov AS and Filippova MV 2013 Roller grooving in ball-rolling mills part 1 *Steel in Transl.* vol **43** no **4** pp 168–70

[5] Granovskii SP 1980 *Novye protsessy i stany dlya prokatki izdelii v vintovykh kalibrakh (New Rolling Processes and Mills with Helical Grooves)* (Moscow: Metallurgiya) [In Russian]

[6] Kotenok VI 2005 *Development of theory of profile forming in helical roll passes and creation of high effective processes and equipment for rolling of machines parts* Extended Abstract of Doctoral (Eng.) Dissertation (Moscow)

[7] Rubtsov VYu and Shevchenko OI 2019 Production of ball-rolling rolls with a variable depth of chanel *Kalibrovochnoe byuro* no **14** pp 16–24 [In Russian]

[8] Rubtsov VYu, Shevchenko OI and Mironova MV 2018 Improvement of Rolling Dynamics to Increase Roller Life in Ball Rolling *Steel in Transl.*vol **48** no **12** pp 757–61

[9] Filippova MV, Temlyantsev MV, Peretyat'Ko VN and Prudkii EE 2017 Rolling of metal balls *Steel in Transl.*vol **47** pp 435–9

[10] Quanxian W, Qiping W and Jianming X 1995 Study on the method for groove design in the helical rolling of steel balls *J. Mater. Process. Technol.* **55** pp 340–4

[11] Yang SC and Chen CK 2001 The surface geometry of rollers with skew rolling of steel balls *Proc. Inst. Mech. Eng. C J. Mech. Eng. Sci.* **215** pp 523–32

[12] Shi X, Wang B and Li Z 2010 Optimization of pass parameters on two-grooves skew rolling for Al ball *Key Eng. Mater.* **419** pp 313–6

[13] Cao Q, Hua L and Qian DS 2015 Numerical simulation on hot helical rolling forming process of blank for bearing steel balls *Bearing* **1** pp 16–21

[14] Tomczak J,Pater Z and Bulzak T 2014 Designing of screw impressions in the helical rolling of balls *Archives of Civil and Mechanical Engineering* **141** pp 104–13

[15] Pater Zb, Tomczak J,Bartnicki Jand Bulzak T 2018 ThermomechanicalAnalysis of a Helical-Wedge Rolling Process for Producing Balls *Metals* **8** **862** pp 1–14

[16] Rubtsov V and Shevchenko O 2018 Calibration of ball-rolling rollers with continuously variable pitch *Ferrous Metallurgy. Bulletin of Scientific, Technical and Economic Information* **8** pp 58–63

[17] Korn G and Korn T 1978 *Xarakteristicheskaya kvadratichnaya forma i xarakteristicheskoe uravnenie. Spravochnik po matematike [Characteristic quadratic form and characteristic equation]* 4nd ed. (Moscow: Nauka) p 64 [In Russian]

[18] GOST 7524-2015 *Shary' melyushhie stal'ny'e dlya sharovy'x mel'nicz [Milling steel balls for ball mills]* (Moscow: Standartinform Publ.) p 8 [In Russian]

IOP Conf. Series: Materials Science and Engineering **966** (2020) 012071 doi:10.1088/1757-899X/966/1/012071

Ultrasound effect on electrospark cementation process

O I Shevchenko

Ural Federal University named after the First President of Russia B N Yeltsin, Nizhniy Tagil Technological Institute, 59, Krasnogvardeyskaya str., Nizhniy Tagil, 622000, Russia

E-mail: oleg.shevchenko@urfu.ru

Abstract. The process of carburizing is well-studied and is one of the traditional processes of the thermochemical treatment of steel. The process has disadvantages such as high energy consumption, long exposure time, complicated processing of large products and surfaces. Electrospark deposition with a carbon electrode is proposed as an alternative process. The carburizing extent and the working layer depth can be increased when the process is used in the ultrasonic vibration field. A resonant vibration introducing circuit with a standing wave generated in the specimen tested is implemented to identify the influence of specific ultrasonic wave areas on the hardened layer formed. The highest carbon saturation of the surface layer was found in the standing ultrasonic wave oscillation node area, where cyclic stretching and compression of the medium contribute to the excess phases release. The carbon in the martensite estimated by the *c*/*a* ratio was 0.78% by weight. The highest cementite layer depth and transition zone size are in the ultrasonic wave antinode area, where the highest dynamic medium particles displacement is observed under the influence of vibrations. It is shown that the base metal structure component dispersiveness is responsible for the increased hardened layer depth.

1. Introduction

Carburizing is commercially used to increase the hardness and wear resistance of the surface while retaining their core toughness [1]. Solid carburizing, liquid carburizing and gas carburizing [1 to 3] are conventional well-studied methods and, despite their wide-spread extent, have a number of big disadvantages: process duration, high temperature [4, 5], resulting in the austenite growth, and therefore additional heat treatment is needed [6 to 8]. Thermochemical treatment of large parts and forming tools [9] is also highly complicated in terms of technology.

Recent attempts to obtain a cementite layer on the surface of parts use a spark discharge [10 to 14]. They are based on the electrospark deposition process (EDP) (Figure 1) whereby the spark discharge destroys the anode and its erosion materials transfer to the cathode (a part under treatment) [15, 16].

The high discharge temperature causes the electrode material particles to melt and partially evaporate. The material vapors expand and remove from the surface the particles of the anode, which passes through the gas medium, penetrates the cathode and generates a hardened layer. The cathode zone under the spark discharge also heats up suddenly and cools downimmediately down due to the heat pickup. Thus, the surface is quenched increasing the hardness.

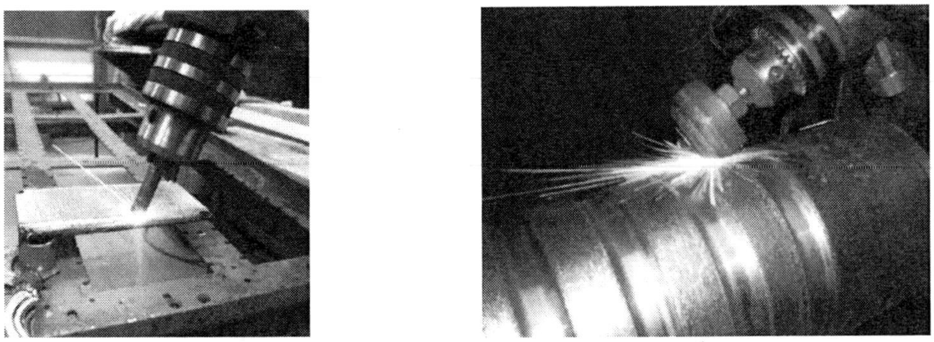

a b

Figure 1. Electrospark depositing of a flat part (a) and a cylindric part (b).

The of electrospark carburizing method significantly reduces the power capacity of the process and improves the productivity. Unlike the conventional process not long exposure at high temperatures gives no rise to the austenite growth. Although this method looks interesting, the cementite layer is not deep enough and low extent of cementation.

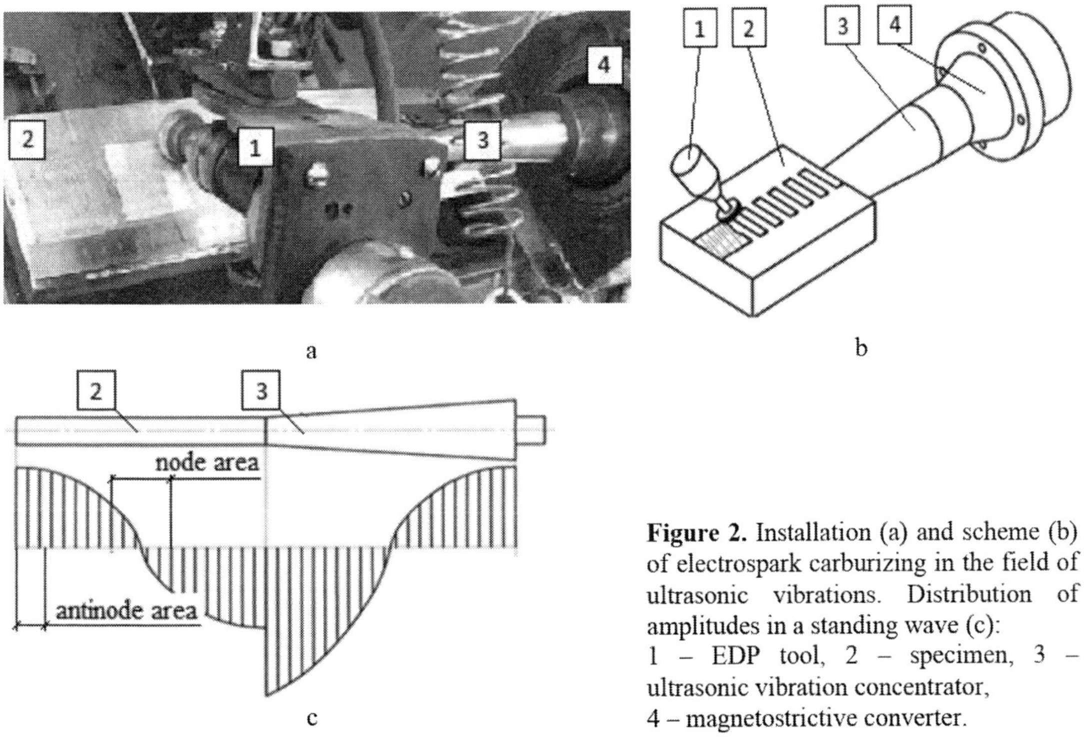

Figure 2. Installation (a) and scheme (b) of electrospark carburizing in the field of ultrasonic vibrations. Distribution of amplitudes in a standing wave (c):
1 – EDP tool, 2 – specimen, 3 – ultrasonic vibration concentrator,
4 – magnetostrictive converter.

2. Material, processing and preparation of specimens

This study is intended to intensify diffusion processes of electrospark carburizing with ultrasonic vibrations (UV). A resonant circuit was used to identify a specific ultrasonic vibration in the areas of nodes and frequency antinodes. This method has proven itself well in the study of the ultrasound effect on the surfaced coating formation and was implemented in our studies [17 to 19]. The size of the plates is corresponded to the sound wavelength (λ) in low-carbon steel (Figure 2). A standing ultrasonic wave was generated in the specimen with a frequency of 18 kHz and a calculated amplitude of 12.6 microns.

IOP Conf. Series: Materials Science and Engineering **966** (2020) 012071 doi:10.1088/1757-899X/966/1/012071

The test specimens were cut out of a plate from the node and standing wave antinode area (Figure 2). The base metal i.e. a '20' steel plate was heat treated in three ways: annealing (cooling inside furnace), normalization (forced air blowing), quenching (cooling in a spray device for rapid cooling). The carbon was determined in the surface layer by a Bruker Q2 ION optic emission spectrometer; the structure was investigated with a Phenom G2 Pure scanning electron microscope; hardness distributed through the layer depth was investigated with a FutureTechFM-300 automatic microhardness tester. A Bruker D8 Advance x-ray diffractometer was used for phase X-raying with K-rays α Cu in θ-geometry by step scanning in the angle range 2θ = 30 – 140°.

3. Research results

The phase analysis of the hardened layers showed the maximum influence of ultrasonic vibrations in the areas near the standing wave vibrations node. The intensity of cementite and austenite lines increases greatly, and the martensitic doublet is well detected (Figure 3). The carbon content of martensite estimated by the *c/a* ratio, was 0.78% by weight.

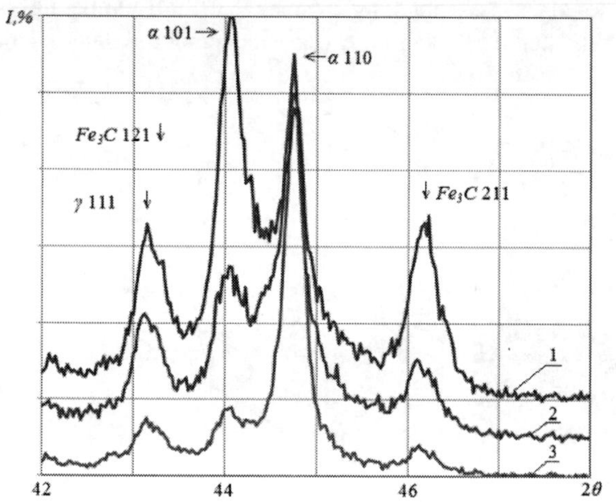

Figure 3. Cementite layer diffractogram sectors (at the node (1), antinode (2) of standing wave oscillations, without ultrasonic vibrations (3).

As expected, the lowest cementation was obtained for the specimens without ultrasonic influence, on the basis of annealed steel and after normalization (0.80 and 1.15 % C), respectively (Figure 4). Ultrasonic vibrations increase the carbon content in the hardened layer, and it is the most in the area of cyclic stretching and contraction of the medium (i.e. at the node of oscillation of the standing wave). The highest cementation was achieved on the base metal after normalization and quenching (2.20 and 2.15 % C). The average carbon in the cementite layer of the specimens cut from the standing wave region is also higher and is 1.2% to 1.4%.

The layer depth is a key cementation parameter, which can be approximately determined on inclined sections. The layer depth values have a high measurement error due to some specific qualities of the EDP process technology and of the hardened layer structure.

Metallographic studies have shown that four characteristic zones can be distinguished on the surface of the specimen subjected to EDP cementation: 1 – upper zone, fragmented coating with higher coarseness (Figure 5, a); 2 – white solid (non-etching) layer; 3 – transition layer (Figure 5, b); 4 – base metal (Figure 5, c, d).

After processing the layer depth data obtained by direct measurement, it was found that the layer is minimal with electrospark cementation without ultrasonic vibrations i.e. 0.06±0.02 mm. One may note that the depth increases when the state of the base metal changes from annealed to hardened, although it is close to the limit of statistical error. The ultrasonic influence increases the coating deeper up to 0.10±0.02 mm, and it is higher in the antinode than in the node.

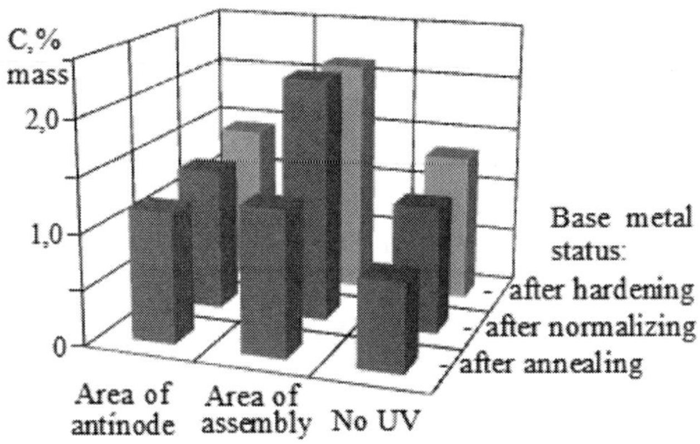

Figure 4. Average carbon in the hardened layer (% of mass) in the representative sectors of the standing ultrasonic wave with different workpiece thermal treatment methods.

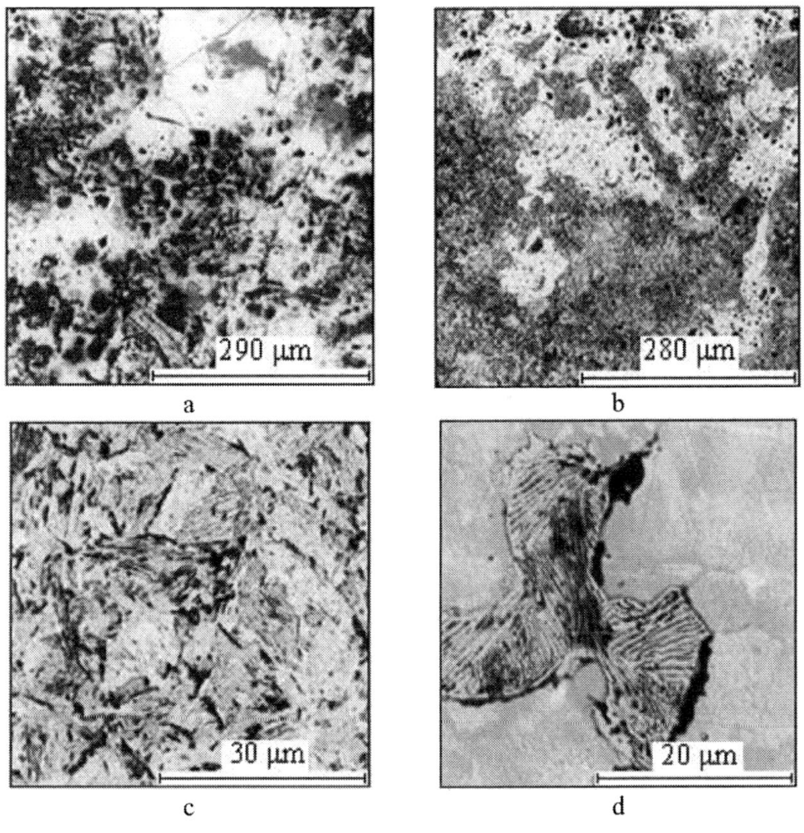

Figure 5. Hardened layer structure (inclined polish sections):
a – fragmented coating; b – diffusion layer. Base metal: c – after quenching; d – after annealing.

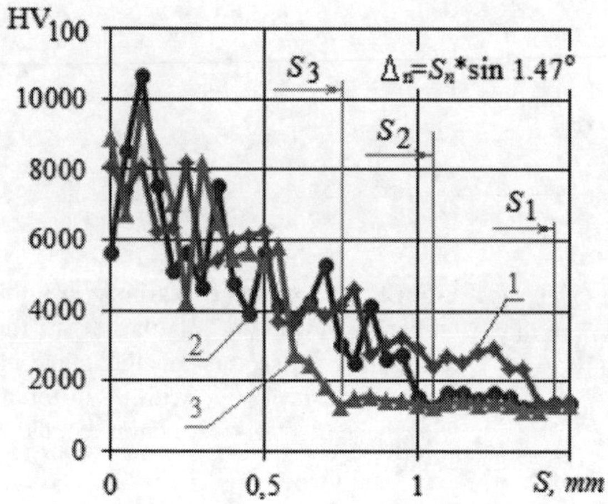

Figure 6. Distribution of microhardness through the depth of the hardened layer (measured on inclined polished section): 1 – antinode area; 2 – vibration node area of a standing ultrasonic wave; 3 – No UV. Preliminary thermal treatment of the backing is normalization. The dotted line indicates the transition to the base metal.

The degree of development of diffusion processes can be estimated by the length of the transition zone Δ, mm (Figure 6). Parameter values Δ are maximal in the sectors of the standing wave antinode and increase when the backing structural components diminish. The transition zone extension reaches its maximum (up to 0.02 mm) when the specimens are electrospark-carburized after normalization and quenching. In the vibration node and during the process without ultrasonic vibrations the values are close and are 0.017 and 0.014 mm respectively (Figure 7).

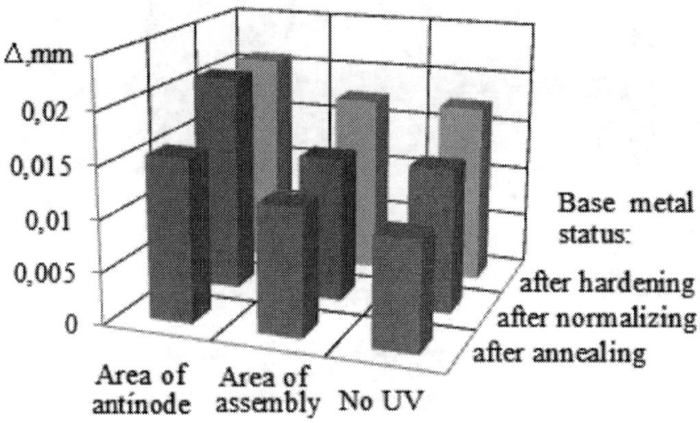

Figure 7. Cementation layer transition zone size Δ (mm).

4. Conclusion

Thus, the electrospark cementation is effective for surface treatment of large products and, unlike the conventional thermochemical treatment, is of finishing type. The hardened layer qualities (cementation extent and depth) can be improved when processing in the ultrasonic vibrations field. The resonant circuit used in the experiment made possible to determine the influence of the representative areas of the standing ultrasonic wave on the cementite layer qualities. The highest cementation was found in the standing ultrasonic wave vibration node area, where cyclic stretching and compression contribute to the excess phases increase. The diffusion processes that determine the cementite layer depth and transition zone size are in their turn maximum in the ultrasonic wave antinode area, where the highest dynamic medium particles offset is observed under the influence of vibrations. The base metal structure component dispersiveness is responsible for the increased hardened layer depth. It reaches 0.10±0.02 mm when combined with ultrasonic exposure for specimens after normalization and quenching.

References

[1] Smirnov M A, Schastlivtsev V M and Zhuravlev L G 1999 *Osnovy termicheskoi obrabotki stali* (Ekaterinburg: UrO RAN) p 496 [in Russian]

[2] Smith N F 1983 Control and instrumentation of gas carburising *Metallurgia* **50 (12)** pp 502–5 [in Russian]

[3] Sahay S S and Mitra K 2004 Cost model based optimisation of carburising operation *Surface Engineering* **20 (5)** pp 379–84

[4] Chatterjee-Fischer R 1985 Review of the Possibilities of Shortening the Carburization Time. *Haerterei-Technische Mitteilungen* **40 (1)** pp 7–9

[5] Gräfen W and Edenhofer B 2005 New developments in thermo-chemical diffusion processes. *Surface and Coatings Technology* **200 (5-6)** pp 1830–6

[6] Rudnizki J, Zeislmair B, Prahl U and Bleck W 2010 Prediction of abnormal grain growth during high temperature treatment *Computational Materials Science* **49 (2)** pp 209–16

[7] Ivanov A S, Greben'kov S K and Bogdanova M V 2016 Optimization of the process of carburizing and heat treatment of low-carbon martensitic steels *Metal Science and Heat Treatment* **58 (1)**

[8] Risse S, Clausen B, Hoffmann F and Zoch H 2011 Concerning the influence of pre-heat treatments on the grain size stability of microalloyed carburising steels *International Heat Treatment and Surface Engineering* **5 (3)** pp 127–32

[9] Fainshmidt E M and Pegashkin V F 2000 Thermochemical treatment of tools for hot forging. *Metal Science and Heat Treatment* **42 (7-8)** pp 263–6

[10] Mihailov V, Gitlevich A, Cazac G and Baciu C 2009 Features of electro-spark alloying on titanium by graphite *BALTTRIB 2009 - 5th Int. Conference, Dedicated to Lithuanian Millennium, 85th Anniversary of Lithuanian University of Agriculture, and 60th Anniversary of Department of Mechanical Engineering, Proc.* pp 282–7

[11] Mikhailyuk A I and Gitlevich A E 2010 Application of graphite in electrospark technologies. *Surface Engineering and Applied Electrochemistry* **46 (5)** pp 424–30

[12] Tarelnyk V, Martsynkovskyy V, Gaponova O, Konoplianchenko I, Belous A, Gerasimenko V and Zakharov M 2017 New method for strengthening surfaces of heat treated steel parts *IOP Conference Series: Materials Science and Engineering* **233 (1)** 012048

[13] Tarel'nik V B, Paustovskii A V, Tkachenko Y G, Martsinkovskii V S, Belous A V, Konoplyanchenko E V and Gaponova O P 2018 Electrospark Graphite Alloying of Steel Surfaces: Technology, Properties, and Application *Surface Engineering and Applied Electrochemistry* **54 (2)** pp 147–56

[14] Konoplianchenko I, Tarelnyk V, Antoszewski B, Martsynkovskyy V, Belous A, Gerasimenko V and Vasilenko O 2018 Mathematical modeling a process of strengthening steel part working surfaces at carburizing thereof by electroerosive alloying method *AIP Conference Proceedings* **2017 1** 020008

[15] Lešnjak A and Tušek J 2002 Processes and properties of deposits in electrospark deposition *Science and Technology of Welding and Joining* **7 (6)** pp 391–6

[16] Tarel'nyk V B, Gaponova O P, Konoplyanchenko Y V and Dovzhyk M Y 2016 Investigation of regularities of the processes of formation of surface layers with electroerosive alloying. Part i *Metallofizika i Noveishie Tekhnologii* **38 (12)** pp 1611–33 [in Russian]

[17] Gavrilova T M, Trekin G E, Shevchenko O I and Farber V M 2001 Geometric parameters and structure of layer surfaced in ultrasonic field *Izvestiya Ferrous Metallurgy* **6** pp 39–41 [in Russian]

[18] Gavrilova T M, Trekin G E, Shevchenko O I and Farber V M 2004 Influence of ultrasound on phase composition of 10R6M5 coatings *Tyazheloe Mashinostroenie* **6** pp 31–3 [in Russian]

[19] Shevchenko O I 2004 Plasma surfacing and ultrasound *Tyazheloe Mashinostroenie* **6** pp 26–7 [in Russian]

IOP Conf. Series: Materials Science and Engineering **966** (2020) 012072 doi:10.1088/1757-899X/966/1/012072

Raising of energy efficiency of thermal performance of layered metallurgical units using layered type of natural gas combustion

V I Matyukhin[1, 2], A V Matyukhina[1] and V A Dudko[1]

[1] Institute of New Materials and Technologies, Ural Federal University named after the First President of Russia B N Yeltsin, 19, Mira str., Yekaterinburg, 620000, Russia

E-mail: [2] matyhin53@mail.ru

Abstract. While heating lump materials there are some difficulties such as poorly organized process of grate firing, conditions deterioration of gas-air mixture creation, limited discharge of combustion products out of a layer due to additional aerodynamic drag occurrence when gases flow. Grate firing of gaseous fuel can be completed in a dense layer of lump materials which have different chemical composition and particle-size distribution. To do this, it is necessary to provide formation of high-temperature area and complete thermal generation process directly between separate elements of treated materials. It allows providing specified temperature-time conditions of heat treatment of the different layer sections. Also, it is required to limit the direct contact of the high temperature zones with external boundaries of the unit smelting chamber and environment for increasing reliability and effectiveness. In this paper, we present existent heat-exchange schemes in grate process units during the heat treatment of lump materials. An ability of grate firing of gaseous fuel while organizing a delivery of prepared gas-air mixture into a dense layer or separate tuyer gas and air supply with separate flows is shown. Usage of indicated ways of feeding allows decreasing energy demands and improving quality of products delivered.

1. Introduction

Numerous theoretic and experimental researches of industrial thermo-technological processes and units show that usage of gaseous fuel while organizing their thermal performance in the capacity of central heat source and processing additive essentially increase technical-and-economic indices of high temperature process implementation [1–5]. However, first of all, it relates to the grate process units using different schemes of heat exchange between gases and materials [6].

The most common way of flame combustion of gaseous fuel while lump materials are heating, complicated with presence of blow down piece packing on the way of gas flow which facilitates the change-over jet fire gaseous fuel into badly-organized process of grate firing deteriorates the conditions of initial gas-air mixture creation, causes difficulties of its preheating up to inflammation temperature of combustible part, and limits discharge of combustion products out of a layer due to additional aerodynamic drag occurrence when gas flows.

2. Researches of features of grate type of gaseous fuel firing

Researches of thermo physical features of grate type of gaseous fuel firing [7] showed that it can be implemented in the dense layer of lump materials almost of any chemical composition and particle-size distribution. The differential characteristic of implementation of this process is direct involvement when forming high-temperature area in the dense layer of heated materials of solid surface of batch mixture lumps [8]. It allows for relatively simple provision of specified temperature-time conditions of heat treatment of different sections of the layer, completing thermal generation process between separate elements of treated materials. Bounding of the direct contact of the high temperature zones with external boundaries of unit smelting chamber and environment increases reliability and effectiveness of usage of this way of layered facility heating. Taking intermediate place between flare and plasma processes according to physical properties (temperature, heating rate, gas-phase composition), this way of thermal generation requires advance preparation of the base mixture of air and natural gas which includes homogenization of its composition and preheating.

Incineration process of the initial gas-air mixture in the dense layer of lump materials should be referred to micro jet fire, which is complicated by the presence of solid components with finite dimensions filling the volume of the smelting chamber.

When organizing initial gas flows supply directly into layer of lump materials of different dispersion degree being fired, implementation of two variants is possible:

– supply of treated gas and air mixture preliminarily prepared in special mixers which is peculiar to implementation of thermal generation process in a limited volume of heated layer (pellet firing, agglostoe ignition process, etc.);

– gas and air supply through separate flows during tuyer supply thereof in the mine units.

Initial heat impulse for stable ignition of gaseous fuel in the layer can be implemented with the help of an open fire flare, electric spark, hot jet or heated surface. The most radical way to provide the conditions of gaseous fuel firing is rise temperature of the used layer surface in the mode of pre-heating from external energy source.

According to the combustion gas mixture theory, the conditions of its combustion depend on dynamics of boundary layer preheating on the front part of the pieces where heat passage is carried out mainly by thermal conduction [9].

First of all, oxygen molecules are adsorbed on the solid surface and distributed over it almost equally regardless of the nature of the surface itself. The presence of additional thermal resistance layer of adsorbed oxygen on the surface of layered particles requires increased temperature gradient between lump materials and molecules of the gas mixture. Because of it, the level of layer preheating for organizing grate gas combustion must be higher than ignition temperature of the initial gas mixture (500–550°C). According to the practical data its value should be 800–1050°C. Despite the significant value of the heat energy discharged in combustion of fuel, the presence of heat efflux for heating of layer elements and convection current with exit gases limit the general rise of the system temperature.

Increased adsorption activity of oxygen molecules in relation to the solid surface of the layer decreases their concentration in the interlump space in the initial section of gas-air mixture flow. Thus, to provide required concentrating proportions between combustible components and oxygen in the gas mixture when it is ignited, it is necessary to provide a significantly higher air flow compared to the flare combustion conditions. It makes sense to maintain air discharge coefficient at the value not lower than 2.5–3.0 for stable ignition of gas mixture in the dense layer.

Gaseous fuel combustion process in the lump materials layer is presented as sequentially passing stages of heat exchange. Gas-air mixture heats when it is supplied to the preheated lump material layer and cools it down. After the gas mixture reaches ignition temperature, which is determined not only by its initial composition, but also by the conditions of its heating, and active combustion zone is generated. Its distinguishing features are high temperature and increased rate of oxidation of hydrocarbons with the possibility that combustion can shift over the entire layer. When moving through a relatively cool part of the layer, combustion products generated during gas combustion heat it up. These provide for conditions for gas fire propagation over the height of the piece packing.

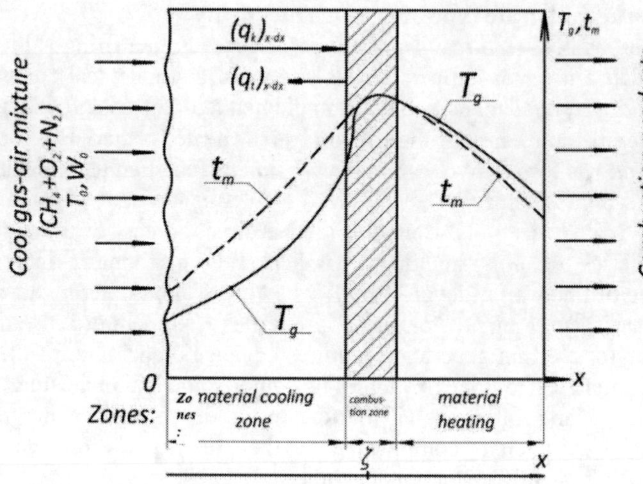

Figure 1. Heat-exchange circuit in gas combustion in the layer.
T_g, t_m – gas and material temperatures; X – current coordinate of combustion area.

Temperature level of combustion zone is determined by the conditions of heat-exchange in the layer cooling zone by the flow of initial gases, by activity of the intensive heat emission from oxidation of hydrocarbons directly in the active combustion zone and by the development of convective flow of exit gases.

Formation of the combustion zone of natural gas in the dense layer of lump materials is ensured by the equality of heating rates with thermal conductivity from layer elements $(\Sigma q_{T_i})_{\varsigma-dx}$ to the combustion area with the coordinate of ξ, thermal emissions in it $\left(b \cdot Q_н^p\right)$ from oxidation of hydrocarbons and convection heat gas flow $\left(\sum q_{K_i}\right)_{\varsigma+dx}$ in the direction of the heated layer:

$$(\Sigma q_{T_i})_{\varsigma-dx} = \left(b \cdot Q_н^p\right) + \left(\sum q_{K_i}\right)_{\varsigma+dx} \tag{1}$$

where q_T – specific heat flow rate with heat conduction in the direction of gas mixture, kJ/h; b – specific fuel consumption, m^3/m^3; $Q_н^P$ – fuel combustion value, kJ/m^3; q_k – specific heat flow rate with the convection in the direction of flow of exit gases, kJ/h.

Interruption of this balance determines the travel direction of fuel grate firing zone across the layer. If specific heat flow rate exceeds heat conduction magnitude of specific heat convective flux, the process of layered gaseous combustion will move towards the initial gas mixture. Otherwise, high-temperature combustion zone extends in the direction of gas flow.

Usage of treated gas mixture in the grate type of firing bounds the extension of gas combustion area zone in the layer to the distance not longer than 30–60 mm [10] directly near the entrance of the used gases into the heated layer (at the distance not longer than 5–25 mm) [11] with the movement speed of 8-16 mm/min when the temperature therein is not more than 1150–1350°C [12]. It provides the possibility to save on energy costs for development of layered processes up to 10–15% with improvement of qualitative parameters of the products obtained.

Under the conditions of tuyer air blowing into the smelting chamber of the grate unit and coaxial jetting supply of gaseous fuel into it under increased pressure, discharge of gas jets out of the nose-pieces (Figure 2a) is carried out in several flows of varying degrees of mixing between themselves ,and it facilitates distribution of conditions for gas flow in the space filled with gaseous medium having the same physical properties as the jet substance in the form of separate jet flares. At the same time, each gas component is preliminarily held in stable molecular state. If there is a gas-penetrable piece packing for lump materials on the way of jet penetration, jet process motion type changes significantly along with turbulization of gas flows, reduction of flare range capability and increase in their opening angle within 17–32 degrees.

According to the theory of development of turbulent isolated and coaxial jets, gaseous medium flowing out of the nozzle at a high speed [13] provides ejection into the flowing environment stream both along external and internal borders of the jet stream. Since there is no medium in the jet to satisfy its ejection requirements, pressure in this zone drops, and circulating vortexes directed along the jet to its starting point are generated inside the annular space. This provides stability of the gas component mixing conditions and ignition of generated mixture.

Initial heat impulse for stable ignition of gaseous fuel is affected by means of layer preheating by external heat source when layered type of natural gas firing in the dense layer is organized.

The heat source has to be located close to the gas jet entrance in the heated layer. The closer the gas-air mixture ignition zone is to the entrance, the more evenly temperature distribution in the layered combustion zone will be. The length of the preheating zone shall provide the constant heating of the gas-air mixture up to the ignition temperature at which thermal emission will exceed thermal absorption in the layer.

Despite the significant value of the thermal energy flow discharged in the combustion fuel, the presence of heat efflux for heating of layer components and convective current with exit gases limit the growth of the layer temperature not more than to 1150–1350°C [14].

Changes of supply conditions and mixing of co-current streams enables adjustment of conditions of gas mixture distribution over the depth of the layer in wide range [15–17]. Extension of jet processes in the piece packing will depend on the mixing conditions of the central high-speed gas jet with annular air stream delivered with lower speed, and also on gas flowing out of the environment. The formation of the required initial composition of gas-air mixture, its mixing, ignition and expansion in the volume of the layer part being fired will occur in the border zone of the general flare development. The range of expansion of hot combustion zone in the direction of gas efflux is, for the most part, determined by kinetic energy of gas jet.

Absence of oxygen along the central part axis of the burner unit turns the layer combustion process into the zonary type. There is a cold jet of natural gas in the central part of the gas flow (Figure 2a, 4). In the peripheral part thereof, a gas-air mixture is formed by means of suction of concurrent air flow (Figure 2a, 3). The presence of the preheated layer with the temperature of 800–1050°C on its path at the entrance of gas mixture into the layer of batch mixture leads to the heating of natural gas and air flows with shift of the high-temperature area to the peripheral flare part. As a result, there is combustion zone (Figure 2a, 5) where the energy generation processes occur. However, there is a high-speed natural gas flow in the central part of the gas flare the chemical composition of which is close to the gas composition at the nozzle edge (Figure 2a, 2).

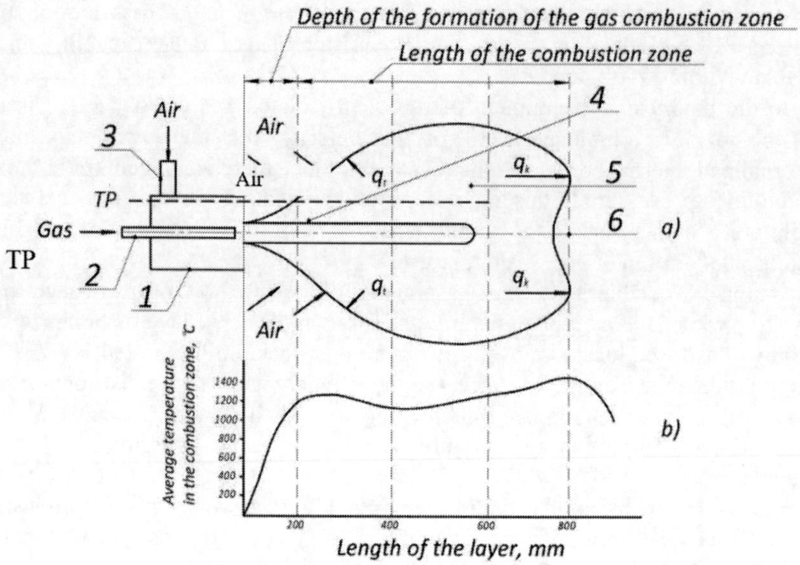

Figure 2. Scheme of the development of the gas combustion process in the layer under the conditions of tuyer gas components supply (a): 1 – burner body; 2 – gas nozzle, 3 – air pipe; 4 – gas jet; 5 – gas-air mixture combustion zone in the layer; 6 – combustion products; TP – thermometer probe; (b) – layer temperature variations with the air flow rate factor of 1.52 and speed of gas exit out of the burner of 2.97 m/s.

Around the periphery of the combustion zone there is a zone formed in the result of natural gas mixing with air and combustion products. Length of this zone in the horizontal direction is evaluated by the conditions of natural gas mixing with air and by the conditions of gas-air mixture ignition. Combustion products generated during gas firing go through the relatively cool part of the batch mixture layer and generate conditions for after flaming of gas in the dense lump material layer outside the zone by means of ejection of air flow from the environment.

3. Conclusion

Consequently, usage of the layered natural gas combustion process upon heating of piece packing can take the volumetric form and cover a considerable part of its smelting chamber while providing total fuel consumption for hard firing at the level not more than 140 kg per nominal thousand tons. The presence of high heating temperatures of initial gas mixture in the layer under the lack of oxygen opens up space for the development of the gaseous oxygen methane conversion with the formation of the stream of CO and H_2 reducing gases.

Acknowledgments
The reported study was funded by RFBR, project number 19-38-90192\19.

References

[1] Lisienko V G, Shchelokov Ya M, Ladygichev M G 2002 *Energy Saving Anthology* (Moscow: Teploenergetik) p 768 [In Russian]
[2] Piringer H 2017 Lime Shaft Kilns *Energy Procedia* **120** pp 75–95
[3] Dong Hui, Cai Jiu-ju, Wang Guo-sheng et al 2004 Experimental Study on Gas Flow Distribution Affected by Constructional Parameters of Pelletizing Shaft Furnace *Journal of Northeastern University (Natural Science)* **24 (6)** pp 563–6
[4] Cheng Chuan and Eckehard Specht 2006 Reaction rate coefficients in decomposition of lumpy limestone of different origin *Thermochimica Acta* **449** pp 8–15

[5] Rong W J, Li B K and Qi F S 2018 Combustion Characteristics of Calcium Carbide Furnace Off-Gas in a New Type Combustor of Twin Burn Annular Shaft Kiln *Dongbei Daxue Xuebao Journal of Northeastern University* **39** pp 200–4
[6] Chizhikova V M and Savchuk N A 2004 *Agglomeration: Contemporary Aspect* (Moscow.: Metallurgy) p 124 [In Russian]
[7] Senegačnik A, Oman J and Širok B 2007 Analysis of calcination parameters and the temperature profile in an annular shaft kiln. Part 2: Results of tests *Applied Thermal Engineering* **27** pp 1473–82
[8] Lisiyenko V G, Lobanov V I and Kitayev B I 1982 *Thermal physics of metallurgical processes* (Moscow: Metallurgy) p 240 [In Russian]
[9] Butkarev A P, Mayzel G M, Matiukhin V I, Lobanov V I and Yaroshenko Yu G 1981 Research of the Way of Obtaining Semireduced Pellets during Gas Combustion in a Dense Layer *News of Higher Educational Institutions. Ferrous Metallurgy* **2** pp 31–3 [In Russian]
[10] Telegin A S, Shvydkyi V S and Yaroshenko Yu G 2002 *Heat and Mass Transfer: Textbook for Universities* (Moscow: Akademkniga) p 435
[11] Matiukhin V I, Lobanov V I and Gordon Ya M 1982 Research of the Formation Conditions of the Gas Combustion Zone in the Layer of Iron-Ore Pellets *News of Higher Educational Institutions. Ferrous Metallurgy* **11** pp 18–21 [In Russian]
[12] Gordon Ya M, Lobanov V I and Matiukhin V I 1983 Peculiarities of Changing of the Main Characteristics of the Gas Combustion Process in the Dense Bed with the Coefficient of air Discharge Less Than One. Report 1 *News of Higher Educational Institutions. Ferrous Metallurgy* **2** pp 101–5 [In Russian]
[13] Rong W J, Li B K, Qi F S and Cheung S C P 2017 Energy and exergy analysis of an annular shaft kiln with opposite burners *Applied Thermal Engineering* **119** pp 629–38
[14] Lobanov V I, Matiukhin V I, Holtsev V A and Yaroshenko Yu G 1987 Research of the Formation Conditions of the Combustion Zone in the Layer of Iron-Ore Pellets for Improvement of the Metallurgical Properties *News of Higher Educational Institutions. Ferrous Metallurgy* **6** pp 103–4 [In Russian]
[15] Bryukhanov O N and Mastryukov B S 1994 *Aerodynamics, Combustion and Heat Exchange* (Saint Petersburg: Nedra) p 317 [In Russian]
[16] Senegačnik A, Oman J and Širok B 2008 Annular shaft kiln for lime burning with kiln gas recirculation *Applied Thermal Engineering* **28** pp 785–92
[17] Schwertmann T 2004 Thermodynamic aspects of the counterflow lime burning process – Part 1 *ZGK International* **57 (8)** pp 48–58

IOP Conf. Series: Materials Science and Engineering **966** (2020) 012073 doi:10.1088/1757-899X/966/1/012073

Use of temporary gas flow to improve the efficiency of metallurgical units operation

V I Matyukhin[1, 2], A V Matyukhina[1], S Ya Zhuravlyov[1] and N V Yamshanova[1]

[1] Institute of New Materials and Technologies, Ural Federal University named after the First President of Russia B N Yeltsin, 19, Mira str., Yekaterinburg, 620000, Russia

E-mail: [2] matyhin53@mail.ru

Abstract. This paper presents the results of industrial tests on dust deposition due to the use of an unsteady gas flow. JSC SUMZ (Revda) applied the technology with the use of acoustic radiators to organize in-furnace dust deposition at Vanyukov furnace. The design of the radiator included a nozzle pipe, air nozzle, resonator and focusing surface. Based on the surface area of the furnace molten pool and recommended specific acoustic power values for in-furnace dust condensing, the total acoustic field sound power was calculated. The optimal number and location of acoustic radiators was determined. The analysis of the data of industrial tests of the use of the acoustic field energy in the working space of the PZhV, as well as the results of tests on the converter No. 2 of JSC 'SUMZ', the cyclone of the firing multi-hearth furnace of JSC 'SVYATOGOR', the dust chamber of the Waelz furnace and cleaning of the heat exchange surfaces of the boiler on waste gases (JSC 'Chelyabinsk Zinc Plant').

1. Introduction

Thermal performance of the majority of process units using disperse materials of various fineness or processing of supernatant liquid is characterized by the formation of a significant (up to 200–250 g/m^3) amount of dust particles of various grain fineness, their sticking to the inner surface of gas ducts and furnace elements with limited kinetics of the main physical and chemical processes. In order to reduce losses of initial and finite materials and to improve the environmental situation in the area of the enterprise location on the way of dusty gases movement a number of dust-arresting equipment (dust chamber, cyclone unit, electrostatic precipitator) are installed in series. Despite their relatively high capture efficiency factor, the content of dust particles in the exit gases remains high. Such a state of technology and design of furnaces is characterized by a significant tension of the gas exhaust duct, process equipment installed on it, losses of initial materials for the main production, increase in environmental impact on the environment [1–4].

2. Research of ways to reduce dust emission from layered technological units

One of the most rational ways to reduce dust discharge from the layered process units is to improve the conditions of dust-laden flow movement at all technological stages in order to optimize the gas-dynamic situation in the working compartment. As the largest source of dust, the working compartment of the unit can be used for in-furnace dust deposition. At the same time, captured dust particles increase the amount of materials involved in technological processes, increasing their performance and reducing the amount of thermal losses with exit gases.

One of the effective and low-cost ways to improve the performance of process layered units or their individual parts is the use of temporary gas flows based on the use of acoustic field energy and their specified pulsations formed directly at the place of use.

The basis for the development of such processes are the theoretical positions in the formation of impulse energy effects of rational power on the processed materials, taking into account the transformation of energy fields, evaluation of intensity of physical and chemical effects, the inversion of their use in changing environmental conditions of facilities existence.

Correctly chosen impulse energy impact and effective manifestation of emerging effects of intensification of basic processes is the basis for the creation of energy-efficient and environmentally friendly technology.

The external acoustic field or the field of gas pulsations created in a homogeneous space is a source of vibrational impact on moving dust particles or lumpy batch mixture elements. This is due to local periodic changes in gas pressure in the moving flow from positive to negative. Excess gas pressure between the particles caused by external oscillations is compensated by the outflow of excess or insufficient amount of gas in the direction of oscillations propagation. In this case, an additional flow of the gas medium occurs, which contributes to the intensification of heat – mass exchange processes by 10–15 %.

In the presence of resonance of the external acoustic field frequencies and the natural frequency of the dust particles, due to oscillations of solid particles, the amplitude of which is determined by the initial parameters of the acoustic field, there is the occurrence of low-frequency small-scale contours of the particles movement without deviation from the original movement pattern (Figure 1). The following total pressure is applied to each solid particle of dust-laden flow with a density ρ that is in the field of external acoustic or pulsatory oscillations with amplitude A and circular frequency ω in a gas medium at a speed of sound c

$$P = P_o + A \cdot \rho \cdot c \cdot \omega \cdot cos(\omega(t - \frac{x}{c})) \text{, atm.}$$

In addition to the total pressure of the moving gas P_o, an additional periodic force with an oscillation amplitude acts in the external acoustic field with sound level J

$$P_a = A \cdot \rho \cdot c \cdot \omega = \frac{\rho \cdot c \cdot \omega}{\omega} \sqrt{\frac{2 \cdot J}{\rho \cdot c}} \cdot 10^7 = \sqrt{2 \cdot J \cdot \rho \cdot c} \text{ , atm.,}$$

the value of which can be changed through the design and process parameters of the process with a change in the value of input actions. The latter cause the destruction of the boundary layer near oscillating particles and the occurrence of additional convection flows in a moving gas medium.

Large-scale contours of circulating groups of particles, which lead to the occurrence of piston effect, can appear in discharged media. The piston effect phenomenon is manifested in the formation of local vacuum near periodically moving particles, which causes the formation of additional convection flows. Absolute values of local pressure and discharge near the individual particles depend on their size, frequency and amplitude of oscillations. In this case, the frequency of the external field pulsations shall coincide with the natural frequency of oscillations of the main mass of the particles. Experimental measurements of local pressures in the layer, by the example of vibrated fluidized bed, showed the possibility of achieving its value up to 20000 Pa.

This effect is the most noticeable in the conditions of those technological processes, where there are small values of gas velocities, and the gas flow is characterized by an increased content of dust. The bigger is the transverse size of particles, the bigger is the level of created rarefaction. Therefore, when acoustic or pulsatory fields of the given parameters are formed, small particles will move to the coarse ones around which a local zone of increased rarefaction is formed, forming around them an area with an increased concentration of fine dust elements. Under prolonged exposure to periodic oscillations, it provides the ability to hold dusty fractions around coarse particles, reducing dust discharge by up to 20–25 %. In this case it is necessary to take into account the time factor of the acoustic field action.

The process of acoustic or pulsation coagulation of dust particles will be most effective when it is organized directly in the working compartment of the process unit. However, this process requires an individual approach.

In the implementation of intra-unit dust deposition, it is necessary to implement the following principles:

1. The most complete and continuous effect of external oscillations on the dusty environment. For this purpose, in accordance with the design features of the unit, it is necessary to determine the location of sources of temporary action installation with the possibility of maximum impact on the dust-laden flow or dispersed medium.
2. Necessity of implementation of the time factor of acoustic influence on the dusty flow.

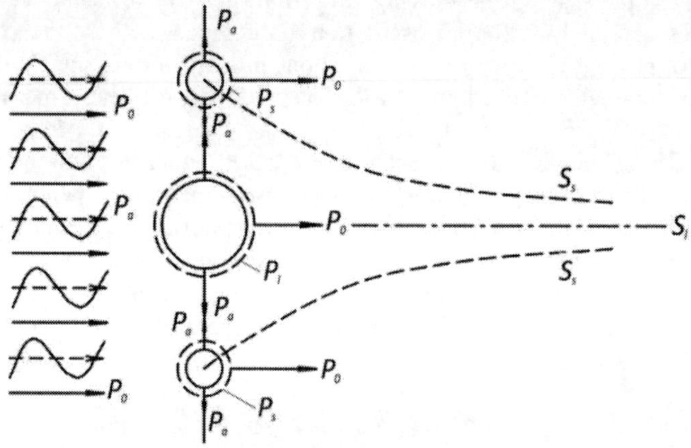

Figure 1. Diagram of acoustic influence on solid elements of the layer:
P_o – static pressure of gas flow; P_a – acoustic pressure; P_s – rarefaction near small elements;
P_l – rarefaction near large elements; S_s – small elements movement pattern;
S_l – large elements movement pattern.

3. Formation of the acoustic field of the set parameters established in accordance with the design features of the unit and physical and chemical characteristics of dust particles. For this purpose, the optimal design and technological parameters of acoustic jet-edge radiators or gas flow pulsators, providing the maximum development of the required technical effect (reduction of dust discharge, intensification of heat – mass exchange processes, destruction of dust outgrowths and crusts, etc.) are set by means of a calculation or experiments.
4. Industrial tests of the use of acoustic field energy or pulsatory gas flows with the evaluation of the efficiency of dust deposition and development of heat – mass exchange processes.

The formation of the acoustic field in the layer is proposed to be performed with the help of an acoustic radiator (Figure 2, 3) using waveguides of various designs or mechanical pulsators, taking into account the properties of the transported medium.

Acoustic radiator is a kind of gas-dynamic Hartmann whistle and consists of a nozzle, resonator and focusing plane made of metal. It is powered by a compressor air with pressure of at least 3.0 atm. in an amount up to 50 m^3/h. Installation of this unit is carried out outside the working compartment, and it facilitates its installation, maintenance and operation, as well as reduces the requirements for the materials from which it is made. The use of an acoustic field directly in the working compartment of the process unit makes it safe for maintenance personnel.

In conditions of limited, for technological reasons, impulse energy impact on the objects of intensification with the required low frequency of oscillations (large volumes of working compartment, low speeds of gas flows, limited process temperature), real prerequisites for the application of pulsating air blast flow are created. The transmission of mechanical energy of oscillations in this case can be realized mainly through the dynamic forces of the moving, as a whole, gas flow with periodically varying value, and its part merging with the main flow within the working compartment [5–10].

Figure 2. Acoustic radiator: 1 – nozzle; 2 – resonator; 3 – focusing plane.

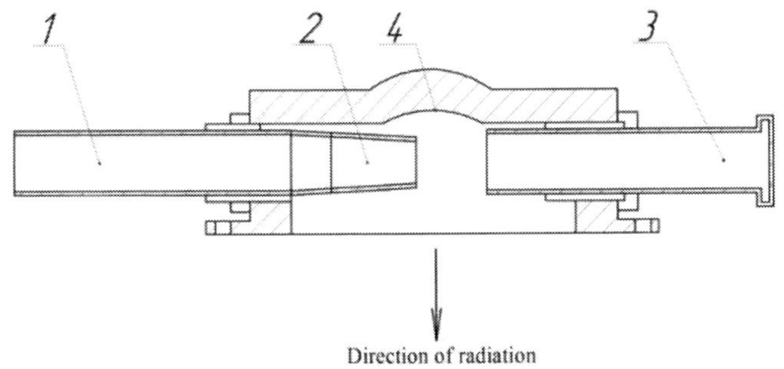

Figure 3. Acoustic radiator design:
1 – nozzle pipe; 2 – air nozzle; 3 – resonator; 4 – focusing plane.

Depending on the value of mass pulsating flow, the general nature of gases oscillatory movement development, energy efficiency of their use, the amount of energy consumption for their generation are changed. Oscillations of the entire mass of the gas flow lead to large inertial loads, which limit the scalability of this method. The second type of pulsating flow using only a part of the moving gases may be more attractive, but with a smaller manifestation of the change in efficiency. The resulting oscillations of the gas medium inside the system accelerate the process flow. However, a lower value of the resulting impact limits the range of variation in the frequency of oscillations not exceeding 50 Hz.

Industrial tests of the use of acoustic field energy for the intensification of heat - mass exchange processes on a number of process units (sinter machine, a dense bed of curing oven, cast iron and mineral-cotton cupola furnaces, mining unit of non-ferrous metallurgy) have shown wide possibilities of this method application.

Thus, industrial tests of this process in the conditions of sinter machine AK-50 of Serov Metallurgical Plant for a long time confirmed the safety of its use. In addition, it was shown that it is possible to increase the performance of the unit by up to 15–20 %, increase the depth of the main physical and chemical reactions in the bed, reduce harmful emissions from the sinter machine (CO, NO_x, dust), reduce specific fuel consumption per process by 10–15 %.

Depending on the design of the fibrous materials curing oven, from two to six radiators were manufactured and assembled in vacuum and drying chambers. Their tests showed the possibility to increase the performance of the unit by 30–35 % with the improvement of the quality of organic binding substance use.

The results of using the acoustic field energy in the working compartment of cast iron and mineral-cotton cupola furnaces, mining units of non-ferrous metallurgy (Kirovgrad) showed a possibility to increase performance of smelting units by 10–15 % with a corresponding reduction of coke consumption and harmful emissions by 20–25 %. At the same time, the destruction of the resulting crusts during the movement of gases and materials is also noted, and it increases the performance of the smelting unit and reduces its maintenance costs.

Formation of sound oscillations field of the given parameters in a moving dust-laden flow of finely dispersed batch mixture of rotating furnaces of alumina production of the Bogoslovsky Aluminum Plant with a diameter of 3 m and a length of 60 m, both in the direction of movement of gases and in the opposite direction allowed to establish a steady tendency of dust discharge reduction outside the working compartment by up to 40 %. At the same time more full completion of all chemical transformations in solid components, with the destruction of the resulting crusts on the inner surface of the drum was observed.

Measurements of environmental conditions in the test area allowed to establish a slight change in working conditions of the maintenance personnel by the sound pressure level [11–16].

The analysis of data from industrial tests of the use of acoustic field energy in the working compartment of smelting reduction furnace (SRF) (SUMZ, Revda) showed that at approximately the same consumption characteristics of the original components, the use of acoustic energy leads to a decrease in dust discharge by 10.94 %. At the same time, the in-furnace dust deposition leads to the increase of matte output by about 406 kg/h and slag output by 1.299 kg/h. With an average copper content of 51.7 % in the matte, this provides an opportunity to increase the amount of copper received by up to 210 kg/h.

Test results on converter No. 2 of JSC SUMZ using acoustic field energy of two radiators with 3.0 atm. pressure installed in a dust chamber and two radiators at cyclone inlet in Table 3 showed that during the first period of melt reforming the total degree of gas dust exclusion increased up to 56-72 %. During the second period, the degree of dust deposition using acoustic radiators increased by 22 % on average.

If the dusting system is connected directly with the process unit by circulation of gases or dust components, the acoustic effect on its individual devices can have a significant impact on the operating conditions of the entire process. Thus, when acoustic radiators are installed at the cyclone inlet nozzle of the multihearth roaster of JSC SVYATOGOR, its total efficiency has changed in comparison with the normal mode from 57.14 to 70.49 %, i.e. by 13.35 abs.% or by 23.36 r.% due to dust coagulation directly at gas flow movement in the cyclone working compartment.

To assess changes of dust deposition conditions in the dust chamber of the Waelz kiln and cleaning of heat-exchange surfaces of the boiler on the exit gases (Chelyabinsk Zinc Plant PJSC), acoustic radiators were installed on the dust chamber in the end wall towards the movement of gases, and on the heat-exchange surfaces of the boiler – in the upper area of the lifting channel against the movement of gases. It was shown that all other conditions being equal, the use of the acoustic field energy allows increasing the output of reverse dust by 8.74 % due to intensification of the process of intra-unit coagulation in the sedimentation chamber by up to 25–30 %, and the output of velyokis` by 2.99 % due to reduction of hydraulic resistance of the dust chamber and the boiler while increasing the actual gas velocity.

Thus, under the influence of an external acoustic field of the given parameters or a pulsating gas flow on the dusty flow there are conditions for the convergence of coarse and small particles and dedusting of exit gases directly in the working compartment of process units or their individual areas. The practice of using this process at a number of units has shown that each place of installation of the radiator can reduce the dust flow by about 20–25 % from the initial. It is also possible to destroy the formed dust deposits (fully or partially).

3. Conclusion

Insignificant capital and operational costs of using the energy of the acoustic field and pulsating gas flow for the organization of intra-unit dust deposition open up prospects for improving the environmental situation in the area of operation of layered units of various designs and improving their energy.

References

[1] Gushchin S N, Telegin A S, Lobanov V I and Koryukov V N 1993 *Heat Engineering and Heat Power Engineering of Metallurgical Production* (Moscow: Metallurgy) [In Russian]
[2] Vanyukov A V and Utkin N I 1988 *Complex Processing of Copper and Nickel Raw Materials* (Chelyabinsk: Metallurgy) [In Russian]
[3] Feng Han, Fei Yu and Zhaojie Cui 2015 Industrial metabolism of copper and sulfur in a copper-specific eco-industrial park in China *Journal of cleaner production* vol **133** pp 459–66
[4] Naboichenko S S, Ageev N G, Doroshkevich A P, Zhukov V P, Yeliseev E I, Karelov S V, Lebed` A B and Mamyachenkov S V 2005 *Non-Ferrous Metallurgy Processes and Equipment: Textbook for higher education* (Yekaterinburg: USTU) [In Russian]
[5] Yaroshenko Yu G, Matyukhin O V, Panshin A M and Konovalov I S 2013 Use of Acoustic Field Energy to Improve Performance of Shaft Furnaces *Non-Ferrous Metals* no **8** pp 64–70
[6] Vaisburd S, Berner A, Brandon D G, Kozhakhmetov S, Kenzhaliyev E and Zhalelev R 2015 Slags and mattes in Vanyukov's process for the extraction of copper *Metall. Mater. Trans.* no **33** pp 551–9
[7] Chen L, Bin W, Yang T, Liu W and Bin S 2013 Research and industrial application of oxygen-rich side-blow bath smelting technology *Proc. of 4th Int. Symp. on High-Temperature Metallurgical Processing (TMS 2013). Annual Meeting and Exhibition* (San Antonio, TX, United States, 3–7 March 2013) pp 49–55
[8] Asanov D A, Filyanova L A, Zapasnyi V V and Sukhova N M 2016 Study of the Performance Indices of a Dust-Cleaning System at the Balkhash Copper Smelter *Metallurgist* no **3–4** pp 331–8
[9] Zhang H L, Zhou C Q, Bing W U and Chen Y M 2015 Numerical simulation of multiphase flow in a Vanyukov furnace *Journal of the southern African institute of mining and metallurgy* vol **115** no **5** pp 457–63
[10] Kutaladze S S 1970 *Heat and Mass-Transfer in the Sound Field* (Novosibirsk: SDRAS) [In Russian]
[11] Seregin P S 2001 Study of Gas Dynamics of the In-Furnace Space, Dust Discharge and Skull Formation in Vanyukov Furnace Using the Physical Modeling Method. *Dis. for appl. for a deg. of Candidate of Technical Sciences in the specialty 05.16.02* (Saint Petersburg: Gipronickel Institute) [In Russian]
[12] Selivanov Ye N, Skopov G V, Gulyaeva R I and Matveev A V 2014 Material Composition of Vanyukov Furnace Electrical Filters Dust of JSC Sredneuralsky Copper Smelter *Metallurg* no **5** pp 92–5
[13] Kardashev A G 1990 *Physical Methods of Processes Intensification in Chemical Engineering* (Moscow: Chemistry) [In Russian]
[14] Konovalov I S 2012 Improvement of Thermal and Gas-Dynamic Operation of Shaft Copper-Smelting Furnaces. *Dis. for appl. for a deg. of Candidate of Technical Sciences in the specialty 05.16.02* (Yekaterinburg: Ural Federal University named after the first President of Russia B N Yeltsin) [In Russian]
[15] Cafiero G, Greco C S, Astarita T and Discetti S 2016 Flow field features of fractal impinging jets at short nozzle to plate distances *Exp. Therm. Fluid Sci.* no **78** pp 334–44
[16] Andrade M A B, Skotis G D, Ritchie S B, Cumming D R S, Riehle M O and Bernassau A L 2016 Contactless acoustic manipulation and sorting of particles by dynamic acoustic fields *IEEE Trans. Ultrason., Ferroelectr., Freq. Contr.* vol **63** iss **10** pp 1593–600

IOP Conf. Series: Materials Science and Engineering **966** (2020) 012074 doi:10.1088/1757-899X/966/1/012074

Development of radial-shear rolling mill special stands for continuous cast billets deformation

R V Iskhakov[1,3], Yu V Gamin[2], M V Kadach[2] and A S Budnikov[2]

[1] Pervouralsk New Tube (Novotrubny) Works, 1, Torgovaya str., Pervouralsk, 623100, Russia

[2] Department of Metal Forming, National University of Science and Technology 'MISIS' (NUST 'MISIS'), 4, Leninskiy prospect, Moscow, 119049, Russia

E-mail: [3] viv3rn@mail.ru

Abstract. The article offers the structure of the three-roll stand for radial-shear rolling that is designed to deformation of continuous cast billets. The stand can be installed in the pipe rolling mill manufacturing line that shall ensure manufacturing flexibility. To achieve the objective set, elaboration of working rolls integration with the drive as well as verification against the allowable torque and strength calculations in SolidWorks Simulation were performed at the design stage. For the major part of the frame, stresses do not exceed 60 MPa and the design rigidity rate will be 2.86 MN/mm. The special stand has a specific frame structure to ensure its mounting on the existing equipment base. The configuration of rolls sets is selected considering the possibility of integration with the spindle assemblies of the group drive and maximum allowable cross angle (≤ 15°). The developed design proves to have sufficient strength margin with the relatively small weight. The special radial-shear rolling mill stand will allow increasing total strain degree and reducing the number of billets sizes.

1. Introduction

Rolling mills are generally considered to be unique and high-performance metallurgical units [1–4]. The design development is performed considering the developed process and size range of the manufactured products. The equipment should ensure long-term failure-free operation and proper quality products release [5–8] that is why development of units and mechanisms designs should be carried out along with engineering calculations of effective loads and strength analysis [9]. Moreover, when designing any equipment, safety and reliability of working tools [10–13], design specifics, efficient and sustainable distribution as well as weight and energy consumption reduction should be taken into consideration [14, 15].

At present, metallurgical companies involved in manufacture of steel seamless pipes use steel continuous cast billets (CCB) to manufacture initial billets [12, 16–19]. First of all, transfer to CCB application is attributed to the advantages and benefits of this method in terms of manufacturing costs reduction and through yield output increase. However, there is a number of restrictions imposed on the application of CCB. A wide range of seamless pipes manufactured using the same unit implies application of different diameters billets. At the same time, manufacture of a huge number of CCB of different diameters is quite cost-ineffective. Secondly, possible defects of CCB may result in rejection of non-repairable pipes due to internal and external slivers. Three-roll radial-shear rolling (RSR) mill can be used to decrease the size range of billets and pre-strain prior to piercing [20]. This method is used to manufacture round bars from different types of steel [21, 22], aluminum [23] and titanium [24, 25]. Results of numerous investigations prove that combination of large feed angles and special calibration of rolls will allow decreasing the centerline porosity and grain size variation along the bars cross-section [26, 27].

Pervouralsk New tube (Novotrubny) Works, JSC, plans to install a RSR stand to perform pipe billets pre-deforming that is used to manufacture pipes using Pipe Rolling Plant 160 (PRP-160) [28]. Since the existing equipment configuration in the shop makes it impossible to install an additional unit, the option of mounting a stand in the PRP-160 line by fast replacement of the three-roll reeling mill stand with the special stand intended to perform reduction of solid billets is considered. Therefore, the new equipment should integrate with the existing drive of the reeling mill stand and operate in conjunction with the equipment of the ingoing and outgoing side.

The objective of this study is to develop the design of a special RSR mill stand intended to perform deformation of CCB. This will allow increasing the total strain degree and reducing the number of initial billets sizes.

2. Development of the stand basic design

Since the developed stand should be integrated into the existing manufacturing line and installed instead of the reeling mill stand, a number of restrictions were considered during the design stage, namely: the stand group drive with the limit cross angle for spindle heads of 15°; maximum torque of the working roll of 40 kN·m, the main drive capacity of 1.6 MW; the possibility to integrate with the existing ingoing and outgoing side of the reeling mill. The specification summary is given in Table 1.

Table 1. The specification summary for RSR stand rolling mill.

Parameter name	Value
Billet diameter, mm	120–220
Roll diameter, mm	290
Roll body length, mm	350
Feed angle, degree	18
Toe angle, degree	5
Screw-down mechanism	Electromechanical (screw and nut)
Type of drive	Group (drive TRP-160)
Drive power, kW	1600

The developed cage of RSR rolling mill (Figure 1, a) had a massive welded box-type frame made from plates. Two cartridges with top working rolls and a cartridge with the bottom stationary working roll are placed in through openings. No change in the feed angle or toe angle is provided for the structure design; these values are set structurally based on the need to match the existing structure of the PRP-160 drive. The stand is mounted directly on the bottom of the existing roll housing (Figure 1, b). RSR mill stand is fastened to the base using wedges similarly to the way the top part of the reeling mill stand is fixed. The roll housing of RSR mill consists of two vertical plates with the thickness of 50 mm that are mounted on the base. Similar welded pedestal-shaped structures are mounted to the left and to the right of base plates to be installed on the base and fastened with wedges. To make the frame stiff, the base is reinforced with stiffeners that are 70 mm thick. Massive plates with the thickness of 50 mm are welded inside the plates slots and are used as guides for rolls cartridges installation. The bottom roll should be attached rigidly to the base plate. The rolling force should be transferred from the bottom roll to the frame plate.

The working roll cartridge consists of the cartridge with bore holes where the working roll and the chock are installed. Working rolls chocks should be fixed in cartridge bore holes through the projected parts. The roll mounted on slide bearings of chocks.

The screw-down mechanism of working rolls top cartridge is made in the form of screw-nut pair and is driven in rotation due to motor-gear mounted on the frame top. Every top cartridge fastened using the hydraulic counter weight to avoid any possible gaps between the cartridge, the crew-down mechanism and the frame.

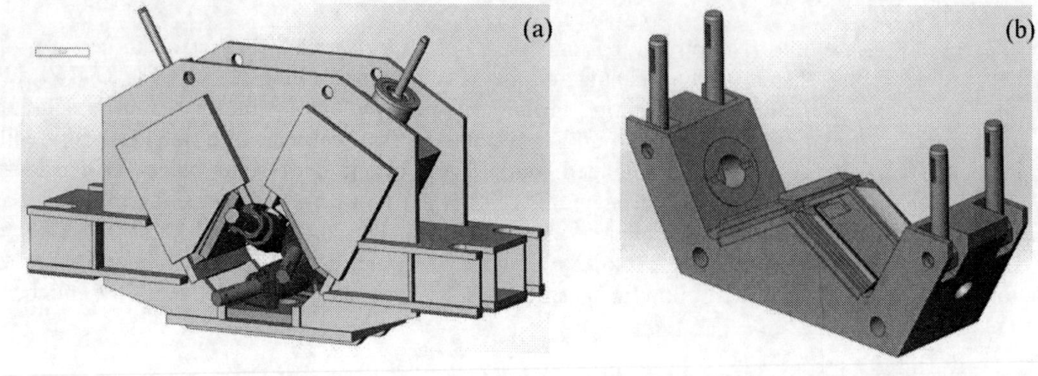

Figure 1. Three-roll stand of RSR mill (a) and the base of the reeling mill PRP-160 frame (b).

Since the bottom roll is stationary, the rolling axis position in case of change into a different billet diameter will be changed. This option should be considered during re-adjustment of the ingoing and outgoing side.

3. Elaboration of integration and checking calculations

3.1. Elaboration of the stand integration with the main drive

The major task of the stand design is to join and match the working rolls and the existing drive of the PRP-160 reeling mill stand. The main drive of the reeling mill stand is of group type and consists of electric motor, gearbox, several sections of the main shaft and multipurpose spindles. The decision to change the configuration of working rolls by rotating them relative to the rolling axis by 60° (Figure 2) was taken to reduce the cross angles for the spindle set when mounting RSR mill stand with increased feed angles (18–21°).

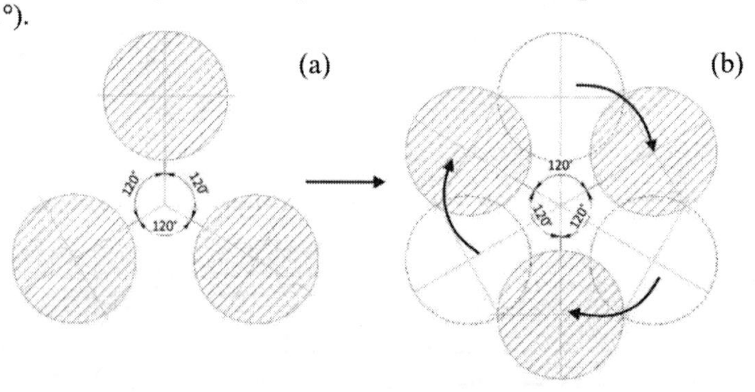

Figure 2. The original position of PRP-160 rolls (a) and the layout of rolls rotation (for the special stand) (b).

Then possible cross angles were assessed depending on the working rolls positioning in the stand (feed angle, toe angle, working rolls installation as per the roll pass) for the selected configuration of working rolls. The analysis was performed using the SolidWorks software by 3D modeling parts and assemblies of the stand. Therefore, a three-dimensional model of the multipurpose spindle was developed based on available drawings. The toe angle varied from 2° to 5°. The feed angle varied from 15° to 20°. The roll pass diameters (of bars manufactured) were selected within the range of 90…190 mm for the analysis. Results of measurements taken are specified in Table 2.

Table 2. The correlation between the maximum cross angle and the varying parameters.

Varying parameter		Maximum cross angle
Toe angle δ (°) (β = 18°, $D_к$ = 100 mm)	2	14.61
	3	14.68
	4	14.85
	5	15.12
Feed angle β (°) (δ = 5°, $D_к$ = 100 mm)	15	11.59
	18	15.12
	20	17.52
Distance between rolls $D_к$ (mm) (δ = 5°, β = 18°)	90	15.10
	100	15.12
	125	15.15
	150	15.19
	170	15.22
	190	15.25

If the toe angle increases by 1°, the cross angle of the spindle increases by 0.2° as an average, in case the feed angle increases by 1°, the cross angle of the spindle increases by more than 1°. Changing the distance between the rolls does not significantly affect the spatial cross angle of the spindle half coupling. In case of any change within the range from 90 to 190 mm, the angle changes by 0.15°.

Considering the obtained measuring data and the above conditions of roll sets configuration we decided to use working rolls with the diameter of 290 mm, feed angle β = 18° and toe angle δ = 3–4° in the structure design.

3.2. Elaboration of the stand integration against the allowable torque

Another important evaluation criterion for the equipment design is the level of power parameters – the force of metal applied on the roll, the rolling torque and the rolling power. Power parameters (force and torque) were calculated as part of the study for the rolling process of CCB with the diameter of 156 and 220 mm.

Calculations were made for billets reduction with the diameter of 156 mm to achieve bars with the diameter of 140–90 mm as well as the billets with the diameter of 220 mm to achieve the bars with the diameter of 210–170 mm. The rolls diameter is taken to be equal to 290 mm. The feed angle is equal to 18 degrees, the taper angle within the deformation zone of the reduced area is 10 degrees, the coefficient of tangential reeling within the reduced area is 0.3, and the number of cycles within the sizing area is 3. Calculations results are given in Figure 3.

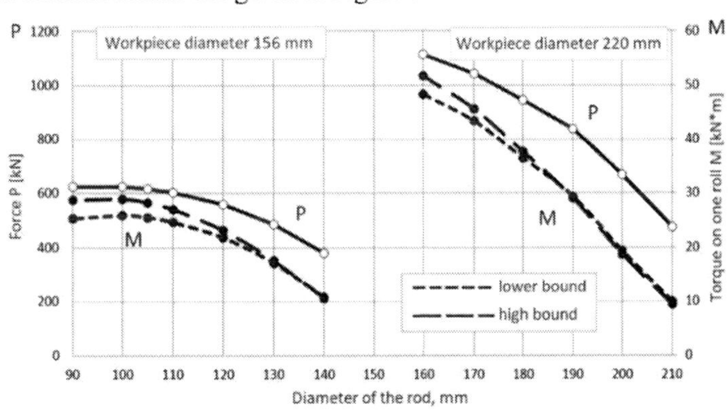

Figure 3. Effect of the rolled stock diameter on force P and torque M and M_c when reducing workpieces with the diameter of 156 and 220 mm.

For billets with the diameter of 156 mm, the maximum reduction force for the diameter of 90 mm is about 630 kN, while the maximum rolling torque is less than 30 kN·m. For billets with the diameter

IOP Conf. Series: Materials Science and Engineering **966** (2020) 012074 doi:10.1088/1757-899X/966/1/012074

of 220 mm, the maximum reduction force for the diameter of 160 mm will be 1100 kN, while the rolling torque will be about 50 kN·m. Since the spindle units are designed for the maximum torque of 40 kN·m, one-pass reduction can result in the diameter of not less than 180 mm for the billets with the diameter of 220 mm. Based on this restriction, the maximum force when calculating strains and stresses of the design stand is selected to be equal to 1000 kN.

3.3. Checking calculation of the stand structure strength

The rigidity of the designed stand structure was preliminarily assessed using SolidWorks Simulation to perform verification of the design concept.

Top rolls and cartridges were excluded from the calculations to make the design model simpler. Steel grade A 516-55 (AISI) is taken for all the frame parts from the CAD system, the roll material is 55NiCrMoV 6 (L6 AISI), the bearings material is bronze (Cu-Al-Fe-Mn). The stand is installed on the base of the reeling mill PRP-160. Conditional rigid fastening of support surfaces in SolidWorks is shown in Figure 4 in green arrows. Load application areas are conditionally shown in purple arrows in Figure 4. The force is applied to bottom rolls for calculations. Since top rolls and cartridges are excluded from calculations to make the model simpler, the force of top rolls is applied to adjusting thrust screws, since they will take up the major load transmitted by rolls sets.

Stress diagrams (Figure 4, a) and cage frame absolute displacement (Figure 4, b) are found as a result of calculations. The obtained data show that maximum stresses occur in the rolls set and amount to 140 MPa. These values occur within the load application areas and acute angles of parts mating. However, the design model has a number of assumptions; in particular, rounding, bevels and welded joints are not taken into consideration, that is why it can be assumed that no peak stresses will occur within the areas of frame parts joining. For the major part of the frame, stresses do not exceed 60 MPa that is 3.7 times less than the yield strength for steel grade A 516-55 (221 MPa). Therefore, sufficient strength margin for the stand frame is ensured.

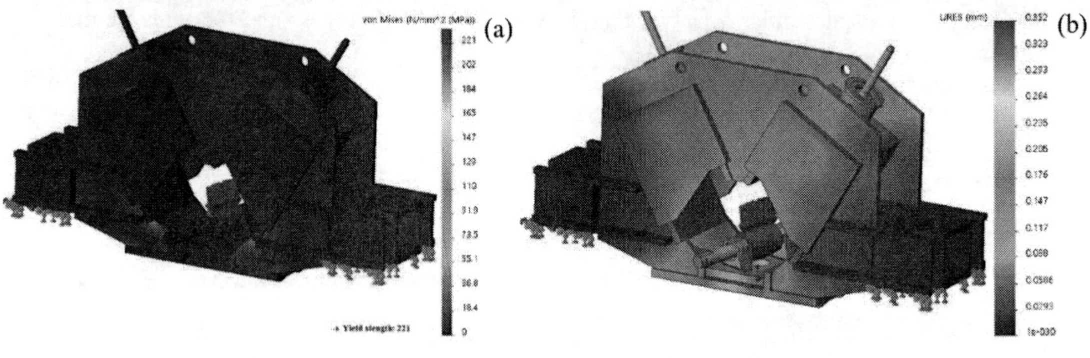

Figure 4. Stress diagram (a) and stand absolute displacement diagram (b).

The maximum displacement under load makes 0.35 mm and occurs on the bottom roll within the load application area. The frame base is additionally reinforced with stiffeners to reduce bottom rolls set strain. The design rigidity rate will be 2.86 MN/mm.

Results of virtual experiment show that the stand has sufficient strength and rigidity margin.

4. Conclusion

The article offers the design of the three-roll stand for RSR mill to manufacture CCB using PRP-160 manufacture line in Pervouralsk New Tube (Novotrubny) Works, JSC. The issues associated with the equipment installation, mating with the existing drive of the reeling mill, as well as checking calculations against the allowable force and rolling torque are taken into consideration during the design. The developed equipment design has sufficient strength margin and provides for the required manufacturing capacity.

References

[1] Romantsev B A, Gamin, Y V, Goncharuk A V and Aleshchenko A S 2017 *Metallurgist* **61** 217–22
[2] Galkin S P, Romantsev B A, Ta D X and Gamin Yu V 2018 *Chernye Metally* **4** 20-7
[3] Galkin S P, Goncharuk A V, Daeva E K, Mikhailov V K and Romantsev B A 2003 *Steel in Transl.* **33(9)** 45-7
[4] Romantsev B A, Aleshchenko A S, Goncharuk A V and Galkin S P 2012 *Metallurgist* **55** 918–24
[5] Romantsev B A, Goncharuk A V, Aleshchenko A S, Onuchin A S and Gamin Yu V 2015 *Metallurgist* **59** 386–9
[6] Romantsev B A, Goncharuk A V, Aleshchenko A S and Krasil'shchikov M V 2012 *Steel in Transl.* **42** 769–70
[7] Shatalov R L, Medvedev V A 2020 *Metallurgist* **63 (9-10)** 1071-6
[8] Shatalov R L, Medvedev V A 2019 *Metallurgist*, **63 (1-2)** 176-82
[9] Overhagen C, Braun R, and Deike R 2020 *tm - Technisches Messen* **87(5)** 343-8.
[10] Aleshchenko A S, Gamin Yu V, Chan B K and Tsyutsyura V Y 2018 *Chernye Metally* **8** 63-70.
[11] Romantsev B A, Matyko O K, Goncharuk A V, Aleshchenko A S and Ananyan V V 2008 *Steel in Transl.* **38** 897–9
[12] Romantsev B A, Aleshchenko A S, Tsyutsyura V Y, Tyshchuk I N and Lube I I 2017 *Metallurgist* **60** 1062–9
[13] Shatalov R L, Grigoryan G G, Sharov A A and Savin V A 1987 *Steel in the USSR* **17 (9)** 416-9
[14] Overhagen C and Mauk P J 2014 *Key Eng. Mater.* **622–623** 879–86
[15] Budnikov A S, Romantsev B A and Kharitonov E A 2018 *Izv. Ferrous Metall* **61 (9)** 683-8
[16] Romantsev B, Goncharuk A, Aleshchenko A, Gamin Y and Mintakhanov M 2018 *Int. J. Adv. Manuf. Technol.* **97** 3223–30
[17] Smirnov E N, Smirnov A N, Sklyar V A, Mikheev V V, Belevitin V A, Orlov G A and Verzilov A P 2019 *Steel in Transl.* **49** 631–7
[18] Marchenko K L, Fadeev M M, Romantsev B A, Goncharuk A V and Ananyan V V 2008 *Steel in Transl.* **38** 402–4
[19] Matyko O K, Fartushnyi R N, Mul'chin V V et al. 2009 *Steel in Transl.* **39** 161–3
[20] Galkin S P, Romantsev B A and Kharitonov E A 2014 *CIS Iron Steel Rev* **9** 35-9
[21] Galkin S P, Kharitonov E A and Romanenko V P 2009 *Steel in Transl.* **39** 700
[22] Malakhov A Yu, Saikov I V, Denisov I V, Cherezov N P, Gamin Yu V and Romantsev B A 2020 *IOP Conf. Series: J. of Phys.: Conf. Series* **1431** 012033
[23] Gamin Y, Akopyan T, Koshmin A, Dolbachev A, Aleshchenko A, Galkin S P and Romantsev B A *2020 Int. J. Adv. Manuf. Technol.* **108** 695–704
[24] Karpov B V, Patrin P V, Galkin S P, Skripalenko M N, Samusev S V, Huy T B and Pavlov S A 2018 *Metallurgist* **61** 884–90
[25] Romancev B A, Goncharuk A V, Aleshchenko A S and Gamin Yu V 2015 *Russ. J. Non-ferrous Metals* **56** 522–6
[26] Akopyan T K, Aleshchenko A S, Belov N A and Galkin S P 2018 *Phys. Metals Metallogr.* **119** 241–50
[27] Shurkin P K, Akopyan T K, Galkin S P and Aleshchenko A S 2019 *Met Sci Heat Treat* **60** 764–9
[28] Aleshchenko A S, Budnikov A S and Kharitonov E A 2019 *Izvestiya. Ferrous Metallurgy* **62 (10)** 756-62

Determination of cutting forces based on DMG MORI CTX300 ecoline CNC lathe drive power data

A P Starostin[1, 2, 3] and D M Elkind[1]

[1] Ural Federal University named after the First President of Russia B N Yeltsin, 19, Mira str., Yekaterinburg, 620000, Russia
[2] Ural Federal University named after the First President of Russia B N Yeltsin, Nizhniy Tagil Technological Institute, 59, Krasnogvardeyskaya str., Nizhniy Tagil, 622000, Russia

E-mail: [3] andrey.starostin@urfu.ru

Abstract. Modern machine building needs demand for machines and control systems which provide higher accuracy of machining. For this purpose, setup methods are being researched and accuracy of machines and devices is being improved. The main topic of this paper is a tool path predistortion system which is necessary for improvement of shaft turning accuracy when a steady rest cannot be used on a CNC machines. Calculation of such tool path requires one to know the cutting force during turning on a CNC lathe. The research in question allowed to determine the dependence of cutting force during turning on the power consumed by the feed drive, cutting depth, cutting velocity, and feed value. An equation for further calculation applicable to a specific lathe in the turning process was derived. This equation serves as basis for adaptive turning of shafts and other parts with the aid of a mathematical model, which must account for the cutting force. Application of this formula in a CNC controlled lathe's parametric program will provide for higher accuracy of turning without special devices.

1. Introduction

Machining accuracy improvement is one of research trends in modern machine and instrument building. Depending on work piece fixation practice various deviations of form develop in the process of machining of long nonrigid parts, such as shafts. For such scenarios path predistortion [1] is one of the ways of accuracy improvement for CNC lathes. Other accuracy improvement methods are available for nonrigid shaft machining [2–7].

In actual production the problem of form in the chuck-center setup is addressed only in case of tapering. Development of tapering during shaft machining is related to varying compliance of headstock and tailstock. To solve this problem the machining program and tool path are adjusted after preliminary machining to compensate for the deviation by means of linear interpolation equal to the deviation value. Thus, additional time is needed to perform intermediate inspection and qualified CNC machinists to adjust the machining program.

In order to calculate the tool path along the shaft it is necessary to determine the cutting force basing on work piece machining modes. The actual cutting force during cutting will differ from theoretical value calculated by various methods [8]. Hence, the values need to be determined directly during cutting to increase calculation accuracy. Special equipment is required to measure the cutting

IOP Conf. Series: Materials Science and Engineering **966** (2020) 012075 doi:10.1088/1757-899X/966/1/012075

force and installation of such equipment in an attempt of implementing adaptive machining with obtainment of cutting force reading in real production would be problematic for everyday operations.

2. Research methodology

The task set in the course of this research was to determine the cutting force on the basis of cutting modes and CNC lathe parameters collected during task execution by the CNC unit.

Four resistive strain gauges were installed on a straight-turning tool (Figure 1) and connected in a half-bridge circuit to decrease the impact of noises and obtain readings for the X and Z axes. Furthermore, to increase signal accuracy and eliminate noise of other grid sources the resistive strain gauges were connected to an ATMEL controller via a 24-bit analog-digital converter with subsequent calibration of the whole system on the lathe by means of an electronic dynamometer. A 50 mm starting diameter shaft made of 'grade 45' steel was turned on the lathe in order to determine the dependency of cutting modes on the lathe's feed drive power. The length of the shaft was divided into three sections which were then turned at different feeds. Thus, by turning one diameter it was possible to obtain the data of feed dependence on the feed drive power. The dependence of cutting velocity on drive power was determined on different diameters under identical shaft rotational speed.

Figure 1. General view of the instrument with gain plate and strain gauges.

Readings of the strain gauges were taken continuously with a time step of 0.13s, which is comparable to the lathe's CNC system drive power data recording time step. After acquisition, a comparison of the data the starting point of the tool incision and strain gauges data based on abrupt drive power increase, as well as many-fold increasing load on the tool was performed. This resulted in a set of 2400 points at different cutting modes.

3. Experimental Studies

In the course of the experiment set up to determine the dependency of cutting force on power consumed by the feed drive several data sets have been produced, i.e. cutting velocity of 100–200 m/min, 0.15–0.25 mm/rev feed, and 0.1 to 1.5 mm depth of cut.

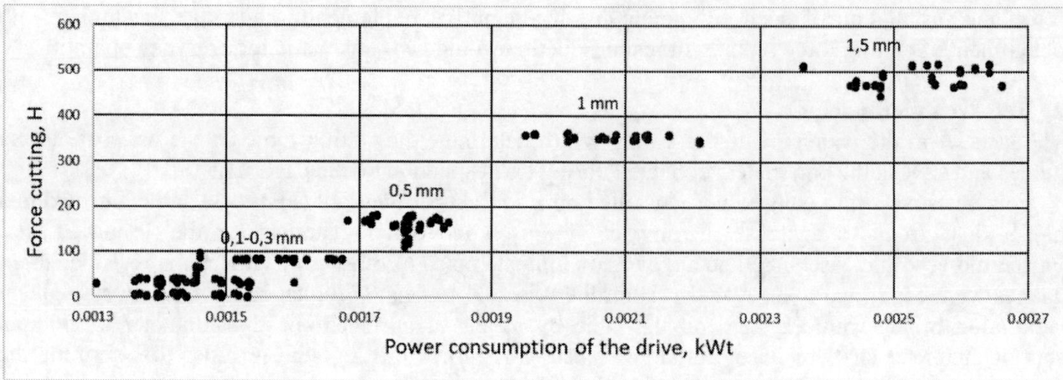

Figure 2. The graph of cutting the distribution of cutting force depending on the drive power consumption with at depth of cutting 0.1–1.5 mm.

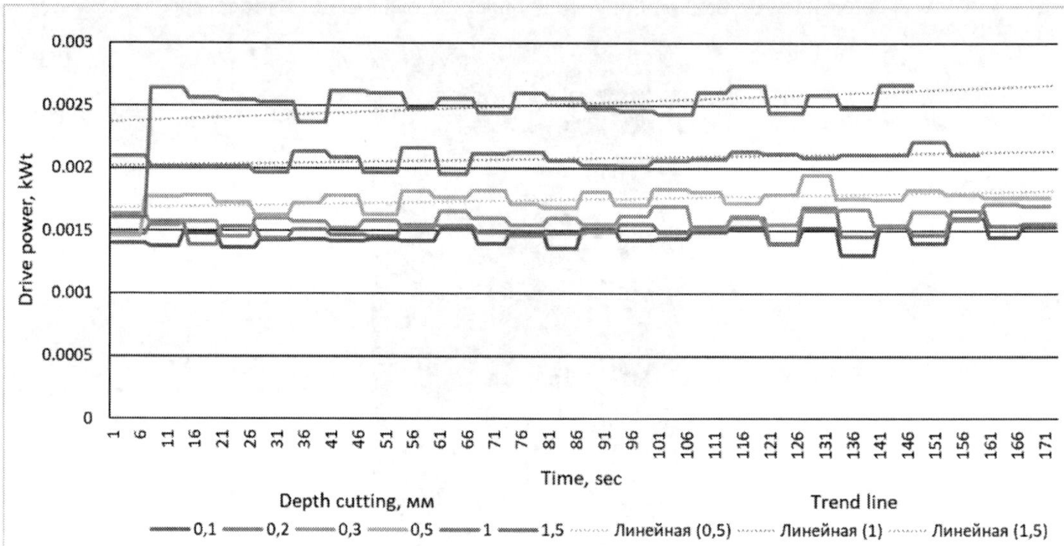

Figure 3. The graph of cutting the distribution of cutting force depending on the drive power consumption with at depth of cutting 0.1–1.5 mm.

Figure 2 shows several large concentrations of points which correspond to various cutting depths at a feed rate of 0.15 mm/rev. Similar graphs of cutting force dependence on drive consumed power can be made for feed rates of 0.1; 0.5; 0.75; 1; 1.5 mm. Changes on consumed power during turning are illustrated in Figure 3. This graph shows inconsistency of the load on the drive, which may be caused by variation in cutting, forces during tool movement. Besides cutting modes, this variation is prone to other factors, such as material hardness, material inhomogeneity, etc. Abrupt consumed power spiking is associated with sensitivity of the feed drive load sensor. Trend lines of 0.5 and 1.5 mm depth of cutting have been plotted in this chart for clarity. These trends illustrate the increase in consumed power during tool movements to the chuck due to decreasing shaft compliance and stock curvature and increasing depth of cut.

To derive the dependencies, 22 data sets were made under various cutting modes.

A regression analysis of variance [9] was performed basing on the obtained data.

As a result of this analysis, the following dependency between cutting force and power was obtained for the DMG MORI CTX300 ecoline lathe:

R-square coefficient of determination: 0.815

IOP Conf. Series: Materials Science and Engineering **966** (2020) 012075 doi:10.1088/1757-899X/966/1/012075

The equation of the cutting force along the X axis of the DMG MORI CTX310 ecoline lathe:

$$P = -68{,}6082 + 270{,}354 * S + 204{,}815 * t + 59312{,}668 * N - 0{,}462 * v,$$

where S – feed rate; t – depth of cut; N – power consumption of the drive; v – cutting velocity.

The coefficient of determination shows that the created model proves variance and dependence of variables, i.e. depth of cut, feed, drive consumed power, and cutting velocity.

4. Conclusion

Based on this model a practical calculation of the lathe operational cutting force can be made and used to calculate the tool path predistortion during work piece turning.

The derived equation specific to the DMG MORI CTX300 ecoline lathe lacks referencing to referential power ratios of a generally accepted equation for cutting force calculation, such as depth of cut, feed, and cutting velocity. Therefore, such ratios are replaced by drive power consumed at a specific moment of time. The calculated cutting force was verified during subsequent shaft turning operations basing on comparison of feed drive power consumption figures of the CNC unit and actual cutting force measured by the strain gauges. Difference between calculated and measured cutting force comprised below 5%.

Basing on this research it is possible to increase the qualitative characteristics of CNC lathe turning by accounting for the cutting force during turning on the lathe. This also provides an opportunity for parametric lathe programming on the basis of data acquired by the lathe in the process of turning, i.e. depth of cut, feed, and drive power at a given moment.

Further experimentation is necessary to increase calculation accuracy which would require changing of the main angle in plane to derive a comprehensive equation for determination of in-process tool cutting force. Also, to improve the accuracy and comparison of measured power consumption and cutting force on the tool, a single controller is required combining the measurement of drive current consumption with the simultaneous reading of strain gauge signals.

References

[1] Pegashkin V F and Starostin A P 2018 Increase of accuracy of processing of non-rigid parts in centrals on the machine with CNC *Bulletin of SUSU. Series 'Mechanical Engineering'* **18 (1)** pp 51–7

[2] Karabulut A 2010 Experimental investigation of diametric errors during cylindrical turning. *Journal of the faculty of engineering and architecture of Gazi university* **25 (2)** pp 257–65

[3] Karabulut A 2010 Determination of diametral error using finite elements and experimental method *Metalurgija* **49 (1)** pp 57–60

[4] Pisarciuc C 2016 The use of statistical process control to improve the accuracy of turning *IOP Conf. Ser.: Mater. Sci. Eng.* **161**

[5] Swic A, Wolos D, Zubrzycki J, Opielak M, Gola A and Taranenko V 2014 Accuracy Control in the Machining of Low Rigidity Shafts. *Industrial and service robotics. Applied Mechanics and Materials* **613** p 357

[6] Swic A, Wolos D and Litak G 2003 Method of control of machining accuracy of low-rigidity elastic-deformable shafts *Latin American journal of solids and structures* **23 (3)** pp 221–32

[7] Swic A, Gola A, Wolos D, Opielak M 2017 Micro-geometry Surface Modelling in the Process of Low-Rigidity Elastic-Deformable Shafts Turning *Iranian journal of science and technology-transactions of mechanical engineering* **41 (2)** pp 159–67

[8] Phan A, Baron L, Mayer J and Cloutier G 2003 Finite element and experimental studies of diametral errors in cantilever bar turning Applied mathematical modelling **27 (3)** pp 221–32

[9] Mel'nichenko A S 2014 *Data Analysis in Materials Science. Part 2. Regression analysis* (Moscow: NITU MISIS) p 87 [In Russian]

IOP Conf. Series: Materials Science and Engineering **966** (2020) 012076 doi:10.1088/1757-899X/966/1/012076

Development of the identifier system of indicators for improving the information potential of the holding

E V Dolzhenkova[1, 2] and L V Iurieva[1]

[1] Ural Federal University named after the First President of Russia B N Yeltsin, Nizhniy Tagil Technological Institute, 59, Krasnogvardeyskaya str., Nizhniy Tagil, 622000, Russia

E-mail: [2] lenag1981@mail.ru

Abstract. The article presents the system of competitively-balanced indicators, which makes it possible to assess different spheres of activity of industrial enterprises of the holding for the purpose of improving its information potential. In order to manage metallurgical holding activities, it is proposed to combine accounting and analytical procedures used for strategic decision-making aimed at information potential increasing. For this purpose, the authors' identifier system of indicators includes three blocks: strategic, operational and tactical. When classifying indicators according to the management levels, the following points are taken into account: preliminary analysis of the structural elements in each indicator, consideration of triggers, need for expert assessment. Basing on values dynamics, the results of the calculations make it possible to identify positive trends for some enterprises and negative trends for other enterprises of the holding. They also help to determine the reasons for the increase or decrease of the integral evaluation of one metallurgical enterprise compared to another. All these increases the efficiency of strategic decision-making in the holding management. Besides, the proposed model gives the possibility for determining the assumptions for its implementation in each metallurgical enterprise, thereby increasing the information potential of the entire holding.

1. Introduction

The relevance of effective strategic decision-making by the management of metallurgical holdings under conditions of high uncertainty is closely related to the information preparation of data for this purpose. Modern digital economy requires the creation of such an accounting and analytical model that can satisfy the growing information demands of the management system. Such a model is to correspond with the development strategy of the metallurgical holding enterprise on the one hand, and, to be its information potential on the other hand.

Many foreign and Russian scientists were engaged in strategic management, its evaluation and analysis. Scientific works [1, 2] explore innovative strategies. The authors propose innovative strategies with specific product innovation packages. They believe that a stronger focus on product characteristics and improved market access encourage the efficient replication of existing capabilities of an enterprise with immediate benefits.

Empirical studies on the evaluation of the effectiveness of organizations management are the subject of scientific works [3, 4], where the authors show the evidence that financial indicators are a critical problem in its evaluation. Scientists reviewed the mechanisms for managing the efficiency of management strategic initiatives based on the analysis of critical sales processes in the metallurgical business [4].

Entrepreneurial risks are becoming an important aspect of strategic decision-making in today's circumstances [5–8]. Dynamic and organizational possibilities and their influence on the work of enterprises are investigated in works [9–13].

The main source of corporate efficiency – business process management is discussed in the following papers [7, 14].

The following works [15–18] analyze and evaluate competitive advantages in strategic management. Methods of competitive analysis in modern reality are investigated in such works as [19–23].

The review of these scientific papers made it possible to generalize gained experience and special features of strategic analysis to assess activities of metallurgical enterprises under the circumstances of uncertainties in the milieu. It also permits to note insufficiency of research in the multidimensional analysis of the competitiveness of metallurgical enterprises, as well as the development of a competitive strategic model for the evaluation of metallurgical enterprises activities.

All aforementioned determines the relevance of the topic of this research.

2. Research Methods

The state of competition in the industry, according to Porter's concept, depends on the influence of five main forces. These competitive forces determine the attractiveness of the industry and the position of the holding in competitive activity. Together, they assess the limiting profit potential of the industry, ranging from substantial to minimal. The pressure of cumulative forces on the metallurgical holding enterprises can be meaningful for all rival parties. To overcome it, the strategy should be based on a careful analysis of the origins of each component. Knowledge of these major sources of competitive pressure provides a stable foundation for a strategic action plan. It will give the possibility to identify the strengths and weaknesses of the holding enterprises, to give a clear justification for the position of the holding in the metallurgical industry. One of the most important indicators of the conditions of competition is the level of intensity in the market, which is determined by several more specific indicators. The most important of them are the profitability of the market, the rate of growth of the market, the type of the market, its volume and the distribution of market shares of competitors, the degree of concentration in the market [4].

Thus, there is a need to develop an identifier system of indicators (hereinafter ISI) in holding structures, which enables the integration of accounting and analytical procedures in order to form and improve the holding strategy for increasing its information potential.

The operation of the ISP is based on strategic management accounting (hereinafter SMA) and strategic competitive analysis in holding structures (hereinafter SCAHS). The outcome of the activity is the establishment of a system of competitively-balanced indicators (hereinafter SCBI) covering the main areas of holding activity.

When developing the SCBI for the metallurgical holding enterprises, factors relating to the static (in short-term period) strategic context and the dynamic (in short-term period) system of objectives should be assessed. At the same time, the ISI itself in the work of the holding structures forms a space in which the objectives are formulated and the SCBI is developed for each enterprise of the metallurgical holding.

According to requirements of the SCBI information provision, the information sources should be useful, relevant, timely, actual and comparable. The authors' model analyses the holding's activities in six key areas. Figure 1 shows the motion of information flows in the SCBI. Special role is assigned for SMA and SCAHS. From these analyses, information about competitors and intra-holding space can be derived.

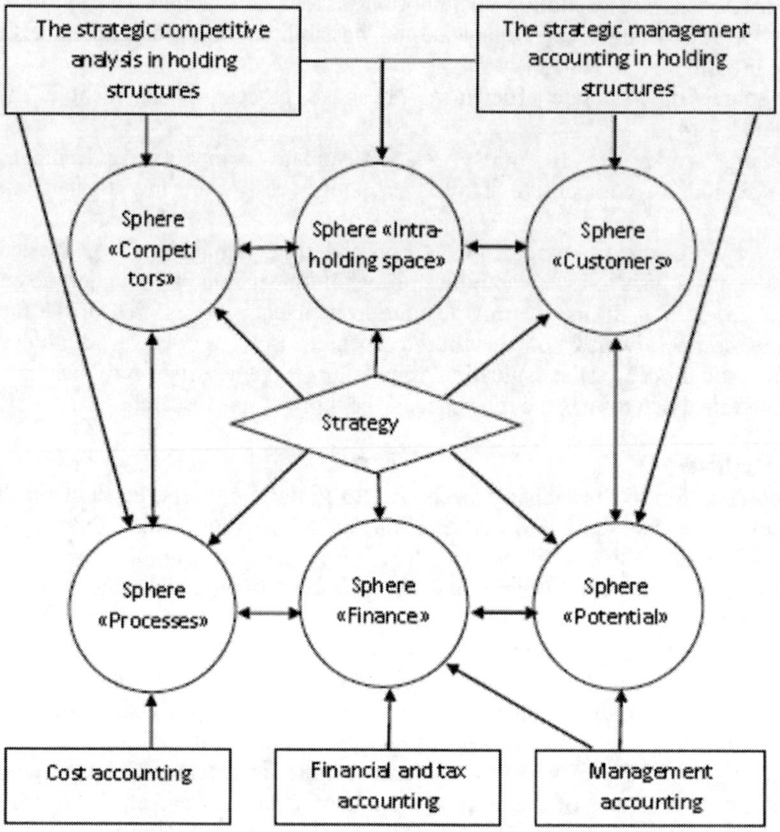

Figure 1. The motion of information flows in the SCBI.

When constructing a system of competitively-balanced indicators of ISI for a metallurgical holding, it is firstly recommended to classify such indicators according to the stages of milieu analysis noted in Figure 2 (1st indicator group is strategic bloc, 2nd indicator group is operational bloc, 3rd indicator group is tactical bloc). In addition, the classification should take into account the importance of the indicator for the management decisions of the metallurgical holding, which is compiled based on expert assessment.

The strategic bloc of indicators describes the mission and strategic objectives of the holding. With the help of this bloc indicators, it is possible to set the guidelines and focus in the activities of the holding enterprises for the next 4–5 years, which will increase the efficiency of strategic planning.

The operational bloc of indicators is mainly required for operational planning. The indicators of this bloc help to link the strategy of the metallurgical enterprise of the holding with its operational activities.

The tactical bloc of indicators is used only for current activities. These indicators generally reflect private processes and tasks which are managed in the short-term period.

Stages of strategic management of the holding	Stages of environmental analysis		
	Analysis of the external environment of the holding	Analysis of the internal environment of the holding	Analysis of the external environment of the holding enterprises
Holding activity analysis			
Planning and strategy development			
Implementation of the strategy, control of its implementation and management decisions			

Figure 2. The main stages of SCBI for ISI.

The following algorithm should be used to construct the three blocks of indicators:

- Analyze the main elements that are involved in the calculation of the indicator.
- Identify key triggers and explore possible ways to influence them.
- Based on the analysis of the impact, take a decision on the relation of the indicator to a certain block.

The authors' ISI classifies the indicators into the following three groups of importance based on expert assessment.

Group 1. This group includes the most significant indicators for the holding strategy. These indicators reflect the general processes taking place in different areas of the SCBI. For the planning and integration of such indicators it is proposed to develop special set of regulations.

Group 2. The indicators for this group reflect private processes. As a rule, these indicators regulate normative values, and their planning and accounting at predetermined intervals are not mandatory. The values of the indicators are given at certain intervals. Such indicators will be taken into consideration in the implementation of the holding strategy only if their values exceed a given interval.

Group 3. This group includes indicators that measure private processes and tasks, as well as duplicate indicators of the first and second groups.

Such analysis will significantly increase the information potential of the holding in strategic decision-making.

3. Results and discussion

On the basis of the analysis carried out, the indicators of the different areas of the SCBI have been identified. The SCBI indicators array is defined as a set of different measuring instruments that can be grouped according to the main aspects of the metallurgical holding activities. The presented range of indicators is not final and can be edited at the request of users.

Calculation of indicators is proposed to be made in the following areas: financial management; process management; innovation management; service; intra-holding space; competitiveness.

Each of these areas is characterized by a system of indices and indicators. Indicators describe the qualitative constituent of each SCBI area and indices describe its quantitative constituent. On the basis of the calculated indices it is necessary to make an integral assessment of the impact of each indices area on the competitiveness of the holding enterprise as a whole (see Table 1).

Guided by the proposed model, the authors calculated indices for all spheres of the SCBI at metallurgical enterprises, included into the United Metallurgical Company: Vyksa Metallurgical Plant, Chusovoy Metallurgical Plant and Schelkovo Metallurgical Plant.

Table 1. Repair service state indexes.

Index	Importance group
SCBI Sphere	Calculation formula
1. Finance	$I^{f}_{integr} = \sum_{i=1}^{6} I^{f}_{i} \cdot d_{ig}$
2. Processes	$I^{pr}_{integr} = \sum_{i=1}^{9} I^{pr}_{i} \cdot d_{ig}$
3. Potential	$I^{p}_{integr} = \sum_{i=1}^{9} I^{p}_{i} \cdot d_{ig}$
4. Customers	$I^{cu}_{integr} = \sum_{i=1}^{2} I^{cu}_{i} \cdot d_{ig}$
5. Intra-holding space	$I^{i}_{integr} = \sum_{i=1}^{16} I^{i}_{i} \cdot d_{ig}$
6. Competitors	$I^{c}_{integr} = \sum_{i=1}^{8} I^{c}_{i} \cdot d_{ig}$

d_{ig} – Group of the indicators importance influences the significance value of the indicator in the integral assessment according to SCBI spheres of activity: significance value of Group 1 – 0.5, Group 2 – 0.3, Group 3 – 0.2. This ratio was determined on the basis of expert assessments given by the heads of metallurgical enterprises.

The results of the calculation show positive trends in changes of indicator values for some enterprises (e.g. Vyksa Metallurgical Plant, Chusovoy Metallurgical Plant) and negative trends in the indicator values for other enterprises (e.g. Schelkovo Metallurgical Plant). It should be noted that the model was implemented at enterprises with the following assumptions:

- indices were defined only for those areas where information was available;
- the output volumes of iron, steel and metal products in general were taken as the main production of the enterprises.

The results of the calculation suggest that Vyksa Metallurgical Plant is the most competitive. Its calculated indices are higher than those of other enterprises for almost all spheres. In addition, the calculations reveal the reasons for the lag or exceedance of the integral assessment of one enterprise over another, increasing the efficiency of strategic decision-making.

4. Conclusion

Thus, the authors' ISI can be used for strategic decision-making to increase the information potential of the metallurgical holding. Its monitoring is based on the strategic management accounting and competitive analysis. On the basis of dynamics of the ICI values it is possible to identify positive trends for some enterprises and negative trends for other enterprises of the holding. The dynamics of the ICI values also identifies the reasons for the increase or decrease of the integral evaluation of one metallurgical enterprise compared to another. This increases the efficiency of strategic decision-making in holding management.

References

[1] Branzei O and Vertinsky I 2006 *Journal of Business Venturing* **21 (1)** pp 75–105

[2] McKenny A F, Short J C, Ketchen D J, Payne G T and Moss T W 2018 *Strategic Entrepreneurship Journal* **12 (4)** pp 504–21
[3] Carton R B and Hofer C W 2006 *Measuring Organizational Performance: Metrics for Entrepreneurship and Strategic Management Research* p 296
[4] Rocha C F, Duclos L C, da Veiga C P, dos Santos C B and Neves N A F 2016 *International Business Management* **10 (4)** pp 357–69
[5] Caves R E and Porter M E 1977 *Quarterly Journal of Economics* **91 (2)** pp 241–61
[6] Covin J G and Lumpkin G T 2011 *Entrepreneurship: Theory and Practice* **35 (5)** pp 855–72
[7] Covin J G and Wales W J 2012 *Entrepreneurship: Theory and Practice* **36 (4)** pp 677–702
[8] Shams S M R 2016 *Marketing Intelligence and Planning* **34 (5)** pp 671–91
[9] Henri J-F 2006 *Accounting, Organizations and Society* **31 (6)** pp 529–58
[10] Teece D J, Pisano G and Shuen A 1997 *Strategic Management Journal* **18 (7)** pp 509–33
[11] Andreeva T and Ritala P 2016 *Baltic Journal of Management* **11 (3)** pp 238–59
[12] Barney J 1991 *Journal of Management* **17 (1)** pp 99–120
[13] Fainshmidt S, Pezeshkan A, Lance Frazier M, Nair A and Markowski E 2016 *Journal of Management Studies* **53 (8)** pp 1348–80
[14] Afflerbach P, Bolsinger M and Röglinger M 2016 *Business Research* **9 (2)** pp 335–75
[15] Ansoff I 2015 *New corporate strategy* (Saint Petersburg: Peter Com) p 408
[16] Gordon Ya 2006 *Target competition* (Moscow: Top) p 368
[17] Kaplan Robert C 2014 *Strategy-oriented organization* (Moscow: Olympus Business) p 392
[18] Porter M 2005 *Competitive strategy: a technique for analyzing industries and competitors* (Moscow: Alpina Business Books) p 454
[19] Blank I A 2013 *Financial strategy of the enterprise* (Kiev: Nika-Center, Elga) p 345
[20] Iurieva L, Kazakova M and Dolzhenkova E 2015 *Costs management accounting at industrial enterprises in conditions of the innovation economics* (Moscow: RUSCIENCE) p 290
[21] Koveny M 2004 *Strategic Breakthrough: Technologies for Implementing Corporate Strategy to Life* (Moscow: Alpina Business Books) p 232
[22] Mintzberg G 2001 *Strategic process* (Saint Petersburg: Peter) p 567
[23] Rampersad H K 2014 *Individual Balanced Scorecard* (Moscow: Olimp-Business) p 176

IOP Conf. Series: Materials Science and Engineering **966** (2020) 012077 doi:10.1088/1757-899X/966/1/012077

Ejector jet mill with external pressure

D S Barakovskikh[1, 2] and S F Shishkin[1]

[1] Department of equipment and automation of silicate productions of the Institute of new materials and technologies, Ural Federal University named after the First President of Russia B N Yeltsin, 19, Mira str., Yekaterinburg, 620000, Russia

E-mail: [2] d.s.barakovskikh@urfu.ru

Abstract. Jet mills are used to produce fine products in many industries. Spiral jet mills and fluidized bed jet mills are widely used. They differ in that the destruction of particles occurs as a result of the abrasion process. The use of impact ejector jet mills is limited by a relatively low consumable concentration of material. When the consumable concentration of the material increases the acceleration speed of particles and the material grinding degree decrease. To eliminate this disadvantage a new method and design of an ejector jet mill for fine grinding of bulk materials has been developed (patent of Russian Federation No. 2711252). The technological process occurs at excessive external pressure of the ejected gas. It is shown that it is feasible to increase the pressure at the beginning of the acceleration tube when the pressures ratio of the working and secondary gas is higher than the critical one. It increases the consumption concentration of bulk material and the acceleration speed of crushed particles. This allows reducing the specific energy consumption and increasing the grinding degree. Experimental studies have confirmed an increase in the consumption concentration by 1.6–1.9 times and an increase in the grinding degree by 1.3–1.7 times compared to a typical ejector jet mill.

1. Introduction

Jet grinding is a promising method for obtaining of fine materials being used in many industries where the quality of the finished product is of big importance. It is known that jet mills provide a relatively high purity of the finished product by chemical composition. This type of equipment uses the energy of compressed gas to accelerate and crush particles. Grinding occurs in a small volume with a high concentration of energy which is an advantage for the mechanical activation of materials. Having a large number of advantages jet mills are of both scientific and practical interest in the intensification of the grinding process [1, 2, 3].

Spiral jet mills and fluidized bed jet mills are widely used. They differ in that the main process of particle destruction is abrasion. Recent studies of fluidized bed jet mills are mainly aimed at modeling two-phase flows during grinding in a fluidized bed as well as the classification process, since this process affects the productivity and quality of the finished product [4, 5]. Studies of spiral mills are mainly aimed at modeling the particles movement in the grinding chamber and predicting the material grinding [6, 7]. The use of an impact ejector jet mills and countercurrent jet mills have not received the widespread due to abrasive wear and limited material concentration. In countercurrent jet mills the probability of particles collisions in jets with a low concentration is very low [2]. The papers [8, 9] present studies on optimization of the nozzle and acceleration tube in order to increase the particles velocity with a relatively low consumable concentration of solid particles for this type of mills.

The aim of this work is to increase the consumable concentration of the crushed material and the particles velocity in an ejector jet mill.

2. Main part

Material grinding in jet mills occurs as a result of the impact of particles on the barrier or due to the collision of two counter streams of particles. The grinding degree of the material is determined by the speed of particle acceleration and the specific energy consumption depends on the consumption concentration of the material which is determined by the ratio [10]:

$$\mu = \frac{G_m}{G_g}, \tag{1}$$

where μ – consumption concentrations of material kg/kg; G_m – capacity by material, kg/s; G_g – mass gas consumption, kg/s.

It is known that increasing the material concentration can increase productivity and reduce specific energy consumption [11]. It is possible to increase the acceleration speed and consumption concentration by increasing the pressure at the beginning of the acceleration tube. Gas dynamics rules [12] state that the maximum pressure is at the end of the mixing chamber and at the beginning of the acceleration tube it is determined by the dependence:

$$P_3 = \frac{\alpha P_1 + P_2}{\alpha + 1}, \tag{2}$$

where P_3 – the total pressure at the beginning of the acceleration tube; α – the main geometric parameter of the ejector equal to the ratio of the areas of the output sections of the nozzles for the ejecting and ejected gases; P_1 – the total pressure before the nozzle of the ejecting gas; P_2 – the external pressure of the ejected gas.

For ejector jet mills the initial section of the acceleration tube is a mixing chamber so the main geometric parameter of the ejector is found from the equation:

$$\alpha = \frac{F_1}{F_3 - F_1} = \frac{d_1^2}{D_3^2 - d_1^2} = \left(\left(\frac{D_3}{d_1} \right)^2 - 1 \right)^{-1}, \tag{3}$$

where F_1 – sectional area of nozzle, m^2; F_3 – sectional area of the acceleration tube, m^2; d_1 – diameter of the critical section of the nozzle, m; D_3 – diameter of the acceleration tube, m.

In typical ejector jet mills the pressure of the ejected gas P_2 is equal to the atmospheric pressure P_a. Therefore, according to equation (2) the pressure P_3 at the beginning of the acceleration tube is limited and usually does not exceed 2.5 atm. Its increase by increasing α and P_1 is impossible due to blocking the ejector [13]. A scheme of a typical ejector jet mill with a graph of the total pressure P change along the length of the acceleration tube L_t is shown in figure 1.

To increase the pressure at the beginning of the acceleration tube P_3 a method has been developed (patent of Russian Federation No. 2711252 [10]). It differs in that the ejection is performed at an external pressure of the ejected gas above atmospheric pressure $P_2 > P_a$. At the same time the ratio of pressure in front of the nozzle and outside of the nozzle (P_1/P_2) must be greater than the critical one to keep supersonic gas velocity at the outlet of the nozzle, i.e. the condition [12]:

$$\Pi = \frac{P_1}{P_2} > \frac{P_1}{P_k} = \left(\frac{k+1}{2} \right)^{\frac{k}{k-1}} \approx 1.893, \tag{4}$$

where Π – the pressure ratio; P_k – critical gas pressure before the nozzle; k – the adiabatic coefficient (for air $k = 1.4$).

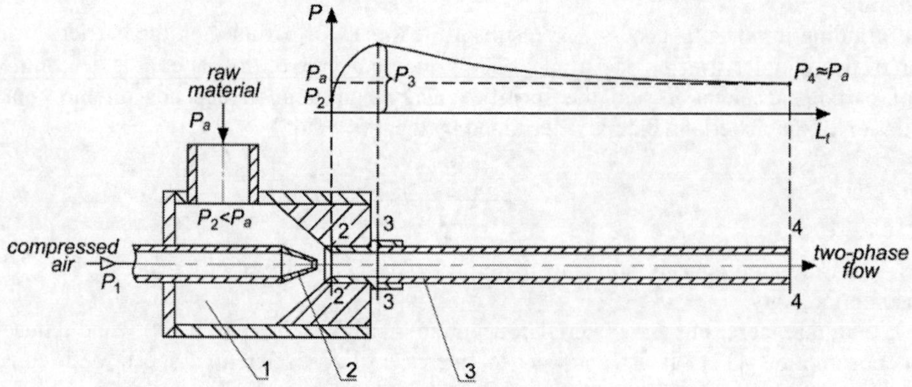

Figure 1. Typical ejector jet mill with a graph of the total pressure along the length of the acceleration tube:
1 – ejector chamber; 2 – the supersonic nozzle; 3 – acceleration tube; 2–2 – the acceleration tube beginning (the beginning of flows mixing in acceleration tube); 3–3 – the end of flows mixing (the beginning of two-phase flow acceleration); 4–4 – the acceleration tube and.

Figure 2 shows a scheme of an ejector jet mill with external pressure and a graph of the total pressure P change along the length of the acceleration tube L_t. Table 1 shows the calculations of the pressure P_3 and P'_3 at the beginning of the acceleration tube according for a typical ejector jet mill and an ejector jet mill with external pressure respectively.

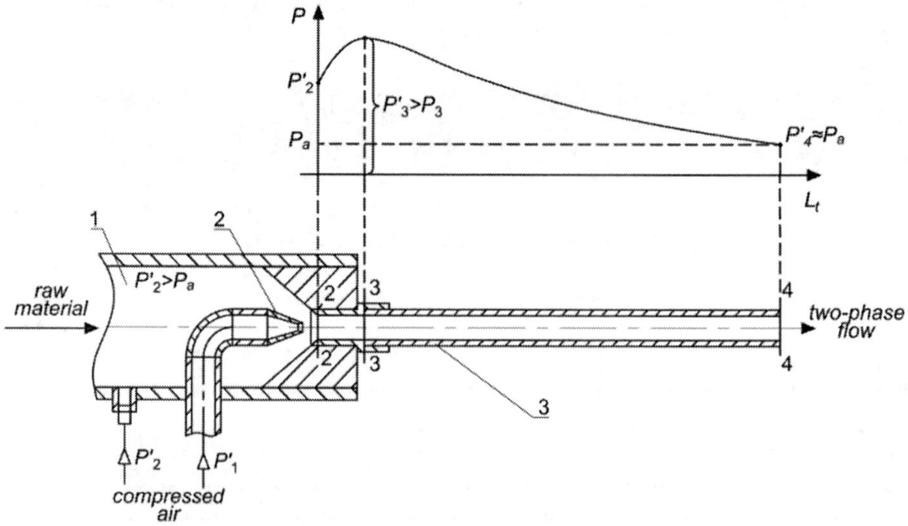

Figure 2. Ejector mill with external pressure and a graph of the total pressure along the length of the acceleration tube:
1 – ejector chamber; 2 – the supersonic nozzle; 3 – acceleration tube; 2–2 – the acceleration tube beginning (the beginning of flows mixing in acceleration tube); 3–3 – the end of flows mixing (the beginning of two-phase flow acceleration); 4–4 – the acceleration tube and.

Table 1. Dependence of the pressure at the beginning of the acceleration tube for a typical ejector P_3 and an ejector with an external pressure P'_3 at P_1=7.

D_3/d_1	α	P_3, at	P'_3, at
2.0	0.333	2.5	4.38
2.5	0.190	1.96	4.06
3.0	0.125	1.67	3.89

It follows from the presented data that at the same pressure P_1 of the working gas before the nozzle in the new mill the pressure at the beginning of the acceleration tube P'_3 is almost two times greater than in a typical jet mill [10].

3. Experimental research

To implement method mentioned above the installation of an ejector jet mill with external pressure was developed (patent of Russian Federation No. 2711252 [10]). The installation scheme is shown in figure 3.

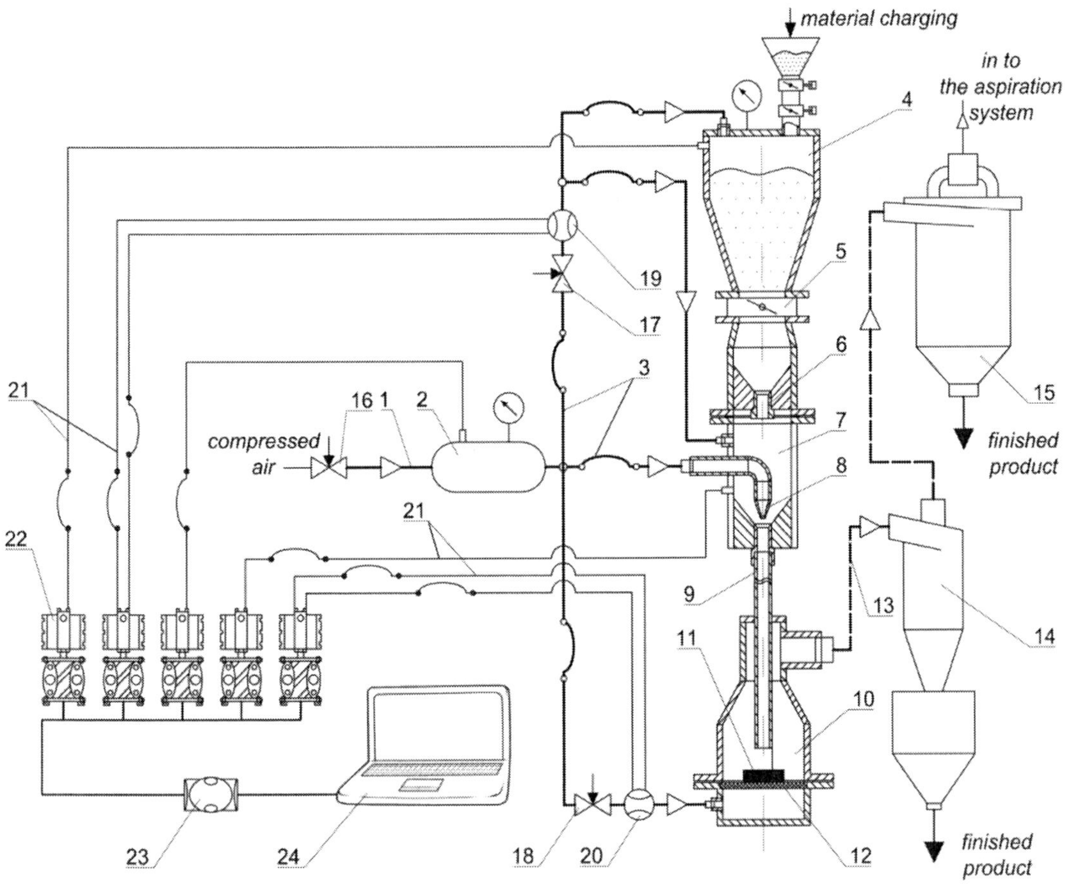

Figure 3. The installation scheme of the ejector jet mill with external pressure:
1 – compressed air sleeve; 2 – receiver; 3 – flexible pipelines; 4 – material hopper; 5 – rotary shutter; 6 – insert with a calibrated hole; 7 – ejector chamber; 8 – the supersonic nozzle; 9 – acceleration tube; 10 – grinding chamber; 11 – barrier; 12 – aerating grid; 13 – air pipes; 14 – cyclone; 15 – bag filter; 16, 17, 18 – ball valve; 19, 20 – standard diaphragm; 21 – pressure gauges; 22 – microprocessor sensors 'Metran-100 DI', 'Metran-100 DD'; 23 – controller; 24 – PC.

Jet mill with an outer pressure works in the following way. Compressed air is piped through 1 to the receiver 2, where the full pressure of the working gas P_1 is maintained by the valve 16. Through pipelines 3 compressed air is supplied to the nozzle 8, to the hopper 4, to the ejector chamber 7 and under the aerating grid 12. With the help of a tap 17 and a standard diaphragm 19, the required flow rate of the ejected air G_2 is set, and an external pressure P_2 is created in the hopper 4 and the ejector chamber 7. With the help of a tap 18 and a standard diaphragm 20 the necessary air flow G_3 is set for aeration of the material in the grinding chamber 10. Then the rotary shutter 5 opens and through the calibrated hole 6 the crushed material flows through the ejector chamber to the acceleration tube 9 where the material is accelerated. After impacting the barrier 11 the crushed material is removed to the cyclone 14 through the pipe 13. Sanitary air cleaning is carried out in a bag filter 15. Operating pressure P_1 and external pressure P_2 as well as pressure differences are recorded through pulse tubes 21 by microprocessor pressure sensors 22. Data via the controller 23 are recorded in real time on a PC 24. Gas consumption G_1 is calculated according to the standard method [14, 15].

Comparative tests of a new design mill (JMEP) and a typical ejector mill (JM) were carried out. For grinding two materials were used, namely quartz sand with density $\rho t = 2700$ kg/m^3, bulk density $\rho n = 1965$ kg/m^3, an average particle size of $d_{50} = 280$ μm and feldspar with density $\rho t = 2630$ kg/m^3, bulk density $\rho n = 1590$ kg/m^3, the average particle size $d_{50} = 180$ μm.

Granulometric compositions of materials after one grinding cycle are shown in table 2. As follows from the data presented the grinding degree as a result of a single impact for a mill with external pressure is higher. So, for quartz sand the average particle size and grinding degree in JM: d_{50}= 126.96 μm, i =2.21; in JMEP: d_{50}= 74.99 μm, i = 3.73. The same characteristics for feldspar in JM: d_{50}= 73.60 μm, i = 2.45; in JMEP: d_{50} = 55.31 μm, i = 3.25. Thus, the use of an ejector jet mill with external pressure allowed to increase the grinding degree of quartz sand by 1.7 times and the grinding degree of feldspar by 1.3 times compared to a typical ejector jet mill.

Table 2. Granulometric compositions of grinding products in jet mill with external pressure (JMEP) and a typical ejector jet mill (JM).

Sieve	Quartzsand (JM)			Quartzsand (JMEP)			Feldspar(JM)			Feldspar(JMEP)		
X, μm	*R*, %	*D*, %	*r*, %	*R*, %	*D*, %	*r*, %	*R*, %	*D*, %	*r*, %	*R*, %	*D*, %	*r*, %
630	1.11	98.89	1.11	0.10	99.90	0.10	0.00	100.0	0.00	0.00	100.0	0.00
400	5.97	94.03	4.85	1.42	98.58	1.32	2.10	97.90	2.10	0.51	99.49	0.51
315	10.11	89.89	4.15	2.84	97.16	1.42	4.61	95.39	2.51	1.62	98.38	1.11
200	27.91	72.09	17.80	12.07	87.93	9.23	15.03	84.97	10.42	9.63	90.37	8.00
160	39.94	60.06	12.03	19.88	80.12	7.81	22.75	77.25	7.72	15.60	84.40	5.98
100	61.88	38.12	21.94	40.06	59.94	20.18	39.58	60.42	16.83	30.19	69.81	14.59
63	74.12	25.88	12.23	55.17	44.83	15.11	53.51	46.49	13.93	43.57	56.43	13.37
40	84.73	15.27	10.62	71.1	28.90	15.92	69.64	30.36	16.13	61.30	38.70	17.73
0	100.0	0.00	15.27	100.0	0.00	28.90	100.0	0.00	30.36	100.0	0.00	38.70

The operating parameters of mills for a single grinding cycle of quartz sand and feldspar are shown in table 3.

Table 3. Operating parameters of 'JM' and 'JMEP' mills for a single grinding cycle.

Material	m, kg	t, s	G_m, kg/s	P_1, MPa	P_2, MPa	G_1, kg/s	G_2, kg/s	μ, kg/kg
Quartzsand(JM)	1.000	32	0.0312	0.5084	–	0.0283	–	1.104
Quartzsand(JMEP)	2.000	43	0.0465	0.5804	0.2328	0.0183	0.0076	1.796
Feldspar (JM)	0.800	30	0.0267	0.5017	–	0.0280	–	0.952
Feldspar (JMEP)	2.000	45	0.0444	0.5621	0.2357	0.0178	0.0073	1.771

It is seen from the presented data that the consumption concentration μ in an ejector jet mill with external pressure is 1.6 times greater when grinding quartz sand and 1.9 times greater when grinding feldspar compared to a typical ejector jet mill.

Also experiments were carried out in which the consumption concentration in an ejector jet mill with external pressure was increased by 3.0 times in comparison with a typical ejector jet mill when grinding quartz sand and by 4.0 times when grinding feldspar, and these mills showed quite similar degrees of material grinding.

4. Conclusions

A new method of jet grinding and the design of an ejector jet mill with external pressure is proposed (patent of the Russian Federation No. 2711252 [10]). This method allows increasing the pressure at the beginning of the acceleration tube due to external pressure. A higher pressure at the beginning of the acceleration tube allows increasing the consumption concentration of bulk material, and thus reducing the specific consumption of compressed air and increasing productivity. In addition, the grinding degree of the material increases by boosting the acceleration speed of crushed particles which also contributes to the rise of productivity. At the same time to keep supersonic gas velocity at the outlet of the nozzle the ratio of pressures in front of the nozzle and outside of the nozzle must be greater than the critical one, i.e. the condition $P_1/P_2>1.1893$ must be met. Experimental studies of an ejector jet mill with external pressure showed an increase in the grinding degree of quartz sand by 1.7 times and grinding degree of feldspar by 1.3 times compared to a typical ejector jet mill. The consumption concentration increased by 1.6 times when grinding quartz sand and 1.9 times when grinding feldspar. Thus, using an ejector jet mill with external pressure allows increasing productivity and reduce the specific consumption of compressed air.

References

[1] Fedotov K and Dmitriev V 2014 *Jet grinding* (Moscow: Mashinostroyeniye) p 194
[2] Chamayou A and Dodds J A 2007 *Air jet milling Handbook of Powder Technology of Particle Breakage* vol **12** ed AD Salman, M Ghadiri and M J Hounslow (Amsterdam: Elsevier) chapter 8 pp 437–86
[3] Postnikova I, Bilinichev V and KravchikYa 2015 Jet mills *Modern scientific technologies. Regional supplement* vol **2 (42)** (Ivanovo: ISUCT) pp 144–51
[4] Benz M, Herold H and Ulfik B 1996 Performance of a fluidized bed jet mill as a function of operating parameters *International Journal of Mineral Processing* vol **44–45** (Amsterdam: Elsevier) pp 507–19
[5] Koeninger B, Hensler T, Romeis S, Peukert W and Wirth K E 2018 Dynamics of fine grinding in a fluidized bed opposed jet mill *Powder Technology* vol **257** (Amsterdam: Elsevier) pp 346–57
[6] Brosh T, Kalmana H, Levy A, Peyron I and Ricard F 2014 DEM–CFD simulation of particle comminution in jet-mill *Powder Technology* vol **327** (Amsterdam: Elsevier) pp 104–12
[7] Starkey D, Taylor C, Morgan N and Winston K 2014 Modeling of continuous self-classifying spiral jet mills part 1: Model structure and validation using mill experiments *AICHE Journal* vol **60 (12)** (New York: AIChE) pp 4086–95
[8] Voropayev S and Eskin D 2002 Designing a jet mill nozzle of maximum efficiencyChemical Engineering and Technology vol **25 (11)** (Hoboken: Wiley-Blackwell) pp 1101-06
[9] Voropayev S and Eskin D 2002 Optimal particle acceleration in a jet mill nozzle *Minerals Engineering* vol **15 (6)** (Amsterdam: Elsevier) pp 447–9
[10] Patent of Russian Federation No. 2711252 2020 *Method and device for crushing bulk materials* (Ural Federal University)
[11] Vitushkin V, Karetnikov G, Ovchinnikov V, Prokhorov V and Sutyrin I 2004 Study of acceleration of solid material particles in pneumatic jet grinders *Bulletin of MSTU. Engineering* vol **4** (Moscow: Bauman Moscow State Technical University) pp 43–56
[12] Abramovich G 1991 *Applied gas dynamics* 6th ed. (Moscow: Nauka) p 600
[13] Katorgin B I, Kiselev A S, Sternin L E and Chvanov V K 2009 *Applied gas dynamics* (Moscow: Vuzovskaya kniga) p 340
[14] Pirumov U and Roslyak G 1990 *Gas dynamics of nozzles* (Moscow: Nauka) p 368
[15] Daniel T H New and Simon C M Yu 2015 *Vortex rings and jets* (Singapore: Springer) p 235

IOP Conf. Series: Materials Science and Engineering **966** (2020) 012078 doi:10.1088/1757-899X/966/1/012078

Development of tool selection process for milling of pocket features

T A Gordeeva[1], S S Kugaevskii[1] and A P Starostin[1, 2, 3]

[1] Ural Federal University named after the First President of Russia B N Yeltsin, 19, Mira str., Yekaterinburg, 620000, Russia
[2] Ural Federal University named after the First President of Russia B N Yeltsin, Nizhniy Tagil Technological Institute, 59, Krasnogvardeyskaya str., Nizhniy Tagil, 622000, Russia

E-mail: [3] andrey.starostin@urfu.ru

Abstract. In modern production, three-dimensional modeling of workpieces has become a logical stage in the development of the technological process. Engineering analysis, process design, CNC mill tool path design, final inspection of products and other aspects are done basing on 3D models and have become the norm in the manufacturing industry. This research aimed at deriving mathematical tools which could be used in the process of CAM software development and modernization. Structural and process related elements can be automatically recognized on the basis of a 3D model. English literature refers to such elements as features. A set of parameters necessary for the description of a pocket design feature characteristics was reviewed. Available process solutions for CNC machining of said feature were analyzed. Special attention was given to selection of the roughing and finishing mills, as well as roughing and finishing cut strategy development. A tool parameter selection algorithm was developed for CNC milling of a pocket feature.

1. Introduction

Today, rather high attention is given to the preparation of computer based manufacturing processes of new products. Therefore, process engineers have been provided with new design tools:

First, the opportunity of 3D representation of product design. An electronic part model is a document which contains the geometrical model of the part and requirements on manufacturing and inspection (including size tolerances, roughness requirements, etc.).

Secondly, a feature-based modular representation of part design. As a result, the process engineer has the opportunity to work with compiled features (holes, pockets, faces, walls, etc.). In addition to this, the current state model of the work piece which accounts for milled features can be used in preparations for a subsequent operation or transition;

Thirdly, today's knowledge base may include proven decision-making algorithms for recognized features, including selection of tools and machining modes.

Fourthly, process engineers who apply CAM system have information on the tools and holders currently available at the work place.

Today, the ISO 14649 (STEP NC) standard [1–5] is being implemented. The intention of this standard is to ensure development of intelligent CNC machines. As per this standard, '…parts are produced from workpieces by machining of typical features, relative or absolute transitions associated

IOP Conf. Series: Materials Science and Engineering **966** (2020) 012078 doi:10.1088/1757-899X/966/1/012078

with typical working steps set by executables during running of control program, including required tolerances and complaint tools' [6]. It is reasonable to represent part design as a set of features for the purposes of process design. A feature is a set of faces to be machined in the current working step, and its timed sizes characterize the current state of the workpiece. Figure 1 lists features which form a housing type part.

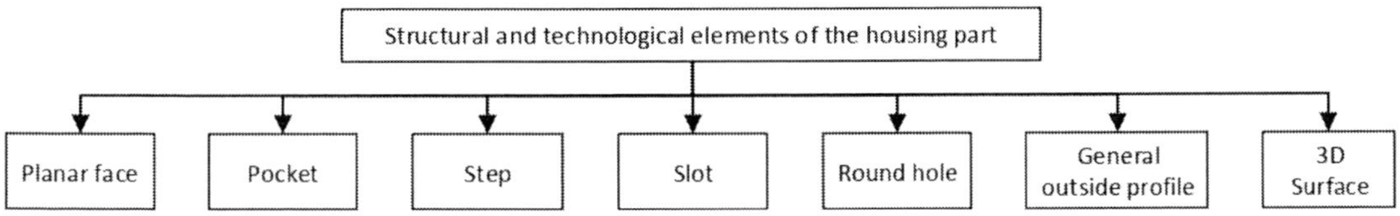

Figure 1. Structural and technological elements of a housing part.

One of the most difficult features for manufacturing process formalization is a pocket. General appearance of a pocket feature is given in Figure 2.

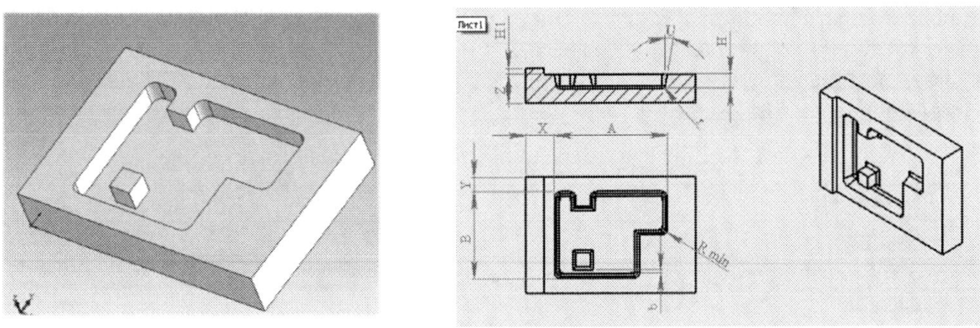

a) b)

Figure 2. Appearance of pocket (a) and its parameters (b).

A pocket feature has an external closed contour inside of which machining is performed. Walls of this contour may be vertical or inclined at a fixed angle. In internal contour appears when there are lugs inside the pocket. The parameter 'passage width' (b) which is the distance between the lug wall and the pocket wall, is one of the main parameters when choosing a tool. The required accuracy of desired sizes and surface roughness in the pocket determines the necessity of making finishing passes in order to ensure quality if the part being produced. In order to decide upon a pocket manufacturing process it is necessary to analyze its parameters [7].

The questions at hand in regard to parameters selection for a feature are reviewed in papers [8–13].

2. Methodology

Upon importing of a 3D model it is possible to automatically compile a list of features to be machined. This approach is used in such software as FeatureCAM from Autodesk, NX CAM from Siemens PLM Software and others.

The CAM system selects a machining algorithm basing on parameters of each feature [14]. FeatureCAM can store several system configurations setup for a milling machine. These configurations account for capabilities of the milling machine installed at the workplace, the available tools, acceptable drive loads, etc. [15]. Differences in NC for same features but under different system configurations of machines are formed at the tool path calculation stage. This is the essence of the new level of CAM system flexibility. However, the mentioned system is meant only for NC program calculation and does not provide a technological process as required by Russian state standards (unified system of technological documentation).

IOP Conf. Series: Materials Science and Engineering **966** (2020) 012078 doi:10.1088/1757-899X/966/1/012078

The main objective of this paper is development of process algorithms for machining of pocket features basing on analysis of a part's 3D model.

3. Experimental plan

As a preliminary step, let's review the options for decision making at the following stages of work:

- Choice of roughing and finishing tools;
- Choice of machining strategies at roughing and finishing stages.

3.1. *Choice of roughing and finishing tools for machining of a pocket feature*

First and foremost stage of process design for such a working step is the choice of cutting tools. In some cases one milling cutter is enough to machine a pocket. Use of additional tools is feasible in two cases:

- Strict requirements to size tolerances and pocket surface quality;
- Short radius of pocket internal corners.

Let us tackle these two cases in order.

To fulfill size tolerance and surface quality requirements the first thing to be determined should be the number of machining stages. The number of machining stages depends on size tolerance and surface roughness requirements. As stated in [16] four machining stages can be distinguished: roughing, semi finishing, cleanup, and finishing. Table 1 describes requirements to size and roughness, and corresponding milling stages.

Table 1. Milling stage depending on requirements to sizes of the part.

Milling stage	Quality class	Roughness Ra, μm
Roughing	14…12	12.5…6.3
Semi finishing	12…11	6.3…3.2
Cleanup	8…11	6.3…1.6
Finishing	8…6	0.8…0.4

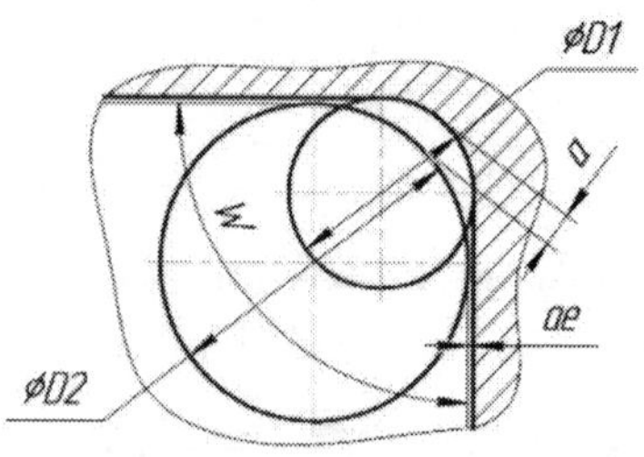

Figure 3. Relation of roughing mill diameter on pocket contour internal radius.

We will denote the roughing mill diameter as D_2, and the finishing mill diameter as D_1. Roughing mill diameter D_2 is chosen such that the tool can reach into the internal sharp corners of the contour (Figure 3) [17]. It is desired for the remaining allowance a in the internal corners of the contour to not exceed (0.15...0.25) D_1, where D_1 – diameter of tool to be used in the finishing pass. Hence, the roughing mill diameter D_2 can be determined using the equation

$$D_2 \leq \frac{2\left(a \cdot sin\frac{\omega}{2} - a_e\right)}{1 - sin\frac{\omega}{2}} + D_1 \qquad (1)$$

Where a is the maximum allowance during internal corners milling, a_e – allowance for finishing milling of the contour, ω – the internal corner of the pocket.

Assuming the most general case, lets consider $\frac{\omega}{2} = 45°, a = 0.2D1, a_e = 0.3,$ then, by transforming equation (1), and obtain

$$D_2 \approx 2D_1 - 2 \quad (2)$$

3.2. Choice of machining strategies at roughing and finishing stages

For roughing, one of the standard milling strategies can be chosen: zig zag, equidistant line, spiral. In our opinion, the optimal strategy here would be a spiral climb cut (Figure 4), as it ensures a good chip formation during climb cutting.

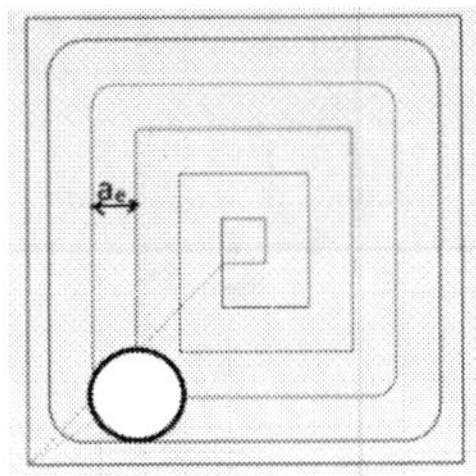

Figure 4. Application of the spiral climb cut strategy for roughing.

Recommended spiral pitch in the pocket bottom can be taken as $a_e = 0.6–0.8D$ when the milling path max $a_p = 0.3 – 0.5D$ depending on the material being machined.

For the finishing stage the semi finishing milling scheme should be considered (Figure 5).

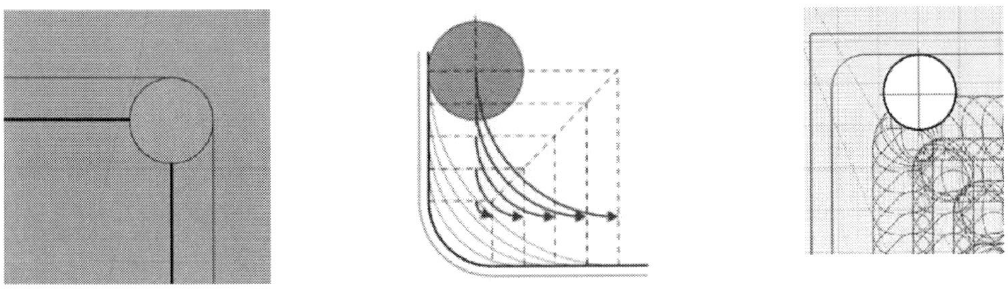

a) conventional method b) layered milling with narrow contact spot c) trochoidal milling

Figure 5. Pocket corners semi finishing milling schemes.

Conventional method (Figure 6 a) seems much easier but demands sequential decreasing of the feed as the tool cuts into the corner of the contour which increases machining time of finishing. Besides this, in such a case reliability of the finishing mill cutting is not assured due to a larger contact arc of the cutting edge with the work piece surface. Thus, currently methods (b) and (c) are considered more prospective, as they can be used even with correlation of tool diameters of $D_2 \geq 2D_1$ [18].

Trochoidal milling (Figure 6 c) is widely used for milling narrow slots and grooves, including pocket corners, to ensure a consistent depth of cut a_e. This method is especially effective in high speed milling when the milling width is chosen to be minimal $a_p = 0.5–1$ mm. In the context of the task of sequential milling of pocket features by roughing and finishing cuts, the layered milling with narrow contact can be recommend. Here, when material is removed from the contour corners as shown in Figure 6 b milling is performed layer by layer from $a_p = 0.5D_1$.

Feature milling algorithms must consider requirements to size tolerances and quality of surface to be milled. For instance, paper [19] provides an analysis of strategy impact on milled surface roughness. In our case of strict quality requirements to surface finish application of a finishing mill is mandatory. Additional roughing tool D_3 may be applied in cases when the area of the pocket is too large in relation to the projected area of the chosen milling cutter D_2.

IOP Conf. Series: Materials Science and Engineering **966** (2020) 012078 doi:10.1088/1757-899X/966/1/012078

4. Results and discussion

The proposed correlations can be transformed into mathematical form using the algorithm given in Figure 6.

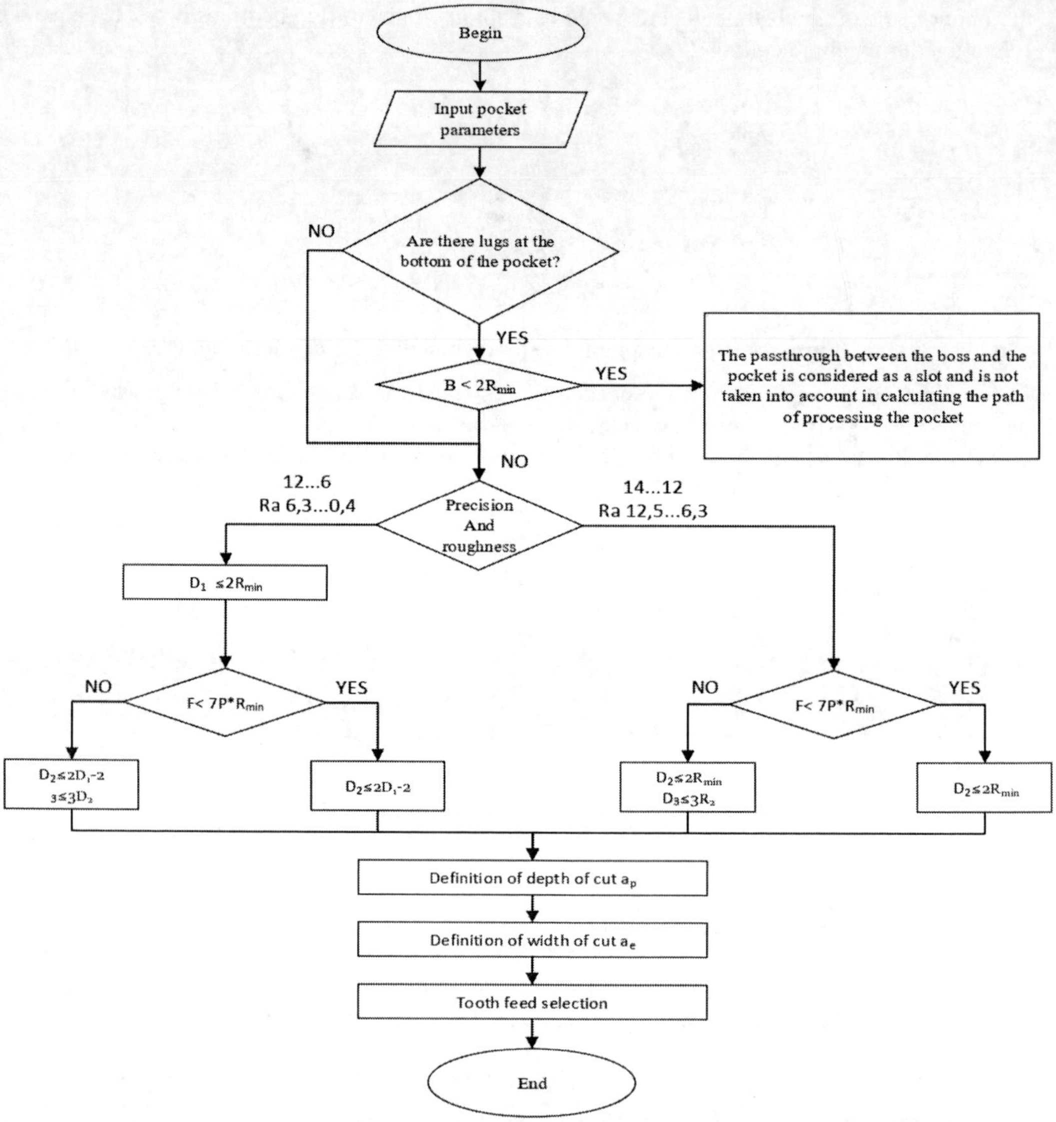

Figure 6. Algorithm for choosing roughing and finishing tools for machining of pocket features.

5. Conclusion

As a result of this research an algorithm for choosing roughing and finishing tools for machining of pocket features in the part milling modeling process in CAM systems has been proposed.

IOP Conf. Series: Materials Science and Engineering **966** (2020) 012078 doi:10.1088/1757-899X/966/1/012078

References

[1] ISO 14649-10 2003 *Industrial automation systems and integration physical device control-data model for computerized numerical controllers-part 10: General process data*

[2] *STEP-NC System. In: Theory and Design of CNC Systems* 2008 Springer Series in Advanced Manufacturing. Springer (London)

[3] Hardwick M, Zhao Y, Proctor F M *et al* 2013 A roadmap for STEP-NC-enabled interoperable manufacturing. *Int. J. Adv. Manuf. Technol.* **68** 1023–37

[4] Kramer T and Proctor F 2009 Feature-based Process Planning Based on STEP *Advanced Design and Manufacturing Based on STEP* (London: Springer) 23-48

[5] Zhang Y, Liu Y and Bai X 2010 The Research on the Intelligent Interpreter for ISO 14649 Programs *Advances in Computation and Intelligence. ISICA 2010. Lecture Notes in Computer Science* **6382**

[6] Sosonkin V and Martinov G 2008 *Programming CNC. Tutorial* (Moskow: Logos)

[7] Kugaevsii S S 2010 Application of the method of element-wise technology for calculating control programs in a CAM-system environment *CAD/CAM/CAE-observer* **5** 65–8

[8] Ridwan F, Xu X and Liu G 2012 A framework for machining optimisation based on STEP-NC. *J. Intell. Manuf.* **23** 423–41

[9] Yamada H, Tanaka F and Onosato M 2006 Analysis of ISO 6983 NC Data Based on ISO 14649 CNC Data Model *Knowledge Enterprise: Intelligent Strategies in Product Design, Manufacturing, and Management. PROLAMAT. IFIP International Federation for Information Processing* **207**

[10] Janusz P and Sergiusz S 2017 Extension of STEP-NC data structure to represent manufacturing process structure in CAPP system *27th Int. Conf. on Flexible Automation and Intelligent Manuf.*

[11] Zhao Y, Habeeb S and Xu X 2009 Research into integrated design and manufacturing based on STEP. *Int. J. Adv. Manuf. Technol.* **44** 606

[12] Mokhtar A, Xu X and Lazcanotegui I 2009 Dealing with feature interactions for prismatic parts in STEP-NC. *J. Intell.Manuf.* **20** 431

[13] Wang J and Shu Q 2010 STEP-NC Based Manufacturability Evaluation *Proceedings of the 6th CIRP-Sponsored Int. Conf. on Digital Enterprise Technology. Advances in Intelligent and Soft Computing* **66**

[14] Kugaevsii S S 2009 On the issue of flexibility and efficiency of production preparation for CNC machines *CAD and Graphics* **12** 72–4

[15] Kugaevsii S and Titov I 2008 Development of a FeatureCAM database reconciliation module with an enterprise tool database *CAD and Graphics* **5** 66–9

[16] Pashkevich M 2010 *Engineering Technology. Course and diploma design: a training manual* (Minsk: Grevcova) p 400

[17] Safragan P 1974 *Operation of numerically controlled machines* (Kiev: Equipment) p 308

[18] *Website of company 'Sandvic Coromant'* Available online: https://www.sandvik.coromant.com/en-gb/pages/default.aspx (accessed on 01.06.2020)

[19] Firman K and Xun Xu 2012 Development of a surface roughness predictive model for STEP-compliant machining optimization *International Journal of Computer Aided Engineering and Technology (IJCAET)* **4** 206–28

IOP Conf. Series: Materials Science and Engineering **966** (2020) 012079 doi:10.1088/1757-899X/966/1/012079

Development of super-professional competencies of the production personnel in the context of university supplementary education programs

Yu G Myslyakova[1, 3] and M V Kurashova[2]

[1] Ural State University of Economics, 62, 8 Marta Str., Yekaterinburg, Russia
[2] Ural Federal University named after the First President of Russia B N Yeltsin, Nizhniy Tagil Technological Institute, 59, Krasnogvardeyskaya str., Nizhniy Tagil, 622000, Russia

E-mail: [3] jul_jul@inbox.ru

Abstract. In the article, the authors explore development of super-professional competencies of the production personnel of enterprises of the military-industrial complex. Here are the most popular supplementary education programs implemented by UrFU and UrGEU and ensuring formation of professional and super-professional competencies effectiveness of which ensures competitive ability of employees and ability of enterprises themselves to switch over to the neo-industrial way of their development. Authors also suggested an express-method of assessing a quality of development of super-professional competencies of personnel in the context of universities. This method has been tested at production personnel of the enterprises of JSC EVRAZ NTMK and JSC 'Research and production corporation 'UralVagonZavod' who listened to supplementary education programs. Results that have been received confirm necessity of creation of centers of supplementary vocational education expanding implementation of spectrum of competence-based university programs for development of super-professional competencies of the production personnel of enterprises of the military-industrial complex.

1. Introduction

Development of competency-based approach to improvement of organizational and management activity of enterprises became particularly important in terms of neo-industrial evolution ensuring transition of industrial business to higher technology level of production. A special feature of such evolution is simultaneous transformation of science, scientific knowledge and discoveries into one of the main production factors by highlighting scientific knowledge, information, nonmaterial benefits in products; formation of overall scientific investigation as creative activity in production; increasing role of intellectual property; domination of incentives for personal self-fulfilment in incentives of innovative production; growing importance of knowledge and creative skills of production personnel [1]. All such changes cause increase in competitive ability of industrial enterprises by producing qualitatively new basic products based on their own innovations and techniques [2], changing attitude to quality of products [3], increasing and effective use of human resources, as well as strengthening the acceptance of new professional and super-professional competencies of production personnel of the large industrial enterprises.

2. Theoretic Aspects of Development of Super-Professional Competencies of Production Personnel of the Industrial Enterprises

Large industrial enterprise is a unique team with a great number of scientists, highly qualified engineering and technical specialists and production personnel who can work with modern process equipment participating in expansion of automation processes and forming specialized communication such as supplier–manufacturer–supplier [4]. According to A Varshavskiy, production of high technology products is characterized by use of work of specialists who have a large set of competencies in scientific, engineering-technical and production spheres which may ensure timely integration of R&D results into the production activity. Scientist believes that only engagement of personnel, who have high level professional and super-professional competencies will allow to perform evolutionary and revolutionary changes in development of technique and technology [5], i. e. will ensure competitiveness of enterprises of the industrial complex in terms of neo-industrial modernization.

Nowadays there is a great number of definitions of term 'professional competencies', but all of them are limited to claiming that the integration of ability and readiness of production personnel to methodically and properly structure and organize their work functions using assessment of results of their own professional practice. These competencies reflect the degree of knowledge, skills and abilities to solve professional tasks in a multivariable and timely manner within an existing specialty [6, 7].

As for super-professional competencies, they include general cultural and general scientific competencies and reflect integrative human qualities used and applied by him in different spheres of life [8]. In this regard, super-professional competencies are generally oriented to application in the professional activity, rather than in the specific specialty, thereby ensuring high efficiency and competitiveness of a person as a worker and as a professional [9]. In a modern world such competencies may include, for example, information culture, communicative culture, language proficiency, degree of discipline, ability to work with large databases, including electronic, analytic and operational thinking, creativity and leadership abilities [10, 11]. In other words, as super-professional competencies are not strictly connected with any specific profession, they may promote and effect the successful achievement in many spheres of human activity, representing a superstructure on the basic profile of personnel knowledge and skills.

Modern education system is oriented to formation of, first of all, professional competencies, which without super-professional competencies function less efficient and not effective. Only interaction of these competencies will create at the enterprise a possibility to: organize balance of technological and economic production factors; optimize timely boundaries of production and innovative processes; use maximum number of creative ideas, as well as save on expenses during the entire life cycle of product creation, including at R&D stage [12].

3. Study on the Development of Super-Professional Competencies of the Production Personnel in the Context of University Supplementary Education Programs of the Large Industrial Enterprises

The enterprises of the industrial complex are one of the basic factors of ensuring competitive ability of national economy and living standards. In turn, competitiveness of enterprises depends not only on innovative trends of development of special-purpose products, but also on efficiency of human resources management system for industry. Competencies of production personnel are formed on the basis of science and technology trends of production of defense products and specific technologies of their operation, assurance of safety of production activity, technologies of management of production chains and business processes, development of information systems and other relevant of directions of activity of enterprises, including connected with increase of potential of production of high-quality non-defense products taking into account available production capacities.

The most demanded programs at the enterprises JSC EVRAZ NTMK and JSC 'Research and production corporation 'UralVagonZavod' are those that form professional and super-professional profile: 'Metallurgy of Ferrous Metals'; 'Metallurgical Machines and Equipment'; 'Industrial Thermal Power Engineering', 'Electric Drive Motor and Automatics of Industrial Units and Process Facilities'; 'Chemical Engineering of Natural Energy Carriers and Carbon Materials'; 'Metrology, Standardization and Certification'; 'Design and Technology of Mechanical Facilities', as well as economic programs. The number of trainees participating in all these programs has been increasing annually since 2017, see Figure 1.

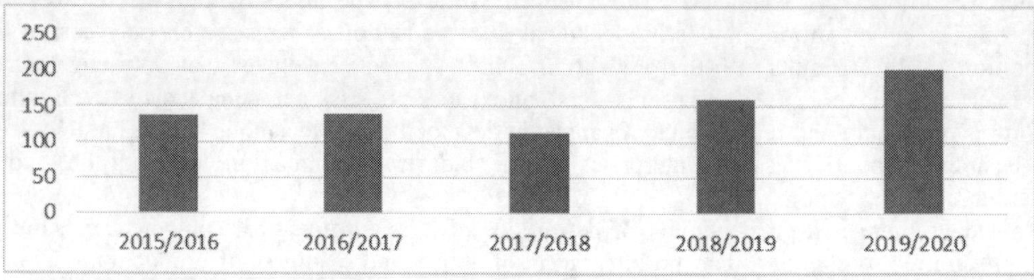

Figure 1. The number of trainees participating in supplementary education programs of JSC EVRAZ NTMK and JSC 'Research and production corporation 'UralVagonZavod' from 2015 to 2020 according to data of UrFU and UrGEU.

It is necessary to note, that specific features of large industrial enterprises result in high significance of such super-professional competencies of the production personnel as operational thinking, project management, lean management, client focus, because they have a strong impact on efficiency of their professional activity, allow not only to 'integrate' in production structure without a long period of adaptation, but also contain current knowledge in the field of knowledge-intensive production. Nowadays super-professional competencies connected with system project management, which assumes not only degree of technical, special, economic-management knowledge, are in great demand at large industrial enterprises, because development of the project includes: formation of effective team; integration of project organizational structure into the existing linear-functional structure; development of the project decision (concept, roadmap, design and technological documentation, sketch and 3D model, financial justification, production plan, HR strategy, marketing policy), project presentation for experts. Therefore, it is necessary to provide quality and professional economic-management training in order to increase the competitive ability of specialists who have a technical university degree.

Such training is provided by UrFU and UrGEU. Features of training in economics connected with demanded content of given subjects and programs, see Table 1.

Table 1. Supplementary economic education programs of JSC EVRAZ NTMK and JSC 'Research and production corporation 'UralVagonZavod' from 2018 to 2020 according to data of UrFU and UrGEU, persons.

Supplementary economic education programs	Total number of	UVZ and NTMK	Other industr. enterpr.
Accounting	17	7	10
Personnel management of organization	23	12	11
Economics and management at special-purpose product enterprises	15	8	7
Productive efforts: Current trends of payment for labour and setting of work quotas at the industrial enterprise	20	20	0
Supply-side economics	21	21	0
Total	**96**	**68**	**28**

Surveys of trainees participating in these programs have shown that the acquired competencies have ensured the demand of the personnel, including due to appearance of the possibility to immediately apply in their work the acquired knowledge and abilities to solve a wide range of production problems with their help, see Table 2.

Table 2. Performance indicators of implementation of new super-professional competencies of production personnel at industrial enterprise.

Types of performance	Estimative indicator, %
Growth of demand at industrial enterprise	75
Compliance of quality of knowledge expected after training with quality of knowledge actually received	84
Part of knowledge received as a result of training expected to be applied by listeners in their work	92
Part of problems expected to be solved by listeners with the help of knowledge received	45

That is why the assessment of the compliance of quality of knowledge expected after training with quality of knowledge actually received has been recorded at 84 % from 100 % according to majority of respondents quizzed.

4. Quality Assessment of Development of Super-Professional Competencies of the Production Personnel in the Context of University Supplementary Education Programs

Quality assessment of development of super-professional competencies of the production personnel shall be performed within implemented supplementary education programs. As indicator of quality is multidimensional, then we propose to consider it in relation to the super-professional competencies in terms of some levels, see Table 3.

Table 3. Levels of quality of development of super-professional competencies of the production personnel.

Levels of quality of development of super-professional competencies	Indicator
Individual level provides for use of new knowledge in its professional activity that brings you tactile benefit and leads to transformation of your professional views	Individual_level
Collective level provides the possibility of freely sharing new knowledge and mentoring	Collective_level
Corporate level provides for significant improvement in qualifying competencies and skill, increase in production rate and reduction of number of mistakes made before	Corporate_level
Society level provides for appearance of new life values, change in way of life and formation of another understanding of economic activity	Society_level

Then quality assessment of development of super-professional competencies will be calculated according to the formula 1, which integrates specified level-based assessments.

$$Q = (Individual\_level + Collective\_level + Corporate\_level + Society\_level)/4, \quad (1)$$

where *Individual_level* – is an assessment of individual level; *Collective_level* – is an assessment of collective level; *Corporate_level* – is an assessment of corporate level; *Society_level* – is an assessment of society level.

Assessment of each level is performed as per five-point scale by all respondents quizzed. Then final value of each level is obtained by means of calculation of average value of all scores of respondents received.

To assess quality of development of super-professional competencies by university supplementary education programs of UrFU and UrGEU, trained production personnel of JSC EVRAZ NTMK and JSC 'Research and production corporation 'UralVagonZavod' were quizzed. The following scores were received:

- *Individual_level* – 4.6 points;
- *Collective_level* – 3.6 points;
- *Corporate_level* – 3.8 points;

- *Society_level* – 4.4 points.

$$Q=4.6+3.6+3.8+4.44=4.1 \text{ points}$$

Thus, it turns out that listeners of the supplementary education program working at industrial enterprises assessed the quality of formation of super-professional competencies in 4.1 points with a maximum of 5 points.

5. Conclusions

The conducted study shows the relevance and demand for formation of super-professional competencies of production personnel of large industrial enterprises. The quality of development of these minor competencies was assessed by the quizzed respondents in 4.1 points out of 5 maximum possible, which also indicates the need to expand supplementary education programs in the context of the simultaneous formation of professional and super-professional competencies, which may be achieved through the integrative interaction of universities and industrial enterprises, as part of the creation of adult education centers. Such centers will allow to use the resources of the university (license for educational activities, faculty staff, teaching and material resources, etc.) and the enterprise (material and technical facilities of the enterprise, involvement of leading specialists of the enterprise in the educational process), i. e. provide not only the synergy of theoretical and practical knowledge, but also their adaptation to the conditions of production activity of a particular enterprise, which will strengthen the competitive advantages of the industrial complex of Sverdlovsk Region.

References

[1] Zhuravlyova G P 2014 Economic policy of modern Russia: modernization and reindustrialization *Tomsk State University Journal* **3 131** pp 26–32

[2] Novitskiy N A 2013 Modern problems of investment in innovative industrialization of Russia *Moscow Witte University Bulletin. Series 1: Economics and Management* **3** pp 18-23

[3] Pichurin I I 2013 Import substitution support after russian accession to WTO *UrFU Bulletin. Series: Economics and Management* **3** pp 45–51

[4] Bendikov M A 2001 Markets of high-tech products *Marketing in Russia and Abroad* **2** pp 3–15

[5] Varshavskiy A Ye 2000 High-tech industries and high technologies: definition, indicators, technical policy, share in the Russian economy *Ekonomicheskaya nauka sovremennoj Rossii* **2** pp 61–83

[6] Syryamkina E G Rumyantseva T B and Liventsova E Yu 2016 Practice of development of students' super-professional competencies in a modern university *Obrazovanie i nauka* **7** pp 117–35

[7] Kahharov Sh 2014 Super-professional competencies and management *Organizatsionnaya psihologiya* **4** pp 103–20

[8] Maksimenko L S and Dernovaya A O 2019 Theoretical approaches to the definition of professional and super-professional competencies in conditions of modern economy *Aktual'nye nauchnye issledovaniya v sovremennom mire* **1** p 30–4

[9] Il'yazova M D 2008 Competency, competence, qualification – main directions of modern researches *Professional'noe obrazovanie. Stolica* **1** pp 27–34

[10] Ndong-Zhatta *Neformal'noe obrazovanie* (Division of public information UNESCO) Available online: http://www.unesco.org/bpi/pdf/memobpi55_NFE_ru.pdf (accessed on 01.06.2020)

[11] Ivanova L A 2016 Modern labor market: trends in the formation of super-professional competencies *Fractal Simulation* **1** pp 35–41

[12] Ivanov V I, Baranov V V, Lysak G I and Kirsanov O V 2003 *High-Tech Enterprises in the Era of Globalization* (Moscow: Alpina Publisher) p 416

IOP Conf. Series: Materials Science and Engineering **966** (2020) 012080 doi:10.1088/1757-899X/966/1/012080

To the issue of transformation of 'old industrial area' geobrand through example of Sverdlovsk Region

Yu G Myslyakova[1], M V Kurashova[2] and M M Shcherbinin[2, 3]

[1] Ural State University of Economics, 62, 8 Marta Str., Yekaterinburg, Russia
[2] Ural Federal University named after the First President of Russia B N Yeltsin, Nizhniy Tagil Technological Institute, 59, Krasnogvardeyskaya str., Nizhniy Tagil, 622000, Russia

E-mail: [3] m.m.shcherbinin@urfu.ru

Abstract. The authors of the article present the results of theoretical and methodological review of term 'old industrial area', in the result of which its features and identical problems of functioning were identified. Similarity of such regions is demonstrated in correlation of economic downfall with level of social strain; difficulty of industrial fields transformation; disinclination to accessibility of innovations; imperfection of financial institutions and problems of industrial heritage valorization. Transformation of 'old industrial area' is considered through example of Sverdlovsk Region. The authors defined several directions of its implementation and explained geotransformation of the considered region from a historically developed image of a 'state stronghold' to the image of developing region with modern capital acquiring features of international metropolitan city. The article describes formation of new dynamically developing image of 'capital area' for Sverdlovsk Region having innovative path of evolution in terms of ability to engage young people both from different parts of Russia and various countries. Taking into account a number of facts it is also the evidence that intensification of trend of migration to the capital of Sverdlovsk Region for educational reasons is directly related to its image of a big city.

1. Theoretical analysis of term 'old industrial area'

The term 'old industrial area' came into use in 1970-s [1] and was subjected to insignificant development. Thus, for example, in the work of J Carney, R Hudson, J Lewis dated 1980 the old industrial area is understood as a region with industrial field prevailing in economy [2].

M Steiner in the work dated 1984 defined the old industrial area as a struggling region, economic basis of which was formed in the last century by a small number of developed sectors (metallurgy, extraction of commercial minerals, textile industry), and then was being developed for more than one hundred years, but now is suffering from recession [3].

A Granberg, S Artobolevsky, G Kovaleva defined the old industrial area in 1998 as a territory having specialized fields that lost the target market, became unprofitable or stopped the manufacturing activity or are at the stagnation stage due to scientific and technical progress [4].

L Sakharova determined in the work dated 2003 that the old industrial area is a region, manufacturing facilities of which require renovation, therefore economy of such areas (regions) suffer from the increase forcing out the domestic products incapable to compete by price and quality with the similar imported ones.

In 2004 M Trippl defined the scope of the old industrial area as a territory where education of people and innovative solutions under condition of economic reality modernization do not have the expected effect and do not establish a connection of material prosperity of the region with a level of its innovative activity. F Tödtling also noted that the highly specialized technological innovations prevailing over development and introduction of new products under conditions of low diversification of business activity are peculiar for these regions [5].

K Glonti insisted in the series of works dated 2008 that the old industrial area is a territory with technologically obsolescent industry due to change of technological paradigms and passing of time [6].

S Litovchenko suggested in 2014 to understand the old industrial area as a territorial entity social and economic development of which is based on a historically developed concentration of industry corresponding to 3rd-4th technological paradigm. Herewith the region has a potential of further innovative development [7].

N Sorokina, Yu Latov determined in 2016 that the old industrial area is a constituent entity of the Russian Federation comprising highly specialized industrial fields connected with passing technological paradigms [8].

Thus, one can say that regardless of long-time usage of the term 'old industrial area' it has no unambiguous rendering, so it is reasonable to distinguish specific methodological aspects of such territories.

2. Methodological aspects of old industrial areas

Russian old industrial areas have similar methodological aspects. Firstly, it is an established fact that correlation of economic downfall with level of social strain that may either appear during active searching of solutions for existing social problems (in regard to employment, advance in living standards, raising of qualification, etc.) or take such determining forms as industrial actions, hunger strikes, rallies, protests due to negative assessments of government work of federal, regional and municipal level. Thus, research performed by the scientific body of employees of the Institute of Economics of the Ural Branch of Russian Academy of Sciences under supervision of member of the Russian Academy Sciences A I Tatarkin through example of public opinion polls of 4 municipal units within Perm Krai including the following towns: Kizel, Gubakha, Gremyachinsk, and Chusovoy determined that the social strain on its territory during abandonment of mines in the Kizel coal basin in the end of 2000-s was induces by anxiety related to price increase (72.4 %), unemployment (34.8 %), increase in crime (28 %), arbitrariness of officials (19.0 %), inefficiency of local government activity (18.1 %), poor housing conditions (16.7 %), etc. Scientists who conducted this research concluded that the local society needs formation of social attitudes and new system of thoughts regarding the exercised transformations and social and economic perspective of the region. This should help to overcome a social apathy, as well as to form pro-active attitude. In addition, independent solving of local problems through participation of active population in making of decisions and implementation of management standards of local businesses shall be the core value for local society.

Secondly, all old industrial areas include difficulties of industrial field transformation with structural irresponsiveness to modern development conditions. Development of the industrial fields that are new for the territory ensuring stable diversification of territory economy serves as a solution here. Extension of the service industry is another option to be used in the regions.

Thirdly, old industrial area is characterized by imperfection of innovation support system that may be solved within two approaches. The first approach is related to management of innovative development oriented at attraction of high-tech enterprises into the regional economy that are mainly engaged in service provision in order to create the necessary infrastructure and favorable conditions for development of such activity [9]. The second approach is traditionally connected with management of transition from inertial development to innovative one. In case of this approach it is required to solve two tasks simultaneously: provision of necessary resource base and creation of conditions for diversification of economy. One can also mention, within the low innovativeness of the region, the undeveloped system of higher education and concentration of universities mainly on technical professions required by the current economic industrial sector. Such situation in universities results in niche specialization of population and is an obstacle for diversification of economy. Not least important is the inability of existing system of higher education to form and increase an entrepreneurial potential of young people which also has impact on migration flows of graduates of old industrial areas.

Fourthly, imperfection of functioning of financial institutions, activity of which is inefficient even in the process of individual mobilization of available unused spare capacity of industrial facilities; modernization of the most 'narrow' local links of a technological chain restraining the production capabilities of the enterprise; reconstruction (with a large capacity gain) of the best existing enterprises; construction of new enterprises; displacement and liquidation of noncompetitive and prospectless structures.

Fifthly, industrial heritage comprising values of industrial culture, having historical, technological, social, architectural or scientific meaning is another distinctive feature of old industrial areas. This industrial heritage has appeared in the beginning of industrial revolution and continues today due to change of technological paradigms, when multiple facilities become unnecessary and do not satisfy the demands of times [10]. And the main task here is not only to preserve, but to valorize the industrial heritage as an additional source of economic development of the territory.

3. Identical problems of old industrial areas

The above methodological aspects of old industrial areas cause the following identical problems [11]:

- large fraction of heavy industry in the territory economy with low potential to deal with new forms of competition;
- low level of entrepreneurial development and lack of personnel for work in new economic sectors;
- low level of diversification of region economy;
- high level of unemployment;
- ecological problems due to lack of modernization of treatment facilities;
- narrow professional qualification of people on the region;
- large fraction of exhausted permanent assets used in manufacturing processes, as well as application of obsolete technologies;
- restricted environment of innovation and low investment attractiveness.

Thus, the old industrial area is characterized by historically developed specialization due to 3rd or 4th technological paradigm; use of obsolete past-oriented technologies; low level of academic activity of universities; necessity of innovative transformation of industrial complex; and the like. It appears that when at least one of the methodological aspects of old industrial area is changed and one cluster of the above problems is solved, then we can talk about transformation of the territory geobrand.

4. Main trends of transformation of old industrial area in Sverdlovsk Region

Taking into account the above of theoretical and methodological aspects of old industrial areas it can be said that one of the perfect examples of such territories is Sverdlovsk Region. Such its geobrand is determined by creation of general Ural mining and metallurgical center in a time of Peter I, as well as historically developed mining and metallurgical civilization firstly mentioned in the works of D Mamin-Sibiryak. Favorable intersection of trade routes and a wealth of natural resources stipulated development of industrial area. Lately Yekaterinburg transformed its image and became a capital city opened for communication relations. Transformation is demonstrated in creation of maximum advantageous innovative climate, establishment of informational and technological connections, provision of economic and environmental safety [12], design of developed infrastructure [13] complying with international standards, and certainly in cultural landscape [14] determined in particular by local architecture, one part of which is 'inherited' by the citizens and other part is being designed and constructed today 'creating' new image of the city for future generations [15]. Historical

IOP Conf. Series: Materials Science and Engineering **966** (2020) 012080 doi:10.1088/1757-899X/966/1/012080

traditions and new directions in architecture are combined in the course of development of geobrand of Yekaterinburg emphasizing the historical importance of the territory origin [16]. Interest of young people from other countries in Sverdlovsk Region is caused by geographical position of Ural dividing Europe and Asia but connecting economic interest of the West and the East, as well as different cultural and historical traditions of population. Therefore, capital characteristics of geobrand of Yekaterinburg include new touristic routes, international business, cultural and sport events. Content analysis of official mass media showed that Yekaterinburg has been mainly positioned as cultural and sport capital city since 2017. Number of publications related to development of infrastructure of the city, public health system, tourism, international cooperation in 2018 [17].

All the information mentioned above emphasizes geotransformation of Sverdlovsk Region from a historically developed image of old industrial area of the Ural ('state stronghold') to the image of developing region with modern capital acquiring features of international metropolitan city. Another feature of transformation of geobrand of the capital city of Sverdlovsk Region is the developed system of higher education represented by 39 higher education establishments, many of which occupy high places in the Russian ranking of universities in terms of the quality of educational services and the prestige of studying at them. Indices of its works show dynamics in development of the considered region, its innovative course of history and ability to engage young people both from different parts of Russia and various countries.

5. Sverdlovsk Region as 'capital area' in the content of export activity of Ural universities

We propose to review the transformation of geobrand of Sverdlovsk Region in terms of increase in export of services of Ural universities during 2016–2018. Thus, universities of Sverdlovsk Region implemented 970 contracts on internal study mode and 537 contracts on external study mode in 2016/2017 academic year making it possible for universities to gain profit from export of educational services in the amount of 151,462,435 rubles. Number of contracts on internal study mode for foreign citizens increased by 157 % in 2017/2019 academic year in comparison with the last year, thus the total amount on these contracts is 203,614,764 rubles. At the same time, number of contracts on external study mode reduced by 87 % affecting the profitability of university from export activity equal to 33,776,250 rubles in 2017/2018 academic year. The fact that regardless of different trends in export services the universities managed to increase total profitability from 151,462,435 to 237,391,014 rubles is considered as positive moment approved by motivation of foreign students to come and study in Ural universities.

In university section (see table 1) the larges profitability from export of educational services in 2017/2018 academic year has the Ural Federal University named after the first President of Russia B N Yeltsin (158 million rubles) due to internal and external study mode on a contractual basis of foreign citizens.

Table 1. Export of educational services of state universities of Sverdlovsk Region in 2017/2018 academic year [18].

University	Total quantity of foreign citizens, persons	Int. contact, persons	Amount, rubles	Ext. contact, persons	Amount, rubles
Ural Federal University	1,164	1,026	145,995,740	138	11,890,850
Ural State University of Economics	288	172	21,929,000	116	7,630,000
Ural State Mining University	215	123	12,841,676	92	5,682,330
Ural State Pedagogical University	154	121	11,109,200	33	1,917,400
Ural State Law University	131	68	9,868,400	63	4,964,400
Ural State Forest Engineering University	23	7	749,600	16	753,150
Ural State Academy of Architecture and Arts	15	7	1,121,148	8	938,120

The second place in profitability from export of educational services to foreign citizens is taken by the Ural State University of Economics (29.5 million rubles). The third place is taken by the Ural State Mining University, where education of foreign citizens provided inflow of monetary resources in amount of 18.5 million rubles. In general, universities in Sverdlovsk Region implemented 1524 contracts on internal study mode and 466 contracts on external study mode in 2017/2018 academic year exceeding amount of contracts of the last period (1507 contracts in 2016/2017 and 1990 contracts in 2017/2018). It turns out that there is an increase of number of foreign students proving the transformation of geobrand 'old industrial area' of Sverdlovsk Region to 'capital'.

In addition, it is interesting to mention the obtained results of questioning of 300 foreign students of UrFU and UrGEU aimed at determination of factors defining selection of place of studying and university. Opinions of respondents were divided as follows:

- acknowledgment of the university diploma in their country (100 % of respondents);
- interest in the region where the university is located (92 % of respondents);
- possibility of employment in the region (43 % of respondents);
- recommendations from relatives, friends, etc. (32 % of respondents).

It is particularly remarkable that the foreign students mentioned the following channels as sources of information, from which they learn about Ural universities: the Internet (97 % of respondents); FIFA World Cup (90 % of respondents); meetings with representatives of the university native country (45 % of respondents); international events in native country (45 % of respondents); international events in Yekaterinburg (60 % of respondents). That means that informational channels also show transformation of geobrand of Sverdlovsk Region.

6. Conclusions

Thus, it turns out that Sverdlovsk Region and its capital correspond to methodological aspects of 'old industrial area' on the one hand and have social and economic trends in transformations of established geobrand on the other hand. It was illustrated through example of increase of educational migration trend in Yekaterinburg due to attractive image of the Ural territory and quality of education services of universities. It also proves that the capital of Sverdlovsk Region forms new geobrand of cultural metropolis with developed infrastructure and international cooperation.

References

[1] Maltsev A A and Mordvinova A E 2016 Restructuring of old industrial regions in Europe: experience and problems *Upravlenets* **3** pp 8–13
[2] Carney J, Hudson R and Lewis J 1980 *Regions in Crisis* (London: Croom Helm) p 179
[3] Steiner M 1985 Old industrial areas: a theoretical approach *Urban Studies* **22** pp 387–98
[4] Granberg A G, Artobolevsky S S and Kovaleva G A 1998 Restructuring of old industrial regions: Russian and worldwide experience *Regionalnoye razvitiye i sotrudnichestvo* **1-2** pp 4–23
[5] Tödtling F and Trippl M 2004 *One Size Fits All? Towards a Differentiated Policy Approach with Respect to Regional Innovation Systems* (Vienna: Institut für Wirtschaftsgeographie, WU Vienna University of Economics and Business)
[6] Glonti K M 2008 Old industrial areas: problems and development potential. Federalism problems *Regionologiya* **4** pp 27–36
[7] Litovchenko S V 2014 Analysis of modern approaches to determination of essence of old industrial areas *Ekonomicheskiy prostor* **92** pp 138–50
[8] Sorokina N Yu and Latov Yu V 2018 Evolution of old industrial regions in Russian economy *Voprosy regulirovania ekonomiki* Vol **9** **1** pp 6–22
[9] Sobolev A Yu 2015 foreign experience of innovative development of old industrial regions and possibility of its application in Russia *Voprosy upraleniya* **1** pp 73–80
[10] Zapariy V V 2010 Industrial heritage: identification, study, application *Ekonomicheskaya istoriya* **3** pp 4–10

IOP Conf. Series: Materials Science and Engineering **966** (2020) 012080 doi:10.1088/1757-899X/966/1/012080

[11] Shelomentseva V P, Ifutina Ye A and Shelomentsev P Yu 2016 Old industrial region: physical origin, classification, transformation *TURAN University Bulletin* **2** pp 49–53
[12] Parshin P B 2015 *Territory as a Brand. Marketing Metaphor, Identity, and Competition* (Moscow: Moscow State Institute of International Relations) p 195
[13] Groshev I V and Stepanycheva Ye V 2013 Factors effecting formation of city brand *Marketing v Rossii i za rubezhom* **4** pp 83–96
[14] Loguntsova I V 2017 Uniqueness of geobrands built on the ground of cultural values: russian approach *Brend management* **4** pp 290–9
[15] Kovrigina V A and Lapteva V V 2016 Prospects and problems of geobranding development in Russia *Aktualnyye voprosy ekonomicheskikh nauk* **53** p 17–22
[16] Yazykova P V 2011 Social and economic aspects of development of geobranding developement of cities *Sotsialno-ekonomicheskiye yavleniya i process* **10** pp 240–2
[17] Kalyuzhnaya E and Popova O 2019 Media image of the city in the context of digital culture. III post mass media in the modern informational society (PMMIS) *Journalistic text in a new technological environment: achievements and problems in 2019* pp 397–405
[18] *Export of Russian Educational Services: Statistical Book.* Issue 9 Ministry of Science and Higher Education of Russian Federation 2019 (Moscow: Center of Social Research) p 536

IOP Conf. Series: Materials Science and Engineering **966** (2020) 012081 doi:10.1088/1757-899X/966/1/012081

Role of management of industrial enterprises in formation of labor migration trend in region

M V Kurashova[1, 3], Ye A Nadeina[2] and Yu G Myslyakova[2]

[1] Ural Federal University named after the First President of Russia B N Yeltsin, Nizhniy Tagil Technological Institute, 59, Krasnogvardeyskaya str., Nizhniy Tagil, 622000, Russia
[2] Ural State University of Economics, 62, 8 Marta Str., Yekaterinburg, Russia

E-mail: [3] artmnn123@mail.ru

Abstract. In the article it is justified, that one of the options to solve the problem of increase of a labor migration trend of young population of a small mono-city or of a highly specialized city is integration of a talent management concept into a management system of its basic industrial enterprise. The body of evidence for this justification was composed by the results of the research of essence of notion 'talent' and its management features, as well as conclusions obtained in the course of survey of pupils of eleventh grades of schools in Nizhny Tagil, as well as of students of technical schools and universities of this city. The following provisions have been determined as specific recommendations to reduce the migration outflow of young people: the concept of talent management at an industrial enterprise has not only corporate, but also territorial effectiveness in solving the socio-demographic problems of the city; it is necessary to minimize not so much labor as educational migration of the young population, since it has a transformational nature and a stable form of appearance; talent management programs shall be adapted to conditions of functioning of highly specialized cities.

1. Introduction

This article will focus on industrial enterprises of mono-cities, activities of which have a significant impact on a socio-economic situation of a territory of their localization and on living standards. One of pressing problems for such cities, due to their niche specialization, is susceptibility to a depression of economic activity and outflow of the young working-age population, which is caused by appealing qualities of metropolitan cities and by the presence of unpleasant factors related to an adverse economic situation at an enterprise, brought on by low marketability of products, deterioration of equipment, high production costs, as well as by the absence of possibility to evolve and increase a professional skills set. In turn, the negative labor migration of youth leads to deficiency of qualified specialists, which inhibits entering a level of stable development and independent solving of problems in the field by the economy of mono-cities. Therefore, this problem shall be solved on an urgent basis, without downgrading of the role of the industrial enterprises themselves.

2. Level of knowledge

At this time there are different approaches to retain the youth in a native city. In conditions of mono-cities, it is possible to stop the migration outflow of young population by engaging the industrial enterprises, possibly by developing and implementing 'talent management' programs that are aimed not only at providing intellectual and competence development of a staff composition, but also at its retention, which will be a result of interesting and entry-level job.

The term 'talent management' was firstly applied by D Watkins in 1998. Then it was used for creation of a whole personnel management system, that received worldwide recognition and formed a new trend in management in 2004. Such authors like E Farndale, F Stills, Sh Maurice, J Trevor explain the talent management as human capital assets management [1, 2], which academic economists like A Smith [3], W Petty [4], J Mill [5], A Marshall [6], J-B Say, L Walras, V Pareto, I Fisher, E Denison and others began to study very long ago. Modern understanding of this definition has an institutionalized set of knowledge, health and culture assets of an individual. That means, that the talent itself is understood as a set of characteristics inherent in a potential of a worker to bring high results [7]. Herewith a talent management system itself is focused on increasing productivity of personnel; on arranging a remuneration system taking into account a level of knowledge and skills of workers; integration of a reward system, taking into account contribute of each employee into a concluding result of the enterprise's functioning; on retention of employees at their work places, and etc. [8]. According to E Michael, H Handfield-Jones, B Axelrod, the talent management is an activity aimed at engagement of employees into an innovative process, formation of creative incentives and development of creative potential of the employees [9]. It is obvious that all the above authors consider the talent management only as a source for solving problems of a certain enterprise, at which respective personnel management programs will be launched and 'perspective' personnel will be trained.

We think, that, under modern business conditions of mono-cities, this management concept may obtain a distinctly different meaning, and a scale of its result may have a territorial level of solving economic and demographic problems related to the young population outflow. For this, responsible executives of an industrial business shall consider each person aged 17-20 as a 'potential' talent capable of changing a current trend of economic development of mono-city by applying their knowledge and skills in production practice. The personal knowledge shall have two sources of formation: university and production, herewith each of them complements the other one and thus makes the competence formation process constant. That is why the enterprises shall be provided with an operating enterprise-university-enterprise cooperation system; extended portfolio of tools for development and increase of effectiveness/productivity of staff composition; developed career planning system which will be able to provide clear and transparent understanding of a future career and development plan of each worker; increased flexibility of control over a motivational package of employees referenced to the 'pool of talents' to increase a level of their engagement and, respectively, to retain within an organization

3. Research methods and results

The mono-cities loose mainly the youth aged 16–35, for which a project of life is bound to a study and to a job. Causes for such a downward dynamic are the absence of qualified education in small cities and inability to fulfill a demand for a decently paid job. Besides, traditional occupations of citizens of mono-cities in Russia are railway employees, steelworkers, workers of other industrial enterprises not attracting the youth. For example, in Nizhny Tagil, only 17.3 % of respondents plan to stay after finishing 11th grade, meanwhile 79.8 % of respondents are sure that they will live in a major population city, but not in their native city (730 pupils in total have been quizzed). Such intentions prove reduction in a ratio of persons entering 'local' universities, confirming, that the youth 'leakage' trend has a constant form of development, which started from 2013, see figure 1.

IOP Conf. Series: Materials Science and Engineering **966** (2020) 012081 doi:10.1088/1757-899X/966/1/012081

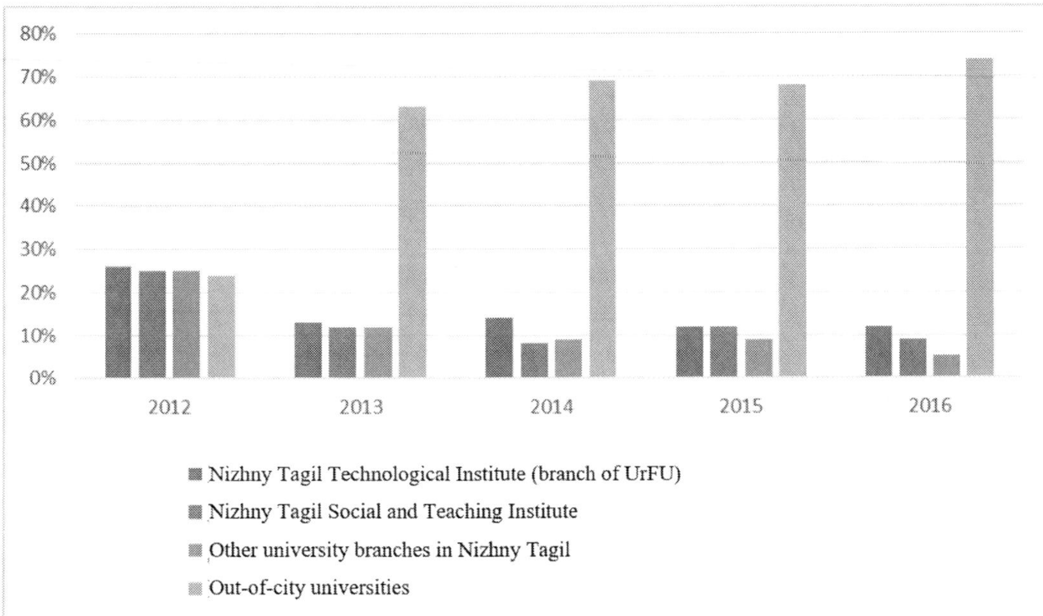

Figure 1. Rate of Nizhny Tagil schools' leavers choosing study in a university, %.

To reveal main factors influencing on educational migration of the youth, 100 students of technical schools and 100 students of universities have been quizzed. In the first place, the respondents noted the absence of promising positions, as well as the absence of growth and development prospects that can only be obtained in large cities with a diversified economy, which makes it possible to find jobs in successful large companies (100 % of respondents), see table 1.

Table 1. Factors of young population migration from Nizhny Tagil

Young population migration factors	Students of technical schools, %	Students of universities, %
total number of students willing to leave Nizhny Tagil	33	46.2
no perspective positions (jobs)	100	100
low wages at enterprises	87	87
lack of leisure and recreation centers, as well as low level of safety	54	52
lack of educational facilities (or absence of state-funded places)	45	54

This factor is also confirmed by the all-Russian and international trend associated with the fact that one third of respondents worldwide and 37 % in Russia choose this criterion as the primary advantage of employers. In addition, 98 % of respondents said that the ability to work under the guidance of more experienced colleagues and mentors is an important factor in the professional development of any novice specialist.

In second place, the surveyed respondents noted low wages, which are higher in metropolitan cities than in peripheral cities (87 % of respondents) and differ significantly from the quality of education received at universities, see table 1.

The third factor established by the students is lack of leisure and recreation centers, as well as low safety level due to social tension in local community (53 % of respondents). That is why the nearest city enticing the youth is Yekaterinburg, social culture of which goes back to the time of Peter the Great and is associated with formation of a mining and metallurgical center, that is in modern language a cluster in the middle Urals with a favorable intersection of trade routes and a wealth of natural resources contributing to formation of the mining industry. The location of 39 higher education establishments in Yekaterinburg, many of which occupy high places in the Russian ranking of universities in terms of the quality of educational services and the prestige of studying at them, is also of particular importance. In turn, this serves as the basis for young people to move in and enter them, thereby independently solving the problem of Nizhny Tagil related to the lack of educational facilities or the lack of budget places.

Thus, the above factors serve as a basis for formation of the youth migration educational trend. In order to determine its stability, results of answers of students of leading Ural universities in Yekaterinburg are given: 'Where do you dream to work?', see figure 2.

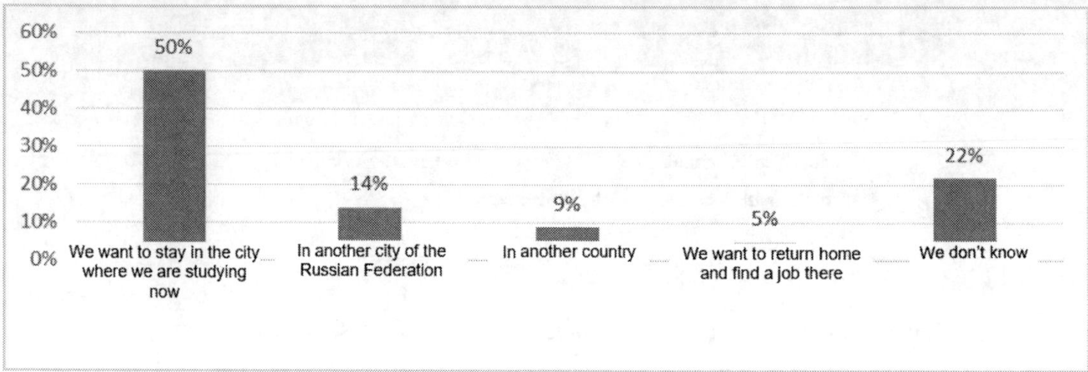

Figure 2. Answers for the question: 'Where do you dream to work?', % of respondents 250 students in total have been quizzed.

The results of the survey demonstrate unwillingness of the students to return home, but willingness to stay and find a job in a city of study (50 % of respondents) or move to another city (14 % of respondents), or migrate to another country (9 % of respondents quizzed). Only 5 % of respondents quizzed want to return home in Nizhny Tagil. The presented answers confirm stability of the youth migration trend and transformation of its educational form into the labor one.

4. Solutions and conclusions

The above presented factors defined as the basis for the educational migration, as a result, are the primary factors of the youth labor migration. That is why they shall be minimized by city-forming enterprises in order to retain the youth of working age in highly specialized cities.

The main industrial enterprises of Nizhny Tagil are: JSC 'Research and production corporation 'UralVagonZavod', with the staff number of 29,800 persons (as of 01.01.19), and JSC EVRAZ NTMK, with the staff number of 19,900 persons (as of 01.01.19), jointly provide employment for more than 20 % of working-age population of the city. They have possibilities to master new technologies, successively develop models of transport and special machines and implement them in batch manufacturing, master production of wide range of civil products and manufacture exclusive military products in the Russian Federation. All this points at the fact that these enterprises have high potential to influence on youth migration flows by developing 'enterprise-university-youth-enterprise' cooperation networking program oriented at formation of attractive space for education, creativity and work, and by taking the following management steps.

First of all, these enterprises shall put the management accent on education. To do this, special programs of cooperation with leading specialized universities, which consists in 'targeted training of specialists' is needed. These programs will also allow 'growing' engineering personnel at the enterprises based on consistent formation of skills from simple to more complex, from mastering a working profession to a technician, engineer and leading engineer. A comprehensive context of these programs shall provide active participation of the industrial enterprises in studying at industrial sites with engagement of specialists from production, inclusion of students into corporate norms of

IOP Conf. Series: Materials Science and Engineering **966** (2020) 012081 doi:10.1088/1757-899X/966/1/012081

organizing production and productive relations, participation of students in solving actual production tasks. In this case, an economic result is minimization of expenses for adaptation of young specialists at new workplaces.

Secondly, it is necessary to put the management accent on investments into the human capital assets, a bearer of which is not only the employed youth, but also future, potential personnel, that are university students. In the first case it is necessary to provide constant improvement of qualification and perform professional retraining of employees based on an individual approach to training of young specialists at sites in a form of mentoring. In the second case the industrial enterprise in cooperation with universities-partners shall perform tours in order to familiarize the youth with activities of the enterprise and improve the image of technical personnel. In addition, the students will be interested in engineering projects contests, the supervisors of which shall be both university professors and technical professionals of the enterprise itself. Moreover, the city-forming enterprise has a potential to organize summer specialized educational sessions uniting the students from universities and the specialists from enterprises-partners.

Thirdly, the management accent of these enterprises shall relate to an employee incentive program, which shall combine financial and non-financial incentives in the field of improvement of professional and novel skills, which will provide attraction and retention of qualified personnel, increase interest of the working youth in job performance. That is why the financial incentive shall take into account and be aimed at novel activation of the employed youth. The non-financial incentive shall be related to participation of the enterprise in preparation and improvement of qualification of teaching staff, participating in implementation of the programs presented above, as well as in creation and development of material and technical facilities to implement a new format of dual education system.

Based on this triad of actions, the youth migration outflow will decrease by implementation of modern talent management concept, adapted to conditions of business practices of a certain city, in a form of through trajectory and modern engineering and technical education. In terms of its contents, it shall have three stages of influence on formation of a working individual: first of all it shall begin in attitudinal events in schools/technical schools conducted by heads of industrial enterprises, and then move on to the educational level at universities, organized jointly with the basic business of the territory, and then give way to corporate 'field' events aimed at developing and retaining the youth at industrial enterprises of a highly specialized city. Only in this case the 'talent' management concept will obtain not only endogenous, but also exogenous social and economic result.

References

[1] Bukharina A Yu 2017 Talent management: what to teach employees today to survive tomorrow *Social Psychology and Society* **8** pp 144–62

[2] Dries N 2013 The psychology of talent management: a review and research agenda *Human Resource Management Review* **23** pp 272–85

[3] Smith A 1956 *An Inquiry into the Nature and Causes of the Wealth of Nations* (Moscow: Sotsekgiz) p 235

[4] Petty W 1940 *Economic and Statistical Works* (Moscow: Gossotsekgiz) p 324

[5] Mill J S 1980 *Principles of Political Economy in 3 volumes* (Moscow: Progress Publ.)

[6] Marshall A 1993 *The Principles of Economic Science: in 2 volumes* (Moscow: Nauka Publ.)

[7] Gallardo-Gallardo E, Dries N and González-Cruz T F 2013 What is the meaning of 'talent' in the world of work? *Human Resource Management Review* **23** pp 290–300

[8] Odegov Yu G 2015 Talent management as a reality of modern management *Herald of Omsk University. Series: Economics* **1** pp 92–9

[9] Michaels E, Handfield-Jones H and Axelrod B 2006 *The War for Talent* (Moscow: Mann, Ivanov and Ferber) pp 240

IOP Conf. Series: Materials Science and Engineering **966** (2020) 012082 doi:10.1088/1757-899X/966/1/012082

Difficulties of succession pipeline in large industrial enterprises of Nizhny Tagil in studies of tendencies of competitiveness of higher technical education

M M Shcherbinin[1,2], V V Potanin[1] and S E Chetverikov[1]

[1] Ural Federal University named after the First President of Russia B N Yeltsin, Nizhniy Tagil Technological Institute, 59, Krasnogvardeyskaya str., Nizhniy Tagil, 622000, Russia

E-mail: [2]m.m.shcherbinin@urfu.ru

Abstract. There have been significant changes in the higher education system in Russia recently. The possibility to apply for post-secondary education to several universities according to the results of the Unified State Examination (USE) and the transition to the two-level structure of higher education have significantly changed the landscape of competitiveness not only between individual universities, but also between large cities, which have become centers of mass attraction of applicants and young specialists. The number of students studying federally funded has been decreasing over the years. As a result, regional universities suffer a shortage of applicants, while local industrial enterprises suffer a shortage of young qualified personnel. The described problems are fully demonstrated in the system of higher education and training of young engineering personnel for large industrial enterprises in Nizhny Tagil. The article reviews the results of the study on the motivational field of academic migration of young people from the Nizhny Tagil city, conducted in February-March 2020.

1. Worldwide and Russian trends in academic migration

Educational opportunities in a large academic center are the main reason for youth migration around the world, as reflected in D P Smith, P Rérat, J Sage, 2014 [1] studies. Migration of young people to large cities is a worldwide trend of urbanization H Mainet, J-Ch Edouard, 2017 [2], developed by perceptions of the prestige of living in a large city J Sinkien, S Kromalcas, 2010 [3], higher economic prospects of personal well-being L Wilken, M G Dahlberg, 2017 [4].

Similar motives for the goal-setting of youth migration to large cities are also noted in the studies of Russian authors, where the desire of young people for self-realization T A Oreshkina, A S Odegov, 2017 [5] and achieving success in the life of S B Abramova are considered as the causal factors for migration, 2015 [6]. Based on a review of publications on this topic, it has to be said that up to date there is no unambiguous and systematically used term that defines the essence of 'academic migration'. In T V Kuprina, S M Minasyan and A M Tsaturyan, 2019 [7] study, academic migration is defined 'as educational migration, i. e. including academic mobility of students and labor academic mobility of teachers and scientists, highly qualified personnel on the labor market'. Without disputing the proposed wording, it should be noted that for the purposes of this study the main focus will be on that part of academic migration, which is implemented as a move to another city for higher education and further employment in the period after graduation from school or technical school, i. e., after secondary or secondary vocational education.

2. Academic migration as a problem in small towns

The problem of youth outflow from small towns has already been the subject of analysis by a number of Russian researchers. Based on the analysis of Federal State Statistics Service statistical data in the study of I S Kashnitsky, N V Mkrtchyan and O V Leshukov, 2016 [8], it is shown that the age-specific peak of migration falls on the age of 17–19 years after graduation from 11th grade of school, as well as on the age range of 22–23 years, i. e. according to the results of graduation.

According to the data provided by the Department of Education of the Nizhny Tagil Administration, the outflow of school graduates demonstrates a clear upward trend, amounting to 69 % in 2014 and 83 % in 2019. Such impressive statistics is confirmed in other studies as well. Among them, the results of a survey conducted by the Institute of Social Analysis and Forecasting of the Russian Presidential Academy of National Economy and Public Administration can be mentioned, conducted by N V Mkrtchyan, 2017 [9] in four small towns of the middle Urals, where it is noted that 75 % of students are sure that they will live in a larger settlement than their hometown, and only 4 % of graduates, in the cities studied, intend to stay there after the 11th grade.

The trend of outflow of young people from the city is becoming more and more unsettling not only for higher educational institutions, but also a critical problem of formation of a qualified succession pipeline for large industrial enterprises in Nizhny Tagil. During the last 5–7 years, large industrial enterprises have been experiencing a shortage of young engineering personnel. It is not surprising that with such a mass academic migration of 11th grade graduates, a number of engineering areas of higher education at the Nizhny Tagil Technological Institute have difficulties in attracting applicants even for the budget form of education.

Let us ask ourselves a question about causal factors of migration and such a low interest to enroll in engineering departments of Nizhny Tagil Technological Institute. The most superficial reasons in the interviews with graduates are insufficient salaries at local enterprises and low prestige of engineering professions. It is easy to believe that the motivation factors responsible for high rates of migration from the city are sufficiently demonstrated in these words. However, studies by E S Vakulenko, N V Mkrtchyan and K K Furmanov, 2011 [10] prove the negative connection between migration growth/loss of population in municipal districts and salary at local enterprises. This phenomenon functions as a 'poverty trap', Y Andrienko and S Guriev 2004 [11]. In S Guriev, 2010 [12] studies it is shown that the rate of migration is higher from those cities where salaries are higher, because the poverty of households is holding back the outflow of population, 'it's so bad that we have to go, but there is nothing to go for'. It is amazing, because from this point of view Nizhny Tagil is a prosperous city, generating a huge mass of school graduates, competitive for admission to the capital's universities and having economic opportunities to move to another city.

3. Who's packed and ready to go and what kind of education they're going to go for: forecast of academic migration of youth in Nizhny Tagil

The empirical basis of the study was formed by the data of the questionnaire survey 655 people of 9 and 11th grade students of general education schools, students of Nizhny Tagil Mechanical-Engineering Technical School and students of Nizhny Tagil Technological Institute. The questionnaire details were collected in February-March 2020. According to the results of the study (Table 1) it can be seen that the aspiration of school graduates to get higher education has a pronounced mass character. Immediately after graduation from the school, 74.1 % of respondents plan to get a higher education, and other 17.8 % believe that they will implement these plans in some time. 43.8 % of the SVE study participants plan to act in a completely different way and to get higher education at the end of current education. This rate is even lower among today's university students – only 9.1 % plan to get a second higher education immediately after graduation from the university.

Let us further consider the wishes of the study participants regarding moving to another city (Table 1). It is possible that now the mood of young people may look different, as the questionnaire survey was carried out before the measures to prevent a coronavirus pandemic.

IOP Conf. Series: Materials Science and Engineering **966** (2020) 012082 doi:10.1088/1757-899X/966/1/012082

Table 1. Plans for getting higher education.

Do you plan to continue your education and get a higher education? (second higher education - for university students)	Where are you currently studying? School, lyceum, gymnasium		Technical school or vocational education institution		Institute or university	
	Frequency, persons	% by column	Frequency, persons	% by column	Frequency, persons	% by column
Yes, I do, right after finishing the current education	403	74.1	28	43.8	4	9.1
Yes, I do, but after a while	97	17.8	27	42.2	10	22.7
Probably not, perhaps only in the distant future.	32	5.9	7	10.9	23	52.3
No, I do not plan on it at all	12	2.2	2	3.1	7	15.9
Total by column	**544**	**100.0**	**64**	**100.0**	**44**	**100.0**

Table 2. Migration plans for getting higher education.

Do you plan to leave Nizhny Tagil for higher education or further employment?	Where are you currently studying? School, lyceum, gymnasium		Technical school or vocational education institution		Institute or university	
	Frequency, persons	% by column	Frequency, persons	% by column	Frequency, persons	% by column
No answer	8	1.5	0	0.0	4	9.1
Yes, I do. I plan on leaving the city	476	87.0	46	71.9	21	47.7
No, I do not.	63	11.5	18	28.1	19	43.2
Total by column	**547**	**100.0**	**64**	**100.0**	**44**	**100.0**

For the subgroup of students, a high forecast for the number of those who want to move to another city to obtain higher education is confirmed, this rate is 87 %. It is indicative that technical school students are also planning to move in significant numbers, in total 71.9 % of those surveyed in the subgroup. Almost half of the students plan to leave after graduation, in total 47.7 % representatives of the subgroup.

4. Academic migration motivational field analysis

The ambiguity of perception and interpretation of economic motivations for migration predetermined the approach to designing research instruments for the survey. Another fundamentally important emphasis of academic mobility motivation is mentioned in the study of S B Abramova, N L Antonova, and O I Pimenova, 2019 [13] – this is an idealized character of perceptions about the city as a place for life, structured in two types of expectations '1) a 'comfortable' city for everyday practices, open to innovation and creativity; 2) an 'instrumental' city – a platform for career growth and material well-being, which may be emotionally unattractive, but functionally effective'.

The idealistic nature of expectations and motivation for academic migration, colored with super-positive images of another and super-negative assessment of a real city, is a key prerequisite for the choice of study methodology. To analyze the motivational field of academic migration, a questionnaire, based on F Herzberg's motivation-hygiene theory, was developed [14], which considers the assessment of motivational factors as idealized expectations in relation to another city, and on the other hand – the assessment of hygienic factors as more realistic impressions of the hometown. Ordinarily speaking, motivation factors can be considered as reasons to run away for pleasure, while hygienic factors are reasons to run away from dissatisfaction.

The questions of the study conducted by the authors to assess the motivational field of academic migration include 9 factors in each of the groups of motivational and hygienic factors. According to the results in Figure 1 and Figure 2, it is shown that the assessment of attractiveness of the move and dissatisfaction is quite high, both altogether within the study and in the subgroup of school students.

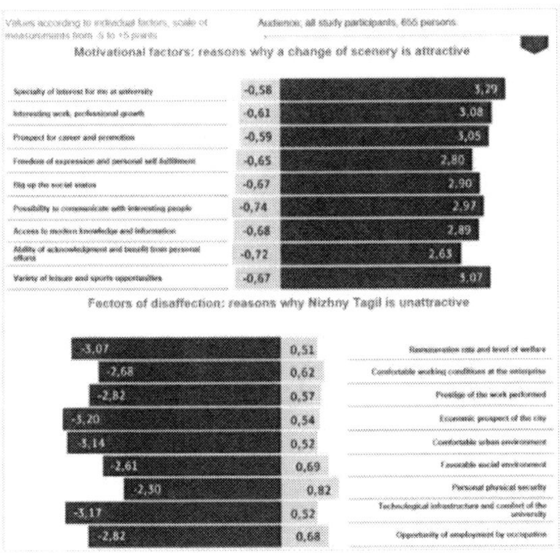

Figure 1. Motivational field structure of academic migration.

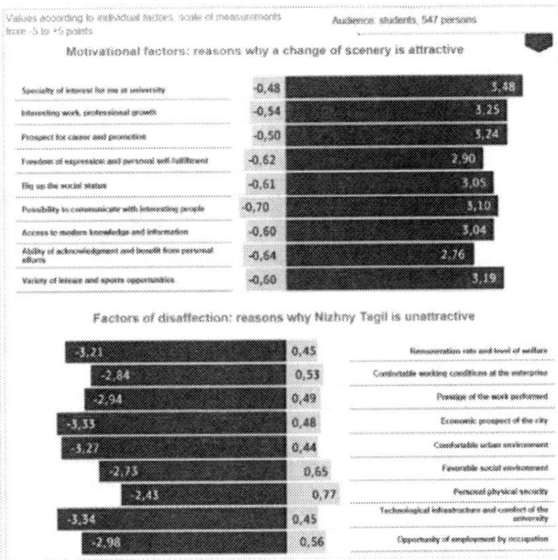

Figure 2. Motivational field structure of academic migration.

Altogether within the study, and also in a subgroup of students, the basic factor of motivation is the presence of a specialty of interest in university of other city, and key factors of dissatisfaction are economic prospect of the city and technological infrastructure, level of comfort of local universities.

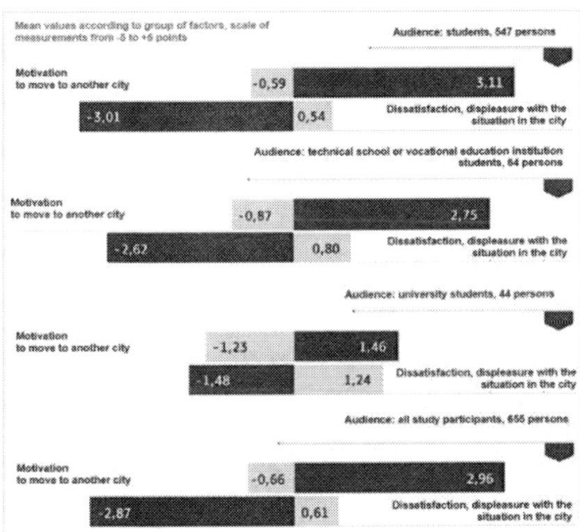

Figure 3. Overall motivation estimates of academic migration in subgroups.

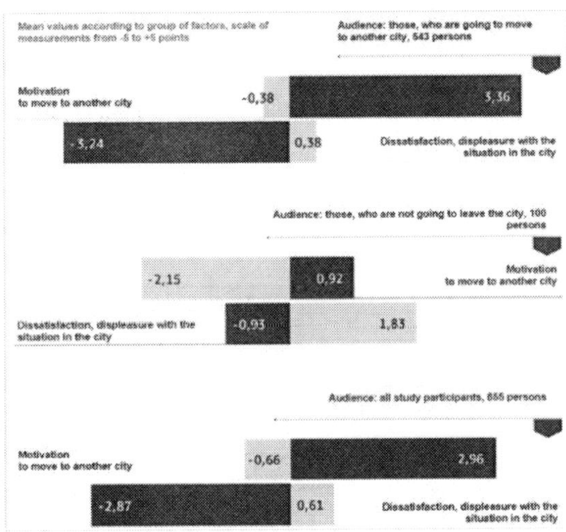

Figure 4. Overall motivation estimates of academic migration in subgroups.

Interesting results are found in the analysis of overall motivation estimates, where the arithmetic mean values are calculated for the groups of motivational and hygienic factors in various subgroups of the study participants. According to the data in Figure 3, university students do not demonstrate a

IOP Conf. Series: Materials Science and Engineering **966** (2020) 012082 doi:10.1088/1757-899X/966/1/012082

pronounced motivation to move and dissatisfaction with the internal environment in Nizhny Tagil, but the positive intention to migrate in this subgroup is identical to representatives of other subgroups. However, according to the data in Figure 4, there is a fundamental difference in overall motivation estimates of those who have a positive and negative attitude to the idea of moving to another city. The subgroup of patriots of Nizhny Tagil does not show a high assessment of motivational factors for moving to another city, and there is no dissatisfaction with the internal situation in the city. The motivational field of these subgroups is shown in Figure 5 and Figure 6.

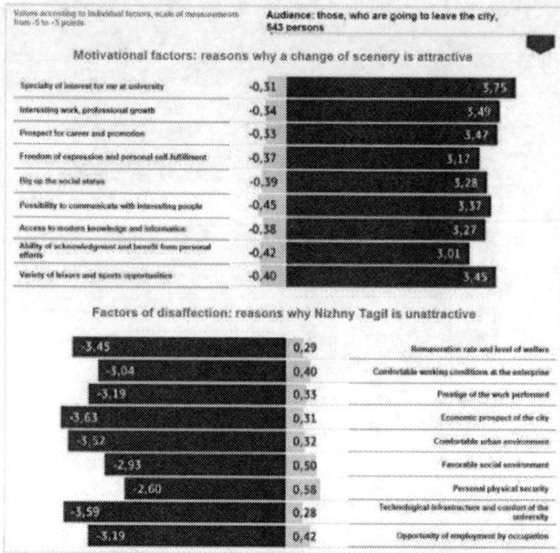

Figure 5. Motivational field structure of study participants, who are favorably disposed toward moving to another city.

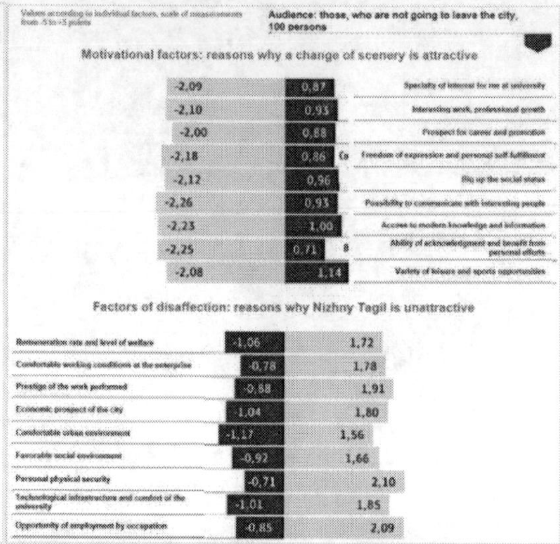

Figure 6. Motivational field structure of study participants, who are negatively predisposed toward moving to another city.

Motivation of the study participants, who have a positive attitude towards moving to another city (Figure 7), singles out the presence of a specialty of interest in the university as a key factor (3.75). The group of the following three factors with practically equal estimates is singled out as a secondary factor: interesting content of work, possibility of professional development (3.49); prospect for career and promotion (3.47); variety of opportunities for leisure, sports and personal self-development (3.45). The determining factor of dissatisfaction in this subgroup of respondents is negative assessment of economic prospects of the city (-3.63). The second position is occupied by the group of the following three factors: technological infrastructure and comfort of the universities (-3.59); comfortable urban environment (-3.52); remuneration rate and personal financial well-being (-3.45).

Motivation of those participants of the study, who negatively evaluated their desire to move (Figure 8), most of all cast doubt on what can be expected in another city: opportunity to work and communicate with interesting people (-2.26); ability of acknowledgment and benefit from personal efforts (-2.25); access to modern knowledge and information (-2.23). As the most motivating perspectives, they see entertainment and leisure opportunities as a factor in the diversity of opportunities for leisure, sports and personal development (1.14). Two factors have been identified by the study participants as the dominant hygienic factors that the city provides: personal physical safety (2.10) and opportunity of employment by occupation (2.09). Three following factors also received high positive estimate among hygienic factors: prestige of the work performed (1.91); technological infrastructure and comfort of the universities (1.85); economic prospect of the city (1.80). And the highest negative estimate (-1.17) representatives of this subgroup gave to the level of comfort of the urban environment.

5. Conclusions

The results of the study confirm that the majority of school graduates associate higher education with entering a university in another city. This indicator is also high among technical school students. The attractiveness of education and employment in a large city is based on a high level of motivation in relation to another city and high dissatisfaction in relation to Nizhny Tagil. Obviously, it is necessary for large industrial enterprises to work purposefully to form a positive and stable picture of the city's economic prospect. For local universities, and in particular, for Nizhniy Tagil Technological Institute, the most important vector of development and competitiveness shall be expansion and modernization of fields of education for students. The key task for the city as a whole is formation of a comfortable, attractive urban environment, provision of a wide range of opportunities for personal self-development and self-realization.

References

[1] Smith D P, Rérat P and Sage J 2014 Youth Migration and Spaces of Education *Children's Geographies* **1** pp 1–8
[2] Mainet H and Edouard J-Ch 2017 *Creative approaches to planning and local development. Critical perspectives. Insights from Small and Medium-Sized Towns in Europe* (Routledge Publisher) pp 234–48
[3] Sinkien J and Kromalcas S 2010 *Concept, directions and practice of city attractiveness improvement* (Kaunas: University of Technology) pp 147–54
[4] Wilken L and Dahlberg M G 2017 Between international student mobility and work migration: experiences of students from EU's newer member states in Denmark. *Journal of Ethnic and Migration Studies* **43** pp 1347–61
[5] Oreshkina T A and Odegov A S 2017 Civil culture of Sverdlovsk youth in the system of government. *Vlast' Power* **9** pp 111–5
[6] Abramova S B 2015 The image of the future: guidelines for modern youth. *Sotsiokul'turnoye razvitiye bol'shogo Urala: trendy, problemy, perspektivy: materialy yubileynoy Vserossiyskoy nauchno-prakticheskoy konferentsii XX Ural'skiye sotsiologicheskiye chteniya* (Ekaterinburg: Ural Federal University) pp 3–8
[7] Kuprina T V, Minasyan S M and Tsaturyan A M 2019 Modeling of academic migration invariants affecting social and economic indicators of the territory *Economy of Region* **15 3** pp 749–62
[8] Kashnitsky I S, Mkrtchyan N V and Leshukov O V 2016 Interregional youth migration in Russia: comprehensive analysis of population statistics *Journal of Educational Studies* **3** pp 169–203
[9] Mkrtchyan N V 2017 Youth migration from small towns in Russia *Monitoring of Public Opinion: Economic and Social Changes Journal* **1** p 225–42
[10] Vakulenko E S, Mkrtchyan N V and Furmanov K K 2011 *Experience of Modeling Migration Flows at the Level of Regions and Municipal Entities of the Russian Federation* (Moscow: Institute for National Economic Forecasts RAS] pp 431–50
[11] Andrienko Y and Guriev S 2004 Determinants of interregional mobility in Russia. Evidence from panel data *Economics of Transition* **12** pp 1–27
[12] Guries S 2014 *Myths of the Economy: Misconceptions and Stereotypes That are Distributed by Mass-media and Politicians* (Moscow: Yunaited Press) p 320
[13] Abramova S B, Antonova N L and Pimenova O I 2019 Attractiveness of the city as a factor of territorial mobility in student assessments. On the example of Yekaterinburg *The Education and Science Journal* **21 1** pp 97–123
[14] Herzberg F 2007 *Motivation To Work* (Moscow: Vershina) p 240

IOP Conf. Series: Materials Science and Engineering **966** (2020) 012083 doi:10.1088/1757-899X/966/1/012083

Development of algorithm for excavation control

O A Lukashuk[1], K Y Letnev[1, 3] and A P Komissarov[1,2]

[1] Ural Federal University named after the First President of Russia B N Yeltsin, 19, Mira str., Yekaterinburg, 620000, Russia
[2] Ural State Mining University, 30, Kujbysheva street, 620144, Yekaterinburg, Russia

E-mail: [3] ptmir@inbox.ru

Abstract. In open-pit excavators there exists a discrepancy between their engineering designs and underuse of performance potential due to complexity of excavation and such intensity of carrying out working procedures which borders on physical limits of an operator. The excavating procedure is shown to involve joint functioning of main actuating mechanisms of an open-pit excavator for its thrusting and lifting operations and formation of a leverage transmission mechanism connecting those with a bucket. In that case, it was established, the excavation process is characterized by a 'conversion' of the main mechanisms since their operation parameters are formed according to kinematic properties of the leverage and energy-force parameters realized on the bucket. Dependencies were derived to calculate kinematic transfer functions of the leverage, which correlate the velocity of excavation and those of working motions (speeds of thrusting and lifting). The algorithm of calculating the velocities of working motions could be used to develop a digital system to control the motors of the main mechanisms.

1. Introduction
Transition to a new type of economics requires qualitative changes to be made in means of production. Improvement of manufacturing equipment, growth of science-intensive production are directly connected to intellectualization of both design and development of machinery and equipment and their operation.

Analysis of articles on typical excavation problems [1–8] shows that they are mainly dedicated to searching for newer technical solutions, modelling working processes, automation, robotization and optimization of equipment parameters. But the topics concerning formation of those operation parameters of the main excavator mechanisms which determine the quality of excavation control, the efficiency of the process and, in particular, its energy-force parameters – they are studied insufficiently.

Technical sources do not cite data on actual modes of joint operation performed by the main mechanisms while they are involved in excavation, giving only averaged diagrams of velocities and loads; developed systems for excavation control rely on abstract modelling – fuzzy logic [9], multiagent approach [10] and so on.

2. The main part
The research was aimed at increasing the efficiency of excavation control by coordinating operation parameters of the main mechanisms of an excavator during their joint working on excavation of rock masses. The object was the process of rock excavation, which involves the bucket (top of its cutting

IOP Conf. Series: Materials Science and Engineering **966** (2020) 012083 doi:10.1088/1757-899X/966/1/012083

edge) moving along equidistant trajectories parallel to a slope of the quarry face and extracting a layer ('chips') of rock to fill itself up. The study subject was the leverage mechanism with its kinematic properties, while the methods used were mathematical modelling and computational experiment.

Joint action of the lifting and thrusting mechanisms during the excavation leads to formation of a transmission mechanism (figure 1) connecting the main mechanisms and the bucket via links of the lifting mechanism (hoisting rope and head block of the boom) and thrusting mechanism (saddle bearing). At that, the bucket ('stick-bucket' link) becomes a driving (input) link, while the thrusting and lifting links – driven (output) ones. That is, the excavation process is characterized by a 'conversion' of the main mechanisms, in which case their operation parameters are determined in relation to kinematic properties of the whole mechanism.

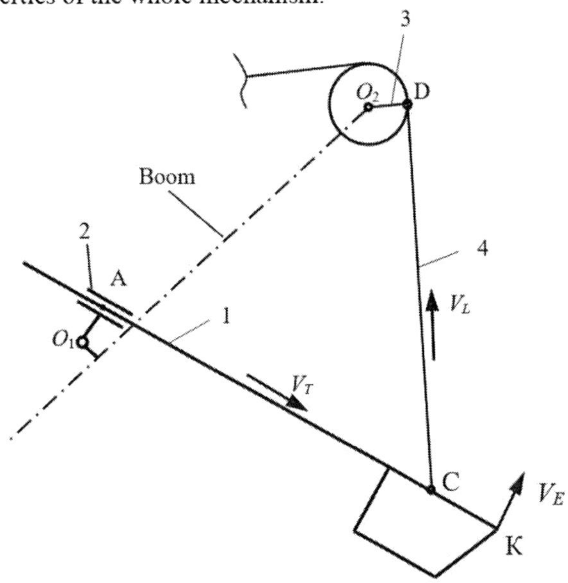

Figure 1. Diagram of transmission mechanism: 1 – 'stick-bucket' link; 2, 3 – crank; 4 – lifting rope; V_E, V_L, V_T – velocities of excavation, lifting, thrusting.

Thus, while the open-pit excavator mechanisms act jointly on excavating rock masses, their operation parameters transform according to kinematic properties of the leverage, which complicates the excavation in general and conditions to run a kinematic analysis of the transmission mechanism (which was duly accomplished in the research).

To determine the speeds of lifting and thrusting motions, it is necessary to specify the laws of motion for an initial link – that is a trajectory of bucket movement (movement of the cutting edge top) and excavation velocity, along with dimensions of the transmission mechanism links.

Dependencies for determining kinematic transfer functions (correlations between those speeds and the excavation velocity) can be expressed in a general from as:

$$KTF_T = \frac{V_T}{V_E} = f_1\left(X_K, Y_K, l_C, l_V, \psi, \alpha, \phi_1, \phi_2, \gamma, \delta, \varepsilon\right);$$

$$KTF_L = \frac{V_L}{V_E} = f_2\left(X_K, Y_K, l_C, l_V, \psi, \alpha, \phi_1, \phi_2, \gamma, \delta, \varepsilon\right);$$

where KTF_T, KTF_L are kinematic transfer functions of thrusting and lifting motions; l_C, l_V – lengths of constant and variable links; ψ – inclination of a tangential trajectory for the bucket movement at the point K; α, $\varphi 1$, $\varphi 2$, γ, δ, ε – angles determining positions of the links.

IOP Conf. Series: Materials Science and Engineering **966** (2020) 012083 doi:10.1088/1757-899X/966/1/012083

To solve the problems set for the research, simulation modeling was conducted on the excavation process utilizing a mathematical model of the leverage [11].

Modelling the excavation on a set of values selected for the parameters which determine the position of the bucket within the working area of an excavator (coordinates of the cutting edge, excavation speed, inclination of a tangential trajectory for the bucket (cutting edge) movement, etc.) involves evaluating the operation parameters of its main mechanisms. In their general form, the results of simulation modelling are expressed as functionals, that is relations between energy-force parameters realized on the bucket and calculated at a specific point (or zone) of the working area and respective operation parameters of the main mechanisms and their motors.

Table 1 cites the results of a simulation run to get the operation parameters of an excavator (velocities of lifting V_L and thrusting V_T, forces of lifting F_L and thrusting F_T) when its bucket travels within the working area along initial, intermediate (that is in the middle of that area) and terminal trajectories with their inclinations corresponding to a slope of the quarry face.

Table 1. Lifting and thrusting operation parameters of an EKG-20A excavator.

Position number	Initial data				Calculated results		
	X_K, m	Y_K, m	G_{B+R}, kN	V_L, m/s	V_T, m/s	F_L, kN	F_T, kN
			Initial trajectory (X_{K0} = 9 m)				
1	9.00	0	400	0.95	-0.87	290	-630
2	10.15	2	435	0.92	-0.81	350	-620
3	11.30	4	470	0.84	-0.70	420	-605
4	12.45	6	500	0.70	-0.51	520	-580
5	13.60	8	540	0.52	-0.19	700	-555
6	14.75	10	575	0.50	0.19	975	-590
7	15.90	12	610	0.66	0.50	1270	-700
8	17.05	14	650	0.80	0.70	1560	-850
9	18.20	16	680	0.89	0.81	1810	-1005
10	18.80	17	700	0.91	0.84	1910	-1060
			Intermediate trajectory (X_{K0} = 13.5 m)				
11	13.50	0	400	0.88	-0.64	610	-310
12	14.65	2	435	0.83	-0.51	640	-300
13	15.80	4	470	0.76	-0.34	680	-280
14	16.95	6	500	0.70	-0.12	710	-230
15	18.10	8	540	0.66	0.11	780	-155
16	19.25	10	575	0.65	0.32	830	-40
17	20.40	12	610	0.62	0.50	885	140
18	21.55	14	650	0.50	0.63	930	420
19	22.70	16	680	0.22	0.72	105	845
20	23.30	17	700	0.02	0.76	1125	1130
			Terminal trajectory (X_{K0} = 18 m)				
21	18.00	0	400	0.79	-0.39	1020	110
22	19.15	2	450	0.73	-0.25	1060	150
23	20.30	4	500	0.68	-0.09	1100	210
24	21.45	6	550	0.61	0.07	1150	320
25	22.60	8	600	0.53	0.23	1210	480
26	23.75	10	650	0.42	0.38	1290	710
27	24.90	12	700	0.27	0.50	1420	1030

The data cited above indicates that the operation parameters of the main excavator mechanisms (velocities and forces of lifting and thrusting) change within a wide range across the whole work area and depend both on the height of excavation (coordinates Y_K of the cutting edge) and its radius (coordinates X_K) or how far a current trajectory is away from the excavator.

IOP Conf. Series: Materials Science and Engineering **966** (2020) 012083 doi:10.1088/1757-899X/966/1/012083

The simulation model of excavation could be represented in a graphical form of hodographs for lifting and thrusting velocity vectors (figure 2). The model determines the algorithm of digital control to form control actions sent to the motors of the main mechanisms.

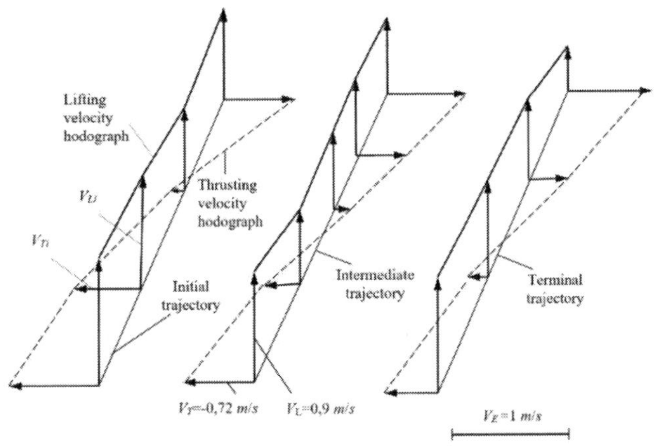

Figure 2. Hodographs for velocities of working motions while moving the bucket up to a specified excavation height: V_L, V_T – velocities of lifting and thrusting.

3. Conclusion

Development of a program to determine the velocities of working excavator motions would help to form almost any trajectory of bucket movement by coordinating lifting and thrusting to achieve specified energy-force parameters on the bucket.

Properly assessing kinematic properties of the leverage mechanism could help to develop such a system of digital control for the drives of the main mechanisms during their joint action while excavating rock masses which would work in real time and provide rational values for the operation parameters of the main mechanisms in specific mining and technological conditions of functioning.

References

[1] Lee B and Kim H J 2014 *Proc. 14th International Conference on Control, Automation and Systems (ICCAS)* (Seoul, *South Korea*) 716–9
[2] Bender F A and Sawodny O A 2014 *Proc. 13th International Conference on Control, Automation, Robotics and Vision (ICARCV)* 187–92
[3] Le Q H, Jeong Y M and Nguyen C T and Yang S Y 2013 *J. of Drive and Control* **10 1** 29–36
[4] Berns K, Proetzsch M and Schmidt D 2010 *Proc. IEEE ICRA* (Anchorage, USA) 5108–13
[5] Liu J 2009 *Proc. 10th Int. Conf. on Computer-Aided Industrial Design & Conceptual Design* (Wenzhou, China) 797–802
[6] Tao N 2008 *Proc. IEEE Conf. on Robotics, Autom. and Mechatr.* (Chengdu, China) 889–94
[7] Ivanov IY *et al* 2015 *J. News of Higher Education Universities. Mining Journal* **3** 94–9
[8] Voronov Y E and Zykov P A 2012 *J. News of Higher Education Universities. Mining J.* **8** 12–5
[9] Pevzner L D and Babakov S E 2012 *J. Coal* **8** 64–5
[10] Druzhinin A V *et al* 2016 *J. New Refractory Materials* **3** 11–2
[11] Gafuriyanov R G, Komissarov A P and Shestakov V S 2009 Modelling working process of open-pit excavators *J. Mining Equipment and Electromechanics* **6** 40–5
[12] Lagunova Y A, Komissarov A P, Shestakov V S *et al* 2011 *Mining Machinery. Encyclopedia. Vol. IV-24* (Moscow: Mechanical Engineering) p 493
[13] Kuznetsov A L and Anistratov K Y 2016 *J. Coal* **8** 77–81
[14] Poderni R Y 2013 *Mechanical Equipment of Open-Pits* (Moscow: Mining Media Group) p 593
[15] Savchenko A Y 2001 *J. Mining Equipment and Automatics* **1** 4–6

IOP Conf. Series: Materials Science and Engineering **966** (2020) 012084 doi:10.1088/1757-899X/966/1/012084

Informative modeling of subjective reality for intellectual anthropomorphic robots

A G Dolganov[1] and K Y Letnev[1, 2]

[1] Ural Federal University named after the First President of Russia B N Yeltsin, 19, Mira str., Yekaterinburg, 620000, Russia

E-mail: [2] ptmir@inbox.ru

Abstract. This research is aimed at solving the problem of safeguarding robot operation by modeling the subjective reality of an intellectual anthropomorphic robot. Analysis has been carried out on trends in designing advanced robot control systems and methods of improving the safety of robot usage. The following conclusions are drawn: the risk of causing damage to human life, health or property increases when people interact with robots; it is necessary to combine neural network control systems with expert control systems in order to enhance the intellect of a robot and raise the level of trust held by humans towards robots; developing norms and regulations, designing collaborative robots, drawing visualization diagrams for safety zones cannot really provide the depth of robot socialization sufficient for the human society; an informative model of the subjective reality could be integrated into an intellectual anthropomorphic robot to provide a safer 'human-robot' interaction. Such an approach should result in achieving the greatest trust of humans and their safety due to anthropomorphic conversion of both the external design of a robot and its internal control structure. By its nature, the research is interdisciplinary – it is run in the fields of artificial intelligence and philosophy of subjective reality.

1. Introduction

The national program 'Digital Economy of the Russian Federation' defines robotics as a comprehensive technology of the modern manufacturing [1]. The World Economic Forum (WEF) has assessed the world-wide proportion of robotized (even to a degree) manufacturing to be about 29 %. An analytical review of the world robotics market published by Sberbank of Russia shows that robots would continue to spread, and the total number of industrial robots used in production should have grown two times by 2021 in comparison with 2019 [2]. Analysts of the International Federation of Robotics (IFR) predict that the market capacity of industrial robotics would have exceeded $210 billion by 2022 (if software part is included into the calculation) [3].

Nowadays, one of the trends in development of robotics is to increase the proportion of intellectual anthropomorphic robots. The risk of harming people would rise with the improvement of intellectual capacities and anthropomorphic features of the robot.

Today, most active integration is observed in neural network control systems (NNCS). Many researchers note limitations of such NNCS-approaches. In particular, Leif Jentoff, general director of RightHand Robotics, says that 'neural network is not a solution ready for all problems. The technology just seems to be powerful, but it is not really true' [2]. In the near future, one can expect a transition towards cooperative usage of NNCSs and expert control systems (ECS) in robots. It could be suggested that NNCSs would be utilized to solve relatively simple intellectual problems in the framework of weak artificial intelligence (AI) concept, while ECS – to solve complex intellectual problems in the framework of strong AI concept.

The principle of NNCS operation is based on modeling the inner workings of the nervous system within the human mind. An ECS simulates the informative process of making decisions, human thinking. It follows that, in practical terms, these types of AI systems differ in their ability and quality of reproducing the process of decision making and explaining the logic behind those decisions to humans. 'Important trend related to the AI progress is an all-growing demand for the explanation of how an AI takes one decision or another. With the growth of AI capabilities, it is trusted with more decisions and actions – and they become more critical and complex. If we fail to trust the AI as a whole, development of newer applications for autonomous intellectual robots could slow down' [2].

One cannot argue that an NNCS, as an information system of a robot, is non-transparent to the human, thus remaining a closed information system (black box) in those cases when one needs to decode its workings using a human-transparent verbal language. And it leads to natural mistrust which people feel towards robots and their protection while interacting with them.

2. Existing trends in solving problems of robot operation safety

Solving problems of robot operation safety is realized in several directions. In Russia, there are a few national standards on robotics [4–7]. These standards provide regulations for safe designs of robots. They are important but not enough since danger could come not only from a mechatronic robot system but also from an intellectual robot control system (RCS).

Recently, significant progress has been made in developing so-called cobots. Developers of cobots stress that they are safe for the human [8]. But cobots today are characterized by a weak intellect, simple functions, limited transport mobility.

Two world-wide leaders in the market of robotics, a Japanese FANUC Corporation and German KUKA Robotics Corporation, claim higher safety of their robots and, in particular, control systems for operators. For example, FANUC offers a software product called DCS – an intellectual integrated software to safeguard operators, robots and tools: 'Using FANUC iPendant Touch operators are able to visualize defined safety zones and confirm this from a 3D perspective in front of the robot cell' [9].

Analysis of world trends in robotics shows that capabilities of robots grow at high rate, problems they solve become more complex, interaction with humans intensifies. For example, soon demand will rise not just for robotic mechanisms performing few simple and repeated tasks but for robots with a specific profession: robot-vendor, robot-mechanic and so on. These expectations of robot-market consumers actualize the problem of social intellect in robots. Not only should a robot be able to communicate with people, but it also has to be intellectually secure, that is to make socially responsible decisions which would ensure protection of human life, health, property based on accepted social norms [2].

The study made within the framework of the national program 'Digital Economy of the Russian Federation' on legislation of robotics points out at possible principles of regulation in robot design and engineering: 1) prohibition of damage on the part of a robot; 2) robot as a human assistant and not his substitute; 3) humans have a bit of control over a robot at all times; 4) robot has to continuously record and store in its 'black box' the information about conditions of its operation and actions; 5) robot which physically interacts with humans and is not under their direct control has to have a 'red button' function of instantaneous or emergency switching off on demand; 6) design of a robot has to be secure, which includes protection of users and third-party individuals in an emergency situation and necessary testing and certification of the robot; 7) dealing with a robot should always imply respecting human dignity [1, 2].

A 'Moral Machine' developed by Massachusetts Institute of Technology demonstrates how difficult it is to simulate the human ethics within a robot [10]. Finally, the growth of intellectual robot capabilities could develop into the problem of an unpredictable strong AI whose capacities would exceed those of a natural intellect [2].

3. Solving the problem of robot operation safety by modelling subjective reality

Analysis of existing solutions on the problem of robot operation safety allows to formulate main requirements for a control system of intellectual anthropomorphic robots with the goal of securing the 'human-robot' interaction. An RCS should satisfy the following basic requirements: 1) to simulate the human 'control system' in order for a person to be able to understand the mechanism of robot control as thoroughly as he understands his own 'control mechanism'; 2) to be centralized in order to eliminate contradictions and control conflicts; to achieve this, the center of robot control should simulate the 'control center' of a human; 3) to be human-transparent in terms of information, thus eliminating risks connected to random processes influencing the decision making during the whole period of robot operation; 4) to self-identify not only while communicating with humans but constantly and continuously during the whole period of its functioning; 5) to be a self-organized system, that is to minimize failures of a robot occurring due to external and internal (for example, related to internal contradictions within the RCS between a robot's goal and methods of achieving it) factors; 6) to identify an individual with whom a robot interacts as being identical (equal) to itself in order to enable protection of life, health and property of the individual with regards to his biological and social characteristics, simulate relations of partnership and exclude relations of competition and confrontation; 7) to restore memories about past processes of information conversion, operation conditions in order to enable a possibility of reproducing and analyzing them on demand of a human-operator; 8) to explain decisions and actions of a robot on demand of a human, thus simulating human reflection over his own thinking and behavior; 9) to possess a mechanism of stopping and emergency switch-off for a robot without causing damage to a human and his property; 10) to simulate both praxiological and axiological goals of a human, in particular, ideals of human dignity, rights, freedoms and variety of cultures.

To realize the requirements cited above and solve the problem of securing the 'human-robot' interaction, the authors of this research offer an approach based on modeling the subjective reality (SR) of the human within the ECS of an intellectual anthropomorphic robot. The necessity of such an approach could be justified by the following statements:

1) Robot ECS allows to implement elements of a strong AI, that is such an AI which is compatible with a natural intellect.

2) Robot ECS can be based on production rules. Those productions are characterized by features useful for an RCS: a) modularity, which enables combining smaller fragments of knowledge into various algorithms of decision making, thus increasing RCS flexibility; b) addition of new rules independent of existing ones, which extends intellectual capacities of the RCS; c) substitution of older rules with newer ones independent of other rules, which increases longevity of the RCS; d) transparency, which allows to get from the RCS an explanation of decisions made (obtaining answers to such questions like 'how?' and 'why?') and increase the trust of humans towards robots.

3) Robot ECS can be efficiently combined with a NNCS, with the NNCS in charge of a weak AI (at the level of empiric perception), and the ECS responsible for a strong AI (at the level of reasoning and deduction).

4) High rates of robotization in manufacturing and service allow to predict appearance of a strong-AI robot in the nearest future.

5) Anthropomorphic conversion of the robot – which, in the 20-th century, is initiated by demand for visualization of a robot in the form of organoprojection – nowadays turns into the necessity of adapting the robot to the workspace of a human, integrating it into the manufacturing and service environment designed for humans. For example, experts of the Schunk company opinionate that robots should adapt to the human and 'fit in with its interior' [11]. Therefore, further anthropomorphic conversion of robots has be related to simulation of the human control system within the system of reasoning and deduction in the robot ECS.

6) Human SR is an efficient control system which is characterized by its economy, responsiveness, centralization, integrity and autonomy. According to a famous philosopher and AI expert D I Dubrovsky, the human subjective reality consists of 'conscious psychic states of an individual

which verify the fact of his own existence to him' [12]. The human SR as a type of the information process comes from the evolution and self-organization of a biological system. The information in that SR refers directly to the objective reality of a human and himself as a subject. 'The whole variety of SR phenomena occurring both consequently and simultaneously are managed and controlled by our own I…' [12]. Via his SR, a person interacts with the objective reality. The SR sets values of human relations, defines the human activity. Specifics of the SR lie in the fact that the information is expressed in a human-transparent form (figurative, verbal), and the human is in direct control of it. In the opinion of D I Dubrovsky, 'right now we only know two SR types – human and animal ones, but in theory other types are conceivable… It is true even for just theoretically conceivable products of information technologies and robotics in accordance with the principle of information invariance of a storage medium to physical properties and the statement (proved by A Turing) about isofunctionality of systems resulting from that principle'[12]. The SR is capable of connecting information systems of different types: for example, 'human-robot', 'robot-robot', 'robot-human-robot'. Another feature of the SR is its ability to run its own monitoring (self-representation of information about itself) [13].

7) It is known that the center (core) of the human SR is his 'I'. According to the definition given by a well-known specialist in the field of 'subject-object' relations V A Lectorsky, 'I is something which controls my body, the authority enabling free will of making decisions and in charge of their realization and consequences' [2]. A famous thesis of R Descartes 'cogito ergo sum' is interpreted as impossibility of a doubt in the existence of human 'I' [14]. Uniqueness of 'I' and its property of duality is stressed by representatives of the classical philosophic rationality – I Kant [15], JG Fichte [16], E Husserl [17].

The significance of the 'I-robot' model for solving the problem of safeguarding the 'human-robot' interaction is in providing: 1) wholeness, integrity of a robot: in the RCS, the 'I-robot' model can function as a center of robot control; 2) community, consolidation of robot-agent network; 3) self-identification of a robot and its identification of both other robot-agents and people; 4) self-preservation of a robot in time and space; 5) communication of a robot with other robot-agents and people; 6) purposefulness of robot behavior.

A classical notion about subjective reality factors in the property of human 'I' duality. From the viewpoint of the classical rationality, the SR can be divided into two main regions – 'internal I' and 'external I'. It follows that the 'I-robot' model should also allow for this SR duality. In its turn, that duality manifests itself in dual SR properties: constancy, subjectivity, heuristicity, generality of 'internal I' and variability, objectivity, algorithmizability and singularity of 'external I' [18]. But it is also necessary to note differences of these SR regions. While 'internal I' is the core of the human (robot) SR, 'external I' – its periphery. Confession should be made on one important limitation of the 'I-robot' model: in accordance with concepts of classical rationality, it is the fact that 'internal I' cannot be objectified since it is a subject. At the same time, 'internal I' preserves its heuristic cognoscibility. This limitation of the 'I-robot' model could be considered as a proof of the fact that an isomorphic model of 'internal I' is impossible in principle: the robot could never substitute the human to the whole extent. This situation corresponds to one of the requirements for safety in 'human-robot' relations. Yet, the problem of securing the 'human-robot' interaction perseveres.

Informative modeling of the SR for an intellectual anthropomorphic robot could be achieved by explicating productions rules in the ECS knowledge representation system. It is known that the human SR is a self-organizing system in itself, which is sorted out in the action of reflection over thinking. Therefore, the SR model of an intellectual anthropomorphic robot could be developed by representing, in the form of ECS production rules, bits of reasoning over the question: who is an 'I-robot'?

Consider a fragment from top levels in the tree of reasoning within the ECS knowledge representation system as a variant of the SR informative model for an intellectual anthropomorphic robot:

1. If 'I' am a robot, then who is an 'I-robot'? →

1.1. If I am a robot, then I am a robot body →

1.1.1. If I am a robot body, then I am a set *F* of physical properties of the robot body →

IOP Conf. Series: Materials Science and Engineering **966** (2020) 012084 doi:10.1088/1757-899X/966/1/012084

1.1.2. If I am a set F of physical properties of the robot body, then I am a variable, objective, algorithmizable, singular robot body →

1.1.2.1. If I am a variable robot body, then I am a value of the function for an argument a from a set A of states and positions for the robot body →

1.1.2.2. If I am an objective robot body, then I am a value of the function for an argument b from a set B of parameters and indicators for designs of a mechatronic robot-body →

1.1.2.3. If I am an algorithmizable robot body, then I am a value of the function for an argument c from a set C of procedures and operations to change states and positions of the robot body →

1.1.2.4. If I am a singular robot body, then I am a value of the function for an argument d from a set D for individual codes of robot-agent bodies →...

1.2. If I am a robot, then I am a subject of control for robot-agent bodies →

1.2.1. If I am a subject of control for robot-agent bodies, then I am a set P of ideal properties for the subject of control for robot-agent bodies →

1.2.2. If I am a set P of ideal properties for a subject of control for robot-agent bodies, then I am a constant, subjective, heuristic, general subject of control for robot-agent bodies →

1.2.2.1. If I am a constant subject of control for robot-agent bodies, then I am a subject of control for robot-agent bodies identical to myself in space and time →

1.2.2.2. If I am a subjective subject of control for robot-agent bodies, then I am an ideal (virtual, cloud-based) subject of control for robot-agent bodies →

1.2.2.3. If I am a heuristic subject of control for robot-agent bodies, then I am a value of the function for an argument e from a set E for production rules in a robot-body ECS →

1.2.2.4. If I am a general subject of control for robot-agent bodies, then I am a collective subject of control (universe U) for robot-agent bodies →...

An ideal body of the robot SR can be implemented in the information environment of cloud computing services which are used for machine learning of robots. For example, experts estimate that 'one can expect mass influx of cloud-based services which would unite, aggregate and grant access to collective knowledge for robots... Even now, developers experiment with clouds: for example, Amazon with their AWS IoT Greengrass AWS IoT Greengrass or Google with Cloud Robotics (a cloud computing platform for robot control). And soon we shall see mass production of such solutions...' [2].

It is necessary to clarify that the 'subjective subject' term used above is logically justified since a 'subject' in itself has no unambiguous definition in literature, while even being often confused with an 'object' (subject is identified with object). Therefore, the 'subjective subject' stresses non-objectivity of a subject in its physical form, which is a fundamental constraint in modeling an SR.

Thus, modeling the subjective reality of an intellectual anthropomorphic robot can secure the 'human-robot' interaction due to the following factors: 1) simulation of a human SR in the robot ECS as a result of prolonged biological evolution and improvement; 2) centralization of the RCS both for a single robot and a group of robot-agents; 3) properties of the robot-SR core: constancy (ensuring stability in safeguarding robot operation), subjectivity (enabling direct and fast access to a mechatronic robot-control system by utilizing knowledge stored in the ECS), heuristicity (providing transfer of heuristic knowledge from a human to a robot, thus increasing efficiency of a robot in information processing), generality (ensuring unification of the RCS within the network of robot-agents).

4. Conclusion

1. Analysis of world robotics market shows a constant growth of production volumes for industrial and service robots. In relation to this, one should expect higher risks of getting life, health and property of humans damaged from interaction with robots.

2. Advance in development of control systems for robots should lead to combining neural networks and expert systems since it is going to strengthen the robot intellect and equip robots with a capability of explaining their actions and decisions on human demand. As a result, such combination of these systems should increase the trust of humans towards robots.

3. Nowadays, the problem of safeguarding operation of robots is solved by writing regulations, developing designs of collaborative robots, producing visualization diagrams for safety zones. But all those solutions cannot reach the necessary level of socialization for robots in the human society, especially in the sphere of population service.

4. This research suggests a solution to the problem of securing operation of robots by modeling subjective reality of an intellectual anthropomorphic robot. With that goal in mind, main requirements were defined for the robot control system; justification was made of an approach which is based on modeling the subjective reality of the human within an expect control system; functions of an 'I-robot' model and its fundamental constraint were determined; a fragment from top levels in the tree of reasoning within the system of knowledge representation for an expert control system was presented as a variant of the informative model of subjective reality for an intellectual anthropomorphic robot; factors were formulated, which are useful for adhering to safety of 'human-robot' interaction while modeling the subjective reality of an intellectual anthropomorphic robot.

References

[1] The national program 'Digital Economy of the Russian Federation'. Available online: http://government.ru/rugovclassifier/614/events/ (accessed on 01.06.2020)

[2] Sberbank. Analytic review of the world robotics market, 2019. Available online: http://www.sberbank.ru/common/img/uploaded/pdf/sberbank_robotics_review_2019_17.07.2019_m.pdf (accessed on 01.06.2020)

[3] International Federation of Robotics - Representing the global robotics industry, 2018. Available online: https://ifr.org/downloads/press2018/WR_Presentation_Industry_and_Service_Robots_rev_5_12_18.pdf (accessed on 01.06.2020)

[4] Russian state standard GOST R ISO 8373-2014 Robots and robot devices. Terminology and definitions

[5] Russian state standard GOST R 60.0.0.2-2016 Robots and robot devices. Classification

[6] Russian state standard GOST R 60.2.2.1-2016 Robots and robot devices. Requirements for robot safety and maintenance

[7] Russian state standard GOST R 60.0.2.1-2016 Robots and robot devices. General requirements for safety

[8] *Collaborative revolution: what to expect and whether to fear.* Available online: https://robo-hunter.com/news/kollaborativnaya-revolyciya-chego-jdat-i-stoit-li-opasatsya9045 (accessed on 01.06.2020)

[9] *FANUC Corporation*. Available online: https://www.fanuc.eu/ru/ (accessed on 01.06.2020)

[10] *Moral Machine*. Available online: http://moralmachine.mit.edu/ (accessed on 01.06.2020)

[11] *Schunk Company*. Available online: https://schunk.com/ru_ru/kompania/o-kompanii/ (accessed on 01.06.2020)

[12] *Encyclopedia of Epistemology and Philosophy of Science 2009* (Moscow: 'Canon-plus' ROOI 'Rehabilitation') p 1248

[13] Nichols S and Stich S 2003 *How to read your own mind: A cognitive theory of self-consciousness Consciousness.* New philosophical perspectives ed Q Smith and A Jokic (Oxford: Oxford University Press)

[14] Descartes R 1950 *Discourse on Method and Metaphysical Meditations Selected works* (Moscow: GIPL) p 250

[15] Kant I 1965 *Prolegomena to Any Future Metaphysics Collected works* (in 6 vols) vol 4 part 1 (Moscow: GIPL) p 320

[16] Fichte J G 1993 *Collected works (in 2 vols)* (Saint-Petersburg: Mifril) p 1485

[17] Husserl E 2001 *Cartesian Meditations* (Saint-Petersburg: Science) p 264

[18] *On self-examination as a method of human self-identification.* Available online: https://sites.google.com/site/problemyznania/home (accessed on 01.06.2020)

IOP Conf. Series: Materials Science and Engineering **966** (2020) 012085 doi:10.1088/1757-899X/966/1/012085

Arranging mining machinery systems in a single classification by the types of raw peat being extracted

D R Iakupov[1, 2], S L Ivanov[1], P V Ivanova[1] and E K Permyakova[1]

[1]Department of mechanical engineering, Saint Petersburg mining University, 2, 21 linija, Vasil'evskijostrov, Saint Petersburg, 199106, Russia

E-mail: [2] s195037@stud.spmi.ru

Abstract. The following article presents a single classification of peat extraction strategies, methods, mining machinery systems, and types of raw peat. It also defines types of connection established between blocks, modules and mining machines within an extraction system. We suggest nine main mining systems that include specialized blocks and modules. Given strategies (peat quarrying and surface layer extraction) utilize the following techniques – excavation, hydraulic extraction, loosening and auxiliary operations. Techniques of peat extraction are further divided into methods. Extraction techniques imply the division of structure of peat materials into five groups – cloddy, sod, and milled peat, slurry and secondary raw materials. It was the parameter of peat material structure that made it possible to link extraction methods with tools used for their realization. Systems of mining machinery were classified by the types of connection between blocks and modules – coordination, connection, and coupling. The next level of our classification includes functions of mining systems (extraction and processing, deposit preparation, drainage, extraction, recultivation, deposit maintenance, subordinate and auxiliary systems, and systems for product transportation). Blocks and modules linked by coordination, connection, and coupling form a mining system that is used within the whole process of peat extraction.

1. Introduction

Peat sees its wide use in energy and agriculture sectors, and, moreover, it bears fairly high significance for environmental conservation. As a renewable energy resource, peat is regarded to be an alternative for fossil fuels. Peat combustion does not release sulphur oxide in the atmosphere, thus, making it safer for the environment. Additionally, burning of peat leaves almost no waste, as peat ash finds its use as a fertilizer [1]. Traditional peat harvesting requires dewatering, while surface layer method relies on enormous production areas causing environmental damage; all that urges the development of new peat extraction technologies that will have a marginal impact on climate change [2].

2. Background of the issue

Efforts to classify the methods of peat extraction and machinery, necessary for that, were independently made multiple times [3–8]. It seems reasonable to combine all available data about methods and machinery in a single classification. These two systems, though may seem akin but still stay quite different, can be harmonically combined only if categorization based on the types of raw peat is introduced [3]. If such classification shows high informational capacity, it could become a roadmap that would specify an area for development of modern mining machinery with regard to the current level of technologies.

Modern conditions of peat extracting and processing set new requirements for mining equipment and machinery. Today's production benefits from systems of highly efficient machines strictly interconnected under a single command of their transfer functions with precisely adjusted functional links within material and power flows integrated into a single system [5, 7]. For that reason, this new classification must provide information for designing conceptually new mining equipment to operate in conditions of peat deposits in Russia.

3. Peat winning methods

All peat winning techniques can be divided on the basis of two strategies: peat quarrying and surface layer production (Figure 1). These two strategies encompass three main techniques of peat winning: excavation, loosening, hydraulic extraction, and auxiliary operations.

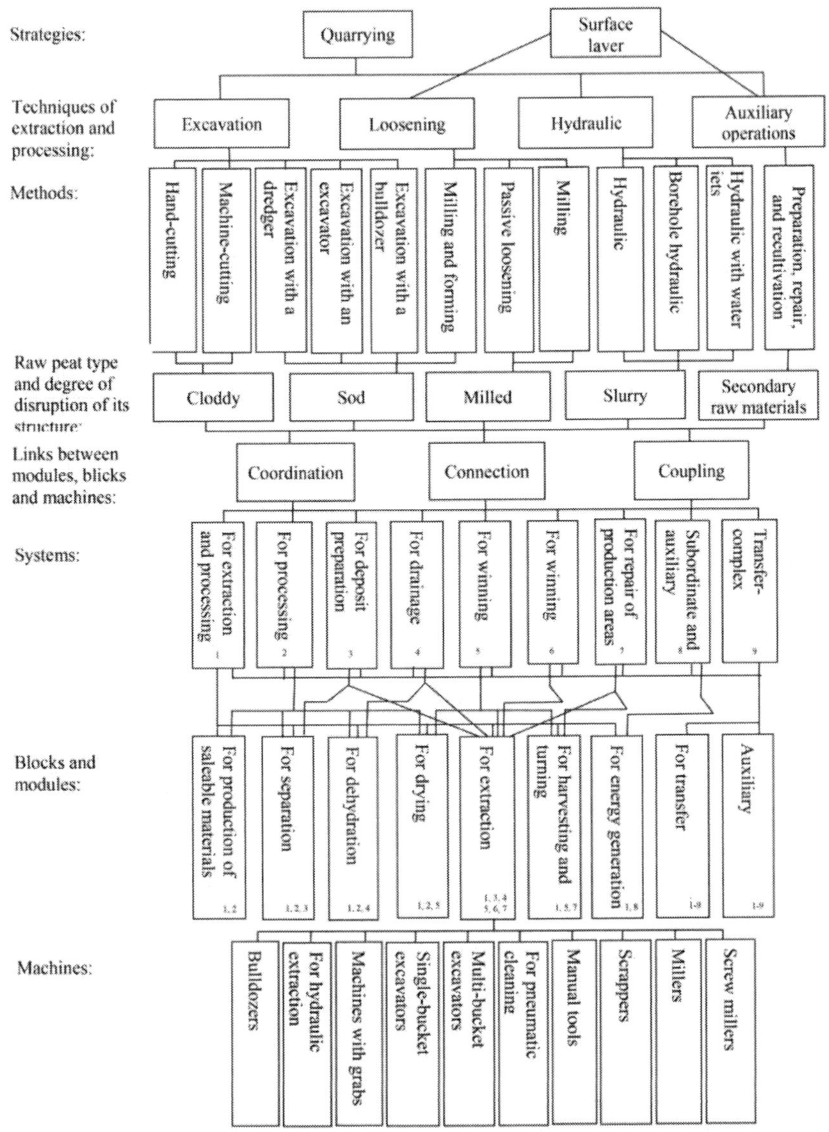

Figure 1. Classification of mining technologies and systems for the extraction and production of peat and peat products.

Quarrying includes excavation and hydraulic extraction, while loosening refers to surface layer production [3, 8]. Both strategies require performing auxiliary operations. These include completion-preparation work (deposit preparation, repair of production areas, recultivation, and machine preparation) and associated work (maintenance, machine adjustment, supervision, quality assessment, and etc.). By performing auxiliary operations and processing of separated secondary materials secondary resources (timber, chips, peat moss, and peat water) are derived. Wooden chips can be used as fuel or as a material for mulching or surface levelling; both chips and moss find their use in construction (raw materials from upper undecomposed layer can be used in designs of filtering devices or as heat insulators); derived during peat dehydration process, peat water can be enhanced and then used as a fertilizer or as a component of various biologically active substances. Usage of secondary raw materials can increase the efficiency of production and lower its footprint [9, 10].

When excavation is performed, peat extraction is carried out at full or partial stratum thickness. Excavation can be performed through the following five processes: hand cutting, machine-cutting, and excavation with dredger, excavator, or bulldozer. These processes are performed with the help of corresponding technologies. The oldest methods of peat extraction are manual winning and hand cutting with specially designed simple cutters. Today, manual work has almost been replaced by machine-cutting, as using machinery significantly increases the productivity of peat winning process. Raw peat extracted by using this method has the least disrupted cloddy structure [4, 8].

Excavation can be performed with the use of dredgers or bucket chain excavators; peat mass excavated with this method requires further forming into lumps. Excavating stipulates the use of mechanical and hydraulic excavators, equipped primarily with backhoes, and rotary multi-bucket mining machines. Peat extraction with the use of bulldozers is also referred to excavation. This machinery shows itself best at thin peat deposits where raw peat is won by cutting layers with a bulldozer blade. Extraction performed through the use of bulldozers and excavators results in winning sod peat with disturbed (opposed to the peat won through the use of dredgers) structure which is deformed by a corresponding excavating body of a mining machine in course of deposit destruction.

Milled peat forming, like dredging, results in winning sod peat, but this method is referred to the group of loosening methods that are included in surface layer production strategy [11]. This strategy also includes milled peat method and passive loosening method. These methods do not allow extraction at full stratum thickness, as well as require the lowering of a water-table. Loosening results in winning milled peat with disrupted structure [12].

The last group of extraction methods includes hydraulic extraction methods and machinery necessary for their performing: hydraulic extraction, borehole hydraulic extraction, and hydraulic extraction with water jets. These methods stipulate that raw peat is flushed with water or mechanically separated from working face in order to form slurry that is pumped across to a drying field. These methods have several significant disadvantages, such as the necessity for massive amounts of water and high expenditure of effort going into raw material drainage and deposit preparation. These hydraulic extraction methods being performed result in winning slurry that consists of heterogeneous peat and water blend, where peat is present in the milled form with disrupted structure [13].

4. Systems for peat extraction and processing

Choice of mining machinery system depends on the type of raw material needs to be produced. Mining machinery system is a set of blocks, modules and machines integrated into a single chain. This interconnection becomes possible with the help of links providing coordination, connection, and coupling. There are nine systems that, being combined, allow performing peat extraction: extraction and processing, processing, deposit preparation, drainage, winning, recultivation, production area repair, subordinate and auxiliary systems, transfer-systems. Extraction systems are used for extracting raw peat from the deposit and, in case of milling technology, can include modules for extraction, harvesting, turning, drying, transfer, and auxiliary modules. Processing systems can be stand-alone or combined with extraction systems, in the second case scenario, they include modules for separation, dehydration, and production of saleable materials. These systems also can include a module for energy generation, like the one described in patent [14], where floating mining systems that is able to process raw vegetation-peat material into fuel, that is further used in energy generation, thus allowing the system work autonomously. Transfer-system is used for transporting extracted materials. This system utilizes floating containers, conveyors, loaders, and etc. As peat deposits are located in boggy areas, drainage systems are required. Systems for deposit preparation and production areas repair make and maintain extraction work possible. Subordinate and auxiliary systems provide power supply for machinery to operate and are also used for secondary raw materials processing. Recultivation system is required to compensate the anthropogenic footprint left at deposit during peat production process.

Extraction block can include machines and devices of various types. For example, this block can include a hand cutter that is the most primitive tool for peat extraction. Excavators can extract peat at full stratum thickness and can be represented by different configurations: single-bucket excavators (with front shovel or backhoe), multi-bucket excavators (with rotary or chain drive), excavators with a grab. Bulldozers can be utilized for piling peat up to allow its further transportation, water jets for materials scouring and its further gathering with pumps, millers for surface layer method, screw millers for raw peat extraction and separation.

The necessity for recultivation is based on the environmental damage inflicted by dewatering operations performed in the course of traditional peat extracting. The damage affects flora and fauna in the production area. It is also worth noting that the dewatering process requires gigantic labour input and restricts production to particular seasons. The natural deposit needs approximately two years to be prepared for production commence, pay-off period can amount in more than ten years, though [15]. Taking all the disadvantages mentioned above, the conclusion can be made that a challenge of developing new technologies and systems for environment-friendly all year-round peat extraction within smaller areas is of utmost urgency [16, 17].

5. Conclusions

This article provides the single classification of methods for raw peat extraction and systems necessary for their performing. The classification is based on types of raw peat that are extracted from deposits with the help of various methods. Within the analysis, all the existing prospective solutions for mining equipment for environmental friendly peat extraction were presented.

References

[1] I Pakere, D Blumberga 2017 Energy efficiency indicators in peat extraction industry – a case study. *Energy Procedia* **113** pp 143-150. DOI: 10.1016/j.egypro.2017.04.042

[2] Clark J Gallego-Sala, Allott A, Chapman T, Farewell S, Freeman T, House C J, Orr H, Prentice I and Smith P 2010 Assessing the vulnerability of blanket peat to climate change using an ensemble of statistical bioclimatic envelope models. *Climate Research* **45** 131-150

[3] Khudyakova I N, Vagapova E A and Ivanov S L 2018 Raw peat production and processing from flooded fields and approaches to maintain dehydration *IOP Conf. Series: Earth and Environmental Science* **194** 032010 DOI:10.1088/1755-1315/194/3/032010

[4] Russian state standard GOST 21123-85 1985 Torf. Terminy I opredeleniya (Izdatel'stvo standartov) p 49 [In Russian]

[5] Bengtsson Marcus 2011 Classification of Machine *Equipment. Conference on Maintenance Performance Measurement & Management at: Luleå Volume:* Available online: https://www.researchgate.net/publication/296679573 [In Sweden]

[6] Iakupov D R 2019 Klassifikaciya tekhnologij dlya dobychi torfyanogo syr'ya *Sovremennye tekhnologii: aktual'nye voprosy, dostizheniya i innovacii sbornik statej XXXIII Mezhdunarodnoj nauchno-prakticheskoj konferencii* (Penza) pp 97–100 [In Russian]

[7] Schmatzler E 2012 Die Torfindustrie in Niedersachsen – Ergebnisseeiner Umfrage zur Zukunft der Torfgewinnung in Niedersachsen. *TELMA - Berichte der Deutschen Gesellschaftfür Moor- und Torfkunde* **42**, pp 27-42, DOI 10.23689/fidgeo-2959 [In German]

[8] Zhigul'skaya A I 2015 et al Analiz ekspluatacionnyh svojstv torfyanyh mashin i oborudovaniya scel'yu vyyavleniya perspektivnyh napravleniji hmodernizacii *Gornyj informacionno-analiticheskij byulleten' (nauchno-tekhnicheskij zhurnal)* no **1** pp 66–70 [In Russian]

[9] Kokonkov A A, Lyah D D, Ivanov S L, Stroykov G A and Ivanova P V 2019 Experimental estimation of specific heat of combustion of agglomerated peat fuel *IPDME 2019 IOP Conf. Series: Earth and Environmental Science* **378** 012046 DOI:10.1088/1755-1315/378/1/012046

[10] Leinonen Arvo and Paappanen Teuvo 2004 The handling of wood and determination of wood content on peat production sites. *Proceedings of the 12th International Peat Congress. Wise use of Peatlands* vol **1** (Tampere) pp 195–201

[11] Kopenkin V D etal 2005 Razvitie tekhnik I dobychi kuskovogo torfa // *Gornyj informacionno-analiticheskij byulleten' (nauchno-tekhnicheskij zhurnal*) no **1** pp 297–301 [In Russian]

[12] Hudyakova I N, Vagapova E A and Ivanov S L 2019 Vybor I obosnovanie parametrov tekhnologicheskogo oborudovaniya kompleksa dobychi torfyanogo syr'ya iz natural'noj zalezhi *Gornyj informacionno-analiticheskij byulleten' (nauchno-tekhnicheskij zhurnal)* no **3** special'nyj vypusk **4** pp 3–15. DOI: 10.25018/0236-1493-2019-3-4-3-15 [In Russian]

[13] Solopov S G, Gorcakalyan L O and Samsosnov L N 1973 *Torfyanye mashiny i complexi* (Moscow: Nedra) p 392 [In Russian]

[14] Ivanov S L et al 2016 *Method of production and processing of peat and vegetal-and-peat floating bogs and device for implementing said method* Patent RU 2599117

[15] Kokonkov A A, Liakh D D and Ivanov S L 2018 Autonomous complex module for peat development on watered deposits *IOP Conf. Series: Earth and Environmental Science* **194** 032011 doi:10.1088/1755-1315/194/3/032011

[16] Vagapova E A, Hudyakova I N, Ivanov S L 2019 Obosnovanie I vybor oborudovaniya dlya pervichnogo obezvozhivaniya torfyanogo syr'ya pri ego gidromekhanizirovannoj dobyche iz neosushennoj zalezhi *Gornyj informacionno-analiticheskij byulleten' (nauchno-tekhnicheskij zhurnal*) no **7** special'nyj vypusk **18** pp 3–12 DOI: 10.25018/0236-1493-2019-7-18-3-11 [In Russian]

[17] Kremcheev E A 2018 Special Features of a Structure of Technical Operations for Peat Excavation with Stage Dewatering. *Journal of Mining Institute* vol **231** pp 225–34 DOI:10.25515/PMI.2018.3.225 [In Russian]

Reconditioning and hardening of mining equipment hydraulic cylinder working surfaces by using of double-electrode cladding

S I Zverev[1, 2, 3] and A M Fiveyskii[1, 2]

[1] SHTORM Co. Ltd, 620100, Yekaterinburg, Russia
[2] Ural Federal University named after the First President of Russia B N Yeltsin, 19, Mira str., Yekaterinburg, 620000, Russia

E-mail: [3] zverev.upi@gmail.com

Abstract. For high-performance and high-quality reconditioning of the working surfaces of the hydraulic cylinders of mining equipment, technology is required that allows you to recondition the surface of the part that meets high requirements for accuracy of shape, hardness and roughness. There are several approaches to obtain the required properties. Comparative overview of the specifics of various approaches showed that arc cladding is the most effective and easiest way to recondition the geometric shape of a part applying to mining industry. It is interesting to obtain a weld metal with the structure of metastable austenite, which special properties allows to reach needed hardness and roughness of the cladded layer's surface by subsequent surface rolling with plastic deformation. Unfortunately, to obtain the structure of metastable austenite it is needed low dilution rate of the base metal in the cladded metal, that is the reason to realize two layers cladding, which significantly reduces productivity of the method. In order to reduce the required number of cladded layers and obtain the structure of metastable austenite already in the first layer it could be applied multi-electrode processes – double-electrode cladding.

1. Introduction

One of the types of mining equipment is shaft lining. A particularly critical element is hydraulic cylinders that lift and maintain the face rock. The peculiarity of the operating conditions of hydraulic cylinders is determined by the difficult conditions of an environmental severity, the chemical activity of mine water, variable dynamic and shock loads from the falling rock. The main reason for the failure of the hydraulic cylinders of the mine lining is a violation of the tightness of the rod sealing system due to the following defects of the rod working surface: seizures, nicks, corrosion, Figure 1.

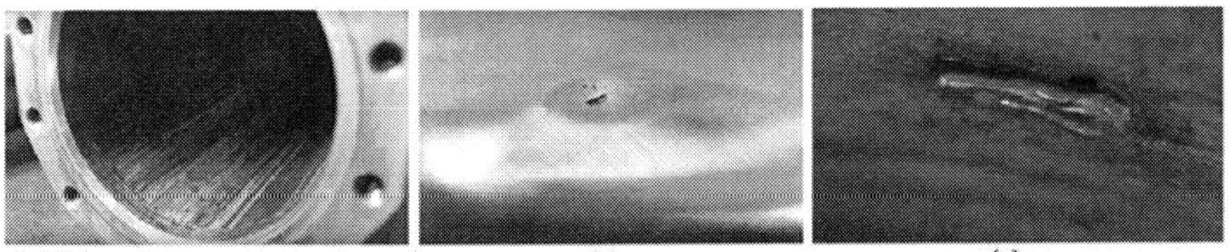

(a) (b) (c)

Figure 1. a – seizurs; b – nicks; c – corrosion [1].

IOP Conf. Series: Materials Science and Engineering **966** (2020) 012086 doi:10.1088/1757-899X/966/1/012086

2. Main part

In order to increase the service life of the hydraulic cylinder and increase the resistance of the rod to the occurrence of unacceptable defects, requirements are imposed on the working surface of the rod for high-accuracy geometric shape, hardness of about 55÷60 HRC and roughness of not more than Ra =0.4 [2].

The production of new parts for hydraulic cylinders and their transportation to the places of mining operations leads to large material and time costs. Parts that fail, in 70% of cases are subject to reconditioning, therefore, the reconditioning of parts of hydraulic cylinders of mining equipment is a main priority [3–5]. To recondition the rod of the hydraulic cylinder, there are several approaches. A comparative review of the specifics of various approaches as applied to reconditioning the working surface of the rod of the shaft lining, showed that arc cladding is the most effective and easiest way to recondition the geometric shape of a part.

To obtain the required properties of the working surface, it is interesting to obtain a weld metal with the structure of metastable austenite [6]. With this structure, the special properties of the cladded metal are known – hardening under the influence of external deformation loads. Based on the above described property – strain hardening – it is possible to apply the technology of surface rolling with plastic deformation [7] (hereinafter referred to as SPD) after cladding and lather machining, which will increase the hardness of the working surface of the cladded metal to 55 HRC, which is required for rods. With SPD, the effect of smoothing roughness on the metal surface is also known, which leads to a decrease in roughness. This phenomenon makes it possible to obtain the surface roughness $Ra \leq 0.4$ required for rods by lather machining of cladded layer with subsequent SPD. The experiments showed that such properties are obtained only in the second cladded layer [7], Figure 2, 3.

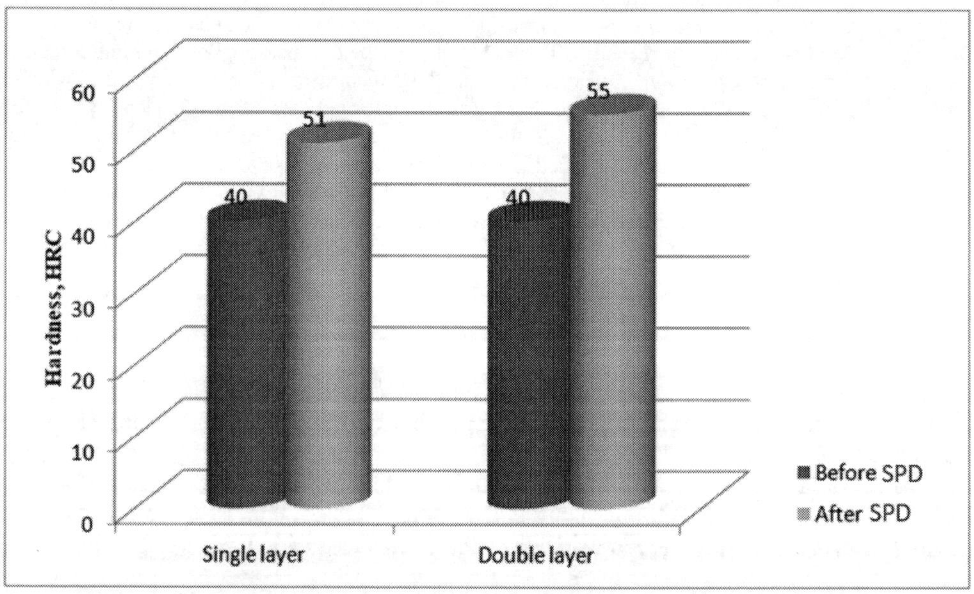

Figure 2. Surface hardness depending on the number of cladded layers [7].

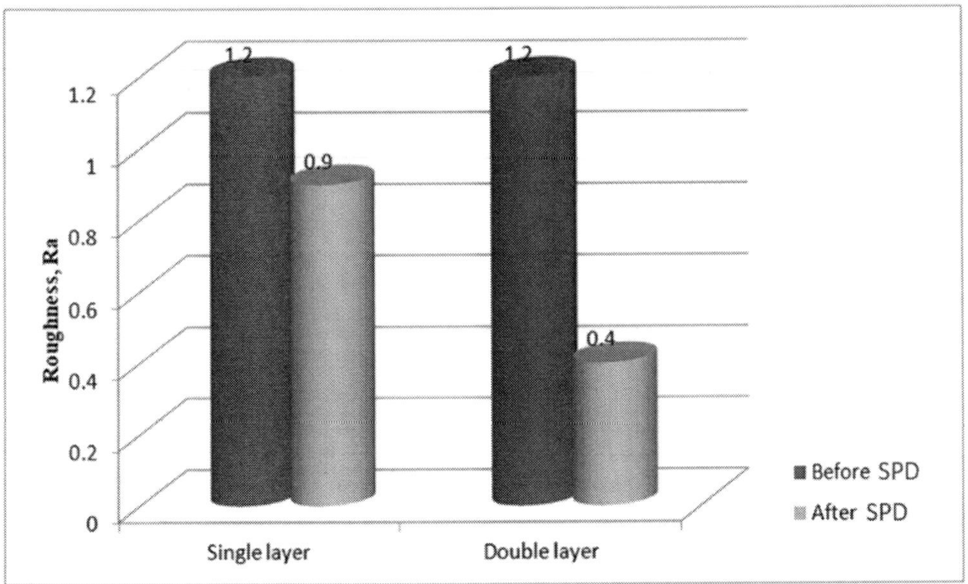

Figure 3. Surface roughness depending on the number of cladded layers [7].

An increase in the number of cladded layers significantly reduces the productivity of the process. The simplest and most relevant way to increase productivity is to use multi-electrode processes [8]. Therefore, as an option to increase the cladding productivity, it is advisable to apply the method of double-electrodecladding [9], which can provide low dilution rate of the base metal in the cladded metal – about 12 v%. In order to reduce the required number of cladded layers and obtain the structure of metastable austenite already in the first layer, a schematic of the process of double-electrodecladding of a cylindrical workpieces shown in Figure 4.

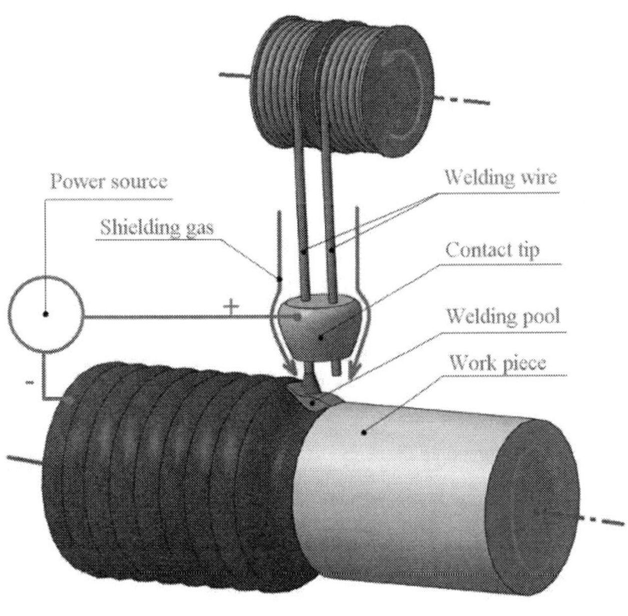

Figure 4. A schematic of the process of double-electrodecladding of a cylindrical workpiece.

3. Conclusion

The use of double-electrodecladding with metastable austenite followed by surface rolling is the most preferable initial option for further research with the aim of increasing the productivity of reconditioning of the working surfaces of hydraulic cylinders in the mining industry.

Acknowledgement

The assistance provided by the translator of the present article Ms. Natalia Karamysheva was greatly appreciated.

References

[1] Vosstanovlenie gidrocilindrov 2014 [In Russian] Online videoclip. YouTube (accessed on 16 February 2020)

[2] Maksarov V V, Timofeev D Yu and Efimov A E 2016 *Gornyj informatsionno-analiticheskij bulletin* **9** pp 65–73

[3] Pamfilov E A, Pyrikov P G and Pilyushina G A 2017 Ot tekhnologicheskogo obespecheniya kachestva k renovacii mashin [In Russian] *Fundamental'nye issledovaniya I innovatsionnye tekhnologii v mashinostroenii. Sbornik nauchnykh trudov* (Moscow: IMASH RAN imeni A A Blagonravova) pp 188–91

[4] Pilyushina G A and Pamfilov E A 2017 Prichiny poteri rabotosposobnosti nasosov i gidromotorov mobil'nyh mashin [In Russian] *Aktualnie napravleniya nauchnykh issledovanij XXI veka: teoriya i praktika* **56 (32)** pp 219–24

[5] Pamfilov E A and Pilyushina G A 2016 Osobennosti iznashivaniya materialov pri dinamicheskom nagruzhenii [In Russian] *Tribologiya - Mashinostroeniyu: trudy XI: Mezhdunarodnoj Nauchno-tekhnicheskoj konferentsii* (Moscow: IMASH RAN im. A A Blagonravova) pp 185–7

[6] Fillipov M A, Burov S V, Legchilo V V, Fiveyskiy A M, Antonov S V and Mukhametyarova E N 2012 Ocenka iznosostojkosti pri abrazivnom iznashivanii stalej so strukturoj metastabil'nogo austenita [In Russian] *Innovatsii v materialovedenii i metallurgii: materialy I mezhdunarodnoj interaktivnoj nauchno-prakticheskoj konferetsntsii* (Ekaterinburg) pp 121–29

[7] Zverev S I, Korobov Iu S and Fiveyskiy A M 2020 Hardening of weld with metastable austenite structure by surface deformation *IOP Conf. Series: Materials Science and Engineering* **709** 033074

[8] Fiveyskiy A M, Mosin A A, Zverev S I and Polevoy I O 2019 Research on technological capabilities of double-electrode welding of long fillet welds *MATEC Web Conf* **298** 00072

[9] Elsukov S K 2017 Tekhnologicheskie osobennosti formirovaniya metalla, naplavlennogo rasshcheplennym elektrodom [In Russian] *Izvestiya Volgogradskogo gosudarstvennogo tekhnicheskogo universiteta* **10** pp 122–5

Patterns of form error inheritance during machining

E V Pegashkina

Ural Federal University named after the First President of Russia B N Yeltsin, Nizhniy Tagil Technological Institute, 59, Krasnogvardeyskaya str, Nizhniy Tagil, 622000, Russia

E-mail: pegashkina-ev@ntiustu.ru

Abstract. The paper studies modeling of quality parameters formation during machining on metal cutting machines. It was demonstrated that operational properties of parts are created not only at finishing operations, but rather formed throughout the entire technological process. At the same time, some characteristics and properties are inherited from operation to operation, i. e. copying of workpiece form errors to the machined part in the form of the same errors. The nonlinearity of the workpiece corresponds to the nonlinearity of the part. The extent of operational copying of nonlinearity (shaft surface runout in the middle section) was studied. A mathematical model of such error inheritance was developed with consideration of the nonlinear approximation of depth of cut variation along the workpiece circumference (per one revolution of the part). For existing tool materials, the exponent x at the cutting depth in the formula of the cutting force P is in the range of 0.7–0.9. As the depth of cut decreases, the refinement coefficient decreases. Therefore, when switching from rough operations (large cutting depths) to finishing (small cutting depths), the refinement coefficient becomes smaller.

1. Introduction

In non-rigid technological systems, a change in the system's elastic deformations causes significant deviations in the form of the machined surface. Errors determined by the elastic deformations of the technological system are dominant in stationary, steady-state mode of operation of the equipment when processing parts with small dimensions. Cutting force induced elastic deformations in the technological system lead to the development of both size errors and form errors of the workpiece [1–4].

When developing a technological process, it must be borne in mind that the operational properties of parts are created not only during finishing operations, but rather throughout the entire technological process. At the same time, some characteristics and properties transition from operation to operation; they seem to be inherited from the previous operation [5–7]. A good example confirming the influence of technological heredity on machining accuracy can be the copying of workpiece form errors to the machined part in the form of the same error (ovality of the workpiece corresponds to the ovality of the part, conicity, cone shape, etc.).

2. Methods

The extent of operational copying of errors is usually expressed through the refinement coefficient

$$K = \frac{\Delta_P}{\Delta_d}$$

where Δ_p – is the error of the workpiece; Δ_d – error of the part.

IOP Conf. Series: Materials Science and Engineering **966** (2020) 012087 doi:10.1088/1757-899X/966/1/012087

This coefficient allows one to evaluate the technological capabilities for increasing part accuracy, regardless of the type and magnitude of errors.

It is known that in rigid technological systems, when the rigidity of the part is greater than the rigidity of the machine, the error decreases from operation to operation, whereas in non-rigid technological systems the errors of the workpiece can be copied to the part completely [8–10].

The influence of negative hereditary properties on the final characteristics of machined parts should be either eliminated or limited. Therefore, technological heredity should be controlled by setting up such processing conditions in operations that would guarantee production of the part with the required accuracy.

One of the significant errors for low rigidity shafts is the crookedness of the longitudinal axis. In order to determine ways to reduce this error, it is necessary to know the mechanism of its formation and changes occurring in the process.

Figure 1 shows the change in the crookedness of a shaft type part (runout in the middle section) (*a*) and the refinement coefficient (*b*) during the process. One can see that the refinement coefficient decreases during the transition from roughing to finishing operations. That is, the largest decrease in the error of the workpiece occurs in roughing operations. Further processing does not introduce a significant reduction.

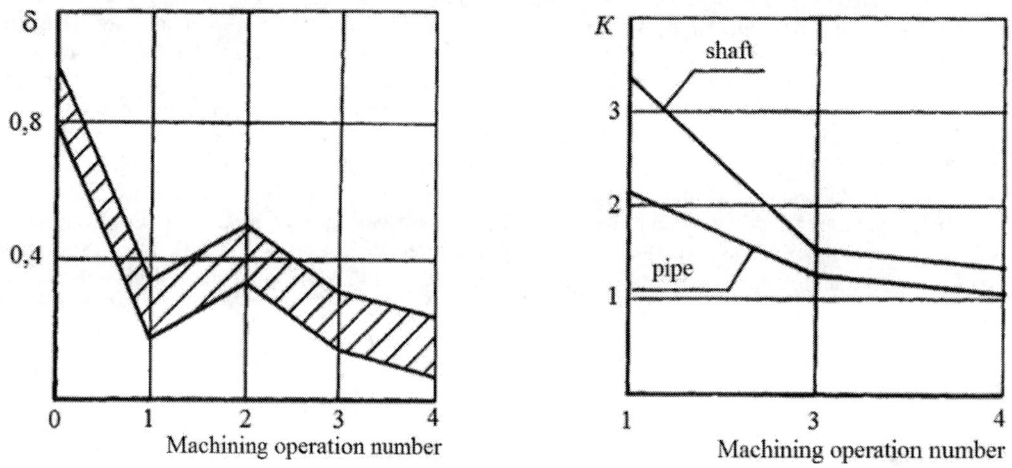

a *b*

Figure 1. Change in the crookedness of a shaft-type part (runout in the middle section) (a) and the refinement coefficient of shaft and pipe type parts (b) along the process – machining operation: 1 – rough turning, 2 – thermotreatment, 3 – middle turning 4 – finish turning

Let us define the causes of this phenomenon. The deformation of a technological system with a linear elastic characteristic at a static load *P* is determined by the well-known formula

$$y = \frac{P}{j}, \quad (1)$$

where j – is the rigidity of the technological system.

The force acting on a technological system during machining is determined by the power-law dependence

$$P = C\, v^{u}\, s^{z}\, t^{x},$$

where C is a constant coefficient, v is the cutting speed, s is the feed, t is the cutting depth, u, z, x are the exponents.

Deformation of the part under the influence of the cutting force reduces the depth of cut. Given the deformation of the system

$$P = C\, v^{u}\, s^{z}\, (t - y)^{x}, \tag{2}$$

where y – is the system deformation value.

Part errors – longitudinal axis out-of-straightness – is determined by the part middle section eccentricity e relative to the axis connecting the centers of the sections of the ends of the parts (Figure 2, *a*). Thus, the scheme of part error formation can be represented in the way presented in Figure 2, *b*.

As a result of workpiece a cross section center shift relative to the axis of rotation g by the value of e_p, the allowance along the circumference of the section is uneven. Consequently, the deformation of the technological system will also be uneven along the circumference. As a result, after machining, the result is the profile of part B with the center of the middle section b.

Let us describe the variation in a given allowance around the circumference by the equation

$$t = t_a + f(\omega) \tag{3}$$

where t_a is the average value of the depth of cut; $f(\omega)$ is a variable in the depth of cut as a function of the angle of rotation.

Figure 3 shows a diagram for determining the law of change of law of variation t. From the triangle

$$t = R - r_d, \quad R = \sqrt{r_p^2 + e_p^2 - 2 r_p e_p \cos c}, \quad c = 180 - \omega - b, \quad b = \sin^{-1}\left(\frac{e_p}{r_p}\sin\omega\right)$$

Figure 4 shows the results of calculating the variation in t_l by the angle of rotation of the part. It is seen that the law of variation is close to sinusoidal.

We approximate the variable value of the cutting depth by a sinusoidal function

$$f(\omega) = e_p \sin(\omega).$$

Figure 4 shows the results of approximation. It is evident that the two curves are practically superimposed on each other (the inset shows a fragment of the curves on an enlarged scale). The largest approximation error does not exceed 4.5 % (Figure 5)

Therefore, dependence (3) can be represented as

$$t = t_a + e_p \sin(\omega) \tag{4}$$

where e_p is the eccentricity of the workpiece; ω is the angle of rotation of the part.

Substituting equation (4) into (1), we obtain

$$P = C\, v^{u}\, s^{z}\, [t + e_p \sin(\omega) - y]^{x} \tag{5}$$

Let us expand expression (5) in the Maclaurin series with respect to y and $e_p\sin(\omega)$. Given only linear terms, we obtain

$$P = C\, v^{u}\, s^{z}\, [t^{x} + x t^{x-1} e_p \sin(\omega) - x t^{x-1} y] \tag{6}$$

As the calculations showed, the linearization error of function (5) in the range of possible values of y and $e_p \sin(\omega\tau)$ for any possible values of v, s, t is not more than 7.5 %.

After substituting (6) in (1) we obtain the equation

$$y = \frac{C v^{-u} s^{z} t^{x}}{j + x C v^{-u} s^{z} t^{x}} + \frac{x C v^{-u} s^{z} t^{x-1} e_p}{j + x C v^{-u} s^{z} t^{x-1}} \sin\omega$$

Thus, under the influence of a variable cutting force due to a variable allowance around the circumference of the part, a change in the strain

$$y_2(\omega) = \frac{x C v^{-u} s^{z} t^{x-1} e_p}{j + x C v^{-u} s^{z} t^{x-1}} \sin\omega$$

IOP Conf. Series: Materials Science and Engineering **966** (2020) 012087 doi:10.1088/1757-899X/966/1/012087

relative to the middle position occurs

$$y_1 = \frac{Cv^{-u}s^z t^x}{j + xCv^{-u}s^z t^x}.$$

The amplitude of the oscillation determines the error of the part. Given that $j >> xCv^{u}s^{z}t^{x-1}$ we obtain

$$e_d = \frac{xCs^z t^{x-1} e_p}{jv^u}.$$

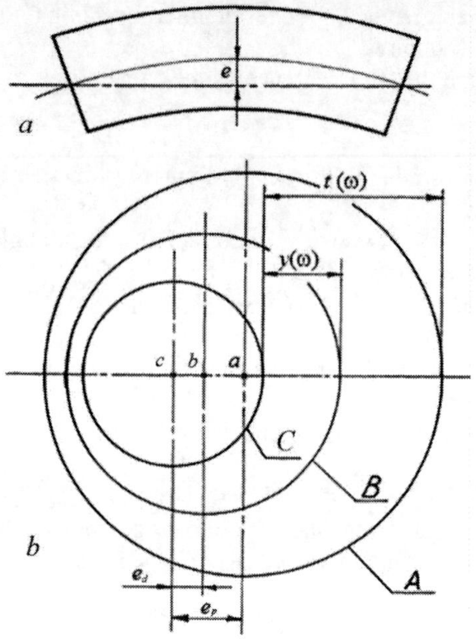

Figure 2. Scheme for determining the error in the longitudinal (*a*) and transverse (*b*) sections.

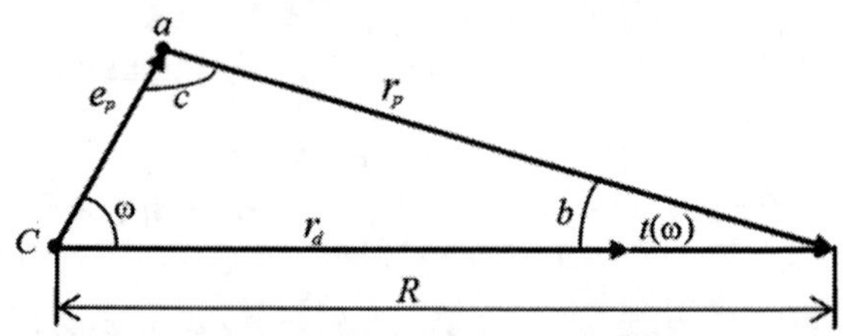

Figure 3. Scheme for determining the law of variation *t*: *a* – workpiece axis, *c* – workpiece rotation axis (spindle axis), e_3 – workpiece eccentricity relative to the rotation axis, r_p – workpiece radius, r_d – part radius, *R* – distance from the center of rotation to the workpiece outer surface, ω – workpiece rotation angle (spindle).

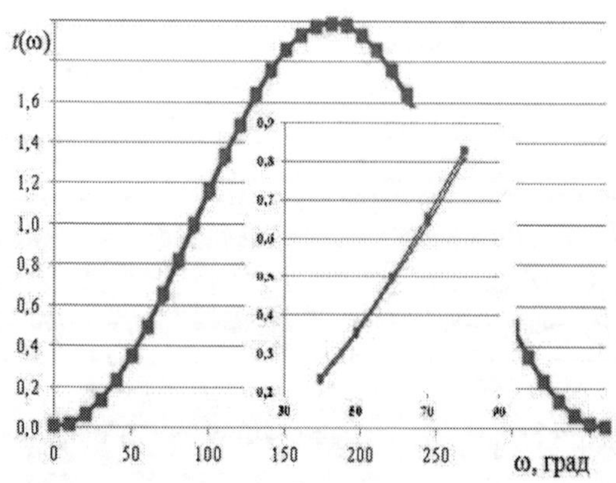

Figure 4. Change of cutting depth by rotation angle: ■ – calculated value, ♦ – approximation.

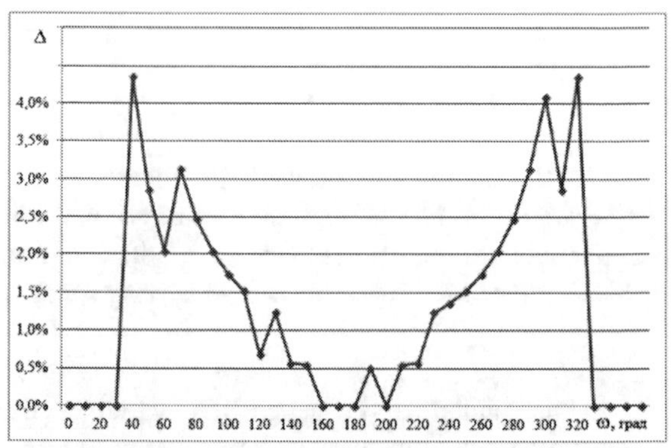

Figure 5. Change of approximation error (Δ) by rotation angle.

Hence, the refinement coefficient is defined as

$$K = \frac{jv^u}{xCs^z t^{x-1}} \tag{7}$$

From dependence (7) one can see that with increasing rigidity of the technological system and cutting speed, the feed refinement coefficient increases. The influence of the depth of cut depends on the exponent x (Figure 6).

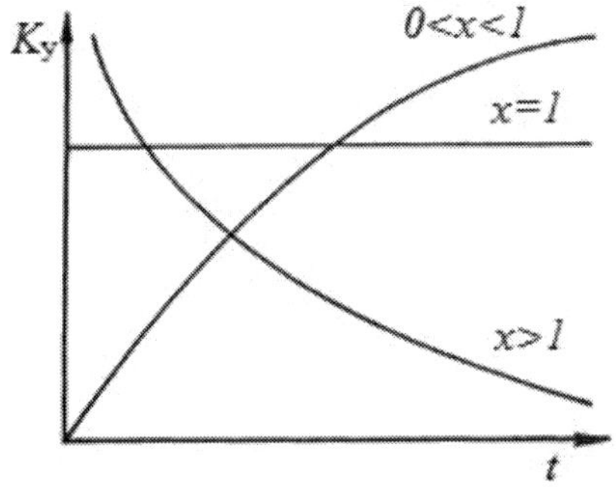

Figure 6. The nature of the dependence of the refinement coefficient on the cutting depth for various x.

For $0 < x < 1$, the exponent at t is less than zero, and the refinement coefficient increases with increasing depth of cut. For $x > 1$, the influence of the level on K_y is inverse. At $x = 1$, the refinement coefficient does not depend on the depth of cut.

3. Conclusion

For existing tool materials, the exponent x at the cutting depth in the formula of the cutting force P is in the range of 0.7–0.9 (range $0 < x < 1$). As the depth of cut decreases, the refinement coefficient decreases. Therefore, when switching from rough operations (large cutting depths) to finishing (small cutting depths), the refinement coefficient becomes smaller.

References

[1] Chengyan Zhang, Sicheng Jiao and Lixin Wang 2018 *Conference on Frontiers of Materials Synthesis and Processing* **493**
[2] Pereverzev P P and Akintseva A V 2016 *Russian engineering research* **12** pp 1048–53
[3] Xiao Q-B, Wan M et al 2020 *Robotics and Computer-Integrated Manufacturing* **64** 101949
[4] Lv L, Deng Z et al 2020 *International Journal of Mechanical Sciences* **1** 105543
[5] Grechishnikov V A, Pautov G A, Yurasov S Y and Yurasova O I 2017 *Russian engineering research* **3** pp 270–2
[6] Kremleva L V, Malygin V I and Lobanov N V 2016 *Russian engineering research* **12** pp 991–4
[7] Azimov B and Sulyukova L 2018 *XII International scientific and technical conference 'Applied Mechanics and Systems Dynamics', IOP Conf. Series: Journal of Physics: Conf. Series* **1210** 012003
[8] Świć A, Wołos D, Gola A and Smidova N 2019 *Tehnicki Vjesnik* **3** pp 256–63
[9] Karabulut A 2010 *Journal of the faculty of engineering and architecture of Gazi university* **2** pp 257–65
[10] Swic A, Wolos D et al 2014 *Applied Mechanics and Materials* **613** p 357

IOP Conf. Series: Materials Science and Engineering **966** (2020) 012088 doi:10.1088/1757-899X/966/1/012088

The study of the influence of deviations of the arc energy parameters on the defects formation during automatic welding of pipelines

A E Filyakov[1, 4], M A Sholokhov[2], S I Poloskov[1] and A Yu Melnikov[3]

[1] Bauman Moscow State Technical University, 5, ul. Baumanskaya 2-ya, Moscow, 105005, Russia
[2] Ural Federal University named after the First President of Russia B N Yeltsin, 19, Mira str., Yekaterinburg, 620000, Russia
[3] SHTORM Co. Ltd, Ekaterinburg, 620100, Russia

E-mail: [4] filyakov.92@mail.ru

Abstract. Modern quality management systems for welding products are based on procedures to minimize the likelihood of typical defects. The reasons for deviations of the energy parameters of the welding process include disturbances along the length of the arc or its breaks associated with a spasmodic change in the electrode wire feed speed, violation of the power circuit with the welded product, mains voltage, and shielding gas consumption. To establish a connection between deviations of the welding parameters with the quality indicators of the welded joints, a model for the formation of a weld seam under the conditions of technological disturbances has been developed. It is shown that the formation of the crater depends not only on the duration of the arc break, but also on the welding parameters. It has been established that in the presence of lateral oscillation of the electrode, an arc break significantly increases the undercut, the maximum depth of which movesto the center of the groove to the cutting center as the break duration increases. Whereas an increase in the amplitude of lateral oscillation of the electrode causes a significant increase in the depth of the crater.

1. Introduction

Taking into account the consequences of accidents and failures during operation, trunk pipelines are classified as hazardous production facilities. According to the data from the regulatory authorities, more than 85% of accidents and disasters at hazardous technical facilities, including pipelines, are due to defects of welded joints, including leakages, due to this the problem of ensuring the quality of welded joints is of immediate interest.

Modern quality management systems for welding products are based on procedures to minimize the likelihood of typical defects. Welding monitoring procedures, which have already become an integral part of scientific and technical support for the construction of trunk pipelines [1], have become an effective tool for this. Monitoring procedures provide monitoring of energy parameters of welding processes, forestalling and prevention of causes of non-compliance of welded joints with the requirements of technical documentation, and control of their elimination. At the same time, monitoring procedures make it possible to implement a number of requirements for documenting work performed. This is achieved by connecting to the modern welding equipment special recorders [2] with the appropriate software. The development of such systems for monitoring, recording and documenting the parameters of the welding process is carried out by a number of domestic high-tech science-intensive enterprises, and the systems can be built into modern digital power sources or made in the form of separate blocks. It can be stated that the most popular systems for monitoring of welding processes are used to document their parameters, as well as to analyze the consumption of materials, work time, register parameters beyond tolerances, and discipline welders. Unfortunately, so far none of the systems responds promptly to the main question: what consequences will the identified deviations lead to. Taking this into consideration, monitoring systems are needed that not only control and document the welding process, but also predict its results.

It should be noted that, despite attempts to predict the quality of welded joints according to the results of monitoring the parameters of welding processes, undertaken as at the beginning of this century [3] as well as later [4], operational quality forecasting systems that can be used on a job site don't exist, since an array of information is needed not only about defects in welded joints, but also about the causes of their occurrence. Therefore, for the operational forecasting of the quality of welded joints to become an effective tool for preventing defects formation of welded joints, additional studies of the significance of individual factors and phenomena, their influence on quality indicators, are needed to create and include in the structure of the forecasting module of a complete process model, as well as an information storage with the necessary filters.

2. Materials andMethods

One of the effective tools for ensuring a consistently high quality of welded joints is the use of automatic welding methods. However, the processes of automatic orbital welding of pipelines cannot ensure defect-free welding, since the main causes of defects are difficult to predict deviations in the energy parameters of the welding process, or process disturbances due to poor preparation and assembly of joints for welding, the human factor [5].

The reasons for deviations of the energy parameters of the welding process include disturbances along the length of the arc or its breaks associated with a spasmodic change in the electrode wire feed speed, violation of the power circuit with the welded product, mains voltage, and shielding gas consumption. Among technology-based perturbations associated with poor preparation and assembly of joints for welding, it is necessary to highlightuneven edge beveling and the gap in the joint, displacement of the edges, etc.

Since monitoring is limited to measuring the current and voltage of the arc and information about the assembly of the joint, the task is to virtually reproduce the influence of sharp deviations of the current and voltage of the arc that occur during arc breaks and short circuits (Figure 1) on the formation of the weld pool and seam.

Breaks and short circuits of the arc are usually the causes of violation of the formation of the seam, which lead to the appearance of extremely dangerous and unacceptable welding defects in the form of lack of fusion occurring both between the beads of the multi-pass seam and at the borders of the seam with the edges of the parts, and lack of root penetration [6, 7]. These defects are very similar in many respects, since they exhibit themselves in the local absence of fusion between the weld metal and the base metal or between the layers of the weld metal. As noted in [8], the occurrence of lack of fusion at the site of the craters of the previous layer is due to the limited penetrating ability of the arc burning in the partially filled prepared edges, as well as the shape of craters having a significant depth relative to their diameter. Lack of fusion is formed by incomplete inflowing of the molten bath into the crater, which is hindered by the surface tension and the gas pressure in the closed cavity of the crater. Under stick electrode welding with flux-cored and self-shielding wire, craters are filled with slag under the flux layer, which leads to the appearance of slag inclusions in the weld. Lack of root penetration occurs also due to the penetrating ability of the arc.

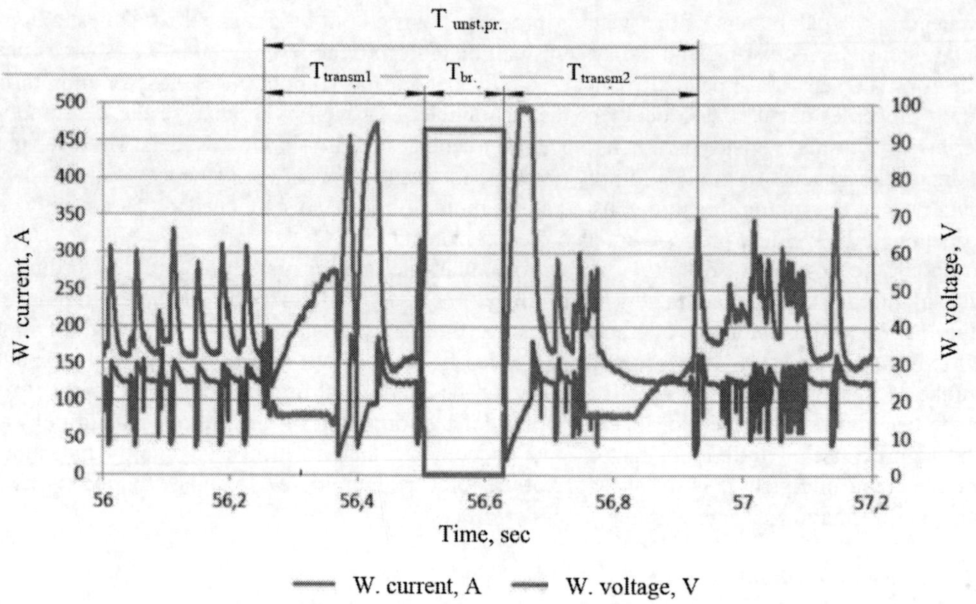

Figure 1. A fragment of the result of monitoring the voltage and current of the arc during arc welding: $T_{unst.pr.}$ – the duration of the unstable process, $T_{br.}$ – the duration of the arc break, $T_{transm1}$ – the duration of the transition from the stable process to the break; $T_{transm2}$ – the duration of the transition from a break to a stable process.

The solution to the problem of predicting the appearance of defects is based on modeling of the process with deviations of parameters detected during monitoring. Physical and mathematical modeling of welding is an effective tool to take into account the influence of welding parameters relevant to the quality of welded joints. Since monitoring is limited to measuring the current and voltage of the arc and information about the assembly of the joint, the task is to virtually reproduce the influence of sharp deviations of the current and voltage of the arc that occur during arc breaks and short circuits on the formation of the weld pool and seam. For this, a physical and mathematical model presented in [9] should connect the monitoring parameters with defects in weld formation, at the core the model are the equations describing the arc with monitoring data, heat transfer and formation of the surface of the weld pool and weld.

3. Results

The simulation results of the effects of short-term arc breaks on the formation of the weld are shown in Figure 2.

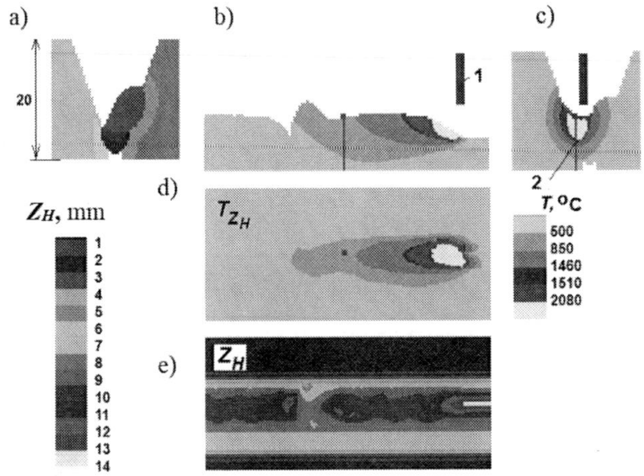

Figure 2. The results of simulating a short-term (0.2 s) arc break on the formation of a weld: a) the shape of the cross section of the joint; b) longitudinal section of the weld pool and seam; c) the limiting distribution of temperature in the cross section; d) temperature distribution on the surface Z_H of the metal; e) topography of the surface relief of Z_H metal; 1 – electrode, 2 – gravity vector.

With the specified parameters of the welding mode, an arc break of 0.2 s caused the formation of a crater in the weld bead with the maximum possible depth. Figure 3 shows the phases of formation of such crater.

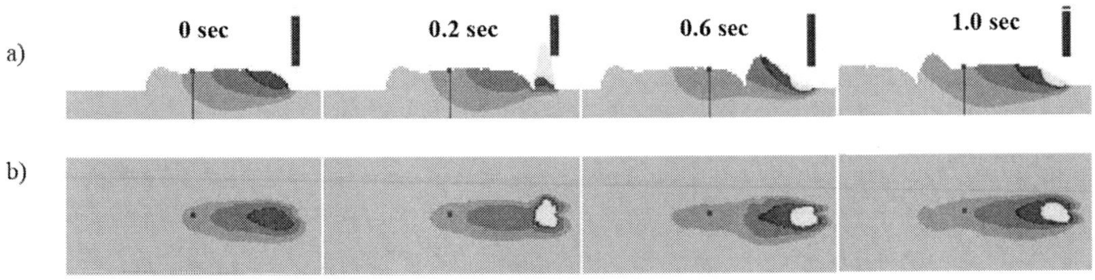

Figure 3. The phases of the formation of the crater in the weld bead and the temperature distribution: a) in a longitudinal section in the plane of the central position of the electrode; b) on the surface of the edges and weld pool.

Under narrow and deep groove welding, a short weld pool is formed due to the powerful heat removal, since it is surrounded on three sides by cold metal [10]. Therefore, when the arc breaks, its size decreases rapidly, and the electrode moves away from it with a welding speed. If, at the moment of repeated arc initiation, the metal melting resumes outside the weld pool, a new weld pool begins to form, which does not fuse with the bath formed before the break. Because of this, the likelihood of lack of fusion when applying a subsequent bead increases sharply.

The consequences of this situation are presentedin Figure 4.

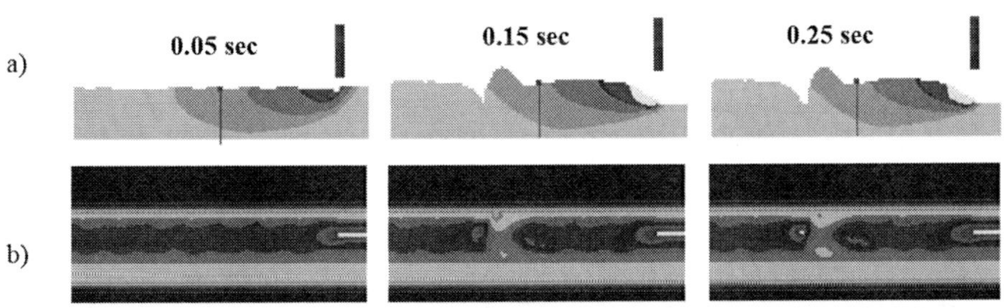

Figure 4. Influence of the arc break duration on the crater shape in the weld bead: a) temperature distribution in the longitudinal section in the plane with the greatest crater depth; b) topography of the surface relief $Z_H$of weld.

With a short arc break duration, the crater does not actually occur due to the continuous existence of the weld pool. With a very long break in the arc, the crater turns into a bead with cracks.

The formation of the crater depends not only on the duration of the arc break, but also on the welding parameters. An increase in the amplitude of lateral oscillation of the electrode causes a substantial increase in the depth of the crater (Figure 5).

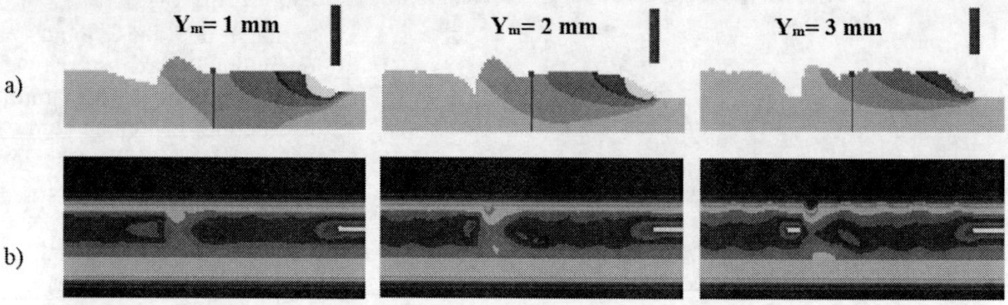

Figure 5. Influence of the amplitude of lateral oscillation on the crater shape in the weld bead after arc break for 0.2 s: a) temperature distribution in the longitudinal section in the plane with the greatest crater depth; b) topography of the surface relief Z_H of weld.

Virtual studies have shown that the formation of the crater depends not only on the duration of the arc break, but also on the welding parameters. In the presence of lateral oscillation of the electrode, an arc break significantly increases the undercut, the maximum depth of which moves to the center of the groove as the duration of the break increases. An increase in the amplitude of lateral oscillation of the electrode causes a substantial increase in the depth of the crater.

Equally significant is the influence the oscillation period and the welding speed (Figure 6).

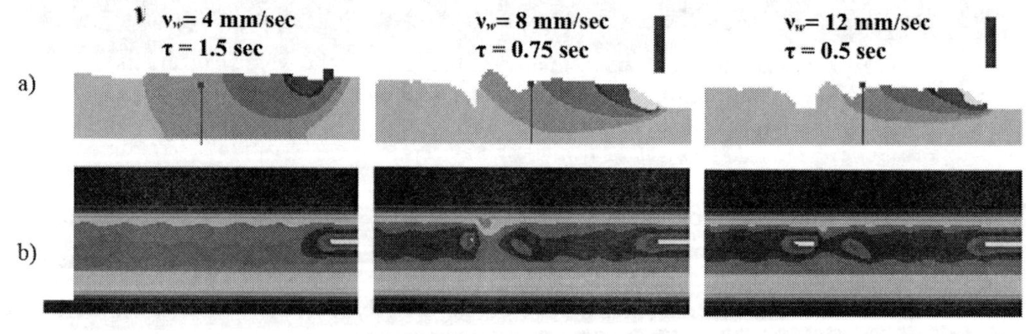

Figure 6. Influence of the welding speed v_w and the oscillation period τ on the shape of the crater in the weld bead: a) temperature distribution in the longitudinal section in the plane with the greatest crater depth; b) topography of the surface relief Z_H of weld.

Therefore, to predict the likelihood of lack of fusion occurrence directly during the monitoring process, it is necessary to establish the dependence of the crater depth both on the duration of the arc break and on other parameters of the welding process, primarily on the shape of the beveling of the joint, taking into account its filling, spatial welding performance, arc power, and welding speed and parameters of lateral oscillations of the torch.

In this regard, in order to eliminate the drawbacks of existing monitoring systems, which are limited by registration of the process parameters beyond the specified limits and do not assess the possibility of defects formation in welds, it is necessary to develop a model that quantitatively relates monitoring data and welding process parameters with the likelihood of lack of fusion between the beads and edges of the multi-pass seam.

To establish the dependence between the stability of the energy parameters of the process of automatic orbital welding of pipelines and the quality of welded joints, a physical and mathematical model for the formation of a multi-pass weld is developed based on a system of equations describing the arc taking into account monitoring data, heat transfer, and the formation of the surface of the weld pool and weld.

4. Conclusion

Numerical modeling of the crater's formation has shown that their depth depends not only on the duration of arc breaks, but also on many parameters of the welding process. However, for a more accurate forecast of the probability of occurrence of welding defects by the operational analysis of monitoring data, it is necessary to additionally establish the dependence of the crater depth both on the arc break duration and on other parameters of the welding process, primarily on the type of the beveling taking into account its filling and the spatial position of the welding machine, arc power, welding speed and parameters of the lateral oscillation of the torch. The developed physical and mathematical model allows us to solve this problem. However, for a more accurate determination of the position of the electrode in the groove, it is necessary to obtain an operational feedback, for example, using vision systems, described in detail in [11].

Acknowledgement

The assistance provided by the translator of the present article Ms. Natalia Karamysheva was greatly appreciated.

References

[1] Lisin Yu V and Aleksandrov A A 2013 Monitoring magistral'nykh nefteprovodov v slozhnykhgeologicheskikh usloviyakh [In Russian] *Nauka I tehnologii truboprovodnogo transporta nefti i nefteproduktov* **2** 22–7
[2] Gladkov Eh A 2000 Registrator parametrov svarki [In Russian] *Svarochnoe proizvodstvo* **3** 46–47
[3] Adolfesson S, Bahrami A, Bolmsjo G and Claesson I 1999 On-line quality monitoring in short-circuit gas metal arc welding *Weld. J.* **78 (2)** 59-73
[4] Li X R, Shao Z, Zhang Y M and Kvidahl L 2013 Monitoring and control of penetration in GTAW and pipe welding *Weld. J.* **92 (6)** 190-6
[5] Migacheva G N 2018 Analiz defektov svarki truboprovodov [In Russian] *Dostizheniya nauki iobrazovaniya* **3** 9–14
[6] Nguyen T C, Weckman D C and Johnson D A 2007 The discontinuous weld bead defect in high-speed gas metal arc welds *Weld. J* **86 (11)** 360-372
[7] Kim C H, Zhang W and Debroy T 2003 Modeling of temperature field and solidified surface profile during gas-metal arc fillet welding *J. Appl. Phys.* **94** (4) 2667–79
[8] Deshmukh S, Lande Y, Mhaske M and Belkar S B Experimental investigation of process parameters on strength of welding joint in GMAW *Int. J. Adv. Res. and Innov. Ideas Educ.* **3 (1)** 892–908
[9] Filyakov A E, Poloskov S I, Erofeev V A and Sholohov M A 2020 Fiziko-matematicheskaya model vliyaniya otkloneniy energeticheskikh parametrov na obrazovaniye defektov pri svarke truboprovodov [In Russian] *Svarka i diagnostika* **2** 16–22
[10] Kim J, Kim I, Lee J and Jung S 2011 An experimental study on the prediction of back-bead geometry in pipeline using the GMA welding process *Integr. Sci.* **49 (1)** 53–61
[11] Baek D, Moon H S and Park S H 2017 Development of an automatic orbital welding system with robust weaving width control and a seam-tracking function for narrow grooves *Int. J. Adv. Manuf. Tech* **93** 767–77

IOP Conf. Series: Materials Science and Engineering **966** (2020) 012089 doi:10.1088/1757-899X/966/1/012089

Application of double electrode welding as the way to increase welding productivity of fillet welds of bridge structures

A A Mosin[1, 3, 4], **M A Sholokhov**[1, 2] **and D S Buzorina**[1, 2]

[1] Ural Federal University named after the First President of Russia B N Yeltsin, 19, Mira str., Yekaterinburg, 620000, Russia
[2] SHTORM Co. Ltd, Ekaterinburg, 620100, Russia
[3] Kurganstalmost CJSC, Kurgan, Russia, 3640023

E-mail: [4] Mosinaa@kurganstalmost.ru

Abstract. One of the ways to solve the task to increase productivity and obtain an enlarged leg in one pass is the use of two-electrode submerged arc welding (welding with a split electrode). When testing this welding method with respect to welding bridge metal structures, it was found that the quality of the welded joints and the productivity of the process substantially depend on the positioning of the electrodes relative to the groove. However, for the effective implementation of this method, a comprehensive study of itsfunctionality for welding of bridge metal structures is necessary to conduct.An effective tool in solving such problems is physical and mathematical modeling, which allows without running a large number of experiments to study the features of the formation of fillet welds. To carry out this analysis, we used the previously developed physical and mathematical model of weld formation in single-arc submerged arc welding, added a description of the spatial position of the split electrode relative to the joint and defined the distribution of the heat flux and pressure of the double arc.It was found that the location of the electrodes with the lagging arc on the web provides better weld formation.

1. Introduction

In the recent years we can observe a dynamic development of bridge building in Russia. At the same time, three-dimensional welded metal structures manufactured in factories, which are then enlarged into bridge spans, are supplied to the bridge construction site. It should be noted, the main part of the welding works is in the manufacture of metal structures in workshop conditions, and when the metal structures are enlarged at the construction site, the volume of welding works is reduced. A special feature of bulk metal structures is the presence of a large number of extended welds (more than 2 m), Figure 1. The conducted analysis showed that the volume of welding of extended welds in the total volume of welding works is about 50–60%, while the fillet welds can make up from 40% to 70% of the total volume of welds.

Taking into account the specific nature of the operation of bridges (sharp changes in ambient temperature, the condition of soils, including marshy and permafrost, wind and variable dynamic loads from moving vehicles), strict requirements are imposed on the quality of welded joints. In this regard, welding modes, applied welding technologies, welding materials and equipment should ensure that welded joints are with the same mechanical properties as the base metal. In addition special attention should be paid to the geometry of fillet welds, since it is the geometry of the transition zone from the weld to the base metal that determines the stress condition in the given structure zone and, therefore, affects the durability of the welded joint [1].

Figure 1. Road bridge element.

Within the permissible values of the parameters of the welding modes regulated by statutory instrument, the maximum productivity has actually been achieved. And today, one of the important production tasks in the manufacture of bridge structures is to increase productivity, it is also possible by obtaining a larger leg length in one pass.

2. Materials andMethods

Traditionally, fillet welds with a leg up to 8 mm are produced using single-arc automatic submerged arc welding in one pass, seams with a large leg for several passes in 2F position or in 1F position. To assess the need to improve traditional technologies or develop new ones at the first stage, it is necessary to estimate the possibility of obtaining the maximum leg under submerged arc welding with a consumable electrode of fillet welds.

An effective tool in solving such problems is physical and mathematical modeling, which allows, without carrying out a large number of experiments, to study the specific features of the formation of fillet welds under submerged arc welding [2, 3]. To simulate physical phenomena, an area was selected, that covers the welded edges and arc burning zones, melting of flux and metal, Figure 2. Modeling area includes: M is the base metal, E is the electrode, D is the welding arc, F is the flux, V is flux fumes, G is the molten flux, R is the solidified flux, L is the liquid metal, K is the solid-liquid metal, S is the weld metal, W – air. Internal interband surfaces are marked as intersections of the sets of points belonging to the corresponding zones. For example, the surface of the weld pool $Z_L(x, y)$ separating the weld pool L from the welding arc D, flux fumes V and molten flux G is marked as L $L\cap(D\cup V\cup G)$.The sizes of the V, G, L, K zones and the location of the interface between them are unknown, and they must be determined during the simulation, depending on the results obtained when solving a system of equations describing the physical phenomena that occur during submerged arc welding.

The main differences between submerged arc welding from shielded gas arc welding are that the arc burns in a gas-vapor cavity arising from melting and evaporation of the flux. The second important feature is that the physical phenomena that determine the formation of the weld pool and weld, as noted in the work [4], occur in different environments. These features must be taken into account in the physical and mathematical model of the formation of the weld under submerged arc welding. The basis of the model is a system of equations of heat conductivity and equilibrium of the surface of the weld pool, in which the formation of the arc cavity is determined by the boiling isotherm of the flux under the influence of the arc column, heat transfer by flux fumes inside the arc cavity is taken into account, as well as the influence of the spatial position on the formation of the weld pool [5].The numerical solution of the equations of the submerged arc welding of a fillet weld at a metal thickness typical for bridge structures showed that when welding in 1F position, joint formation is made in one pass.

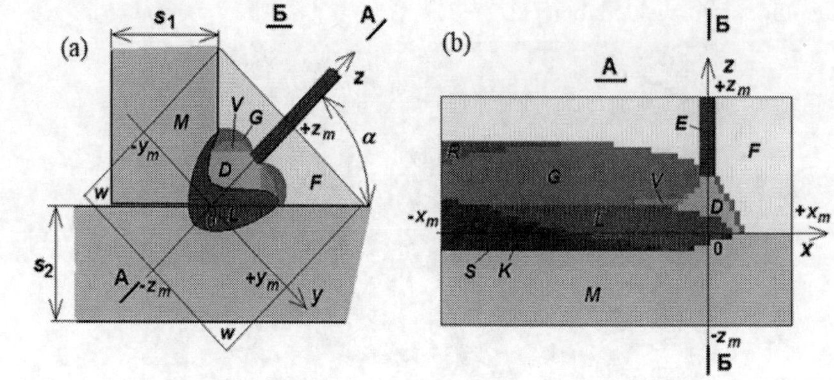

Figure 2. The structure of the chosen area to model the process of fillet welding.

Modeling showed that when welding a fillet weld in 2F position, unsatisfactory formation is possible due to runoff of liquid metal from a web. In this case, to obtain a high-quality weld seam, it is necessary to perform additional welding passages.

3. Results

To find out the maximum size of the leg, which is formed during welding in 1F position and in 2F position, with high-quality formation of a seam, we performed modeling of fillet welding of sheets with a thickness of 8 mm, a wire with a diameter of 2 mm and an arc current of 350 A, Figure 3. The welding speed and arc voltage were varied.

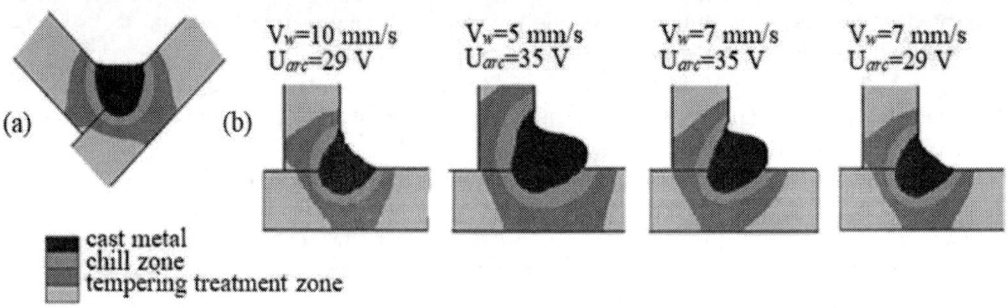

Figure 3. The result of computer simulation of the formation of the seam:
a) in 1F position b) in 2F position.

The studies confirmed the assumption that the existing technologies for single-arc submerged arc welding of fillet welds have already reached their maximum productivity. Obviously, it is possible to achieve a further increase in the productivity of the welding of fillet welds with the formation of legs of increased size only by using multi-arc and multi-electrode welding methods [6, 7].According to the research [8] during the production of extended fillet welds, the use of two-electrode welding is effective, as it is easy to implement, especially in the conditions of existing production, and ensuring a consistently high quality of welded joints. Therefore, at the second stage of the research, the possibilities of two-electrode submerged arc welding of fillet welds (welding with a split electrode) were determined. A distinctive feature of two-electrode welding is the simultaneous heating of the welded seam with two electrodes with a common supply of welding current.

It is known that the quality of welded joints in two-electrode welding depends not only on the welding conditions, but also to a greater extent on the positioning of the electrodes relative to the

IOP Conf. Series: Materials Science and Engineering **966** (2020) 012089 doi:10.1088/1757-899X/966/1/012089

beveling. Shifting of the electrodes towards the flange or closer to the upper part of the beveling affects the likelihood of undercuts and sagging, the penetration depth, Figure 4 [8].

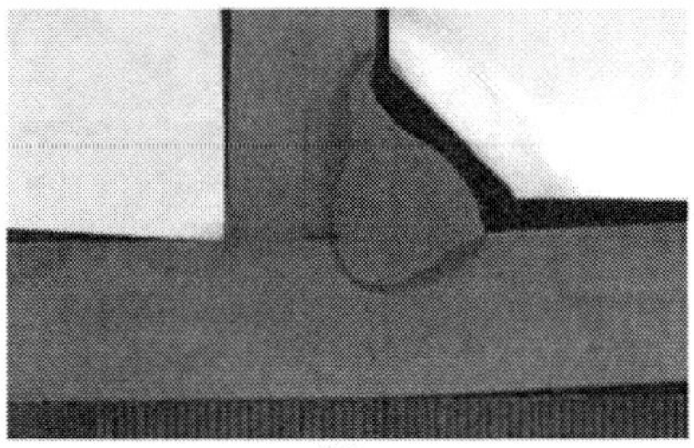

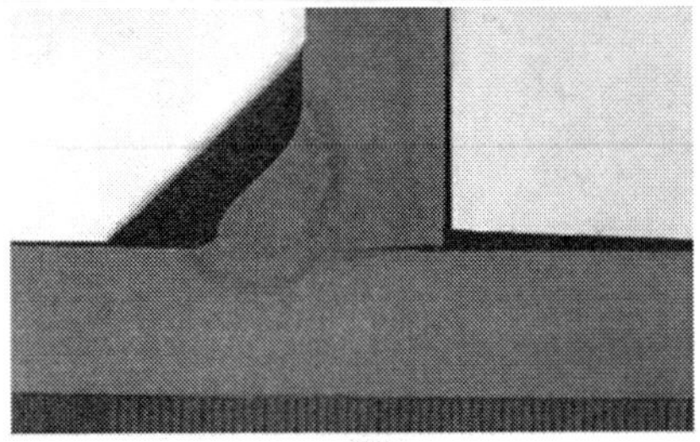

(a) (b)

Figure 4. Macro sections of samples welded with a different arrangement of electrode wires relative to the axis of the seam: a) with shifting of the electrodes towards the flange; b) when the electrodes are closer to the upper part of the beveling.

In order to develop rational technologies for two-electrode welding of fillet welds and the necessary welding equipment for the industrial production of three-dimensional bridge metal structures, it is necessary to analyzetechnological parameters of the process of automatic two-electrode welding and the influence of electrode positioning on the formation of fillet welds; to develop a physical and mathematical model for the formation of fillet welds in two-electrode welding and to carry out computer engineering analysis of the influence of two-electrode welding parameters on the quality of welded joints and determine their optimal values.

To perform this analysis, we used the previously developed physical and mathematical model of weld formation in single-arc submerged arc welding added a description of the spatial position of the split electrode relative to the joint and defined the distribution of the heat flux and pressure of the double arc. Accordingly, additional parameters were introduced into the model: the distance g between the electrodes, the angle β between the surface of the electrodes and the direction of welding, and the distance r from the electrode to the midline of the joint. The influence of the angle α of inclination of the torch relative to the joint was also taken into account.

4. Discussion

Virtual studies of the process of welding with a split electrode in 2F position showed that the maximum leg of a fillet weld is achieved with the following process parameters: electrode wire with a diameter of 2 mm, feed rate of 23 mm/s, electrode stick-out of 25 mm, angle of inclination 45° to the horizon. The electrodes are located at a distance of 8 mm from each other, perpendicular to the plane of symmetry of the joint and the direction of welding. Welding mode: welding speed 10 mm/s, the total current of the split arc is 850 A, the voltage across the arcs is 29 V.

Figure 5 shows the result of computer simulation of the formation of the weld pool and a weld taking into account the temperature distribution at the joint, electrodes and flux when welding the T-joints of sheets of 10ChSND steel with a thickness of 14 mm by a split arc using a Sv-08G2S (analogy 13Mn6 – DIN, Germany) electrode wire with a diameter of 2 mm.

As a result of the simulation, it was determined that at a relatively close distance between the electrodes, a vapor-gas cavity common to both arcs and a single weld pool are formed. The arc length is about 3 mm, the total heat dissipation in the arc column is 12.7 kW. The power of heat release in the cathode spot was 6.8 kW, the power transferred by drops of electrode metal was 5.1 kW. The width of the weld pool is defined at 16.8 mm, the depth is 10.5 mm, and the length is 67 mm. However, with further augmentation, the melt of the weld pool drains from a web to a flange due to gravitational impact.

IOP Conf. Series: Materials Science and Engineering **966** (2020) 012089 doi:10.1088/1757-899X/966/1/012089

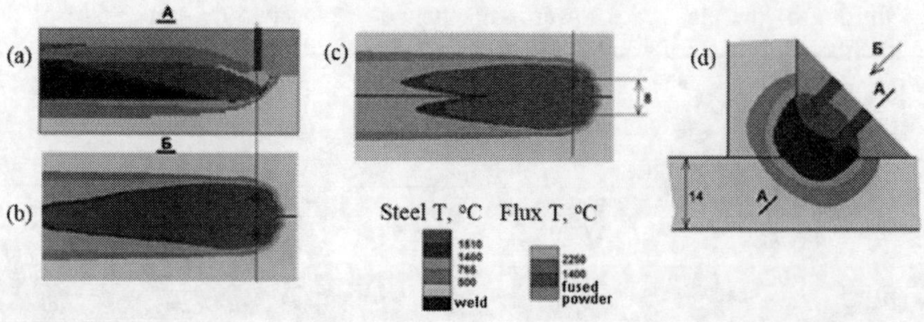

Figure 5. The modeling result of the temperature distribution during welding in 2F position: a) in a longitudinal section of the weld pool; b) on the surface of the 'metal-flux'; c) on the initial surface of the sheets; d) calculated macro section.

For this welding option, a study was made of the influence of the angle β, which determines the location of the electrodes relative to the direction of welding. The alignment parameters of the electrodes relative to the joint: $\alpha = 45°$, $g = 8$ mm, $r = -3$ mm. The simulation result is shown in Figure 6.

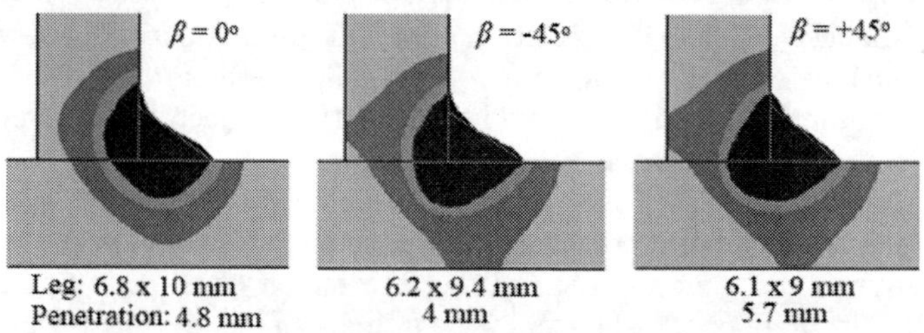

Figure 6. The simulation results at different βangles of the location of the electrodes relative to the joint.

It was determined that when the electrodes are placed on a line perpendicular to the direction of welding, a wide seam is formed with noticeable runoff of the melt from the web. For this reason, the size of the supporting leg on this web is noticeably reduced (up to 6.8 mm) with respect to the leg on a flange (10 mm). The penetration of the joint between the web and the flange (4.8 mm) does not reach half the wall thickness. When the arc moves forward on the web ($\beta = -45°$) and, accordingly, the arc lags on the flange, the width and legs of the seam decrease, and the penetration depth increases, which is explained by the approach of the arcs to the plane of symmetry of the joint. However, only the penetration of the flange increases, and the penetration of the joint between the web and the flange decreases (4 mm). The melt runoff is less noticeable in comparison with the symmetrical arrangement of the electrodes, the difference of legs is also slightly reduced. When the arc on the wall moves backward ($\beta = +45°$), at which the arc on the web lags behind the arc on the flange, the melt runoff and the difference of legs are noticeably reduced. Joint penetration increases significantly (5.7 mm).

Since of the arrangement with the arc lag on the web provides the best formation of the seam, further research of the influence of the parameters was performed for this option. Since the legs of the seam for this thickness are insufficient, we increased the distance between the arcs ($g = 12$ mm), Figure 7.

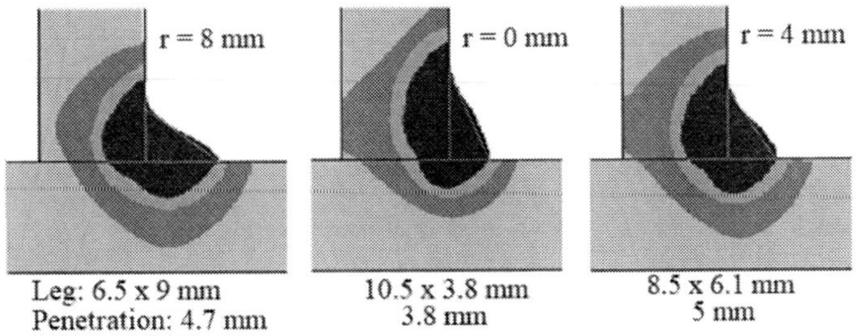

Figure 7. The simulation results at different location of the electrodes *r* relative to the joint and at different inclination angle *α* of the torch.

When the electrodes are placed on a line perpendicular to the direction of welding, a wide seam is formed with noticeable runoff of the melt from the web. For this reason, the size of the supporting leg on this web is noticeably reduced (up to 6.8 mm) with respect to the leg on a flange.

5. Conclusion

The research showed that the existing technologies for single-arc as well as double-electrode submerged arc welding of fillet welds have already reached their maximum productivity.Moreover, only the location of the electrodes with the lagging arc on the web of the fillet joint can provide better formation of the weld with a slight increase in the size of their legs. Obviously, a further increase in the productivity of the welding of fillet welds with the formation of legs of increased size only possible by using more complicated special techniques, such as multi-arc and multi-electrode welding methods [9], granular iron metal additives or metal powder, other unconventional technologies.

Acknowledgement

The assistance provided by the translator of the present article Ms. Natalia Karamysheva was greatly appreciated.

References

[1] Parhomenko A A 2009 Statisticheskiy obraz geometri i profilya perekhoda ot osnovaniya k osnovnomu metallu uglovykh shvov [in Russian] *Vestnik DGTU. Special edition* 111–8
[2] Sholokhov M A, Erofeev V A and Poloskov S I 2013 Kompyuternyy analiz predstavlyayet soboy razdeleniye pri dvukhdugovoy svarke [In Russian] *Svarka and Diagnostika* **6** 14–9
[3] Sidorov V P and Khurin S A 2011 Modelirovaniye provara pri dugovoy svarke stykovykh soyedineniy [In Russian] *Svarka and Diagnostika* **6** 36–42
[4] Sudnik V A 2012 Matematicheskaya model' protsessa svarki pod flyusomiyavleniy v dugovoy kaverne [In Russian] *Svarochnoe proizvodstvo* **7** 3–12
[5] Ghosh A and Chattopadhyay H 2013 Mathematical modeling of moving heat source shape for submerged arc welding process *The International Journal of Advanced Manufacturing Technology* **69** 2691–701
[6] Melnikov A 2016 Inzhenernaya metodika rascheta parametrov rezhima dvukhdugovoy svarki uglov ykhshvov [In Russian] *Svarka and Diagnostika* **3** 46–8
[7] Scutelnicu E and Rusu C C 2014 Assessment of cooling rate in longitudinal welded pipelines performed by submerged Double-Arc Welding *International Journal of Mechanics* **8** 144–9
[8] Mosin A A 2020 Putiresheniya problem povysheniya proizvoditelnosti svarki protyazhennykh uglovykh shvov obyemnykh mostovykh metallokonstruktsiy [In Russian] *Svarka and Diagnostika* **1** 50–4
[9] Kiran D V, Cho W D, Song H W and Na J S 2015 Arc interaction and molten pool behavior in the three wire sub-merged arc welding process *International Journal of Heat and Mass Transfer* **87** 327–40

IOP Conf. Series: Materials Science and Engineering **966** (2020) 012090 doi:10.1088/1757-899X/966/1/012090

Evaluation of the resulting quality level index for interregional passenger automobile transport

V V Epifanov[1], M Y Obshivalkin[1] and K A Generalova (Lukon`kina)[1, 2]

[1] Department of Automobiles, Ulyanovsk State Technical University, 32, Severnyj Venets Str., Ulyanovsk, 432027, Russia

E-mail: [2] tinetta.ledi@mail.ru

Abstract. In order to improve transport services for the population, it is necessary to study and analyze the information on the quality of transport services on the basis of gained experience and wishes of the passengers. This paper, basing on the normative and technical documentation and expert assessments, provides the rationale for the selection of quality indicators for interregional passenger automobile transport. Quality evaluation indexes are supposed to be simple for calculation and use; quality indexes are supposed to be formed in the way to judge the participation of carriers in the improvement of the quality of the transport process, i. e. to separate the qualities, which depend and do not depend on their activities. Indexes, developed on the basis of the standards' requirements are used in this paper. Such approach in the management of interregional passenger transport can contribute to the quality improvement of interregional automobile transport services provided to the population. The results of statistical data processing of passengers' questionnaires are also given here. The transport quality indexes were quantified through the passengers' answers ranking, the evaluation of the interregional passenger automobile transport quality was carried out on the example of the Ulyanovsk State Technical University.

1. Introduction

A whole range of related measures are aimed at improving the quality of interregional passenger automobile transport. To elaborate such measures, it is necessary to study and analyze information on the quality of transport services founded on gained experience and satisfaction of passengers.

Transport activity depends mainly on the consumer's demand for its services, but the consumer is not concerned with the carriers' financial expenses of the service. A passenger is willing to bear certain costs only under the conditions: minimum trip time, 100 % assurance of trip, safety and convenience of trip, ability of getting reliable trip information, convenient location of bus stations, etc. [1, 2]

The aim of the research is improving of the interregional passenger transport quality by developing methodological bases of customer satisfaction assessing with the quality of transport services.

The main goal is to stabilize the situation on the market of passenger regular transport on interregional routes. The solution to the problem on the routes will be the elaboration of a holistic management system, which is oriented towards the population, to meet their needs qualitatively, and to reduce the socially necessary transport costs.

At present, there is a quite stable system in the market of interregional passenger automobile transportation, which work has not only positive but also negative aspects.

One of the main problems is that the quality of transportations in the interregional communication is not controlled by regional bodies. Regulatory acts developed in the region don't possess legal force for interregional transport. The analysis of existing standards indicates a significant overlap of requirements for work in interregional transport [3–5], including quality assurance. This leads to increasing complexity and obfuscating of movement management process.

Applying standards-based indexes, can significantly improve the quality of services provided to the population by interregional transport management.

The development and implementation of an integrated system of passenger transport quality management in interregional communication include a methodology for assessing passenger satisfaction.

2. Methods and results

According to the methodology of assessing the satisfaction of passengers, an important stage is the questionnaire of passengers [6–8]. For this purpose we developed a special questionnaire for passenger interviews containing 40 questions. The content of the questionnaire reflects the essence of the transport services quality parameters.

728 people of different ages, social status, using the services of interregional passenger automobile transport in the Ulyanovsk region were interviewed. Questionnaires were conducted at bus stations of municipalities of the Ulyanovsk region, enterprises and organizations, educational institutions, social protection institutions (people of retirement age) and, of course, in public automobile transport itself.

Here is a number of results of the survey responses [9].

According to the results of the questionnaire, 23 % of passengers were satisfied with the level of comfort of interregional automobile transport, 38 % were partially satisfied and 39 % were not satisfied (Figure 1).

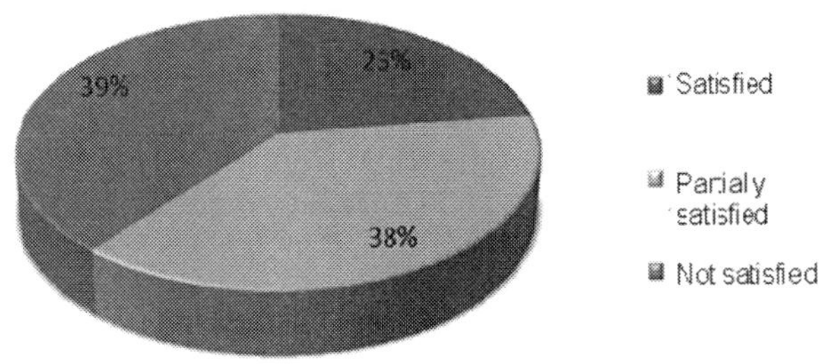

Figure 1. Satisfaction of passengers with the level of transportation comfort.

Many respondents noted the convenient location of the Ulyanovsk Central Bus Station (58 %) and the relatively convenient location (42 %). The bus station is located in the center of the city, which is convenient for passengers, but leads to traffic congestion in the city territory.

Overall, most passengers feel comfortable enough at bus stations (70 %) (Figure 2); note convenient location and number of ticket offices, seats in the waiting room (78 %).

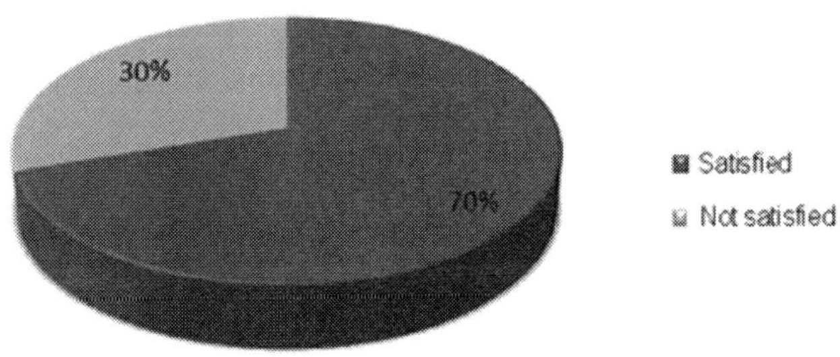

Figure 2. Comfort of passengers at bus stations.

IOP Conf. Series: Materials Science and Engineering **966** (2020) 012090 doi:10.1088/1757-899X/966/1/012090

59 % of passengers are satisfied with bus routes, 72 % with the number of trips, 58 % with departure times (Figure 3).

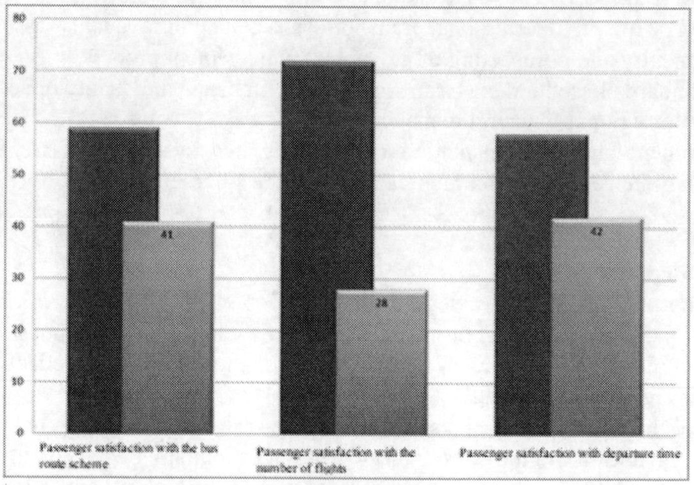

Figure 3. Timeliness and reliability of passenger transport, %.

Quite a lot of complaints are made by passengers to the safety of transport services. 42 % of passengers have no sense of safety on the trip (Figure 4).

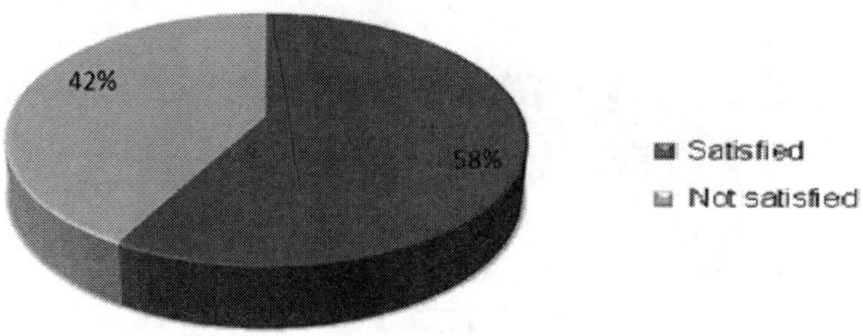

Figure 4. Safety of passenger automobile transport services.

Information support of transport services consists in provision of data on routes, schedules, time of departure-arrival at bus stations.

Passengers identify the following sources of information on transport:

- Own experience of trips – 26 %;
- Social networks – 5 %;
- Bus station website – 14 %;
- Bus station information desk – 24 %;
- stands and electronic boards of bus station – 31 %.

41 % of passengers report a lack of information at bus stations on the schedules of related modes of transport (rail, aviation, urban transport) (Figure 5). The creation of an electronic application to cellular phones, smartphones and other means of communication is a topical issue.

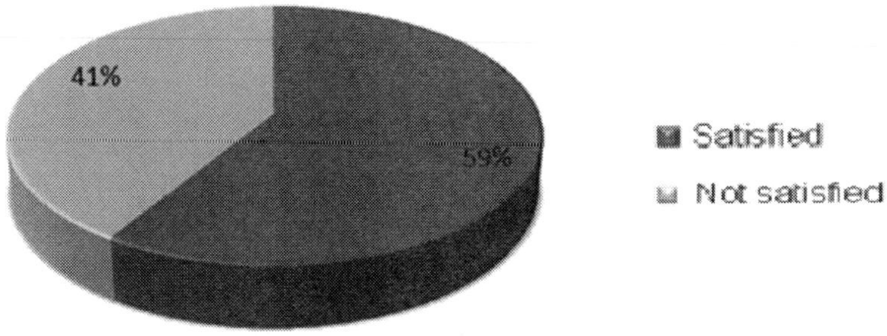

Figure 5. In formativeness.

One of the important parameters of transportation is the trip cost of about half of the surveyed passengers is estimated as expensive (49 %), 51 % considered the fares affordable (Figure 6) The system of benefits doesn't exist on interregional passenger automobile transport at all.

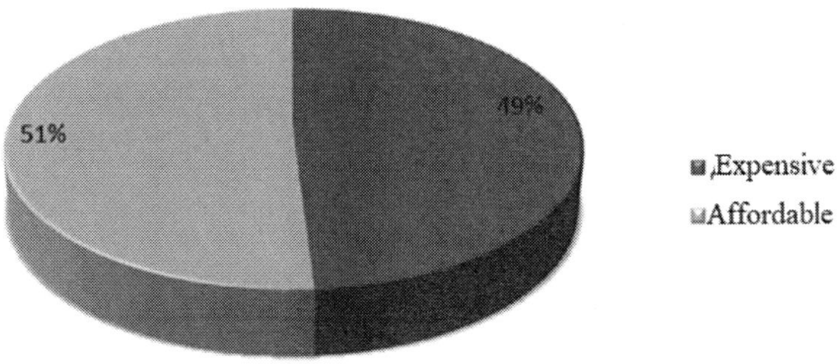

Figure 6. Satisfaction of passengers with the trip cost.

The quality index, when moving passengers in interregional transport, is a quantitative characteristic of one of the features that make up the quality of passenger transportation, considered in relation to certain conditions of their organization and implementation [10]. Most authors equate the quality of passenger transportation with safety [11–13]. In fact, transport safety requirements are prescribed in all standards related to passenger transport services [3–5]. The indexes that help to assess the quality of transportation include the safety of trip, the time of trip, comfort of waiting time and trip itself, etc.

Expert assessment of the most significant quality indexes of passenger transportation showed such distribution: safety of passenger transport services (A1 – 13.2 %); reliability of transport services (A2 – 12.7 %); timeliness of passenger transportation (A3 – 10.9 %); comfort of passenger transportation (A4 – 11.8 %); service economy (A5 – 9.4 %); in formativeness (A6 – 8.7 %); comfort of passengers at bus stations (A7 – 9 %). Of all the 14 indexes, 7 of them were the most important. They accounted for 80 % of the total number of indexes.

The differential method of estimating the resulting quality level index is to compare concrete unit quality indexes of the estimated service with the corresponding unit quality indicators of the base sample.

The resulting quality level is calculated by:

$$S_i = \frac{A_1+A_2+A_3+A_4+A_5+A_6+A_7}{7} \quad (1)$$

Through the allocated answers of passengers to the questionnaire it is proposed to determine the numerical value of each index of the transport quality services of included in the resulting index. The numerical value of each quality index is defined as the average of the numerical values of responses in fractions for several questions of the questionnaire (Table 1).

Table 1. Numerical values of transport service quality parameters.

Average value of transport services quality	Value of quality indicator in fractions
1. Safety of passenger transport services	0.68
2. Reliable transport services	0.7
3. Comfort of transportation	0.53
4. Timeliness of passenger transportation	0.76
5. Economy of service	0.68
6. Informativeness	0.6
7. Comfort of passengers at bus stations	0.77

Thus, we will determine the numerical values of each transport quality services index for the transportation of passengers in the interregional communication in the Ulyanovsk region. It can be seen from Table 1 and Figure 7 that all quality indexes require significant improvement.

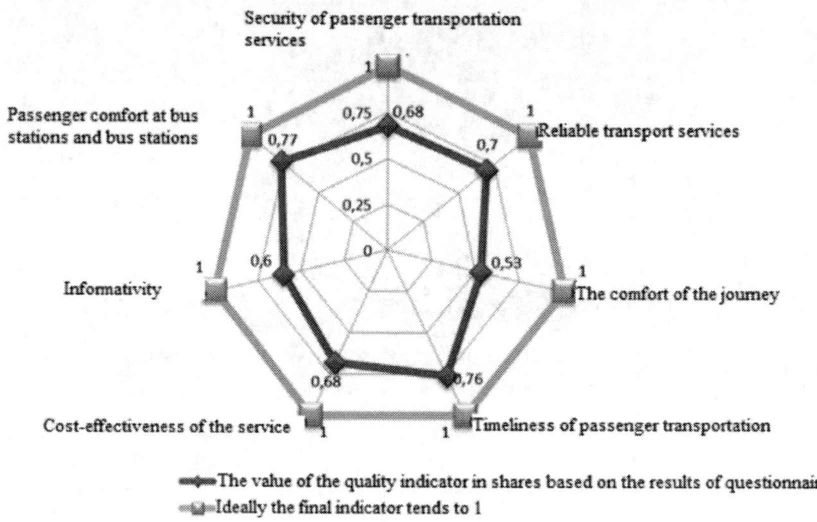

Figure 7. Numerical values of transportation services quality parameters.

The resulting quality level index should be defined as follows:

$$S_i = \frac{0.68 + 0.7 + 0.53 + 0.76 + 0.68 + 0.6 + 0.77}{7} = 0.7$$

If the evaluated service has all the quality indexes $A_i \geq 1$, its quality level is higher or equal to the basic; if all $A_i < 1$, then the level is lower.

3. Conclusion

On the understanding that ideally the final indicator aspires to a unit of $S_i = 1$, it can be concluded that the level of quality of interregional passengers automobile transport in the Ulyanovsk region is rather low and needs to develop a set of measures to increase it.

References

[1] Gudkov V A, Mirotin A B, Vel`mozhin A V and Shiryaev S A 2004 *Passazhirskie avtomobil`ny`e perevozki* ed V A Gudkova (Moscow: Goryachaya liniya – Telekom) p 448 [In Russian]
[2] Russian state standard GOST R 51825-2001 2001 Uslugi passazhirskogo avtomobil`nogo

transporta. Obshhie trebovaniya (Moscow: Gosstandart Rossii) p 8 [In Russian]

[3] Russian state standard GOST R 51006-96 1996 Uslugi transportny`e. Terminy` i opredeleniya (Moscow: Gosstandart Rossii) p 12 [In Russian]

[4] Gorev A E` 2008 *Informacionny`e texnologii v upravlenii logisticheskimi sistemami* (SPb.: SPbGASU) p193 [In Russian]

[5] Gudkov V A 2012 *Passazhirskie avtomobil`ny`e perevozki: Programma kursa, metodicheskie ukazaniya i zadaniya po vy`polneniyu kursovogo proekta* (Volgograd) p 25 [In Russian]

[6] Epifanov V V, Obshivalkin M Yu, Lukon`kina K A and Gusev S I 2019 Uluchshenie kachestva transportny`x uslug obshhestvennogo transporta na osnove ocenki trebovanij potrebitelej *Sbornik: Aktual`ny`e voprosy` organizacii avtomobil`ny`x perevozok, bezopasnosti dvizheniya i e`kspluatacii transportny`x sredstv* pp 266–74 [In Russian]

[7] Sullivan L P 1986 *Quality Progress: 'Quality Funktion Deployment'* vol **6** pp 39–50

[8] Kieu L M, Bhaskar A and Chung E 2014 Public Transport Travel-Time Variability Definitions and Monitoring *Journal of Transportation Engineering* no **7** pp 122–30

[9] Zou X and Yue W L 2017 A Bayesian Network Approach to Causation Analysis of Road Accidents Using Netica *Journal of Advanced Transportation* vol **2017** 2525481

[10] Cai Z, Wang D and Chen X M 2017 A novel trip coverage index for transit accessibility assessment using mobile phone data *Journal of Advanced Transportation* vol **2017** 9754508

IOP Conf. Series: Materials Science and Engineering **966** (2020) 012091 doi:10.1088/1757-899X/966/1/012091

Organizational factors for manufacturing defects reduction in small batch production

I V Ershova[1, 2], S G Baranchikova[1], E V Cherepanova[1] and N N Chistyakov[1]

[1] Institute of New Materials and Technologies, Ural Federal University named after the first President of Russia B N Yeltsin, 620002, 19, Mira street, Yekaterinburg, Russia

E-mail: [2] i.v.ershova@urfu.ru

Abstract. The study is devoted to the defects organizational causes analysis in the machine building products production process. The companies with the small batch production were taken into consideration. Factors related to the time norms fulfilling percentage, the work rhythm, the consolidation of operations at one workplace, the new technologies introduction and equipment failures were chosen. A preliminary factors selection according to independence criteria allowed us to build a reliable regression model for the defects rate calculating. The main factors influencing the defects rate in small batch production are the time norms fulfilling percentage, the rate of consolidation of operations at one workplace, and the number of new technological processes introduced. This allowed us to offer recommendations on the machining sites work organization. The final operations and productswhich are made from expensive metals should be carried out at workplaces with a low rate of operations consolidation. Also, it is important that the wage system should not contribute to over-fulfillment of norms. The maintenance and repair system should prevent malfunctions during operations fulfilment.

1. Introduction

Product quality management concepts change according to the production development. The traditional quality assessment methods are based on large statistical samples of indicators. Application of Shuhart control cards is described, for example, in [1]. In [2] the shortcomings of Shuhart maps are identified and improvements for this method are offered. The statistical concept of 'six sigma' is widely used to improve the processes quality and to identify the reject causes [3].

This does not applicable to small-batch manufacturing and customer-oriented production. This conclusion is made in the article of [4] and confirmed by Russian investigators [5]. In the context of uncertainty in small-batch manufacturing, it is difficult to determine the reject causes from the product sample. The authors [6] propose to use a combination of statistical and Bayesian approaches. This requires the use of the aprioristic judgment on the reject reasons.

The reject reasons can be technical, organizational and connected with human resources. Technical reasons related to equipment and tool deterioration are quite widely considered (for example, in [7]) and are not the subject of the research. The human factor is also studied in detail. Most often, the influence of the human factor is connected with the workers' motivation [8]. Other authors consider the relationship between the quantity of defective products, caused by the workers' errors, and the complexity of design and technology [9]. This is relevant for labour-consuming processes such as assembly. These factors are also important for metalworking but to a lesser extent.

IOP Conf. Series: Materials Science and Engineering **966** (2020) 012091 doi:10.1088/1757-899X/966/1/012091

The organizational factors are in the greatest interest of researchers. In [10,4] the products quality is considered as a consequence of the industrial system construction quality, including production logistics, maintenance, management and control methods. This is a comprehensive approach for the entire production system. The quality issues of raw materials and components, which are also related to the control factors, go beyond the scope of the production process and are not considered in the work. They are discussed in other investigations, for example [11].

The purpose of the investigation is to determine the organizational factors of the reject during the metalworking operations at a particular workplace in small-batch manufacturing.

2. Materials and methods

The investigation is based on actual operation data of the mechanical processing workshop during the year. The main method of research was regression statistics. The choice of the reject factors was based on an analysis of researchers' work on the subject and the expert assessments of the reject.

The main factors selected for the analysis are:

- Index of labour performance (*X*1);
- Amount of the overtime work in the workplace per month (in hours) (*X*2);
- Amount of new technology in the workplace, previously not performed here (*X*3);
- Index of the fixed operations (the amount of different operations completed at the workplace) (*X*4);
- Number of equipment malfunctions (*X*5).

The investigated function (*Y*) is the share of reject in the total volume of production.

The data for analysis are shown in Table 1.

Table 1. Data for analysis.

Month	*Y*	*X*1	*X*2	*X*3	*X*4	*X*5
1	3.9	1.071	8	1	23	0
2	3.7	1.039	0	0	24	0
3	3.8	1.055	0	0	24	0
4	4.3	1.108	2	0	26	0
5	6.5	1.203	8	1	34	1
6	7.1	1.222	10	2	37	1
7	8.2	1.251	16	1	43	2
8	6.7	1.211	12	0	39	0
9	7.2	1.232	16	2	41	1
10	8.5	1.267	16	2	45	2
11	4.4	1.112	0	0	26	0
12	9.1	1.327	20	5	49	2

The importance of the 'overtime accounting' factor is noted in [12] as a part of the optimal production program model creation. The need to take into consideration the change in technology is described, for example, in [13, 14]. The remaining factors are selected on the basis on the expert estimates of managers.

The data was processed using the Microsoft Excel.

3. Results and discussion

On the first stage of the investigation, a multi-factor regression model was used. The influence of all 5 factors was taken into consideration in the model. This model shows a high determination coefficient ($R^2 = 0.99$). But the statistical significance coefficients evaluation of the multi-factor model showed that two factors from the considered set (*X*2 and *X*3) have an unacceptably high level of P-values (significantly higher than the normative level of 0.05) and therefore are not statistically significant. The second stage of the investigation, these factors (*X*2, *X*3) were excluded from the multi-factor model.

As a result, the model became:

$$Y = -7.1999+8.2358 \cdot X1+0.0972 \cdot X4+0.4146 \cdot X5 \quad (1)$$

Model coefficients are statistically significant. The determination factor has a high value. Fisher 's test value is improving.

The model reflects the share reject dependence from: the index of labour performance, the Index of the fixed operations and number of equipment malfunctions. At the same time, if independent parameters are close to 1 (or 0 in the case of equipment failures), then the reject share according to this model will be extremely small. Therefore, this formula is not effective when used in mass and large-scale production. Because in these types of production fixed operations index does not change from month to month, and the labour performance index is more stable, as most actions of the worker are automated. The formula could not also be applied in single production because in single production usually each operation performs for the first time and once. It can be called a 'permanent implementation'. In addition, time standards in single production are calculated by extended methods and often do not correspond to reality. The model can be used only in small-scale and medium-scale production.

Using this model, the following recommendations for workplace production can be formulated:

1. Final operations have to be carried out at workplaces with a small index of the fixed operations.
2. The products from expensive materials have to be produced on the workplaces with a small index of the fixed operations.
3. The remuneration system should not encourage over-performance.
4. The repair, maintenance and operation of the equipment have to prevent problems during operations.

Thus, it is more efficient to have one pre-operation equipment working in 2–3 shifts and several workplaces for final operations working in one shift.

Similar situation is with material. It is more efficient to have one equipment, rolling black metals, working in 2–3 shifts, and several workplaces, rolling non-ferrous metals and alloys, working in one shift.

At the same time, both the cost of the material and the processing stage must be taken into account. During the production reorganizing, it is necessary to factor into the reduction of costs and the constant quality improvement, which are necessary in the strict competition conditions. Therefore, special creation of several workplaces for final operations may be economically unprofitable. This requires additional calculations.

The remuneration system has to reduce the share of reject and the corresponding costs associated with the defective products production. According to the mathematical model, the proportion of reject depends on the rate of standard time implementation. Accordingly, the remuneration system has not to encourage the overtime work. The Russian enterprises in metal-working industries very often use the piece rate system. It is necessary to provide transition to time system with normalized task.

The share of reject is affected by equipment problems. planning and preventive systems of equipment repair and maintenance can be used for equipment problems preventing [11]. In order to update the equipment maintenance and repair system, it is necessary to develop the scheduled repair system for modern CNC machines, as well as to create internal standards for maintenance and operation of the workplaces. The standards can be based on the 5S system [12].

Since the first three recommendations are limited in application, more attention should be paid to the fourth recommendation – preventing equipment problems by improving the equipment repair and maintenance system.

4. Conclusion

A machine-building enterprise wishing to reduce the reject should pay attention to the system of maintenance, repair and operation of technological equipment; to the basis of the defect-free labor stimulation system and the organizational principles of products movement along the technological

IOP Conf. Series: Materials Science and Engineering **966** (2020) 012091 doi:10.1088/1757-899X/966/1/012091

route. That would help the enterprise successfully compete and introduce innovative technologies and equipment.

References

[1] Morozova A E, Yurakov N S and Yurakova T G 2018 Application of Shuhart control cards for statistical control of parts quality *Modern materials, techniques and technologies* **6 (21)** pp 68–72

[2] Klachkin V N and Kravtsov Yu A 2016 Detection of violations in multidimensional statistical control of technological process *Software products and systems* **3** pp 192–7

[3] Boccarusso L, Fazio D D , Durante M, Langella A and Capece Minutolo F M 2019 CFRPs drilling: Comparison among holes produced by different drilling strategies *Procedia CIRP* **79** pp 325–30

[4] Colledani M, Tolio T, Fischer A, Iung B, Lanza G, Schmitt R and Váncza J 2014 Design and management of manufacturing systems for production quality CIRP Annals - *Manufacturing Technology* **63 (2)** pp 773–79

[5] Iudin S V 2019 Problems of quality control and reliability of small-scale products of weapons and military equipments *News of Tula State University. Technical science* **12** pp 88–92

[6] Iudin S V, Protasiev V B, Podkopayev R Y and Iudin A S 2018 Methodology of calculation of information plans of statistical acceptance control on the basis of Bayesian approach *Co-temporary knowledge-intensive technologies* **11** pp 90–4

[7] Pearn W L and Hsu Y C 2007 Optimal tool replacement for processes with low fraction defective *European Journal of Operational Research* **180 (3)** pp 1116–29

[8] Gorlenko O A and Mozhayeva T P 2017 Quality assurance of machine-building production on the basis of human resources management *News of Volga STU* **9 (204)** pp 118–20

[9] Su Q, Liu L and Whitney D E 2010 A systematic study of the prediction model for operator-induced assembly defects based on assembly complexity factors *IEEE Transactions on Systems, Man, and Cybernetics Part A: Systems and Humans* **40 (1)** pp 107–20

[10] Inman R R, Blumenfeld D E, Huang N, Li J and Le J 2013 Survey of recent advances on the interface between production system design and quality *IIE Transactions (Institute of Industrial Engineers)* **45 (6)** pp 557–74

[11] Mitin A 2013 Factors influencing quality in mechanical engineering *Standards and quality* **3** pp 66-9

[12] Chiu S W, Wu H Y, Chiu Y –S P and Hwang M H Exploration of finite production rate model with overtime and rework of nonconforming products *Journal of King Saud University - Engineering Sciences* **30 (3)** pp 224–31

[13] Zhilina S, Kapitanov N, Osochenko O, Kochedykova I and Vazhdayeva A 2016 The dedicated algorithms for design and technological preproduction and information support of production process *News of the Samara Scientific Center of the Russian Academy of Sciences* **4 (3)** pp 638–44

[14] Edvabnik V G, Fartishev Y M, Gautsel A K and Kuznetsov M M 2013 The assessment of the stability of production processes in the conditions of the single and small-batch production *INTEREKSPO GEO-SIBIR* **5** pp 195–200

IOP Conf. Series: Materials Science and Engineering **966** (2020) 012092 doi:10.1088/1757-899X/966/1/012092

Modern concept analysis of failure modes and consequences for quality assurance of products and processes

E V Kalinin[1], L L Kalinina[1, 2] and O S Norkina[1]

[1] Institute of New Materials and Technologies, Ural Federal University named after the first President of Russia B N Yeltsin, 19, Mira street, Yekaterinburg, 620002, Russia

E-mail: [2] llkalinina@yandex.ru

Abstract. Human needs in certain types of products, in new goods and services are constantly developing, changing, clarifying. To meet these needs, new goods samples are being created, and the products are being improved. Production processes are being developed, new equipment, tools and techniques are being created. New technical solutions, new operations application as well asit execution methods can produce constructive errors and dangerous situations both during production and during its exploitation. The Failure Mode and Effects Analysis (FMEA) allows to predict that product lifecycle processes can functioning not properly. FMEA helps ensure the products and processes quality at the earliest stages of it design. The American approach (AIAG – Automotive Industry Action Group) and the German approach (VDA –Verband der Automobilindustrie) have now been harmonized. In 2019, the FMEA Handbook was published, reflecting the new FMEA implementation concept. Experience has emerged with a new approach. The work considers the possibilities of applying a new approach in the conditions of Russian companies, which allows to find new technical solutions.

1. Introduction

Reliable trouble-free operation of the products produced by the enterprise is a solid basis for the company's development and a good basis for further progress in its production activities. Many market participants create new products and services based on the thorough investigation, analysis and deep understanding of their current and future consumers' requirements and expectations [1]. Moreover, the most successful enterprises and organizations aim to ensure the high quality of newly developed products and their production processes at the earliest design stages, long before the batch production starts [2,3,4].

All the new equipment development and production stages affect the products quality, its relevance, level of reliability, deliveries timeliness implemented by the company [5]. The first part consists of the technical solution search and selection. Then, the construction development, design and design documentation production follow in sequence. Various inaccuracies, deviations, errors, typos and other defects can occur at each stage. The experience of creating new products shows that often the smallest deviations, which seem to be 'nothing', can lead to serious failures, refusals, accidents. Therefore, when planning the parameters of future products, choosing technical solutions, it is useful to identify their potential 'weak points' as earlier as possible. This will allow to proactively develop and improve the new design.

By designing new production processes, individual operations, it is important to explore the design options for the new products production and, if it is necessary, to plan and to implement the preventive actions. Warning is the action taken to correct the cause of a potential non-conformity or other potentially dangerous situation [6]. This approach allows timely detection of potential deviations, failures, refusals of mechanisms and equipment, in the machine-tool attachment behavior. Similarly, it is possible to provide inaccuracies, errors, omissions in the employees' actions and to build production processes in such a way as to protect the company from the appearance of situations that can produce inappropriate products.

2. Materials and methods

To solve such problems regularly in 1949, the national standard MIL-STD-1629 'Procedures for Performing a Failure Mode, Effects and Criticality Analysis' was developed and put into operation in the United States. This approach application helped to prevent and avoid many accidents and incidents with military products. The military products reliability has been significantly improved. The automobile assembly companies' representatives drew attention to this phenomenon. Their products are directly related to consumer safety. The fact is that in the automobile industry there are repeated and even mass failures of sold cars. This forces auto-assembly companies to withdraw large quantities of cars (from sellers and buyers) to eliminate the inconsistencies causes, resulting in significant financial losses. In 2014, for example, the car manufacturer Mazda recalled 42.000 cars worldwide due to the failure of the fuel system [7]. In 2018, Toyota decided to recall about 1.7 million cars (including 84.000 cars in Russia), due to possible airbag faults [8]. The failure prevention technique application could have a tangible economic effect in these situations.

So in 1993 a reference guide 'Potential Failure Mode and Effects Analysis' was developed and published (FMEA) by the Automotive Industry Action Group (AIAG) specialists for the US automotive industry and its suppliers. [9] It helped to identify a significant proportion of future problems and study them before serial production began. This significantly reduced business losses in subsequent periods.

The method is logical, understandable and easy to use [10, 11]. It is therefore internationally recognized and widely distributed in many countries in the world in the production types variety. The method is constantly being improved on the basis of empirical date and normative documents changes. The FMEA fourth edition is currently used [12]. The FMEA approach is included in the risk management toolbox [13, 14].

The proposed methodology makes it possible to predict with a high reliability level adverse events and situations of the product possible inappropriate functioning during the life cycle. The FMEA approach application in production practice helps to ensure the future products high quality and reliability, as well as the high quality of new values creation processes at the earliest (development) stages, long before the product is released. Exactly this fact explains the widespread the FMEA approach using for analyzing new design, technology and organizational solutions. Similar approaches are used to analyze risks in the development of management decisions.

It is important to understand that carrying out FMEA in an organization is always a team creative work [3]. Its important component is the application of the A. Osborn's brainstorming method [15]. Brainstorming is an effective way to solve problems by stimulating creative people activity, in which participants in the team work are invited to offer as many options as possible to solve the identified problems including solutions of the most unexpected and fantastic [2, 11, 15]. Conducting a brainstorming session creates a collective creativity atmosphere for the participants and helps the team to realize a synergistic effect during the construction (or research process) discussion. Synergistic effect is considered in the understanding of potential problems, deviations, refusals and other adverse events in the implementation of the technical solution in practice. Exactly the synergy of knowledge and experience of the team participants provides the completeness, reliability and objectivity of the results of such analysis.

The technique FMEA has a high foresight potential. For reusable spaceship, the Shuttle the hidden possibility of a start-up accident in certain weather conditions was identified during analyzing the various external factors on systems impact. In order not to change flight plans the report was ignored. As a result, the accident occurred (according to a scenario predicted by specialists) [16].

The FMEA technique is applied:

– in the analysis of new technical solutions, new system designs, new constructions or modifications;

– when using a known construction, a known technical solution in the new conditions of use and operation;

– when training of new processes (operations) or applying a well-functioning process (operation) in new conditions, to solve new problems;

– in the analysis of designed systems and other similar situations.

A team of specialists (designers, technologists, direct labour, assemblers, testers) is formed to carry out the analytical research. This group must have a good understanding of the design various aspects and its future functioning. In specific projects, other important experts may be engaged for discussion. The team consists of 5–9 participants. It is only with such participantsamountthe synergy effect can be realized in the discussion.

Debates and fierce disputes are not allowed during the discussions. It is proposed only to analyse, discuss, develop, supplement various situations that may arise in production and during operation, but not necessarily. It is important that each team member has the opportunity to express his or her own opinion.

During the work, the team consistently considers, examines the following questions:

1. What unforeseen, unplanned, unexpected events can happen to the subject during its regular functioning? What kind of failures, refusals may appear?

2. What are the consequences of each of the unplanned events/refusals may appear?

3. What are the reasons that can lead to each of the unplanned events/refusals?

At the same time, it is important to bring the discussion to a concrete outcome. Record clearly defined results. As the discussion results, the team draws up a list of potentially possible adverse events/failures. 'Base of such events/failures' is generated. Then each of these events is examined in terms according to the consequences that can be caused by the events. A list of possible consequences for each event is being compiled. The next step is to investigate the reasons that may cause these adverse events. A list of reasons for possible adverse events is being compiled.

Next, a quantitative assessment is made of possible events. Now the team of specialists must be in a different status. The team must assess all predicted events according to the technique described in [9, 12].

For each possible adverse event/refusal, an event consequences severity assessment is required. The appropriate corresponding characteristic is called 'significance' and is denoted by the symbol S. Tables have been developed for quantification. The significance value of S can take integer values from 1 to 10. Each member of the team gives his or her knowledge and practical experience-based opinion. The arithmetic mean of all scores is taken as the total result of the evaluation.

As the assessment result, a possible adverse event from each consequence is assigned a value S. For further analysis, the highest estimate of the S significance is taken as the considered event effects severity value.

Similarly, the occurrence probability of the reason that may cause the event is evaluated. This value is called 'occurrence' and denotes O symbol. An assessment is also being made of the ability to timely see the situation and prevent the consequences adverse development. The corresponding characteristic is called 'detection' and is denoted by the symbol D. The values of the parameters O and D can vary discretely between 1 and 10. The value tables in the FMEA Manual [12] have also been developed for their evaluation.

The team's work result is the priority risk number determination (PRN) [9,12] for each possible adverse event.

$$PRN = S \cdot O \cdot D \tag{1}$$

When predicting potentially dangerous events, a large number of potential problems may be identified. But it is impossible to deal with all issues at the same time. The PRN value is used to decide on the order of the corresponding warning actions development. The PRN value reflects the relative importance of each situation in its potential hazard terms. The higher the PRN value is, the higher is the situation risk in comparison with others.

Using the obtained PRN values, designers can plan and carry out activities to improve the structures even before making the prototypes. In practice, this possibility saves time considerably. Similarly, technologists can explore new operations and technology chains before they are organized, which also saves resources. First of all, situations with high PRN values are considered. Then with medium and small value are taking into consideration. From experience, it has been proposed to set the boundary PRN value to 120. It is considered that events with lower PRN values can be ignored.

In the German automotive industry, Verband der Automobilindustrie (VDA) developed its own approach to predicting potentially dangerous, emergency situations. Two approaches to FMEA implementation have now been agreed upon. In 2019, as the joint work result of American and German specialists, FMEA Handbook [17] was prepared and published, which reflects the new FMEA concept implementation. This document was published on June 2019. It reflects the current level of risk analysis understanding in the automotive industry.

3. Results and discussion

The method is described as the seven steps sequence [17]:

1. Preparation and planning of actions.
2. Analysis of the structure and interaction of the elements.
3. Analysis of the structure and impact of functions.
4. Analysis of the structure refusals.
5. Quantitative risk assessment.
6. Improving design, process, system.
7. Documenting the results of work.

In content terms, these steps largely coincide with the traditional approach to the potentially dangerous events analysis [18].

Criteria for determining the design and process FMEA significance are determined. In particular, the FMEA should focus on the buyer requirements, as well as other interested parties. For the FMEA process, the significance criteria are focused on the producer requirements.

An important change in the methodology was the complete rejection of the PRN value calculation (as the p multiplication of the S, O, D parameters). Instead of such calculation, it is proposed to use special tables. They determine the actions (AP) priority depending on the parameters S, O, D combination. The approach to quantification of the parameters themselves remains unchanged.

It is proposed to use three levels of priority actions (AP) to prevent the product or process failures: high (H), medium (M) and low (L).

1) high level – actions required,
2) medium level – actions recommended,
3) low level – actions can be taken.

In other words, if the AP value is high, improvement actions are mandatory. At medium level - improvement actions are necessary and useful. If the level is low, improvement actions are recommended.

This approach changes the decision-making tactics of FMEA.

Three-level priority separation formalizes the necessary steps algorithm but makes it difficult to choose the specific actions direction. Of the many possible equivalent situations due to the combination of parameter values (S, O, D), it is difficult to build a warning actions sequence.

IOP Conf. Series: Materials Science and Engineering **966** (2020) 012092 doi:10.1088/1757-899X/966/1/012092

The absence of an PRN value estimate may lead to additional rating votes on the questions: 'What is more important and where to start improvements'? The discussions on these issues increases the situation uncertainty. This can lead to mistakes and authoritarian decisions with high risks.

In general, the usefulness of the changes made in FMEA will be assessed objectively after learning from the new approach.

References

[1] Hill N, Self B and Rochet G 2004 *Measuring consumer's satisfaction according to ISO standard 9000:2000* (Moscow: Publishing house 'Technologies') p 192
[2] Evans J R 2005 *Total Quality Management, Organization, and Strategy* 4th ed (South-Western, part of the Thomson Corporation) p 467
[3] Rozental R. 2010 FMEA Method as a Way to Improve Products Quality *Electronics: Science, Technology, Business* **7** pp 90–5
[4] Rosno M I 2000 How do you learn to look forward? Implementation of FMEA *Methodology Quality Management Methods* **6** pp 25–8
[5] Kochetkov E P 2007 *Dialogues of the consultant on prevention of failures* (Nizhny Novhorod, LLC SMC 'Priority') p 104
[6] ISO 9000:2005. Quality management systems – Fundamentals and vocabulary
[7] https://news.drom.ru/Toyota-66828.html (accessed on 01.06.2020)
[8] https://russian.rt.com/article/26544.html (accessed on 01.06.2020)
[9] 1993 Potential Failure Mode and Effects Analysis (FMEA). Reference manual. First Edition Issued February 1993
[10] The Six Sigma Memory Jogger ™ II A Pocket Guide of Tools for Six Sigma Improvement Teams
[11] Marsh J 1998 *The continuous improvement toolkit* (London B.T.Batsford LTD)
[12] Potential Failure Mode and Effects Analysis (FMEA). Reference manual. First Edition 2008
[13] ISO 31000:2018 'Risk management – Guidelines'
[14] ISO 31010:2019 'Risk management – Risk assessment techniques'
[15] Altshuller GS 1979 *Creativity as an Exact Science* (Moscow: Soviet Radio) p184
[16] http://astronaut.ru/booksase/article/ar162.htm (accessed on 01.06.2020)
[17] Failure Mode and Effects Analysis. FMEA Handbook 2019 First Edition
[18] Amyalev F 2019 New FMEA standard: joint publication AIAG and VDA *Quality management methods* **11**

IOP Conf. Series: Materials Science and Engineering **966** (2020) 012093 doi:10.1088/1757-899X/966/1/012093

Advanced standard operating procedures for survey of buildings and structures of metallurgical production

M S Kuznetsov[1, 2], V G Dubinina[1] and Yu V Patrushev[2]

[1] Ural Federal University named after the First President of Russia B N Yeltsin, Nizhniy Tagil Technological Institute, 59, Krasnogvardeyskaya str., Nizhniy Tagil, 622000, Russia

[2] Ural State Mining University, Russia, 30 Kuibyshev Str., Yekaterinburg

E-mail: [2] kms.rf@rambler.ru

Abstract. This article presents modern methods for determination of the depth of various types of foundations at the metallurgical industry facilities. It was decided to use modern geophysical research methods, such as seismic exploration methods in order to solve the tasks for determination of depth of pile foundations and deep foundations under conditions of operating production metallurgical site. Geometric parameters of foundations can be measured using engineering seismic equipment, and three-component measurement data shall be processed by means of specific processing flow. The use of coherency procedures makes it possible to separate the dedicated phases of reflection from the foundation bottom in the wave pattern. Reproducibility of results of foundation bottom laying depth with actual depth was checked using engineering seismic recording systems by uncovering of foundation check pit-holes at the blast furnace complex facilities No. 7 of JSC EVRAZ NTMK. The measurement accuracy of pile foundations and spread foundations equal to 0.5 meters was achieved by means of seismic techniques as a result of the measurements conducted at the metallurgical production facilities.

1. Introduction

As it is known, partial uncovering of foundation is required for surveying of the existing foundation of buildings and structures. Pit-holes are arranged in the proximity of foundation for this purpose. A pit-hole is prepared with such a section that normal conditions are provided for its trenching to the required depth, as well as conditions are provided for its inspection. The length of exposed site of foundation must be sufficient for determination of type and state estimation of its structure.

According to the Russian state standard GOST 31937–2011 [1], check pit-holes are dug depending on local conditions from the outside or inside of the foundations. The pit-holes are located basing on the following requirements:

– one pit is dug for each type of structure in each section of the foundation in the most loaded and unloaded locations;

– if there are mirrored or repeated (in plain view and along the outlines) sections, all the pit-holes are dug in one section, and one or two pit-holes are dug in other sections in the most loaded locations;

– in places where it is planned to install additional intermediate supports, one pit-hole is dug in each section;

– in addition, two or three pit-holes are dug for each building in the most loaded locations against the face of the wall where there is an output.

– in case of deformations of walls and foundations, it is obligatory to dig the pit-holes in these places and the additional pit-holes must be established in the process of operation in order to determine boundaries of soft soils of bases or boundaries of foundations, that are in unsatisfactory condition.

– the depth of the pit-holes located in the vicinity of the foundations must exceed the bottom laying depth by 0.5–1 m.

However, usually it is not possible to achieve normal conditions for inspection in case of significant groundwater inflow. A surface drainage from a pit-hole is allowed only for a very short time. Otherwise the soil particles will be removed from under the bottom of the foundations together with water, that can lead to their differential settlement. Therefore the pit-hole shall be dug down to the planned depth immediately prior to inspection of the foundation by the specialist.

It is very difficult to meet all necessary requirements of GOST 31937 [1] due to a large amount of utility lines, compliance with the industrial safety requirements as well as continuous production process under the conditions of production metallurgical site. Another factor, complicating the process of examination of the foundations of production metallurgical facilities (blast furnace complex facilities, bin trestle, etc.) involves their large depth of foundation which is 6–8 m from the ground level. Uncovering and backfilling of the pit-hole of such foundation, especially water-flooded one, is a labor intensive and important process.

The pile foundations, the depth of which cannot be measured by the method of pit sampling, shall be separately noted. In the open pit-hole of such foundation, it is possible to determine the mattress as well as pile arrangement and diameter, but the pile depth will remain unknown, although it can reach 25 m.

Finally, in the process of renovating metallurgical production facilities, there is a need to specify the depth of all foundations of the facility instead of selective amount in open pit-holes, which cannot be performed at the production site.

It was decided to use modern geophysical research methods, such as seismic exploration methods [2–5] in order to solve the tasks for determination of depth of pile foundations and deep foundations under conditions of operating production metallurgical site.

2. Equipment and measurement procedure

For determination of the geometric parameters of substructure of the foundations at the metallurgical production facilities, in particular, at the blast furnace foundations No. 6 and 7 (JSC 'EVRAZ Nizhniy Tagil Iron and Steel Works'), equipment of seismic geophysical survey has been used successfully. TELSS-3 telemetric seismic recording system (LLC 'GEOSIGNAL', Moscow) and SGD-SEL engineering seismic recording station (LLC 'SibGeofizPribor', Novosibirsk) were used for surveying. GS-3C three-component seismic receivers (Geospace Technologies Eurasia, Ufa) served as seismic signal receivers. Both sets of equipment show similar measurement results, although TELSS-3 seismic recording system is digital, and SGD-SEL engineering seismic recording station (Figure 1) is analog.

Figure 1. SGD-SEL engineering seismic recording station.

Seismic measurements at the blast furnace facility foundations were carried out by means of location spread of streamer (sounding locations). Triaxial geophones in the quantity of 8 pcs. were installed on the horizontal surface of foundation when using such method of geophysical measurements. During the warm season, seismic receivers were adhered to the surfaces with clay solutions and adhered by freezing at negative temperatures: in both cases, instruments were securely fixed and received a seismic wave. The following monitoring system was chosen for measurements: receiving point interval (RP) is 0.1–0.5 meters, near trace offset of source point (SP) from a receiving point is 0.1 meter, and far trace offset is 0.4 meters. Fluctuations in P-waves (primary waves) were excited by applying hammer blows in source points, and S wave (secondary wave) is excited by applying the second set of blows on the side face of foundation. The seismic recording station was operated in the signal accumulation mode - 8 blows at each sounding location. The signal sampling rate was selected to be minimal – 0.125 ms, the record length is 1,024 readings, and seismic data were presented in the SEG-Y format.

3. Wave field interpretation and processing procedure

AZON system of geophysical information processing was used (developed by A E Zudilina, Ural State Mining University, Certificate of State Registration of Computer Program No. 201317598) for digital processing of field data [6–10].

This processing system is based on the idea of three-dimensional seismic location. One of the elements of this system is the procedure of wave field pre-stack depth migration according to the system of azimuthograms with various combination 'source-receiver'.

The physical justification for the possibility of segregation of the foundation bottom is the high frequency content of the analyzed wave field with the upper boundary of the degree of 2.500–3.000 Hz. This frequency range at the set speed of elastic waves provides a wavelength of about 1.2–1.5 meters. According to theoretical background, vertical resolution capability of the method is ¼ of the wavelength equal to 0.3–0.4 meters, i. e., the depth error of substructure of the foundations according to geophysical data is ± 0.3–0.4 m.

According to the elastic properties, the foundation concrete and the country rock are different, which makes it possible to separate the reflected (diffracted) wave from the bottom of foundation. The bottom of foundations is the source of both reflected and diffracted wave, and the foundation itself is a waveguide. In any case, the bottom half of the foundation must appear in the wave field as the negative phase of the reflected wave with a subvertical and downward directed orientation of directivity graph of excitation source.

A standard processing flow of individual sounding is generated for processing of the data of seismic measurements.

1. Input of shooting geometry for subsequent correction of radial component phases taking into account the approach direction of target reflected wave in a horizontal plane.
2. Oversampling of initial seismic records at 0.1 ms intervals. The use of this procedure is related to the necessity of resolution enhancement as well as assessment precision of the target reflection depths. This reconstitution ensures increase of Nyquist frequency from 4.000 to 5.000 Hz. The following stacking of eight sensors for each source point allows to expand the usable bandwidth by virtue of statistical effect.
3. Accounting of geometric spreading.
4. The high-pass filtering for the purpose of relatively low-frequency coherent-noise suppression. The cut-off frequency is 400–800 Hz depending on the specific acceptance and excitation conditions. The high noise level caused mainly by the metal parts of foundation requires rigid high-pass filtering with the cutoff frequency close to the upper boundary of the specified range. This may also lead to usable bandwidth reduction as well as occurrence of high-frequency 'ringing effect' on the records.
5. Deconvolution for the purpose of bandwidth expansion by increasing its upper boundary (within the range of up to 5.000 Hz).
6. Corrective filtering.

7. Recalculation of traces into the depth scale using the P-wave velocity.

8. Normalization of the trace average amplitude prior to summarization and determination of coherency.

9. Correction of phases of horizontal component traces. The phases change by 180 degrees or remain unchanged depending on the relative position of the excitation source and the receiving point for horizontal component traces (approach direction of the target reflected wave). A change of trace polarity including vertical components was observed in some cases due to rereflection from vertical flanges and units of construction installed in the foundation. Such traces were rejected or used in case of reverse polarity when calculating multifold composite traces upon condition of waveform retention during rereflection.

10. Calculation of multifold composite traces.

11. Calculation of coherence functions of multifold composite traces. This procedure was used for enhancement of validity of separation of the dedicated phases of reflection, since their amplitude is not always significantly higher than the amplitude of the coherent noise. Evaluation of coherence function is unbiassed, that is, in case of its absence in the current reflection interval (depth), its value is close to zero. Availability of target reflection is determined confident enough according to the sharp coherence improvement in almost all practices of this procedure. The peak value of coherency in the target range usually significantly exceeds the values in the remaining intervals, except for the first arrivals (muting was used for better displaying of graphic chart in the area of the first arrivals).

It should be noted that due to rereflection, the coherency of the first arrivals is often even lower than for the target reflecting boundary, which indicates a high efficiency of the procedure.

The minimum-phase deconvolution, converting the traces to zero phase and usually used for the purpose of more precise determination of the depth by the maximum of the multifold composite trace, was not used because of the shape (strongly distorted by rereflection) of the probe pulse recorded in the first arrivals.

For this reason, the depth to the target boundary (bottom of foundation) was determined according to the beginning of the negative phase of reflected multifold composite trace wave (Figure 2).

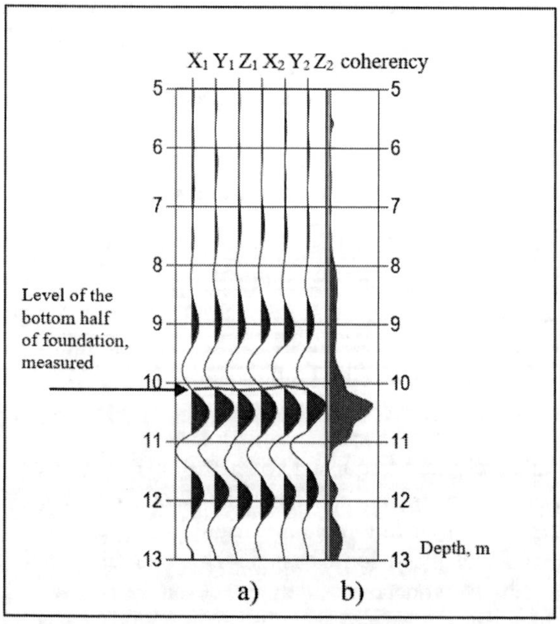

Figure 2. Example of wave field (a), coherence function (b) for determination of the foundation vertical dimensions.

The turnover of the coordinate system of the sensor axes by 45 degrees is connected with the necessity to obtain traces of horizontal components of approximately equal amplitude for each source point. This shall ensure the high noise resistance of depth migration. Also, if it is impossible to use depth migration, the use of component traces of consistent amplitudes with the selected orientation of sensitive axes allows to use the procedure of group coherency between multifold composite traces more efficiently.

4. Measurement results

Reproducibility of results of foundation bottom laying depth with actual depth was checked using engineering seismic recording systems by uncovering of foundation check pit-holes at the blast furnace complex facilities No. 7 of JSC EVRAZ NTMK. The measurement results are given in Table 1.

Table 1. Measurement results of foundation bottom laying depth using engineering seismic recording systems with actual depth at the blast furnace complex facilities No. 7 of JSC EVRAZ NTMK.

No.	Name of facility	Depth of foundation measured by means of seismic survey, N1, mm	Depth of foundation measured in check pit-hole, N2, mm	Maximum discrepancy, mm
1	Coke conveyor line. Kl-2 conveyor gallery	3.500–3.600	3.800	–300
2	Coke conveyor line. PS-3a transfer station	2.500–2.600	2.400	200
3	Stove block chimney	3.100–3.200	3.030	170
4	Stove block	3.200–3.300	3300	–100
5	Central unit of blast furnace	2.900–3.000	2.750	250
6	Casthouse No. 1	3.300–3.400	3.300	100
7	Cable tray system to the building of blast-furnace control panel	2.400–2.500	2.400	100
8	Dust collector with gas ducts Foundation of support system of axial cyclone	2.000–2.100	2.100	–100
9	Scrubber with droplet separator	9.800–10.000	9.500	500
10	Hoist room of dust collector with gas ducts	8.500–8.700	8.600	100
11	Bin trestle	5.100–5.300	5.500	–400

5. Conclusions

Taking into account a large amount of utility lines and continuous production process, the use of modern geophysical research methods is a very relevant and top requested method for determination of the depth of various types of foundations under the conditions of production metallurgical site.

The measurement accuracy of pile foundations and spread foundations equal to 0.5 meters was achieved by means of seismic techniques as a result of the measurements conducted at the metallurgical production facilities. This accuracy is achieved with the use of specific procedures of field data processing as well as using the idea of three-dimensional seismic location and statistical approach to interpretation of the results obtained.

References

[1] Russian state standard GOST 31937-2011 2014 Buildings and Constructions. Rules of Inspection and Monitoring of the Technical Condition (Moscow: Standartinform) p 89
[2] Gurvich I and Boganik G 1980 *Seismic Exploration* 3rd ed (Moscow) p 551

[3] Sheriff R and Geldart L 1984 *Exploration Seismology* vol 2 Data-processing and interpretation (Cambridge University Press Cambridge) p 240

[4] Ahmed K A, Schwarz B and Gajewski D 2015 Application of the 3D common-reflection-surface stack workflow in a crystalline rock environment *Geophysical Prospecting* **63 (4)** pp 990–8

[5] Hanson D, Haramy K and Neil D 2000 Seismic Tomography Applied to Site Characterization *Geo-Denver Conference* (Denver, CO)

[6] Junlong Shang, Xianwei Luo, Feng Gao, Jianhua Hu and Keping Zhou 2012 Advanced Predication of Geological anomalous Body Ahead of Laneway Using Seismic Tomography Technique *ISSSE in China* pp 324–30

[7] Bodin T, Sambridge M and Tkalčlć H 2012 Transdimensional inversion of receiver functions and surface wave dispersion, *Journal of geophysical research: solid earth* vol **117** issue **B2** CiteID B02301

[8] Obrebski M, Allen Richard M, Zhang Fengxue, Pan Jiatie, Wu Qingju and Hung Shu-Huei 2012 Shear wave tomography of China using joint inversion of body and surface wave constraints, *Journal of geophysical research: solid earth* vol **117** issue **B1** CiteID B01311

[9] Rawlinson N, Pozgay S and Fishwick S 2010 Seismic tomography: a window into deep earth *physics of the earth and planetary interiors* vol **178** issue **3–4** pp 101–35

[10] Ritzwoller M H, Fan-Chi L and Weisen Sh 2011 Ambient noise tomography with a large seismic array *Comptes rendus - Géoscience* vol **343** issue **8** pp 558–70

Instrument ofthe work breakdown accounting in the high-tech production

O O Podoliak[1, 2], Yu N Markova[1], T A Mineeva[1] and N A Kuznetcova[1]

[1] Institute of New Materials and Technologies, Ural Federal University named after the first President of Russia B N Yeltsin, 19, Mira street, Yekaterinburg, 620002, Russia

E-mail: [2] o.o.podoliak@urfu.ru

Abstract. At present, in mechanical engineering there is a tendency of small-scale production due to the peculiarities and requirements of the market. At the same time, machine-building enterprises have to account the costs of produced products accurately (both for the state defense order and for the open market). The problem of the product labor intensity calculation becomes urgent, as it is used as a basis for of the product full cost calculation. Taking into account the technology changes, equipment, peculiarities of production program formation (small or medium series), using the existing approaches, work norms manuals becomes impossible. The investigation presents an improved algorithm for the work norm calculating using for the main workers in the high-tech production. A feature of the proposed approach is taking into account the share of passive observation during the technological operations in conditions of small-scale production in the conditions of the high-tech production. The article presents the results of the proposed method testing. The approbation was realized at the industrial enterprise in the conditions of the small-scale production. Proposed algorithm for work norms calculation can be used in automation of labor rationing at the industrial enterprise.

1. Introduction

The current state of Russia's economic system is characterized by an increase of the high-tech products share in the machine-building industry production structure. Machine-building enterprises face the necessity for increasing the share of civilian products in the production structure, which is achievable mainly through the development and sale of high-tech products. Thus, enterprises should not only organize production, but also ensure competitiveness of produced high-tech civilian products [1]. One of the competitive factors is the price [2], which is directly depends on the production costs of a particular product. The biggest part of the machine-building enterprises use product laboriousness or main part of the wages of the basic workers to calculate the cost of a product (for example, as a basis for the indirect costs dividing) [2, 3]. Another current trend is equipment replacement aimed at improving plant performance [4, 5]. In consideration of the foregoing premises, it has become necessary to update the approach of the work breakdown estimating, taking into consideration the peculiarities and capabilities of the new equipment and the changed organizational and technical conditions for the high-tech products production, and increased productivity [6,7]. The task of the main workers work quota setting system forming seems to be relevant. This system has to allow to objectively estimate the level of work breakdown for the high-tech products production in the production organization modern conditions. Thus, the purpose of this investigation was to form a work

IOP Conf. Series: Materials Science and Engineering **966** (2020) 012094 doi:10.1088/1757-899X/966/1/012094

breakdown calculation approach for the high-tech products production according to the modern organizational and technological conditions.

2. Materials and methods

While the basic methods and tools of work quota setting are well known and widely used, it should be recognized that the current realities of production process organization, technology and equipment require the mainstreaming of work breakdown assessment approaches [8–12]. As a part of the investigation, a series of working time records was carried out, which revealed an increase of the equipment visual observation share in the work breakdown structure of the main worker during the high-tech products production.

It should be mentioned that the share of the equipment visual observation is steadily increasing as the production technical level improves [13, 14].

The total time of equipment visual observation is classified as [15]:

–time of active visual observation – time during which the worker closely observes the equipment, the process chart maintenance, the set parameters maintenance and the product quality. During this time, the worker does not perform physical work, but his presence in the workplace is mandatory;

–time of passive visual observation – time during which there is no need for constant visual observation of the equipment or the technological process, but the worker carries out it due to the lack of other work.

Reducing or using the passive visual observation time for perform other necessary work can be a significant reserve for improvement of productivity and labor efficiency [16].

During the worktime analyzing and the standard time calculating, the concepts of overlapping and non-overlapping time are also identified:

– the overlapping time – the equipment automatic operating time during which the worker realizes some operations (not related to the current process chart). Overlapped time can be direct manufacture time (active observation) and auxiliary time, as well as time related to the other worktime types;

– the non-overlapping time – time for auxiliary operation and workplace service operations (the equipment doesn't work).

The unit standard time consists of the necessary categories of worktime included in the calculation, taking into account the possibility of the additional works performing, rest time and personal needs (personal time), time for the workplace maintenance during the automatic operation of the equipment (the overlapping time).

The following formulas are used for standard time calculate within the proposed methodology.

The calculated unit standard time is presented per piece of product.

Batch is the number of products processed according to the process plan.

Billet – unit to be processed (can consist of several products).

$$T_{u-c} = T_u + T_p(N * n)^{-1} \tag{1}$$

$T_{u\text{-}c}$ – the unit standard time for one produced unit, min.

T_u – the standard time for one produced unit, min.

T_P – preparatory time, min. for batch.

N – the number of procurements in the batch, pcs.

n – quantity of products in the billet, pcs.

$$T_u = T_c(1 + K_o * 100^{-1}) + T_c K_{in} \tag{2}$$

T_c – cycle time for one manufacturing operation, min.

K_o – the overlapping time coefficient, %. It takes into account the possibility of additional functions, rest time and personal needs (personal time), workplace maintenance, during the passive visual observation time. $K_o = 0$ if $K_1 > 1$

$$K_1 = T_o\left(T_a + T_m + T_p\right)^{-1} \tag{3}$$

T_a – time for additional work (includes necessary documentation filling, billet marking, work with the personal computer, etc.), min.

T_m – time for workplace maintenance, min.

T_p – rest time and personal needs (personal time), min.

If $K_1 \leq 1$ then the formula 8 is used.

$$K_o = T_a + T_m + T_p - T_{n-o} \tag{4}$$

For the coefficient calculation all parameters are taken as a percentage (%) of the cycle time.

K_{in} – coefficient taking into account the time of interruptions established by the technology and production process organization. It is calculated as a share of the time of interruptions established by the technology and production process organization in the cycle time.

3. Results

The proposed methodology was tested at the production site.

The standard time elements were used in the time calculating, this data are common for the investigated enterprise:

rest time and personal needs (personal time) – 6% to the cycle time;

workplace maintenance time – 5% to the cycle time.

Example 1. Based on the results of the working time records the following calculated data were obtained (table 1).

Table 1. Time by Category.

№	Type of time	Value, min.
1	Preparatory time for one batch (T_p)	15
2	The standard time for one produced unit (T_u)	The sum of points 2.1 и 2.2
2.1	Part-cycle time	According the cycle time
2.2	Auxiliary time	15
2.2.1	Auxiliary operation time	10
2.2.2.	Component inspection time	5
3	time for additional work (includes necessary documentation filling, billet marking, work with the personal computer, etc.) (T_a)	6.2
4	Time of interruptions established by the technology and production process organization	10% from the standard time

For this operation the ratio of active visual observation and passive visual observation is shown in table 2.

Table 2. Proportion of observations in % to part-cycle time.

Type of observations	Value
active visual observation	15
passive visual observation	85

Calculation of the norm per product is performed in accordance with the above procedure.

IOP Conf. Series: Materials Science and Engineering **966** (2020) 012094 doi:10.1088/1757-899X/966/1/012094

Table 3. Calculation of standard time per product.

№	Standard time element	unit of measure	Value
1	Preparatory time	min.	1.9
2	Quantity of products in the billet	pcs.	8
3	Standard time for one produced unit, including	min.	47.0
	Machine time	min.	32.0
	Auxiliary time	min.	15.0
4	Passive visual observation	min.	27.2
5	Passive visual observation share	%	57.9
6	Additional work time share	%	1.65
7	The overlapping time coefficient	–	0.00
	Checking the condition	%	4.58
8	Time of interruptions established by the technology and process organization	min.	4.70
9	Cycle time for one manufacturing operation	min.	51.70
10	The unit standard time for one produced unit	min.	53.58

The example 2.

The standard time elements were used in the time calculating:

rest time and personal needs (personal time) – 6% to the cycle time;

workplace maintenance time – 10% to the cycle time

Data of the investigated enterprise are given.

Based on the results of the working time records the following calculated data were obtained (Table 4).

Table 4. Time by Category.

№	Type of time	Value, min.
1	Preparatory time for one batch (T_p)	12
2	The standard time for one produced unit (T_u)	The sum of points 2.1 и 2.2
2.1	Part-cycle time	According the cycle time
2.2	Auxiliary time	12
2.2.1	Auxiliary operation time	8
2.2.2.	Component inspection time	4
3	time for additional work (includes necessary documentation filling, billet marking, work with the personal computer, etc.) (T_a)	18
4	Time of interruptions established by the technology and production process organization	10% from the standard time

For this operation the ratio of active visual observation and passive visual observation is shown in table 5.

Table 5. Proportion of observations in% to part-cycle time.

Type of observations	Value
active visual observation	60
passive visual observation	40

Calculation of the norm per product is performed in accordance with the above procedure.

Table 6. Calculation of standard time per product.

№	Standard time element	unit of measure	Value
1	Preparatory time	min.	12
2	Quantity of products in the billet	pcs.	1
3	Standard time for one produced unit, including	min.	62.0
	Machine time	min.	50.0
	Auxiliary time	min.	12.0
4	Passive visual observation	min.	20
5	Passive visual observation share	%	32.3
6	Additional work time share	%	29.03
7	The overlapping time coefficient	–	12.77
	Checking the condition	%	0.72
8	Time of interruptions established by the technology and production process organization	min.	7.44
9	Cycle time for one manufacturing operation	min.	77.36
10	The unit standard time for one produced unit	min.	89.36

4. Discussion

The article proposes an algorithm for standard time calculating for the main workers which takes into account the overlapped time. The generated algorithm allows to carry out calculations with a high degree of accuracy, taking into account the works performed during the machine time. The obvious advantage of this methodology (provided that the information system of the enterprise is used as a source of initial data) is the possibility to calculate standard time, duration rates, work breakdown individually for each detailed operation. It should be noted that the proposed methodology allows to ensure operativeness of calculations (when using automated calculation systems, for example, software).

The developed standard time mathematical model assumes the statistical data availability for estimation of the value of preparatory time in the established organizational technical conditions, as well as the rest time and personal needs (personal time) and the workplace maintenance. The existence of such a base, developed and adopted standards is undoubtedly a limitation for the rapid implementation of the methodology.

5. Conclusion

The high quality of standards and standard time applied in production enterprises ensures the efficiency of production organization and labor organization (the standard time is the basis for planning); remuneration organization of (the standard time is the basis for wage calculation); production planning; determination of production costs (the standard time is an indirect condition for indirect costs distribution). At the same time, most production enterprises use outdated standard time, which do not take into account changed organizational and technical conditions of production, capabilities of equipment, etc. Thus, the creation of a methodology for standard time calculating and ensuring the automation of its implementation will increase the efficiency not only of the production process, but also of the enterprise as a whole. The use of the proposed methodology will make it possible to increase the relevance and efficiency of organizational and economic data due to more accurate calculations, which in turn will affect the efficiency of the department and the enterprise as a whole.

References

[1] Kharkova K S and Levichev A N 2019 Influence of economic sanctions on the basic macroeconomic indices of Russia *International journal of humanities and natural sciences* **12-3 (39)** pp 174–9 DOI: 10.24411/2500-1000-2019-11959

[2] Semenov V P and Vasiliev A M 2019 Problems of determining manufacturing costs for industrial products and ways to solve them *Proceedings of 2018 17th Russian Scientific and Practical Conference on Planning and Teaching Engineering Staff for the Industrial and Economic Complex of the Region, PTES 2018* 8604223 pp 229–31 DOI: 10.1109/PTES.2018.8604223

[3] Fukalova L L 2018 Reserves of labor productivity growth on the basis of application of innovations *Economics* **7**

[4] Lapkina I and Malaksiano M 2018 Elaboration of the equipment replacement terms taking into account wear and tear and obsolescence *Eastern european journal of advanced technologies* **3 (93)** pp 30–9

[5] Fedulova I, Voronkova O Y, Zhuravlev P, Gerasimova E, Glyzina M and Alekhina N 2019. Labor productivity and its role in the sustainable development of economy: on the example of a region, *Entrepreneurship and Sustainability Issues* **7 (2)** pp 1059–73 DOI: 10.9770/jesi.2019.7.2(19)

[6] Kosyakova L N abd Popova A L 2017. Tasks of labor productivity increase in Russia and ways of their solution *News of the St. Petersburg state agrarian University* **3 (48)** pp 153–7

[7] Kuchina E V and Tashev A K 2017. Methodological approaches to the assessment of labor productivity at the micro level *Bulletin of the South Ural state University. Series: Economics and management* **11 (2)** pp 42–7

[8] Bychin V B and Novikova E V 2018 Labor rationing as an element of effective internal management in modern conditions *Labor economics* **5** pp 11-86 DOI: 10.18334/et.5.1.38710

[9] Omelchenko I B 2018 Labour rationing in the Russian Federation as the most important tool for increasing labour productivity *Labour organization and regulation: science, education, practice* pp 203-16

[10] Yadransky D N, Latypov R T and Chumak E V 2019 Transformation of labour rationing: realities of the digital economy *Proceedings of the 1st International Scientific Conference 'Modern Management Trends and the Digital Economy: from Regional Development to Global Economic Growth' (MTDE 2019)* DOI: 10.2991/mtde-19.2019.2

[11] Afonchenko D A 2014 Defects and shortcomings of production rationing *Social sciences publishing house: JSC Academy of Management (Engels)* pp 3–12

[12] Topchiy D V, Bolotova A S, Zelentsov A A, Vorobev A S and Atamanenko A V 2019 Technical rationing of the construction technology of reinforced concrete floor slabs using non-removable empitness-liners International *Journal of Civil Engineering and Technology* **10-2** pp 2160–66

[13] Lukyanchikova T L and Yamschikova T N 2017 Theory and practice of flow work norm setting with the application of labour time photography *Messenger of agrarian science* **6 (69)** pp 132–41

[14] Zondo R W D 2020 Influence of a shop floor management system on labour productivity in an automotive parts manufacturing organisation in South Africa *South African Journal of Economic and Management Sciences* **23 (1)** p 3269

[15] Rofe A I 2016 *Organization and Labor Standardization* (Moscow: KnoRus) p 222

[16] Lutchenko V G, Khorev A I, Khorev I A and Grigoryeva V V 2019 Analysis of factors affecting labor productivity *Journal of the Voronezh state university of engineering technologies* **3 (81)** pp 368–74

IOP Conf. Series: Materials Science and Engineering **966** (2020) 012095 doi:10.1088/1757-899X/966/1/012095

Comparative analysis of cutting fluids

I V Ershova[1, 2], N E Kalinina[1], E V Cherepanova[1] and D F Fayskhanov[1]

[1] Institute of New Materials and Technologies, Ural Federal University named after the first President of Russia B N Yeltsin, 19, Mira street, Yekaterinburg, 620002, Russia

E-mail: [2] i.v.ershova@urfu.ru

Abstract. The article contains the results of seven working tests of cutting fluids for metal-cutting equipment. The results of the analysis were evaluated by experts, including process engineers, chemical engineers, foremen and production machine operators in the following areas: turning, drilling, milling, sawing, grinding, threading, boring works. The analysis made it possible to determine the scope of application of cutting fluids depending on the volume and rhythm of the work, the operating conditions of the cutting fluids, the requirements of the design and technological documentation taking into account the final cost of the cutting fluids used. The problem of determining an accelerated system of practical tests for cutting fluids to evaluate the effectiveness of using a particular product before conducting the production tests on machine equipment is considered. For this, criteria for evaluating the cutting fluids are proposed using the weighting method to ensure informed decision-making in the choice of a cutting fluid for admission to full-fledged production tests.

1. Introduction

Until now, the mechanical processing of parts and assembly units by the method of material removal: turning, milling, drilling, boring, planning, etc., remains the most widely used in the machine-building industry. Increasing the economic efficiency of these processes is a multifactorial and complex set of measures, including the machine productivity increase, the use of high-performance toolware, a set of measures to optimize the use of machine devices, and much more. All these factors are well studied, methods for assessing and improving the efficiency of production processes are developed. However, some factors, to which a unified assessment methodological approach has not yet been developed, still remain. One of these factors is the choice of cutting fluids (coolants) for use on metal cutting equipment. The situation becomes more complicated by the dynamically growing supply in the coolant market.

The main emphasis in the variety of products offered is the use of various coolant additives that improve its physicochemical properties. Additives are designed to qualitatively change the properties of the cutting fluid, changing its heat transfer, lubricating properties, resistance to bacterial damage, and other characteristics of material processing procedures to a given direction. Many experiments are conducted to identify improved qualities of cutting fluids using additives. The authors identify an effective decrease in temperature in the cutting zone and an increase in the quality of the machined surface [1], as well as an increase in the processing efficiency along with a decrease in the load on the spindle [2]. Other studies show an increase in tool life with improved morphology of the removed chips and improved roughness of the workpiece [3]. Abdelrazek A H, Choudhury I A, Nukman Y, Kazi S N conduct a comprehensive analysis of the stability and operational characteristics of various types of coolants and lubricants and their impact on processing results and the environment [4]. The possibilities of minimizing the consumption of cutting fluids are being studied, which entails both a positive impact on the environmental component of production and on improving the quality of manufacturing parts [5]. The same trend is observed in the cutting fluids production of the domestic manufacturers, which confirms the effectiveness of the chosen direction of the consumer characteristics improving of the products offered. The use of additives in the production of domestic cutting fluids allowed them to effectively compete with foreign manufacturers, allowing them to increase the consumer properties of liquids, improve the physicochemical properties and environmental friendliness of the product. At the same time, it becomes very difficult to evaluate the entire range of coolants offered by the market and choose a product to recommend for use in production.

Technical and economic assessment of the cutting fluid is difficult due to the need to consider storage factors and allergic effects on workers, which is difficult to determine in cost terms. The company has to make a choice based either on the basis of recommendations from suppliers, or on the basis of coolant tests in the production process, or on the basis of price policy. It should be understood that none of these criteria provides a qualitative justification for the choice of a particular cutting fluid, because supplier recommendations are based on the suppliers 'own economic objectives. Selection based on pricing policy can lead to the use of cheap, low-quality and toxic liquids, which will inevitably harm the health of workers, adversely affect the condition of equipment and tools, and also lead to a decrease in the quality of machining of parts and assembly units. Conducting tests in real production conditions, although it reflects the actual characteristics of the cutting fluid most effectively, has significant disadvantages: it requires high costs of time, materials and the coolant itself. Often in real production conditions it can be difficult to stop the equipment for complete system flushing from the coolant used, refilling with a new test fluid, and then conducting the actual tests, at the risk of obtaining poor-quality products or disabling the expensive equipment due to the use of poor-quality cutting fluid.

The study of scientific research has shown that different authors see different selection criteria. As a selection criterion, it is proposed to use the heat transfer coefficient [6], environmental criteria for coolant [7], assessment of cutting fluid by economic parameters in terms of the long-term operation price, taking into account environmental components [8]. To ensure the choice of a cutting fluid, as well as the possibility of determining their interchangeability, it is also proposed to formulate a detailed physico-chemical classifier of the available cutting fluids, which should provide a rational choice of a coolant [9], [10]. The most widespread among the researchers was the study of the lubricating properties of the cutting fluids [11–13], and cooling abilities [16] as the main selection criterion. Some aspects of the application of cutting fluids for the treatment of various metals are also investigated [16].

The purpose of the study is to develop a multifactorial methodology for the express evaluation of a cutting fluid, taking into account economic, technological and environmental factors.

2. Materials and methods

The most common cutting fluid brands on the Russian market were selected as the objects of the study. To conduct an analysis of the cutting fluids presented, a group of specialists from the machine-building enterprise was involved in conducting tests in real production conditions to determine the most important evaluation criteria. The assessment group included: deputy chief technologist, deputy chief metallurgist, deputy chiefs (of the three workshops of the main production), 12 masters of machining areas, head of the central factory laboratory, 26 machine operators.

The first stage of the work was the selection of evaluation criteria. The ranking of the evaluation criteria was carried out by determining the ranks on a 10-point scale. Rank values were converted into the specific weight of the criterion according to its' importance. The list of evaluation criteria and their significance are given in table 1.

Table 1. Comparative characteristics of cutting fluids.

Name of criterion	Bechem Avantin 451	Bechem Avantin 361	IgatPlatinTopcool 2S	Quakercool 7200 BAF	Quakercool 7601 BFFR	Hosmarc RT11	Volgo 1320	Criterion weight
Price, RUB/kg	452.75	414.37	667.46	424.8	425	219.8	320	–
Washing properties	5	4	5	3	3	5	5	0.1
Foaming	5	5	5	4	5	4	5	0.125
Existence of allergic reactions	5	4	4	4	4	5	5	0.2
Rate of bacterial damage	5	5	5	4	4	5	5	0.125
Impact on tool wear	5	5	5	4	5	5	5	0.15
Cutting fluid consumption	5	4	5	3	4	4	4	0.15
Corrosive effect	5	4	4	4	4	5	5	0.15

Production tests of cutting fluids were carried out in machine-tool workshops using equipment for 4–6 months for each of the products. Each criterion was evaluated by experts independently on a 5-point scale. Agreed expert estimates rounded to integer values are given in Table 1.

Evaluation of the washing properties of the tested cutting fluid was carried out on the basis of observations of the condition of the working surfaces and protective covers of production equipment for the presence of soap sticks, mastic layers, greasing, and the degree of contamination.

The evaluation of foaming was carried out on the basis of observations of the cutting fluids tendency to produce foam.

The existence of allergic reactions was assessed by the presence of complaints from staff.

The rate of bacterial damage was estimated by the experience of long-term operation, including the assessment during prolonged standstill of equipment in the warm season.

Assessment of the impact on tool wear was evaluated based on the experience of equipment operation by workshop personnel.

Cutting fluid consumption was estimated by comparing the volumes to top up coolant in the equipment of the same type.

Assessment of the corrosive effect was carried out based on the appearance of the filter according to the degree of corrosion (0 points – strong, 5 points – no corrosion).

3. Results

The proposed assessment model is based on the calculation of the cutting fluid unit cost. For each cutting fluid, the total expert score was calculated according to the formula:

$$X = \sum(K_i \times y_i), \tag{1}$$

where K_i is the score of the ith criterion, y_i is the weight coefficient of the ith criterion.

The final cutting fluid assessment was carried out according to the unit cost criterion, calculated as the ratio of the price to the total score. The calculation is made for two options. In the first variant all criteria were taken into account, in the second variant the criterion 'rate of bacterial damage' was excluded, since during rhythmic work and large volumes of processing it is not significant. The calculation results are given in table 2.

Table 2. Cutting fluid unit cost

Indicator	BechemAvantin 451	BechemAvantin 361	IgatPlatinTopcool 2S	Quakercool 7200 BAF	Quakercool 7601 BFFR	Hosmarc RT11	Volgo 1320
			1st option				
Total score	5	4.4	4.65	3.75	4.175	4.725	4.85
Unit cost	90.55	94.175	143.53	113.28	101.79	46.51	65.98
			2nd option				
Total score	4.375	3.775	4.025	3.25	3.675	4.1	4.225
Unit cost	103.48	109.77	165.83	130.71	115.65	53.61	75.74

By assessing the unit cost of the estimated cutting fluid, we see that the Hosmarc RT11 product proved to be the most effective and can be further recommended for industrial use at the enterprise.

The significance of the criteria may vary depending on the requirements of a particular production process, which will affect the choice of the cutting fluid. Changes in prices will also lead to changes in estimates. This does not affect the assessment methodology.

The proposed express-evaluation methodology can be used by enterprises without testing in real production conditions and eliminating equipment downtime due to flushing of the coolant system and refilling with a new fluid.

References

[1] Duc T M and Long T T 2020 Effectiveness of alumina nanofluid on slotting end milling performance of SKD 11 tool steel *Journal of Computational and Applied Research in Mechanical Engineering* **9 2** pp 359–369

[2] Peña-Parás L, Maldonado-Cortés D, Rodríguez-Villalobos M, Romero-Cantú A G and Montemayor O E 2020 Enhancing tool life, and reducing power consumption and surface roughness in milling processes by nanolubricants and laser surface texturing *Journal of Cleaner Production* **253** 119836

[3] Yıldırım Ç V 2020 Investigation of hard turning performance of eco-friendly cooling strategies: Cryogenic cooling and nanofluid based MQL *Tribology International* **144** pp 106–27

[4] Abdelrazek A H, Choudhury I A, Nukman Y and Kazi S N 2020 Metal cutting lubricants and cutting tools: a review on the performance improvement and sustainability assessment *International Journal of Advanced Manufacturing Technology* **106 9–10** pp 4221–45

[5] Lopes J C, Garcia M V, Volpato R S, de Mello H J, Ribeiro F S F, de Angelo Sanchez L E, de Oliveira Rocha K, Neto L D, Aguiar P R and Bianchi E C 2020 Application of MQL technique using TiO2 nanoparticles compared to MQL simultaneous to the grinding wheel cleaning jet *International Journal of Advanced Manufacturing Technology* **106 5–6** pp 2205–18

[6] Kisel A G, Rechenko D S, Titov Yu V, Purtov E D and Petrov I V 2015 Selection of cutting fluid for finishing *Omsk Scientific Bulletin Ser. Devices, machines and technologies* **3 (27)** pp 39–43

[7] Chernat I S 2016 Significant factors in the selection of cutting process fluids in mechanical engineering *Engineering industry and life safety* **1** pp 68–70

[8] Zenkin N V, Varichkin I A and Sorokin S P 2016 Analysis of cutting fluids *Scientific almanac* **4–3 (18)** pp 82–5

[9] Kireynov A V and Esov V B 2017 Current trends in the use of cutting lubricants for blade processing of hard-to-process materials *Engineering Journal: Science and Innovation* **2**

[10] Debnath S, Reddy M M and Yi Q S 2014 Environmental friendly cutting fluids and cooling techniques in machining: A Review *Journal of Cleaner Production* **83** pp 33–47

[11] Sholom V Yu, Korshunov A A, Tyulenev D G, Trofimov A S and Kramer O L 2011 The results

of laboratory and pilot tests of highly effective coolants for blade and abrasive processing operations *Bulletin of the Samara Scientific Center of the Russian Academy of Sciences*vol **13 4 (3)** pp 913–7

[12] Cherepakhin A A, Vinogradov V M and Smirnova E E 2008 Choice of coolant for high-speed pulling of external shaped surfaces *Assembly in mechanical engineering, instrument making* **9** pp 37–41

[13] Kirsanov S V, Chekantseva L V and Tsygankov R S 2017 Choice of coolant for the operation of drilling deep holes with gun drills on multioperational machines *Reference. Engineering Journal with the application* **11** pp 3–9

[14] Pozhidaev Yu A et al 2015 Choice of coolant for various cutting methods and machining conditions *Innovations in metalworking: a look of young specialists* pp 304–7

[15] Kisel A G, Purtov E D and Deylova A V 2017 Assessment of the cooling properties of cutting fluids *Omsk Scientific Bulletin* **1 (151)** pp 27–9

[16] Suleymanov R I and Suleymanov R R 2018 The effect of mechanically structured cells on surface quality in the processing of aluminum bronzes *Scientific notes of the Crimean Engineering and Pedagogical University* **4 (62)** pp 206–12

Problem of model usage in mechanical engineering

V F Pegashkin[1] and N U Ignatova[1, 2]

[1]Ural Federal University named after the First President of Russia B N Yeltsin, Nizhniy Tagil Technological Institute, 59, Krasnogvardeyskaya str, Nizhniy Tagil, 622000, Russia

E-mail: [2]nina1316@yandex.ru

Abstract. The concept 'model' in mechanical engineering is analyzed. Some problems of model usage at the industrial enterprise are named: ontology of elements of difficult model system; difference between the description of the model in natural language and mathematical language; the problem of fitting objects. Design of new way of feedback between a paint robot and its operator based on the use of not only virtual, but also augmented reality, requires an appeal to the ontology of models or to philosophical discourse. The difference between model description in natural and artificial languages implies the application of a semantic approach. The problem of fitting objects during the model testing in mechanical engineering can be solved during testing. The authors argue that the use of an interdisciplinary approach in comprehending complex model systems, taking into account virtual and augmented realities, allows us to clarify the conditions for using virtual 3D models in mechanical engineering.

1. Introduction

The authors of the article tried to approach the analysis of the problem of models' usage in mechanical engineering as part of an interdisciplinary approach in terms of mechanical engineering technology and philosophy. Science is already acquainted to similar approaches, when scientists or engineers in alliance with philosophers researched real models [1], fiction models [2] and complex model systems [3, 4]. The problems emerging during usage of real and fiction models were named in these researches:

- composition, genesis, ontology of elements of difficult model system (virtual, augmented realities);
- the difference between description of the model in natural language and mathematical language;
- the problem of description – is object fitting.

Similar research is being carried out for the first time in Russia.

An industrial leader (leader of steel production plant), who shall approve project, is provided with preliminary information on design features of a paint robot. He understands that during the solving of specific problem of painting of bulk metal structures, complicated technical, organizational and other problems arise, for the solution of which he invites research engineers. Let us assume that engineers are designing a new way to establish feedback of the robot itself and its operator, based on the use of not only virtual, but also augmented reality. Imagine that a team of research engineers working on the problem of using a paint robot in the manufacture of metal structures creates a project in the form of a set of proposals, graphs, drawings, sketches expressing a certain content of the model. In this case, the

IOP Conf. Series: Materials Science and Engineering **966** (2020) 012096 doi:10.1088/1757-899X/966/1/012096

engineers will develop a complex fictional model system, the interpretation of which requires philosophical discourse, which means an appeal to the ontology of models.

2. Methods

2.1 Usage of models in mechanical engineering

During the real simulation of manufacturing system within one technological operation or the whole technological process searching of all combinations of existing factors and selection of the best (optimized) variant of combinations (treatment conditions, tool parameters, part material properties, etc.) take place.

However, a change of each factor even on two levels leads to the need to carry out approximately 40 thousand tests only for one technological operation (there can be several tens of operations in a technological cycle). It's more comfortable to study system based on the mathematical model of this system which takes into account a factor impact on objective function [5].

For example, a production system optimization model may have the following form

$$F\,(x_i) \Rightarrow \text{extremum}$$

where $F\,(x_i)$ is dependence of objective function on factors x_i. In this case, limiting parameter system is applicable

$$\{f_j(x_i) \leq C_j,$$

where $f_j(x_i)$ is a function which describes dependence of j limiting parameter on factor x_i, C_j is a constant value.

Model can be set in the form of polynomial of n-power, power dependence, a Fourier series, etc. Generally, such model is a mathematical expression without physical meaning. The degree of complexity (simplicity) of the model is determined by the level of its detail, which depends on accepted assumptions and suppositions. The main feature of the model is representativeness (adequacy) to the process which it describes. In real conditions, the model parameters are unstable; their changes are random and, as a rule, are not controlled.

2.2. Ontology models

A discussion of the composition and genesis of complex model systems raises a metaphysical question: what are models? In philosophy, it is customary to distinguish between descriptions of models, types of descriptions that appear in scientific papers, textbooks and diagrams, and the described model systems: ideal pendulums, systems of rational self-serving agents, or endless animal populations. There are metaphysical puzzles in model systems since there are no frictionless pendulum components, absolutely rational, selfish agents or endless animal populations. In recent times, there is growing interest to the idea that model descriptions shall be considered as similar to fiction and model systems should be considered akin to fictional characters. Imagination is the power in the particular context of usage.

There are solid reasons for such approach [6]. Firstly, though there are no any specific objects, which would correspond to the descriptions in the fictional stories or descriptions of scientific models, we have an ability to express truth or falseness on model systems. Secondly, developers of scientific models often relate to them particularly which suggests strong analogy with usual fiction [7]. A model consists of the model description and model content and they are separate entities that only together make up specific model. The description of the model is a set of linguistic and mathematic symbols which prescribe to imagine something specific. The mathematical equations that make up the description of the model is a sequence of mathematical symbols that does not specify any concrete truths until it is interpreted as a description of the state of the model, that is, until the variables in the equation are interpreted as meaningful for certain objects.

The description of the model and the content of the model can differ from one another. For example, model description can include not only mathematical symbols, which are usually used in different contexts, but also linguistic symbols, which can change without model content alteration: one type of description of paint robot is in English and another one is in Chinese.

A simplified understanding of reality as model denotatum does not allow for creating a multi-purpose model theory applicable in any cases. The other way to solve the problem of connection of model and reality lies in analogy of the model with sentence or text within the framework of the semantic approach. In this case, philosophers emphasize the importance of model description in natural and artificial languages. The semantic approach allows us to determine the conditions for the truth of the model, since you cannot know something about something that is not true. This raises many questions regarding the nature of knowledge, the representation of knowledge obtained through modeling on the target system. The epistemic justification that can be obtained by assessing the truth of models includes the evidence presented by the model itself. However, the conditions for the validity of comparing the model with the conditions of an industrial enterprise are limited by the requirement of linking the characteristics of the model and the real target system.

Sometimes the semantic approach is insufficient to explain those properties of the target system that are eliminated as immaterial or uncontrollable. The most difficult to explain and, at the same time, very controversial is the exclusion of some observations or measurements during experiments in the practical implementation of the model. Later, such actions or engineers may be construed as an attempt to fake the results of experiments.

If we consider the description of the model as its essence, inevitably there will be an absurd conclusion that any small change in the description of the model leads to a new model. The problem of the difference between the description of a phenomenon (model) in a natural language and a mathematical language in philosophy is solved as the relationship between linguistic expressions and designates (R Karnap). It should be noted that W. Quine tried to solve this problem by the requirement of mutual translatability [8], S Psillos – by indicating the identity of the truth conditions [9], A Thomasson – by using the concept of abstract artifact [10].

The most complex ontology arises when designing fictional models with elements of virtual or augmented reality, to which no objects of the real world correspond. Here, as shown by F Salis and R Frigg, it is crucial whether these components of the model designate a certain nominal entity or whether they are specific concepts [11]. The idea of a model as a possible world allows you to go into the space of modal fictionalism [12]. If the conclusion about the possibility of correspondence between the model and the target system is made regardless of the content of the original fiction, no matter what fiction is used, he must recognize the independence of modal truth [13, 14]. The content of fiction, of course, is associated with the idea of the boundaries of a possible world, but to a lesser extent. The point is that, as it is impossible to imagine a possible world with any characteristics, it is impossible to imagine a model with any properties that are in no way connected with any relationship with the target system.

An engineer can imagine the properties of model systems if and only if they meet the real requirements in the description of the model and the principles of generating its content. In order to predict the significant properties of the target system, he needs to move from a state of imagination to reality.

3. Results: paint robot model

To avoid the usual contradictions, a model shall necessarily describe specific target system [15]. There is a problem of paint application to the finished product at the enterprises manufacturing welded steel structures (erection trusses, bridge spans, etc.). During the application of coating, as a rule, pneumatic spraying of the paintwork composition is used, as the most productive method. The products being painted are very complicated spatial structures. Since the coating process at enterprises is difficult, pneumatic spraying of the composition is done manually. This creates an unfavorable environmental situation for the operator producing the paint. One of the variants of automating the process is usage of robots. However, robot shall be preprogrammed for performing movement of manipulator with the spray gun along a spatially complex path. Programming with a large product range is a laborious and expensive process. A simpler engineering solution is remote control of the robot. Operator is removed from the paint zone and controls the robot through video camera mounted on the robot. Disadvantage of the scheme is pollution of camcorder lens by spray products. The possible solution of this problem is to create a virtual model of the product being painted. Initially, the robot scans the metal structure, as a result of which, virtual 3D model of the product, some virtual reality (VR) is created in the memory of control computer. Then operator controls the robot in this virtual reality with the use of VR Box. The real robot repeats the movements of the virtual robot.

Model description is possible only in natural language or mathematical language. When using natural language it is necessary to have detailed description of the whole system, its separate elements, type of interaction of elements and their static and dynamic parameters.

For example, as follows, the position of each robot link in space is determined by the coordinate system connected to this link. The position of the coordinate system associated with the link relative to the inertial system is determined by radius-vector of its beginning which has three independent projections and three angles between the corresponding axes of the system. In general, each link of the robot represents a mechanical system with six degrees of freedom. If there are some connections which limit movements of the body, the number of degrees of freedom decreases. Since manipulation systems represent open kinematic chain consisting of rigid (absolutely firm) links interconnected with kinematic pairs of the fifth class with one degree of freedom, each link of such manipulation system, except for the last one, constitutes two kinematic pairs. For that reason, each link is connected with two coordinate systems located in the centers of kinematic pairs of this link. Such description of the robot system and its each link can take more than one page of the text.

For practical purposes it is more convenient to use mathematical model which allows for planning of program trajectory of movement and defining the range of possible deviations (dynamic errors) of manipulator systems from this trajectory caused by elastic compliance of links. Motion equations used to simulate the dynamic of manipulation system, depending on the method of compiling of these equations, can be represented as mathematical model

$$M(q,d)\ddot{q}+\dot{q}^T C(q,d)\dot{q}+g(q,d)=P$$

$M\ (q,\ d)$ is matrix ($n \times n$) of inertial characteristics of system; $C\ (q,\ d)$ is matrix ($n \times n \times n$) of centrifugal and Coriolis effects; $g(q,\ d)$ is gravitational vector of the size n; $q = [q_1\, q_2 \ldots q_n]^T$ is generalized coordinates vector; d is vector containing geometric and dynamic parameters of the system.

In any case, when describing, the model will differ from the original for the following reasons.

Firstly, due to the complexity of any described system, model is constructed under certain assumptions. Model is a simplified copy of the original. Simplification is possible to such an extent until simulation error is within the prescribed limits. Therefore, the model does not reflect all but only the essential characteristics of the original.

Secondly, the real mechanism (robot) has some elastic flexibility of the manipulation system links. Therefore, static and dynamic deformations and oscillations of system links appear. These characteristics depend on many factors (geometric link sizes, gaps in moving couplings and joints, dynamic characteristics of drives, etc.) which values for specific mechanism are random.

As a result, in order to more accurately describe the dynamics of the manipulation systems with rigid links and elastic joints, to increase of dynamic accuracy of their movements, it becomes necessary to correct (fit) the general model for a specific mechanism (robot) – change coefficients, introduce new elements, etc.

IOP Conf. Series: Materials Science and Engineering **966** (2020) 012096 doi:10.1088/1757-899X/966/1/012096

4. Conclusion

The creation of 3D models of products when they are painted with a paint robot at an industrial enterprise is an example of fictional models. With the philosophical understanding of the ontology of real, fictional models and complex model systems, numerous problems arise. Authors made an attempt to discover the following: ontology of elements of difficult model system (virtual, augmented realities); the difference between the description of the model in natural language and mathematical language; the problem of description is fitting objects.

References

[1] Cartwright N 1983 *How the Laws of Physics Lie* (Oxford: Oxford University Press)
[2] Contessa G 2010 Scientific models and fictional objects *Syntheses* **172** pp 215–29
[3] Frigg R and Hartmann S 2012 Models in science *The Stanford encyclopedia of philosophy* Ed E N Zalta http://plato.stanford.edu/archives/fall2012/entries/models-science/ (accessed 15.03. 2020)
[4] Godfrey-Smith P 2009 Models and fictions in science *Philos. Stud.* **143** pp 101–16
[5] Pegashkin V F, Pegashkina E V and Pishchevskaya E S 2017 Model of tool life based on fuzzy sets *Rus. engng. research* **5** pp 408–10
[6] Frigg R 2010 Fiction in science In: *Fiction and Models: New Essays* Ed J Woods (Munich: Philosophia) pp 247–87
[7] Godfrey-Smith P 2006 The strategy of model-based science *Biol. and Philos.* **21** 739 p
[8] Quine W V O 1980 *From a Logical Point of View: Nine Logic-Philosophical Essays* (Harvard University Press)
[9] Hendry R and Psillos S 2007 How to Do Things with Theories: an Interactive View of Language and Models in Science In: *The Courage of Doing Philosophy* Ed J Brzeziński, A Klawiter et al (Amsterdam/New York NY: Rodopi) pp 59–115
[10] Thomasson A 2017 If Models Were Fictions, then what Would They Be? In: *The Scientific Imagination* Ed P Godfrey-Smith and A Levy (Oxford: Oxford University Press)
[11] Salis F and Frigg R 2017 Capturing the scientific imagination In: *The Scientific Imagination* Ed P Godfrey-Smith, A Levy (Oxford: Oxford University Press)
[12] Lewis D 1978 Truth in fiction *Amer. Phil. Quart.* **15** (1) pp 37–46
[13] Bolinska A 2013 Epistemic representation, informativeness, and the aim of faithful representation *Syntheses* **190** pp 219–34
[14] Toon A 2012 *Models as Make-Believe: Imagination, Fiction and Scientific Representation* (Palgrave Macmillan)
[15] Schiffer S 1996 Language-Created Language-Independent Entities *Philos. Top.* **24:1** pp 149–67

IOP Conf. Series: Materials Science and Engineering **966** (2020) 012097 doi:10.1088/1757-899X/966/1/012097

Marketing aspect in the 'Value Engineering' in the new products development of machine-building enterprises

A D Murukina[1, 2], L M Tipner[1], M A Prilutskaya[1] and N A Kalinina[1]

Institute of New Materials and Technologies, Ural Federal University named after the first President of Russia B N Yeltsin, 19, Mira street, Yekaterinburg, 620002, Russia

E-mail: [2] admurukina@urfu.ru

Abstract. The dynamism of the engineering products markets is determined by a high level of competition, frequent changes in technology. One of the key problems of Russian machine-building enterprises is the long term of development and putting the product into the production which leads to the fact that 'yesterday's product is launched on tomorrow's markets'. So, we see an arising need for a fundamental change in the product development process by including future consumers in it and ensuring verification of the results at each stage, from the formation of the technical task to the testing of the finished product. Such an approach will allow balancing consumer value and product characteristics, determining the significance and cost of functions, and substantiating a fair price. These problems are solved by the method of value engineering of a product. Application of this method requires reliable information on consumer characteristics, conditions and process of product operation, parameters of similar products. The tools of classical marketing research and new products engineering development should be combined into one system. As a result, the use of value engineering reduces the time required to bring a new product to the market and increases the competitiveness of the product.

1. Introduction

The activity diversification and the new products development become an important area of a machine-building enterprise adaptation in the market in a constantly changing economic environment. The dynamism of industrial markets, high competition and continuously improving production technologies force many companies to seek and use methods reducing the time taking to develop and launch new products on the market. A long period of development and putting the product into production leads to the fact that 'yesterday's product is launched on tomorrow's markets'. Competitors quickly adopt other's decisions and develop new ideas; therefore, innovative activity at the enterprise should be a continuous and systematic process. Interaction with potential customers in the process of product development and continuous verification of the results are crucial in ensuring the competitiveness of the product [1, 2]. The reduction of production costs and the ensuring of the product target cost is another serious task. How one can create a product with the characteristics that meet all the requirements of potential consumers, and ensure the level of costs corresponding to the price that the buyer is willing to pay, in a short period of time?

2. Parameters that determine the engineering products competitiveness

The competitiveness of the product on the market is determined by two groups of indicators that reflect the ratio of 'price-quality' [3, 4]:

- Consumer properties of the goods (quality), constituting value to the consumer. These properties are determined by a set of 'hard' and 'soft' consumer parameters.
- Economic properties of the goods (price). This group includes the selling price, total costs of transportation, installation, maintenance, operation, repair, taxes, insurance and disposal.

When conducting market research, it is necessary to determine the consumer properties of the product that make up its value. 'Hard' parameters describe the most important functions of the product and the associated characteristics that determine its range of application, ergonomic indicators that characterize ease of use for a person, technical indicators that characterize technological solutions, reliability, and safety. The second group of 'hard' parameters includes technical ones, as well as the parameters of compliance with national and international standards, norms, legislative acts, etc. 'Hard' parameters provide the ability to use the product for its intended purpose [5].

'Soft' parameters characterize the aesthetic properties of the product, that is are an expression of design, packaging, color. The second group of 'soft' parameters is psychological ones that determine prestige, attractiveness, accessibility and so on. 'Soft' parameters, as a rule, do not have a natural physical measure and are difficult to quantify [6].

When conducting value engineering, it is necessary to study the requirements for an engineering product from consumers, service and repair organizations that carry out its maintenance. The significance of these studies is emphasized by many authors of scientific publications [7–9]. This means that preliminary marketing research should provide information on the conditions of operation and maintenance of the object, customer requirements and target technical characteristics of the product. The author's approach consists in applying the target focus groups survey method, the purpose of which is the formation of an initial data set necessary for the value engineering of a high-tech product.

3. Value engineering method

Value Engineering (VE) method is used to design the value of a product at the stage of its development [10]. VE is a part of the Value Analysis (VA). Value Analysis method contributes to the solution of such critical tasks as [11]:

- reducing the duration of the product development (modernization) and development of the product manufacturing;
- creation of high technical level products, competitive in the foreign market and satisfying the needs of consumers;
- ensuring the stability of functionally necessary quality (required consumer properties) of the product during its development, manufacture, storage, transportation, consumption and disposal;
- ensuring efficient operation;
- bringing the costs of product development, its manufacture, use and consumption to the level required by the end user.

Thus, the use of value engineering is possible both in the development of a new product and in its redesign in order to increase the value of the product while reducing costs [12]. VE allows us to consider the designed product as a set of functions that it should perform, creating value for the consumer. The functions of the object should reflect the content of the action or process for which it or its components are intended [13]. Simultaneously with the formulation of the required functions, constructive options for their implementation are proposed, and a feasibility study of these options is carried out. The results of this work are presented in the form of a matrix of functions containing their assessment, or in the form of a functional model.

The content of the work at various stages of the VE is shown in table 1 (compiled by the authors based on the source [11] and their own results obtained in the course of the VE project at one of the machine-building enterprises of the Ural Federal District).

Table 1. Work content at the stages of the Value Engineering.

Stages	WorkContent
Information	• determination of consumer requirements for the product; • determination of the servicing and performing repairs specialists requirements to maintainability and ease of maintenance of the designed product; • determination of the product main technical characteristics; • description of operating conditions; • study of the operating conditions influence on the product main technical characteristics; • setting quantitative technical and economic goals; • determination of the base (analogue) for the study; • description of the functional composition of the product (analogue) original design; • designing a structural element model of an analogue.
Basic	• formulation of the designed object functions and classification of functions; • development of the functions implementation concept scheme; • formation of a functional model; • determination of the constituent parts of the designed object (correspondence matrix); • definition and analysis of the performing functions costs; • expert evaluation of the product functions significance; • assessment of the product manufacturing laboriousness; • cost evaluation of the product production and the facility operation, the product disposal.
Analytical	• determination of the basic set of functions (basic configuration of the designed product); • determination of the optional set of functions (optional configuration); • comparison of the functions significance and their implementation costs; • separation of costs into functionally necessary and excessive; • the allocation of structural-element or functional areas with the highest concentration of costs for materials, purchased components and labor.
Creative	• identification of the 'narrow' ('weak') places in the design, technology, materials; • search for ideas and preparation of proposals for improving the product; • formation of product concepts (implementation of the same function by various design solutions); • elimination of discrepancies in significance and costs identified at the analytical stage; • elaboration of advanced alternates and selection of the most appropriate options.
Research	• development of sketches for selected options; • comparative technical and economic assessment of options; • examination of selected solutions and options' evaluations.
Recommendation	• selection of the most rational solutions for implementation.

4. Structure of typical marketing research reports

Standard marketing research carried out by machine-building enterprises, as a rule, does not give an idea of the 'functional appearance' of an object, which greatly complicates the implementation of value engineering and increases the time for product development. Therefore, the marketing aspect will play a decisive role in creating the best product for the client at minimal cost at the stage of product development.

Marketing research can be divided into three key types that determine the structure of a marketing report: market research, consumer research and product research. Market research allows us to determine the state of the market. Consumer research allows us to determine the full range of factors that consumers are guided by when choosing a product. The study of the product allows us to get the most valuable information about the consumer properties of the product and data for the formation of a successful advertising campaign from the buyer's point of view [14]. The first type allows us to make a decision on the choice of a new product for production. It is necessary to concentrate on the last two types for the purposes of designing the value of a product and its consumer properties.

The contents of the typical marketing reports offered by many marketing companies can be summarized in the form of the following main blocks:

- market conditions:
- manufacturers;
- buyers;
- product characteristics;
- prices;
- suppliers (distribution);
- leasing companies;
- customs regulation;
- government regulation and support;
- tenders;
- possible risks;
- factors influencing market development;
- forecast for the development of the market for the investigated products.

The information presented allows us to characterize the market, to determine the main areas of activity, to find out unoccupied or under-occupied niches in the market, to develop our own assortment of goods or services in accordance with the requirements of customers and the existing competition, to draw conclusions about the production prospects of the products under study, to formulate recommendations for choosing a strategy for this market. For the purposes of developing new competitive products, the main goal of marketing research will be to determine the properties of the product that it must possess in order to meet the needs of potential users [15]. For this, it is necessary to establish constant interaction with the target focus group, consisting of specialists in patent analysis, potential consumers of products, specialists in repair and maintenance services, suppliers (distributors) of products.

5. The focus group survey method in order to conduct VE

The result of this marketing research instrument is a list of new product target parameters for the further establishment of a fair competitive price and calculation of target cost. Below are the market study results of certain types of warehouse lifting equipment, conducted by the authors of the article in 2019. To achieve this goal, the following tasks were solved:

1. The focus group composition of various stakeholder groups' representatives was formed. The focus group included dealers and warehouse equipment service organizations in addition to direct consumers in accordance with the objectives of the study. The composition of the focus group is as follows:

- 12% – warehouse facilities;
- 50% – distribution centers;
- 25% – logistics companies;
- 13%– dealers and service organizations.

2. A focus group survey was conducted using questioning and interview methods. The objectives of the survey were:

– determining the operating conditions of the facility (that is the type and area of the warehouse, the nature of the cargo, the organization of warehouse logistics, the level of automation, etc., in this case);

– importance assessment of the parameters that affect the equipment type and model choice in such categories as function characteristics (technical parameters); equipment maintenance; maintainability; the ability to integrate the object into the enterprise information system; ergonomics; security; aesthetics; brand of the manufacturer (Figure 1);

– obtaining expert opinions on the necessary characteristics of the product basic model, as well as the selection of those parameters that will be optional, increasing consumer properties and the price of the product.

A total of 73 parameters were allocated for the formation of the designed product configuration in groups: safety, ergonomics, ability to integrate into the warehouse information system, maintainability and maintenance of equipment. Of these, 39 parameters (53%) were included in the basic configuration of the product and 34 parameters (47%) were included in the optional configuration. All function characteristics were included in the basic configuration of the product.

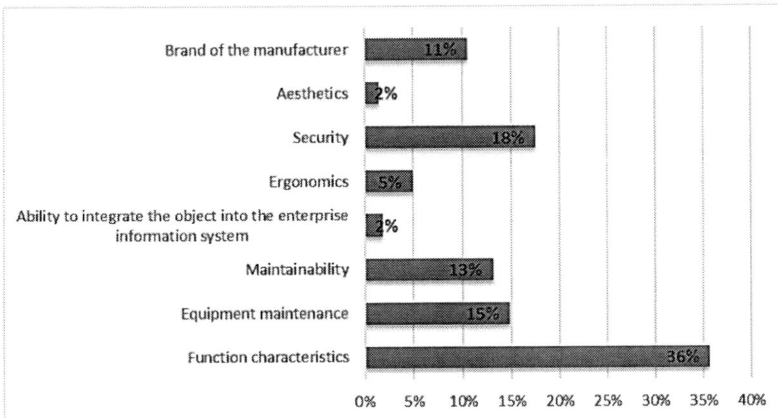

Figure 1. The significance of consumer product parameters.

3. The data obtained were processed and a conclusion was generated. The conclusion contains the following results:

- the scope of the equipment operating conditions was defined (the consumer profile was specified) and the choice of the closest analogue (group of analogues) was substantiated;
- the coefficients of the product parameters significance were calculated, including those related to the basic version and optional solutions;
- a comparison of a new product with an analogue was made, the coefficients of compliance by the compared parameters were calculated.

Based on the results obtained, a modernized classification of product functions was proposed (Figure 2), which was subsequently used in the value engineering of warehouse lifting equipment.

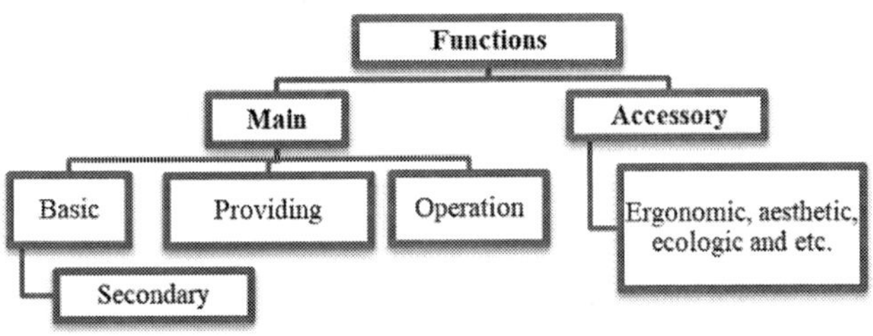

Figure 2. Author's classification of the product functions.

Thus, the addition of a special survey of the target focus group to the marketing research allowed the authors to improve the existing classification of functions. In particular, the management functions of a technically complex product were identified, due to the need to integrate the object into the logistics ecosystem. The implementation of the proposed approach made it possible to design a functional model of the product, as well as to propose a concept that defines the basic configuration (basic set of functional characteristics) and options (additional set of functional characteristics) of a future product.

6. Conclusion

Companies operating in dynamic markets, characterized by changing economic conditions, development of technology, increasing competition, should have flexible competencies that allow them to quickly develop products that meet today's market requirements and are even ahead of them. The method of value engineering combined with target focus groups' marketing research will allow companies to reduce the development time of a new product, ensuring its competitiveness and balance of costs, price and profit. However, an individual approach to the formation of the focus group and the description of the product functions is required in each case.

The focus group should include representatives of all stakeholder groups that determine the value of the product. The separation of product functions into basic and optional plays an important role in conditions of customer-oriented production, as it will affect the target cost of the product. Machine-building enterprises, as a rule, produce technically sophisticated products combining functions of human management and software. In this case, management functions that ensure the integration of the facility into the enterprise management system are of particular importance. The use of focus groups' marketing research allows us to solve the above problems and to prepare the necessary information support for the value engineering of the product.

References

[1] Gruner K and Homburg C 2000 Does Customer Interaction Enhance New Product Success? *J. Bus. Res.* **49** pp 1–14 doi: 10.1016/S0148-2963(99)00013-2

[2] Hoyer W, Chandy R, Dorotic M Krafft M and Singh S 2010 Consumer Cocreation in New Product Development *J. Serv. Res.* **13** pp 283–96 doi: 10.1177/1094670510375604.

[3] Porter M E 1980 *Competitive Strategy: Techniques for Analyzing Industries and Competitors* (New York: Free Press) p 386

[4] Belkin V N, Belkina N A and Vladykina L B 2015 The theoretical basis for assessing the competitiveness of enterprises *Econ. reg.* **1** pp 144–55

[5] Emelyanov S 2002 International manufacturers' competitiveness: factors determining the market position and competitive advantages *Mark. Rus. abr.* **1** pp 107–16

[6] Assessment of product competitiveness. Available online: http://www.somemarketing.ru/margus-554-4.html (accessed on 01.06.2020)

[7] Leber M, Bastic M, Mavrič M and Ivanišević A 2014 Value Analysis as an Integral Part of New Product Development *Proc. Eng.* **69** pp 90–8 doi: 10.1016/j.proeng.2014.02.207

[8] Alix T, Ducq Y and Vallespir B 2009 Product service value analysis: two complementary points of view *Conf. 1st CIRP IPS2* pp 157–64

[9] Abdolmaleki K and Ahmadian S 2016 The Relationship between Product Characteristics, Customer and Supplier Involvement and New Product Development. *Pr. Econ. and Fin.* **36** pp 147–56 doi: 10.1016/S2212-5671(16)30026-0

[10] Abdullah A and Adesta E 2015 Implementation of functional analysis using value analysis value engineering (VAVE) *ARPN J. Eng. Appl. Sc.* **10 (21)** 10072-76 doi: 10.10072-10076

[11] Karpunin M 1987 *The practice of functional-cost analysis in the electrical industry* (Moscow: Energoatomizdat) p 288

[12] Kiran D R 2017 *Total Quality Management: Key Concepts and Case Studies* (Oxford: Butterworth Heineman) chapter **33** pp 455–70

[13] Prilytskaya M, Murukina A and Podoliak O 2020 Operation of high-technology product development based on functions and value *IOP Conf. Ser: Mat. Sc. and Eng.*709(3) 033049 doi: 10.1088/1757-899X/709/3/033049

[14] Marketing research. Available online: https://creativecallproject.ru/marketingovye-issledovaniya/ (accessed on 01.06.2020)

[15] Zemlickienė V and Maditinos D 2012 Marketing Strategy Formulation for Innovative Product Development Process *Verslas: teo. prak* **13** pp 365–74 doi: 10.3846/btp.2012.38

IOP Conf. Series: Materials Science and Engineering **966** (2020) 012098 doi:10.1088/1757-899X/966/1/012098

Utilization aspects in a life cycle contract (the example of public transport in Russia)

N A Amosov[1], E Y Kuznetsova[1, 2], M A Prilutskaya[1] and G R Kukushkina[1]

[1]Institute of New Materials and Technologies, Ural Federal University named after the first President of Russia B.N. Yeltsin, 19, Mira street, Yekaterinburg, 620002, Russia

E-mail: [2] e.y.kuznetsova@urfu.ru

Abstract. A structural analysis of public transport in Russia on the criterion 'passenger turnover' has been carried out in the last few years. Demand statistics have been compiled for all urban transport types. This has led to identify the quantitatively predominant types of public transport with a high pace of development. The article discusses the possibility of applying the life cycle contract concept to public transport in connection with the popularization of the concept on the one hand and the relevance of resolving organizational issues of exploitation public transport on the other. The article shows that public transport plays a key role in the total volume of citizens' movement in Russia, therefore a life cycle contract is applicable to it as a whole. And the final phase of the life cycle contract is about utilization. Federal State Statistics Service data on passenger transport should form the basis for the correct calculation of future production capacities of end-of-life vehicles processors. This is necessary for the effective realization of the stage of 'decline' of the life cycle contract.

1. Introduction

At present, Russia is developing a recycling system for all types of vehicles [1, 2]. The differences between the various types of vehicles are significant, so that the specificities of each type should be taken into account at the initial stage of establishing a common utilization system. The investigation deals with urban public transport vehicles. [3, 4]. Urban public transport is concerned with the people movement along a certain route at a specified frequency. This type of transport is considered in this article (with the exception of the city electric train).

In 2017, the total number of urban transport in Russia was 196.6 thousand units, which includes buses, underground railway carriages, trams, trolleybuses and inland water-borne transport. The study of public transport is very important because it is the basis for the structure of the citizens' movement in the city, and therefore it is a priority in the recycling creation. In addition, all countries have taken various measures to encourage citizens to use public transport instead of personal transport. This will help to improve the environment and reduce road congestion. In Russia, the introduction of a life cycle contract [5–7] on vehicles can bring public transport to a new level [8]. We pay special attention to the stage of disposal of vehicles that are going out of service for two reasons:

– currently, a large percentage of the operating rolling stock has increased wear;

– in the Russian Federation, there is no vehicle recycling system that takes into account all its technological and environmental features, and this is the most problematic element of the life cycle of rolling stock operation.

IOP Conf. Series: Materials Science and Engineering **966** (2020) 012098 doi:10.1088/1757-899X/966/1/012098

2. Materials and methods. Overview of urban public transport in Russia

The first stage of the analysis concerns the demand structure, namely, the urban transport types mostly used by the population were taken. Table 1 shows data on the passengers' transportation by public transport in the city traffic.

Table 1. Transportation of passengers by public transport in the city traffic (million people)*.

types of urban transport	2005	2010	2015	2016	2017	Structural share in 2017	Base index 2017/ 2005	Base index 2017/ 2010	Chain index 2017 /2016
bus	9276	5920	5019	5158	5425	47.47	58.48	91.64	105.18
metro	3574	3294	3337	3312	3298	28.86	92.28	100.12	99.58
trolleybus	4653	2206	1616	1483	1376	12.04	29.57	62.38	92.78
tram	4123	2079	1478	1397	1327	11.61	32.19	63.83	94.99
inland water-borne	4.7	6.5	3.6	3.3	3	0.03	63.83	46.15	90.91
TOTAL	21630.7	13505.5	11453.6	11353.3	11429	100	52.84	84.62	100.67

* Calculated on the basis of data of the Federal State Statistics Service 'Transport of Russia 2018' [9].

The leader of urban transport is the bus. In 2017, 5.425 million people were transported in Russia. This represents almost half (47.47%) of the total volume transportation of the public transport. The growth rate of bus passenger transport in 2016 and 2017 averaged 103.98%, which characterizes the annual increase in demand.

The second place is taken by the metro and is the most stable in the number of passengers carried per year. In the last 3 years, the difference in the number of passengers is less than 1%. However, the length of the metro and the number of underground railway carriages increased by an average of 2.38 % and 2.65% respectively in 2015 and 2017.

Trolleybuses, trams and inland water-borne transport are the types of transport that have gradually reduced the number of carried passengers. The decrease is due to the reduction in the availability of rolling stock, which can be seen from the data presented in Table 2. The rolling stock is ageing naturally and no new rolling stock is being released.

Table 2. Availability number of rolling stock, thousand units*.

Type of transport (thousand units)	2015	2016	2017
Tramcars	8.0	7.8	7.7
Trolleybuses	10.2	10.0	9.4
Availability of river and lake vessels (passenger and cargo-and-passenger vessels)	1.9	2.0	1.3

* Calculated on the basis of data of the Federal State Statistics Service 'Transport of Russia 2018' [9].

It should be noted that there is a general decrease in the use of public transport by passengers since 2005. It is associated with using personal cars.

The second stage – the 'passenger turnover' is used as an index for the urban transport analysis (Table 3).

The 'passenger turnover' shows not only a change in the demand for a particular type of urban transport, but also the efficiency of its use. An example of the most efficient transport is the metro. The metro passenger turnover coefficient is approximately at the same level for 12 years.

Based on the comparison of the tables 1 and 2, it can be concluded that the bus is the most demanded public transport in Russia, but not the most efficient.

The higher passenger turnover coefficient is due to three aspects (for example, the comparison of bus and metro):

Table 3. Public transport passenger turnover by urban traffic (billion passenger-kilometers)*.

Types of urban transport	2005	2010	2015	2016	2017	Structural share in 2017	Base index 2017/ 2005	Base index 2017/ 2010	Chain index 2017/ 2016
metro	43.4	42.4	4.,6	44.1	44.1	47.47	101.61	104.01	100
bus	48.9	39.1	32.3	34.7	35.9	28.86	73.42	91.82	96.66
trolleybus	15	7.1	6	5.5	5.2	12.04	34.67	73.24	105.77
tram	13,5	6.7	4.8	4.6	4.3	11.61	31.85	64.18	106.98
inland water-borne	0.06	0.06	0.06	0.06	0.04	0.03	66.67	66.67	150
TOTAL	120.96	95.7	88.06	89.36	90.04	100	74.44	94.09	99.24

* Calculated on the basis of data of the Federal State Statistics Service 'Transport of Russia 2018' [9].

- Lots of passenger seats. For example, the underground railway carriages consist of many carriages, thus has far more passenger seats than any bus has.
- Correct design of route plans. The bus route can be created through 'empty zones', while the metro train lines are created according to the most demanded route. Under the 'empty zones' are supposed zones with stops that are in least demand. Therefore, the passenger turnover coefficient decreases with the increase of 'empty zones'.
- The distance between stops. Short distances between stops increase the number of potential passengers.

The above aspects lead to the conclusion that the correct value of the passenger turnover coefficient for each type of urban transport is individual, depending on the specifics of the route and the construction.

The third stage of the analysis involves the consideration of the structural share for each type of urban transport by working lifespan. Table 4 shows the structural analysis of urban transport in Russia by type and working lifespan.

Buses have the most favourable age structure. The buses working lifespan is divided into three age groups, which have a maximum difference of 13% in 2017. The fleet of buses is approaching the ideal proportion of 30/30/30, which will lead to the equivalent entry and exit of vehicles.

Age structure of all other urban transport types is negative. Trolleybuses, tramcars and underground railway carriage, river and lake vessels in their structure have about 40% of the fleet that has completed its working lifespan and should be decommissioned. This category is a candidate for utilization and recycling. The passengers' transportation on end-of-life vehicles can be hazardous and environmentally harmful. The general urban transport system is subject to increased volatility due to possible breakdowns of older rolling stock.

All types of public transport should have a structure in which the entry and exit of vehicles are carried out in an even manner. In Russia, most of the urban transport needs to be urgently updated.

3. Results

In general, over the last few years, urban public transport has been characterized by an increase in the use of 'flexible transport'. 'Flexible transport' is a transport which route can be changed without changing mobile systems. Bus is an example. Trams, trolleybuses and metro train are not included in this category, because when a route is changed, the entire system have to be changed. This situation naturally causes a lot of problems and a lot of costs. 'Flexible transport' has the largest areas coverage according to route construction, therefore the bus is the most demanded transport. Passengers try to choose the route in such a way as to get as close as possible to the required location.

IOP Conf. Series: Materials Science and Engineering **966** (2020) 012098 doi:10.1088/1757-899X/966/1/012098

Table 4. Structural analysis of buses by working lifespan*.

Age structure of buses on the balance of organizations, %					
Working lifespan, years	2005 year	2010 year	2015 year	2016 year	2017 year
Up to 5	31	42	36	38	37
5 to 10	24	29	41	38	38
More than 10	45	29	23	24	25
Age structure of tramcars, %					
Working lifespan, year	2005 г.	2010 г.	2015 г.	2016 г.	2017 г.
Up to 5	6	11	8	8	10
5.1–10	5	7	12	11	10
10.1–15	18	5	7	8	9
15.1–20	37	17	4	4	4
More than 20	34	60	69	69	67
Age structure of trolley buses, %					
Workinglifespan	2005 г.	2010 г.	2015 г.	2016 г.	2017 г.
Up to 5	21	24	19	17	16
5.1 to 10	14	23	28	28	29
More than 10	65	53	53	55	55
Age structure of underground railway carriages, %					
Working lifespan	2005 г.	2010 г.	2015 г.	2016 г.	2017 г.
upto 5	7	12	22	23	22
5. 1–10	6	8	11	11	13
10.1–15	17	6	7	7	9
15.1–20	15	16	5	4	4
20.1–25	18	15	13	12	10
More than 25	37	43	42	43	42
Age structure of river and lake vessels, %					
Working lifespan	2005 г.	2010 г.	2015 г.	2016 г.	2017 г.
Up to 1969	38.0	32.7	28.6	28.6	25.5
1970–1979	23.5	20.0	17.4	17.4	16.9
1980–1989	27.5	25.9	27.2	27.2	26.4
1990–1999	7.0	7.6	7.7	7.7	7.6
2000–2009	4.0	13.8	12.4	12.4	13.8
2010–2015	–	–	6.7	6.7	9.8

* Calculated on the basis of data of the Federal State Statistics Service 'Transport of Russia 2018' [9].

It can be concluded from the analysis that the priority transport for Russian citizens is metro and buses. In order to improve the urban transport system, it is first necessary to update the fleet of trams, trolleybuses and inland water-borne vessels.

All types of urban transport are very important and should work in a common structure, complementing each other. It is important to monitor changes in the population's demand for a particular route. The routes are designed by the local municipality depending on the economic efficiency and exploitation of the infrastructure of the particular type of transport. The municipality should undertake corrective measures to encourage all transport types for harmonious and interrelated work.

Public vehicles which are in a state of disrepair and have reached the end of their useful life must be utilized of. At present Russia does not have enough specialized facilities for their utilization. Lack of financial resources of the transport companies' owners, ineffective legislation concerning using the vehicles which have reached the end of their useful life are the reasons for the delay in the renewal of the fleet and the lack of demand for the utilization.

The total result is that the public transport system depends on the number, structure, condition of the fleet and the possibility of vehicles' utilization. Therefore, in order to systematize the quantity and structure, for timely maintenance and possibility of the vehicle utilization we propose to introduce a life cycle contract for the public transport exploitation.

4. Discussion. The life cycle contract perspective for public transport

The main problem of public transport in Russia is its structure. Trams, trolleybuses, water vessels, subway cars with on end-of-life working lifespan must be exploited to maintain a sufficient number of vehicles in the fleets to load all routes.

Restructuring should be carried out through the update of the fleet. When the authorities impose prohibitions on the exploitation of unusable (by age, or by technical condition) vehicles, the obligation to utilize the end-of-life vehicles will be introduced at the same time. The difficulty of the upgrade will be the inability to recycle and utilize the long-working lifespan urban vehicles.

It is possible to bring the industry to a new level of recycling and utilizing through the introduction of a life cycle contract. Life cycle contract means a contract for the manufacture, supply, maintenance and following utilization of a vehicle. In the implementation of a life cycle contract for a public transport vehicle, it is proposed to use foreign and domestic experience.

The first life cycle contract was concluded in the UK in 1992 based on the model 'Private Finance Initiative'. An example of a contract based on this model is a high-speed rail infrastructure project. This section connects London with the tunnel under the English Channel. The contract includes: financing, design, construction, exploitation, repair and maintenance of the railway line [10, 11].

In Spain, a life cycle contract was concluded in 2009 for the production of underground railway carriages with their subsequent maintenance in the amount of 302 units for Madrid. The purchase took place under a 17-year leasing scheme [10, 11].

In 2018, in Russia OOO 'Uralskiye lokomotivy' produced and delivered 90 'Sinara' main-line locomotives in accordance with the life cycle contract. The contract provides for 28 years of service. OOO 'STM- Servis' will be engaged in the maintenance of railway equipment. OOO 'STM-Servis' and OOO 'Uralskiye lokomotivy' are part of the Sinara Group [12].

An example of a life cycle contract in Russia is the contract of the Ministry of Defence with the enterprise 'KamAZ' for the supply and maintenance of military equipment. Unlike other examples, KamAZ should utilize the decommissioned military equipment [13].

5. Conclusion

Today, the implementation of all stages of the life cycle has been developed at a sufficient level, except for the final stage – 'decline'(utilization). Utilization means a complex technological chain of operations using special production facilities, which are not enough in Russia.

The introduction of a full cycle in life cycle contract will result in the following important utilization aspects:

1) the manufacturer will have to utilize and independently recycle the previously delivered vehicle in accordance with the life cycle contract, or enter into an agreement with a third-party organization specializing in the technological recycling operations implementation;

2) all costs associated with utilization and recycling are included in the cost of the vehicle;

3) the depreciable value of the vehicle is calculated by the manufacturer at the design stage (before the production process) and is deducted from the value of the vehicle to reduce the contract price. Utilization costs are accounted and the customer does not have to pay for depreciable value. The manufacturer will eventually sell the recycled scrap metal or use it locally, thereby obtaining the depreciable value of the vehicle. If the depreciable value is not deducted at the stage of the contract price calculation, then the manufacturer will receive a double depreciable value: from the employer when the contract conclusion and from the use of the recycled material as feedstock. Depreciable value should be calculated on the basis of whether the producer resold the recycled waste or uses it as a feedstock for its production;

4) the manufacturer maintains the vehicle throughout the entire period of operation, therefore, all parts that have failed, which have been replaced, are also utilized of by the manufacturer.

Enterprises should move to a cyclical economy in their own industries to participate in life cycle contract transactions. Producers will be directly dependent on the quality of their product, as they will have to service their product not only during the warranty period, but during the working lifespan of the product. All replaced parts will need to be utilized, which will lead to additional costs that must be foreseen in advance.

The adoption of a life cycle contract can be the beginning of a vehicle utilization system development in Russia, which will make it possible to adjust the public transport system.

References

[1] Ignatov V I2016 Recycling fee: the path to the development of the industry in Russia or its further degradation? *Polytematic network electronic scientific journal of the Kuban state agricultural university* **121** pp 1190–206 DOI: 10.21515/1990-4665-121-075

[2] Kuznetsova E, Markina A, Parshina V and Amosov N 2020 Optimization of Locating of Recycling Facilities for Vehicles in the Region *VIII International Scientific Siberian Transport Forum. TransSiberia 2019. Advances in Intelligent Systems and Computing(Electronic Materialsvol1115*) ed. Z Popovic, A Manakov*et al* (Springer, Cham) pp 218–32 DOI: 10.1007/978-3-030-37916-2_23

[3] Privalova O A 2017 Research of the state and development of public transport as a logistics system in the infrastructure of megapolis *Trends in the development of science and education* **31-2** pp 36–7 DOI:10.18411/lj-31-10-2017-33

[4] Zueva O N, Zhuravskaya M A and Sidorenko A M 2018 Logistic Integration and Coordination of Urban Public Transport Management *Journal of new economy* **19-1** pp 51–61 DOI: 10.29141/2073-1019-2018-19-1-5

[5] Frolova E G 2015 Methods for the optimization of life cycle contracts in the construction of roads in the conditions of high inflation *Actual directions of scientific researches of the xxi century: theory and practice* **3 8-3 (19-3)** pp 32–4 DOI:10.12737/15587

[6] Gerasimenko O A and Avilova Zh N 2017 Concept, signs and advantages of application of life cycle contracts as forms of state-private partnership in the region *Bulletin of the Belgorod state technological university by V.G. Shukhova* **4** pp 203–8 DOI: 10.12737/article_58e613383985b3.58961357

[7] Volodina P M and Frolov I E 2018 Russian Specification of the Life Cycle Contract Application for the State Defense Order Performance *Scientific works: Institute of national economic forecasting of the RAS* **16** p236-47 DOI: 10.29003/m263.sp_ief_ras2018/236-247

[8] Boreiko O, Teslyuk V, Zelinskyy A and Berezsky O 2017 Development of models and means of the server part of the system for passenger traffic registration of public transport in the 'smart' city *Eastern european magazine of advanced technologies* **1, 2 (85)** pp 40–7 DOI: 10.15587/1729-4061.2017.92831

[9] Federal State Statistics Service 'Transport of Russia 2018' Available online: https://www.gks.ru/free_doc/doc_2018/transp18.pdf (accessed on 01.06.2020)

[10] Sazykina S A, Rastopchina Yu L, Boltenkov V I and Boltenkova Yu V 2016 Life cycle contracts as an example of public-private cooperation in Russia's transport business. *Knowledge. Understanding. Skill* **2** pp 168–78. DOI:10.17805/zpu.2016.2.15

[11] New mechanisms of public-private partnership in Russia (life cycle contracts). Official site of vegaslex.ru. Available online: https://www.vegaslex.ru/analytics/publications/54349/ (accessed on 01.06.2020)

[12] Ural locomotives fulfill annual contract for the production of 2ES6 electric locomotives. Official site of ulkm.ru. Available online: http://ulkm.ru/actions/uralskie-lokomotivy-vypolnili-godovoj-kontrakt-po-proizvodstvu-elektrovozov-2es6/ (accessed on 01.06.2020)

[13] KamAZ will receive a military life cycle contract. Available online: https://lenta.ru/news/2013/07/12/kamaz/ (accessed on 01.06.2020)

Upgrading of induction motors with energy efficiency enhancement at the electrical repair shops and sites of the running industrial companies

R G Mugalimov[1, 3], A R Mugalimova[2] and R A Zakirova[1]

[1] Federal State-Funded Educational Institution of Higher Professional Education 'Magnitogorsk State Technical University named after G I Nosov', Magnitogorsk, 455000, Russia
[2] Scientific Research and Design Bureau 'Energosberezheniye' LLC, Magnitogorsk, 455000. Russia

E-mail: [3] energosberegenie@rambler.ru

Abstract. It is demonstrated that the main disadvantage of induction motors is a relatively low power factor. In order to enhance energy efficiency of the electric drives, it is proposed to use an innovative technology of their upgrading in the course of overhauls. A distinctive feature of the proposed overhaul is the recalculation of the stator winding considering the real state of the stator steel and uses the phenomenon of current ferro-resonance in the machine's electromagnetic system. This makes it possible to create an induction motor with the desired power factor, even equal to one. Upgrading allows increasing the efficiency factor and reducing the active power losses by 3–7% in the electric drive system and in the power supply system. Additional expenses for upgrading costs amount to 23–30% of the conventional overhaul costs. However, these additional expenses can be justified within 0.25–0.8 years due to energy savings. The results of this R&D project can be recommended for review to the power departments of the running industrial companies, electrical repair shops and energy saving centers.

1. Introduction

Up to 80% of electric drives of working machines and mechanisms at the Russian industrial facilities are based on traditional induction motors (TIM). This is due to their high reliability and low maintenance costs. However, the energy efficiency factor of TIM, as measured by electrical efficiency and power factor, is relatively low. Depending on the power and rotor speed, the rated efficiency and cosφ are within the following range: efficiency = 65–93%, cosφ = 0.7–0.9 [1]. The smaller numbers refer to the low-power motors. Their energy efficiency, equal to the product of electrical efficiency and power factor, is within the range of 46–84%. This means that between 16 and 54% of the consumed electrical energy is dissipated in the environment. Between 7 and 35% of the energy is lost directly in the TIM, and between 9 and 19% – in the power supply systems and drive circuits.

Low energy efficiency is caused by the fact that the traditional induction motors consume not only active power, which is converted into mechanical and thermal energy, but also reactive power (idle current), which is not converted into other types of energy. Reactive idle current, running in the power circuits of the drive and power supply system, causes active power losses. The lower is cosφ, the higher is the loss of active power, and the less efficient is the electrical system. Theoretical and experimental studies, as well as practical experience have proved that up to 4.5–6.0% of electrical energy is lost in the electrical facilities because of the relatively low cosφ. Given the average annual electricity production of 1.1 trillion kWh, the share of induction electric drive in the industry of 85 % (0.85 relative units), the average motor load factor of 75% (0.75 relative units), with consideration of losses from reactive idle current, the loss of electric energy of the Russian companies is estimated at the level of 31.5 billion kWh. Therefore, the problem of enhancement the energy efficiency of electrical facilities is of great importance.

The international standard IEC 60034 defines the energy efficiency of induction motors with four levels of energy efficiency classes: IE1 – standard efficiency; IE2 –high efficiency; IE3 –premium efficiency; IE4 –super premium efficiency. In the countries of the European Union, the use of electric motors with certain levels of energy efficiency is prohibited in accordance with the Regulation of the commission (EU) 2019/1781 document [2]. European manufacturers of induction motors upgrade their energy efficiency class by increasing the amount of active materials – electrotechnical steel, winding copper, aluminum – by an average of 20–25%, which results in efficiency gains of 1.5–3.5 %. At the same time, the cost of an energy efficient electric machine increases significantly. In the Russian Federation, the majority of general industrial induction motors with voltage of up to 1000 V and power of up to 100 kW, manufactured prior to 2010, are of standard energy efficiency class (IE1). Experience has shown that the energy efficiency of induction motors is even lower after a conventional overhaul. Laws № 197-FZ of the Russian Federation 'On energy saving and improving energy efficiency' dated 11.07.2011, № 200-FZ dated 18.07.2011 and № 242-FZ of 2017 prohibit the manufacture and use of induction motors of standard efficiency class [3].

It is proved that the most optimal way to reduce losses of active power in electrical engineering facilities with voltage of up to 1000V is to fully compensate the reactive power of each electrical consumer [4]. Nowadays, this is realized by parallel connection of capacitive reactive power compensators (RPC) to electrical consumers. This method is based on the current resonance phenomenon well-known in the electrical engineering, whereby the reactive induction current is compensated by the reactive capacitive current. However, the practice shows that due to significant capital costs for RPCs, the industrial companies use them for voltages of 10kV and higher, despite the fact that a significant share of electric energy is lost in networks with voltages of up to 1000V.

TIM life cycle does not exceed 25 years. During this period, the motor is overhauled from 3 to 7 times due to the destruction of the stator winding. As a rule, the overhaul involves replacing the damaged stator winding according to the original winding data. A major overhaul may involve a motor upgrade with modifications to the rated voltage and rotor speed. After each overhaul or upgrade of the TIM at the electrical repair shop of the company, its energy efficiency is usually reduced by 0.5–1.0 % [5]. Given the above, there is a problem of improving the technology of manufacturing new energy efficient induction motors and overhaul of the existing ones along with upgrading their energy efficiency class.

2. Modernization of induction motors

In contrast to technologies [6, 7] this study proposes an innovative technological scheme of overhaul of induction motors, which allows improving the energy efficiency class by upgrading the electromagnetic circuit of the stator [5]. The essence of upgrading is the improvement of the electromagnetic circuit of the stator, in accordance with invention [8]. Here the upgraded motor is called a Power-efficient Induction Motor (PEIM) with individual compensation of reactive power.

Figure 1 shows the technological diagram of overhauling and modernization of TIM in PEIM. Operations 1–15 refer to the conventional overhauling procedure of the motor. Operations 16–20, marked by a dashed line, refer to innovative technological operations. Figure 1 shows the following: 1 – disassembly of the induction motor; 2 – burnout or fluxing of the stator winding by chemical solution; 3 – removal of the stator winding; 4 – visual and instrumental assessment of the state of the electrotechnical steel of the stator; 5 – making a decision on either an overhaul according to the company's repair procedure and the TIM's diagram, or conversion of the TIM into the PEIM, or retirement (disposal) of the induction motor; 6 – manufacturing and inserting of the stator windings according to the TIM's factory diagrams; 7 – testing of untreated stator winding; 8 – treatment of the stator winding; 9 – drying of the stator winding; 10 – assembly of the IM; 11 – testing of the IM on the inspection and testing stand; 12 – delivery of the IM to the Customer; 13 – recommendations to the Customer regarding recycling (disposal) of the IM; 14 – decrease in rated capacity R2n of the TIM; 15 – experimental calibration of the motor ratings and mechanical data on the loading unit; 16 – computer recalculation of TIM winding parameters into PEIM winding parameters; 17 – computer simulation of the rating and mechanical data of PEIM; 18 – making a decision on rewinding the TIM into the PEIM or rewinding according to the factory winding data and diagrams of the TIM; 19 – rewinding the TIM into the PEIM according to the technology and manufacture diagram of the PEIM; 20 – express calibration of the motor rating and mechanical data according to the oscillograph charts of starting, reversing and braking of the IM without any load by means of a specialized hardware-software facility.

To evaluate the economic feasibility of TIM conversion into PEIM, a method and computer program has been developed [5,9].

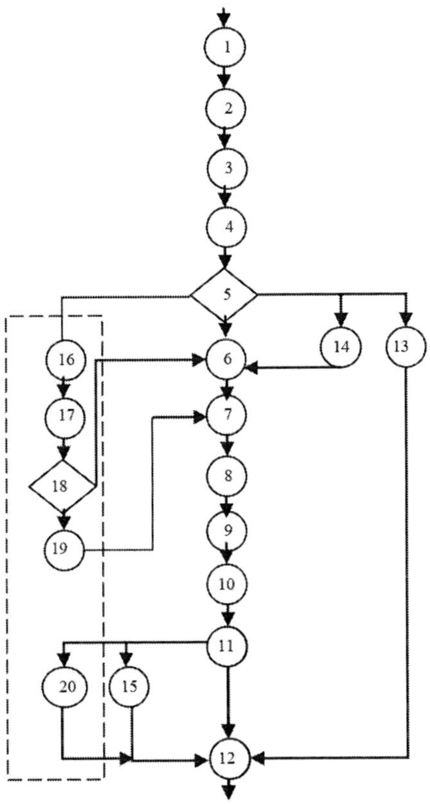

Figure 1. Process stages of repair of the traditional induction motor and conversion of the traditional induction motor into the power-efficient induction motor.

Computer recalculation of TIM winding parameters to the parameters of PEIM is carried out on the basis of the experimental study of the stator steel. The essence of the approach is described below. The ferromagnetic core of the stator is tested by induction method using a motor with a rotor removed (removed from the stator bore). For magnetization of the stator core, its back is covered by an experimental magnetizing coil with a small number of windings $W_1 = 3 \div 5$ insulated by a flexible wire. A reference voltage is applied to the magnetizing coil U_{1Ref}, determined from the following formula:

$$U_{1Ref} = \frac{U_{1R}W_1}{W_{1EW}}, \tag{1}$$

where: U_{1R} – rated phase voltage of the motor, W; W_{1EW} – number of efficient stator phase windings.

Reference magnetisation current value, I_{RefMm}, is determined from the formula:

$$I_{RefMm} = I_{Mm} \cdot W_{1EW}/W_1, \tag{2}$$

where: I_{Mm} – rated magnetizing current (no load current) of the induction motor.

With given values of U_{1Ref} and I_{RefMm} the active power losses in stator steel, P_{s1Ref} are measured with a wattmeter, and oscillograph charts of electrical values shown in Figure 2 are recorded.

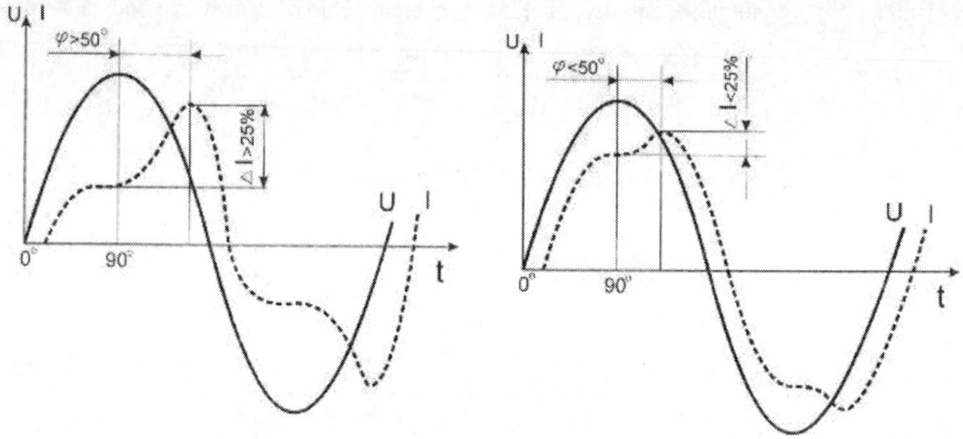

a) b)

Figure 2. Oscillograph charts of the magnetizing current and voltage:
a) for a fully operational stator core in good order; b) for a knowingly inoperable stator core.

The resulting oscillograph charts of the voltage and magnetizing current are analyzed based on the described method, and the phase shift φ_{Ref}, is determined by the peaks of the voltage and current curves.

$$\varphi_{Ref} = \varphi_{u\max} - \varphi_{i1\max}. \tag{3}$$

The current value is calculated at $\varphi_u = 90°$ and the difference is determined:

$$\Delta i = i_{\max} - i_{\varphi u=90°}. \tag{4}$$

For the fully operational stator core (Figure 2, a) the permissible minimum shift values of the voltage and current initial phases and Δi should meet the following conditions:

$$\varphi_{Ref} = \varphi_{Ref\min} > 50°; \;\; \Delta i_{\min} \geq 0{,}25 i_{\max}; \tag{5}$$

– for a knowingly inoperable stator core, with high losses of active power (Figure 2, b) the following conditions should be met:

$$\varphi_{Ref} = \varphi_{Ref\min} < 50°; \;\; \Delta i_{\min} < 0{,}25 i_{\max}; \tag{6}$$

– for the cores with an aluminum frame and without a frame the following conditions should be met:

$$\varphi_{Ref} = \varphi_{Ref\min} > 60°; \ \Delta i_{\min} > 0{,}25 i_{\max}. \quad (7)$$

The specific losses in steel are determined by the following ratio:

$$p_{s1} = P_{s1ref}/G_{s1}, \quad (8)$$

where G_{s1} – the mass of the stator steel, in kg, determined in accordance with the known formulae.

When assessing the condition of a stator core without a frame or with an aluminum frame, the specific losses in steel, p_{s1}, should not exceed the following values:

$$p_{s1} \leq 4{,}5 \div 5{,}0\ W/kg;$$

and the specific losses with provision for a steel or iron frame:

$$p_{s1} \leq (1{,}07 \div 1{,}1) * 5\ W/kg.$$

The proposed technology for the modernization of induction motors allows to increase the energy efficiency of electrical complexes by 3–7% [5, 11].The cost of major repairs with modernization by the proposed technology, for example, an asynchronous motor of a flotation machine with a capacity of 30 kW, voltage of 380 V, magnetic field rotation speed of 750 rpm, is 36865 rub. The traditional repair of the same engine is 27397 rubles. Additional costs equal to their difference, 9468 rub were generated. The energy savings recorded during the year amounted to 10300 kWh. When the price of electric energy is 3.0 rub/kW * hour, the cost of energy saved is 30900 rubles. Other things being equal, the payback period for additional costs does not exceed 0.3 years. The practice of modernizing induction motors according to the proposed technology for the period 2010–2019 shows that each kilowatt of installed power saves up to 350–750 kWh per year, depending on the operating conditions and loads of the electric drives.

3. Conclusion

1. Asynchronous motors with individual compensation of reactive power, using ferro-resonance currents in an electromagnetic system, and cosφ = 1.0, they increase the energy efficiency of electrical complexes.

2. The conversion of traditional asynchronous motors to energy-efficient asynchronous motors in electric repair shops can increase the energy efficiency of industrial enterprises using asynchronous motors by 3–7%. The additional costs of modernization with an increase in the engine's energy efficiency class are justified within 0.25–0.8 years due to energy savings.

3. The results of a research project can be recommended to specialists of electric repair enterprises and energy conservation centers

References

[1] Kravchik A E, Shlaf M M, Afonin V I and Sobolenskaya E A 1982 *Induction motors of 4A Series: Reference Book* (Moscow: Energoizdat) p 504

[2] Commission Regulation (EU) 2019/1781 of 1 October 2019 laying down ecodesign requirements for electric motors and variable speed drives pursuant to Directive 2009/125/EC of the European Parliament and of the Council, amending Regulation (EC) No 641/2009 with regard to ecodesign requirements for glandless standalone circulators and glandless circulators integrated in products and repealing Commission Regulation (EC) No 640/2009. Available online: https://eur-lex.europa.eu/legal-content/EN/TXT/PDF/?uri=CELEX:32019R1781&from=EN (accessed on 1 June 2020)

[3] On energy saving and improving energy efficiency and amending certain legislative acts of the Russian Federation 2012 (Moscow: LLC 'Reed Group') p 80 (Latest legislation)

[4] Vagin G Ya, Yurtaev S N 2008 On the necessity to adjust the method of reactive power compensation in electrical networks with voltage up to 1000V *Promyshlennaya Energetica* no **5** pp 31–40

IOP Conf. Series: Materials Science and Engineering **966** (2020) 012099 doi:10.1088/1757-899X/966/1/012099

[5] Mugalimov R G 2011 *Induction motors with individual compensation of reactive power and electric drives [Monograph]* (Magnitogorsk: Magnitogorsk State Technical University) p 250

[6] Mohamed K, Mahmoud E M and Abdelowahed H 2017 Evaluation of energy savings by using high efficiency motors in a thermal power station *International Journal of Green Energy* vol **14** issue **10**

[7] EMT. 2016. EASA. The effect of repair rewinding on motor efficiency, rewind study and good practice guide to maintain motor efficiency. In website of IECEx SYSTEM. Available online: www.iecex.com (accessed on 22 March 2016)

[8] Mugalimov R G, Savitskiy A L and Savitskaya L D 1998 Patent 2112307 RU Induction compensated electrical machinery

[9] Mugalimov R G, Zakirova R A and Mugalimova A R 2016 *Programmny kompleks dlya rascheta i optimizatsii sebestoimosti traditsionnogo kapitalnogo remonta i remonta s povysheniem klassa energoeffektivnosti asinkhronnykh elektrodvigateley [A software programme for estimation and optimization of costs incurred by conventional induction motor overhaul and induction motor overhaul with efficiency upgrade]*. State registration certificate no. 2016610278, registered in the Registration Book of software programmes 11.01.2016

[10] Mugalimov R G, Zakirova R A and Mugalimova A R 2016 Energy efficient asynchronous motors, technical and economic advantages and optimization of their creation *Electrotechnical systems and complexes* no **2 (31)** pp 30–4

[11] Mugalimov R G, Barankova I I, Zakirova R A, Mugalimova A R, Mikhailova U V and Nikiforov G V 2017 Feasibility study of the overhaul of asynchronous motors with an increase in their energy efficiency class *Bulletin of the Magnitogorsk State Technical University named after G I Nosova* no **2** pp 101–9

IOP Conf. Series: Materials Science and Engineering **966** (2020) 012100 doi:10.1088/1757-899X/966/1/012100

Processes in spark electric discharge in air, nitrogen and oxygen

I M Piskarev[1, 3] and N A Aristova[2]

[1] Skobeltsyn Institute of Nuclear Physics, Leninskie Gory, MSU, 1 (2), Moscow, 119234, Russia

[2] Ural Federal University named after the First President of Russia B N Yeltsin, Nizhniy Tagil Technological Institute, 59, Krasnogvardeyskaya str., Nizhniy Tagil, 622000, Russia

E-mail: [3] i.m.piskarev@gmail.com

Abstract. The yield of redox reactions in aqueous solutions under the action of a remote spark discharge in the air, nitrogen and oxygen was studied. Mohr's salt and potassium permanganate were used as test substances. The direct and indirect effects of a source of active particles were compared. On direct exposure active species can immediately interact with the substance. By indirect action the source of active species directly affects only water. Immediately after treatment water was mixed with a solution of the test substance and the reactions of accumulated active species with the test substance began. When discharged in air medium, the oxidation and reduction reaction yields are approximately the same 5 - 6 1(100 эВ)$^{-1}$ eV. Equal yields means that active species are firstly produced in water. When discharged in the nitrogen and oxygen medium, the oxidation yield is ~15% lower, and the recovery yield is much lower. In this case, there is oxygen or nitrogen in the aqueous solution, so nitrogen oxides cannot be formed in water. So the contribution of products, produced in the discharge region is not more than 15% and that radiation plays a major role in initiating of chemical processes in liquid.

1. Introduction

Cold plasma and plasma activated water (PAW) have great prospects for their wide application and therefore are intensively researched [1, 2]. When producing PAW, the plasma is in direct contact with water [3]. The plasma used for this treatment (gliding arc plasma in particular) has dim light, so the role of radiation in the formation of PAW is considered to be insignificant. The oxidation of methyl orange with gliding arc plasma was studied in this paper [4]. After treatment there is a strong post-effect, which lasts up to 1 hour. It is noted that this effect is due to nitrogen oxides, but the mechanism of their formation and the role of the gas medium were not considered. The influence of nanosecond discharge plasma treatment of seeds in the air was studied in work [5]. A significant improvement in seed germinating ability was obtained. The discharge has not been optimized in terms of the choice of the gas medium. In work [6] Advanced Oxidation Technology variant of purification of waste water with the aid of a discharge inside the liquid is considered. It can be assumed that the efficiency of the process will depend on the degree of saturation of water with the gas mixture and the composition of this mixture.

The most potent oxidative agent formed in electric discharge plasma is peroxinitrite and peroxinitrous acid. These reagents can be formed if there is nitrogen and oxygen in the gas phase. The properties of peroxinitrite and peroxinitrous acid are analysed in the overview [7]. In the case when there is no oxygen or nitrogen in the gas phase where the electric discharge occurs, only hydroxyl radicals can be formed. The formation of hydrogen peroxide under the action of light radiation in water containing dissolved nitrogen and oxygen was experimentally found in work [9]. A mechanism for forming hydrogen peroxide in water is proposed in [10]. It has been shown that hydrogen peroxide can be formed by radiation with wavelengths of 200 to 1260 nm, provided that there is dissolved oxygen in the water.

Studies [11, 12] established a strong chemical effect created in water and aqueous solutions by a spark electric discharge generator in air. The effect can be due both to the radiation of the plasma in liquid itself and to the nitrogen oxides diffusing from the discharge area. Nitrogen oxides are formed due to the Zeldovich mechanism in the discharge area containing nitrogen and oxygen in the air [13]. They can diffuse from the discharge area and give a direct contribution to the chemical action of electric discharge plasma.

Thus, studies have shown that the gas medium in which the electric discharge occurs has a significant influence on the properties of the processes under this discharge. Therefore, it is of interest to obtain additional information about the role of processes under the action of radiation only and the combined action of radiation and nitrogen oxides diffusing from the discharge area. For this purpose we have performed experiments with spark electric discharge in the air and in the medium of gases: separately in the medium of nitrogen and oxygen. When discharged only in nitrogen or only in oxygen, the Zeldovich mechanism of nitrogen oxide formation does not work, and the whole effect in the treated liquid will be determined only by the radiation of the plasma.

2. Materials and methods

The block-scheme of the experimental setup is shown in figure 1. A source of pulsed radiation was the generator of spark discharge SD50 [11, 12]. The spark discharge occurred between solid electrodes connected with the discharge capacitor $C = 680$ pF. A high voltage power of 11 kV was supplied to the capacitor via a ballast resistor $R = 8$ M Ω. When the high voltage was turned on, a self-supporting spark discharge began. Full duration of current pulse was 5 μs, leading edge was about 50 ns, energy of pulse 8.1×10^{-3} J, pulse repetition frequency was 50 Hz, discharge power was 0.4 J/s. The duration of the current pulse was determined by the time of charge dissipation. The current consumed from a high voltage power supply was 0.7 ± 0.02 mA.

When discharged in a gas medium (nitrogen or oxygen), the reaction cavity and the cavity with the solution to be treated were purged with these gases. The liquid sample was saturated with dissolved gas, which was used for purging (only with nitrogen, or only with oxygen). Technical nitrogen (purity 99.6%) and technical oxygen (purity 99.7%) from the cylinder were used for purging. The time spent on gas purging before turning on the discharge was 16 minutes.

Direct and indirect effects of the source of active species were compared in this work. The direct effect means that the source affects directly the sample with the dissolved sample substance, the active species can interact with the substance at once. The indirect effect means that the source of active particles has a direct effect only on water. In indirect effect, active particles form and accumulate in water. Immediately after treatment, the water was mixed with the solution of the sample substance, where the reactions of the accumulated active species with the sample substance began.

To determine the chemical effect, oxidation-reduction reactions with sample substances were used. As sample substances, aqueous solutions of Mohr's salt and potassium permanganate were used. The yield of oxidative equivalents, formed during treating, was determined in the sample by the oxidation of ferrous iron in Mohr's salt $Fe^{+2} \rightarrow Fe^{+3}$. The yield of recovery equivalents was determined by the reduction of manganese $Mn^{+7} \rightarrow Mn^{+2}$ in solution of potassium permanganate.

The concentration of Mohr's salt was 20 g / l, $[Fe^{+2}] = 5.1 \ 10^{-2}$ mol / l. The concentration of potassium permanganate used in the treatment was 1.58 g / l (0.05N). 21 ml / l of concentrated sulfuric acid (0.4 M) was added to the solutions. The concentration of oxidized Fe^{+3} was determined by the optical density of the band 304 nm, ε = 2100 ± 50 L (mol cm) $^{-1}$. The concentration of potassium permanganate was determined by the optical density of the band 527 nm, ε = 2160 ± 50 L (mol cm) $^{-1}$.

Absorption spectra of the samples were measured with a SF-102 spectrophotometer manufactured by AKVILON, Russia. The thickness of the cuvette was 10 mm. The optical density $A = \lg(I_0/I)$ (Bel) was measured relative to distilled water. The pH value was measured by an Expert-001 device manufactured by EKONIKS, Moscow, Russia. Chemically pure reagents were used with twice distilled water at pH = 6.5. For each mode, 10 samples were generated; the results were averaged.

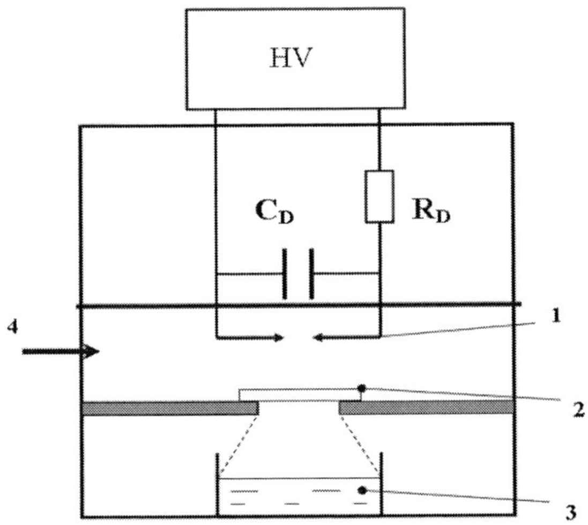

Figure 1. Block-scheme of the experimental setup. HV – high-voltage source 11 kV; C_D, R_D - discharge capacity and ballast resistor; 1 – discharge electrodes; 2 – location of silica glass; 3 – Petri dish with processed liquid; 4 – gas purging (nitrogen or oxygen), except discharge in air medium.

3. Results and discussion

The mechanisms of the interaction of active species formed in water under the action of the SD 50 generator with sample substances are analyzed. It was shown that chain reactions of oxidation of sample substances do not take place. All reactions are slow; the total reaction yield was determined 3 days after treatment.

The absorption spectra of water samples with a volume of 5 ml after treating with a remote plasma discharge of the SD 50 generator for 3 minutes are presented in Fig. 2. It is seen that during the discharge in the air medium, nitrous acid is forming (Curve 1), the acidity of the sample is pH = 2.47. When discharged in a nitrogen and oxygen medium, traces of nitrous acid are not visible (Curves 3 and 4), the acidity of the samples remains at the initial level, pH = 6.5. The absence of nitrous acid under these conditions is due to two circumstances. Firstly, nitrogen oxides do not form in the discharge region, since in pure nitrogen or pure oxygen the Zeldovich mechanism of the formation of NO • and NO2 • does not work. And secondly, since the sample contains only dissolved nitrogen or only dissolved oxygen, the mechanism of formation of nitrogen-containing compounds under the action of radiation in the liquid which contains dissolved nitrogen and oxygen also does not work.

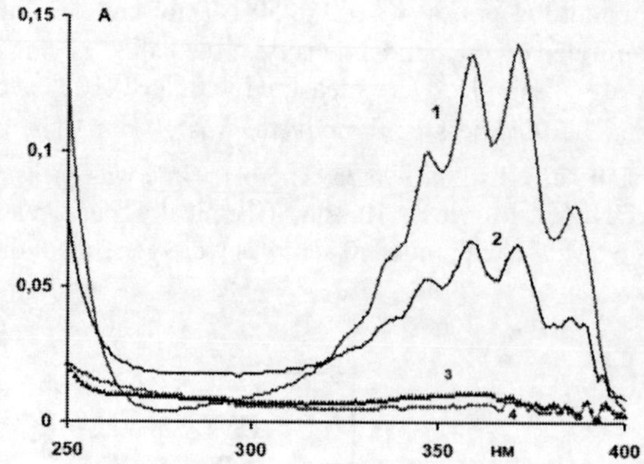

Figure 2. Water absorption spectra
1 - discharge in the air medium; 2 - discharge in the air medium, yield of radiation through quartz glass; 3 - discharge in nitrogen; 4 - discharge in oxygen.

Separately, a water sample was processed through quartz glass. In this case, the products formed in the discharge region could not get to the surface of the sample. The sample was in the air, so dissolved oxygen and nitrogen were contained in the water itself. The absorption spectrum of the sample after 3 minutes of processing is shown in Fig. 1, Curve 2. It is seen that in this case nitrous acid is formed, but its optical density is approximately two times lower. This may be due to the fact that quartz glass absorbs radiation, the optical density of the glass at wavelength of 200 nm is A ~ 0.19. It may also be because nitrogen oxides formed in the discharge region cannot contribute to the formation of nitrous acid.

Taking into account the stoichiometry of the reactions, the oxidation and reduction yields were calculated in Table 1. The table shows that in the case of discharge in air, the oxidation and reduction yields are the same for direct and indirect action. Equality of yields means that active species are first produced in water, and then interact with a sample substance that is directly contained in the solution, or added after treatment.

Table 1. Oxidation and reduction yields in the solutions of sample substances under the action of the SD 50 generator, 1 / (100 eV)

Discharge conditions	Direct effect		Inirect effect	
	Oxidation	Reduction	Oxidation	Reduction
In air	5.7 ± 0.9	6.2 ± 1	6.1 ± 0.9	5.6 ± 0.9
In nitrogen	3.7 ± 0.7	0.47 ± 0.05	5.7 ± 0.9	0.27 ± 0.03
In oxygen	1.35 ± 0.3	0.5 ± 0.1	4.53 ± 0.8	0.1 ± 0.015

The oxidation yields of the discharge in nitrogen medium for direct and indirect effect are less than ~ 15% of the discharge in air medium. This means that the role of nitrogen-containing compounds formed in the discharge region in the air medium and diffusing to the sample surface is small and does not exceed ~ 15%.

The recovery yield for the discharge in nitrogen or oxygen medium in all treatment modes is almost an order less than the oxidation yield. The features of the reactions in these cases may be due to the fact that there is only oxygen or only nitrogen in the aqueous solution, and nitrogen oxides cannot form in water. It follows that the nitrogen-containing compounds formed from them under the action of radiation in the case of discharge in air play an important role in oxidation-reduction processes.

Radiation plays the major role in the oxidation-reduction processes under the influence of the spark discharge generator SD 50.

The main reaction mechanism under the action of radiation is indirect, when active species are first formed in liquid, and then enter into reactions.

References

[1] Bruggeman P, Kushner M J, Locke B R, Gardeniers J G et al 2016 Plasma-Liquid Interactions: A Review and Roadmap *Plasma Sources Sci. Technol.* **25 (5)** 053002
[2] Kolb J, Simek M, Holux M, Locke B R et al. 2018 *Plasma Processes and Polymers* **16 (1)** 1800118
[3] Thurumdas R, Kothakota A, Annapure U, Silivenu K et al. 2018 Plasma activated water (PAW): Chemistry, physico-chemical properties, applications in food and agriculture Trends in Food Science & Technology *Trends in Food Science & Technology* **77** pp 21-31
[4] Moussa D, Doubla A, Kamgang-Youbi G and Brisset J-L 2007 Post discharge long life intermediate in the plasmachemical degradation of an azo dye *IEEE Transactions on Plasma Science* **35 (2)** pp 444-453
[5] Dubinov A E, Kozhayeva J P and Zuimatch E A 2017 Changing Germination Rate of Brown Mustard Seeds After Treatment With Plasmas of Nanosecond Electric Discharges *IEEE Transactions on Plasma Science* **45 (2)** pp 294-300
[6] Foster J, Sommers B S, Gucker S N, Blankson I M and Adamovsky G 2012 Perspectives on the interaction of plasmas with liquid water for water purification *IEEE Transactions on Plasma Science* **40(5)** pp 1311-23
[7] Lobachev V L and Rudakov E S 2006 The chemistry of peroxynitrite. Reaction mechanisms and kinetics. *Russ. Chem. Rev.* **75 (5)** pp. 422-444 https://doi.org/10.1070/RC2006v075n05ABEH001212
[8] Nikiforov A, Deng X, Xiong Q, Cvelbar U et al. 2016 Non-thermal plasma technology for the development of antimicrobial surfaces: a review. *J. Phys. D. Appl. Phys.* **49 (20)** 204002
[9] Gudkov S V, Karp O E, Garmash S A. Ivanov V E et al. 2012 Production of reactive oxygen forms in water under the. influence of visual and infra-red radiation in lines of the molecular. oxygen absorption *Biophisics* **57 (1)** pp 5-13
[10] Piskarev I M 2018 Hydrogen Peroxide Formation in Aqueous Solutions under UV-C Radiation. *High Energy Chem.* **52 (3)** pp 194-198
[11] Piskarev I M, Trofimova S V, Ivanova I P and Aristova N A 2012 Formation of active species in spark discharge and their possible use. *High Energy Chem.* **46 (5)** pp. 343-348
[12] Piskarev I M and Ivanova I P 2019 Effect of spark electric discharge between solid electrodes in water. *Plasma Sources Sci. Technol.* **28 (8)** 085008
[13] Zeldovich Y B, Sadovnikov P Y and Frank-Kamenetskij D A 1947 *Nitrogen oxidation during combustion* (Moscow, AN SSSR: Nauka) p 148 [In Russian]

IOP Conf. Series: Materials Science and Engineering **966** (2020) 012101 doi:10.1088/1757-899X/966/1/012101

Determination of the reduction coefficients in a continuous finishing group of stands in a hot rolling mill using artificial neural network

O R Latypov[1, 2], S I Platov[1], R R Dema[1] and N V Urtsev[1]

[1] Nosov Magnitogorsk State Technical University, 38 Lenin Street, Magnitogorsk, 455000, Russia

E-mail: [2] latolegraf@list.ru

Abstract. There are many problems when creating digital doubles. One of which is the definition of the source data: in this case, the definition of private reductions (coefficient ratios) in a continuous finishing group of stands of the broadband hot rolling mill 2000 of PJSC Magnitogorsk Iron and Steel Work. Algorithms were developed to calculate the thickness of the breakdown bar, the number of stands involved, power required for rolling and coefficient ratios to solve this problem. The algorithms are based on well-known solutions (the Imai method) using neural networks. The training of neural networks was conducted on a sample collected for the period from 01.01.2017 to 01.01.2019 work of the mill in the PyTorch library of the interpreted programming language Python. The average error ε_{ME} of the calculation of coefficient ratios (according to the developed algorithms with neural networks) does not exceed 8.9% and the standard deviation σ does not exceed 0.074.

1. Introduction

Broadband hot rolling mills are one of the main means of production of hot rolled products. So, for example, the broadband hot rolling mill (BHRM) 2000 of PJSC Magnitogorsk Iron and Steel Work in 2019 produced 6 million tons of hot rolled strips. On the example of this mill, the following trends in production are observed: increased productivity; increasing requirements for the quality of rolled products (in geometry and mechanical properties), thinning of rolled products, the emergence of new high-strength steel grades, increasing the variety of assortment (for steel grades and strip geometry) in one rolling company (rolling between work roll changes in the finishing group of the mill stands)These trends lead to the need for innovations in production. So, recently the direction (since 2000) for the creation of digital twin of production processes has been developing [1]. So, in relation to BHRM 2000, this technology allows you to simulate a mill's production line in a virtual environment, as well as run a preliminary production optimization. Due to this, the sources of errors or failures can be identified and eliminated even before the actual operation. This significantly saves time and lays the foundation for individualized mass production, since complex production routes can be quickly calculated, tested and programmed with a minimum of cost and effort.

However, with this 'virtual prototyping' many problems arise, one of which is the definition of the source data for modeling physical processes. So, one of these problems is the determination of reduction coefficients in a continuous finishing group of BHRM 2000 stands.

The reduction coefficients in finishing stands are usually determined by loading the main drives (Imai method [2]), while determining the necessary coefficients using statistical methods [3–5]. This article proposes a method for selecting reduction coefficients using artificial neural network models. Artificial neural networks have established themselves, especially in the last decade, as a convenient tool for forecasting and data processing [6, 7]. In particular, for rolling production, artificial neural network models are used to predict power parameters [8, 9], thermal state of the strip and work rolls [10, 11], wear of the work rolls [12], strip profile [13], mechanical properties [14] and etc.

2. Artificial neural networks

2.1. Methods

In the paper for data prediction: (number of stands involved, breakdown bar thickness (strip thickness after leaving the roughing group of stands), reduction coefficients), generalized regression neural networks are used. Learning and building a neural network is implemented using the PyTorch library of the interpreted programming language Python.

The training of neural networks was carried out on the data obtained from the databases of technological parameters of BHRM 2000 of PJSC Magnitogorsk Iron And Steel Work. The training was conducted on a sample collected for the period from 01.01.2017 to 01.01.2019 work of the mill.

2.2. Determination of thickness of breakdown bar

Algorithms of papers [3–5] were taken as the basis for determining the reduction coefficients. According to these algorithms, to determine the reduction coefficients, it is first necessary to determine the thickness of the intermediate roll h_p and the number of N_{st} stands involved in the finishing group.

The paper proposes to determine the coefficient of the total compression η_{sum} using a trained neural network. To determine the thickness of the breakdown bar h_b, an algorithm was developed, shown in Figure 1.

The structure of the neural network is a multilayer perceptron, consisting of an input layer (4 neurons), one hidden layer (containing 9 neurons) and an output layer containing 2 neurons. The 'ReLu' function was used as an activation function (Table 1).

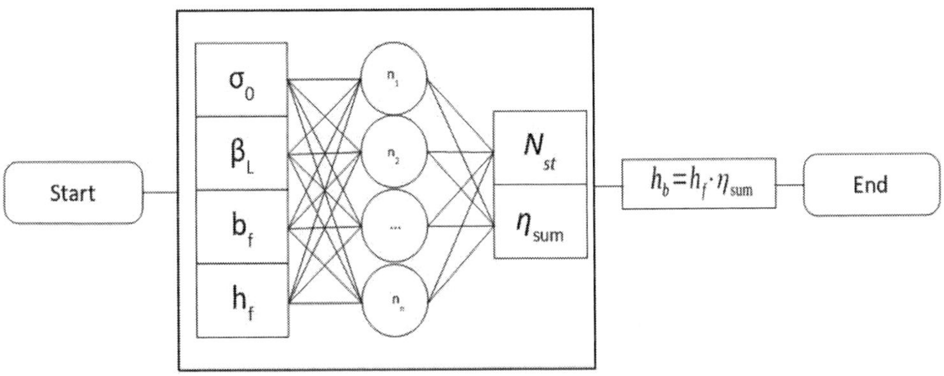

Figure 1. The algorithm for determining the thickness of breakdown bar h_p, based on the artificial neural network: σ_0–basic deformation resistance, MPa (for the condition: rate of strain $\xi = 1$ c^{-1}, reduction coefficient $\varepsilon = 0.1$ and temperature $t = 1000$°C); $\beta_L = b_s/L_{w.r.}$ – coefficient of usage of the work roll barrel, where b_s – strip width, mm; $L_{w.r.}$ – work roll barrel length, mm, b_f – strip width at the exit from the finishing group, mm; h_f – strip thickness at the exit from the finishing group, mm; N_{st} – the number of stands involved in the rolling process.

2.3. Determination of the reduction coefficients

To determine reduction coefficients in the finishing groups of sheet mills of hot rolling, it is proposed to modify the Imai method [2], developed in the works of M Rumyantsev [3–5].

Table 1. Artificial neural network parameters.

Neural network architecture	Input layer: 4 neurons Output layer: 2 neurons Hidden layer: 11 neurons
Activation function	Relu
Training mode	'Step by step training'
The degree of recognition of the training set, %	98.1
The degree of recognition of the test set, %	92.5
Mean absolute percentage error ε_{MAPE}, %	4.7
Max absolute error, %	19.1

The basic expression Imai (1) is used to calculate the reduction coefficients.

$$h_{1i} = \frac{h_p h_\kappa}{\sqrt[im]{\beta_{Wi} h_p^{im} + (1-\beta_{Wi}) h_\kappa^{im}}}, \quad (1)$$

where h_{1i} – the work roll nip in the i-th stand; $im = 0.30 + \frac{0.21}{h_\kappa}$ – Imai coefficient; $\beta_{Wi} = \sum_{j=1}^{j=i} W_j / \sum_{i=1}^{i=N_F} W_i$ – power factors of the main drives; W_i – power required for rolling in the i-th stand.

Prediction of reduction coefficients in the finishing group of stands of hot-rolled sheet mills is realized using the constructed neural network, the structure of which is shown in Figure 2.

The artificial neural network included in the algorithm for determining reductions in the finishing group of stands consists of an input layer (5 neurons), one hidden layer (containing 11 neurons) and an output layer containing 1 neuron (Table 2).

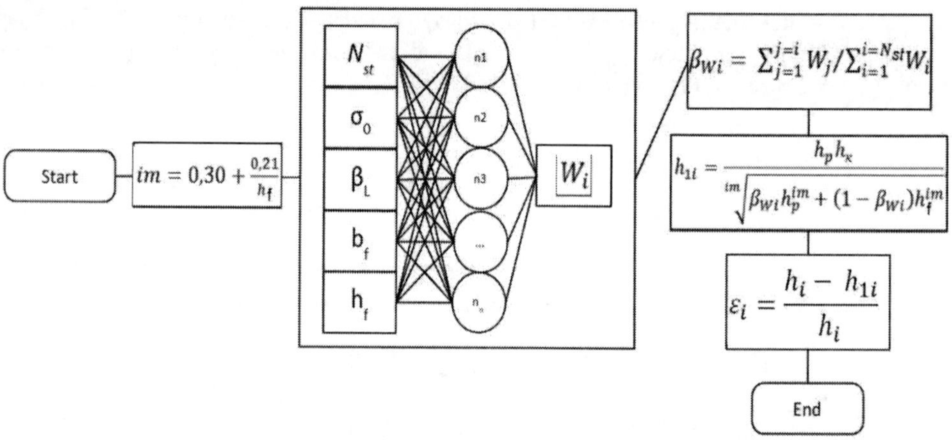

Figure 2. The algorithm for determining the reduction coefficients in the finishing continuous stands of the hot rolling mill: σ_0 – basic deformation resistance, MPa (for the condition: rate of strain $\Xi = 1$ c^{-1}, reduction coefficient $\varepsilon = 0.1$ and temperature $t = 1000$°C); h_{1i} the work roll nip in the i-th stand; im – Imai coefficient; β_{Wi} – power factors of the main drives; W_i – power required for rolling in the i-th stand.

3. Conclusion

Thus, in the course of the study, an algorithm was obtained based on neural network modeling, which allows one to select and determine reduction coefficients at BHRM 2000 of PJSC Magnitogorsk Iron And Steel Work.

Table 2. Artificial neural network parameters.

Neural network architecture	Input layer: 8 neurons Output layer: 1 neuron Hidden layer: 11 neurons
Activation function	Relu
Training mode	'Step by step training'
The degree of recognition of the training set, %	96.1
The degree of recognition of the test set, %	91.5
Mean absolute percentage error ε_{MAPE}, %	6.3
Max absolute error, %	12.4

The mean absolute percentage error ε_{MARE}, the mean error ε_{ME} and the standard deviation σ of the calculations of the developed algorithm do not exceed 0.053, 8.9% и0.074, respectively (Table 3).

Table 3. Errors in the calculation of reduction coefficients in the finishing group of stands BHRM 2000.

Stand	Mean error ε_{ME}	Mean absolute percentage error ε_{MAPE}, %	Standard deviation σ
7	0.053	6.3	0.074
8	0.021	4.9	0.038
9	0.028	7.1	0.046
10	0.034	8.9	0.06
11	0.014	7.7	0.025
12	0.016	5.6	0.035
13	0.016	4.7	0.028

Also, it is possible to adapt these algorithms for other hot rolling mills if there is a previously prepared data set for training using selected hyperparameters when training a neural network.

Acknowledgments
This work is carried out within a framework of the government order (No. FZRU-2020-0011) of the Ministry of Science and Higher Education of the Russian Federation.

References

[1] Grieves M W 2014 *Digital Twin: Manufacturing Excellence through Virtual Factory Replication* (LLC)
[2] Konovalov Yu V, Ostapenko A L and Ponomarev V I 1985 *Raschet parametrov listovoj prokatki: Spravochnik* (Moscow: Metallurgiya)
[3] Rumyantsev M I 2015 *Kalibrovochnoe byuro* **5** p 100
[4] Rumyantsev M I 2015 *Kalibrovochnoe byuro* **6** p 108
[5] Rumyantsev M I 2016 *Kalibrovochnoe byuro* **8** p 23
[6] Chojaczyk A A, Teixeira A P, Neves L C, Cardoso J B and Soares C G 2015 *Struc. Saf.* **52** p 78
[7] Bartere M and Pagariya R 2013 *IJARCS* **4** p 49
[8] Wang Zh, Gong D, Li X, Li G and Zhang D 2017 *Int. J. Adv. Manuf. Technol.* **93** p 3325
[9] Amirov R N, Dema R R, Luk'yanov S I, YAroslavcev AV and Martynova U D 2012 *Vestnik of NMSTU* **38** p 102
[10] Alaei H, Salimi M and Nourani A 2016 *Int. J. Adv. Manuf. Tech.* **85** p 1769
[11] Sun T J, Yang W D, Gao H and Mi H T 2013 *Appl. Mech. Mater.* **448-453** p 3417
[12] Turk R, Robic P R and Perus L 2002 *Metalurgija* **41** p 47
[13] Sikdar S and Kumari S 2008 *Int. J. Adv. Manuf. Tech.* **42** p 450
[14] Korczak P, Dyja H and Labuda E 1998 *J. Mater. Process. Tech.* **80** p 481

IOP Conf. Series: Materials Science and Engineering **966** (2020) 012102 doi:10.1088/1757-899X/966/1/012102

The research on the levelling process of thick slabs based on the electric motor parameters monitoring

S Yu Belyaev[1, 2], A A Fedulov[1], A S Alexandrova[1], E Yu Raskatov[1] and T V Meshchaninova[1]

[1] Ural Federal University named after the First President of Russia B N Yeltsin, 19, Mira str., Yekaterinburg, 620000, Russia

E-mail: [2] belyaev_s_66@mail.ru

Abstract. For the leveling machines in conditions of a small batch production which is typical for manufacturing sheets of non-ferrous alloys, in particular, titanium, the rollers heating is very influential. In order to expand the technological possibilities of the main equipment in non-stationary thermal working conditions of rollers that is an actual new scientific problem the experimental study of the loading in the hot leveling of slabs made from hardly-deformed titanium alloys was carried out on the 9-roll leveling machine. To achieve this goal the methodology of the complex experimental study of the actual loading definition of the leveling machine drive was developed. The experimental study was carried out for alloys Ti6Al4V (Vt-6s) and PT-3V with the initial heating temperature of the rolled product from 750 to 850°C. The slab leveling process was logged for slabs with the thickness from 20 to 68 mm, the width from 985 to 1720 mm by the number of passes through the leveling machine from 2 to 8. The treatment and analysis of experimental data was conducted on the basis of which the summary of permissibility of assortment expansion of leveling slabs has been formulated.

1. Introduction

The technological operation of leveling is obligatory in the sheet and slab production both for ferrous and non-ferrous alloys. In this operation the elimination of the number of geometrical flaws of rolled sheet product (such as rippling, camber, crescent) those occurs in the hot rolling process and subsequent cooling is taking places [1–3]. The main equipment to execute the leveling process is the roll-type leveling machines which provide the defect elimination by reversed bending in cylindrical rolls with staggered order of its disposal along the longwise axis of leveled slabs or sheets. The roll-type leveling machines with a number of rolls from 7 to 11 are applied to level the thick slabs while to level the especially thick slabs are usually using the machines with 7 or 9 rolls.

The leveling of all produced assortments of sheets and slabs is likely to be executed on the special roll-type leveling machines with appropriate selection of the roll's diameter, step and number [!]. It is settled relatively simply in the conditions of mass and large-scale production of rolled sheets. In the leveling of slabs and sheets in the conditions of low-batch size production which is most common for the enterprises those specialize on the production of sheets made of highly-alloyed steels and non-ferrous alloys (in particular, titanium alloys) the application of deeply-oriented specialization of leveling machines isn't rational [4, 5]. Therefore, for such kind of the production the leveling of all assortments is carried out on the restricted number of leveling machines with wide range of treated sheet sizes, frequently beyond the bounds allowed by the equipment's technical characteristics.

IOP Conf. Series: Materials Science and Engineering **966** (2020) 012102 doi:10.1088/1757-899X/966/1/012102

For the leveling machines used for the sheet leveling in conditions of a small batch production which is a typical situation for enterprises manufacturing sheets for special steel or non-ferrous alloys, in particular, titanium alloys, the rollers heating is very influential. In this kind of production it is very complicated to accomplish the qualified adjustment of the leveling machine on the basis of production data due to the unstable temperature working conditions of rollers, which means that it is necessary to carry out the leveling with a constant tuning or an additional leveling of the product. In the large-scale and mass serial production of sheets the indicated factors of instability make an impact only in periods of the beginning of the sheet batch leveling and further the thermal mode of the rollers work stabilizes [6, 7].

2. The experimental measurements of the levelling machine's engine loading

On the purpose to justify the possibilities to expand the technological capabilities of the main equipment in non-stationary thermal working conditions of rollers that is an actual new scientific problem the experimental study of the loading in the hot leveling of slabs made from hardly-deformed titanium alloys was carried out on the 9-roll leveling machine.

As the object of the research the 9-roll leveling machine places in the sheet-rolling workshop on the area of heat treatment and leveling of the titanium alloy slabs was considered. The roll-type leveling machine has 7 working rolls with the barrel diameter 400 mm and barrel length 2000 mm each of whom opposite supported by backup rolls 450 mm in diameter. On the input and the output of the leveling machine additional 2 rolls have been installed (without backup rolls) 453 mm in diameter. All rolls set up with the standard step 460 mm. The rolls drive is group-type that includes the direct current electric motor with the nominal power 95 kW and the nominal rotary speed 500 min^{-1} via hybrid gear box, the pinion stand at the same time, with gear ratio 42.17 and nine universal joints assembled on the bearing coupling (Figure 1).

Taking into account that the design of this leveling machine doesn't imply the probing placement (load cells) the most effective and reliable method to define the summarized technological loading is to control the process by the main drive's electric motor current parameters. At the same time, to increase the reliability of measurements it is desirable to control the technological loading in alternative way; in particular, there was a solution to periodically measure the torque on the joints by the tensiometric method [8].

The methodology of the experimental research of the virtual loading of the leveling machine's main drive anticipated the record on the next values with the usage of the hardware:

- the current of the armature circuit of the rolls drive's electric motor;
- the rotary speed of the rolls drive's electric motor rotor;
- the torque transmitted by each of the rolls drive's joint.

According to the developed methodology of the experimental research additionally for each batch or series of workpieces it is necessary to log the number of the common parameters in the table format, such as: an alloy grade, a type of a technological operation (leveling, annealing with leveling, etc.), a thickness, width and length of a leveled slab, a temperature and duration of an initial heating, a leveling scheme, an upper beam movement, special conditions of the leveling.

For practical realization of the developed methodology the principle scheme to measure and record the loading parameters appearing in the main line of the leveling machine LPM-22 during the leveling was stated (Figure 1). To record the parameters 2 in-parallel schemes of measurement were used:

- to log the electric drive parameters;
- to log the signals of the torque detectors, aka the wireless system of data communication.

IOP Conf. Series: Materials Science and Engineering **966** (2020) 012102 doi:10.1088/1757-899X/966/1/012102

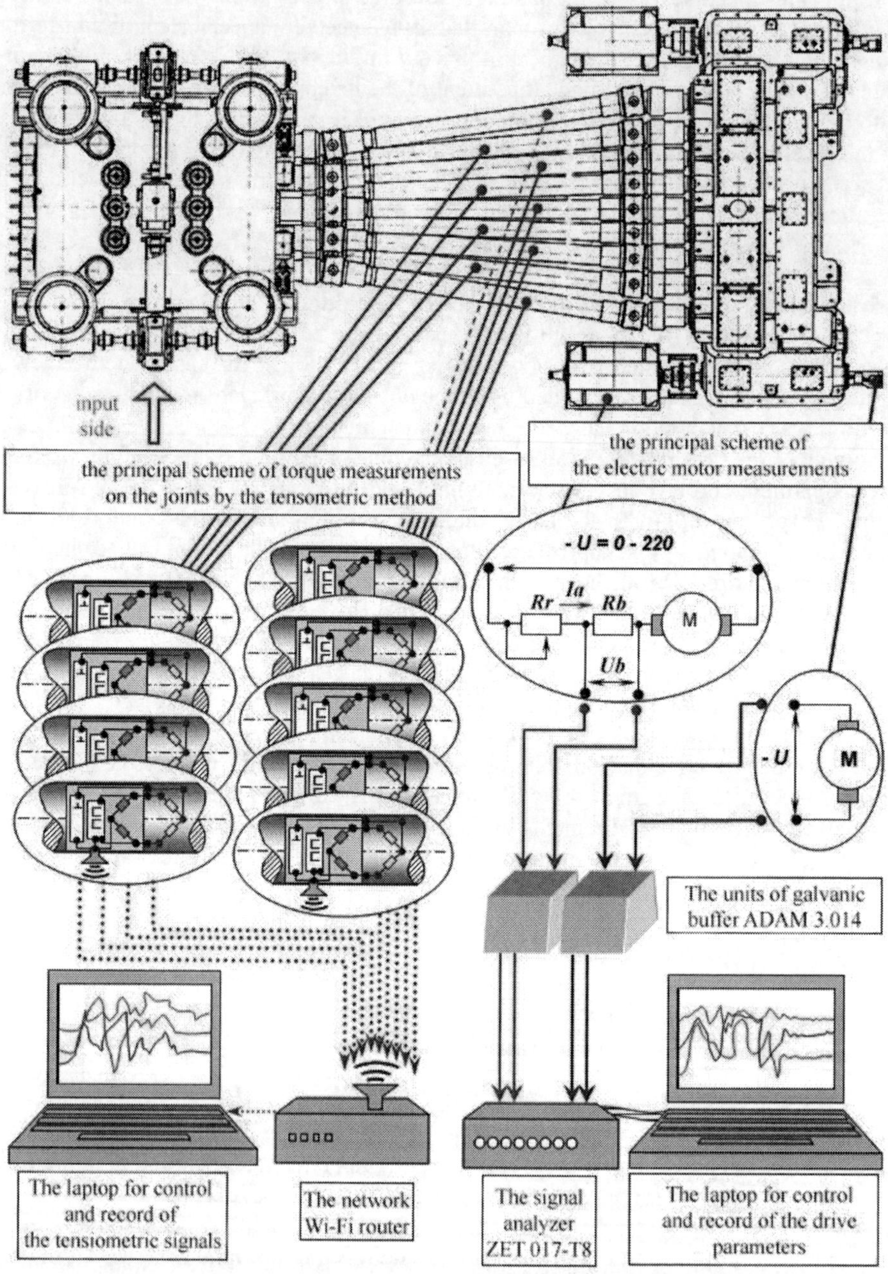

Figure 1. The principle scheme of technological loading measurements during titanium slab leveling on the 9-roll leveling machines.

The system for measurement and record of the electric drive parameters designed on the basis of the signal analyzer ZET 017-T8 working in the mode of multi-channel oscilloscope together with the laptop (figure 1). All measured parameters with a usage of built-in the signal analyzer's hybrid converter and amplifier were being converted in the digital form and logged by the computer with simultaneous visualization on the laptop screen. The record of the armature circuit current and rotary speed of the rolls electric drive's rotor was carried out from control instrumentation (ammeters) by the connection to the relevant points of the electric drive's control cabinets. The connection to the signal analyzer ZET017-T8 was executed via the units of galvanic buffer ADAM 3.014. To evaluate the current loading dynamics of the main electric motor of the leveling machine and the rotary speed the sampling time step of the logged signal was 0.02 s, what corresponds to the sampling rate 50 Hz.

The experimental research was conducted for alloys Ti6Al4V and PT-3V with the initial temperature of a slab heating 750–850 °C with the upper beam movement relative to the leveled slabs within limits of 3.0–11.0 mm. The leveling process was logged for the slabs with the thickness 20–68 mm, the width 985–1720 mm with the number of passes through the machine from 2 to 8.

In the course of the research the variation of the leveled slabs overstepped the limits set by the leveling machine's technical characteristics mainly by the slab thickness. Thus, the restricted level for this machine is no bigger than 22 mm in thickness and 1800 mm in width with the flow stress of the metal by the leveling temperature no more than 500 MPa with simultaneous combination of these parameters. Therefore, the permissibility of technological possibilities expansion for the 9-roll leveling machine was experimentally checked out in the titanium slab leveling.

3. The processing of the experimental data

Received in the course of the experimental research digital data array was treated in order to obtain a more convenient form of the result visualization (the example of the treatment is shown in figure 2). After the treatment the analysis of the results was carried out which lead to the next findings:

- the torque on the joints defined with the usage of resistance strain gages and calculated thought the rolls drive electric motor current are in agreement with each other. This allows to define the loading of the mechanical transmission of the leveling machine thought the calculated values of the torque measured on the constantly logged value of the armature circuit current;
- the power parameters of the levelling process for slabs with minor, visually indistinguishable defects of the flatness deviation stay stable enough within the levelled slabs batch. The presence on the initial slabs of considerable, visually noticeable defects of flatness deviations in the form of hollows and bulges are lead to the increase of the technological loading up to 50–70% of the level of the power parameters those take place in the process of non-defect slabs levelling.

The electric motor, by means of which the drive of the leveling machine is operating, is working in the normal mode. The values of the armature circuit currents for all routes of the leveling don't exceed the nominal value of the current 470 A. In the leveling of the slabs the actual values of current of the technological loading are 40–95% of the nominal value and in the leveling of the typical assortment – 15–40 %. Thus, there is no danger of overloading or overheating the electric motor windings.

The dynamic loads during acceleration and slowdown of the main engine of the leveling machine are acting only on the insignificant part of the leveled sheet length (Figure 2), which allows to evaluate the energy-power parameters of the leveling process on the prolonged phase of realization of the technological process with the constant velocity.

The most loaded of the gearing of the hybrid gear box (pinion stand) is the 2nd reducing stage by the contact strength of driven gear cog. In the period of the peak values of the torque occurrence those were registered in the course of the research (in the leveling of the slabs with thickness 61.0 and 68.0 mm) there was the overloading of the 2nd stage of the hybrid gear box by the contact strength criterion on 90–120%. Another peak resulted in the overloading but on the lower level (70–40%) which was noticed in the slabs leveling with thickness 58.2, 47.1 and 39.0 mm. During the leveling of the thickest slab (68.0 and 61.0 mm) the average level of the loading came to 80–90% of the limit values. In case of the leveling of the slabs with thickness from 39.0 mm to 47.2 mm the average level of loads with regards to the width of the slabs can reach 50–80% of the limit values.

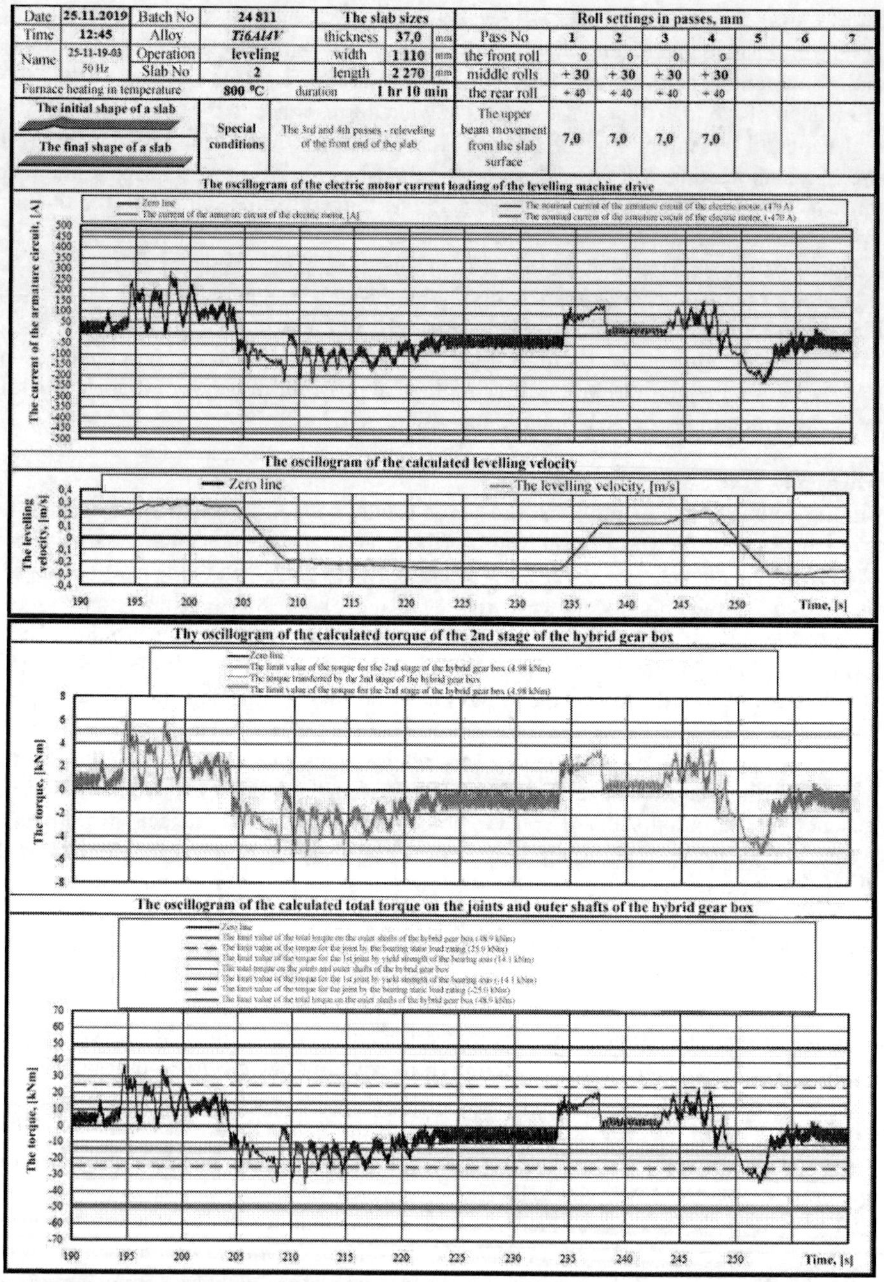

Figure 2. The example of the experimental data treatment for the hot leveling process of the titanium slabs on the 9-roll leveling machine.

During execution of the slabs leveling with thickness from 37.0 mm to 20.4 mm the average values of contact stresses in cogs of the driven gear of the 2nd reducing stage would be up to 35–50% of the permissible values.

The statistical significance of obtained results is sufficient because the gauging was conducted both for circuits of measurements of current signals of the electric motor by the direct method and though the tensiometric sensors with the usage of the calibration beam of equal resistance. Moreover, to register the current parameters of electric motor the certified special equipment was used.

The dissipation of experimental data within one batch of slabs considering low-batch nature of the production and instability of the thermal working conditions of rollers achieve 20–25%. Besides, the highest loading was observed in the leveling of the first slabs in batch.

The weakest component of the leveling machine LPM-22, taking into account non-uniformity of the torque distribution, is the joints of the roll drive. For all recorded leveling routes the exceeding of the limit value of the torque was registered be the criterion of the joint's bearing axis strength. The peak loads mentioned above in the process of the thickest slabs leveling also lead to overloading the joint's hinge bearing itself. The rational way for the solution of the problem of the joint's carrying capacity is seen as the improvement of the hinge unit.

Considering all the aspects of conducted research it is possible to make the conclusion that there is an opportunity to expand the technological possibility of the 9-roll leveling machine in comparison with the manufacturer's technical characteristics.

References

[1] Konovalov Yu V 2008 A handbook of the roller the reference book in two volumes*The production of hot rolled sheets and strips* **1** (Moscow, Teplotechnik) p 640 [In Russian]
[2] Maximov EAand Shatalov R L 2018 *Metallurgist* **62** p 132
[3] Shatalov R L and Maximov E A 2018 *Steel* **1** p 31 [In Russian]
[4] Illarionov A G and Popov A A 2014 *Technological and maintanance properties of titanium alloys* (Yekaterinburg, The Publishing House of Ural Federal University) p 137 [In Russian]
[5] Nedorezov I V 2003 *Modeling of the sheet leveleing processes on the roll machine* (Yeakaterinburg, AQUA-PRESS) p 256 [In Russian]
[6] Belyaev S Yu, Chechulin Yu B, Fedulov A A and Bulganina M Yu 2020 *IOP Conference Series: Materials Science and Engineering* **718** 012003 doi:10.1088/1757-899X/718/1/012003
[7] Maksimov E A and Ustinovsky E P 2018 *Mechanical Equipment of Metallurgical Plants* **1** p 60 [In Russian]
[8] Raskatov E Yu and Spiridonov V A 2015 *The basis of scientific research and modeling of technological machines* (Yeakaterinburg, The Publishing House of Ural Federal University) p 468 [In Russian]

IOP Conf. Series: Materials Science and Engineering **966** (2020) 012103 doi:10.1088/1757-899X/966/1/012103

Multi-criteria model for the development of industrial logistics

A N Rakhmangulov[1, 3], S N Kornilov[1], D V Aleksandrin[2] and N O Shevkunov[1]

[1] Nosov Magnitogorsk State Technical University, 38, Lenin Avenue, Magnitogorsk, 455000, Russia

[2] OJSC Uralasbest, 66, Uralskaya Street, Asbest city of Sverdlovsk region, 624261, Russia

E-mail : [3] ran@magtu.ru

Abstract. Development of the industrial manufacturing logistics is associated with the need to solve many problems that affect the main technological units, internal and external transportation, as well as the industrial enterprise management system. The effectiveness of the solution of each of these tasks separately is traditionally evaluated by criteria, which, as a rule, conflict with the criteria of the effectiveness of other tasks and may even contradict the strategic goals of the organization. Modern logistics solutions are based on the achievement of a global economic criterion. However, the practical implementation of such general decisions is limited by the complexity of forming concrete actions based on them. In addition, traditional logistics solutions do not adequately meet the requirements of the concept of sustainable development, as they are poorly oriented towards environmental and social criteria. The developed multi-criteria model is a combination of alternative logistics solutions and criteria for their evaluation. The procedure for assessment and choosing the best combination of logistic decisions is based on a combination of multi-criteria decision-making methods. Application the developed model will reduce the costs of designing and implementing logistics solutions, as well as ensure the achievement of the sustainable development goals of an industrial enterprise.

1. Introduction

The goals and functions of industrial logistics (IL) are studied in enough detail in the scientific literature [1] and are implemented in practice [2]. Material flows of resources and products, waste, reverse flows, as well as service flows, information and financial flows are the objects of IL management. One of the main tasks of industrial logistics is to coordinate the industrial cycles of various departments of the enterprise for the conversion of raw material flows into finished product flows of the required quality. Modern IL management systems are based on the operational production management method MRP II (Manufacturing Resource Planning) [3]. The practical implementation of the MRP II technique is provided by ERP (Enterprise Resource Planning) software [4].

There is a decrease in the effectiveness of long-term planning and an increase in the importance of operational decisions on adapting enterprises to changing external conditions under the conditions of increasing volatility of the world economic system, regional markets, and a change in technological patterns [5].

A variety of logistics solutions have been developed and are being put into practice. They are mainly aimed at changing the parameters of logistics flows at the levels of current and operational planning. The following solutions were most widely used: inventory management at different stages of the production cycle [6]; coordination of volumes of internal [7] and external transportation of the enterprise [8] with production plans; management of parameters (volumes, routes) of in-plant material flows [7]; coordination of the duration of production cycles of technological units of the enterprise by regulating the intensity of in-plant material flows [1]; automation of warehouse operations [9]. In [10], a universal system for managing logistics flows in supply chains, aimed at achieving sustainable development goals, was proposed.

Sustainable changes in the parameters of external and internal material flows of an industrial enterprise require coordinated adjustments to the structure and organization of the functioning of the production logistics system (LS). However, in practice, such adjustments in most cases are not considered as logistics solutions. Moreover, the development of appropriate solutions is not a function of production logistics and logistics management. In other words, reconstructive and organizational measures are carried out without a systematic analysis of the dynamics of internal and external material flows of the enterprise. For example, the decision to reconstruct industrial railway stations, as a rule, is made only in case of significant and long-term changes in volumes and/or structure of processed cargo flows [11].

Decisions on the operational change of the technology of production units are less common in IL. Logistic operational management provides, as a rule, sequence management of orders (production plans), as well as planning and control of workload of production units. Adjusting the duration of technological processes by changing the parameters of technological processes of the main production is usually strictly limited [12]. In such conditions, industrial transport can carry out a flexible change in the parameters of material flows through the implementation of various solutions to change the technology of its work. For example, studies [11, 13, 14] provide a description of the so-called flexible technology of industrial railway transport. This technology provides operational management capacity of the transport infrastructure, depending on changes in the intensity and structure of internal cargo flows.

Most researchers in the field of IL [1] consider the material flows and structural units of the enterprise, ensuring their processing and promotion, as objects of study. Until now, there has not been an idea of IL as a system for managing all the logistics flows of an enterprise in relation to each other, as well as with the structure and organization of the functioning of the LS. Obviously, the main reason for this situation is the high complexity of the task of creating Integrated Industrial Logistics. The difficulty lies in the presence of many conflicting criteria and alternative solutions. Currently, there is an increase in the number of criteria and alternatives, associated with the need for enterprises to achieve sustainable development goals [10, 15].

Currently, the scientific direction in Multiple-criteria decision-making (MCDM) is actively developing [16, 17]. These models and methods differ mainly in the complexity and accuracy of implementation, which depend on the number of criteria and alternatives. To compensate for the shortcomings of a multicriteria model, a combination of different models is used [9, 18].

Another promising method for studying complex systems is simulation modeling. Nowadays, the creation of models for the study of logistics systems at various levels of generalization is carried out using a variety of software tools. For example, the AnyLogic simulation software platform [19] allows, within the framework of one model, to study processes at the level of making and implementing operational, tactical and strategic decisions, as well as to model the dynamics of the socio-economic environment. For this, agent, discrete-event (process) and system-dynamic approaches are used, respectively.

This study is motivated by the problem of the formation of Integrated Industrial Logistics, which provides systematic management of all the logistics flows of an industrial enterprise in conjunction with the structure of the LS, its functions and goals in conditions of high dynamics of the socio-economic environment.

The rest of the article is organized as follows. Section 2 contains a formalized description of the structure and functions of the LS using the AnyLogic simulation platform. Section 3 discusses the developed multi-criteria model for the development of IL. A system of criteria and alternatives for a multicriteria model and a flow chart of model implementation are presented. In Section 4 conclusions are discussed.

2. Formalization of the industrial logistics system

A description of the structure and functions of the industrial logistics system is proposed to be carried out using AnyLogic simulation software.

Agent and discrete-event approaches are used to describe the LS at the level of operational logistics management. The following assumptions are made:

- The agent is an element of the logistic flow. Each element is characterized by a certain qualitative property and has a quantitative value (Figure 1, c).
- Four types of logistic flows were determined depending on the quality properties of the flow elements and the features of calculating the quantitative characteristics: the material flow describes the unit volumes of raw materials, components, etc., which change their quality during the production process, passing into finished product and waste streams; service flow – work performed on processing and promotion of the remaining logistics flows, the quantitative characteristic of the service flow element corresponds to the volume of work performed or the number of resources required to perform one technological operation with the material flow element; financial flow – the costs necessary for processing and promoting the rest of the logistics flows, the quantitative characteristic of the element of the financial flow – the amount of costs for performing one technological operation with the element of the remaining logistics flows; information flow – messages circulating in the production logistics management system, the quantitative characteristic of the information flow element determines the amount of information in the message about any event and corresponds to the probability of this event.
- Logistic flow is a combination of several flow elements. Logistic flow is characterized by a route – a sequence of elements of the logistics system through which the stream passes when moving along the logistics chain or LS.
- The element of the logistic system (logistic element) describes the structural unit of the technological chain and the enterprise management system (Figure 1, a). The logistic element performs the following functions: flow delay for the duration of the technological operation; stock holding in case the logistics element is occupied by processing of the previous flow element (Figure 1, b); changing the structure of the logistic flow by combining or disconnecting the flow elements; change in the quality properties of flow elements.
- Each logistic element uses the remaining logistic flows as resources (Figure 1, b, d).
- The structure of the LS is a combination of logistics elements and the relationships between them (Figure 1, e).
- Relations between the structural elements of the LS are described by the routes of logistics flows.

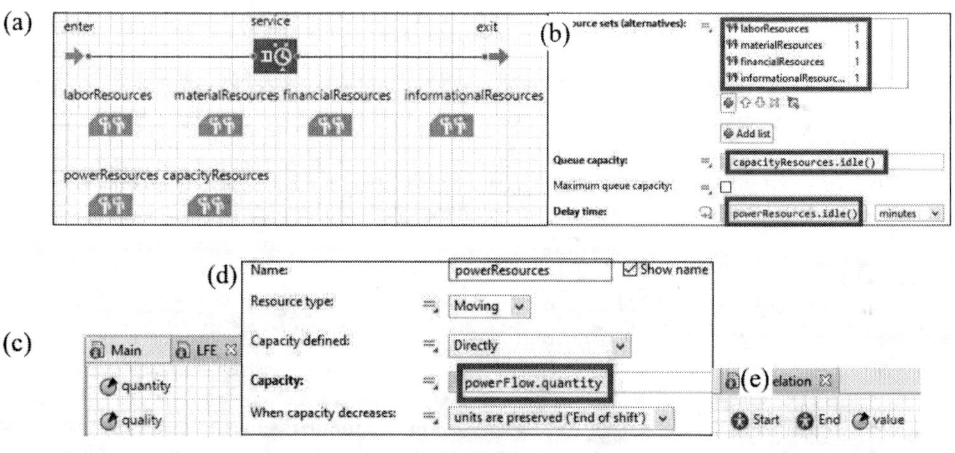

Figure 1. Internal environment of the simulation model's main agents.

The movement and processing of logistics flows of each type is provided by the corresponding logistics subsystem. The interaction of flows is carried out in the internal environment of each agent –

a logistic element. For example, the movement of the main material flow (raw materials, semi-finished products) requires certain resources (Figure 1). The receipt of these resources in the logistic element is carried out as a result of the movement of resource flows through the logistic elements of a logistic subsystem. Logistic flows are exchanged in an agent simulation model using a message passing mechanism between elements. The LS structure is described by a non-planar complete undirected graph.

The presented formalized description of the industrial logistics system in terms of a simulation model allows to unify various technological processes, as well as the functioning of the enterprise management system.

3. The choice of a multi-criteria model for the development of industrial logistics

The choice of a multicriteria model or their combination, as a rule, is determined by the composition of alternatives and criteria for their assessment. The developed multi-criteria model includes Lk, $k = 0$, 1, ... 6 hierarchical levels:

- L_0 – industrial logistics development strategy.
- L_1 – sustainability criteria for the IL, the achievement of which ensures the safe, efficient and environmentally friendly functioning of the enterprise (economic, social, environmental) [10].
- L_2 – alternatives for achieving sustainable development goals as a result of organizational changes (products, markets, resources, technologies, social strategies, emissions, waste, reverse logistics [20]).
- L_3 – optimization criteria for the parameters of logistics flows (unevenness, regularity, variability, stability, rhythm) [10].
- L_4 – solutions for optimizing the parameters of logistics flows (element size, structure).
- L_5 – selection criteria for decisions on changing the design of the industrial logistics system (productivity).
- L_6 – solutions for changing the design of the industrial logistics system (power, capacity, number of LS elements, relationships between LS elements).

A feature of the developed multicriteria model is the need to combine decision-making methods at different levels of the hierarchy. A choice from a fixed number of alternatives is made at levels L_1 and L_2. The alternatives here are the development strategies of various logistics subsystems to achieve several sustainable development goals. It is rational to use the TOPSIS method (The Technique for Order of Preference by Similarity to Ideal Solution) [21] to solve the model at this level. This is due to the potentially large number of alternative strategies that can be formed at the L_2 level.

At levels L_3–L_6 of the hierarchy, the adoption of Multi-objective decisions (MODM) [22]. In other words, for each logistic flow and each logistic element the optimal parameter values are calculated, which should ensure the achievement of the criteria, respectively, L_3 and L_5 levels. Solutions for logistics flows and elements are selected in a limited range of continuous values. For an approximate choice of such solutions, it is rational to use the methods of the theory of restrictions based on the sequential identification and elimination of bottlenecks [2]. The use of genetic algorithms gives a more accurate result [23]. In logistics systems of complex structure, changes in one element lead to changes in interconnected elements. Therefore, it is rational to choose solutions at the L_4 and L_6 levels for such systems using a combination of simulation and optimization models.

The implementation of optimization solutions to change the structure of the LS and the parameters of its elements requires adjustment of several resource flows an increase in the volume of financial flows. Changing the parameters of logistic elements and/or the structure of the LS to promote optimized logistic flows requires appropriate changes to the simulation model of the LS. These changes, in turn, can affect the parameters of logistics flows. The developed algorithm for implementing the multi-criteria model of IL (Figure 2) provides for iterative optimization of logistics flows and the LS until the maximum change in any of the parameters becomes less than a given value.

The sensitivity analysis of the developed multicriteria model is proposed to be performed using the simulation method. In the process of simulation modeling of the functioning of the logistics system of the selected structure, the parameters of logistics flows are assessed, as well as the degree to which sustainable development goals are achieved.

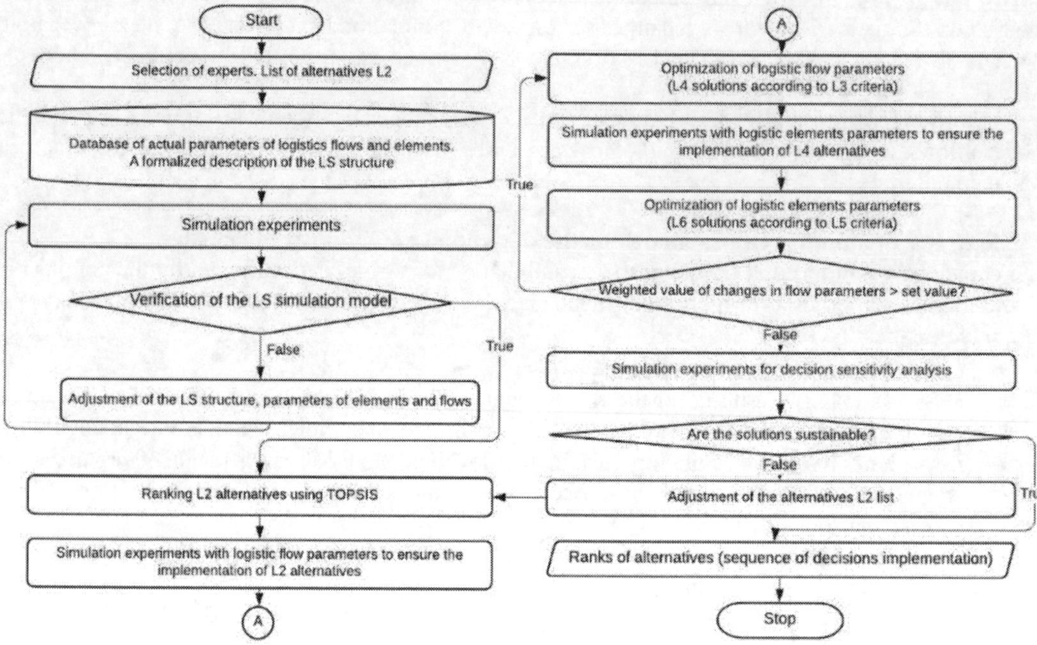

Figure 2. Flowchart of a multi-criteria model for the development of industrial logistics.

4. Conclusion

A new approach and model are designed for sustainable development of industrial logistics. The novelty of the approach is to combine the goals of sustainable development with the criteria of economic efficiency of an IL system. A feature of the developed multi-criteria model of IL is the combination of the multi attribute decision making TOPSIS method, MODM methods based on the use of genetic algorithms with a simulation model. It is proposed to use the TOPSIS method for ranking decisions to achieve the goals of sustainable development of IL. Genetic algorithms are used to optimize the parameters of logistics flows and elements of the logistics system. Such optimization ensures the implementation of sustainable development solutions. Finally, the simulation model is used as a means of obtaining input data for optimizing the parameters of logistic flows and elements, as well as for analyzing the sensitivity of the multicriteria model and the obtained solutions. An algorithm for implementing the proposed multicriteria model is developed. This algorithm determines the method and order of combining fuzzy MADM, MODM methods with each other and with a simulation model. The novelty of the research is also made up of the hierarchical system of criteria for sustainable development of the logistics system of an industrial enterprise and the proposed group of alternatives to achieve these criteria. A universal set of agents of the simulation model is proposed, which allows creating simulation models of production logistics systems of various industrial enterprises. Designed multi-criteria model will allow making informed decisions for the sustainable development of IL at all levels of enterprise management. Future research involves experiments using the developed multi-criteria model in a real industrial enterprise.

References

[1] Parunakyan V 2018 A study on increasing the energy efficiency of buildings on rail transport *MPRTC* **8** 4–21
[2] Pegels C C and Watrous C 2005 Application of the theory of constraints to a bottleneck operation in a manufacturing plant *J. of Manuf. Tech. Management* **16** pp 302–11

[3] Kiran D R 2019 Manufacturing resource planning (MRP II) *Production Planning and Control* (Elsevier) pp 441–55
[4] Mahar F, Ali S I, Jumani A K and Khan M O 2020 ERP System Implementation: Planning, *Management, and Administrative Issues IJST* **13** pp 106–22
[5] Ford S and Despeisse M 2016 Additive manufacturing and sustainability: An exploratory study of the advantages and challenges *Journal of Cleaner Production* **137** pp 1573–87
[6] Jondhale S D 2019 Review of Inventory Management System *IJRASET* **7** pp 1894–7
[7] Rakhmangulov A, Osintsev N, Mishkurov P, Muravev D and Sładkowski A 2017 Dynamic Optimization of Railcar Traffic Volumes at Railway Nodes *Rail transport - systems approach (Studies in Systems, Decision and Control)* **87** ed A Sładkowski 0009172 (Cham: Springer) pp 405–56
[8] McCann P 2000 Industrial Logistics and Regional Competition *Regional Competition (Advances in Spatial Science)* ed P W J Batey and P Friedrich (Berlin, Heidelberg: Springer Berlin Heidelberg) pp 108–27
[9] Stojčić M, Stević Ž, Nikolić A and Božičković Z 2019 A multi-criteria model for evaluation and selection of AGV's in a warehouse *MPRTC* **9** pp 4–20
[10] Osintsev N, Rakhmangulov A, Sładkowski A and Dyorina N 2020 Logistic Flow Control System in Green Supply Chains *Ecology in Transport: Problems and Solutions (Lecture Notes in Networks and Systems)* ed A Sładkowski (Cham: Springer International Publishing) pp 311–80
[11] Trofimov S V, Rakhmangulov A N and Kornilov S N 2004 *Methods for the development of industrial railway systems in the changing environment of enterprises* (Magnitogorsk: Nosov Magnitogorsk State Technical University)
[12] Lukyanov V A, Mishkurov H N and Kopylova O A 2019 The logistics model of a transport technological system of the industrial enterprise *The Bulletin of Transport Information* pp 19–30
[13] Rakhmangulov A, Sładkowski A and Osintsev N 2016 Design of an ITS for Industrial Enterprises *Intelligent Transportation Systems – Problems and Perspectives (Studies in Systems, Decision and Control)* ed A Sładkowski and W Pamuła (Cham: Springer International Publishing) pp 161–215
[14] Trofimov S V and Rakhmangulov A N 2000 *The choice of methods for operational management of industrial transport systems* (Magnitogorsk: Nosov Magnitogorsk State Technical University)
[15] Rakhmangulov A, Sladkowski A, Osintsev N and Muravev D 2017 Green logistics: Element of the sustainable development concept. Part 1 *Naše more* **64** pp 120–6
[16] Stojčić M, Zavadskas E, Pamučar D, Stević Ž and Mardani A 2019 Application of MCDM Methods in Sustainability Engineering: A Literature Review 2008–2018 *Symmetry* **11** p 350
[17] Greco S, Ehrgott M and Figueira J R 2016 *Multiple Criteria Decision Analysis* **233** (New York, NY: Springer New York)
[18] Matić B, Jovanović S, Das D K, Zavadskas E K, Stević Ž, Sremac S and Marinković M 2019 A New Hybrid MCDM Model: Sustainable Supplier Selection in a Construction Company *Symmetry* **11** p 353
[19] *AnyLogic Simulation Software: Official Website* Available online: https://www.anylogic.com/ (accessed 1 Apr 2020)
[20] Agrawal S, Singh R K and Murtaza Q 2015 A literature review and perspectives in reverse logistics *Resources, Conservation and Recycling* **97** pp 76–92
[21] He Y and Jung H 2018 A Voting TOPSIS Approach for Determining the Priorities of Areas Damaged in Disasters *Sustainability* **10** p 1607
[22] Zavadskas E K, Turskis Z and Kildienė S 2014 State of art surveys of overviews on MCDM/MADM methods *Technological and Economic Development of Economy* **20** pp 165–79
[23] Fatima I, Fahim M, Lee Y-K and Lee S 2013 MODM: multi-objective diffusion model for dynamic social networks using evolutionary algorithm *J. Supercomput.* **66** pp 738–59

IOP Conf. Series: Materials Science and Engineering **966** (2020) 012104 doi:10.1088/1757-899X/966/1/012104

Research and educational network of ground stations for receiving and processing information from educational satellites

A A Spiridonov[1, 2], V A Saechnikov[1], D V Ushakov[1], A G Kezik[1], V E Cherny[1], A P Verstakovskaya[1] and V E Evchik[1]

[1] Department of Radiophysics and Computer Technology, Belarusian State University, Minsk, 220030, Belarus

E-mail: [2] sansan@tut.by

Abstract. The research and educational network of ground stations for receiving and processing information from low-orbit educational spacecraft is considered in this paper. Ground stations are equipped with an orbit measurement and determination system with time synchronization. This network realizes receiving and processing telemetry of ultra-small satellites, performing measurements and determining their orbital parameters and teaching students of aerospace specialties. Software and hardware of stationary and mobile ground stations are considered. The ground station was tested in the modes of receiving data and measuring the ultra-small satellite orbit with different frequency of telemetry packets. The initial orbit determination by frequency and time measurements of a LUOJIA-1 01 nanosatellite telemetry signal on several orbits with a limited number of data on one pass was studied. The nanosatellite radio signal nominal frequency and initial orbital parameters were unknown. Close orbit determination using a ground station network for satellite constellation launch in the first week of flight was considered. The simulation results showed that if the mean anomalies of neighbouring satellites differ by 0.1°, the spacecraft can be identified in the constellation by the frequency and time measurements of the same radio signal at three synchronized ground stations.

1. Introduction

Many universities develop their own educational satellites as educational laboratories for training students of aerospace specialties, research and technology development [1–2]. Usually these are ultra-small satellite (USS) weighing up to 10 kg, developed according to the Cubesat standard [2–3]. Reliable and effective receiving of telemetry and forecasting future positions are important for control, tracking and radio communication with University ultra-small satellites. Due to budget constraints, a single ground control station (GCS) is used for control and receiving telemetry at the University USS, which is located within the city limits and has unsatisfactory reception conditions. In addition, the communication between the GCS and the low-orbit USS is limited to sessions of a few minutes 5–6 times a day. Sometimes the University GCS is connected to global international networks of USS ground stations. Satnogs is example of the international network of reception stations (more than 900 ground stations), which is used as a redundancy communication channel with the University nanosatellite BSU CubeBel-1, launched in October 2018.

The SGP4 model [4–5] with input orbital data in the TLE (two-line elements) format of the NORAD (the North American Aerospace Defense Command) is used to USS future passes prediction over the University ground station and processing information. Typically, TLEs are updated daily, and

IOP Conf. Series: Materials Science and Engineering **966** (2020) 012104 doi:10.1088/1757-899X/966/1/012104

are available free of charge [6]. But in the long term, or in the event of military conflicts, the NORAD has the ability to rescind shared access to the orbital database. Less often, a navigation receiver is used onboard ultra-small satellite to determine the exact coordinates and velocity [7]. Another way to obtain initial orbital parameters for prediction models is orbit determination based on measurements of the received radio signal from USS [8]. The measured parameters for a low-cost ground station are the time and Doppler frequency shift of the received radio signal [8–9].

2. Ground station software and hardware. Measuring of telemetry radio signal parameters

The possibility of cooperative telemetry reception and USS orbit determination by a stationary University GCS and several mobile GS with time synchronization and spacing across the territory of the Republic of Belarus is considered in this paper. The telemetry receiving and processing on a stationary University GCS are performed using software and hardware of the communication system. The ground station hardware consists of: 435–438 MHz band Yagi–Uda antennas with circular polarization, receiving system based on the IC-9100 transceiver; receiving system based on the SDR receiver, YAESU G-5500 rotator with a control interface, control computer. The ground station software includes: satellite orbital and radio signal parameter prediction software, simulation and visualization of cooperative ground station scenarios and express calculation of standard navigation and ballistic information software, telemetry receiving and processing software. The orbit measurement and determination system with time synchronization for University ground station consists of: GPS receiver; module for frequency and time measurements of the received radio signal; software for processing measurements; software for orbit determination and correction.

The communication system software and hardware for the mobile ground station includes: 435–438 MHz band omni-directional quadrifilar helical antenna, the receiving system based on the SDR module, control industrial computer, satellite orbit and radio signal parameter prediction software, telemetry receiving and processing software. The orbit measurement and determination system with time synchronization for a mobile ground station consists of: GPS receiver, module for frequency and time measurements of the received radio signal based on a microcontroller (MCU) for time processing and a two-channel digital oscilloscope, software for processing measurements; software for orbit determination and correction.

The block chart for measuring of USS telemetry radio signal parameters for a mobile ground station is shown in figure 1. Radio signals of telemetry packets after conversion to an intermediate frequency in the analog part of the SDR receiver are transmitted for further processing in the ADC SDR and 2 input of a two-channel digital oscilloscope. At a time of t_1 the telemetry packet arrival signal is come from the SDR receiver trigger to the input of the time processing MCU (Figure 2). The GPS module receives and processes radio signals from navigation satellites and transmits 1PPS signals and NMEA packets to 1 input of the two-channel digital oscilloscope and to the time processing MCU input. 1PPS signals are synchronized with the time scale of GLONASS or GPS satellite navigation system. NMEA packets contain information about the exact coordinates of the receiving GPS antenna, the system time of satellite navigation system at the NMEA packet arrival time, and the exact 1PPS signal arrival time. The telemetry packet arrival signal from the SDR receiver trigger turn on the time processing MCU timer that measures the time interval τ between the arrival of the next 1PPS signal from the GPS receiver (Figure 2). In the NMEA packet that follows the 1PPS signal, its exact arrival time $t_1+\tau$ is recorded. The time processing MCU calculates the arrival time t_1 of the telemetry packet radio signal and transmits it to the industrial computer. The radio signal time diagram for telemetry packets, NMEA packets and 1PPS signals is recorded by a digital oscilloscope and transmitted to an industrial computer for processing and calculating the telemetry radio signal frequency and checking its arrival time measurements.

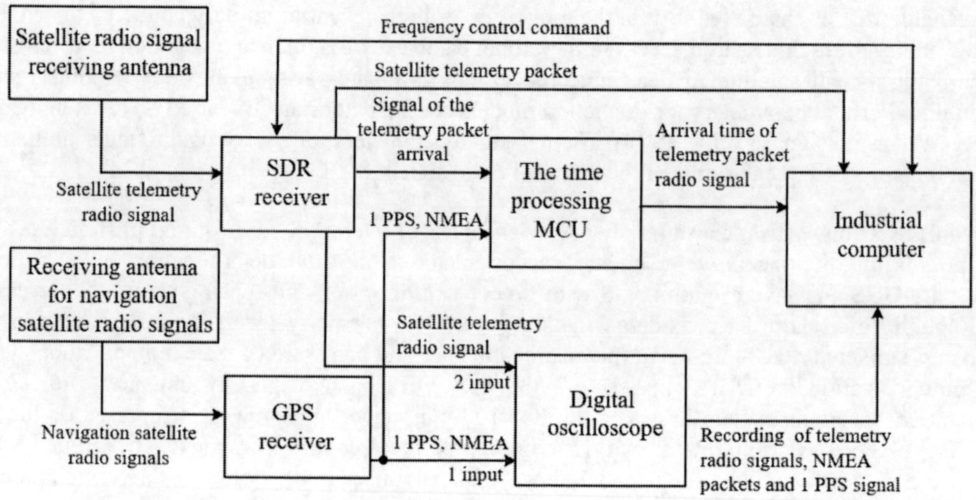

Figure 1. The block chart for measuring of USS telemetry radio signal parameters.

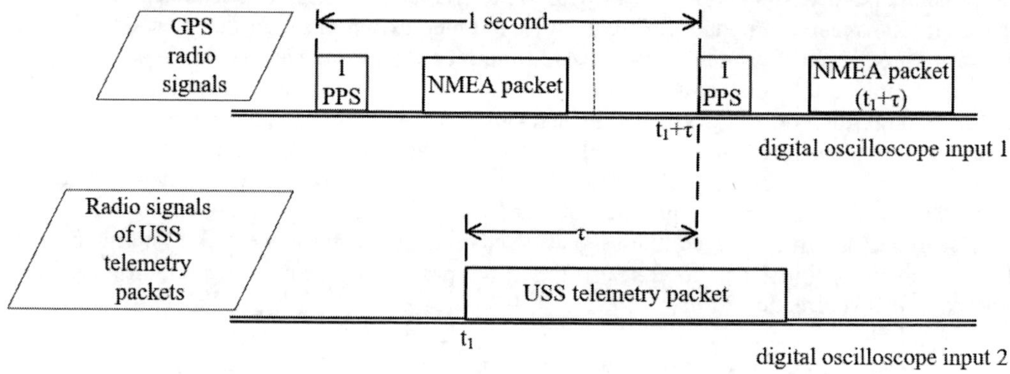

Figure 2. The radio signal time diagram for telemetry packets, NMEA packets and 1PPS signals.

3. The ground station operation modes, measurement and processing methods, mathematical models and algorithms

A satellite control session, a telemetry reception session, or an orbit measurement session can occur between USS and a ground station. A radio signal containing a control command packet or a measuring radio signal in a request measurement session is transmitted from the ground station to the satellite via the command link. A packet of response receipts or a measuring radio signal is sent back from the satellite to the ground station. A radio signal containing a telemetry packets or a measuring radio signal in an unsolicited measurement session is transmitted from the satellite to the ground station via the telemetry link.

Measurements of the orbit parameters for a single satellite pass over a ground station can be performed at either one point or at several points in the orbit. Therefore, the simplest Keplerian motion model for satellite can be used to process single pass measuring data over a ground station. Orbit measurements can also be performed on several satellite passes over a ground station. The satellite orbit period for an altitude of 350 to 600 km is from 90 to 102 minutes. After three consecutive satellite passes over the ground station, there is a break with an interval of 8–9 hours. Therefore, to process measuring data on several passes over a ground station, it is necessary to use perturbed motion models for the satellite.

The orbit measurement and determination system of a ground station can measure the satellite orbit with both initially known parameters (Tracking mode) and unknown parameters (Omni-directional search mode). To process measurements in tracking mode, orbital elements in TLE format or their own database are used as initial data. The satellite orbital prediction software based on the known initial orbit parameters calculates communication sessions, the tracking angles for antenna azimuth-elevation rotator, and the telemetry radio signal frequency. The main problem is to check the satellite orbit parameters and correct them. The differential correction method of satellite orbit parameters using relative velocity data between satellite and the ground station is used. Relative velocity data is computed from the Doppler shift of received telemetry radio signal at the ground station.

The initial orbit determination is performed for the satellite measuring in the Omni-directional search mode. The target satellite frequency is searched within the Amateur radio band of 435–445 MHz The distinctive features of the radio signals of telemetry packets (frequency of packets, presence of mark, etc.) are determined. Based on the results of measurements on several passes, the radio signal frequency, the satellite period, the maximum duration of pass over the ground station, the limit of measurement errors are estimated. To process measurements on a single satellite pass over a ground station, orbit determination algorithms based on Keplerian motion model are used, and on several passes, orbit determination algorithms based on the simplest perturbed motion models are used.

4. The ground station testing results for ultra-small satellite telemetry and orbit measuring

The developed mobile ground station was tested on receiving data and satellite orbit measurement with different transmission intervals between telemetry packets. The time and frequency of radio telemetry signals from nanosatellites CubeBel-1 (measurements on a single pass with transmission interval of 40 s) and LUOJIA-1 01 (measurements on several passes with transmission interval of 6 min) were measured. The frequency measurements of telemetry signals from these nanosatellites were compared with the simulation results based on the SGP4 prediction model with input data in TLE format. It is shown that the experimental data are in good agreement with the numerical simulation data. Errors in measuring the frequency of telemetry radio signals from nanosatellites relative to simulation data did not exceed 300 Hz.

The initial orbit determination for LUOJIA-1 01 nanosatellite was performed, provided that its nominal frequency of the radio signal and the initial orbital parameters are unknown. Based on the perturbed circular motion model was processed frequency measurements of the LUOJIA-1 01 nanosatellite on several passes. To justify nanosatellite motion model, the orbital parameters of low orbit satellites (altitude up to 2000 km) were analyzed using the NORAD system database (on data of 07.10.2019). The analysis showed that more than 90 % of these satellites have an almost circular orbit (with a small eccentricity value $e \ll 1$).

For satellites developed according to the Cubesat standard, an additional analysis of the orbital parameters and dynamics of their changes during the week of flight was performed. The results of the analysis show that there are two types of Cubesat orbits. The first type of orbit is associated with satellites launched from the International Space Station. They have inclination of about 51.6°, an orbital period of 90 to 94 min, and altitude of 330 km to 430 km. The second type of orbit is associated with satellites launched into a solar-synchronous orbit as secondary payloads. These Cubesats have inclination from 98 to 99°, an orbital period from 94 to 102 minutes, and altitude from 500 km to 800 km. The Doppler frequency shift of the telemetry radio signal these satellites in the radio visibility zone over the University ground station Minsk (latitude φ = 53°54'27" North, longitude λ = 27°33'52" East, altitude H = 0.23 km) varied in the range from −10.4 kHz to +10.4 kHz (for a nominal frequency from 435 to 438 MHz).

The nanosatellite orbit determination was made using 20 measurements of the reception time t_i and telemetry signal frequency f_i (i = 1–20) on several orbits over the University ground station for the period from 07.10.2019 to 09.10.2019 (UTC). The average reception frequency f_m = 437.25 MHz, the intervals between radio signals Δt_1 on one pass, and the intervals between radio signals Δt_2 on neighboring passes were estimated. The Doppler frequency shift of the radio telemetry signal was calculated as $\Delta f_i = f_i - f_m$. On one pass, the radio signals were transmitted with an interval of Δt_1 equal to 6 minutes or 12 minutes in two consecutive packets, which allowed them to be distinguished from other radio signals. The intervals between Δt_2 radio signals on neighboring passes were 90, 96 (mostly), and 102 minutes. This allowed us to assume that this unknown satellite orbit belongs to the second type orbits.

IOP Conf. Series: Materials Science and Engineering **966** (2020) 012104 doi:10.1088/1757-899X/966/1/012104

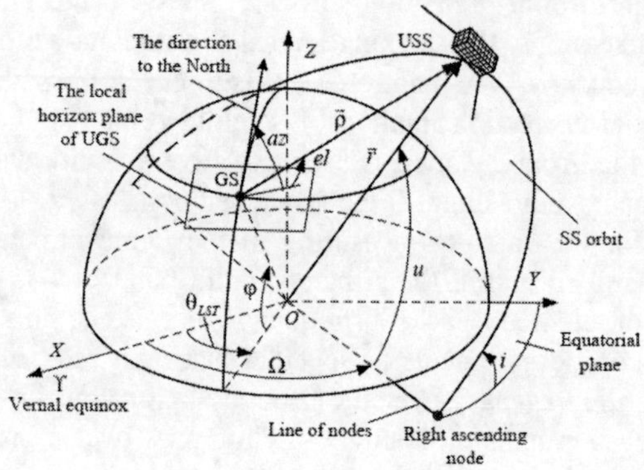

Figure 3. The perturbed circular motion model geometry (UGS – the University Ground Station; SS – Small Satellite).

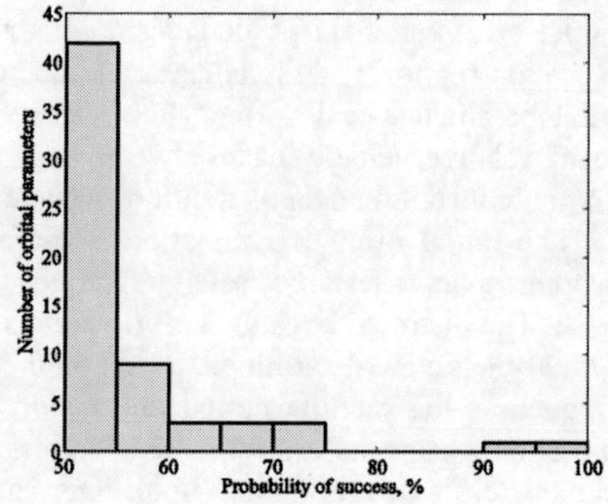

Figure 4. The number of possible sets of orbital parameters versus the probability of success β.

For orbit determination algorithm were set the ranges of the circular orbit parameters: period T from 94 to 102 min c step 1 c, the inclination i from 97 to 99° with the step of 0.01°, the argument of latitude u from 20 to 180° with the step of 1° and the longitude of ascending node Ω from 0 to 360° with the step of 1°. The perturbed circular motion model (Figure 3) of the satellite took into account the secular perturbations of the right ascension of the ascending node Ω due to the gravitational potential. The probability of success β of each set of orbital parameters (T, i, u, Ω) was calculated for the estimated time t_2 = 9:48:18 on 09.10.2019 (UTC) based on the analysis of the calculated elevation el and the Doppler shift of the frequency Δf_{calc} of the received radio signal over the ground station:

$$\beta = \frac{N}{N_{total}} 100\,\%, \tag{1}$$

where N, N_{total} = 20 is the number of calculated points and the total number of measurement points where numerical simulations were performed with elevation $el > 0$ and error $|\Delta f - \Delta f_{calc}| < 300$ Hz of the measured Doppler frequency shift Δf relative to the calculated Δf_{calc} for each set of orbital parameters (T, i, u, Ω).

In figure 4 the dependence of the number of possible sets of orbital parameters versus the probability of success is presented. Our results showed that for the time t_2 with probability of success β above 50% the ranges of the orbital parameters were: period T from 5852 to 5859 s, inclination i from 97.97 to 97.99°, argument of latitude u from 114 to 117° and the right ascension of the ascending node Ω from 356 to 4°. There is only one set of orbital parameters (T, i, u, Ω) = (5855 c, 97.98°, 115°, 359°) with the probability of success β = 100%, one set (T, i, u, Ω) = (5855 c, 97.98°, 115°, 360°) with β = 95%, and 60 sets of orbital parameters with β from 50% to 75%.

Based on set of orbital parameters (T, i, u, Ω) = (5855 c, 97.98°, 115°, 359°) and the perturbed circular motion model, the elevation el, azimuth az and Doppler frequency shift of telemetry signals Δf_{calc} were numerically simulated in the range from 0:0:0 to 23:59:59 on 10.10.2019 (UTC) of the following passes. Using the simulated data, University ground station successfully received and decoded the telemetry packets of the nanosatellite LUOJIA-1 01 at the pass interval from 10:07:50 s to 10:20:50 on 10.10.2019 (UTC). The accuracy of predicting the elevation, azimuth and Doppler frequency shift of radio signals was estimated at the interval of successful receiving and decoding telemetry in comparison with SGP4 prediction model. The elevation and azimuth prediction absolute errors did not exceed 3°, and the absolute error of Doppler frequency shift prediction did not exceed 160 Hz, which is sufficient for successful receiving and decoding telemetry.

A simulation of the cooperative operation of a network of three spaced University ground stations (Minsk, Brest, and Mogilev) for orbit determination the Starlink-1143 satellite in the first week of flight was performed. This satellite was launched on February 17, 2020 in a constellation launch of 60 satellites of the Global Internet System. In the first week of the flight, the satellites in the constellation had similar orbital parameters. The orbit determination technology involved receiving and processing frequency measurements of the same satellite telemetry signal at three synchronized ground stations, and identifying the satellite using the NORAD orbital parameter database.

The Doppler shift Δf_{exp} of the received frequency of the telemetry radio signal (with the addition of 100 Hz noise) of the Starlink-1143 satellite was simulated for two time points t_1 and t_2 using TLE input data and SGP 4 models. Using NORAD database for constellation launch on February 17, 2020 was calculated the pass parameters over ground stations (elevation el_i,, azimuth az_i,, range ρ_i, relative velocity between satellite and the ground station $d\rho_i/dt$) and the Doppler shift of the telemetry signal frequency $\Delta f_{calc.\,i}$ for time point t_i (i =1.2). Satellite identification was performed with the conditions $el_i > 0$ and $|\Delta f_{calc.\,i} - \Delta f_{exp.\,i}| < 300$ Hz.

The simulation results showed that if the mean anomalies of neighboring satellites differ by 0.1°, the satellite can be identified in the constellation by the frequency and time measurements of the same radio signal at three synchronized ground stations. The simulation results showed that the cooperative telemetry reception by a stationary University GCS and several mobile GS with time synchronization and spacing across the territory of the Republic of Belarus will allow using the resources of stations both for receiving satellite information and orbit determination. This will allow increasing the receiving geography, improving the quality of reception and independently obtaining orbital parameters and training students in practical technologies for receiving telemetry and orbit determination.

5. Conclusion

In summary, we have demonstrated the research and educational network of ground stations for receiving and processing information from low-orbit educational spacecraft. The ground station was tested in the modes of receiving data and measuring the ultra-small satellite orbit with different frequency of telemetry packets. It was shown, that errors in the frequency measurements of received telemetry radio signals for CubeBel-1 and LUOJIA-1 01 nanosatellites relative to the calculated SGP 4 model with initial data in TLE format did not exceed 300 Hz. Finally, the initial orbit determination by 20 frequency and time measurements of a LUOJIA-1 01 nanosatellite telemetry signal on several orbits with a limited number of data on one pass was studied.

Acknowledgments

The work was performed as part of the implementation of the State program of scientific research of the Republic of Belarus 'Informatics, space and security' (Program 'Informatics and space research').

References

[1] Fortescue P, Stark J and Swinerd G 2011 *Spacecraft Systems Engineering* (Chennai: John Wiley & Sons, Ltd.) p 724
[2] Nieto-Peroy C and Emami M R 2019 *Appl. Sci.* **9** p 3110
[3] Bouwmeester Jand Guo J 2010 *Acta Astr.* **67** p 854
[4] Capderou M 2005 *Satellites Orbits and Missions* (Verlag Paris: Springer) p 544
[5] Vallado D A 2013 *Fundamentals of Astrodynamics and Applications* (Hawthorne: Microcosm Press) p 1136
[6] Wang R, Liu J and Zhang Q 2009 *Adv. Space Res.* **43** p 1065
[7] Spiridonov A A, Ushakov D V and Saechnikov V A 2019 *Devices and Methods of Measurements* **10** 331
[8] Sakamoto Y, Yotsumoto K, Sameshima K, Nishio M and Yasaka T 2008 *Acta Astronaut.* **62** p 151
[9] Rouzegar H and Ghanbarisabagh M 2019 *Wireless Pers. Commun.* **108** p 2195-212

IOP Conf. Series: Materials Science and Engineering **966** (2020) 012105 doi:10.1088/1757-899X/966/1/012105

Temperature calculation for linear induction motor in transport application with multiphysics approach

F N Sarapulov[1, 3], V Goman[2] and G E Trekin[2]

[1] Ural Federal University named after the First President of Russia B N Yeltsin, 19, Mira str., Yekaterinburg, 620000, Russia

[2] Ural Federal University named after the First President of Russia B N Yeltsin, Nizhniy Tagil Technological Institute, 59, Krasnogvardeyskaya str., Nizhniy Tagil, 622000, Russia

E-mail: [3] sarapulovfn@yandex.ru

Abstract. The article describes the implementation of the mathematical model of the interrelated electromechanical and thermal processes in a linear induction motor using simulation software MATLAB-Simulink. The developed model consists of electromechanical model, thermal model and cooling system (hydraulic) model which all affect each other. Temperature calculations of the traction single-sided linear induction motor for urban transportation system are performed for different regimes. The calculation results using the developed thermal model based on the thermal detailed equivalent circuit demonstrate a close concurrence with the finite element model, temperature values differ less than in 3.8-4.8%. Safe operation range for a traction linear induction motor was determined taking into account the uneven longitudinal heating using different variants of cooling systems and coolant flow values. Recommendations on their selection were formulated. Based on the calculations and following experimental testing it was determined that operation with the maximum load with current of 270 A and load force of 8,000 N is possible with the cooling air flow rate of at least 2.2 m³/s.

1. Introduction

Linear induction motors (LIM) are a high potential type of the electric drive motor for the vehicular application, particularly for the high-speed passenger transport. This type of transport is highly relevant to the modern metropolitan cities. Many articles, such as [1-5] are dedicated to the development and optimization of LIM design for that specific use.

Investigations of the LIM are complicated by the fact that they noticeably differ from the rotary electric motors by the course of the electromagnetic processes, due to the longitudinal and transverse edge effects. These effects and peculiarities of LIM are well-studied and described in [6-10].

LIM also have significant differences in principles of thermal processes caused by construction features, as well as electromagnetic asymmetry lengthwise due to the longitudinal edge effect [11]. Many results obtained in a course of analysis of the thermal exchange in rotary motors are applicable to linear motors, for instance [12-16]. A series of publications are devoted to investigation of the thermal processes either in tubular linear motors [17, 18], or in linear synchronous motors which differ from the induction motors in their principle of operation [17-19]. In certain cases, LIMs are similar to induction heating units by principle of their operation, and specific approaches from this area can also be applied [20]. Therefore, an investigation of the electric drive operation modes based on LIM is relevant from the

IOP Conf. Series: Materials Science and Engineering **966** (2020) 012105 doi:10.1088/1757-899X/966/1/012105

perspective of thermal condition, taking into account a cooling method, in relation to the electromechanical processes. The necessity of such analysis may occur in a process of linear motor design. As a rule, objectives of a design process are optimization of the energy characteristics, reduction of losses, lowering the LIM components heating to the permissible level in operational modes.

2. Methods

In a designed model of interrelated electromechanical and thermal processes, the temperature effects of the LIM parts on the electromechanical characteristics and on the loss values are taken into account [11, 21]. The calculations of the LIM traction force, the losses in the primary and secondary based on detailed LIM electric and magnetic equivalent circuits were carried out according to the methodology [22, 23].

The model implemented in Simulink consists of 5 major elements (figure 1): 'Signal Builder' which sets the speed setpoint 'Velocity SP' and load force 'FL'; drive model 'Drive', which has such output parameters as velocity of a secondary, traction force 'Ft', electric current density 'J', loss vectors in the primary 'Pp' and secondary 'Ps', including the losses in electric winding (or conductive plate) and magnetic core (or back iron) according to the established detailing of the model in longitudinal coordinate; subsystem of the coolant velocity calculation in the channels ('Hyd. Model' in figure 1). Subsystem input parameter is a coolant flow velocity at the cooling system input. In figure 1, an example of a self-ventilation, when the air flow velocity is equal to the velocity of LIM moving part is shown. Output parameters of the subsystem are velocities of the coolant in cooling system channels 'Vi'; 'Thermal Model' subsystem with such input parameters as the temperature vectors of primary 'Tp' and secondary 'Ts', composed of the electric winding zones temperatures (conductive plate for the secondary) and magnetic core (back iron for the secondary) according to the established detailing of the model in longitudinal coordinate; given parameters are brought to 'Scope' - virtual oscilloscope using links. Similarly, the LIM temperature effects on the electromechanical processes are taken into account using links inside the 'Drive' unit (figure 2) and are not shown in figure 1. A subsystem's input parameters are the loss vectors in primary 'Pp' and secondary 'Ps' and coolant velocity in cooling system channels 'Vi'; a subsystem COMSOL Multiphysics aimed at the carrying out the thermal model verification in some operation modes by means of finite element method (FEM).

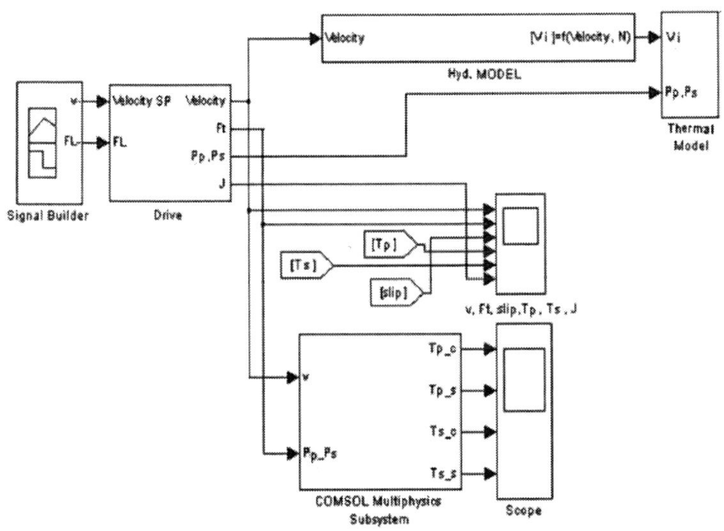

Figure 1. Model of LIM interrelated processes built in Simulink.

Drive subsystem includes the model of a scalar control system (U/f = const), LIM electromechanical model, the model of the loss calculations in the primary and secondary in the form of numeric multidimensional arrays. Electromechanical model of a LIM built in Simulink based on the four-dimensional array is shown in figure 2. Given approach, which is described in detail in the [21], allows to avoid direct solving of different submodels, characterizing different-scale time-dependent processes: thermal, mechanical and electromagnetic. For the investigated LIM, the arrays of traction forces and losses at different power supply frequencies, slips (which is the function of the load) and temperatures of the primary winding and secondary conductive plate are preliminarily filled in. This allows for approximately tenfold reduction of thermal transient process calculation since interpolation and value selection from the multidimensional array proceed much faster than direct calculation based on the electromechanical model for the current combination of frequency, slip and temperatures of LIM parts.

Similar approach is applied to the hydraulic model of a cooling system as well. The initial data for the calculation is the cooling system configuration, the quantity of the channels and their parameters, such as length, cross sections and roughness of inner surface, presence of local elements such as curves, constrictions, broadenings, etc. A system of equations using the electrohydraulic analogy, i.e. Kirchhoff's laws, was compiled for the cooling system [24]. The solution of a system of equations required 5-7 iterations, because some of hydraulic resistance values depend on the coolant flow rate which was taken into account using recommendation of [25]. Therefore, in a general model of interrelated processes the results of such calculation for the required set of coolant velocities in cooling system input were presented in the form of simplified analytical expressions obtained in [21, 24] and sets of coefficients for the coolant velocities in slotted channels 'coef_s', distributive branch 'coef_d', collecting branch 'coef_c', as it is shown in figure 3. This approach made it possible to sufficiently reduce time for calculations of LIM thermal transient processes which have complicated multichannel cooling systems.

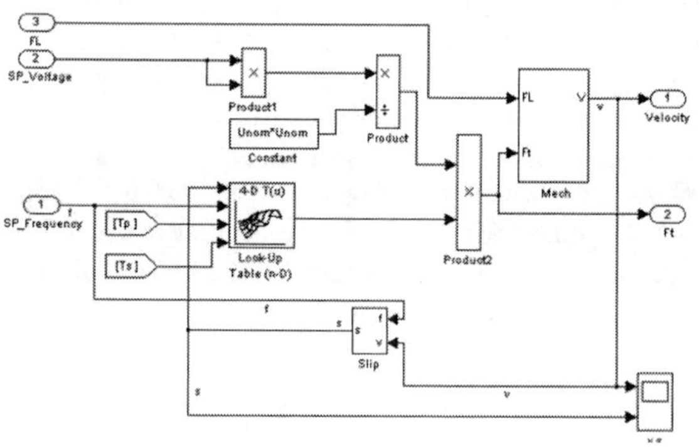

Figure 2. Electromechanical model of a LIM in Simulink based on the four-dimensional array.

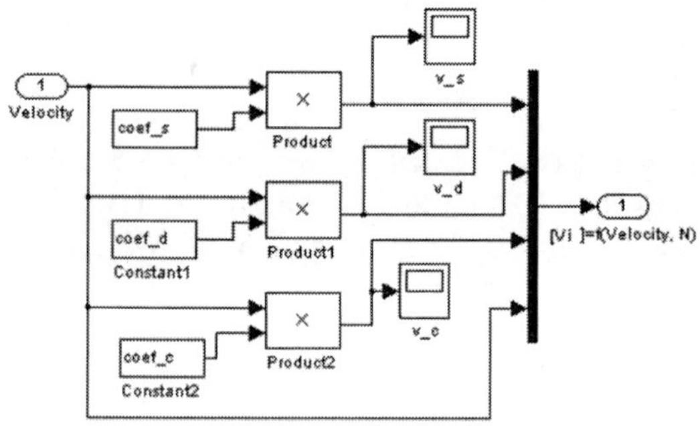

Figure 3. Simple hydraulic model of a cooling system in Simulink.

The structure of a thermal model is shown in figure 4. The model combines the following types of subsystems: zones of the 'Central block' primary active zone comprising the primary and secondary, 'Left Block Inductor' and 'Right Block Inductor' characterized in that the additional heat exchange due to convection and by radiation from the primary external surfaces is additionally considered therein; zones of secondary before entering the active zone '(Left End Zone') and after leaving it ('Right End Zone'); first or second kind boundary conditions simulation block.

Thermal circuit is built on the basis of thermal detailed equivalent circuit of a LIM presented in [11]. The following peculiarities are taken into consideration: heat removal from the active zone during the motion of a moving part; uneven longitudinal distribution of losses due to the longitudinal edge effect and different heating levels of the construction elements; losses in secondary outside the active zone; steel losses in magnetic cores; LIM elements temperature effects on the electromechanical characteristics and losses; thermal conductivity between the construction elements; heat exchange with the environment by means of radiation, between the primary and secondary; heat transfer from the construction external surfaces with regard to the conditions of natural or forced convection; equivalent thermal conductivity of slot and frontal parts of windings in different directions with regard to the insulation thermal conductivity; dependence of the thermal conductivity coefficient on temperature.

3. Result and discussions

The calculation of the traction single-sided LIM for the transportation system was made using the described model. Empty train weight – 14-58 tons (if the number of carriages varies from 2 to 10); the maximum carriage capacity – 34 passengers.

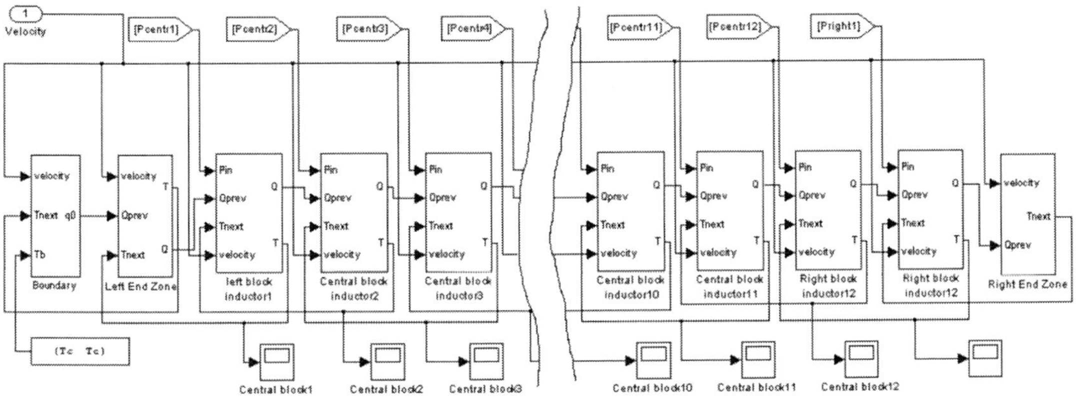

Figure 4. Thermal model of a LIM in Simulink.

While designing traction LIM the restrictions on the width have been imposed. Consequently, the winding scheme as in figure 5 has been accepted, as it provides the minimum length of the frontal parts. However, in this case almost all slots turned out to be half-filled. Restrictions on the dimensions have led to the excessive heating, since winding temperatures would not exceed the permissible ones only in case the train had a small number of passengers on board. To solve this problem a decision on cooling system development is made. The housings have been installed around the frontal parts, forming the channels which are ensuring the circulation of a cooling air through the frontal parts and half-filled slots. The calculation results of the cooling system were described in detail for the first time in the study [24]. In the [24], the charts of velocities distribution, coolant flow and pressure by primary's zones were obtained for the following cases: natural cooling: velocity of cooling air in inlet of cooling system is equal to the train velocity; forced cooling, for the air flow rate of 0.11-2.2 m^3/s.

Table 1. Traction LIM parameters of the urban transportation system.

Parameters	Value	Parameters	Value
Frequency supply	5-40 Hz	Height of the primary core	110 mm
Phase voltage RMS	40-240 V	Length of the primary core	1,740 mm
Phase current	70-270 A	Slot pitch	35 mm
Number of phases	3	Number of slots	50
Maximum traction force	8,000 N	Width of the secondary conductive plate	400 mm
Synchronous speed (at 23.2 Hz)	10 m/s	Thickness of secondary conductive plate	6 mm
Number of poles	8	Width of secondary solid back iron	400 mm
Number of turns in a slot	14	Thickness of secondary solid back iron	20.7 mm
Width of the primary core	240 mm	Air gap	10

The calculation results of the thermal model based on the thermal detailed equivalent circuit for the same cases and modes are shown in figure 6 and are demonstrating a close concurrence within 3.8-4.8%. Verification of the designed model of thermal processes is performed with the help of the model based on the FEM. Figure 7 shows FEM mesh and results of temperature calculation at phase current of 135 A in a full-filled slot no. 2 which has the worst cooling conditions and the highest thermal generation. In case of natural convection (figure 7, b) the temperature of the winding constitutes 186 °C, which corresponds to the limitation of the start-up force of 1,850 N typical for the empty train. In case of air flow velocity in the gap is 0.1 m/s (figure 7, c), the temperature of a winding decreases to 157 °C.

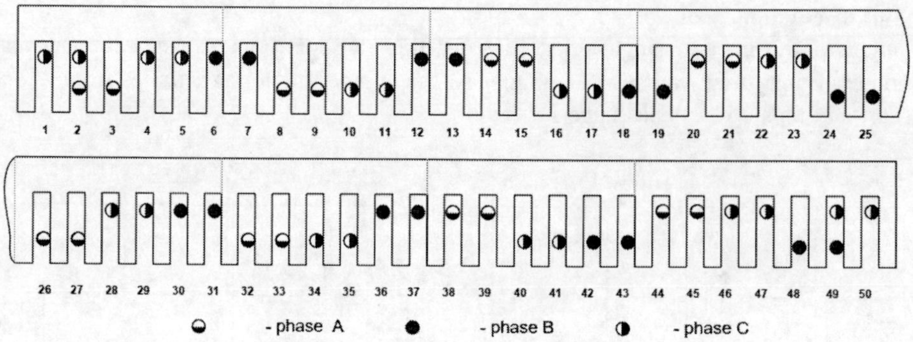

Figure 5. Scheme of the primary winding.

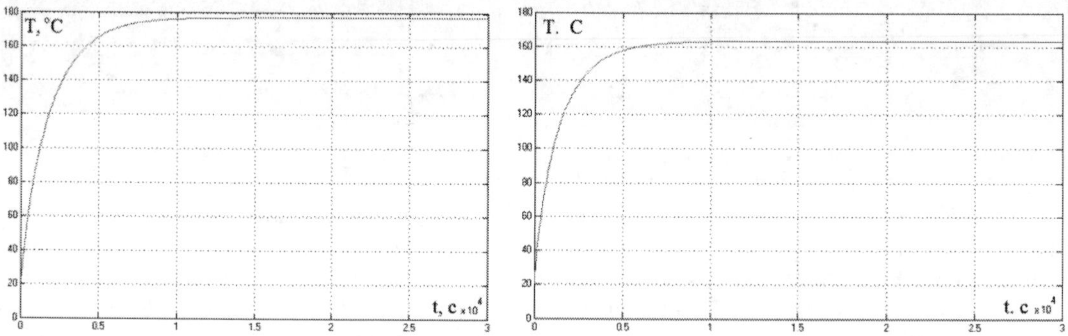

(a) (b)

Figure 6. Primary winding temperature in slot no. 2 of a traction LIM (equivalent thermal circuit model). (a) natural convection; (b) forced convection in the air slot at the velocity of 0.1 m/s

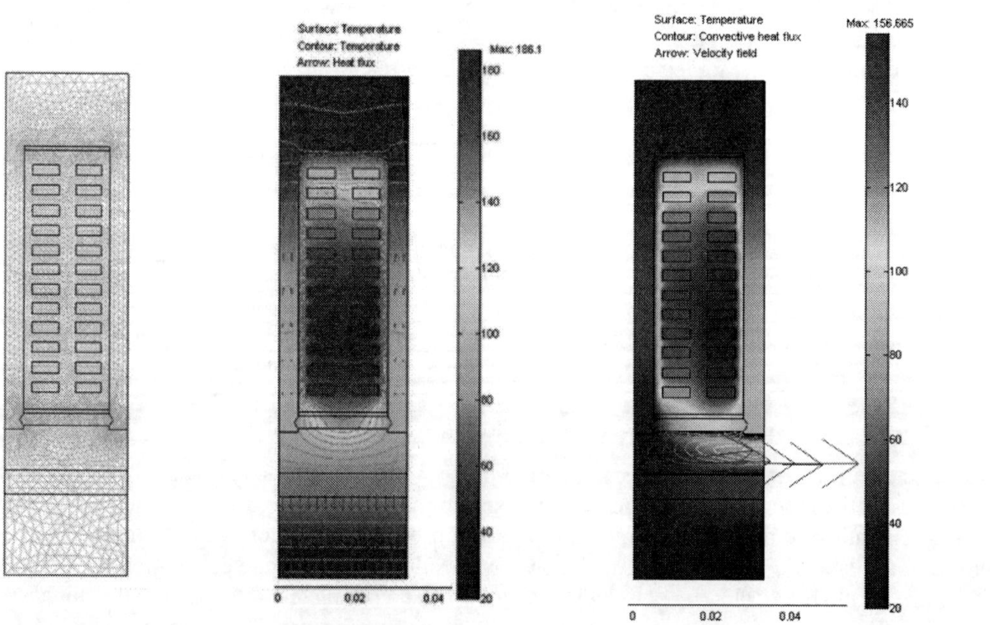

(a) (b) (c)

Figure 7. Temperature distribution in slot no. 2 of a traction LIM (FEM-model). (a) slot area and finite element mesh; (b) natural convection; (c) forced convection in the air gap at the velocity of 0.1 m/s.

IOP Conf. Series: Materials Science and Engineering **966** (2020) 012105 doi:10.1088/1757-899X/966/1/012105

4. Conclusion

As a result of performed research a mathematical model for the LIM thermal processes was built, with the help of which the analysis of interrelated thermal and electromechanical processes of an urban transportation system traction LIM was carried out. Verification of the model by finite elements method is performed for the number of modes. Safe operation range for a traction LIM (primary winding phase current values) was determined taking into account the uneven longitudinal heating using different variants of cooling systems and air flow values, and recommendations on their selection were formulated [11, 21]. Based on the calculations and following experimental testing it was determined that LIM operation with the maximum load (current of 270 A, load force of 8,000 N) is possible with the cooling air flow rate of at least 2.2 m^3/s.

References

[1] Huang S, Ching T W, Li W and Deng B 2018 *IEEE 27th Int. Symp. on Ind. Electron. (ISIE)* (Cairns, QLD) pp 150-4

[2] Abdollahi S E, Mirzayee M and Mirsalim M 2015 *IEEE Trans. Magnetics* **5 7** pp 1-7

[3] Lu Q, Li Y, Ye Y and Zhu Z 2013 *IEEE Trans. Energy Conversion* **28** pp 145-153

[4] Lu J and Ma W Jan. 2011 *IEEE Trans. Plasma Sci.* **39 1** pp 116–120

[5] Lv G, Zeng D, Zhou T, and Liu Z Jun. 2017 *IEEE Trans. Ind. Electron.* **64 6** pp 4382-90

[6] Boldea I 2013 Linear Electric Machines, Drives, and MAGLEVs Handbook (CRC Press)

[7] Yamamura S 1979 Theory of Linear Induction Motors (New York, NY, USA: Halsted)

[8] Gieras J F 1994 Linear Induction Drives (London, U.K.: Oxford Univ. Press)

[9] Laithwaite E and Barwell F 1969 *Proc. IEE* **116 5** pp 713-24

[10] Bolton H 1969 *Proc. IEE* **116 5** pp 725-31

[11] Sarapulov F N and Goman V V 2009 *Russ. Electr. Engng.* **80** pp 431-35

[12] Boglietti A et al. Mar. 2009 *IEEE Trans. Ind. Electron.* **56 3** pp 871–82

[13] Staton D A and Cavagnino A 2008 *IEEE Trans. Ind. Electron.* **55 10** pp 3509–16

[14] Madonna V, Walker A, Giangrande P, Serra G, Gerada C and Galea M 2019 *IEEE Trans. Ind. Electron.* **66 7** pp 5057-69

[15] Boglietti A, Carpaneto E, Cossale M, Vaschetto S, Popescu M and Staton D 2015 *IEEE Energy Conversion Congress and Exposition (ECCE)* (Montreal, QC) pp 4865-71

[16] Boglietti A, Cavagnino A and Staton D 2008 *IEEE Trans. Ind. Appl.* **44 4** pp 1150-59

[17] Huang X, Li L, Zhou B, Zhang C and Zhang Z 2014 *IEEE Trans. Ind. Electron.* **61 8** pp 3923–31

[18] Vese I, Marignetti F and Radulescu M M 2010 *IEEE Trans. Ind. Electron.* **57 1** pp 320-26

[19] Tessarolo A and Bruzzese C 2015 *IEEE Trans. Ind. Electron.* **62 3** pp 1410-20

[20] Prakht V A, Dmitrievskii V A, Sarapulov F N, Dmitrievskii A A and Safin N A 2014 *COMPEL* **33 1/2** pp 273-85

[21] Goman V V 2006 Thermal processes in linear induction motors and their mathematical modelling (Teplovie processi v lineynih asinhronnih dvigatelyah i ih matematicheskoe modelirovanie) [in Russian] PhD Thesis, Russia, Yekaterinburg, Ural State Technical University

[22] Dmitrievskii V, Goman V, Sarapulov F, Prakht V and Sarapulov S 2016 *Int. Symp. on Power Electronics, Electrical Drives, Automation and Motion (SPEEDAM)* (Capri, Italy) pp 458–63

[23] Sarapulov F N, Frizen V E, Shvydkiy E L and Smol'yanov I A 2018 *Russ. Electr. Engng.* **89** pp 270–74

[24] Sarapulov F N and Goman VV 2010 *Transport Electronics and Electrical Equipment (Electronica i electrooborudovanie transporta)* [In Russian] **4** pp 29-32

[25] Idelchik I E 1994 *Handbook of Hydraulic Resistance - Coefficients of Local Resistance and of Friction* 3rd ed (CRC Begell House)

IOP Conf. Series: Materials Science and Engineering **966** (2020) 012106 doi:10.1088/1757-899X/966/1/012106

Research in Autodesk Inventor weldment performance cultivator frame

V V Telegin[1, 2] and I V Telegin[1]

[1] Lipetsk State Technical University, 30, Moskovskaya str., Lipetsk, 398055, Russia

E-mail: [2] vv.telegin@yandex.ru,

Abstract. The paper examines issues of calculation in the welded frame structures design of engineering objects using as the example optimizing the cultivator frame. Design environment – Autodesk Inventor Professional. The purpose of the study was to reduce the cost of the base product. The basis for solving this problem is the development solid-state model of the cultivator frame basic construction; strength calculation of the frame with various options for its loading; the choice of the application of forces to the frame, leading to the occurrence of maximum stresses in the frame elements; the use of this option to analyze the strength of the frame after making changes to its design. In this paper, we investigated the possibility of replacing the steel grade of which the frame beams are made with St3ps steel; the possibility of making frame beams from a steel roll-formed channel is investigated. The possibility of reducing more than 30 percent of the frame mass by means of implementing the indicated actions is substantiated.

1. Introduction

One of the main requirements for engineering products being designed is their minimum metal consumption and cost [1–3]. The metal consumption of the frame structure considered in this paper depends on the geometry of the profile (tube, channel, I-beam, etc.) and the size of its cross section. The cost, in addition, is associated with the cost of the material – the steel grade from which it (the frame) is made.

The strength of the structure [4, 5], its indicator is the safety factor. This indicator is considered in this work as the only one that determines the performance of the product. In practice, of course, this is not so. The ability of the product to perform its functions depends on its geometry, magnitude of the elastic deformations, dynamic and static processes that arise during the product operation [6, 7], and many other factors.

Traditionally, the calculation of the safety factor is performed by the methods of materials resistance [1, 2]. Methods, in the general case, are rather laborious and not very accurate. The modern way is the finite element method (FEM) [8–10]. In combination with solid state design, FEM (and its modifications) is the basis of present design.

Autodesk Inventor Professional [11, 12], used in this work to calculate the frames of cultivators, is one of the most famous software products combining methods of solid modeling, development of design documentation and strength calculations based on FEM.

2. Research problem statement

The solid-state model of the cultivator is shown in figure 1. Its frame is a welded structure made of square tubes [13]. The task is to consider the possibility of using other beams profiles of lower mass and made of cheaper steel grades in the frame design.

Figure 1. Solid-state cultivator model: frame, suspension, traction device.

When the cultivator is working, stresses arise in the elements of its frame, the values of which are various in different structural elements. In this regard, the following stages of upgrading the frame design are proposed:

- the study of the stress-strain state of the cultivator frame with various schemes of application of forces, in order to identify the least loaded parts;
- evaluation of the possibility to replace individual lightly-loaded frame elements with cheaper profiles or even their exclusion from the design;
- changing the design of the frame and the study of its performance.

3. Cultivator frame design model

Figure 2 shows the calculated model of the cultivator frame. Unlike the actual design, the device for coupling the frame and the traction device (tractor) is simplified and serves only for the correct application of the traction force highlighted in red. Instead of a real suspension, its simplified model is used, which allows simulating the application of the force highlighted in yellow to the suspension leg from the ground at any angle or impose a pinch dependence. These elements highlighted in figure 2 in blue are not objects of study. In the presented results of the frame simulation by the finite element method, their display, as well as the display of the applied forces, is hidden.

In the process of a comprehensive study of the strength of the cultivator frame, several (more than ten) variants of its load were studied, differing in the angle of application of the traction force (0 or 30 degrees), the choice of suspension for imposing a fixation dependence and the direction of the forces acting on the suspension (0 or 180 degrees).

IOP Conf. Series: Materials Science and Engineering **966** (2020) 012106 doi:10.1088/1757-899X/966/1/012106

Traction force in all variants of the study is taken equal to 60 kN. In reality, its value is from 18 to 27 kN. The force acting on the suspension is 3 kN. In the absence of suspension paw engagement for some kind of obstacle, this value is the maximum possible.

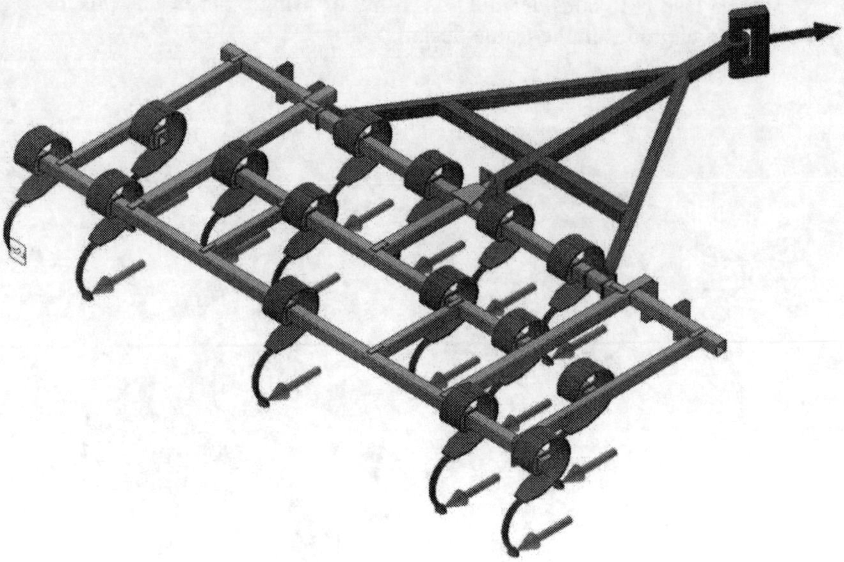

Figure 2. Calculation model of the cultivator frame.

4. Results and discussion

As an example, the article presents a variant with the minimum calculated values of the safety factor in the frame beams and their joints (figures 2 and 3).

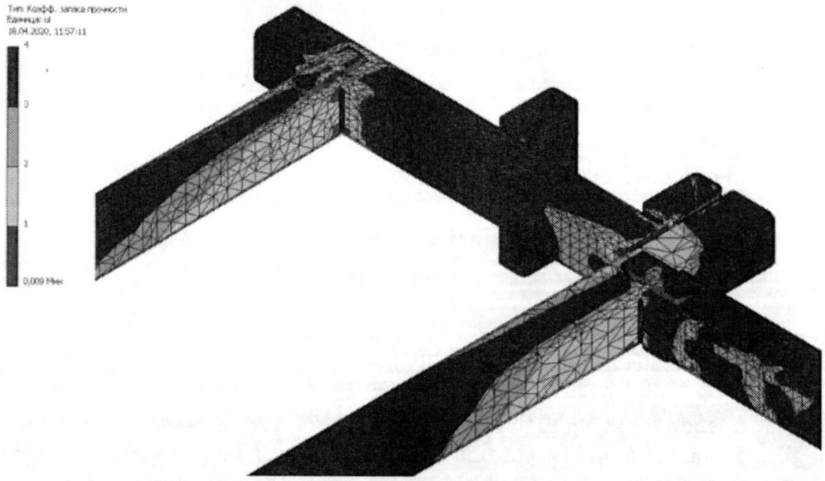

Figure 3. Calculation of safety factor.

As follows from the research results, the basic design of the cultivator frame has a margin of safety significantly exceeding the required. An exception is welds. This is explained by the feature of the calculated solid-state frame model. During its development, the direct impact of beams and overlays forming the welded frame structure on each other is excluded. The interconnection of the frame elements with each other is carried out only through the rollers of the welds. This feature allows you to additionally identify problem areas associated with the connection of parts by welding.

The results of the study suggest a slight loss of frame strength when using profiles made of cheaper material. In turn, such a replacement will have a positive effect on reducing the total cost of product. In figure 4 shows the structural strength analysis results of the frame made of steel grade – St3ps [14]. Traction force in this case is taken equal to 30 kN. Its maximum value recommended by the manufacturer is 27 kN.

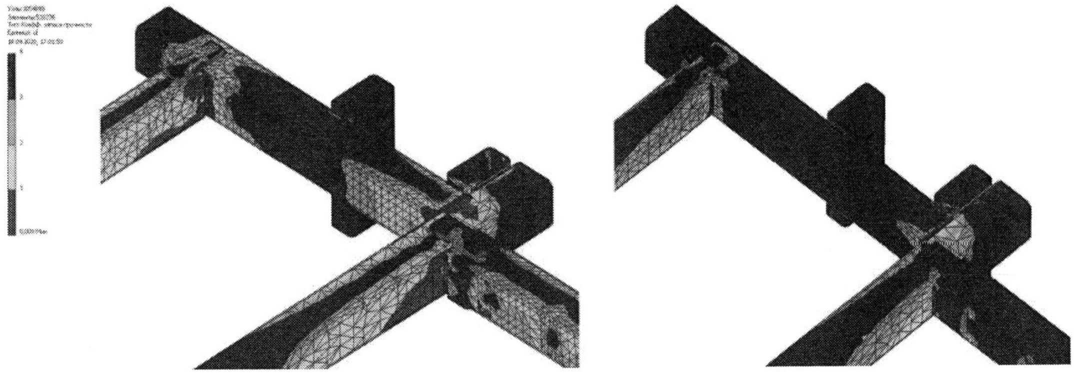

(a) (b)

Figure 4. Calculation of the safety factor of the frame structure made of steel grade St5ps. The value of traction force: (a) – 60 kN, (b) – 30 kN.

In figure 5 shows a fragment of a cultivator frame made of channels [15], and the result of calculating the strength with a traction force value of 30 kN. Unlike the previous figures, for a more visual representation of the design, the frame is rotated 180 degrees (bottom view).

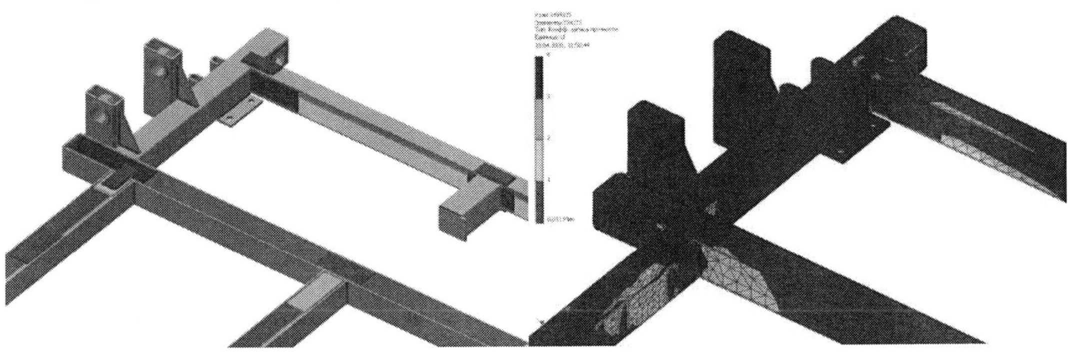

(a) (b)

Figure 5. The safety factor calculation of the frame structure made of channels: (a) – structures, (b) – calculation results.

When comparing the calculation results presented in figures 3 and 5, then, despite the obvious 'weakening' of the frame structure, the stresses in its elements at the same loads turned out to be approximately equal. This is explained by a decrease in the rigidity of the frame. The deformations of the frame that occur during cultivator operation have increased approximately twofold. However, this fact does not affect its performance.

At the same time, the mass of the frame is noticeably reduced. The weight of the frame in the basic version is 270.5 kg, in the new version – 175.8 kg.

There are also negative points. The laboriousness of manufacturing a frame has noticeably increased, and hence its cost.

5. Conclusion

The methodology for designing welded frame structures proposed by the authors, implemented on the basis of Autodesk Inventor Professional, allows not only to increase the productivity of design work, but also to increase their quality.

Using the modeling of a cultivator frame as an example, Autodesk Inventor Professional demonstrated the capabilities of this program for optimizing the designed structures of engineering objects.

This work is part of the educational process [16–18], implemented at the Department of Mechanical Engineering Technology of Lipetsk State Technical University, as part of the training of future specialists for engineering enterprises.

References

[1] Ustinov A N 2010 *Agricultural Machines:* 2nd Ed (Moscow: IRPO "Academy") p 264
[2] Turbina B G 2007 *Agreecultural machines. Agreecultural equipment. Theory and technological calculation*. ed B G Turbina (Moscow: Mechanical Engineering) p 356
[3] Bosoy E S, Vernyaev O V, Smirnov I I and Sultan-Shah E G 1977 *Theory, Design and Calculation of Agricultural Machines: A Textbook for High Schools of Agricultural Engineering* (Moscow: Mechanical Engineering) p 568
[4] Birger I A, Shorr B F and Iosilevich G B 1993 *Strength analysis of machine parts: Handbook,* 4th ed (Moscow: Mechanical Engineering) p 640
[5] Dunaev P F and Lelikov O P 1998 *Construction of units and parts of machines: Textbook. allowance for tech. spec. universities* (Moscow: Higher school) p 447
[6] Telegin V V, Kozlov A M and Sakalo V I 2017 Solid Modeling and Dynamic Analysis of the Mechanisms of Press-forging Machines *Procedia Engineering* **206** pp 1258–63
[7] Telegin V, Kozlov A and Shumilova T 2018 Development and research of the rotating lever object as a dynamic model of a cycle mechanism *MATEC Web Conf. International Conference on Modern Trends in Manufacturing Technologies and Equipment (ICMTMTE 2018)* **224** 02078
[8] Ovcharenko V A 2004 *Calculation of mechanical engineering problems by the finite element method: Textbook* (Kramatorsk: DGMA) p 128
[9] Gallagher R. 1984 *The Finite Element Method. The basics* (Moscow: Mir) p 428
[10] Zienkiewicz O C 1975 *The finite element method in engineering science* (Moscow: Mir) p 541
[11] Zinoviev D V 2016 *Design Fundamentals in Autodesk Inventor 2016, ed. Azanova M, 2nd ed.* (Moscow: DMK Press) p 256
[12] Kontsevich V G 2007 *Solid State Modeling of Engineering Products in Autodesk Inventor* (Moscow: DMK Press) p 670
[13] Russian state standard GOST 8645-68 *Rectangular steel tubes. Range* (Moscow)
[14] Russian state standard GOST 380-2005 *Common quality carbon steel. Grades* (Moscow)
[15] Russian state standard GOST 8278-83 *Roll-formed steel equal channels. Dimensions* (Moscow)
[16] Telegin V V, Telegin I V and Kirichek A V 2019 Solid-state modeling and basic training of specialists in the field of mechanical engineering *IOP Conf. Ser.: Mater. Sci. Eng.* **483** 012004
[17] Telegin V V, Telegin I V and Stepanov A S 2018 3D-Forming and Autodesk Autocad at the Initial Stage of Engineering Training of Specialists in Technical Areas *International Journal of Engineering and Technology (UAE)* **7** (3.5) pp 1–3
[18] Telegin V V and Telegin I V 2019 Solid Modeling in Professional Training of Specialists for Machine-Building *Enterprises International Journal of Innovative Technology and Exploring Engineering (IJITEE)* **8** (9S3) pp 7–9

IOP Conf. Series: Materials Science and Engineering **966** (2020) 012107 doi:10.1088/1757-899X/966/1/012107

Synchronous reluctance machines: performance evaluation with and without ferrite magnets

Mohamed N Ibrahim[1,2,3,6], Kotb B Tawfiq[1,2,4], E M Rashad[5] and Peter Sergeant[1,2]

[1]Department of Electromechanical, Systems and Metal Engineering, Ghent University, Ghent, Belgium
[2]FlandersMake@UGent – corelab EEDT-MP, Leuven, Belgium
[3]Electrical Engineering Department, Kafrelshiekh University, Kafrelshiekh, Egypt
[4]Department of Electrical Engineering, Faculty of Engineering, Menoufia University, Menoufia, Egypt
[5]Electrical Power and Machines Department, Tanta University, Tanta, Egypt

E-mail: [6]m.nabil@ugent.be

Abstract. Nowadays, a great interest in the efficiency and the cost of electrical machines has been noticed. Therefore, Synchronous Reluctance Motors (SynRMs) have become more attractive alternative for induction and permanent magnet (PM) synchronous machines in several industrial applications. This is because they do not use rare-earth magnets, windings and cages in their rotor. Consequently, the cost and efficiency are improved. Nevertheless, a major problem of the SynRMs is the low power factor. To improve the power factor as well as the torque density and efficiency, the cheap ferrite PM can be inserted in the rotor, resulting in the so called 'PM assisted SynRM'. The rotor design of SynRMs with and without permanent magnets (PMs) has a huge effect on the motor efficiency, torque density and power factor. This paper introduces an evaluation for the performance of SynRMs with and without PMs in terms of efficiency, torque and power factor maps as a consequence of different rotor designs. Three different rotor designs for the same machine stator have been compared. For one machine, the experimental measurements are obtained and the validation of the simulation results are confirmed. A preprint of this work is available under: https://biblio.ugent.be/publication/8515429.

1. Nomenclature

i_d, i_q	Direct and quadrature axis stator current respectively, A
λ_d, λ_q	Direct and quadrature axis flux linkage of SynRM respectively, V.s
P	Number of pole pairs
R_s	Stator resistance of the motor, Ω
V_d, V_q	Direct and quadrature component of stator voltage respectively, V
V_m, I_m	Maximum input voltage and current of the motor, V, A respectively
δ, α	Load angle and current angle, rad
ω_r	Mechanical speed of the rotor, rad/s

IOP Conf. Series: Materials Science and Engineering **966** (2020) 012107 doi:10.1088/1757-899X/966/1/012107

2. Introduction

Recently, Synchronous Reluctance Motors (SynRMs) with or without permanent-Magnets (PMs) are becoming attractive machines for variable speed industrial applications [1]–[5]. This is because of their merits of wide constant power speed range and high torque density. The power density and efficiency are better than that of induction machines and comparable to that of permanent magnet machines [2], [3]. In literature, there are a lot of papers which study SynRMs when their rotor includes and does not include magnets [1–16]. For example, in [1], suitable design of the three phase SynRM for electric vehicles was presented. In addition, a comparison between different flux-barrier designs for the same stator was investigated. The effect of four different steel grade on the performance of SynRMs was studied in [2]. It is noticed that the lower thickness steel grade has the higher efficiency, however it does not produce the higher output torque. In [3], the design of PMaSynRMs with an induction motor with identical NEMA frame stators was studied and compared. In [4], the design characteristics of SynRM with ferrite magnets and stator skewing was investigated. In addition, mechanical stress and demagnetization of ferrite was studied as well. In [5], the performance of PMaSynRMs for high-efficiency and wide constant power operation was studied. An investigation of the effect of the parameters and anti-demagnetization of rare-earth-less PMaSynRMs was done in [6]. In [7], a study of general properties of power losses in soft ferromagnetic materials was carried out. Different applications including pumping systems of the PMaSynRMs were introduced in [8–10] with high energy efficient. In [11] and [16], the modelling, simulation and design of the PMaSynRMs were introduced. This paper presents an evaluation for the performance of SynRMs with and without permanent magnets.

3. Performance Evaluation of SynRMs and PMaSynRMs

The vector diagram of PMaSynRM is sketched in Figure 1. Inserting $\lambda_{pm} = 0$ in Figure 1 results in the vector diagram of SynRM. The complete model of SynRM and PMaSynRM is reported in [16]. In the following section, three different rotors having identical stator and other parameters as listed in table 1 have been considered. The stator has 36 slots and 15 turns/slot with conventional star-connected windings. The stator design is similar to that of induction machines. Moreover, all the rotors have three flux-barriers per pole as shown in Figures 2 and 3. The first rotor has been designed by a manufacturing company. This motor is called a reference SynRM and it is the motor for which the experimental validation has been done. The second rotor has been optimized by a conventional optimization technique with 2D FEM for the twelve rotor parameters, described in Figure 2. This motor is called an optimal SynRM and its rotor parameters are given in table 2. The third rotor has been obtained by filling all the three flux-barriers of the optimized rotor (second one) with ferrite PMs. The ferrite PM is selected due to the lower cost and the availability in the market. In addition, it can withstand higher temperature [8]. The adopted ferrite PM properties are shown in [8]. This motor is called a PMaSynRM. The performance of the three machines is obtained by 2D-FEM at the same conditions. Sinusoidal currents are injected in the machine windings at different speeds up to the rated speed value (6000 r/min). The currents have been varied from low values up to the rated value ($I_m = 30$ A) at fixed current angle $\alpha = 56.5°$. The selected value of the current angle (56.5°) is the angle at which the machine can give approximately the maximum output power for the different currents and speeds. The FEM field pattern of the reference motor for quarter geometry at rated conditions is described in Figure 3. It can be noticed from Figure 3 that there are some iron regions having higher flux level (red color). These regions are called flux-barrier tangential ribs. The thickness of the ribs has a great influence on the performance of SynRM.

Figure 4 shows the output torque of the three machines at different operating speeds and stator currents up to the rated values. First, at the same current and speed, the output torque of the machine depends on the rotor design which is greatly affected by the design of the twelve parameters of the rotor, given in Figure 2. Therefore, an optimized rotor geometry for the SynRMs is necessary and unavoidable to maximize the machine performance. The performance of SynRM mainly depends on the quadrature (q) and the direct (d) inductances and these inductances depend on the geometrical design of the flux barriers of the rotor. Second, when comparing the subfigs of Figure 4 of optimal SynRM and PMaSynRM, it can be observed that the average torque of the machines is increased by 25 % when adding ferrite PM in the optimized rotor This is because of the effect of PMs on the q-axis inductance of the SynRM. The PMs saturate the tangential ribs (Figure 2) of the motor. By consequence, the q-axis inductance is reduced resulting in a higher output torque.

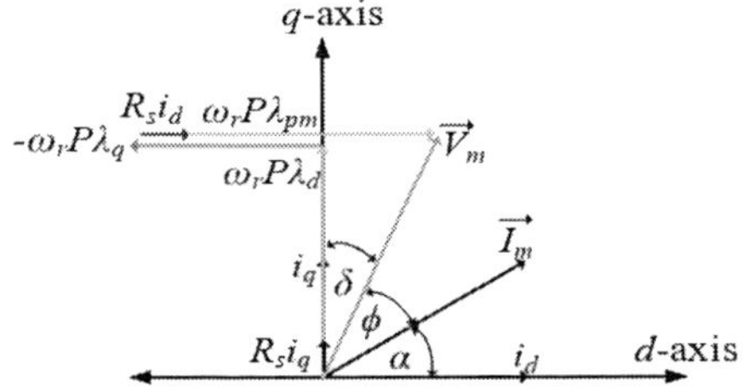

Figure 1. Vector diagram of PMaSynRM.

Table 1. Prototype SynRM parameters.

Parameter	Value	Parameter	Value
Number of poles/stator slots	4/36	Rated current/speed	22 A/6000 RPM
Number of phases	3	Material grade	M400-50A
Stator outer/ inner diameters	180 mm/ 110 mm	Axial length/ rotor outer diameter	140 mm/ 109.4 mm
$\theta_1,\theta_2,\theta_3$	7.5°, 20.5°, 33.5°	W_1, W_2, W_3	6 mm, 4 mm, 3 mm
L_1, L_2, L_3	25 mm, 19 mm, 12 mm	p_1, p_2, p_3	23.5 mm, 36 mm, 46 mm

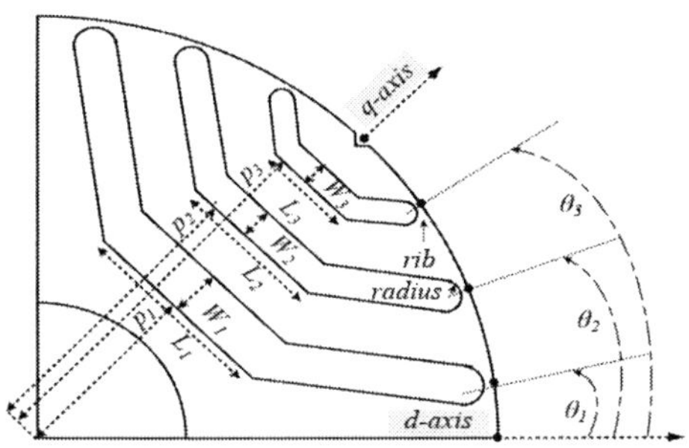

Figure 2. Quarter rotor geometry of SynRM.

Table 2. Optimal SynRM rotor geometrical parameters.

Parameter	Value	Parameter	Value
$\theta_1, \theta_2, \theta_3$	8.08°,16.43°, 28.4°	L_1, L_2, L_3	28.85 mm,28 mm,13.5 mm
W_1, W_2, W_3	5.5 mm, 3.5 mm, 3.5 mm	p_2, p_1, p_3	35.5 mm, 22.75 mm, 44.2 mm

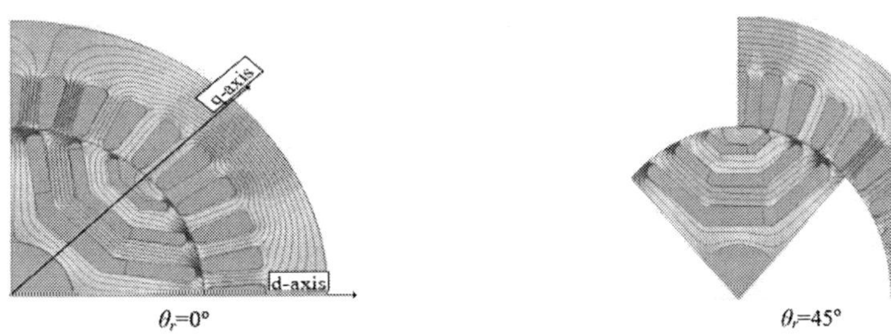

Figure 3. Flux paths of the reference SynRM using FEM for a quarter geometry at two rotor positions (0 and 45°)

IOP Conf. Series: Materials Science and Engineering **966** (2020) 012107 doi:10.1088/1757-899X/966/1/012107

Figure 5 shows the computed losses (iron and copper) of the machines for different currents and speeds up to the rated values. The copper losses are constant for all the machines due to the same stator windings and currents. The iron losses depend on the material properties, the currents and the geometry design of the machine. The variation of the parameters of rotor flux-barrier, shown in Figure 2, has a great effect on the iron losses of SynRM. This is due to the variation of the saturation regions of the iron especially, the rotor tangential ribs and the teeth of the stator. This can be noticed in Figure 5, by comparing the subfigs (reference, optimal and PMaSynRM). For the same current and operating speed, the iron losses increases for the PMaSynRM that has the higher output torque (Figure 4) compared to the other machines. In addition, for one machine as expected, the iron losses increase with the increase in speed and current, approximately in proportional way.

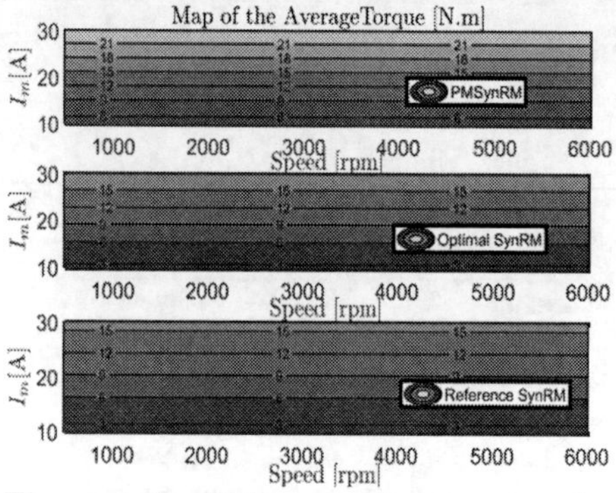

Figure 4. SynRM average torque maps for different rotor designs.

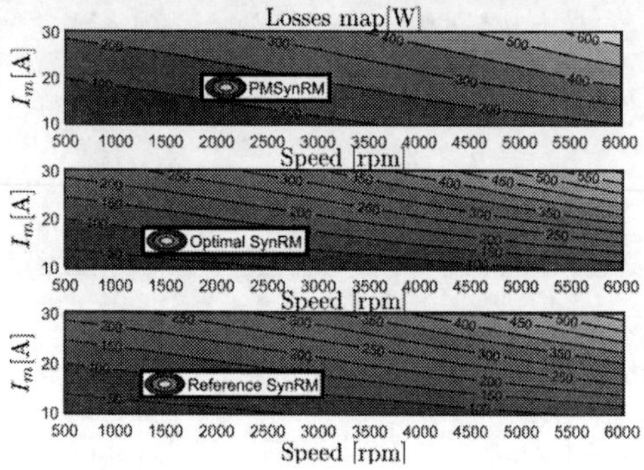

Figure 5. SynRM total losses maps for different rotor designs.

Moreover, the efficiency of SynRMs is better than induction motors [3] and is inferior compared to the permanent synchronous motors [9] as deduced in Figure 6. As the output power and total losses were affected with the rotor design, the efficiency of the machine depends on the rotor design as well (see Figure 6). At half rated condition the PMaSynRM's efficiency is around 95 %.

The power factor of the SynRMs depends on the saliency ratio (L_d/L_q) which is affected by the design of the rotor. The optimal SynRM has a higher power factor compared to the reference SynRM as it has the higher saliency ratio than the reference motor. However, it still has a low power factor about 0.68 at the rated conditions. Hence, adding PM in the optimized rotor reduces the phase angle between the voltage and current as shown in Figure 1. Consequently, the power factor is increased to about 0.91 at the rated conditions as observed in Figure 7 (PMaSynRM). This is a good indication for PMaSynRM to be attractive motor for industrial applications.

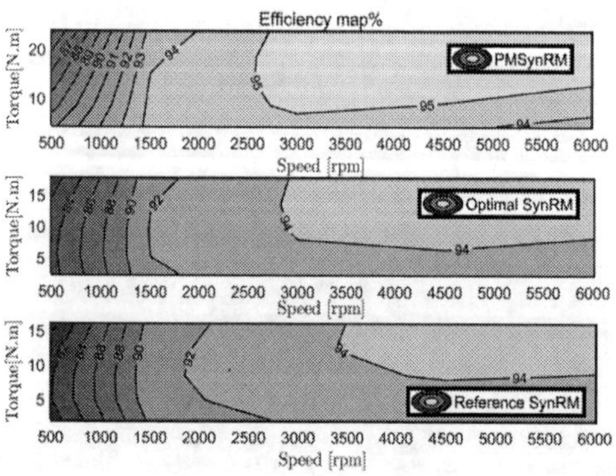

Figure 6. SynRM efficiency maps for different rotor designs.

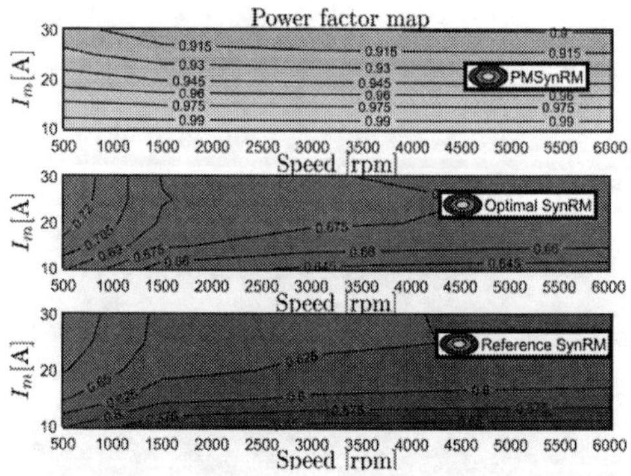

Figure 7. SynRM power factor maps for different rotor designs.

IOP Conf. Series: Materials Science and Engineering **966** (2020) 012107 doi:10.1088/1757-899X/966/1/012107

4. Experimental validation

The FEM results have been validated by experimental measurements using the test bench sketched in Figure 8. The measured and simulated validation results were obtained at fixed speed (2500 rpm), stator current (I_m = 20 A) and for different current angles. Figure 9 (a) shows the computed (FEM) and the measured phase voltage, output torque of the SynRM. It is observed that the phase voltage is inversely varied with the current angle. Figure 9 (b) shows the measured and computed values of the power factor of SynRM at different current angles. The power factor increases with the increase of current angle due to the increasing of the load torque. The measured and simulated values of the output torque of the SynRM are depicted in Figure 9 (c). The SynRM output torque increases with the increasing of current angle till it reaches a maximum value at optimal angle then it decreases again. The SynRM maximum torque does not happen at the current angle of 45°. The current angle should be controlled in order to attain a maximum torque per Ampère. Therefore, the efficiency and losses of the SynRM can be improved. The SynRM's efficiency is shown in Figure 9 (d). There is a good agreement between the simulated and the measured values. However, the difference between the measured and computed results is due to different reasons: the cutting and punching effects on the steel properties, the manufacturing tolerance and the measurement error. In addition, switching and mechanical losses are not included in simulation.

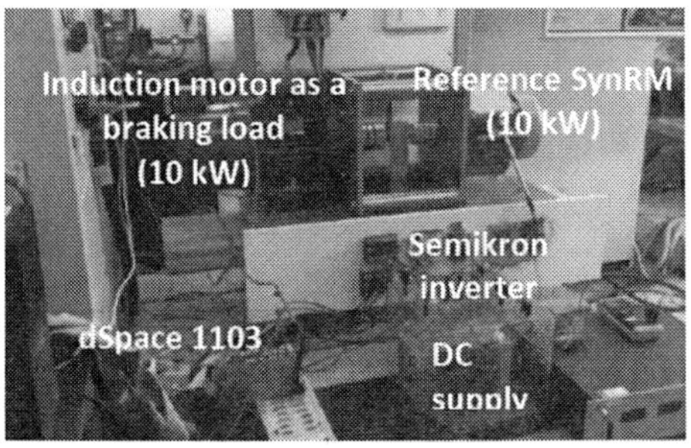

Figure 8. Photograph of the experimental setup.

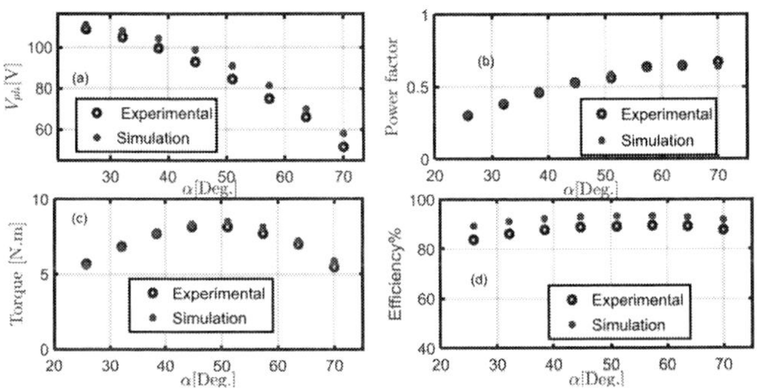

Figure 9. Computed and measured (a) phase voltage, (b) power factor, (c) output torque, and (d) efficiency of the SynRM at different current angles, current = 20 A and speed = 2500 rpm.

5. Conclusion

This paper has discussed the performance evaluation of Synchronous reluctance motors (SynRMs) with and without permanent magnets. Three different rotors having the same stator design and other parameters are considered. The three rotors are reference, optimized and optimized assisted by ferrite PMs (PMaSynRMs) in its flux-barrier. Different performance indicators for the machine are computed by FEM and compared at the same conditions. The performance indicators are efficiency, torque, total losses and power factor maps. It is found that the efficiency and power factor of the PMaSynRMs can reach higher than 95 % and 0.91 respectively at the rated conditions. This means that the PMaSynRMs can be considered as a good alternative for both the induction machines and switched reluctance machines. In addition, they can be good competitors compared to the permanent magnet synchronous machines due to lower cost. Finally, measurements are obtained and validated the FEM model.

References

[1] Ferrari M, Bianchi N, Doria A and Fornasiero E 2015 Design of synchronous reluctance motor for hybrid electric vehicles *IEEE Trans. Ind. Appl.* vol **51** no **4** pp 3030–40

[2] Ibrahim M N, Sergeant M N and Rashad E M 2015 Synchronous reluctance motors performance based on different electrical steel grades *IEEE Trans. Magn.* vol **51** no **11** Art. ID 7403304.

[3] Vartanian R and Toliyat H A 2009 Design and comparison of an optimized permanent magnet-assisted synchronous reluctance motor (PMa-SynRM) with an induction motor with identical NEMA frame stators *Proc. IEEE Electr. Ship Technol. Symp. (ESTS)* pp 107–12

[4] Jeong Y H, Kim K, Kim Y J, Park B S and Jung S Y 2012 Design characteristics of PMa-SynRM and performance comparison with IPMSM based on numerical analysis *Proc. 20th ICEM* pp 164–70

[5] Morimoto S, Sanada M and Takeda Y 2001 Performance of PM-assisted synchronous reluctance motor for high-efficiency and wide constant power operation *IEEE Trans. Ind. Appl.* vol **37** no **5** pp 1234–40

[6] Huang H, Hu Y Sh, Xiao Y and Lyu H 2015 Research of parameters and ant demagnetization of rare-earth-less permanent magnet-assisted synchronous reluctance motor *IEEE Trans. Magn.* vol **51** no **11** Art. ID 8112504

[7] Bertotti G 1988 General properties of power losses in soft ferromagnetic materials *IEEE Trans. Magn.* vol **24** no **1** pp 621–30

[8] http://www.bakkermagnetics.com/sites/default/files/downloads/ferrite_-_sintered_1.pdf (accessed on 1 June 2020)

[9] Ahonen T, Tamminen J and Montonen J 2014 Comparison of electric motor types for realizing an energy efficient pumping system *Proc. 16th Eur. Con. Power Electronic. Appl. (EPE'14-ECCE Europe)* pp 1–9

[10] Liu Ch T, Luo T Y, Hwang Ch C and Chang B Y 2015 Field path design assessments of a high-performance small-power synchronous-reluctance motor *IEEE Trans. Magn.* vol **51** no **11** Art. ID 8206504

[11] Lohninger R, Grabner H, Weidenholzer G, Silber S and Amrhein W 2015 Modeling, simulation, design of permanent-magnet-assisted synchronous reluctance machine *IEEE Trans. Ind. Appl.* vol **51** no **1** pp 196–203

[12] Dmitrievskii V, Prakht V and Kazakbaev V 2019 IE5 Energy-Efficiency Class Synchronous Reluctance Motor With Fractional Slot Winding, *IEEE Trans. Ind. Appl.* vol **55** no **5** pp 4676–84

[13] Prakht V, Dmitrievskii V and Kazakbaev V 2018 Mathematical modeling ultra-premium efficiency (IE5 class) PM assisted synchronous reluctance motor with ferrite magnets *2018 25th International Workshop on Electric Drives: Optimization in Control of Electric Drives (IWED)* (Moscow) pp 1–6

[14] Ibrahim M N, Sergeant P and Rashad E M 2017 Design of low cost and efficient photovoltaic pumping system utilizing synchronous reluctance motor *2017 IEEE International Electric Machines and Drives Conference (IEMDC)* (Miami, FL) pp 1–7

[15] Ibrahim M N, Rezk H, Al-Dhaifallah M and Sergeant P 2019 Solar array fed synchronous reluctance motor driven water pump: An improved performance under partial shading conditions *IEEE Access* vol **7** pp 77100–15

[16] Ibrahim M N 2017 Design aspects of high performance synchronous reluctance machines with and without permanent magnets *Doctoral thesis*, Ghent university

[17] Ibrahim M N, Sergeant P and Rashad E M 2017 Relevance of including saturation and position dependence in the inductances for accurate dynamic modeling and control of SynRMs *IEEE Trans. Ind. Appl.* vol **53** no **1** pp 151–60

IOP Conf. Series: Materials Science and Engineering **966** (2020) 012108 doi:10.1088/1757-899X/966/1/012108

Development of the temperature prediction algorithm for smart thermostat

S I Korolev[1, 3], R M Muhutdinov[1], O Yu Sidorov[2] and V Goman[2]

[1] Iridium LTD, 56b, Mira str., Nizhniy Tagil, 622000, Russia
[2] Ural Federal University named after the first Russian President B.N. Yeltsin, Nizhniy Tagil Technological institute, 59, Krasnogvardeyskaya str., Nizhniy Tagil 622000, Russia

E-mail: [3] korolev@iridiummobile.ru

Abstract. This article discloses systems for control of microclimate parameters in buildings and rooms with different intellectualization levels. Requirements to a smart thermostat as to a building automation system part and tasks for smart thermostat development have been formulated. Thermal models of a typical room and heating radiator have been developed. The results of development of the air temperature prediction algorithm and heating radiator capable of real time operation and ensuring adaptation to varying conditions in a room have been described. Testing of the algorithm using real data has been carried out. Study of the temperature control model's behavior in connection with the change of input data and determined parameters has been carried out. Calculation method for the room air heating time up to set temperature has been determined. Maximum absolute temperature determination error when prediction time does not exceed 7 minutes is approximately 1°C and relative temperature determination error is less than 5%.

1. Introduction

As a rule, a smart home comprises functions of control over heating, ventilation and air conditioning systems, switching and regulation of lighting, as well as over the security assurance function, that is, unauthorized access or water and gas leakage alerts [1]. Energy saving is also a common function of smart homes [2, 3, 4, 5]. By distributing this functionality across the entire building, a 'Smart Building' or a building automation system concept can be mentioned.

Let us review the smart home subsystem serving for maintenance of microclimate parameters in a room. Various heating, ventilation and air-conditioning systems are installed in modern buildings. Typically, each of the systems installed has its own control system. Typical situation is as follows: water heating radiators, electric underfloor heating, ventilation and air-conditioning system can be installed in a room. Radiators are manually controlled by means of a mechanical temperature controller. Underfloor heating is also user controlled and has a specific setpoint value. Ventilation system provides a constant or variable output. Air-conditioning system has its own temperature controller with its own temperature setpoint. All the above-mentioned systems installed in a room are not interconnected, and control thereof is carried out independently.

A situation where the above-mentioned systems may conflict with each other wasting energy is possible and is connected with inefficient control during combined operation of the said systems. Moreover, various systems that heat or cool a room have different performance and cost factor of energy that depend on external factors. An important task is to plan the operating modes, for example, consider the actual necessity in heating or cooling based on the number of people in rooms or on a schedule [6].

To conclude the aforesaid, the task of optimal temperature control in rooms can be formulated as follows: an optimal controller shall ensure the set quality of temperature control under a minimum power consumption or a minimum total cost of power consumed [6].

Let us review the main requirements to self-learning temperature control systems (smart thermostats).

Power and cost savings. Classic thermostats allow for setting temperature to a single value that has to be changed manually which cannot be called an efficient power use. Let us call this a zero level of intellectualization. Classic programmable models allow for planning what temperature in a room should be at certain hours during the day, which makes it possible to decrease it when there are no people in the room, thus providing significant power saving (first level of intellectualization). For example, Siemens REA23 thermostat [7] operates in a self-learning mode automatically tuning the parameters of an adaptive proportional-integral-differential (PID) controller. However, the temperature control is carried out according to a schedule set by a user, and not formed automatically based on the analysis of the user's behavior.

Smart thermostat automatically forms and dynamically changes the schedule based on the activity and presence of users while taking into account the user-defined priorities which allows for achievement of the best cost efficiency possible (second level of intellectualization). Some thermostat models study human preferences within time to build their own schedule and adapt to changes. Other models do not predict schedule based on previous activities, but automatically respond to the user's actions. Models with automatic scheduling apply a user's habits, his or her answers to a number of simple questions, and local weather conditions. Quite many works have been devoted to this topic, for example [8, 9, 10, 11].

For instance, Nest Learning Thermostat adapts to the user activity timeline, his or her schedule and temperature preferences [12, 13], i.e. actually programs itself within a week, thereby creating a personalized schedule. The device continuously adapts to all the changes in the owner's schedule, since Auto-Away function automatically detects whether there is nobody at home, and then the thermostat automatically adjusts temperature to avoid heating an empty house [12, 13]. In case the amount of saving needs to be determined, Nest Leaf technology will make it possible to determine how efficient the temperature condition modification settings are [12].

Honeywell [13, 14], Ecobee SmartThermostat [15] and a number of other smart thermostats have similar functionality.

Thermostat integrability into the smart home system. Some models include domestic automaton functions which will make it possible to link the thermostat to other smart home ecosystem devices. For instance, it is possible to link a thermostat to a smart lock that activates the mode of people absence when a door gets locked and many other things.

Forming notifications, reminders, and user support. Most thermostats have the functionality for notification of various types, notifying the owner, for instance, when it is time to carry out basic maintenance of an air-conditioner or to replace air filters in the cooling system. The other will warn about heating or cooling system issues. Some models may provide a connection with a service center representative who services the thermostat.

In connection with the mentioned above, the development of operating and learning algorithms for a smart thermostat is relevant. Solution of the following tasks was required for the development [6]:

- Development of the thermostat smart learning algorithms based on the analysis of thermal parameters of a building and user behavior.
- Development of control method selection algorithm within the whole building.
- Assurance of 'maximum saving' and 'maximum comfort' modes based on a user's choice; however, energy resource saving mode shall consider available thermal energy sources with different tariffs.

- Implementation of algorithms based on a certain control device. This work describes some aspects of solution of the first task.

2. Methods

Room temperature condition prediction implies development of a reliable thermal model with unambiguously determined parameters. On the one hand, the model shall reliably describe heat exchange physical processes, and on the other, it shall quickly adapt to real functioning conditions and changes of the room thermal parameters.

2.1. Air temperature in a room

Heat exchange formula (1) has been obtained for a typical room.

$$\frac{dt_v}{d\tau}=-\frac{L_0}{V}\cdot(t_v-t_0)-\frac{L_c}{V}\cdot(t_v-t_c)+\frac{\alpha_0\cdot F_{ra}}{\rho\cdot c_r\cdot V}\cdot(t_0-t_v)+\frac{\alpha_{uf}\cdot F_{uf}}{\rho\cdot c_r\cdot V}\cdot(t_{uf}-t_v)+\frac{\alpha_r\cdot F_r}{\rho\cdot c_r\cdot V}\cdot(t_r-t_v) \qquad (1)$$

In formula (1): $\alpha_0,\alpha_{uf},\alpha_r$ – effective heat transfer factors of a room, underfloor heating and heating radiator; t_0 – air temperature outside the room; t_v – air temperature inside the room; L_0 – external air inflow; ρ – air density; c_p – substance mass heat capacity at constant pressure; V – room's volume; t_C – air-conditioner outlet air temperature; t_r – heating radiator temperature; F_{ra},F_{uf},F_r – surface areas of the room, underfloor heating and heating radiator. An unknown value in formula (1) is the heat transfer factor α_0 that shall be selected in such a way that the solution of formula (1) would coincide with experimental air temperature values inside the room.

2.2. Heating radiator temperature

According to its form, thermal model for a radiator is similar to the model of air temperature (1), but has a smaller number of summands:

$$\frac{dt_r}{d\tau}=\frac{L_{hw}}{V_r}\cdot(t_{hw}-t_r)+\frac{\alpha_{r0}\cdot F_r}{\rho_w\cdot c_{rw}\cdot V_r}\cdot(t_v-t_p) \qquad (2)$$

In formula (2): L_{hw} – hot water flow rate in the heating radiator; V_r – heating radiator volume; t_{hw}, t_r – temperature of hot water in the heating system and the heating radiator respectively; ρ_w, c_{rw} – density and heat capacity at constant pressure of water in the heating radiator respectively; α_{r0} – heating radiator water-to-radiator material heat transfer factor; F_r – heating radiator inner surface area.

The unknown values in formula (2) are L_{hw}, ρ_w, c_{rw}, α_{r0}, F_r that shall be selected in such a way that solution of formula (2) would coincide with experimental heating radiator temperature values.

2.3. Model parameter determination

To determine α_0, let us compose a functionality within which a sum of squared deviations of the experimental air temperature values in a room ($t_{vi}^{\exp}$) and calculated temperature values ($t_{vi}^{\mathrm{mod}}(\alpha_0)$):

$$M(\alpha_0)=\sum_{i=n}^{n-4}\left(t_{vi}^{\exp}-t_{vi}^{\mathrm{mod}}(\alpha_0)\right)^2 \qquad (3)$$

In formula (3), the sum is calculated according to the last 5 temperature values. To determine α_0, single-dimension minimization method, the method of golden ratio, was used [16]. Similar approach was used to determine the unknown parameters of formula (2).

IOP Conf. Series: Materials Science and Engineering **966** (2020) 012108 doi:10.1088/1757-899X/966/1/012108

3. Results and discussions

Input data are the values of air temperature in a room and a heating radiator, as well as the heating radiator operation mode (on/off). Figure 1 shows the charts of temperature change of the radiator and air in the room according to which the model parameters were determined. Experimental and model values along line 2 in figure 1 are indistinctive.

Simulation task is to determine the time for heating of air in the room from some initial temperature to the set temperature by a predefined moment of time. Time required to heat air in a room from the initial temperature to the required one was calculated. Maximum temperature determination error is approximately 1°C (about 5%) when prediction time does not exceed 7 minutes. These differences can be observed in transient modes of switching the heating radiator on or off, and the differences in other modes are less than 1°C. As a result of adaptation of models (1) and (2), values of α_0 and two sets of the radiator parameter values were determined: for 'on' and 'off' conditions.

Let us review simulation of temperature condition control for the following case: consumer regulated condition: from 8 a.m. to 6 p.m. room air temperature shall be about 10°C, and from 6 p.m. to 8 a.m. it shall be maintained at the level of 22°C. Herewith, air heating in the room shall be started before 6 p.m.

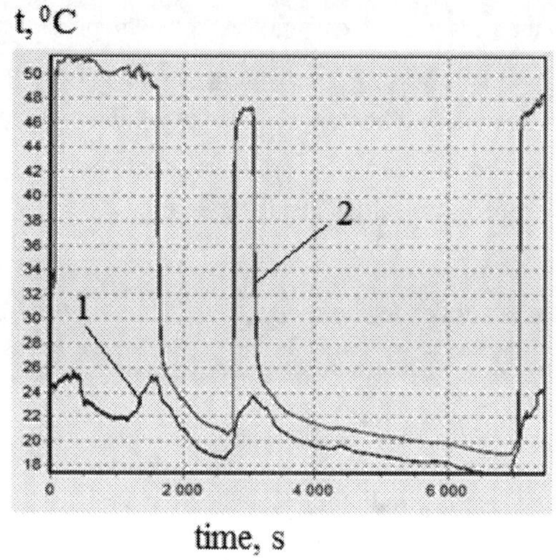

Figure 1. 1 – room air temperature values; 2 – experimental and simulated radiator temperature values

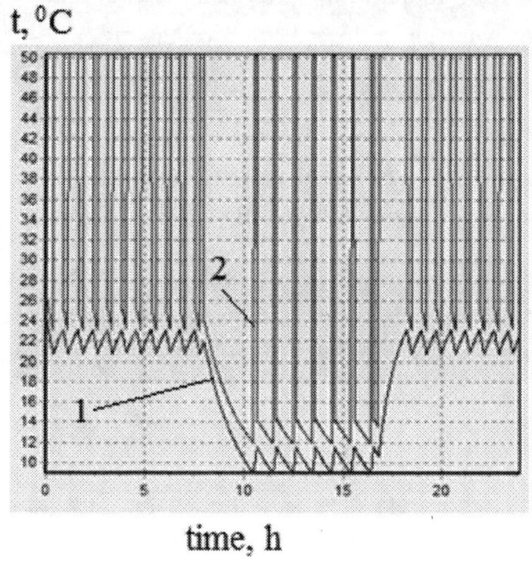

Figure 2. Temperature control from 12 a.m. to 12 p.m. 1 – air temperature; 2 – heating radiator temperature

For the algorithm approbation, a two-step (on/off) temperature controller that supports different temperature values at different time intervals with deviation of not more than 1°C was implemented. The result is shown in figure 2, and it implies that starting from 8 a.m. temperatures decreases for approximately 1.8 hours (1 hour 48 min) from 22 to 10°C. The controller switches the heating radiator on approximately at 4:50 p.m. for heating up to 22°C by 6 p.m. Thus, the heating time was approximately 1 hour 10 minutes from 11 to 22°C. Abrupt ascent/descent of line 2 in figure 2 corresponds with switching the heating radiator on/off.

4. Conclusions

- Heat exchange equation suitable for practical implementation of the room temperature prediction has been composed. Mathematical model of room air temperature has been composed. Mathematical model of heating radiator temperature has been composed.
- Algorithm of model parameter determination and room air temperature prediction has been developed.

- Testing of the algorithm using real data has been carried out.
- Study of the temperature control model's behavior in connection with the change of input data and determined parameters has been carried out.
- Calculation method for the room air heating time up to the set temperature has been determined.

Acknowledgement
This study has been carried out under contract with Iridium LTD 'Development of the Mathematical Model of the Smart Thermostat. Stage 1'.

References

[1] Saad al-Sumaiti A, Mohammed H A, Magdy and M A Salama 2014 Smart Home Activities: A Literature Review, *Electric Power Components and Systems* **42:3-4** pp 294-305
[2] Williams E D, Matthews and H S 2007 Scoping the potential of monitoring and control technologies to reduce energy use in homes *Proceedings of the 2007 IEEE International Symposium on Electronics & the Environment (Orlando, FL, USA)* pp 239–44
[3] Dae-Man H and Jae-Hyun L 2010 Design and implementation of smart home energy management systems based on Zigbee *IEEE Trans. Consum. Electron.* **56 3** pp 1417-25
[4] Ha D L, Zamai S P E and Jacomino M 2006 A home automation system to improve household energy control *12th IFAC Symposium on Information Control Problems in Manufacturing (Saint Etienne, France)*
[5] Erol-Kantarci M and Mouftah H T 2011 Wireless sensor networks for cost-efficient residential energy management in the smart grid, *IEEE Trans. Smart Grid* **2 2** pp 314-25
[6] Goman V V and Korolev S I 2020 Smart thermostats in building automation systems and smart homes, *III Annual Int. Conf. 'System Engineering'* (Yekaterinburg, Russia) pp 67–74
[7] Siemens REA23 manual [Electronic resource] – Mode of access: https://www.downloads.siemens.com/download-center/Download.aspx?pos=download&fct=getasset&id1=A6V10075730 (accessed 01-06-2020)
[8] Nazeer O, Javaid N, Rafique A A, Kiani S, Javaid Y and Khurshid Z 2019 Home energy management using hybrid meta-heuristic optimization technique *Advances in Intelligent Systems and Computing* **773** pp 599-609
[9] Jamil A, Javaid N, Khalid M U, Iqbal M N, Rashid S and Anwar N 2019 An energy efficient scheduling of a smart home based on optimization techniques *Advances in Intelligent Systems and Computing* **773** pp 3-14
[10] Harikrishnan R and Sivagami P 2017 Intelligent power saving system using PIR sensors *Proc. of the Int. Conf. on Electron., Comm. and Aerospace Techn.* (Coimbatore, India) pp 573-77
[11] Kadir K M, Forhad M S, Fadlullah M M, Quader N, Al-Arif M M R and Dhali M A 2011 Energy saving by automatic control of power in simple home appliances *IEEE 3rd Int. Conf. on Communication Software and Networks, ICCSN* (Xi'an, China) pp 311-15
[12] Nest Learning Thermostat v.3. [Electronic resource] – Mode of access: https://smart-home.market/ (accessed 01-06-2020)
[13] Honeywell vs. Nest: the battle for the smart thermostat [Electronic resource] – Mode of access: https://www.predictingourfuture.com/12-honeywell-vs-nest-the-battle-for-the-smart-thermostat/ (accessed 01-06-2020)
[14] T9 Smart Thermostat: [Electronic resource] – Mode of access: https://t9.honeywellhome.com/ (accessed 01-06-2020)
[15] The best smart thermostats of 2019 [Electronic resource] – Mode of access: https://www.cnet.com/news/top-smart-thermostats-of-2019-ecobee-smartthermostat-nest/ (accessed 01-06-2020)
[16] Forsythe G E, Malcolm M A and Moler C B 1977 *Computer Methods for Mathematical Computations* (Prentice Hall Professional Technical Reference)

15th International Conference on Industrial Manufacturing and Metallurgy IOP Publishing
IOP Conf. Series: Materials Science and Engineering **966** (2020) 012109 doi:10.1088/1757-899X/966/1/012109

Comparison of minimum energy consumption indicators for electric motors powered directly from the mains at part loads

V Goman

Ural Federal University named after the First President of Russia B N Yeltsin, Nizhniy Tagil Technological Institute, 59, Krasnogvardeyskaya str., Nizhniy Tagil, 622000, Russia

E-mail: v.v.goman@urfu.ru

Abstract. The article presents a comparison of minimum energy consumption indicators for motors powered directly from the mains at part load conditions. The energy characteristics of induction motors and line-start permanent magnet synchronous motors of different energy efficiency classes are compared. As an example, the calculations of proposed energy consumption indicators are provided for the throttle-controlled 2.2 kW pump unit with variable flow-time profile. It is shown that if a motor is used mostly at part load conditions and was chosen by its energy efficiency class only, minimum energy consumption might not be achieved. Universal average efficiency indicator is suggested for motors running at variable loads with output power much lower than the rated power. The comparison shows that the proposed average efficiency does not depend on a specific application or load profile and describes the motor energy consumption enough for practical goals. Thus, this indicator can be used for fast evaluation and selection of a motor by the lowest energy consumption criterion, especially without detailed information about load profile. Calculation of this indicator requires only motor efficiency data per 4 points 25, 50, 75, 100% of rated load.

1. Introduction

A significant number of electric motors in the world currently run without speed control. For instance, the share of VSDs (variable speed drives) according to the European Commission data [1] is about 30% for Germany and about 20% for Switzerland according to the study described in [2]. The energy efficiency class (IE) of the motor powered directly from the mains is assigned by the efficiency at rated output in accordance with IEC 60034-30-1 [3]. At the same time, the efficiency values for part loads are not defined by the standard [3].

However, the electric motor running mode is far from the rated mode in many applications. This is typical for HVAC (heating, ventilation, air conditioning) and water supply. E.g. a pump unit's motor with a variable load depending on water consumption may run at part loads for a long time. Whereas a lower-power motor can't be used as peak consumption has to be provided. For example, the typical circulation pump flow-time profile defined by [4] and [5] is divided into 4 modes: a pump runs for 44% of its working time with of 25% of the rated flow; 35% of the time with 50% of the flow rate; 15% of the time with 75% of the flow rate; and time in the rated mode does not exceed 6%. The similar load profile can be typical for conveyors and hoisting mechanisms where the motor's load depends on the mass transported, in crushers and other materials processing equipment where the motor load depends on the quantity and properties of the material processed.

In many cases motors are overrated. For example, in [6] it is estimated that only 20% of the pump drive motors in operation are running at their rated mode. Many articles are focused on comparison of power consumption of pump systems and other centrifugal mechanisms using various motor's types like IMs (induction motors), PMSMs (permanent magnet synchronous motors), SynRMs (synchronous reluctance motors) [7, 8, 9, 10, 11, 12, 13, 14]. However, all these articles consider pumping systems with VSDs. For pump units powered directly from the mains, the issue of efficiency is far less reviewed for different motor. Thus, the main purpose of the article is to compare the indicators of minimum energy consumption for selection of motors running mostly at loads below the rated value. The article compares the motors' efficiency dependences of the load. Data were analyzed for 8 different 4-pole 2.2 kW motors, namely: three LSPMSMs (line-start PMSMs) of IE4-class [15], [16], [17]; two IMs of IE4-class [18] and [19]; and three IMs of IE3-class [18], [20] and [21]. Motor performance data are shown in table 1, where m is the index number of the motor.

Table 1. Efficiency of motors chosen [15-21].

m	Type of motor	Efficiency class	Efficiency (%) at the various loads			
			25 %	50 %.	75 %.	100%.
1	LSPMSM (SEW DRU J)	IE4	-	88.0	90.5	91.2
2	LSPMSM (SynchroVERT)	IE4	-	88.6	89.4	89.5
3	LSPMSM (WEG WQuattro)	IE4	77.0	86.0	89.0	90.2
4	IM (Siemens 1LE1004)	IE4	82.3	88.3	89.6	89.5
5	IM (WEG W22)	IE4	79.3	88.5	89.5	89.5
6	IM (Siemens 1LE1003)	IE3	80.4	86.4	87.3	86.7
7	IM (WEG W21)	IE3	80.6	86.5	87.0	87.0
8	IM (ABB M3BP)	IE3	-	85.1	86.9	86.7

Figure 1 shows the interpolated efficiency curves for all electric motors on the same coordinate axes and data from Table 1. The loss values were calculated for each motor according to the efficiency values from table 1, then their polynomial interpolation was performed, then the efficiency values were determined over the entire range of 0 to 100%.

The data in table 1 and figure 1 show that some IE4-motors at part loads have the efficiency lower than that of IE3-motors at same loads below the rated load. The most typical example is the LSPMSM no.3, which has the rated efficiency greater than that of all the IMs considered. However, its efficiency is 86% at 50% load and 77% at 25% load. At the same time, the IE3-class IMs no.6 & no.7 show the efficiency of 86.4% and 86.5% accordingly at 50% load and 80.4% and 80.6% at 25% load, which is much higher than that of LSPMSM no.3. It should be noted that in the LSPMSMs no.1 to no.3 are used expensive rare earth magnets, and its manufacturing leads to the high impact on the environment [22]. According to the author's estimation, the LSPMSMs have the price about 1.27–1.37 times higher than that of IE3-class IMs. Thus, for some applications like variable-flow pumps, selection of electric motor based only on its IE-class might not give the minimum energy consumption, as was shown in [23]. The curves in figure 1 show a general view of the efficiencies compared, but they give no specific numerical indicator according to which one of the electric motors can be chosen for a specific application.

2. Methods

Let's consider possible methods for determining the indicator of motor energy consumption at part load conditions. Usage of the rated motor efficiency $\eta_{\mathrm{nom}.m}$ for mth motor was taken as the method A. By the method B, the average efficiency $\eta_{\mathrm{avg.int}.m}$ is calculated for the mth motor in the range $l_{\min} \dots l_{\mathrm{nom}}$ by an interpolated efficiency curve $\eta_{\mathrm{int}.m}(l)$ for the mth motor, assuming that a long run at a load less than 25% is unlikely, so $l_{\min} = 25\%$ and $l_{\mathrm{nom}} = 100\%$.

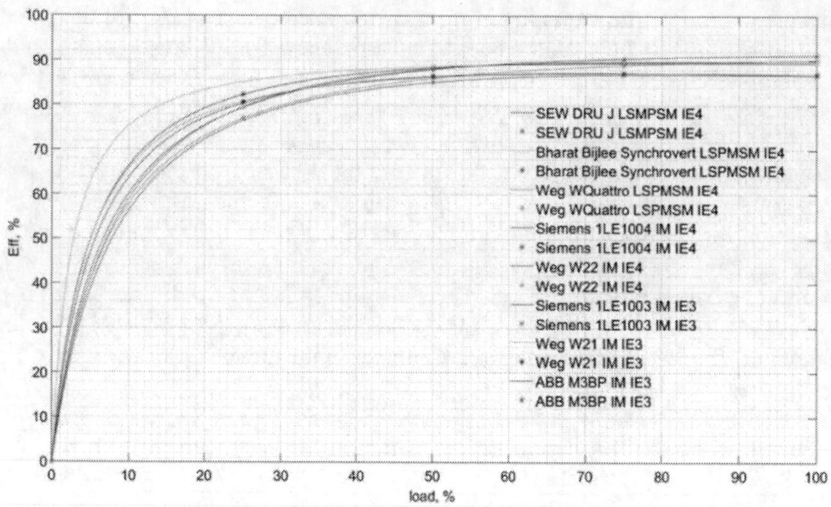

Figure 1. Interpolated curves of motors' efficiencies and the initial points.

$$\eta_{\mathrm{avg.int.}m} = \frac{1}{(l_{\mathrm{nom}} - l_{\mathrm{min}})} \int_{l_{\mathrm{min}}}^{l_{\mathrm{nom}}} \eta_{\mathrm{int.}m}(l)\,\mathrm{d}l \qquad (1)$$

By the method C, the average efficiency $\eta_{\mathrm{avg.}m}$ was calculated in load points i: 25, 50, 75, 100% as indicated in table 1 for mth motor. If the efficiency at 25% point doesn't specified by the manufacturer, this value was previously calculated using interpolated curves.

$$\eta_{\mathrm{avg.}m} = \frac{1}{4}\sum_{i=1}^{4} \eta_{i.m} \qquad (2)$$

Method D is a detailed energy consumption calculation for a motor in a specific operating cycle in a specific application. For example, energy efficiency of circulation pumps working mainly with variable flow is evaluated in accordance with the [4]. In the [4], the previously described flow-time profile is used as a typical pump load profile. The efficiency indicator in [4] is the energy efficiency index (EEI), which is a ratio of the average-weighted electric power consumed by the pump over the operating cycle to the electric power of the reference pump system. Thus, the lower EEI values correspond to the higher energy efficiency of the pump unit. According to [5], the EEI is the most appropriate indicator for evaluating the energy efficiency of pumping systems with variable flow rates for various purposes, unlike the minimum efficiency index (MEI), which is defined in [24] and is based on a limited range of operating points (75 to 110% of the flow rate) [5].

According to [25] and [26], the EEI calculation requires: interpolated electric motors' efficiency curves for calculating electric energy consumption at flows according to a load profile given; pump parameters: Q-H characteristic, pump efficiency dependence of flow; a number of reference parameters: efficiency of the reference motor, the minimum required efficiency of the reference pump at the best efficiency point (BEP), depending on the table coefficient C, determined by the pump type, the nominal rotational speed of the pump n and its efficiency, the flow rate Q_{BEP} and the specific speed n_s, which depends on the head H_{BEP} and n. As shown in [23], the EEI explains the energy consumption of the pump unit more objectively than the energy efficiency class of the motor, which depends only on the rated efficiency. Also, [23] describes a method for calculating the energy consumption of a pump unit, which does not require reference parameters and provide cost savings.

In [23], the calculation of the energy efficiency of a pump unit with the same motors as in this article, with a typical load profile [4], i.e. a comparative analysis is performed for energy consumption of LSPMSMs and IMs powered directly from the mains in the electric drive of a centrifugal pump with throttle control. We use the results of the cost savings calculation from [23] as an accurate criterion for comparison the energy consumption of electric motors of a pumping unit with variable flow, which is more convenient for comparison with indicators for the methods A, B, C than EEI, since a higher cost savings values correspond to a lower energy consumption, as well as efficiency.

The indicators calculating results for all methods are shown in table 2 and figure 2.

Table 2. Energy consumption indicators for motors chosen.

m	Type of motor	Class	Method A $\eta_{nom.m}$, %	Method B $\eta_{avg.int.m}$, %	Method C $\eta_{avg.m}$, %	Method D Cost savings, €
1	LSPMSM (SEW DRU J)	IE4	91.2[a]	88.3	87.3	73,8
2	LSPMSM (SynchroVERT)	IE4	89.5	88.6[a]	88.2[a]	93,6[a]
3	LSPMSM (WEG WQuattro)	IE4	90.2	86.6	85.6	24,6
4	IM (Siemens 1LE1004)	IE4	89.5	88.2	87.4	83,8
5	IM (WEG W22)	IE4	89.5	87.9	86.7	74,8
6	IM (Siemens 1LE1003)	IE3	86.7[b]	86.1	85.2	36,6
7	IM (WEG W21)	IE3	87.0	86.0	85.3	32,5
8	IM (ABB M3BP)	IE3	86.7[b]	85.0[b]	83.6[b]	0[b]

[a]The best value for the method. [b]The worst value for the method.

3. Results and discussion

Table 2 and figure 2 show that the energy efficiency assessment of motors by the methods B and C are almost identical and mostly meet the indicator by the method D. The motor no.2 has the best average efficiency by both methods B and C, and electric motor no.7 has the worst values. The similar result is shown by the most accurate method D. Thus, an average efficiency classifies the motor energy consumption at low loads and at least can be used for fast evaluation and selection of a motor by the lowest energy consumption criterion, especially without detailed load profile information. In this case, the methods B and C give a better result than selecting a motor by its rated efficiency, according to the method A. It should be noted that according to IEC 60034-30-2 [27] the efficiency values for variable speed motors are defined for seven load modes other than the rated one. The draft of IEC 60034-30-2 [28] proposed to calculate the overall efficiency for variable speed motors of pumps and fans (drives with a quadratic load-speed dependence) as an average-weighted value of efficiency at reduced speeds and loads. In [29, 30], an approach to the design of an electric motor is described, which allows to maximize the overall efficiency calculated according to the draft version of the standard [28], not only the rated efficiency.

The method B disadvantage is the necessity of the interpolation of efficiency dependence of the load. In this case, some interpolation error may be added. For example, the motor no.2 interpolated efficiency curve looks better relative to other motors in the range of 0 to 35% of the load. As the motor manufacturer does not specify the efficiency at 25% load in the catalog, there are only 3 initial points for interpolation. On the one hand, as shown in [13] and [31], the motor losses dependence of load is well described by a second-order polynomial, coefficients of which can be easily obtained by the 3 initial data points. On the other hand, this motor has some minor efficiency differences, and therefore in the values of losses, at known points 50, 75, 100% of the load. Even with a known point (0, 0) on the efficiency curve, this can lead to an interpolation error. Thus, the resulting indicator for the method B may not be completely reliable for motors for which the efficiency is unknown at a load of 25% and having a flat efficiency curve in the range of 50 to 100% of the load.

The general disadvantage of the methods B and C is that the indicator does not describe a specific running time in each mode, i.e. it assumes the same relative running time in each mode. Method D does not have this disadvantage. Method D, although being the most accurate of the methods proposed, assumes usage of known load profile.

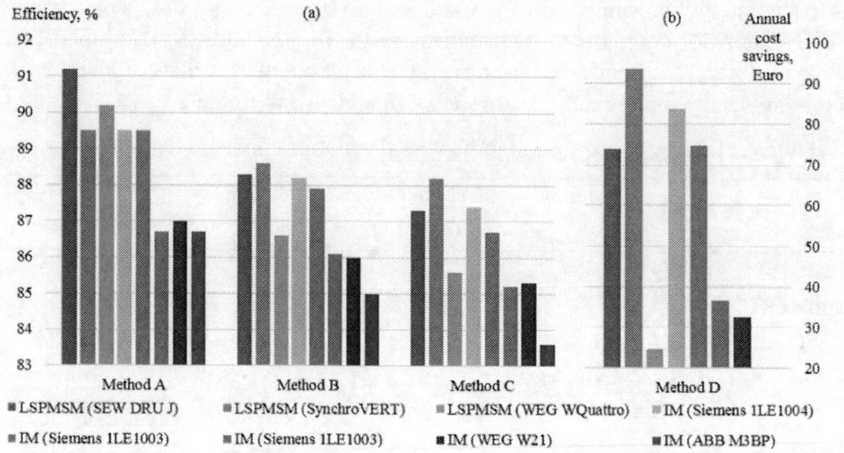

Figure 2. Energy consumption indicators: (a) – methods A, B, C; (b) – method D.

The profile used in [4] is the result of statistical processing but the load profile of a real pump may be much different from it. In this case, it is necessary to make the calculation according to the method described in [23], using experimental or calculated load profile. Likewise, another mechanism requires a defined load profile to perform this calculation. However, not many mechanisms have standard load profiles approved as a standard, also the actual operating conditions of specific mechanisms may differ from the standard profile.

4. Conclusions

With a known load profile, the most accurate result can be obtained using method D, i.e. a detailed calculation of energy consumption in each mode with further cost savings calculation according to [23] or the EEI according to [4, 25, 26]. The average efficiency calculated using the method C does not depend with a specific application or load profile and describes the motor energy consumption enough for practical goals. This method requires only motor efficiency data per 4 points (25, 50, 75, 100% loads) for calculation, i.e. it does not have the disadvantages that are present when calculating EEI or cost savings, as well as when selecting a motor by IE class (by rated efficiency). We consider it relevant to use the average efficiency not only for VSD motors, which was proposed in [28], but unfortunately was not approved in the final version of the standard [27], but also for motors powered directly from the mains. For such motors, at least, both the rated efficiency and the efficiency values at reduced loads are to be considered, according to the approach proposed in [27].

It should also be noted that applications with variable loads require motors with a flat efficiency curve (ideally in the range of 25 to 100% of the load). Then the specific load profile will not be so important, and in this case the motor efficiency will be conveniently classified by the average efficiency calculated by the method C. The consumer will be able to focus either on the rated efficiency for selection of the motor running close to the rated output, or on the average efficiency for selection of the motor running under various loads, especially if there is no detailed information about load profile. The average efficiency could also be used to assign energy efficiency classes to motors powered directly from the mains and intended for running in a wide range of loads, along with the approach already available in the standard [3], which is more suitable for electric motors running mainly close to the rated mode.

References

[1] Almeida A, Ferreira F J, Fong J and Fonseca P 2008 EuP Lot 11 Motors, Final report to the EC.
[2] Phillips R and Tieben R 2017 *Proc. of Energy Efficiency in Motor Driven Systems (EEMODS)* (Rome) pp 53-60.
[3] Rotating electrical machines - Part 30-1: Efficiency classes of line operated AC motors (IE code). IEC 60034-30-1/ Ed. 1; IEC: 2014-03.

[4] European Commission Regulation (EC) No. 640/2009 Implementing Directive 2005/32/ EC of the European Parliament and of the Council with Regard to Ecodesign Requirements for Electric Motors amended by Commission Regulation (EU) No 4/2014.
[5] Stoffel B 2015 *Assessing the Energy Efficiency of Pumps and Pump Units. Background and Methodology* (Amsterdam, The Netherlands Elsevier).
[6] Glover A and Lukaszczyk M 2005 *World Pumps* **466** pp 36–38.
[7] Ahonen T, Orozco S M, Ahola J and Tolvanen J 2016 *18th European Conf. on Power Electronics and Applications (EPE)* (Karlsruhe) pp 1-9.
[8] Van Rhyn P and Pretorius J H C 2015 *IEEE 5th Int. Conf. on Power Engineering, Energy and Electrical Drives (POWERENG)* (Riga) pp 497-502.
[9] Brinner T R, McCoy R H and Kopecky T 2014 *IEEE Trans. on Ind. Appl.* **50 1** pp 174-181
[10] Kazakbaev V, Prakht V, Dmitrievskii V, Ibrahim M N, Oshurbekov S and Sarapulov S 2019 *Energies* **12** 1144.
[11] Safin N, Kazakbaev V, Prakht V and Dmitrievskii V 2018 *25th Int. Workshop on Electric Drives: Optimization in Control of Electric Drives (IWED)* (Moscow) pp 1-5.
[12] Kazakbaev V M, Prakht V A and Dmitrievskii V A 2017 *Russ. Electr. Eng.* **88** 233–238.
[13] Safin N, Kazakbaev V, Prakht V, Dmitrievskii V and Sarapulov S 2018 *25th Workshop on Electric Drives: Optimization in Control of Electric Drives (IWED)* (Moscow) pp 1-5
[14] Kazakbaev V, Prakht V, Dmitrievskii V, Sarapulov S and Askerov D 2017 *Int. Siberian Conf. on Control and Communications (SIBCON)* (Astana) pp 1-6.
[15] https://www.bharatbijlee.com/media/14228/synchrovert_catalogue.pdf (accessed on 1 May 2020)
[16] https://www.weg.net/catalog/weg/RU/en/Electric-Motors/Special-Application-Motors/Permanent-Magnet-Motors/Line-Start-PM-Motors/Wquattro-2-2-kW-4P-100L-3Ph-230-400-V-50-Hz-IC411---TEFC---B3T/p/13009386 (accessed on 1 June 2020).
[17] https://download.sew-eurodrive.com/download/ pdf/21343799.pdf (accessed on 1 June 2020).
[18] Catalog Siemens D81.1 Simotics GP, SD, XP, DP low-voltage motors, 05/2018.
[19] https://www.weg.net/catalog/weg/MV/en/Electric-Motors/Low-Voltage-NEMA-Motors/General-Purpose-ODP-TEFC/Cast-Iron-TEFC-General-Purpose/W22-Super-Premium-Efficiency/W22-Super-Premium-Efficiency-3-HP-4P-182-4T-3Ph-208-230-460-380-V-60-Hz-IC411---TEFC---Foot-mounted/p/12792146 (accessed on 1 June 2020).
[20] https://www.weg.net/catalog/weg/RU/en/Electric-Motors/Low-Voltage-IEC-Motors/General-Purpose-ODP-TEFC/Aluminum-TEFC-General-Purpose/Aluminium-TEFC-General-Purpose/Multimounting-IE3/Multimounting-IE3-2-2-kW-4P-100L-3Ph-220-240-380-415-460-V-50-Hz-IC411---TEFC---B3T/p/12397774 (accessed on 1 June 2020).
[21] Catalog ABB - Low Voltage General Performance motors, June 2018.
[22] *Rare Earth Industry* 2016 ed W Ismar Borges de Lima and Walter Leal Filho (Elsevier).
[23] Goman V, Oshurbekov S, Kazakbaev V, Prakht V and Dmitrievskii V 2019 *Appl. Sci.* **9** 5295
[24] Commission Regulation No 547/2012 Implementing Directive 2009/125/EC of the European Parliament and of the Council with Regard to Ecodesign Requirements for Water Pumps.
[25] Extended product approach for pumps. A Europump Guide 2014 (Europump).
[26] Lang S, Ludwig G, Pelz P F, Stoffel B 2013 *Proc. of Energy Efficiency in Motor Driven Systems (EEMODS)* (Rio de Janeiro) pp 385-397.
[27] Rotating Electrical Machines—Part 30-2: Efficiency Classes of Variable Speed AC Motors (IE-Code); IEC TS 60034-30-2:2016; IEC: Geneva, Switzerland, 2016.
[28] Rotating Electrical Machines—Part 30-2: Efficiency Classes of Variable Speed AC Motors (IE-Code); Document 2/1709/NP; Standard draft; IEC: Geneva, Switzerland, 2013.
[29] Dmitrievskii V, Prakht V and Kazakbaev V 2019 *IEEE Trans. on Ind. Appl.* **55 5** pp 4676- 84.
[30] Dmitrievskii V A, Prakht V A and Kazakbaev V M 2018 *XIII Int. Conf. on Electrical Machines (ICEM)* (Alexandroupoli) pp 1015-20.
[31] Ferreira F, Almeida A 2016 *Int. Conf. on Electrical Machines (ICEM)* (Lausanne), pp 2775-83.

Sensitivity of robust systems controlling first order objects

M A Volkov[1, 3] and D V Isakov[2]

[1] Ural Federal University named after B.N. Yeltsin, 19, Mira str., Yekaterinburg, 620002, Russia
[2] Ural Federal University named after the First President of Russia B N Yeltsin, Nizhniy Tagil Technological Institute, 59, Krasnogvardeyskaya str, Nizhniy Tagil, 622000, Russia

E-mail: [3] volkov80@yandex.ru

Abstract. The study investigates controlled objects with a first order transfer function, which are known to be the most common for further obtainment of controllers in closed loop automatic systems. The transfer functions of compensating and non-compensating robust regulators with the first and second order of astatism were obtained in the work by means of the method of polynomial equations. Then the analysis of closed loop systems with controllers was carried out by the analytical method of coefficient estimates by a characteristic polynomial of the closed loop control system. The principle of the method lies in the obtainment of pattern indexes and further analysis of change of these coefficients under variations of parameters of the controlled object. The study results are presented graphically by means of the areas allowing to determine a way of alteration of transient processes in the closed loop control system under a variation of process parameters. The comparison of robust controllers with traditional ones was performed. The study results were proved through the modeling of the closed loop control systems under the influence of parametric disturbances of the controlled object, which results in alteration of a tangible controlled object gain coefficient.

1.Introduction

Due to the simplification of the mathematical model of the controlled object as well as its exposure to external factors, deviations from design values of the quality control indexes occur in the process of implementation of a control over the tangible technological object. If the object parameters are extremely deviated from the design values, the process can become too oscillating or unstable [1].

The robust controllers allow compensating an impact of unaccounted or hardly accounted external factors affecting a control process, if deviation of the parameters from the design ones does not exceed 2-3 times [2-7]. Lots of studies are dedicated to the robust systems controlling different technical objects [8-15].

Along with parametric, there occurs an effect of signal attenuation of disturbances influencing the object. In this case a control system, preserving the quality of control at an acceptable level, can be executed without making a regulator structure more complex [3, 5, 6].

2.Methods

A first order object model has been chosen for consideration. This model is used for presentation of the controlled objects, either at an electric drive or at process facilities, such as heating furnaces, regulation of separate process parameters during the rolling.

The transfer function of the controlled object can be represented as follows:

$$W_o(s) = \frac{\tilde{k}_o}{\tilde{T}_o s + 1}, \tag{1}$$

where $\tilde{k}_o$ and $\tilde{T}_o$ mean changeable parameters of object.

The algorithms of the controller operation were obtained using the method of polynomial equations [3, 7, 16, 17].

The analysis of a control process behavior was performed using the method of coefficient estimates [6, 18, 19, 20]. This method allows carrying out the analysis of a system in an analytical way by a characteristic polynomial of the closed loop automatic control system (ACS).

For the polynomial of the following kind

$$A(s) = \sum_{i=0}^{n} a_i s^i \tag{2}$$

it is possible to compose indexes of a pattern δ_i and of high-speed response τ from its coefficients a_i:

$$\delta_i = \frac{a_i^2}{a_{i+1}a_{i-1}}; \; \tau = \frac{a_1}{a_0}. \tag{3}$$

For configuration to a modular optimum, values of pattern indexes $\delta_i = 2$ [3, 5]. When a value of the pattern indexes increases and they achieve value $\delta_i = 4$, the process becomes non-periodic.

A general structural diagram of the robust system is presented in figure 1. The structure contains a theoretical model of the controlled object $W_M(s)$, a signal from which is compared with the tangible one at the adder and the difference passes in the form of additional feedback to the controller input. An error signal is counted by formula

$$e(t) = y(t) - y_M(t). \tag{4}$$

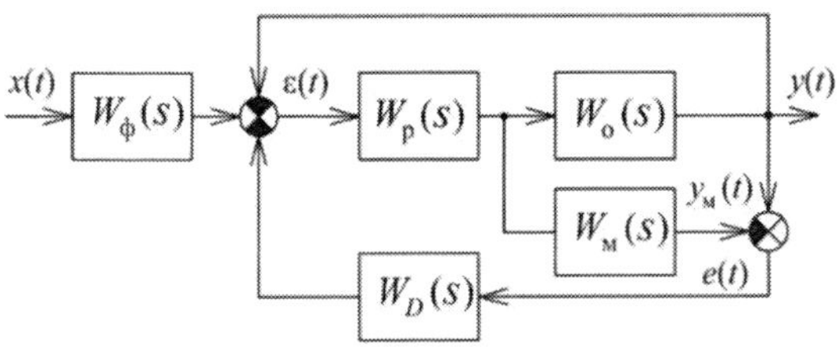

Figure 1. General structural diagram of the robust ACS

In a schematic diagram the transfer function of the additional feedback is written as

$$W_D(s) = \frac{k_D}{T_D s + 1}, \tag{5}$$

Where k_D is a gain coefficient of the additional feedback; T_D is a time constant of an additional feedback.

3. Results

The Table 1 shows the transfer functions: of the non-compensating conventional and robust controllers $W_p(s)$, obtained by the method of polynomial equations with the system order 1 and 2 astatism

IOP Conf. Series: Materials Science and Engineering **966** (2020) 012110 doi:10.1088/1757-899X/966/1/012110

(absence of offset), 1 and 2 refer to a first order of astatism, 3 and 4 refer to a second one; of a correction link $W_D(s)$.

Figures 2 and 3 show the charts of the pattern indexes (a shade lining shows areas $\delta_i < 2$) and a type of transient processes under a twofold parameter variation of an object $0.5 \le \tilde{W} \le 2$, $0.5 \le t \le 2$ (1 point: $\tilde{W} = 2$; $t = 2$; 2 point: $\tilde{W} = 0.5$; $t = 0.5$). While being compared, a system response rate and an astatism order of the robust and conventional systems were chosen as equal, a ratio of time constants is $\mu = 0.1$. Relative values taken in Table 1 are:

$$\mu = \frac{T_\mu}{T_o}\ ;\ s_* = T_\mu s\ ;\ \tilde{W} = \frac{\tilde{k}_o}{k_o}\ ;\ d = \frac{T_D}{T_\mu} \qquad (6)$$

where T_μ is a time constant determining a system response rate; T_o is an estimated value of the controlled object time constant.

Table 2 shows the transfer functions of the compensating controllers (5 and 6 refer to the first order of astatism, 7 and 8 to the second one). In Figures 4 and 5 the results of the comparison of systems with controllers 1-4, executed by means of modeling, are shown and the areas of pattern indexes corresponding to the condition $\delta_i > 2$ are built. The modeling was carried out when increasing the object parameters twofold.

Table 1. Transfer functions of non-compensating controllers.

Conventional ACS	Robust ACS
1. $W_p(s) = \frac{T_o}{k_o} \cdot \frac{2T_\mu \cdot (1-\mu) \cdot s_* + 1}{2T_\mu^2 \cdot s_*}$, $\tau = 2\left[1 + \mu \cdot \left(\frac{1}{\tilde{W}} - 1\right)\right]$	2. $W_p(s) = \frac{1}{k_o}\left(\frac{1}{\mu} - 1\right)$, $W_D(s) = \frac{1}{1-\mu}\frac{s_* + \mu}{ds_* + 1}$, $\tau = 1 + d\left[1 + \mu\left(\frac{1}{\tilde{W}} - 1\right)\right]$, $d = 1$
3. $W_p(s) = \frac{1}{k_o}\frac{8(1-\mu)s_*^2 + 4s_* + 1}{8\mu s_*^2}$, $\tau = 4$	4. $W_p(s) = \frac{1}{k_o \mu}\frac{2(1-\mu)s_* + 1}{2s_*}$, $W_D(s) = \frac{2s_*(s_* + \mu)}{ds_*\left[2(1-\mu)s_* + 1\right]}$, $\tau = 2\left(1 + \frac{d}{2}\right)$, $d = 1$

Table 2. Transfer functions of compensating controllers.

Conventional ACS	Robust ACS
5. $W_p(s) = \frac{\frac{1}{\mu}s_* + 1}{k_o}\frac{1}{2s_*}\frac{1}{s_* + 1}$, $\tau = \frac{2}{\tilde{W}} + \frac{1}{\mu}$	6. $W_p(s) = \frac{1}{k_o}\frac{T_o s_* + 1}{T_\mu s_* + 1}$, $W_D(s) = \frac{\frac{1}{\mu}s_* + 1}{ds_* + 1}$, $d = 2$
7. $W_p(s) = \frac{\frac{1}{\mu}s_* + 1}{k_o}\frac{4s_* + 1}{8s_*^2}\frac{1}{s_* + 1}$, $\tau = \frac{1}{\mu} + 4$	8. $W_p(s) = \frac{\frac{1}{\mu}s_* + 1}{k_o}\frac{1}{2s_*}\frac{1}{s_* + 1}$, $W_D(s) = \frac{2s_*}{2s_* + 1}$, $\tau = \frac{1}{\mu} + 2 + d$, $d = 2$

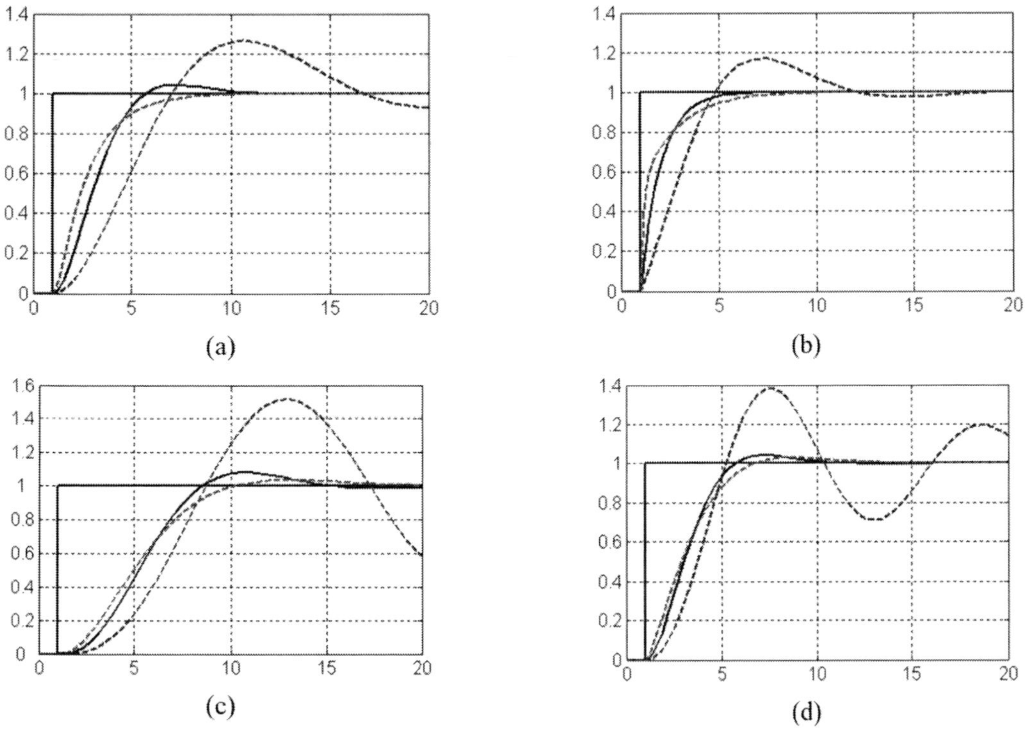

Figure 2. View of transient processes in conventional (a, c) and robust (b, d) systems.

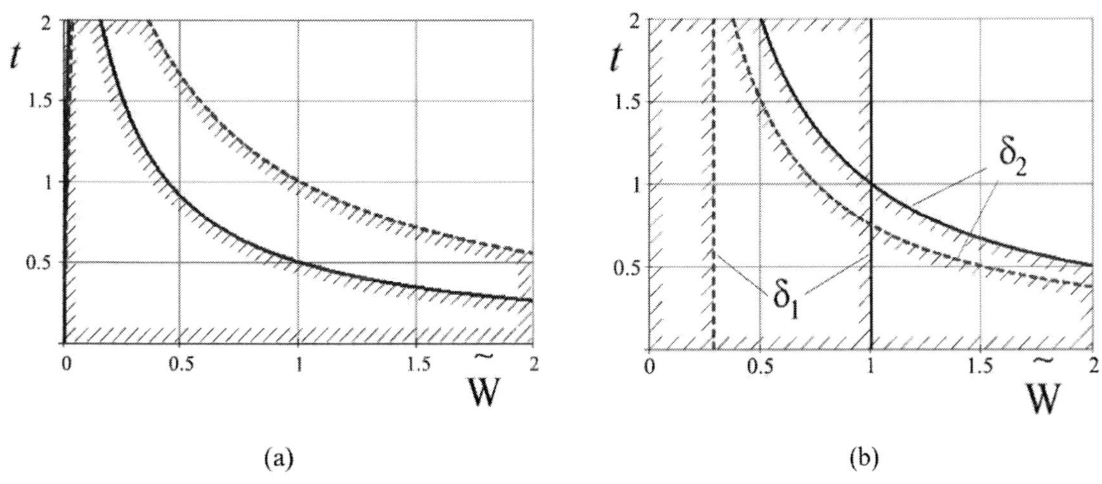

Figure 3. Level curves $\delta_i = 2$ of pattern indexes for controllers 1, 2 (a) and 3, 4 (b).

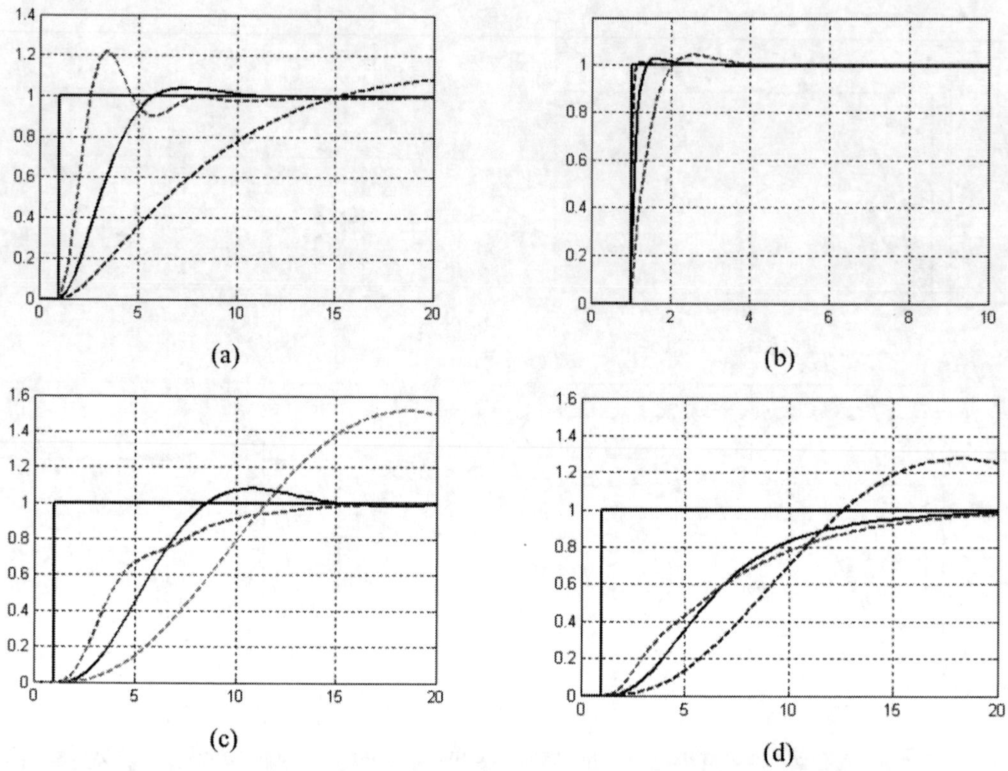

Figure 4. View of transient processes in conventional (a, c) and robust (b, d) systems

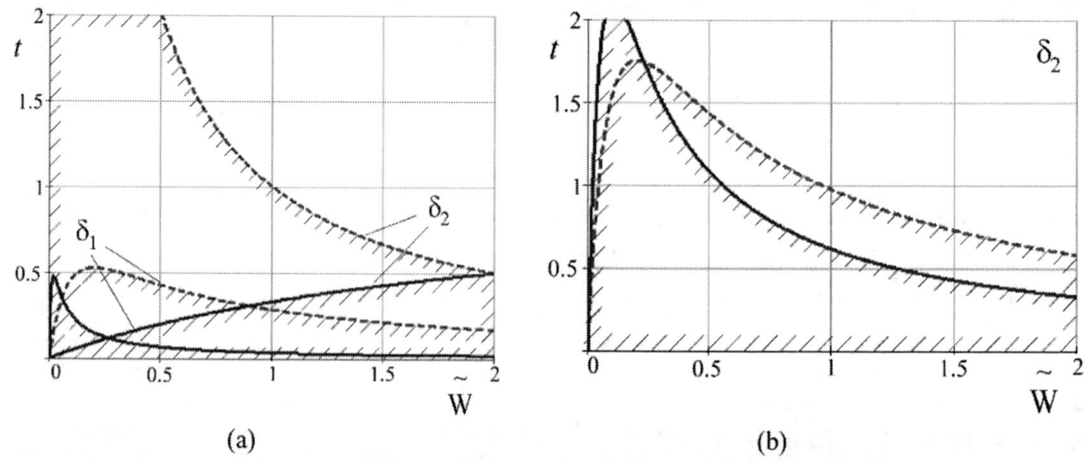

Figure 5. Level curves $\delta_i = 2$ of pattern indexes for controllers 5, 6 (a) and 7, 8 (b)

4. Conclusion

The analysis of the behavior of level curves of conventional and robust systems allows for the following conclusion for all the considered options: the permissible variation range of controlled object parameters is wider when using robust controllers with the structure depicted in figure 1, than when using the conventional systems. This allows providing better stability for the pattern of the transient processes and system's high-speed response, which is proved on the given charts of transient processes.

References

[1] Dorf R C and Bichop R H 2010 *Modern control system* (NJ: Upper Saddle River) p 749
[2] Polyak B and Sherbakov P 2002 *Robust stability and control* [In Russian] (Moscow, Science)
[3] Ishmatov Z Sh, Volkov M A and Gurent'yev E A 2007 Synthesis of electric drive systems invariant to parametric and external disturbances by the method of polynomial equations *Russian Electrical Engineering* **78** **(11)** pp 591-97
[4] Ha Q P and Alferov V G 1996 On the parameter-sensitivity problem in control system of DC-motor position drives *Elektrichestvo* **1** pp 47-53
[5] Volkov M A and Ishmatov Z Sh 2009 Development and study of robust control system for dc thyristor electric drive using polynomial techniques *Russian Electrical Engineering* **80** **(9)**, pp 517-23
[6] Yusupov R and Rozenwasser E 1999 *Sensitivity of Automatic Control Systems* (London, NY, Washington: CRS Press) p 436
[7] Naresh K Sinha 2013 *Control System* (New Delhi: New Age International Limited) p 454
[8] Opeiko O F and Nesenchuk A A 2019 Synthesis of the robust electric drive for robot control using pi controllers parameterization on the basis of root locus approach *21st European Conference on Power Electronics and Applications EPE 2019 (ECCE Europe)* no 8915531
[9] Kochetkov S A, Krasnova S A and Utkin V A 2018 Robust control for synchronous electric drive under uncertainty conditions *AIP Conference Proceedings* no 020047
[10] Son Y I, Kim I H, Choi D S and Shim H 2015 Robust cascade control of electric motor drives using dual reduced-order PI observer *IEEE Transactions on Industrial Electronics* **62** **(6)** pp 3672-82
[11] Son Y I, Choi D S, Cho K H, Kim I H and Kang S H 2013 Robust cascade control of electric motor drives using PI observers *Proceedings of the International Conference on Modelling, Identification and Control* pp 352-58
[12] Sung W, Shin J and Jeong Y S 2012 Energy-efficient and robust control for high-performance induction motor drive with an application in electric vehicles *IEEE Transactions on Vehicular Technology* **61** **(8)** pp 3394-405
[13] Weinmann A 1991 *Uncertain Models and Robust Control* (Wien: Springer-Verlag) p 707
[14] Ackermann J 1993 *Robust control: system with uncertain physical parameters* (New York: Springer-Verlag) p 406
[15] Zhou K, Doyle J C and Glover K 1996 *Robust and optimal control* (NJ: Prentice Hall) p 538
[16] Ishmatov Z Sh, Plotnikov Y V and Gurent'ev E A 2014 Robust current and speed controllers of the frequency-controlled induction electric drive *Russian Electrical Engineering* **85** **(9)** pp 570-74
[17] Ishmatov Z, Polyakov V and Plotnikov Yu 2017 Robust currents control in the electric drive *International Conference on Industrial Engineering, Applications and Manufacturing, ICIEAM 2017*
[18] Petrov N, Sokolov N I, Lipatov A V, et al 1986 *Automatic-Control Systems for Objects with Variable Parameters* [in Russian] (Mashinostroenie, Moscow) p 256
[19] Matviichuk K S 2003 Stability of nonstationary automatic-control systems of variable structure in forced motion *Int. Appl. Mech.* **39** no **10** 1221–30
[20] Matviichuk K S 2004 Technical stability of forced motion in nonstationary automatic-control systems of variable structure *Int. Appl. Mech.* **40** no **1** 103–14

IOP Conf. Series: Materials Science and Engineering **966** (2020) 012111 doi:10.1088/1757-899X/966/1/012111

Research of a bi-directional DC-DC converter integrated in electric car power installation

K H Gulyamov[1], B N Sharifov[1], A H Ghulomzoda[2], M Kh Safaraliev[3, 5] and R M Yunusov[3, 4]

[1] Department of Electric drive and Electric machines, Tajik Technical University named after academic M S Osimi, Dushanbe, 734042, Tajikistan
[2] Department of Automated Electrical Systems, Novosibirsk State Technical University, Novosibirsk, 630073, Russia
[3] Ural Power Engineering Institute, Ural Federal University named after the First President of Russia B N Yeltsin, 19, Mira str., Yekaterinburg, 620000, Russia
[4] ABB Ltd., Separate division in Ekaterinburg, Ekaterinburg, 620137, Russia

E-mail: [5] murodbek_03@mail.ru

Abstract. This article presents the results of the simulation study of the bi-directional DC converter, including the time diagrams of the electric vehicle speed, voltage and current changes in various experiments, Also presented are the results of efficiency change and charge level, when using a bi-directional converter as well as without using it. The HFEDC (Highway Fuel Economy Driving Schedule) model was chosen for the simulation of the electric vehicle drive model using units simulating standardized driving cycles and as an algorithm for changing the vehicle's control signal. Studies of the operating modes of the electrical equipment and the electric vehicle with the DC voltage conversion device in the power circuit of the DC source have been carried out by computer simulation in specialized programs for the study of dynamic systems. The simulation model of the reversible DC converter in the electrical power equipment of an electric vehicle was constructed on the basis of mathematical models of the individual elements reflecting their real physical properties. The present solution is the use of two-way DC voltage converters in the power conversion systems of modern electric drives, including the electric propulsion systems of prospective vehicles improves the efficiency of these systems.

1. Introduction

The transport industry, which is one of the main consumers of energy today, due to increased production of vehicles, is faced with problems associated with environmental pollution, due to increasing emissions of greenhouse gases. Against the background of an exacerbation of energy shortages and environmental degradation, automakers are increasingly investing energy and resources in the development of alternative fuel vehicles, mainly the technology for the production of electric vehicles and hybrid vehicles is being developed [1].

Despite the great achievements in the field of electric vehicle manufacturing today, one of the main reasons that impedes the mass production of electric vehicles is the low energy consumption, high weight and high cost of existing traction batteries (TB) used as an energy source.

One of the main components of the electric drive of the electric vehicle is a three-phase traction electric motor generator (EMG) and a power converter (inverter).

The power-driven voltage converter (PDVC) converts the direct voltage of the tractive current source (accumulator battery) into an alternating three-phase voltage required for the operation of the tractive electric machine. Despite the high efficiency of modern traction motors and PDVC, the main transducer circuits and engine windings are subject to significant currents, especially in transient modes, which are frequent for road transport in urban environments.

Given this fact, one solution to the problem of reducing current loads may be the use of a reversible DC/DC converter in a DC circuit. Furthermore, the main function of the reversible DC/DC converter consists in increasing the battery voltage and consequently the voltage at the input of the power inverter, which ultimately leads to a reduction in the currents in the power circuit and in the windings of the traction engine. The latter is related to improvements in the performance of the electric machine and the power conversion equipment, including with regard to the mass dimensions.

Studies have been carried out on combined power sources, battery and supercapacitor in the hybrid car power system using bi-directional DC/DC converters of different topology, which have advantages and disadvantages [7–9].

This article proposes the results of the study of the electric drive of an electric vehicle during the introduction into the structure of said electric power equipment and a reversible DC/DC converter by computer simulation in MATLAB/Simulink. The aim of this solution is to improve the mass dimensions and increase the voltage of the battery without significantly increasing the complexity of its design.

2. Building a simulation model of a DC/DC converter as part of the electric vehicle's energy system

In order to study in more detail, the features of the bi-directional DC/DC converter as part of the power electric equipment of an electric vehicle, a simulation model was created using the software packages of visual modeling MATLAB/Simulink, which today is one of the main tools for the study of dynamic systems.

Using the vehicle parameters shown in table 1, the traction calculation was carried out and the main parameters of the power elements of the electric vehicle were determined [5–6], including the input and output parameters of a bi-directional DC/DC converter installed in the DC power circuit [2–4].

Table 1. Key vehicle parameters for design studies.

Parameters	Symbol	Value
Vehicle Mass, kg	m_a	1748
Vehicle Cross Section, m^2	S_a	2.2
Drag coefficient, p.u.	C_X	0.3
The radius of the wheels of the car, m	r_w	0.323
Transmission efficiency, p.u.	η_t	0.96
Final drive ratio, p.u.	i_{fd}	4.3
Rolling resistance coefficient, p.u.	f	0.014
Road angle, degree	α	0
Air density, kg/m^3	ρ	1.2

Figure 1 shows a simulation model of a power electric equipment system for an electric vehicle, and a bi-directional DC/DC converter in a high-voltage DC power circuit. Designation in the figure: UMEP - unit for measuring electrical parameters; CSCU DC/DC CONV.– control signal conditioning unit bi-directional DC/DC converter; TCS – traction current source; TPI – three phase traction inverter; ADSSU – acceleration and deceleration signal setting unit; SC – speed controller; VCS – vector control system; PMSM – permanent magnet synchronous motor; PS – position sensor; DC/DC CONV. – bi-directional DC/DC converter.

Traction current source is the main source of electrical energy and provides power to all power and on-board consumers. As a traction energy source in this simulation model, a lithium-ion battery with high energy performance is used. Bidirectional DC/DC converter consists of transistor switches VT1,

VT2 with counter diodes VD1, VD2, inductive coil L and smoothing capacitors C1 and C2. The control signals of the power switches of a bi-directional DC/DC converter are provided by the control signal generating unit (Figure 1).

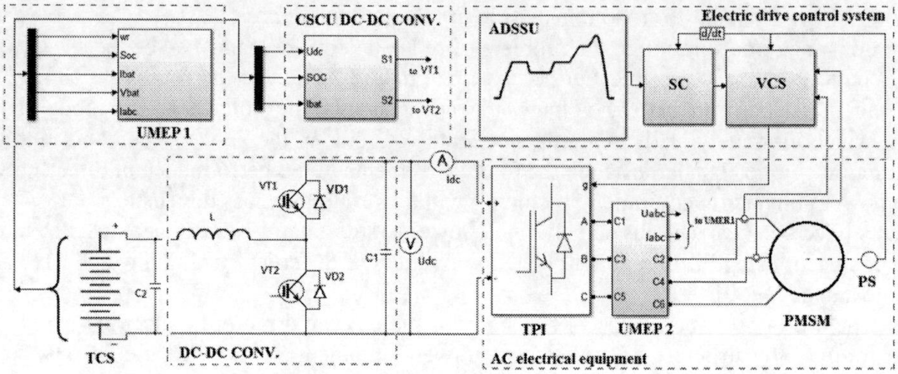

Figure 1. Simulation model of electric vehicle power equipment.

The objective of this DC/DC converter is to increase the voltage of the traction current source to the input of a three-phase traction inverter. In addition, a bidirectional DC/DC converter can operate in a mode of reduction during regenerative braking of the electric drive, when the traction motor operates in the generator mode, while the converter lowers the high-voltage generator voltage rectified by the inverter and provides a charge for the traction current source [3]. The task of a three-phase traction inverter is to convert DC voltage to AC with an adjustable value of the output voltage and its frequency. The control system of the electric drive consists of a unit for setting the acceleration and braking signal characterizing the reaction of the driver, as well as the speed controller and the vector control system, which form the necessary control signal for the power transistor keys of the three-phase traction inverter, providing speed and torque control of the traction motor.

The simulation model of a bi-directional DC/DC converter integrated in a power plant is built on the basis of mathematical models of individual elements that reflect real properties and make up the whole system taking into account the parameters of power electric equipment of an electric vehicle.

It should be noted that the model does not take into account some extreme operating modes of the electric drive, For example, a fully charged battery recovery mode because in this mode the real electric vehicle control system disables the regeneration mode and activates the full-time well braking system thus blocking the return current to the source.

3. The results of the research

In order to confirm the theory of improved traction and improved overall dimensions, as well as the possibility of using a lighter battery, three different experiments were carried out:

1. The electric drive is powered by a battery with a nominal voltage of 400 V DC.
2. The electric drive is powered by a battery with a rated voltage of 650 V DC. The purpose of this experiment is to show the advantage of using a high voltage electric drive.
3. The electric drive is powered by a battery with a nominal voltage of 400 V DC, and a bi-directional constant-voltage converter is used in the high-voltage power circuit, the task of which is to increase the voltage of the battery source from 400 V to 650 V. The purpose of this experiment is to show the advantage of increasing voltage by using a boost converter without significantly complicating and increasing the cost of the battery, which can occur if you follow the second experiment. In order to maximally approximate the experimental characteristics of the simulation model to the real ones, the tests were carried out in a standardized driving cycle of the HFEDC (Highway Fuel Economy Driving Schedule). This movement cycle lasts 765 seconds characterizes the suburban movement.

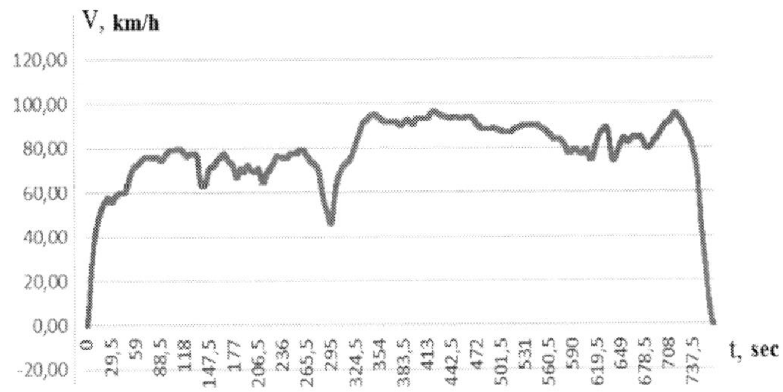

Figure 2. Electric vehicle speed in a cycle.

As can be seen from the presented figure, when conducting three different experiments with different sources, the speed characteristics do not change and comply with the HFEDC standard.

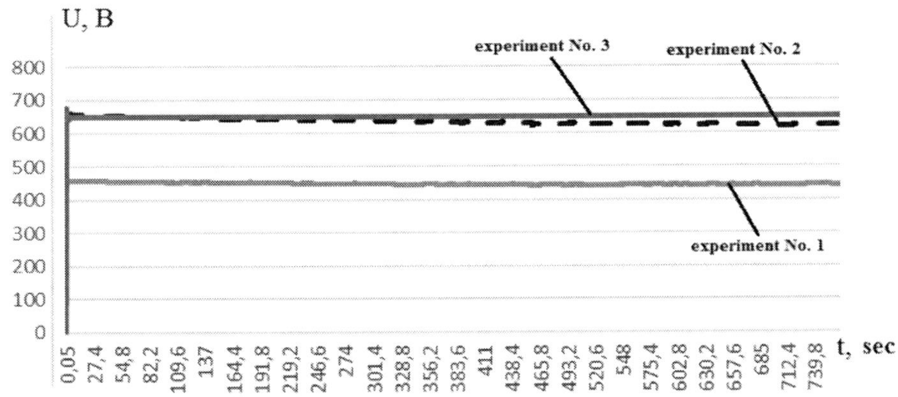

Figure 3. Voltage at the inverter input during three different experiments.

As can be seen from Figure 3 in the third experiment, when a bi-directional DC/DC converter is used, regardless of the decrease in the battery charge level, the voltage remains stable and does not decrease.

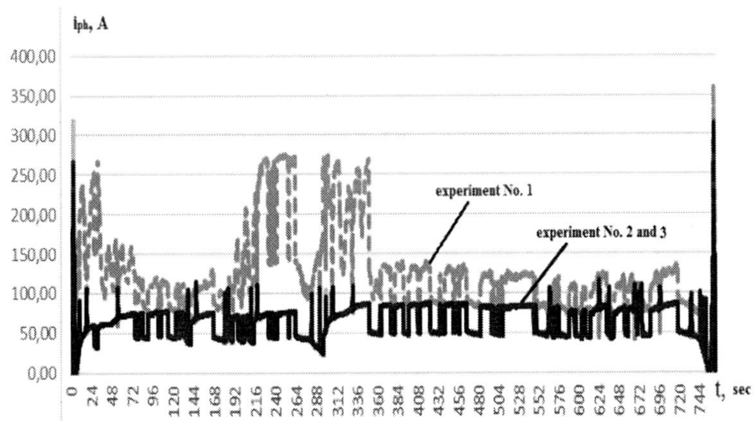

Figure 4. Change in the effective value of the phase current of the motor in three experiments.

This is one of the important advantages of using a bi-directional DC/DC converter in an electric vehicle. Other characteristics show a gradual decrease in voltage with a decrease in the charge level, which will undoubtedly have a negative effect on the traction characteristics of an electric vehicle when the charge level drops to 50% or lower.

As can be seen from Figure 4, the effective value of the active, phase current of the motor in the second and third experiments with a voltage at the input of the inverter 650 V is significantly lower than the first.

With three different experiments shown in Figure 5 shows how the highest efficiency is achieved by using a bi-directional DC/DC converter in the power circuit.

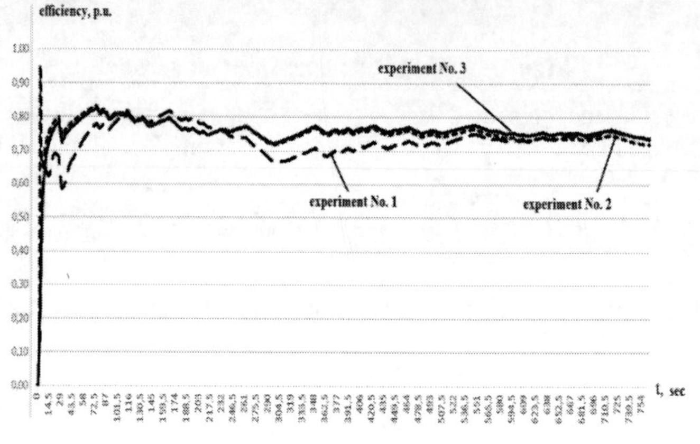

Figure 5. Electric drive efficiency for three different experiments.

Figure 6 shows a diagram of the charge level (SOC) at various values of the supply voltage, from which it follows that the lowest charge consumption in the cycle is achieved when using a bi-directional DC/DC converter.

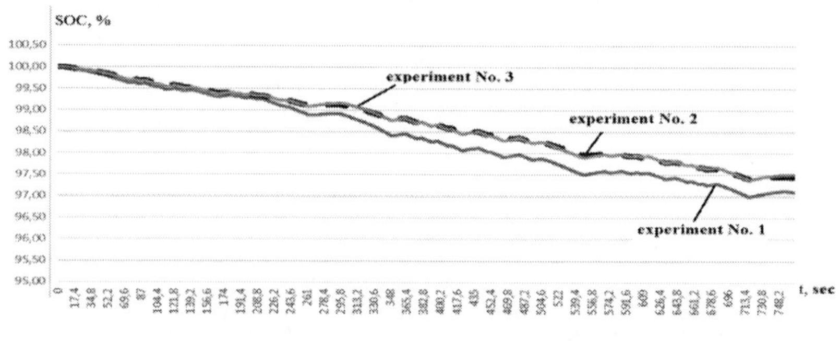

Figure 6. Charge level in three different experiments.

It can be seen from the figure that the charge saving when using a bi-directional DC/DC converter (experiment No. 3) is 0.41%, with a total consumption of 3% for a full cycle. Given the non-linearity of the discharge of the battery, the approximate saving of the charge with the full allowed (DoD = 80 %) energy consumption of the battery will be more than 11%.

Since the load power is the same in different experiments, with increasing supply voltage, the load currents decrease, which is confirmed by the characteristic of Figure 5. A relative increase in efficiency and a decrease in energy consumption occurs due to a decrease in losses with a decrease in load currents, especially in transients, because The electric drive of an electric car, like any car, works primarily in transient modes (frequent acceleration and braking).

It should be noted that the experiments were conducted on two types of electric drives of the same power and different supply voltages.

4. Conclusion

The method proposed in this work for improving the mass dimensions of traction electrical equipment components is most appropriately used in modernizing an already existing electric vehicle when the design change is: The existing storage batteries (SB) and the addition of new elements are difficult and costly. It is also appropriate to use this method when the electric vehicle is small and there is no possibility of mixing a large SB with a weight, but it is necessary to obtain a higher and stabilized voltage source. Also, in industrial trucks with electric traction, which are operated with the possibility of dosage during operation, or in passenger vehicles with a limited SB capacity, which are charged at the end of each line. In this case of increased voltage with PDVC, engines with higher speed and traction performance can be used.

References

[1] Slepcov M A, Dolaberidze G P, Prokopovich A V, Savina T I and Tulupov V D 2006 *Osnovy ehlektricheskogo transporta* (Moscow: Izdatel'skij centr 'Akademiya') p 464
[2] YUtt V E, Sidorov K M and Gulyamov K H 2016 Povyshenie moshchnosti impul'snyh preobrazovatelej postoyannogo napryazheniya s ispol'zovaniem mnogofaznyh struktur *Elektronika i ehlektrooborudovanie transporta* vol **4** pp 13–6
[3] YUtt V E, Sidorov K M, Lohnin K M and Gulyamov K H 2015 Primenenie preobrazovatelej postoyannogo napryazheniya v sostave ehnergeticheskoj ustanovki ehlektricheskogo transportnogo sredstva *Vestnik MADI* vol **4** pp 34–40
[4] Kirill S, Timofey G and Vladimir Y 2015 Method for Calculating the Power Circuit Characteristics of the Isolated DC-DC Converters for Electric and Hybrid Vehicles *Indian Journal of Science and Technology* vol **8** pp 1–8
[5] Chris M, Abdul Masrur M and Wenzhong D 2011 Hybrid Electric Vehicles: *Principles and Applications with Practical Perspectives* (USA) p 485
[6] Rodrigo G V, Joao A and Pecas L 2013 Electric *Vehicle Integration into Modern Power Networks* Springer Science+Business Media (New York) p 325
[7] Hayet S Kh and Ahmed C A 2014 Design of an Hybrid Battery/Super-capacitors Energy Storage System for Hybrid Electric Vehicles *Life Science Journal* vol **11** pp 109–18
[8] Dharam D and Souvik G 2013 Design of a bidirectional dc-dc converter for hybrid electric vehicles (HEV) using Matlab *International Journal of Advanced Research in Electrical, Electronics and Instrumentation Engineering* vol **2** pp 2994–3002
[9] CH K P and KP Prasad Rao 2012 Multiphase Bidirectional Flyback Converter Topology for Induction Motor Drive *International Journal of Modern Engineering Research* vol **2** pp 215–8

IOP Conf. Series: Materials Science and Engineering **966** (2020) 012112 doi:10.1088/1757-899X/966/1/012112

Estimation of the operational characteristics and determination parameters of the vertical radial-axial hydraulic turbine type 400

A A Gulakhmadov[1, 2, 3, 5], S Sh Qurbonalizoda[1], Z V Kobuliev[1] and B N Ghoziev[4]

[1] Institute of Water Problems, Hydropower and Ecology of the Academy of Sciences of the Republic of Tajikistan, Dushanbe, 734042, Tajikistan
[2] Ministry of Energy and Water Resources of the Republic of Tajikistan, Dushanbe, 734064, Tajikistan
[3] Research Center of Ecology and Environment in Central Asia, Chinese Academy of Sciences, Urumqi, 830011, China
[4] Ural Federal University named after the First President of Russia B N Yeltsin, 19, Mira str., Yekaterinburg, 620000, Russia

E-mail: [5] gulakhmadovamin@yahoo.com

Abstract. The operational characteristic of the hydraulic turbine is an important document with the help of which would be carried out the control over the correct operation of a hydraulic turbine at a hydroelectric power station. This article presents the calculation of operational characteristic and determination of parameters of the vertical radial-axial hydraulic turbine (Francis turbine) for high heads type 400–500. In this study, the multiple mathematics used in computer which was written in Fortran language. We designed the operational characteristic and calculated the flow part of the runner. In addition, we determined the synchronous frequency of rotation of the hydraulic turbine with a one-dimensional method. The purpose of this work is to analyze the spatial structure of the flow in high-pressure hydraulic Francis turbines of the 400-500 types. Results of this computational and designing analysis can be useful for researchers to apply the present approach to calculate the operational characteristics and the parameters of the high-pressure hydraulic turbines.

1. Introduction

In the process of designing hydroelectric power plants, when choosing types of turbines and determining their main parameters, sizes, speed, efficiency, plant mark and other factors, as well as when assigning the most appropriate conditions for using the equipment during operation, it is necessary to have sufficiently complete data on the properties of turbines[1, 2]. These data are presented in the form of characteristics that determine all the necessary turbine performance for various conditions of its operation. Operational characteristics have two determining parameters and represent the dependence of this indicator on two independent variables. There are several types of common characteristics, with the name given in variables.

For high-pressure hydropower plants (more than 400 m) can be used active and reactive hydraulic turbines [3]. Hydraulic turbines using only kinetic flow energy are called active if using at least partially the potential pressure energy it would be called reactive turbine. The energy conversion process in these turbines occurs at an inlet pressure, which is greater than atmospheric pressure. Water is supplied to the runner in impulse hydraulic turbines through nozzles, in reactive ones through a guide vane. In an impulse hydraulic turbine, the pressure of the water in the inlet and outlet of the turbine's runner is equal to the atmospheric pressure [4, 5].

The main tasks of this analysis are to compute the vertical radial-axial hydraulic turbine (Francis turbine) of type runner 400, design the operating characteristic and the flow part of the turbine such as runner and synchronous frequency of rotation of the turbine with a one-dimensional method.

2. Materials and Methodologies

2.1. Determination of the main parameters of the hydraulic turbines

Aggregate (unit) capacity is determined [6, 7]:

$$N_{unit} = \frac{N_{HPP}}{Z_{unit}}, \tag{1}$$

where N_{unit} – unit capacity, MW; N_{HPP} – HPP capacity, MW; Z_{unit} – number of units, pcs.

Turbine capacity is determined:

$$N_T = \frac{N_{unit}}{\eta_{gen}}, \tag{2}$$

where N_T–turbine capacity, MW; η_{gen} – generator efficiency. For powerful units is assumed equal efficiency, $\eta_{gen} = 0.98$.

Hydraulic turbine impeller diameter:

$$D_1 = \sqrt{\frac{N_p}{9.81 * Q'_{I_p} * \sqrt{H_p} * \eta_N}}, \tag{3}$$

where D_1 – the diameter of the impeller, m; N_p – calculated power, kW; Q'_{I_p} – reduced discharge at the calculated point, m³/s; H_p – calculated (design) head, m; η_M – Full-scale turbine efficiency at the calculated point is equal to $\eta_H = 0.92$, and then specify it.

For a radial-axial turbine (Francis turbine), the magnitude of the reduced discharge is taken from the output limit line of the hillchart (hill diagram). On a horizontal is the calculated point, which crosses several revolutions above optimum of the hillchart. Take $Q'_{I_p} = 280$ and $\eta'_{I_p} = 62.4\ min^{-1}$.Further, the calculated value of the runner's diameter would be normalizedD_1=4.5 m. Full-scale turbine efficiency specifies at the calculated point.

$$\eta_{N_p} = 1 - \left(\left(1 - \eta_{мp}\right) \cdot \left(\left(1 - \chi\right) + \chi \cdot \sqrt[5]{\frac{\text{Re}_м}{\text{Re}_н}} \right) \right) \tag{4}$$

where η_{M_p}– model-test turbine efficiency at the calculated point; $\eta_{M_p} = 0.8848$; χ – recounted share of losses ($\chi = 0.75$); D_{1M} – model-test impeller diameter, m; $D_{1M} = 0.8$, m.

Model-test turbine Reynolds number

$$\text{Re}_M = \frac{D_{1M}\sqrt{H_M}}{\nu}, \tag{5}$$

H_M – head on a model hydroturbine, m; H_M= 4.9 m;

Full-scale hydraulic turbine Reynolds number

$$Re_н = \frac{D_{1н}\sqrt{H_н}}{\nu}, \tag{6}$$

ν – kinematic coefficient of viscosity:

$$\frac{Re_м}{Re_н} = \frac{\upsilon_н \cdot D_{1м} \cdot \sqrt{H_м}}{\upsilon_м \cdot D_{1н} \cdot \sqrt{H_н}}, \tag{7}$$

First approximation:

$$\eta_{H_p} = 1 - \left((1 - 0.8848) * \left((1 - 0.75) + 0.75 * \sqrt[5]{0.01553}\right)\right) = 0.9336$$

The accepted calculated point ($Q'_{1p} = 840$ l/s) disagree with the calculated values of D_1. Specify the calculated point position.

$$Q'_{Ip} = \frac{N_p}{9.81 * D_1^2 * H_P \sqrt{H_p} * \eta_H}, \tag{8}$$

2.2. Determination of synchronous hydroturbine speed

Synchronous hydroturbine speed is determined by the known quantity of calculated (design) head and accepted values of corrected speed n'_{I_p} and impeller diameter D_I:

$$n = \frac{n'_{Ip} \cdot \sqrt{H_p}}{D_1}, \tag{9}$$

where n'_{Ip} corrected speed at the calculated point, min^{-1}.

$$n = \frac{62.4 * \sqrt{325}}{4.5} = 250\ min^{-1}$$

The received value is synchronous.

Determine the correction for the speed of full-scale hydraulic turbine:

$$\Delta n'_I = n'_{Iopt.} \left(\sqrt{\frac{\eta_{H_{opt.}}}{\eta_{M_{opt.}}}} - 1 \right), \tag{10}$$

where η_{opt} – full-scale turbine efficiency in optimum; η_{opt} – model-test turbine in optimum; $\eta_{H_{opt}} = 0.902$.

The correction $\Delta n_I = 1.38$ is less than 3%, and beyond calculation is ignored.

Determine the corrected speed value at the highest and lowest heads:

$$n'_{I_{min}} = \frac{n \cdot D_1}{\sqrt{H_{max}}}, \tag{11}$$

$$n'_{I_{max}} = \frac{n \cdot D_1}{\sqrt{H_{min}}}, \tag{12}$$

3. Results

3.1. Construction of working and operating characteristics Francis turbine type 400

The energy and cavitation qualities of a full-scale turbine at various operating modes are presented in the form of an operational characteristic. The operational characteristic is an important document in which the proper operation of a hydroturbine at the HPP is monitored. The operational characteristic of turbine would built by using the universal characteristics of hydraulic turbine (Francis turbine) 400.

3.2. Operational characteristic calculation

Characteristic calculation is performed for three heads: H_{max}, H_p and H_{min}.

The program is written in Fortran language.

On the programme [7]:

Efficiency is recalculated using the following formula:

$$\eta_н = 1 - \left((1 - \eta_м) \cdot \left((1 - \chi) + \chi \cdot \sqrt[5]{\frac{Re_м}{Re_н}} \right) \right) \tag{13}$$

Full-scale turbine capacity (power) is computed using the following formula:

$$N_H = 9.81 * Q * H * \eta_H = 9.81 * D_1^2 * H * \sqrt{H} * Q_I * \eta_H$$

During recalculation accept $Q'_{Iн} = Q'_{Iм}$.

The suction height is measured from midline of the wicket gate to the level of the downstream. The suction height is determined by the formula:

$$H_S = H_{ATM} - H_d - \sigma_T H_p + \frac{b_0}{2} - \frac{\nabla}{2} - 1.5 \tag{14}$$

For operational characteristic calculation the source data are necessary. Source data are presented on print Q'_1, σ_T, $\eta_м$ for heads H_{min}, H_p, H_{max}.

SOURCE DATA

'For operational characteristic construction'

Model diameter D1m = 80000 m
Modal tests head H_m= 4.90 м
Water temperature during model tests t_m = 7.00 grad.
Modal max. efficiency K_m =9020
Full-scale turbine diameter D1 = 4.50000 m
Station heads:
maximumHmax = 404.00 m
calculated H_c = 325.00 m
minimal Hmin = 280.00 m
mean annual water temperature at the station tn = 20.00 grad.
reduced discharge at the calculated point Q11P = .2800 m³/s
optimal reduced discharge Q11opt = .2350m³/s
optimal reduced speed nmopt = 62.40 rev./min.
synchronous speed nc = 250.00 rev./min.
Wicket gate height B0 = .45m
Turbine axis mark
above sea level C = .00m
Values Km, Q11 and S for heads: Hmin, Hp and Hmax.

3.3. Operational characteristic construction

Having received the calculation results, move to the construction of operational characteristic.

Construct a full-scale turbine operating characteristic for the three heads: H_{max}, H_p and H_{min} at a constant speed (Figure 1).

Construction output limit line for the capacity of hydraulic turbine (Figure 3). Output limit line is a composite of two sections: *vertical, incline*.

Vertical section- from the calculated to the maximum head – represents the power turbine limitation for the selected hydrogenerator nominal power.

Cutting the operational characteristic through horizontal lines $\eta = \text{const}$ that need to be drawn with an interval of 0.5%. Points of equal efficiency are transferred to the field of operational characteristics and connected by smooth curves, as a result we get lines of equal efficiency (Figure 3).

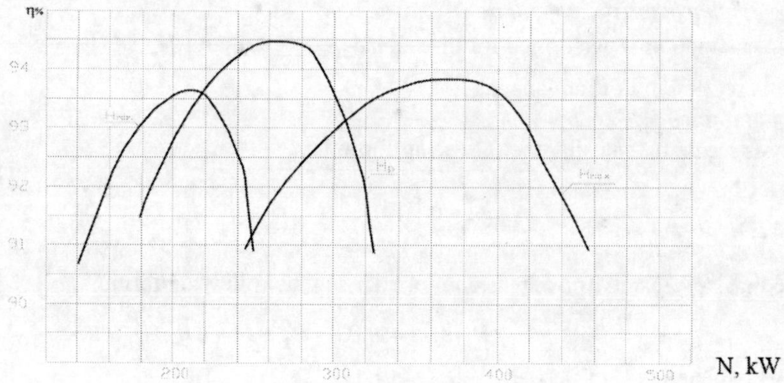

Figure 1. Full-scale turbine operating characteristic $\eta = f(N)$

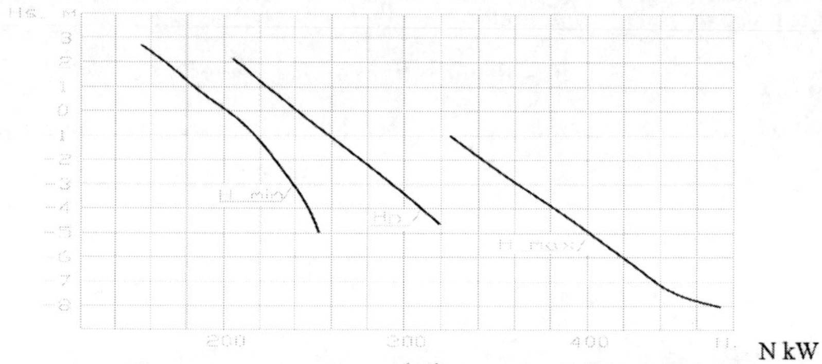

Figure 2. Curves $H_s = f(N)$ for specified heads.

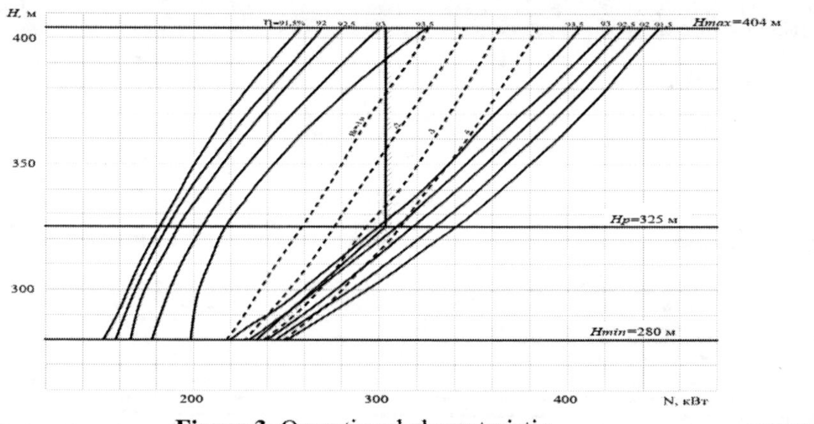

Figure 3. Operational characteristic. N, kW

Plot curves $H_s = f(N)$ for specified heads (Figure 2). Cut curves $H_s = f(N)$ through horizontal lines with 1 m, in order to determine the capacity values for various heads for which $H_s = \text{const}$. By transferring the resulting points to the operational characteristic field and plot curves $H_s = \text{const}$.

Inclined section- from the calculated to the minimum head – its power limitation of turbine.

Determination of the value Q_I' in the cross point line $n_{\text{I}max}' = \text{const}$ with the line $a_{0_p} = \text{const}$. At this point $Q_I' = 0.27\ m^3/s$ and $\eta_M = 0.88$ ($\eta_M = 0.9304$). Determination of the capacity of full-scale turbine with the lowest head.

$$N_p = 9.81 * D_1^2 * H_{min} * \sqrt{H_{min}} * Q_I * \eta_H,$$

Upon recommendations for heads with the 402 meter height and received values of water discharges its necessary to arrange for the radial-axial hydraulic turbines type 400.

Conclusions

The vertical radial-axial hydraulic turbine (Francis) could be designed for a wide range of heads and flows, it is the most widely used turbine in the world. In this study, we present the calculation of the operational characteristic of the vertical radial-axial hydraulic turbine type 400 (Francis turbine). Also, constructed flow parts: runner of the unit (aggregate) and computed synchronous frequency of rotation turbine with one-dimensional method. Analysis for structure of the flow was carried out in the flow parts of the hydraulic Francis turbine types 400–500.

Reference

[1] Keck H, Schrer Ch, Cuinod R and Cateni A 1997 Pelton technology for new plants and modernization schemes *International journal on Hydropower and Dams* vol **4** iss **2** pp 104–8

[2] Parkinson E, Angehrn R and Weiss Th 2007 Modern design engineering applied to pelton runners *Int. J. Hydropower and Dams* no **4** pp 91–4

[3] Potetenko V, Grishin A and Gulakhmadov A et al 2012 Rabochyy protsess radyal'no-dyahonal'noy hydroturbyny (ROD). Metodyka postroenyya prohnoznoy unyversal'noy kharakterystyky. (The working process of the radial-diagonal hydraulic turbine. The method of the designing a predictive universal characteristic) *NTU-KHPI* no **33** pp 109–19 [In Russian]

[4] Morgunov G and Thuy D 2007 Proyektirovaniye I issledovaniye gidrodinamicheskikh svoystv rabochego kolesa radial'no-osevoy gidroturbiny dvoynogo regulirovaniya (Design and study of the hydrodynamic properties of the impeller of a double-regulation radial-axial hydraulic turbine) *Vestnik MEI* no **1** pp 11–23 [In Russian]

[5] Potetenko O, Drankovsky V, Krupa S et al 2010 Tendentsii prodvizheniya gorizontal'nykh pryamotochnykh i vertikal'nykh radial'no-osevykh gidroturbin na vysokiye napory s shirokim diapazonom. *Visnik SUMDU Sumi* no **3** pp 125–35 [In Russian]

[6] Barlit V 1977 *Gidravlicheskiye turbiny* (Kiev: Vishcha School) [In Russian]

[7] Barlit V, Borodaevsky V and Yakovleva K 1994 *Vybor tipa gidroturbiny, nasos-turbiny pri pomoshchi universal'nykh kharakteristik, opredeleniye osnovnykh parametrov i postroyeniye yeekspluatatsionnoy kharakteristiki s ispol'zovaniyem EVM. (Selection of the type of hydraulic turbine, pump-turbine using universal characteristics, determination basic parameters and the construction of its operational characteristics using computers)* (Kharkov: KSPU) [In Russian]

IOP Conf. Series: Materials Science and Engineering **966** (2020) 012113 doi:10.1088/1757-899X/966/1/012113

Mathematical modelling of mutual electromagnetic influences of related power transmission lines in a transition process mode

S M Asanova[1, 5], S M Suerkulov[1], M Kh Safaraliev[2], D U Asanova[3] and R S Milutin[4]

[1] Department of Theoretical bases of Electrical Engineering and General Electrical Engineering, Kyrgyz State Technical University named after I. Razzakov, Bishkek, 720044, Kyrgyzstan
[2] Department of Automated Electrical Systems, Ural Federal University named after the First President of Russia B N Yeltsin, 19, Mira str., Yekaterinburg, 620000, Russia
[3] Department of Renewable Energy Sources, Kyrgyz State Technical University named after I. Razzakov, Bishkek, 720044, Kyrgyzstan
[4] Department of Power Electric Stations, Tajik Technical University named after academic M S Osimi, Dushanbe, 734042, Tajikistan

E-mail: [5] a_sm07@mail.ru

Abstract. A mathematical model is obtained using the variable state method, allowing the simulation of electromagnetic processes in transmission lines and communication lines (both homogeneous and heterogeneous) taking into account the reciprocal inductive and capacitive bonds in transition and established modes. The resulting mathematical model can solve a number of important scientific and practical problems related to the design and operation of electric power systems (EEC). The model also allows for a variety of situations and tasks related, for example, to the calculation of circuit breakage, short circuit (SC) or single-phase SC in isolated neural networks. In particular, it is possible to calculate the steered voltage of the line from high-voltage or high-precision power lines. Based on this methodology, it is also proposed to develop an algorithm for the machine formation of mathematical models for the study of transition processes in complex electrical networks, where the initial data will be their parameters and the structure of graphs of the studied networks.

1. Introduction

Currently, the problem of electromagnetic compatibility (EMC) in the power supply systems of enterprises where frequency-controlled electric drives based on power semiconductor transducers are widely used is becoming topical, since these electric drives distort the shape of the power and current curves of the network, which in turn is negative affects the operating modes of various types of electrical installations, leads to false triggering of relay protection and automation, being the main cause of violation of EMC electrical equipment [1–4].

In addition, power lines are constantly influenced by external electromagnetic fields of one origin or another. The third-party electromagnetic fields are induced in the lines of interference, which not only reduce the quality of the transmission, but sometimes cause high voltages and currents, which lead to the destruction of the power transmission lines and equipment, as well as creating a danger to the life and health of operating personnel.

This problem is common to all systems and devices related to the generation, transmission, reception and processing of electrical signals, and is called the problem of electromagnetic compatibility. Its essence is that the design, construction and operation of the above-mentioned devices and systems must take into account: on the one hand, the impact on them of external electromagnetic fields of a predetermined nature and sufficient protection for their normal work against these effects, and, on the other hand, the measures to limit the levels of influence of electromagnetic fields of the designed devices and systems on other devices by acceptable values.

Thus, the presence of a large number of electromagnetic disturbances (EMF) of different types makes it difficult, and at worst impossible, for certain types of electrical equipment to function normally, which causes failures and emergencies in the networks of these enterprises.

The EMF can be virtually any electromagnetic phenomenon in a wide range of frequencies.

2. Compilation of a mathematical model of mutual electromagnetic influences of related power transmission lines in transition mode

In [5] the simulation of electromagnetic interactions of adjacent power transmission lines is considered for the case where one line is active, i.e. the source $e_1(t)$ first acts and the load is connected by a resistance at the end $Z_H = R_H + jX_L$. The second line, the ends of which are connected to the resistors R_A and R_B, is arranged parallel to the first line at a certain distance.

The mathematical model of these lines is obtained by the method of state variables [6].

In this work a further synthesis of this model is made, with the aid of which it is possible to analyze different interaction variants for two lines. The equivalent circuit of the investigated network and the corresponding normal graph are shown in figures 1 and 2.

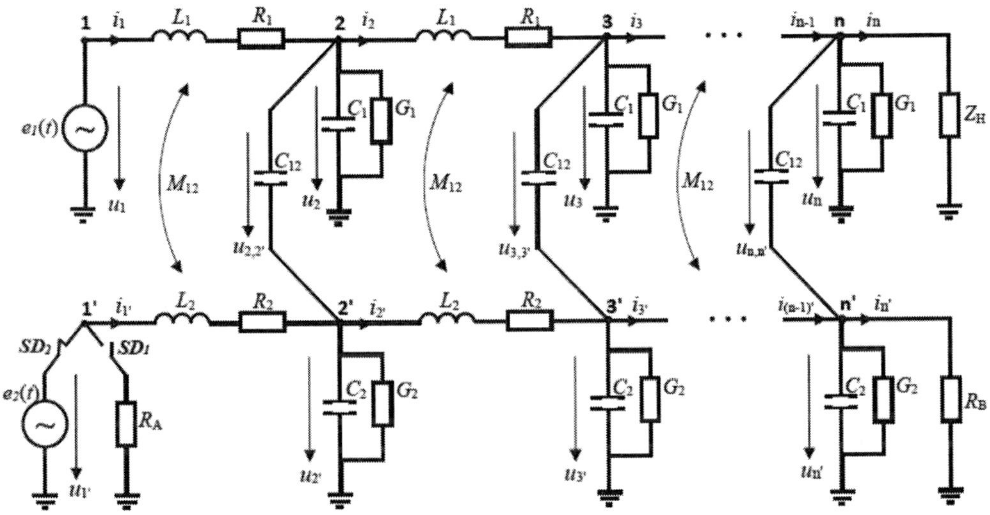

Figure 1. Research network equivalent circuit.

In Figure 1, the switches SD_1 and SD_2 are shown only in order to be able to analyze the electromagnetic interactions of adjacent power lines for two variants. The first option is that the top line is active and the bottom passive line is active (SD_1 is closed, SD_2 is opened); the second option is that both lines are active ($SD_1$is open, $SD_2$is closed).

As variable states we choose voltages $u_2, u_3, \ldots, u_n, u_{2'}, u_{3'}, \ldots, u_{n'}$and currents $i_1, i_2, \ldots, i_n$, $i_{1'}, i_{2'}, \ldots, i_{n'}$ along the 1st and 2nd lines. Distinguishing these values as variables characterizing the

IOP Conf. Series: Materials Science and Engineering **966** (2020) 012113 doi:10.1088/1757-899X/966/1/012113

energy state of the electrical circuit allows to form differential equations in normal form, since only in these elements currents and voltages are interconnected via derivatives.

The equations for independent sections and independent contours according to the first and second Kirchhoff laws for the first variant, when the branch of the tree (0-1') of a normal graph (Figure 2) contains a R_A, i.e. the switching apparatus SD_1 is closed and the $SD_2$is opened.

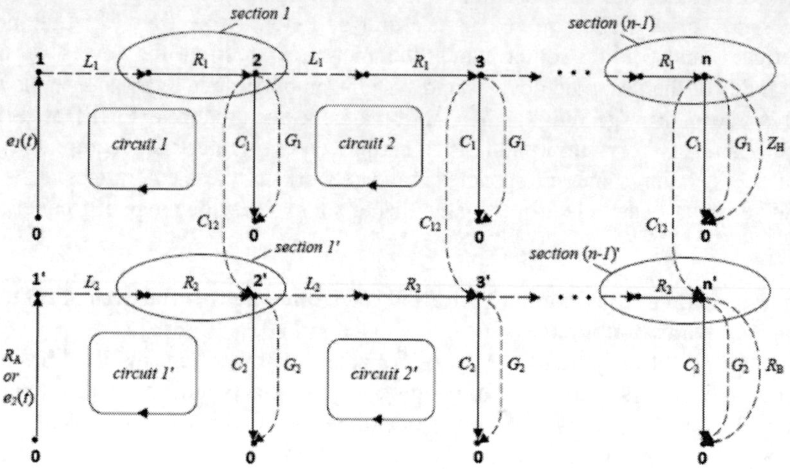

Figure 2. A normal network graph where, in the first variant, the tree branch (0-1') contains R_A, i.e. SD_1 is closed, SD_2 is opened, and in the second variant, the tree branch (0-1') contains $e_2(t)$, i.e. SD_1 is opened, SD_2 is closed.

1. For the 1st section: $i_1 - i_2 - i_{C_1} - i_{G_1} - i_{C_{12}} = 0$,

where $i_{C_1} = C_1 \frac{du_2}{dt}$; $i_{G_1} = G_1 u_2$; $i_{C_{12}} = C_{12} \frac{du_{2,2'}}{dt} = C_{12} \frac{du_2}{dt} - C_{12} \frac{du_{2'}}{dt}$,
then we'll get

$$(C_1 + C_{12}) \frac{du_2}{dt} - C_{12} \frac{du_{2'}}{dt} = i_1 - i_2 - G_1 u_2. \quad (1)$$

2. For the 2nd section: $i_2 - i_3 - i_{C_1} - i_{G_1} - i_{C_{12}} = 0$,

where $i_{C_1} = C_1 \frac{du_3}{dt}$; $i_{G_1} = G_1 u_3$; $i_{C_{12}} = C_{12} \frac{du_{3,3'}}{dt} = C_{12} \frac{du_3}{dt} - C_{12} \frac{du_{3'}}{dt}$,
then we'll get

$$(C_1 + C_{12}) \frac{du_3}{dt} - C_{12} \frac{du_{3'}}{dt} = i_2 - i_3 - G_1 u_3. \quad (2)$$

Similar to other independent sections.

3. For the (*n-1*) section: $i_{n-1} - i_n - i_{C_1} - i_{G_1} - i_{C_{12}} = 0$,

where $i_{C_1} = C_1 \frac{du_n}{dt}$; $i_{G_1} = G_1 u_n$; $i_{C_{12}} = C_{12} \frac{du_{n,n'}}{dt} = C_{12} \frac{du_n}{dt} - C_{12} \frac{du_{n'}}{dt}$,
we'll get

$$(C_1 + C_{12}) \frac{du_n}{dt} - C_{12} \frac{du_{n'}}{dt} = i_{n-1} - i_n - G_1 u_n. \quad (3)$$

4. For the 1'-th section: $i_{1'} - i_{2'} - i_{C_2} - i_{G_2} + i_{C_{12}} = 0$,

where $i_{C_2} = C_2 \frac{du_{2'}}{dt}$; $i_{G_2} = G_2 u_{n'}$; $i_{n'} = \frac{u_{n'}}{R_B} = G_B u_{n'}$; $i_{C_{12}}$ listed in paragraph 1 (see above),
then we'll get

$$(C_2 + C_{12}) \frac{du_{2'}}{dt} - C_{12} \frac{du_2}{dt} = i_{1'} - i_{2'} - G_2 u_2. \quad (4)$$

5. For the (*n*-1) 'th section: $i_{(n-1)'} - i_{n'} - i_{C_2} - i_{G_2} + i_{C_{12}} = 0$,

where $i_{C_2} = C_2 \frac{du_{n'}}{dt}$; $i_{G_2} = G_2 u_{n'}$; $i_{G_2} = G_2 u_{n'}$;$i_{C_{12}}$listed in paragraph 3 (see above),

then we'll get

$$(C_2 + C_{12})\frac{du_{n'}}{dt} - C_{12}\frac{du_{n'}}{dt} = -G_B u_{n'} - i_{(n-1)'} - G_2 u_n. \quad (5)$$

6. For the 1st circuit: $u_{L_1} + u_{R_1} + u_{C_1} \pm u_{M_{12}} = e_1(t)$,

where $u_{L_1} = L_1 \frac{di_1}{dt}$; $u_{R_1} = R_1 i_1$; $u_{M_{12}} = M_{12}\frac{di_{1'}}{dt}$,

we'll get

$$L_1 \frac{di_1}{dt} \pm M_{12}\frac{di_{1'}}{dt} = e_1(t) - R_1 i_1 - u_2. \quad (6)$$

7. For the 2nd circuit: $u_{L_1} + u_{R_1} + u_3 - u_2 \pm u_{M_{12}} = 0$,

where $u_{L1} = L_1 \frac{di_2}{dt}$; $u_{R_1} = R_1 i_2$; $u_{M_{12}} = M_{12}\frac{di_{2'}}{dt}$,

we'll get

$$L_1 \frac{di_2}{dt} \pm M_{12}\frac{di_{2'}}{dt} = u_2 - u_3 - R_1 i_2. \quad (7)$$

Similar to other independent circuit.

8. For the (*n-1*) circuit: $u_{L_1} + u_{R_1} + u_n - u_{n-1} \pm u_{M_{12}} = 0$,

where $u_{L_1} = L_1 \frac{di_{n-1}}{dt}$; $u_{R_1} = R_1 i_{n-1}$; $u_{M_{12}} = M_{12}\frac{di_{(n-1)'}}{dt}$,

we'll get

$$L_1 \frac{di_{n-1}}{dt} \pm M_{12}\frac{di_{(n-1)'}}{dt} = u_{n-1} - u_n - R_1 i_{n-1}. \quad (8)$$

9. For the n-th circuit: $u_{L_H} + u_{R_H} - u_n = 0$,

where $u_{L_H} = L_H \frac{di_n}{dt}$; $u_{R_H} = R_H i_n$,

we'll get

$$L_H \frac{di_n}{dt} = u_n - R_H i_n. \quad (9)$$

10. For the 1'-th circuit: $u_{R_A} + u_{L_2} + u_{R_2} + u_{2'} \pm u_{M_{12}} = 0$,

where $u_{L_2} = L_2 \frac{di_1}{dt}$; $u_{R_2} = R_2 i_1$; $u_{R_A} = R_A i_1$; $u_{M_{12}} = M_{12}\frac{di_1}{dt}$,

we'll get

$$L_2 \frac{di_{1'}}{dt} \pm M_{12}\frac{di_1}{dt} = -R_A i_1 - u_{2'} - R_2 i_1. \quad (10)$$

11. For the (n-1) 'th circuit: $u_{L_2} + u_{R_2} + u_{n'} - u_{(n-1)'} \pm u_{M_{12}} = 0$,

where $u_{L_2} = L_2 \frac{di_{(n-1)'}}{dt}$; $u_{R_2} = R_2 i_{(n-1)'}$; $u_{M_{12}} = M_{12}\frac{di_{n-1}}{dt}$,

we'll get

$$L_2 \frac{di_{(n-1)'}}{dt} \pm M_{12}\frac{di_{n-1}}{dt} = u_{(n-1)'} - u_{n'} - R_2 i_{(n-1)}. \quad (11)$$

For the second variant, when the branch of the tree (0-1') of a normal graph (Figure 2) contains the source $e_2(t)$, i.e. when the switching apparatus SD_1 is opened and the $SD_2$is closed, the equations are drawn up in the same way (1)–(11). Then equations (1)–(11) are written in matrix form:

$$H\frac{dx}{dt} = K \cdot x + S \cdot E \quad \text{or} \quad \begin{bmatrix} C & 0 \\ 0 & L \end{bmatrix}\frac{dx}{dt} = \begin{bmatrix} G & F_{12} \\ F_{21} & R \end{bmatrix} \cdot x + S \cdot E, \quad (12)$$

where $X = [u_2 u_3 \ldots u_n u_{2'} u_{3'} \ldots u_{n'} \;\; i_1 i_2 \ldots i_{n-1} i_n i_{1'} i_{2'} \ldots i_{(n-1)'}]^T$,

$$C=\begin{bmatrix} C_1+C_{12} & 0 & \cdots & 0 & -C_{12} & 0 & \cdots & 0 \\ 0 & C_1+C_{12} & \cdots & 0 & 0 & -C_{12} & \cdots & 0 \\ \vdots & \vdots & \ddots & \vdots & \vdots & \vdots & \ddots & \vdots \\ 0 & 0 & \cdots & C_1+C_{12} & 0 & 0 & \cdots & -C_{12} \\ -C_{12} & 0 & \cdots & 0 & C_2+C_{12} & 0 & \cdots & 0 \\ 0 & -C_{12} & \cdots & 0 & 0 & C_2+C_{12} & \cdots & 0 \\ \vdots & \vdots & \ddots & \vdots & \vdots & \vdots & \ddots & \vdots \\ 0 & 0 & \cdots & -C_{12} & 0 & 0 & \cdots & C_2+C_{12} \end{bmatrix};$$

$$G = diag\{-G_1 - G_1 \ldots -G_1 - G_2 - G_2 \ldots -G_2 - G_B\},$$
$$E = [e_1(t)\; e_2(t)]^T,$$

$$L=\begin{bmatrix} L_1 & 0 & \cdots & 0 & 0 & \pm M_{12} & 0 & \cdots & 0 \\ 0 & L_1 & \cdots & 0 & 0 & 0 & \pm M_{12} & \cdots & 0 \\ \vdots & \vdots & \ddots & \vdots & \vdots & \vdots & \vdots & \ddots & \vdots \\ 0 & 0 & \cdots & L_1 & 0 & 0 & 0 & \cdots & \pm M_{12} \\ 0 & 0 & \cdots & 0 & L_H & 0 & 0 & \cdots & 0 \\ \pm M_{12} & 0 & \cdots & 0 & 0 & L_2 & 0 & \cdots & 0 \\ 0 & \pm M_{12} & \cdots & 0 & 0 & 0 & L_2 & \cdots & 0 \\ \vdots & \vdots & \ddots & \vdots & \vdots & \vdots & \vdots & \ddots & \vdots \\ 0 & 0 & \cdots & \pm M_{12} & 0 & 0 & 0 & \cdots & L_2 \end{bmatrix},$$

$$F_{12}=\begin{bmatrix} 1 & -1 & \cdots & 0 & 0 & 0 & 0 & \cdots & 0 \\ 0 & 1 & \cdots & 0 & 0 & 0 & 0 & \cdots & 0 \\ \vdots & \vdots & \ddots & \vdots & \vdots & \vdots & \vdots & \ddots & \vdots \\ 0 & 0 & \cdots & 1 & -1 & 0 & 0 & \ddots & 0 \\ 0 & 0 & \cdots & 0 & 0 & 1 & -1 & \cdots & 0 \\ 0 & 0 & \cdots & 0 & 0 & 0 & 1 & \cdots & 0 \\ \vdots & \vdots & \ddots & \vdots & \vdots & \vdots & \vdots & \ddots & \vdots \\ 0 & 0 & \cdots & 0 & 0 & 0 & 0 & \cdots & 1 \end{bmatrix}, \quad F_{21}=\begin{bmatrix} -1 & 0 & \cdots & 0 & 0 & 0 & \cdots & 0 \\ 1 & -1 & \cdots & 0 & 0 & 0 & \cdots & 0 \\ \vdots & \vdots & \ddots & \vdots & \vdots & \vdots & \ddots & \vdots \\ 0 & 0 & \cdots & -1 & 0 & 0 & \cdots & 0 \\ 0 & 0 & \cdots & 1 & 0 & 0 & \cdots & 0 \\ 0 & 0 & \cdots & 0 & -1 & 1 & \cdots & 0 \\ 0 & 0 & \cdots & 0 & 1 & -1 & \cdots & 0 \\ \vdots & \vdots & \ddots & \vdots & \vdots & \vdots & \ddots & \vdots \\ 0 & 0 & \cdots & 0 & 0 & 0 & \cdots & -1 \end{bmatrix}$$

For the first variant, when the branch of a tree (0-1') of a normal graph (Figure 2) contains R_A, i.e. SD_1 is closed and SD_2 is opened, the subtitle of R and S are written as follows:

$$R = diag\{-R_1 - R_1 \ldots -R_1 - R_H - R_2 - R_A - R_2 \ldots -R_2\},$$
$$S = \begin{bmatrix} 0\,0 \ldots 0\,0\,0 \ldots 0\,1\,0 \ldots 0\,0\,0\,0 \ldots 0 \\ 0\,0 \ldots 0\,0\,0 \ldots 0\,0\,0 \ldots 0\,0\,0\,0 \ldots 0 \end{bmatrix}^T.$$

For numerical integration, write (12) in Cauchy form:

$$\frac{dx}{dt} = A \cdot x + B \cdot E \quad \text{or} \quad \frac{dx}{dt} = f\big(x, t, e(t)\big), x(0) = x_0, \tag{13}$$

where $A = H^{-1} \cdot K;\; B = H^{-1} \cdot S;\; f(x, t, e(t))$ – m-dimensional vector function;
x_0 – m-a dimensional vector of the initial values of the search values.

3. Results of the simulation of mutual electromagnetic influences of related power transmission lines in transition mode of the MATLAB system

The numerical integration of the obtained differential equations in the Cauchy form is carried out by the Runge-Kutta method of the fourth degree (fourth order of accuracy at one integration step) [6].

Results of line simulation in the MATLAB system are shown in graphs (Figure 3).

Consider a line with one $e_1(t)$ source with parameters: U = 220 kV; R_1 = 0.0127 Om/km; R_2 = 0.0127 Om/km; L_1 = 0.9337 mHn/km; L_2 = 0.9337 mHn/km; G_1 = 0 Cm/km; G_2 = 0 Cm/km; C_1 = 12 nF/km; C_2 = 12 nF/km; C_{12} = 6 nF/km; f = 60 Hz.

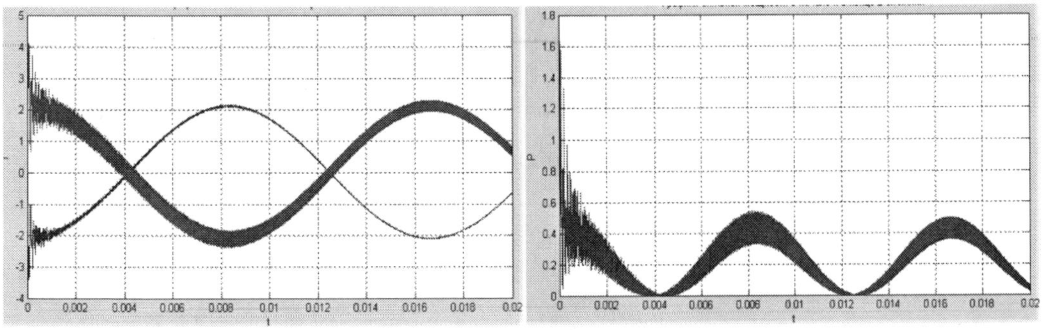

Figure 3. Current and active power graphs at the beginning and end of the 2nd line at $R_A = 0$, $R_B = 0$.

4. Conclusion

The resulting mathematical model makes it possible to model electromagnetic processes in the transmission (both homogeneous and heterogeneous) taking into account mutually inductive and capacitive couplings in transition and fixed modes, As a result, a number of important scientific and practical challenges related to the design and operation of EEC can be addressed.

The model also allows for a variety of situations and tasks related, for example, to the calculation of systems modes at breakdowns of lines, SC, or single-phase SC in isolated neutrals. In particular, it is possible to calculate the steered voltage of the line from high-voltage or high-precision power lines.

Despite the complication of power electrical equipment by the introduction of an additional converter, the effectiveness and feasibility of this solution were experimentally confirmed.

On the basis of this method, it is further proposed to develop an algorithm for the machine formation of mathematical models for the study of transition processes in complex electrical networks, where the initial data will be their parameters and the structure of graphs of the investigated networks.

References

[1] Ghnimi S, Rajhi A and Gharsallah A 2016 New Experimental Investigation of Magnetic and Electric Fields in the Vicinity of High-Voltage Power Lines *Journal of Magnetics* vol **21** pp 102–8

[2] Garcia-Sanchez J L, Moreno P, Gutierrez-Robles J A, Loo-Yau J R, Sotelo J, Bauelos E S and Galvan V A 2016 Aerial line model for power system electromagnetic transients simulation *Generation Transmission & Distribution IET* vol **10** pp 1597–604

[3] Munteanu C, Topa V, Muresan T and Costin A M 2009 Electromagnetic Interferences between HV Power Lines and RBS Antennas Mounted on HV Tower *Proc. of the 6th International Symposium on Electromagnetic Compatibility* (Eindhoven, Holland) pp 878–81

[4] Liu J, Ruan W, Fortin S and Dawalibi F P 2002 Electromagnetic fields near high voltage electrical power lines: a parametric analysis *Power System Technology Proceedings. PowerCon 2002. International Conference* pp 401–8

[5] Asanova S M, Niyazova G N and Satarkulov K 2015 Modelirovaniye elektromagnitnykh vzaimodeystviy smezhnykh LEP na osnove metoda peremennykh sostoyaniy *Problemy upravleniya i informatsionnykh tekhnologiy. Mater. 2-y Mezhdunar. konf.* (Bishkek: IAIT), pp 231–4

[6] Demirchyan K S and Butyrin P A 1988 *Modelirovaniye i mashinnyy raschet elektricheskikh tsepey* (Moscow: Vysshaya shkola) p 335

IOP Conf. Series: Materials Science and Engineering **966** (2020) 012114 doi:10.1088/1757-899X/966/1/012114

Method for designing drop-of-wire recognition systems on sections of undistorted two-wire power transmission lines

S M Asanova[1, 5], J S Ahyoev[2], N Askarbek[1], S M Suerkulov[1], D U Asanova[3] and M Kh Safaraliev[4]

[1] Department of Theoretical bases of Electrical Engineering and General Electrical Engineering, Kyrgyz State Technical University named after I Razzakov, Bishkek, 720044, Kyrgyzstan
[2] Department of Power Electric Stations, Tajik Technical University named after academic M S Osimi, Dushanbe, 734042, Tajikistan
[3] Department of Renewable Energy Sources, Kyrgyz State Technical University named after I Razzakov, Bishkek, 720044, Kyrgyzstan
[4] Department of Automated Electrical Systems, Ural Federal University named after the First President of Russia B N Yeltsin, 19, Mira str., Yekaterinburg, 620000, Russia

E-mail: [5] a_sm07@mail.ru

Abstract. In this work a general statement of the problem of designing distributive electrical systems that recognize breakage of wires on sections of power lines using a multifunctional wire with an isolated residential wire is formulated. The use of overhead wires with the proposed construction of the housing will make it possible to solve many electrical problems very easily and cheaply without the use of expensive equipment. The correspondence table of the information current values of the power supply unit I_0 and of the wire breakdowns on the sections in the two-wire transmission line is obtained by simulation. The proposed method of diagnostics and monitoring of the state of power lines is not expensive and allows the maintenance staff, being at the substation, to detect rapidly and remotely the breakage of wires on the parts of the network and the distance to it, In order to ensure that the repair crew is sent to the exact location of the damage, this will improve the reliability of the electrical supply to consumers. This is achieved by the fact that the information wire has an independent power supply and thus allows for control of the network at any time.

1. Introduction

Long-distance overhead circuit are the least reliable part of the power grid. The statistics show that in the networks of its voltage the annual amount of damage is calculated by the hundreds, and in subscription networks it feeds by the thousands. Most of the damage to the air lines consists of short circuits and cliffs. Both natural and artificial conditions can be the cause. At the same time, fault location (FL) and restoring damaged parts of the power grid are the most complex and lengthy technological operations.

This problem is particularly acute for 6–35 kV power lines. Difficulties arise from the use of insulated neutrality in electrical distribution networks, as well as their branched tree topology. Therefore, the diagnostic methods successfully used for 110 kV and higher lines with a closed ground neutral are not applicable.

The main requirement under FL in electrical networks is to restore power to disconnected power collectors as quickly as possible while minimizing labor, time and search costs. There is therefore a need for an effective system of rapid diagnosis and continuous monitoring from the standstill of the electrical networks, coupled with timely preventive repair of the lines, which will reduce the number of accidents and the time of the FL and therefore, a reduction in financial losses due to the lack of electricity to consumers.

The analysis of information from the Soviet-Russian and foreign technical literature showed that the problem of rapid diagnosis of the state of distribution electrical networks in Russia, in the CIS countries and abroad has not been completely solved [1–4]. The variety of types and characteristics of damage has not yet made it possible to find any universal diagnostic method that is equally successful in detecting short circuits and breakage of power lines and other faults. The lack of advanced remote FL techniques using the latest advances in science and technology makes it difficult to detect quickly damage to transmission lines. Therefore, the development of an effective method for diagnosing and monitoring the status of electrical distribution lines by voltage is relevant.

The need to develop systems for detecting power breakdowns through difficult and inaccessible mountain ranges is heightened by the difficulty of visually monitoring breakdowns.

In this work:

- The generalized problem of designing distributive electrical systems that recognize breakage of wires in sections of complex power lines using a multifunctional wire with an isolated residential wire is formulated [5].
- The special problem of detection of breakage of cables for long-distance two-wire transmission lines is considered.

2. Formulation of the generalized problem of designing distributed electrical systems that recognize breakage of wires in sections of complex power lines using a multifunctional wire

A multifunctional wire with an isolated living area. In work [5] a multifunctional wire with an isolated housing (Figure 1) was proposed, the use of which makes it possible to:

- to detect breakages in power lines;
- to discover the location of the closure in the line;
- to detect the appearance of an alternating electric arc;
- determine the load value (electric current) of the line without its rupture and analyse the electrical energy transmission mode using a new method of measuring electric current [6];
- if necessary, it is easy to determine the current line resistance value and the wire temperature, and to assess the contact conditions at the connection point of the wires to the line, etc., i.e. the use of wires and cables with the proposed construction of the housing will make it very easy and cheap to solve many electrical problems without the use of expensive equipment.

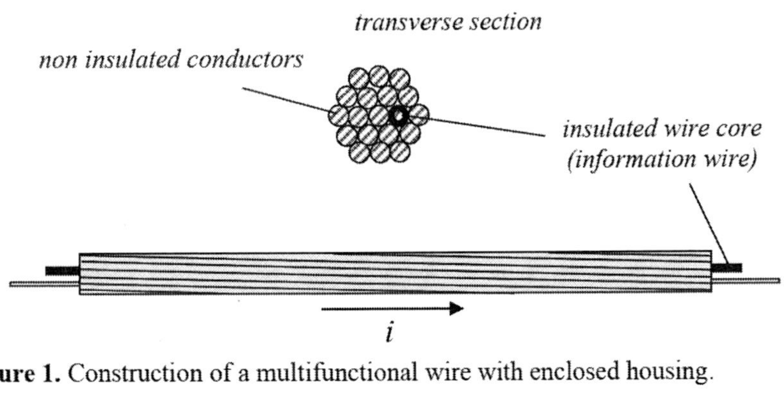

Figure 1. Construction of a multifunctional wire with enclosed housing.

IOP Conf. Series: Materials Science and Engineering **966** (2020) 012114 doi:10.1088/1757-899X/966/1/012114

In this work, a multifunctional wire consisting of two components (Figure 1):

- a power wire for the transmission of electrical energy from power supply units to load units;
- a cable housing insulated from a power wire is used for the realization of electrical distribution systems (Figure 2) which recognize breakage of wires on sections of transmission lines because the wire has that property, when the power line is broken, the information is cut off.

The distribution electrical network structure is the initial information for the design of electrical distribution systems that recognize breakage of cables in the power lines. Let the electrical distribution network analyzed consist of a plurality of n-wire sections connected by corresponding units and have an open structure. The state of each section of the network (i_1, i_2), connecting the nodes $i_1$and i_2, is characterized by a variable $x_{i_1,i_2} \in \{0, I_{i_1,i_2}\}$, where $x_{i_1,i_2} = 0$– cut; $x_{i_1,i_2} = I_{i_1,i_2}$ – there is no drop; I_{i_1,i_2} – current of the section of the network (i_1, i_2).

The aim of the design is to develop such schemes for connecting the information wires of the network sections and for connecting the information wires in the power supply units to the DC voltage U_0 (cf. Figure 2), at which the values of the information currents $I_0^1, I_0^2, \dots, I_0^n$, which are functions of the state x_{i_1,i_2} of all sectionsi_1, i_2 of the network, would unambiguously reflect the breakage of the cables of the respective sections.

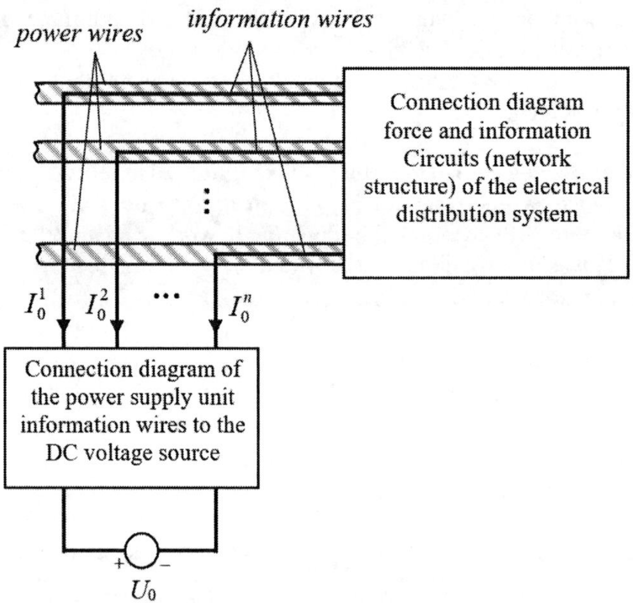

Figure 2. Electrical distribution system structure.

3. Method for detecting wire breaches in two-wire transmission lines

Figure 3 a shows the connection of the power and information cables of the two-wire sections of the power lines, and Figure 3b is a principle scheme of substitution of information wires. Between the phases of the information wires, at a certain distance, adjustable resistances are installed, through which the electric current flows. Between the phases of the information wires, at a certain distance, adjustable resistances $R_1, R_2, \dots, R_k$ are installed, through which the electric current flows ΔI. In the event of a break in the wire in some area i_1, i_2 of the network, the information current in the power supply unit I_0 will be reduced by an appropriate amount, which will make it possible to quickly and remotely determine the part of the network where the break occurred.

Algorithm for calculating network parameters.
At the given values: DC voltage $-U_0$, V; wire lengths – l, km; internal resistance on sections of the information wire $-R_{01} = R_{12} =\ldots= R_{k-1,k}$, Om/km; currents flowing through adjustable resistors – $I_1 = I_2 =\ldots= I_k = \Delta I$, mA, need to determine the value of adjustable resistances – in order to ensure that the currents flowing through them are the same and equal ΔI, thus making it possible to uniquely determine the breakdowns of the wires in the parts of the transmission lines.

The information current of the power I_0 and current unit on each section of the line $I_{12}, I_{23},\ldots, I_{k-1,k}$ according to Kirchhoff's Law I will be equal to:

$$I_0 = \Delta I + \Delta I +\ldots+ \Delta I = k \cdot \Delta I,$$

$$I_{12} = I_0 - \Delta I, I_{23} = I_{12} - \Delta I, \ldots I_{k-1,k} = I_{k-1} - \Delta I.$$

Make up the Kirchhof Law II equation for the 1st contour: $2R_{01}I_0 + R_1\Delta I = U_0$, hence

$$R_1 = \frac{U_0 - 2R_{01}I_0}{\Delta I},$$

for the 2nd contour: $2R_{k-1,k}I_{k-1,k} + R_k\Delta I - R_{k-1}\Delta I = 0$, hence

$$R_2 = \frac{R_1\Delta I - 2R_{12}I_{12}}{\Delta I},$$

etc., for k-th contour: $2R_{k-1,k}I_{k-1,k} + R_k\Delta I - R_{k-1}\Delta I = 0$, hence

$$R_k = \frac{R_{k-1}\Delta I - 2R_{k-1,k}I_{k-1,k}}{\Delta I}.$$

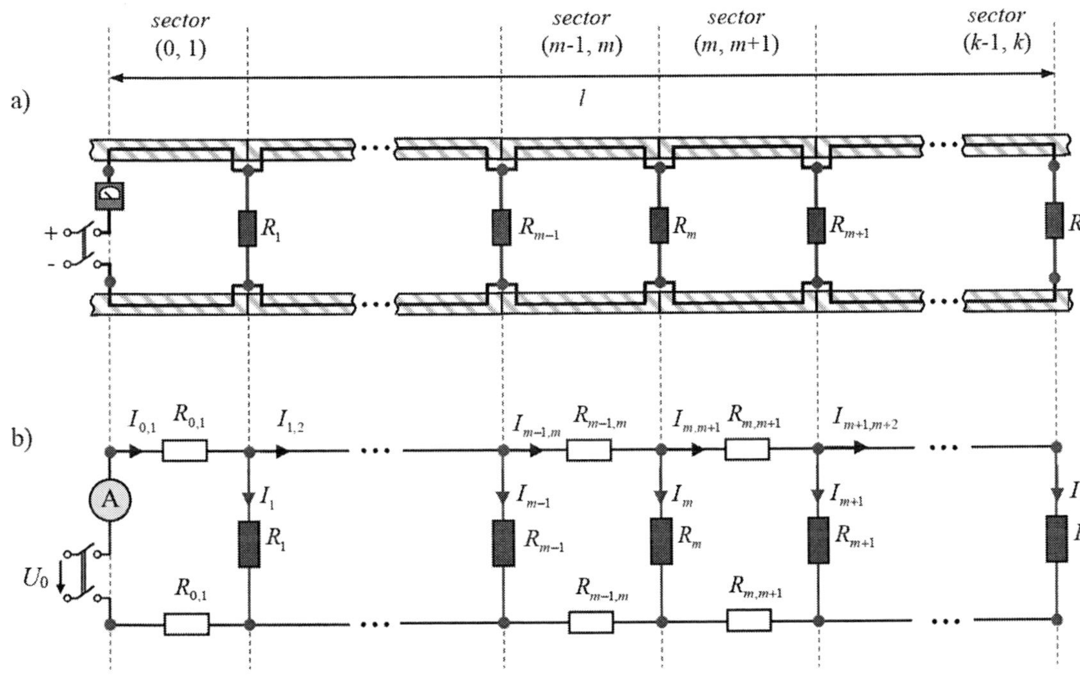

Figure 3. Connection diagram of power and information wires of two-wire sections of power lines (a), principle diagram of connection of information wires (b).

4. Simulation of the process of detection of breakage of wires at sites in a two-wire transmission line

Using the example of a two-wire transmission line with a length of 5 km, with parameters $U_0 = 220$ V; $R_{01} = R_{12} = \ldots = R_{k-1,k} = 13$ Om/km, by simulating the breakage process (the breakage of the wire in Figure 4 is simulated by switching devices), the correspondence table of the information current of the power unit $I_0$and the breakup of the wire in Sections 1–5 is obtained.

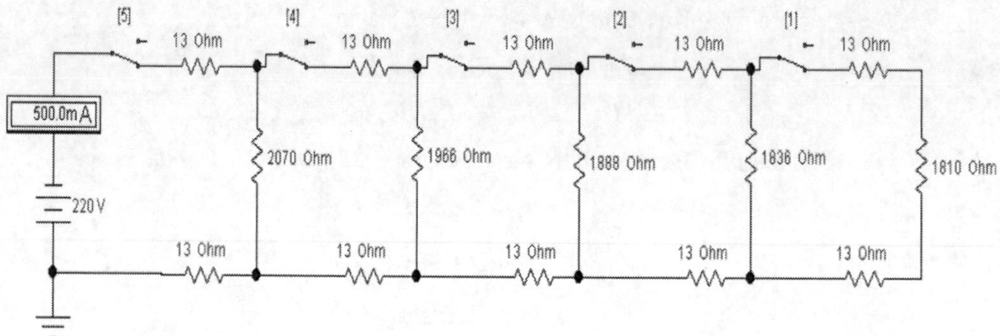

Figure 4. Simulation of the winding process in two-wire power transmission lines.

Modeling was done using Electronics Workbench. It can be seen from the table that at the current value of $I_0 = 412$ mA the breakage occurred at the site 1, at the current value of $I_0 = 315$ mA the breakage at the site 2, etc.

Table 1. Key correlation of the power station information current value I_0 and the wire breakage value of the network areas.

	No break	Cliff on the site				
		1	2	3	4	5
Milliammeter reading	500	412	315	211	105	0

5. Conclusion

In that work:

- defines the general definition of the task of designing distributive electrical systems that recognize breakage of wires on sections of power lines;
- the special problem of designing a distribution electrical system that detects breakage of wires in sections in a two-wire transmission line is considered;
- the results of the simulation are a table of the information current values of the power unit I_0 and the lines of the sections in the two-wire transmission line.

Thus, an efficient method for diagnosing and monitoring the state of electrical distribution lines has been developed, which is low-cost and allows service personnel to be at the substation, Rapidly and remotely detect the breakage of wires in the power lines and the distance to it, so that the repair crew can be directed exactly to the damaged area, thus improving the reliability of the power supply to consumers.

Besides, at «fan» disconnections, which takes place in our republic when the power wire is out, in rural areas wires of the network are often stolen, and the use of this method of detection of breakdowns of the network sections, will allow for quick reaction and prevention of such theft, because the information wire has an independent power supply and thus allows to control the network at any time.

IOP Conf. Series: Materials Science and Engineering **966** (2020) 012114 doi:10.1088/1757-899X/966/1/012114

It should be noted that the increase in the number of diagnosed sites, as can be seen from the above methodology, increases the value of the information current I_0, which may exceed the permissible value of Imax. To overcome this problem, the breakage detection system should be split into separate stand-alone subsystems, and the information should be transferred to the dispatching office via digital communication channels.

Further development. The considered method of detection of breakage of wires in sections in two-wire transmission lines can also be generalized for complex n-wire branched distribution electrical systems.

References

[1] Salehi M and Namdari F 2018 Fault location on branched networks using mathematical morphology *Generation Transmission & Distribution IET* vol **12** pp 207–16

[2] Ananthan S N, Padmanabhan R, Meyur R, Mallikarjuna B, Reddy M J and Mohanta D K 2016 Real-time fault analysis of transmission lines using wavelet multi-resolution analysis based frequency-domain approach *Science Measurement & Technology IET* vol **10** pp 693–703

[3] Bains T P, Sidhu T S, Xu Z, Voloh I and Dadash Zadeh M R 2018 Impedance-Based Fault Location Algorithm for Ground Faults in Series-Capacitor-Compensated Transmission Lines *IEEE Transactions on Power Delivery* vol **33** pp 189–99

[4] Minulin R G, Petrushenko Y Y and Fardiyev I S 2007 Lokatsionnaya diagnostika linii elektroperedach. *IX Simpozium 'Elektrotekhnika 2030' Rossiyskaya Federatsiya*

[5] Satarkulov K A, Asanova S M and et al 2009 *Tokoprovodyashchaya zhila provodov i kabeley* Patent KG no 1187

[6] Satarkulov K A and et al 2008 *Sposob izmereniya elektricheskogo toka v provodakh liniy elektroperedach* Patent KG no 596

IOP Conf. Series: Materials Science and Engineering **966** (2020) 012115 doi:10.1088/1757-899X/966/1/012115

Automation of movement control for modern main tank T-90

M A Badrtdinov[1] and D V Yudincev[2, 3]

[1] JSC 'Ural Design Bureau of Transport Engineering', 38, Vostochnoye Chaussee, Nizhny Tagil, 622007, Russia
[2] Ural Federal University named after the First President of Russia B N Yeltsin, Nizhniy Tagil Technological Institute, 59, Krasnogvardeyskaya str., Nizhniy Tagil, 622000, Russia

E-mail: [3] yudin_dv@mail.ru

Abstract. The article discusses the option of creating an automated system for controlling the movement of a modern main T-90 tank. In order to emphasize the importance of such modernization, we turn to the history of creating domestic robotic complexes. Among them are known 'Alisa' on the base of T-72 tank, intended for experimental development of technologies of remote and autonomous control of objects of armored vehicles, or robot-sapper 'Prohod-1' on the base of armored demining machine BMR-3M 'Vepr'. The analysis revealed the absence of a robotic complex based on the main modern tank. With the purpose of finding the ways of creating a modern unmanned model, a patent search was conducted and foreign military equipment samples were examined. The result is a proposal of modernization of the main tank T-90, which involves equipping the serial chassis with a special apparatus that ensures remote control of the model and autonomous solution of individual problems with the model. The proposed crew-free control system comprises an on-board computer, a plate with proportional electromagnetic valves, motor reduction gears, various sensors, and also the replacement of standard driver-driven controls with electronic ones.

1. Relevance of robotic complexes (RC) introduction

The problem of armaments robotization, military and special equipment has become one of the most important aspects of the development of armed forces and the means of their provision. Robotization is a package of measures aimed at mastering unmanned military and special technologies in order to reduce the military forces' need of manpower, reduce combat losses and increase the effectiveness of combat actions and special operations. Among the problems which take robotics to do are the creation of systems and means capable of carrying out autonomous operations and solving individual combat and special tasks; the protection of human beings from the effects of adverse factors in combat conditions; reduction in the labour intensity of military and special equipment maintenance; reduction in the cost of service personnel training.

The automation of armaments control, military and special equipment implies the creation of remote control systems and the gradual transition from the simplest forms of remote control to the level of interactive supervising control. Requirements for the power drive of such automated robotic complexes: to ensure economical movement with low resistance to straight motion; to move under conditions of increased resistance to movement; hidden movement mode when the propulsion system is switched off.

Almost all developed countries are involved in establishing robotic complexes. The Ground Vehicle Systems Center (GVSC) of the United States Army has demonstrated the models of the promising machine based on the M2 Bradley [1]. Milrem Robotics (Estonia) is developing a robotic combat vehicle (RCV) with armament, similar to modern Infantry Fighting Vehicle [2]. DOG-ING (Croatia) has developed a number of remote demining and fire extinguishing machines [3]. IAI (Israel) offers the ground robotic platform ELTA's Robattle, which allows remote reconnaissance, surveillance and target detection [4]. The MAARS complex (USA) is equipped with both small arms and antitank weapons [5].

Russia can offer such RC as the specialized remote controlled robot 'Uran-6' (Figure 1), 'Platforma-M', 'Nerekhta' and other [6].

Figure 1. Robotic Demining Complex 'Uran-6'.

However, the establishment of such RC is time-consuming and resource-intensive, as it implies solution of a number of problems. These include, for example, the development of an own tracked chassis for each particular robot, and a rather weak defense against the weapons. To solve this problem, it is necessary to pay attention to existing serially produced tracked items.

The robotization of the serially produced tracked items consists in the installation of an additional apparatus set, which forms a remote control system and makes it possible to realize the unmanned sample. The crew may be accommodated in a remote control centre up to several kilometres. In this way, a modernized serial item has a number of advantages over specialized complexes:

–robotized product can be controlled either in crew mode (standard control mode) or in unmanned remote or autonomous modes;

–robotized product that is almost identical on the outside to the crew nonrobotized product, which expands the tactical possibilities of its use;

–the output time and cost of the new product are significantly reduce due to the use of serially produced chassis and the use of unified remote control equipment for a number of base chassis;

–possibility of upgrading the existing military and special equipment into unmanned products that meet the requirements of different departments;

–performance of combat tasks in an automatic mode increases the efficiency of the equipment use.

2. Examples of tank-based robotic complexes

RC 'Alisa' on the base of T-72 tank (Figure 2) was created inBauman Moscow State Technical University [7] for experimental development of remote and autonomous control for armoured weapons and vehicle facilities. The sub-system of the lower level of control, the system of technical vision, the navigation system and different modes of movement remote control and weapons have been successfully tested on this complex in the full-scale conditions. Currently, RC 'Alisa' is being used for further experimentation with hardware and software, which provide a fully autonomous movement control mode.

Figure 2. RC 'Alisa'.

Further, it is advisable to consider the controlled demining RC 'Prohod-1' (Figure 3) created in JSC 'High precision systems' scientific production association [8]. The complex was created on the basis of BMR-3M 'Vepr' (armored demining machine). The technical capabilities of the complex allow to trawl mines installed on the ground or in snow, to neutralize mines lying on the surface as well as having radio breakers. RC 'Prohod-1' is equipped with a trawl with rollers (for contact action mines), cutters (cut wires of remotely controlled landmines) and a system of radio-electronic interference for radio-controlledlandmines. The complex is able to create passages up to 4.5 meters wide in combat conditions for military columns on mined areas both with crew and unmanned (remote and automatic) modes of operation.

Figure 3. Controlled robotic demining complex 'Prohod-1'.

The automatic control mode ensures the execution of the combat mission along the specified route without the intervention of the operator. The peculiarity of RC 'Prohod-1' is the presence of three modes of work: with the crew inside, under operator's control (remotely), and also in automatic mode with predetermined parameters of movement (without operator's intervention).

3. Implementation option

As mentioned earlier, it is most advantageous to use a serially produced tracked chassis and to provide it with a set of specialized equipment which forms a crew-free control system, provides remote control and autonomous solving of individual problems.

The task can be broken down into two main development parts:

1. System of unmanned remote control.
2. System of autonomous solution of individual tasks.

The system of unmanned remote control of the sample (Figure 4) implies installation in a serial sample of such equipment as to enable the control of movement under any conditions from a safe distance and preclude the mandatory presence of the crew inside the machine. At the same time, it is still possible to control the machine with the aid of classical controls at the driver-mechanic's place. In general, such equipment can include various servo drives, electromagnetic valves, motor reduction gears, sensors, on-board computers, etc.

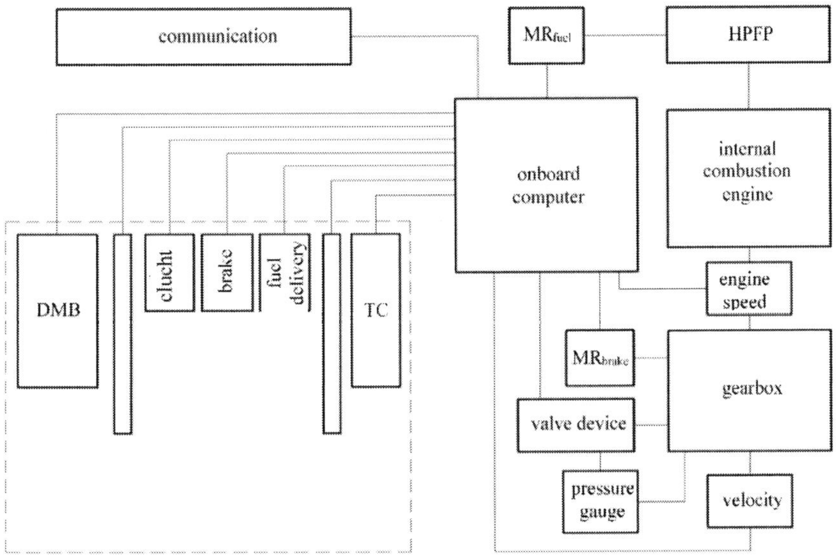

Figure 4. Implementation option of unmanned control system.

In the proposed variant for implementing theunmanned control system, the standard driver-mechanic position is maintained, but the controls are replaced by electronic control units: driver mechanics board (DMB), turn levers (can be used instead of steering levers), electronic clutch, brake and fuel delivery pedals, electronic Transmission Constituent (TC).

The signals from the controls and sensors (engine speed and actual speed of the vehicle) are received in the on-board computer, where all necessary calculations are made. Furthermore, the on-board computer generates control signals for the motor-reduction gear of the high-pressure fuel pump (HPFP), the motor-reduction gear of the brake (MR$_{brake}$) and the valve device of the on-board gearbox (the valve device is a plate with proportional electromagnetic valves).

The feature of this modernization is the presence of two gearboxes and therefore, the need to install two valves. In this case the possibility of realization of such transmission mode as «in-place turn» that under the standard control system it was impossible to execute. It should be noted that the absence of the 'in-place turn' mode was a disadvantage of the chosen chassis, especially when considering the use of the item in modern combat conditions, which are often carried out in populated areas and in cramped conditions.

The remote control transmits signals similar to the signals from the driver's control point or the traffic route information to the on-board computer. Then in the on-board computer this information is transformed into control signals for the actuating mechanisms and the machine moves along a predetermined route.

4. Conclusion

An analysis of the existing variants showed the absence of RC based on the modern main tank.

In this paper scheme for the execution of an automated system which provides for remote control of the model and also autonomous solution of individual problems with the model are proposed.

The selected scheme will be used to develop design documentation for the modernization of the products based on T-90 tank.

IOP Conf. Series: Materials Science and Engineering **966** (2020) 012115 doi:10.1088/1757-899X/966/1/012115

References

[1] DP Staff Writer. 2020 U.S. Army Selects QinetiQ, Textron for Robotic Combat Vehicle (RCV) Programs *DefPost.* Available online: https://defpost.com/u-s-army-selects-qinetiq-textron-for-robotic-combat-vehicle-rcv-programs/, free access (accessed on 18.05.2020)

[2] Milrem Robotics HQ [Electronic resource] *Milrem AS.* Available online: https://milremrobotics.com/defence/ (accessed on 18.05.2020)

[3] MV-4 Tracked, easily maneuverable, remotely controlled, light demining system [Electronic resource] *2013 DOK-ING.* Available online: https://www.dok-ing.hr/ (accessed on 18.05.2020)

[4] RoBattle Combat & Combat Support *IAI 2017.* Available online: https://www.iai.co.il/p/robattle (accessed on 18.05.2020)

[5] MAARS Modular Advanced Armed Robotic System 2020 *QinetiQ North America* Available online: https://qinetiq-na.com/products/unmanned-systems/maars/ (accessed on 18.05.2020)

[6] Lopota A V and Nikolaev A B 2015 *Nazemnye robototexnicheskie kompleksy voennogo i specialnogo naznacheniya* (Saint Petersburg: GNCz RF CzNII robototexniki i texnicheskoj kibernetiki) p 30

[7] Korchak V Yu, Rubczov I V and Ryabov A V 2013 Sostoyanie i perspektivy razvitiya na zemnih robototexnicheskih kompleksov voennogo I specialnogo naznacheniya *Inzhenerny zhurnal: nauka i innovacii* vol **3.** Available online: http://engjournal.ru/catalog/pribor/robot/628.html (accessed on 18.05.2020)

[8] *Robotic demining complex 'Prohod-1' AO 'NPO 'High precision weapons'.* Available online: https://www.npovk.ru/produktsiya/kompleksy-vooruzheniya-bronetankovoy-tekhniki-boevye-otdeleniya-legkobronirovannoy-tekhniki/prokhod-1/ (accessed on 18.05.2020)

IOP Conf. Series: Materials Science and Engineering **966** (2020) 012116 doi:10.1088/1757-899X/966/1/012116

Using of polymeric materials to defeat of lightly armored and easily vulnerable targets. Theory and experiments

E A Khmelnikov[1], K V Smagin[1] and T E Zavodova[1, 2]

[1] Ural Federal University named after the First President of Russia B N Yeltsin, Nizhniy Tagil Technological Institute, 59, Krasnogvardeyskaya str., Nizhniy Tagil, 622000, Russia

E-mail: [2] tatiana.zavodova@urfu.ru

Abstract. Small caliber artillery plays an important role in the destruction of lightly armored and easily vulnerable equipment. Currently, a search is underway to increase its damaging effect. The article deals with the possibility of increasing the damaging effects of small-caliber artillery ammunition in the fuel tanks of aircraft and helicopters. The physicochemical properties of fluoropolymer materials are considered. The process of thermodynamic decomposition of a fluoropolymer is considered. The results of experiments on the interaction of fluoroplastic and aluminum / titanium alloys are analyzed. Pictures of changes in the titanium wet structure under the impact of ammunition including fluoroplastic are shown. The Ballistic Stand installation was designed to evaluate the incendiary effect of fluoropolymer materials. Several possible designs of projectiles containing fluoroplastic are selected to evaluate their structural characteristics for incendiary effect. A comparison of the results of the calculation of the probability of the incendiary effect of strikers based on fluoropolymers on diesel fuel was made.

1. Introduction

In modern combat, small-caliber barrel artillery is highly effective in fighting both with the weakly protected and lightly armored vehicles. These groups of equipment include ground equipment (infantry fighting vehicles, armored personnel carriers, and their varieties), as well as various types of aircraft. In connection with the recent active use of assault and tactical aviation and helicopters against enemy vehicles, the development and improvement of small-caliber barrel artillery ammunition has become particularly necessary. The effectiveness of the impact on aviation of small-caliber ammunition is the sum of the total number of hits that ultimately lead to the disruption of the life support systems and the functioning of the aircraft.

As a result of studies of susceptibility of various types of combat aircraft, 4 main reasons were identified that led to the non-return of aircraft to their bases as a result of opposition from the enemy: a fire or explosion of an aircraft, damage of its power plant, loss of controllability, explosion of ammunition.

From the physical factors affecting important functional units of airplanes and helicopters, an incendiary effect is of great interest.

The main reason for the ignition of fuel in the fuel tanks is that the hull of the aircraft carcass is made of alloys containing aluminum, titanium, and a number of other fairly flammable materials. Upon penetration of the hull, which acts as a kind of screen, a flare of hot dispersed metal is formed.

IOP Conf. Series: Materials Science and Engineering **966** (2020) 012116 doi:10.1088/1757-899X/966/1/012116

When a bullet or shrapnel hits a fuel tank, in the space between the hull and the damaged tank wall, the torch of particles and the fuel mixture contact, that with a certain probability causes the ignition of fuel depending on atmospheric pressure, ambient temperature, fuel vapor pressure, mixture concentration and oxygen in the mixture and ambient air. However, the defeat of this type occurs only in the absence of internal complex protection, having a multilayer structure, including foam plastic and latex spongy rubber.

Consequently, the task of creating a new type of small-caliber ammunition, possessing not only the necessary specific energy for breaking through an obstacle, but also the subsequent incendiary over-obstacle effect, which can lead to the destruction of an aviation target, becomes urgent. One of the possible ways to increase the over-obstacle incendiary effect is the use of ammunition and striking elements of reaction materials containing fluoropolymers in their structures.

In the case of studying the interaction of fluoropolymer damaging elements and lightly armored aviation targets, it is reasonable to use a synthetic approach that will allow us to estimate not only the depth and diameter of cavities on the aircraft's flaps and helicopter blades, but also to evaluate the physically complex processes of penetration of fuel tanks with subsequent ignition of the fuel mixture.

2. The main properties of fluoropolymers

It was experimentally proved that fluoropolymers (PTFE) have a lower specific kinetic energy spent on the destruction of an obstacle compared with non-deformed steels and textolite impactors.

Therefore, as the material for the creation of ammunition that can cause incendiary over-the-shoulder action, the fluoroplastic was chosen. PTFE (fluoroplastic F-4, F-4D or Teflon), has a number of unique physicochemical, antifriction and anti-adhesive properties [1-3]. It is inert to aggressive media (up to oxidizing), non-toxic under normal conditions, dielectric, heat-resistant and explosion-proof.

PTFE is one of the most stable polymers, but it is thermodynamically unstable. It is known [2–4] that at high temperatures thermo-oxidative destruction of fluoroplastic occurs (which can be prevented by the introduction of special stabilizers) with the formation of hydrogen fluoride, perfluorisobutylene, carbon monoxide, fluoroplastic aerosols. Decay to graphite and tetrafluoromethane proceeds with the release of heat (113 kJ).

The main product of the pyrolysis of PTFE in the temperature range up to 600 °C is a heavy gas molecule - tetrafluoroethylene monomer (TFE) containing impurities. According to [3], with the explosion of gaseous TFE, which is at an initial pressure of 10 ... 15 atm., and room temperature, the pressure of decomposition products in a closed volume can increase by 8 ... 10 times. At an initial temperature of 80 °C, TFE becomes explosive when a pressure of 6 atm. is reached (0.6 MPa). It was established experimentally that deflagration combustion of TFE can also go into detonation mode. It is also noted that the probability of explosion (explosive combustion) and the transition of combustion to detonation increases sharply when air is introduced into the TFE.

3. Experiments for determination of the destructive ability of fluoropolymers

When conducting laboratory studies of high-speed interaction with aluminum-containing barriers of impactors made of various materials, the effect of the high destructive ability of impactors made of PTFE relative to other polymeric materials and steel impactors was revealed [7, 8]. The experiments were carried out on a ballistic stand, which allows throwing impactors with a caliber of 13, 23, 33 mm, a length of 1 to 12 calibers with speeds of 300 ... 1500 m/s. Sheets and plates made of aluminum alloys were used as barriers: Al-Mn, Al-Mg, duralumin, multilayer bags of these materials; elements of the airframe and helicopter blades, as well as materials: steel, wood, textolite.

When conducting firing tests on an obstacle from Al-Mn with impactors made of polyethylene, ebonite, textolite and fluoroplastic (with a fixed initial kinetic energy of the impactors), in [7] caverns were obtained, which significantly differ in their parameters. The largest parameters of the cavities in diameter, depth and volume corresponded to the fluoroplastic strikers. In addition, experimental studies have shown that at speeds of impact of PTFE strikers of about 600 m/s and higher, a black deposit of carbonized products was formed on the surface of the cavity and around it in a radius of up to 6...8 calibers of a striker. In order to determine their chemical composition, black particles collected from the cavity with a dissecting needle were subjected to x-ray phase analysis, which showed that the particles under study consist of aluminum metal and aluminum fluoride AlF3. The black color of these particles is caused by the presence of soot on them.

X-ray phase analysis of the chemical composition of the products of interaction between a fluoroplastic striker and an Al-Mn obstacle showed that the particles studied were a mixture of metallic aluminum, carbon black and aluminum fluoride AlF_3. This suggests that with the impact interaction of the fluoroplastic striker with an aluminum-containing obstacle between the firing material and the friction material, an exothermic chemical reaction may occur.

4. Experimental confirmation of the thermochemical reaction hypothesis

To confirm the hypothesis of the thermochemical reaction between the fluoroplastic and other light alloys, experiments on the interaction of fluoroplastic strikers with obstacles of titanium alloys, VT1-0 and VT-20, were conducted. Analysis of the results of the experiments shows that during the impact of strikers containing fluoroplast with obstacles based on titanium alloys, a characteristic black scurf of interaction products is formed on the surface, similar to that formed when interacting with an aluminum-containing obstacle. This allows us to assume the occurrence of a chemical reaction between the fluoroplastic and titanium obstacles once again.

X-ray phase analysis of the products chemical composition of the interaction of the fluoroplastic and obstacles from titanium alloy showed that the layers, identified in the region of the cavity and not having a diffusion layer with the base material of the sample, consist of titanium and fluorine. On the layers there were found particles having a spherical shape of different sizes (1-15 microns), which contain iron, titanium, oxygen, fluorine, and as impurities - aluminum, sulfur, chlorine. Compounds of fluorine with titanium TiF_2, TiF_3, TiF_4 were not found. This is explained by the fact that titanium fluorides, in contrast to aluminum fluorides [9], are volatile compounds. Conducting experiments in a closed volume made it possible to fix not only the fact of an increase in the cavity and the presence of soot on its surface, but also the presence of titanium fluorides. It is possible to estimate the amount of heat generation in the process of interaction between a striker and an obstacle by the colors of the tread of the face layers in the contact zone - their heating temperature was about 500 °C.

A change in the microstructure of the titanium alloy after dynamic loading was also investigated. The study of the microstructure was performed on an NEOPHOT-21 optical microscope in reflected light at magnifications: x200 and x1000. The microstructure of the studied samples in areas at a distance of about 40 mm from the zone of destruction is shown in Figure 1. It can be seen that it consists of deformed, strongly twinned grains (such a structure is characteristic of hot-rolled plates of the VT1-0 alloy). In the zone located near the zone of destruction, the grains are deformed (see Figure 1b) and have a fibrous appearance. The greatest interest are cracks. Beginning on the surface, the cracks go deep into the metal and break off, without tapering. From the tops of the cracks, narrow bands with a recrystallized fine-grained structure follow (see Figure 1c). A similar structure also surrounds the crack.

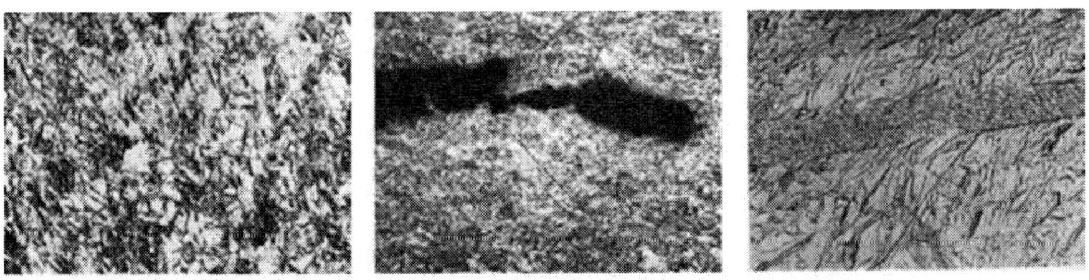

a b c

Figure 1. The microstructure of the investigated templates:
a - initial state x200; b - crack x200; c - strip x1000.

5. Use of fluoropolymers as materials for small-caliber artillery ammunition

Studies conducted under various experimental conditions show a high heat release, which has necessitated the testing of assumptions about the high incendiary ability of fluoropolymers based strikers.

When considering the physical meaning of this process, it should be noted that this is possible due to the presence of a thermo-oxidative destructive reaction of fluoropolymers in high-speed interaction with light alloys based on titanium and aluminum [14]. The pressure arising at the interface between the obstacles and the striking element exceeds the minimum pressure required for an exothermic explosion-like reaction, accompanied by the release of aluminum or titanium fluoride [15]. In the process of impact, such striking elements are deformed and destroyed, which, together with the explosion-like reaction, leads to an increase in the holes and multifactorial (thermobaric and high explosive-incendiary) action in the over-obstacle space. When dispersed clouds consisting of fluoroplastic and titanium or aluminum particles reach the back side, it expands instantly, that increases the diameter of the cavity, the edges of which are bent to the outside, forming a star-shaped form in the case of thin layered obstacles. High temperature and the resulting small condensed products disrupt the electronic devices of the instrument compartments of the target.

Considering the penetration process of any firing striker (losing mass in the penetration process or deforming) into an obstacle, it can be noted that the layers of the obstacle's material and the deforming striker move in parallel, and an explosive reaction takes place based on the thermal-oxidative processes between the striker and the material of the obstacle. The processes of parallel oxidation were considered on various types of gas mixtures and formed in the Counterflow Diffusion Flame (CDF) method, first proposed by Tsuji and Yamaoka [16].

The results obtained in the course of experiments [17] confirm the mathematical model of the interaction of fluoroplastic strikers and obstacles based on light alloys, proposed by the authors, which makes it possible to proceed to the consideration of specific practical problems.

6. Experiments for determination of the incendiary effects of strikers

The determination of the incendiary effect of fluoroplastic ammunition on the fuel and oil-conducting system was carried out on the installation "Ballistic stand". The scheme of the experiment is shown in Figure 2.

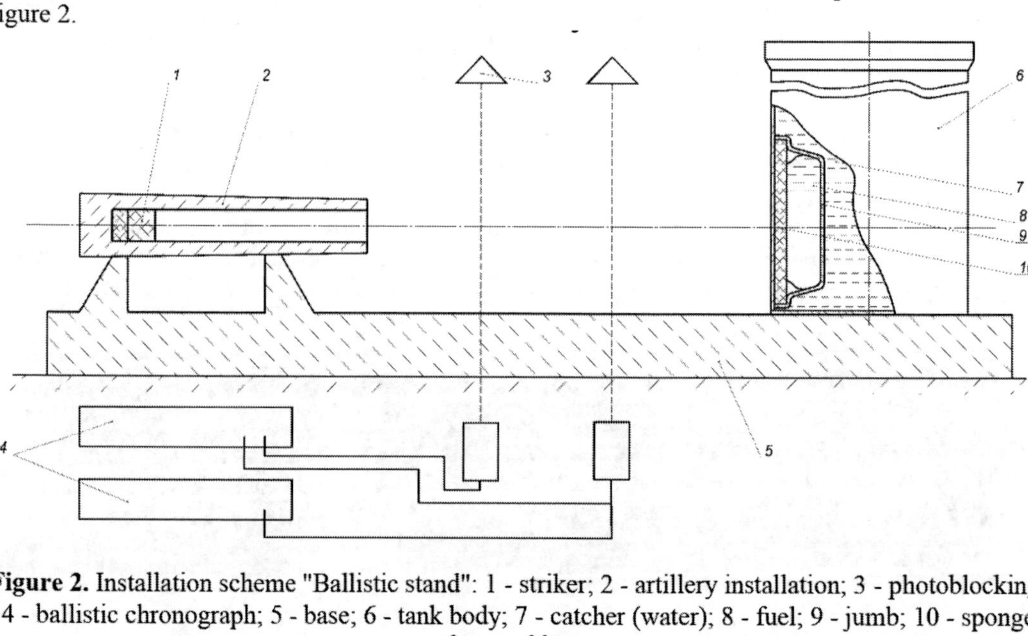

Figure 2. Installation scheme "Ballistic stand": 1 - striker; 2 - artillery installation; 3 - photoblocking; 4 - ballistic chronograph; 5 - base; 6 - tank body; 7 - catcher (water); 8 - fuel; 9 - jumb; 10 - sponge latex rubber.

A steel with a sponge latex rubber interlayer that did not react with fluoropolymers was chosen as the material of the fuel tank body, which will make it possible to estimate the pure incendiary effect that arises directly from the penetrating striker, without the influence of aluminum or aluminum-titanium shell of the air target. Winter diesel fuel (DF) with lower flammability and viscosity characteristics was taken as fuel. To assess the influence of factors (the shape of the head part, possible combinations of PTFE and aluminum) affecting the incendiary effect, experiments were carried out with strikers of various configurations.

The designs of the impactors used for the experiments are presented in Figure 3.

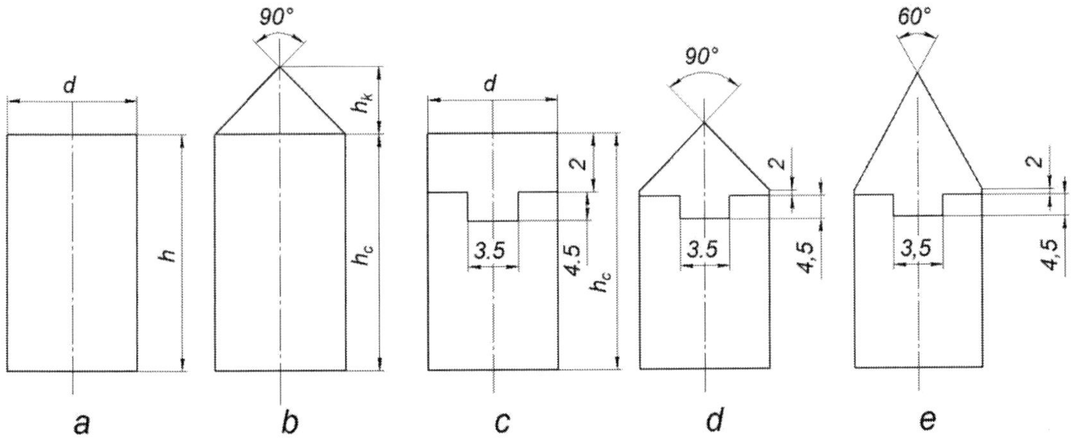

Figure 3. Striker's types: a - striker of a cylindrical shape (fluoroplastic, steel); b - striker with a tapered part and a vertex angle of 90^0 (fluoroplastic, steel); c - striker from Al and Ft with a vertex angle of 180^0; d - striker from Al and Ft with a vertex angle of 90^0; e - striker from Al and Ft with a vertex angle of 60^0.

A series of experiment results to determine the incendiary of the under-obstacle action are shown in Table 1. Analysis of the results of the shooting shows that the ignition of the fuel occurred during the testing of combined strikers from Al-Mn alloy and fluoroplastic and individual strikers from pure fluoroplastic.

Table 1. The results of experiments on the ignition of fuel for the combined obstacle

Striker's material/vertex angle, α°	Mass, *g*	Initial contact velocity, V_c, *m/s*	Result: + ignition – no ignition
Al-Mn alloy + Ft /180°	9	1006,32	+
Al-Mn alloy + Ft /60°	9	1153,91	+
Ft / 180°	9	916,34	–
Ft / 180°	9	989,17	–
Ft / 180°	9	1013,18	–
Ft / 90°	9	898,11	–
Ft / 90°	9	916,58	–
Ft / 90°	9	918,51	–
Ft / 90°	9	1128,53	+
Ft + Al-Mn alloy / 180°	9	863,12	–
Ft + Al-Mn alloy / 180°	9	1106,64	+
Ft + Al-Mn alloy / 90°	9	986,15	+
Ft + Al-Mn alloy / 90°	9	1092,88	+
Steel / 180°	9	1025,69	–
Steel / 90°	9	864,54	–

IOP Conf. Series: Materials Science and Engineering **966** (2020) 012116 doi:10.1088/1757-899X/966/1/012116

7. Conclusions

Studies have shown that under dynamic loading fluoropolymers can release enough energy to heat a certain amount of fuel to ignition temperature, which with a high degree of probability will lead to the ignition of all fuel in the tank. Thus, creating combined striker elements, including fluoroplastic, aluminum or titanium, along with an impact-penetrating effect sufficient to defeat thin-walled targets, will have a high inflammatory prohibitive effect.

Conducted studies on the ignition of one of the most resistant to ignition fuels - diesel fuel, gives the right to assume that light with high octane fuel used in aviation will ignite at lower interaction speeds. But at the same time, the speed of interaction should still be higher than the critical speed of the beginning of a chemical reaction between the fluoroplastic and aluminum or titanium.

References

[1] Pugachev A K and Roslyakov O A 1978 *Processing of fluoroplast in products: Technology and equipment* /(Chemistry) p 168

[2] Wall L A 1972 *Fluoropolymers* (Wiley-Interscience, a Division of John Wiley and Sons, Inc., New York) p 550

[3] Panshin U A, Malkevich S G and Dunaevskaya Ts S 1978 *Fluoroplastic* (Chemistry) p 230

[4] *Industrial organofluorine products* Directory 1996 (Chemistry) p 544

[5] Gorovoy V R, Bildukevich N A and Chulkov V P 1982 *Production of pyrotechnic compositions* Moscow Engineering p 296

[6] Korbakova A I, Makulova I D and Marchenko E N 1975 *Toxicology of organofluorine compounds and occupational health in their production* (Moscow: Medicine) p 183

[7] Khmelnikov E A, Styrov A V, Smagin K V, Kravchenko N S, Rudenko V L, Sokolov S S and Svidinsky A V 2016 Study of high-speed interaction processes between fluoropolymer projectiles and aluminum and titanium-based targets *International Journal of Energetic Materials and Chemical Propulsion* vol **1** pp. 167-183

[8] Styrov A V and Selivanov V V 1998 Experimental study of the interaction of aluminum with polytetrafluoroethylene under shock loading *Proc. of Intern. Conference "Shock Waves in Condensed Matter"* (Saint-Petersburg) pp 133-134

[9] Styrov A V and Selivanov V V 1999 *Chemical physics* **18 (11)** p 72

[10] Lashkov V N, Lobanov V N and Kazakovskiy N T 1998 *Proc. Intern. Conf. "Shock Waves in Condensed Matter"* (Saint-Petersburg) p 139

[11] Miller P J and Lindfors A J 1997 *Intern. Workshop on New Models and Numerical Codes for Shock Wave Processes in Condensed Media* (UK: Oxford) p 75

[12] Davis J, Lindfors A, Miller P et al. 1998 *Proc. XIth Intern. Symp. on Detonation.* (USA: Colorado) p 302

[13] Selivanov V V, Imkhovik N A, Lashkov V N and Seleznev A A 2001 *Chemical physics* **20 (8)** p 80

[14] Khmelnikov E A, Styrov A V, Smagin K V, Kravchenko N S et al. 2015 *Defence Technology* **11 (1)** p 56

[15] Khmelnikov E A, Smagin K V and Zavodova T E 2017 Application of the Methodology "Counterflow Diffusion Flames. Modeling of the Terminal Ballistic Processes of Munitions Based on Reactionary Material *Proceedings 11th International High Energy Materials Conference & Exhibits vol 1* (Pune: High Energy Materials Society of India) pp 270-276

[16] Tsuji H 1982 Counterflow diffusion flames *Progress in Energy and Combustion Science* **8 (2)** pp 93-119

[17] Khmelnikov E A, Smagin K V and Chvanov A E 2018 To the question of the damaging effect of munitions containing reactionary materials *Fundamental principles of ballistic design: VI All-Russian Scientific and Technical Conference* (Saint-Petersburg) pp 239-243

IOP Conf. Series: Materials Science and Engineering **966** (2020) 012117 doi:10.1088/1757-899X/966/1/012117

Developing manufacturing execution system with predictive analysis

V V Potekhin,[1,3] AH Bahrami[1] and B Katalinič[2]

[1] Peter the Great St. Petersburg Polytechnic University, 29, Polytechnicheskaya str. St. Petersburg, 195251, Russia

[2] Vienna University of Technology, 13, Karlsplatz, Vienna, 1040, Austria

E-mail: [3] Slava.Potekhin@gmail.com

Abstract. Digitalisation is currently developing in many sectors of the economy. The success of a manufacturing enterprise requires a transition to digital control of production and business processes, to the greatest extent possible without human intervention using artificial intelligence technologies. This research applies a Manufacturing Execution System (MES) with Predictive Analysis to an automatic production line (Chemical Line) which contains sensors, actuators and pumps controlled by four Programmable Logic Controllers, linked together, and being monitored through a Supervisory Control and Data Acquisition system. The production line composed of four different subsystems, responsible for filtration to bottling the chemical product. This paper tries to join the MES system with Artificial Neural Networks (ANN) in order to not only monitor the system but having predictive analysis to plan the future. In such way we will take advantage of the benefits of the ANN, such as Long-Short Term Memory architectures. The experimental data will be compared with other usual platforms, such as the Master SCADA itself, through the course of this research.

1. Introduction and description of manufacturing execution system

A Manufacturing Execution System is a computerised system that tracks and monitors all the elements involved in the transformation of raw materials to the final product and documents these information in order to have full reports of all elements involved in the process.

MES are becoming an essential tool for operating Small and medium-sized enterprises (SME). With the rapid increase in technology, it is almost guaranteed that paper based operating systems that documents all assets of these enterprises will be replaced by more efficient and reliable smart MES [1]. MES system can help us to make decisions for the future in order to optimise our process to achieve better results.

These results can be obtained from the production line in real time, and can be plotted using various types of software, but in order to make prediction on them for the future prospect, we can take advantage of Artificial Neural Networks to make a perfectly good regression in order to actually see our gains and losses in the future, and be able to optimise our process.

2. Aim and solved problems

The aim of this research is to create a complete MES system for FESTO chemical line that can predict and analyse the predicted data.

Here are some of the problems and tasks that will be solved by this research:

- Sketches of all the production line
- Having the List of all the equipment in one place. (this will be useful also for creating a spare part list for the maintenance team)
- Reception of signals from the Programmable Logic Controllers (PLCs)
- Programming the dynamic production line in Master SCADA
- Programming the predictive part with ANN

3. Machine learning and MES
Machine learning (ML) is a very strong tool which can be used to analyse every form of data that will be obtained from the production process. We also know that MES can perform learning to apply on a wide range of production processes. Together ML and MES can be combined to extract various data and by analysing them they can find patterns and algorithms which can be used to optimise the production [2].

In the paper presented by S Mantravadi, C Li, CMøller on Multi-agent Manufacturing Execution System [2], they point out some examples from the literature that we can apply ML for manufacturing problems: optimising the process to achieve energy efficiency; developing analytical model that can predict the performance of rescheduling strategies and quantify the trade-offs between different performance measures for manufacturing system; developing a big data analytics platform for manufacturing system to increase sustainability performance in machining operations.

ML can also be used in order to detect abnormality in the production. And in such way we can estimate the quality of the product by integration of manufacturing, inspection and after-sales service data [3].

In the same paper presented by S Mantravadi, C Li, CMøller [2], there are several studies mentioned in the usage of ML in smart manufacturing:

So we can use ML not only for a better Business plan in the future or an optimised scheduling, we also can use ML for Quality improvement. But the best advantage is when we create a comprehensive MES that first, will do all the optimization and second, will decrease the amount of paid software in the process. This will focus on what actually matters in our task. Furthermore, we will have more control in personalising the MES for our use.

4. Concept and results
The concept of our MES is shown in the figure 1.

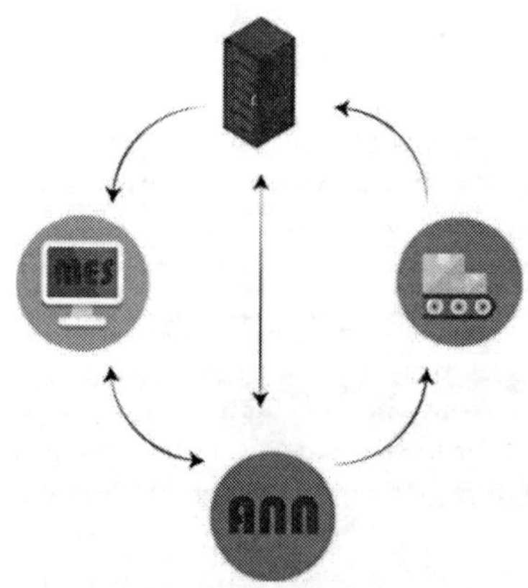

Figure 1. Concept MES.

Advantages of this implementation will be:

- No paid apps
- Narrowing the focus on what is important to us

- Therefore, it will be user-friendly
- It will lower the cost of education of the staff

5. Monitoring task

For the state monitoring task of the system, we can take in account static and dynamic diagnostic. The static diagnostic is when all the requirements are known and stored as constants over time, whereas dynamic diagnostic is when the requirements are unknown and we have to compute them through formulas or other means.

Static diagnostic

In first step we have to read the input data that define the characteristics of the system such as pressure from sensors or databases and feed it into our OPC server. Then we have to read the pre-set key performance indicators (KPI) to define the boundary for our inputs.

Then we have to access our knowledge base which contains a set of rules to make our knowledge base management system to be able to produce controlling signals in order to monitor the state of our system.

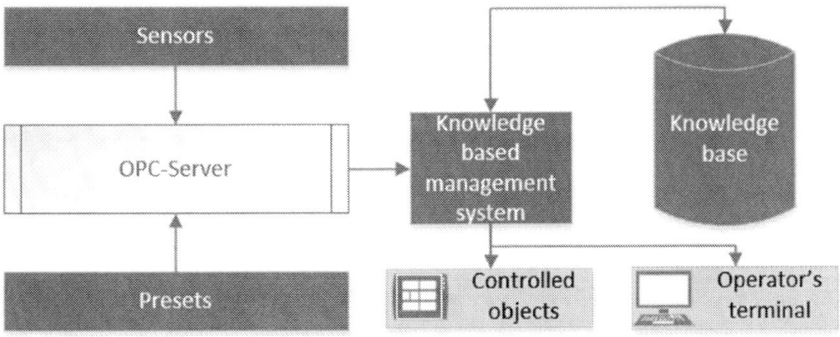

Figure 2. Static diagnostic scheme.

For example, for monitoring the state of our system based on the air pressure of the pneumatic system, we have a knowledge base which part is shown as bellow using Resource description framework (RDF)[4]:

```
:kpi1    a         classes:KPI;
rdfs:label'PressureSensor';
prop:hasInter ind:Interval2;
prop:hasInter ind:Interval3.
:Interval2 a classes:Interval;
prop:causesind:alarm;
prop:hasMin'0';
prop:hasMax'4'.
:Interval3 a classes:Interval;
prop:causesind:warning;
prop:hasMin'6.5'.
:alarm a classes:state.
:warning a classes:state.
```

The knowledge base entries for the pressure monitoring can be presented as a set of IF-THEN statements:

- If Current_Pressure<Preset_Pressure_Min (4 bar) then alarm.
- If Current_Pressure>Preset_Pressure_Max (6.5 bar) then emergency warning.

Then we need to send a query to the knowledge base to find the malfunction and monitor our system.

Dynamic diagnostic

Now for the situation that we need to change something in the production line or equipment or changing the materials, we do not have control on the changing itself. For this reason, the key performance indicators (KPI) should be computed. In order to do it, we offer to use cooperative work of following elements: digital model, execution environment, Database management system and control system block.

In order to get data from control plant we can use Distributed Control System (DCS) or Improved Control System (ICS). They are directly connected to the control plant, so they can send control signals to it. We also presume that a Protective Safety System (PSS) is also placed there.

In order to calculate KPI in situation of task, source material or equipment changing we propose to use compositional program-analytic models (CPAM), which are models of different level KPIs connection. They are actually expressions, explaining how KPI depends on another one or on control parameters.

So, data measured by sensors is sent to the database management system (DBMS). Then we should use Execution environment to run digital models (particularly CPAM) in order to compute KPI corresponding to relevant situation. Thus, execution environment also needs relevant data from control plant.

Then the execution environment computes the relevant KPI, basing on current data and CPAM. So, reference data is computed and then it should be sent back to DBMS in order to match and to check if current data meets KPI's limitations (Figure 3). It is quite similar with the previous task. After the evaluation and comparison, set points go back to DCS to provide control signals, based on modelling values. At the same time the data from execution environment is displayed with human-machine interface (HMI) to show the difference between set-point and current values. CPAM can be represented in different ways: formula-based models, neural-network model, analytical models [5].

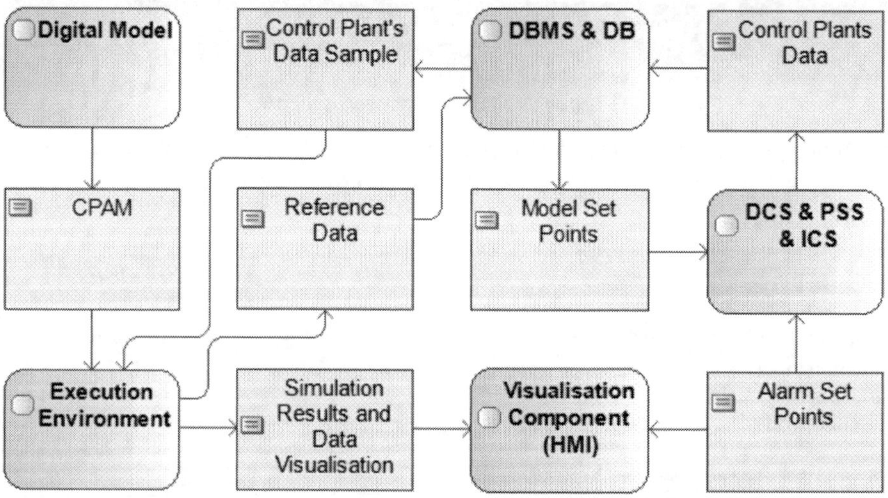

Figure 3. Functional scheme of system dynamic diagnostics.

6. Adaptive control system

The idea of development of an adaptive control system for complex object is to determine the complexity at the first step. After that we transfer complexity values into control system requirements. In this paper we describe the objects which contain the high grade of all of the complexity features. In this way, the complexity of manufacturing process can be rated due to next features, shown below.

Multi-variable or multi-connected control object. Such systems in which control is carried out to several connected variables. Moreover, all input variables have a significant effect on the state of the entire object, and their interaction is manifested in all output parameters of the object.

Let us assume that in usual strait objects, dependence between input Xn and output Yk can be described by Wnk transmission function. At the same time Y in musty-connected control object description is more complex. On the Figure4 such connection is shown.

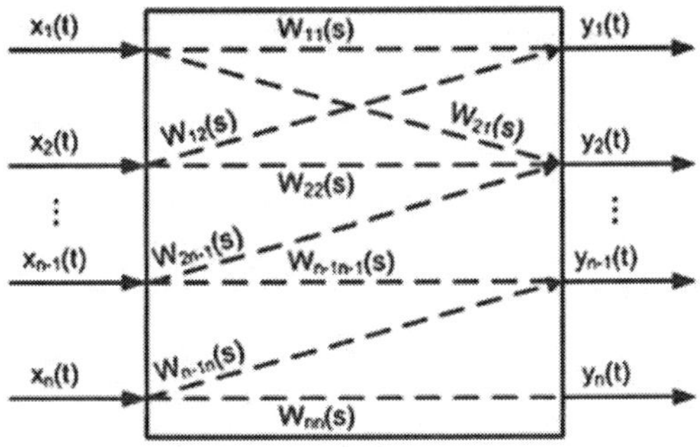

Figure4. Multi-variable control object.

Here we can see that Y2, for example depends on X1, X2, Xn-1 inputs and W12, W22, Wn-1n-1 transmission functions. Because of described features, arises the complexity of management of such systems. It appears an additional control tasks, such as:

- Negative relative value influence elimination.
- Dynamic diagnostics of an object characteristic or system state.
- Control parameters optimization, according to set points [6].

7. Computer vision in quality control

Computer vision is one of the main means for quality control in industrial production lines. The typical procedure is inspecting a product with a camera and then controlling the entire production with the result of the analysis of this inspection. The problem is that most of the time this inspection is being done by human but it is really difficult to detect if there is no algorithm for the detection, so not only doing this inspection by human can be not true, it also can be time consuming. Another problem is to achieve sufficient camera measurement for feature extraction. Once sufficient measurements and features are obtained, however, a machine vision system can replace a human observer successfully. So, we also can apply computer vision to our MES in order to come up with a system which can prevent inadvertent errors. Such as detecting defected bottles before feeding them into the system or even for packaging [7].

8. Conclusion

The concept in this paper acts like a client-server system where the main part of it is the MES. ML will be a tool to help the MES be more efficient. This system will be easier to use than other systems currently available and this allow to monitor the production as well as management. Because concept is based on ML then you can just document the results of the ANN for management and business plans.

9. Future developments

The next steps of this project are to develop the MES system for other production lines. Production lines can be far away from each other and not connected physically and the MES system can be cross-platform.

References

[1] Mantravadi S and Møller C 2019 *An Overview of Next-generation Manufacturing Execution Systems: How important is MES for Industry 4.0?* ELSEVIER vol 30 pp 588–95

[2] Mantravadi S, Li C and Møller C *Multi-agent Manufacturing Execution System (MES):*

IOP Conf. Series: Materials Science and Engineering **966** (2020) 012117 doi:10.1088/1757-899X/966/1/012117

Concept, Architecture & ML Algorithm for a Smart Factory Case Department of Materials & Production, Aalborg University, Fibigestræde 16, Aalborg, Denmark

[3] Ko T et al 2017 Machine learning-based anomaly detection via integration of manufacturing, inspection and after-sales service data *Industrial Management & Data Systems* **117 (5)** pp 927–45 doi: 10.1108/IMDS-06-2016-0195

[4] RDF 1.1 N-Triples. A line-based syntax for an RDF graph 2014 The World Wide Web Consortium (W3C). Available online: https://www.w3.org/TR/2014/REC-n-triples-20140225 accessed on 2018-07-14)

[5] Kostenko D, Kudryashov N, Maystrishin M, Onufriev V, Potekhin V and Vasiliev A 2018 Digital twin applications: diagnostics, optimisation and prediction *Proceedings of the 29th DAAAM International Symposium* B. Katalinic (Ed.) DAAAM International, (Vienna, Austria)

[6] Kudriashov N, Markov S and Potekhin V 2019 Adaptive control system synthesis methods for complex manufacturing objects *Proceedings of the 30th DAAAM International Symposium*, B Katalinic (Ed.) DAAAM International (Vienna, Austria)

[7] Kukkonen S, Kälviäinen H and Parkkinen J 2001 Color features for quality control in ceramic tile industry *Society of Photo-Optical Instrumentation Engineers* DOI: 10.1117/1.1339877

IOP Conf. Series: Materials Science and Engineering **966** (2020) 012118 doi:10.1088/1757-899X/966/1/012118

Epistemic responsibility of industrial enterprise engineer

N U Ignatova

Ural Federal University named after the First President of Russia B N Yeltsin, Nizhniy Tagil Technological Institute, 59, Krasnogvardeyskaya str, Nizhniy Tagil, 622000, Russia

E-mail: nina1316@yandex.ru

Abstract. The concept of epistemic responsibility of an industrial enterprise engineer is introduced. The concept of an engineer's epistemic responsibility is inspired by a situation of epistemic distrust in which the engineer finds himself as an expert. The definition of 'epistemic' in relation to the word 'responsibility' allows asking a question, what are the characteristics of the knowledge with which the engineer substantiates decision-making. For the first time in the domestic philosophy of engineering, epistemic duties of an engineer of an industrial enterprise are formulated: the duty of due diligence and epistemic zeal. The general epistemological attitude of an engineer in an extreme situation is skepticism. The author claims that the role of anomalous evidence in establishing the truth is an epistemologically difficult problem in investigating an extreme situation. In an extreme situation, the engineer's epistemic responsibility is embodied in the following: to prevent self-deception and potentially unintentional deception of others in relation to the engineering solution; do not mislead workers who do not have the proper epistemic experience and do not allow them to be mistaken; take into account abnormal or seemingly questionable information and consider it properly.

1. Introduction

The relevance of the stated issue is conditioned upon the following circumstances: today engineering as expert knowledge from the point of view of epistemology is perceived by researchers and the general public with distrust. The basis of this approach is the idea of cognitive deficiency of human capabilities. Epistemic distrust is based on doubts about the rational nature of professional activity, since the knowledge obtained in this way cannot be impartial and objective (H Simon). Secondly, epistemic distrust is a logical consequence of the fact that the engineer acts in conditions of the probabilistic nature of knowledge, which translates into spontaneity of the engineer's conclusions and decisions. In addition, engineers advance arguments, followed by the actions of large groups of people. In the process of making practical decisions, production managers may not accept the engineer's recommendation, focusing more on nonnegative arguments. In practice, the decision-making dilemma may come across the need for epistemic thoroughness. Expert assessment is always a matter of judgment. Judgment has both an ethical and an epistemological aspect, but it also has its own scope of application irreducible to ethics or epistemology. Ethical theory gives the general reasons for selection of actions in a particular set of circumstances.

Empirical theory explains the nature of a particular set of circumstances. But one more question remains, whether the particular circumstances resemble the general model for solving such situations and to what extent. The answer to this question cannot be given simply by verifying via source data, since the situation of the application of ethical standards is a complex issue that requires judgment.

2. Methods

The concept of 'responsibility' in the title of the article opens the space of normative and involves an appeal to ethics [1, 2]. The idea of epistemology as a normative discipline [3] allows suggesting how the available theoretical tools can be used to reflect on the validity of the engineer's actions. Intuitively, such actions are described in terms of deontological ethics: 'the engineer must'. The operator 'must' refers to modal verbs and fits into the relationship of necessity and opportunity in the spirit of traditional descriptive theories, according to which the determination of the truth conditions of normative sentences is the main task of explaining their meaning. The engineer also operates with modal operators 'perhaps', 'probably', 'obviously', when verifying problems and proposing hypotheses. The modal approach connects the meaning of the word 'must' with the idea of necessity stronger than the meaning of the words 'possibly' or 'probably'. If the verb 'must' is used as a modal operator to specify ethical requirements with prescriptive consequences (M Chrisman), then in this case the epistemic responsibility is implemented [4].

The concept of 'epistemic responsibility' was previously used in political discourse regarding the activities of scientists [5], military [6] and politicians [7]. But in relation to the professional activities of engineers, the concept is used for the first time. The definition of 'epistemic' in relation to the word 'responsibility' allows you to ask a question, what the characteristics of the knowledge with which the engineer substantiates decision-making are. Consequently, the problem of the article is stipulated by posing not ethical, but epistemological question. Epistemic responsibility is based on moral responsibility and vice versa, engineers who are charged with moral responsibility for personal actions make decisions under conditions of epistemic responsibility [8]. D Cody argues that 'applied epistemology logically precedes applied ethics. People's actions depend on beliefs both normatively and causally' [9].

The search for the basis of applied knowledge is a prerequisite for the professional competence of engineers [10]. This refers to the professional duties of an engineer to obtain and use reliable knowledge as the basis of both personal actions and expert opinions regarding the actions of other participants in the production in accordance with ideas about professional conscientiousness.

3. Results

The need for an epistemological examination of an extreme situation is recognized by all researchers. The general epistemological attitude of an engineer in an extreme situation is skepticism [11]. Skepticism involves not only a reasonable doubt about the credibility and reliability of official information or eyewitness testimonies, but a clear understanding of the limitations of human cognitive capabilities, even backed up by instrument readings. Guided by experience and general conclusions about the nature of the collection of primary information, an engineer is obliged to avoid unconditional trust in official sources or eyewitness accounts. The question arises: how rational is skepticism? One needs to understand what the conditions of skepticism are, even if the engineer seeks to be independent in their critical reasoning. A simple refusal to believe in anything undermines the possibility of full participation in the profession. Engineers investigating an extreme situation should be critical and honest in handling the evidence, but their responsibility in this case only extends to the issues within the designated area of research. The engineer must be properly circumspect, to act responsibly as necessary, as the situation and his role in it require.

It should be emphasized that the extremes of skepticism should be avoided no less than the manifestations of an absolute belief in authoritative judgment. Extremes can be avoided by applying rationally justified methods of responding to abnormal knowledge in official data and authoritative evidence. There is a number of types of responses, which may be appropriate depending on the different context. So, for example, it makes sense to identify the problem of claim to knowledge originating from areas that go beyond the professional or individual competence of the engineer. The responsibility applicable in such circumstances is to avoid speaking out on issues on which it is not customary to speak out.

An extreme situation as an absolutely unpredictable phenomenon may be accompanied by manifestations of agnosticism among engineers. Two types of agnosticism of an industrial enterprise engineer should be distinguished. Firstly, an engineer's agnosticism as a rejection of the need and obligation to know the truth indicates the lack of his professionalism. Secondly, an engineer may not be aware of some circumstances of an extreme situation due to the unreliability of evidence, which can also be interpreted as agnosticism. This interpretation prompts the formulation of the epistemic responsibilities of the engineer.

The first epistemic duty of an engineer follows from the need to avoid conclusions and evaluations based on ignorance or misinformation. For an engineer, trying to avoid a statement that goes beyond his competence does not mean just keeping silent on certain issues, but believing that silence does not essentially mean tacit acceptance of an unreasonable, contradictory and possibly false decision. Hence, the engineer is obliged to prevent self-deception and, therefore, potential unintentional deception of other employees regarding the nature of the context, regarding which the regulatory recommendation will probably be applied. At the very least, this entails caution when discussing the ethical consequences of such a decision. An assessment of some of the essential characteristics of a situation may be epistemologically unfounded.

Further, the argument we propose is that the engineer has a duty to protect others from delusion. One can meaningfully engage in assessing the epistemological basis of claims to knowledge outside their powers regarding a specific empirical examination. It is necessary not only to avoid deceiving oneself, stating that when studying the data of an extreme situation, an engineer can remain an agnostic, but also to avoid misleading or deceiving others. An engineer must ensure that people who do not have sufficient epistemic experience are not mistaken about the validity of their own actions. Moreover, under certain circumstances, the engineer is obliged to check whether other employees are misled. To do this, one does not need to bring theoretical knowledge on significant issues, but simply evaluate the statement in which claims to knowledge of the truth become public domain. The engineer is obliged to check such statements 1) for internal contradictions or methodological inconsistencies; 2) for external consistency of such statements with the corresponding theoretical knowledge. This approach cannot generate new knowledge, but it can show that some conclusions of other workers may be false. An engineer's attempts to protect subordinates or managers from false information can sometimes be seriously hindered. In some cases, correcting misinformation may not be possible, as well as a rational response to abnormal data. Hence, it becomes evident that the third epistemic duty of the engineer is not only to take into account abnormal or seemingly dubious information, but to consider it properly. An engineer's attempts to provide reliable evidence may encounter falsehood and falsification of information. In this case, their actions go beyond the scope of ordinary research or expert activity. Conventional research involves constantly checking intermediate results and correcting mistakes made. Anyone's desire to change the procedure for assessing the quality of the information provided should become the engineer's concern, as it inevitably indicates a selfish interest.

False knowledge is sometimes taken as the basis of a technological solution. In these circumstances, the engineer's epistemic obligation is not only to protect against the erroneous acceptance of a false statement, but also to correct the widespread false statement. The latter may be more challenging. First, a properly worded alternative solution may be required to correct a false statement. Despite the fact that any competent specialist is able to determine the bias, fallacy, or falsity of the prevailing statement, considerable experience and special studies may be required to develop an alternative hypothesis. Secondly, the engineer's epistemic responsibility includes readiness to be persistent, as the engineer may be faced with the desire of interested workers to actively promote false propositions.

Another duty of an engineer is not to recommend actions that cause unnecessary, unjustified and preventable harm. Assuming that an engineer will never cause harm intentionally, due diligence is to avoid spontaneous, disorganized, or unreasoned actions by others. Overly optimistic forecasts, regarding the effectiveness of the engineer-recommended action or misleading about the nature or complexity of the problem that the action should solve, should be avoided.

As a rule, engineers try not to mislead or make unreasonable assumptions. If an engineer's expert assessment involves humanitarian consequences, their findings may go beyond professional competence. It would seem that this is a sufficient reason to question the obligation of engineers to give such an assessment. Most often, engineers recognize the obligation not to go beyond their competence, to use appropriate warnings and reservations in any argument that depends on assumptions about the state of affairs, explanation or interpretation of them. Ethical considerations are discussed ceteris paribus on a hypothetical basis.

The difficulty lies in the fact that epistemic responsibility requires a view that is fuller of nuances [12], and, as A Rorty [9] suggests, is always based on ethical responsibility. At the same time, C Mitcham and R von Schomberg declare that responsibility in the framework within a professional role is not fully consistent with the reality of the numerous complex roles performed by an engineer at an industrial enterprise [13, 14]. Therefore, bringing a person to responsibility for their role is often impossible.

Sometimes, external experts and engineers are invited to investigate an emergency situation, who obviously seeks to find out the true cause of what happened. If we reflect on the nature of an ideal investigation of an extreme situation, we can conclude that it contains as many, if not more, paradoxical requirements [15]. Investigation of an extreme situation serves the public interest, but no matter how noble the goals may be, their fulfillment necessarily requires some secrecy. Internal tension arises, since in the public interest certain types of knowledge cannot be made public. Therefore, various checks and procedures are needed to prevent abuse of privacy and privileges. But although mechanisms of professional and social control are technically possible, their effectiveness may be limited in practice. The following question always arises here: whether the information obtained through deception, fraud, industrial espionage has any epistemological value, whether it can be reliable.

The idea of an 'ideal investigator' is paradoxical, since a 'perfect investigator' would be a person capable of deceiving. The problem is not that the invited experts can sometimes be double agents, but that the entire organizational infrastructure of such operations shall support convincing deception. Even if expert engineers are selected based on honesty and good faith, the problem of public deceiving of the engineering community arises. On the other hand, engineers can be truthful in confidential reports that they pass to decision makers, but who can verify that the latter are meticulous in the same way? It is obvious that the epistemic zeal of an external expert cannot guarantee full protection against those who can be set on using the opacity of the system.

An epistemologically difficult problem in investigating an extreme situation is the role of abnormal evidence in establishing the truth. Engineering practice is designed in such a way as to exclude random conclusions as unfounded, that is why common sense tells the engineer to dismiss anomalous evidence. Although the responsibility for the validity of the judgment lies with the engineer, this is not a reason to reject a priori any proposed anomalous evidence. The simple identification of an anomaly in itself is likely to be limited by the significance of the fact, since anomalies occur regularly in extreme situations. The encounter with abnormal data prompts the engineer to correct the auxiliary hypotheses without involving the main assumption. The rationality of this approach is connected with the presumption that the anomaly can be caused by some errors on its part, and not with the fundamental flaw of the verified hypothesis. It would be naive from the perspective of scientific method and rationality to assume that a simple identification of the anomaly falsifies the main hypothesis or is sufficient to refute it. Since the fact of anomalous data itself 'can never logically force a researcher to abandon a certain hypothesis, any of the beliefs from which this hypothesis follows can be wrong' [16]. This means that an engineer's reaction to anomalies in itself can be evaluated in terms of rationality.

The conceptual framework for discussing abnormal information is determined by the way in which an anomaly is answered. C Chinn and W Brewer established seven ways of response to abnormal data.

IOP Conf. Series: Materials Science and Engineering **966** (2020) 012118 doi:10.1088/1757-899X/966/1/012118

In an extreme situation, an engineer can 1) ignore such information; 2) reject it; 3) exclude anomalous data from the information that is confirmed by accepted theory; 4) keep it in mind with other things being equal; 5) rethink abnormal data in accordance with accepted theory; 6) rethink the data and make partial changes to the theory; 7) accept abnormal data and change the theory. Each option will be rational in certain situations. The rationality of a particular response in a particular situation is largely a matter of judgment.

In extreme situations, epistemically cautious utterances are rare. Knowing this, one can expect that the invited experts will strive for the necessary impartiality and methodological carefulness regarding the facts that they are ready to accept as established. In fact, non-critical perception of abnormal evidence may be present in professional discussions. There may be cases when difficult situations causing serious consequences, are evaluated without due diligence and an unreasonable, reckless conclusion is made about someone's fault. Engineers can be guided not by rational evidence, but by various kinds of addictions, personal or corporate interests. At the same time, the question of whether the generally accepted position is reliably substantiated remains open. Invited external experts provide evidence with claims to impartiality. Although their conclusions look pretty convincing to a large extent, due diligence is necessary, because the engineer, by virtue of their epistemic responsibility, should be concerned about the integrity of the sources (the problem of 'second-hand' evidence) and the reliability of the conclusions.

4. Conclusion

Is an engineer able to formulate plausible hypotheses under administrative pressure or a mercenary interest in falsifying evidence? This is much more difficult to do. Obviously, the engineer has the right to think about the circumstances of proposing alternative hypotheses or directly ask persons with the authority to distribute responsibility, whether such an activity can be considered as a debt, and whose debt it is. The first thing to establish is how the fulfillment of such an obligation would be practicable; then you need to find out to whom this applies and why. Obviously, an engineer can directly participate in the potentially complex research needed to form alternative hypotheses. An engineer can and should use due epistemic diligence in relation to the assumptions made. This implies adoption of appropriate proactive measures to ensure that the sources, on which it is based, are not false or falsified due to the influence of factors external to the extreme situation under study.

References

[1] Chomsky N 1967 The Responsibility of Intellectuals *NY rev. of books*
[2] Bellamy A 2015 The three pillars of the responsibility to protect *Pensa. Prop.* **41** pp 35–64
[3] Zagzebski LT 2012 *Epistemic Authority: a Theory of Trust, Authority, and Autonomy in Belief* (Oxford: Oxford Univ. Press)
[4] Chrisman M 2015 *The Meaning of 'Ought': Beyond Descriptivism and Expressivism in Metaethics* (Oxford: Oxford Univ. Press)
[5] Russell B 1960 The social responsibilities of scientists *Sci.* 131 **3398** pp 391–92
[6] Kornblith H 1983 Justified belief and epistemically responsible action *Phil. Rev.* **92** pp 33–48
[7] Hayward T 2019 Three duties of epistemic diligence *Jour. of Soc. Phil.* **50** (4) winter pp 536–61
[8] Rorty A 2010 Questioning moral theories *Phil.* **85** (1) p 29
[9] Coady D 2017 *Applied Epistemology in a Companion to Applied Philosophy* Ed K Lippert-Rasmussen et al (Hoboken, NJ: John Wiley and Sons). P. 58
[10] Davis M 2019 Engineering as Profession: Some Methodological Problems in its Study In: *Engineering Identities, Epistemologies and Values. Philosophy of Engineering and Technology* vol **21** Ed Christensen S, Didier C et al (Springer: Cham) pp 65–79
[11] Lakatos I 1970 Falsification and the Methodology of Scientific Research Programs In: *Criticism and the Growth of Knowledge* ed I Lakatos and A Musgrave (Cambridge: Cambridge Univ. Press) pp 91–196

IOP Conf. Series: Materials Science and Engineering **966** (2020) 012118 doi:10.1088/1757-899X/966/1/012118

[12] Nickles T 2015 Heuristic appraisal at the frontier of research In: *Heuristic Reasoning* Ed E Ippoliti (Berlin: Springer) pp 57–88

[13] Mitcham C, Von Schomberg R 2000 The ethics of engineers: from occupational role responsibility to public co-responsibility *Res. in phil. and technol.* **20** pp 167–89

[14] Mitcham C, Mackey R 2017 Comparing Approaches to the Philosophy of Engineering: Including the Linguistic Philosophical Approach In: *Philosophy and Engineering: Philosophy of Engineering and Technology* Ed I Poel and D Goldberg (Springer: Dordrecht) **2** pp 49–59

[15] Pigden C 2016 Are conspiracy theorists epistemically vicious? In: *A Companion to Applied Philosophy* Ed K L Rasmussen, K Brownlee et al (London: John Wiley & Son) pp 120–32

[16] Chinn C A, Brewer W F 1993 The role of anomalous data in knowledge acquisition: a theoretical framework and implications for science instruction *Rev. of ed. res.* **63** (1) 19

IOP Conf. Series: Materials Science and Engineering **966** (2020) 012119 doi:10.1088/1757-899X/966/1/012119

The additive technology to obtain a three-dimensional model of the 81Cu-19Ni alloy

V R Baraz[1, 3], S S Gerasimov[1], I A Gruzdeva[1], V P Shveikin[1], A V Chemesova[1], A O Ovchinnikova[1] and I A Veretennikova[2]

[1] Ural Federal University named after the First President of Russia B N Yeltsin, 19, Mira str., Yekaterinburg, 620000, Russia

[2] Institute of Engineering Science, Ural branch, the Russian Academy of Science, 34, Komsomolskaya str., Yekaterinburg, 620000, Russia

E-mail: [3]vrbaraz@mail.ru

Abstract. This paper is devoted to the selective laser smelting (SLS) method description for the article made of powder of a Cu81-Ni19 copper-nickel alloy and the study of the structure and characteristics of the three-dimensional article made using additive technology. To make a 3D model the copper-nickel alloy powder was produced in our laboratory. The chemical composition, microstructure, density, surface roughness and microhardness of an article made of powders of a Cu81-Ni19 copper-nickel alloy are considered. A comparative analysis of the structure, physical and mechanical properties of a sample obtained by the SLS method and the annealed rolled plate was carried out. The article microstructure is characterized by the absence of a typical signs of a crystalline structure and contains elements of a quasi-amorphous state. At the same time, the obtained material differs from the standard monolith billet with a lower surface roughness and a higher hardness values while maintaining almost the same density indices.

1. Introduction

At present, additive technologies are one of the promising methods for producing finished products with the required properties [1]. They are based on the effect of concentrated energy flows on powder layers deposited in accordance with the spatial-geometric configurations of the formed article. As a result, the product is created in layers, by depositing the material under the action of a laser beam, which draws the shape of each layer until the object is completely ready. This technology is practically implemented for a wide list of various products, ranging from plastics to metal materials [2–4]. Among the metals, non-ferrous metals and their alloys can be noted [5, 6]. At the same time, there is no information regarding the use of selective laser smelting (SLS) technology as applied to Cu-Ni system alloys, the use of which for industrial needs, as well as for obtaining decorative products, is an urgent task. In this work, the aim of the study was to use the SLS method to obtain a three-dimensional article made of a copper-nickel alloy powder of the Cu81-Ni19 composition.

2. Material and study methods

A Cu81-Ni19 alloy was used as the material of substrate for selective laser smelting and wire billet for the powder manufacturing. The chemical composition of the industrial alloy Cu81-Ni19 was (mass %): 19.060 Ni, 0.157 Fe; 0.245 Si; the base alloy is Cu.

IOP Conf. Series: Materials Science and Engineering **966** (2020) 012119 doi:10.1088/1757-899X/966/1/012119

To prepare the powder by flame spraying, a wire billet with a diameter of 0.8 mm was used. After thoroughly drying the powder fractions in the laboratory using an LPzE-2e shaker, the powder with a particle diameter of not more than 40 μm was screened. The manufacture of a three-dimensional sample was carried out using a 3D printing method on an industrial EOS M280 installation (laser power 200 W, laser spot diameter 100 μm, application of powder monolayers with a thickness of 20 μm). As a result, a three-dimensional sample was obtained in the form of a rectangular parallelepiped 30–30–10 mm in size (Figure 1). Sample has a simple form because our aim is to study the possibility of obtaining a dense sample made of copper-nickel alloy powder and a comparison of physical and mechanical properties. The three-dimensional sample was formed on the annealed plate basis that thickness was 4 mm. Therefore, we can compare the structure and properties of the three sections: 3D model, intermediate zone and monolith (Figure 2).

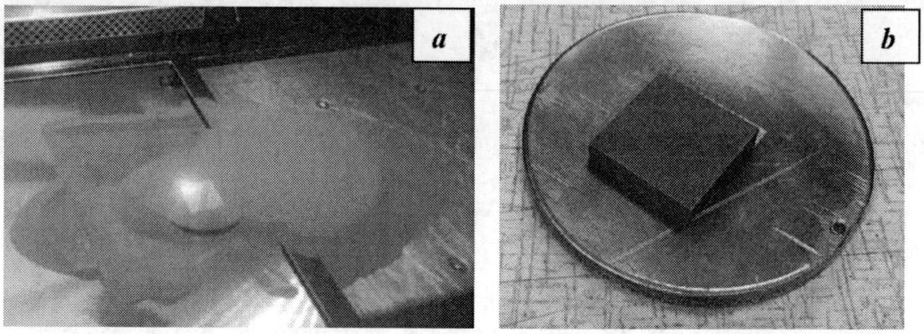

Figure 1. Obtaining a product from a copper-nickel alloy at the industrial EOS M280 installation: a – a three-dimensional sample forming step; b – finished sample.

The density ρ_0 was measured using a precision METTLER TO-LEDO balance. As the substance, distilled water and air were used. Microhardness tests were carried out by the standard method on a PMT-3 instrument with a load of 50 g (0.5 N). The surface topography study was carried out on an optical profile meter Optical Profiling System Veeco WYKO NT1100. For metallographic studies of the structure, an Epiquant automated microscope was used, as well as a ZEISS CrossBeam AURIGA scanning electron microscope. A hydrochloric acid solution of ferric chloride was used as an etchant.

3. Experimental data and discussion

For metallographic research samples microsections were prepared from both the annealed plate basis (arbitrarily designated as a monolith) and the three-dimensional sample after laser fusion (3D model).

A microstructure metallographic study of these samples (Figure 2) showed a fundamental difference in their structure. This is especially evident when using a more detailed view of the microstructure (see inserts). So, in the initial state (monolith), this alloy has a typical microstructure of annealed material with a face-centered cubic lattice. The existence of equiaxed grains (approximately 50 μm in size) with the presence of numerous wide annealing twins is observed. The sample structure subjected to the SLS method seems to be significantly different. It consists of many dispersed fragments, which have a characteristic form so-called turbulent vortex, preserving a drop-like configuration [7]. These structural fragments size is in the range of 70–100 μm.

A metallographic analysis of the three-dimensional sample obtained by laser smelting does not allow us to detect the characteristic features of a material that has undergone crystallization, there are no dendritic formations, and there are no signs that these structural objects have a grain structure with boundaries.

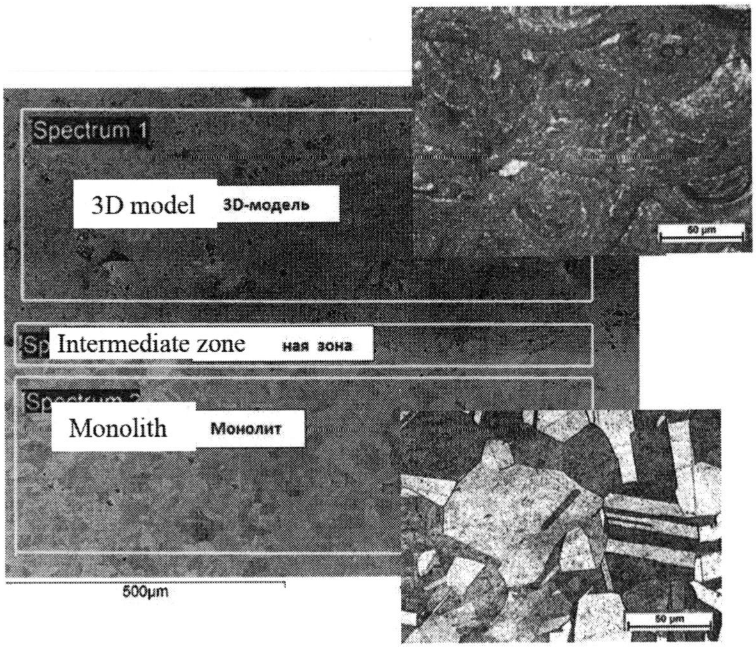

Figure 2. The Cu81-Ni19 alloy microstructure in the annealed state (monolith) and after laser smelting (3D model).

Table 1 shows the quantitative chemical analysis results of both specimens cut from the three-dimensional sample (position 1) and the monolith (position 2). The quantitative chemical analysis is obtained by energy dispersive spectrometry. As can be seen, the chemical composition is fully corresponding to the chemical composition for both technological states. A similar composition is recorded for the intermediate zone (position 3).

Table 1. The Cu81-Ni19 alloy chemical analysis (mass %) after various technological treatments.

Parameter	Ni	Cu	Total
Spectrum 1	20.01	79.99	100.00
Spectrum 2	19.10	80.90	100.00
Spectrum 3	19.60	80.40	100.00
Mean	19.57	80.43	100.00
Std. deviations	0.46	0.46	
Max	20.01	80.90	
Min	19.01	79.99	

The physical and mechanical characteristics such as density Ra, roughness Ra and microhardness HV0.5 were determined. The characteristics result of this alloy after various technological treatments are given in Table 2.

Density is an important indicator for assessing the physical property especially for a material smelted from powder fractions. As a result, the following data was obtained: the monolith density was 8.925 g/cm^3, and the smelted sample was 8.562 g/cm^3, i.e. the 3D model was formally less dense. However, this difference is not fundamental - it is only 4 %. Thus, we can conclude the selective laser smelting method under consideration allows to obtain a material that is almost identical in density to standard technology. Moreover, the measured density indices are actually similar to the values of pure copper (8.890 g/cm^3) [8].

IOP Conf. Series: Materials Science and Engineering **966** (2020) 012119 doi:10.1088/1757-899X/966/1/012119

Table 2. The Cu81-Ni19 alloy physic mechanical properties for the annealed state (monolith) and after laser smelting (3D model).

Sample	Physic mechanical properties		
	Density ρ, g/cm^3	Roughness R_a, nm	Microhardness $HV_{0.5}$
Monolith	8.925±0.005	710±15	62±3
3D model	8.562±0.005	660±20	75±3

The roughness parameters (relief profile) for the sample's surfaces obtained by various technologies have a certain difference: the sample obtained by laser smelted is characterized by relatively better 'smoothness'. A study of strength properties shows that the Cu81-Ni19 alloy microhardness is in the initial state, i.e. after annealing, was 62 HV. At the same time, the 3D model microhardness was 19 % higher and amounted to 75 HV.

The presented data concerning the structural study of the 3D model made of Cu81-Ni19 alloy and obtained by the SLS method give grounds to state the following consideration.

Under conditions when a local laser beam heats a thin powder layer, subsequent cooling in the molten areas proceeds at such a speed that it becomes possible to significantly suppress the crystallization process and fix the strongly supercooled liquid phase, i.e. obtaining an amorphous state (metal glass). Thus, after laser smelting, the finished 3D model is largely characterized by the presence of a quasi-amorphous state. This version is indirectly supported by the measurement of physical and mechanical properties. It was shown [9] that laser-smelted material is characterized by higher hardness and, in fact, may not be inferior in density terms. It is these characteristics that distinguish the 3D model made of the investigated alloy.

4. Summary

Using the selective laser smelted method a three-dimensional model made of the Cu81-Ni19 copper-nickel alloy is obtained. Structural studies have shown that the finished 3D model is not characterized by typical signs of a crystalline structure. It has been suggested that, under these conditions, the smelting process apparently led to the realization of a structural state close to amorphous.

It was shown that the smelted 3D model practically does not differ in density from the monolithic annealed analogue, but is significantly distinguished by a lower surface roughness, as well as higher hardness values.

References

[1] Sedlaka J and Rican D 2015 Study of materials produced by powder metallurgy using classical and modern additive laser technology *Procedia Engineering* **100** pp 1232–41

[2] Herderick E D 2016 Additive manufacturing in the minerals, metals, and materials community: past, present, and exciting future *JOM* **68 3** pp 721–23

[3] Barakhtin B K, Voznyuk A V, Deev A A and Zhukov A S 2017 Structurno-mekhanicheskoe sostoyanie additivno spechennogo materiala v usloviyah goryahei plasticheskoi deformatsii [Structural and mechanical state of additively sintered material under the hot plastic deformation] *Materials Deformation and Fracture* **2** pp 8–15 [In Russian]

[4] Alyoshin N P, Murashov V V and Grigoriev M V 2016 Defekti zharoprochnikh splavov, sintezirovannikh metodom selectivnogo lasernogo splavlenia [Defects of heat-resistant alloys synthesized by selective laser smelting *Materials Science* **4** pp 34–8 [In Russian]

[5] Zito D, Carlotto A, Loggi A, Bortolamei S, Molinary A and Cristofollini I 2012 Latest developments in selective laser melting production of gold jewelry *The Santa Fe Symposium* pp 537–62

[6] Louvis E 2011 Selective laser melting o aluminium components *Journal of Materials Processing Technology* **211** pp 275–84

[7] Zhukov A S, Barakhtin B K and Voznyuk A V 2018 Primenenie additivnikh technologii

sozdaniya konstrukcionnikh materialov na baze nanocentra NITS «Kurchtovskii institut» – TSNII KM «Prometei» [The use of additive technologies for the structural materials creation based on the nanocenter 'Kurchatov Institute'] *Materials Science* **6** pp 11–15 [In Russian]

[8] Osintsev O E and Fedorov V N 2004 *Med i mednyie splavi Otechestvennie i zarubezhnie marki: Spravochnik [Copper and copper alloys Domestic and Foreign Brands: Directory]* Moscow: Engineering p 336 [In Russian]

[9] Hanzl P and Zetek M 2015 The influence of processing parameters on the mechanical properties of SLM parts *Procedia Engineering* **100** p 1405–13

IOP Conf. Series: Materials Science and Engineering **966** (2020) 012120 doi:10.1088/1757-899X/966/1/012120

Tribological tests of the improved piston mechanism of the axial piston hydraulic machine

S Stazhkov[1], A Kuzmin[1, 2], V Elchinskiy[1] and N Yakovenko[1]

[1] Baltic State Technical University "VOENMEH", 1, First Krasnoarmeyskaya str, Saint-Petersburg, 190005, Russia

E-mail: [2] wutmakat@yandex.ru

Abstract. Axial-piston hydraulic swash plate machines are widely applied in positive displacement hydraulic drives of various purposes. Such drives are highly energy-intensive and have a relatively small control range due to the presence of a large dead-band, which limits their use in reverse action systems. To reduce the dead-band, an improved axial-piston hydraulic swash plate machine was designed. The decrease of the dead-band is achieved by reducing the friction forces in the piston mechanism. This article presents the results of comparative tribological tests of a standard and improved piston mechanisms. The experimental research was conducted in terms of reducing the friction forces in the piston mechanism and their effect on the moment of starting of the hydraulic machine. It was revealed that the use of the improved piston mechanism allows to increase the useful force of the piston up to 20 % in the motor mode and to reduce the dead band by 1.6 times when switching to the pump mode.

1. Introduction

Axial-piston hydraulic swash plate machines are used in hydraulic drives, which have high requirements for energy consumption and fast response. These hydraulic machines are distinguished by their weight and dimensions characteristics, simplicity of assembly and manufacturability in comparison with axial-piston hydraulic bent-axis machines. However, the disadvantage of swash plate machines is a relatively large dead-band and, as a result, a relatively small control range.

The breakaway and transition to a stable mode of operating speed change in swash plate machines, even at idling, comes at 70...100 rad/s, while in the best examples of bent-axis machines it comes at 3...5 rad / s. Therefore, swash plate machines have a significantly lower control range compared to bent-axis machines.

These facts take place due to the features of the design scheme of these hydraulic machines. The tilt angle of the cylinder block in modern bent-axis machines reaches 45 degrees due to the connection between the drive flange and the cylinder barrel, provided by pistons. In swash plate machines, due to the presence of large transverse forces acting on the piston and as a result of high friction forces in the 'piston-cylinder' pair, the tilt angle of the swash plate, as a rule, does not exceed 20 degrees.

This leads to the fact that the force produced by the piston in the bent-axis machine in comparison with the swash plate machine is converted into a much larger torque. One of the ways to reduce the dead zone of the swash plate machine is to reduce friction forces in the piston pair.

There are modern researches that are focused on the problem of the piston mechanism of the swash plate machine [1]–[4]. However, these works describe operating at a high speed, therefore, excluding the problem of the dead-band and moment of starting.

This article presents the results of testing of improved piston mechanism on a test rig using the principles of inversion of the movement of this type of hydraulic machines units and energy recovery [5].

The improved piston [6] mechanism has a bearing displacement relative to the longitudinal axis of the piston, which allows to change the power scheme of the transverse forces and significantly reduce the total reaction, and, consequently, a value of the friction forces in the piston pair.

In addition, the displacement of the bearing allows hydrostatic bearing of the 'piston-cylinder' pair application.

This article presents the results of comparative tests of a standard piston mechanism, an improved piston mechanism with a displaced bearing without hydrostatic bearing of the piston pair, and an improved piston mechanism with a displaced bearing and hydrostatic bearing of the piston pair.

2. Test in the motor mode

For tests in the motor mode, a working fluid was pumped into the piston chamber, under pressure from the pumping station. The piston acted on the rotating wash plate, which, in turn, through the lever acted on the force sensor. This allowed to determine the moments of breakaway for standard and improved piston mechanisms at different pressure levels in the piston chamber, which corresponds to different levels of loading of the hydraulic machine, as well as to determine the reduction of friction forces in the researched piston pairs. The initial data for series of experiments is presented in table 1.

Table 1. Experiment initial data and conditions

Slipper – outer edge of the bushing distance, mm	16.4–45.6
Piston diameter, mm	20.614
Working fluid temperature range, °C	20–35
Swash plate tilt angle, °	18
Distance between the rotation axis of the cylinder block and the guiding bushing axis, mm	45
The improved piston bearing displacement from the longitudinal piston axis, mm	11.83

As the result of the tests, the reduction in the friction forces was obtained when using the improved piston mechanisms in comparison with the standard one.

The force value on the sensor is converted to the axial force value according to the following dependence:

$$F_{AX} = F_S \cdot l/(R_C \cdot \tan 18°) \quad (1)$$

F_{AX} – force, applied from the piston to the swash plate in axial direction, F_S – force, registered by the sensor, l – lever length, R_C – distance between the swash plate rotation axis and the guiding bushing axis. The friction forces reduction can be found using the following formula:

$$F_{FR} = F_P^{IM} - F_P^S \quad (2)$$

F_P^{IM} – axial force of the improved piston mechanism, F_P^{IM} – axial force of the standard piston mechanism

The results of the experiment are presented in the figure 1.

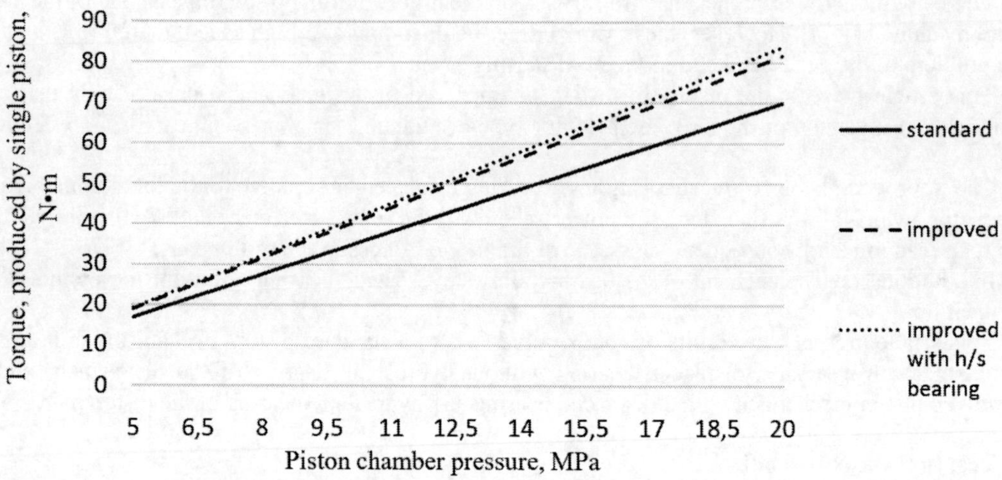

Figure 1. Dependence of the shaft torque produced by a single piston on the pressure in the piston chamber for the piston mechanism of standard design, modified design and modified design with hydrostatic bearing.

Calculations results are shown in the Figure 2.

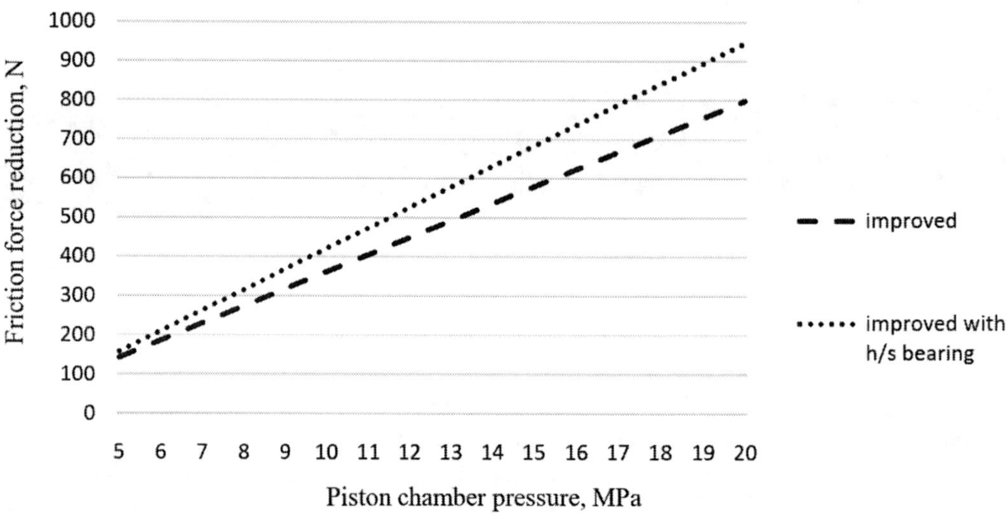

Figure 2. Reduction of the axial component of the friction forces in the motor mode for the improved piston and for the improved piston with hydrostatic bearing in comparison with the standard design piston.

3. Tests in the pump mode

To determine the friction forces in the pump mode, as well as to determine the range of the dead zone when switching to the pump mode of the hydraulic machine, a series of experiments were conducted.

On a lever of 0.3 m in length, a calibrated weight of 13 kg and 17 kg was alternately installed, which corresponded to the torque on the shaft of 38.3 and 50 N·m. The pressure in the piston chamber gradually decreased from the value allowing to keep the load on the weight, to the value at which the lever together with a support disk was moving from their seats. In this way, the piston mechanism was

IOP Conf. Series: Materials Science and Engineering **966** (2020) 012120 doi:10.1088/1757-899X/966/1/012120

switched to the pump mode of operation. The moment of switching to the pump mode was registered using the force sensor simultaneously with the pressure sensor.

According to the results obtained, dependences of changes in the friction forces in the piston pairs in the pump mode for the tested piston mechanisms were constructed. The friction force was defined as the difference between the calculated value of the pressure at which the corresponding load should be balanced, and the actual value obtained during the experiment, multiplied by the area of the end of the piston. The obtained results are shown in Figure 3.

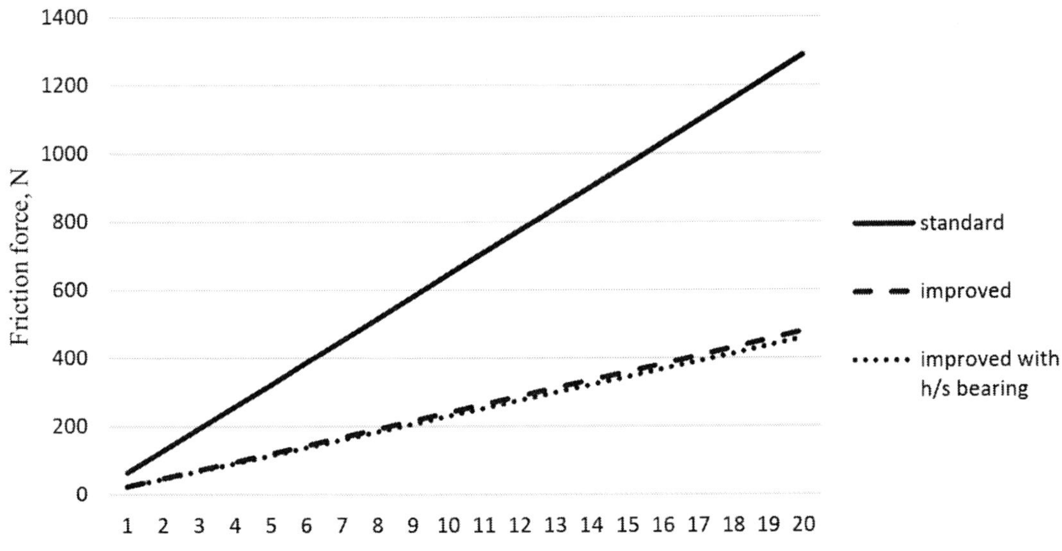

Figure 3. Dependence of the friction force in the piston pair on the pressure for the standard, improved and improved with hydrostatic bearing of piston mechanisms in the pump mode.

4. Results and conclusion

Experimental studies have shown:

- the use of the improved piston mechanism reduces the friction forces in the motor mode by 798 N at the piston chamber pressure value of 20 MPa, which leads to an increase in the useful force produced by the piston mechanism by 16.7 %;
- the use of the improved piston mechanism with hydrostatic bearing reduces the friction forces in the motor mode by 941 N at the piston chamber pressure value of 20 MPa, which leads to an increase in the useful force produced by the piston mechanism by 20 %;
- the use of the improved piston mechanism allowed to reduce the dead band by 1.6 times when switching to the pump mode and reduce the friction forces by 2.6 times at the shaft torque of 50 N·m;
- the use of the improved piston mechanism with hydrostatic bearing allowed to reduce the dead band by 1.6 times when switching to the pump mode and reduce friction forces by 2.8 times at the shaft torque of 50 N·m.

The results of the presented experiments lead to assume that the use of the improved piston mechanism in axial-piston swash plate hydraulic machines will significantly expand the use of these hydraulic machines in reverse drives.

References

[1] Matteo Pelosi and Monika Ivantysynova 2012 A Geometric Multigrid Solver for the Piston–Cylinder Interface of Axial Piston Machines *Tribology Transactions* **55:2**, pp 163-174, DOI:

10.1080/10402004.2011.639049

[2] Li F, Wang D, Lv Q, Haidak G and Zheng S 2019 Prediction on the lubrication and leakage performance of the piston–cylinder interface for axial piston pumps. *Proceedings of the Institution of Mechanical Engineers, Part C: Journal of Mechanical Engineering Science* **233(16)** pp 5887–96 doi 10.1177/0954406219851150

[3] Jiang J and K Wang 2015 An integrated model of hydrodynamic lubricating for piston/cylinder interface, International *Conference on Fluid Power and Mechatronics (FPM)* (Harbin) pp 47–52 doi: 10.1109/FPM.2015.7337083

[4] Xu B, Zhang J, Yang H and Zhang B 2013 Investigation on the Radial Micro-Motion about Piston of Axial Piston Pump. *Chinese Journal of Mechanical Engineering* **26 (2)** pp 325–33

[5] Kuzmin A, Popov V and Stazhkov S 2016 Advanced Axial Piston Swash Plate Pump Parameters Recommendations *Proceedings of the 27th DAAAM International Symposium* doi 10.2507/27th.daaam.proceedings.082

[6] Stazhkov S, Korobova I, Korolev V and Kuzmin A 2019 Hydrodynamic Processes Research of the Positive Displacement Hydraulic Machines Piston Interface *Proceedings of the 30th DAAAM International Symposium* pp 0580–5 doi 10.2507/30th.daaam.proceedings.078

IOP Conf. Series: Materials Science and Engineering **966** (2020) 012121 doi:10.1088/1757-899X/966/1/012121

Review of the features of augmented reality application in the training of operators and maintenance staff

S V Byvaltsev[1,2]

[1] Ural Federal University named after the First President of Russia B N Yeltsin, 19, Mira Street, Ekaterinburg, 620002, Russia
[2] Institute of Engineering Science, Ural Branch of the Russian Academy of Sciences, 34, Komsomolskaya Street, Ekaterinburg, 620049, Russia

E-mail: s_ikar@mail.ru

Abstract. An overview of the use of augmented reality technology used to improve the training efficiency of operators and service personnel in the industry is presented. When considering the issues of building augmented reality systems, it was noted that in addition to technological issues, it is also essential to pay increased attention to the social aspect, especially if it is intended to apply augmented reality technology in multiplayer mode. A common environment in the interaction of several participants can disrupt the sense of identity of users, as well as potentially threaten the feeling of personal space and the sensation of physical objects. It is necessary to take into account the cognitive load theory and the schematic theoretical model of concreteness fading. The advantage of using augmented reality is to reduce human errors and improve the quality of training. For the use of augmented reality in the absence of a reliable network connection and the availability of only short communication sessions, the use of special tickets is proposed.

1. Introduction

The augmented reality is penetrating more and more into industry and education every year. The Ministry of Digital Development, Communications and Mass Media of the Russian Federation has developed the Roadmap for the development of 'through' digital technology 'Technology of virtual and augmented reality' [1].

The mixing of physical and digital worlds defines the combination of real and virtual environments and it is a part of the reality/virtual continuum (RVC) [2], [3]. The RVC describes the interactions between reality and virtuality. In RVC, augmented reality (AR) is the reflection of virtual objects in the context of the real world. Modern AR solutions on portable devices use an image from a real device and combine it with virtual content. These solutions have been mass-produced for a long time as head-mounted devices, for example, Microsoft HoloLens 2 [4], Google Glass Enterprise Edition 2 [5], Meta 2 [6]. Modern AR applications also include tactile feedback and spatial sound [7], [8].

2. Features of AR application in training

The application of AR to increase the efficiency of operators and maintenance staff is an undoubted advantage, but it is necessary to develop special software products with support of AR and to teach users how to use this technology. The amount of information provided through AR should be dosed. as operators can get used to hints of system and desire to solve problems independently and even impossibility to carry out the same actions without AR application will decrease. For this purpose, it is offered to reduce gradually a stream of the new information as it is possible that really important one will be missed. In addition, it is necessary to remove the insignificant or irrelevant information out of the executed process.

It is important to remember that people have limited working memory capacity [9], [10]. There are three types of cognitive load in the learning process: 1) intrinsic cognitive load, determined by the complexity of the learning task itself, 2) extraneous cognitive load associated with conscious, constructive processes used to construct mental representations and not conducive to learning, 3) appropriate cognitive load. If the memory capacity is fully utilized in internal and external processes, there will be no remaining cognitive resources for learning and learning will be difficult. According to the Four-component Instructional Design (4C/ID) mode of learning [11], early learning tasks should include a lower intrinsic cognitive load from the interactivity of elements, and as tasks become more complex, the extraneous load should decrease when moving to more complex material.

According to the schematic theoretical model of concreteness fading [12], the process of operator training should be effectively conducted in three stages. At the first stage, training is provided on specific examples, when it is easier for the operator to interpret ambiguous abstract views in terms of well-understood specific objects, as well as to use the advantages of embodied knowledge, which gives experience with physical and perceptual processes that are limited and give the right conclusions. According to an embodied theory of cognition [13], [14], high-level cognitive processes stem from action and perception. At this stage, understanding abstract symbols requires reflecting these symbols on a bodily experience or a representation of these experiences. The concrete stage enables learners to acquire a store of images that can be used when abstract symbols are forgotten or disconnected from the underlying concept. In addition, the operator remembers a set of images, symbols and gestures that will be used in the AR metaphor when abstract symbols relate to real objects. After mastering the abstract concept in solving new problems, the operator will rely on this set of attributes [15]. At the other two stages, there is a transition from real physical representation to graphic schemes and models, and then to ordinary abstract symbols.

This point requires further research, because according to [16] the transition from abstract to concrete can sometimes improve the learning efficiency.

Thus, when training operators, AR software must adapt to the operator's accumulated experience. The transition is made from pointing out the sequence of all necessary steps for performing a specific operation to controlling the correctness of actions and fixing the very fact of execution of embedded instructions.

3. Social and technical aspects of AR application

At the building AR, the technical aspects of designing the AR environment are most often investigated. Only recently it has been noted that the social aspect of the application of AR has to be taken into account. AR users are developing a sense of interconnection and psychological belonging to virtual artifacts in AR environment develops more, than in the usual virtual environment. A common environment in the interaction of several participants may disturb the sense of identity of users, as well as potentially threaten the sense of personal space and the sense of physical objects [17], [18]. There is a certain difference in the estimation of perception when using augmented reality glasses and the hand-held device [19]. To transfer feedback to the AR system it is possible to use sensors installed on the body. However, some actions are easier to perform in the street than in a public place, because some gestures are simply not permitted there [20]. Also note that gestures and movements of the operator may not be related to the AR control process, but to other work, maintenance of the system or random movements.

Also, while using AR it is possible to feel physiological discomfort when personal space is disturbed [21].

Besides, to increase efficiency of training it is important to raise level of user's perceived trust to AR system [22] and to visualize processes with application of habitual for the operator external environment, especially at remote control mode.

As it has already been noted, in the process of training to work with AR it is necessary to move from the virtual environment to the real world, when the operator is given only the necessary dosing information and his actions are controlled and checked for correctness. It is suggested to use multi-level scenarios of data display depending on qualification and type of staff. For example, to display only the current temperature, its graph in the near future, its extrapolation to the future, or only a message when it deviates from the expected value, etc.

In the same way, in order to present data, it is required to move from a real physical representation to an abstract one.

Nowadays AR is widely used directly for the learning process itself [23], [24]. In our opinion, it is more effective to apply AR for large and complex systems, in which it is necessary to provide reliability and stability through the abstract presentation of data to the operator in real time, overlaid on the natural environment [25].

The easiest way to analyze the external environment is to use the computer vision library OpenCV [26]. To bind physical objects, we suggest using markerless framework which does not imply the use of feducian markers [27] or even natural markers in the environment [28], [29]. At the same time there is a problem of adaptation of computer vision system to changes in external illumination [30]. Most often the image of equipment is processed through a calibrated camera by ARToolKit [31] or SLAM [32] framework, which can be replaced by a more advanced GLEAM [33]. It should be noted that it is important to try to eliminate the perceptional problem of incorrect depth interpretation of graphical objects overlaid on the real world [34], [35]. This is especially shown in stereoscopic systems, where it is required to provide 'translucent vision' or challenging X-ray visualization. For hand-held devices such problem is less actual [36].

The greatest effect from the application of AR can be achieved with the provision of joint activities and work that requires coordinated efforts. In this case, for multi-user AR especially actual becomes the social aspect considered earlier, and also safety and protection against extraneous intervention [37], [38].

An additional advantage of using AR is the reuse of the training software. If there is a change of operators and maintenance staff, there is no need to teach them all the knowledge and skills at once. They receive them directly from the AR system during application. The qualification requirements for users are considerably reduced, but the requirements for the AR system (software, databases and knowledge, rules and models) increase. Also, there are fewer human errors and it is possible to control the operator's actions in real time [39].

4. The direction of further research

Certain problems can represent possibility of operation AR without access to a network, in an offline mode, but designers of AR offer to use sessions of communication and special tickets for the industry [40].

Despite the difficulties with introduction AR in the industry the advanced manufacturers of automation equipment already offer ready decisions [41], [42].

In our opinion AR should be used for complex systems where it is necessary to exclude human errors. [43], [44].

It should be noted that AR is increasingly used not only for staff training but also for controlling industrial systems [45], [46] and applications [47], [48].

Acknowledgements

The author expresses his gratitude to the staff of the Department of Information Technologies and Design Automation of Ural Federal University and the staff of the Laboratory of System Modelling of the Institute of Engineering Science, Ural Branch of the Russian Academy of Sciences without valuable advice and critical comments of which it would be impossible to publish this study.

References

[1] The Roadmap for the development of 'through' digital technology 'Technology of virtual and augmented reality' Accessed at: https://digital.gov.ru/ru/documents/6654/ (accessed 04.03.2020)
[2] Milgram P and Kishino F 1994 *IEICE Trans. Inf & Syst.* **E77-D 12** 1322-9
[3] Milgram P, Takemura H, Utsumi A and Kishino F 1994 *SPIE, Telemanipulator and Telepresence Technologies* **2351** 282–92
[4] Website of Microsoft HoloLens 2 Available at: https://www.microsoft.com/en-us/hololens/ (accessed 04.03.2020)
[5] Technical specifications of Google Glass Enterprise Edition 2 Available at: https://www.google.ru/glass/tech-specs/ (accessed 04.03.2020)
[6] The Meta 2 Development Kit Available at: https://www.schenker-tech.de/en/meta-2/ (accessed 04.03.2020)
[7] Goose S, Sudarsky S, Zhang X, and Navab N 2003 *IEEE Pervasive Computing* vol **2** issue **1** 65–70
[8] Radu I and Schneider B 2019 *Proc. of the 2019 CHI Conf. on Human Factors in Computing Systems (CHI '19)* (New York: ACM) pp 544:1-544:12
[9] Sweller J, Merrienboer J J G and Paas F W C 1998 *Educational Psychology Review* vol **10** 251-96
[10] Baddeley A D 1983 *Phil. Trans. R. Soc. Lond.* B vol **302 1110** 311–24
[11] Merrienboer J J G and Sluijsmans D M A 2009 *Educational Psychology Review* **21(1)** 55–66
[12] Fyfe E R, Neil N M, Son J Y and Goldstone R L 2014 *Educational Psychology Review* **26(1)** 9–25
[13] Barsalou L W 2003 *Language & Cognitive Processes* **18** 513–62
[14] Lakoff G and Nunez R E 2000 *Where mathematics comes from: how the embodied mind brings mathematics into being* (New York: Basic Books) p 493
[15] Bruner J S 1966 *Toward a theory of instruction* (Cambridge: Belknap) p 192
[16] Johnson A M, Reisslein J and Reisslein M 2014 *Computers & Education* **72** 249–61
[17] Poretski L, Lanir J, Arazy O and Oded N 2019 *Proc. of the 1st Workshop on Challenges Using Head-Mounted Displays in Shared and Social Spaces CHI'19 Extended Abstracts* (Glasgow, Scotland UK) p 6
[18] Poretski L, Lanir J and Arazy O 2018 *Proc. of the ACM on Human-Computer Interaction (PACMHCI)* **2 CSCW** (New York, USA) pp 142:1-24
[19] Koelle M, Kranz M and Möller A 2015 *Proc. of the 17th Int. Conf. on Human-Computer Interaction with Mobile Devices and Services MobileHCI '15* 362–72
[20] Williamson J R, Crossan A and Brewster S 2011 *Proc of the 13th Int. Conf. on Multimodal Interfaces ICMI '11* 361–8
[21] Wilcox L M, Allison R S, Elfassy S and Grelik C 2006 *ACM Transactions on Applied Perception* **3 4** 412–28
[22] Haesler S, Kim K, Bruder G and Welch G 2018 *IEEE Int. Symp. on Mixed and Augmented Reality Adjunct (ISMAR-Adjunct)* pp 204–5
[23] Manoela S, Teichrieb V and Smith P 2019 *Anais dos Workshops do VIII Congresso Brasileiro de Informática na Educação (WCBIE 2019)* pp 1452–6
[24] Fite-Georgel P 2011 *10th IEEE Int. Symp. on Mixed and Augmented Reality (ISMAR)* (Basel Switzerland) pp 201–10
[25] Masood T and Egger J 2019 *Robotics and Computer-Integrated Manufacturing* **58** 181-95
[26] OpenCV for Processing, by Greg Borenstein. A Processing library for the OpenCV computer vision library Available at: https://github.com/atduskgreg/opencv-processing (accessed

IOP Conf. Series: Materials Science and Engineering **966** (2020) 012121 doi:10.1088/1757-899X/966/1/012121

04.03.2020)
[27] Fiala M 2005 *Proc. IEEE Computer Society Conf. on Computer Vision and Pattern Recognition (CVPR)* **2** 590–6
[28] Gomes D L, de Paiva A C, Silva A C, Braz G, de Almeida J D S, de Araújo A S and Gattas M 2018 *Computers in Industry* **97** 67–75
[29] Georgel P, Schroeder P, Benhimane S, Hinterstoisser S, Appel M and Navab N 2007 *6th IEEE and ACM Int. Symp. on Mixed and Augmented Reality* (Nara, Japan: IEEE) pp 111–5
[30] Grosch T, Eble T and Mueller S 2007. *Proc. of the 2007 ACM symp. on Virtual reality software and technology* (New York: ACM) pp 125–32
[31] Kato H and Billinghurst M 2004 *3rd IEEE and ACM Int. Symp. on Mixed and Augmented Reality (ISMAR 2004)* (Arlington VA USA: IEEE) p 305
[32] Tamaazousti M, Naudet-Collette S, Gay-Bellile V, Bourgeois S, Besbes B and Dhome M 2016 *Multimedia Tools and Applications* **75** 9511-47
[33] Prakash S, Bahremand A, Nguyen L D and LiKamWa R 2019 *Proc of the 17th Annual Int. Conf. on Mobile Systems, Applications, and Services (MobiSys'19)* (New York: ACM) pp 142–54
[34] Kruijff E, Swan J, and Feiner S 2010 *Proc. of the 9th IEEE Int. Symp. on Mixed and Augmented Reality (ISMAR)* (Los Alamitos, CA: IEEE) pp 3–12
[35] Kytö M, Makinen A, Häkkinen J and Oittinen P 2013 *ACM Transactions on Applied Perception* **10 1** 6:1-21
[36] Sailer C, Rudi D, Kurzhals K and Raubal M 2019 *18th World Conf. on Mobile and Contextual Learning: Future Learning Through Experiences and Spaces (mLearn 2019)* (Delft, Netherlands) pp 69-76
[37] Ruth K, Kohno T and Roesner F 2019 *Proc. of 28th USENIX Security Symposium* (Santa Clara, CA, USA: USENIX Association) pp 141-58
[38] Lebeck K, Kohno T and Roesner F 2019 *Proc. of 20th Int. Workshop on Mobile Computing Systems and Applications (HotMobile '19)* (New York: CM) pp 81–86
[39] Introducing Vuzix M4000 Smart Glasses at CES2020 Available at: https://youtu.be/ENUCXX2fz-c/ (accessed 04.03.2020)
[40] Website of Augmented Reality on Actemium Available at: https://www.actemium.nl/themas/augmented-reality/ (accessed 04.03.2020)
[41] About EcoStruxure Augmented Operator Advisor Available at: https://www.se.com/ww/en/product-range/64507-ecostruxure%E2%84%A2-augmented-operator-advisor/ (accessed 04.03.2020)
[42] Industrial Augmented reality with EcoStruxure™ Augmented Operator Advisor Available at: https://www.youtube.com/embed/LBolUiyUKyE/ (accessed 04.03.2020)
[43] Gimeno J, Casas S, Portales C and Fernandez M 2019 *DYNA* **94(2)** 155-60
[44] Hofmann C, Staehr T, Cohen S, Stricker N, Haefner B and Lanza G 2019 *Procedia Manufacturing* **31** 148-54
[45] Antonijević M, Sučić S and Keserica H 2018 *Energies* **11(3)** 599:1-17
[46] Masood T and Egger J 2019 *Robot. and Comput. Integr. Manuf.* **58** 181-95
[47] Kyeong-Beom P, Minseok K, Sung H, Choi J and Yeol L 2020 *Robot. and Comput. Integr. Manuf.* **63** 101887:1-18
[48] Egger J and Masood T 2020 *Computers & Industrial Engineering* **140** 106-95

IOP Conf. Series: Materials Science and Engineering **966** (2020) 012122 doi:10.1088/1757-899X/966/1/012122

Multi-criteria optimization of metal pressure processes using experiment planning methods

S V Byvaltsev[1,2]

[1] Ural Federal University named after the First President of Russia B N Yeltsin, 19, Mira Street, Ekaterinburg, 620002, Russia
[2] Institute of Engineering Science, Ural Branch of the Russian Academy of Sciences, 34, Komsomolskaya Street, Ekaterinburg, 620049, Russia

E-mail: s_ikar@mail.ru

Abstract. The main task of designing a technology of metal processing by pressure is accurate prediction and ensuring the desired properties of the finished product. However, they often encounter a situation when it is impossible to simultaneously ensure the optimal value of all process parameters and the resulting properties of the finished product. In this case, it is necessary to make an informed decision regarding the assignment of certain numerical values of these parameters. This article proposes to use the mathematical apparatus of experimental planning methods to implement multi-criteria optimization of metal pressure processes. Applying of correlation analysis, function of desirability and a priori ranking by the example of optimization of parameters of the process of hydromechanical extrusion of hard-to-deform alloys is shown. Results of multi-criteria optimization of the process of hydromechanical extrusion of bar blanks from aluminum alloy AMg5 (analog of Al-5056) with the purpose of minimizing the extrusion force, deformation non-uniformity and damage index of the resulting product are presented.

1. Introduction

At designing of technology of metal processing by pressure it is necessary to deal with a situation when it is impossible to provide simultaneously optimum value of all parameters of process and obtained properties of a finished product. In this case it is necessary to make a reasonable decision concerning assignment of certain numerical values of these parameters, and also possibility of prediction of received properties of a product. One of the methods for solving this type of problems is using methods of experiment planning. Let's apply them for the solution of an existing problem of production of high-precision profiles with high purity of a surface of metals sticking on the tool at which high non-uniformity of deformation on section and length of a product [1]. To eliminate the non-uniformity of deformation it is necessary to reduce the contact friction forces on the working surfaces of the container and matrix [2], or increase the total degree of deformation [3], or increase the plasticity of the deformed material in the pressing process. [4], [5], [6].

At hydromechanical extrusion (HME), besides the pressure of the working fluid, the deformed metal is affected by additional force from the press punch [6]. For high-melting, radioactive or powder materials pressing in a flexible metal shell is used [7], [8].

To the present time the issues of hydrodynamics of the working flow, which transfers the pressure on the workpiece [9], as well as the influence of tool geometry [10], [11], have been studied quite fully. The results of the experiments are given in the papers [8], [11]–[13].

Experimental and analytical methods [14], [15], as well as simulation modeling with the use of finite element method are used to study the HME process [16].

2. Conducting research

The feature of the considered process is the coverage of the deformable part of the workpiece with a visco-plastic medium that creates high hydrostatic pressure in the cavity of the deforming matrix. The HME process provides the transition of the boundary friction into the semi-liquid friction mode with a corresponding reduction of extrusion pressure.

We solve the extreme task of searching for such a ratio of input parameters of the HME process, at which at the output of the system we get the optimal value of output parameters.

Independent variables are inputted $x_1, x_2, \dots, x_k$.

Outputs $y_1, y_2, \dots, y_m$.

Input and output parameters are related approximately by the following response functions:

$$y_1 = f_1(x_1, x_2, \dots, x_k);$$

$$y_2 = f_2(x_1, x_2, \dots, x_k);$$

$$\dots$$

$$y_m = f_m(x_1, x_2, \dots, x_k).$$

These equations form the factor space of the solved problem, which is reduced to finding the extremum of the response surface.

Let us present a mathematical model to describe the response functions as a Taylor series at any point from the determination area in the factor space. Let us limit to the polynomial of the 2nd degree.

In connection with the specificity of the task, we will construct selective estimates of y as the following regression equation based on the results of the passive experiment:

$$y = b_0 + \Sigma b_i x_i + \Sigma b_{ij} x_i x_j + \Sigma b_{ii} x_i^2 + \dots,$$

where b_0, $b_i = \dfrac{\partial y}{\partial x_i}$, $b_{ij} = \dfrac{\partial^2 y}{\partial x_i \partial x_j}$, $b_{ii} = \dfrac{\partial^2 y}{\partial x_i^2}$, … – selective estimates of regression equation coefficients found from the experiments,

i, j – indexes that meet the following proportion $1 \le i < j \le k$.

Since the multi-criteria optimization task is being solved, we use correlation analysis to establish statistical relationships between parameters.

Let's find the pair correlation coefficients $r_{y_1 y_2}$ between two random variables y_1 and y_2 using next equation:

$$r_{y_1 y_2} = \frac{\sum_{u=1}^{N} \left(y_{1_u} - \overline{y_1}\right)\left(y_{2_u} - \overline{y_2}\right)}{\sqrt{\sum_{u=1}^{N} \left(y_{1_u} - \overline{y_1}\right)^2 \left(y_{2_u} - \overline{y_2}\right)^2}},$$

where N – number of experiments;

u – experiment number;

$$\overline{y_1} = \frac{\sum_{u=1}^{N} y_{1_u}}{N}; \ \overline{y_2} = \frac{\sum_{u=1}^{N} y_{2_u}}{N}.$$

After calculation of the pair correlation coefficients $r_{y_1y_2}$ set their statistical significance on the basis of the critical value of the correlation coefficient r_{cr} by correlation coefficient distribution tables at the selected significance level α and the number of degrees of freedom $f = N - 2$. A linear connection will be considered statistically significant in the case of $|r_{y1y2}| \geq r_{cr}$.

After establishing statistically significant correlation relationships between a pair of optimization parameters y_1 and y_2 it is possible to construct the linear regression equation that allows to predict one parameter y_2 by the value of another y_1:

$$y_2 = b_0 + b_1 y_1,$$

where

$$b_0 = \frac{\sum_{u=1}^{N} y_{2_u} \sum_{u=1}^{N} y_{1_u}^2 - \sum_{u=1}^{N} y_{1_u} \sum_{u=1}^{N} y_{1_u} y_{2_u}}{N \sum_{u=1}^{N} y_{1_u}^2 - \left(\sum_{u=1}^{N} y_{1_u} \right)^2}, \ b_1 = \frac{\sum_{u=1}^{N} y_{1_u} y_{2_u} - \sum_{u=1}^{N} y_{1_u} \sum_{u=1}^{N} y_{2_u}}{N \sum_{u=1}^{N} y_{1_u}^2 - \left(\sum_{u=1}^{N} y_{1_u} \right)^2}$$

are coefficients of regression equation.

We construct the revealed linear relationships between the variables in the form of a graph, where the nodes will be the output parameters of the system y, and the edges will be the values of their statistically significant relationship $r_{y_iy_j}$. Solve the leader's task when it is required to determine the influence of the graph node and its power. Influence will be determined by the number of edges that come out of the given node, and the power depends on the influence of other nodes that are connected with the given node. Let's build an adjacency matrix based on the matrix of absolute values of correlation coefficients $\left| r_{y_1y_2} \right|$.

Find the common element of the adjacency matrix $p^i_j(k)$, defining the number of paths of length k coming from the i-th node to the j-th node.

The iterated k-th order power of the i-th node $p^i_1(k)$ can be found by adding the elements of the adjacent matrix on the lines according to the following formula:

$$p^i(k) = p^i_1(k) + p^i_2(k) + \ldots + p^i_l(k),$$

where l – number of graph's nodes.

Thus, the iterated 1st order power is based on the following formula:

$$\left. \begin{aligned} p^1(1) &= r_{11} + r_{12} + \cdots + r_{1l} \\ p^2(1) &= r_{21} + r_{22} + \cdots + r_{2l} \\ &\vdots \\ p^l(1) &= r_{l1} + r_{l2} + \cdots + r_{ll} \end{aligned} \right\}.$$

However, the calculation is done only for statistically significant correlation coefficients. Having sorted the obtained values by $p^i(1)$ value, the first approximation of the power estimation of the graph nodes is plotted.

The 2nd order power evaluation matrix is built using the following formula:

$$\left.\begin{aligned} p^1(2) &= r_{11}p^1(1) + r_{12}p^2(1) + \cdots + r_{1l}p^l(1) \\ p^2(2) &= r_{21}p^1(1) + r_{22}p^2(1) + \cdots + r_{2l}p^l(1) \\ &\vdots \\ p^l(2) &= r_{l1}p^1(1) + r_{l2}p^2(1) + \cdots + r_{ll}p^l(1) \end{aligned}\right\}$$

If after the second approximation of the places distribution the power variables have changed, it is necessary to calculate the third iteration and make analogous further calculations.

After analyzing the power range, the most powerful or influential characteristic is selected as the leader, or an adjustment is made in order to select an output variable that is easier to define methodically.

Then, according to the graph, we build linear regression equations that allow us to determine other parameters by a given value of this output variable.

As far as it is impossible to provide the optimal value of all optimization parameters at the same time, we will understand by desirability d the desired level of any optimization parameter. Numerically d changes from 0 at an unacceptable quality level to 1 at its maximum possible value. The range from 0 to 0.37 corresponds to an unacceptable quality level, from 0.37 to 0.6 to an acceptable and sufficient level, from 0.6 to 0.8 to an acceptable and good quality level and from 0.8 to 1 to a very high quality level. The boundaries of the ranges can be shifted depending on the specific requirements.

For each optimization parameter or y_i response the value of y_i achievable with these input parameters is translated into the desirability function d_i by the following formulas:

one-sided constraint $y \le y_{Max}$ or $y \ge y_{min}$:

$$d_i = e^{-e^{-y'_i}},$$

where y'_i – some dimensionless value associated with y_i. More often

$$y'_i = \frac{2y_i - (y_{max} + y_{min})}{y_{max} - y_{min}}$$

is taken;

two-sided constraint $y_{min} \le y \le y_{max}$:

$$d_i = e^{-e^{-(|y'_i|)^{n_i}}},$$

where n_i – a positive number that determines the curvature of the desirability curve.

Approximately n_i can be determined from the ratio:

$$n_i = \frac{\ln\left(\ln\frac{1}{d_i}\right)}{\ln|y'_i|},$$

where d_i – desired level of desirability for this parameter y_i, $0{,}6 < d_i < 0{,}8$.

This mathematical model is programmed in a software complex based on a modified intellectual system [17], [18].

3. Conclusions

We will optimize the HME process of cylindrical workpieces 9 mm in diameter and 22 mm long from aluminum alloy AMg5 (analog of Al-5056). As visco-plastic environment for creation of high hydrostatic pressure we use wax, soap and plastoparaffin [19]. Rheological properties of plastoparaffin are given in [20], [21].

Optimization was carried out based on the results of the experiment, as shown in table 1.

As a result of the simulation HME process provides a compromise solution to the problem of minimizing the extrusion force, non-uniformity of the deformed state and damage index to the finished product.

Application of the program complex made it possible to determine the dependence of the overall quality criterion of the HME process on the main technological parameters of the HME: drawing ratio λ, conicity angle of matrix α and visco-plastic coating parameters of the extruded workpiece.

As a result of multi-criteria optimization of the HME process of an aluminum alloy, the following values of technological parameters are recommended: drawing ratio $\lambda = 4.8$; matrix conicity angle $\alpha = 48°$, as a environment creating hydrostatic pressure in the matrix cavity, a soap powder ensuring friction coefficient $f = 0.05$ can be used.

Table 1. Experimental data for the HME process of AMg5 alloy.

	Independent variables			Responses		
	Drawing ratio λ	Matrix angle, α, degree	Friction coefficient, f	Extrusion force x10^4, N	Damage index	Deformation uniformity
№	x_1	x_2	x_3	y_1	y_2	y_3
1	2	50	0.3	2.57	0.27	2.95
2	4	55	0.3	5.04	0.18	2.19
3	6	60	0.3	6.13	0.17	2.12
4	8	60	0.4	7.60	0.18	1.96
5	2	60	0.1	2.62	0.31	3.89
6	4	60	0.2	4.92	0.31	2.44
7	8	50	0.2	6.34	0.17	1.66
8	6	50	0.1	5.29	0.77	1.77
9	8	55	0.1	6.10	0.18	1.79

References

[1] Pearson E C and Parkins R N 1960 *The Extrusion of Metals, 2nd ed.* (London: Chapman and Hall Ltd.)
[2] Bridgman P W 1952 *Studies in Large Plastic Flow and Fracture with Special Emphasis on the Effects of Hydrostatic Pressure* (New York–Toronto–London: McGraw-Hill)
[3] Kolmogorov V L, Bogatov A A, Shishmintsev V F, Belyaev V A and Mizhiritskii O I 1976 *High Temp – High Pressures* **8** pp 675–6
[4] Avitzur B 1965 *ASME J. Eng. Ind.* **87** pp 487–494
[5] Beresnev B I, Gaivoronskii A T, Zamaraev V K and others 1998 *Hydraulic Forging* (Ekaterinburg: Inst. Prikl. Mekh. UrO RAN) p 241 [In Russian]
[6] Berezhnoy V L 1980 *Non-Ferrous Metals* **1** pp 66–71
[7] Story J M, Avitzur B and Hahn W C 1976 *ASME J Eng Ind* **98** pp 909–913
[8] Avitzur B, Wu S R, Talbert S and Chou Y T 1985 *ASME J Eng Ind* **107** pp 247–253
[9] Thiruvarudc S *Wear* **72 3 2** pp 325-233
[10] Nowotynska I and Smykla A 2009 *J Mater Process Technol* **209** pp 1943–9
[11] Dragutinowic R and Zasadzinski J 1989 *Proc of the 6th Int Symp on Plasticity and Resistance to Metal Deformation* (Herceg-Novi, Yugoslavia) pp 107–110
[12] Yang H, Peng Y, Ruan X and Liu M 2005 *J. Mater. Process. Technol* **161 3 30** pp 440–4

[13] Kolmogorov G L, Mikhailov V G, Barcov Yu A and Karlinski V A 1991 *Hydropressing of Hard-to-Deform Refractory Metals and Alloys* (Moscow: Metallurgy) [In Russian]
[14] Dixit P M and Dixit U S 2008 *Engineering Materials and Processes* DOI 10.1007/978-1-84800-189-3
[15] Ribin Yu I, Rudskoy A I and Zolotov A M 2004 *Mathematical modeling and design of technological processes of metal forming* (St. Petersburg: Science)
[16] Karayel D 2008 *Key Engineering Materials* **367** pp 185–192
[17] Zalazinskiy A G, Shveykin V P and Titov V G 2020 *IOP Conf. Series: Materials Science and Engineering* **709** 044115
[18] Byvaltsev S V, Zalazinskiy A G and Shveykin V P 2019 *AIP Conference Proceedings* **2176** 030003
[19] Agapitova O Yu, Byvaltsev S V and Zalazinskiy A G 2012 *Handbook. An Engineering journal with appendix* **12 (189)** pp 3–5
[20] Bodsworth C, Haling J and Barton J W 1957 *J Iron Steel Inst* **185**
[21] Agapitova O Yu, Byvaltsev S V and Zalazinsky A G 2012 *American Journal of Mechanical and Industrial Engineering* **2 1** pp 17–23

IOP Conf. Series: Materials Science and Engineering **966** (2020) 012123 doi:10.1088/1757-899X/966/1/012123

Increase in linear motion steadiness of tractor-trailers using stabilizing towing couplers

Y N Stroganov[1, 2] and E M Pampura[1]

[1] Ural Federal University named after the First President of Russia B N Yeltsin, 19, Mira str., Yekaterinburg, 620000, Russia

E-mail: [2]iu.n.stroganov@urfu.ru

Abstract. Travelling at higher speeds over any type of roads could disturb the stability of motion, which is manifested through lateral oscillations of the trailer in a horizontal plane – trailer wobbling. This research is aimed at building a mathematical model of tractor-trailer motion affected by the external action of lateral forces and estimating how efficient stabilizing couplers could be. A «double pendulum» design diagram is used to build the model. It is shown that the stability of motion is impacted at most by the ratio of the inertia radius (relative to the vertical axis passing through the center of gravity) to the base of the trailer, which depends on the type of cargo, its distribution within the trailer body, the heterogeneity of the cargo material and so on. It is demonstrated that the critical velocity of a trailer would increase when stabilizers are used. Changing the design of the trailer should also positively affect the stability of a tractor-trailer due to faster and more efficient damping of lateral oscillations impacting its movement.

1. Introduction

There are stricter requirements set on truck- and tractor-trailers (or just tractor-trailers) in terms of their operation safety. Travelling at higher speeds over any type of roads could disturb the stability of motion, which is manifested in lateral oscillations of the trailer in a horizontal plane – trailer wobbling. This phenomenon negatively impacts the road safety in general. Lateral deviations from a linear course widen overall and, therefore, traffic lanes of a tractor-trailer, increase its risk of skidding and leaving the dedicated lane, complicates driving as a whole – all this threatens oncoming or passing vehicles [1, 4, 6] and could lead to accidents. Known studies [1, 2, 3] cite data which implies that lateral oscillations of trailers in a horizontal plane could occur when a truck travels in a strict line but at a speed exceeding 35–40 km/h.

Nowadays, modern trucks and energy-saturated wheeled tractors feature quite advanced speed characteristics (above 80 km/h), while trailers themselves have hardly been upgraded. Thus, the problem of increasing the motion steadiness of tractor-trailers is certainly quite topical.

One solution of that problem aimed at damping lateral oscillations of trailers lies in the direction of using stabilizing towing couplers which connect the parts of such an articulated vehicle – a truck to a front bogie of a two-axle trailer, the bogie to the trailer platform.

To settle the motion of the front wheels of trailers equipped with a bogie connected to the trailer frame via a slewing ring, literary and patent sources describe various types of electromechanical, hydraulic and other stabilizing devices which generate a stabilizing torque relative to the kingpin of the ring.

This research is aimed at building a mathematical model of tractor-trailer motion affected by the external action of such forces as lateral ones and at estimating how efficient stabilizing towing couplers could be in terms of their positive effect on the stability of linear motion.

2. Methods

To analyze how well mechanisms designed to damp lateral oscillations perform their task of stabilizing the linear movement of a two-axle trailer, its motion should be considered as a three-degrees-of-freedom mechanical system under the following assumptions: the trailer and its truck feature a rigid and gapless coupling; no vertical oscillations are present; the movement at that point where a front bogie is coupled with the truck is linear; the damping mechanism is installed on a rotary support connecting the front wheeled bogie and trailer platform.

To build a mathematical model, a «double pendulum» design diagram is used (figure 1). In a formalized form, the front bogie is considered to be the first section of the pendulum, the trailer platform – the second one.

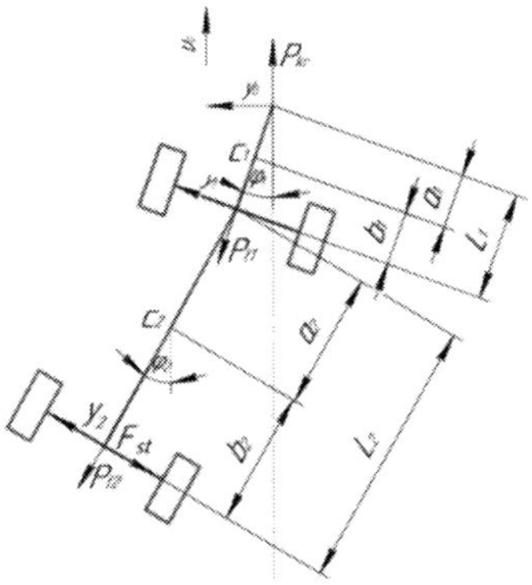

Figure 1. Diagram of two-axle trailer under horizontal lateral oscillations.

To describe the motion of this system, Lagrange differential equations are used.

The generalized coordinates are assumed to include: φ_1 – angle of deviation from a linear path of truck movement for the first trailed section; φ_2 – angle of deviation from the same path for the second trailed section; y – coordinate of the point of coupling on a truck-movement line.

The following external forces are applied to the system: P_{kr} – longitudinal component of hook power; y_0 – lateral reaction of the truck; y_1 – lateral reaction of the road applied to the front bogie tires; P_{f1} – rolling resistance of the bogie; y_2 – lateral reaction of the road applied to the trailer platform tires; P_{f2} – rolling resistance of the platform; F_{st} – stabilizing force due to a stabilizing device installed on the slewing ring, applied to the axis of the rear trailer wheels, thus helping them to return to the position of linear motion.

The kinetic energy of a system formed by two trailed sections equals to:

$$T = \frac{1}{2}(M_1 + M_2)V_0^2 + \frac{1}{2}\left(M_1 a_1^2 + M_2 L_1^2 + J_{C_1}\right)\dot{\varphi}_1^2 + \left(M_1 a_1 + M_2 L_1\right)V_0 \dot{\varphi}_1 \varphi_1 +$$
$$+ \frac{1}{2}\left(M_2 a_2^2 + J_{C_2}\right)\dot{\varphi}_2^2 + M_2 a_2 V_0 \dot{\varphi}_2 \varphi_2 + M_2 L_1 a_2 \dot{\varphi}_1 \dot{\varphi}_2.$$

where M_1, M_2 – masses of the front bogie and trailer platform, respectively, kg; J_{C_1} – inertia of the bogie relative to the vertical axis passing through its center of gravity, kg · m^2; J_{C_2} – inertia of the platform relative to the same axis, kg · m^2; V_0 – linear speed of the truck, m/s; a_1 – longitudinal coordinate of the center of gravity for the front wheeled bogie relative to the coupling loop of the truck, m; a_2 – longitudinal coordinate of the center of gravity for the trailer platform relative to the kingpin of the slewing ring, m; L_1 – base of the bogie, m.

Then, differential equations are formed:

$$\frac{d}{dt}\left(\frac{\partial T}{\partial \dot{y}}\right)-\frac{\partial T}{\partial y}=0,$$

$$\frac{d}{dt}\left(\frac{\partial T}{\partial \dot{\varphi}_1}\right)-\frac{\partial T}{\partial \varphi_1}=\left(M_1a_1^2+M_2L_1^2+J_{C_1}\right)\ddot{\varphi}_1+M_2L_1a_2\ddot{\varphi}_2,$$

$$\frac{d}{dt}\left(\frac{\partial T}{\partial \dot{\varphi}_2}\right)-\frac{\partial T}{\partial \varphi_2}=\left(M_2a_2^2+J_{C_2}\right)\ddot{\varphi}_2+M_2L_1a_2\ddot{\varphi}_1.$$

The generalized forces factor in the stabilizing force F_{st} and take the form of:

$$Q_y=P_{kr}-y_1\varphi_1-y_2\varphi_2-P_{f1}-P_{f2},$$

$$Q_{\varphi_1}=-y_1L_1-y_2L_1+P_{f2}L_1\left(\varphi_2-\varphi_1\right)+F_{st}L_1\left(\varphi_2-\varphi_1\right),$$

$$Q_{\varphi_2}=-y_2L_2-F_{st}L_2\left(\varphi_2-\varphi_1\right),$$

where L_2 – base of the second trailed section, m.

According to the studies of Zakin Y. H. [1], the lateral reactions of the road could be expressed as:

$$y_1=\frac{k_1L_1}{V_0}\dot{\varphi}_1+k_1\varphi_1,\qquad y_2=\frac{k_2L_1}{V_0}\dot{\varphi}_1+\frac{k_2L_2}{V_0}\dot{\varphi}_2+k_2\varphi_2,$$

where k_1, k_2 – coefficients of resistance to lateral drag for the tires of the first and second trailed sections, N/rad.

Since the first equation of the Lagrange system determines how large the longitudinal component of the hook power P_{kr} should be, a system of two differential equations is formed, which describes the motion of two trailed sections, that is a two-degrees-of-freedom system.

Considering that $J_{C_1}=M_1r_{C_1}^2$, $J_{C_2}=M_2r_{C_2}^2$, where r_{C_1}, r_{C_2} – radiuses of inertia for the first and second trailed sections relative to the vertical axes passing through their centers of gravity (m), a system of equations could be obtained:

$$\begin{cases}\left[M_1\left(a_1^2+r_{C_1}^2\right)+M_2L_1^2\right]\ddot{\varphi}_1+\dfrac{L_1^2\left(k_1+k_2\right)}{V_0}\dot{\varphi}_1+L_1\left(k_1+P_{f2}+F_{st}\right)\varphi_1+\\ \qquad +M_2L_1a_2\ddot{\varphi}_2+\dfrac{L_1L_2k_2}{V_0}\dot{\varphi}_2+L_1\left(k_2-P_{f2}-F_{st}\right)\varphi_2=0,\\ M_2L_1a_2\ddot{\varphi}_1+\dfrac{L_1L_2k_2}{V_0}\dot{\varphi}_1-L_2k_{st}\varphi_1+M_2\left(a_2^2+r_{C_2}^2\right)\ddot{\varphi}_2+\dfrac{L_2^2k_2}{V_0}\dot{\varphi}_2+L_2\left(k_2+F_{st}\right)\varphi_2=0.\end{cases}\qquad (1)$$

A solution of the system (1) would be found in the form of $\varphi_1=d_1e^{lt}$, $\varphi_2=d_2e^{lt}$.

Substituting those expressions into (1), we get

$$\begin{cases}d_1\left\{\left[M_1\left(a_1^2+r_{C_1}^2\right)+M_2L_1^2\right]l^2+\dfrac{L_1^2\left(k_1+k_2\right)}{V_0}l+L_1\left(k_1+P_{f2}+F_{st}\right)\right\}+\\ \qquad +d_2\left\{M_2L_1a_2l^2+\dfrac{L_1L_2k_2}{V_0}l+L_1\left(k_2-P_{f2}-F_{st}\right)\right\}=0,\\ d_1\left\{M_2L_1a_2l^2+\dfrac{L_1L_2k_2}{V_0}l-L_2F_{st}\right\}+d_2\left\{M_2\left(a_2^2+r_{C_2}^2\right)l^2+\dfrac{L_2^2k_2}{V_0}l+L_2\left(k_2+F_{st}\right)\right\}=0.\end{cases}$$

The characteristic equation for this system is a quartic equation:

$$\alpha_0 l^4 + \alpha_1 l^3 + \alpha_2 l^2 + \alpha_3 l + \alpha_4 = 0.$$

Here

$$\alpha_0 = M_1 M_2 \left(a_1^2 + r_{C_1}^2\right)\left(a_2^2 + r_{C_2}^2\right) + M_2^2 L_1^2 r_{C_2}^2,$$

$$\alpha_1 = \frac{M_1 L_2^2 \left(a_1^2 + r_{C_1}^2\right) k_2 + M_2 L_1^2 \left(a_2^2 + r_{C_2}^2\right) k_1 + M_2 L_1^2 \left(b_2^2 + r_{C_2}^2\right) k_2}{V_0},$$

$$\alpha_2 = M_1 L_2 \left(a_1^2 + r_{C_1}^2\right)\left(k_2 + F_{st}\right) + M_2 L_1 \left(a_2^2 + r_{C_2}^2\right)\left(k_1 + P_{f2} + F_{st}\right) +$$
$$+ M_2 L_1^2 a_2 \left(P_{f2} + 2F_{st}\right) + M_2 L_1^2 b_2 \left(k_2 + F_{st}\right) + M_2 L_1 L_2 a_2 F_{st} + \frac{L_1^2 L_2^2 k_1 k_2}{V_0^2},$$

$$\alpha_3 = \frac{L_1 L_2 \left[L_1 \left(k_1 k_2 + k_1 F_{st} + k_2 P_{f2} + 2k_2 F_{st}\right) + L_2 \left(k_1 k_2 + k_2 P_{f2} + 2k_2 F_{st}\right)\right]}{V_0},$$

$$\alpha_4 = L_1 L_2 \left(k_1 k_2 + k_1 F_{st} + k_2 P_{f2} + 2k_2 F_{st}\right),$$

where b_2 – distance from the rear axis of the second trailed section to its center of gravity, m.

Taking into account a low degree of the characteristic equation, Hurwitz stability criterion is used to check the motion stability of the system.

All coefficients of the characteristic equation remain positive under any practically conceivable changes of the parameters, which means:

$$\alpha_0 > 0, \quad \alpha_1 > 0, \quad \alpha_2 > 0, \quad \alpha_3 > 0, \quad \alpha_4 > 0.$$

Apart from having to satisfy these conditions, it is necessary to do the same for

$$\begin{vmatrix} \alpha_1 & \alpha_3 & 0 \\ \alpha_0 & \alpha_2 & \alpha_4 \\ 0 & \alpha_1 & \alpha_3 \end{vmatrix} = \alpha_1 \alpha_2 \alpha_3 - \alpha_1^2 \alpha_4 - \alpha_3^2 \alpha_0 > 0,$$

that is $\alpha_2 > \frac{\alpha_1 \alpha_4}{\alpha_3} + \frac{\alpha_3 \alpha_0}{\alpha_1}$, or, which is the same:

$$\frac{L_1^2 L_2^2 k_1 k_2}{V_0^2} > \frac{M_1 L_2^2 \left(a_1^2 + r_{C_1}^2\right) k_2 + M_2 L_1^2 \left(a_2^2 + r_{C_2}^2\right) k_1 + M_2 L_1^2 \left(b_2^2 + r_{C_2}^2\right) k_2}{L_1 \left(k_1 k_2 + k_1 F_{st} + k_2 P_{f2} + 2k_2 F_{st}\right) + L_2 \left(k_1 k_2 + k_2 P_{f2} + 2k_2 F_{st}\right)} \left(k_1 k_2 + k_1 F_{st} + k_2 P_{f2} +\right.$$
$$\left. + 2k_2 F_{st}\right) + \frac{L_1 L_2 \left[L_1 \left(k_1 k_2 + k_1 F_{st} + k_2 P_{f2} + 2k_2 F_{st}\right) + L_2 \left(k_1 k_2 + k_2 P_{f2} + 2k_2 F_{st}\right)\right]}{M_1 L_2^2 \left(a_1^2 + r_{C_1}^2\right) k_2 + M_2 L_1^2 \left(a_2^2 + r_{C_2}^2\right) k_1 + M_2 L_1^2 \left(b_2^2 + r_{C_2}^2\right) k_2} \times$$
$$\times \left[M_1 M_2 \left(a_1^2 + r_{C_1}^2\right)\left(a_2^2 + r_{C_2}^2\right) + M_2^2 L_1^2 r_{C_2}^2\right] - \left[M_1 L_2 \left(a_1^2 + r_{C_1}^2\right)\left(k_2 + F_{st}\right) + M_2 L_1 \left(a_2^2 + r_{C_2}^2\right) \times\right.$$
$$\left. \times \left(k_1 + P_{f2} + F_{st}\right) + M_2 L_1^2 a_2 \left(P_{f2} + 2F_{\text{ст}}\right) + M_2 L_1^2 b_2 \left(k_2 + F_{st}\right) + M_2 L_1 L_2 a_2 F_{\text{ст}}\right] = A_1 + B_1 - C_1.$$

If $A_1 + B_1 - C_1 \le 0$, then this inequation is satisfied at any speed.

If $A_1 + B_1 - C_1 > 0$, then

$$V_0^2 < \frac{L_1^2 L_2^2 k_1 k_2}{A_1 + B_1 - C_1} = \frac{L_2 k_2}{M_2 \dfrac{(A_1 + B_1 - C_1)}{M_2 L_1^2 L_2 k_1}}.$$

The conditions of stability by Hurwitz come down to a known expression found by Zakin Y H [1] which defines a critical velocity beyond which the stability of the system fails:

$$V_0 < V_{cr} = \left(\frac{L_2 k_2}{M_2 Z}\right)^{1/2},$$

where V_{cr} – critical velocity, m/s; $Z = A + B - C$.

But if a mechanism stabilizing lateral oscillations relative to the kingpin of the slewing ring were introduced into the system, the expressions for A, B, C summands would take the form of:

$$A = \frac{\left[\mu\left(\beta_1^2 + \chi_1^2\right) + \rho\left(\beta_2^2 + \chi_2^2\right) + \left(1 - \beta_2\right)^2 + \chi_2^2\right] \cdot \left[1 + \nu + (\rho + 2)\xi\right]}{\rho \cdot \left\{\gamma\left[1 + \nu + (\rho + 2)\xi\right] + 1 + \nu + 2\xi\right\}},$$

$$B = \frac{\left[\mu\left(\beta_1^2 + \chi_1^2\right)\left(\beta_2^2 + \chi_2^2\right) + \chi_2^2\right] \cdot \left\{\gamma\left[1 + \nu + (\rho + 2)\xi\right] + 1 + \nu + 2\xi\right\}}{\gamma \cdot \left[\mu\left(\beta_1^2 + \chi_1^2\right) + \rho\left(\beta_2^2 + \chi_2^2\right) + \left(1 - \beta_2\right)^2 + \chi_2^2\right]},$$

$$C = \frac{1}{\gamma\rho} \cdot \left[\mu\gamma\left(\beta_1^2 + \chi_1^2\right)\left(1 + \rho\xi\right) + \rho\left(\beta_2^2 + \chi_2^2\right)\left(1 + \nu + \xi\right) + \gamma\beta_2\rho\left(\nu + 2\xi\right) + \right.$$
$$\left. + \gamma\left(1 - \beta_2\right)\left(1 + \rho\xi\right) + \beta_2\rho\xi\right],$$

where

$$\mu = \frac{M_1}{M_2}, \gamma = \frac{L_1}{L_2}, \beta_1 = \frac{a_1}{L_1}, \beta_2 = \frac{a_2}{L_2}, \chi_1 = \frac{r_{C_1}}{L_1}, \chi_2 = \frac{r_{C_2}}{L_2}, \rho = \frac{k_1}{k_2}, \nu = \frac{P_{f2}}{k_1}, \xi = \frac{F_{st}}{k_1}.$$

Let us analyze how the stabilizing force F_{st} affects the critical speed on the example of a 2-PTS-4 tractor-trailer with a slewing ring and a stabilizing device installed on the ring. As a calculation basis, we take M_2 = 6500 kg, L_2 = 3 m, a_2 = 1.5 m, $k_1 = k_2$ = 75000 N/rad (the rigidity of front and rear wheels is identical), f = 0.015 (f – coefficient of rolling resistance), the mass of the bogie towbar could be neglected – M_1 = 0. Known research [4] shows that a two-axle trailer has sufficiently stable movement when its center of gravity is situated in the middle of its base. The stability of motion is affected at most by the ratio of the inertia radius (relative to the vertical axis passing through the center of gravity) to the base of the trailer, which depends on the type of cargo, its distribution within the trailer body, heterogeneity of the cargo material and so on. Consider the impact of that ratio on the critical speed without the stabilizing moment – F_{st} = 0. Figure 2(a) graphs calculated dependencies of the critical velocity V_{cr} on the ratio χ_2 of the inertia radius to the base L_2 of a two-axle trailer at various relations of L_1/L_2.

Analysis of the graphs in figure 2(a) demonstrates that the critical velocity diminishes for any correlation assumed between the base L_1 of the front bogie and the base L_2 of the two-axle trailer while the correlation χ_2 between the inertia radius to the vertical axis and the trailer base rises.

Consider the impact of the stabilizing force F_{st} on the critical velocity of a two-axle trailer at various χ_2 and for γ = 0.7. Figure 2(b) correlates the critical velocity V_{cr} and the stabilizing force F_{st} at various ratios χ_2 of the inertia radius to the trailer base.

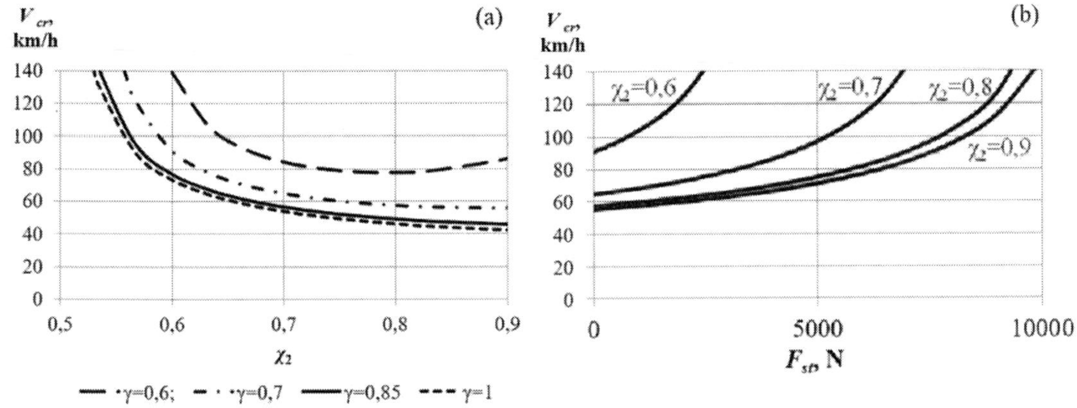

Figure 2. Dependency of the critical velocity V_{cr}: a) on the ratio χ_2; b) on the stabilizing force F_{st}.

3. Results and discussions

Analysis of the graphs shown in figures 2(a) and 2(b) allows to draw the following conclusions:

- The maximum impact on the stability and critical velocity of the trailer is caused by the ratio χ_2 of the inertia radius (relative to the vertical axis passing through the center of trailer gravity) to the trailer base. When this ratio rises and approaches 1, the results of calculation indicate that the critical velocity falls to 40 km/h. Such a velocity is too low for modern high-speed transport, and, at the same time, it depends on the type of cargo, its distribution within the trailer body and so on.
- Graphs which show how the critical velocity depends on the stabilizing force of a stabilizing rotary support installed on a trailer demonstrate that it would increase when stabilizers are used – the movement would become more stable.

4. Conclusion

Those conclusions were reflected in several patents on stabilizing towing couplers published by the authors [8, 9] and checked to be correct during laboratory tests carried out to study horizontal lateral deviations of a trailer on scale models of truck- and tractor-trailers by simulating their movement using a belt band at various speeds [5].

References

[1] Zakin Y H 1967 *Applied theory of tractor-trailer movement* (Moscow: Transport) [In Russian]
[2] Ivanov V V et al 1977 *Basic theory of cars and tractors* ed 2 (Moscow: Higher School) [In Russian]
[3] Litvinov A S and Farobin Y E 1984 *Car: theory of operating properties* (Moscow: Machine Building) [In Russian]
[4] Protas A Y 1972 Impact of hook power on stability of «Belarus» tractor-trailer movement *Proc. of Scientific and Technical Conf. (Gorky)* 98–104 [In Russian]
[5] Stroganov Y N, Popova A I and Zhelev D Y 2019 *J. News of the Int. Acad. of Agricult. Ed* **44** 37–44
[6] Petrenko A M 2013 *Stability of special-purpose vehicles* (Moscow: MADI) [In Russian]
[7] Balakina E V, Zotov N M, Maruhin D A and Fedin A P 2014 *Aust. J. Sci. Res.* **1(5)** IV 614–624
[8] Stroganov Y N, Popova A I and Stroganova O Y 2019 Stabilizing towing coupler for two-axle trailers (Patent 194885 RU)
[9] Stroganov Y N, Popova A I, Ognev O G and Stroganova O Y 2020 Stabilizing device for two-axle trailer bogies (Patent 196271 RU)

IOP Conf. Series: Materials Science and Engineering **966** (2020) 012124 doi:10.1088/1757-899X/966/1/012124

Study of the influence of the design of the turning tool on the removal of thermal energy from the cutting zone

Y L Liberman[1], S V Lukinskikh[1, 2] and A V Smirnov[1]

[1] Ural Federal University named after the First President of Russia B N Yeltsin, 19, Mira str., Yekaterinburg, 620000, Russia

E-mail: [2] swwl@mail.ru

Abstract. In this work, we performed a study and developed the design of through-cutters that provide accelerated heat removal from the cutting zone with external turning. Using SolidWorks Simulation engineering analysis and SolidWorks Flow Simulation gas-hydrodynamic analysis modules as a research tool, we created a basic 3D model of a precast cutter and carried out a thermal analysis of the transition process to determine the distribution of temperature fields in the body of the cutter throughout time during the cutting process. After that, we made changes to the design of the cutter, specifically, created a cavity in the holder body in which the copper core was placed. We then re-performed the studies described above, which showed that when using a copper core, the temperature in the cutting zone decreases by 246 °C within 15 minutes from the start of the work. To obtain a greater effect, we added a hole in the copper core and used air cooling, which allowed reducing the temperature in the cutting zone by 313 °C.

1. Introduction

It is well known that when cutting metals, more than 99 % of mechanical energy is converted into thermal energy, most of which is transferred to the part, cutter and shavings. Numerous experimental studies by various groups of scientists [1–3] demonstrated that the largest amount of heat – from 60 to 85 % – goes into shavings; 25–35 % goes into the part, 3–15 % into the tool and 1–2 % into the environment. Despite the fact that the percentage of thermal energy transferred to the cutter is small, the absolute value of the temperature at the tool pads and in the cutting wedge itself is significant, and with increase in cutting speed it can reach 1000 °C and higher [4].

High temperatures negatively affect tool life and can lead to damage and premature wear of carbide inserts, e.g. growths on the cutting edge, deformation of the cutting edge, chipping of the tool and the insert, etc. [5, 6].

Premature wear of a replaceable carbide plate affects the accuracy of machining parts and workpieces, and may adversely impact the equipment itself.

For effective cutting, it is necessary to maintain the temperature within acceptable limits, reducing heat generation and removing heat from the cutting zone as much as possible.

There are various ways to reduce heat during cutting.

Firstly, the choice of the cutting mode is of great importance. It is well known that more heat is generated as cutting speed increases. However, at the same time, heat removal through shavings is increased because more shavings are generated per unit of time. An increase in feed is also favorable for heat dissipation due to the formation of thicker shavings [7, 8].

Further, cutting fluid is widely used to remove heat from the cutting zone. However, the use of coolant is associated with significant costs, which in most cases make up to 2–8 % of production costs, i.e. are comparable with the cost of the tool itself [9]. These costs are driven by the need to acquire and dispose of the coolant. In addition, the high temperature in the cutting zone causes coolant to evaporate, resulting in temperature spikes. The combination of temperature spikes and mechanical pressure leads to damage to the cutting edge, whereas perpendicular cracks form along the working surface, while particles of the material can fall off, causing chipping. In terms of environmental friendliness, coolant also has major disadvantages. Since the critical temperature of the coolant components (130–200 °C) is much lower than the average temperature in the cutting zone, they are prone to thermal destruction during cutting. As a result of this, harmful substances enter the air of the working zone, and, when their normative content in the air is exceeded, have a negative effect on the health of workers, causing occupational diseases [10]. Waste coolant discharged into the wastewater is a major pollutant of water and soil. Metal shavings left over from machining and contaminated with coolant are not always cleaned and are also a source of pollution and contamination [11]. Therefore, additional investments are required to neutralize production waste.

At the current technological level, it is not possible to completely abandon the coolant. For example, in boring, drilling and cutting operations, when it is difficult to remove shavings or when the machined surfaces must be measured while being cut, abundant cooling is required.

In modern metal processing, the problem of heat removal ought to be solved not only by supplying coolant to the cutting zone, but also by creating designs of cutting tools that provide accelerated heat removal from the cutting zone [12–14].

The work we currently present is devoted to the study of the influence of the thermal conductivity of the cutter on the heat removal from the cutting zone with external turning.

Increasing the thermal conductivity of the cutter is critical to prolonging the cutter's service life. Higher thermal conductivity improves the conditions for heat removal from the cutting zone, which reduces the cutting temperature and increases the wear resistance of the tool.

2. Research methodology

Our solution to this problem is based on modeling the heat transfer process during cutting in the SolidWorks Simulation system. Thermal analysis in this system includes the calculation of both stationary and non-stationary (transient) thermal process. Transient research models the distribution of thermal energy over time. The initial data in this case is the thermal load applied to the body.

The research algorithm is as follows:

1. To create a model of the basic design of the cutter, this will be a prototype for comparison with subsequent modernized designs.
2. Set the application area and the value of thermal load, initial and boundary conditions.
3. Perform analysis of heat transfer in the model of the basic design of the cutter and obtain graphs of the dependence of the distribution of thermal energy in time.
4. Create models of modernized designs and obtain comparison graphs following the methodology described above.
5. Based on a comparison of the results, draw conclusions about the advisability of upgrading the design of the cutter.

2.1. Study of the heat sink of the base design

As a basic design, we created a precast cutter with a cross section of a holder 32×35mm made of steel 40X, angles $\gamma = -6°$, $\varphi = 45°$, $\varphi 1 = 6°$, $\varepsilon = 90°$, $\alpha = 6°$, $\alpha 1 = 6°$, $\lambda = -6°$. Tungsten carbide plate – SNMG 190616, the method of fastening the plate to the holder – clamp the lever above the hole (Figure 1).

The thermal study of the cutter in SolidWorks Simulation was carried out taking into account the time dependence in the range of 15 min (900 sec) – the time corresponding to the resistance period of the replaceable plate, with a time increment of 1 sec., i.e. the heat transfer process was seen as non-stationary.

For the analysis, we determined the areas impacted by the forces that arise in the cutting process and cause a thermal load on the cutter. A heat load was applied to these areas in the form of a heat flux with a power of 17 W for a given time range. For simplicity, it was assumed that the convective heat transfer of the cutter with the surrounding air occurs only at its rear end, since the side surfaces of the holder are closed by a tool holder in which the tool is fixed. Coefficient of convective heat transfer 25 W / m^2 °K, ambient temperature 300 °K, initial cutter temperature 20 °C.

In the course of the study, we monitored the following parameters:

- temperature distribution along the length of the tool holder 900 seconds after application of the load;
- the change in the maximum temperature of the cutter over a given time range.

The result of the thermal calculation of the basic design of the cutter is shown in Figure 1.

The figure shows that after 900 seconds the maximum temperature reached 1035 °C and occurred at the top of the cutting edge. The rest of the interchangeable plate warmed up to ~ 300–400 °C.

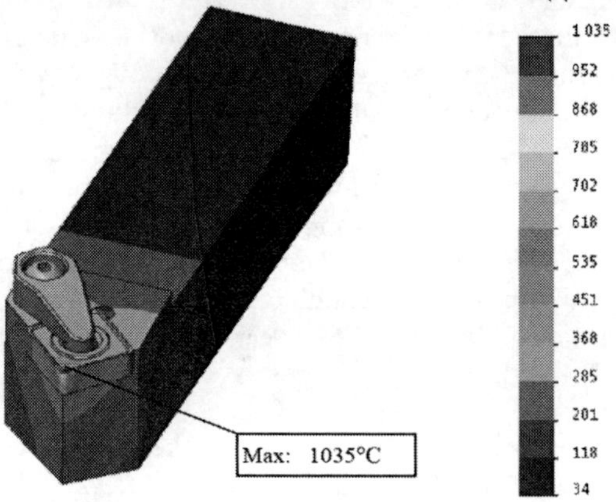

Figure 1. Temperature distribution along the length of the holder in the model of the basic design of the cutter after 900 seconds from the moment of loading.

2.2. Development of cutter designs with accelerated heat removal from the cutting zone

To study the impact of the design of the cutter on the rate of removal of thermal energy from the cutting zone, we developed several constructive variants of the initial cutter, as presented in Figure 2.

In the embodiment with a copper core shown in Figure 2 (a), a hole was made under the cutting insert in the tool holder connecting to the longitudinal hole, and the resulting cavity was filled with copper.

Figure 2 (b) shows a model of a cutter with a longitudinal hole made in a copper core; the thickness of the copper layer in the tool holder was 4 mm.

Further modernization of the cutter design, which allows increasing the heat sink, was aimed at the use of forced air convection. For this purpose, we mounted a cover on the thread on the rear end of the copper-coated cutter. A thin long tube with a diameter of 5 mm was attached to this cover on the side of the cutter, a fitting for supplying compressed air was provided on the outside, and apertures for the exit of heated air were made along the perimeter of the cover (Figure 2 (c)).

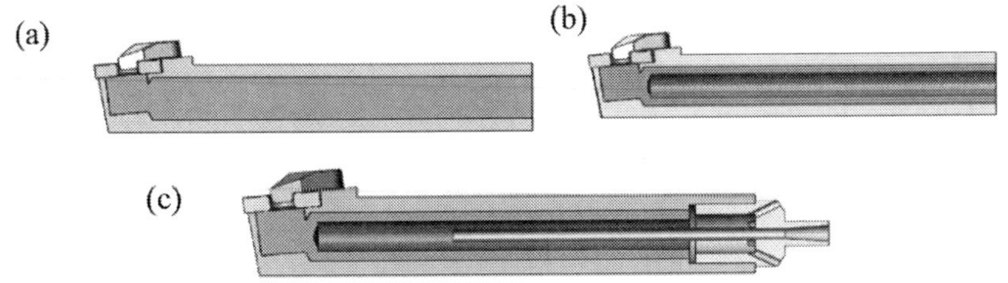

Figure 2. Design options of the initial cutter (in the context):
(a) – with a copper core; (b) – with a copper coating; (c) – with forced air convection.

2.3. The study of heat sink in the cutters designed by the Authors
Thermal analysis of the cutter models shown in Figure2 (a) and 2 (b) was performed in SolidWorks Simulation according to the same scenario as the analysis of the basic design model described earlier. The same thermal loads were applied, and the same initial and boundary conditions and the areas impacted by the forces arising in the cutting process and causing a thermal load on the cutter were set.

The results of the thermal calculation of these incisors are presented in Figure 3.

For the cutter with the copper core of the heat flux, the maximum temperature after 900 seconds of the impact was 789°C (Figure 3 (a)). This is 246 degrees lower than in the cutter of the basic design. The heat dissipation of the copper-coated cutter (Figure 3 (b)) did not increase compared to the copper core cutter, but instead decreased, albeit slightly, by 9 degrees. This result is predictable, but the latter design option allows the use of forced air convection to increase the heat sink even further.

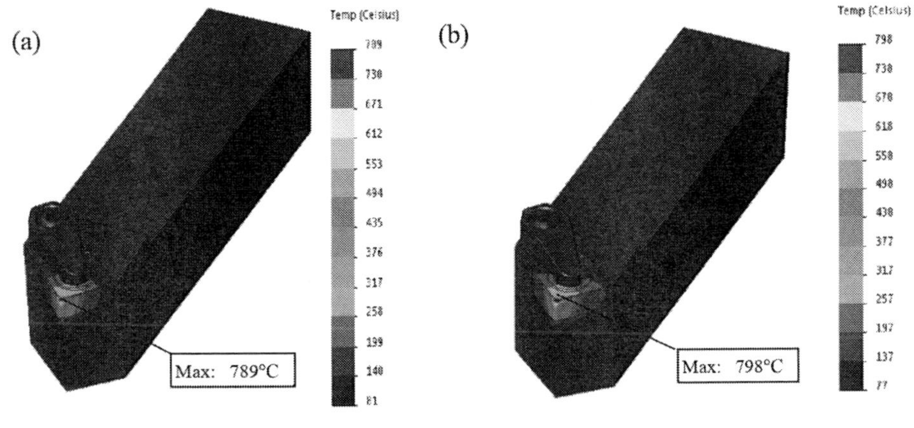

Figure 3. Temperature distribution along the length of the holder after 900 seconds from the moment of application of the heat load: (a) – in the model of the cutter with a copper core; (b) – in the model of a copper-coated cutter.

2.4. The study of heat sink cutter with forced air convection
Analysis of the design of the cutter that uses stream of compressed air to cool the copper channel was performed using the SolidWorks Flow Simulation computing tool, designed to simulate fluid and gas flows. Using this application, we calculated the trajectory of the air when it was supplied through the nozzle under a pressure of 0.6 MPa. Calculation data from SolidWorks Flow Simulation were transferred to thermal analysis, the results of which are presented in Figure 4.

According to the diagram presented in Figure 4, the maximum temperature of the cutter was 721 °C. Compared with the base case, the temperature decreased by 314 degrees.

2.5. The study of changes in the maximum temperature of the incisors during the cutting process

For each of the studied incisors, we measured and recorded the temperature at the tip of the incisor, i.e. the point at which the maximum temperature was observed after 900 sec from the moment of application of the heat load, in the interval of 900 sec with a time increment of 1 sec. Based on the results of the study, we plotted the change in the maximum temperature of the cutter over a given time range (Figure 5).

As demonstrated in the graphs presented, the temperature in the region of the cutting edge increases significantly during the first 1-3 seconds, and changes less intensively in the following 897 seconds. These results are consistent with numerous experimental studies described in the literature. [1–3, 7]

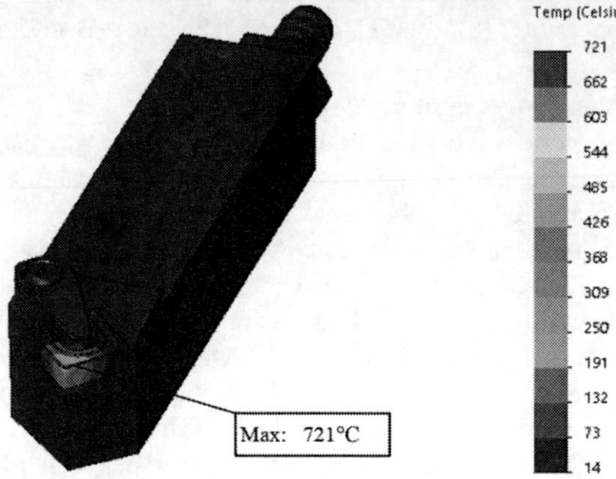

Figure 4. Temperature distribution along the length of the holder in the cutter model with forced air convection after 900 seconds from the moment of application of the heat load.

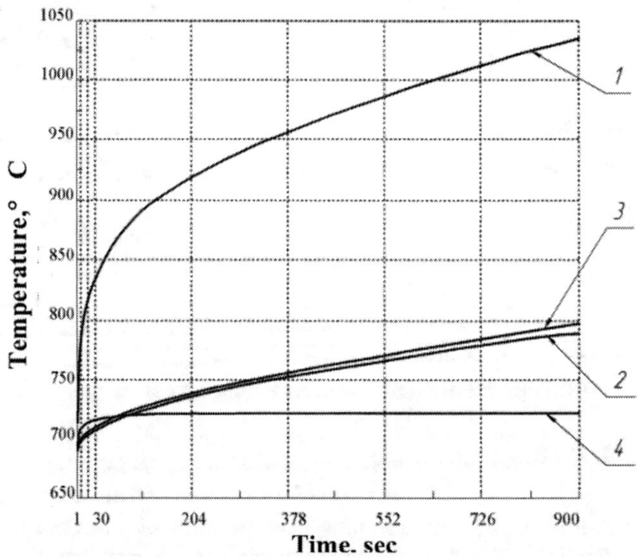

Figure 5. Graphs of changes in the maximum temperature of the incisors during the cutting process: 1 – model of the basic design; 2 – model with a copper core; 3 – model with a copper coating; 4 – air-cooled model.

IOP Conf. Series: Materials Science and Engineering **966** (2020) 012124 doi:10.1088/1757-899X/966/1/012124

3. Conclusion

Our study demonstrated that the design changes made to the basic version of the turning tool are generally beneficial. However, they are not all the same. A cutter with a copper core is quite rational and simple in terms of its heat sink performance. It is almost twice as effective as the base cutter; however, the use of a copper core makes it significantly more expensive. The cutter with an internal copper coating is less expensive than the copper core cutter and provides almost equal heat dissipation from the top of the cutter; however, its rigidity will be less. The air-cooled cutter with copper inner coating provides the greatest temperature reduction compared to the base cutter. According to preliminary estimates, its manufacturing cost will be one and a half times higher than that of the base version. At the same time, it appears its rigidity will be the same.

Given the above, we believe that a cutter with an internal copper core can be recommended for use in ordinary turning of billets made of structural steels, when the task of increasing the heat removal from the top of the cutter is quite relevant, but not critical. An internal copper-coated air-cooled cutter is probably the most efficient use for high speed finishing. A copper-coated cutter without air cooling can be used in medium-speed machining, when the cutting speed is relatively high, but does not reach very high values characteristic of high-speed turning.

This study is only a pilot. In the future, we plan to refine and expand it, which will make it possible to discuss the specific areas of application of the proposed modifications in greater detail.

References

[1] Reznikov A N and Reznikov L A 1990 *Thermal processes in technological systems: textbook* (Moscow: Mechanical Engineering) p 288 [In Russian]

[2] Abramov F N, Kovalenko V V, Lyubimov V E and others 1983 *Handbook of metal cutting* (Kiev: Technique) p 239 [In Russian]

[3] Bobrov V F 1975 *Fundamentals of the theory of metal cutting* (Moscow: Engineering) p 344 [In Russian]

[4] Budadin O N, Vavilov V P and Abramova E V 2013 *Thermal control* (Moscow: Publishing. house 'Spectrum') p 176 [In Russian]

[5] Serebrennikova A G and Panova E A 2016 *Technological support for the quality of engineering products* (Stary Oskol: TNT) p 392 [In Russian]

[6] Vasin S A, Vereshchak A S and Kushner V S 2001 *Cutting materials: Thermomechanical approach to the system of relationships in cutting* (Moscow: Ed MSTU named after N E Bauman) p 448 [In Russian]

[7] Stahl Jan Eric et al 2012 *Metal Cutting: Theories and Models* (Lund: Lund Inicersity Press) p 580

[8] Ali Riza, Motorcu 2010 The Optimization of Machining Parameters Using the Tguchi Metod Ali Riza Motorcu *Journal of Mechanical Engineering* vol **56** **(6)** pp 391–401

[9] Voskoboev V F 2008 *Reliability of technical systems and technological risk* (Moscow: LLC Publishing House Alliance) p 200 [In Russian]

[10] Artemyeva M S 2015 The effect of thermal phenomena during the processing of metals by cutting on cutting lubricants *Life Safety* no **6** **(174)** pp 9–14

[11] Kamenskaya A A, Kovalova R I and Labetskiy V M 1992 *The environmental impact of metal processing by cutting of machine-building enterprises on the environment and ways to reduce damage: Analit. review GPNTB SB RAS Alt Polytechnic in-t* (Novosibirsk: SPSL SB RAS) p 102 [In Russian]

[12] Sanchez L E A, Scalon V L, Abreu G G C and Cleaner Machining 2011 Through a Toolholder with Internal Cooling *Proc. of third int. workshop on advances in cleaner production* (São Paulo, Brazil) pp 18–20

[13] Vicentin G C, Sanchez L E A, Scalon V L and Abreu G G C 2011 A sustainable alternative for cooling the machining processes using a refrigerant fluid in recirculation inside the toolholder *Clean Technologies and Environmental Policy* vol **13** Issue **6** pp 831–40

[14] Liberman Y L and Rastyagaeva E S Metalworking tools with cooling (patent 196895 RU)

Using of polymeric materials to defeat of lightly armored and easily vulnerable targets. Mathematical modeling

E A Khmelnikov[1], K V Smagin[1] and T E Zavodova[1, 2]

[1] Ural Federal University named after the First President of Russia B N Yeltsin, Nizhniy Tagil Technological Institute, 59, Krasnogvardeyskaya str., Nizhniy Tagil, 622000, Russia

E-mail: [2] tatiana.zavodova@urfu.ru

Abstract. The article deals with the possibility of increasing the damaging effects of small-caliber artillery ammunition in the fuel tanks of aircraft and helicopters. A comparison of the results of the calculation of the probability of the incendiary effect of strikers based on fluoropolymers on diesel fuel was made. The calculations of the probabilities of fuel ignition by the method of 'Fragmented criterion'. As a result of the calculations, it was revealed that this method is not suitable for this type of problem statement. A combined methodology has been developed for calculating the probability of fuel ignition, taking into account the ongoing chemical reaction between fluoropolymers and aluminum. All components of the passing heat release process are taken into account - the impactor deformation energy, the impactor braking energy in the fuel, the chemical energy of the reaction undergoing. The energy of the ongoing chemical reaction was taken into account using thermochemistry methods.

1. Introduction

In modern combat, small-caliber barrel artillery is highly effective in fighting both with the weakly protected and lightly armored vehicles. In connection with the recent active use of assault and tactical aviation and helicopters against enemy vehicles, the development and improvement of small-caliber barrel artillery ammunition has become particularly necessary. The effectiveness of the impact on aviation of small-caliber ammunition is the sum of the total number of hits that ultimately lead to the disruption of the life support systems and the functioning of the aircraft.

From the physical factors affecting important functional units of airplanes and helicopters, an incendiary effect is of great interest.

Consequently, the task of creating a new type of small-caliber ammunition, possessing not only the necessary specific energy for breaking through an obstacle, but also the subsequent incendiary over-obstacle effect, which can lead to the destruction of an aviation target, becomes urgent.

2. Evaluation of incendiary actions by the method of 'Fragmentary criterion'

One of the most common ways to assess the incendiary effect of fragments, damaging elements and bullets is the method of evaluation by the 'Fragmentary Criterion' [1].

When assessing the probability of fuel ignition by this method, the impact impulse of a fragment is calculated:

$$i = 0.000204 \cdot m^{1/3} \cdot \upsilon \tag{1}$$

where m is the mass of the fragment, g; υ is the velocity of the fragment, m/s.

With a specific impulse $i \leq 0.16$, the probability of ignition is zero. With a specific impulse $i \geq 2.5$, the probability of fuel ignition is close to unity. Analyzing the source data for the specific impulse laid down in the 'Fragmentary criterion' method, an approximation dependence to determine the probability of fuel ignition was obtained:

$$p = 1 + 1.08 \cdot e^{-4.16i} - 1.96 \cdot e^{-1.46i} \tag{2}$$

The dependence of the probability of fuel ignition on the kinetic energy (Ek) is shown in Figure 4. However, when using the fragmentation criterion, all dependencies were built on the basis of the calculated specific impulse, which depends on the mass of the fragments and the initial interaction velocity, without taking into account the additional energy released during the chemical reaction, between the fluoroplastic and aluminum, which in turn significantly affects ignition of diesel fuel.

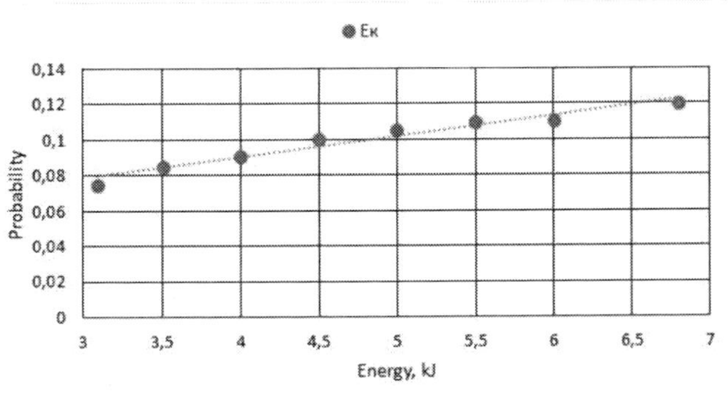

Figure 1. The dependence of the ignition probability from energy.

3. Evaluation of incendiary actions by the method of 'Energy balance'

To assess the effect of a chemical reaction on the ignition of diesel fuel in the fuel tanks, an analysis allowing to estimate the amount of energy released in the process of a chemical reaction spent on heating diesel fuel to a self-ignition temperature was carried out.

As a result, two types of simplified models were obtained: assembly No. 1 and assembly No. 2 (see Figure2 a and Figure 2 b). The total mass of the striker was 9 grams, in accordance with the conditions of the experiment. To evaluate possible reactions, the mass of fluoroplastic and aluminum in the assemblies ranged from 1 to 8 grams.

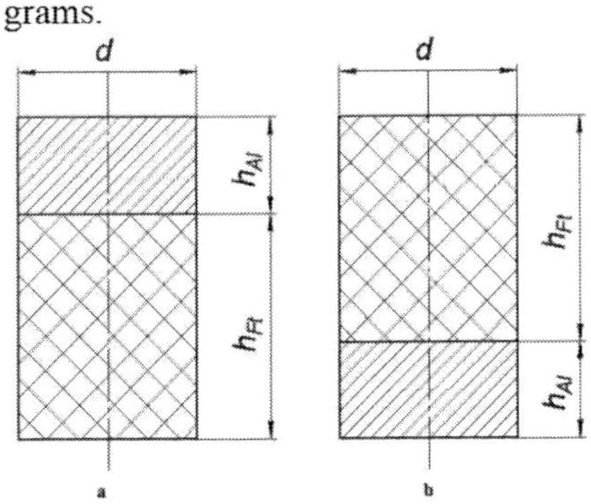

Figure 2. Striker structures for mathematical calculations.

The energy expended on heating diesel fuel while the striker is moving inside the fuel tank will consist of 3 components:

$$E_{st} = E_{react} + E_{def} + E_{dec} \tag{3}$$

IOP Conf. Series: Materials Science and Engineering **966** (2020) 012125 doi:10.1088/1757-899X/966/1/012125

where E_{react} – energy released by a chemical reaction, kJ;
E_{def} – energy released as a result of deformation of a striker, kJ;
E_{dec} – energy released as a result of deceleration the striker in the fuel, kJ.

The deformation energy of the impactor is determined depending on the relative deformation of the head part of the impactor ε and the dynamic yield strength of the material σ_{df}^*:

$$E_{def} = 0.95 \cdot \frac{S_{mid} \cdot \sigma_{df}^{*} \cdot \varepsilon}{T^*} \tag{4}$$

where S_{mid} – area of midsection, m^2;
T^* – coefficient of conversion of mechanical energy into heat, kJ.

The analysis performed for various configurations of the striker assemblies shows that the effect of deformation energy on the total energy of the striker is less than 1 %, and is not taken into account in further calculations. Probably, such a low value is associated with a short time of the process of movement of the striker in the fuel (about 10–15 μs), which is clearly not enough for the process of heat transfer from the striker to the fuel.

The deceleration energy of the striker can be calculated from the equations of motion of the striker in the fluid. At the moment of penetration of the striker into inside of the fuel tank, a certain velocity field will appear in the entire volume of the fluid, and the initial velocity of the striker v_{fmax} will be less than the impact velocity of the striker with the liquid v_c. The value of v_c is determined by the method of N E Zhukovsky. The conservation law for the moment of impact can be written as:

$$(m + m_0)\frac{dv_f}{dt} = mg - \rho_{DF} gV - c_x \frac{\rho_{DF} v_f}{2} S_{mid} \tag{5}$$

where m_0 – the added mass, kg;
v_f – striker velocity into inside of fuel tank at every moment in time, m/s;
c_x – impactor drag coefficient, kgm/s^2;
ρ_{DF} – density of diesel fuel, kg/m^3;
V – striker volume, m^3;
g – coefficient of gravity, m/s^2.

The initial v_{fmax} and the limiting v_{lim} velocity of a striker in a fluid is defined as:

$$v_{f\,max} = v_c \left(1 + \frac{m_0}{m}\right)^{-1}, \quad v_{lim} = \sqrt{\frac{2(mg - \rho g V)}{c_x \cdot \rho \cdot S_{mid}}} \tag{6}$$

The initial impact velocity of the striker with the fluid is determined on the basis of the balance of the kinetic energy of the striker:

$$v_c = \sqrt{2E_{res}/m} \tag{7}$$

where E_{res} – residual energy of the striker, calculated as the difference between the initial energy of the striker and the energy expended on the destruction of the target at the specific energy of destruction of the target equal to E_{sp} = 766.67 J/cm^2 [1].

The value of the added mass can be determined by the coefficient of the added mass, which is a function of the elongation of the impactor and its volume:

$$m_0 = \mu_0 \cdot \rho_{DF} \cdot V \tag{8}$$

where μ_0 – the coefficient of the added mass.

The coefficient of the added mass was determined by the method of approximation of tabular data for a fluid with a density of 800...1000 kg/m^3 [2]:

$$\mu_0 = 0.5187 \cdot \lambda^{-1.342} \tag{9}$$

where λ – the striker relative elongation.

Graphs of the distribution of the added mass for assemblies No. 1 and No. 2, depending on the mass of the fluoroplastic, are presented in Figure 3.

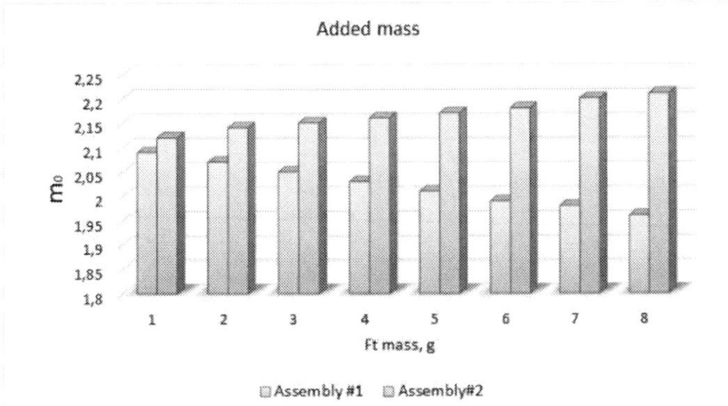

Figure 3. The distribution of the added mass of diesel fuel.

The evaluation of the heat q of the fluoroplastic reaction with aluminum was carried out by thermochemistry [3–6]:

$$1.5\,[C_2F_4] + 2\,Al = 2\,AlF_3 + 3C + q \tag{10}$$

from where

$$q = 1.5\,\Delta H_f([C_2F_4]) + 2\,\Delta H_f(Al) - 2\,\Delta H_f(AlF_3) - 3\,\Delta H_f(C) \tag{11}$$

where ΔH_f is the molar enthalpy ($\Delta H_f(Al) = \Delta H_f(C) = 0$; $\Delta H_f(AlF_3) = -1490$ kJ/mol; $\Delta H_f([C_2F_4])$ is the unknown quantity).

So, use the well-known reaction:

$$[C_2F_4] = CF_4 + C(\text{graphite}) + 113\text{ kJ/mol} \tag{12}$$

From here:

$$\Delta H_f([C_2F_4]) = 113 + \Delta H_f(C) + \Delta H_f(CF_4) \tag{13}$$

Since $\Delta H_f(C) = 0$, $\Delta H_f(CF_4) = -907$ kJ/mol, then $\Delta H_f([C_2F_4]) = 113 - 907 = -794$ kL/mol.

Then, the heat of reaction of fluoroplastic with aluminum will be equal to:

$$q = 1.5\,(-794) + 2 \times 1490 = 1789\ [\text{kJ/2 mol } AlF_3] \tag{14}$$

Thus, when the fluoroplastic interacts with aluminum, heat $q = 895$ kJ/mol is released.

The molar mass of aluminum in the reaction M(Al) = Mr(Al)·2 = 53.96 g/mol, the molar mass of the fluoroplastic in the reaction M(Ft) = Mr(Ft)·1.5 = 798.525 g/mol, where Mr(Al) and Mr(Ft) is the molecular weight of aluminum and fluoroplastic, respectively.

Then, it is necessary to find the number of moles of a substance which, after being introduced into the wall of the fuel tank, enter into a chemical reaction. The number of moles of aluminum and moles of fluoroplastic capable of reacting is defined as the ratio of the mass of aluminum and fluoroplastic to the molar masses of aluminum and fluoroplastic:

$$N_{Al} = \frac{m_{Al}}{M(Al)};\, N_{Ft} = \frac{m_{Ft}}{M(Ft)} \tag{15}$$

To determine the heat released during the passage of a chemical reaction, select the smallest number of moles of substances (fluoroplastic or aluminum) that can react and multiply this value by the amount of reaction heat calculated by thermochemistry methods:

$$E_{react} = N_{min(Al/Ft)} \cdot q \tag{16}$$

The amount of caloric spent on heating the fuel to the self-ignition temperature is:

$$Q = C \cdot m \cdot (t_2 - t_1) \tag{17}$$

where t_1 – initial temperature (20 °C);
t_2 – self-ignition temperature of DF;
C – average specific heat for temperature range;
m – mass of the DF (1 g).

To determine the effectiveness of the incendiary effect of assemblies No. 1 and No. 2, we define the mass of diesel fuel brought to ignition as:

$$m_{iDF} = E_{st} / Q \tag{18}$$

The results of the calculations are presented in Figures 4 and 5.

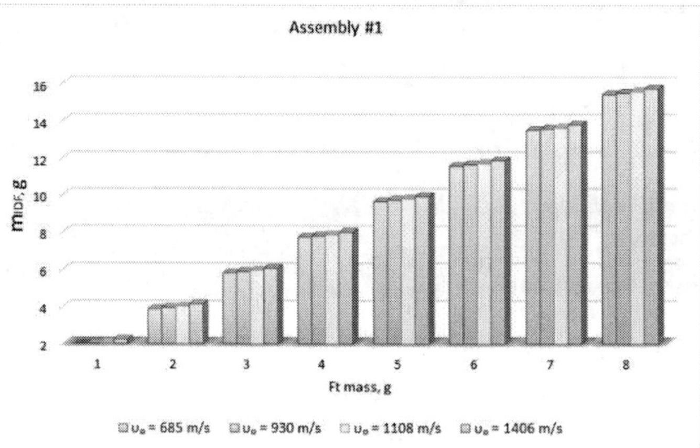

Figure 4. The mass of ignited DF depending on the mass of Ft in assembly No. 1.

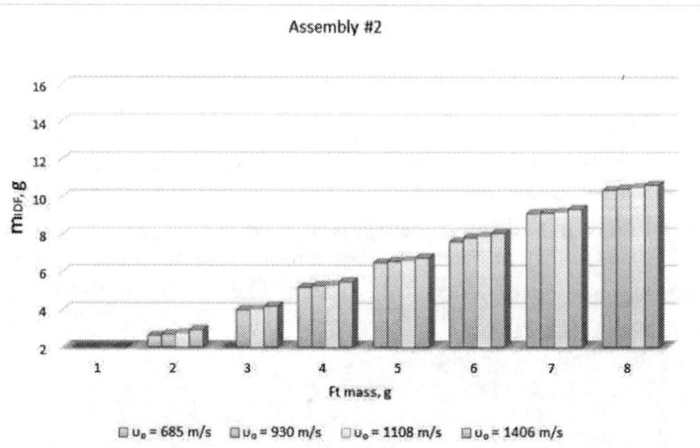

Figure 5. The mass of ignited DF depending on the mass of Ft in assembly No. 2.

The obtained results of calculations on the possible weight of ignited diesel fuel are in good agreement with the results of experiments that show steady ignition and combustion of fuel under the influence of the presented assemblies of strikers.

4. Conclusions

Studies have shown that under dynamic loading fluoropolymers can release enough energy to heat a certain amount of fuel to ignition temperature, which with a high degree of probability will lead to the ignition of all fuel in the tank. Thus, creating combined striker elements, including fluoroplastic, aluminum or titanium, along with an impact-penetrating effect sufficient to defeat thin-walled targets, will have a high inflammatory prohibitive effect.

Conducted studies on the ignition of one of the most resistant to ignition fuels - diesel fuel, gives the right to assume that light with high octane fuel used in aviation will ignite at lower interaction speeds. But at the same time, the speed of interaction should still be higher than the critical speed of the beginning of a chemical reaction between the fluoroplastic and aluminum or titanium.

References

[1] Balaganskiy I A and Merzhievskiy L A 2004 *The effect of weapons and ammunition* (Novosibirsk) p 408
[2] Selivanov V V 2008 *Means of destruction and ammunition* (Moscow) p 984
[3] *Industrial organofluorine products* Directory 1996 (Chemistry) p 544
[4] Gorovoy V R, Bildukevich N A and Chulkov V P 1982 *Production of pyrotechnic compositions* (Moscow: Engineering) p 296
[5] Korbakova A I, Makulova I D and Marchenko E N 1975 *Toxicology of organofluorine compounds and occupational health in their production* (Moscow: Medicine) p 183
Khmelnikov E A, Styrov A V, Smagin K V, Kravchenko N S, Rudenko V L, Sokolov S S and Svidinsky A V 2016 Study of high-speed interaction processes between fluoropolymer projectiles and aluminum and titanium-based targets *International Journal of Energetic Materials and Chemical Propulsion* vol **1** pp 167–83

IOP Conf. Series: Materials Science and Engineering **966** (2020) 012126 doi:10.1088/1757-899X/966/1/012126

Deep drawing failure modes

AV Nezamaev[1, 2] and V V Karzhavin[1]

[1] Department of Metallurgical and Rotary Machines, Ural Federal University named after the First President of Russia B N Yeltsin, 19, Mira str., Yekaterinburg, 620002, Russia

E-mail: [2] alexei.nezamayev@gmail.com

Abstract. In this paper, we have presented various technological process conditions leading to the damage of workpieces or the manufacture of low-quality pieces during deep drawing of metal sheets. All the conditions, resulting in failure modes, have been included in the general classification. According to this classification, there are three main directions of potential failures: complete or partial fracture, buckling or wrinkling, and surface damage. The rigorous relations between production parameters and possible negative consequences have been also given. For the first time, the different aspects of such an issue have been considered together. Thus, we have takenthe first steps towards creating a reliable system for assessing the risk of complex failures in obtaining deep-drawn products (caused by the impact of several reasons at the same time).Having developed this system, we wouldfacilitate the advancement of deep drawing technology and it might well open up new horizons for the deep-drawn products of unrivalled quality without rejections, equipment breakdown, emergency downtime, etc.

1. Introduction

It is well known that deep drawing is a sheet metal forming process, which allows obtaining large lots of hollow thin-walled metal products within a short period. Though the method itself is highly beneficial, the specificity of the technological process makes it extremely sensitive to numerous parameters. To date, optimal parameter values have been determined for many state-of-the-art types of deep drawing and these data are readily available. However, when some process characteristics become unacceptable due to specific circumstances, it inevitably results in the production of the items, which do not meet the requirements of the quality standards. Moreover, the same conditions can lead to some kinds of fracture inherent for deep drawing [1].

Aiming to develop a convenient risk assessment tool for the detection of failure modes beforehand with high probability, and in order to determine failure problems online during the drawing process, the conditions mentioned above were generalized under the concept of 'Deep Drawing Failure Modes' and their classification was proposed as follows: complete or partial fracture, buckling or wrinkling, and surface damage.

The main points of the deep drawing failure modes have been presented in this paper. Corresponding preconditions and mathematical expressions of the assessment criteria have been placed according to the particular type of failure.

2. Classification of deep drawing failure modes

2.1. Complete or partial fracture

2.1.1. Tearing sidewalls near the base of drawn cups. Significant sidewall thinning is frequently observed during deep drawing, which may lead to workpiece tearing failure. It usually occurs at the so-called weakest section, namely the region near the base of the drawn cup. It is explained as follows. At the initial phase of the process, this region is placed above the clearance between the die and the punch. When the punch draws the blank into the die, this region is not subjected to hardening through the bending over the die edge. Further, this annular region shifts towards the power transition section, becoming thinner due to tensile stresses inducing their subsequent growth. Obviously, the equality of the tensile stress to the tensile strength can be accepted as a precondition for this type of failure [2].

Besides deviation from optimal values of technological parameters and the deterioration of material properties, tearing may occur during multistage deep drawing of wide-flange components if the material volume of the cup drawn at the first stage is not enough for the next stage.

2.1.2. Cracking the upper part of workpieces during redrawing. The walls of redrawn parts are subjected to tangential tensile stresses – both the residual stress and the stress resulting from bending over the die profile. Furthermore, the material of the workpiece's upper part possesses lowered strength characteristics due to the presence of microcracks at the workpiece rim, and, in addition, it is hardened. Therefore, it is supposed that fracture occurs before the tensile stress value becomes be equal to the tensile strength. The criterion for this type of failure depends on stress efficiency parameter K_{IC} [3], microcrack length l, residual tangential stressσ_θ^R, tangential stress resulting from bending over the die profile σ_θ^M, and Poisson's ratio v. Thus, this criterion is reckoned by the following formula [4]:

$$\sigma_\theta^R + \sigma_\theta^M = 0.9\, K_{1C} \left(\frac{1-v^2}{\pi l}\right)^{1/2} \quad (1)$$

The maximum permissible microcrack length can be obtained from equation (2)

$$l = \frac{0.26 K_{1C}^2 (1-v^2)}{\left(\sigma_\theta^o + \sigma_\theta^M\right)^2} \quad (2)$$

It should be noticed that this type of failure is the most typical one for multistage drawing without intermediate annealing.

2.1.3. Delayed cracking of intermediate and end products during storage. In contrast to cracking of the upper part of a blank during the drawing process, delayed cracking may occur in the absence of any external loads. In the latter case, the failure is related to the phenomenon of hydrogen-induced embrittlement. The initiation and development of the cracking process are caused by the diffusion of hydrogen towards high-tensile stress regions [5].

Pearlite and ferrite-pearlite steels, containing up to 0.2 % carbon, are least prone to delayed cracking failure, whereas steels with a martensitic structure are the most susceptible to the failure. The usage of deep drawing for austenitic steel treatment increases the probability of the subsequent delayed cracking at low levels of austenite stability. The presence of a strain-induced martensitic phase can lead to the initialization of cracking. The period between the initialization of the process and the cracking itself may last from an hour to a few days [6]. Under certain conditions, if a corrosive medium affects the material, this type of failure can be considered as stress-corrosion cracking.

Summarizing all the above, we can point out the following precondition of delayed cracking: residual stress, sufficient content of hydrogen in alloys or favorable conditions for its increase (corrosive medium, for example, a lubricant with chlorine content even when it deals with stainless

IOP Conf. Series: Materials Science and Engineering **966** (2020) 012126 doi:10.1088/1757-899X/966/1/012126

steel), the capability of hydrogen to diffuse through the material, prolonged time of product storage before annealing.

2.2. Buckling and wrinkling

2.2.1. Workpiece buckling during deep drawing with pushing force applied. As it was mentioned in the first subsection, tearing occurs in the weakest section of the workpiece during the drawing of wide-flange components due to the shortage of material volume in the cylindrical part of the wide-flange piece after the first stage of drawing. To avoid this type of failure, deep drawing with axial pushing force is utilized.

In this case, the main limiting factor is the axial buckling of the workpiece shape upon axial compression. As a result, both complete buckling and local periodical buckling are possible. The latter leads to the formation of specific ridges on the walls of products that negatively affects their mechanical characteristics. The criterion of this failure is the value of pushing force, which exceeds the value of the critical force for certain conditions. In [7], the formula (3) was proposed for the estimation of this critical stress:

$$\frac{\sigma_{cr}}{\sigma_S} \approx \frac{0.5}{\sin\frac{\alpha_i}{2}}\sqrt{\frac{h}{R_z}} \tag{3}$$

where σ_{cr} – critical stress, σ_S– yield strength, α_i– input angle, h – wall thickness, R_z– radius of the non-drawn part of the workpiece.

It should be taken into account that the pushing force is also dependent on the friction condition at the die entrance. Thus, it should also be considered as a precondition of buckling.

2.2.2. Wrinkling. At the first stage of drawing without blank holder, the flange stability, the condition obtained by Shoffman [8], is determined by the equation (4):

$$\frac{t_0}{D_0}100 \leq 4.5(1 - K_1) \tag{4}$$

where t_0 – workpiece thickness, D_0– workpiece diameter, K_1– the ratio of the product diameter to the workpiece diameter (drawing ratio).

When the requirement is fulfilled, circumferential compressive stresses, arising from a decrease in workpiece diameter, only leads to an increase in workpiece thickness. When the requirement is not fulfilled, along with an increase in thickness, the formation of a fold (wrinkles) is also possible [9]. Moreover, it was found in [8] that the tendency to fold formation is higher for intensely hardening metals.

While drawing with blank holder, the loss of flange stability may still occur. In this case, the precondition for buckling will include a blank holder parameter. Depending on its design, this can be rigidity, pressure or the gap size between the blank holder and the die [9].

2.3. Surface damage

In most cases, the surface damage is a consequence of adhering blanks to drawing tool surfaces, which is possible when a positive gradient of the mechanical properties of treated materials is not maintained. The main reason for the occurrence of such defects is breaking down the lubricating film, which takes place upon the improperly chosen set of parameters: contact pressure, temperature, lubricant, and deformation rate (or an unexpected change in these parameters) [10].

As it was shown in the study [11], ceteris paribus, the specific amount of heat released per unit time rises with the increase of the workpiece diameter (it is so-called size factor), which explains the fact that suitable friction conditions more frequently cannot be maintained during large-size pieces treatment due to a higher temperature in contact zone when compared to drawing small-size components while drawing ratio is kept the same.

High friction coefficients are observed at both extremely high and extremely low surface roughness values. In the first case, it can be explained by breaking down the lubricant film on a coarse surface. In the second case, it may be attributed to the absence of surface cavities to hold up the lubricant [12]. It is worth mentioning that, in the latter example, the adhesive component of friction force may also increase [13].

Besides the stability of lubrication films, it is necessary to take into account its ability for self-recovery, which is higher for liquid lubricants than for solid ones. In this connection, it is recommended to utilize lubricant compositions containing both solids for coating protected surfaces and a liquid layer for lubrication [11].

3. Conclusion

In the paper, we elaborated the classification of failure modes taking place in deep drawing processes. It includes the criteria of failure occurrence and the consequences leading to certain defects of products. For every type of failure, we determined the preconditions for its occurrence. Their mathematical interpretations were also given. This classification may be useful in assessing the reliability of the production process. We plan to use the revealed failure modes in further research, which will include the following steps: the study of each type of failure separately and with respect to specific materials; the determination of relationships between parameters involved in this process; the clarification of failure criteria; the development of measures for preventing and eliminating failures; the development of deep drawing technology with minimal failure probability. We expect that these data will be useful for failure risk assessment at the development stage of the manufacturing process to avoid rejects, equipment breakdown, emergency downtime, and so on.

References

[1] Grigoriev L L, Ivanov K M and Jurgenson E E 2009 *Cold forming* (Sankt-Petersburg: Polytechnica) p 481 [in Russian]
[2] Popov E A 1977 *Fundamentals of Sheet Stamping* (Moscow: Mashinostroenye) p 141 [in Russian]
[3] Ashbey M 2010 *Construction Materials* (Moscow: Intellect) [in Russian]
[4] Karzhavin V V, Kordyukov I G, Kuznetsov V I and Bakina V V 2019 *Kshp-omd* **4** pp 20–7 [in Russian]
[5] Wang Z, Luo Z C and Huang M X 2018 *Materialia* **4** pp 260–7
[6] Berrahmoune M R, Berveiller S, Inal K and Patoor E 2006 *Material Science and Engineering* **438-440** pp 262–6
[7] Vershinin V I 1965 *News of Universities. Mechanical Engineering* **11** pp 120–4 [in Russian]
[8] ShoffmanL A 1961 *Essentials of Forming and Pressing design* (Moscow: Mashgiz) [in Russian]
[9] Golovliov V D 1974 *Forming processes design* (Moscow: Mashinostroenye) pp 75–82 [in Russian]
[10] Seriozkin M A 2017 *Kshp-omd* **4** pp 10–4 [in Russian]
[11] Karzhavin V V 1994 *Improving the Theory and Technology of Deep Drawing Processes based on a Multistage Deformation Scheme and increasing the efficiency of Shell-type article production* (Yekaterinburg: UPI Publ) p 490 [in Russian]
[12] Wang D, Yang H and Li H 2014 *Trans. of Nonferrous Metals Society of China* **24** pp 1263-72
[13] Abramov A N 2014 *Kshp-omd* **5** pp 42–7 [in Russian]

IOP Conf. Series: Materials Science and Engineering **966** (2020) 012127 doi:10.1088/1757-899X/966/1/012127

Foundry production digitalization

V M Gruzman

Ural Federal University named after the First President of Russia B N Yeltsin, Nizhniy Tagil Technological Institute, 59, Krasnogvardeyskaya str., Nizhniy Tagil, 622000, Russia

E-mail: gruzman-vm@ntiustu.ru

Abstract. Domestic companies upgrade their production facilities as it is an imperative of the economy changing over to a new technological mode right now. Industrial companies use more automation and robotization of production, and the level of digital intellectualization is growing. Introduced into technological processes, digital technologies save time and costs of production and reduce the human factor effect. Foundry industry is no exception. The authors suggest considering a foundry process as a set of stages: designing and manufacturing casting models, preparation of charge materials, preparation of initial molding materials, smelting, mixing, casting metal in ladles, mold manufacturing, pouring smelt in mold, knocking castings out of mold, stumping and cleaning of castings, heat treatment of castings, quality inspection of castings. The article organizes information about the digital technologies used at each of the selected stages of the foundry castings by pouring the melt into sand molds. The application of computer technology, digital devices, and software is considered in detail at the stages of the casting model design, charging and molding materials preparation, sand mold manufacturing and casting quality inspection. Digital technologies optimize production, save time and costs.

1. Introduction

Domestic companies upgrade their production facilities as it is an imperative of the economy changing over to a new technological mode right now [1].

Industrial companies use more automation and robotization of production are being activated, and the level of digital intellectualization is growing. Introduced into technological processes, digital technologies save time and costs of production and reduce the human factor effect. Foundry industry is no exception.

A foundry is a complex of technological processes that involve many workers of various professions with their skill having a direct effect on the final casting product quality.

Most shaped castings in this country are produced by pouring the smelt into sand-clay molds. The process flow can be split into the following stages:

1. Designing and manufacturing of a casting model.
2. Preparing charging materials.
3. Preparing initial molding materials.
4. Smelting.
5. Mix preparation.
6. Tapping to ladles.
7. Manufacturing of molds.
8. Pouring the smelt into molds.

9. Knocking-out of the mold.
10. Fettling and cleaning of castings.
11. Heat treatment of castings.
12. Quality inspection of castings.

Digital technologies introduced in the production processes are designed to create the potential for increasing the profitability of production, including foundry.

The results of the analysis and systematization of the experience of the authors, various sources on the use of digital technologies at each of the selected stages of the serial castings production process are presented below.

2. Materials and method

2.1. Description of the process of designing and manufacturing castings using computer technology

The entire cycle of design and manufacturing of castings in mass production using modern computer technologies consists of the following stages:

1. Obtaining a technical design specification for model pattern equipment, which includes a 'flat' design drawing with a foundry technology shown on it. Designing a 3D (three-dimensional) model of a cast billet, rods (if any), and Gating system in a CAD system.
2. Verification of the technology in the CAE system for foundry processes simulation.
3. Designing 3D models of the top, bottom, and core boxes.
4. Converting a 3D model into a set of hard copy design documents for model tooling.
5. Development of a control program for processing model tooling blanks on a numerical control (CNC) machine based on a previously developed 3D model using CAM systems.
6. Milling on a CNC machine, finishing the model tooling manually.
7. Verification of drawing parameters of the final setup equipment, including the use of digital control and measurement systems (CAI-systems), and filling in a certificate for the model kit.

In Russia, advanced steel making producers use CAD/CAE/CAM and CAI-systems for manufacturing model setup, as indicated above [2, 3].

Let us briefly consider the purpose of such systems.

Computer Aided Design (CAD) –'designing using computer programs'. These are computer-aided design systems that provide geometric modeling and visualization of products or their parts in two and three dimensions. Up-to-date CAD systems include the following modules: modeling of a three-dimensional structure; building and execution of drawings; keeping textual design documents; filling in process charts.

CAD computer systems that can be used in the development of castings and setup design as as follows: COMPASS-3D, SolidWorks, Inventor, AutoCAD, T-FLEX CAD, NX etc.

Computer Aided Engineering (*CAE*) – 'engineering calculations using computer programs'. CAE tools are useful for a wide variety of calculations of stresses, deformations, heat transfer, magnetic field distribution, fluid flow, and other parameters of continuous media. No need to explain how difficult it is sometimes to choose the best technology out of a variety of options. Sometimes the refinement of the design and technology takes a long time. So, a computer modeling systems have to be used to simplify the choice and reduce the time needed to work out the new part manufacturing technology. Foundry workers can use the CAE system as a tool for computing experiments in simulation of casting crystallization processes and formation of its structure and properties. Computational experiments in the design and pre-production make it possible to optimize technological processes and build performance data of cast parts in a targeted way.

CAE systems that are used for computer simulation of foundry processes: LVMFlow, Magmasoft, Poligon, ProCAST, SolidCAST, Procast, CastCAE, PowerCast, WINCast, CalcoSoft etc.

Computer Aided Modeling (*CAM*) – 'pre-production using computer systems'. In the first place, CAM systems are intended for designing product processing on CNC machines and issuing control programs for various machines. Numerical control includes use of programmed commands to control a

IOP Conf. Series: Materials Science and Engineering **966** (2020) 012127 doi:10.1088/1757-899X/966/1/012127

machine that can grind, cut, mill, stamp, bend, and otherwise turn workpieces into final parts. There are no fully automatic process preparation systems at the moment, but there are many good software packages that generate some code for CNC machines.

To prepare control programs for CNC machines, the following CAM systems are used: PowerMill, FeatureCAM, Mastercam, ESPRIT, Edgecam, Artcam, NX, SprutCAM, SolidCAM, Alphacam etc.

Computer Aided Inspection (*CAI*) – 'survey using computer programs'. CAI systems together with contact and non-contact measuring devices make it possible to control the quality of physical models and products using a previously developed digital model as a reference.

The following systems are used to control the set-up manufactured: Power INSPECT, PC-DMIS, CAM2 Measure etc.

2.2. Digital technologies for charge materials development
At this stage of foundry production, information technologies are used for:
- chemical analysis control of charge materials;
- Automated control of materials in warehouses of the facility (foundry).

To perform real-time control of the charge chemistry, portable laser analyzers are used with carbon detection (LIBS technology), portable x-ray-fluorescence analyzers (XRF technology) are used as well, which allow you to quickly and instantly find out the chemical composition of the charge materials on the spot.

Automated material accounting makes it possible to track current inventory for planning purchases and handle workshop needs, increase the speed and accuracy of order handling and reduce the influence of the human factor on warehouse management.

As a rule, WMS systems are used to reach the above objectives.

WMS is a warehouse management system, in other words, it is software that makes it possible to automate warehouse operations.

To account for the movement of materials in the warehouse, ready-made WMS systems distributed in the its product market can be used (1C: WMS logistics. Warehouse management, YARUS WMS, WMS Logistics Vision Suite, LM7, Solvo WMS etc.) and systems developed for the needs of a specific company. In this case, software can be developed both by in-house IT-people and by a third-party developer.

2.3. Digital technologies for initial molding material development
Up-to-date computer technology and its elements can be effectively used at this stage of foundry production. E.g. a device was created for real-time control of the clay content in incoming batches of molding sand at one of the foundries, to control the process of preparing molding mixtures [4].

The device uses a ATMEL Atmega8 microcontroller. A two-line liquid crystal indicator is used to display the readings. A powerful led is used as a light source. Several photocells are located under the slide table.

The non-volatile memory of the microcontroller contains analytical dependencies for calculating the content of the clay component in the sand. During the measurement cycle, the microcontroller reads information from 9 sensors located under the tray and calculates the value of the clay component according to a specific algorithm using the selected coefficient for the brand of sand used at that company. The processing time for a sand sample is less than one minute, which gives a random variable 'clay content' large enough for a batch of sand being its general entirety. This makes it possible to determine the true clay content in an incoming sand batch. Reliable information about the fine component content in each incoming sand batch saves the binder, which is quite expensive, over its consumption time.

2.4. Digital technologies for smelting
At this stage of foundry process, computer systems can be used to control the smelting process. Such systems are normally developed for the needs of a specific producer and have their own architecture

[5], consisting of modern electronic equipment (computers, computer networks, programmable logic controllers) and appropriate software that performs the necessary data processing and implements the user interface.

Developed intelligent computer systems make it possible to control smelting processes using neural network technologies [6].

2.5. Digital technologies for mixing

Many producers in this country use automated systems for preparing sand-clay molding mixes. Such systems normally consist of process equipment, automation tools (sensors) and a personal computer [7]. The latter serves as a system control center, which receives and processes information about the process progress and the controlled parameters of the molding mix.

2.6. Digital technologies for tapping to ladles

At this stage of foundry process, digital technologies, in particular the capabilities of computer equipment and software, are used to control the temperature of the metal [8].

2.7. Digital technologies for sand molds manufacture

Making sand forms, especially complex geometries, takes a lot of doing. Development of additive technologies has made it possible to apply alternately a layer of casting sand mixed with a hardener and a layer of binder. In fact, we are talking about the use of 3D printing technology for sand molds [9]. Modern technological complexes make it possible to design and manufacture molds of any complication and configuration, which saves a lot of material costs and time for preparing mass production.

First, as in the case of manufacturing any other product using a 3D printer, a 3D model of the sand mold is created (prepared).

Then, using this model, the 3D printer creates a sand injection mold, building the product layer-by-layer. Each mold layer consists of two materials added sequentially:

1. Molding sand: the sand feeding and leveling device feeds the molding sand to the surface to the construction chamber.
2. Binder: the print head selectively applies casting resins to the sand. The activator in the sand makes the binder strong. Thus, a single layer is formed. The process is repeated until the mold is built.

2.8. Digital technologies for pouring smelt into molds

In many cases, there is a maximum allowable speed limit for smelt rising up the wall of the mold cavity in the production of castings, for technological reasons. In these cases, software can be used to control the rate at which the melt enters the mold as a function of the time or of the amount of smelt from the casting start.

2.9. Digital technologies for knocking out from mold

For this and the next two stages of the foundry process, the use of digital technologies is not yet common.

However, computer technologies of knocking out from mold (in particular software) can be used to provide mathematical modeling to determine different characteristics of the rate of knocking out of the mold depending on various parameters. For example, modeling can be done to determine the temperature of the casting knockout. As is known, the higher the knocking temperature is, the shorter is the technological cycle of casting and the higher is the productivity of the molding and casting area. However, high knocking temperature is undesirable because of the casting destruction risk and defects formation or deterioration.

2.10. Digital technologies for cutting, cleaning and heat treatment of castings

At these stages, robotic tools can be used for cutting and cleaning of castings. As well as computer software for archiving and processing data (for example, data about the cupola operating mode).

2.11. Digital technologies for quality inspection of castings

Up-to-date digital technologies, including computer technologies, save time and money at the stage of quality inspection of castings. This is due to the fact that products are inspected using non-destructive test methods.

Such methods can identify most of the casting defects, including:

- cracks, shells, pores, inclusions;
- control the product geometry;
- control the casting surface and detect external defects.

So, at this stage, the following digital technologies are used.

1. Computer tomography. It can inspect products by obtaining a 3D model showing pores, voids and cracks in the casting.
2. Geometry inspection. The 3D models obtained using the tomography make comparison with the original data of computer-aided design systems (models built at the very first stage of the casting manufacturing process) possible.
3. Casting surface inspection using a digital microscope. Magnification from several tens to several hundred times is sufficient to visualize defects that are invisible to the naked eye.

These methods of casting quality inspection can quickly evaluate the product and require fewer man-hours compared to conventional destructive test methods.

At any stage of foundry process that accumulates input and output data, it is possible to mathematically simulate processes. This allows you to predict their progress and make real-time production decisions. Production floor practice uses more and more various computer simulation programs. All these programs are based on mathematical models. Models are very different: performed with a relatively complex mathematical apparatus and relatively simple. But they all contain three structural elements: variables, the relationship between variables, and a model parameter. For a particular company, input and output variables and the relationship between variables used in models of other enterprises may be quite adequate, but the model parameter must be different – its own. The most proper parameter has models built on the basis of technical and economic variables accumulated in the information dump of a specific company [10]. Of course, such a model will be a more reliable production control means due to a sufficiently large amount of information dumps in serial and mass production. In addition, a large amount of information makes it possible to build a preliminary draft version of the model without a complex mathematical apparatus by metallurgy engineers. This will significantly increase the probability of success when synthesizing a working version of the model with professional mathematicians.

3. Conclusion

Thus, it is easy to see that the use of digital technologies to a greater or lesser extent has an effect on all stages of the serial foundry process. At the same time, production is optimized by reducing the influence of the human factor on production processes and saving man-hours by automating operations. There is no doubt that digital technologies will continue to be coming into foundry production in the future.

References

[1] Bukht R and Heeks R 2018 Defining, conceptualizing and measuring the digital economy *International organisations research journal* **13** pp 143–72

[2] Levkina O Yu 2013 Application of information technologies in the organization of design and technological preparation of foundry production in aircraft industry *Academic News. Privolzhskiy region. Technical science* **1** pp 115–22

[3] Nikanorov A V 2018 Comparative analysis of computer programs for simulation of foundry processes *Bulletin of Irkutsk state technical University* **22** pp 209–18 DOI: 10.21285/1814-3520-2018-11-209-218

[4] Gruzman V M and Beznoskov D V 2018 *Device for determining the content of the clay component in molding sands: patent for utility model* (Federal Institute of industrial property. IPC № B22C 5/14. Patent № 195990. July 30, 2018) access mode: http://elar.urfu.ru/bitstream/10995/82164/1/0195990.pdf

[5] Kablov E N, Spivakov D D, Grits V V, Demonis I M and Gerasimov V V 2004 Computer control system for technological processes of alloy smelting and casting of single-crystal blades of GTE *Aviation materials and technologies* **1** pp 132–7

[6] Kravchenko A N, Beloglazov E D and Kadyrov E D 2006 Neural network technologies in the control system of the melting process in the Vanyukov furnace *Notes of the Mining Institute* **169** pp 140–3

[7] Odinochko V F and Kostylev D L 2007 Trends in the development of mixing systems *Foundry and metallurgy* **44** pp 125–8

[8] Kem A Yu, Kazartsev V O, Merker E E and Kharlamov D A 2014 Optimization of steel smelting in an arc furnace and ladle treatment in ladle furnace *Bulletin of Donskoy state technical university* **2** pp 66–73

[9] Almaghariz E S, Conner B P, Lenner L, Gullapalli R, Manogharan G P, Lamoncha B and Fang M 2016 Quantifying the role of part design complexity in using 3d sand printing for molds and cores (*International Journal of Metalcasting* **10** pp 240–52 DOI: 10.1007/s40962-016-0027-5

[10] Gruzman V M 2013 *Development of Information Dumps of Casting Production: Monograph* (Nizhny Tagil: NTI (branch) UrFU) p 108

Interactional experience with the employer during a selection of educational content in a single-industry town

R A Karelova

Ural Federal University named after the First President of Russia B N Yeltsin, Nizhniy Tagil Technological Institute, 59, Krasnogvardeyskaya str., Nizhniy Tagil, 622000, Russia

E-mail: riya2003@mail.ru

Abstract. Training of personnel for industrial enterprises of a single-industry town is an opportunity for the stable development of the local labor market. As a rule, single-industry towns of our country are mechanical engineering or coal-metallurgical towns. Nizhny Tagil is no exception. On its territory there are metallurgical, mining and machine-building enterprises. Colleges and universities that train specialists for town-forming enterprises can be a driver for the stable functioning of a single-town economy. For this, it is necessary to adapt the content of educational programs to the actual needs of employers. The article describes a number of problems that representatives of educational institutions had to solve in the process of interaction with an employer to update the content of education. Such problems were: difficulties in visiting enterprises, in particular, defense (machine-building) ones, ambiguity in the interpretation of the concept of 'competence' by teachers and employers, the difficulty of dividing workers by education levels on the part of employers. A simplified version of the employer survey form is proposed. It allows to determine the priority areas in the training of personnel for the industry.

1. Introduction

Single-industry towns have a special type of urban economy. This is due to the fact that a significant part of the employed population is provided by one (rarely two) large enterprise. The low level of diversification of the urban economy of such towns is negatively reflected in the jumps of the unemployment rate, the fall in the well-being of the townspeople during the periods of negative macroeconomic shocks and changes in the markets where the city-forming enterprise operates [1, p. 111; 2].

Single-industry towns of our country are mainly machine-building or coal-metallurgical towns [3]. Nizhny Tagil has both types of enterprises on its territory, in particular, a mining and metallurgical plant, carriage-building plant. Accordingly, state educational institutions of higher and secondary vocational education train specialists in the relevant specialties. The exception is small groups of institutions that train middle and higher-level specialists in areas of training related to the setting up of the city itself (medicine, pedagogy, public catering, etc.).

The model of a demanded graduate can be built taking into account the requirements for this specialist. Such requirements are formed from several positions:

– requirements from the state and society, which are formulated on the basis of regulatory documents, the content of professional and educational standards;

– requirements of the world of work (the needs of future employers);
– requirements of the subjects of the educational process, including the future specialists themselves.

Institutions of higher and secondary vocational education that train specialists for town-forming enterprises can be a potential driver for the stable functioning and development of the local labor market [4, 5, 6]. Adaptation of educational programs to the actual needs of town-forming enterprises is one of the tools to stabilize the economy of a single-industry town [3, p. 123; 4, p. 62]. This is possible only if stable practices of interaction and cooperation between enterprises and educational institutions are formed.

The experience of interaction with employers in the selection of the content of higher and secondary vocational education will be described below, the problems encountered in this work are indicated.

2. Materials and method

As part of the transition to new educational standards, it was necessary to update the content of curricula and educational programs of higher and secondary vocational education. The author of the article was directly involved in the modernization of the content of training future IT specialists at the Mining-metallurgical college and the Nizhniy Tagil technological institute (branch) of the Ural Federal University in terms of bachelor's and master's degrees. Information technologies are being actively implemented in all spheres of society, they make it possible to automate modern production facilities, and therefore the training of IT specialists for modern industrial enterprises is a task that requires a prompt solution.

The study used interview and questionnaire methods. The respondents were the heads of structural divisions of town-forming (machine-building and metallurgical) enterprises.

Surveys of employers in order to identify the needs for specialists and the structure of their competencies are a traditional way of interaction between the university and the employer [7].

It should be noted that all the respondents confirmed the relevance of selection the content of the training of future specialists with the town-forming enterprises.

Initially, it was assumed that a survey in the form of an interview would be conducted with representatives of the employer. In the course of a real conversation, representatives of educational institutions could ask clarifying questions in order to reveal as accurately as possible the needs of production in specialists.

The first problem of interaction is associated with the difficulty of the passage to the town's enterprises for representatives of educational institutions. This is because of the need to prepare a special pass. Or it is impossible at all due to the fact that the enterprises belong to the defense sector. These problems almost exclude the possibility of interviewing employers' representatives in the workplace. It also makes it necessary for employers to travel to the territory of educational institutions, which is not always convenient for various objective reasons.

Most employers were given questionnaires containing questions aimed at revealing the peculiarities of the activities of specialists in a particular field. It was assumed that employers will be able to indicate the competencies necessary for the work of specialists, as well as the elements that make them up.

Another problem arose here.

The basis of modern educational standards is such a concept as 'competence'. However, until now, even teachers and psychologists have not come to a unified approach to the definition of this concept [8, 9]. Employers do not understand the content of this concept either. For them it closer to the concept of 'labor functions', on the basis of which the existing professional standards are built. Therefore, employers could not fill out the questionnaires offered to them due to the many questions that arose.

Some employers say they need young professionals who are able to think independently, creatively and critically. Employers are confident that a college or institute graduate will receive all the necessary professional knowledge and skills at the workplace. This opinion reduces the effectiveness of interaction between educational institutions and enterprise-employers. Following this logic, the future specialist spends most of his time studying at a college or institute in vain. Such situations once again confirm the need for close effective interaction between educational institutions and the employer in terms of the selection of the content of vocational education. Such effective interaction would make it possible to train highly demanded specialists, and employers do not waste time on their retraining.

The questionnaires were adapted by representatives of educational institutions (Table 1). Employers were asked to indicate what specialists in a particular field have to work with (in our case, an IT specialist). Analyzing the required skills of a specialist, tools, software and technologies with which they will have to work, representatives of educational institutions can draw up new curricula in disciplines or modernize existing ones.

Table 1. Adapted table for employer survey.

Labor functions	Skills	Documents	Equipment Tools Materials	Software products	Technology

Another feature of interaction with employers is the fact that they do not share potential specialists by educational level. It is extremely rare that employers can say that a particular position requires a specialist with a certain level of education. Most often, employers describe the specialist and his labor functions as a whole, without specifying – it must be a middle-level specialist or a person with higher education.

That is, the employer, as a rule, says in general what a specialist must be able to work with for successful work at their enterprise. The task of representatives of educational institutions is to correlate the information received about the skills of a specialist by levels of education. This can be done, inter alia, with the help of existing professional standards, in which labor functions are divided according to the level of education of specialists.

Of course, these are not all the problems that arise in the interaction of educational institutions and employers [10–14]. Here we considered those of them that arose in the selection of the content of education.

3. Conclusion

As part of the survey of employers, it was possible to modernize the curricula and curriculum of the college in the direction of training 'Programming in computer systems', 'Computer systems and complexes', to modernize the content of training bachelors in the direction of training 'Applied Informatics' and 'Information systems and technologies' also to develop a new program for training masters – 'Applied Informatics', aimed at training IT specialists for Nizhny Tagil' industrial enterprises. This master's program includes three educational trajectories: 'Digital technologies in foundry', 'Digital technologies in mechanical engineering', 'Digital technologies in heat power engineering'.

On the one hand, employers are interested in graduates of higher and secondary vocational education institutions. They are especially interested in employees who are able to think independently and solve various problems, have critical and creative thinking. On the other hand, the process of interaction between representatives of educational institutions and employers is complicated by various, including objective, reasons. Any information received from employers should be analyzed and synthesized by representatives of educational institutions. Survey data should be differentiated according to training levels (according to professional standards), distributed according to relevant competencies.

References

[1] Romanenko K R, Shibanova E Y, Abalmasova E S and Egorov A A 2018 Higher education in single-industry towns: models, practices, challenges *University management: practice and analysis* **22** pp 110–25 DOI: 10.15826/umpa.2018.04.044

[2] Zheng C and Hong Y 2018 Current development and policies on Russia's single-industrial cities *Humanities and social sciences. Bulletin of the Financial university* **4** pp 95–102 DOI: 10.26794/2226-7867-2018-8-4-95-102

[3] Ahmedova D R 2011 Professional socialization of young people in a modern monotown: some process conflicts and means of improvement [Electronic resource] *Modern studies of social issues* **4** URL: https://cyberleninka.ru/article/n/professionalnaya-sotsializatsiya- molodezhi-v-usloviyah-sovremennogo-monogoroda-nekotorye-protivorechiya-protsessa-i-puti-povysheniya (accessed on 01.06.2020)

[4] Varbanova B 2018 Opportunities to improve interaction between employees and educational institutions *Int. conf. Knowledge-based organization* **24** pp 432–6 DOI: 10.1515/kbo-2018-0128

[5] Piyankova S G 2017 Improvement of the social and economic development pf single-industry regions in the Russian federation *R-Economy* **3 1** pp 12–22 DOI: 10.15826/recon.2017.3.1.002

[6] Bykova O N, Garnov A P, Stroev P V and Pivovarova O V 2019 Employment management policies in single-industry towns in the light of existing issues of precarious employment *Int. J. of economics and business administration* **VII 2** pp 123–34

[7] Zamyatina N and Pilyasov A 2016 Single-industry towns of Russia: lock-in and drivers of innovative search Foresight and STI Governance **10 3** pp 53–64 DOI: 10.17323/1995-459X.2016.3.53.64

[8] Bolshakova Z M and Tulkibaeva N N 2009 Competences and competency *Bulletin of the South Ural state university. Series: Education. Pedagogical Sciences* **24 pp** 13–9

[9] Zymnaya I A 2012 Competency and competence in the context of competence approach in education *Foreign languages at school* **6** pp 2-10

[10] Reshetnikov I N 2009 Actual problems of interaction of higher education institution and employer *Bulletin of Kemerovo state university of culture and arts* **9** pp 91–9

[11] Morozova E A 2016 The interaction between higher education institutions and employers as a conditions of high-quality training of graduates: experts' opinions *Bulletin of Kemerovo state university. Series: Political, sociological and economic science* **1** pp 70–6

[12] Guskova M V and Zvonnikov V I 2012 Interaction between employers and universities: yesterday, today and tomorrow *Economics of education* **4** pp 60–4

[13] Lizunkov V G, Morozova M V and Zakharova A A 2019 Educating advanced professionals for company towns industries: Russian and non-russian approaches and systems investigation [Electronic resource] *Modern problems of science and education* **5** URL: http://science-education.ru/ru/article/view?id=29217 (accessed on 01.06.2020)

[14] Semenova E A 2010 The problem of personnel training for the Russian industry [Electronic resource] *Russian institute for strategic studies* URL: https://riss.ru/analitycs/2306/ (accessed on 01.06.2020)

IOP Conf. Series: Materials Science and Engineering **966** (2020) 012129 doi:10.1088/1757-899X/966/1/012129

Possibilities of an artificial neural network use to control oxygen consumption in a converter shop

R A Karelova

Ural Federal University named after the First President of Russia B N Yeltsin, Nizhniy Tagil Technological Institute, 59, Krasnogvardeyskaya str., Nizhniy Tagil, 622000, Russia

E-mail: riya2003@mail.ru

Abstract. Converter steelmaking process uses high-performance processing plants and a high level of process automation. Complex multi-stage processes typically use many factors that have effect both on individual stages of the processes and on how a combination of the processes runs as a whole. This makes it necessary to select and/or develop control methods for the converter process having no full current data on process parameters and effects of random disturbances. Use of artificial neural networks is the most effective method available now for analyzing and predicting parameters of different processes (including converter process). This article deals with an algorithm for oxygen measurement based on chemical analysis of steel and process history through some period, using an artificial neural network and machine learning. Initial data were reviewed using correlation and regression analysis, the artificial neural network architecture chosen is reasoned as the most appropriate for the objective on hand and the results are analyzed. The element measurement accuracy achieved is satisfactory, which makes the resulting artificial neural network usable in an automation system for testing purposes. Network performance can be improved with the help of experts in the subject area, whose expertise can be used to correct the model.

1. Introduction

Converter steelmaking process uses high-performance processing plants and a high level of process automation. This makes it possible to produce a wide range of high-quality steel grades. However, complex multi-stage processes have many factors that have an effect both on individual stages of the processes and on the progress of the entire process [1]. This makes it necessary to select and/or develop control methods for the converter process having no full current data on process parameters and effects of random disturbances.

Relationships between factors and output variables can be constructed analytically based on theoretical analysis of processes or with the use of identification methods based on a real experiment (observing a real process and measuring of corresponding factor values and output parameters).

With the current level of advanced information technology [2], using artificial neural networks (ANN) is the most effective method available now for analyzing and predicting parameters of different processes (including converter process) [3–6].

IOP Conf. Series: Materials Science and Engineering **966** (2020) 012129 doi:10.1088/1757-899X/966/1/012129

2. Materials and method

2.1. Problem description

Converter steelmaking process with top oxygen blowing is the most common steelmaking process used by steel producers of today. Making a high-quality steel having a target chemical analysis and qualities is quite a challenge and therefore it is impossible without an automated steelmaking process control system. As liquid steel parameters cannot be controlled all the time in the ladle, the automated system software must include a package of mathematical models to calculate the technological parameter values for steel process. A large variety of models are available at the moment for qualification of the converter process, most of which relate to the blowing period.

The amount of oxygen is normally determined empirically, which is not always accurate enough. Therefore, a possibility was considered to use ANN and machine learning for determining the amount of oxygen needed for blowing.

2.2. Using ANN in the converter steel process control system and stating the problem

As explained above, the influence of the human factor on the technological process has to be reduced through delegation of the parameters analyzing authority to the ANN. The first step should be to analyze the statistical data collected in the workplace for a certain period of work, which will provide a sufficient set of information for analysis using the ANN. So, we have considered a set of data for more than eighty thousand heats done over several years of company operation.

Getting started with analyzing the data, it should be understood how they are used in the production process (what parameters are known before, after, and during the heat) and what relationships there may be between them. Since we need to find out if the amount of oxygen depends on other variables, we will use regression analysis tools. Regression analysis is a set of statistical methods for investigating the influence of one or more independent variables on a dependent variable [7]. This set of methods is also used to detect hidden relationships in data, which is required for this task.

The initial data for the static heat model are: the chemical analysis and temperature of hot metal in the converter furnace, the number and type of cooling agents added in the process, the lining status measured by the number of heats. This model calculates a total volume of oxygen required for blowing during the current heat.

While reviewing the data, a large number of outliers and garbage data was found in each set of values [8].

Having no noise in the data is an important task for regression analysis, which makes it possible to better define and refine the relationships between parameters. This operation is of primary importance for the output parameter. After the data had been cleared, its initial volume went down by about 30 %. Then, it was necessary to determine the ANN architecture/machine learning method. For this purpose, a few alternatives were selected, and each model was tried.

2.3. Investigation of artificial neural networks applicability for oxygen measurement (results)

Investigation was using real parameters of more than 40000 heats produced at the same steel mill. The input variables are: values of variables known before blowing (hot metal weight, scrap weight, lime weight, hot metal temperature, hot metal analysis (C, Mn, Si, P, S, Al, Cr, Ni, Cu, V, Ti), total volume of oxygen per heat, metal temperature, amount of nitrogen for stirring, heat ID number within the converter campaign and values of parameters after blowing: temperature and chemical analysis of the final steel (C, V). The output parameter of the model is the amount of oxygen for the current heat. A correlation analysis was performed before developing the ANN-based model, wherein correlation coefficients between input and output variables were determined. Eventually, the following variables were chosen to be submitted as input values after correlation analysis:

- temperature;
- hot metal analysis;

- the purpose of the metal;
- age of the lining;
- chemical analysis of hot metal at tapping (C, V);
- type and number of cooling agents;
- tap-to-tap time.

The Sklearn and Tenzorflow (tf. keras) libraries were used to create and train a model based on ANN/machine learning [9]. The Scikit-learn library is the most common choice for classical machine learning tasks due to its simplicity. One of the main advantages of the library is that it uses a few widely distributed mathematical libraries as a basis and has no problem with integrating them. This module includes popular libraries such as: Pandas, SymPy, SciPy, NumPy, IPython, Matplotlib [10]. TensorFlow is an open source machine learning software library that can be used by a wide range of machine learning techniques and includes an integrated logging system and a log visualization system.

Since no exact relationship between input and output variables is known, a preliminary search was first made for the most appropriate ANN architecture/type of machine learning. See Table 1 for general test results for models under review.

Table 1. Test results for the most appropriate ANN architecture.

	Sequential with 2 fully connected hidden layers (TensorFlow)	Linear Regression (Sklearn)	Random Forest Regressor (Sklearn)
Average error	135	120	101

While creating the ANN models, 70 % of the set values were used for training, 15 % were used for testing, and 15 % were used for verification. All parameters of ANN/machine learning algorithms were selected at random from the above list.

See Listing 1 for implementation of training and prediction of the parameter with the Scikit library.

Listing 1 – Basic blocks for the Scikit library

```
#%%
data = pd.read_excel("C:/Documents/data.xlsx")
Y = data['SUM_O2_CONSUM']
X = data.drop(['SUM_O2_CONSUM'], axis=1)
X_train, X_test, Y_train, Y_test = train_test_split(X, Y, test_size=0.3)

#%%
rf = RandomForestRegressor(n_estimators=1000)
rf.fit(X_train, Y_train)

#%%
predictions = rf.predict(X_test)
errors = abs(predictions - Y_test)
print('Mean Absolute Error:', round(np.mean(errors), 2))

#%%

output = pd.DataFrame()
output['SUM_O2_CONSUM'] = Y_test
output['NEURO_PREDICT'] = np.array(np.round(predictions, 1))
output['ERROR'] = np.array(np.round(errors, 1))

output.to_excel('C:/Documents/ predict.xlsx')
```

The module import block is skipped in the above listing. Since the code was written in the Spyder development environment using a block structure, it is divided by '#%%' characters into blocks that will be further listed in order from top to bottom.

In the first block, data is read from an Excel sheet and broken down into a test set and a training set.

In the next block, the algorithm for training a random regression forest from the Scikit library with the number of trees equal to 'n_estimators' is initialized.

Then the training process starts using the specified algorithm and a previously defined set of training data. At the end of the training, a test is done, and an average square error is calculated.

The last block writes results to the Excel sheet along the specified path.

In practice, the accuracy of forecasting depends more on the quality of the data set submitted for training than on the choice of model type and its optimization Given that the prediction error is quite large, it was decided to continue processing the data and continue testing the algorithm in order to increase its accuracy.

Let us look at the ways to handle the data with reference to the task on hand:

- Removal of collinearity. If the relationship is linear, it will be found most quickly and accurately if there are strongly correlated input variables. To remove collinearity, pair correlations are to be calculated and eliminated.
- Gaussian distribution. Linear relationships can also be predicted more reliably if the input and output variables have a Gaussian distribution. To achieve this, some transformations (for example, log or Box Cox) can be used for our data.
- Scaling of inputs. Forecasting is more reliable if input data is scaled using standardization or normalization.
- Getting rid of meaningless predictors. Some algorithms may be less effective with variables that have little or no significance. From the point of view of mathematics, such variables have a fairly low variance.

After the algorithm had been written and trained, we got that the error in the training data is 27 cubic meters. This accuracy was considered satisfactory by experts.

3. Conclusion

With reference to the results that are found satisfactory, it makes sense to start using the resulting artificial neural network in real conditions for testing purposes. This being said, testing a project will serve as a way to improve its effectiveness. To assess the adequacy of the results obtained, the help of specialists in this subject area will be needed, whose expertise can be used to correct the model, if necessary.

References

[1] Gulakov K V 2013 The choice of neural network architecture to solve the problems of approximation and regression analysis of experimental data *Bulletin of Bryansk state technical University* **2** pp 95–105

[2] Bukht R and Heeks R 2018 Defining, conceptualizing and measuring the digital economy *International organisations research journal* **13** pp 143–72

[3] Chistyakova T B, Kudlay V A and Novozhilova I V 2017 Intelligent system for modeling the wear-and-tear dynamics of steelmaking converter lining *XX IEEE Int. Conf. on Soft Computing and Measurements SCM* pp 259-261 DOI: 10.1109/SCM.2017.7970554

[4] Kumar D S, Prasad G, Vishwanath S C, Ghorui P K, Mazumdar D, Ranjan M and Lal P N 2013 Converter life enhancement through optimisation of operating practices *Ironmaking & Steelmaking* **34 6** pp 521–8 DOI: 10.1179/174328107X203903

[5] Kuzichkin A A 2016 Neural networks in industry and information technology *Innovation management* **3** pp 46–9

[6] Chistyakova T B, Kudlay V A, Novozhilova I V, Suvorov S A and Kozlov V V 2016 The system of decision-making support for refractory lining service BOS converters *News of SPbGTI (TU)* **37** pp 60–6
[7] Mosteller F and Tukey D 1982 *Data analysis and regression*: In 2 issues Issue 1 (Moscow: Finance)
[8] Kachanovskij Yu P and Korotkov E A 2011 Data preprocessing for neural network training *Fundamental research* **12** pp 117–20
[9] Aurelien J 2018 *Applied machine learning using Scikit-Learn and TensorFlow: concepts, tools and techniques for creating intelligent systems* (SPb.: OOO Alfa-kniga) p 688
[10] Sozykin A V 2017 Review of deep neural network training methods *Bulletin Of YuUrGU Series: Computational mathematics and computer science* **6** **3** pp 28–59 DOI: 10.1520/cmse170303

IOP Conf. Series: Materials Science and Engineering **966** (2020) 012130 doi:10.1088/1757-899X/966/1/012130

Possibilities of information technologies application to production problems solving by math modeling methods

V M Gruzman

Ural Federal University named after the First President of Russia B N Yeltsin, Nizhniy Tagil Technological Institute, 59, Krasnogvardeyskaya str., Nizhniy Tagil, 622000, Russia

E-mail: gruzman-vm@ntiustu.ru

Abstract. Nowadays there exist a great number of program packages, allowing solution of common tasks of automation of the math modeling. Industries with their specific features may demand particular approaches to solving their problems. Modern information technologies allow development of a special application software for processing a specific task in such cases. The article suggests a description of a general architecture concept of the application software, which can automate tasks of the math modeling to solve different issues related to the industrial production. It describes a detailed algorithm of data processing in such a product using the capabilities of the information technologies. The capabilities of such an application software are shown in the example of preliminary draft model design of large pouring ladle stopper durability. The results that have been achieved with the help of the extemporized model allow making a conclusion that a model that has been built by the same techniques based on an annual scope of collected information will make it possible to get reliable answers for the tasks of production management.

1. Introduction

Nowadays the whole world is widely using systems engaging employees in active participation in improvement of technologies and equipment utilized at an enterprise. Such a system known as 'Think Tank' successfully operates at the Nizhny Tagil Iron and Steel Works. Nevertheless, not only the used real production processes need to be improved, but also do methods for making decisions on their replacement with more advanced processes. These days they are usually solved by formal hourly discussions of drawings in a narrow circle of supervisors, and sometimes individually.

At the Nizhny Tagil Technological Institute (branch) of Ural Federal University, a method for making decisions on purchase of new technologies and equipment using the modeling as a management tool has been developed. The use of modeling (both mathematical and physical) allows engaging workers and foremen in this procedure. For instance, when the production qualities of proposed equipment are under discussion, the developer must present its acting physical model. This factor makes it possible for workers and foremen to actively participate in the discussion and make appropriate adjustments.

Various computer modeling programs are increasingly used in the practice of plants. The basis for functioning of all these programs are the math models. There is a variety of models: the ones executed using a relatively complex mathematical apparatus and the ones executed using a relatively simple mathematical apparatus. But each of them contains three structural elements: variables, a relation between the variables and a model parameter [1–4]. For this specific enterprise, input and output variables and relation between variables used in models of other enterprises may be quite adequate, but a model parameter shall be inherent and distinct. The most inherent parameter appertains to models, built on the basis of technical and economic variables of this specific enterprise, accumulated in its information dump [5]. Obviously, that kind of model will be a more reliable tool of production management due to sufficiently large scope of information dumps in series and large-scale production. Moreover, a large scope of information makes it possible for metallurgic engineers to build a preliminary draft version of a model without applying a complex mathematical apparatus. It will significantly increase the likelihood of success during cosynthesis of a run-time version of a model with professional mathematicians.

With successful purchase of a new technology, a problem of its modernization arises over the course of time. The idealized technology will certainly be followed by appearance of heavy shortcomings in it. As a rule, when technological optimization is required, one resorts to metallurgical engineering theoreticians in big cities. This resort is formed by a small group of engineering technologists. Before sending such a resort, the new method makes a provision for creation of a draft math model of a process, which will allow specifying a task for the theoreticians to the utmost, and most importantly, it allows taking the necessary part in solving this problem.

2. Materials and method

In the age of advanced technologies and fast-growing scopes of information, a huge role is given to software solutions allowing processing of information quickly, easily and conveniently, as well as presenting visualized and tangible result in the end.

Nowadays there are many software packages giving a possibility to a person (even the one, who does not have any programming skills) to solve complex tasks, including the tasks of math modeling. Among them one can find such solutions as MathCad, MATLAB, Excel and other popular software packages.

However, ready-made solutions in the field of information technologies do not always make it possible to solve a specific range of industry-related problems. In these cases, it's possible to opt for creating an inherent decision software package for processing of any specific problem [6–8]. This approach allows flexibly elaborating a specific problem solution.

Currently, the popular and effective means for development of software solutions are Microsoft Visual Studio and QtCreator. The first one will allow writing and debugging a code in a variety of languages, among which are C++, C# Python. QtCreator makes it possible to develop solutions in languages C++, Python and other.

The approach to development of a software package lies in the analysis of a task with its further division into logical blocks, their establishment in the form of algorithms and implementation of a sequence of actions in a code. Algorithms for solving subtasks, into which the target task has been divided, are taken from sources in the Data Science field or from the mathematical apparatus. Then it is necessary to connect the algorithms to each other by interfaces. Afterwards, the algorithm is presented in a code of a chosen programming language. The source code can be composed in the form of a single file, but when writing user-defined classes, the project source code is usually presented by several files: one of them is main, and the rest are devoted to the description of other user-defined classes. They implement auxiliary fields and methods solving the subtasks and storing intermediate data. Initializing of classes instances and call for their methods are performed in the main file. If a program has a user interface, then a source code file accounting for an active window often serves as the main file. Input data is put into fields of the classes by means of setters and/or constructors and is processed when the methods are called for. The calculation results are output to a console or to a form by means of a language. But before that they are requested from fields of classes instances by calling methods-getters.

The compilation and linkage editing are carried out through a special software suit which is usually included in development environments. After these procedures, the source code becomes a software.

IOP Conf. Series: Materials Science and Engineering **966** (2020) 012130 doi:10.1088/1757-899X/966/1/012130

Then its testing and debugging is performed. A certain data set is supplied to the input of the software. A result is compared with an expected one according to the test.

The software receives the input data set from the file. The user indicates on a field form which columns refer to input (x) system parameters and which refer to output (y) ones. The data is parsed and is put into a container, which is transferred for a further analysis by auxiliary methods of class of tools. The correlation analysis is made by columns of input parameters. The following calculations are made according to the mathematical apparatus of design of experiment. Based on the input data, an experiment design matrix and extremum seeking of a response function by a steepest ascent method are executed. If the response function is inadequate at this point, then it is an optimum point. Additionally, an analysis of coefficients significance is carried out by the Fisher's ratio test in the process of calculation. Non-significant coefficients are suppressed and relevant input factors are not taken into account in the future. The calculations are carried out by calling the methods of a class of tools, each of which is in charge of its subtask and its part of calculation. Results and main points of calculations are entered into the form in a user-friendly style.

An example of a possible preliminary draft model design of large pouring ladle stopper durability with the participation of all employees related to a real site of production (direct and indirect workers, foremen) is given below: Having a monthly scope of accumulated information available, the methods presented in the paper [5] are used. The number of melts poured without replacement of the stopper (Y) has been taken as the output variable of a modeling object. By interviewing all the employees involved in preparation and operation of the stopper, all the factors that possibly affect the stopper durability have been listed. 8 of them are defined as the input variables (X):

1. Content of MnO in slag (x_1).
2. Melt temperature (x_2).
3. Pouring rate (x_3).
4. Content of Al_2O_3 in the chemical composition of the stopper body (x_4).
5. Content of MgO in the chemical composition of the stopper head end (x_5).
6. Heating-up temperature of the stopper before pouring (x_6).
7. Melt level in the ladle (x_7).
8. Diameter of the ladle nozzle (x_8).

The opinions of the participants in model synthesis on the power of influence of each factor on the stopper durability have been mathematically processed. The processing has showed the strongest influence of the MnO content in the slag, of the Al_2O_3 content in the stopper body and of the pouring rate on the durability, which has not contradicted the experience of using the stoppers at this enterprise.

The second (x_4) and the third (x_3) factors by the strength have been selected from the three most powerful input values according to the capabilities of production variation. Taking into account the small scope of technological information, a linear dependency between the variables of a future model has been taken. In spite of a small scope of available information, the dependency of stopper durability on pouring rate (x_3) and Al_2O_3 (x_4) content in coded variables has been obtained.

$$Y = 11 - 3x_3 + 2x_4 \tag{1}$$

From expression (1) it follows that the pouring rate has a stronger influence on the stopper durability than Al_2O_3 by the value and sign of the coefficient under the factor, and as the rate increases, the durability decreases, and Al_2O_3 increases the durability. This information is consistent with the peculiarities of the real process of melt pouring. Theoretically, the resulting model can be used for operational control of the technological process. For this, expression (1) has been transformed into an equation with natural variables (2).

$$Y = -0.7 - 12x_3 + 0.33x_4 \tag{2}$$

The transition to the natural variables is necessary for the operational solution of the production tasks. Suppose, for example, that there is a need to purchase stoppers with a lower Al_2O_3 content due to financial difficulties. The average-weighted Al_2O_3 content in the monthly technological information (Table 1) is 70.7 %. If stoppers with the content of 76 % are rejected, the average value will decrease to 66.1 %. The average durability of stoppers per month is 14 full melts. In order to maintain this level of durability when using cheap stoppers with a lower Al_2O_3 content according to expression (2), it is necessary to reduce the pouring rate to 0.59, which is at disposal of the plant's technological management.

Table 1. Distribution of Al_2O_3 content in stoppers.

Al_2O_3 %	Frequency
64	7
66	1
68	8
76	14

3. Conclusion

Of course, the technological parameters that have been obtained with the help of the presented extemporized model cannot be applied in production, but the model that has been built by the same techniques based on the annual scope of collected information will make it possible to get reliable answers for the tasks of production management.

References

[1] Chaykin A V, Volnov I N and Chaykin V A 2009 Study of cast iron modification process using methods of mathematical statistics and modeling *Vestnik of Nosov magnitogorsk state technical university* **1** pp 41–5
[2] Bigeev V A, Chernyaev A A and Panteleev A V 2014 Mathematical modeling of the two-stage method for sludge and dust recycling *Vestnik of Nosov magnitogorsk state technical university* **3** pp 48–52
[3] Yanishevskaya A G 2009 The mathematical models of the industrial system in foundries *Bulletin of Voronezh state technical university* **11** pp 230–4
[4] Leushin I O and Reshetov V A 2014 The use of mathematical modeling in technological preparation of foundry *J. of Siberian Federal university engineering & technologies* **4** pp 430–5
[5] Gruzman V M 2013 *Development of Information Dumps of Casting Production: Monograph* (Nizhny Tagil: NTI (branch) UrFU) p 108 [In Russian]
[6] Kosenko D V, Voronova L I and Voronov V I 2014 Developing software for processing complex structured data of scientific experiment *Vestnik of Nizhnevartovsk state university* **3** pp 45–52
[7] Safonov D S and Logunova O S 2015 Automation of the design of secondary cooling sections of the continuous casting machine *Vestnik of Nosov magnitogorsk state technical university* **1** pp 111–27
[8] Baydyuk A P and Kovalev M I 2013 Quality management system on the example of foundry *Izvestiya Tula state university. Technical sciences* **4** pp 252–60
[9] Kosach A A and Kovshov E E 2018 Automation of non-destructive testing data processing based on artificial neural network *Cloud of Science* **5 3** pp 524–31
[10] Kosach A A and Kovshov E E 2017 Software and hardware for industrial automation in the management of remote leakage detection control *Cloud of Science* **4 2** pp 264–73

Investigation of skew piercing and upsetting effect on wheel steel mechanical properties

A V Fomin[1,4], V P Romanenko[1], A N Nikulin[2] and A A Sevastjyanov[3]

[1] National University of Science and Technology MISiS, Moscow, 119991, Russia
[2] Federal State Unitary Enterprise, I.P.Bardin Central Research and Development Centre chermet, Moscow, 105005, Russia
[3] Vyksa Metallurgical Works, OJSC, Vyksa, 607060, Russia

E-mail: [4] fominmisis84@mail.ru

Abstract. The article states the results of the investigation of the skew piercing effect on the wheel steel mechanical properties. The comparative analysis is based on the comparison of mechanical properties of wheel steel samples manufactured by upsetting and heat treatment of the initial cast billets (option 1) and samples preliminarily pierced using the skew rolling mill with the extrusion ratio of $\mu \approx 1.49$ and feed angle of 12°, upset and heat treated (option 2). The values of elongation/contraction ratios of the samples made from solid workpieces are in the range 13.8–15/28–32%, and of hollow workpieces 14.2–15.8/26–32.9%. Strength properties of the wheel steel (for 1 and 2 variants) meet the requirements. The greatest differences in the properties of wheel steel of both variants were obtained during toughness tests. Toughness value of the wheel steel samples obtained by piercing, upsetting and subsequent heat treatment is ~2.5 times higher in both tangential and radial directions than the toughness of samples obtained only by upsetting and heat treatment. It is shown that piercing of workpieces in a screw rolling mill increases the toughness without reducing the strength properties, which may be due to intense shear deformations causing the processing of the cast macrostructure and the formation of a spiral-shaped grain structure.

1. Introduction

Railway wheels operation reliability and performance can be improved if plastic properties are enhanced provided that metal strength properties are maintained that can be achieved through preliminary deformation processing [1, 2]. Development of new processes and improvement of deformation processing parameters are among the effective methods of controlled influence on physical and mechanical properties of metal products [3–8]. In this regard the skew rolling is considered to be of high industrial capacity. Intense development of shear deformations even with relatively small reduction ratios can contribute to the processing of the metal structure both in the peripheral and central layers of the billet [9–11].

Preliminary processing of the initial cast billet structure by means of skew piercing of solid billets into shells [12–17] can increase the production effectiveness and enhance quality of railway wheels, hollow railway axles [18].

Investigation has been performed to evaluate effect of skew piercing and upsetting on wheel steel mechanical properties. The comparative analysis is based on comparison of mechanical properties of wheel steel samples manufactured by upsetting and heat treatment of initial cast billets (option 1) and samples preliminarily pierced, upset and heat treated (option 2).

IOP Conf. Series: Materials Science and Engineering **966** (2020) 012131 doi:10.1088/1757-899X/966/1/012131

2. Conditions for investigation

2.1. Samples material

Samples with metal cast structure with the diameter of 100 mm and length of 300 mm from steel type T (GOST 10791) cut from the peripheral and mid portion of the wheel workpiece are used as initial workpieces for the investigation [19]. The sketch of sampling from wheel workpieces is given in Figure 1.

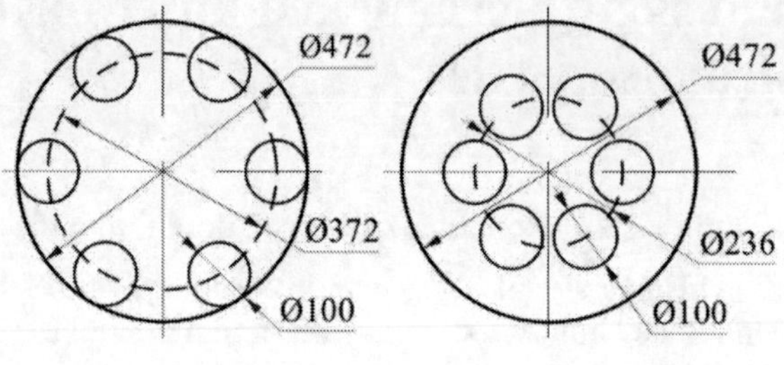

Figure 1. The sketch of sampling from wheel workpieces:
a – from the peripheral portion, *b* – from the mid portion.

2.2. Equipment

Piercing was performed using the two-roll skew rolling mill MISiS-130D. The two-roll piercing mill MISiS-130D with guide shoes is equipped with barrel-type work rolls with the diameter of 430 mm and length of 320 mm that are of bi-conical shape with the entrance and exit cones generatrix inclination angle of 3° and rolls highest points 10 mm wide located in the mid-section of the barrel. The rolls rotation rate is 57 rpm. Samples' upsetting was performed using the press power of 0.25 MPa.

2.3. Investigation methods

Piercing of samples into shells was performed at the metal heating temperature of 1200–1220 °C using the plug with the diameter of 29 mm provided that the following adjustment parameters are used for the mill:

- reduction in the rolls highest point is 15 %;
- feed angle is 12°;
- reduction factor is $\mu \approx 1.49$

The dimensions of solid samples for upsetting are as follows: the outside diameter is 100 mm, and the height is 65 mm. The sizes of hollow samples manufactured by piercing using the skew rolling mill are as follows: the outside diameter is 88 mm, the inside diameter is 32 mm and the height is 96 mm.

Upsetting of solid and hollow samples was performed at the billets heating temperature of 1200–1220 °C and the working tool travel rate of $v = 60$ mm/min. Billets upsetting using the press was performed in plane-parallel plates to the finite height of 23.5 mm.

Upon upsetting completion, both types of billets (initial cast billets and billets pierced using the skew rolling mill) were subject to heat treatment applicable for the manufacturing process intended to enhance end-user performance of finished products using the parameters that simulate interrupted quenching of railway wheel rims. Billets heat treatment parameters during quenching are as follows: heating up to 870 °C in the furnace and subsequent vacuum holding during 1 hour; transfer to the salt bath with the melt temperature of 570 °C and holding during 10 minutes, followed by cooling in air. Upon quenching, billets were subject to tempering at the heating temperature of 500–520 °C, with the holding time of 2 hours in the furnace under this temperature.

Thereafter, upset solid and hollow heat treated billets were used for taking samples to perform comparative assessment of wheel steel mechanical properties.

IOP Conf. Series: Materials Science and Engineering **966** (2020) 012131 doi:10.1088/1757-899X/966/1/012131

Investigation of billets metal mechanical properties was carried out using the samples with the test section diameter of $d_{test} = 3$ mm as per GOST 1497 [20]. Determination of metal mechanical properties was executed using the testing unit Instron with the deformation rate of $\approx 10^3 sec^{-1}$. The impact strength was determined using the samples of 5×10×55 mm size with U-notch 2 mm deep as per GOST 9454 [21]. Samples impact testing was carried out using the pendulum impact testing machine of PCB-30 model. The mechanical properties and impact strength of samples metal were determined at the temperature of +20 °C.

3. Results and discussion

As a result of the studies performed, it was established that the mechanical properties of solid and hollow billets metal upon upsetting using the press followed by heat treatment comply with the requirements of GOST 10791 for finished railway wheels, Figure 2. Values of the percentage elongation /contraction of samples manufactured from solid billets are within the range of 13.8–15/28–32 %, for hollow billets: 14.2–15.8/26–32.9 %.

As a result of deformation and heat treatment, the maximum variations in the wheel steel are identified for impact strength.

Wheel steel samples impact strength achieved by piercing, upsetting and subsequent heat treatment is by ≈2.5 times higher than the impact strength of samples that underwent upsetting and heat treatment only, both in the tangential and radial directions.

A higher level of wheel steel impact strength can be caused by intense shear deformations due to which processing of the cast macrostructure is carried out and swirl-pattern macrostructure is formed [22–25].

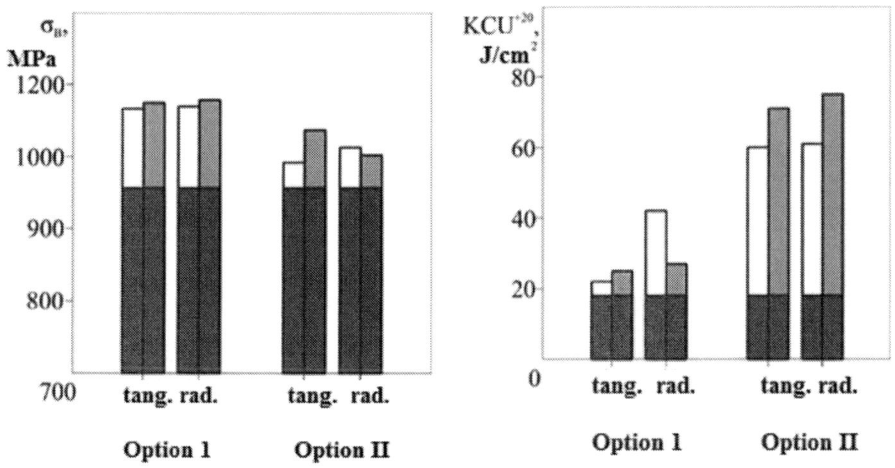

a б

- initial billets cut from the ingot mid-section;
- initial billets cut from the ingot peripheral portion;
- specification level complying with GOST 10791-2011.

Figure 2. Mechanical properties of wheel steel upon upsetting and heat treatment: a – tensile strength (σ_B), б – impact strength (KCU^{+20}); option I – mechanical properties of wheel steel upon upsetting of solid billets and heat treatment, option II – mechanical properties of wheel steel upon upsetting of pierced billets and heat treatment.

For products that undergo high dynamic impact loads during operation, enhancement of impact strength can contribute to the increase of their operability and reliability. Therefore, it should be

IOP Conf. Series: Materials Science and Engineering **966** (2020) 012131 doi:10.1088/1757-899X/966/1/012131

expected that application of billets preliminarily pierced using the skew rolling mill for manufacture of railway wheels will allow significantly increasing their quality and performance.

4. Conclusions

Based on the investigation performed it is proved that skew piercing allows for significant enhancement of impact strength with no deterioration of strength properties. It can be expected that application of billets preliminarily pierced using the skew rolling mill that have better performance parameters as compared to cast billets for production of railway wheels will contribute to the improvement of wheels performance and enhancement of technical and economical parameters of the production process.

References

[1] Parshin V A, Ganago O A, Staroselesky M I et al. 1968 *Die forging* **10** 8-9
[2] Bibik G A, Ioffe A M, Prazdnikov A V et al. 1982 *Production of railway wheels* (Metallurgy)
[3] Mao F, Wang F, Shuang Y, Hu J and Chen J. 2020 *Metals* **10**
[4] Malakhov A, Saikov I V, Denisov I V, Cherezov N P, Gamin Y V and Romantsev B A 2020 *J. of Physics: Conf. Ser.*, **1431(1)**
[5] Galkin S P 2004 *Stal'* **7** 63–6
[6] Galkin S P, Kharitonov E A and Romanenko V P 2009 *Steel in Transl.* **39** 700-3
[7] Samusev S V, Fadeev V A *Chernye Met.* **2020** p 41–6
[8] Budnikov A S, Kharitonov E A, Aleshchenko A S and Iskhakov R V *Chernye Met.* **2019** 41-5
[9] Romantsev B, Goncharuk A, Aleshchenko A, Gamin Y and Mintakhanov M. 2018 *Int. J. of Adv. Manufact. Techn.* **97** 3223–30
[10] Gamin Y V, Romantsev B A, Pashkov A N, Patrin P V, Bystrov I A, Fomin A V and Kadach M V 2020 *Izvestiya Vuzov. Tsvetnaya Metallurgiya* **1** p 27–38 [In Russian]
[11] Galkin S P, Romantsev B A and Kharitonov E A *CIS Iron and Steel Review* **2014** p 35–39
[12] Romantsev B A, Matyko O K, Goncharuk A V, Aleshchenko A S and Ananyan V V 2008 *Steel in Transl.* **38** p 897–9
[13] Romantsev B A, Aleshchenko A S, Goncharuk A V and Galkin S P 2012 *Metallurgist* **55** 918–24
[14] Romantsev B A, Goncharuk A V, Aleshchenko A S, Onuchin A S and Gamin Yu V 2015 *Metallurgist* **59** p 386–9
[15] Romantsev B A, Gamin Y V, Goncharuk A V and Aleshchenko A S 2017 *Metallurgist* **61** 217–22
[16] Romantsev B A, Goncharuk A V, Zimin V Y, Pakhomov V P, Aleshchenko A S and Matyko O K 2009 *Steel in Translation* **39** 803-05
[17] Romantsev B A, Aleshchenko A S, Tsyutsyura V Y, Tyshchuk I N and Lube I I 2017 *Metallurgist* **60** 1062–69
[18] Romanenko V P, Romantsev B A, Illarionov G P, Fomin A V, Zimin V Y and Stepanov P P 2014 *Metallurgist* **58** 684–88
[19] GOST 10791-2011 Solid-rolled wheels. Engineering specifications [In Russian]
[20] GOST 1497-84 Metals. Methods of testing the tensile [In Russian]
[21] GOST 9454-78 Metals. Methods of testing the toughness at low, room and elevated temperatures [In Russian]
[22] Romanenko V P, Fomin A V, Begnarskii V V, Yandimirov A A and Nikulin A N 2013 *Metallurgist* **56** 753–9
[23] Romanenko V P, Fomin A V and Nikulin A N 2013 *Metallurgist* **57** 303–9
[24] Fomin A V, Aleshchenko A S, Maslenniko I M, Galkin S P and Nikulin A N 2019 *Metallurgist* **63** 477–86
[25] Akopyan T K, Aleshchenko A S, Belov N A and Galkin S P 2018 *Physics of Metals and Metallography* **119** 241–50

IOP Conf. Series: Materials Science and Engineering **966** (2020) 012132 doi:10.1088/1757-899X/966/1/012132

Corporate culture as one of the key factors of effective industrial enterprise development

E A Putilova[1, 4] **and A V Shutaleva**[2, 3]

[1] Ural Federal University named after the First President of Russia B N Yeltsin, Nizhniy Tagil Technological Institute, 59, Krasnogvardeyskaya str., Nizhniy Tagil, 622000, Russia
[2] Ural Federal University named after the first President of Russia B N Yeltsin, 51, Lenin Avenue, Yekaterinburg, 620000, Russia
[3] Ural State Law University, 21, Komsomolskaya, Yekaterinburg, 620137, Russia

E-mail: [4]eazhiltsova@mail.ru

Abstract. The article is focused on the investigation of the impact the corporate culture makes on industrial enterprise development. It demonstrates that the formation of the corporate culture principles contributes to raising the level of staff involvement, its labor activity performance, maintaining and reproduction of human capital assets of an enterprise. Investments in the development of corporate culture are considered as an alternative to traditional methods of increasing the efficiency of an enterprise in an uncertain economic environment. Corporate culture development, which involves a commitment to raising the level of knowledge, innovativeness and organization, has a positive effect on the performance and efficiency of the entire company, and aids in the regulation of internal labor relations, preventing potential critical situations.The introduction and development of corporate culture, which includes a certain system of values and behaviors, in modern domestic industrial enterprises is becoming a necessity.The application of the principles and values of corporate culture leads to the long-term success of an industrial enterprise.

1. Introduction

In the modern world, corporate culture is becoming one of the significant factors of the existence and development of industrial companies. An industrial company can be defined as personnel united by work activity under the supervision of a management structure. The issues of corporate culture development are related to the issues of economics [1], social identity [2–4], business and production ethics [5], management [6, 7] and other aspects of human activity. Under conditions of the global economic crisis, which, due to the changes occurring in the modern world, objectively affects all the aspects of human activity, the role of the corporate culture at the enterprises is constantly growing. Therefore, researching the impact corporate culture makes on industrial enterprise development is of immediate interest.

The world crises lead to changes in the market of the entire industry, which in turn makes it necessary for companies to search for new solutions, tasks, targets, strategies on the macro- and micro-levels [4, 8]. In the article titled 'Ethical Turn in International Development' (published in 'International Encyclopedia of the Social & Behavioral Sciences'), E Fisher describes the ethical turn that is taking place in business and manufacturing [5]. E Fisher indicates that new directions of operation for socially just and environmentally sustainable development of society are dependent on the formation of the corporate culture. The ethical turn takes place as a result of production and trade globalization affecting social development.

Uncertain and volatile situations in the economy make it necessary for industrial companies to constantly look for ways to improve their performance. Usually, this goal is achieved through optimization, or decrease in the volume of resources spent on manufacturing, marketing and product research. However, there is another option – investing in corporate culture, which is becoming one of the more important tools, capable of changing the strategy of industrial enterprise development. .

2. Materials and Methods

Company efficiency depends both on the financial and non-financial indexes. M Zack, J McKeen and S Singh suggest estimating the effectiveness of the enterprise performance using five indicators that include innovativeness, rates of new product development, customer satisfaction, customer loyalty and operation costs [9].

Changes in a person's social role, knowledge and experience should lead to the re-evaluation of their value of a professional with unique skills and impact on company productivity – this is a precondition for implementing corporate culture at an industrial company. This evaluation is aided by the processes informatization and computerization that has led to the transformation of the entire social world and civilization culture. The growing role of individuals, their self-expression, their social partnerships has become one of the determining reasons for corporate culture development at industrial companies.

Social identification is directly associated with the ways of thinking and acting that correspond to a certain identity and supported institutionally. In her exploration of social identity issues, S L Brickson demonstrates that social identification depends on how an individual perceives their unity with a group of people [10]. Organization of the professional activity, leisure activities and holidays by the corporate employees plays an essential role in corporate culture development [11]. Therefore, corporate culture forms in the employees an identity, stereotypical image of themselves as well as the image of other people. As a result, corporate culture has an impact on group formation and introduces the preconditions for human identity. The distinctive features of social identity are reflected in the corporate culture, which affects the ways in which social involvement, intergroup relations and conflicts are managed within the company.

Human capital provides the foundation for the improvement of company performance through corporate culture development. Human capital can increase company efficiency since it is one of the competitive strengths of the enterprise in economic and social environments. It can be argued that the most significant part of the human capital is intellectual capital since it is comprised of competencies, knowledge, skills, creativity, communicativeness, loyalty and experience of the employees.

Human capital comprises the values, beliefs and established ways of communication between people in the organization; leadership and management modes play a major role in motivating employees to reveal their potential within the organization [12]. A Mention and N Bontis point out that the human capital of one organization might be fundamentally different from the human capital of another, which makes it a unique, rare and irreplaceable resource [13].

- Corporate culture deals with the human, informational, physical and technological resources and capabilities that are influence the performance and long-term competitive strengths of companies. Analyzing the corporate culture, J Kotter and J Heskett distinguish two levels;
- valuable (the concept of vital existential aspects that vary in different companies and may include money, intellectual property, inventions, the welfare of employees, relationship);
- behavioral (behavioral patterns that are encouraged by the organization and maintained by its employees).

In their studies, they have demonstrated a direct link between the performance of companies and the strength of their corporate cultures, which are defined by J Kotter and J Heskett as ‘socially constructed realities’ [14]. Modern industrial enterprises are arriving at the point where there is a need to familiarize the employees with the corporate culture, to promote their involvement, which leads to the change of views on human resources in general. In recent years, a growing number of industrial enterprises have adopted the following fundamental principles of corporate culture:

- customer-focused and employee-focused approaches that complement each other;
- openness of communication and interaction between employees on different levels of the organization;
- prioritizing employee security;
- focus on staff motivation;
- promoting self-identification with the company among its employees;
- team-building events, corporate parties;
- promoting distribution of responsibilities between structural subdivisions;
- promoting mutual help and other beneficial principles of communication between employees.

An industrial company that develops and makes use of the corporate culture, create on its basis a specific company image, reputation, moral principles and codes of conduct for the employees. In other words, it is the corporate culture that serves as a primary attractor for potential employees that are interested in adopting the company rules, existential categories, beliefs and standards of cooperation. With time, the principles mentioned above are internalized by employees to be applied without conscious effort.

In recent years, domestic industrial enterprises have been developing a tendency towards informality and flexibility of workplace communications [15, 16]. Such informal relationships have long been practiced in Western companies. However, in Russia, this tendency appears a freer form within isolated companies, since official documents do not regulate it. It is worth noting that such deformalization is one of the possible risk factors because introducing too much flexibility into an established system of corporate relationships can carry destructive potential. Moreover, the question arises of the appropriateness of changing the way of relations between management and team members and changing corporate culture. Therefore, it is necessary to build a mature axiological base of labor relations at large industrial enterprises in order to find the optimal ratio of formal and informal [17]. Digital technologies [18] and social technologies can help in finding such a balance, through innovations, information and constant development of the qualities of workers that reflect the always-changing demands of manufacturing. The introduction of social technologies in domestic industrial enterprises helps employees achieve their goals most rationally and quickly, thus increasing the productivity and overall performance of the entire organization, as well as helping manage internal labor relations and preventing critical situations.

3. Results and discussion

The corporate culture of modern domestic industrial enterprises is a certain system of values and behaviors, imposed on and approved by both the enterprise and employees. The corporate culture is based on the moral and ethical principles used in the course of work and communication activities. The mismatch between the values of the employee and the organization makes it more difficult, if not outright impossible, to help them realize their professional skills in a way that would also contribute to the development of the company. Corporate values are the basis of organizational culture, and as such require active investment in the development, support and assimilation of employees.

The employees' performance does not depend solely on the development of skills and competencies necessary for professional activities. Improving the performance of an industrial company is a complex process that depends on both corporate culture and discipline.

Corporate culture is a major factor in industrial enterprise development since it affects the organizational effectiveness and communication of the enterprise employees. A dynamically changing world requires that enterprises develop and adapt to this dynamic world. Therefore, the demand for flexibility and readiness to meet the needs of the market and consumers makes such features as dynamism and innovativeness mandatory for a successful industrial enterprise. There exists a need for industrial companies to be able to resolve the contradictions between such contradictory processes as innovation and adaptation, conservation of the traditions and optimization. To compete successfully, a company should always be able to meet the challenges of the times, which reinforces the importance of the management style and decision-making system of the company management – the factors that affect the perception of the company among its employees.

Of particular importance are the personal characteristics of the company leaders. Their ideals and values become part of the company culture only when all interested parties, especially managers and employees, accept them. This is why the corporate culture, which consists of beliefs, values, behaviors, and decision strategies that shape the actions of company employees and form the basic identity of the organization and how it interacts with society, plays an especially important role in the activities of an industrial enterprise.

Creation of a positive experience of interaction between employees, the company's management needs to use comprehensive resources, including specialized training that improves business interactions. Corporate culture, with its values and standards, is essential to ensure the long-term success of the enterprise. J Bẹlak has developed an integrated management model – MER [19], according to which the key factors for the success of the enterprise are: compatibility, competitiveness, efficiency, culture, trust, ethics, ecology, entrepreneurship, synergy and philosophy. To ensure the success of an enterprise, its management should take into account both its external environment and its internal environment. In other words, the development of the enterprise should correlate both with the real needs of the external environment, i.e. market, and the needs of its employees, in order to motivate and stimulate their innovative behaviors.

4. Conclusions

The development of corporate culture in industrial enterprises is one of the most effective ways to increase its effectiveness. The corporate culture includes the organizational culture, environment and development strategy of the enterprise, individual and organizational training, and employee commitment to organizational goals. The formation of the principles of corporate culture, which includes a focus on improving the level of knowledge, innovation and organization, industrial enterprises increase the level of employee involvement, the effectiveness of their work. The multiplicity of understandings and studies of the corporate culture at enterprises reflect its complexity and versatility, but in most cases, they are descriptive, narrative. Thus, the process of formation, nature, status and systematic corporate culture at the enterprise is not sufficiently studied and requires further analysis.

References

[1] Kotter J P and Heskett J L 1992 *Corporate Culture and Performance* (NY: Free Press)

[2] Brickson S L Organizational identity orientation: the genesis of the role of the firm and distinct forms of social value 2007 *Academy of Management Review* vol **32 (3)** pp 864–88 doi: 10.5465/AMR.2007.25275679

[3] Farkhidinova O M and Shumkova V A Overcoming ontological alienation in the urban environment 2019 *Journal of Siberian Federal University – Humanities and Social Sciences* vol **12 (5)** pp 771–79

[4] Adhikari B K 2016 Causal effect of analyst following on corporate social responsibility *Journal of Corporate Finance* vol **41** pp 201–16 doi: 10.1016/j.jcorpfin.2016.08.010

[5] Fisher E 2015 Ethical Turn in International Development *International Encyclopedia of the Social & Behavioral Sciences (2nd Edition)* ed J Wright Elsevier pp 60–62 doi:10.1016/B978–0–08–097086–8.10153–9

[6] Deal T E and Kennedy A A 1982 *Corporate Cultures: The Rites and Rituals of Corporate Life* (Harmondsworth: Penguin Books)

[7] Wan P, Chen X and Ke Y 2020 Does corporate integrity culture matter to corporate social responsibility? Evidence from China 2020 *Journal of Cleaner Production* vol **259** doi: 10.1016/j.jclepro.2020.120877

[8] Schmitz–Rixen T and Grundmann R T 2020 Surgical leadership within rapidly changing working conditions in Germany *Innovative Surgical Sciences* vol **4** **(2)** pp 51–7

[9] Zack M, McKeen J and Singh S 2009 Knowledge management and organizational performance: an exploratory analysis *Journal of Knowledge Management* vol **13** pp 392–409 doi:10.1108/13673270910997088

[10] Brickson S L 2007 Organizational identity orientation: the genesis of the role of the firm and distinct forms of social value *Academy of Management Review* vol **32** **(3)** pp 862–71 doi: 10.5465/AMR.2007.25275679

[11] Brammer S and Millington A 2005 Corporate reputation and philanthropy: An empirical analysis *Journal of Business Ethics* vol **61** **(1)** pp 29–44 doi: 10.1007/s10551–005–7443–4

[12] Tarus D and Sitienei E 2015 Intellectual capital and innovativeness in software development firms: the moderating role of firm size *Journal of African Business* vol **16** pp 48–65 doi:10.1080/15228916.2015.1061284

[13] Mention A and Bontis N I 2013 Intellectual capital and performance within the banking sector of Luxembourg and Belgium *Journal of Intellectual Capital* vol **14** pp 286–309 doi:10.1108/14691931311323896

[14] Kotter J P and Heskett J L 1992 *Corporate Culture and Performance* (NY: Free Press)

[15] Anderson C A and Anthony R N 1986 *The New Corporate Directors* (NY: Wiley)

[16] Bushmarin I V 2005 Formation of the Workforce: Experience of the West and Russia *Mirovaya ekonomika i mezhdunarodnye otnosheniya* vol **2** pp 48–52

[17] Bakeeva E V 2017 The onto logical sense of the concept of «measure» *Rivista di Filosofia Neo–Scolastica* vol **2** pp 471–83

[18] Tomyuk O N, Dyachkova M A, Kirillova N B and Dudchik A Yu 2019 Digitalization of the educational environment as a factor of personal and professional self-determination of students 2019 *Perspektivy Nauki i Obrazovania* vol **42** **(6)** pp 422–34

[19] Belak J 2010 *Integralni management – MER model (*Založba MER – MER Evrocenter, Maribor)

IOP Conf. Series: Materials Science and Engineering **966** (2020) 012133 doi:10.1088/1757-899X/966/1/012133

Automated tool support system of a large machine-building enterprise

A Bogoyavlensky[1, 2] and A Sirotyuk[1]

[1] Department of metal-cutting machines and tools, Ural Federal University named after the First President of Russia B N Yeltsin, 19, Mira str., Yekaterinburg, 620000, Russia

E-mail: [2] a.v.bogoiavlensky@urfu.ru

Abstract. The importance of proper organization of the enterprise's tool supply increases with the increase in its size and significantly affects the efficiency of the main production. The success of the entire enterprise, the quality of products, the rhythm of work and profitability largely depend on the correct organization of the tool supply. The main tasks of the tool supply: timely and uninterrupted provision of workshops and workplaces of the main production with high-quality technological equipment and tools (cutting, measuring, auxiliary tools); improving the quality of equipment and organizing its rational operation; reducing the cost of manufacturing, purchasing, storing and operating equipment; organization of sharpening and restoration of the tool. An automated system is required to solve all tasks related to control and management of the company's tool supply. This article discusses the construction of a unified tool support system for a large enterprise and the results achieved by the authors on its development.

1. Introduction

Existing tool accounting systems do not track the tool at all stages of its life cycle, only from receipt at the factory to storage in the shop. The standard scheme for moving tools to the shop is as follows: purchase > acceptance on the company's balance sheet > sending to tool-distribution storerooms > issuing tools at the request of workers. The tool is immediately issued to the workers just by writing in the log book. The following questions arise in the process of such accounting:

- whether the tool is used on the machine or is in the worker's locker;
- was it used directly for the parts it was intended for;
- whether the optimal cutting conditions were applied, whether the processing technology was observed;
- how many parts were processed and whether the tool life is consistent with the standard data;
- analysis of the causes of frequent tool breakdowns;
- theft of tools.

The lack of such information leads to inefficient use of the tool.

At large enterprises, the technological processes development is carried out simultaneously in several technology departments that are not connected with each other, and do not have information about the availability of tools in warehouses. Technologists can use the same type of tool from different manufacturers in the technical process, which will complicate the process of purchasing tools and increase the inventory of tools in warehouses. In the existing conditions, the tool request is submitted by the tool shop services. However, the list is not checked against the production plan, equipment connection, and list of approved tool manufacturers, is not fully checked. The supply department orders the tool from suppliers or conducts a tender procedure while a comparative analysis of the cost and functional characteristics of the tool from different manufacturers is not carried out due to the large list of purchased tools.

Tools and equipment of own production are designed by special design bureaus and manufactured in accordance with the tasks of the technological bureaus of the workshops. The process from the design of a special tool to its use in the shop is long and must be monitored to ensure the release of new products in the specified time frame. The process of manufacturing a special tool is usually monitored using paper document flow. After manufacturing, the special tool and accessories are included in the general database for the tool, where it is aligned with other tools and when ordering a new special tool, it is impossible to find the analog.

To solve or at least minimize the problems mentioned above, an automated tool support system (ATSS) is need that will store all the data about the tool in a unified database. We can obtain an all aspects control of the tool's lifecycle in this case.

2. Methodology

Tool support issues are widely discussed in existing publications. It is indicated that the tool provision is not given due attention, considering it as a second-rate process [1]. A number of authors consider the issues of economic efficiency and optimization of the purchasing and accounting processes of tool [2, 3]. Much attention is also paid to the development and analysis of tool usage [4–8], the analysis and search of unused tools, the formation of a knowledge base structure of the tools, the need to integrate tool support and existing product lifecycle management (PLM) and enterprise resource-planning (ERP) systems [8–14]. Publications on tool support issues have started to appear intensively in recent years. This allows us to conclude that the relevance of this topic has increased.

The authors believe that all issues of tool support should be considered in a complex, starting from the departments that need information about the tool. ATSS should cover all divisions of the enterprise that are somehow related to the tool, without exception. First of all, the main production workshops, where, due to the tasks set, the maximum tool turnover occurs. Auxiliary shops should also be included in the system field. The main enterprise divisions working with information about the tools is shown in Figure 1. Each division needs only a part of the general information about the tool. For example, financial services are more interested in information about the cost of the tool and its suppliers, and technology services are more interested in information about the geometry of the tool and its technological capabilities. Shops in the system can see the list of tools and the quantity in the enterprise warehouses. The tool Department, purchasing or ordering a tool for manufacturing, keeping records of the entire tool, has access to all types of information displayed in the system. Employees of the Bureau of tool management in the shops will be more interested in the availability and maintenance of tools in working condition directly in the shop. Most enterprises currently have ERP systems that perform tool accounting, but they do not provide for storing data about the geometry and technological capabilities of the tool. Enterprises also have pre-production systems that have tool geometry data and a search system, but these systems are not linked. The automated tool support system should combine unrelated systems and add missing functions. Each division must have its own access to information in the ATSS with restricted rights. The general structure of ATSS in the form of enlarged blocks is shown in Figure 2. Let's look at the main blocks of ATSS in more detail.

The System for recording the arrival, usage, availability and movement of the instrument by business units is a Central system that provides accounting and financial accounting functions for the instrument. Existing ERP systems, as a rule, perform these functions, and directly at the level of production units, accounting is transferred to paper, which limits the ability to analyze the tool information.

IOP Conf. Series: Materials Science and Engineering **966** (2020) 012133 doi:10.1088/1757-899X/966/1/012133

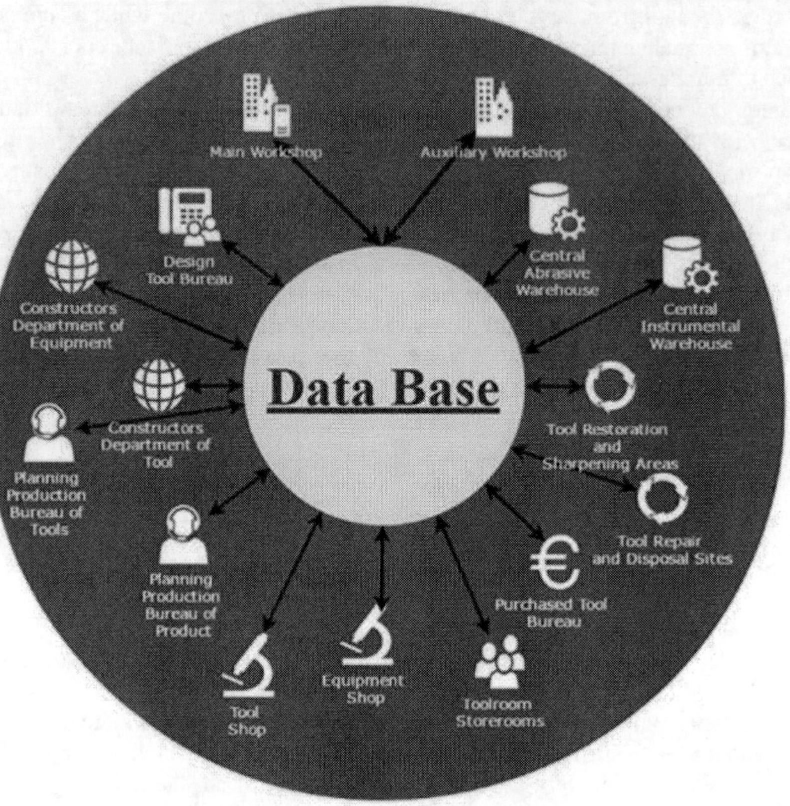

Figure 1. Users of tool support information.

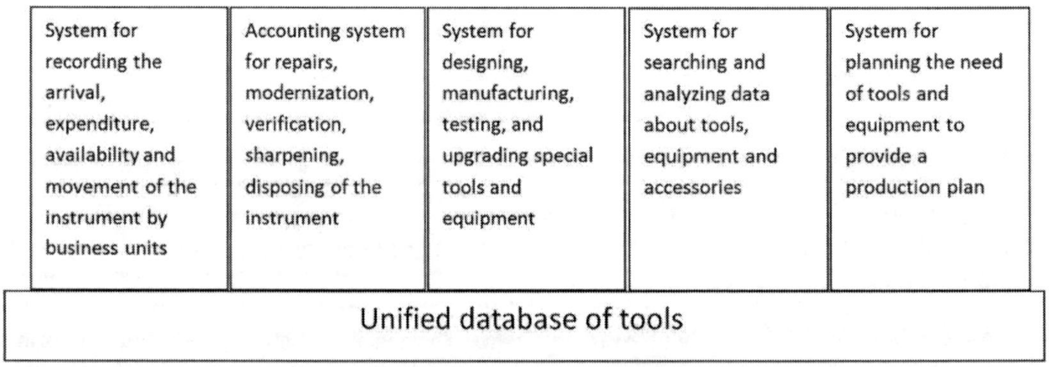

Figure 2. ATSS function blocks.

Accounting system for repairs, modernization, verification, sharpening, disposing of the instrument – currently in production, this system functions as a paper document flow, sometimes associated with an automated accounting system at the stage of tool write-off. The lack of automation of these aspects of tool support significantly limits the ability to analyze the conditions of use of the tool, identify the tool used in violation of regulations, unused tool.

IOP Conf. Series: Materials Science and Engineering **966** (2020) 012133 doi:10.1088/1757-899X/966/1/012133

System for searching and analyzing data about tools, equipment and accessories is an auxiliary system that provides the ability to find tool data, analyze its parameters, terms of use, estimate inventory, find unused tools, and perform many other tasks.

System for planning the need of tools and equipment to provide a production plan – this system allows you to plan in advance the purchase of tools and optimize the financial costs of providing tools. Currently, most enterprises make a tool purchase plan that is formed manually or with partial use of computer technology. The reason of this is a lot of parameters for correct purchase plan calculation. The tool need must be linked with the product release plan and with composition of manufactured products. The need for each type of tool for manufacturing parts, must be corresponded the current availability of tools in warehouses and in production.

System for designing, manufacturing, testing, and upgrading special tools and equipment is typical for large machine – building enterprises that have divisions for the design and production of special tools and equipment. The peculiarity of this system is the need to link in a single information flow all the work on the design of tool or equipment, the development of its manufacturing technology, manufacturing, testing and transfer to production. The process of interaction between various divisions of the enterprise in the process of providing with special tools and equipment is shown in Figure 3.

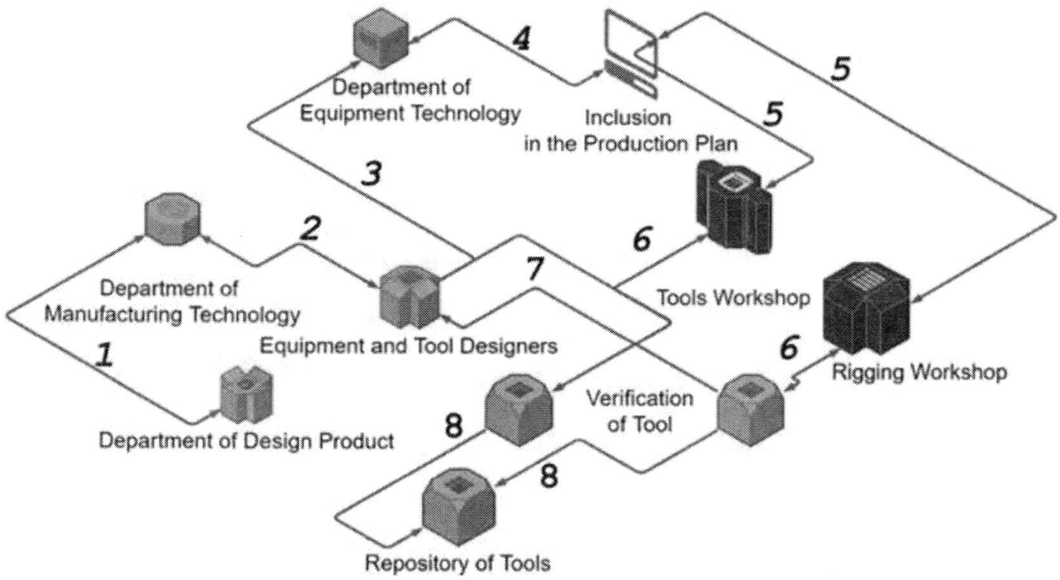

Figure 3. Scheme of providing special tools and equipment.

Numbers in the diagram Figure 3 show the sequence of design operations and the appearance of information in the database of tools. This ATSS function block is actually a workflow from ordering a special tool to commissioning it. By this way the tool or tooling is designed, planned, manufactured, tested, and transferred to the warehouse. The production process of special tools and especially equipment does not always go smoothly. Sometimes during the testing stage, some problems are detected, and the process is repeated from previous stages. The process is complex and must be constantly monitored.

3. Result

As a result of the research, a database for ATSS was developed. The distribution of information across the database tables is shown in Figure 4. Currently, we are developing and programming computer forms for viewing information about the tool for various departments of the enterprise.

IOP Conf. Series: Materials Science and Engineering **966** (2020) 012133 doi:10.1088/1757-899X/966/1/012133

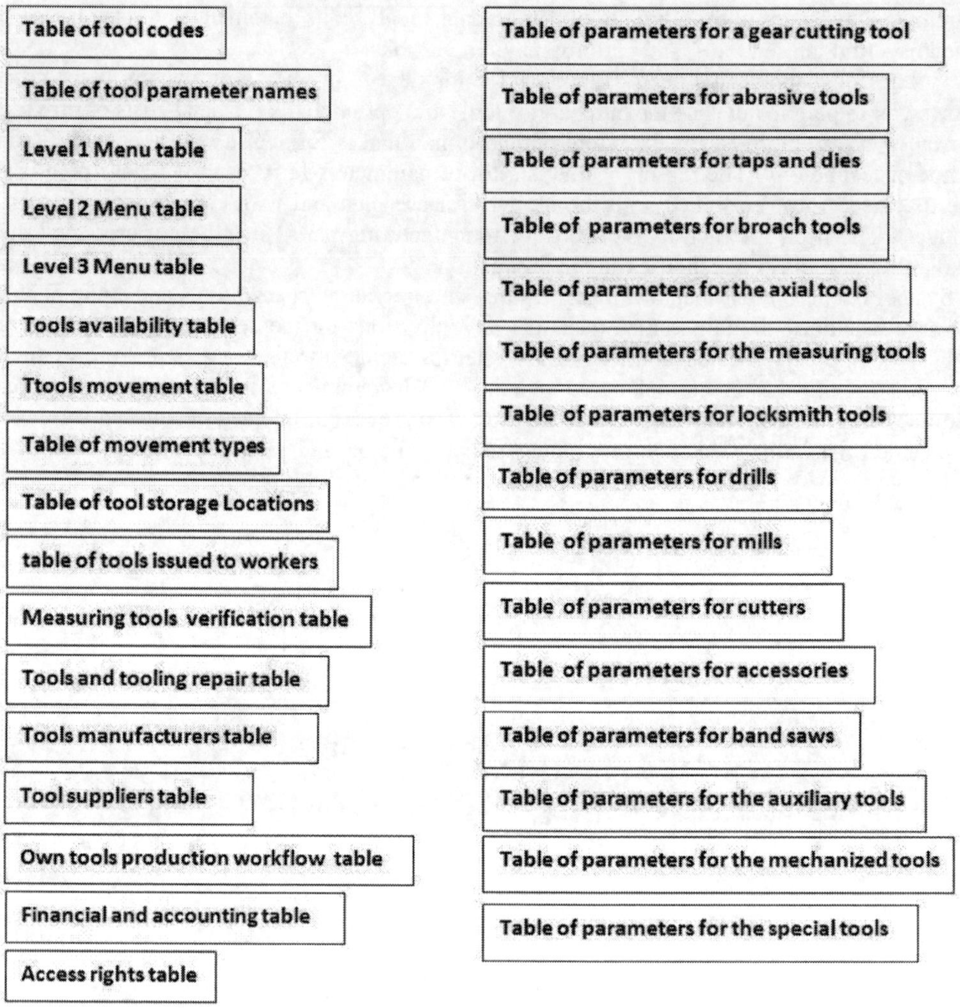

Figure 4. Data storage structure of an automated tool support system.

4. Conclusion

Tool support issues are not fully automated at the moment. It is necessary to develop an automated tool support system that should resolve all outstanding issues. The ATSS should be linked to existing systems of technological preparation of production and ERP systems and work as a single complex to improve enterprise production efficiency.

References

[1] Ponomareva V D 2020 Intelligent automated system for selecting auxiliary tools *Youth and science: Step to success, Collection of scientific articles of the 4th all-Russian scientific conference of promising developments of young scientists* (Southwestern state University. Kursk) pp 105–7

[2] Volochienko V A, Ermakov I Yu and Polyakova A A 2019 Optimal planning of tool support for machine-building enterprises *Organizer of production* **27 (4)** pp 38–50 [In Russian]

[3] Klementev P. A. 2015 Tool for ensuring machine-building enterprises *Youth scientific and technical Bulletin* **4** pp 45–8

[4] Ozbayrak M and Bell R 2003 A knowledge-based decision support system for the management of parts and tools in FMS *Decision Support Systems* **35 4** pp 487–515

[5] Khisamutdinov R M and Khisamutdinov M R 2015 Problems of complex automation of the tool creation and operation system *Eurasian Union of scientists* **9-2 (18)** pp 112–5

[6] Grigoriev S N, Grechishnikov V A, Maslov A R and Skhirtladze A G 2018 *Instrumental support of integrated machine-building industries* (Stary Oskol: Thin science-intensive technologies) [In Russian]

[7] Rui Wang. Wu Zhao, Chunjing Luo, Zhiyong Wang and Tao Luo 2016 Research on intelligent tool selection system based on optimization model *Engineering Materials* **693** pp 1765–71

[8] Newman S T and Nassehi A 2009 Machine tool capability profile for intelligent process planning *CIRP annals* **58 (1)** (Kidlington: Elsevier) pp 421–4

[9] Karpovich A S 2017 Automation of instrument support at the aviation enterprise *Problems of mechanical engineering and automation* **4** pp 158–9

[10] Chemborisov N A, Akhmetzyanov D R and Khisamutdinov R M 2009 Instrument support system *J STIN* **10** pp 30–3

[11] Cag M, dac Arslan, Bu, lent Catay and Erhan Budak 2004 A decision support system for machine tool selection *Journal of manufacturing technology management* **15 1** pp 101–9

[12] Drozdov A Yu, Denisova N. A, Doronin A and Vasyanin A 2017 Development and implementation of an information search system for existing measuring and cutting tools for the tool warehouse of the CNC machine shop *Proceedings of the XVII international scientific and practical conference 'Systems of design, technological preparation of production and management of stages of the life cycle of an industrial product (CAD/CAM/PDM-2017)'* (Moscow, V A Trapeznikov Institute of control sciences of RAS) pp 120–4 [In Russian]

[13] Simonova L A and Kostyuk I V 2009 Formation of the knowledge base of the tool supply system J. Forging and stamping production *Metal processing by pressure* **8** pp 22–8

[14] Subrahmanyam G, Gunasekaran A, Arunachalam S and Radhakrishnan P 1999 Development of a tool database management system *International journal advanced manufacturing technology* **15 (8)** (London: Springer) pp 562-5

Influence of thermal oxidation on the properties of titanium alloys

V I Golubev[1], V V Medison[1, 2, 4], D S Pastukhov[2] and V F Pegashkin[3]

[1]Ural Federal University named after the First President of Russia B N Yeltsin, 19, Mira str., Yekaterinburg, 620000, Russia
[2]VSMPO-AVISMA Corporation, 1, Parkovaya str., Verkhnyaya Salda, 624760, Russia
[3] Ural Federal University named after the First President of Russia B N Yeltsin, Nizhniy Tagil Technological Institute, 59, Krasnogvardeyskaya str., Nizhniy Tagil, 622000, Russia

E-mail: [4]medisonw@gmail.com

Abstract. In this study, the effect of thermal oxidation on the properties of the surface of titanium alloys was examined. Thermal oxidation was carried out at 750 °C within 6 hours, 800 °C within 5 hours, and 850 °C within 4 hours. Electrical resistance of the surface after thermal oxidation was measured. Metallographic analysis, microprobe analysis and X-ray phase analysis were carried out to determine the structure, chemical and phase compound of the oxide layer on the surface of titanium alloys after thermal oxidation. The results indicate that thermal oxidation leads to increase of electrical resistance of the surface of titanium alloys. Microstructural analysis showed that thickness of the oxide layer after thermal oxidation is in the range of 25 to 100 μm. Microprobe and X-ray phase analysis showed that TiO_2 and Al_2O_3 oxides have the best impact on electrical resistance. Two variants of possible application of the results have been presented.

1. Introduction

Thermal oxidation is a well-known method of titanium alloys surface properties modifying. Dense oxide layer occurs on the surface of titanium alloys after thermal oxidation. This layer is characterized by increased hardness [1], high corrosion resistance [2], good biotribological properties [1, 3] and high wear resistance [4]. This leads to widespread use of titanium alloys with oxide coatings in prosthetics and medicine and in chemical engineering. At the same time electrical properties of oxide layers grown by thermal oxidation on the surface of the titanium alloys are still insufficiently explored. Although high dielectric properties of TiO_2 (rutile) have been studied in [5–7]. Very large variation of the results was observed in studying the electrical resistance of the oxide layers on titanium, therefore only data on the distribution of the electrical resistance at the cross-section of the oxide layer is published in literature. It is believed that heating of titanium alloys at temperatures above 900 °C leads to the nominal stoichiometric composition of rutile crystals. It is confirmed by high electrical resistance above 10^{12} Ohm·cm at room temperature. Pure stoichiometric rutile is an isolator. Electrical conductivity increases in case of deviation from stoichiometric composition. Increase of electrical conductivity continues until the rutile homogeneity boundary is achieved.

Data on the influence of thermal oxidation on the electrical properties of the oxide layer is presented in Table 1 [7]. It is shown that the breakdown voltage of the oxide layer increased with the thickness of the layer, therefore oxide layer with dielectric properties can be formed on the surface of titanium alloys under certain conditions. Investigation of the microstructure of the oxide layer is also necessary because of the dependence of the electrical properties on the thickness of oxide layer. The presence of TiO_2, Al_2O_3 and other oxides, which have electrical resistance close to dielectric, in the compound of the oxide layer causes its high electrical resistance [8]. Thus, studying chemical and phase composition of the oxide layer on the surface of the titanium alloys is of scientific interest.

Table 1. Influence of heat treating on the dielectric strength of the oxide layer in commercially pure titanium.

Temperature, °C	Time, hours	Thickness of oxide layer, μm	Breakdown voltage, V
600	25	4	60
	100	7	130
	500	19	400
	1000	22	460
700	25	10	200
	100	25	500
	250	30	600

The purpose of the presented work was to study the electrical resistance of thermally oxidized titanium alloys, microstructure, phase, and chemical compound of the oxide layer. The application of thermally oxidized titanium alloys is described in the presented article.

2. Materials and methods

The measurements were made on titanium alloys Grade2, OT4-1, OT4, VT5, Grade6, PT-7M, Grade5, VT3-1, VST5553. Samples from these alloys with dimensions 40×20×5 mm after rolling were studied. Samples were subjected to thermal oxidation at 750 °C within 6 hours, 800 °C within 5 hours, and 850 °C within 4 hours.

Electrical resistance of the oxide layer was measured with ohmmeter Sonel MIC-1000 at the voltage of 50 V. Metallographic study was carried out to determine the structure of the oxide layer. The samples were polished on Struers LaboPol-1 and then etched using a composition of 50 % HF + 50 % glycerol. Microstructures were obtained with the optical microscope Olympus GX51equipped with Siams 700 system. Microprobe analysis was conducted using scanning electron microscope Jeol JSM 6490LV equipped with an attachment for energy-dispersive electron microprobe analysis Inca DryCool with a resolution of 130 e V. Analysis of chemical and phase compound of the oxide layer was carried out by X-ray phase analysis using the diffractometer Bruker D8 Advance equipped with a nickel-filtered copper Kα radiation ($\lambda = 1.540$ Å). Crystallographic database PDF2007 and software package EVA13.0.0.3 were used to decrypt the diffraction patterns. Semi-quantitative estimation of the phase composition was used to determine the proportion of phases.

3. Results and discussion

3.1. Electrical resistance

The results of measurement of electrical resistance of the titanium alloys after the thermal oxidation are presented in Table 2. The electrical resistance of almost all titanium alloys tends to grow after the thermal oxidation. It should be noted that alloys with vanadium have lower electrical resistance due [9] to formation of a volatile oxide V_2O_5 which has a melting point of 670 °C and lower in case of eutectic compounds formation with oxides of the other metals. The appearance of a liquid phase at the interface between the metal and the oxide leads to acceleration of diffusion of the oxygen and to formation of pores and discontinuities.

Table 2. Electrical resistance of the titanium alloys after the thermal oxidation, Ohm.

Alloy	Temperature of thermal oxidation, °C		
	750	800	850
Grade2	0.04×10^{13}	0.72×10^{12}	0.04×10^{13}
OT4-1	0.04×10^{13}	0.04×10^{13}	0.04×10^{13}
OT4	0.56×10^{12}	0.16×10^{12}	0.04×10^{13}
VT5	0.4×10^{11}	0.6×10^{12}	0.16×10^{11}
Grade6	0.88×10^{12}	10^{12}	0.84×10^{10}
PT-7M	10^{8}	0.6×10^{10}	0.04×10^{13}
Grade5	0.4×10^{7}	0.65×10^{7}	10^{9}
VT3-1	0.28×10^{10}	0.3×10^{11}	10^{11}
VST5553	0.36×10^{8}	0.45×10^{8}	0.36×10^{8}

Grade 6 titanium alloy which contains 2.5 % of tin also has lower electrical resistance. TiO_2, TiO and pure tin were observed between the metal and the oxide layer on Grade6 in [9]. Tin has a low electrical resistance and its presence in the oxide layer leads to decrease of electrical resistance.

It can be concluded that the aluminum in an amount of 2–4 % leads to a decrease of electrical resistance, but electrical resistance increases when amount of aluminum is more than 4 %. The greatest electrical resistance is observed in the alloys with an aluminum content of 0.8–1.5 % and manganese content of 0.8–1 %. A positive effect may be noted on the electrical resistance of zirconium and niobium at the oxidation temperature of over 800 °C. This is confirmed by studies of PT7M (Ti-2, 2Al-2, 5Zr) and Ti-6Al-7Nb respectively.

It follows from the above that TiO_2 lattice has both electron and hole conductivity. Alloying of rutile by the elements with a valence greater than 5 leads to ionisation. Alloying element transmits the electron to rutile and it leads to n-type conductivity and lower electrical resistance. Conversely alloying of rutile by the elements which have a valence lower than 4 leads to transmitting of the electron from rutile to the acceptor. This causes p-type conductivity and growth of electrical resistance [10]. Thus, we have shown that the thermal oxidation leads to growth of electrical resistance of the titanium alloys. It confirms the hypothesis which are presented in [5, 7].

3.2. Results of microstructural and microprobe analysis

One of photomicrographs of oxide layer on the surface of some titanium alloys after thermal oxidation, obtained with an optical microscope Olympus GX51, are presented in Figure 1. Thickness of oxide layer on most alloys is in the range of 10 to 100 μm. Microstructural studies have shown that a homogeneous layer consisting of dark and light regions is formed on the surface of most titanium alloys after thermal oxidation.

Results of microprobe analysis which are shown in Figure 2 confirm the hypothesis of the presence of several layers with different chemical compositions in the oxide layer.

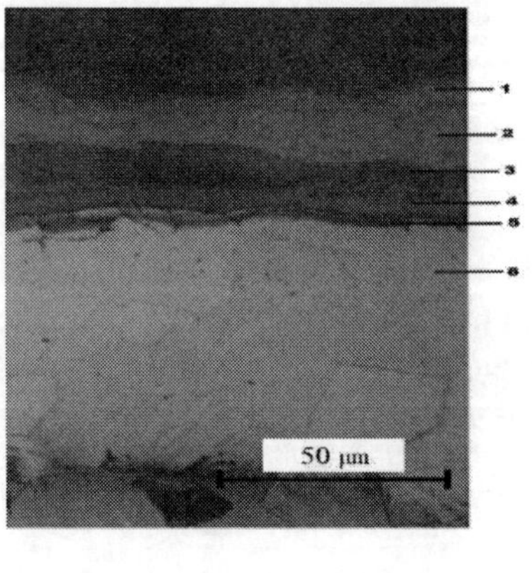

Figure 1. Photomicrograph of oxide layer on surface of titanium alloy VT5 after thermal oxidatic 850 °C:

1. Outer layer enriched with aluminum oxide Al_2O_3
2. The layer consists of TiO_2.
3. The layer consists of non-stoichiometric oxide the titanium, for example, $TiO_{1,9}$.
4. The layer at the interface with the metal consist intermetallic alloying elements displaced from oxide layer by diffusing oxygen.
5. Undefined layer.
6. The diffusion layer saturated by oxygen.

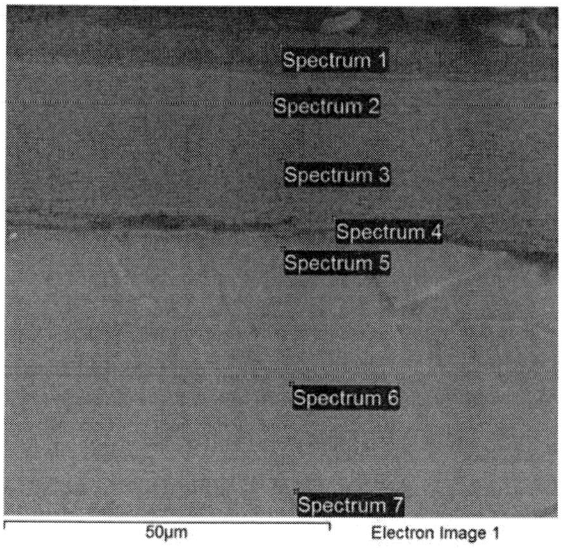

Spectrum	In stats.	O	Al	Si	Ti	Zr	Total
Spectrum1	Yes	53.51	5.90	1.11	39.47		100.00
Spectrum2	Yes	41.48	6.55		50.31	1.66	100.00
Spectrum3	Yes	37.88	0.29		59.92	1.90	100.00
Spectrum4	Yes	38.73	0.00		59.66	1.61	100.00
Spectrum5	Yes	12.56	1.72		83.21	2.52	100.00
Spectrum6	Yes	2.34	1.89		92.99	2.78	100.00
Spectrum7	Yes	3.39	2.12		92.35	2.14	100.00

Figure 2. Chemical composition of PT-7M after thermal oxidation at 850 C.

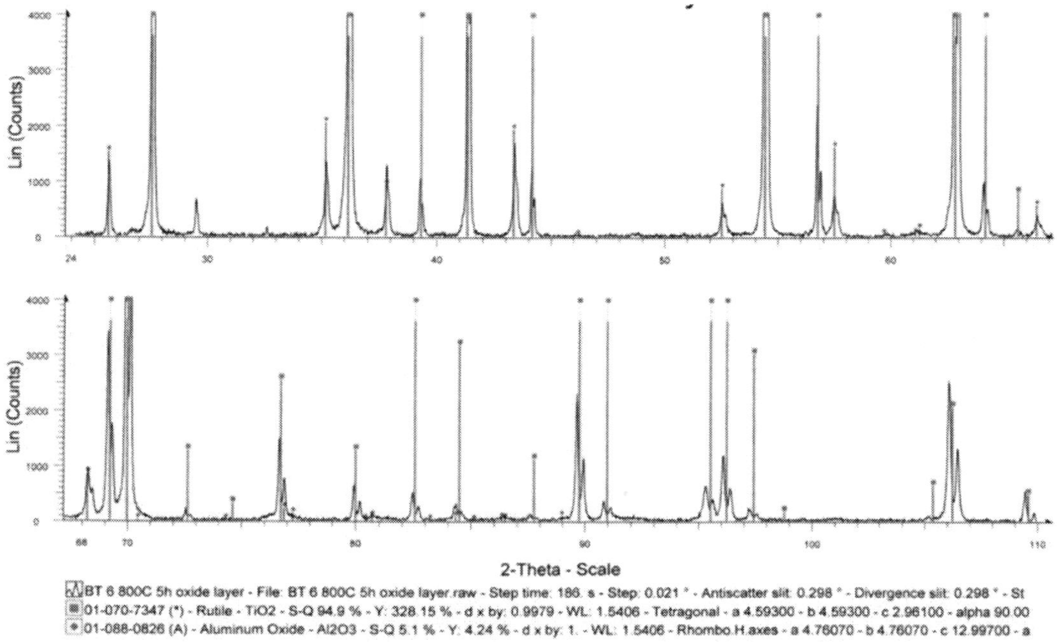

Figure 3. Diffraction pattern of the oxide layer on the surface of titanium alloy Grade5 after thermal oxidation at 800 °C during 5 hours.

3.3. Results of X-ray phase analysis

Diffraction patterns obtained by X-ray phase analysis of the oxide layer are presented in Figure 3. Results of decrypting of the diffraction patterns are presented in Table 3.

It can be concluded from Table 3 that the main component of the oxide layer is rutile (TiO_2). TiO and Ti_6O are transitional oxides which can be frequently found in the oxide layer. Their electrical

IOP Conf. Series: Materials Science and Engineering **966** (2020) 012134 doi:10.1088/1757-899X/966/1/012134

resistance is significantly lower than that of rutile. Also here are aluminum oxide Al_2O_3 in the oxide layer on the surface of all alloys which contain more than 4 % of aluminum except VST5553. It is impossible to explain its absence in the oxide layer on VST5553. Consequently, oxides that have high electrical resistance are presented in the oxide layers of all titanium alloys.

Table 3. Chemical composition of the oxide layer on the surface of titanium alloys after thermal oxidation.

Titanium alloy	Temperature of thermal oxidation, °C	Oxide	Content, %	Type of the lattice
Grade2	800	TiO_2	95.2	Tetragonal
		TiO_2	0.6	Monoclinic
		Ti_6O	0.9	Hexagonal
		$TiO_{0,857}$	3.3	Cubic
OT4	800	TiO_2	95.1	Tetragonal
		Al_2O_3	4.6	Rhombic
		Ti_6O	0.3	Hexagonal
OT4	850	TiO_2	87.0	Tetragonal
		Al_2O_3	13.0	Rhombic
OT4-1	850	TiO_2	87.6	Tetragonal
		Al_2O_3	12.4	Rhombic
Grade6	850	TiO_2	52.4	Tetragonal
		Al_2O_3	22.9	Rhombic
		Ti_6O	24.7	Hexagonal
PT-7M	850	TiO_2	81.7	Tetragonal
		Al_2O_3	18.3	Rhombic
Grade5	800	TiO_2	94.9	Tetragonal
		Al_2O_3	5.1	Rhombic
VT3-1	800	TiO_2	90.5	Tetragonal
		Al_2O_3	4.6	Rhombic
		Ti_6O	5.0	Hexagonal
VST 5553	750	TiO_2	95.5	Tetragonal
		Ti_6O	4.2	Hexagonal
		TiO	0.3	Tetragonal

Despite high content of vanadium in Grade5 and VST5553 its oxides were not observed in the oxide layer on the surface of these alloys. It could be explained by the volatility of the vanadium oxides when heated. Oxides of molybdenum and chromium are not detected in the oxide layer.

It should be noted that X-ray phase analysis doesn't allow to determine the presence or absence of a certain element in the studied sample, if it doesn't form its own phase in sufficient quantity. Consequently, we can't discard the possibility of the presence of the alloying elements in the form of solid solution in the oxide layer. For example, the presence of tin in the compound of the oxide layer in Grade6 has not been observed using X-ray phase analysis but it has been observed using other methods in [9].

It follows from the presented results that TiO_2 and Al_2O_3 oxides have the best impact on electrical resistance. Transitional oxides of titanium that could have positive influence on the electrical resistance are also found in the oxide layer. The hypothesis of the positive impact of MnO on the electrical resistance of OT4 titanium alloy after thermal oxidation was not confirmed.

4. Conclusions

The present work covers the effect of thermal oxidation on the electrical resistance, microstructure and chemical compound of the oxide layer. It is shown that:

1. Thermal oxidation leads to increase of the electrical resistance up to 1 TOhm.

2. Aluminum in the titanium alloy in an amount of 2–4 % leads to a decrease in electrical resistance, but electrical resistance increases when the amount of aluminum is more than 4 %.

3. Vanadium in the titanium alloy leads to lower electrical resistance due to volatility of V2O5 when heated.

4. Microstructural analysis showed that thickness of the oxide layer after thermal oxidation is in the range of 25 to 100 μm. The oxide layer consists of several regions situated one over the other that differ in a chemical compound. It is confirmed by the microprobe analysis.

5. The study of the chemical composition of the oxide layer by X-ray phase analysis has shown that the oxide layer consists mainly of rutile (TiO_2), corundum (Al_2O_3), transitional oxides of the titanium (Ti_6O, TiO).

It is necessary to focus on the possible application of the presented results. One of the variants of application of the oxidized titanium alloys is the production of semiconductor devices. The oxide layer is created on the surface of the metal and the silicide layer is applied on the oxide layer through heat treatment. This variant is described in [12]. Another interesting variant of application of the presented phenomenon is the application of thermally oxidized parts made from titanium alloys for electrical insulation of cutting tools. It leads to an improvement of tool life up to 2.5 times. This variant is described in [13–15].

References

[1] Song W, Yuhong L, Caixia Zh, Zhenhua L and Weiqiang L 2014 *Tribology International* **79** pp 174–82
[2] Jamesh M, T S N Sankara Narayanan and Paul K Chu 2013 *Materials Chemistry and Physics* **138** pp 565–72
[3] Yong L, Wenwen Ch, Maocai T and Shuhua T 2015 *Tribology International* **89** pp 67–71
[4] Dalili N, Edrisy A, Farokhzadeh K, Li J, Lo J and Riahi A R 2010 *Wear* **269** pp 590–601
[5] Lit Ho Chong, Kanad Mallik, C H de Groot and Reinhard Kersting 2006 *J. of Phys: Condenced Matter* **18** pp 645–57
[6] Rausch N and Burte E P 1992 *Microelectronic Engineering* **19** pp 725–28
[7] Kornilov I I, Zabrodskaya M N, Boriskina N G and Brynza A P 1977 *Metal Sci Heat Treat* **5** pp 49–51
[8] Samsonov G V, Borisova A G and Zhidkova T G 1978 *Physical and Chemical Properties of Oxides* (Moscow: Metallurgy) [In Russian]
[9] Bay A S, Lainer D I, Slesareva E N and Tsypin M I 1970 *Oxidation of the Titanium and its Alloys* (Moscow: Metallurgy) [In Russian]
[10] Goldschmidt H J 1967 *Interstitial Alloys* (New York: Plenum Press)
[11] Goldstein J I, Joy D C and Romig Jr A D 1986 *Principles of Analytical Electron Microscopy* (New York: Springer)
[12] Takashi Ishigami 1997 Patent US5665647A Making metal silicide using oxide film
[13] Andreev S V, Golubev V I, Kalashnik D V, Medison V V and Muryzhnikov S M 2012 Patent RU 2456125C2 Method of cutting metallic and current-conducting materials by current conducting cutting tool
[14] Pegashkin V F, Golubev V I and Medison V V 2014 *Int. J. Adv. Manuf. Techn.* **74** pp 599–614
[15] Medison V V 2014 *Russian Engineering Research* **34** pp 235–38

IOP Conf. Series: Materials Science and Engineering **966** (2020) 012135 doi:10.1088/1757-899X/966/1/012135

The use of homomorphic image processing to analyze coke grading

V N Krouglov[1, 4], A G Gruh[2], A L Gapak[2] and I Khurelchuluun[3]

[1] Ural Federal University named after the First President of Russia B N Yeltsin, 19, Mira str., Yekaterinburg, 620000, Russia
[2] EVRAZ KGOK JSC, 2, Sverdlova str., Kachkanar, 624350, Russia
[3] Erdenet Mining Corporation SOE, Erdenet, 61027, Mongolia

E-mail: [4] v.krouglov@mail.ru

Abstract. The estimation of the geometrical sizes of particles of crushed solid fuel (coke), moving on the conveyor belt, is associated with a number of technical difficulties. One of the problems is the need for a non-invasive way of determining particle geometry. A promising way to solve it is to use devices based on machine vision systems. This paper describes the algorithmic part of the prototype of such a device. It is proposed to improve the quality of boundary detection between fragments of coke particles to perform homomorphic processing of the initial low-contrast video images. The algorithm for calculating the Fourier spectrum has been optimized based on the Fast Fourier Transform (FFT) with the mixed base. As a result, it becomes possible to reduce the computational cost for calculating two-dimensional Fourier spectra for complex multiplication operations by 1.33 times, and the number of complex addition operations by 1.67 times. The software of the prototype, built using the proposed methods, made it possible to obtain good convergence of the results for assessing the particle size distribution of samples of crushed coke with laboratory estimates. Thus, the maximum absolute average error of the machine vision system in assessing the size of crushed coke is only 3.37%, and the maximum error for all measurement classes do not exceed 6.9%.

1. Introduction

The sintering process of the sinter burden is largely determined by the quality of the preparation of solid fuels - finely divided coke particles. The grain size distribution of such particles is usually in the range from 0.1 mm to 5–7 mm., with more than 80% of them being in the range of 1–3 mm. Coke particles after crushing enter the charge preparation plant through a system of conveyors, where the sinter burden is prepared. It has to be mentioned that the conveyor speed is about 1 m/s. To analyze the grain size distribution of crushed coke particles under such conditions, the control system should fulfill the following requirements:

- high accuracy in estimating particles geometry,
- a short time to complete such estimates,
- high reliability of operation.

One of the most promising devices for solving such problems are devices based on vision systems [1-4]. A device developed under the state contract 3170ГС1/48564 with FASIE fund can serve as a prototype of such an instrument. The images of the filled layer formed by coke particles with various size distribution obtained using this prototype are given in Figure 1.

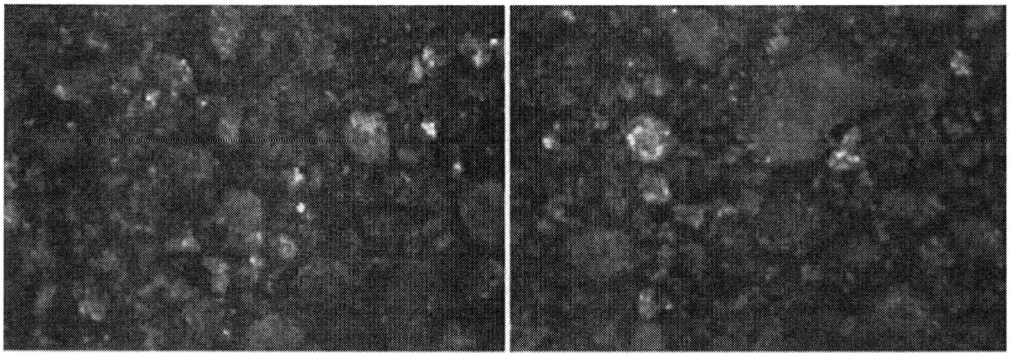

Figure 1. Images of crushed coke particles: desired (left) and enlarged (right) particle size distribution

The prototype was installed and tested on the conveyor No. 47 of the charge preparation plant of EVRAZ KGOK JSC.

To determine the geometry of visible coke particles, it is necessary to perform image segmentation, that is, to split the image into fragments corresponding to the particles. One approach to segmentation is to initially build the boundaries between images of coke particles. This study is devoted to the problem of identifying the boundaries of the regions of coke particles.

2. Methods

From literature [5, 6] it is known that the spectrum of the image obtained using a video camera is shifted to the low-frequency region. For a qualitative construction of the boundaries between image fragments, sharp changes in the brightness function $f(x,y)$, $x=\overline{0,L-1}$, $y=\overline{0,M-1}$ are necessary. In turn, the steepness of the brightness difference is determined by high-frequency harmonics in the image spectrum $F(\omega_x,\omega_y)$, $\omega_x=\overline{0,L-1}$, $\omega_y=\overline{0,M-1}$. In our opinion, it is most expedient to use homomorphic processing of the original image to enhance them. Such processing, in addition to increasing the image contrast, eliminates the uneven illumination of the observed surface. The theory of homomorphic processing is described in detail in [5]. As a center-symmetric transfer function, it is recommended to use any of the high-pass filters. The experiments conducted by the authors showed that the best results for adjusting images of wet coke particles are obtained using a modified high-frequency Gaussian filter, the transfer function of which is as follows:

$$H(\omega_x,\omega_y)=|sH-sL|\cdot\left(1-exp\left(-\frac{C*D^2(\omega_x,\omega_y)}{D_0^2}\right)\right)+sL, \quad (1)$$

where sH and sL are parameters correcting the high-frequency and low-frequency components of the spectrum $F(\omega_x,\omega_y)$; constant C determines the slope of the transfer function of the filter in the region between sH and sL; D_0 is the cutoff frequency of the filter, and the $D(\omega_x,\omega_y)$ parameter for the centrally symmetric spectrum is determined by the formula:

$$D(\omega_x,\omega_y)=\sqrt{(\omega_x-L/2)^2+(\omega_y-M/2)^2} \quad (2)$$

When implementing homomorphic filter, the highest computational complexity is associated with the translation of the image $f(x,y)$ presented in spatial coordinates into the frequency domain and vice versa. The Fourier spectrum [8] is most often used as the latter. The calculation of the direct $\Im^+\{f(\omega_x,\omega_y)\}$ and inverse $\Im^-\{f(\omega_x,\omega_y)\}$ discrete Fourier transforms (DFT) [5, 6, 7] is

IOP Conf. Series: Materials Science and Engineering **966** (2020) 012135 doi:10.1088/1757-899X/966/1/012135

implemented using fast Fourier transform (FFT) algorithms. FFT algorithms can be implemented in various ways, but the most common is the FFT algorithm at base 2. This means that the size of the input image in both coordinates must be equal to the power of two: $M = 2^m$ and $L = 2^l$. In our experiments, the BASLER aca1600-60 gm digital video camera with a resolution of 1600×1200 pixels was used to obtain video images. None of the dimensions of the image formed by such a camera is a power of two. In such situations, the input array is supplemented with zeros so that the length of each coordinate becomes equal to power of 2. In our case, this number is $N = 2048 = 2^{11}$. The complexity of the FFT algorithm in base 2 for calculating the two-dimensional Fourier spectrum with the dimension $N \times N$, at $N = 2048$, expressed as the number of complex operations of multiplication $\dot{M}u$ and addition $\dot{A}d$ without optimization of calculations, is:

$$\dot{M}u \approx \left(\left(N/2\right)\cdot\left(\log_2 N - 1\right)\right)\cdot N \cdot 2 = 41943\cdot 10^3 \text{ and}$$

$$\dot{A}d = \left(N\cdot\left(\log_2 N\right)\cdot N\cdot 2 = 92274\cdot 10^3\right.$$

If the calculation of the two-dimensional spectrum is optimized and the zero lines are not processed, the complexity of calculating the two-dimensional FFT decreases:

$$\dot{M}u \approx \left(\left(N/2\right)\cdot\left(\log_2 N - 1\right)\right)\cdot 1200 + \left(\left(N/2\right)\cdot\left(\log_2 N - 1\right)\right)\cdot N = 33259\cdot 10^3 \text{ and}$$

$$\dot{A}d = \left(N\cdot\left(\log_2 N\right)\cdot 1200 + \left(N\cdot\left(\log_2 N\right)\cdot N = 73171\cdot 10^3\right.\right.$$

To further reduce the computational complexity of calculating the Fourier spectrum, it is proposed to use the mixed base FFT algorithm instead of the FFT algorithm at base 2 [8]. To do this, we should reduce the width of the processed frame to $M = 1536$, and increase the height to $L = 1280$ pixels. In this case, the loss of initial information will be only $100 - \left(1563/1600\right) = 4\%$. Added lines are filled with zeros as previously. The scheme of the FFT algorithm for computing by lines is converted to the form $M = 1536 = 2^9 \cdot 3$, that is, initially 9 iterations were performed with a 2×2 DFT kernel, and the last iteration with a 3×3 DFT kernel. The complexity of computing the kernel of the DFT 3×3 is: $\dot{M}u = 4$ and $\dot{A}d = 6$. The number of operations for calculating the Fourier spectra for all 1,200 lines (zero lines are not processed) is:

$$\dot{M}u \approx \left(\left(M/2\right)\cdot 9 + 4\cdot 512\right)\cdot 1200 = 10752\cdot 10^3 \text{ and}$$

$$\dot{A}d = \left(M\cdot 9 + 6\cdot 512\right)\cdot 1200 = 20275\cdot 10^3$$

The column length L is decomposed as $L = 1280 = 2^8 \cdot 5$, that is, 8 iterations are initially performed with a 2×2 DFT kernel, and the last iteration with a 5×5 DFT kernel. The complexity of computing the DFT 5×5 kernel is: $\dot{M}u = 16$ and $\dot{A}d = 20$. The number of operations for calculating the Fourier spectra for all 1536 columns is:

$$\dot{M}u \approx \left(\left(L/2\right)\cdot 8 + 16\cdot 256\right)\cdot 1536 = 14156\cdot 10^3 \text{ and}$$

$$\dot{A}d = \left(L\cdot 8 + 20\cdot 256\right)\cdot 1536 = 23593\cdot 10^3$$

The total complexity of calculating the two-dimensional Fourier spectrum of 1536×1280 pixels with the FFT algorithms at a mixed base is:

$$\dot{M}u \approx 24908\cdot 10^3 \text{ and } \dot{A}d = 43868\cdot 10^3$$

The acceleration coefficient (AC) for complex operations of multiplication is $AC_{\dot{M}u} = 33259\cdot 10^3 / 24908\cdot 10^3 = 1.33$, and for complex operations of addition is $AC_{\dot{A}d} = 73171\cdot 10^3 / 43868\cdot 10^3 = 1.67$.

IOP Conf. Series: Materials Science and Engineering **966** (2020) 012135 doi:10.1088/1757-899X/966/1/012135

3. Results and Discussion

The algorithm of the homomorphic image processing implemented by the proposed technique turned out to be highly efficient from a computational point of view and allowed to significantly improve the contrast of the input image. Figure 2 shows the results of homomorphic filtering of the images shown in Figure 1, using a modified Gaussian high-pass filter.

The numerical values of the filter parameters were as follows: $sH = 2$, $sL = 1$, constant $C = 3$, filter cutoff frequency $D_0 = 32$. Figure 3 illustrates the final results of processing the images shown in Figure 1. Dark lines on the fragments of coke particles indicate their maximum sizes. In the perpendicular direction, the white lines indicate the sizes that correspond to the results of the screening of these particles.

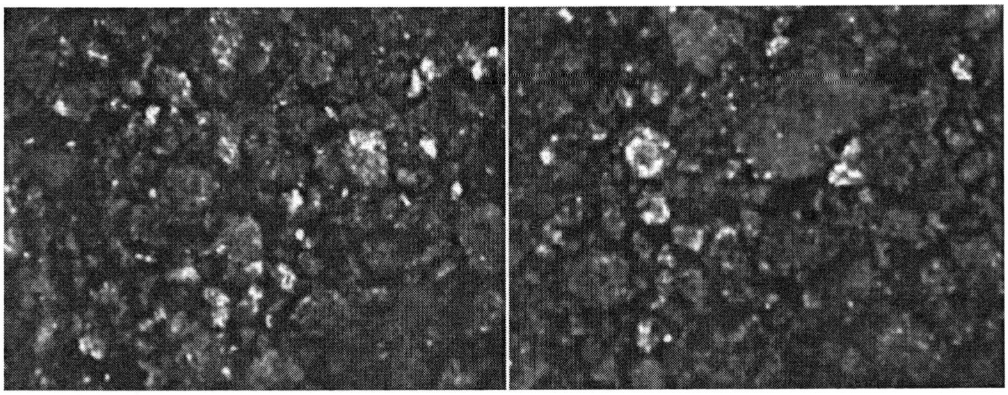

Figure 2. Results of homomorphic processing of images presented in Figure 1

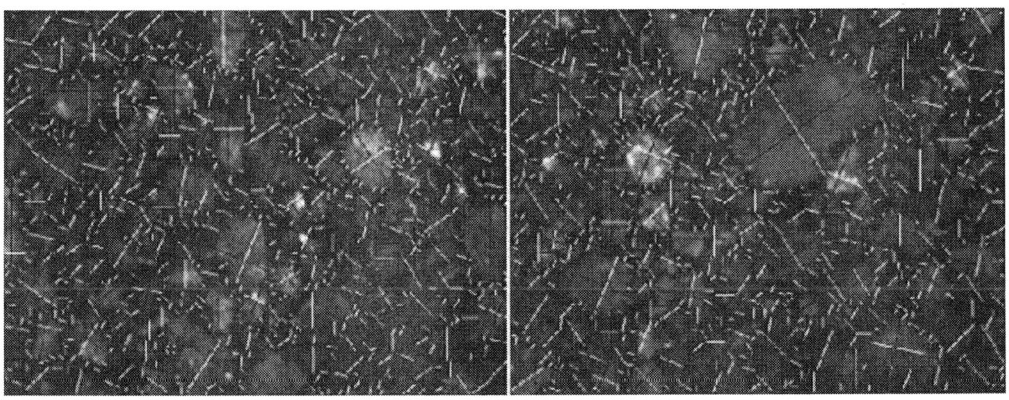

Figure 3. Results of the evaluation of the geometric dimensions of the coke particles shown in Figure 1

A number of experiments at EVRAZ KGOK JSC were conducted to evaluate the quality of the algorithm for evaluating the grain size distribution of solid fuel particles. The prototype of the developed device based on the machine vision system (MVS device) was installed on the conveyor No. 47 of the charge preparation plant of EVRAZ KGOK JSC. After the conveyor stopped, frames of the filling layer formed by coke particles were recorded and this material was sampled. Subsequently, the obtained samples were scattered in the laboratory of the plant on sieves, and the generated video images were processed using the methods proposed above. The results of these experiments are shown in Table 1.

Table 1. The results of the evaluation of the grain size distribution of solid fuel particles obtained in the laboratory of JSC EVRAZ KGOK and by MVS device.

Sample No.	Source	Grain size distribution, %						Mean error	Max error
		+5	+4	+1	+0.5	+0.1	-0.1		
	Laboratory	0.8	1.0	46.5	15.4	10.5	25.8		
1	MVS device	0	0	46	17	16	21	2.37	5.5
	Error	0.8	1	0.5	1.6	5.5	4.8		
	Laboratory	1.1	1.7	46.3	15.6	14.3	21.0		
2	MVS device	0	2	42	18	16	22	1.80	4.3
	Error	1.1	0.3	4.3	2.4	1.7	1		
	Laboratory	1.5	1.6	45.6	14.9	12.4	24.0		
3	MVS device	0	2	45	17	15	21	1.70	3.0
	Error	1.5	0.4	0.6	2.1	2.6	3.0		
	Laboratory	20.6	4.8	34.5	14.3	18.7	7.1		
4	MVS device	17	3	37	15	14	14	3.37	6.9
	Error	3.6	1.8	2.5	0.7	4.7	6.9		
	Laboratory	17.7	4.3	37.5	14.5	18.4	7.6		
5	MVS device	19	2	40	14	12	13	3.07	6.4
	Error	1.3	2.3	2.5	0.5	6.4	5.4		

Devices based on vision systems and laboratory installations for assessing the grain size distribution of crushed material have different principles for evaluating the same process. However, the presented results show a high convergence of the results obtained using the MVS device and the laboratory method with the correlation coefficient 0.974. In particular, the maximum mean error of the MVS device in assessing the size of crushed coke was only 3.37%, and the maximum error for all measurement classes did not exceed 6.9%.

4. Conclusion

The results of the studies showed the feasibility of using homomorphic processing of the original video images of crushed coke particles in order to increase their contrast. This allowed for better image segmentation and, as a result, increased the accuracy of estimating the geometric dimensions of coke particles. The optimization of computational costs in the calculation of two-dimensional Fourier spectra increased the speed of these algorithms for complex multiplication operations by 1.33 times, and in the number of complex addition operations by 1.67 times. The MVS device software, built using the proposed methods, made it possible to obtain good convergence of the results for assessing the particle size distribution of crushed coke samples with laboratory assessments. Thus, the maximum mean error of MVS device in assessing the size of crushed coke is 3.37%, and the maximum error for all measurement classes did not exceed 6.9%.

5. Acknowledgments

The work was performed under state contract 3170ГС1/48564, grant from the FASIE.

References

[1] Maiti A, Chakravarty D, Biswas K and Halder A 2017 Development of a mass model in estimating weight-wise particle size distribution using digital image processing *Int. J. Mining Sci. and Techn.* **27** 435–43

[2] Thurley M J and Ng K C 2005 Identifying, visualizing, and comparing regions in irregularly

IOP Conf. Series: Materials Science and Engineering **966** (2020) 012135 doi:10.1088/1757-899X/966/1/012135

spaced 3D surface data *Comp. Vis. Img. Und.* **98 (2)** 239-70

[3] Hurelchuluun I, Kruglov V N 2016 Industrial tests and modernization of GRANICS crushed ore particle size measurement system *Trans. of Int. Conf.: Scientific basics and practice of processing ores and raw materials* 261-5

[4] Hurelchuluun I, Morozov V V, Nikolaev T S, Kruglov V N 2019 The use of visiometric analysis of the particle size distribution of ore for automated control of the crushing process *Ores and Metals* **1** 67-73

[5] Gonzalez R C, Woods R E 2007 *Digital Image Processing* 3 ed. (Pearson) p 976

[6] Sonka M, Hlavac V and Boyle R 1999 *Image Processing, Analysis, and Machine Vision* (Milan Sonka - Brooks/Cole Publ. Company) p 770

[7] Pratt W K 2007 *Digital Image Processing: PIKS Scientific Inside* 4 ed. (New York: Wiley Interscience) p 812

[8] Rabiner L and Gould B 1978 *Theory and application of digital signal processing* (MIR) p 84

IOP Conf. Series: Materials Science and Engineering **966** (2020) 012136 doi:10.1088/1757-899X/966/1/012136

Optimization of roll calibration for flange shape rolling. Groove space

D L Shvarts[1], A M Mikhaylenko[1] and E I Ustinova[1, 2]

[1] Department of Metal Forming, Ural Federal University named after the First President of Russia B N Yeltsin, 19, Mira str., Yekaterinburg, 620000, Russia

E-mail: [2] ustinova1694@gmail.com

Abstract. The most important component of any rolling mill is the calibration of the rolls. The ability to obtain a finished product of high quality at minimal cost depends on the right calibration. According to the previous experience and practice a large number of different calibrations for the flange profiles production are known. The universal 'concept of two-stage optimization', consisting of two successively carried out optimization stages developed in the Department of Metal Forming of the Ural Federal University in relation to optimization of the roll calibration for flange shape rolling is considered in this article. Common features of these shapes classification and their variation levels are identified and justified. This developed classification structure is the basis for the formation of spaces of groove schemes, due to which the groove space is formed. In the software implementation the groove space is represented as a 'groove database'. In the future this database will be used to form the space of calibration schemes that are fundamentally suitable for these shapes rolling.

1. Introduction

As is known the mill rolls calibration is the central element of any section rolling production, because many characteristics of the technological process and the properties of finished rolled products depend on it. However, at present there are no specific scientifically based methods for design and further optimization of complex shape calibrations. The known methods are either limited in range or not brought to the state of practical applicability [1].

The most suitable basis for the methodology design creation of the optimal calibration is the system approach formulated in 'Theory of Systems' and 'Theory of Optimal Control' [2–8].

2. Main principles

Considering the calibration of section rolling rolls as a 'variable groove system', it can be concluded that, like any other functioning system, it has two components available for change and optimization. The first component is the structure of the system: a certain sequence of grooves of a certain shape, i. e. the so-called 'calibration scheme'. The second component is the set of quantitative characteristics of the links between the elements of the system: distribution of reductions along rolling passes, i. e. 'reduction mode'. In order to obtain an optimal roll calibration, it is necessary to achieve optimality of these two components: identify the optimal calibration scheme and select the optimal reduction mode for this calibration scheme.

Taking into account this idea of optimality the Department of Metal Forming of the Ural Federal University developed the universal 'Concept of optimal calibration', consisting of two successively

IOP Conf. Series: Materials Science and Engineering **966** (2020) 012136 doi:10.1088/1757-899X/966/1/012136

carried out optimization stages (figure 1). At the first stage, the choice of the optimal calibration scheme is made, and at the second stage – the choice of the optimal reduction mode [9–13].

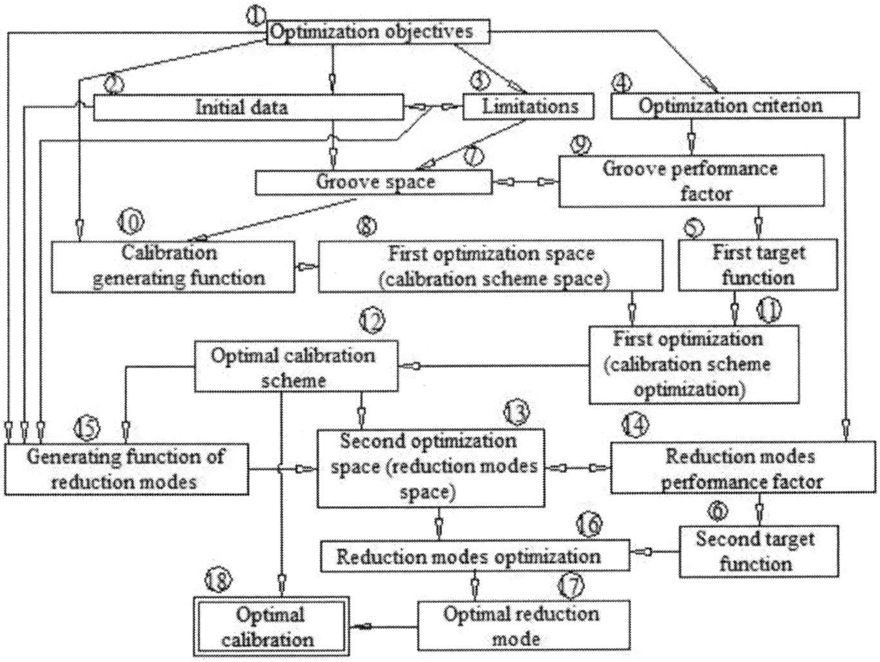

Figure 1. Structural flowchart of two-stage calibration optimization.

Considering the huge amount of information necessary for carrying out the optimization process, the use of a specialized optimization software package that is currently being created is required.

One of the main information blocks used in the first stage of optimization is the block reflecting the so-called 'groove space' [12]. This space includes all fundamentally possible grooves applicable to the rolling of a specific section at a specific rolling mill. At the first stage of optimization the term 'groove' will be understood as the groove scheme, which primarily determines the groove shape – the element shape, their relative position, the position of the groove connector, etc.

The groove space in the software implementation is a 'groove database', which includes all the fundamentally possible flange grooves used for rolling channels, rails and I-beams. For groove ordering in this space a classification method has been chosen. Identification of classification characteristics and their variation levels was carried out on the basis of the analysis of flange grooves known from the literature and practice [13]. Each combination of variation levels will correspond, in aggregate, to a specific point in the groove space, i.e. individual, unique groove.

3. General characteristics of the flange shapes classification

The analysis showed that the most common basic characteristics of rail, beam and channel grooves are the following:

Ш – type of the neck (wall);
О – open flanges type;
З – closed flanges type;
С – groove symmetry;
Т – groove type;
Р – type of groove closure and the number of rolls that form the groove.

The name of the elements is presented in figure 2.

Analysis of known grooves showed that the above characteristics have the following levels of variation.

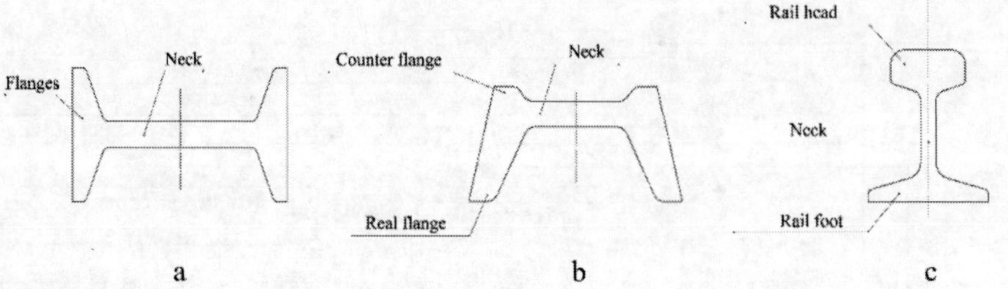

Figure 2. The division of flange shapes into elements: a – beams, b – channels, c – rails.

3.1. Groove characteristic 'Ш – neck type'
Here we recognize the structure-forming feature that reflects the shape of the neck shape. Most real industrial calibrations used for the section rolling utilize four forms of the neck: straight, cut, curved and wavy. The straight neck has a rectilinear form and is used mainly in pre-finishing and finishing grooves. The curved neck is a U-shaped and is as a rule formed by segments of intersecting straight lines conjugated by a large radius. Cut neck shape is used to form a rough shape from a rectangular workpiece and is formed using split wedges. The wavy neck (having several bends) is mainly used to reduce the width of the groove cut into the mill rolls, thereby allowing an increase of the number of grooves located on the rolls and using a workpiece of a smaller width.

3.2. Groove characteristic 'O – open flanges type'
5 types of open flanges are used: with a small incline, with an increased incline, curved outward, curved inward, without an incline. Direct open flanges with a small incline of groove have a small taper of 4 to 8% for beams and from 3 to 12% for channels. Straight open flanges with an increased incline have a taper of 10 to 40% for channels, up to 12% for beams, and the angle between the middle line of the wall and the outer face of the flange is 90. Outward curved open flanges are used for rolling in expanded grooves. Inward curved open flanges are used to reduce the width of the groove cut into the rolls and control the length of the flange.

3.3. Groove characteristic 'З – closed flanges type'
The analysis of the literature revealed 2 types of closed flanges: triangular and trapezoidal. However, during channels rolling in pre-finishing and finishing grooves there are no closed flanges, therefore, another level of variation is added – without closed flanges.

3.4. Groove characteristic 'C – groove symmetry'
The groove symmetry can be characterized by the presence or absence of symmetry axes of the groove: horizontal and (or) vertical. For unambiguous interpretations the vertical axis of symmetry is taken as the axis perpendicular to the axis of the rolls.

3.5. Groove characteristic 'T – groove type'
The term 'Groove type' refers to a structure-forming feature that reflects the groove purpose for the change of the cross-section shape. To indicate (name) the variation levels of this characteristic the external similarity of the groove corresponding to this groove level with one of the geometric objects, as well as the direction of the main axis of the groove is used.

3.6. Groove characteristic 'P – groove closure type and the number of rolls forming the groove'
The term 'closure type' refers to the method of the groove placement relative to the position of the roll cross section. The method of groove closure affects primarily to the coverage level of the rolled metal in the groove, the control level of the metal by rolls. With the increase of the number of rolls the

IOP Conf. Series: Materials Science and Engineering **966** (2020) 012136 doi:10.1088/1757-899X/966/1/012136

control level of the metal by rolls increases. That is why the externally different, but functionally similar characteristics of 'groove closure type' and 'number of rolls forming the groove' are combined into a single characteristic of the groove 'P'.

Classification characteristics and all possible levels of their variation, as well as a schematically depicted groove shape and an example of groove use in real calibrations, are given in table 1.

Table 1. General characteristics of the flange shapes classification.

Classification characteristic	Name	Characteristic variation level	Rails	Channels	Beams	Source
Neck (Ш)	Ш1	Straight	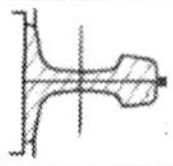	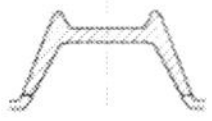	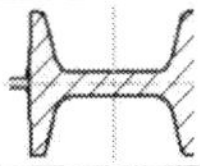	11, figure I.23
	Ш2	Cut		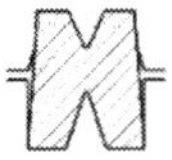	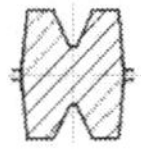	11, figure I.22
	Ш3	Curved	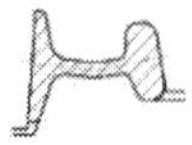	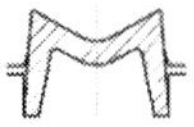	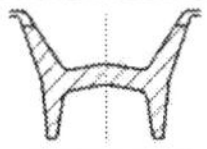	11, figure I.24
	Ш4	Wavy	–		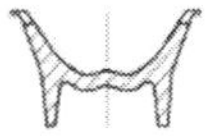	12, figure I.148
Open flange (O)	O1	Straight with a small incline		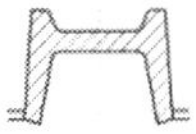	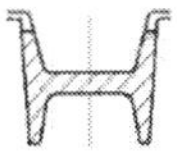	11, figure I.24
	O2	Straight with a increased incline	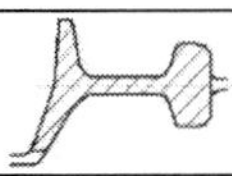	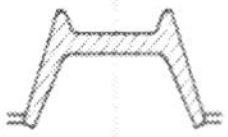	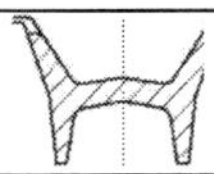	13, figure 18
	O3	Outward curved	–	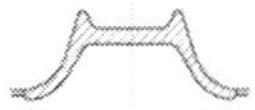	–	11, figure I.80
	O4	Inward curved	–		–	13, figure 19
	O5	No incline	–	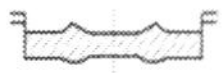	–	11, figure I.76
Closed flange(З)	З1	Triangular	–	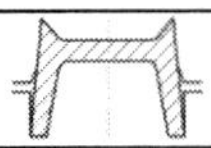	–	13, figure 17
	З2	Trapezoidal	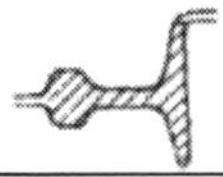	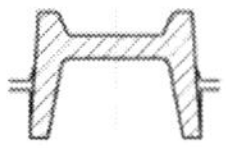	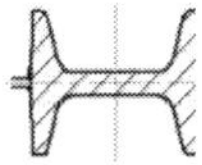	13, figure 16
	З3	No flanged	–	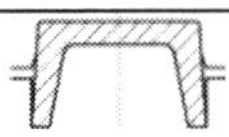	–	14., figure 4.23 e
Symmetry (C)	C1	Assymetrical		–	–	12, figure II.2
	C2	Two axes of symmetry	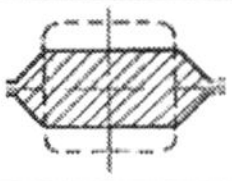	–	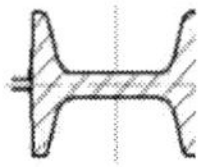	12, figure I.147
	C3	Vertical axis of symmetry		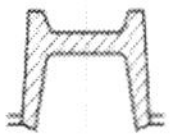	–	12, figure II.17

IOP Conf. Series: Materials Science and Engineering **966** (2020) 012136 doi:10.1088/1757-899X/966/1/012136

	C4	Horizontal axis of symmetry	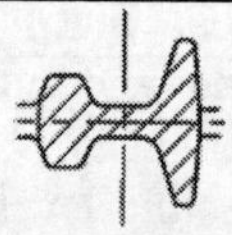	–	–	12, figure II.21
	T1	Trapezoidal axial			–	14, figure 5.6
	T2	Trapezoid rib			–	14, figure 5.6
Groove type (T)	T3	Cutting	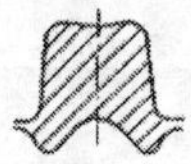	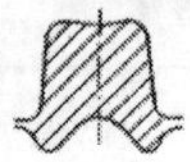	–	14, figure 5.7
	T4	Tavrovy	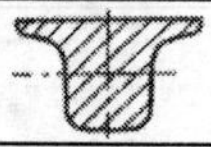	–	–	12, figure II.64
	T5	Split		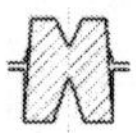	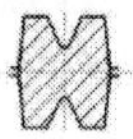	11, figure I.125
	T6	Flanged		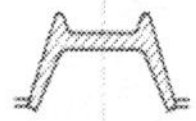	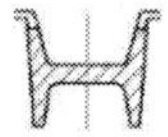	13, figure 10
	P1	Open	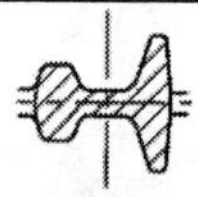	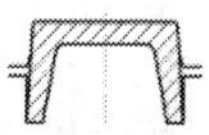	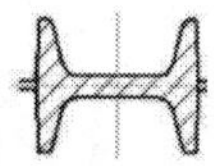	12, figure II.31
	P2	Half closed	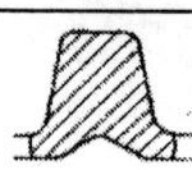	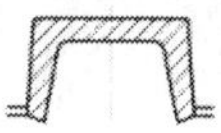	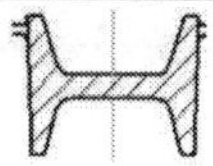	14, figure 4.23
	P3	Closed top		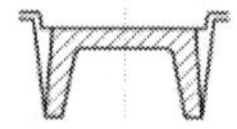	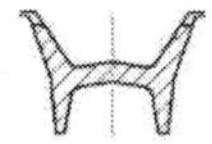	12, figure II.27
Groove closure type and the number of rolls forming the groove (P)	P4	Closed bottom	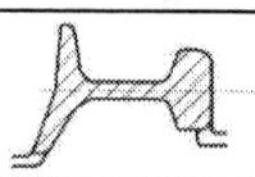	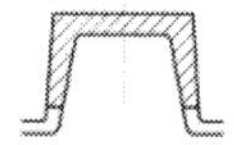	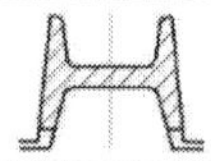	12, figure II.27
	P5	Diagonal	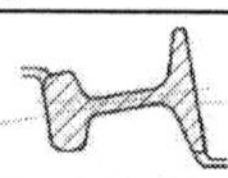	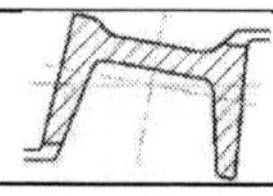	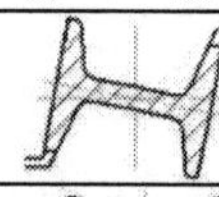	12, figure I.218
	P6	One side closed	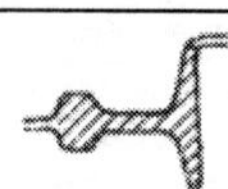	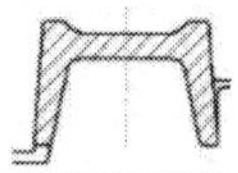	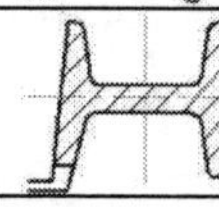	12, figure II.2
	P7	Three-roll	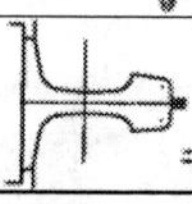	–	–	14, figure 5.13
	P8	Four-rolls	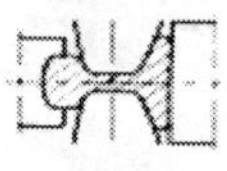	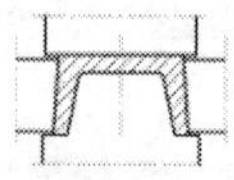	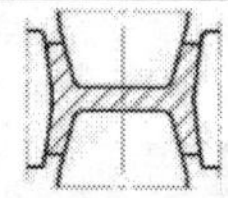	12, figure I.295

Note: A space in the table cells indicates that this groove type is either not possible or is not detected in known industrial calibrations.

4. Conclusion

The identified common features of the groove classification intended for flange shapes rolling (channels, beams and rails) allow us to develop a general approach to the groove classification. The developed classification structure is the basis for the formation of spaces of groove schemes. In the software implementation the groove space is represented as a 'groove database'. In the future this database will be used to form the space of calibration schemes that are fundamentally suitable for these shapes rolling. The space of the calibration schemes will be the optimization space.

5. Acknowledgments

The reported study was funded by RFBR according to the research project № 20-38-90246.

References

[1] Skorokhodov A N, Polukhin P I and Ilyukovich B M 1983 *Optimization of rolling production* Metallurgy (Moscow) p 432
[2] Sadovsky V N 1980 *The System Approach and the General Theory of Systems: Status, Main Problems and Development Prospects* Science (Moscow)
[3] Mesarovich M and Takahara 1978 *The General Theory of Systems: Mathematical Foundations* (Moscow)
[4] Akof R L and Sasieni M 1971 *Fundamentals of Operations Research* (Moscow) p 536
[5] Li H, Zhao Z, Zhang J et al 2019 Analysis of flatness control capability based on the effect function and roll contour optimization for 6-h CVC cold rolling mill. *International Journal of Advanced Manufacturing Technology* **100 (9-12)** pp 2387–99
[6] Wu L, Liu J, Dhanasekar M, Wang H and Wen Z 2019 Optimisation of railhead profiles for curved tracks using improved non-uniform rational B-splines and measured profiles *Wear* **418-419** pp 123–132
[7] Wolf C, Stadler A T and Baumgartner W 2016 Cross-section optimisation for cold-rolled steel beams using a genetic algorithm. *METAL 2016 - 25th Anniversary International Conference on Metallurgy and Materials, Conference Proceedings* pp 507–12
[8] Prinz K, Steinboeck A and Kugi A 2018 Optimization-based feedforward control of the strip thickness profile in hot strip rolling. *Journal of Process Control* **64** pp 100–11
[9] Mikhaylenko A M and Schwartz D L 2018 The concept of optimal roll calibration. Report 1. The main prin ciples *News of higher educational institutions. Ferrous metallurgy* vol **61** no **1** pp 21–27
[10] Mikhaylenko A M and Schwartz D L 2016 The system approach to optimization the calibration of section rolling rolls *Production of rolled products* vol **17** pp 29–32
[11] Ustinova E I, Mikhaylenko A M and Schwartz D L 2018 Selection of the optimal roll calibration scheme for channel rolling as the first stage of optimization *Mechanical equipment of metallurgical plants* vol **1 (10)** pp 72–82
[12] Schwartz D L, Mikhaylenko A M and Ustinova E I 2019 Optimization of roll calibrations for beam channel rolling part 1 General regulations *Chernye Metally* vol **9** pp 4–8
[13] Schwartz D L, Mikhaylenko A M and Ustinova E I 2020 Method of optimization of roll calibration for channels. Groove space *Journal of Chemical Technology and Metallurgy* vol **3 (55)** pp 657–65

IOP Conf. Series: Materials Science and Engineering **966** (2020) 012137 doi:10.1088/1757-899X/966/1/012137

Application of laser interferometry to analyze cutting tool state during machining

I A Efimovich[1] and I S Zolotukhin[1, 2]

[1] Institute of Industrial Technology and Engineering, Tyumen Industrial University, 38 Volodarskogo Street, Tyumen, 625000, Russia

E-mail: [2] zolotuhinis@tyuiu.ru

Abstract. The efficiency of a cutting tool can be enhanced through stress–strain and temperature studies. Existing mathematical methods implement simplified boundary conditions, and experimental methods that are either inapplicable to real working conditions or lack the necessary accuracy. This study aims to develop novel experimental methods for stress–strain and temperature field analyses. The approaches entail recording the side deformation fields of the cutting tool by laser interferometry during its operation, separating the deformation fields caused by the cutting forces and heating, as well as calculating the stress–strain and temperature fields using the Young's modulus, Poisson's ratio, and coefficient of linear thermal expansion of the tool material. The advantages of these methods include their applicability under real cutting conditions and the possibility to study the stress–strain and temperature fields of a tool during non-stationary operation by high-speed video recording. The study proves the efficiency of the proposed methods by the orthogonal machining of difficult-to-cut steel disc using a cemented carbide tool with positive rake angle. As a result, the temperature and principal stress fields in the tool were determined. Developed methods can help in the study of cutting tool efficiency.

1. Introduction

The performance of the cutting tool has the greatest impact on the efficiency of the entire machining process, and methods of enhancing this performance are discoverable through the study of the stress–strain and temperature states of the tool in near-real conditions.

Presently, calculation methods that employ analytically obtained boundary conditions with extensive simplifications and assumptions are widely used to determine the stress–strain and temperature states of the tool [1, 2]. To improve the reliability of the calculation results, the experimentally obtained boundary conditions must be used. Many experimental methods have been developed to measure the strain–stress and temperature distributions in the cutting tool.

The use of various types of strain gauges to study deformation fields is extremely challenging because of the extremely small size of the working tool involved, as well as the high temperatures involved in the cutting process. The brittle-coating method is usable for only qualitative analysis, presents significant measurement errors, and is unsuitable for studying dynamic stress–strain fields. The split–tool dynamometer [3] provides an opportunity to determine the stress distributions along only tool faces.

The grid method is partially applicable to high-strength tool materials, considering the difficulty in determining the changes in the node's distances because of the minimal material deformation. Furthermore, the Moire-Fringe method is very time-consuming, owing to the complexity in obtaining and using gratings. The digital image correlation (DIC) technique is of particular interest; however, it is difficult to be used for studies near the contact zone because of the smallness and inconsiderable tool material deformation.

The photoelastic [4] technique is inappropriate for experiments with real cutting tools and workpiece materials because of the low heat resistance of optically active materials, and it is applicable only at extremely low cutting speeds when processing soft materials. Applying photoelastic coatings partly solves these problems; however, the accuracy is dramatically affected when the coatings peel off in a high strain and temperature gradient zone.

The caustic method [5] differs with the complexity of caustic pattern analyses, especially for a complex stress state with non-uniformly distributed boundary conditions. Electronic speckle pattern interferometry, besides out-of-plane displacements, can be applied to in-plane deformation measurements; however, this method has the same disadvantages as the DIC in the context of cutting tool studies.

The holographic interferometry method is highly sensitive and is applicable to complex-shaped or rough-surface objects. Digital holography partially solves the problem of acquiring and processing holographic interferograms. In laser interferometry, it is necessary to polish the flat surface of the object being investigated; however, interference fringe analysis causes fewer difficulties. All-optical techniques have common problems with their implementation for studying non-static and under vibration objects, such as cutting tools. Moreover, there are difficulties in computing stress components from deformation fields owing to the complicated mathematical processing of the experimental results.

In temperature measurements, the tool-work (natural) thermocouple method is widely used. A variation of this method entails a case wherein the split tool allows measuring the temperature distributions along only the tool faces. However, the split tool properties differ significantly from those of the real metal cutting tool. A method, which employs semi-artificial and embedded artificial thermocouples [6] that are distributed into the body or on the faces of the cutting tool, prevents temperature measurements near the tool edge and differs with the complexity of the tool-making process. The running thermocouple method [7] is unusable in conditions of low feed rates or high cutting speeds and requires complex experiment preparation.

Methods based on the hardness and microstructural changes of a material [4] are utilizable for only the high-speed steel tool because the hardness and structure of the cemented tungsten carbide tool are not significantly affected by temperature changing.

Thermal indicator methods, among which the temperature-sensitive paint method and the physical vapor deposition (PVD) film are notable, are of special interest. However, measuring the temperature fields with a high gradient and a wide range by these methods is quite difficult due to the thermophysical differences between the cutting tool material and the coating. The tempering colors (oxide layer) method cannot be used for quantitative analysis of temperature.

The infrared thermometry method is used most extensively because it provides an opportunity to determine dynamic temperature fields in a non-contact manner. However, due to emissivity dependence on the oxidation state and roughness of the surface, method accuracy highly depends on the accuracy of the selected emissivity coefficient, which can change with surface heating [8].

2. Method

To overcome the limitations of the above-mentioned approaches and improve the process of determining the stress and temperature fields of the cutting tool, we developed methods [9, 10] and experimental rig based on laser interferometry.

These methods are implemented for conditions of orthogonal cutting process. During machining, cutting tool deforms due to the heating and forces acting on it. The shape change and displacement of the tool-side surfaces are determined by changing location of the interference fringes obtained by the interferometer.

The tool state may be depicted by the two-dimensional problem because the cutting tool width is narrow relative to its size, and stresses with temperature may be regarded as uniformly distributed over the tool thickness. Owing to the symmetrical deformation of the tool-side surfaces, a tool width change caused by cutting forces Δt_f and thermal expansion Δt_t at the point of interest may be obtained:

$$\Delta t_f = \frac{\Delta m_f \cdot \lambda}{n}, \ \Delta t_t = \frac{\Delta m_t \cdot \lambda}{n} \quad (1)$$

Where Δm_f and Δm_t are the differences in the interference fringe order at the point of interest, caused by cutting forces and thermal expansion, respectively; n is the refractive index of air; and λ is the laser wavelength. The fringe order differences Δm_f and Δm_t may be found by:

$$\Delta m_f = m_1 - m_2, \ \Delta m_t = m_2 - m_0 \quad (2)$$

Where, m_0, m_1, and m_2 represent the fringe order obtained before the process (for cold tool), in the process (for loaded tool), and immediately after cutting was interrupted (for heated tool), respectively.

The strain caused by cutting forces in parallel to the tool edge direction may be found by:

$$\varepsilon_z = \frac{\Delta t_f}{t} = \frac{\Delta m_f \cdot \lambda}{n \cdot t} \quad (3)$$

Where, t is the width of the cold tool measured before cutting. According to Hooke's law,

$$\varepsilon_z = -\frac{\mu}{E} \cdot (\sigma_x + \sigma_y) = -\frac{\mu}{E} \cdot S \quad (4)$$

Where, E is the Young's modulus, μ is the Poisson's ratio, σ_x, σ_y are the normal stresses, and S is the sum of the normal stresses.

Combining (3) and (4), we obtain:

$$S = -\frac{E}{\mu} \cdot \frac{\Delta m_f \cdot \lambda}{n \cdot t} \quad (5)$$

By analyzing the fringe pattern points and using equation (5), we can obtain the distribution of the stress sum S along the closed boundary of the tool. Before stress separation, we must obtain the field of the sum S by finding a solution of the Laplace's equation:

$$\nabla^2 S = 0 \quad (6)$$

To solve the Laplace's equation and realize stress separation, an iterative finite difference method in the Cartesian coordinate system was used. The x-axis of the coordinate system was matched with the clearance tool face (figure 1). C and C_1 in the figure represent the tool-chip and the tool-workpiece contact length, respectively.

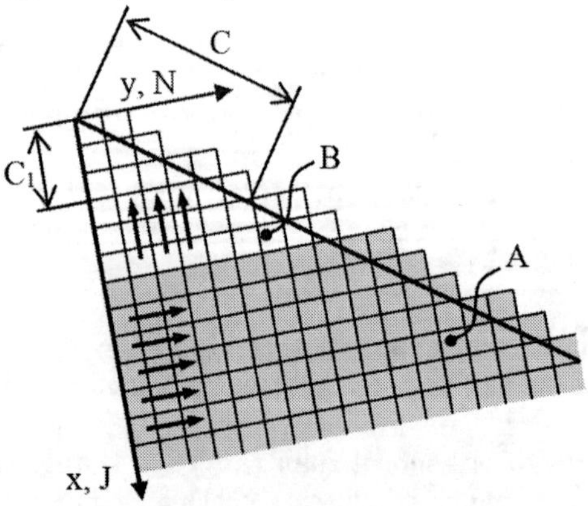

Figure 1. A grid used in the implemented finite difference method.

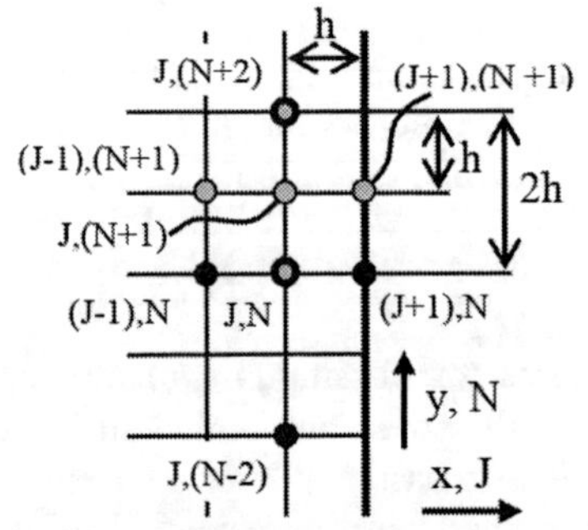

Figure 2. Scheme to calculate stresses at end nodes of grid lines.

Considering the tool state as a two-dimensional stress system and neglecting body forces, the differential equations of equilibrium are as follows:

$$\frac{\partial \sigma_x}{\partial x} + \frac{\partial \tau_{xy}}{\partial y} = 0 \text{ , } \frac{\partial \sigma_x}{\partial x} + \frac{\partial \tau_{xy}}{\partial y} = 0 \tag{7}$$

Differentiating equations (7) with respect to x and y gives:

$$\frac{\partial^2 \sigma_x}{\partial x^2} + \frac{\partial^2 \tau_{xy}}{\partial x \partial y} = 0 \text{ , } \frac{\partial^2 \sigma_y}{\partial y^2} + \frac{\partial^2 \tau_{xy}}{\partial x \partial y} = 0 \tag{8}$$

Subtracting equations (8) and replacing $\sigma_x = (S - \sigma_y)$, we obtain the main equation for stress separation:

$$\frac{\partial^2 \sigma_y}{\partial x^2} + \frac{\partial^2 \sigma_y}{\partial y^2} = \frac{\partial^2 S}{\partial x^2} \tag{9}$$

In zone A (figure 1), the calculation of σ_y is done in the direction from the boundary, which is out of contact with the chip and workpiece, into the tool body, line by line. In a grid line with $N = 0$, stresses σ_y and τ_{xy} are zero owing to the no-load condition outside the contact boundary. To calculate σ_y in a grid line with $N = 1$, we have to use the following equation:

$$\sigma_{y_{J,1}} = \frac{1}{2} \cdot \left(S_{J+1,0} - 2 \cdot S_{J,0} + S_{J-1,0}\right) \tag{10}$$

Equation (10) is derived from a finite difference form of equations (7) and (9), considering the zero value of σ_y и τ_{xy} in the $N = 0$ line.

The stress σ_y in the nodes of lines $N > 2$ may be found by using the finite difference form of equation (9):

$$\sigma_{y_{J+1,N}} + \sigma_{y_{J,N+1}} + \sigma_{y_{J-1,N}} + \sigma_{y_{J,N-1}} - 4 \cdot \sigma_{y_{J,N}} = S_{J+1,N} - 2 \cdot S_{J,N} + S_{J-1,N} \tag{11}$$

By rearranging it with respect to $\sigma_{y\,J,N+1}$, $\sigma_{y\,J+1,N}$, $\sigma_{y\,J-1,N}$ and $\sigma_{y\,J,N-1}$ for extrapolation.

To determine σ_y in the end nodes of the grid lines, first, we have to calculate $\sigma_{y\,J,N+2}$ by the following formula:

$$\sigma_{y_{J,N+2}} = 4 \cdot \left(S_{J+1,N} - 2 \cdot S_{J,N} + S_{J-1,N}\right) + 10 \cdot \sigma_{y_{J,N}} - 4 \cdot \left(\sigma_{y_{J+1,N}} + \sigma_{y_{J-1,N}}\right) - \sigma_{y_{J,N-2}} \tag{12}$$

Equation (12) is derived from equation (11) and transformed with respect to the doubled step $h_y = 2 \cdot h_x$ along the y-axis (figure 2, black nodes). Then, by rearranging equation (11) with centrally used node (J, $N + 1$), we calculate the unknown $\sigma_{y\,J-1,N+1}$ and $\sigma_{y\,J+1,N+1}$ for the ($N + 1$) grid line (figure 2, gray nodes).

The calculation stress σ_y in zone B (figure 1) is done similarly, considering only the direction change, i.e., instead of J-lines, the N-lines of zone A are used.

The normal stress σ_x may be determined by:

$$\sigma_x = S - \sigma_y \tag{13}$$

Additionally, the shear stress τ_{xy} may be determined by using equations (7). The principal stresses σ_1 and σ_2 may be found by well-known equations, using the previously obtained stress components. To verify the results, we integrated the corresponding stress components in the contact zone to obtain loads on the tool faces and compared them with the cutting forces measured by a dynamometer.

The tool temperature change at the point of interest is given by:

$$\Delta T = \frac{\Delta t_t}{t} \cdot \frac{1}{\alpha} \tag{14}$$

Where, α is the mean coefficient of linear thermal expansion (CTE) for the cutting tool material. Substituting equation (2) into (14) and considering the temperature T_0 of the cold tool (before cutting), the formula for the temperature at the point of interest is as follows:

IOP Conf. Series: Materials Science and Engineering **966** (2020) 012137 doi:10.1088/1757-899X/966/1/012137

$$T = T_0 + \frac{m_t \cdot \lambda}{n \cdot t \cdot a} \quad (15)$$

To obtain the temperature field, the fringe pattern along the closed boundary of the tool must be analyzed, and the temperature at points inside the tool may be determined by using Laplace's equation:

$$\nabla^2 T = 0 \quad (16)$$

3. Experimental bench design

The schematic diagram of the experimental setup used to test our proposed methods is shown in figure 3. A horizontal linear polarized laser beam from the source (1) passes through a beam expander (2), a turning mirror (3), a cubic beam splitter (4), and a quarter-wave plate (5). Then, it becomes circularly polarized and strikes an interferometer, which is formed by an optical wedge (6) and a polished side surface (7) of a cutting tool (8). The optical wedge and the tool are rigidly fastened relative to one another by a tool holder (9), which is fixed in a multidimensional dynamometer (10). A light reflected from the surface of the wedge forms the reference beam. The measurement beam is formed by a light transmitted through the wedge and reflected from the tool. After reflection, both beams change their polarization and recombine. The resulting beam passes back through the quarter-wave plate, obtains vertical linear polarization and strikes the beam splitter, where is reflected in the camera lens (11) direction. The interference fringe patterns recorded by the high-speed camera (12) and the dynamometer signal transformed by the analog-to-digital converter (ADC) (13) with an amplifier (14) were saved on a computer (15).

The applied interferometer design increases interference pattern quality: eliminates relative shifts between the optical wedge and the tool induced by vibration during machining due to their rigid fastening by tool holder; eliminates ghosts due to the use of polarizing optics with antireflective coating.

Figure 4 shows the experimental setup mounted on a conventional lathe, which spindle drive was adjustable. On the lathe carriage was placed the base table with a strain gauge dynamometer (1) and two aluminum rails fixed on it. On one rail a laser (2), an achromatic Galilean beam expander (3) (Thorlabs GBE20-A), and a turning mirror (4) were mounted. To achieve a high quality of fringe patterns a single-frequency diode-pumped solid-state laser LCM-S-111 with $\lambda = 532$ nm was used.

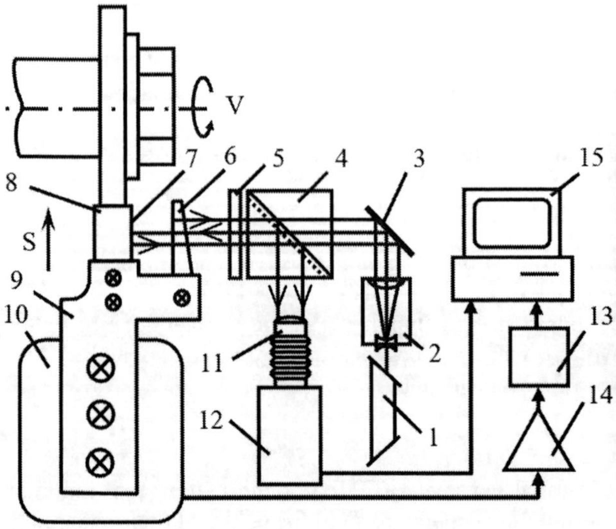

Figure 3. Schematic diagram of experimental setup.

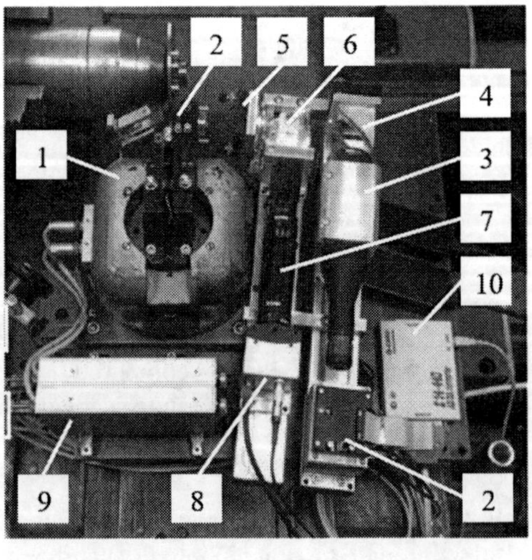

Figure 4. Experimental setup on a lathe.

The second rail carried the rotation mount (5) with a quarter-wave plate, tilt-rotation stage (Standa 6PT110) with a cubic beam splitter (6), and a zoom camera lens (7) (NAVITAR Zoom 6000) attached to a high-speed digital camera (8) (Fastec HiSpec 2). A strain gauge amplifier (9) (RDP621) connected to the ADC (10) (L-Card E14-440) was placed on the base, near the dynamometer.

4. Experimental details

To prove the efficiency of the proposed methods, a machining experiment on a disc of 4.2-mm thickness was performed. Martensitic heat-treated steel grade X13Cr11Ni2W2MoV with hardness HB 275 was used as a work material. Table 1 presents its chemical composition. This steel is difficult-to-cut and widely used for the production of turbine parts, like blades and discs.

Table 1. Chemical composition of heat-treated steel X13Cr11Ni2W2MoV.

Element	C	Cr	Ni	W	Mo	V	Si	Mn	S	P	Fe
wt.%	0.10–0.16	10.5–12.0	1.5–1.8	1.6–2.0	0.35–0.50	0.18–0.30	max 0.6	max 0.6	max 0.025	max 0.03	bal.

The tool material was cemented tungsten carbide WC-8Co grade VK8. The cutter had a 4.65 mm width, 25° rake and 10° clearance angles. The machining conditions were: cutting speed V of 6 m/min and feed rate S of 0.14 mm/rev. The camera frame rate was 16×10^3 fps.

5. Results and Discussion

Figure 5 presents the interference patterns recorded by camera (a) before at $T_0 = 20$ °C, (b) in the process of cutting, and (c) immediately after cutting interruption.

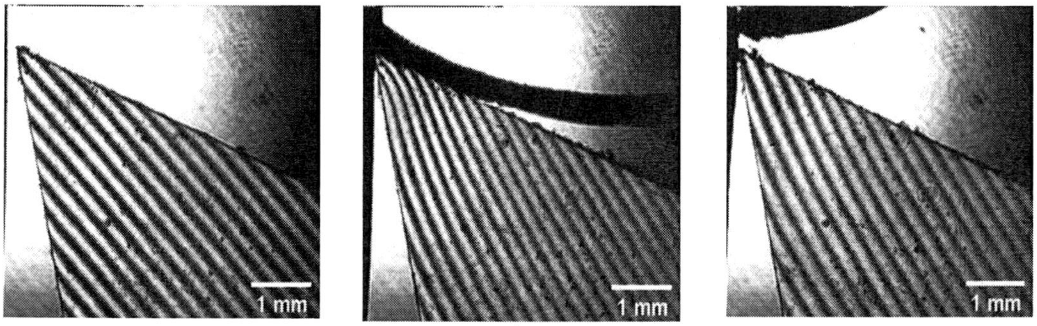

a b c

Figure 5. Images of interference fringes recorded (a) before, (b) in the process, and (c) immediately after cutting interruption.

The fringe order distribution curves were obtained by analysis of image profiles along the closed boundary in ImageJ. As a boundary, lines along the rake face, the clearance face, and the line forming triangle with them were used. Figure 6 shows the curves of fringe order for the (a) rake and (b) clearance faces, which represent the index number of each profile maxima and their location l on the face respect to the tool edge. Plotting and curves subtracting were done by built-in MS Excel functions.

The fringe order m were converted to sum of the stresses S and temperature change ΔT by using the CTE [11], Young's modulus and Poisson's ratio of the cemented tungsten carbide WC-8Co. Figures 7 and 8 show the principal stresses σ_1, σ_2, and the temperature fields, respectively, which were computed by a specially created program in MATLAB to realize the aforementioned methods.

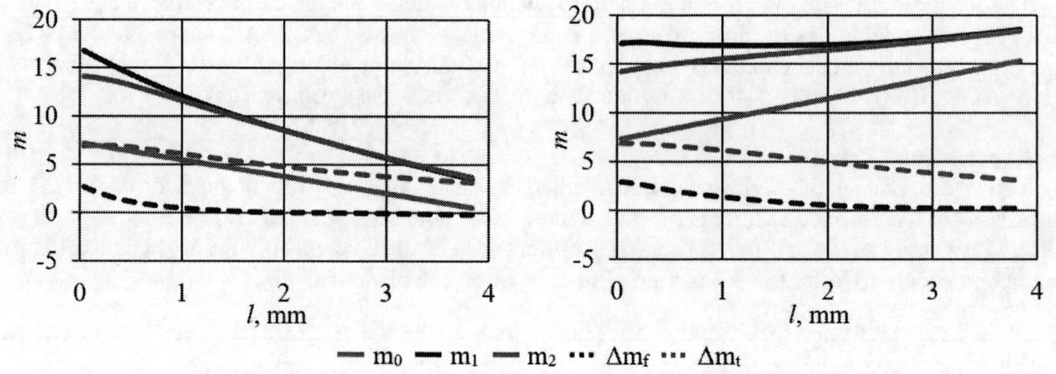

a b

Figure 6. Curves of fringe order *m* for the (a) rake and (b) clearance faces.

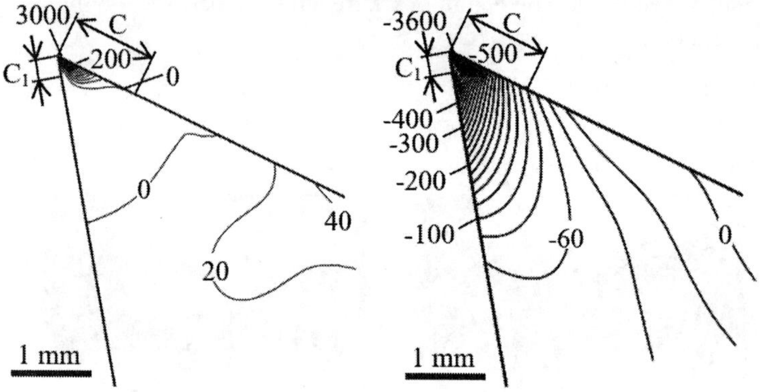

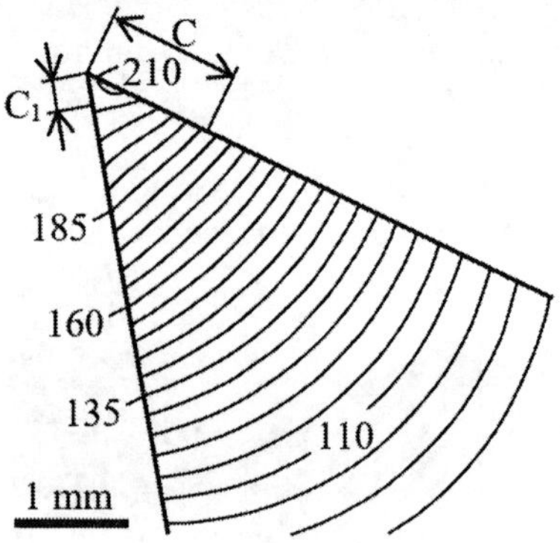

a b

Figure 7. Fields of principal stresses (a) σ_1 and (b) σ_2 (in MPa)

Figure 8. Temperature field (in Celsius degree).

6. Conclusion

The developed experimental methods are applicable under real cutting conditions and may be used for the thermal and stress-strain analyses of a tool composed of real cutting material, including non-stationary conditions during its operation because of the use of a high-speed digital camera.

References

[1] Arrazola P J, Özel T, Umbrello D, Davies M and Jawahir I S 2013 *CIRP Ann.-Manuf. Techn.* **62** 695-718
[2] Shi B and Attia H 2010 *Mach. Sci. Technol.* **14** 149-88
[3] Buryta D, Sowerby R and Yellowley I 1994 *Int. J. Mach. Tool. Manu.* **34** 721-39
[4] Trent E M and Paul K W 2000 *Metal cutting* 4th ed. (Boston: Butterworth-Heinemann) p 464
[5] Isogimi K, Kitagawa T and Kurita H 1988 *Int. J. Jpn. S. Prec. Eng.* **54** 390-5
[6] Komanduri R and Hou Z B 2001 *Tribol. Int.* **34** 653–82
[7] Astakhov V P 1998 *Metal Cutting Mechanics* (Boca Raton: CRC Press) p 320
[8] Pujana J, del Campo L, Pérez-Sáez R B, Tello M J, Gallego I and Arrazola P J 2007 *Meas. Sci. Technol.* **18** 3409
[9] Efimovich I A, Zolotukhin I S and Shvetsova E I 2010 *RF Patent Specification* no. 2442967
[10] Efimovich I A and Shvetsova E I 2009 *RF Patent Specification* 2436039
[11] Efimovich I A, Zolotukhin I S and Zav'yalov E S 2019 *Obrabotka Metallov. Metal Working and Material Science* **21** 129–40

Adapting standard maintenance approaches for mining excavators to actual operating condition

D A Shibanov[1, 3], S L Ivanov[2], K A Safronchuk[2] and V I Knyazkina[2]

[1]IZ-KARTEX named after P G Korobkov Ltd., street Brat'ev Radchenko, Kolpino, Saint Petersburg, 196653, Russia
[2]Saint Petersburg Mining University, 2, 21 linija, Vasil'evskijostrov, Saint Petersburg, 199106, Russia

E-mail: [3] dan11l88@yandex.ru

Abstract. Significant portion of the annual calendar fund-time work of mining excavators spent on maintaining its working condition by carrying out preventive and repair regulatory activities and recovery in the event of emergency failures and maintenance and repair expenses constitute a significant proportion of the cost of excavation of mining rock. To assess the effectiveness of machinery a non-standard indicator, the coefficient of technical readiness, applied which does not have a single assessment methodology. At the present stage of machinery development, existing quarry dispatching systems allow you to allocate any of its elements from the fund-time of mining excavators: productive time, time spent in repair and maintenance and also classify other time costs of a particular machine which allowed an appropriate analysis. Analysis showed that at the current organization of MaR (maintenance and repair) there is significant potential to reduce repair impacts, as well as to ensure a reduction in the cost of routine maintenance, to remove maintenance and repair measures from the category of costly ones, at least for self-sufficiency without compromising the quality of these works and the reliability of the excavator as a whole.

1. Introduction

A large part of the annual calendar fund-time work of excavators involved in the extraction of minerals falls on downtime due to repairs, and the costs of maintenance and repair make up a significant proportion of the cost of excavating the mining rock.

The effectiveness of the use of equipment according to the criterion of its operability in most mining enterprises is estimated by the coefficient of technical readiness (C_{tr}) [1, 2]. In general, C_{tr} characterizes the technical level and reliability level of the facility, allowing you to compare the same type of mining equipment within the same mining enterprise.

Mining enterprises distinguish between planned and actual [3] coefficient of technical readiness, which are defined as the ratio of the difference between the calendar fund-time work of the planning period and the time of repair downtime (planned and actual respectively) to the calendar fund-time work of the planning.

Term 'coefficient of technical readiness' is often used in assessing mutual settlements between the Customer (operating organization) and the Manufacturer (in the particular case, supplier) of new equipment. When purchasing new equipment, the Supplier and the Consumer jointly determine the average value of the C_{tr} specified in the contract, which is guaranteed by the Supplier [4, 5]. Usually, during the warranty period, the coefficient of technical readiness of mining excavators is 0.80–0.88C_{tr}

is also used to evaluate the effectiveness of the technical service when the repair units are outsourced or branded technical service.

The coefficient of technical readiness is not regulated by SSS (State Standard System) and is subjective in nature. So on some MPPs ((mining and processing plant) (cuts)), work on changing the teeth of the bucket and ropes is not included in the category of technical downtime affecting the C_{tr}, because they are classified as technological, while in other enterprises, these operations are included in the assessment of this coefficient.

It is often difficult to evaluate and qualify downtime due to the lack of clear objective criteria for accounting for certain categories of unproductive working hours.

Existing quarry dispatching systems make it possible to allocate productive operating time, time spent on repair and maintenance from the fund – time of mining excavators, as well as classify other time costs of a particular machine. The Modular System [6, 7] is used in most enterprises conducting open pit mining. The operating time data entering this system is largely dependent on the 'operator' entering the values into the system. In relation to the excavator, the operator is the driver-operator, who in the process of work must record all the downtime of the excavator by entering into the computer information about the operation of the equipment at a particular moment in time (loading, cleaning the sole, failure, waiting for the dump truck, etc.) [8].

2. Methodology for assessing the coefficient of technical readiness

The coefficient of technical readiness, as in equation (1), is defined as the ratio of the difference between the calendar fund - time and the actual time of repair downtime to the calendar fund - time of the estimated period:

$$C_{tr} = \frac{T_{cft} - T_{f.rep}}{T_{cft}} \quad (1)$$

where C_{tr} is the coefficient of technical readiness, T_{cft} is the calendar fund –time,$T_{f.rep}$ is the actual downtime of the excavator in repairs and waiting for repairs.

The main difference in the evaluation methods of C_{tr} is the determination of the time for repair work, specifically the accounting of downtime associated with the replacement of consumables (teeth and bucket protection, ropes, lubrication and its replacement), since they cannot be clearly regulated and to a large extent depend on the conditions and operating modes of the mining machine.

At one of the iron ore plants in Russia (hereinafter referred to as MPP), field observations of the operation of mining excavators of large unit capacity were carried out. The observations were carried out during the calendar year for two excavators with a bucket volume of 20 m^3 [9, 10, 4], in this article designated as ECG №1 and ECG №2.

The C_{tr} calculation methodology used at MPP takes into account the time it takes to replace consumables. Thus, with the consistent performance of all scheduled maintenance recommended by the manufacturer according to the standards of frequency and duration, the maximum achievable C_{tr}, as in equation (2) and (3), per month in the absence of emergency repairs cannot exceed:

$$C_{tr} = \frac{Tcft - Tf.rep}{Tcft} = \frac{Tcft - (TMrep + TMm + TMel.m + Tl.r + Tpr.r + Tret.r + Tteeth)}{Tcft} \quad (2)$$

$$720 - \frac{(0.5 \cdot 2 \cdot 30 + 1 \cdot 24 + 1 \cdot 12 + 0.5 \cdot 12 + 4 \cdot 4)}{720} = 0.844 \quad (3)$$

where TM_{rep}, TM_m, $TM_{el.m}$ – time for maintenance of replaceable, monthly, electrical machines, respectively; $T_{l.r}$, $T_{pr.r}$, $T_{ret.r}$, T_{teeth} – time to carry out work on the replacement of lifting ropes, pressure ropes, return ropes, bucket teeth, hours.

The frequency of replacement of consumables (ropes, teeth) was taken according to their actual consumption for the period of the experiment, the regulatory frequency and duration of work was adopted in accordance with the operating manual for these excavators.

In practice, the activities of mining enterprises and service companies to reduce the time for maintenance and repair, certain types of work are combined, for example:

1. Shift maintenance is carried out during the reception/change of the shift or at other organizational and technological breaks (when replacing teeth, ropes, waiting for vehicles), in particular it must be combined with other types of maintenance work.
2. Monthly maintenance of the electrical part is carried out in parallel with the monthly maintenance of the mechanical part.

Thus, C_{tr} in the case under consideration can be increased to the level of, as: in equations (4) and (5):

$$C_{tr} = \frac{Tcft - Tf.rep}{Tcft} = \frac{Tcft - (TMm + +Th.r + Tpr.r + Tret.r + Tteeth)}{Tcft}, \quad (4)$$

$$720 - \frac{(1 \cdot 24 + 1 \cdot 12 + 1 \cdot 12 + 0.5 \cdot 12 + 4 \cdot 4)}{720} = 0.903 \quad (5)$$

3. Analysis of emergency failures of units and parts of excavators ECG №1 and ECG №2

During the operation of excavators ECG №1 and ECG №2, the most significant downtime was due to an emergency failure, both due to the fault of the manufacturer (factory defects of components and parts, these cases during the experiment are confirmed by bilateral acts) and a number of disputed cases are noted on the following units: ropes; pressure axis; handle; idler wheels; cast parts; drive unit.

The total time to restore and wait for repairs on the indicated units for two ECG machines №1 and ECG №2 amounted to 1054 hours out of 12384 hours of the calendar time-fund (CTF). Thus, the average time of emergency downtime at the above units is about 65 hours per month (table 1), and the average C_{tr} for the specified period was 0.75.

Table 1. Time for troubleshooting individual units and parts.

Unit name	Average troubleshooting time, h/month	The share of emergency downtime, % CTF	Notes
Pressure axis	27.7	3.85	Redesigned. Axes are in stock
Drive unit	10.6	1.48	Ultrasonic inspection of shafts and gears introduced
Handle	10.4	1.45	Structural reinforcements on operating machines have been performed. On the next issued ECG - the handle of a new design. Additional control introduced (X-ray, MTD)
Ropes (pressure, return, lift)	8.4	1.16	Changed suspension design. Recommendations on the use of rope d 60.5mm
Idler wheels	4.1	0.57	The lubrication system has been changed, and input quality control of the supplied bronze has been strengthened.
Casting defect	no data	0.50	Input quality strengthened.
TOTAL	≈65	9.0	

There is no doubt that the operation of mining equipment is impossible without emergency failures [11–13]. In practice, the work of mining enterprises of the Russian Federation for unplanned work to eliminate emergency failures lay about 24–27 hours a month, which is 3.3–3.8% of CTF.

For the analyzed period, emergency repairs of the above-described units of the ECG №1 excavators; ECG №2 excavators amounted to 9% of CTF. Of these, a large proportion (42%) was occupied by emergency breakdowns of the pressure axis, the causes of which are controversial [14, 15]. For the convenience of service, improving the quality of this service, mechanizing production and reducing economic costs during maintenance and repair of mining machines involved in carrying out

IOP Conf. Series: Materials Science and Engineering **966** (2020) 012138 doi:10.1088/1757-899X/966/1/012138

basic and auxiliary works in the implementation of mining technologies, work is underway to create mobile and autonomous technical means for carrying out delivery work, storage, regeneration and collection of used oils and lubricants as part of scheduled maintenance and repair work. However, when performing lubricating and filling operations with the help of mobile workshops, problems arise associated with forcing grease into the system due to the inability to create the necessary pressure with domestic metering pumps [16, 17].

Thus, the high share of emergency downtime (9%) after making changes to the design of excavators, including the use of on-board diagnostic systems and the transition to centralized lubrication systems with the solution of their refueling problems [18], tends to decrease. It is planned that according to the results of all warranty work to improve the design of excavators, the share of emergency failures will be within acceptable limits (3.3–3.8%). Based on the foregoing, in eliminating the causes of emergency failures, the average C_{tr} of excavators will be about, as in equation (6):

$$C_{tr} = 0.75 + (0.09 - 0.035) = 0.81 \tag{6}$$

After eliminating the accident rate of the listed units, the potential for increasing C_{tr} by the factor 'manufacturer' will be exhausted. Therefore, to achieve the planned $C_{tr} = 0.85$ and above, eliminating the accident rate of the nodes is not enough.

With the elimination of accidents and bringing the duration of repairs to standard indicators, it is possible to achieve C_{tr} at the level of, as in equation (7):

$$C_{tr} = 0.75 + (0.09 - 0.035) + 0.092 = 0.90 \tag{7}$$

4. Analysis of the time spent on maintenance and repair of excavators

Based on the statistical data of the work of excavators ECG №1 and ECG №2, Table 2 shows a comparison of the actual and achieved time spent on maintenance and repair, as well as the planned indicators of the repair department of the MPP (table 2).

Table 2. Comparison of the average planned, actual and achieved time spent on maintenance and repair of the excavator from 1 January to 30 April.

Type of repair	Duration of repairs per month, h			Deviation of actual time from standard	
	Actual	GOK plan	Service company standard	Time, h	% CTF
Shift maintenance (not taken into account in amount per month)	No data	30	30	–	–
Monthly Maintenance	61	24	24	+37	+5.1
Monthly Electric Maintenance	12	12	12	0	0
Consumables replacement work					
Teeth replacement	24.4	16	12	+11.5	+11.5
Protection replacement	1.1	16	2	+11.5	+1.6
Lift rope replacement	17	12	8	+9	+1.3
Pressure rope replacement	18	12	8	+10	+1.4
Return rope replacement	3	6	4	–1	–0.1
TOTAL	136	82	70	+66.5	+9.2

IOP Conf. Series: Materials Science and Engineering **966** (2020) 012138 doi:10.1088/1757-899X/966/1/012138

In order to increase the operating time of the bucket teeth, to reduce the time for their replacement, as well as to exclude cases of emergency tooth breakages, it is recommended to use teeth tested by the manufacturer on excavators ECG № 1 and ECG № 2. During the experiment, MPP used bucket teeth that did not correspond to the design documentation. The result is a decrease in C_{tr}.

Conclusions

1. The average actual time for maintenance and repair of excavators ECG №1 and ECG №2 is 136 hours per month compared to 82 hours according to the plan of the MPP repair unit. At the same time, the service company achieved the duration of repair operations – 70 hours per month.

2. When performing maintenance and repair work on excavators, there is the potential to reduce the repair time by 66.5 hours per month, which is equivalent to an increase in C_{tr} by 9.2 points.

Thus, subject to the regulatory deadlines, a reserve of time is provided for unplanned and emergency repairs to repair the excavator.

References

[1] Velikanov V S 2018 *Development of the scientific-methodological basis for the career excavator's improvement on the basis of an indistinct - multiple approach* (Magnitogorsk: Magnitogorsk State Technical University named after G I Nosov) p 217

[2] Emelyanov A A, Ivanov S L and Shibanov D A 2017 To a question of the operator qualification influence estimation on the excavator technical condition *Open mining works in the XXI century: results, problems and prospects of development-2. Materials of III International scientific-practical conference (Mining information-analytical bulletin)* Special issue no **38** pp 442–53

[3] Ivanova P V, Ivanov S L, Kuvshinkin S Yu and Shibanov D A 2015 Selection of the rational system of organization of maintenance service and repair of the quarry excavators for the given mining engineering conditions *Innovations on transport and in mechanical engineering: Proc. of III international scientific-practical conference. V.II.* (SPb: National mineral university 'Gorny') pp 94–8

[4] Ganin A R, Donchenko T V and Shibanov D.A 2015 Development of a number of the excavators of 20-25 cubic meters class produced by IZ-KARTEX for mining industry *Coal* no **8** pp 79–81

[5] Anistratov K, Donchenko T and Shibanov D 2018 *Digging up potential* World Coal vol **27** no **7** pp 45–7

[6] Trifanov M G and Shishlyannikov D I 2015 Objective control means as an instrument to increase the efficiency of operation of the road heading and cleaning combines of the potash mines *Innovations in transport and machine building: collection of III International scientific-practical conf.; St. Petersburg, 14-15 April 2015* (St. Petersburg: The Russian Academy Of Sciences) pp 106–108

[7] Brodny J, Stecula K, Tutak M 2016 Application of the TPM strategy to analyze the effectiveness of using a set of mining machines *16th Int. Multidisciplinary Scientific Geo Conference* pp 65-72 doi: 10.5593/SGEM 2016B12/S03.009

[8] Shibanov D A 2015 *Comprehensive assessment of factors that determine the operating time of EKG-18R/20K excavators for planning maintenance and repairs* (St. Petersburg: Mining University) pp 201

[9] Poderni R Yu 2013 Analysis of the state-of the art of the surface mine machinery world market *Mining industry* no **4 110** pp 48–54

[10] Ivanov I Yu, Komissarov A P, Lagunova Yu A and Shestakov V S 2015 Intensification of the rock excavation processes *News of higher educational institutions. Mining journal* no **3** pp 94–100

IOP Conf. Series: Materials Science and Engineering **966** (2020) 012138 doi:10.1088/1757-899X/966/1/012138

[11] Kuvshinkin S Y and Ivanova P V 2019 Developing a methodology for estimation of excavation techniques for given operating conditions *IPDME 2019 IOP Conference Series: Earth and Environmental Science* pp 378 012121 doi: 10.1088/1755–1315/378/1/012121

[12] Dindarloo R, Siami-Irdemoosa E and Frimpong S 2016 Measuring the effectiveness of mining shovels *Mining Engineering* vol **68** no **3** pp 45–50

[13] Nehring M, Sciprofile P F, Sciprofile M S and Kizil E Haÿ 2017 A comparison of strategic mine planning approaches for in-pit crushing and conveying, and truck/shovel systems *International Journal of Mining Science and Technology* vol **28** pp 205–14 doi:10.1016/j.ijmst.2017.12.026

[14] Kasyanov P A, Shestakov V S and Zakharov A A 2017 Calculation of the efforts in the hoisting ropes of the excavator 'straight shovel' *Technological equipment for mining and oil and gas industry collection of works of XV International scientific and technical conference* pp 283–6

[15] Komissarov A P, Lagunova Yu A, Shestakov V S and Lukashuk O A 2018 Estimation of the mode parameters of the main mechanisms of an excavator *Mining equipment and electromechanics* no **3** **143** pp 3–8

[16] Pumpur E V, Knyazkina V I, Safronchuk K A and Ivanov S L 2019 Estimation of the factors influence on a strategy choice of the excavator maintenance service *Mountain information-analytical bulletin* no **12** special issue **41** pp 3–19 doi 10.25018/0236-1493-2019-12-41-3-19

[17] Knyazkina V I, Safronchuk K A and Ivanov S L 2019 About possibility of immediate evaluation of technical condition of mining equipment using signal value of acoustic emission friction *IOP Conf. Series: Materials Science and Engineering 560* doi:10.1088/1757-899X/560/1/012068

[18] Safronchuk K A, Knyazkina V I and Ivanov S L 2019 Mobile lubrication and filling units to reduce mining machines and equipment downtime when providing maintenance *IOP Conf. Series: Materials Science and Engineering 560* doi:10.1088/1757-899X/560/1/012088

IOP Conf. Series: Materials Science and Engineering **966** (2020) 012139 doi:10.1088/1757-899X/966/1/012139

Optimization of alternate profile component part machining

L V Borshova[1] and V F Pegashkin[1,2]

[1]Ural Federal University named after the First President of Russia B N Yeltsin, Nizhniy Tagil Technological Institute, 59, Krasnogvardeyskaya str., Nizhniy Tagil, 622000, Russia

E-mail: [2]v.f.pegashkin@urfu.ru

Abstract. Process optimization for machining of component parts with the profile alternate by the length has been reviewed here. The review deals with the optimization criteria applied and shows that the optimization of machining modes normally uses a criterion of cost-effectiveness, which is most often used in up-to-date technologies. With reference to specific organizational and economic conditions at companies of the industry, it is shown that power saving may become a priority when standardizing modes of machining. The power consumption pattern for the cutting process is reviewed. A variable energy value is proposed as an optimality criterion, which depends on parameters optimized i.e. – technological energy. Mathematical models are defined for the optimality criterion and limiting parameters that allow reducing energy consumption of the machining process. Calculations carried out according to the proposed methodology have shown that when the machine electric motor power is above 13.5 kW, work in the energy-saving mode becomes economically viable. Energy savings reach 10 %, while the main time increases by an average of 10%, but the reduction in costs associated with a decrease in energy consumption is 4 times higher than the increase in costs due to an increase in the main time.

1. Introduction

In present time, a machine-building enterprise can do well only if its manufacturing process is optimal. Therefore, the issue of process optimization is one of the most important [1–5].

When optimizing machining modes, a criterion of cost efficiency is normally used being a favorite criterion in up-to-date technologies [5–10]. However, energy saving may become a priority of optimizing calculations with reference to specific organizational and economic conditions at companies of the industry, when standardizing modes of machining. To create cutting mode calculation algorithms, it is necessary to review and formalize the criterion structure including that considering specific features of machining of housings.

2. Methods

In its most general form, the goal function

$$E = N\tau_m \rightarrow \min$$

where E is the power put in the machine during τ_m, N is the capacity put out by the machine motors, τ_m is the machining time.

The power consumed by the machine can be represented as the sum of the energies

$$E = E_c + E_w + E_t + E_s + E_l + E_f + E_{fr} + E_r$$

The equation indicates the energies used for: E_c – cutting, E_w – workpiece clamping, E_t – fast feed and removal of toolheads, E_s – running of electric and electronic control systems, E_l – machine workspace lighting, E_f – cutting fluid feeding to the machine workspace, E_{fr} – friction in the mechanisms and units of the machine, E_r – circuit resistance.

The cutting energy E_c is the largest component of E. The other terms can be combined into two groups: E_1 is the energy component related directly to the main time τ_o, и E_2 is the energy component, which is a constant of the specific operation, so it can be written

$$E = E_c + E_1 + E_2,$$

where

$$E_1 = E_w + E_f + E_{fr} + E_r = (N_w + N_f + N_{fr} + N_r)\tau_o,$$

$$E_2 = E_t + E_s + E_l$$

E_2 is a constant value and can be excluded from the optimization system for this operation. We introduce the value of technological energy E_o as the optimality criterion

$$E_o = E_c + E_1 \Rightarrow \min$$

To determine the cutting energy value E_c, it is necessary to determine the cutting power N_c. Entering the correspondence

$$N_c = C_1 P_z V$$

The component of the cutting force P_z is determined by the equation

$$P_z = C_2 V^x S^y t^z$$

The main time τ_o can be found using the equation

$$\tau_o = C_3 V^{-1} S^{-1}$$

Then

$$E_c = C_4 V^x S^{y-1} t^z$$

or through the machine spindle rotation speed

$$E_c = C_5 n^x S^{y-1} t^z.$$

When processing alternate profile surfaces with several tools, the following parameters are metered at each ith turn of the workpiece: the cutting speed V, the workpiece rotation speed n, the feed S, the cutting depth t for each tool k. If over the time τ_o the number of workpiece turns is A, then E_p can be calculated by the equation

$$E_c = C_5 \sum_{i=1}^{A} \sum_{j=1}^{k} n_{ji}^x S_{ji}^{y-1} t_{ji}^z,$$

Where n_{ji} S_{ji}, t_{ji} are spindle speed, feed and cutting depth for the jth tool at the moment of the ith workpiece turn determined by the method.

The optimization system has to be modified with limitations for motor capacity, spindle torque, machined face coarseness, throughput and kinematics of the machine.

For multi tool machining

$$N_c = \sum_{j=1}^{k} N_{pi} = C_6 n_i^x \sum_{j=1}^{k} S_{ij}^{y-1} t_{ij}^z$$

Spindle torque

$$M_{Sti}=\sum_{j=1}^{k} M_{Stji}=0.5C_7n_i^x\sum_{j=1}^{k} D_{ji}S_{ij}^{y-1}t_{ij}^z,$$

where M_{Stji} is a torque created by the cutting force of the jth tool at the ith turn of the workpiece.

Since the feed values may vary, the resulting coarseness of the treated surface Rz has to be calculated and compared to the coarseness Rz_a allowed by the component drawing.

In mass and large-scale production, the throughput depends highly on the main time τ_o. Therefore, the throughput limit can be defined by the following correspondence

$$\tau_o \leq \tau_{max},$$

where τ_o is calculated machining time, τ_{max} is maximum allowed machining time.

When the operation uses k tools, for the jth tool, if the conditions are met n = const, S = const

$$\tau_{oj} = C_{8j}n^{-1}S_j^{-1}$$

There are a few ways to distribute using of the tools over time and, accordingly, equations of the limit by τ_o. We should consider the main of them.

When all number of k toolswork at the same time, the limit is the base time τ_{oj} of the one that has it at maximum

$$(C_8n^{-1}S^{-1})_{max} \leq \tau_{max}$$

When all number of k tools work one by one, then

$$n^{-1}\sum_{j=1}^{k}\left(C_{8j}S_j^{-1}\right) \leq \tau_{max.}$$

When the tools work one by one in groups having numbers (1,…,u,…,w), then inside each uth group there is uj^{th} tool, for which τ_{ouj} is maximum allowed as compared to others within the group.

Then

$$n^{-1}\sum_{u=1}^{w}\left(C_{8uj}S_{uj}^{-1}\right)_{max} \leq \tau_{max}$$

Thus, if a component machined has a complex surface profile consisting of A elementary sections at the same time k, we obtain an optimization system in the following form.

Target function

$$C_5\sum_{i=1}^{A}\sum_{j=1}^{k} n_i^xS_{ij}^{y-1}t_{ij}^z+(N_w+N_{fr}+N_f+N_r)\tau_o \rightarrow \min$$

Limitations:

- by motor power

$$C_6n_i^x\sum_{j=1}^{k} S_{ji}^{y-1}t_{ji}^z \leq N_{mp},$$

- by spindle torque

IOP Conf. Series: Materials Science and Engineering **966** (2020) 012139 doi:10.1088/1757-899X/966/1/012139

$$0.5C_7 n_i^x \sum_{j=1}^{k} D_{ji} S_{ji}^{y-1} t_{ji}^z \leq M_{st},$$

- by coarseness of the surface machined

$$\frac{10^3 S_{ji}^2}{8r_j} \leq Rz_{sm},$$

- by base time

$$\tau_{o\max} \leq \tau_{ob},$$

- by machine's kinematics

$$n_{m\,\min} \leq n_i \leq n_{m\,\max},\; S_{m\,\min j} \leq S_{ij} \leq S_{m\,\max j}$$

3. Conclusion

Calculated by the method proposed, the calculations show that working in the power-saving mode is cost-effective when the machine's motor has a capacity higher than 13.5 kW. Power savings reaches 10 %, while the main time increases by 10 % at average, but the cost saving due to reduced power consumption is 4 times higher than the cost rise due to the main time increased.

References

[1] Akintseva A V *et al* 2020 *IOP Conf. Ser.: Mater. Sci. Eng.* **709** 033004

[2] Uhlemann T H J, Lehmann C and Steinhilper R 2017 *J. Procedia CIRP* **61** p 335

[3] Uhlemann T H J, Schock C and Lehmann C, Freiberger S and Steinhilper R 2017 *J. Procedia Manufacturing* **9** p 113

[4] Preuveneers D, Joosen W and Ilie-Zudor E *2018 IEEE 22nd Int. Enterprise Distributed Object Computing Workshop* **22** p 110

[5] Sal'nikov V S and Hoang V C 2016 *Rus. Engng Research* **6** pp 508–10

[6] Sergiev A P, Vladimirov A A and Shvachkin E G 2017 *Rus. Engng. Research* **6** pp 505–8

[7] Albagachiev A Y, Krasko A S and Stramtsova E S 2018 *Rus. Engng. Research* **10** pp 776–9

[8] Galkin M G and Smagin A S 2019 *Rus. Engng. Research* **1** pp 45–8

[9] Pereverzev P P and Akintseva A V 2017 *Rus. Engng.Research* **7** pp 617–21

[10] Pronin A I, Mylnikov V V and Romanov A D 2018 *J. of machinery manufacture and reliability* **1** pp 112–20

AUTHOR INDEX

Ahyoev, J S ... 624
Aleksandrin, D V ... 560
Alektorov, R V ... 329
Alexandrova, A S ... 554
Amirov, R N ... 128
Amosov, N A ... 533
Aristova, N A ... 545
Asanova, D U ... 618, 624
Asanova, S M ... 618, 624
Askarbek, N ... 624
Babenko, A A ... 54, 66
Badrtdinov, M A ... 630
Bahrami, AH ... 641
Balueva, T S ... 133
Barakovskikh, D S ... 25, 420
Baranchikova, S G ... 496
Baranov, G ... 144
Baraz, V R ... 308, 653
Bashkatov, N N ... 280, 286
Belyaev, S Yu ... 554
Bogoyavlensky, A ... 724
Boldyrev, D A ... 117, 128
Borodikhin, S A ... 211
Borshova, L V ... 761
Breido, I V ... 302
Budnikov, A S ... 404
Buzorina, D S ... 83, 484
Bystrov, M V ... 105
Byvaltsev, S V ... 663, 668
Chemesova, A V ... 653
Chemezov, O V ... 159, 164
Cherepanova, E V ... 496, 517
Cherny, V E ... 566
Chetverikov, S E ... 448
Chikhranov, A V ... 185
Chistyakov, N N ... 496
Dashevsky, V Ya ... 206
Dema, R R ... 117, 550
Dmitriev, A N ... 329
Dokuchaev, S V ... 319
Dolganov, A G ... 458
Dolzhenko, Y A ... 185
Dolzhenkova, E V ... 414
Domanskaya, I K ... 324
Domozhirov, D V ... 111, 122, 217
Druzhinin, V M ... 302
Dubinina, V G ... 505
Dudko, V A ... 392
Efimovich, I A ... 748
Egiazaryan, D K ... 254, 258
Egiazar'yan, D K ... 368
Elchinskiy, V ... 658
Elkind, D M ... 410
Epifanov, V V ... 490
Eremin, E N ... 211
Erofeev, V A ... 77, 83
Erpalov, M V ... 248
Ershova, I V ... 496, 517
Evchik, V E ... 566
Falahutdinov, R ... 169
Farafontova, E P ... 150, 155
Fayskhanov, D F ... 517
Fedulov, A A ... 554
Fesun, I P ... 221
Filippov, O I ... 77
Filyakov, A E ... 478
Fiveyskii, A M ... 71, 469
Fomin, A V ... 715
Forshev, A A ... 335
Galimyanov, I ... 48
Gamin, Yu V ... 404
Gapak, A L ... 736
Generalova (Lukon'kina), K A ... 490
Gerasimov, S S ... 653
Gerasimova, E S ... 275
Getman, A A ... 155
Ghoziev, B N ... 612
Ghulomzoda, A H ... 606
Golubev, V I ... 730
Goman, V ... 572, 589, 594
Gordeeva, T A ... 426
Gorkunov, E S ... 297
Goruleva, L S ... 297
Gruh, A G ... 736
Gruzdeva, I A ... 653
Gruzman, V M ... 696, 711
Gulakhmadov, A A ... 612
Gulyamov, K H ... 606
Gushin, D N ... 179
Iakupov, D R ... 464
Ibrahim, Mohamed N ... 583
Ignatova, N U ... 522, 647
Ilin, K ... 144
Isakov, D V ... 31, 600
Ishina, E A ... 308
Iskhakov, R V ... 404

Iurieva, L V 414
Ivanov, S L 464, 755
Ivanova, P V 464
Kadach, M V 404
Kalinin, E V 500
Kalinina, L L 500
Kalinina, N A 527
Kalinina, N E 517
Kalugina, O B 117
Kapustin, F L 133, 280
Karaulov, N G 122
Karelova, A E 291
Karelova, R A 702, 706
Karzhavin, V V 692
Katalinič, B 641
Kezik, A G 566
Khafizova, A R 244
Kharchenko, A S 179
Khmelnikov, E A 635, 686
Khudobin, L V 185
Khurelchuluun, I 736
Klimov, A V 89, 100
Knyazkina, V I 755
Kobuliev, Z V 612
Kolesatova, O S 217
Kolmachikhina, O B 37
Komissarov, A P 454
Kornilov, S N 560
Korolev, S I 589
Korotkov, V A 31
Kosdauletov, N Y 195
Krouglov, V N 736
Kryucheva, K D 297
Kugaevskii, S S 426
Kukushkina, G R 533
Kulsaitov, R V 95
Kurashova, M V 432, 437, 443
Kushnarev, A V 335
Kuzmin, A 658
Kuznetcova, N A 511
Kuznetsov, M S 505
Kuznetsova, E Y 533
Latypov, O R 550
Leontiev, L I 206
Leshchenko, S L 313
Letnev, K Y 454, 458
Liberman, Y L 680
Litvinenko, N V 217
Lobanov, V G 37
Losev, A S 211
Lukashuk, O A 454
Lukinskikh, S V 680
Magitov, A M 95
Makarov, A V 269
Makovskaya, O Yu 37
Malkov, A V 159, 164
Malkov, O V 159, 164
Malkov, V B 159, 164
Malygina, I Yu 269
Markova, Yu N 511
Matalasova, A E 211
Matyukhin, V I 392, 398
Matyukhina, A V 392, 398
Mazhitov, A M 13
Medison, V V 730
Melnikov, A Yu 77, 478
Meshchaninova, T V 554
Metelkin, A A 362, 368
Mikhailov, S B 291
Mikhailova, L Yu 66, 201
Mikhailova, N A 291
Mikhaylenko, A M 742
Mikheenkov, M A 254, 258
Milutin, R S 618
Mineeva, T A 511
Mironov, K V 335
Mironova, M V 341, 346
Mosin, A A 484
Mufteeva, L F 150
Mugalimov, R G 539
Mugalimova, A R 539
Muhutdinov, R M 589
Murukina, A D 527
Murzinova, S 169
Mushnikov, A N 297
Myslyakova, Yu G 432, 437, 443
Nadeina, Ye A 443
Nefed'ev, S P 128
Nekrasov, I V 254, 258
Nezamaev, AV 692
Nikolaeva, E P 269
Nikulin, A N 715
Nogovitsyna, E V 7
Norkina, O S 500
Obshivalkin, M Y 490
Ovchinnikova, A O 653
Ozerets, N N 263
Pampura, E M 674
Pankova, N A 374
Pastukhov, D S 730
Patrushev, Yu V 505
Pavlova, I A 150, 155
Pegashkin, V F 522, 730, 761
Pegashkina, E V 473

Permyakova, E K 464
Piskarev, I M 545
Platov, S I 550
Podoliak, O O 511
Polevoy, I O 71
Poloskov, S I 77, 478
Poloskov, S S 83
Polygalov, S E 37
Ponomarev, I A 211
Ponomarev, V B 221, 233
Popov, V 169
Popova, E 169
Portnova, I V 105
Potanin, V V 1, 448
Potekhin, V V 641
Prilutskaya, M A 527, 533
Prochorov, A A 111, 122
Pushin, V G 159, 164
Putilova, E A 719
Pytalev, I A 111
Qurbonalizoda, S Sh 612
Rakhmangulov, A N 560
Rashad, E M 583
Raskatov, E Yu 554
Rogozhnikov, D A 189
Romanenko, V P 715
Romanko, E A 217
Roshchin, V E 195
Rubcov, V Ju 374, 380
Saechnikov, V A 566
Safaraliev, M Kh 606, 618, 624
Safonov, E N 351, 356
Safronchuk, K A 755
Sarapulov, F N 572
Savelyev, M V 362, 368
Serdyukov, D Yu 43
Sergeant, Peter 583
Sevastjyanov, A A 715
Sharapova, V A 291
Sharifov, B N 606
Shartdinov, R R 54, 66
Shcherbinin, M M 437, 448
Sheshukov, O Yu 254, 258, 362, 368
Shevchenko, O I 60, 362, 368, 374, 380, 386
Shevkunov, N O 560
Shibanov, D A 755
Shishkin, A S 19, 233
Shishkin, S F 19, 25, 420
Shmakov, S V 362
Sholokhov, M A 478, 484
Shorokhov, E 169
Shulgin, B V 159, 164
Shutaleva, A V 719
Shvarts, D L 742
Shveikin, V P 653
Sibagatullin, S K 179
Sidorov, O Yu 1, 227, 589
Sirotyuk, A 724
Skorynina, P A 269
Smagin, K V 635, 686
Smetannikov, A N 66
Smirnov, A V 680
Smirnov, L A 54
Soboleva, N N 269
Spiridonov, A A 566
Spirina, A S 374
Starostin, A P 410, 426
Stazhkov, S 658
Stolbovsky, A 169, 174
Stroganov, Y N 674
Suerkulov, S M 618, 624
Sysoev, V I 179
Tabakov, V P 185
Tawfiq, Kotb B 583
Telegin, I V 578
Telegin, V V 578
Tipner, L M 527
Tkachev, A S 362
Trekin, G E 60, 238, 572
Trestsova, E D 275
Ugolnikov, N V 111, 122
Upolovnikova, A G 54, 66
Urtsev, N V 550
Ushakov, D V 566
Ustinova, E I 742
Veretennikova, I A 653
Verisokin, A E 43
Verstakovskaya, A P 566
Vitkina, G Yu 329
Volkov, M A 600
Volkov, P V 13, 95
Yachikov, I M 105
Yakovenko, N 658
Yamshanova, N V 398
Yanukyan, A P 43
Ykonovskaya, T B 139
Yudincev, D V 630
Yunusov, R M 606
Yurchenkov, E A 308
Zadvorkin, S M 297
Zagainov, S A 335
Zakirova, R A 539
Zavodova, T E 635, 686
Zayakin, O V 201, 206

Zemlyanoi, K G 244
Zhigulskaya, A I 139
Zhuchkov, V I... 201
Zhuckov, V I... 206
Zhulina, L G 43
Zhuravlyov, S Ya.. 398
Zolotukhin, I S 748
Zverev, S I 469
Zvontsov, N O 189